# Building
# Materials
## Testing and Sustainability

### N. SUBRAMANIAN

*Consulting Engineer*
*Maryland, USA*

## OXFORD
### UNIVERSITY PRESS

# OXFORD
## UNIVERSITY PRESS

Oxford University Press is a department of the University of Oxford.
It furthers the University's objective of excellence in research, scholarship,
and education by publishing worldwide. Oxford is a registered trade mark of
Oxford University Press in the UK and in certain other countries.

Published in India by
Oxford University Press
22 Workspace, 2nd Floor, 1/22 Asaf Ali Road, New Delhi 110002, India

ISBN-13: 978-0-19-949721-8
ISBN-10: 0-19-949721-4

Typeset in TimesLTStd
by B2K Infotech Private Limited, Erode
Printed at Manipal Technologies Limited, Manipal

*This book is dedicated to
all the ladies who crossed my life,
and taught me
love, compassion, and patience!*

# PREFACE

Ancient civilization lived in caves. After fire and tools were discovered, they started to use natural products such as stones, bricks, and wood to build habitats, which provided protection from the natural environment. Bricks have been found in Egypt as early as 14,000 BC. By 200 BC, the Romans successfully used concrete in the majority of their construction (though the present day concrete construction is much different from theirs). In the industrial age (between the mid-1700s and 1840), several manufacturing processes to produce cement (Joseph Aspdin, 1824), concrete, macadam roads (John Loudon McAdam, c. 1815), mass production of steel from molten pig iron by the Bessemer Process (Henry Bessemer, 1856), method of converting pig iron into wrought iron (Henry Cort, 1784), etc., were invented, resulting in the extensive use of concrete and steel. Now, concrete is the most widely used substance on earth, after water. We now use a variety of materials in our buildings and constructions ranging from stones, bricks, cement, reinforced concrete, steel, stainless steel, wood and wood products, glass, fibreglass, aluminium, plastics, ceramics, gypsum, paints and varnishes, bitumen, to copper, zinc, and aluminium alloys.

The exploding population growth (3 billion in 1960 to an estimated 7.7 billion in 2019) and the resulting urbanization has resulted in greater energy use [according to BP's Statistical Review of World Energy, world primary energy consumption reached 157 terawatt-hour (TWh) in 2017]; it has to be noted that materials such as cement and steel require large amounts of energy as their production requires them to be heated to temperatures exceeding 1300°C in kilns. This causes emission of huge quantities of greenhouse gases like $CO_2$ and also large amounts of industrial waste by-products like fly ash, which are harmful to the environment. In addition, indiscriminant mining of natural resources for the production of building materials has resulted in severe adverse effects on the environment, including loss of biodiversity, erosion, contamination of surface and ground water, and soil. Several (rocky) mountains are disappearing and river beds and even beaches (which took numerous years to form) are being denuded of sands. As these natural resources are limited, there is a shortage of coarse aggregates and sand in several parts of the world, necessitating the use of alternate materials or recycling of used materials and even the use of industrial wastes like fly ash. Several other materials such as plastics and lead are harmful to the environment or to the heath of human beings and have to be handled carefully. In addition, some materials like paints, varnishes, or wood impregnated with preservative chemicals may off-gas volatile organic compounds (VOC), which may affect the health of people living in air-conditioned interiors.

Almost all materials that are used in the construction have to follow the norms stipulated in the national codes (Bureau of Indian Standards has numerous codes on building materials). In addition, as buildings account for about one-third of worldwide energy consumption and are one of the largest contributors to GCG emissions, several countries including India have recently developed *Energy Conservation Building Codes*. They usually contain mandatory and voluntary provisions on insulation, thermal and solar properties of the building envelope, heating, ventilation and air conditioning, and also hot water supply systems, lighting, and electrical power.

From the above discussions, it is clear that all those engaged in the design and construction of buildings should have a sound knowledge about the manufacture, energy required in the manufacture, properties, effect on health, environmental friendliness, recyclability, sustainability, etc., of the materials they are using and also the norms prescribed for these materials. This book has been written to fulfill these needs. This text is based on several latest Indian Standard codes on building materials. SI units have been used throughout the book.

## About the Book

*Building Materials, Testing, and Sustainability* is a comprehensive text book designed to meet the requirements of the undergraduate students of civil engineering, based on the recent AICTE syllabus on *Materials, Testing & Evaluation*. This book will also be useful to students of architecture (who usually specify the materials), students of diploma courses in civil engineering and will serve as an invaluable reference to postgraduate students and practising engineers, as well as researchers.

Each chapter starts with an introduction about the origin and development of the material discussed in that chapter, the manufacture, properties, uses, advantages and disadvantages, and sustainability of that material. After presenting the subject, multiple-choice questions (MCQ) and review questions are provided at the end of each chapter, which will help the students assimilate the topics presented in the chapters. Answers to the MCQ are given at the end of each chapter, so that the students can check their answers. Several interesting case studies are also included as part of every chapter.

## Key Features

The following features in the book make it stand out among the other books in this area:

- It covers traditional materials to the most modern materials such as plastics, gypsum, and ceramics.
- Each chapter covers a brief history, composition, classification, manufacture, properties, advantages and disadvantages, use in buildings, environmental effects, sustainability, etc., of the material discussed in that chapter.
- Several Indian codes are available which stipulate norms for these materials. Most of these codes are cited in the book and important extracts are provided. The list of several codes, which are not cited, are also included in the references of each chapter and included in the online resource centre (ORC). In addition, a bibliography is provided in the book which lists the important references, for further study and research.
- Several topics, which are normally not found in other books, such as different types of brick kilns, substitutes for bricks, green cement, industrial by-products that can be used to replace cement, green substitutes for coarse aggregates and sand, green mortars, green and special concrete, mix design of concrete, controlled permeability formwork (CPF), industrial timber products, various reinforcing bars, structural insulated panel (SIP), sustainability of various materials, green building rating systems, bamboo, nano-materials, composite materials and concrete canvas, health effects and precautions to be taken while handling certain materials, etc., are discussed.
- 30 interesting practical case studies are provided.
- Students and engineers will find the separate chapter on testing and evaluation of these materials to be useful.
- A rich pedagogy provides the required rigour for students to excel in this subject in the examinations: Over 750 review questions and 440 multiple-choice questions (with answers) to test the understanding of the students; over 300 illustrative figures and photographs and 200 tables to supplement the text; more than 1440 references, which include relevant Indian and American codes.
- Provides most updated information in this subject covering the state-of-the-art trends and developments.

## Online Resources

The following resources are available to support the faculty and students using this text:

**For faculty**
- Lecture PPTs

**For students**
- List of references

## Using the Book

The text is divided into 25 chapters and completely covers the undergraduate (UG) curriculum of most of the universities. The teacher adopting this book is requested to exercise discretion to select portions of the text to be presented for a particular course. It is suggested that portions of Chapters 2–8, 11–13, 15–18, and 25 may be taught and Chapters 1, 9–10, 14, and 19–24 may be left for self-study.

Although relevant information from some important codes of practice has been included in the text, readers are advised to buy and refer to the latest codes published by the Bureau of Indian Standards, New Delhi. It is recommended that readers should use the book along with the latest codes/publications released by the Bureau of Indian Standards, for better clarity.

## Contents and Coverage

The text is divided into 25 chapters.

*Chapter 1* deals with the general information on physical, mechanical, thermal, and other properties of materials. It also gives some indications on sustainable (healthy and ecological) materials. Introduction to various green building systems is given and a comparison of structural steel, reinforced and pre-stressed concrete, and wood, which are the major materials used in construction, is provided. A discussion on Alternative Building Materials and building codes is also included.

Various aspects of stone including durability, deterioration, preservation, selection and uses of stones are discussed in *Chapter 2*. It also has brief introduction to stone masonry and a comparison of stone and brick masonry.

Bricks, which are usually used for constructing the walls of buildings, are the subject of *Chapter 3*. In India, the current brick manufacturing is through the use of highly polluting, energy inefficient, and uneconomical kilns. Hence various types of kilns are discussed and the vertical shaft brick kiln/tunnel kiln is suggested. Qualities of good bricks, properties, characteristics, etc., of bricks are provided. Several substitutes for bricks are also suggested.

*Chapter 4* deals with lime, which is a green material, and was used in olden days; this chapter discusses its manufacture, types, classifications, uses, and precautions while handling it. The manufacture, chemical composition, properties, and hydration of various types of cement, which is an important ingredient of concrete, are explained in *Chapter 5* along with the various pozzolana/green cement replacement materials. *Chapter 6* deals with the characteristics and properties of coarse and fine aggregates, which are mainly used in concrete, and include topics on grading of aggregates, alkali–aggregate reaction, and green substitutes for coarse aggregates and sand. *Chapter 7* is concerned with mortars and plasters, which are used in building masonry and providing protective coating on walls and ceilings, respectively.

Various aspects of concrete and RCC, which is used extensively in India in buildings, bridges, dams, etc., are covered in *Chapter 8*. Various special concretes such as ready mixed concrete, high-strength/high-performance concrete, self-compacting concrete (which is the material of the future), structural light weight concrete, foamed concrete, fibre reinforced concrete, ultra-high performance concrete, polymer concrete, geploymer concrete, prestressed concrete, precast concrete, decorative concrete, etc., are discussed in *Chapter 9*. As concrete contributes to about 5–6% of global emissions of greenhouse gases, the methods to reduce these emissions by the use of industrial by-products such as fly ash, GGBS, etc., (which also improve the properties of concrete) are also described in Chapter 9.

*Chapter 10* describes gypsum, a fire retardant and sustainable material, requiring low energy in its production, and mostly used in wall panels. The technology developed at IITM, using glass fibre reinforced gypsum panels, could reduce the cost and time required to build houses.

Wood is the most sustainable and renewable building material having a low level of embodied energy, and wood products can be carbon negative. In order to use it we should maintain sustainable forestry. *Chapter 11* explains the classifications, defects, conservation, seasoning and preservation, properties, selection and testing of timber. Several wood products such as layered timber composites, parallel laminates, particle composites, fibre composites, and timber-concrete composites are now available; their use and properties are briefly explained along with cork and linoleum.

*Chapter 12* deals with the various forms of iron (pig, wrought, cast) and *Chapter 13* describes steel, which is the second most used structural material. Steel could be alloyed with other elements (mainly with carbon) to improve the properties. It has equal strength in tension and compression, but its main drawback is its corrosion. Steel is used in various forms from structural sections to bolts, nuts, and nails. Hot rolled and cold-formed steel sections are available. Rebars used in concrete and prestressed concrete are usually made of high strength steel. Several techniques have been developed to mitigate the corrosion problem including the development of stainless steel and weathering steel. The production of steel is also energy intensive and requires high temperatures (up to 1650°C), but is considered sustainable due to its 100% recyclability. *Chapter 14* is concerned with non-ferrous materials such as aluminium, copper, zinc, lead, etc., and its alloys like brass and bronze. Steel is galvanized using zinc.

Glass, which is obtained from silica sand, lime, soda, and alumina, and mainly used in windows and curtain walls of multi-storey buildings, is discussed in *Chapter 15*. Common types (sheet, plate, float, and extra clear) and special types of glass (safety, translucent, etched, tinted, reflective/coated, insulated, double-glazed, glass blocks) are described. Clay roofing tiles of various types and ceramic products are discussed in *Chapter 16*. *Chapter 17* discusses plastics, which were invented in the 1800s and have revolutionized the construction industry and being used in a variety of applications. This chapter describes its classification, methods of production, properties, and uses. Most of the plastics are not recyclable and are not bio-degradable and hence have to be used with caution.

Details about paints and varnishes are provided in *Chapter 18* and asphalt, bitumen, and tar in *Chapter 19*. *Chapters 20–22* deal with thermal and sound insulating materials and waterproofing materials, respectively. Miscellaneous and recent materials are discussed in *Chapter 23*. A brief description of deformation and fracture of materials is provided in *Chapter 24*. *Chapter 25* describes the various tests that are performed on some of the important materials, in order to evaluate them.

Though care has been taken to present error-free material, some errors might have crept in inadvertently. I would highly appreciate if these errors are brought to the attention of publishers. Any suggestions for improvement are also welcome.

## Acknowledgements

I was greatly influenced during the preparation of this book by several books, papers, and websites on building materials, sustainability and testing. Although sufficient effort was taken to acknowledge the source of images, tables, etc., I wish to apologize for the use of any phrase, image, or illustrations used in this book inadvertently without acknowledgement.

I thank the following organizations/publishers for permitting to reproduce material from their publications: Ms Angela R. Matthews of American Concrete Institute (ACI), Ms Rachel Jordan of

American Institute of Steel Construction (AISC), Mr Toru Kawai, Executive Director of Japan Concrete Institute (JCI), The US Green Building Council, Dr Steven H. Kosmatka, former Vice President, and Guiyun Wang of Portland Cement Association (PCA), USA, Mr Chris Shaw of UK, The Indian Concrete Journal, Mr Craig Donnelly of the Scottish Government, Er Robert Jackson of Fast+Epp, of Vancouver, BC, Canada, Dr Warren South, Director – Research and Technical Services of Cement Concrete & Aggregates, Australia, Ms Rousalina Barkhamatova, Secretary, Industrial Minerals Association Europe (IMA-Europe), Brussels, Belgium, Mr Opinder Saggi of Allied Concrete Ltd and AML Ltd, Auckland, New Zealand, Dr Mohammed Nadeem of Ambuja Cements, Mr Drew Burns, Executive Director of Slag Cement Association, Dr M.C. Nataraja, Professor of JSS Science and Technology University, Mysore, Prof. Devdas Menon of IITM, Er Alok Bhowmick of B&S Engineering Consultants Pvt. Ltd, Mr Samuel Sundar Singh of The Indian Green Building Council (IGBC), and M/s Tata Steel.

I thank the following organizations/persons for giving permission to use photographs of testing equipment/buildings/materials: Er Dar Adil of IITD, Mr Mohammed Razal of Bangalore Tile Company, Bangalore, Prof. Manu Santhanam of IITM, Mr Robin Bailey of Humboldt Mfg Co., Elgin, IL, Prof. Jake Hiller of Michigan Technological University, USA, Dr Ajoy Kumar Mullick, former Director General of the National Council for Cement & Building Materials and Consultant, New Delhi, Mr Garry Watkins of Concrete Canvas Ltd, UK, Mr Giorgio Maestroni and Ms Elena Agazzi of Matest S.P.A. Unipersonale, Italy, Ms Sadia Parveen of ELE International, UK, Ms Aarti Bhargava of Aimil Ltd, New Delhi, Mr Jean Pol Grandmont (Flickr), Er Sebastian Kaminski of Arup, UK, Prof. M. Sekar of Anna University, Mr Lori Maloney of JC Steele & Sons, Inc., Mr John Lamond (Controls USA, Inc.) and Ms Paola Bettinelli (Controls S.p.A.) of Controls Group, Milan, Italy, M/s Elsevier (Fig. 5.4), Mr Nemish Sheth, Director of Asona India Pvt. Ltd Mumbai, Ms Sara DeWaay of the University of Oregon Libraries, Ms Alison of Findlay–Evans Waterproofing & Liquid Rubber Melbourne, Australia, Dr Narayan V. Nayak, Advisor, Gammon India Ltd, Dr V.S. Parameswaran, Former Director of SERC, Dr G. Mohan Ganesh of Vellore Institute of Technology, Vellore, Er S.H. Gajendran, Er E.S. Jayakumar, Superintending Engineer (Civil), BSNL Civil Wing, Kerala, Er Rahul Leslie, Deputy Director of Kerala PWD, Cochin, and Er M. Karthik.

I thank all those who assisted me in the preparation of this book. First and foremost I thank Dr Col. Nallathambi, Managing Director, Sakthi Civil & Structural Consultancy Pvt. Ltd, Chennai for his help in writing Chapters 20–22. My sincere thanks are due to Prof. B. Vijaya Rangan, Emeritus Professor of Civil Engineering, Curtin University of Technology, Perth, Australia, Prof. Dr P. Suryanarayana, Professor & Former Dean (Retd), Maulana Azad National Institute of Technology, Bhopal, Dr Ch. T. Madhavi, Professor & Head, SRM Institute of Science and Technology, Ramapuram Campus, Dr Abdul Rashid Dar, Professor and former Director, National Institute of Technology Srinagar, J&K, India, and Dr G. Mohan Ganesh, of VIT, for going through the chapters patiently and offering useful comments, and Er Hemal Mistry and Er Dar Adil of IITM for their help in locating the literature. Finally I express my gratitude to Er Kumar Abhishek Singh of New Delhi, Er Vivek Abhyankar of L&T Mumbai, Er Siddique, and Er Ganesh of Nagpur for their encouragement.

I will be failing in my duty if I do not acknowledge the help and wonderful assistance I received from Ms S. Chithra at all the stages of this book-writing project. Lastly, I acknowledge the excellent support and coordination provided by the editorial team of Oxford University Press India.

**N. Subramanian**

# BRIEF CONTENTS

# DETAILED CONTENTS

# LIST OF ACRONYMS

## A

| | | |
|---|---|---|
| AAC | : | Autoclaved aerated concrete |
| AAR | : | Alkali-aggregate reaction |
| ACA | : | Ammoniacal copper arsenate |
| ACC | : | Autoclaved cellular concrete |
| ACI | : | American Concrete Institute |
| ACM | : | Aluminium composite material |
| ACV | : | Aggregate crushing value |
| ABR | : | Abrasion-resistant iron |
| ABS | : | Acrylonitrile butadiene styrene |
| ACR | : | Alkali-carbonate reaction |
| ACRR | : | Alkali-carbonate rock reaction |
| ACZA | : | Ammoniacal copper zinc arsenate |
| AFG | : | Austenitic flake graphite iron |
| AFRP | : | Aramid fibre reinforced polymer |
| AHL | : | Artificial hydraulic lime |
| AIBN | : | Azobis-iso-butyro-nitrile |
| AIV | : | Aggregate impact value |
| $Al_2O_3$ | : | Alumina |
| ALC | : | Autoclaved lightweight concrete |
| AMBT | : | Accelerated mortar bar test |
| ANSI | : | American National Standards Institute |
| APP | : | Atactic polypropylene |
| ASG | : | Austenitic spheroidal or nodular graphite iron |
| ASHRAE | : | American Society of Heating, Refrigerating, and Air-Conditioning Engineers |
| ASR | : | Alkali-silica reaction |
| ASTM | : | American Society for Testing and Materials |

## B

| | | |
|---|---|---|
| bcc | : | Body-centred cubic |
| BF | : | Blast furnace |
| BFS | : | Blast furnace slag |
| BIS | : | Bureau of Indian Standards |
| BM | : | Black-heart malleable iron |
| BOF | : | Basic oxygen furnace |
| BOS | : | Basic oxygen steel making |
| BBIT | : | Butyl-benzisothiazolinone |
| BREEAM | : | Building Research Establishment Environmental Assessment Method |
| BTK | : | Bull's trench kiln |

| | | |
|---|---|---|
| BUR | : | Built-up roof |
| BWP | : | Boiling water proof |
| BWR | : | Boiling water resistant |

## C

| | | |
|---|---|---|
| C&D waste | : | Construction and demolition waste |
| $C_3A$ | : | Tricalcium aluminate |
| $C_2S$ | : | Dicalcium silicate |
| $C_3S$ | : | Tricalcium silicate |
| $C_4AF$ | : | Tetracalciumalumino ferrite |
| CAC | : | Calcium aluminate cement |
| $CaCO_3$ | : | Calcium carbonate |
| CAE | : | Carboxylic acid with acrylic ether |
| CaO | : | Calcium oxide (also calcia and lime) |
| $Ca(OH)_2$ | : | Calcium hydroxide |
| CaS | : | Calcium sulphide |
| CASBEE | : | Comprehensive Assessment System for Building Environmental Efficiency |
| CCA | : | Copper-chrome-arsenate |
| CCAA | : | Cement Concrete & Aggregates Australia |
| CDK | : | Cross-draught kiln |
| CDS | : | Cold-drawn seamless |
| CEB | : | Compressed earth blocks |
| CEW | : | Cold-drawn electric-resistance welded |
| CFC | : | Chlorofluorocarbon |
| CFRP | : | Carbon fibre reinforced polymer |
| CFS | : | Cold-formed steel |
| CH | : | Calcium hydroxide |
| CHS | : | Calcium silicate hydrate |
| CLT | : | Cross-laminated timber |
| CMAC | : | Cold mix asphalt concrete |
| CMC | : | Carbon-matrix composites |
| CMIT | : | Chloromethylisothiazolinone |
| CMU | : | Concrete masonry unit |
| CPT | : | Concrete prism test |
| CO | : | Carbon monoxide |
| $CO_2$ | : | Carbon dioxide |
| CPF | : | Controlled permeability formwork |
| CPO | : | Concrete production operations |
| CPVC | : | Chlorinated poly vinyl chloride (also critical pigment volume concentration) |
| CR | : | Cold rolled |
| CRM | : | Crumb rubber modifier |

| | | |
|---|---|---|
| CRMB | : | Crumb rubber modified bitumen |
| CS | : | Steel castings |
| CSC | : | Corrosion-resistant steel castings |
| CSCEC | : | China State Construction and Engineering Corporation |
| CSH | : | Heat-resistant steel castings |
| CSPE | : | Chlorosulfonated polyethylene |
| CTD | : | Cold twisted deformed (also Torsteel) |
| CTOD | : | Crack tip opening displacement |

### D

| | | |
|---|---|---|
| DC | : | Double charge |
| DCOI | : | Dichloro-octyl-isothiazolone |
| DDK | : | Down-draught kiln |
| DPC | : | Damp-proof course |

### E

| | | |
|---|---|---|
| EAF | : | Electric arc furnace |
| EDTA | : | Ethylene diamine tetra acetic acid |
| EMC | : | Equilibrium moisture content |
| ENM | : | Engineered nanomaterials |
| EPD | : | Electrophoretic deposition |
| EPDM | : | Ethylene propylene diene monomer |
| EPS | : | Expanded polystyrene |
| ERW | : | Electric-resistance welded |
| ESCS | : | Expanded shale, clay, and slate |
| ESCSI | : | Expanded shale, clay and slate institute |
| ETFE | : | Ethyl tetra fluoro ethylene |
| EVA | : | Ethyl vinyl acetate |

### F

| | | |
|---|---|---|
| Fal-G | : | Fly ash-lime-gypsum |
| FCBTK | : | Fixed chimney bull's trench kiln |
| fcc | : | Face-centred cubic |
| $Fe_2O_3$ | : | Iron oxide |
| FG | : | Grey iron |
| FGD | : | Flue gas desulphurization |
| FHWA | : | The Federal Highway Administration, USA |
| FM | : | Fineness modulus |
| FRC | : | Fibre reinforced concrete |
| FRP | : | Fibre-reinforced plastic/polymer |
| FSP | : | Fibre saturation point |

### G

| | | |
|---|---|---|
| GBFS | : | Granulated Blast Furnace Slag |
| GFRG | : | Glass fibre reinforced gypsum |
| GFRP | : | Glass-fibre reinforced polymer/plastic |
| GGBS | : | Ground Granulated Blast Furnace Slag |
| GHG | : | Greenhouse gases |
| GLT | : | Glued Laminated Timber (also Glulam) |
| GRP | : | Glass-reinforced plastic |
| GVT | : | Glazed vitrified tile |

### H

| | | |
|---|---|---|
| HCFC | : | Hydro-chlorofluorocarbon |
| hcp | : | Hexagonal close-packed |
| HDPE or PE | : | High-density polyethylene |
| HFC | : | Hydro-fluorocarbon |
| HFW | : | Hot-finished welded |
| HFS | : | Hot-finished seamless |
| HMAC | : | Hot-mix asphalt concrete |
| HPC | : | High-performance concrete |
| HPI | : | Hydrophobic pore-blocking ingredient |
| HPL | : | High pressure decorative laminates |
| HPS | : | High-performance steel |
| HR | : | Hot rolled |
| HRM | : | High reactivity metakaolin |
| HSC | : | High strength concrete |
| HSLA | : | High-strength low-alloy |
| HSS | : | Hollow steel section |
| HVFA | : | High volume fly ash |
| HWRA | : | High-range water-reducing admixtures |
| HYSD | : | High yield strength deformed steel |

### I

| | | |
|---|---|---|
| IBC | : | International building code |
| IF | : | Induction furnace |
| IIT | : | Indian Institute of Technology |
| IR | : | Insoluble residue |
| ISA or ISEA | : | Indian Standard Equal Angles |
| ISDT | : | Indian Standard Rolled Deep Legged Tee Bars |
| ISFl | : | Indian Standard Flat |
| ISHB | : | Indian Standard Heavyweight Beam/column |
| ISHT | : | Indian Standard Slit Tee Bars from H-section |
| ISJB | : | Indian Standard Junior Beam |
| ISJC | : | Indian Standard Junior Channel |
| ISJT | : | Indian Standard Junior Tee section |
| ISLB | : | Indian Standard Lightweight Beam |
| ISLC | : | Indian Standard Lightweight Channel |
| ISLT | : | Indian Standard Slit Lightweight Tee Section |

| ISLT, ISMT, and ISHT | : | Indian Standard Lightweight/Heavy-weight/Heavy Tubular section |
|---|---|---|
| ISMB | : | Indian Standard Medium-weight Beam |
| ISMC | : | Indian Standard Medium-weight Channel |
| ISMCP | : | Indian Standard Medium-weight Channel Parallel Flange |
| ISMT | : | Indian Standard Slit Medium-weight Tee Section |
| ISNT | : | Indian Standard Rolled Normal Tee Section |
| ISPL | : | Indian Standard Plate |
| ISRO | : | Indian Standard Rolled Round bar |
| ISSC | : | Indian Standard Column Section |
| ISSQ | : | Indian Standard Rolled Square bar |
| ISST | : | Indian Standard Strip |
| ISWB | : | Indian Standard Wide-flange Beams |

## L

| LCA | : | Life-cycle assessment |
|---|---|---|
| LCC | : | Lightweight cellular concrete |
| LCCA | : | Lifecycle Cost Analysis |
| LD | : | Linz and Donawitz |
| LDPE | : | Low-density polyethylene |
| LEED | : | Leadership in Energy and Environmental Design |
| LFC | : | Lightweight foamed concrete |
| LMC | : | Latex-modified concrete (also PMC) |
| LMF | : | Ladle metallurgical furnace |
| LOI | : | Loss of ignition |
| Low-e | : | Low-emissivity |
| LP | : | Lime pozzolana |
| LSF | : | Lime saturation factor |
| LSL | : | Laminated strand lumber |
| LVL | : | Laminated veneer lumber |

## M

| MA | : | Mastic asphalt/bitumen |
|---|---|---|
| MB | : | Modified bitumen |
| MCE | : | Multi-polycarboxylate ethers |
| MDF | : | Medium density fibreboard |
| MEE | : | Moisture-excluding effectiveness |
| MEK | : | Methyl ethyl ketone |
| MF | : | Melamine-formaldehyde or Formica |
| MgO | : | Magnesium oxide |
| $Mg(OH)_2$ | : | Magnesium hydroxide |
| MII | : | Mixed-in inhibitor |
| MIT | : | Methylchloroisothiazolinone |

| MLS | : | Modified lignosulphonates |
|---|---|---|
| MM | : | Mortar mix |
| MMA | : | Methyl methacrylate |
| MMC | : | Metal-matrix composites |
| MS | : | Mild steel |
| M-sand | : | Manufactured sand |
| MR | : | Moisture resistant |
| MT | : | Million tonnes |
| MUF | : | Melamine-urea-formaldehyde |

## N

| NAAC | : | Non-autoclaved aerated concrete |
|---|---|---|
| NaOH | : | Sodium hydroxide (also known as caustic soda) |
| NBC | : | National Building Code of India |
| NDT | : | Non-destructive tests |
| NHL | : | Natural Hydraulic Lime |
| NiTi | : | Nickel-titanium |
| Nitinol | : | Nickel Titanium Naval Ordnance Laboratories |
| NLT | : | Nail-Laminated Timber |
| NM | : | Nanomaterials |
| NMDC | : | National Mineral Development Corporation |
| NOAA | : | National Oceanic and Atmospheric Administration |
| NRC | : | Noise reduction coefficient |
| NRMB | : | Natural rubber modified bitumen |
| NRMCA | : | National Ready Mixed Concrete Association |
| NSC | : | Normal strength concrete |
| Nylon | : | Polyamide |

## O

| OAW | : | Oxy-acetylene welded |
|---|---|---|
| OBD | : | Oil-bound distemper |
| OIT | : | Octylisothiazolinone |
| OPC | : | Ordinary Portland Cement |
| OSB | : | Oriented strand board |
| OSL | : | Oriented strand lumber |

## P

| PAE | : | Poly acrylic ester |
|---|---|---|
| PAN | : | Poly acrylamide |
| PB | : | Polymer-based |
| PC | : | Portland cement/Polymer concrete/Polycarbonate |
| PCA | : | The Portland Cement Association |

| | | |
|---|---|---|
| PCE | : | Polycarboxylate ether |
| PE | : | Polyethylene |
| PET or | | |
| PETE | : | Polyethylene terephthalate |
| PEVA | : | Polyethylene-vinyl acetate |
| PEX | : | Cross-linked polyethylene |
| PF | : | Phenol-formaldehyde |
| PIC | : | Polymer impregnated concrete |
| PIR | : | Polyisocyanurate |
| PLC | : | Portland limestone cement |
| PM | : | Pearlitic malleable iron |
| PMB | : | Polymer-modified bitumen |
| PMB(E) | : | Elastomeric thermoplastic based bitumen |
| PMB(P) | : | Plastomeric thermoplastic based bitumen |
| PMC | : | Polymer-modified concrete |
| PMM | : | Polymer modified mortar |
| PMMA | : | Acrylic-Polymethyl-Methacrylate |
| POLYISO | : | Polyisocyanurate |
| PP | : | Polypropylene |
| PPC | : | Portland pozzolana cement |
| PPCC | : | Polymer-Portland-cement concrete (also PMC) |
| PPR | : | Polypropylene random |
| PR | : | Parallel flow regenerative |
| PRF | : | Phenol-resorcinol-formaldehyde |
| PSA | : | Pressure sensitive adhesive |
| PSC | : | Portland slag cement |
| PSD | : | Particle size distribution |
| PSL | : | Parallel strand lumber |
| PS | : | Polystyrene |
| PTFE | : | Polyfluoroethylene (also known as Teflon) |
| PUR | : | Preformed rigid polyurethane |
| PVA | : | Poly vinyl acrylic |
| PVAl | : | Polyvinyl alcohol |
| PVB | : | Polyvinyl butyral |
| PVC | : | Polyvinyl chloride (also known as pigment volume concentration) |

## R

| | | |
|---|---|---|
| RA | : | Recycled aggregates |
| RAP | : | Recycled asphalt pavement |
| RAS | : | Recycled asphalt shingles |
| RCA | : | Recycled concrete aggregate |
| RC | : | Reinforced concrete |
| RC/MC/SC | : | Rapid/Medium/Slow curing |
| RCC | : | Reinforced cement concrete/Roller-compacted concrete |

| | | |
|---|---|---|
| RH | : | Relative humidity |
| RHA | : | Rice husk ash |
| RHS | : | Rectangular hollow sections |
| RMC | : | Ready mixed concrete |
| R-Value | : | Thermal resistance value |

## S

| | | |
|---|---|---|
| SAA | : | Sound absorption average |
| SAE | : | Polystyrene-acrylic ester |
| SCL | : | Structural composite lumber |
| SBR | : | Styrene butadiene rubber |
| SBS | : | Styrene-butadiene styrene |
| SCC | : | Self-compacting concrete |
| SCEB | : | Stabilized compressed earth block |
| SCMs | : | Supplementary cementitious materials |
| SFLA | : | Sintered fly ash lightweight aggregates |
| SFS | : | Steel furnace slag |
| SG | : | Spheroidal or nodular graphite iron |
| SHGC | : | Solar heat gain coefficient |
| SHS | : | Square hollow sections |
| SIP | : | Structural insulating panel |
| $SiO_2$ | : | Silica |
| SLWC | : | Structural lightweight concrete |
| SMA | : | Stone-mastic asphalt or shape-memory alloy |
| SMB | : | Stabilized mud block |
| SMF | : | Sulphonated melamine formaldehyde condensates |
| SNF | : | Sulphonated naphthalene formaldehyde condensates |
| SPF | : | Sprayedpolyurethane foam |
| SPL | : | Sound pressure level |
| SPUR | : | Silyl-modified polyurethanes |
| SRC | : | Sulphate resisting Portland cement |
| SRI | : | Sound reduction index or solar reflectance index |
| SS | : | Stainless steel |
| SSD | : | Saturated surface-dry |
| STC | : | Sound transmission class |
| STPE | : | Silyl-terminated polyether (also MS Polymer) |

## T

| | | |
|---|---|---|
| TCC | : | Timber-concrete composites |
| TFV | : | Ten percent fines value |
| TL | : | Transmission loss |
| TMT | : | Thermo-mechanically treated |

| | | |
|---|---|---|
| TMTCRS | : | Thermo-mechanically treated corrosion resistant steel |
| TPO | : | Thermoplastic polyolefin |

## U

| | | |
|---|---|---|
| UDK | : | Up-draught kiln |
| UF | : | Urea formaldehyde |
| UHPC | : | Ultra-high performance concrete |
| UPVC | : | Unplasticized polyvinyl chloride |
| USBR | : | United States Bureau of Reclamation |
| USEPA | : | United States Environmental Protection Agency |
| USGBC | : | United States Green Building Council |
| UTM | : | Universal testing machines |
| UV | : | Ultraviolet light |
| U-Value | : | Reciprocal of thermal resistance |

## V

| | | |
|---|---|---|
| VA/VeoVa | : | Vinyl acetate and versatate copolymer |
| VAC | : | Vinyl acetate copolymers |

| | | |
|---|---|---|
| VG | : | Viscosity grade |
| VOCs | : | Volatile organic compounds |
| VSBK | : | Vertical shaft brick kiln |
| VSI Crusher | : | Vertical shaft impactor (VSI) |
| VT | : | Visible transmittance |

## W

| | | |
|---|---|---|
| W/c ratio | : | Water-cement ratio |
| W/cm ratio | : | Water-cementitious ratio |
| WM | : | White heart malleable iron |
| WMA | : | Warm mix asphalt |
| WR | : | Weather resistant or weathering steel |
| WRA | : | Water-reducing admixtures |
| WWF | : | Welded wire fabrics |

## X

| | | |
|---|---|---|
| XPS | : | Extruded polystyrene |

# CHAPTER 1
# PROPERTIES OF BUILDING MATERIALS

## 1.1 Introduction

In the early societies, human beings lived in caves and almost certainly rested in the shade of trees. Gradually, they learnt to use naturally occurring materials such as stone, timber, mud, and biomass (e.g., leaves, grasses, and natural fibres) to construct houses, which was then followed by brick making, rope making, and glass and metal work (see Table 1.1). From these early beginnings, the modern materials–manufacturing industries developed.

The principal modern building materials are bricks, cement, concrete (e.g., mass, reinforced, and prestressed), timber, structural steel (in rolled and fabricated sections), aluminium, glass, paints, plastics, and ceramic products. All these materials have particular advantages in certain applications and hence the construction of a building may include various materials; for example, a commercial multi-storey

**Table 1.1** Historical developments in building materials

| Material | Period |
|---|---|
| Mud, stones, and thatch | 400,000 BC |
| Sun-dried bricks | 7500 BC |
| Timber | 5000 BC |
| Fired bricks | 2000 BC (First appeared in the Middle East) |
| Lime | 3000 BC |
| Glass | 2500 BC (The Egyptians provided the first examples with silica and calcium.) |
| Iron | 1350 BC |
| Lime–pozzolana cement | 300 BC–AD 476 (Romans and Greeks) |
| Aluminium | In 1808, it was discovered by English chemist Sir Humphrey Davy. In addition, the smelting process was discovered by Charles Martin Hall of the USA in 1886. |
| Portland cement | Invented by Joseph Aspdin in 1824 |
| Natural rubber | A method of processing invented by Charles Goodyear in 1839 |
| Steel | In 1855, Sir Henry Bessemer invents the Bessemer process for the mass production of steel |
| Plastics | In 1862, by Alexander Parkes at the 1862 Great International Exhibition in London |
| Stainless steel | Invented by Henry Brearly in 1913 |

building may have concrete/steel frame, infilled, and partition walls made of bricks or concrete blocks, wooden doors, and double-glazed windows of aluminium. The architect or design engineer has to think about the various alternatives and suggest a suitable material, which will satisfy economic, aesthetic, functional, and ecological/sustainability requirements.

Since construction is the largest consumer of natural materials, India was one of the first few countries in the world to add a specific provision on environment protection in its Constitution, through the 42nd Amendment, during 1976. The Environment (Protection) Act, 1986 was introduced as an umbrella legislation that provides a holistic framework for the protection and improvement to the environment.

## 1.2 Functions of a Building

The major requirement of any building is to shelter its occupants from the environmental conditions and offer a pleasant, comfortable, and healthy indoor environment. Thus, a well-designed building will provide good weather resistance (adequate resistance to rain and wind penetration), ample lighting, proper ventilation, adequate thermal insulation (to prevent heat loss to the environment in cool areas and heat gain in hot areas), low noise levels (to prevent airborne sound and impact sound from outside and prevent the passage of sound from one inner space to another inner space), have adequate strength and stability, privacy, security, and fire resistance (for the occupants to exit safely in the case of major fires), proper ingress and egress, good appearance, durability, and have reasonable cost. These requirements can be achieved by the proper selection of building materials and products. Proper site location, shape, orientation, and vegetation around the building may also help in obtaining better lighting and thermal performance (Reid, 1984).

## 1.3 Choosing Materials for Construction

Building materials may be categorized based on the source of availability as natural and manufactured. They may also be classified as traditional and modern. The traditional building materials are generally the naturally occurring substances such as earth (e.g., clay, sand), stone and rocks, lime, and wood logs. Whereas, modern building materials include many artificial or manufactured synthetic and composite products such as bricks, lightweight concrete blocks, concrete, metals, glass, ceramics, plastics, and petroleum-based paints. The manufacture of building materials is an established industry, and the use of these materials is generally segmented into specific specialty trades such as concrete, masonry, carpentry, plumbing, roofing, electrical, mechanical, insulation, and HVAC (heating, ventilation, and air conditioning). In the countries like the USA, in order to work in these trades, one has to get certified by passing the Law and Business and the Trade Examinations.

For a material to be considered suitable for construction, it should have some essential engineering properties. These properties are broadly classified as follows:

| | |
|---|---|
| 1. Physical properties | 5. Optical properties |
| 2. Mechanical properties | 6. Acoustical properties |
| 3. Thermal properties | 7. Physiochemical properties |
| 4. Chemical properties | 8. Metallurgical properties |

These properties of building materials are responsible for its quality and capacity and will be useful while deciding the use of the material in different applications. Physical and mechanical properties, such as strength, porosity, etc., are generally considered while selecting a material for a particular use. Chemical properties are considered when the material is used in aggressive environments. Some of these properties are discussed in Sections 1.4–1.7. Before using the materials, they should be tested and evaluated to ascertain whether the properties assumed in the design are actually available. For this purpose, samples are taken randomly (the number of samples tested is based on statistical analysis) and tested. The test methods, required equipment, and selection criteria are discussed in Chapter 25.

## 1.4 Physical Properties

The physical properties generally include the shape, size, density/bulk density, specific gravity, and porosity of the material.

**Density ($\rho$)**    *Density* of a homogeneous material is defined as the mass per unit volume and expressed in kg/m$^3$. Density is also called the unit weight of substance.

$$\rho = \frac{m}{V} \qquad (1.1)$$

where $m$ = mass (kg) and $V$ = volume (m$^3$).

**Bulk density ($\rho_b$)**    This is defined as the mass per unit volume of material in its natural state (including pores and voids), and expressed in kg/m$^3$. It is calculated as:

$$\rho_b = \frac{m}{V_b} \qquad (1.2)$$

where $m$ = mass of specimen (kg) and $V_b$ = volume of specimen in its natural state (m$^3$).

For most of the materials, bulk density will be less than their densities; however, for liquids and materials like glass and dense stone materials, density and bulk density will not differ much. Bulk density represents the degree of compactness of material. Bulk density of a material depends upon the packing of particles, particle shape and size, moisture content, and grading. For example, in coarse aggregates, a higher bulk density indicates fewer voids that are to be filled by sand and cement in concrete.

Properties like strength and heat conductivity are greatly affected by their bulk density. Density and bulk density of some building materials are compared in Table 1.2.

**Table 1.2** Comparison of density and bulk density of some building materials

| Material | Density (kg/m³) | Bulk density (kg/m³) |
|---|---|---|
| Brick | 1920–2400 | 1600–1800 |
| Granite | 2600–2900 | 2500–2700 |
| Wood (teak) | 1500–1600 | 630–720 |
| Steel | 7750–8050 | 7850 |
| Concrete | 2400 | 2080–2400 |

**Density index ($\rho_0$)**    This is the ratio of bulk density of a material to its density and is expressed as:

$$\rho_0 = \frac{\text{Bulk density}}{\text{density}} = \frac{\rho_b}{\rho} \qquad (1.3)$$

It indicates the degree to which the volume of a material is filled with solid matter. For natural building materials, $\rho_0$ will be less than 1.0 because natural materials are not absolutely dense.

**Specific weight ($\gamma$)**    This is also called *unit weigh*. This is the weight per unit volume of material in kN/m$^3$, and is expressed as:

$$\gamma = \rho g \qquad (1.4)$$

where, $\rho$ = density of the material (kg/m$^3$) and $g$ = acceleration due to gravity (m/s$^2$).

Unit weight is used to determine the dead load of a structure in structural design. The unit weight of water is 9.81 kN/m$^3$ at 4°C.

**Specific gravity or relative density ($G_s$)** This is the ratio of the density of a substance to the density of a reference substance, and is a dimensionless quantity. The reference substance is usually specified as water at 4°C. At this temperature, the density of water will be the highest at 981 kg/m$^3$ (approximately taken as 1000 kg/m$^3$). Hence, the specific gravity may be expressed as:

$$G_s = \frac{\gamma_s}{\gamma_w} = \frac{\rho_s g}{\rho_w g} = \frac{\rho_s}{\rho_w} \tag{1.5}$$

**True or absolute specific gravity ($G_a$)** When both the permeable and impermeable voids are excluded to determine the true volume of solids, the specific gravity is called true or absolute specific gravity and can be expressed as:

$$G_a = \frac{(\rho_s)_a}{\rho_w} \tag{1.6}$$

The absolute specific gravity is not much used in practical applications.

**Porosity ($n$)** This is a measure of the void or empty spaces in a material, and is expressed as a ratio of the volume of voids ($V_v$) to the total volume ($V$), between 0 and 1, or as a percentage between 0 and 100%.

$$n = \frac{V_v}{V} \tag{1.7}$$

Porous materials absorb more moisture. Porosity influences many properties like thermal conductivity, strength, bulk density, and durability. Porosity reduces the resistance to freezing, thawing, and abrasion. Rocks usually have porosity of less than 20%. Dense materials, which have low porosity, have to be used when high mechanical strength is required, whereas walls of buildings are commonly built with materials like bricks, which have considerable porosity.

**Void ratio ($e$)** This is defined as the ratio of volume of voids ($V_v$) to the volume of solids ($V_s$) in the material.

$$e = \frac{V_v}{V_s} \tag{1.8}$$

The following relationship exists between void ratio and the porosity.

$$n = \frac{e}{1+e} \tag{1.9}$$

**Durability** This is the ability of a material to perform its required function over a lengthy period under normal conditions of use without excessive expenditure on maintenance or repair.

It has to be noted that the strength and durability are two separate aspects: neither guarantees the other. Durability may be affected by a number of parameters. For example, the durability of reinforced concrete is affected by the following (Subramanian, 2013):

1. Environment
2. Temperature or humidity gradients
3. Abrasion and chemical attack
4. Permeability of concrete to the ingress of water, oxygen, carbon dioxide, chloride, sulphate and other deleterious substances

5. Alkali-aggregate reaction (chemical attack within the concrete)
6. Chemical decomposition of hydrated cement
7. Corrosion of reinforcement
8. Concrete cover to the embedded steel
9. The quality and type of constituent materials
10. Cement content and water/cement ratio
11. Degree of compaction and curing of concrete
12. Shape and size of member
13. The presence of cracks

## 1.5 Mechanical Properties

The important mechanical properties of building materials are strength (compressive, tensile, bending, and impact), elasticity, plasticity, ductility, hardness, toughness, malleability, brittleness, fatigue, impact strength, abrasion resistance, creep resistance, and stiffness/flexibility.

**Strength** This is the ability of a material to resist stresses caused by the external forces (i.e., tension, compression, bending, torsion, and impact), without failure or fracture. It is of importance to note that materials such as stones and concrete have high compressive strength but low tensile, bending, and impact strengths (about 1/5 to 1/50$^{th}$ of compressive strength).

*Ultimate strength* This is the minimum guaranteed ultimate tensile strength at which a metal would fail, is obtained from a tensile test on a standard specimen, as described in Section 25.20.1 of Chapter 25. The stress–strain curve obtained from this test for various materials is shown in Fig. 1.1.

*Compressive strength* Compressive strength of concrete is found by testing standard cylinders (150 mm diameter and 300 mm long) or cubes (150 mm size) in compression testing machines, usually on the 28$^{th}$ day of casting them—cylinders have lower resistance than cubes of the same cross-sectional area. Properties such as modulus of elasticity, tensile strength, shear strength, and bond strength are usually expressed in terms of the compressive strength (see Subramanian, 2013, for more details).

*Bending strength* Tests of building strength on concrete are performed on small beams supported at their ends and subjected to one or two concentrated loads, which are gradually increased until failure takes place.

**Hardness** This is a measure of the resistance of the material to indentations and scratching. Several methods are available to determine the hardness of steel and other metals. In all these methods, an 'indenter' is forced on the surface of the specimen. On removal, the size of indentation is measured using a microscope. Based on the size of the indentation, the hardness of the specimen is determined.

**Elasticity** This is the ability of a material to regain its original shape and size after removal of the external load. Ideally, elastic materials obey the Hooke's law, which states that, within elastic limits, stress is directly proportional to strain. The ratio of unit stress to unit deformation is termed as the *modulus of elasticity*. A large value of it represents a material with very small deformation.

**Plasticity** This is the ability of a material to change its shape under load without cracking and retain its shape after the load is removed. Some of the examples of plastic materials are steel, copper, and hot bitumen.

**Ductility** This is the ability of a material (such as a metal or reinforced concrete) to undergo plastic deformation without fracture and is required for materials to resist earthquake loads. The lack of ductility is often termed as *brittleness*. Cast iron, stone, and brick and plain concrete are comparatively brittle materials. *Malleability* is a similar property, and is the ability of material to deform under pressure (compressive stress), without rupture. If malleable, a material may be flattened by hammering or rolling. Copper is the most malleable building material.

**Creep** This is defined as the deformation of structure under sustained load. More materials will creep or flow to some extent and eventually fail under a sustained stress less than the short-time ultimate strength. For example, when load is applied on a concrete specimen, it shows an instantaneous deformation followed by a slow increase of deformation over a period of time. Creep and long-time strength, at atmospheric temperatures, must sometimes be taken into account while designing the members of nonferrous metals and while selecting allowable stresses for wood, plastics, and concrete.

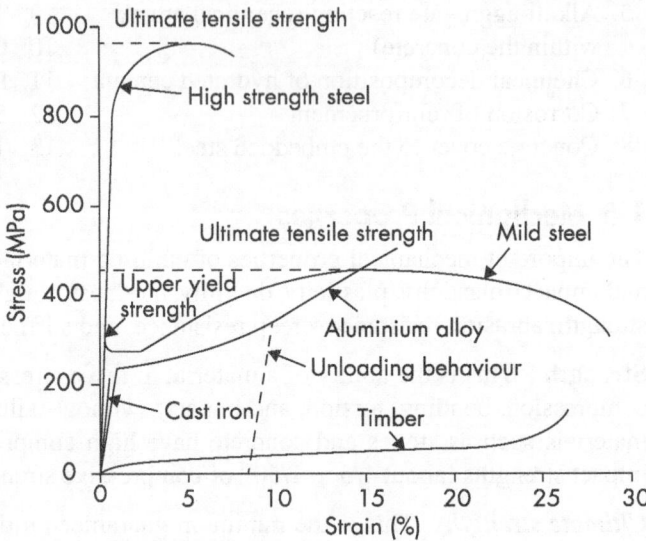

**Fig. 1.1** Typical tensile stress–strain curves for some structural materials

**Stiffness** This is a measure of the resistance offered by a member to deformation ($k = F/d$), and has a unit of Newton per meter (N/m). Stiff materials have high modulus of elasticity, hence, will result in small deformations for a given load. Stiffness is the reverse of flexibility.

**Fatigue strength** This is the highest stress that a material can withstand for a given number of cycles without breaking. A material has a tendency to fail at a lesser stress level when subjected to repeated loading, such as those occurring in steel bridges and cranes. Welding details also affect the fatigue strength.

**Impact strength** This is the capacity of the material to withstand a suddenly applied load, and expressed in terms of energy. It is often measured with the Izod impact strength test or the Charpy impact test, both of which measure the impact energy required to fracture a sample. It thus indicates the *toughness* of the material, which is the ability of a material to absorb energy when impacted. Stainless steels and titanium alloys are tough, whereas glass and ceramics are very *fragile* (opposite of tough). Hardness and toughness have an inverse relationship. For a particular solid, the toughness decreases when hardness increases.

## 1.6 Thermal Properties

The important thermal properties are thermal conductivity, thermal capacity, and fire resistance.

**Thermal conductivity** This is also called heat conductivity. This is the ability of a material to conduct heat, and is measured in Watts per metre-Kelvin [W/(m·K)]. It is influenced by the nature of material, its structure, porosity, character of pores, moisture content, and mean temperature at which heat exchange takes place. Higher the thermal conductivity, faster will be the heat transfer—it is usually measured by the $U$-value. Materials with large size pores have high heat conductivity because the air inside the pores enhances heat transfer. Moist materials have higher heat conductivity than dry materials. The lower the $U$-value of a material, the better will be its ability to resist heat conduction. This property is important while selecting insulating materials for walls of heated buildings, which should have negligible thermal conductivity. The reciprocal of thermal conductivity is called thermal resistivity. Designated as $R$ ($R$-value), thermal resistance indicates how effective any material is as an insulator. A higher $R$-value indicates a better insulating performance. It has to be noted that $U = 1/R$ and $R = 1/U$.

**Thermal capacity**   This may be defined as the ability of a material to store heat per unit volume, and is measured by the product of density and specific heat, with units of Joule per Kelvin (J/K). When a material shows greater thermal capacity, it shows that it can store more heat in a given volume for every degree of increase in temperature. Specific heat is a measure of the amount of heat required to raise the temperature of given mass of material by 1° [measured in J/(kg·K)]. It takes less energy input to raise the temperature of a low specific heat material than that of a high specific heat material. For example, it requires one calorie of heat energy to increase the temperature of water by 1°C. As water has a high heat capacity, it is sometimes used as thermal mass in buildings. Generally, materials with higher thermal capacity can reduce heat flow from the outside to the inside of buildings by storing the heat within the material. Thus, by using a material of adequate thermal capacity, the heat (produced by the Sun) entering a wall during the daytime could be stored within the wall itself for several hours, and conveniently made to flow out during the cool night hours. Table 1.3 shows the typical thermal properties of some building materials.

**Table 1.3** Thermal properties of some building materials

| Material | Modulus of elasticity, $E$ (GPa) | Poisson's ratio | Coefficient of thermal expansion, $\alpha$ ($\times 10^{-6}/°C$) | Thermal conductivity, $\lambda_a$ (W/mK) | Specific heat capacity, $c_a$ (J/kg · K) |
|---|---|---|---|---|---|
| Aluminium alloys | 72 | 0.33 | 23.5 | 56–205 | 900 |
| Concrete (M30) | 27.4 | 0.15–0.25 | 11 | 0.8–1.40 | 840–880 |
| Copper | 118 | 0.33 | 17.6 | 385 | 386 |
| Glass | 70 | 0.24 | 9.0 | 0.8–1.0 | 840 |
| Iron (grey cast) | 90 | 0.26 | 12.1 | 79.5 | 448 |
| Structural steel | 200–207 | 0.30 | 11–12 | 54 | 425 |
| Stainless steel | 193 | 0.30 | 16 | 14.6 | 450 |

**Fire resistance**   This is the ability of a material to resist the action of high temperature without any appreciable deformation and loss of strength. While a fire-resistant material is one that is designed to resist burning and withstand heat, fire-retardant materials are designed to burn slowly. Some examples of fire-retardant materials are: fire-retardant treated wood, brick, concrete, mineral wool, gypsum boards, and intumescent paint. Steel suffers considerable deformation and loss of strength under the action of high temperature. *Refractoriness* denotes the ability of a material to withstand the prolonged action of high temperature of about 1580°C without melting or losing shape.

## 1.7 Other Properties

Chemical properties include corrosion resistance, chemical composition, acidity, or alkalinity. Optical properties include the colour, light reflection, and light transmission. Acoustical properties comprise sound absorption, transmission, and reflection. Physiochemical properties include hygroscopicity and water absorption. Useful in construction, metallurgical properties are metal fusibility, weldability, hardening, and tempering. Some of these properties are briefly described here.

**Chemical resistance**   This is the capacity of any material to resist the deteriorating action of substance such as harmful acids, alkalis, seawater, and gases. For example, natural stones such as limestone, marble, and dolomite are affected even by weak acids, wood has low resistance to acids and alkalis, and bitumen disintegrates due to alkaline liquids.

**Hygroscopicity**  It is the property of a material to absorb water vapour from air, resulting in volume change (shrinkage or swelling). It is influenced by air temperature and relative humidity, and types, number and size of pores. Wood, concrete, brick, plaster, and several engineering polymers are hygroscopic, including nylon, polycarbonate, cellulose, and poly (methyl methacrylate). Hydrophobic materials are the opposites of hygroscopic materials and repel water; their typical examples include glass, metals, and plastics.

**Water absorption**  This denotes the ability of the material to absorb and retain water. It is expressed as percentage in weight or of the volume of dry material:

$$W_w = \frac{M_1 - M}{M} \times 100 \tag{1.10a}$$

$$W_v = \frac{M_1 - M}{V} \times 100 \tag{1.10b}$$

where $M_1$ = mass of saturated material (g), $M$ = mass of dry material (g), and $V$ = volume of material including the pores ($mm^3$). Water absorption by volume is always less than 100%, whereas that by the weight of porous material may exceed 100%.

The properties of building materials can change considerably when saturated. The water resistance of a material is expressed by the coefficient of softening, which is the ratio of compressive strength in water-saturated condition to that in dry condition. For example, the coefficient of softening for clay is zero, as it soaks readily in water. However, for materials like glass and metals, which are not affected by water, it is taken as one. In locations exposed to moisture (like roofing or foundations), materials with the coefficient of softening less than 0.8 should not be used.

**Corrosion resistance**  Formation of rust (iron oxide) in metals, when they are subjected to atmosphere, is called *corrosion*, and is a recurring problem in steel structures and steel reinforcements. To mitigate corrosion in steel structures, several methods are applied such as treatment of the environment to render it non-corrosive, coating/painting systems (surface preparation, such as sand blasting, plays an important role in the durability), galvanizing, thermal (metal) spraying, cathodic protection, and use of corrosion-resistant structural steels (e.g., weathering steel and stainless steel). For reinforcements in concrete, cathodic protection, corrosion-inhibiting admixtures, or the use of thermo-mechanically treated corrosion-resistant steel bars (TMT-CRS bars), fusion-bonded epoxy-coated rebars, galvanized rebars, stainless steel rebars, fibre-reinforced polymer bars, and basalt bars may be used.

Metallurgical, acoustical, and optical properties are discussed in detail in the relevant chapters. Properties of some materials used in building construction are given in Table 1.4.

**Table 1.4** Properties of some materials used in building construction

| Material | Density, kg/m³ | Stiffness (E), GPa | Poisson's ratio | Strength[1], MPa | Fracture toughness, MN/m³/² |
|---|---|---|---|---|---|
| Clay bricks | 1480–2400 | 14–18 | – | 5–110 (compression) | 0.6 |
| Timber[2] | 170–980 (dry) | 3–21 | 0.25–0.49 | 10–80 (tension) 15–90 (compression) | 7–13 |
| Structural steel | 7850 | 195–205 | 0.3 | 235–960 | 50–100 |
| Cast iron | 6900–7800 | 170 | 0.26 | 220–1000 | 20–50 |
| High strength steel | 7850 | 205 | 0.3 | 500–1900 | 50–125 |
| Aluminium alloys | 2700 | 70 | 0.33 | 80–505 | 25–35 |

*(Contd)*

**Table 1.4** *(Contd)*

| Material | Density, kg/m³ | Stiffness (E), GPa | Poisson's ratio | Strength[1], MPa | Fracture toughness, MN/m³/² |
|---|---|---|---|---|---|
| Concrete (M20–M100) | 1800–2500 | 22–50 | 0.15–0.25 | 2–12 (tension) 20–100 (compression) | 0.8–1.2 |
| Glass | 2500 | 70 | 0.24 | 40–70[3] | 0.8 |
| Rubber | 830–910 | 0.1–1 | 0.5 | 15–30 | – |
| Polyethylene (high-density) | 960 | 1.1 | 0.4–0.45 | 20–30 | 2.5 |
| Glass fibre composite | 1400–2000 | 6–50 | – | 40–1250[4] | 7–23 |
| Carbon fibre composite | 1500–1600 | 70–200 | – | 600–2000[4] | 6–80 |
| Epoxy resin | 1100–1400 | 2.6–3 | – | 30–100 | 0.5–1 |

1. In tension unless stated, yield or proof strength for metals, ultimate strength for other materials
2. Based on tests on small specimens, loaded parallel to grain
3. Modulus of rupture
4. In the longitudinal direction

## 1.8 Sustainable Materials

The definition of sustainability, suggested by the then Prime Minister of Norway, Gro Bruntland, in 1987, is *meeting the needs of the present without compromising the ability of future generations to meet their needs.* Sustainability is thus is a realization that today's population is merely borrowing resources and environmental conditions from future generations. The greatest threats to the sustainable development on earth are as follows:

1. Population growth and urbanization
2. Energy use and global warming
3. Excessive waste generation and the subsequent pollution of soil, air, and water
4. Transportation in cities
5. Limited supply of resources

Many of them are interrelated and discussed in the work of Subramanian (2016).

The materials we use for construction affect the environment. Their production and transportation deplete natural resources, consume considerable energy, and pollute the environment. Several building materials, and the energy needed to produce them, are becoming scarce. If the present trends continue, some of the common raw materials and energy sources (like oil and natural gas) will be exhausted within about the next century. As per www.msci.org, the grades of mined copper and iron ore are declining and the natural reserves of lead, molybdenum, chromium, nickel, copper, zinc, tin, and radium are depleting. There is an urgent need to use alternate materials—for example, using M-sand in the place of natural sand—to preserve natural resources.

A reclassification of all building materials and products based on sustainability, and to meet criteria pertaining to personal health and health of environment, is necessary. Traditional materials like clay, lime, and stone are still abound, and timber (especially softwoods) can be replenished by properly managed forestation. In addition, these materials can be easily reused or recycled; they produce little or no pollution and are reabsorbed into the natural cycles of environment, when they are discarded. Recycling materials like steel and aluminium also preserves the natural resources and saves considerable energy.

### 1.8.1 Healthy Materials

In order to satisfy the criteria of being healthy to human beings, Pearson (1998) has suggested the following:

1. The materials should be clean and should not contain pollutants or toxins, emit no biologically harmful vapours, dust, particles or be odours either during manufacture or usage. They should also be resistant to bacteria, viruses, moulds, and other harmful microorganisms.
2. They should not be radioactive and must not emit any harmful levels of radiation.
3. They should not be electromagnetic and should not allow the conduction or built-up of static electricity or emit harmful electric fields of any type.
4. They should have good sound-reduction properties and should not themselves produce any noise.

### 1.8.2 Ecological Materials

For materials to be environment friendly, they should satisfy the following criteria (Pearson, 1998):

1. They should be renewable and abundant, and coming from diverse natural sources. Their production should have a low impact on the environment.
2. They should be non-polluting and should not emit harmful vapours, particles, or toxins into the environment, either during manufacture or usage.
3. They should be energy efficient, and use low energy in production, transport and should generally be available locally (see also Section 1.8).
4. They should be durable and easy to maintain and repair—additionally, it is better if they are tried and tested over several generations, as in the case of natural materials.
5. They should produce less waste during production and be capable of being reused and recycled, so that the vast amount of energy spent on processing raw materials could be saved.

**Fig 1.2** Green Seal and EU Ecolabel

Many countries now have a system of labelling environment-friendly products. In the USA, since 1989, a non-profit organization called Green Seal identifies products and services that have less impact on our health and environment and awards a 'Green Seal' as shown in Fig. 1.2 (www .greenseal.org). Similar 28 international eco-labelling programmes exist, including Germany's Blue Angel, the European Union's Ecolabel, and the Nordic Swan, and all are members of the Global Eco-labelling Network (GEN).

### 1.9 Green Building Rating Systems

To promote the design and construction practices that reduce the negative environmental impacts of buildings and improve occupant health and well-being, the US Green Building Council (USGBC), a Washington D.C.-based non-profit coalition of building industry leaders, developed the LEED® green building rating system in 1993. In the USA and in a number of other countries around the world, LEED® certification is the recognized standard for measuring building sustainability. Similar assessment systems are available in other countries also. Some of these are

1. the British green building rating system developed by Building Research Establishment (BRE) in 1992 called the Building Research Establishment Environmental Assessment Method (BREEAM),
2. Griha in India (info@grihaindia.org),

3. the comprehensive Assessment System for Building Environmental Efficiency (CASBEE) of Japan,
4. Green Star of Australia, and
5. Green Globes, which is a web-based, interactive learning tool developed from BREEAM to the needs of US commercial buildings.

All these systems are designed to encourage the construction of green buildings, which will minimize the disruption of local ecosystems; ensure the efficient use of water, energy, and other natural resources; and ensure a healthy indoor environment. However, they differ in terminology, structure, assessment of performance, points assigned to different performance criteria, and documentation required for the certification. These systems, while voluntary in nature, continue to gain recognition. It is interesting to note that adoption of these systems also results in economic incentives, as owners and renters are increasingly demanding facilities with high green building ratings.

## 1.9.1 LEED-NC

From 1994 to 2009, LEED® grew from one standard for new construction to a comprehensive system of six interrelated standards covering all aspects of the development and construction process: LEED-NC, for New Construction; LEED-EB, for Existing Buildings; LEED-CI, for Commercial Interiors; LEED-H, for Homes; LEED-CS, for Core and Shell projects; and LEED-ND for Neighbourhood Development (Kibert, 2005). LEED-NC, which was originally developed for office buildings but is now being used for all types of buildings except single family homes, is briefly discussed here.

LEED-NC 2009 is structured with eight prerequisites and a maximum of 110 points. These points are divided into the following seven major categories:

1. Energy and atmosphere (35 maximum points)
2. Indoor environmental quality (15 points)
3. Sustainable sites (26 points)
4. Materials and resources (14 points)
5. Water efficiency (10 points)
6. Innovation and design process (6 points)
7. Regional priority (4 points)

A building is LEED® certified if it obtains at least 40–49 points. Silver, gold, and platinum levels are awarded for 50–59, 60–79, and greater than 80 points, respectively (see Table 1.5). Note that LEED is continuously evolving and improving. The recent update to the rating system LEED-NC 4.0 was launched in 2014. LEED-NC 4.0 has the following six main credit categories (Subramanian, 2017):

1. Location & Transportation (LT)
2. Sustainable Sites (SS)
3. Water Efficiency (WE)
4. Energy & Atmosphere (EA)
5. Materials & Resources (MR)
6. Indoor Environmental Quality (EQ)
7. Innovation and Design Process
8. Regional Priority

**Table 1.5** Overview of LEED-V4 categories and credits

| Integrative Process: 1 Possible Point | |
|---|---|
| Location and Transportation: 16 Possible Points | Required |
| Credit 1 Sensitive Land Protection | 1 |
| Credit 2 High Priority Site | 2 |
| Credit 3 Surrounding Density and Diverse Uses | 5 |
| Credit 4 Access to Quality Transit | 5 |
| Credit 5 Bicycle Facilities | 1 |
| Credit 6 Reduced Parking Footprint | 1 |
| Credit 7 Green Vehicles | 1 |

*(Contd)*

**Table 1.5** (*Contd*)

| Sustainable Sites: 10 Possible Points | |
|---|---|
| Prerequisite 1 Construction Activity: Pollution Prevention | Required |
| Credit 1 Site Assessment | 1 |
| Credit 2 Site Development - Protect or Restore Habitat | 2 |
| Credit 3 Open Space | 1 |
| Credit 4 Rainwater Management | 3 |
| Credit 5 Heat Island Reduction | 2 |
| Credit 6 Light Pollution Reduction | 1 |
| **Water Efficiency: 11 Possible Points** | |
| Prerequisite 1 Outdoor Water Use Reduction | Required |
| Prerequisite 2 Indoor Water Use Reduction | Required |
| Prerequisite 3 Building-Level Water Metering | Required |
| Credit 1 Outdoor Water Use Reduction | 2 |
| Credit 2 Indoor Water Use Reduction | 6 |
| Credit 3 Cooling Tower Water Use | 2 |
| Credit 4 Water Metering | 1 |
| **Energy and Atmosphere: 33 Possible Points** | |
| Prerequisite 1 Fundamental Commissioning and Verification | Required |
| Prerequisite 2 Minimum Energy Performance | Required |
| Prerequisite 3 Building-level Energy Metering | Required |
| Prerequisite 4 Fundamental Refrigerant Management | Required |
| Credit 1 Enhanced Commissioning | 6 |
| Credit 2 Optimize Energy Performance | 18 |
| Credit 3 Advanced Energy Metering | 1 |
| Credit 4 Demand Response | 2 |
| Credit 5 Renewable Energy Production | 3 |
| Credit 6 Enhanced Refrigerant Management | 1 |
| Credit 6 Green Power and Carbon Offsets | 2 |
| **Materials and Resources: 13 Possible Points** | |
| Prerequisite 1 Storage & Collection of Recyclables | Required |
| Prerequisite 2 Construction & Demolition Waste Management Planning | Required |
| Credit 1 Building Life-Cycle Impact Reduction | 5 |
| Credit 2 Building Product Disclosure & Optimization - Environmental Product Declarations | 2 |
| Credit 3 Building Product Disclosure and Optimization - Sourcing of Raw Materials | 2 |
| Credit 4 Building Product Disclosure & Optimization - Material Ingredients | 2 |
| Credit 5 Construction & Demolition Waste Management | 2 |
| **Indoor Environmental Quality: 15 Possible Points** | |
| Prerequisite 1 Minimum IAQ Performance | Required |
| Prerequisite 2 Environmental Tobacco Smoke (ETS) Control | Required |

(*Contd*)

**Table 1.5** (Contd)

| | |
|---|---|
| Credit 1 Enhanced Indoor Air Quality Strategies | 2 |
| Credit 2 Low-emitting Materials | 3 |
| Credit 3 Construction Indoor Air Quality Management Plan | 1 |
| Credit 4 Indoor Air Quality Assessment | 2 |
| Credit 5 Thermal Comfort | 1 |
| Credit 6 Interior Lighting | 2 |
| Credit 7 Daylight | 3 |
| Credit 8 Quality Views | 1 |
| Credit 9 Acoustic Performance | 1 |
| **Innovation and Design Process: 6 Possible Points** | |
| Credit 1 Innovation in Design | 5 |
| Credit 2 LEED® Accredited Professional | 1 |
| **Regional Priority: 4 Possible Points** | |
| Credit 1 Regional Priority | 1–4 |

**Project Total:** 100 base points; 6 possible Innovation in Design and 4 Regional Priority points
*Certified:* 40–49 Points; *Silver:* 50–59 Points; *Gold:* 60–79 Points; *Platinum:* > 80 Points

(*Source*: www.usgbc.org)

Figure 1.3 shows the views of a LEED Platinum certified building in Rockville, MD, USA.

**Fig. 1.3** Views of the LEED Platinum certified building in Rockville, MD, USA

A part of the Confederation of Indian Industry (CII), the Indian Green Building Council's (IGBC) Green Building rating systems were launched in 2003. CII–Sohrabji Godrej Green Business Centre building in Hyderabad was the first to receive the prestigious Platinum rated green building rating in India (see Fig. 1.4). Since then, the rating systems have been successfully applied to more than 4025 buildings, with a foot-print of 4.50 billion square feet. It is given under the following 16 different categories (https://igbc.in):

1. IGBC Green New Buildings
2. IGBC Existing Buildings
3. IGBC Green Homes
4. IGBC Green Residential Societies
5. IGBC Green Healthcare
6. IGBC Green Schools
7. IGBC Green Factory Buildings
8. IGBC Green Data Centres
9. IGBC Green Campus
10. IGBC Green Villages

11. IGBC Green Townships
12. IGBC Green Cities
13. IGBC Green SEZ
14. IGBC Green Landscapes

15. IGBC Green Mass Rapid Transit System
16. IGBC Green Existing Mass Rapid Transit System

The task of selecting building materials and products, for a high-performance green building, is the most difficult and challenging task for any design team. Several tools are available for this process and one best tool is the *life-cycle assessment* (LCA). LCA provides information about the resources, emissions, and other impacts resulting from the life cycle of material use. Hence, one must consider the impact of the material from extraction to disposal. One such LCA programme is Building for Environmental and Economic Sustainability software (BEES-NIST). Ideally, the material cycle should be a closed looped and waste free. Thus, the following rules apply while selecting the materials for green construction:

**Fig. 1.4** Platinum-rated CII–Sohrabji Godrej Green Business Centre, Hyderabad (*Source*: CII–IGBC)

1. They should consume least energy to manufacture.
2. They should not involve long-distance transportation (for the raw materials as well as finished product).
3. The natural resources and raw materials do not affect the environment.
4. They must be easy to recycle and safe to dispose in landfills.
5. Materials should be harmless in production and use.
6. Materials dissipated from recycling must be harmless.
7. They should have long life and durability.
8. Buildings must be de-constructible.
9. Building components must be easy to disassemble.

It may be difficult to identify a material that obeys all the aforementioned rules. Especially, the last rule of disassembly has not been considered in traditional building materials, except prefabricated steel structures. Disassembly also discourages the use of composite materials. It has been shown that by using concrete, one can earn 37–62 LEED® points (see Table 1.6).

**Table 1.6** Summary of possible points to increase LEED ratings of buildings

| Category | Total number of points | Points earned using concrete |
|---|---|---|
| Sustainable sites | 26 | 12 |
| Water efficiency | 10 | 10 |
| Energy and atmosphere | 35 | 1–19 |
| Materials and resources | 14 | 9 |
| Indoor environmental quality | 15 | 2 |
| Innovation credits | 6 | 2–6 |
| Regional priority | 4 | 1–4 |
| **Total** | **110** | **37–62** |

(*Source*: Adapted from RMC-Guide, 2010)

Green buildings adopt various strategies for water management: using low flow or ultra-low flow plumbing fixtures, electronic controls and fixtures, substitution of alternative water sources (rainwater, reclaimed water, and grey water) for potable water, rainwater harvesting, xeriscaping, and use of other technologies and approaches that result in the reduction of potable water consumption (Kibert, 2005).

## 1.10 Embodied Energy and Energy Efficiency

Energy from fossil fuels is becoming scarce and the amount used in the production and transportation is high. As mentioned earlier, the best materials are those which are energy efficient (using low energy in production, transport, and use), need minimum processing and are available locally. *Embodied energy* refers to the total energy consumed in the acquisition and processing of materials, including manufacturing, transportation, and final installation. Products with greater embodied energy usually have higher environmental impact due to the greenhouse gas emissions associated with their energy consumption. However, a true indicator of environmental impact will be obtained, only when the embodied energy is divided by the number of times the product is used or recycled. Thus, aluminium may have low embodied energy per time in use, as it is very durable. Similarly, recycled aluminium and steel have less than 10–20% of embodied energy, compared to original steel or aluminium made from ores.

Typical embodied energy of some common building materials is shown in Table 1.7. From this, it is seen that the embodied energy of locally grown and reclaimed timber is low (Haseltine, 1975). Clay used as adobe or unbaked brick is another example of material requiring low energy. In contrast, synthetic and processed products such as plastics, aluminium, steel, glass, and oven-fired bricks and clay tiles, have higher embodied energy (Pearson, 1998). In addition, the materials used should be good energy conservers with high insulation value that should retain heat in winter and keep the building cool in summer.

**Table 1.7** Embodied energy of common building materials (http://www.yourhome.gov.au)

| Material | Embodied energy (MJ/kg) |
|---|---|
| | **Very high energy** |
| Aluminium | 227.0 |
| Carpet (synthetic) | 148.0 |
| Polystyrene insulation | 117.0 |
| Synthetic rubber | 110.0 |

*(Contd)*

**Table 1.7** (*Contd*)

| Material | Embodied energy (MJ/kg) |
|---|---|
| **Very high energy** | |
| Linoleum | 116.0 |
| Stainless steel | 100+ |
| Plastics (general) | 90.0 |
| PVC (polyvinyl chloride) | 80.0 |
| Copper | 70–100 |
| Acrylic paint | 61.5 |
| Zinc | 51.0 |
| **High energy** | |
| Steel | 32.0 |
| Galvanized steel | 38.0 |
| Hardboard | 24.2 |
| Glass | 12.7 |
| Glue-laminated timber | 11.0 |
| Plywood | 10.4 |
| Cement | 5.6 |
| Steel (recycled) | 8.9 |
| Aluminium (recycled) | 8.1 |
| Particleboard | 8.0 |
| **Medium energy** | |
| Autoclaved aerated concrete (AAC) | 3.6 |
| Kiln dried sawn softwood | 3.4 |
| Lime | 3-5 |
| Gypsum plaster | 2.9 |
| Clay bricks and tiles | 2.5 |
| Lumber | 2.5 |
| Kiln dried sawn hardwood | 2.0 |
| Precast steam-cured concrete | 2.0 |
| Concrete (30 MPa) | 1.3 |
| Concrete blocks | 1.5 |
| **Low energy** | |
| Stabilized earth | 0.7 |
| Fly ash, rice husk ash (RHA), volcanic ash | <0.5 |
| Sand, aggregate | 0.1 |

These figures should be used with caution because:

1. The actual embodied energy of a material manufactured and used in one location may be very different from the same material transported by road to another location.
2. Though materials like stainless steel have high embodied energy value, they are recycled many times, reducing their life cycle impact.

As discussed in Section 1.8, it is not enough to consider the energy requirement of material during the production stage; *a life-cycle assessment approach* is necessary to determine its environmental impact. Thus, all stages in the life of a product should be analysed, that is, raw material acquisition, manufacture, transportation, installation, use, recycling, and waste management. There is also a *Life-cycle Cost Analysis* (LCCA), which deals with the cost impact of a product or material, but does not deal with the environmental impact. Both have important roles to play in sustainability assessment of a material/product.

## 1.11 Comparison of some Major Building Materials

Though the various building materials are covered individually in the subsequent chapters, the four major building materials, which are used extensively, are briefly discussed here.

**Masonry** Masonry may be defined as the work done by masons using individual units, which are often laid in and bound together by mortar; the term masonry can also refer to the units themselves. The common materials of masonry construction are brick, stone, concrete block, glass block, and cob (made of soil, water, straw, and sometimes with lime).

Some important features include:

1. Different types of bricks and concrete blocks exist (see Chapter 3). The process of manufacturing of bricks from clay involves preparation of clay, moulding, and then drying and burning of bricks in kilns, at temperature of about 1100°C, to give them their final hardness and appearance.
2. The vertical shaft brick kiln (VSBK) technology, which consumes less fuel and energy and emits lower suspended particulate matter (SPM), as compared with Bull's Trench Kiln (BTK), is suitable for medium sized kilns. VSBK technology was introduced during 1996; at present, there are more than 40 operational VSBKs in India (www.teriin.org).
3. Masonry is mainly used for load bearing walls and walls taking in-plane or transverse loads. It is durable, fire resistant, and aesthetically pleasing. It can be used for buildings with moderate heights, that is, up to 20 storeys. (Unfortunately, the bricks produced in India do not have uniform quality and the bricks produced in south India have low strength. Hence, buildings with load-bearing brick masonry are built only up to three to four floors).

**Reinforced and prestressed concrete** Concrete is a composite material made of two or more sizes of aggregates (generally with gravel and sand), with a binding medium of Portland cement and water. After mixing, it is placed into moulds (called formwork) with proper compaction, and after proper curing, the cement hydrates and eventually hardens into a stone-like material. The following are some of its features:

1. As concrete is weak in tension, it is usually strengthened with steel bars (known as reinforcement or rebars). This strengthened concrete is called as *reinforced concrete* or RCC.
2. Currently, concrete is the predominant building material in India and several other countries. Reinforced concrete framed or shear wall construction, if properly poured and cured, is very durable and fire resistant.
3. Since reinforced concrete can be cast to any required shape, it is used for a variety of constructions including tall buildings and floors and foundation of all types of buildings.
4. Substitution of cement by several wastes such as fly ash, ground-granulated blast furnace slag, silica fume, reactive rice-husk ash, etc., can lead to significant reductions in the amount of cement needed to make concrete, hence reduces emissions of $CO_2$ and consumption of energy and raw materials, and results in reduced landfill/disposal burdens.

5. Recent advancements such as self-compacting concrete (SCC), though having more binder content, reduce manpower, as SCC is easier to place even in congested structural members, has reduced noise level, and is environment friendly.

*Prestressed concrete* is a concrete in which high strength reinforcing steel bars are stretched and anchored to compress it, thus increasing its resistance to stress. It is used for floor construction of large span structures and in buildings, bridges, and towers.

Some of the shortcomings are as follows:

1. In India, though concrete is used extensively in all types of construction, except by a small number of big companies, quality control is not exercised during mixing of concrete.
2. Moreover, the curing of concrete is mostly ignored or not done properly for the code prescribed duration.
3. In addition, the steel reinforcements (especially the smaller diameter rods) available in the market are produced by re-rollers and do not possess the required ductility and strength.
4. Since concrete can be mixed and poured to any required shape, it is misused by several small contractors, who do not give much importance to design or detailing.

The aforementioned factors have led to the deterioration of several concrete structures all over the country and resulted in the failure of several concrete structures in the recent earthquakes. Since prestressed concrete is used in major constructions, as well as used by major contracting companies, the quality of prestressed concrete in India is up to the standards.

**Structural steel**  Most metals used for construction purposes are alloys. For example, structural steel is a primary alloy of iron and carbon (0.10–0.25%). The properties of steel vary widely, depending on its alloying elements. Some of its features are as follows:

1. Steel is made using the basic oxygen steel making (BOS) process or the electric arc method.
2. Raw materials such as iron ore, scrap steel, coke, limestone, and dolomite are charged in a blast furnace, and heated up to 1600°C, and oxygen of greater than 99.5% purity is blown into the mix.
3. The liquid steel is solidified into large blocks called ingots and then rolled into semi-finished products and then into plates, structural shapes, bars, etc.
4. The main advantages of steel are strength, speed of erection, prefabrication, and demountability.
5. They are used in load-bearing frames in buildings, and as members in trusses, bridges, and space frames. Steel, however, requires fire and corrosion protection.
6. Steel is also used in conjunction with concrete in composite constructions and in combined frame and shear-wall constructions.
7. In many cases, the fabrication of steel members is done in the workshop and the members are then transported to the site and assembled.
8. Tolerances specified for steel fabrication and erections are small compared to reinforced concrete structures.
9. Different steel sections are joined by welding or bolting. Welding or tightening of high strength friction grip bolts requires proper training.
10. Due to these factors, steel structures are often handled by trained persons and assembled with proper care, resulting in structures with better quality.
11. Compared with concrete, steel offers much better compressive and tensile strength, and enables lighter constructions. Unlike masonry or reinforced concrete, steel can be easily recycled.

**Wood**  This imparts natural, human warmth that steel and concrete lack. Due to this, wood has long been used for housing (up to three floors) and for historical structures in western countries such as the

USA, the UK, Germany, France, and Japan where there is cold climate. However, with the development of wood composites—thin, pressed sheets—combined with joints and steel frames have changed the scene. Glued laminated wood has been used in a number of large span structures. Prominent wood composite structures are Tacoma Dome and North Michigan University stadium in the United States and Odate Jukai Dome in Japan. All these domes have diameters in the range of 160–180 m. Since wood is a natural product, it is environment friendly, though the resins used in glued laminated wood many contain harmful chemicals. However, not all woods can be used for constructions and quality wood is in short supply in India and hence wood is used in India only for doors and windows. (Nowadays, even they are replaced with aluminium, steel, ferrocement, or plastic doors and windows.)

Important structural properties of steel, concrete, and wood are compared in Table 1.8.

**Table 1.8** Important properties of steel, concrete, and wood

| Item | Mild steel | Concrete* M20 grade | Wood |
|---|---|---|---|
| Unit mass, kg/m$^3$ | 7850 (100) | 2400 (31)[#] | 290–900 (4–11) |
| Maximum stress in MPa Compression Tension Shear | 250 (100) 250 (100) 144 (100) | 20 (8) 3.13 (1) 2.8 (1.9) | 5.2–23[+] (2–9) 2.5–13.8 (1–5) 0.6–2.6 (0.4–1.8) |
| Young's modulus, MPa | $2 \times 10^5$ (100) | 22,360 (11) | 4600–18,000 (2–9) |
| Coefficient of linear expansion $\times 10^{-6}$/°C | 12 | 10 to 14 | 4.5 |
| Poisson's ratio | 0.3 | 0.20 | 0.20 |

\* Characteristic compressive strength of 150 mm cubes at 28 days
[+] Parallel to grain
[#] Relative value when compared to steel

## 1.12 Alternative Building Materials

A number of industrial, agricultural, and mining wastes are used in the production of alternative building materials (Venkatarama Reddy, 2004). The industrial wastes include: fly ash, phosphogypsum, blast furnace slag, silica fume, alumina red mud, slate and marble waste, glass powder, paper-mill pulp, sludge, and discarded tires. Fly ash is used in the production of Portland-pozzolana cement, lime-fly ash bricks, fly ash-lime-gypsum (Fal-G) concrete, fly ash-lime cellular concrete, and sintered fly ash lightweight aggregate. However, the response by the Indian building community to the use of fly ash is poor, as only 5% of generated fly ash is used in India. It has to be noted that not all the available fly ash is suitable for use in concrete and other products.

Calcium silicate bricks are manufactured from a mixture of sand and/or siliceous waste and a small portion of lime, which is mechanically pressed and autoclaved. These bricks can also solve the problem of waste disposal.

The agricultural wastes that could be used in building products such as roofing units, thermal insulating materials, and walling boards are bagasse, jute stalks, groundnut hulls, hemp, flax, reed, natural wool, expanded cork, and straw bales. Rice husk ash, saw dust, cork granules, and coconut kernel are used as substitutes in concrete.

In addition, several substitutes for teak, rosewood, and white cedar, such as secondary species of timber, poly-vinyl chloride (PVC), mild steel and galvanized steel, aluminium, precast concrete, ferrocement, particle boards, fibre boards, fibre glass, and glass reinforced gypsum composite boards, have been tried. The use of some of these waste products and other substitutes as building materials are discussed in the respective chapters (3, 5–7, 9, and 11).

## 1.13 Building Codes and By-Laws

A building code contains a set of rules that knowledgeable people recommend for others to follow. It is not a law, but can be adopted into law. National building codes contain administrative regulations, development control rules, and general building requirements; fire safety requirements, stipulation regarding materials, structural design and construction, building and plumbing services, landscape development, guidelines for sustainability, asset and facility management, and other aspects of buildings to ensure safety and health for people living in or around buildings (see SP 7-2016, SP 21:2005, SP 41-1987, IS 3362:1977, IS 3792:1978, and IS 6060:1971). They are mandatory in nature and serve to protect buildings against fire, earthquake, noise, structural failures, and other hazards (IS 1641:1988, IS 1950:1962, and IS 2526:1963). Building by-laws usually contain provisions for parking, peripheral open spaces including set-backs, disaster management, and fire safety. They may also contain green building and sustainability provisions, rainwater harvesting, wastewater reuse, and recycle and installation of solar roof-top photovoltaic (PV) norms. Recent by-laws also have guidelines for mitigating electromagnetic radiations (Model Building Bye-Laws, 2016).

In addition, the Energy Conservation Building Code (ECBC) was launched by the Ministry of Power, Government of India in May 2007, as a first step towards promoting energy efficiency in the building sector (ECBC, 2006). The ECBC provides design norms for:

1. Building envelope, including thermal performance requirements for walls, roofs, and windows;
2. Lighting system, including day lighting, and lamps and luminary performance requirements;
3. HVAC system, including energy performance of chillers and air distribution systems;
4. Electrical system; and
5. Water heating and pumping systems, including requirements for solar hot-water systems.

## SUMMARY

- The major requirement of any building is to shelter its occupants from the environment and offer a pleasant, comfortable, and healthy indoor environment.
- Building materials are categorized based on the source of availability as natural (e.g., clay, sand, stone, lime, and wood logs) and manufactured (e.g., concrete, metals, glass, ceramics, and plastics).
- The different properties which are of concern are physical, mechanical, thermal, chemical, optical, acoustical, physiochemical, and metallurgical.
- The physical properties include the shape, size, density/bulk density, specific gravity, and porosity of the material.
- The important mechanical properties are strength, elasticity, plasticity, ductility, hardness, toughness, malleability, brittleness, fatigue, impact strength, abrasion resistance, creep resistance, and stiffness/flexibility.
- Production and transportation of building materials deplete natural resources, consume considerable energy, and pollute the environment. Several materials are also becoming scarce.
- In order to have sustainable development, it is important to preserve these materials, by reducing their use (by using alternative materials), recycling them after use, or reusing them. In addition, the materials used should be environment friendly and not affect the health of the occupants. A number of industrial, agricultural, and mining wastes are now used to provide alternative building materials.
- Several rating systems have been developed, such as LEED, BREEAM, Green Globes, and CASBEE. For a building to be LEED® certified, it should obtain at least 40–49 points.
- Embodied energy is the total energy consumed in the mining, manufacturing, transportation, and final installation. Products with greater embodied energy have higher environmental impact.
- A building code, like the NBC of India, contains a set of rules that should be followed to have better functioning.

## EXERCISES

### Multiple-choice Questions

1. Density of concrete is about
   (a) 2000 kg/m$^3$
   (b) 2400 kg/m$^3$
   (c) 3200 kg/m$^3$
   (d) 7000 kg/m$^3$

2. The modulus of elasticity of steel is about
   (a) 150 GPa
   (b) 200 GPa
   (c) 220 GPa
   (d) 240 GPa

3. Match the following:
   (a) BREEAM
   (b) LEED
   (c) CASBEE
   (d) Green Star
   (e) IGBC
   (i) The USA
   (ii) India
   (iii) Australia
   (iv) Japan
   (v) The UK

4. As per LEED-NC, the combination of energy and atmosphere has the following possible points:
   (a) 26
   (b) 10
   (c) 35
   (d) 5

5. As per LEED-NC, sustainable sites has the following possible points:
   (a) 26
   (b) 10
   (c) 35
   (d) 5

6. To get a Gold rating in LEED-NC, one has to obtain the following points:
   (a) 40–49
   (b) 50–59
   (c) 60–79
   (d) 80–100

7. To get a Silver rating in LEED-NC, one has to obtain the following points:
   (a) 40–49
   (b) 50–59
   (c) 60–79
   (d) 80–100

8. Which of the following materials has the maximum value of embodied energy?
   (a) Steel
   (b) Copper
   (c) Aluminium
   (d) Zinc

9. Which of the following materials has the lowest value of embodied energy?
   (a) Recycled steel
   (b) Concrete
   (c) Recycled aluminium
   (d) Cement

### Review Questions

1. What are the functions of a building?
2. How are building materials classified?
3. How are the properties of building materials classified?
4. List and define the different physical properties of building materials.
5. Define the following:
   (a) Density
   (b) Bulk density
   (c) Density index
   (d) Specific weight
   (e) Porosity
   (f) Void ratio
6. List the mechanical properties of building materials.
7. What are the thermal properties of materials?
8. What are the chemical, optical, acoustical, physiochemical, and metallurgical properties of materials?
9. What are the factors influencing the choice of a building material?
10. Write short notes on the following:
    (a) Ductility
    (b) Thermal conductivity
    (c) Selection of building materials
    (d) Fire-resistant and fire-retarding materials
    (e) Corrosion resistance
    (f) Durability
11. Define sustainability. Why is it important to consider sustainability while selecting a building material?
12. When is a product or material considered sustainable? Illustrate with an example.
13. What are healthy materials?
14. When are materials considered as ecological materials?
15. List any three green-labelling systems.
16. Name any three green building rating systems
17. Describe the LEED-NC rating system briefly.
18. What is embodied energy? Is it the true indicator of environmental impact?

**19.** Briefly describe four important building materials.

**20.** What are the alternative or substitute building materials. Name five such materials and describe two of them briefly.

**21.** Why is it important to have standards for building materials?

## ANSWERS

**Multiple-choice Questions**

| | | |
|---|---|---|
| **1.** (b) | **2.** (b) | **3.** (a)–(v), (b)–(i), (c)–(iv), (d)–(iii), (e)–(ii) |
| **4.** (c) | **5.** (a) | **6.** (c) |
| **7.** (b) | **8.** (c) | **9.** (b) |

# CHAPTER 2
# BUILDING STONES

## 2.1 Introduction

*Dimension stone* or simply *building stone* can be defined as a natural rock material quarried for the purpose of obtaining blocks or slabs that are worked into a specific size and shape for the purpose of using them in masonry, tiles, and other construction. Rocks and minerals have been forming since 4.2 billion years, and the process continues even today, at the earth's surface, in the crust (generally, in hills), on the ocean floor, and in the mantle deep below. Rock-forming minerals (which are naturally occurring, solid inorganic substances) grow or cement together to form rocks. *Feldspar* is one of the most abundant minerals found in many types of rocks. Other examples of rock-forming minerals are quartz, olivine, mica, dolomite, and magnetite. Before the extensive use of concrete, stones were preferred for the construction of bridge piers, seaside and harbour walls, and walls of buildings. They are still used where they are freely available. When foundations are liable to be flooded, stonework is preferred compared to brickwork, as submerged bricks may disintegrate with time.

Most of the prehistoric forts, monuments, and temples, which were built all over the world and have sustained themselves even today, were built out of stone. Old roads with heavy traffic were also paved with stone. The Taj Mahal of India, the Coliseum in Rome, Italy, the Great Pyramid in Giza, Egypt, and the Great Wall of China are a few examples, which stand as a testimony to the durability of stones. They show that stone is more permanent than other building materials, such as brick or wood. Figure 2.1 shows the 169 m tall Washington monument in Washington D.C., USA, which is the tallest stone structure in the world. Its construction was completed in 1885. It is made of marble, granite, and bluestone gneiss.

In India, stone quarries are mainly located in the states of in Rajasthan, Madhya Pradesh, Andhra Pradesh, and few locations in Gujarat, Odisha, Karnataka, Tamil Nadu, and the Andaman and Nicobar Islands. Granite resources are largely located in south India and marble deposits in western India (Rajasthan and Gujarat). World granite and marble production was estimated at 142 MT in 2013. The top five stone-producing countries in 2016, in the descending order by tonnage, were China, Turkey, India, Iran, and Italy; they accounted for about 72% of the world's granite and marble production (https://minerals.usgs.gov).

**Fig. 2.1** The 169 m tall Washington monument is the world's tallest stone structure

The various types of rocks from which building stones are usually derived are granite, basalt, limestone, slate, shale, sandstone, travertine, and marble. Stone is used as building and decorative stone, and to produce aggregates, which are used mainly in concrete. Stone has been extensively used in load-carrying units as well as to enhance the beauty and elegance of any structure. The governing factors of using stone for structural purposes are its cost, adaptability to sculptural treatment, ornamental value, and durability. Limestone is used to produce lime and cement. Traditionally, building stone has been used in places such hilly regions, where it is freely available. At these places, as stones occur naturally and need not be manufactured, stone masonry may work out to be cheaper than brickwork.

As a building material, stone has gradually lost its importance due to the increased use of reinforced cement concrete (RCC). Other major factors that prevent its wider use are its time-consuming dressing operations, its non-availability in plains, and the difficulty of its transportation due to its weight. More-over, the strength of the structural elements built with stones cannot be rationally analysed.

## 2.2 Classification of Rocks

Rocks are composed of mineral grains (e.g., olivine, feldspar, and pyroxene) that grow or cement together. Rocks are seldom made of homogenous materials. The rock-forming minerals are relatively few in comparison to the overall number of known minerals. In general, a rock is made up of an infinite number of crystals of one or more minerals. A mono-mineral rock contains only one major mineral. For example, rocks, such as marble or quartzite, mostly contain calcite/dolomite and quartz, respectively. Granite consists of the minerals, such as feldspar, quartz, and mica. The rocks may be classified according to their geological formation, physical characteristics, chemical composition, and hardness.

**Geological classification** Geologists classify rocks into three types on the basis of how they are formed – *igneous*, *sedimentary,* and *metamorphic*. It has to be noted that over thousands of years, rocks can change from one type to another – from igneous to sedimentary and to metamorphic and back to igneous – as described below. This process is called the *rock cycle* (see Fig. 2.2).

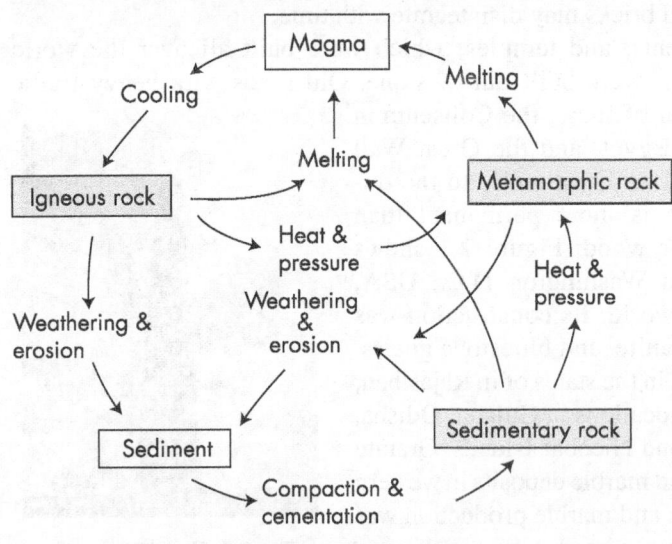

**Fig. 2.2** The rock cycle

1. Igneous rocks can change into sedimentary rocks, metamorphic rocks, or even different igneous rocks. The process of change involves melting and crystallization.
2. Sedimentary rocks can change into metamorphic rocks, igneous rocks, or even different sedimentary rocks. The process of change involves weathering, erosion, deposition, compaction and cementation.
3. Metamorphic rocks can change into sedimentary rocks or igneous rocks. They can also change into different metamorphic rocks; however, this is not shown in Fig. 2.2. The metamorphic change is driven by heat and pressure.

It has to be noted that silicates are the major rock-forming minerals of igneous, metamorphic, and many sedimentary rocks.

**Physical classification**   As per physical characteristics, rocks may be classified as (a) stratified, (b) un-stratified, and (c) foliated. *Stratified rocks* occur in most sedimentary rocks and in some igneous rocks, and show distinct layers (ranging from several millimetres to many metres in thickness, and vary greatly in shape), along which the rocks can be split. The examples are sandstone, limestone, shale, slate, and marble. *Unstratified rocks* do not show any stratification and cannot be easily split into thin layers. They possess crystalline and compact (dense) grains. The examples of such rocks are granite, basalt, and trap. In *foliated rocks,* the grains may line up in distinct, repetitive layers, which have different a texture or colour, giving the rock a distinct wavy appearance. These rocks have a tendency to split along a definite direction only. The direction need not be parallel to each other as in the case of stratified rocks. Most of the metamorphic rocks, such as gneiss, phyllite, and schist, have a foliated structure, except for quartzite and marble.

**Chemical classification**   According to chemical characteristics, rocks may be classified as (a) argilla-ceous, (b) siliceous, and (c) calcareous. *Argillaceous rocks* contain clay like components. These rocks are hard and brittle, for example, shale, siltstone, loess, and mudstone. *Siliceous rocks* are made mostly from silicates. These rocks are very hard and durable, for example, granite, basalt, trap, quartzite, and gneiss. *Calcareous rocks* contain a high proportion of calcium carbonate in the form of calcite or arago-nite and are sensitive to acidic solutions, for example, limestone, marble, and travertine. More than 98% of the rocks in the world are formed from a combination of merely eight elements as shown in Table 2.1.

**Table 2.1** Different elements in rocks

| Element | % of all rocks | Element | % of all rocks |
|---|---|---|---|
| Oxygen (O) | 46.6 | Sodium (Na) | 2.8 |
| Silicon (Si) | 27.7 | Potassium (K) | 2.6 |
| Aluminium (Al) | 8.1 | Magnesium (Mg) | 2.1 |
| Iron (Fe) | 5.0 | Others | 1.5 |
| Calcium (Ca) | 3.6 | | |

**Hardness classification**   Based on hardness, stones are classified as soft, hard, and very hard. The hardness is determined by the *Mohs's scale,* invented by German mineralogist *Friedrich Mohs* in 1812. It measures the hardness based on how easily it scratches. The scale consists of 10 minerals arranged from 1 to 10. The higher the number, the harder will be the mineral. Grades 1 and 2 can be scratched by a fingernail, whereas a steel knife is required to scratch Grades 3 to 6. The minerals that have a Mohs's hardness of 7 (e.g., quartz) or more can scratch glass. A comparison of Mohs's and equivalent Vickers or Brinell hardness is given in Table 2.2 (Richards, 1961).

**Table 2.2** Mohs's scale of hardness

| Mineral | Mohs's hardness | Equivalent Vickers or Brinell hardness |
|---|---|---|
| Talc | 1 | 22 |
| Gypsum | 2 | 35 |
| Calcite | 3 | 65 |
| Fluorite | 4 | 90 |
| Apatite | 5 | 175 |
| Feldspar | 6 | 290 |
| Quartz | 7 | 490 |

*(Contd)*

**Table 2.2** *(Contd)*

| Mineral | Mohs's hardness | Equivalent Vickers or Brinell hardness |
| --- | --- | --- |
| Topaz | 8 | 950 |
| Corundum | 9 | 1,600 |
| Diamond | 10 | 2,500 |

## 2.2.1 Igneous Rocks

*Igneous rocks* or *magmatic rock* are of volcanic origin (The word 'Igneous' comes from the Latin word *'ignis'* meaning 'fire'). They are formed when hot, molten magma inside the Earth is forced up as volcanic eruptions and cools above or below the surface, thus forming solid rocks. Depending on where the cooling process took place, they may be distinguished as (a) volcanic (extrusive) rocks, (b) plutonic (intrusive) rocks, and (b) gangue. Igneous rocks are generally hard, acid-resistant, frost and wear-resistant, and durable (e.g., granite and basalt). The texture of the rock is greatly influenced by the rate of cooling of the magma. Volcanic and plutonic rocks can be identified on the basis of texture. The principal constituents of magma are quartz, mica, and feldspar.

1. *Volcanic rocks*, which are solidified on the earth's surface, have almost no visible minerals (grain sizes are too small to be seen with the naked eye), and are uniform in appearance and composition (e.g., andesite, basalt, rhyolite, and trap). They are non-crystalline and have a glassy texture.
2. *Plutonic rocks* are those that have cooled down gradually and solidified under constant, high pressure, deep within the Earth's crust. They are more coarse-grained and have a crystalline surface (evenly distributed three-dimensional arrangement of minerals), with crystals that are clearly visible (e.g., diorite, granite, gabbro, peridotite, and syenite).
3. *Gangues* are formed in cracks in the Earth's crust (solidify at a relatively shallow depth), where they are cooled slowly and under high, constant pressure. Texture and composition of gangues are quite regular, with finely grained crystalline structure, but there are important variations between the stones themselves (e.g., dolerite, diabase, and porphyry).

## 2.2.2 Sedimentary Rocks

Up to 80 to 90% of the rocks on the Earth's surface are *sedimentary rocks* only. These rocks form on land from pre-existing rocks or pieces of once-living organisms. They also form from deposits that accumulate on the Earth's surface and/or carried by wind or water to the sea where they are buried, get consolidated under pressure and/or heat, and form the layers of rock. Chemical agents also contribute to the cementing of the deposits. The main feature of sedimentary rocks is that they have horizontal or parallel consecutive layers. The thickness and the composition of these layers may vary. The layers are clearly visible as they usually vary in colour, composition, size of the grains, and texture. Sedimentary rocks tend to break more easily along the surface of these layers. They are more uniform, fine grained, and compact.

The three types of sedimentary rocks are: *clastic sedimentary rocks* (86% of sedimentary rocks belong to this category, and the rest are chemical and biochemical), such as breccia, conglomerate, sandstone, siltstone, and shale, are formed from mechanical weathering debris; *chemical sedimentary rocks*, such as rock salt, rock gypsum, iron ore, chert, flint, some dolomites, and some limestones, are formed when dissolved materials precipitate from solution; and *organic (biochemical) sedimentary rocks,* such as coal, some dolomites, and some limestones (e.g., coquina and chalk), form from the accumulation of plant or animal debris.

## 2.2.3 Metamorphic Rocks

*Metamorphic rocks* are previously formed igneous and sedimentary rocks, which undergo changes due to the action of heat or pressure or both. For example, granite becomes gneiss, limestone changes to marble, sandstone becomes quartzite, and mudstone transforms to slate. The pressure causes tabular or sheet-like minerals (e.g., mica) in the rock to grow parallel to each other and perpendicular to the direction of pressure, generating a mineral alignment termed as *foliation*. There are two the basic types of metamorphic rocks: *foliated metamorphic rocks,* such as slate, gneiss, schist, and phyllite, and *non-foliated metamorphic rocks,* such as hornfels, marble, quartzite, and serpentinite.

The geological classification of rocks and their sub-classifications and examples are summarized in Table 2.3 for an easy reference.

**Table 2.3** Geological classification of rocks

| Classification | Sub-classification | Examples |
|---|---|---|
| Igneous rocks | Volcanic (extrusive) rocks | Andesite, basalt, rhyolite, and trap |
| | Plutonic (intrusive) rocks | Diorite, granite, gabbro, peridotite, and syenite |
| | Gangues | Dolerite, diabase, and porphyry |
| Sedimentary rocks | *Clastic sedimentary rocks* | Breccia, conglomerate, sandstone, siltstone, and shale |
| | *Chemical sedimentary rocks* | Rock salt, rock gypsum, iron ore, chert, flint, and some limestones |
| | *Organic (biochemical) sedimentary rocks* | Coal, coquina, chalk, and some dolomites |
| Metamorphic rocks | Foliated metamorphic rocks | Slate, gneiss, schist, and phyllite |
| | Non-foliated metamorphic rocks | Hornfels, marble, quartzite, and serpentinite |

## 2.3 Common Stones used in India

A wide spectrum of stones, such as granite, basalt, sandstone, limestone, marble, slate, and quartzite, are available in the different parts of India. Several indigenous machinery and tool manufacturers are also available to cater to the demands of stone architecture. Traditionally, stones have been used to build temples, forts, and mausoleums like the Taj Mahal. They continue to be used in some modern buildings, such as the Presidential House, the Parliament House, and the Supreme Court, which use high-quality sandstone from Rajasthan. Other famous rock-cut structures include the Elephanta Caves, the Sanchi Stupa, the Khajuraho Temples, the Konark Temple, and the Bahá'í House of Worship in New Delhi. India possesses a wide spectrum of stones, which include granite, basalt, sandstone, limestone, marble, slate, and quartzite, found all over the country. India also has an indigenous resource of machinery and tool manufacturers, which cater well to the demands of this sector.

To select and utilize stones for their satisfactory performance, it is important to know their properties. The strength of rock depends on its mineral constituents, which form the basis of the classification and identification of rocks. Identification of stones may be done in accordance with IS 1123. Some of the characteristics of rocks are given in Table 2.4. Figure 2.3 shows the textures of some typical building stones (https://flexiblelearning.auckland.ac.nz). More details about them and other rocks may be found in IS 1123. In general, igneous rocks, such as granite, basalt, and syenite, are good to be used as masonry stones. Granite is now the most widely used stone for buildings, monuments, and bridges. Limestone is the second most used building stone.

**Table 2.4** Characteristics of building stones

| Type | Physical properties: Colour, texture, and mineral content | Average engineering properties | Availability and uses |
|---|---|---|---|
| **Class: Igneous rocks** | | | |
| Granite | Colour: White to light grey, pink, and red; crystalline, medium-to-coarse grained; essentially quartz and feldspar with mica | Sp. Gr. 2.63–2.75, Comp. strength: 80–150 MPa Porosity: 0.4–4% Resistance to abrasion: 44–88% Modulus of elasticity: 20–60 GPa | Occur throughout the country; used for bridge piers, dams, river walls, monumental buildings, pavements, and kerbs; in addition, used as decorative stone, coarse aggregate and road metal |
| Basalt | Colour: Dark grey to black; medium-to-fine grained, dense, and compact Pyroxene, plagioclase, olivine, and magnetite | Sp. Gr. 2.60–3.00, Comp. strength: 150–200 MPa Porosity: 0.1–1% Resistance to abrasion: 15–19% Modulus of elasticity: 60–100 GPa | Forms when lava cools and solidifies on the Earth's surface; occurs as 'trap' rocks in western India Used for bridge piers, dams, river walls, masonry works, pavements and kerbs, and monumental buildings |
| Syenite | Colour: Grey, pink, and red. Crystalline, medium-to-coarse grained, and massive Potassium feldspar, plagioclase, biotite, amphibole, and pyroxene | Sp. Gr. 2.60–2.80, Comp. strength: 35–50 MPa Porosity: 1.38–1.54% Modulus of elasticity: 60–80 GPa | Less abundant than granite; uses: same as granite; looks similar to granite but, unlike granite, it contains little, if any, quartz |
| **Class: Sedimentary rocks** | | | |
| Sandstone | Colour: cream, grey, red, brown, and even and dark grey Stratified, fine-to-coarse grained. Mainly quartz with feldspar and dark minerals in a siliceous, calcareous, and argillaceous cement | Sp. Gr. 1.85–2.70, Comp. strength: 50–70 MPa Porosity: 5–25% Resistance to abrasion: 2–29% Modulus of elasticity: 5–80 GPa | Commonly available; used for masonry work, dams, bridge piers, river walls, buildings, pavements, and kerbs; famous building to use is Angkor Wat, Cambodia |
| Limestone and dolomite | Colour: White, grey, pink, red, blue, buff brown, green, yellow, and black Stratified, fine-to-medium grained; Mainly calcite with varying amounts of magnesium carbonate | Sp. Gr. 2.14–2.80 (limestone), 2.50–2.80 (dolomite) Comp. strength: 50–60 MPa Porosity: 5–20% Resistance to abrasion: 1.3–24% Modulus of elasticity: 10–80 GPa | Commonly available; used as slabs and tiles in any building; large-sized blocks (Porbandar stone) as ornamental stone. On burning limestone, it produces lime, which is used to make cement; in addition, used as raw material in glass manufacture. |
| Laterite | Colour: Brownish red, yellow, brown, grey and mottled colours Porous, oolitic, and pisolitic with cavities; at times, stratified. A mixture of hydrated oxides of iron and aluminium with manganese dioxide, titanium oxide and free silica. | Sp. Gr. 2.36–3.04 (depends on heavy iron oxides present) Comp. strength: 2–3 MPa (IS 3620 suggest a minimum of 3.5 MPa) Porosity: 20–40% Modulus of elasticity: Wet: 0.25 to 0.47 GPa, Dry: 0.4 – 0.6 GPa. | Available in Malabar region of Kerala; can be easily cut into brick-shaped blocks for building. Freshly quarried laterite is soft and porous, but when exposed to atmosphere it hardens and makes a very tough material; used as blocks in walls should be plastered from outside. |

*(Contd)*

**Table 2.4** (*Contd*)

| Type | Physical properties: Colour, texture, and mineral content | Average engineering properties | Availability and uses |
|------|-----------|-----------|-----------|
| **Class: Metamorphic rocks** | | | |
| Quartzite | Colour: White, grey, yellowish & brownish grey, buff. Fine-to-coarse grained often granular and banded. Up to 90% is quartz with feldspar and mica in small amounts. | Sp. Gr. 2.55–2.65 Comp. strength: 150–300 MPa Porosity: 0.2–0.6% Modulus of elasticity: 26 to 93 GPa. | Widely available in India; used as blocks and slabs for building stone, paving blocks and as aggregates. |
| Marble | Colour: White, pink, red, green, blue, grey, and black. Fine-to-coarse grained, massive crystalline and granular. Mainly calcite and dolomite or mixture of the two with some impurities | Sp. Gr. 2.6–2.7 Comp. strength: 35–70 MPa Porosity: 0.5–2% Resistance to abrasion: 2–4% Modulus of elasticity: 50 to 70 GPa. | Available in Rajasthan, Gujarat, Madhya Pradesh, Udaipur, and Jaipur; used as blocks, slabs and tiles in monuments, temples, and buildings; coloured marbles as ornamental stones. |
| Slate | Colour: Dark grey, black, greenish grey, purple grey. Fine grained, fissile along planes of original bedding. Composed of mudstones, shale, tuff, and clay minerals | Sp. Gr. 2.6–2.7 Comp. strength: 75–200 MPa Porosity: 0.1–0.5% Modulus of elasticity: 2.3 to 10 GPa. | Available in many parts of India; easily cut into slabs of desired dimensions; used as slabs and roofing tiles in buildings and pavements; in addition, used for making chalkboards. |

| Phyllite | Marble | Quartzite | Manhattan Schist | Slate | Limestone (Crystalline) | Sandstone | Granite |

**Fig. 2.3** Textures of typical building stones (*Source*: Wikipedia)

More details about the various types of rocks may be found in Siegesmund and Snethlage (2014).

## 2.4 Quarrying, Seasoning, and Dressing of Stones

Stones are obtained from the underground by using digging, blasting, or cutting. Collectively, these processes are known as *quarrying* and the open excavation from which the stone is obtained is called a *quarry*. Based on the method of excavation, quarries can be divided in two broad categories: (a) vertical quarries and (b) horizontal quarries. In the vertical quarry, the stone is obtained from the walls of the quarry. Whereas, in the horizontal quarry, stones are obtained from the bed/floor of the quarry. Quarrying often involves both machines and manual work. Traditionally, hand chisels and hammers were used to quarry stones. Several present-day quarries use mining machinery: drilling machines for drilling, diamond wire machines and chainsaws for cutting, hydraulic cushions for splitting, cranes for lifting big blocks, and dumpers and trucks for transport. After quarrying, the stones are subjected to the following: cutting/sawing, dressing, surface grinding and polishing, and edge-cutting/trimming.

### 2.4.1 Quarrying Methods

The method used for the quarrying of stones depends on the type of stone, its intended use, and the geological conditions at the site of quarries. The three methods of quarrying stones are as follows:

1. Quarrying with plug and feather
2. Quarrying by channelling
3. Quarrying by blasting with explosives

The first two methods are the oldest methods for quarrying. They have been, generally, replaced with other more efficient methods, such as line drilling and sawing. Each of the aforementioned three methods is briefly discussed here.

**Quarrying with plug and feather** It involves the drilling of a series of holes of about 100 mm diameter up to a depth of about 200–250 mm along a line with a spacing of about 250 mm, using either hand tools or pneumatic drills. Two steel flat wedges (with its upper end curved outwards and known as *feathers*) and a conical steel plug are placed in each hole (as shown in Fig. 2.4). The plug is then driven evenly with a sledge, gradually increasing the pressure until the stone splits along the drilled line. The blocks thus separated are lifted up and stored. This method is suitable for soft rocks, such as marble, limestone, and sandstone. Other old methods of using hand tools include digging and excavating (hand tools, such as chisels and shovels, are used to dig and excavate soft rocks such as laterite), and heating (used when rocks occur in layers; the surface is heated by fire, which separates the different layers of the rocks).

Line drilling and sawing are more modern techniques for quarrying (Langer, 2001). Line drilling (also called slot drilling) consists of drilling a series of overlapping holes using a drill that is mounted on a quarry bar or frame that aligns the holes and holds the drill in position. Sawing can be accomplished with a variety of saws including wire saws, belt saws, and chainsaws. The introduction of synthetic diamond tools during the 1960s revolutionized stone working.

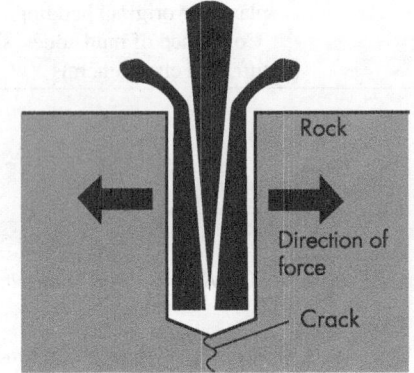

**Fig. 2.4** The three-piece metal tool set for splitting rocks

**Quarrying by channelling** In order to use this method, one of the faces of the solid rock to be quarried should have an exposed face. In this method, special machines called channellers, driven by steam, compressed air, or electricity, are used to cut long, narrow channels into the rock. They cut the stone with a cutting edge that traverses back and forth along the seam of the rock bed until a deep cut is made. The machines can cut 50 to 75 mm width channels up to 24 m in length and 2,400 to 3,000 mm depth. This process consists of the following steps:

1. Channels of sufficient depth are cut along the three sides of the block to be removed.
2. Horizontal holes are then driven beneath the block from the exposed face.
3. Wedges are then inserted and driven into the horizontal holes to make the block break loose.
4. The block is lifted from its bed to be cut further in to the slabs of required thickness.

The use of channelling is extensive in soft rock quarries, such as those containing limestone, marble, and sandstone.

**Quarrying by blasting with explosives** This method is the most commonly used for manufacturing stone aggregates. In this method, explosives capable of blasting away the larger portions of rocks are used. The stone then gets split with the use of wedges or by the plug-and-feather method, or is crushed by a heavy steel ball weighing several tons. Holes are drilled deep enough into the rock that it will break. The drilled holes are partially filled with explosives, which are then detonated. Most quarries first separate the larger masses of rock and then divide the rock mass into smaller blocks of desired sizes. To prevent the stone from shattering, lighter gunpowder is preferred to fracture dimension-stone. In the production of crushed-stone, more powerful dynamites and explosives are used.

In most of the quarries, large blocks (5–20 m long, 5–12 m high, and 4–30 m in depth) are first extracted (see Fig. 2.5). To extract this, first vertical planes normal to the bench face are cut using the diamond saw machines or diamond wire.

The remaining two planes are commonly cut by drilling and blasting. The various steps of the drilling and blasting operation are as below.

1. Boreholes of appropriate diameter (often 28–34 mm) and depth (about 450 mm) are made using jumpers along the lines at required spacing, so that the required size of the blocks could be produced. In hard rocks, the holes may be drilled with pneumatic drilling machine. The drilled holes are cleaned and dried well using a dry cloth.
2. The required amount of explosive, such as gunpowder (a mixture of potassium nitrate [$KNO_3$, also known as saltpeter], charcoal, and sulphur mixed in the ratio of 15:3:2, respectively, by weight) or dynamite (available from the government-managed agencies) is placed at the bottom of bore hole with a cotton fuse wire placed in its position, keeping the top part of the hole free. The quantity of gunpowder used will depend upon the spacing of the boreholes. The cotton fuse wire, which ignites the explosive, consists of a black powder core in a textile tube, covered with a waterproofing agent, and having an outer wrapper of tough textile or plastic. About 1 m of the fuse wire is kept projecting out of the blast hole. It has a slow burning rate of 5–10 mm/s to help the person firing it to run safely away from the explosion. It has to be noted that gunpowder blasting is not possible in a damp environment or under water.

   *Dynamite*, invented by Swedish chemist and engineer Alfred Nobel and patented in 1867, is the first safely manageable explosive, which is stronger than gunpowder. When using dynamite, a detonator is used. The detonator is a device with a copper cylinder (about 6 mm in diameter and 25–50 mm in length) closed at one end and projecting fuse at the other end. The detonator contains an easy-to-ignite primary explosive, which will activate the larger explosion of dynamite. Modern, electronic detonators provide the precise control necessary to produce accurate and consistent blasting results. One end of the fuse wire is connected to the detonator, which is lowered into the hole. The other end of the fuse is kept above the hole as in the case of gunpowder firing. (It has to be noted that dynamite is several times costlier than the gunpowder and stocking such explosives without license is a crime.)
3. The blast hole is filled with the layers of sandy clay with each layer rammed hard with a brass tamping rod.
4. The charge placed inside the blast holes is fired either directly using fire or electric spark. In the case of dynamite, the detonators are activated by electric spark.

When weakness plane is horizontal in granite quarries, lifters (horizontal blastholes) are made with large diameter blastholes (see Fig. 2.5). Most quarries separate the larger masses of rock first and then divide the rock mass into the smaller blocks of desired sizes. The secondary cutting (subdivision of the large primary blocks) is done by drilling and blasting along successive rows of blastholes, as shown in Fig. 2.5. Further subdivision into commercial blocks may be done either by drilling and blasting or using wedges. In softer rocks, such as marble, most often, the diamond saw and disk cutters are used for this phase.

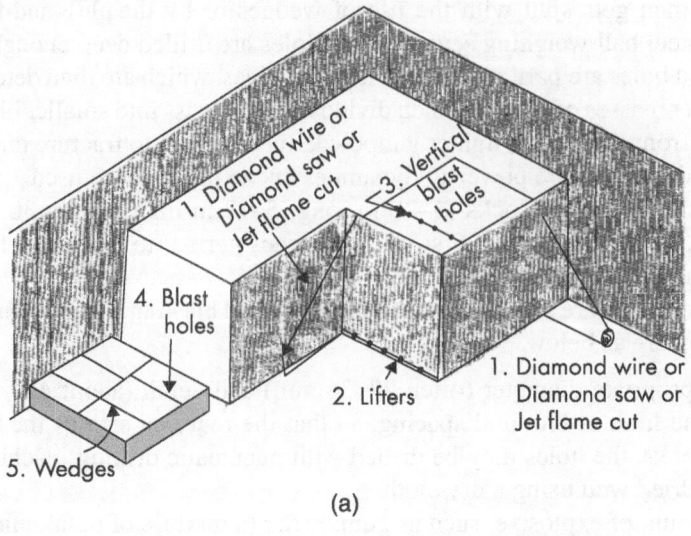

**Fig. 2.5** (a) Dimension stone quarrying sequence (b) Stone quarry in Soignies, Hainaut (province), Belgium (*Credit:* © Jean-Pol GRANDMONT)

Some new blasting and drilling techniques to improve the productivity of quarrying dimension stones are presented by Sanchidrian et al. (1996).

## 2.4.2 Seasoning of Stones

The stones obtained from quarry will contain interstitial moisture in the pores, called quarry water or sap. The process of removing moisture from pores, before using the stone in constructions, is called *seasoning*. Seasoning is performed by placing the stones in sundried condition for 6–12 months, with protection against rain. Seasoning is very essential for laterite stones. When the quarry sap evaporates, it leaves a crystalline film on the faces of the stone and makes them weather resistant. Seasoning may also make stones harder and more durable against the action of frost. Dressing of stones before seasoning improves their weather resistance. It is better to do dressing or carving as early as possible after quarrying. In addition to seasoning, for the stone to weather well, it should be laid with its bedding (lamination) horizontal, as it was first laid down by nature in the quarry.

### 2.4.3 Dressing of Stones

The process of giving a proper size, shape, and finish to the roughly broken blocks obtained from the quarry is called the *dressing of stone*. The various objectives of dressing are as below.

1. It is done to reduce the size of the big blocks so that they are easily handled. This reduction in size is generally carried out at the quarry itself because that saves a lot of transportation cost.
2. As stones are used in buildings at different locations, such as foundations, walls, arches, or flooring that may require differently shaped stones, the reduced blocks are processed to give proper shape to stones, so that they can be used in these locations. This operation can be done using profiling saws at the quarry and at the site.
3. Stones are used not only because of their high strength, hardness, and durability, but also because of their aesthetic value. Stone surfaces can be made very decorative in order to have an appealing appearance, which will last for several years. More intricate work is carried out on the dimension stones to obtain an appealing finish, using polishing machines or by hand. Polished surfaces are obtained by grinding the cut face with successively finer grades of abrasives. Sandstone and limestone are available in honed finish. Marble, harder limestone, and granite can be polished.

Manually, skilled stone-smiths work on stones with chisels, hammers, and abrasives to give different texture and appealing finish. There are 10 to 15 types of dressings, which are popular. Stone that is used as it comes from quarry is called the quarry face or if from the natural break, seam face. Either face may be finished with a variety of ways: axed finish, boasted finish, circular finish, combed finish, chisel drafting, furrowed finish, hammer-dressed finish, punched finish, moulded finish, plain finish, polished finish, reticulated finish, rubbed finish, sand-blasted finish, sunk finish, tooled finish, and vermiculated finish (https://theconstructor.org/). *Honed surface* is produced by grinding the surface with high-grit material to a uniform specification, without producing a reflective surface as in polished finish. Some of the dressed stone surfaces are shown in Fig. 2.6. More information may be found at www.designingbuildings.co.uk/wiki/Stone_dressing and http://stonemtg.com.au/stone-as-a-building-material.html.

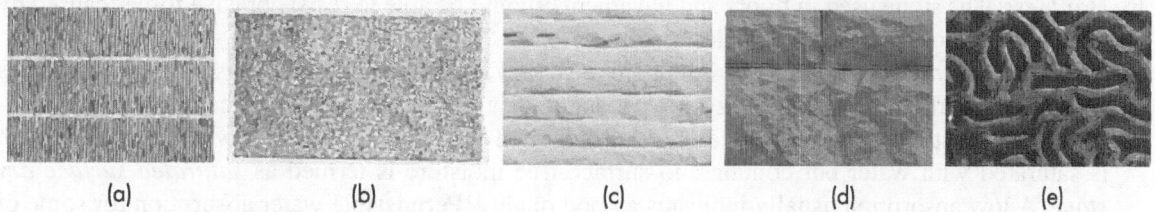

| (a) | (b) | (c) | (d) | (e) |

**Fig. 2.6** Some dressed stone surfaces (a) Boasted/droved finish (b) Chisel drafting (c) Tooled finish (d) Quarry faced finish (e) Vermiculated finish

The fine edge finish with 90° angle in the vertices of tiles or slabs are obtained by machine cutting/sawing, hand chiselling, or hand dressing, and chamfering/bevelling. These fine edges simplify the work of masons, as the stones fit perfectly at site. Chamfered edge slabs are used in kitchen/table tops. Chamfering of the edges requires sawing and polishing.

## 2.5 Requirements of Good Building Stones

To select and utilize stones for their satisfactory performance, it is necessary to know the various properties, which can be determined according to relevant Indian Standards. Identification of stones may be done in accordance with IS 1123. The following are the requirements of good building stones:

1. *Appearance and texture*: For stones used in facing works, appearance is a primary requirement. The colour of the stone and its ability to receive polish are important factors. Texture of stone is indicated by the arrangement, size, and shape of the grains in the stone. Fine-grain stones are suitable for dressing. Facing stones should have pleasing texture and be free from cavities and cracks. For facing work, stones with compact, fine, and crystalline texture, having a light colour, are preferred.

2. *Cost*: Cost is an important consideration in selecting any building material. Proximity of a quarry to the building site will bring down the cost of transportation.

3. *Compactness*: Freshly quarried compact (dense) stones will show a clear and bright surface, with well-cemented particles. The compactness may be tested by striking the stone with a hammer; a clear metallic sound will indicate the good compactness. The compact and hard stones are considered to be durable.

4. *Dressing*: This property depends on the nature of work. It should be easy to dress to the required texture so that the cost of dressing is reduced. Stones sawed to the required dimensions are more durable than those hammered and broken to size.

5. *Durability*: Stones selected should be durable and able to withstand the disintegrating action of the environmental forces, such as wind, rain, and temperature changes. Stones which have crystalline structure, are homogenous, and have good cementing medium are more durable (see also Section 2.6).

6. *Fire resistance*: Igneous rocks are not stable beyond a temperature of about 575°C, due to quartz disintegrating into smaller particles. However, limestone is better at resisting fire up to about 800°C. For better fire resistance, stones should be free from calcium carbonate, oxides of iron, and minerals, which have different coefficients of thermal expansion.

7. *General structure*: Igneous stones with unstratified structure are always hard, durable, and strong. However, they can be easily cut into desired shape and size. Hence, they are used in foundations, heavy structures, and as coarse aggregates in concrete. Stones with foliated structure are usually weak and hence not used in masonry construction. Stones of stratified rocks are commonly used as building units.

8. *Hardness*: The stone used in floors and pavements should be able to resist abrasive forces caused by the movement of people and vehicles over them.

9. *Porosity and water absorption*: Building stone should not be porous. A porous stone disintegrates when the absorbed rainwater freezes, expands, and causes cracking. The percentage of water absorbed by the stone when immersed in water is termed as the *absorption of stone*. The stone which is saturated with water but contains no surface-free moisture is termed as *saturated surface dry stone*. A low absorption usually indicates a good quality. Permissible water absorption for some of the stones is given in Table 2.5. The porosity of commonly used rocks ranges from 0.1 to 20% (see Table 2.4). Greater the porosity, greater will be chances of its disintegration. The resulting permeability and water absorption may affect the bond between the stone and mortar in stone masonry, and the resistance of masonry to freezing and thawing.

**Table 2.5** Permissible water absorption by volume in 24 h

| Type of stone | Water absorption (% not greater than) | Type of stone | Water absorption (% not greater than) |
|---|---|---|---|
| Gneiss | 1 | Sandstone | 10 |
| Granite | 1 | Shale | 10 |
| Limestone | 10 | Slate | 1 |
| Quartzite | 3 | Trap | 6 |

10. *Resistance to wear*: For a good building stone, the percentage wear should be equal to or less than 3%. However, for use as coarse aggregates, a much higher value is required.

11. *Seasoning*: Good stones should be free from the quarry sap. For example, laterite stones are dressed as soon as they are quarried, and seasoned for about 6–12 months. This process gets rid of quarry sap. Iron compounds contained in the laterite stones get oxidized and the stones attain the necessary strength.

12. *Specific gravity*: The specific gravity of a stone may reveal its quality and properties. The higher the value of specific gravity of a stone, the harder and stronger it will be. In general, heavier stones are compact, strong, and are less porous and hence are used in the construction of dams, retaining walls, and docks. The specific gravity of good building stone is between 2.4 and 2.8 (see Table 2.4).

13. *Strength*: The stone should be strong to resist the loads acting on it. The strength of rocks depends on its mineral constituents, which form the basis of classification and identification of rocks. In general, stones have good compressive strength in the range of 50–200 N/mm$^2$ (see Tables 2.4 and 2.6), and hence it is usually not a problem. However, as their tensile strength is limited, they have to be checked for strength, when they are used as bending members, such as lintels.

14. *Thermal movement*: Thermal movements, *per se*, are not a big problem. However, due to temperature, joints in coping and parapets may open-out causing rainwater to enter. Marble slabs show a distinct distortion when subjected to heat. An exposure to heat on one side of marble slab may cause that side to expand, resulting in warping of the slab. On cooling, the slab will not go back to its original shape.

15. *Toughness*: The measure of impact that a stone can withstand is defined as *toughness*. The stone aggregates used in the road constructions should be tough enough to resist the vibratory or moving loads. An impact test value of 19 is considered good and below 13 is considered poor.

16. *Weathering*: The resistance of stone against the wear and tear due to natural agencies should be high. They should also weather well with age.

17. *Weight*: It is an indication of the porosity and density. For the stability of structures, such as dams, and retaining walls, heavier stones are required; whereas for arches, vaults, and domes, lighter stones may be used.

It has to be noted that not a single stone can satisfy all the aforementioned requirements of a good building stone. Hence, the architect/site engineer should study the properties required for the indented work and select a suitable stone. A comparison of the properties of some common stones is provided in Table 2.6.

**Table 2.6** Comparison of the properties of some common stones

| Property | Types of stone | | | | | |
|---|---|---|---|---|---|---|
| | Granite | Marble | Limestone | Sandstone | Slate | Quartzite |
| Specific weight (kN/m$^3$) | 24–29 | 27.5–29 | 19–27 | 19.5–25.5 | 24–29 | 25–26 |
| Compressive ultimate strength (MPa) | 100–200 | 68–150 | 27–135 | 21–135 | 70–100 | 95–100 |
| Coefficient of expansion (/°C) | $11 \times 10^{-6}$ | $4 \times 10^{-6}$ | $4 \times 10^{-6}$ | $12 \times 10^{-6}$ | $11 \times 10^{-6}$ | $11 \times 10^{-6}$ |
| Thermal conductivity (W/m°C) | 3.0 | 2.5 | 1.5 | 1.5 | 1.9 | 3.0 |
| Effect of chemicals | Resistant to most chemicals | Attacked by acids | Attacked by acids | Resistant to most acids | Resistant to most acids | Resistant to most acids |
| Resistance to effect of soluble salts | Poor to good | Good | Poor to very good | Poor to good | Good | Good |

*(Contd)*

**Table 2.6** (*Contd*)

| Property | Types of stone | | | | | |
|---|---|---|---|---|---|---|
| | Granite | Marble | Limestone | Sandstone | Slate | Quartzite |
| Ease of working | Hard | Fairly hard | Easy to hard | Hard | Hard | Hard |
| Ease of cleaning | Difficult | Difficult | Easy | Difficult | Difficult | Difficult |
| Resistant to frost | Good to excellent | Good to excellent | Poor to very good | Poor to very good | Good to excellent | Good to excellent |

## Case Study | Angkor Wat

*Photo Courtesy:* Er Gajendran

Angkor Wat, located in present-day Cambodia, is the largest religious structure built by Suryavarman II, who ruled the Khmer Empire from 1112 to 1152. It is a World Heritage Site. This monument has been built out of 5 to 10 million sandstone blocks with a maximum weight of 1.5 tons each. The sandstone used for the building of Angkor Wat is Mesozoic sandstone quarried in the Phnom Kulen Mountains, about 40 km away from the temple. The foundations and internal parts of the temple contain laterite blocks behind the sandstone surface. Virtually, all of its surfaces, columns, lintels, and even roofs are carved. The masonry was laid without joint mortar.

## 2.6 Durability of Stones

Durability of stones is one of the important criteria while selecting them for use in constructions. In general, stones have a good service life. The National Association of Home Builders expect natural stone, particularly granite, marble, and slate, to last approximately 100 years with proper maintenance (NAHB, 2007). Stone structures, hundreds of years old, in many parts of the world, are still in excellent condition and show little sign of deterioration. However, several factors during the quarrying, seasoning, finishing, and setting of stones may affect their durability (www.stonemtg.com.au). In addition, environmental pollution and weathering may influence the service expectancy of stones.

**Quarrying**  Stones taken from the exposed faces and the top ledges of a quarry are less hard and less durable than unexposed stones. The excessive use of explosives in quarrying shatters the cohesion of the particles in the stone and thus causes cracks and flaws that can make the stone more permeable to moisture. Hence, stone cut out by quarrying machinery is preferable than blasted or wedged out stones. In general, fine-grained, compact stones will absorb less moisture and suffer less from frost than a more open-textured stone. In addition, highly absorbent stones should not be quarried in freezing weather, as there may be tendency for the rock to break.

**Seasoning**  If the stone is not seasoned, the quarry water will be alternately frozen and thawed during its lifetime, making it to break.

**Dressing**  The methods employed in dressing the stone also affect its life. Minute fissures are produced by impact, which will render the stone more susceptible to the atmosphere. Hence, instead of hammering, sawing the stone to the required dimensions will result in more durable stones. Repeated hammering during cutting may injure stones.

**Finishing**  The type of finish given to the exposed faces of a rock may affect its durability. Thus, a smooth or polished surface may be more durable because it will facilitate rapid discharge of rainwater.

**Setting**  Stratified stones should be set with their layers horizontal to avoid water entering between the layers. Stones placed under cornices, belt courses, and window sills are liable to early decay, as rainwater slowly falls or drips on them. To prevent this, the under surface of projecting stones should be provided with a narrow groove, called a *drip*, extending its whole length. The drip will interrupt the water that flows from the upper surface down to the underside face, and make it to fall immediately on to the ground. When mixing stones, one has to be careful. For example, when a sandstone layer is placed under a layer of limestone, the chemicals washed away due to rain from limestone may cause the sandstone to decay. In addition, an unequal expansion of different rocks, due to their coefficient of thermal expansion, may also affect the durability.

---

**Case Study**  **Bowing of marble slabs in buildings**

Researchers were puzzled over several decades for the reason behind the bowing of thin marble veneer cladding on modern buildings, such as the Amoco building in Chicago, the Grande Arch de la Defénse in Paris, and Alvar Aalto's Finlandia city hall in Helsinki. Substantial recent research has found that marble type and its fabrics, stress relief, thermal expansion, and expansion due to moisture and temperature, and the climate are the main causes of this problematic phenomenon (Siegesmund, Ruedrich, & Koch, 2008).

---

A porous stone is less durable than a dense stone, since the former is less resistant to freezing. More information on the durability of stones may be found in the works of Siegesmund and Snethlage, 2014.

## 2.7 Deterioration of Stones

The disintegration or decay of rocks to form sediments, due to the Earth's atmosphere, and biological organisms is called *weathering*, and may be classified as physical, chemical, or biological. The physical or natural agents are air (in the form of wind), water (in the form of rain and ice), and temperature (in the form of cold and heat). The chemical agents are the various acids formed due to the atmospheric pollutants. The biological agents are vegetable growth, bacteria, and insects. Along with air pollution, soluble salts represent one of the most important causes of stone decay (Doehne & Price, 2010). The actions of these agents are briefly described below.

**Rain**  Rainwater acts both physically and chemically on stones. The physical action is due to the erosive and transportation powers of rain combined with wind. Rain may also produce the following effects: (a) oxidation of iron particles that may be present in stones, (b) dissolution of water-soluble particles, and (c) hydration of minerals that may be present in the stones. As granites are least porous, as compared to the most porous sandstones, they are adopted in wet places and in foundations.

**Physical action**   Alternate wetting by rain and drying by the Sun causes internal stresses in the stones and their consequent disintegration. Heavy winds, especially in desert regions, can carry sand and dust, eroding the surface of stones and removing small particles similar to the action of sand blasting. Subsequent rain may wash these small particles away. Disintegration of stones may also result from alternate freezing and thawing in cold regions.

**Chemical agents**   In industrial areas, acidic rainwater reacts with the constituents of stones leading to its deterioration. The 'traditional' pollutants of sulphur oxides, nitrogen oxides, and carbon dioxide dissolve in water to form sulphuric acid, nitric acid, and carbonic acid, respectively, and react with calcareous materials. Limestone, marble, sandstones, granite containing feldspar, and lime mortars are the most vulnerable to such acidic pollution. However, this is a universal problem for all types of stones. This kind of attack is noticeable in exposed areas of buildings and statues, as roughened surfaces, damaged material, and loss of carved details. Even sheltered areas show blackened crusts (composed of gypsum) that spall off in some places, revealing crumbling stone beneath (Doehne & Price, 2010).

**Salts**   The sources of salt are from air pollution, soils, wind from the sea or the desert, deicing salt, unsuitable cleaning materials, incompatible building materials, and garden fertilizers. The growth of salt crystals within the pores of a stone can generate stresses that are sufficient to overcome the stone's tensile strength and turn the stone to a powder. Salt solutions enhance the dissolution of calcite and the alteration of biotite, quartz, and feldspars. The deterioration of many of the world's greatest monuments can be attributed to salts, from Angkor Wat to Venice, and from Petra to the Great Sphinx of Giza (Doehne & Price, 2010). Salt damage can also take place indoors, through the hygroscopic action of the salts.

**Decomposition**   Due to the action of chemically active water, the alkaline silicate of alumina in stones will disintegrate. In addition, water dissolves and removes hydrated silicate and carbonate of alkaline materials, leaving behind only a hydrated silicate of alumina (kaolinite).

**Oxidation and hydration**   Pyrite, magnetite, and iron carbonate oxidize due to oxygen in the air/water and may cause discoloration of the stone. Since such oxidation is accompanied by a change in volume, the surrounding structure may get weakened.

**Biological agents**   Growth of vegetation (such as seedlings of banyan trees from bird droppings) in the joints and cracks of stones, and worms or bacteria can cause decay. Roots of plants/trees that grow exert pressure on stones and may crack them. Certain category of worms bore holes in stones, thus weakening them. It has to be noted that biological growths on stone are both a blessing and blight. Colourful lichens and creepers, such as ivy, can contribute an air of age and romance to a monument, and their removal can leave the stone looking stark and denuded.

**Nature of mortar used**   The binding material, such as cement/lime mortar, used in the construction work may sometimes have ingredients that may react chemically with the mineral contents of stones, causing them to disintegrate.

**Temperature variation**   Large variations of temperature and alternate heating and cooling can cause expansion and contraction and induce differential movements and internal tensile stresses, which can lead to slow and gradual disintegration of stones.

## 2.8 Preservation of Stones

The disintegration or decay of rocks discussed in Section 2.7 can be prevented by adopting some measures taken before or during the construction of stonework and preservation methods taken up after the stonework is completed. Preventing damage can embrace a very wide range of topics: legislation to protect individual buildings and monuments, pollution control, traffic control, control of groundwater, visitor management, and disaster planning.

### 2.8.1 Precautions During Construction

The following measures may be taken while selecting the rocks and during construction:

**Selection of proper stones**   It is important to pay attention while selecting the stone for a particular project. Compact stones with dense crystalline texture should be selected to have high resistance to weathering actions (see also Section 2.9). In industrial areas, limestone and calcareous sandstones should not be used for external walls.

**Size of units**   The size of selected stones should be as large as possible to minimize the number of joints.

**Seasoning**   The stones should be well seasoned and washed clean before they are used.

**Laying**   The construction should strictly follow the required specifications. The stones should be laid in such a manner that the natural bed is at right angles to the applied load (*Natural bed* is the plane of bed on which the sedimentary stone was originally deposited. It may not necessarily be horizontal.). In such a position, stones offer maximum resistance to crushing and disintegration by frost and rainwater. If the stones are used with their natural beds parallel to the direction of load, the stones will split at much lesser loads and will be attacked more severely by frost and rainwater.

**Masonry joints**   All the joints should be completely filled with the cement mortar, made using pure silica sand.

**External rendering**   All exposed stones should be finished with external rendering like pointing; otherwise, they should be plastered with cement mortar, which will, however, spoil the rich look of stones. Compared to rough stones, polished stones should be preferred for external walls.

### 2.8.2 Preservation of Completed Stonework

Preventive conservation measures of more immediate effect are usually concerned with keeping water out of the stone and with controlling the relative humidity and temperature of the air around the stone. Several other methods that have been used to preserve stones are listed below (Doehne & Price, 2010).

**Cleaning**   For a better service life, the stone surfaces should be periodically dusted and washed. A wide range of techniques are available for cleaning stone, ranging from those intended for use on large facades to those for use in finely carved and delicate sculptures. Using lasers to clean stone has now become popular and large-scale commercial laser cleaning has become common in the past 20 years. Spraying the stone surface with water just before laser cleaning can enhance the effectiveness of this treatment.

**Latex poultice method**   Known commercially as ArteMundit, it was originally developed as an improvement to the Mora poultice by Eddy De Witte in 1992. It is a spray-on film containing ethylene diamine tetra acetic acid (EDTA) and other additives.

**Desalination of masonry**   It is usually attempted through the use of poultices, which may consist of a range of materials (e.g., clay, sand, and paper pulp). Where calcium sulphate is to be removed, additional materials, such as EDTA and its sodium salts, sodium bicarbonate, ammonium bicarbonate, and ammonium carbonate, may be added in order to increase its solubility. Desalination efforts often need to be coupled with efforts to reduce the supply of salts, such as the maintenance or installation of a damp-proof course at the base of the building foundation.

**Consolidants** Where stone is severely weakened by decay, some form of consolidation may be necessary to restore some strength. Consolidants are usually applied to the surface of the stone by brush, spray, pipette, or immersion, and are drawn into the stone by capillary action. The alkoxysilanes have been the most widely used stone consolidants over the past 30 years. The majority of materials that have been tried as stone consolidants have been organic polymers, but several inorganic materials, such as calcium hydroxide (slaked lime) and barium hydroxide, have also been used (their mode of operation is however different). Barium hydroxide prevents deterioration due to calcium sulphate ($CaSO_4$), by the following reaction:

$$Ba(OH)_2 + CaSO_4 = BaSO_4 + Ca(OH)_2 \tag{2.1}$$

The resulting barium sulphate ($BaSO_4$) is insoluble and the calcium hydroxide [$Ca(OH)_2$] absorbs carbon dioxide and gives strength to the stone.

**Linseed oil** Naturally occurring compounds, such as linseed oil and cactus juice, and synthetic polymers have been applied to the external face of stone. It has to be noted that the application of boiled linseed oil will discolour the stone, though it may last longer than 1 year applicability of raw linseed oil.

**Surface coatings** A range of materials that have been applied in the past as surface coating of stone include: protective water repellents, emulsions, anti-graffiti coatings, salt inhibitors, protective oxalate layers, sacrificial lime coatings, colloidal silica, biocides, and bioremediation treatments. The property that has been most sought in surface coatings is water repellency. Water repellency has been provided largely by alkoxysilanes, silicones, and fluoropolymers. The emulsions that have been tried include acrylics, silicones, silanes, and fluorinated polyurethanes.

**Painting** It may preserve the stone but changes the colour of the stone. Colourless paints are also available now. If applied under pressure, it can fill the pores in the stone. The selected paint should not react with the stone. Paraffin is also used, alone or dissolved in naphtha, as a paint medium. However, it changes the colour of the stone.

**Other methods** Biological cleaning, for example, by using the anaerobic sulphur-reducing bacterium, has been used but has not been well researched. Protective shelters and canopies have also been used to protect important monuments from the direct action of rain. Though epoxy resins have also been used, there have been some notable failures. Applying coal tar is a crude method, which, in addition, to changing the colour may affect some stones. A mixture of alum and soft soap dissolved in water has been used as a protective coating but it does not offer a long-term solution.

More information on the various preservation methods, their effectiveness, and problems may be found in the works of Ashurst and Dimes (1998) and Snethlage (2008).

## 2.9 Selection and Applications of Stones

Natural stones, which possess high resistance to atmospheric agents, high mechanical strength, and pleasant colours, are widely used in construction as slabs, for walls and foundations of buildings, as facing slabs and stones for exterior and interior walls of buildings, for roads, side-walks, piers, and other structures, where mechanical strength, durability, and aesthetics are required. The selection of the nature and quality of stone in a particular application is governed by the purpose, availability, colour, grain texture and pattern, and surface finish of the stone. Prevailing environmental conditions, durability (essentially based on mineral composition and hardness and past performance), strength, the ability of the stone to take a polish, and the cost of stone are the other important selection criteria.

Of the total amount of stone quarried for the building industry, about 75% are used directly for concrete constructions and road making. The rest of them find applications in the manufacture of cement and other binding materials, and in the chemical industry, as metallurgical fluxes. Although a variety of igneous, metamorphic, and sedimentary rocks are used, the principal rock types are granite, limestone, marble, sandstone, and slate. Suitability and application of the various types of stones for different purposes and situations are briefly discussed below.

**Stones for masonry**   Any type of stone can be used in rough work, such as random rubble masonry. However, for ornamental work where dressing of stone is required to produce different finishes (for example, ashlar masonry), limestones and sandstones are preferred, as they are dressed easily compared to granites. Rubble units can be used for foundations, floors, walls, and in all cases with or without mortar (see also Section 2.10).

**Stones for important structures**   For temples, bridges, retaining walls, piers and abutments, docks and harbours, break-waters, and other marine structures, the stone should be very hard, heavy, strong, and durable. Granite and gneiss are recommended for this purpose.

**Stone for pavements and roads**   Generally, hard-stones of any type (e.g., granite and basalt) can be used for paving walkways and driveways (see Fig. 2.7). Many countries, such as Argentina, Mexico, Puerto Rico, the Philippines, and Uruguay, are well known for their many cobblestone streets, which are still operational and are in good condition. They are still maintained and repaired by the traditional manner, that is, by placing and arranging granite stones by hand. Cut cobble stone is also used to the reinforce slopes of earthworks and banks of water basins. Curb stones are used to separate roadways form sidewalks in streets, on bridges, and in tunnels of mixed traffic.

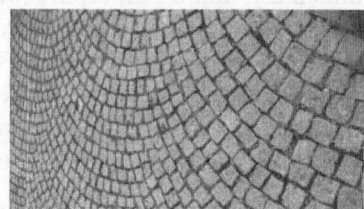

**Fig. 2.7** Stone pathway at St Peter's Abbey, Salzburg, built in the year 696 AD

**Stones for flooring**   In flooring, steps, and doors sills, where there will be a regular flow of traffic, stone which is hard, dense, durable, and has good wear and abrasion resistance should be used. Marble, granite, serpentinite, slates, limestone, and sandstone with good silica bond can be used in such places. Stones are also used for heavy-duty flooring where mechanical strength, durability, and decorative requirements are of concern. It is now possible to produce large slabs for flooring even from hard rocks, such as granite. Marble and Kotastone (a fine-grained variety of limestone quarried at Kota District, Rajasthan) are very popular for fireplaces, bar-tops, and bathrooms, as they can be polished and are available in a wide range of colours. Cuddapah stones (black limestone, quarried at Betamcharla, Cuddapa District, Andhra Pradesh) are popular for kitchen platforms and shelves. Stones used for slabs should have compressive strength not less than 5 MPa and the coefficient of softening between 0.7 and 0.9. See also Section 2.13.

**Stones for facing work**   The facing stones should be attractive and durable. Impervious stones, such as granite, basalt, and marble, are often used for facing work as they are durable, available in attractive colours, and do not change in colour with time, especially in industrial atmospheres. However, pervious stones like limestone are also used for this purpose in the form of thin slabs (veneers). Facing stone may be 300 mm or more in width and up to 4 m in length. Face stones and facing slabs for tunnels and bridges are given grooved or fluted finishes.

**Stone for road metal and aggregates for concrete**   For these uses, stones should be hard, tough, durable, and have high resistance to abrasion. Aggregates of moderate strength from limestone are used in normal strength concretes with strength less than 40 MPa. For high strength concretes and prestressed concrete, aggregates made from hard igneous rocks, such as granite and basalt, and sometimes quartzite are used. Crushed stone particles and manufactured sand made from these stone are used as fine aggregates in concretes.

**Stone for ballast** For railway ballast, the stone should be hard, dense, durable, tough, and easily workable. Sandstone, compact limestone, trap, and quartzite are commonly used.

**Stones for ornamental work and statues** For pillars, balustrade, pedestals, statues, and door and window sill, stones such as granite, marble, and compact limestone are recommend because they can take good polish.

**Building items** Elements of stairs, landings, parapets, and guard rails are manufactured from granite, marble, and limestone, and are given various finishes, depending on the kind of rock, which are similar to facing slabs. Pedestal slabs and stones for making doorways, cornices, corner, and window-sill slabs are made from the same material as the facing slabs and finished in a great variety of ways.

**Foundations and walls below ground level** Foundations are made from sawn stones of igneous, sedimentary, and metamorphic rocks. Impermeable stone, such as granite, can be used as damp-proof course and external cladding of walls. Underground structures and bridges are built of slabs and stones of igneous and sedimentary rocks. For the stability of structures, such as dams and retaining walls, heavier stones are required; whereas, for arches, vaults, and domes, light stones, such as limestone or sandstone, should be chosen. Tunnels, above-water parts of bridges, and dams are built of granite, diorite, and basalt with compressive strengths not less than 100 MPa.

**Stones for fire proof construction** Fire resistance of stone is not good especially if exposed to a stream of water while it is hot. For this reason, many building codes restrict the use of stone in commercial buildings. Some types of stones will weaken or crumble when exposed to intense heat. Granite disintegrates under fire. Limestone crumbles easily. Sandstones with fine grains can resist fire moderately. Marble resists fire better.

**Other uses** Slate tiles are used in roofing. Natural stone veneers over concrete masonry units or cast-in-situ concrete walls are widely used to give the appearance of stone masonry. Limestone is used as a raw material for the production of lime and cement.

## Case Study Discolouration of the marble surface of the Taj Mahal

Damage to stone by air pollution is an important problem in parts of central Europe, China, India, Russia, and other industrialized regions. According to an Indo-US study conducted in 2015 (by the Georgia Institute of Technology and University of Wisconsin in the USA and the Indian Institute of Technology, Kanpur, and the Archaeological Survey of India (ASI)], the white colour of the marble surface of the Taj Mahal, the 366-year-old mausoleum, is slowly turning brownish-yellow because of air pollution. Results indicated that the discolouration is due to the black carbon (emitted by vehicles and other machines that burn fossil fuels), and brown carbon (released through burning of biomass and garbage). Though scrubbers were installed to reduce $SO_2$ near the Taj Mahal, a lack of water, power outages, and the corresponding use of diesel generators were found to reduce the effectiveness of the scrubbers and decrease air quality near the site. This outlines the importance of considering infrastructure development to monument health (Doehne & Price, 2010). Since 2008, the ASI has been trying to fight the yellowing of the monument by giving it a clay-pack treatment using the lime-rich Fuller's earth (Multani Mitti) to clean the marble surface. Researchers are now keen on studying the efficacy of this method and finding ways of improving it.

If a large quantity of stones are required for a project, where the strength is the consideration, it is preferable to inspect the quarry from which the stones are obtained and make laboratory tests to determine its properties (see Chapter 25 for tests conducted on stones). An inspection of long-standing buildings (if any) made with the same material, at the selected location, is also desirable.

## 2.10 Stone Masonry

Stones can be used in different types of masonry as shown in Fig. 2.8. In general, igneous rocks, such as granite, syenite, charnockite (a type of metamorphic rock), and basalt, are very good for masonry work. Stone masonry is made of stone units bonded together with suitable mortar. Stone masonry may be broadly classified into the following two types: (a) rubble masonry and (b) ashlar masonry. Both are described briefly in this section.

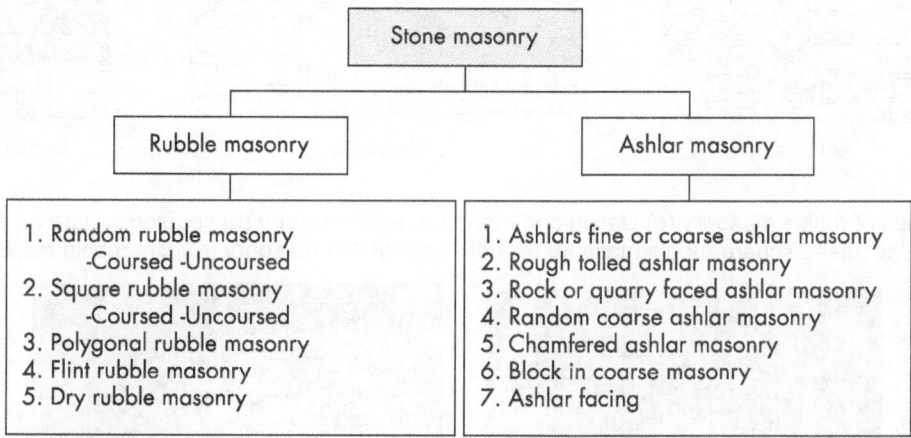

**Fig. 2.8** Types of stone masonry

### 2.10.1 Rubble Masonry

*Rubble* is stone in its natural shape, either as individual stones found in nature or irregular shaped pieces from a quarry. The stone masonry in which either undressed or roughly dressed stone is laid in a suitable mortar is called *rubble masonry*. In this masonry, the joints are not of uniform thickness. Rubble masonry is further sub-divided into the following three types: (i) random rubble masonry, (ii) squared rubble masonry, and (iii) dry rubble masonry. Squared stone is random sized stone that has been roughly dressed and squared to allow for narrow mortar joints. It has to be noted that the sides of these stones are not necessarily all perpendicular and parallel. Some of these types are shown in Fig. 2.9. The rubble masonry in which either undressed or hammer dressed stones are used is called *random rubble masonry* (see Fig. 2.10). When stones of different sizes and shapes are laid without forming courses, it is known as *un-coursed random rubble masonry* (used in walls of low height). When stones are laid in the layers of equal height, it is called *coursed random rubble masonry* (used for the construction of buildings and boundary walls). The rubble masonry in which the face stones are squared on all joints, and beds by hammer/chisel dressing before their actual laying, is called *squared rubble masonry*. *Coursed square rubble masonry* is used for the construction of public/ residential buildings, hospitals, and schools. *Uncoursed square rubble masonry* is used for constructing ordinary buildings in hilly areas, where a good variety of stones are cheaply available. The rubble masonry in which stones are laid without using any mortar is called *dry rubble masonry* and used to construct walls of height less than 3–6 m.

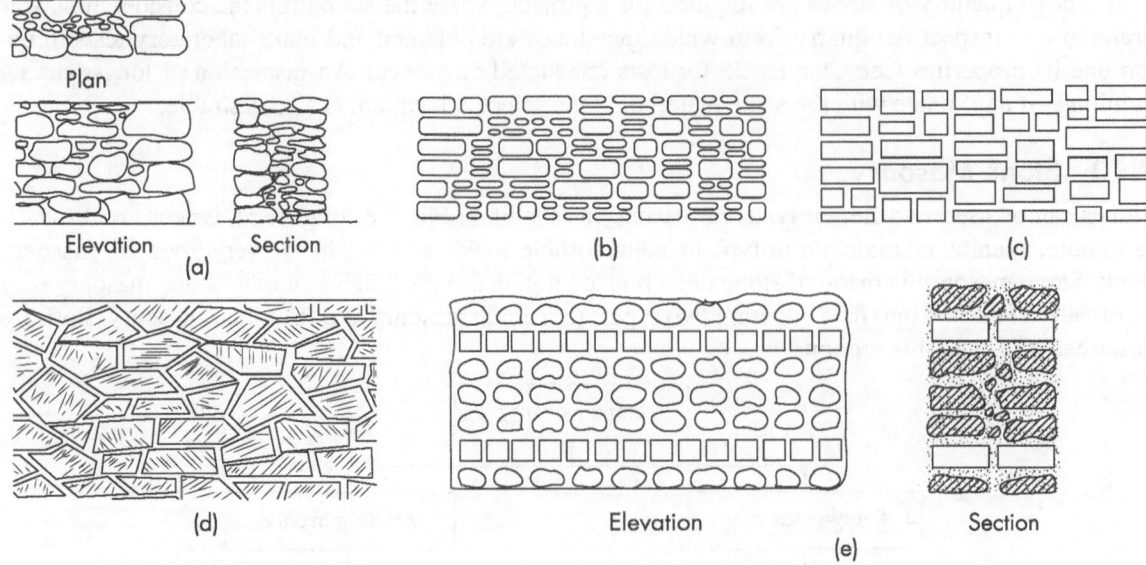

**Fig. 2.9** Types of rubble masonry (a) Uncoursed random rubble masonry (b) coursed square rubble masonry (c) Uncoursed square rubble masonry (d) Polygonal rubble masonry (e) Flint rubble masonry

**Fig. 2.10** (a) Random rubble masonry wall in Kerala (*Photo courtesy*: E.S. Jayakumar); (b) Dry rubble masonry in Maryland, USA

### 2.10.2 Ashlar Masonry

*Ashlar stone* is stone cut to the highest standards. The stone masonry, in which finely dressed stones are laid in cement or lime mortar, is known as *ashlar masonry*. In this masonry, the courses are of uniform height, all the joints are regular, thin, and have uniform thickness. All the stones in ashlar masonry need not be of the same size, but regular units are repeated to have horizontal and vertical mortar joints. This type of masonry is much costly as it requires dressing of stones. This masonry is used for heavy structures, architectural buildings, high piers, retaining walls,

**Fig. 2.11** Ashlar rough-tooled masonry

and abutments of bridges. In *ashlar course masonry*, stone blocks of same height are used in each course, with every stone fine tooled on all sides and having uniform thickness of mortar. Rough-tooled ashlar masonry is shown in Fig. 2.11. We may also have *random coarse ashlar*, which involves stone blocks laid with deliberately discontinuous courses and, therefore, discontinuous joints both vertically and horizontally. The dry ashlar masonry, used in *Inca architecture* of Cusco and Machu Picchu, is famous. *Quarry faced ashlar masonry* is similar to rough-tooled type except that there is a chisel-drafted margin left rough on the face of the stones. *Chamfered ashlar masonry* is similar to quarry faced except that the edges are bevelled to 45°. *Block-in course masonry* is a class of ashlar masonry, which is intermediate between rubble and ashlar, where all stones are squared and properly dressed and the construction resembles rough-tooled ashlar masonry. In *ashlar facing*, the facing alone is built in ashlar and the rest of the thickness in rubble.

### 2.10.3 Stone Veneers

For their majestic appearance, stones are often used in large and important buildings, especially for facing work. Wall faces are covered with thin polished tiles or slabs cut and from stones like sandstones, marble, and granite (see Fig. 2.12). Such veneering (also called as stone wall cladding) provides decorative and maintenance-free finish. A large number of factories are available in India to produce such thin tiles with varied finishes, such as natural, honed, polished, antique, tumbled, and split face. India is also exporting large blocks of stones to countries like the USA and Japan for producing such stone veneers.

**Fig. 2.12** Stone veneering over walls and concrete columns

### 2.10.4 Advantages and Disadvantages of Stone Masonry

The following are the advantages of using stone masonry:

1. Stones, the main material required for the construction of stone masonry and stone-retaining walls, are generally available in a large quantity in hilly regions. Their extraction generally requires low investment and energy input.
2. Stones have high strength and are durable. Most of them require minimal maintenance.
3. Most stones are impermeable and, hence, provide good rain protection.
4. Due to their high thermal capacity, stones are climatically appropriate in arid zones.

However, the use of stone in masonry may have the following disadvantages:

1. Atmospheric pollution may discolour certain stones. For example, sulphur compounds in polluted areas may dissolve in rainwater to produce sulphuric acid, which may react with carbonates in limestone causing blisters. Hence, limestones and calcareous sandstones should be used with caution in industrial towns.
2. Thermal movements may cause damage in stones, especially when fixed rigidly to other materials having different thermal properties (e.g., concrete). Providing movement joints can mitigate this problem.
3. In coastal zones, there may be problems of efflorescence and spalling of stones due to salts. This problem may be overcome by avoiding the use of surface treatments that seal in salts and cleaning the surface of stones periodically.

4. As mentioned earlier, limestones may disintegrate due to exposure to rainwater and prolonged wetting and drying may affect certain sandstones. Detailing of construction to allow the evaporation or drainage of water may be helpful in such situations.

5. Stone masonry is heavy and must be built on properly designed foundations to avoid potential settlement and subsequent cracking. It is better to avoid stone masonry in earthquake zones. Several traditional un-coursed random stone masonry walls have collapsed during past earthquakes killing numerous people. Earthquake resistance of stone masonry may be improved by restricting the thickness of walls to 450 mm, using 1:6 or richer cement–sand mortar or 1:3 lime–sand mortar in joints, using through stones (each extending over full thickness of walls) every 600 mm along the height and at a maximum spacing of 1.2 m along the length, providing RCC plinth, lintel, roof, and gable bands, and limiting the height of walls in each storey to 3 m and unsupported walls between cross walls to 5 m (refer to Bothara & Brzev, 2011; Murty, 2005, for more details).

In addition to the aforementioned, quarrying of stone will result in changes to geomorphology and conversion of land use, with the associated change in visual scene. The major impact may be accompanied by loss of habitat, noise, dust, vibrations, chemical spills, erosion, sedimentation, and dereliction of the mined site (Lad & Samant, 2014; Langer, 2001). However, it is possible to control, mitigate, and keep these impacts at tolerable levels, or restrict them to the immediate vicinity of quarry. By enclosing equipment and removing dust using vacuums, the impacts of noise and dust can be mitigated (Langer, 2001). Engineering activities associated with quarrying can directly change the course of surface water. Sinkholes created by quarrying can intercept surface water flow. Sinkholes can also result in catastrophic subsidence.

## 2.11 Stone Masonry vs Brick Masonry

A comparison of stone masonry and brick masonry (which will be discussed in Chapter 3) is provided in Table 2.7. Both materials have some advantages and disadvantages. It may be easier to obtain bricks of the required type, texture, size, and colour; whereas, stones may have distinctive colour, texture, and quality, depending on their origin. Moreover, working with stones is a slow and laborious process, and requires much more skill than that for brick. All stones are best considered thermal mass materials (that absorb heat and cold readily, and then hold them for quite a while, before letting them out) rather than insulators. In hot climates, thick stones keep the inside of houses cool, but do not effectively trap heat – an insulation layer is often required (preferably, on the outside of the shell, to block the heat from entering the stone). Brick walls benefit from the energy efficiency of thermal mass. The mass of brick walls, especially when combined with insulation, takes longer to heat on warm days and retain heat longer on cooler days. The environmental impact of stone comes from their quarries, while that of bricks comes from the resources taken from the land, the energy required in brick kilns, and their emissions. Cost wise, bricks may be much cheaper than stones, and can be constructed with semiskilled labour. Transportation costs also play a big role, if the materials are not available locally.

**Table 2.7** Comparison of brick and stone masonry

| Property | Stone | Brick |
| --- | --- | --- |
| Type of material | Natural | Man-made |
| Weight | Heavy | Light |
| Shaping and dressing | Expensive and requires man-power and machines | Produced to the desired size and shape |
| Labour required | Skilled | Semiskilled |

*(Contd)*

**Table 2.7** (*Contd*)

| Property | Stone | Brick |
|---|---|---|
| Construction speed | Slow, laborious job, and requires a lot of skill and patience | Fairly fast |
| Cost | Cheap in hilly regions where stones are available | Generally cheaper than stones |
| Strength | Generally high | Depends on the clay used and workmanship; mostly, lower than stones |
| Aesthetics | Natural elegance | If executed properly may present a good appearance (see Fig. 3.1 of Chapter 3) |
| Thermal insulation | Poor | Good |
| Porosity | Less; can be used in wet areas without plastering | High-requires plastering in wet areas |
| Weather resistance | Good | Good but requires plastering in abnormal conditions |
| Environmental impact | Through quarries | Through brick kilns |

## 2.12  Artificial Stones

*Artificial stone*, also called *engineered/cast stone*, is a synthetic building material manufactured to look like natural stone. Some examples of artificial stones include cement concrete, terrazzo, mosaic, and artificial marble. It is used in places where durable natural stone is not available at reasonable cost. Artificial stone is made with white and/or grey cement and natural sand, carefully selected and well-graded aggregates of crushed stone, and suitable colouring pigments to achieve the desired colour and surface finish. Cement and aggregates are mixed in the proportion of 1:3. The prepared material is cast in moulds of the required size and shape and machine pressed. When the bocks are set hard, they are removed from the moulds and the surfaces hardened by keeping them immersed in soda silicate solution for a specified period. The resulting product is designed to have much better physical properties than natural stones. The artificial marble/granite manufactured using vacuum pressing with high pressure may even be stronger than the natural stone. Some products, such as mosaic, are also polished, after laying them over concrete floors. Artificial stone products can be used, as a substitute for natural stone, in all applications. They are widely used for flooring, in walls as veneers, and as counter tops. They can be used for both interior and exterior decorations. The following are the advantages the artificial stones:

1. They can be cast as per the architectural requirements. Stones of any shape and size with features, such as grooves, can be produced easily. Many natural stone textures can be produced in a variety of colours.
2. They are economical and weigh less than full thickness stones and, hence, easier to handle and install. It also makes them easier and less costly to transport.
3. Placing them is more convenient than natural stones as the natural bed is absent, and hence needs less supervision. (Natural stones should be placed with their bed in compression.)
4. As there will not be any cavity or defect as in natural stones, they are more durable.
5. They can be made even in places where deposits of natural stones are not available, as broken stones can be easily transported.
6. The uneven coloration of natural stones is eliminated. However, UV exposure may cause the pigments to fade. Cleaners, deicers, chemicals, and slush may discolour the product.

## 2.13 Stone Flooring

Few building materials match the beauty of a natural stone floor. Due to its nature of no-two-pieces-look-alike and durability, stone flooring has been a favourite, especially in places of heavy traffic such as commercial, institutional, and public buildings. Some of the natural stones used in India for flooring include: (a) marble, (b) granite, (c) Cuddapah slabs, (d) Kota stone, (e) Shahabad stone (a type of limestone quarried at Gulbarga and Belgaum Districts of Karnataka), (f) sandstone, and (g) mosaic. Slate, travertine, and sandstone can also be used.

The standard sizes that are available in India are 300 mm × 300 mm, 300 mm × 300 mm, and 300 mm × 300 mm with the thickness of 20–40 mm. The top surface of the floors can be polished.

**Marble stone flooring** Marble presents an elegant, classic veined appearance and hence used since ancient times. Its properties are given in Table 2.4. Marble is relatively soft, and often polished to a high sheen that shows off its subtle beauty. However, due to its softness, it is prone to scratching and staining. Hence, professional sealing after installation is recommended. Marble is easily recognized for its soft grain patterns and creamy colours that include white (Makrana white and Abu white), grey, pink, brown, green (with grey and white veins), and black. Makrana white marbles were considered to be the best variety but have become extinct due to excessive mining. It has to be noted that the tools and techniques involved in stone floor installation can make the work area hazardous.

Polished marble is slippery when wet and, thus, not the best choice for kitchens and baths. Honed marble has more surface texture and is a good alternative, but doesn't have the dynamic colour of polished marble. It is very difficult to get marble of uniform good quality in large quantities. Good marble will have a specific gravity of 2.6–2.7, Mohs's hardness greater than 3, and moisture absorption (after immersion in water for 24 h) of less than 0.4%. Porosity is the main defect of many verities, which result in discolouration of the stone when oil or ink spills on it.

**Granite floors** Granite is an igneous rock formed under extreme pressure, making it one of the hardest flooring materials. It's virtually impervious to staining and scratching. It comes in many colours – local varieties of granite flooring may be less expensive due to lower shipping costs.

**Slate** It is a metamorphic rock that forms in layers, and, hence, easy to split into thin sheets that are cut into tiles. Slate floor tiles can be left with their naturally rough surfaces or honed smooth. The rougher surfaces offer good traction for entryways, sunrooms, patios, and kitchens. Slate is available in blends of dark green, grey, and brown. Slate comes in various densities; denser slates cost more. The possibilities of slate to chip or flake (called *spalling*) are reduced when it is denser. Some amount of spalling is common in un-honed slate.

**Travertine** It is a sedimentary rock and one of the softer varieties of stone flooring. Over time, travertine will develop a soft, mellow patina. It comes in a wide range of shades and hues, from light cream to dark rust. Polished travertine resembles marble; tumbled travertine is characterized by pits, holes, and rounded edges that give it an antique look. Travertine flooring is porous and reacts quickly to acidic substances; thus, even the spilled orange juice may cause a stain. Regular application of a stone sealer helps to protect travertine flooring.

**Limestone** It is a cousin of travertine, although it is somewhat harder and, over time, will stand up better to everyday use. It has striking grain patterns that, when cut into long, narrow tiles, resemble wood planks. Limestone is porous and, hence, the flooring should be protected with a quality stone sealer to prevent stains, and resealed every few years.

**Terrazo flooring** Popularly called as *mosaic flooring*, they were once very popular in India, and are still used in some places. Unless these tiles are made with proper constituents and are well cured, they tend to get pitted. The flooring also requires polishing after it is laid, and, hence, proper time planning is necessary while adopting this flooring in a building. Terrazo is basically a decorative concrete in which the aggregates are white or coloured marble/ceramics chips (limestone, cuddapah stone, or other stone chips of similar characteristics and hardness can also be used), and marble powder or dolomite powder, or a mixture of the two. The binder used is cement (grey, white, or coloured). The cause of failure of mosaic tiles is often due to the addition of hard stone chips like quartzite. If coloured cement is obtained locally by mixing colouring oxides, it should first be mixed thoroughly in a ball mill to get a uniform colour. The pigment added should be less than 10% of the weight of cement. The mixture of marble powder and cement is usually in the ratio of 1:3 by mass.

First, the cement and marble powder are mixed thoroughly and then the marble chips are added to it in the ratio as shown in Table 2.8. They are again mixed thoroughly to get a homogeneous mix. Sufficient water to make the mix workable should be added.

**Table 2.8** Mix proportions for terrazo work

| Size of aggregates | Cement–marble powder to aggregate ratio |
| --- | --- |
| Grades 0 and 1 | 1:1.75 |
| Grades 2 and 3 | 1:1.50 |
| Grades 4 and 5 | 1:1.25 |
| Mixed aggregates | 1:1.50 |

The tiles can be made either by the dry process or the wet process. The dry process is simpler and preferred by manufacturers. The tiles, made using this process, consist of two layers – the *wearing layer* of chips and concrete mix and the *backing layer* of cement mortar (cement and sand in the ratio of 1:3, with a moisture content of about 12–18% only). The wet process consists of one wearing and two backing layers. The production consists of a steel mould fitted with a removable bottom plate, which is first filled with the wearing mix to about 5–6 mm depth. Over this layer, the moist backing mixture is spread to the overall thickness of the mould. Then, the mould is covered with a top plate and the mixture is subjected to a pressure of about 14 N/mm$^2$, using a hydraulic press. The pressure compresses the mixture to the required thickness. After withdrawing from the press, the tile is removed from the mould and stacked in a wooden rack, standing on their edges. They remain in this position of air curing for 8 to 12 h. Then, the rack is submerged in a water tank for 48 h. They are stored in a curing shed with water spraying for at least 2 weeks. They are removed and polished using a polishing machine (first polishing) before delivery to the construction site. Terrazo tiles should conform to IS 1237. As per this, double layer tiles are produced with the following dimensions (length × breadth × thickness): 200 × 200 × 15 mm, 250 × 250 × 16 mm, 300 × 300 × 20 mm, and 400 × 400 × 25 mm. As per IS 1237, the average percentage of water absorption of the produced tiles should not exceed 10. After the tiles are installed in the flooring of a building, on a bed of cement mortar and set, they are given two or three more rounds of polishing, using heavy machinery.

## SUMMARY

- Stone is an important and sustainable material for building construction.
- Geologists classify rocks into igneous, sedimentary, and metamorphic rocks.
- As per physical characteristics, they are classified as stratified, unstratified, and foliated.
- According to chemical characteristics, rocks may be classified as argillaceous, siliceous, and calcareous.

- The hardness of rocks is determined by the Mohs's scale, which measures it based on how easily it scratches.
- Igneous rocks are of volcanic origin and distinguished as volcanic rocks, plutonic rocks, and gangue.
- The three types of sedimentary rocks, which are found on land are clastic, chemical, and organic.
- Metamorphic rocks are formed when igneous and sedimentary rocks are subjected to heat, pressure, or both. They are classified as foliated and non-foliated.
- Common stones in India are granite, basalt, sandstone, limestone, marble, slate, and quartzite.
- Stones are obtained by quarrying, using plug and feather, channelling, or blasting with explosives. The process of removing moisture from pores is called seasoning. Proper size, shape, and finish are given to the cut rocks by dressing.
- Several factors during the quarrying, seasoning, finishing, and setting of stones may affect their durability.
- Stones are used in several applications, such as masonry (rubble and ashlar), dams, bridges, retaining walls, pavements, flooring, in foundations, and as road metal and aggregates.
- Examples of artificial stones are cement concrete, terrazzo, mosaic, and artificial marble.

## EXERCISES

### Multiple-choice Questions

1. Match the following:
   - (a) Stratified rocks
   - (i) contain clay-like components
   - (b) Argillaceous rock
   - (ii) repetitive layers having different texture/colour
   - (c) Foliated rocks
   - (iii) have a high proportion of calcium carbonate
   - (d) Calcareous rock
   - (iv) have distinct layers

2. Which of the following is unstratified?
   - (a) Sandstone
   - (c) Shale
   - (b) Limestone
   - (d) Granite

3. Marble is a
   - (a) Stratified rock
   - (c) Foliated rock
   - (b) Unstratified rock
   - (d) Argillaceous rock

4. Gneiss is a
   - (a) Siliceous rock
   - (c) Calcareous rock
   - (b) Argillaceous rock
   - (d) None of the above

5. Which of the following statements is correct for basalt?
   - (a) It is an igneous rock.
   - (b) It has the compressive strength of 50 MPa.
   - (c) It is porous.
   - (d) It is used in roof tiles.

6. Which of the following is the hardest mineral?
   - (a) Quartz
   - (c) Corundum
   - (b) Topaz
   - (d) Feldspar

7. Which of the following is the softest mineral?
   - (a) Calcite
   - (c) Fluorite
   - (b) Gypsum
   - (d) Talc

8. Sandstone is
   - (a) Igneous rock
   - (b) Clastic sedimentary rock
   - (c) Metamorphic rock
   - (d) Organic sedimentary rocks

9. Marble is
   - (a) Igneous rock
   - (b) Clastic sedimentary rock
   - (c) Non-foliated metamorphic rock
   - (d) Foliated metamorphic rock

10. Which of the following pairs is correctly matched?
    - (a) Sandstone
    - (i) Igneous rock
    - (b) Shale
    - (ii) Sedimentary rock
    - (c) Basalt
    - (iii) Metamorphic rock
    - (d) Quartzite
    - (iv) Argillaceous rock

11. Granite, after metamorphism, transforms to
    - (a) Quartzite
    - (a) Schist
    - (a) Gneiss
    - (a) Slate

12. Limestone, after metamorphism, changes to
    - (a) Marble
    - (c) Slate
    - (b) Quartzite
    - (d) Phyallite

13. Marble is usually quarried by
    - (a) Blasting
    - (c) Heating
    - (b) Excavating
    - (d) Plug and feather

14. Which of the following method is used for the manufacture of stone aggregates?
    - (a) Blasting
    - (c) Channelling
    - (b) Excavating
    - (d) Plug and feather

15. Which of following stones has the least percentage of water absorption by volume?
    - (a) Trap
    - (b) Slate
    - (d) Limestone
    - (c) Quartzite

16. Which of the following stones has the maximum percentage of water absorption by volume?
    - (a) Shale
    - (c) Slate
    - (b) Granite
    - (d) Quartzite

17. The specific gravity of good building stones lies between
    - (a) 1.8 and 2.4
    - (c) 2.8 and 3.5
    - (b) 2.4 and 2.8
    - (d) 3.5 and 4.5

18. Stones have compressive strength in the range of
    (a) 15–25 MPa          (c) 50–200 MPa
    (b) 25–50 MPa          (d) >500 MPa

19. The consolidant used to protect stones is
    (a) Barium hydroxide
    (b) Linseed oil
    (c) Ammonium bicarbonate
    (d) Sodium bicarbonate

20. The most suitable stone for building piers is
    (a) Granite            (c) Sandstone
    (b) Limestone          (d) Marble

21. Identify the correct match from the following pairs.
    (a) Breakwater         (i) Marble
    (b) Arches             (ii) Basalt
    (c) Road metal         (iii) Granite
    (d) Facing work        (iv) Sandstone (coarse grained)

22. Select the correct match from the following pairs.
    (a) Quartzite          (i) Ornamental work
    (b) Marble             (ii) Ballast
    (c) Slate              (iii) Walls
    (d) Laterite           (iv) Roofing

23. Which of the following stones has better fire resistance?
    (a) Granite
    (b) Limestone
    (c) Basalt
    (d) Marble

24. When stone is placed in masonry along its natural bed, the applied load will act
    (a) parallel to it.
    (b) normal to it.
    (c) at 45° to it.
    (d) at 30° to it.

## Review Questions

1. What are the geological, physical, and chemical classifications of rocks?
2. Explain the process of rock cycle.
3. Is stone a homogenous material? Name at least five minerals that are found in rocks.
4. What is Mohs's scale? How is it used to determine the hardness of minerals?
5. What are igneous rocks? Describe the different types of igneous rocks by giving at least two examples of each.
6. Discuss the formation, mode of occurrence, and types of sedimentary rocks.
7. How are metamorphic rocks formed? What are the two types of metamorphic rocks?
8. Describe the important characteristic features and uses of the followings rocks in civil engineering works: granite, basalt, syenite, sandstone, limestone, laterite, quartzite, marble, and slate.
9. What are vertical and horizontal quarries?
10. List the various methods used in quarrying stones.
11. What are the types of explosives generally used for blasting the rocks?
12. Describe the process of blasting rocks. State the precautions to be exercised.
13. What are the requirements of good building stones?
14. Briefly describe the following:
    (a) Seasoning of stone, (b) Dressing of stone, and (c) Preservation of stone
15. What factors affect the durability during quarrying, seasoning, finishing, and setting of stones?
16. What are the factors that result in the deterioration of stones?
17. What is the natural bed of stone? Why is it necessary to set a stone along its natural bed?
18. What are the precautions to be taken during construction to preserve stones?
19. What preservation methods are adopted on completed stonework?
20. What are consolidants? Describe the action of barium hydroxide used as a consolidant.
21. What considerations would guide you in selecting stone for use in the following situations:
    (a) Face work of a building      (b) Masonry      (c) Marine structures      (d) Flooring
    (e) Ornamental work              (f) Aggregates in concrete
22. List five construction works where stones are used. In addition, mention the type of stones that are suitable in each of these constructions.
23. What are the two types of stone masonry?
24. Differentiate between rubble and ashlar masonry.
25. What are stone veneers? Explain their use.
26. What are the advantages and disadvantages of stone masonry?
27. Compare and contrast stone masonry with brick masonry.
28. What are artificial stones?
29. List the advantages of artificial stones.

30. What are the different types of stone floorings?
31. What are the advantages of using marble for flooring? What are the locations where marble flooring is not used?
32. What is terrazzo? Explain the manufacturing of terrazzo tiles.

## ANSWERS

**Multiple-choice Questions**

| | | |
|---|---|---|
| **1.** (a)–(iv) (b)–(i) (c)–(ii) (d)–(iii) | **2.** (d) | **3.** (a) |
| **4.** (a) | **5.** (a) | **6.** (c) |
| **7.** (d) | **8.** (b) | **9.** (c) |
| **10.** (a)–(ii) (b)–(iv) (c)–(i) (d)–(iii) | **11.** (b) | **12.** (a) |
| **13.** (d) | **14.** (a) | **15.** (b) |
| **16.** (a) | **17.** (b) | **18.** (c) |
| **19.** (a) | **20.** (a) | **21.** (a)–(iii) (b)–(iv) (c)–(ii) (d)–(i) |
| **22.** (a)–(ii) (b)–(i) (c)–(iv) (d)–(iii) | **23.** (d) | **24.** (b) |

# CHAPTER 3
# BRICKS AND BRICK SUBSTITUTES

## 3.1 Introduction

Clay products are one of the most important classes of building materials. They include bricks, tiles, ceramics, terracotta, earthenware, and stoneware. They are basically made by moulding, drying/burning prepared clay. This chapter deals with bricks. Other types of clay materials are discussed in Chapter 16. Brick is one of the oldest manufactured building materials in the world. Hand-moulded clay bricks have been found in the lower layers of Nile deposits in Egypt, dating as far as 14,000 BC. The first sun-dried bricks were made in Mesopotamia, in the ancient city of Ur (Iraq), around 4000 BC (Christine, 2004). Starting from 5000 BC, clay bricks were fired to have durability. From the Middle East, the art of brick making spread West to present Egypt and East to Persia and India. The Romans used fired bricks and were responsible for its introduction and use in England (Christine, 2004).

Today, brick is one of the most used masonry units. Bricks are produced in numerous classes, types, patterns, textures, colours, and sizes. They are produced in bulk quantities and used in a variety of applications. According to an estimate, the global annual production of bricks was about 1500 billion bricks in 2004, out of which about 87% is from Asia (Baum, 2007). Brick production is highly concentrated in the following four countries: China (54%), India (11%), Pakistan (8%), and Bangladesh (4%). In India, the brick production is about 200 billion bricks per year, consuming about 35–40 million tonnes of coal per year (2016 estimate).

The main raw material for bricks is clayey soils, containing 50–60% silica and 20–30% alumina with small quantities of other substances, such as lime, manganese, sulphur, phosphates, and oxide of iron (which gives the colour); the composition of the soil depends on the locality from which the soil originates. These substances, if not present in the clay, are added during mixing. The clayey soil is usually obtained from open pits with may affect the environment by disrupting drainage, vegetation, and wildlife habitat (Hendry & Khalaf, 2001).

The two basic categories of bricks are: fired and non-fired bricks. *Non-fired bricks*, also known as mud-bricks, were sun-dried and used in olden days to construct temporary structures. Thus, there was no chemical change and the bricks only lost the water added during mixing. Such sun-dried bricks may contain mechanical binders, such as straw. Non-fired bricks crumble when placed in water. However, the present-day fired or *burnt bricks* are heated to high temperatures in kilns, resulting in some chemical changes and hence are more durable and resistant to water. The transition from the traditional hand-moulding method of production to a mechanized form of mass-production took place during the first half of the 19th century. A successful brick-making machine was patented by Henry Clayton, at the Atlas Works in Middlesex, England, during 1855. This machine was capable of producing up to 25,000 bricks daily with minimal supervision. Another machine was patented by Richard A. VerValen of New York during 1852.

The term *brick* is used to signify solid clay masonry units. In general, clay bricks are small, rectangular blocks made of fired clay. Most bricks produced today have standard nominal modular sizes and shapes. Bricks are manufactured by grinding or crushing the clay in mills and mixing it with water to make it plastic. The plastic clay is then moulded, textured, dried, and finally fired in kilns (the various types of kilns exist) at about 800–1300°C. Bricks are usually laid in courses and different patterns known as bonds (e.g., most popular Flemish Bond, oldest English bond, Stretcher Bond, Rat Trap Bond, etc.). Such a collection of bricks, which are bonded together by mortar of different kinds, is known as *brickwork*, and is used to construct the different kinds of walls. One of the remarkable structures to be built with both, burnt and sun-dried bricks, during 210 BC is the Great Wall of China. Several structures exist all over the world in all kinds of buildings, testifying the long life of bricks (Fig. 3.1). It is estimated that about 65% of the bricks produced are used in residential buildings and the rest in commercial, industrial, and institutional buildings. Several alternates are being used in place of bricks, incorporating waste materials like fly ash, which include concrete blocks.

(a)          (b)

**Fig. 3.1** Historic brick building (note the brick arches in both the buildings) (a) Harpers Ferry, MD, USA, built in 1848 using lime mortar (b) College of Engineering, Guindy, Chennai, built in 1920 (*Courtesy*: Prof. M. Sekar)

## 3.2 Properties of Brick Clay

The clay that is used in brick making should have enough plasticity such that it can be easily moulded. Though better mouldability is obtained with soils having high clay content, they may result in higher shrinkage and cracking. Laboratory tests such as liquid limit, plastic limit, and shrinkage limit could be done to determine the suitability of soil for brick making. IS 2117:1991 suggests a quick field test to determine the suitability of soil for brick making. As per this, the soil has to be ground to a fine powder and mixed with sufficient water. A lump of this soil is taken in hand and a ball of 80 mm diameter is formed. Then, the ball is kept in the Sun. If the ball is deformed on drying and crumbles easily by the application of light pressure, it is inferred that the sand content is excessive. On the other hand, if the ball is hard but shows cracks on the surface, the sand content is not sufficient.

### 3.2.1 Types of Clay

Clays occur in the following three principal forms, and have similar chemical compositions but different physical characteristics (TN 9 of BIA, 2006).

**Surface clays**   They are found near the surface of the earth, typically at river bottoms.

**Shales**   These are clays that have been subjected to high geologic pressures and vary in hardness from slate to partially decomposed rock.

**Fire clays**   These are usually mined at greater depths below the surface than other clays. Generally, fire clays contain fewer metallic oxides than shales and surface clays, and have more uniform chemical and physical properties.

Surface and fire clays differ from shales in physical structure but are similar in chemical composition. The manufacturer minimizes the variations in chemical composition and physical properties of naturally available clays by mixing clays from different sources and different locations in the pit. Still bricks from different batches from the same manufacturer will have slightly different properties. Naturally, bricks from different manufacturers, which may have the same appearance, may differ in other properties.

### 3.2.2 Composition of Good Brick Clay

The raw material required for making bricks is pure clay (a recyclable natural resource), sand, and water, which when mixed form a plastic viscous mass. This could be moulded, dried, and, when heated red hot to about 800–1300°C, acquires hardness and strength. In order to produce better quality bricks, the clay soil should be (1) plastic when mixed with water, (2) have enough tensile strength to keep its shape, and (3) the clay particles must fuse together. Approximately, 3 m³ of clayey soil and 600 L of water are required to make 1000 bricks.

Sand can be added to the clay soil to make minor adjustments to the quality. Sand is also used as a stabilizer in the mixture. Sand may also be used in the brick-making process to keep the bricks from sticking to the moulds. As per IS 2117:1991, the mixture of clay should be in the following proportion: clay 20–30% by mass, silt 20–35% by mass, and sand 35–50% by mass. In addition, the total content of clay and silt should be greater than 50%.

As per IS 2117:1991, the plasticity index of brick material should be in the range of 15–25. Additives such as fly ash, rice husk ash, and basalt stone dust may be added to modify the shaping, drying, and firing behaviour of clay. Such additions prevent agricultural lands being quarried for raw material and able to utilize large quantities of waste materials that are produced as by-products of industrial goods (see Section 3.10.1 for more discussion on fly ash bricks).

The functions of the various ingredients present in brick material are as follows:

**Alumina ($Al_2O_3$)**   A good brick material should contain 20–30% of alumina. Alumina imparts plasticity to the clay so that it can be easily moulded. When excess alumina is present in the clay, with inadequate quantity of sand, the raw bricks shrink and warp during drying /burning and become too hard when burnt.

**Silica ($SiO_2$)**   A good brick material should contain about 50–60% of silica. Silica exists in clay either as free or combined form. As mentioned earlier, sand is mechanically mixed with clay and in combined form it exists as a chemical composite of alumina (for example, in kaolin group of clay as $Al_2O_3 \cdot 2SiO_2 \cdot 2H_2O$, and in montmorillonite group as $Al_2O_3 \cdot 4SiO_2 \cdot nH_2O$). Presence of silica prevents cracking, shrinking, and warping of raw green bricks and thus imparts uniform shape to the bricks. When the brick material has lime and oxides of iron, silica fuses at lower temperatures in the kiln. Proper proportioning of silica in the brick material is important for the durability of bricks. Excess silica can destroy the co-hesion between the particles of brick material and the resultant bricks may become brittle.

**Lime (CaO)**  As per IS 2117:1991, the total lime and magnesia in the case of alluvial soil should be less than 1% and in other cases should be less than 15%. A small quantity of lime, not exceeding 5%, is desirable in finely powdered state, to prevent shrinkage of raw bricks (higher size particles of lime, even in the size of a pin-head may cause flaking of bricks). It has to be noted that the presence of sand alone is infusible. However, in the presence of lime, sand slightly fuses at the high kiln temperature. Such fused sand works as a hard cementing material for brick particles. However, excess lime causes bricks to melt and, hence, they may not retain their shape. Moreover, lime is converted into quicklime after burning and this quicklime slakes and expands in the presence of moisture and exerts enough pressure to break small chips off the surface of the brick. These chips, called 'lime pops' usually occur within the first year of manufacture of brick, and only affect the aesthetic appearance, and does not affect the strength and durability (www.brick.com). Similar to lime, manganese oxides also result in 'manganese pops', which look like lime pops, except the centre is normally black or dark brown, instead of white as in the case of lime pops.

**Iron oxides (Fe$_2$O$_3$)**  A small quantity of oxides of iron of less than 6–7% is desirable in good brick material. Like lime, it acts as a flex to fuse the silica during burning. It also imparts red colour to the bricks. Non-calcareous clays typically contain silicate of alumina, feldspar, and iron oxides, and will burn to a brown, pink, or red colour, depending on the iron oxide content. When excess iron oxide is present, it will make the bricks dark blue or blackish. When less quantity of iron oxide is present, it will result in yellow-coloured bricks. Calcareous clays have 15% calcium carbonate and, hence, result in yellow- or cream-coloured bricks.

**Magnesium oxide (MgO)**  A small quantity of magnesia (less than 1%) in brick material imparts a yellow tint to the bricks and may decrease the shrinkage. However, the excess of magnesia may lead to the decay of bricks.

### 3.2.3 Harmful Ingredients in Brick Clay

The following are some of the ingredients, which are undesirable in the brick material:

**Lime**  As mentioned already, the excess quantity of lime (>10%) is undesirable in brick material.

**Iron pyrites**  If iron pyrites are present in the brick material, they will be oxidized during burning and result in the bricks being crystallized and disintegrated.

**Alkalis**  They may be present in the form of soda and potash. During the burning process in the kilns, these alkalis act as a flux and cause bricks to fuse, twist, and warp. As a result, the bricks are melted and may lose their shape. Further, when bricks are used in the masonry, the remaining alkalis will absorb moisture from the atmosphere and result in damp conditions. This moisture, when evaporated, leaves behind grey or white deposits, called *efflorescence*, on the wall surface, affecting the appearance of the brick wall. Hence, alkalis should be limited to 10% by volume.

**Pebbles**  The presence of pebbles or grits of any kind is undesirable in the brick material because they will not allow the clay to be mixed uniformly and thoroughly, resulting in weak and porous bricks. In addition, the brick containing pebbles will not break regularly as desired.

**Vegetation and organic matter**  The presence of vegetation and organic matter in the brick material assists in burning. However, when they are not burnt completely, the bricks become porous, due to the entrapment of gases that are produced during the burning of the carbonaceous matter.

## 3.3 Manufacture of Bricks

In general, the manufacturing processes of clay blocks, facing bricks, and clay roof tiles are quite similar. They consist of the following steps:

1. Extraction (mining) and storage of raw materials
2. Preparation (crushing, blending, grinding, and screening)
3. Forming or moulding
4. Drying
5. Colouring (required for clay roof tiles only)
6. Firing
7. Packing and distribution

**Extraction and storage of raw materials**   The most important raw material for bricks and roof tiles is pure clay, which is a recyclable natural resource. Clay is extracted from different clay pits using modern equipment, and transported to plant storage areas. Since the clay pits are usually located close to production facilities, transportation is usually minimal. Sufficient quantity of raw materials, required for many days of plant operation, are often stored for the continuous production of bricks. If clay is in large lumps, it undergoes preliminary crushing before storage. In areas where black cotton soils occur, a more elaborate method of processing, as developed by the Central Building Research Institute (CBRI), Roorkee and as given in Clause 7.1 of IS 2117:1991 should be followed.

IS 2117:1991 suggests that additives, such as fly ash, sandy loam, rice husk ash, and stone dust, be spread over plain ground surface on volume basis, mixed with the excavated, puddled, watered soil mass, and left over for weathering for at least 1 month, if necessary. The weathering may help to develop a homogeneous soil mass, especially when the soil is obtained from different sources, and will also eliminate the impurities, which will get oxidized.

**Preparation**   The following steps are involved in preparation.

1. From storage, clays from different pits and locations are usually blended to produce more uniform raw materials and get uniform colour. Then, they pass through the grinding and milling process to reduce it to relatively smaller pieces.
2. Materials, such as gravel, coarse sand (practical size more than 2 mm), lime, and kanker particles, and vegetable matter, should be removed. Then, the material is processed through an inclined vibrating screen to control particle size, and the coarse particles are returned to the grinder for regrinding.
3. Fineness of the clay influences not only the external appearance of the finished brick, but also physical characteristics, such as compressive strength and water absorption.
4. The next step in preparation is called *tempering*, in which a homogeneous, plastic clay mass is produced by adding water (the quantity added depends on the soil, and the production method being used – this may range from 1/4 to 1/3 by mass of the soil).
5. Other materials, such as lime, pulverized fuel ash, or crushed clinker, may be added to act as fuel, while pigments may also be incorporated to produce specific colours.
6. There are two different types of *extrusion systems* for mixing the brick material: (1) *De-airing machines* (referred to as a *pug sealer*), which remove air from clay by passing into a vacuum after pugging (Fig. 3.2), and (2) *Non de-airing pug-mills*, which simply consolidate and extrude the clay without de-airing it.
7. Both of them can be used with dies to extrude shapes. Both systems contain some kind of auguring machine, rotating slowly inside a tub. The auger has blades arranged in a spiral pattern along a stiff strong central shaft, so that it chops and pushes the clay along inside the tub.

8. This central shaft may be rotated by diesel or electric motor. The tub has a hopper at the input end where the blended clay, with the required amount water, is fed (the amount of water used depends on the method of forming).

9. The tub has a tapering, closed barrel at the other end, with a hole having half the diameter of the main barrel. Squashed and kneaded clay emerges from this end into a long cylinder; it is squashed smoothly together as it comes out the smaller hole.

**Fig. 3.2** Mid-sized 45A Extruder from JC Steele & Sons, Inc., Statesville, NC, USA; note the auger blades

10. At this point in a de-airing process, material falls into a second machine, commonly referred to as an *extruder*. Extruders have a fairly long barrel arranged into a collecting section and a pressure building section, conveying the mixture towards the die and building the pressure needed to force the mixture through the die.

11. An airtight lid fits closely in the chamber between the two machines, and a pipe leads off to a vacuum pump, which runs all the time when the machine is in use.

12. The vacuum pump sucks out as much air as possible from the clay as it passes along the barrel, to give a denser bubble-free product.

13. The yield may be in the range of 1500–15,000 bricks, depending on the size of pug-mill. After pugging, de-airing, and pressurizing, the plastic clay mass is ready for forming through the exit of the extruder die in a green state and the required final product shape.

Note that steps 8–13 belong to the *machine moulding process*.

**Forming or moulding the brick**    Basically, there are three methods of forming: (1) Stiff-mud process, (2) Soft-mud process, and (3) Dry-mud process.

**Stiff-mud process**    The stiff-mud process has the following features.

1. In this process, which is also called the *extrusion process*, water in the range of 10–15% by weight is mixed into the clay to produce plasticity.

2. If a non-de-airing pug-mill was used for pugging, the plastic clay goes through a de-airing machine to remove air pockets and bubbles, which increase workability and strength.

3. As mentioned earlier, the clay is then forced by an auger through an extrusion die, producing a continuous column of clay of desired size and shape. As the clay column leaves the die, textures or surface coatings may be applied.

4. The clay columns pass through an automatic cutter, which cuts individual bricks of proper length by the use of taut wire. Hence, they are called *wire-cut bricks*.

5. Cutter spacing and die sizes must be carefully calculated to compensate for normal shrinkage that may occur during drying and firing.

6. In this process, the shrinkage of clay is low and the size of bricks is easier to control. A further advantage of this process is that drying time is relatively short. The bricks have at least one shallow frog. Most structural bricks are produced by this process, as it results in hard and dense bricks.

7. In this process, holes or other perforations can be formed in the bricks, using proper dies, in order to reduce the weight of the brick. These holes must be less than 25% of the volume of the brick.

**Soft-mud process**    The soft-mud or moulded process is used for making bricks only and suitable for clays containing too much water to be extruded by the stiff-mud process. During tempering, 20–30%

water is added to the clay. The bricks are formed using moulds (the oldest method of production). To prevent clay from sticking, the moulds are lubricated with either sand or water.

Bricks made with sanded moulds are called *sand-struck bricks* and bricks made with water-lubricated moulds are called *water-struck bricks*. Sand-struck bricks usually have sharper corners. Brick may be produced in this manner by machine or hand. Due to the high drying shrinkage of such wet mixes and the plasticity of the unfired brick, the size and shape of such bricks are fairly variable. Therefore, they are not suitable for use in thin mortar beds.

**Dry-press process**  This process is particularly suited to clays of very low plasticity. Clay is mixed with a minimal amount of water (up to 10%). Dry mix is fed into machines, which form bricks in steel moulds under high pressure (3.4 to 10.3 MPa) using hydraulic or compressed air rams. Brick formed in this process has the least amount of shrinkage and warpage and, hence, will have most precise size. They are called as *pressed bricks*. These different forming processes are compared in Table 3.1.

**Table 3.1** Comparison of different forming processes of brick

| Forming process | Water content in raw mix (%) | Used for producing |
| --- | --- | --- |
| Stiff-mud or extrusion process | 10–15 | Engineering bricks, facing bricks, and other brick products with very accurate dimensions |
| Soft-mud process | 20–30 | Common bricks, in which the size and shape are fairly variable |
| Dry-press process | <10 | Engineering bricks and special types of bricks |

**Different types of moulding**  As mentioned earlier, the moulding of bricks can be done by hand or using machines. Hand moulding of brick, which is a part of soft-mud process, is extensively used in India. This could be done either on the ground (*ground moulding*) or on the table (*table moulding*).

**Ground moulding**  This method is adopted when a large and levelled area of land is available. This process has the following features.

1. In this process, the ground is levelled, plastered smooth, and sprinkled over with sand.
2. Bricks moulded by this method may have their lower faces objectionably rough; to overcome this, moulding blocks or boards may be used at the base of the mould.
3. The mould used may be made of wood or steel; a typical wooden mould is shown in Fig. 3.3. As the bricks shrink during drying, moulds are made larger than the size of burned bricks.
4. The exact percentage increase in dimension should be determined by actual experiments – usually, it varies from 8 to 15%. These bricks may have frog (*Frog* is an indentation provided in a face of the brick, indicating the trade mark of the manufacturer. Bricks are laid in masonry with frog up.
5. Importantly, frog provides a key for the mortar and holds the bricks on top firmly in place. The process consists of shaping a lump of well-pugged brick material, having a volume slightly more than that of the brick, by hand.
6. It is then rolled in sand and with a jerk dashed into the mould, to completely fill it. Moulders then blow and press the clay with their fists and press it at the corners and edges of the mould with their thumb.
7. A metal plate with a sharp edge, known as *strike*, is then used to scrap off the excess clay at the top of the mould and the top surface levelled.
8. After the brick has been moulded, the mould is given a gentle stroke to release the mould and the mould is lifted to release the brick, to dry on the ground. The aforementioned process is repeated for other bricks.

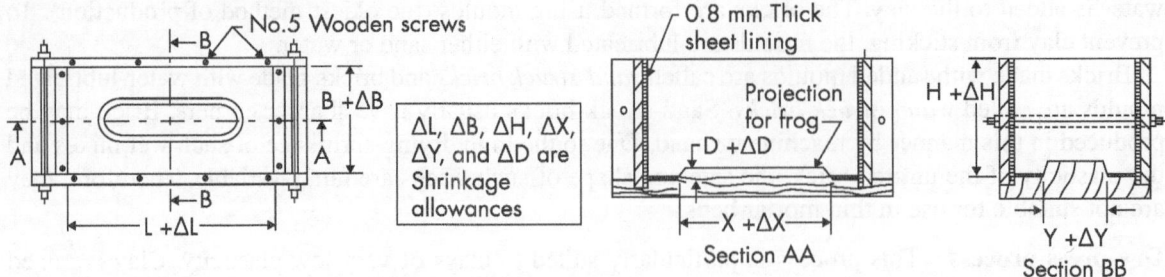

**Fig. 3.3** Typical wooden mould used in hand-moulding

**Table moulding**  In the table moulding, the bricks are moulded on stock boards, pushed on the mould-ing table of size 2 m × 1 m (see Fig. 3.4). Stock boards have the projection for forming the frog. The process of filling clay in the mould is the same as that of ground moulding. A thin board, called *pallet,* is then placed over the mould. The mould containing the brick is then smartly lifted off the stock board and inverted so that the clay filled mould rests on the pallet (IS 2117:1991). The mould is then given a gentle blow and lifted, leaving the brick on the pallet. One more pallet is placed on the brick and it is carried to drying site between the two pallets. It is allowed to dry on side. A skilled brick-maker can produce up to 100 bricks per hour.

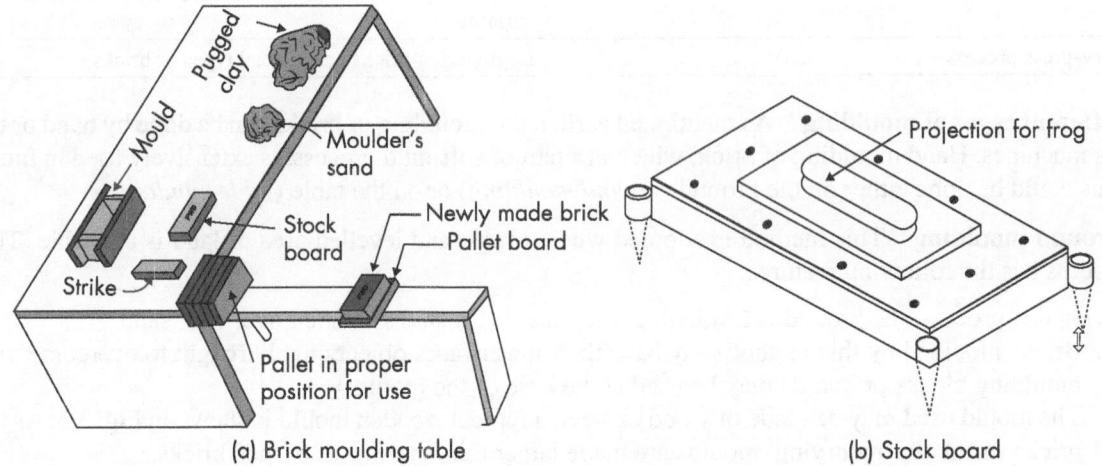

**(a) Brick moulding table**    **(b) Stock board**

**Fig. 3.4** Table moulding of bricks

**Machine moulding**  In this process, the clay bricks are shaped in hand-operated or power-driven machines, such as a hand-screwpress, a soft-mud moulding machine, an extruder (shown in Fig. 3.2) or a semi-dry press. The CBRI, Roorkee, has developed a portable hand-operated C-Brick moulding machine, based on vibro-compaction technique, which could be used to manufacture bricks at low cost.

**Drying**  Wet brick from moulding or forming machines may contain 7–30% moisture, depending upon the forming method. Drying causes shrinkage, and should be allowed for when bricks are being formed, in order to produce bricks or tiles of the required size. Drying is accomplished by either (a) by storing bricks in open drying areas or (b) in drying chambers of kilns.

In natural drying of bricks under the Sun in an open area, the bricks are stacked with sufficient gap between them to provide access to air on all sides of the bricks. (The recommended method of stacking for drying is shown in Fig. 2 of IS 2117:1991.) To avoid the rapid drying, the bricks are covered using a protective covering (e.g., plastic sheeting). If it is likely to rain, drying should be done under a roof. However traditionally, such natural drying of bricks is done only during the dry season.

If the bricks are dried in drying chambers of kilns, they are subjected to temperatures ranging from 38°C to 204°C, and the drying time varies from 24 to 48 h, depending on the type of clay. Although heat may be generated specifically for drying chambers, it usually is supplied from the exhaust heat of kilns to maximize thermal efficiency. In all cases, the heat and humidity are carefully regulated to avoid rapid shrinkage, which will cause cracking in the brick. After drying, the moisture content of bricks drops to below 2%.

**Colouring (in clay roof tiles only)**   Colouring of clay roof tiles may be necessary when they do not remain natural red after firing. To colour them, they are either *engobed* or glazed. The surfaces of engobed clay roof tiles can be made to appear as matt, matt-glossy, or glossy. Engobes (clay slurries) may be applied by dipping, dousing, centrifugal casting, or spraying on the unfired clay tiles. Glazes are made up of silica, metal oxides (acting as a flux), alumina, colourants, and opacifiers that make each glaze unique. Glaze may be applied before or after drying the tiles in kilns and it closes all pores and makes the tiles extremely water-resistant.

**Firing**   After drying, the products are transferred to a kiln, where they are fired over a period of 36 to 150 h, depending on the type of kiln, type of clay, type of glaze, and other variables. There are several types of kilns used by manufacturers (see Section 3.4). Bricks must be set in kilns in the prescribed pattern to allow the free circulation of hot kiln gases (for example, see Fig. 3 of IS 2117:1991). During firing, the components of clay, such as silica, alumina, and oxides of iron, are fused together giving brick the strength and durability. Firing may be divided into the following six general stages.

1. Water-smoking (evaporation of free water)
2. Dehydration
3. Oxidation
4. Vitrification
5. Flashing
6. Cooling

All, except flashing and cooling, are associated with rising temperatures in the kiln.

During firing, the bricks go through a number of property changes. Although the actual temperatures will differ with clay or shale, residual moisture is driven off (called *water-smoking*) at temperatures up to about 205°C, dehydration from about 150 to 980°C, oxidation from 540 to 980°C, and vitrification from 870 to 1315°C – the brick colours are formed at this stage (TN 9 of BIA, 2006). Carbonaceous material, which may be present in the clay or added as fuel, is ignited at temperatures between 200°C and 900°C, which results in a more open, lower density structure. Near the end of the burning process, the bricks may be *flashed* to produce colour variations. This is done by injecting natural gas at the appropriate time. When this extra fuel burns, patches and colour variations are formed throughout the stack of brick. Recent kiln and firing technologies have reduced the required firing time by up to two-thirds in recent years. Additionally, the residual heat from the cooling process is recovered and recycled in the drying process. It has to be noted that the energy requirements for brick-making occur mainly during the drying and firing stages of brick making. Cooling takes place over a 48-to-72-h period, depending on the type of kiln. Cooling must be carefully controlled as the rate of cooling has a direct effect on colour; rapid cooling will also cause cracking in the brick.

**Vitrification**   Silica, alumina, boron, and phosphorous, in the presence of some flux-forming metal oxides and fired at 870 to 1315°C, will fuse into a hard, glassy non-crystalline material. This process is called *vitrification*. As the firing temperature increases, the degree of vitrification also increases. Hence, the temperature at which bricks are fired is an important factor in the manufacture of bricks. If under-fired, the bonding between the clay particles will be poor and the brick will be weak. If the temperature is too high, the bricks will melt or slump. Vitrification does not have to be complete, and does not actually occur in many of the small traditional brick-making plants around the world. However, even part vitrification gives sufficient strength and significantly reduces the permeability of bricks.

**Packing and dispatch**    After the fired products have cooled down, some quality inspection is done in laboratories and the bricks are unloaded from the kiln. They are sorted, graded, packaged, and taken to storage yards or loaded on trucks for shipment.

## 3.4 Different Types of Brick Kilns

Kilns first started in pits; walls were added later. A chimney stack was incorporated to improve the air flow or draught of the kiln, and to help burn the fuel more completely. Over time, kilns have been modified with the varying degrees of efficiency and cost. The different types of brick kiln technologies include the following (Akinshipe & Kornelius, 2017).

1. Clamp kiln
2. Up-draught kiln (UDK)
3. Down-draught kiln (DDK)
4. Bull's trench kiln (BTK), which can have a movable or fixed chimney
5. Zigzag kiln
6. Hoffman (annular) kiln
7. Vertical shaft brick kiln (VSBK)
8. Tunnel kiln.

Less popular types of kilns include the Habla kiln (an energy-efficient variant to the Zigzag kiln invented in Germany), the igloo or beehive kiln (used in Zimbabwe), the Kondagaon kiln, and the Bhadrawati kiln (Akinshipe & Kornelius, 2017). Other less common kilns include the modified clamp kilns, such as the Scove kiln, and the Scotch kiln.

### 3.4.1 Classification of Brick Kilns

There are two types of classification of brick kilns: (1) based on the production process of kilns and (2) based on the flow of air in the kilns.

**Classification based on production process**    The kilns for firing bricks are usually classified based on the production process into two basic types based on the production process: (1) Intermittent kilns and (2) Continuous kilns (see Fig. 3.5).

In *intermittent kilns*, bricks are fired in batches and can be further sub-divided into two categories: *intermittent kilns without chimney*, which do not have any stack/chimney to guide the flue gases (e.g., clamp, scove, and scotch kilns), and *intermittent kilns with chimney*, which have a stack/chimney to create draught for releasing the flue gases at a higher level in the atmosphere (e.g., DDK and climbing kilns). In both types of intermittent kilns, bricks and fuel are stacked in layers and the entire batch is fired at once. After the fire dies down, the bricks are allowed to cool. The kiln is then emptied, refilled again with raw bricks, and a new fire is started for the next batch of bricks. Thus, the heat contained in the hot flue gases, in the fired bricks, and the kiln structure is lost. They have low fuel efficiency and pollute the atmosphere.

In *continuous kilns,* fire will always burn and bricks are warmed, fired, and cooled simultaneously in different parts of the kiln. The continuous kilns are further sub-divided into two categories: moving-fire kilns and moving-ware kilns. In a *moving-fire kiln*, the fire progressively moves round a closed kiln circuit, with the assistance or help of a chimney or suction fan, while the bricks remain stationary (e.g., the Hoffman kiln, fixed chimney bull's trench kiln (FCBTK), and the habla-zigzag kiln). Whereas, in a *moving-ware kiln*, fire remains stationary, while the bricks and air move in counter-current paths (e.g., the tunnel kiln and the vertical shaft brick kiln). These kilns widely vary in efficiency, emissions, and productivity.

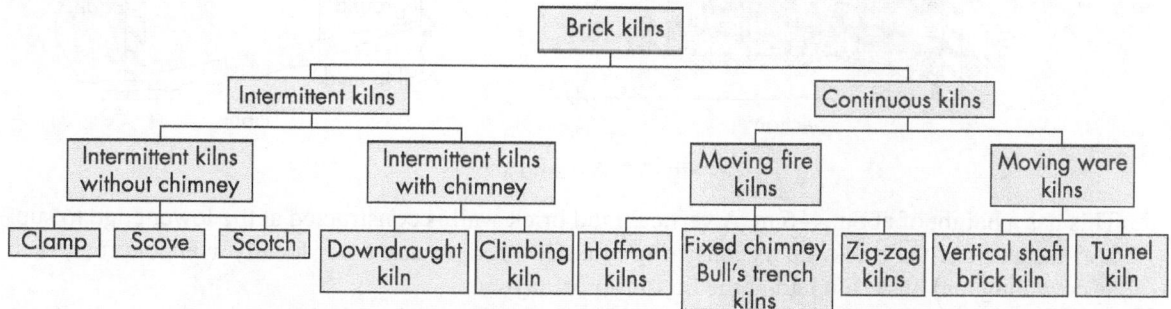

**Fig. 3.5** Classification of brick kilns based on production process

**Classification based on air flow** The direction of air flow in the kiln, with respect to the brick stacking, helps identifying the brick kilns as UDKs, DDKs, and cross-draught kilns (CDK). In an UDK, air enters the kiln from below, and is heated by the fire and moves upwards through the brick stacking, thereby transferring the heat to the bricks. As the upward movement of heated air is a natural, it does not require a stack or fan to cause the air flow. Clamps and VSBK are examples of UDKs. In a DDK, first the air is heated with fire and the hot air is then made to enter the kiln from the top and pass through the brick stacking, with the help of draught created by a chimney. In such kilns, the bricks are not in direct contact with the fire. The DDK is an example of this category of kilns. In the CDK, air is made to flow horizontally through the brick stacking. This type of air movement is made possible either by the natural draught created by a chimney or the forced draught through the use of fans. Since the flow of air is horizontal, these kilns are also called horizontal draught kilns. Examples of CDKs include the Hoffman kiln, fixed chimney bull's trench kiln, and the tunnel kiln (www.breathelife2030.org).

**Fuels used in brick kilns** In brick kilns, generally solid fuels, such as coal, wood, and sawdust; agricultural residue, such as mustard stalk, rice husk, and coffee husk; industrial wastes and by-products, such as used rubber tyres, and pet-coke, are used. Apart from solid fuels, bricks can also be fired using natural gas, diesel, and biogas. Relevant fuel properties, the calculation of specific energy consumption, and instruments required for the energy monitoring of brick kilns are discussed in Kumar and Maithel (2016).

Only the brick kiln technologies that have been used in India are briefly discussed below. The details of other technologies may be found from the references provided in the ORC.

### 3.4.2 Clamp Kiln

The *clamp kiln* is an intermittent kiln, which is primitive and traditional, lacking a permanent structure. It was invented by the Egyptians around 4000 BC. The clamp kiln is one of the most commonly used brick-firing technique in developing countries, including India (25–40%) and South Africa (68–85%). It has the following features.

1. This is generally trapezoidal in plan whose shorter edge among the parallel sides is below the ground with the surface rising constantly at about 15° to reach the other parallel edge over the ground (see Fig. 3.6).

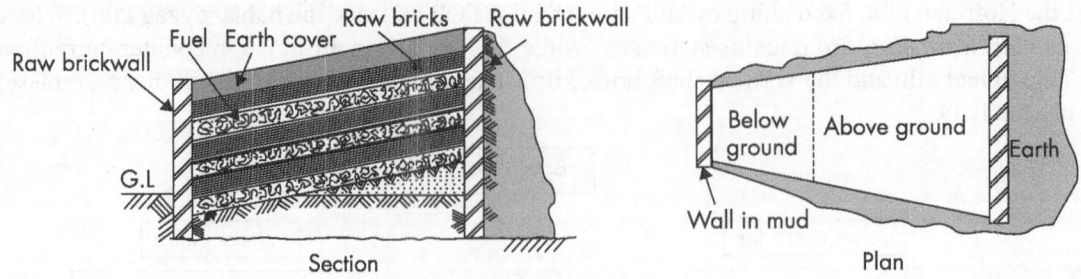

**Fig. 3.6** A clamp kiln

2. This has a height of about 4–6 m. A vertical mud brick wall is constructed at the lower edge to support the stack of the brick. Fuels consisting of coal, wood, or any other locally available fuel, such as cow dung and husk, are laid at the bottom-most layers.

3. Another layer of about 4–5 rows of bricks is laid and then again a fuel layer is laid over it. Some space is left between bricks for the free circulation of hot gasses.

4. The alternate layers of bricks and fuel is covered at the top and at sides with earth in order to preserve the heat. When the bottom layer of the kiln packed with fuel is ignited, it sets the bricks on fire one layer at a time until the whole kiln is ablaze.

5. Each clamp kiln can burn about 25,000–100,000 bricks, and may take about 3 months to burn and cool the bricks.

6. Clamp kilns are labour intensive and often operated in clusters.

7. They burn fuel inefficiently and are highly polluting. However, they are simple to build and, hence, could be located close to a clay source, minimizing the cost of transportation.

8. The bricks at the bottom are over-burnt while at the top are under-burnt. The bricks may also lose their shape, as they descend downwards once the fuel layer is burnt.

9. About 20–30% of bricks produced using clamp kilns are of poor quality. In addition, clamps cannot be operated during monsoon season.

### 3.4.3 Down-draught Kiln

In early 19th century, the product quality and energy efficiency of the kilns were improved. Hence, the UDK and the DDK were invented. The main advantage of this kiln is that the fuel and fuel residue do not come into contact with the kiln charge and, hence, no pollutants are deposited on the surface of the products. The DDK has the following features.

1. The DDK is an intermittent kiln in which the bricks are fired in batches. It has a permanently built structure consisting of rectangular firing chamber that is connected to a chimney through an underground flue duct, and a barrel-vaulted roof.

2. The inner surface of the kiln is constructed with refractory bricks. It has fireboxes at the bottom of the chamber on both sides where the fuel is burnt.

3. Usually, two chambers are connected to a single chimney and fired alternately. Each chamber has a capacity of firing 20,000–40,000 bricks in a batch.

4. The bricks stacked in the chamber/kiln are not in direct contact with the flames. The hot gases from the burning fuel are deflected to the roof of the kiln, which are then drawn downwards by the chimney draught through the green bricks to fire them, and then are out through the chimney stack.

5. Continuous feeding of fuel (e.g., by coal, gas or oil, firewood, and twigs and branches) for about 30 h ensures a uniform heat distribution in the kiln until the target temperature is attained.

6. This target temperature is maintained for a specific period until the fire subsides, thereby ensuring better thermal performance and lesser heat loss.

7. The kiln cools down in 2–3 days. The total time required for a batch from loading and firing of green bricks to cooling and unloading of fired bricks is around 7–10 days.

8. Owing to a uniform heat distribution in the DDK, the percentage of good quality bricks is higher than clamp kilns. However, the DDK has limited heat-recovery features.

9. Other kilns with similar configuration to the DDK are the UDK and the CDK, differing in the direction of the heat flow. It is estimated that there are about 300 DDKs with a capacity of 0.24 billion bricks per year, which is about 1% of the 250 billion bricks produced in India.

### 3.4.4 Vertical Shaft Brick Kiln

The vertical shaft brick kiln (VSBK) was invented in China in 1958 as a modification to the traditional updraft intermittent kiln. The VSBK is a continuous, updraft kiln with a stationary fire and a moving-brick arrangement. Figure 3.7 shows the schematic diagram of the working principle of VSBK. The kiln works in the form of a 'counter-current heat exchanger', since heat exchange takes place between the continuously flowing updraft air and the intermittently moving bricks downward. A VSBK consists of a long, rectangular, vertical shaft through which green bricks and fuel are lowered from top to bottom in batches (the shaft is made of fire-resisting refractory bricks). The height of the shaft is about 6 to 10 m and the cross-section of the shaft ranges from $1.0 \times 1.5$ m to $1.75 \times 3.75$ m. The kiln may consist of two or more shafts. The shafts are enclosed by an outer wall made up of bricks and the gap between the shaft and outer kiln wall is filled with insulating materials like clay, fly ash, rice husk, or glass wool. Each batch typically contains four layers of bricks, set in two distinct pre-determined patterns, called a normal batch or a zigzag batch. The fired bricks are unloaded from the bottom of the shaft.

The shaft has three distinct sections (see Fig. 3.7): the top *preheating zone* – where the incoming green bricks are preheated by the upward moving flue gases; the middle *firing zone* – where fuel combustion occurs; and the lower *cooling zone* – where the fired bricks are cooled down by the cold ambient in-coming air entering the shaft. Peep-hole pipes are fixed in the outer wall of the VSBK shaft to monitor the condition of the fire inside the shaft and the temperature of the different batches. Ramp, steps, conveyors, or a loading gantry are used to transport the green bricks to the loading platform from the ground. The loading platform, at the top of the shaft, is usually covered by a roof and used to store green bricks before stacking them into the shaft.

Green bricks are loaded from the top of the shaft in batches. Once the bricks are stacked at the top zone, they pass through preheating, firing, and cooling zones before reaching the bottom of the shaft where the burnt bricks are unloaded. The fuel may be present as internal fuel and external fuel. The internal fuel is added with the soil prior to brick moulding. The different internal fuels used in the VSBK are coal powder, fly ash, and biomass fuels, such as rice husk and bagasse. Only coal is used as external fuel in the VSBK. The external fuel is added along with the green brick batches from the top of the shaft. The VSBK has the following procedure of brick making.

1. Air for combustion enters the shaft from the bottom as shown in Fig. 3.7. It gets preheated by the hot fired bricks in the brick cooling zone at the bottom, before reaching the firing zone.

2. After combustion, the hot flue gases, preheat the green bricks in the preheating zone before exiting the kiln through the flue ducts and chimneys at temperatures of 60–130°C. Typically, two chimneys are provided at the diagonally opposite corners of each shaft.

3. Dampers (adjustable valves) are provided in the chimney to control the draught and the exit of flue gases from the chimney. The stack of bricks rest on I or square-shaped support bars (which can be removed or inserted), which in turn are supported by a pair of horizontal beams across the arches in the unloading tunnel.

4. A lid cover is provided at the shaft top to cover the shaft, which prevents smoke and heat loss from the shaft top.

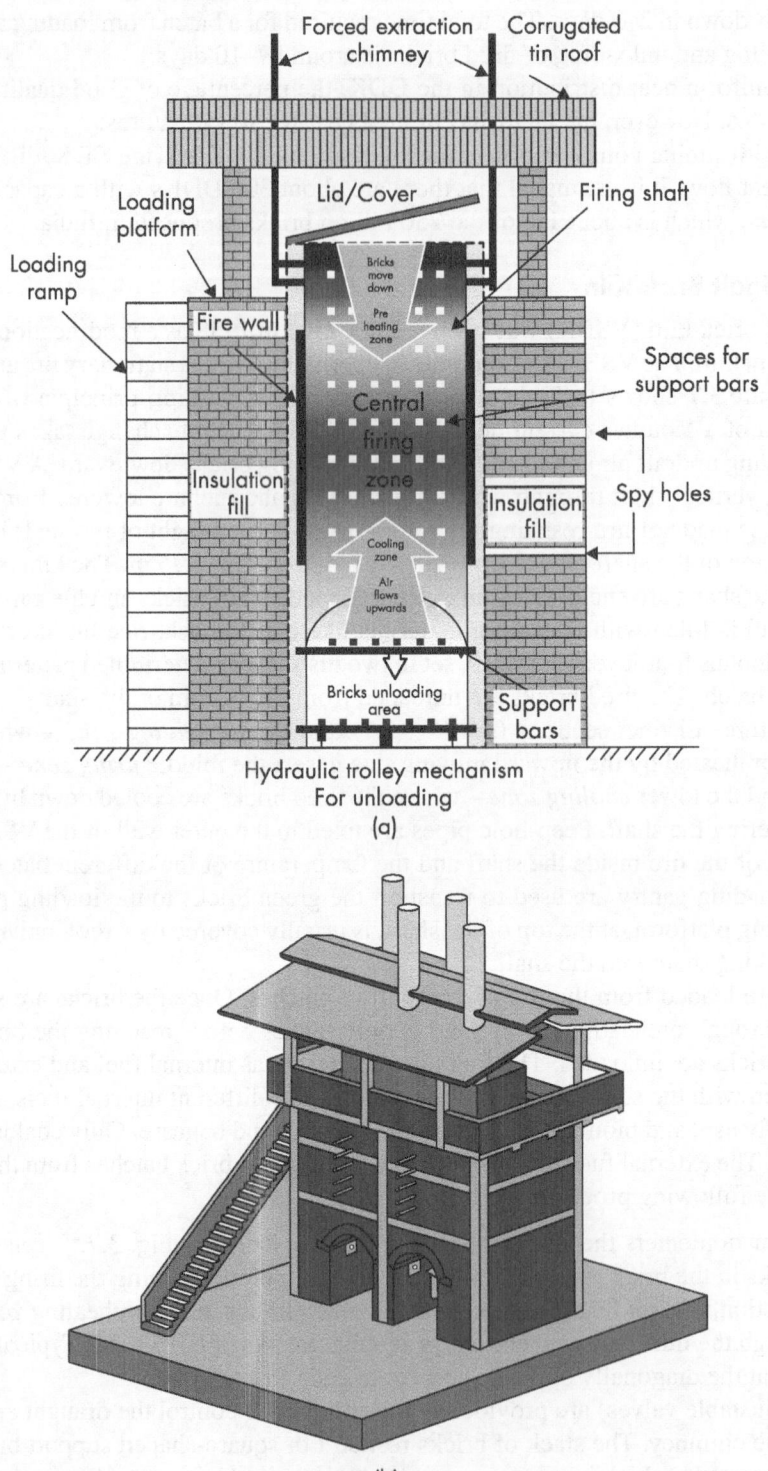

**Fig. 3.7** Schematic of the vertical shaft brick kiln (VSBK) (a) Typical cross-section (b) Isometric view

5. A hydraulic trolley mechanism or a single-screw jack system is used for unloading the fired bricks. Brick unloading is carried out in batches from the bottom using a trolley-every 2–3 h; when one batch is unloaded at the bottom, a batch of green bricks is loaded at the top. At an instant, there will be 8 to 12 batches in a kiln.

The right soil quality is one of the most important factors in the VSBK technology to determine the quality of fired bricks. In general, the VSBK requires clayey soil with relatively higher plasticity. Bricks used in the VSBKs are made from coarse type of clay, which contains particles of fine and coarse sand, providing a more open structure. This helps the bricks to dry more easily by allowing the water to evaporate and increase the resistance of bricks to thermal shock during the water smoking period of firing. The VSBK is a natural draft kiln, requiring no electricity for the supply of combustion air, which enters from the bottom of the kiln.

The VSBK technology may require about 105 tonnes of coal per million bricks, compared to the 160 tonnes required by the BTK. For a shaft size of $1 \times 2$ m, the VSBK has a production capacity of about 8000 bricks per day with a specific energy consumption of 0.8–0.9 MJ/kg of fired bricks. Compressive strengths of 12.5 to 30 MPa for extruded bricks and 7.0 to 11.5 MPa for hand-moulded bricks can be obtained. More information on the construction of the VSBK can be found in Müller et al., (2013). The practical steps to operate a VSBK are provided in Prajapati and Maity (2013).

**Advantages of VSBK**   The following are the advantages of VSBK:

1. The VSBK technology has two major advantages – first, it is energy efficient and consumes less fuel and has lower SPM (suspended particulate matter) emissions; second, it saves about 25–40% of coal when compared to a FCBTK and reduce pollution by 70% when compared to the FCBTK.
2. It provides cleaner working environment, with lesser dust and smoke, thus minimizing health hazards.
3. It also requires less space when compared to other brick making technologies. With a roof protection, the kiln can be operated throughout the year.
4. It does not require external power source. In addition, the bricks can be produced according to the demand – if the demand is high, all the shafts can be operated, if not, only a limited number can be operated to meet the demand.
5. Once constructed, the kiln requires very little maintenance, thus emerging as a very economically viable option.
6. Owing to proper combustion of fuels, efficient heat transfer, and minimal heal losses, the VSBK is one of the most efficient brick kiln technologies.
7. It has low operating costs, and hence is suitable for firing bricks of high quality and specifications.
8. The VSBK is suitable for firing solid bricks; however, it can also be used to fire bricks with perforations.

It is to be noted that fast heating and cooling of the bricks and high load in the lower section of the brick stacking can cause cracks and damage the bricks in the VSBK. These factors may be prominent in the case of hand-moulded bricks, which have low density and low compressive strength. However, extruded and machine moulded bricks result in low percentages of breakages. The VSBK is ideally suited in areas with good clay soils. The VSBK firing is sensitive to minor changes in coal amount, stacking patterns, or unloading frequency. Any unintentional changes of these parameters may affect the quality of bricks.

In 1996, the first pilot VSBK was established at Datia in Madhya Pradesh. This and other three pilot projects in Kerala, Maharashtra, and Orissa demonstrated the energy efficiency, strong environmental performance, and sustainable nature of this technology. According to The Energy and Resources Institute (TERI), currently there are more than 40 VSBKs operational in India.

### 3.4.5 Tunnel Kiln

*Tunnel kiln* is a continuous moving-ware kiln technology, developed around mid-19th century in Germany and widely adopted in several countries after the Second World War (in India, however, there are only a few [less than 10] tunnel brick kilns). A tunnel kiln is a 60–150 m long horizontal tunnel. The clay products are set on 'kiln cars' and moved continuously or at fixed intervals through the tunnel, like a small railroad train. The kiln cars pass through a long stationary firing zone, situated at the central part of the tunnel, which may extend up to 8 kiln cars. The air supply and extraction systems are provided at several points along the kiln structure. The bricks and the air move in opposite directions and the temperature is regulated at 900–1300°C. Generally, green bricks are produced by mixing powdered fuel with clay. The fuel (granulated/pulverized coal) is fed into the firing zone of the kiln through feed holes provided in the kiln roof. The duration of the firing cycle can range from 30 to 72 h.

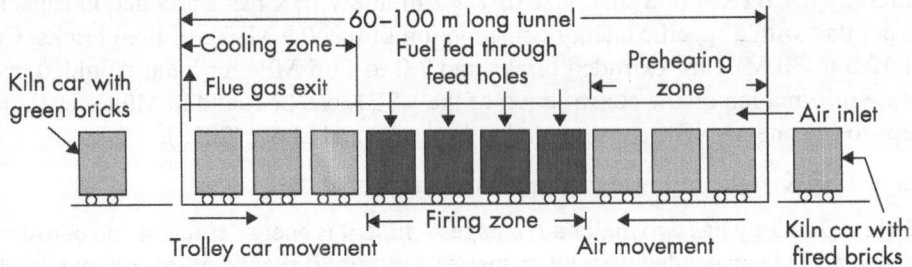

**Fig. 3.8** Schematic of the tunnel kiln technology

There are three distinct zones in the tunnel kiln (see Fig 3.8): (1) the firing zone where the fuel is fed and combustion occurs, (2) the preheating zone (in front of the firing zone), and (3) the cooling zone (at the exit end of the kiln). There is counter current heat transfer in the kiln between the bricks and the air. Cold air enters from the exit end of the tunnel kiln and gets heated while cooling the fired bricks. After combustion, the hot flue gases travel towards the tunnel entrance end and part of the heat is used to dry the green bricks entering the kiln. Hot gases are extracted from the tunnel kiln at several points along the length of the kiln and supplied to the drying chamber. In some of the kilns, there is also provision of a hot air generator to supplement the requirement of hot air for drying. The flue gases from the drying part of the tunnel are released in the atmosphere through a chimney.

Although tunnel kilns consume more electricity and require high capital costs, the man power required to operate them is comparatively less. They may be used to produce 42,000–50,000 bricks per day and 3–5 days are required for the complete drying and firing of the bricks. A variety of bricks can be fired using the tunnel kilns that meet specific demands in terms of size, shapes, and colour. A modified version of the tunnel kiln, called the *roller kiln*, can fire bricks at a short duration of 3–8 h.

**Advantages of tunnel kiln** The following are the advantages:

1. The tunnel kiln is considered to be the most advanced brick making technology. The main advantages of tunnel kiln technology lie in its ability to fire a wide variety of clay products, better control over the firing process, high quality of the products, and large production volume.
2. It has a mechanized and highly automated process, which results in minimal emissions. Tunnel kiln emits about 80% lower particulate matter and negligible black carbon as compared to the FCBTK.
3. Tunnel kiln can be used to fire all types of products such as solid bricks, hollow/perforated bricks, and roof and floor tiles.

This is mainly because of better combustion and use of internal fuel. The emission of $CO_2$ is higher in the case of tunnel kiln, probably due to incomplete combustion of internal fuel (See Table 3.2).

Moreover, tunnel kiln is expensive to construct, and requires reliable power/electricity supply. In addition, the average specific energy consumption (SEC) in a tunnel kiln is 1.4 MJ/kg of fired bricks, which is slightly higher as compared to other continuous kiln technologies. This is mainly because the SEC also includes the energy utilized for the drying of bricks in the tunnel dryer.

**Pollution due to brick kilns**    Pollutants associated with clay brick firing include (Akinshipe & Kornelius, 2017): Particulate matter (PM), Sulphur dioxide ($SO_2$), Sulphur trioxide ($SO_3$), Nitrogen oxides ($NO_x$), (including nitrogen dioxide [$NO_2$] and nitrogen monoxide [NO]), Carbon monoxide (CO), Carbon dioxide ($CO_2$), Metals (including Cooper [Cu], Chromium [Cr], Lead [Pb], Nickel [Ni], Zinc [Zn], Cadmium [Cd], Iron [Fe], and Manganese [Mn]), Fluorides and Organic compounds (including methane, ethane, volatile organic compounds [VOCs], persistent organic compounds [POPs], and some hazardous air pollutants such as hydrogen chloride [HCl] and hydrogen cyanide [HCN]). Based on a study, Akinshipe and Kornelius (2017) concluded that the VSBK and the zigzag kilns have the lowest potential for atmospheric pollution; while the BTK and the DDK have the highest potential for atmospheric pollution. It is estimated that in India, 74% of total brick production is done through the BTKs and 21% through clamps. Based on the direction from the Central Pollution Control Board in 1996 (with a deadline on June 2001) several MCBTKs were converted to the FCBTKs. It has also strict emission limits for particulate matter (in milligrams per cubic meter): for the BTKs: 750–1000, the DDKs: 1200, and the VSKs: 250 mg/m³ (www.cpcb.nic.in).

A comparison of the different kiln technologies is provided in Table 3.2. From this table, it is seen that the tunnel kiln is expensive but emits less pollution. The next best is the VSBK.

**Table 3.2** Comparison of different kiln technologies

| Parameters | Types of kiln | | | | | | |
|---|---|---|---|---|---|---|---|
| | Clamp | Down-draught kiln (DDK) | Fixed chimney Bull's trench kiln (FCBTK) | Natural draft zigzag | Vertical shaft brick kiln (VSBK) | Tunnel kiln | Hybrid Hoffman kiln |
| Air emission of (g/kg) fired brick | | | | | | | |
| $CO_2$ | NA | 282.4 | 131 | 105 | 70.5 | 166.3 | 100 |
| Black carbon | NA | 0.29 | 0.13 | 0.01 | 0.001 | 0.0 | NA |
| Particulate Matter | 1.91 | 1.56 | 1.18 | 0.22 | 0.15 | 0.24 | 0.29 |
| CO | NA | 5.78 | 2.0 | 0.29 | 1.84 | 3.31 | NA |
| Fuel and energy MJ/kg | 1.7–4.2 | 2.97 | 1.30 | 1.06 | 0.8–1.1 | 1.4–2.0 | 1.20 |
| Capital cost (US$) | NA | 20,000–30,000 | 50,000–80,000 | 50,000–80,000 | 60,000–80,000 | 1,000,000 | 600,000–650,000 |
| Production capacity | 10,000–20,000 /batch | 20,000–40,000 /batch | 3–8 million per year | 3–8 million per year | 1.5–3 million per year | 15–18 million per year | 15–18 million per year |
| Type of products | Only solid bricks | All types | All types | All types | Solid and perforated bricks | All types | All types |
| Good quality product (%) | About 50 | About 85 | About 60 | About 85 | About 90 | About 95 | About 90 |

## 3.5 Classification of Bricks

There are different types of classifications described as follows.

**According to use**   According to the way in which they are used, bricks are classified as (a) Common bricks, (b) Engineering bricks, (c) Facing bricks, (d) Fire bricks, and (e) Paving bricks (see Section 3.7.1 for the description of these types of bricks).

**According to Indian code**   According to the IS 1077:1992, bricks may be classified according to their compressive strength, water absorption, and limit of efflorescence, as given in Table 3.3.

**Table 3.3** Indian code classification of bricks according to strength

| Class designation | Average compressive strength not less than(N/mm²) | Water absorption (%) | Efflorescence |
|---|---|---|---|
| 35 | 35.0 | 15 | Slight |
| 30 | 30.0 | 15 | Slight |
| 25 | 25.0 | 15 | Slight |
| 20 | 20.0 | 15 | Slight |
| 17.5 | 17.5 | 15 | Slight |
| 15 | 15.0 | 15 | Slight |
| 12.5 | 12.5 | 20 | Moderate |
| 10 | 10.0 | 20 | Moderate |
| 7.5 | 7.5 | 20 | Moderate |
| 5 | 5.0 | 20 | Moderate |
| 3.5 | 3.5 | 20 | Moderate |

**According to physical requirements**   In some specifications, clay bricks are classified as Class I, Class II, and Class III, according to their properties (see Table 3.4). No good brick should disintegrate when immersed in water for a long period; such disintegration may indicate the lack of adequate burning.

**Table 3.4** Classification of bricks based on physical requirements

| | Class I bricks | Class II bricks | Class III bricks |
|---|---|---|---|
| General requirement | Bricks to be table moulded and have standard shape. The surface and edges of the bricks should be sharp, square, smooth, and straight. It should comply with all the qualities of good brick. | May be ground moulded, burnt in kilns, and have uniform colour. Surface of bricks may be somewhat rough and shape also may be slightly irregular. | May be ground moulded and burnt in clamps. Need not be hard and may have rough surfaces with irregular and distorted edges. May give dull sound when struck together |
| Water absorption after immersing in cold water for 24 h | Not more than 20% by weight | Not more than 22% by weight | Not more than 25% by weight |
| Efflorescence | Slight | Slight | Moderate |
| Applications | Used for superior work of permanent nature | Commonly used in places where brick work is provided with a coat of plaster | Used for unimportant and temporary structures and at places where rainfall is not heavy |

**Note:** Over-burnt bricks are classified as Class VI and used as aggregate for concrete in foundation, floors, and roads.

**According to method of manufacture**   As already mentioned, based on the method of manufacture, bricks are classified as: (a) Hand made: Water-stricken or Sand-stricken, (b) Wire-cut, (c) Stiff-mud, and (d) dry pressed. Generally, factory made (wire-cut) bricks in India give a strength of about 17 N/mm$^2$ and hand-made bricks only 3–5 N/mm$^2$ when dry.

## 3.6 Qualities of Good Bricks

The most important properties of brick are (a) durability, (b) colour, (c) texture, (d) size variation, (e) compressive strength, and (f) water absorption. Engineering bricks, which are used for the construction of important structures, should possess the following qualities.

1. They should have a compressive strength greater than 5.5 N/mm$^2$.
2. They should be table-mounted/wire-cut bricks, well burned in kilns, free from cracks and should have sharp and square edges. The colour should be uniform bright red or copper colour.
3. They should have uniform shape and standard size.
4. The bricks should give a clear metallic ringing sound when struck with each other without breaking.
5. They when broken or fractured should show a homogeneous and uniform compact texture free from voids.
6. They should not absorb water more than 20% by weight for first-class bricks and 22% by weight for second-class bricks, when soaked in cold water for a period of 24 h.
7. They should be sufficiently hard. No impression should be left on a brick surface, when it is scratched with finger nails.
8. They should not break into pieces when dropped flat on hard ground from a height of about 1 m.
9. The bricks should have low thermal conductivity and they should be sound-proof.
10. The bricks, when soaked in water for 24 h, should not show deposits of white salts when allowed to dry in shade.

## 3.7 Architectural Characteristics of Bricks

Architectural requirements of bricks include shape, size, colour, texture, and weight of bricks.

### 3.7.1 Shape and Types of Bricks

Bricks are made in a wide range of shapes in order to suit the requirements of the location where they are to be used. Conventional bricks are rectangular in shape, which are small and light enough to be picked by mason using one hand. They may also have indentations called *frog* on one of the longer faces (see Fig. 3.9). Typical common burnt bricks and hollow bricks are shown in Fig. 3.10.

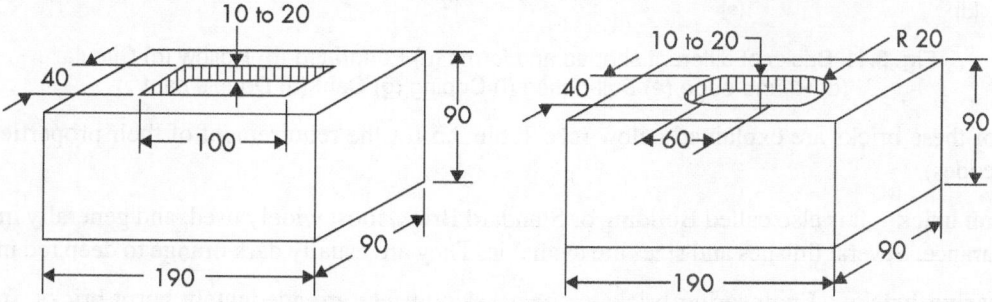

**Fig. 3.9** Shape and size of frog in ordinary bricks

**Fig. 3.10** (a) Common burnt bricks (b) Hollow bricks

Some special forms/types of bricks are also available to satisfy structural considerations or aesthetic purposes. Some of these specially moulded bricks avoid the cumbersome process of cutting and rounding the rectangular bricks to the desired shape. Some of these bricks are shown in Fig. 3.11. (IS 6165:1992 provides dimensions of these specially shaped clay bricks.)

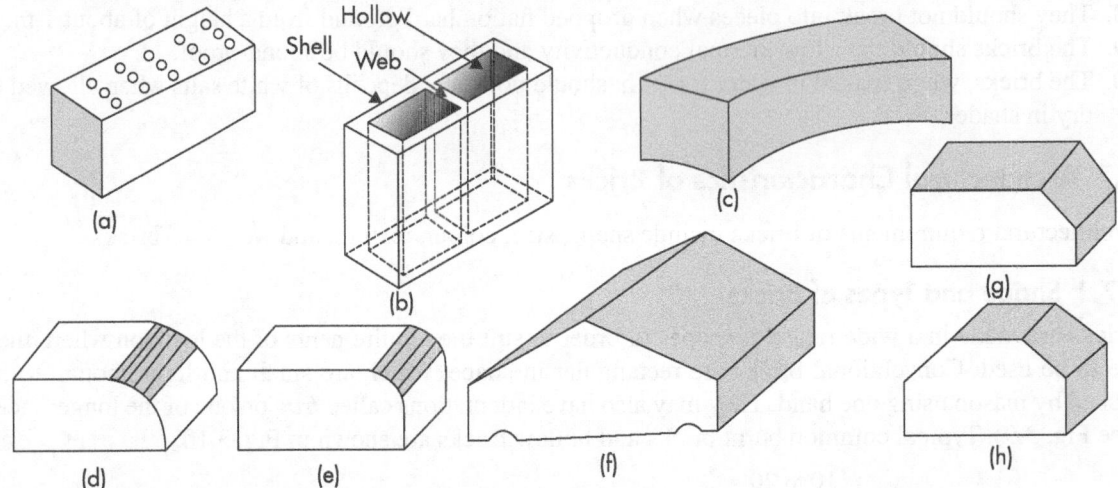

**Fig. 3.11** Bricks of different shapes and forms (a) Perforated (b) Hollow (c) Circular
(d) Round-edge (e) Bull-nosed (f) Coping (g) Cant (h) Double cant

A few of these bricks are explained below (see Table 3.5 for the requirement of their properties as per Indian codes).

**Common brick**    It is also called Building or Standard Brick, most widely used, and generally imperfect in appearance. Several finishes and sizes are available. They are usually dark orange to deep red in colour.

**Engineering bricks**    Engineering bricks or *heavy duty bricks* are adequately burnt bricks, free from cracks and other flaws, and have high durability, high compressive strength and low water absorption

(IS 2180:1988). These bricks provide enhanced technical performance. They are used in heavy engineering works, such as bridge structures, industrial foundations, multi-storeyed buildings, and in damp-proof course. The density of bricks should not be less than 2.5 g/cm$^3$. They are usually red in colour although blue engineering bricks are also available.

**Burnt clay perforated bricks**  These bricks [Fig. 3.11(a)] are standard size bricks with perforations running through their thickness. Perforated bricks are easy to burn and their light weight (60% lighter than conventional bricks) reduces the weight of walls, and the size foundations. As per IS 2222:1991, the area of perforation should be from 30–45% of the total area of the corresponding face of the bricks. The area of each perforation should not exceed 500 mm$^2$. The shorter side of rectangular perforation should not be less than 20 mm and the diameter of circular perforations should not be less than 25 mm. A range of perforated bricks/blocks are produced in India by M/s. Wienerberger India under the brand name *Porotherm Smart Bricks*. These blocks are available in three sizes (400 × 200 × 200 mm, 400 × 150 × 200 mm, and 400 × 100 × 200 mm) and in half-block versions. Porotherm Thermobricks, with specially formulated insulating material has a 'U' Value of 0.6 W/(m²K), and hence provides superior thermal insulation (45% more than conventional bricks), improves indoor comfort and energy efficiency. Porotherm Vertically Perforated Clay Hollow Bricks (Porotherm VP), with compressive strengths ≥ 7 N/mm$^2$, can be used in load bearing walls up to ground plus three floors. These products have a fire rating of F240 (240 minutes) and are credited by the Indian Green Building Council for ratings.

**Burnt clay hollow bricks**  These bricks [Figs 3.10(b) and 3.11(b)] contain hollows with ribs. They are usually made from tile clay as ribs have to be thin and strong. The following three types are produced as per IS 3952:2013: Type A – Bricks with both faces keyed for plastering or rendering, Type B – Bricks with both faces smooth and suitable for use without plastering or rendering on either side, and Type C – Bricks with one face keyed and one face smooth. They are available in three sizes (190 × 190 × 90 mm, 290 × 90 × 90 mm, and 290 × 140 × 90 mm). The thickness of any shell should not be less than 11 mm and that of any web not less than 8 mm. The burnt clay hallow bricks are about one-third the weight of normal bricks and provide thermal and sound insulation, but are not suitable where concentrated loads are expected. Horizontally perforated (Porotherm HP) clay hollow blocks are primarily used as infill masonry for non-load bearing wall construction. M/s. Wienerberger India has also developed a 'Dryfix System' adhesive that creates perfect bonding between the blocks specially when used with Porotherm HP, enhancing the strength of the walls.

**Circular bricks**  These bricks [Fig. 3.11(c)] have internal and external faces curved to meet the requirement of the particular curve and radius of the wall. These bricks are used in structures with curved surfaces, such as wells and towers. Bull nose and round-edged bricks [Figs 3.11(d) and (e)] have rounded edges and are used in copings where rounded corners are preferred.

**Coping bricks**  The top most brick course of parapets may be made with coping bricks [Fig. 3.11(f)]. They are throated on the underside to throw off rain water.

**Cant bricks**  For door and window jambs, cant brick, also called *splay brick*, is most suitable. The double cant brick is used for octagonal pillars [Figs 3.11(g) and (h)].

**Facing bricks**  These bricks are made from clay, shale, fire clay, or a mixture of the aforementioned. They are thoroughly burnt to produce uniform colour and to have plane rectangular faces and sharp, straight, and right-angled edges (IS 2691:1988). They are fired at high temperatures to ensure that they are fire resistant and will not contribute to the spread of fire. Facing bricks have nicely finished surface and textures and are used in the exterior faces of brickwork without any plaster or surface treatment (see Fig. 3.12). They should have sufficient resistance to penetration by rain and all weather conditions. They are used in countries such as the USA, in the face of walls to provide attractive appearance, and energy efficiency. *Glazed facing bricks* are bricks sprayed with a liquid ceramic glazing before being fired and fired at high temperature to provide glaze to the brick. Finishes may appear dull, satin, or glossy. They are used in kitchens and bathrooms, as they are easy to clean.

**Fig. 3.12** Building with facing bricks in Maryland, USA

**Fire/Refractory bricks** These (also called as *fire clay bricks* or *refractory bricks*) are made from special clays called *fire-clays* with a large amount of alumina (aluminium oxide content as high as 50–80%), with correspondingly less silica, flint, and feldspar. They are fired in kilns, at extremely high temperatures, until they are vitrified. It is easily identified by its yellow colour, and for special purposes may also be glazed. Fire-bricks are very heavy/dense, light in colour, soft in texture, have low porosity, and are resistant to heat. There are two standard sizes of fire-bricks: $230 \times 115 \times 75$ mm and $230 \times 115 \times 64$ mm. They are used in lining furnaces, kilns, incinerators, fireplaces, and chimneys, where these are exposed to high temperature (up to 1450°C). They usually have a low thermal conductivity and great energy efficiency. Fire-bricks should be laid with fire-resistant mortar and not with cement mortar. (Fire-resistant mortar consists of one part of alumina cement and two parts of finely crushed powder of fire-bricks.) They are available as acid-resistant, basic, and neutral refractory bricks.

**Acid-resistant bricks** These are made using clay or shale of suitable composition with low lime and iron content, feldspar, and sand, and vitrified at high temperatures in ceramic kilns. These bricks are used with chemical resistant mortars in chemical and allied industries. As per IS 4860:1968, they are available in the size of $230 \times 114 \times 64$ mm and in two grades: Class I and Class II. The required properties for these two grades of acid-resistant bricks along with other types of bricks are given in Table 3.5. In addition, the loss in weight due to acid attack for Type I bricks should not exceed 1.5% and for Class II bricks 4%. Though sewers carrying domestic sewage may be lined with common building bricks, those dealing with industrial effluents have to use acid-resistant bricks in accordance with IS 4860. *Sewer bricks* should conform to IS 4885:1985.

**Table 3.5** Required properties of different types of bricks as per IS codes

| Type of brick | Required properties | | | | IS code |
|---|---|---|---|---|---|
| | Minimum average compressive strength (N/mm²) | Maximum water absorption (%), after 24 h cold water immersion | Degree of efflorescence | Maximum warpage (mm) | |
| Engineering bricks (Heavy duty bricks) | 40–45 | 10 | Nil | – | IS 2180:1988 |
| Burnt clay perforated bricks | 7 | 15 | Slight | 3 | IS 2222:1991 |

*(Contd)*

**Table 3.5** (*Contd*)

| Type of brick | Required properties | | | | IS code |
|---|---|---|---|---|---|
| | Minimum average compressive strength (N/mm²) | Maximum water absorption (%), after 24 h cold water immersion | Degree of efflorescence | Maximum warpage (mm) | |
| Burnt clay hollow bricks | 3.5 | 20 | Slight | – | IS 3952:2013 |
| Facing bricks | 10 | 15 | Nil | 2.5 | IS 2691:1988 |
| Paving bricks | 40 | 5 | Nil | | IS 3583:1988 |
| Burnt clay soling bricks | 5 | 20 | Slight | – | IS 5779:1986 |
| Acid resistant brick Class I | 70 | 2 | – | 2.5 | IS 4860:1968 |
| Class II | 50 | 4 | – | 5 | IS 4860:1968 |

**Paving bricks** These are hard bricks and specially made for paving the surface of patios, drive ways, and streets, as shown in Fig. 3.13. These bricks are usually made from shale, fire clay or a mixture of the two. As per IS 3583:1988, these bricks should be free from cracks and flaws and nodules of free lime, and have smooth rectangular faces with sharp corners. They are machine moulded and are burnt in a continuous kiln, at much higher temperatures to ensure high degree of vitrification and uniform colour and texture. They should not have frogs. They are unaffected by weather and have high resistance to wear from traffic.

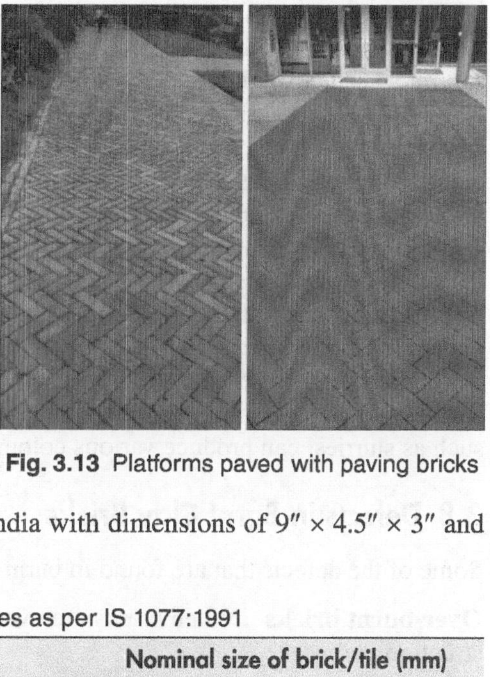

### 3.7.2 Size and Weight of Bricks

The standard and nominal size of common building bricks as per IS 1077:1991 are given in Table 3.6. However, hand moulded bricks are available in many parts of India with dimensions of 9″ × 4.5″ × 3″ and known as *field bricks*.

**Fig. 3.13** Platforms paved with paving bricks

**Table 3.6** Standard size of bricks and tiles as per IS 1077:1991

| Type of brick/tile | Standard size | | | Nominal size of brick/tile (mm) |
|---|---|---|---|---|
| | Length (mm) | Width (mm) | Height (mm) | |
| Modular brick | 190 (200) | 90 (100) | 90 (100) | 200 × 100 × 100 |
| Modular tile brick | 190 (200) | 90 (100) | 40 (40) | 200 × 100 × 40 |
| Non-modular brick | 225 | 110 | 70 | 229 × 114 × 70 |
| Non-modular tile | 225 | 110 | 40 | 229 × 114 × 40 |

**Note:** The dimensions given in brackets are nominal sizes, including the thickness of mortar.

Bricks of 90 mm height are moulded with a frog, 10 to 20 mm deep on one of its flat sides. The shape and size of frog as per IS 1077:1992 is shown in Fig. 3.9.

**Tolerances** The dimensions of modular bricks should conform to the following limits of tolerances per 20 numbers of modular size bricks (IS 1077:1992):

Length from 3720 to 3880 mm (3800 ± 80 mm)

Width from 1760 to 1840 mm (1800 ± 40 mm)

Height from 1760 to 1840 mm (1800 ± 40 mm) (For 90 mm high bricks)

760 to 840 mm (800 ± 40 mm) (For 40 mm high bricks)

**Weight** The deadweight of brickwork is required in the design of buildings. The average weight of a brick is about 3 to 3.5 kg depending on its density. As per IS :875 (Part 1) – 1987 common burnt clay bricks may weigh from 15.70 to 18.85 kN/m³, engineering bricks 21.20 kN/m³, pressed bricks 17.25–18.05 kN/m³, refractory bricks 17.25 to 19.60 kN/m³, heavy-duty bricks (IS 2180:1988) 24.50 kN/m³, sand-lime bricks 20.40 kN/m³, and sand-cement bricks 18.05 kN/m³. The deadweight of brick wall is taken as 18.85–23.55 kN/m³, depending on the type of brick used. Lightweight hollow bricks will weigh only one-third the weight of common bricks.

### 3.7.3 Colour and Texture of Bricks

Brick colour is determined by the mineralogy of its clay, kiln temperature and atmosphere, and, to a small degree, the kiln fuel. Pure clay bricks burn to form white bricks because of the highly refractory nature of the material. As discussed in Section 3.3.2, when clay containing iron oxides is fired, it will exhibit a shade of red due to the formation of ferrous oxide. Depending on the iron oxide content, the colour of brick after firing will be cream or buff (light yellow-brown colour), pink, or red. When excess iron oxide is present, it will make the bricks dark blue or blackish. Calcareous clays result in yellow or cream coloured bricks. Bricks with high lime content will have white colour. Pigments can be added to the clay mix. For example, the addition of manganese dioxide to a cream base creates a grey brick, and when added to red clay base creates a brown brick. As discussed in Section 3.3, near the end of the burning process, the bricks may be flashed with excess fuel to produce blue-black bricks. Given the same raw material and manufacturing method, darker colours are associated with higher firing temperatures, lower absorption values, and higher compressive strengths (TN 9 of BIA, 2006).

Manufacturers sometimes add a glass based material known as *frit* to create a spattered surface effect. Salt, clay particles, carbon, or even small pebbles can be added to the surface. Other surface additives, such as slurries, can produce various colours and styles.

### 3.8 Defects in Burnt Clay Bricks

Some of the defects that are found in burnt clay bricks are given here.

**Over-burnt bricks** When bricks are over-burnt, their shape is lost. Such bricks are not used for construction purposes.

**Under-burnt bricks** Under-burnt bricks results in a higher degree of water absorption and less compressive strength. Such bricks are not recommended for construction purposes.

**Efflorescence** As already discussed, efflorescence is caused by salts, resulting in a crystalline white deposit on the surface, causing unsightly appearance. This can be minimized by selecting proper brick material for manufacturing, preventing moisture to come into contact with the masonry, by providing waterproof coping, and by providing damp-proof course. As per IS 3495-Part 3, efflorescence is considered as

1. Nil- When there is no perceptible deposit of salts on the brick.
2. Slight- When the brick is not covered by salt for not more than 10% of the area.

3. Moderate- When there is heavy deposit of salt covering up to 50% of the area of brick but there is no powdering or flaking of the surface.
4. Heavy- When there is heavy deposit of salt covering more than 50% of the area of brick and there is powdering or flaking of the surface.
5. Serious- When there is heavy deposit of salt accompanied by powdering and/or flaking of the surface and this deposition increases with repeated wetting.

**Lime blowing/lime bursting**   Lime-blowing is caused by limestone lumps in clay that re-hydrate and expand after firing. To avoid lime-blowing, brick material should not have particle size greater than 2 mm; firing the bricks in kilns to a higher temperature (say, 1050°C) could also solve this problem but will increase the cost of brick manufacturing.

**Oversize in thickness**   This defect is often observed with wire-cut bricks. During wire-cutting, a block of clay is forced through a row of wires; the force on the wire may cause some movement, which may change the dimension of the opening, resulting in oversize bricks.

**Bloating (Large crack with bulge in surface)**   It can occur in bricks when the clay is heated too quickly and the surfaces are vitrified before the chemical reactions take place inside the brick. Although bloated bricks will have enough strength, due to their bloated shape, they are not used in visible areas and may be used in locations such as footings.

**Multiple surface cracks in random directions**   Such cracks may occur due to lumps of drier material shrinking differently than the rest of the brick, as a result of differential drying. To prevent this from happening, the clay material should be mixed thoroughly before moulding. This defect can also occur due the presence of pebbles in the clay mix.

**Spots**   Due to the presence of iron sulphide, dark red spots may occur on the surfaces of the brick. Although these spots are not harmful, such bricks are not used in exposed masonry due to aesthetic considerations.

**Lamination crack**   Lamination may occur in the sand-moulded process when a clay brick, covered with sand, is mixed into another piece. When stresses occur during drying or firing, the film of sand may separate the two pieces, resulting in lamination. This defect may happen even with oil moulding. Properly trained workers can easily prevent this defect.

## 3.9 Advantages and Uses of Bricks

Some of the advantages of bricks, as compared to other building materials, are as follows (www.thebalance.com).

**Sustainable**   Bricks are made primarily from clay and shale, which are abundant natural resources. In addition, they are usually made locally, thus reducing transportation costs.

**Minimal wastage**   Poor quality bricks and even broken bricks can be used as fillers in wall and floor cavities, as aggregates in concrete, or finely ground and used as pozzolana (surkhi), or as grogs in brick making.

**Easy to handle**   All the bricks are of uniform size and shape, and hence can be laid in any desired pattern. In addition, they are light in weight and small in size. Hence, they can be easily handled by masons by hand.

**Easy to construct**   The art of bricklaying can be understood very easily, even by unskilled masons. On the other hand, stone masonry construction requires highly skilled masons.

**Aesthetic**   Bricks offer natural and a variety of colours, including various textures.

**Porosity**   The ability to release and absorb moisture, without significant dimensional change, may be useful to regulate temperature and humidity inside structures.

**Strength**   Bricks offer excellent high compressive strength.

**Fire resistance**   Fire resistance of bricks is usually good. In fact, bricks are used to encase steel columns to protect them from fire. The 150 mm thick solid clay or shale brick wall can provide maximum fire protection rating of 4 h (see BIA TN-16: 2008 for more details).

**Sound insulation**   The brick sound insulation is normally 45 dB for a 115 mm brick thickness and 50 dB for 230 mm thick brick.

**Insulation**   Bricks can exhibit above-normal thermal insulation when compared to other building materials. Bricks can help regulate and maintain constant interior temperatures of a structure due to their ability to absorb and slowly release heat. Thus, bricks can reduce peak energy loads, resulting in more than 30% of energy saving compared to wood. Perforated/hollow bricks provide additional insulation.

**Wear resistant**   A brick is so strong that its composition provides excellent wear resistance.

**Emission free**   Bricks used in brick walls do not emit any gases during their life-time.

**Special bricks**   Special purpose bricks with high densities and compressive strength can be produced; refractory bricks with high heat resistance are used for lining kilns and furnaces; Acid resistant bricks and tiles can withstand chemical attack.

**No surface protection needed**   In addition, good quality bricks (e.g., facing bricks) and tiles do not require any surface protection, such as plastering or painting.

**Durability**   Brick is extremely durable and one of the best durable structural building materials.

Owing to these advantages, bricks are used in foundations, load-bearing walls, partition walls, and filler walls of buildings and pavements throughout the world. Construction of brick-walls and the different arrangements in which bricks are used in walls (called *bonds*) are usually covered in books on building construction. In addition, bricks are used in the following: (a) ornamental work (arches and domes), (b) for protecting steel columns from fire, (c) for lining sewer lines, and (d) in roofs having slopes in the range of 18°30' (1:3) to 45° (1:1), tiles of various shapes and sizes are used. Hollow clay blocks are used in composite reinforced concrete beam, and roof slab construction.

## 3.10 Substitutes for Bricks

It has been found that in countries like India and China, the demand for bricks for construction is increasing. (Even in countries like the USA, where various other siding materials are used, demand for bricks, blocks, and pavers is forecast to rise 8.8% per annum from a low 2013 base to $8.9 billion in 2018.) As bricks are made from clay excavated from the Earth, availability of good brick-earth is becoming scarce. Hence, it has become necessary to produce bricks using industrial wastes and other raw materials. The different bricks that use such wastes or other materials include: (1) Fly ash bricks, (2) Sand-lime bricks or Calcium silicate bricks, (3) Compressed Earth Blocks, and (4) Concrete blocks. The term 'block' is used to differentiate it from the 'fired' brick. These products are briefly described in the following sections.

### 3.10.1 Fly Ash Bricks

The following three kinds of fly ash bricks are possible: (1) Burnt clay fly ash bricks, (2) Pulverized fuel ash lime bricks, and (3) Fly ash bricks

Pulverized fuel ash or simply *fly ash* is a residue resulting from the combustion of ground/powdered/crushed bituminous coal or sub-bituminous coal (lignite) in thermal power plants. About 80% of the total ash is fine-grained and emitted through chimneys along with flue gases and collected by electro static precipitators or filter bags. This fly ash is available in large quantities from thermal power plants, as a waste material, and can be added to alluvial, red, black, or marine clays. It is important to use fly ash, which is similar in composition to brick earths and should conform to grade 1 or grade 2 as per IS 3812(Part 1): 2013. The silicates in fly ash will help to increase the strength while firing.

**Burnt clay fly ash bricks** The specifications for *burnt clay fly ash bricks* are provided in IS 13757:1993. The desirable characteristics of fly ash used as an additive to the soil mass, depending on the type of fly ash, as per IS 15648:2006 are given in Table 3.7. Burnt clay fly ash bricks are primarily composed of cement, clay, fly ash, and water. Fly ash has to be mixed in optimum proportion depending on the properties of soils used in the brick. The manufacturing process of burnt clay fly ash bricks is similar to that of fired clay bricks. Bricks should have minimum compressive strength for the respective class ($3.5–30$ N/mm$^2$), when tested according to IS 3495. Water absorption of bricks should be less than 20% by weight up to class 12.5 and 15% for higher classes when bricks are immersed in cold water for 24 h. Efflorescence should not be more than 'moderate' for class up to 12.5 and 'slight' for higher classes.

**Table 3.7** Desirable characteristics of fly ash used for bricks

| S. no. | Characteristics | Desired level | |
|---|---|---|---|
| | | Siliceous fly ash (Grade 1) | Calcareous fly ash (Grade II) |
| 1 | Fineness - Specific surface in m$^2$/kg by Blaine's permeability method; minimum | 250 | 200 |
| 2 | Particles retained on 45 μ IS sieve (wet sieving) in percent; maximum | 40 | 45 |
| 3 | Average compressive strength in N/mm$^2$; minimum | 3.5 | 3.0 |
| 4 | Loss on ignition in percent by mass; maximum | 5 | 5 |
| 5 | Silicon dioxide ($SiO_2$) plus aluminium oxide ($Al2O_3$) plus iron oxide ($Fe_2O_3$) in percent by mass; minimum | 70 | 50 |
| 6 | Silicon dioxide ($SiO_2$) in percent by mass; minimum | 35 | 25 |
| 7 | Reactive silica in percent by mass; minimum | 20 | 15 |
| 8 | Magnesium oxide (MgO) in percent by mass, maximum | 5 | 5 |

The advantages of these bricks include the following:

1. Reduction of drying shrinkage and drying losses
2. Reduction of firing time and coal consumption
3. Higher strength compared with normal clay bricks
4. Production of more number of bricks (up to 40%) with the same quantity of clay soil,

5. Uniform shape and size, rougher texture, with sharper edges and corners of bricks; due to these characteristics, they require less mortar in the joints of brickwork

6. Environment friendly, as agricultural lands are conserved and fly ash is disposed in a very efficient, useful, and profitable way

**Pulverized fuel ash lime bricks and FAL-G bricks** These are made using fly ash (in major quantity, conforming to IS 15648:2006), lime, and an accelerator acting as a catalyst. Pulverized fuel ash-lime bricks are generally manufactured by inter-grinding or blending various raw materials, which are then moulded into bricks and subjected to curing cycles at different temperatures and pressures. Crushed bottom fuel ash or sand may also be used in the composition as a coarser material to control water absorption in the final product. Pulverized fuel ash reacts with lime in the presence of moisture to form a calcium-silicate hydrate which acts as a binder. Thus, pulverized fuel ash-lime brick is a chemically bonded brick and hence does not require firing. The specifications for burnt clay fly ash bricks are provided in IS 12894:2002. The minimum compressive strength, water absorption, and efflorescence requirements of this kind of brick are similar to that of burnt clay fly ash bricks. These bricks are suitable for use in masonry construction just like common burnt clay bricks. In addition to the advantages offered by burnt clay fly ash bricks, these bricks have the following advantages.

1. As no burning of bricks is involved, saves energy and there is also no pollution. In addition to helping the disposal of fly ash (which is a waste product), it also conserves top soil.
2. The bricks can be made locally, even at the job site.
3. These bricks have better mechanical properties and more durability.
4. They are about 10–20% cheaper than burnt clay bricks.

*FaL-G (Fly ash-Lime-Gypsum) process* was developed and patented by Dr. Bhanumathidas and Kalidas and introduced in India during 1991 (see www.fal-g.com). In this process, 60% fly ash, 30% sand, and 10% Portland cement are used. In the lime route, the composition is fly ash (62%), slaked lime (8%), anhydrite gypsum (5%), and stone dust/sand (25%). These ingredients are manually fed into a pan mixer where required amount of water is added and intimately mixed. The mixture is then used in brick making machines, and the bricks are dried and water cured for 14 days. The calcium aluminate is converted into calcium alumino-sulphate resulting in a product having high early strength. FaL-G bricks can be produced with compressive strength of 10–35 N/mm$^2$, and water absorption of 8–15% (Bhanumathidas and Kalidas, 2003). It is notable that FaL-G bricks do not need any pressure and are cured at ambient temperature of 20–40°C. By avoiding both pressing and heating chambers, high economy is achieved and this resulted in the proliferation of 18,000 manufacturing units as of 2016.

**Fly ash bricks** *These* are made using Class C or Class F fly ash and water. They are compressed at 28 MPa and cured for 24 h in a 66°C steam bath, then toughened with an air entrainment agent. Owing to the high concentration of calcium oxide in Class C fly ash, the brick is described as 'self-cementing'. Use of fly ash to make the bricks results in the elimination of the whole processes of mining, transporting, mixing and grinding, and firing that are necessary in the case of the clay- and shale-based bricks. The manufacturing method saves energy, reduces pollution, and costs 20% less than traditional clay brick manufacturing.

### 3.10.2 Sand-lime Bricks and Calcium Silicate Bricks

These are made using sand, lime, and water. Lime and selected sands and water are mixed, compressed in moulds under high pressure and autoclaved to produce a white coloured brick. These bricks have a compressive strength of 10 MPa. They provide good acoustic insulation and excellent fire resistance. However, their production is limited; they are used in the USA and Germany. IS 3115:1992 stipulates the specification for lime based bricks.

*Calcium silicate bricks* were developed and patented in England by Van Derburg in 1866. It was further developed and patented by Dr. Wilhelm Michaëlis, who succeeded in producing a real 'artificial sandstone' in 1880. The specification for these bricks are provided by IS 4139:1989. The essential steps in the process of making sand-lime brick are (www.mortar.org.uk).

1. Hydrated or quicklime is added to siliceous sand or crushed flint, or manufactured sand in the ratio 1 (lime):10 (sand), and the two are mixed thoroughly; additionally, pigments may be added to give any desired colour to the bricks.
2. Water is then added and the mix is once again mixed thoroughly to bring the material to such a consistency that it will hold together when moulded.
3. The damp mixture is pressed into the desired shape.
4. Green bricks are then loaded into trucks, which are moved into curing chambers called *autoclaves*, which are similar to very large pressure cookers.
5. After the green bricks are loaded inside the autoclaves, the ends are closed and steam at a temperature of about 175°C is injected into the sealed chamber. This increases the pressure inside the chamber to 0.8–1.2 N/mm². This results in lime and sand combining chemically, to form calcium silicate, which acts as a bonding agent to hold the sand particles together.
6. Bricks remain in the autoclave for 7 to 10 h and are then unloaded and allowed to cool. The calcium silicate reacts with the atmosphere to form calcium carbonate.

Generally, calcium silicate bricks have a comparable weight to that of a clay bricks. They have negligible soluble salt content and hence are not prone to efflorescence. Based on their average compressive strength, the calcium silicate bricks are classified, as per IS 4139:1989, as in Table 3.8. The average drying shrinkage of calcium silicate bricks, when tested as per IS 4139:1989, should not be greater than that given in Table 3.8.

**Table 3.8** Classification of calcium silicate bricks

| Class designation | Average compressive strength, N/mm² | | Drying shrinkage (% of wet length) |
| --- | --- | --- | --- |
| | Greater than | Less than | |
| 7.5 | 7.5 | 10 | 0.06 |
| 10 | 10 | 15 | 0.06 |
| 15 | 15 | 20 | 0.04 |
| 20 | 20 | – | 0.04 |

Calcium-silicate (sand-lime) bricks may be used for masonry construction just like common burnt clay bricks, and may also be used as facing bricks. Notable uses of these bricks in London include Battersea Power Station and the RIBA building in Portland Place. This type of brick is common in other European countries like Sweden. These bricks use less energy and do not produce air pollutants associated with the firing of clay bricks. Hence, they are environmentally greener than clay bricks. Calcium-silicate bricks are also manufactured in Canada and the USA, and meet the criteria set forth in ASTM C 73–14 Standard Specification for Calcium Silicate Brick (Sand-Lime Brick).

## 3.10.3 Compressed Earth Blocks

*Compressed earth blocks* (CEB), also known as *stabilized mud blocks* (SMB) and *stabilized compressed earth blocks* (SCEB), were developed in the 1950s. They are made by pressing a mixture of soil (having liquid limit below 25%, clay content up to 20%, a minimum sand content of 35%, and plastic index between 8.5 and 10.5%) and stabilizer-like cement (5%) or lime (20%) or a combination of cement (4%) and fly ash (5%), using a manual or motorized machine at suitable moisture content. Where laterite soil is available, it can be used for making CEBs. The CEBs have advantages, such as low cost, low energy requirement (15 times less energy than fired bricks), comparable strength, and durability. More information may be obtained from IS 1725:2013 and Orekanti (2013).

### 3.10.4 Concrete Blocks

The first hollow concrete block was designed in 1890 by Harmon S. Palmer in the USA. These early blocks were usually cast by hand, and the average output was about 10 blocks per person per hour. Today, concrete block manufacturing is a highly automated process that can produce up to 2000 blocks per hour.

*Concrete block*, also called as *concrete masonry unit* (CMU) is a large rectangular brick made of concrete. Concrete blocks are made using a mixture of cement, fine and coarse aggregate (gravel), and are pale grey in colour. They are available in different shapes, such as stretcher, corner, double corner or pier, jamb, header, bull nose, and partition block, and concrete floor units. Some of these blocks are shown in Figs 3.14 and 3.15. Special 'architectural' blocks with a number of surface textures are also available. They include split face (the manufacture results in a rough, stone-like texture on one face of the block, as shown on Fig. 3.16, giving the architectural appearance of a cut-and-dressed stone; no two split face units are exactly alike), rustic face (produced by shot blasting the surface with small pellets), chiselled face, fluted face, and gemstone face (produced by grinding the surface and then applying a sealer, which gives added sheen). Hollow blocks construction facilitates provisions for concealing electrical conduit, water, and soil pipes.

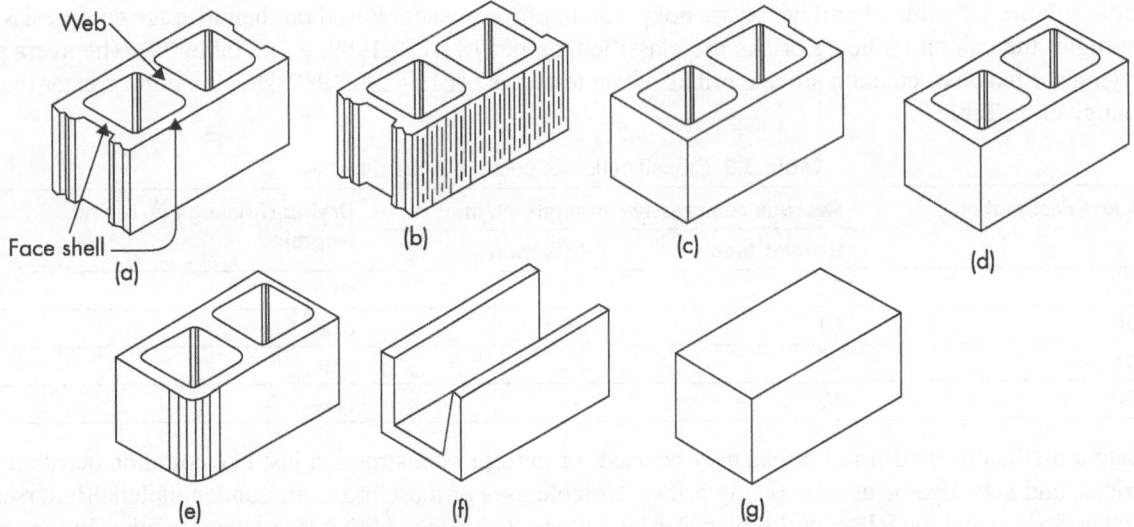

**Fig. 3.14** Typical shapes of concrete blocks (a) Stretcher (b) Wire cut/striated (c) Corner block (d) Double corner (e) Bull-nose (f) Beam or lintel block (g) Solid block

**Classification of concrete blocks** The concrete blocks are classified as per IS 2185 into four types: (1) Hollow, (2) Solid, and (3) Cellular concrete, and (4) Cellular concrete using preformed foam. The specifications for hollow and solid concrete blocks are provided in Part 1; for hollow and solid lightweight concrete blocks in Part 2; cellular concrete blocks in Part 3; and cellular concrete using

**Fig. 3.15** Solid and hollow concrete blocks

preformed foam in Part 4 of IS 2185:2005, respectively. Hollow blocks are those which have one or more large cavities, which either pass through the block (open cavity) or do not effectively pass through the block (closed cavity) and having the solid material between 50 and 75% of the total volume of the block. A block, which has solid material not less than 75% of the total volume of-the block, is considered solid block.

The physical requirements of these blocks for the three Grades A, B, and C, as specified in IS 2185 are given in Table 3.9. Thus, Grade A can have eight different strengths varying from

**Fig. 3.16** Wall of a building built with split face concrete block

3.5 to 15.0 N/mm² on the 28th day and have a minimum density of 1500 kg/m³. The density is calculated by dividing the mass of a block by the overall volume, including holes of cavities and end recesses. In addition, the water absorption should not be more than 10% by mass, drying shrinkage and moisture movement of the dried blocks on immersion in water should not exceed 0.06% and 0.09%, respectively.

**Table 3.9** Physical requirements of concrete blocks as per IS 2185 (Part 1):2005

| Type | Grade | Minimum density (kg/m³) | Minimum average compressive strength (N/mm²) at 28 days | Minimum compressive strength (N/mm²) at 28 days |
|---|---|---|---|---|
| Hollow (open and closed cavity) concrete blocks | A (3.5, 4.5, 5.5, 7.0, 8.5, 10.0, 12.5, and 15.0) | 1500 | 3.5, 4.5, 5.5, 7.0, 8.5, 10.0, 12.5, and 15.0, respectively | 2.8, 3.6, 4.4,5.6, 7.0, 8.0, 10.0, and 12.0, respectively |
| | B (3.0 and 5.0) | 1100 to 1500 | 3.0 and 5.0, respectively | 2.8 and 4.0, respectively |
| Solid load bearing units | C (4.0 and 5.0) | 1800 | 4.0 and 5.0, respectively | 3.2 and 4.0, respectively |

The dry loose bulk density of lightweight aggregates (conforming to IS 9142) used in light-weight concrete blocks should be less than 1120 kg/m³ for fine aggregates, 880 kg/m³ for coarse aggregate, and 1100 kg/m³ for combined aggregates. For more details, reference should be made to IS 2185 (Part 2):2005.

*Autoclaved cellular (aerated) concrete blocks* are covered in Section 9.6.1 of Chapter 9 (Also see IS 2185 (Part 3):2005). Two grades – Grade 1 with strengths of 2.0, 4.0, 5.0, 6.0, and 7.0 N/mm² and Grade 2 with strengths of 1.5, 3.0, 4.0, 5.0, and 6.0 N/mm² (with density varying from 450 to 1000 kg/m³, and thermal conductivity in air dry condition varying from 0.21 to 0.42 W/(mK) – are specified in IS 2185(Part 3). These blocks require steam-curing and autoclaving for 14–18 h at 700 kPa pressure and 185°C. The aerated cells of the cellular concrete blocks are formed by generating a gas with the mix by chemical action, prior to hardening, using suitable chemical foaming agents and mixing devices. An appendix is provided in IS 2185(Part 3), which gives the details of the manufacture of these blocks.

Foamed concrete, produced using preformed stable foam, in the form of blocks or poured *in situ* is used for thermal insulation over flat roofs or for cold storage walls, in addition to the other uses of ordinary concrete blocks. Low-density *preformed foamed cellular lightweight concrete blocks* are covered in IS 6598:1972 and medium- and high-density blocks, used in partitions and load bearing walls, are covered in IS 2185(Part 4):2008 (also see Section 9.6.2 of Chapter 9).

**Paver blocks**   The blocks are solid concrete blocks of various shapes, sizes, textures, and colours, made for exterior ground-paving on side-walks, drive ways, and parking lots. They are more durable than bricks but are less expensive. Some manufacturers design their pavers to inter-lock, assuring solid surface. Typical applications of these paver blocks in Chennai are shown in Fig. 3.17. The specifications of these concrete paver blocks may be found in IS 15658:2006 and IS 10360:1982.

**Materials used**   The concrete blocks are made with cement, coarse, and fine aggregates, with the minimum amount of water. Cement complying with any of the following Indian Standards may be used: 33/43/53 grade Ordinary Portland Cement (OPC)/Portland slag cement/Portland Pozzolana cement/Supersulphated cement/Rapid hardening Portland cement/White Portland cement or Hydrophobic Portland. When 33/43/53 grade OPC is used, the cement may be replaced by fly ash (conforming to IS 3812 (Part 1)) up to 25%. The aggregates used in blocks should con-form to IS 383. Fine aggregates may be replaced by fly ash conforming to IS 3812 (Part 2) up to 20%. Potable water, free of matter harmful to concrete or reinforcement, or matter likely to cause efflorescence should be used. Admixtures such as accelerating, water reducing, air-entraining, and super plasticizer (as per IS 9103) as well as waterproofing agents conforming to IS 2645 may be used. In addition, pigments may be added to give the blocks a uniform colour throughout. The surface of the blocks may also be coated with a baked-on glaze to give a decorative effect or provide protection against chemical attack. The glazes are usually made with a thermosetting resinous binder, silica sand, and colour pigments.

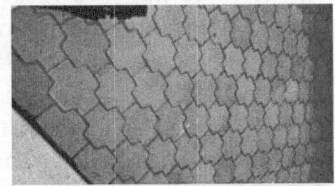

**Fig. 3.17** Application of paver blocks in Chennai

When volcanic cinders or granulated coal are used instead of sand and gravel, the block is called a *cinder block*, which will have dark grey colour, a medium-to-coarse surface texture, and typically weighs 11.8–15.0 kg. Cinder blocks have good strength, good sound-proof, and higher thermal insulat-ing properties than regular concrete blocks. *Precast concrete stone masonry blocks* (IS 12440:1988) are solid blocks made with concrete and stone spalls ranging in size from 50 to 250 mm.

Normal weight units are made with such aggregates as sand, crushed stones (with density 2000 kg/m³), and air-cooled slag. By replacing sand and gravel with expanded clay, shale, or slate, lightweight con-crete blocks are obtained. Expanded clay, shale, and slate are produced by crushing the raw materials and heating them to about 1093°C. At this temperature, small quantities of trapped organic material are com-busted resulting in rapid generation of gases, which bloat the size of blocks. These light-weight blocks, weighing about 10.0–12.7 kg, are used in non-load-bearing walls and partitions. Lightweight blocks are also made using expanded blast furnace slag, and natural volcanic materials, such as pumice and scoria.

**Production of concrete blocks**   The concrete mix used for blocks should not be richer than one part by volume of cement to six parts by the volume of combined aggregates before mixing. It is suggest-ed that the fineness modulus of the combined aggregates may be between 3.6 and 4.0. Owing to the thin web and faces of the concrete blocks, small-sized coarse aggregates of size 6 to 12 mm (which are not generally used for conventional concrete) are used. In general, the concrete mixture used for blocks has a higher percentage of sand (60–80%) and a lower percentage of gravel (20–40%) and water than the concrete mixtures used for general construction purposes. This produces a very dry, stiff mixture that holds its shape when it is removed from the block mould (Frasson Jr., et al., 2012). This 'dry-mix' necessitates the use of *vibro-compression machines*, which are special compaction devices that simultaneously apply compression and vibration to eliminate air voids when moulding the blocks (Frasson Jr., et al., 2012). The formed units are removed from the moulds and cured in a water tank (water in tank changed every 4 days) or yard for at least 14 days. After curing, the blocks should be dried for about 4 weeks before using them in walls. The blocks should be stacked with voids in the horizontal direction to facilitate easy drying. Alternatively, they can be stream-cured and dried. This process allows the initial drying shrinkage of the blocks to take place before they are used. It is important to prevent the use of freshly made or uncured concrete blocks in the construction.

**Dimensions and tolerenaces**   As per IS 2185 (Part 1):2005, the nominal dimensions (including mortar thickness of 10 mm) of concrete blocks are as follows (Fig. 3.14).

1. Length: 400, 500, or 600 mm
2. Height: 200 or 100 mm
3. Width: 50, 75, 100, 150, 200, 250, or 300 mm

In addition, blocks are also made in half lengths of 200, 250, and 300 mm to correspond to the full length. The maximum variation allowed is ±5 mm in length and ±3 mm in height and width. These strict tolerances could easily be achieved in machine made blocks. Full-length and half-length U-blocks are also available for constructing bands and lintels. In load bearing-walls, blocks with a minimum width of 200 mm are used; in filler or parapet walls, a minimum thickness of 100 mm may be used. The hollow load-bearing concrete block of standard 400 × 200 × 200 mm size will weigh 17–26 kg when made with normal weight aggregates. The two outer plates of the hollow concrete block are termed as face shells. These are connected together by webs (see Fig. 3.14). Table 3.10 gives the thicknesses of face-shell and the webs prescribed in IS 2185 (Part 1):2005. The core moulds could be so designed to have face shells and webs that are flared and tapered or straight tapered, with the former providing wider surface for mortar.

**Table 3.10** Minimum face-shell and web thickness

| Block width (mm) | Face-shell thickness (mm) | Web thickness (mm) |
| --- | --- | --- |
| Over 200 | 35 | 30 |
| 150–200 | 30 | 25 |
| 100–150 | 25 | 25 |
| Less than 100 | 25 | 25 |

**Use of hollow concrete blocks**   Concrete masonry units are used for interior and exterior load-bearing and non-load bearing walls (both below and above grade), fire safe walls around stairwells, elevators, and enclosures, fire walls and curtain walls, partitions and-panel walls, as backing for brick, stone, and other facings, for piers and columns, for retaining walls, garden walls, chimneys and fire places, as fillers in concrete joist floor construction, and as fireproofing over structural members. A wall built with split face concrete blocks is shown in Fig. 3.16.

Concrete masonry units manufactured from lightweight aggregate concrete are used for both load bearing and non-load bearing internal walls, partition and panel walls, inner leaf of cavity walls or as backing to brick masonry, and for external load bearing walls. They can be used as panel walls in steel or reinforced concrete frame construction, when protected from weather by rendering or by some other efficient treatment.

**Advantages of concrete blocks**   Concrete masonry has several advantages as listed below, such as economy, durability, strength and structural stability, fire resistance, insulation, and sound absorption.

1. They are cost effective than burnt clay bricks. Though they cost about 1.75 to 2 times more than ordinary and table-moulded fired clay bricks, they have 4.5 times the volume (a concrete block of size 390 × 190 × 190 mm can replace eight bricks).

2. As concrete blocks are regular in size, they require less mortar in joints. Owing to the large size of blocks, the number of joints in the masonry is also less. Fewer joints result in considerable savings in mortar as compared to normal masonry constructions, and in increasing the strength of the wall. The strength of mortar used need not be greater than the strength of blocks.

3. As the blocks are made with concrete, which is well compacted by vibration, they are much stronger than bricks. Proper curing increases compressive strength of the blocks.

4. Blocks are light and hence easy to handle. Owing to their lightness, the loads transferred to the foundations are much less than stone/brick masonry. This may be an important consideration in locations where soil has low bearing capacity.

5. The hollows in the blocks result in great saving of material.

6. The units are relatively large and true in size and shape. This result in rapid construction of walls per man-hour compared with bricks, resulting in saving time and labour.

7. Owing to the true plane surfaces obtained, plastering the walls of unimportant buildings situated in low rainfall areas is not necessary. Even when plastering is done, the quantity of mortar required for satisfactory coverage is significantly small.

8. The air in hollow of the block does not allow outside heat or cold inside the house. Thus, it keeps houses cool in summer and warm in winter. This thermal insulation results in reduced air-conditioning load, resulting in energy savings. Owing to the hollow space, the resulting wall will have much greater insulation properties against sound and dampness also.

9. There is no efflorescence due to salts. Hence, great saving in the maintenance of final finishes to the walls.

10. Environment friendly, as waste products like fly ash can be used as one of the raw materials. Moreover, clay bricks require excavation of earth from fertile areas and require fuel and energy for firing, which are not required in the manufacture of concrete blocks. The only energy intensive ingredient is cement; but only, a small quantity is required.

11. Maintenance cost of the hollow concrete block is less than that of brick masonry.

12. Concrete masonry has an attractive appearance and is readily adaptable to any style of architecture. It lends itself to a wide variety of surface finishes for both exterior and interior walls. The surfaces may also be finished with cement plaster. Concrete masonry units provide a strong mechanical key, uniting the concrete masonry backing and the plaster finish in a strong permanent bond [IS 2185 (Part 1):2005].

13. No additional formwork or any special construction machinery is needed for reinforcing the hollow block masonry, if required.

14. However, in walls made with hollow blocks, it is difficult to fix fittings like washbasins. Hence, such portions of walls are to be made with brickwork. Unless they are reinforced, their performance is poor during earthquakes or cyclones. Hence, some countries restrict their use in severe cyclonic or earthquake zones.

## 3.10.5 Plastic Blocks

The company Conceptos Plásticos (Plastic Concepts) in Colombia developed in 2011, LEGO-like bricks, which are made from recycled plastics and can be used to build houses up to two stories high in 5 days. Using an extrusion process, the plastic is melted and emptied into a final mould, creating a 3 kg brick, similar to the usual clay bricks with the same dimensions. When assembled under pressure, the bricks insulate heat and have additives that retard combustion. Additionally, they provide thermal and acoustic comfort and earthquake-resistance (www.archdaily.com).

## 3.11 Sustainability and Brick

Brick is made primarily from clay and shale, which are abundant natural resources. In addition, most brick manufacturing facilities are located near the clay and shale mining sites thus reducing transportation. Since fired brick is inert, it can safely encapsulate many waste materials. During the brick-making process, post-industrial and postconsumer waste products can be added to the clay or shale. Waste products such as bottom ash and fly ash from coal-fired generators, stone dust, glass, ceramic tile, and waste brick are included by some brick plants. Burn-out materials, such as sawdust and rice husk, are added to produce lower-weight brick units with fewer raw materials. Reclaimed industrial metallic oxides can also be used as colorants in brick. Brick manufacturing is an efficient process. Only about 5% of the mined material is lost in the production of bricks. Only about 3.5% of bricks go as scrap, which is used again in the manufacturing process or used as a filling material. Thus, raw material wastage is minimal. Water used in the production of bricks and heat generated in the kilns are also recycled and reused.

The brick making industry contributes significantly to global warming, especially in developing countries where inefficient brick kilns are used. In Asia, the brick industry is one of the most significant sources of pollution, and is the single largest emitter of both $CO_2$ and black carbon (soot). This is mainly due to the use of coal and scavenged fuels (such as plastic trash, used motor oil, battery cases, dung, and discarded tires) in outdated kiln technologies, which burn these fuels inefficiently resulting in incomplete combustion. Thus, the industry produces devastating effect on the environment that affects the health of kiln workers and the surrounding population. The brick industry in developing countries is generally associated with poor and marginalized workers. Other problems of the brick industry include land degradation, reduced crop yields, compromised food security, deforestation, poor work practices, respiratory illness, and bonded and child labour (Schmidt, 2013). Rapid urbanization and construction are increasing the demand for brick production and further usage of inefficient brick kilns. However, many brick manufacturers reclaim pit and mined areas when operations are completed by re-establishing vegetation or creating new ponds or lakes. Overburden and topsoil are also replaced in some cases, so that the resulting property can be reused for other purposes.

However, brick manufacturing in countries like the USA, which use tunnel kiln technology, is more energy efficient now than ever. In 2012, the average energy needed to manufacture fired bricks was in the range of 1.86–2.09 MJ/kg of brick produced, which is 30% less than the 2007 value. The improvements made in manufacturing include providing more energy-efficient kilns, increasing production of brick that require less energy to make, and installing energy-efficient lighting. Many plants also use scrubbers to control kiln emissions, and recycle the waste lime/limestone material generated by the scrubbers. Many manufacturers are using methane gas from landfills or sawdust to partially or totally fire their kilns, and one plant in the USA even received LEED Gold certification.

Brick can assist in getting up to 38 points out of the possible 110 points of the LEED-NC (2009) and 24.5 points out of the 136 points of LEED for homes (2008). For more information on how these points may be obtained, refer TN 48 of the *Brick Industry Association*. For example, the few areas in which bricks can contribute to a project being awarded a LEED™ certification under LEED-NC 2009 are shown here.

## Category: Sustainable sites

*Credit 6 Storm Water Design (1–2 points):* Permeable pavements utilizing flexible brick pavements allow for water to filter back into the ground.

*Credit 7 Heat Island Effect: Non-Roof (1 point):* Light-coloured brick pavements have a 'solar reflectance index' of at least 29.

## Category: Energy and atmosphere

*Credit 1 Optimized Energy Performance (1–19 points):* Brick is an energy-efficient material with insulating value and high thermal mass. It can also be used in passive solar construction by utilizing its thermal lag to reduce peak energy loads.

## Category: Materials and resources

*Credit 1 Building Reuse (1–4 points):* Brick has a useful life of more than 100 years. Brick walls and non-structural elements can be reused.

*Credit 2 Construction Waste Management (1–2 points):* Salvaged brick can be used in road construction and in some way in buildings.

*Credit 3 Materials Reuse (1–2 points):* Brick masonry is among the most commonly salvaged building material.

*Credit 4 Recycled Content (1–2 points):* Brick can be made with recycled or industrial wastes, such as fly ash, that are rendered harmless when the brick is fired.

*Credit 5 Regional Materials (1–2 points):* The raw materials of brick, clay, and shale are abundant and most of the time available locally, making bricks efficient to transport and distribute.

## Category: Indoor environmental quality

*Credit 4 Low-Emitting Materials – Flooring Systems (1 point):* The use of brick floors avoids carpets and adhesives – leading to the avoidance of VOCs.

## Category: Innovation in design

*Credit 1 Innovation in Design (1–5 points):* Brick can provide superior acoustic comfort. Brick interior walls do not require any paint thus avoiding VOCs. Brick also has Life Cycle Assessment advantage.

## Category: Regional priority

*Credit 1 Regional Priority (1–4 points):* Brick can help achieve credits in this category.

*Note:* Points shown here refer to total points allowed in credit, not the points contributed solely by brick.

## SUMMARY

- Bricks are important construction materials. These were used as early as 14,000 BC.
- Bricks are produced from surface clays, shales, and fire clays. A good brick material should contain 20–30% of alumina, 50–60% of silica, small quantities of finely powdered lime, and iron and magnesium oxides; the last two constituents providing colour to the brick, when fired.
- The manufacturing process of bricks and other clay products consists of the following steps: (a) extraction and storage of raw materials; (b) preparation (crushing, blending, grinding, and screening); (c) forming; (d) drying; (e) firing; and (f) packing and distribution.
- Moulding of bricks can be done by (a) stiff-mud process, (b) soft-mud process, and (c) dry-mud process.
- Several types of kilns can be used for firing bricks, which include (a) clamp kiln; (b) the UDK; (c) the DDK; (d) Bull's (movable/fixed chimney)trench kiln; (e) zigzag kiln; (f) Hoffman kiln; (g) VSBK; and (h) tunnel kiln. Bricks may be classified based on their compressive strength, according to physical requirements, and based on the method of manufacture.

- The most important properties of bricks are: (a) durability; (b) colour; (c) texture; (d) size variation; (e) compressive strength; and (f) water absorption. Architectural characteristics of bricks include their shape, type, size, weight, colour, and texture.
- The main types of bricks are engineering bricks, perforated bricks, hollow bricks, facing bricks, refractory bricks, paving bricks, and acid-resistant bricks. Modular bricks are normally made in sizes of 190 × 90 × 90 mm or 190 × 90 × 40 mm, and weigh about 3.0–3.5 kg.
- The properties of bricks that may affect their performance during their service life are (a) compressive strength; (b) water absorption; (c) efflorescence; (d) dimensional tolerance; (e) hardness; (f) soundness; and (g) structure.
- There are several substitutes for bricks which include: (a) fly ash bricks; (b) sand-lime bricks or calcium silicate bricks; (c) compressed earth blocks; and (d) concrete blocks. The use of concrete blocks has many advantages over fired bricks.
- As bricks are made with a natural material such as clay, they are more sustainable.

# EXERCISES

## Multiple-choice Questions

1. Which of the following is necessary for a good brick material?
   - (a) About 30% alumina
   - (b) About 30% silica
   - (c) 10% of iron oxides
   - (d) About 15% magnesia

2. Match the list on the right with the list on the left.
   - (a) Silica            (i) <5%
   - (b) Powdered lime      (ii) 20–30%
   - (c) Alumina           (iii) <10%
   - (d) Alkalis           (iv) 50–60%

3. Match the list on the right with the list on the left.
   - (a) Silica            (i) Imparts colour
   - (b) Alumina           (ii) Fuses sand at high kiln temperature
   - (c) Lime             (iii) Prevents shrinkage and cracking
   - (d) Ferric oxide      (iv) Imparts plasticity

4. Which of the following is harmful in the clay used for making bricks?
   - (a) Iron oxide
   - (b) Iron pyrite
   - (c) Silica
   - (d) Magnesia

5. What should be the water content in stiff mud process?
   - (a) <10%            (c) 20–30%
   - (b) 10–15%          (d) >25%

6. How much percentage the size of moulds should be larger than actual size of bricks, to account for shrinkage?
   - (a) 2–5%            (c) 8–15%
   - (b) 6–10%           (d) 15–20%

7. When is frog provided?
   - (a) 90 mm high bricks only
   - (b) 40 mm high bricks only
   - (c) Wire-cut bricks only
   - (d) None of these

8. What is the most important purpose of frog in a brick?
   - (a) To show the trade mark of manufacturer
   - (b) Reduce the weight of brick
   - (c) Form keyed joint between brick and mortar
   - (d) All of these

9. In which of the following operations of brick manufacturing, pug sealer/pug mill is used?
   - (a) Weathering         (c) Tempering
   - (b) Blending          (d) Burning

10. After drying in the drying chambers of kilns, the moisture content of bricks drops to
   - (a) Below 2%           (c) About 6%
   - (b) Below 5%           (d) None of the above

11. What is the temperature range in which bricks are burnt?
   - (a) 500–600°C          (c) 900–1300°C
   - (b) 700–800°C          (d) >1500°C

12. Vitrification of bricks takes place at about
   - (a) 205°C             (c) 540–870°C
   - (b) 150–540°C          (d) 870–1315°C

13. Which of the following is an intermittent kiln?
   - (a) Clamp kiln
   - (b) Tunnel kiln
   - (c) Hoffman kiln
   - (d) Vertical shaft brick kiln

14. Which of the following is a continuous kiln?
   - (a) Clamp kiln
   - (b) Scove kiln
   - (c) Downdraught kiln
   - (d) Vertical shaft brick kiln

15. Which of the following is the advantage of vertical shaft brick kiln?
   - (a) It saves about 25–40% of coal compared to the FCBTK.
   - (b) It reduces pollution by 70% compared to the FCBTK.
   - (c) Bricks can be produced according to the demand.
   - (d) All of these

16. Water absorption of Class I bricks is less than
    (a) 20%                    (c) 25%
    (b) 22%                    (d) 30%

17. As per IS 2222, for perforated bricks, the ratio of area of perforation to the total area should be
    (a) 10–20%                 (c) 25–35%
    (b) 15–25%                 (d) 30–45%

18. The IS classification of bricks is based on
    (a) Compressive strength
    (b) Water absorption
    (c) Dimensional tolerance
    (d) None of these

19. Water absorption for Class II bricks should not be more than
    (a) 12%                    (c) 20%
    (b) 15%                    (d) 25%

20. Size of modular bricks is
    (a) $9'' \times 4.5'' \times 3''$
    (b) $190 \times 90 \times 90$ mm
    (c) $230 \times 110 \times 70$ mm
    (d) None of the above

21. Weight of modular brick is approximately
    (a) 2 to 2.5 kg            (c) 4 kg
    (b) 3 to 3.5 kg            (d) 5 kg

22. Efflorescence of bricks is due to
    (a) Soluble salts present in clay for making bricks
    (b) High porosity of bricks
    (c) High silt content in brick earth
    (d) Excessive burning of bricks

23. What is efflorescence?
    (a) The appearance of white patches on the brick surface due to insoluble salts
    (b) Swelling of brick
    (c) Deformation of brick
    (d) Impurities that are seen on the surface after burning

24. When efflorescence is considered moderate?
    (a) Heavy deposit of salt covers 25% of the area of brick but there is no powdering/flaking of the surface
    (b) Heavy deposit of salt covers 30% of the area of brick and there is no powdering/flaking of the surface
    (c) Heavy deposit of salt covers 50% of the area of brick but there is powdering of the surface
    (d) Heavy deposit of salt covers 50% of the area of brick but there is no powdering/flaking of the surface

25. What is effect of using fly ash as an additive with clay?
    (a) Higher strength
    (b) Sharper edges and corner of bricks
    (c) Reduced drying shrinkage
    (d) All of these

26. Grade A hollow concrete blocks, as per IS 2185 (Part 1), have a minimum density of
    (a) 1100 kg/m³             (c) 1100–1500 kg/m³
    (b) 1500 kg/m³             (d) 1800 kg/m³

27. Concrete blocks are produced with
    (a) 20–40% of sand and 60–80% of gravel
    (b) 60–80% of sand and 20–40% of gravel
    (c) 50% of sand and 50% of gravel
    (d) None of these

28. Before using them, concrete blocks should be
    (a) Cured for 10 days and then dried for 2 weeks
    (b) Cured for 10 days and then dried for 4 weeks
    (c) Cured for 14 days and then dried for 2 weeks
    (d) Cured for 14 days and then dried for 4 weeks

29. Out of the possible 110 points of LEED–NC, brick can assist to get
    (a) Up to 20 points
    (b) Up to 25 points
    (c) Up to 30 points
    (d) Up to 38 points

## Review Questions

1. What are the constituents of good brick earth? Is it possible to make good bricks using black cotton soil?
2. What are the harmful ingredients in brick earth?
3. Name the different operations involved in the manufacture of brick.
4. Why is tempering essential in brick making?
5. What are the two types of extrusion systems for mixing the brick material? Explain them briefly.
6. Describe briefly the process of forming the green bricks. What are the three methods of forming bricks? Which method is used in India?
7. How does hand moulding differs from table moulding?
8. What is meant by drying of bricks? What are the different methods of drying bricks? Which method is preferable in winter seasons?
9. What are the six different stages of firing bricks in a kiln?
10. Explain the operations involved while firing bricks in a kiln.
11. How do intermittent kilns differ from continuous kilns?
12. Name any five kilns that are used for firing bricks.

13. Describe clamp kiln and state why it is harmful to the atmosphere.
14. Write short notes on
    (a) Down-draught kiln        (b) Vertical shaft brick kiln    (c) Tunnel kiln
15. Explain briefly the features of the vertical shaft brick kiln with a neat sketch.
16. What are the advantages of the vertical shaft brick kiln?
17. Explain briefly the concept of tunnel kiln. What are the advantages and drawbacks? Why is it not used extensively in India?
18. What are the different ways in which bricks are classified?
19. What are engineering bricks and how are they differing from common bricks?
20. How are bricks classified as per IS 1077?
21. What are Class I bricks? How are they differentiated from Class II and Class III bricks?
22. What are the differences between water-stricken, sand-stricken, and wire-cut bricks?
23. What are the six most important properties of bricks?
24. List any five qualities of engineering bricks, to be used in important structures.
25. What is a frog? What is its function? Can we have frog in wire-cut bricks?
26. Which type of brick will you recommend in bridges? Why?
27. What are the rules, as per IS 2222, to consider a brick as perforated brick? What are the advantages of using such bricks?
28. How will you differentiate between hollow and perforated bricks?
29. What are facing bricks? What are the characteristics that differentiate them from common bricks?
30. What are refractory bricks? Do they require special composition and firing to have fire-resistant properties?
31. What are the dimensions of modular brick as per IS 1077? How their dimensions differ from that of field bricks that are commonly available in India?
32. What are the properties of bricks that affect their performance during their service life?
33. Describe at least five common defects in burnt clay bricks.
34. List at least five advantages of bricks, as compared to other building materials.
35. Where are bricks used in building construction?
36. List at least three substitutes for burnt clay bricks.
37. What are the three kinds of fly ash bricks? What are the advantages of fly ash bricks?
38. What are FAL-G bricks? What are the additional advantages of these bricks?
39. Can we use fly ash alone in bricks? How are they made? What are the advantages?
40. Write shot notes on (a) Calcium silicate bricks (b) Compressed earth blocks (c) Concrete blocks
41. What are the four types of classification of concrete blocks as per IS 2185?
42. What are the materials used for producing concrete blocks? How are lightweight concrete blocks produced?
43. How are concrete blocks produced?
44. Compare the sizes of concrete blocks with that of clay bricks.
45. What are the uses and advantages of hollow concrete blocks?
46. Why are bricks considered sustainable building materials?

# ANSWERS

## Multiple-choice Questions

|  |  |  |
|---|---|---|
| 1. (a) | 2. (a)–(iv) (b)–(i) (c)–(ii) (d)–(iii) | 3. (a)–(iii) (b)–(iv) (c)–(ii) (d)–(i) |
| 4. (b) | 5. (b) | 6. (c) |
| 7. (a) | 8. (c) | 9. (b) |
| 10. (a) | 11. (c) | 12. (d) |
| 13. (a) | 14. (d) | 15. (d) |
| 16. (a) | 17. (d) | 18. (a) |
| 19. (d) | 20. (b) | 21. (b) |
| 22. (a) | 23. (a) | 24. (d) |
| 25. (d) | 26. (b) | 27. (b) |
| 28. (d) | 29. (d) |  |

# CHAPTER 4
# BUILDING LIME

## 4.1 Introduction

Throughout the world, limestone is a naturally occurring mineral. It is a general name given to a wide variety of sedimentary rocks, which are composed primarily of calcium carbonate ($CaCO_3$). About 10% of all sedimentary rocks, such as limestone, chalk, marble, and travertine, are limestone. White chalk is a pure limestone whereas *kankar* is an impure limestone. *Lime*, i.e., calcium oxide (CaO), is obtained from limestone by heating these stones to a temperature of about 1000°C, in kilns. This process is called *calcination*. It is also obtained by the calcination of shell, coral, and other calcareous substances. Lime has been used as the chief cementing material in building construction for both mortars and plasters, for several centuries. Use of lime as a binder in construction dates back to 6,000 BC. A terrazzo floor excavated in Catalhüyük, Turkey, laid with lime mortar is considered to be belonging to 7,500–5,700 BC (Elsen, 2006). Romans developed producing lime using kilns. Indians used lime for construction from very ancient days. The Qutub Shahis (1518–1687) often used lime mortar for constructing many monuments in and around the city of Hyderabad, including the Charminar. The durability of lime mortars is proven by more than 2,000-year old lime structures that withstood the ravages of time. Owing to the varied geological character of India, several types of limestone are available here, which can be burnt to obtain building limes. Owing to the variability of limestone from place to place, the resultant lime also varies in quality. Major uses of lime are metallurgical (steel, copper, gold, aluminium, and silver), environmental (flue gas desulphurization, water softening and pH control, sewage-sludge stabilization, hazardous waste treatment, and acid neutralization), for construction purposes (soil stabilization, bitumen additive, and masonry lime), and agricultural (soil conditioning and pH control).

Though Portland cement has almost replaced lime in modern structures, lime is used where it is available locally and perhaps during the period of shortage of ordinary Portland cement. Notably, along with durability, lime provides a cheap alternative to cement. Three varieties of lime, viz., quicklime, hydrated lime, and hydraulic lime, could be manufactured. *Quicklime* is not a stable product and hence is hydrated or 'slaked' with the sufficient quantity of water, as soon as it is produced, to produce *hydrated lime* (i.e., *slaked lime*). To be specific, the term lime in construction always refers to slaked lime and not quicklime. *Hydraulic lime* (which can be used like cement) can be manufactured from caliche (also known as hardpan, calcrete, *kankar* [in India], or duricrust), which is a sedimentary rock consisting of a mixture of $CaCO_3$, clay, and silica. Caliche has been used to build some of the Mayan buildings in the Yucatán Peninsula in Mexico.

In 2014, global lime production, including captive lime, was estimated to be 360 MT. (Captive lime is the lime produced for internal consumption in integrated plants [in sugar, pulp, and steel industries]) (https://minerals.usgs.gov/). More than 65% is produced in China and 4.5% in India. The cement and lime production processes are similar: both use fuel-fired kilns to process raw materials at high temperatures, and both are energy-intensive, continuous-production industrial processes.

The manufacturing of lime, its different forms and classification, storing, different uses, and precautions while working with lime are discussed in this chapter. Though the production of lime involves the emission of carbon dioxide ($CO_2$), it is considered a green product, as it reabsorbs $CO_2$ from the atmosphere to harden and forms $CaCO_3$, which has cementing properties.

## 4.2 Constituents of Limestone or Varieties of Lime

The chemical composition of limestone varies greatly from region to region as well as between different deposits in the same region. Therefore, the end product from each natural deposit is different. The production of limestone in India during 2012–2013 was about 279.7 MT. Typically, limestone is composed of $CaCO_3$ and the mineral dolomite [calcium and magnesium carbonate ($MgCO_3$)], silica ($SiO_2$), alumina ($Al_2O_3$), iron (Fe), sulphur (S), and other trace elements]. Limestone is generally classified into the following types (www.lime.org):

1. High calcium: derived from limestone containing 90–95% $CaCO_3$ and 0–5% magnesium carbonate
2. Magnesian: derived from limestone containing 5–35% magnesium carbonate
3. Dolomitic: derived from limestone containing 35–46% magnesium carbonate

In addition, we could have shell lime, which is very pure lime obtained by the calcination of shells of sea animals and corals (Coral and shell lime also have over 95% $CaCO_3$ content). Some naturally occurring limestone, called *kankar* in India, is an impure lime and occurs in the form of nodules and compact blocks. It may contain 5–25% of clay and is commonly used for making hydraulic lime, which hardens in the presence of water-like cement. While more pure form of lime is needed for chemical and industrial use, lime with impurities may be desirable for use in building construction. Thus, lime for construction purposes may have up to 15% insoluble residue content, and up to 20% silica, alumina, and ferric oxide content.

## 4.3 Lime Cycle

The *quicklime* that is produced by burning limestone in kilns should be slaked soon after it is drawn from the kiln. Otherwise, it tends to react with the $CO_2$ and humidity present in the atmosphere and becomes air slaked, and returns to its original form of $CaCO_3$. This process is called *lime cycle* (Fig. 4.1). Consequently, the lime loses its properties and becomes unsuitable for sound construction. The hydration of lime is accompanied by an increase in volume of about 1.7 to 2.5 times (it depends on the volume of water used and the type of limestone).

The dolomitic lime 'cycle' is more complicated than high-calcium limestone, in that the magnesium and calcium compounds hydrate and carbonate in a

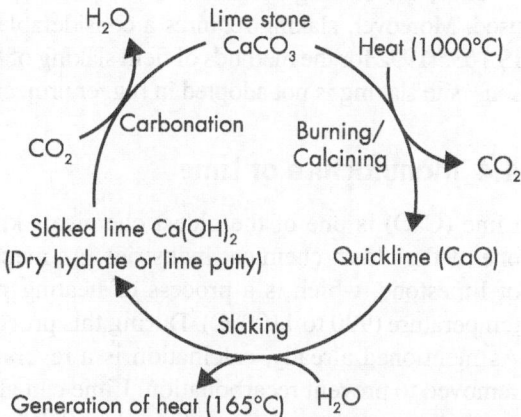

**Fig. 4.1** Lime cycle (for high-calcium limestone)

different manner, apparently leading to a wider variety of phases (Fig. 4.2). Dolomitic limestone is calcined to produce both quicklime and magnesium oxide (MgO), resulting in *hydraulic lime*. The two oxides of dolomitic lime hydrate to calcium hydroxide [Ca(OH)$_2$] and magnesium hydroxide [Mg(OH)$_2$], although the hydration of MgO is much slower than that of quicklime. The carbonation of Ca(OH)$_2$ to calcite (CaCO$_3$) is simple; however, the carbonation of Mg(OH)$_2$ to magnesite (MgCO$_3$) occurs at a much slower rate, and an assortment of hydroxycarbonate compounds, in varying ratios, may also occur, as shown in Fig. 4.2. The magnesium compounds may be significantly reactive to acid rain, in industrial locations, leading to the formation of magnesium sulphate salts. According to Hartshorn (2012), these magnesium sulphate salts may result in sulphate attack and damage lime mortar when they are dry and expand due to recrystallization.

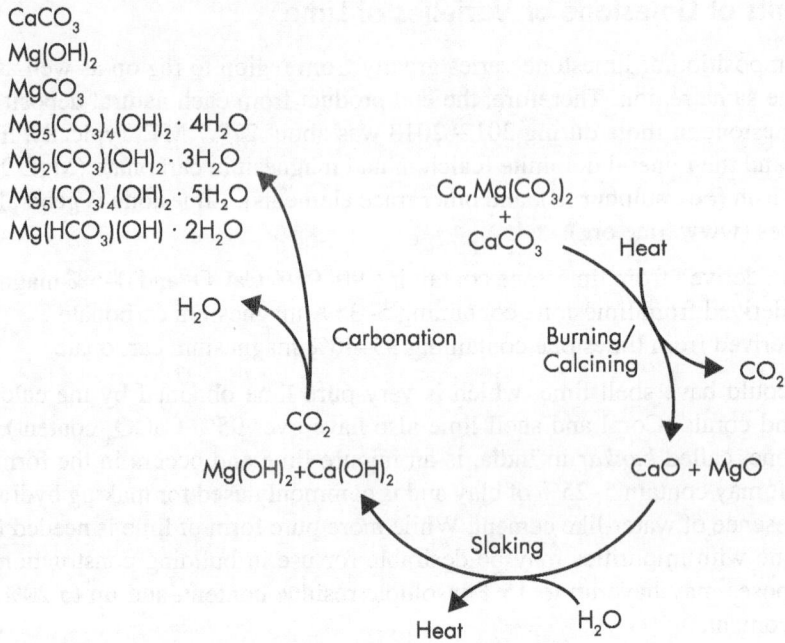

**Fig. 4.2** Lime cycle for dolomitic lime

Thus, the slaking of limestone requires great skill and knowledge of the type of limestone being used. Moreover, slaking requires a considerable space, which is not always readily available at sites (see IS 1635:1992 for the methods of field slaking of building lime and preparation of putty). Owing to these reasons, site slaking is not adopted in bigger projects and slaking is done in factories (see also, Section 4.5.2).

## 4.4 Manufacture of Lime

Lime (CaO) is one of the oldest chemicals known to man and the process of lime production is also one of the oldest chemical industries. As mentioned earlier, quicklime is derived from the calcination of limestone, which is a process of heating pieces of limestone, in various types of kilns, to a high temperature (900 to 1100°C). During this process, CO$_2$ is released from CaCO$_3$ (see also, Section 4.4.4). As mentioned already, calcination is a reversible chemical reaction; hence, the emitted CO$_2$ must be removed to prevent recarbonation. Lime can also be produced from aragonite, chalk, coral, marble, and seashells; these sources contain only calcite or aragonite (CaCO$_3$ with an orthorhombic rather than a rhombohedral crystal structure). *Quicklime* coming out of the kilns is also called as *lump-lime*. When the quicklime has high-calcium oxide component, it is called as *fat lime*. In some lime plants, the resulting lime is reacted (slaked) with water to form hydrated lime (see Section 4.5.2). The various steps in the production of lime are explained, with reference to Fig. 4.3, as below.

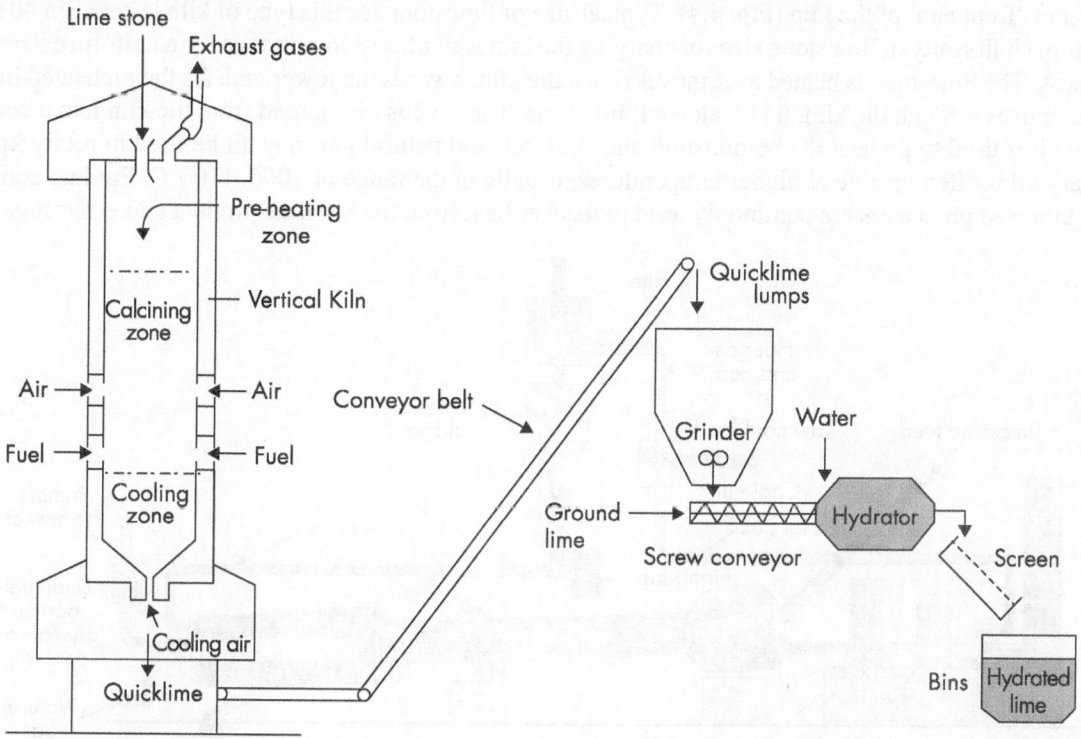

**Fig. 4.3** Flow diagram of manufacturing of lime with vertical kiln

## 4.4.1 Preparation of Raw Materials

The first two steps of manufacturing of lime are concerned with the preparation of raw materials. Raw limestone is excavated from quarries or mines using explosives or mechanical rock-breaking hammers. The extracted limestone is selected according to its chemical composition, as different sources require different calcination temperatures in the kilns.

In order to obtain uniform burning and facilitate even gas flow in the kilns, the excavated limestone are crushed to a small uniform size of about 50 to 80 mm diameter (about the size of fist of human beings). Energy required to quarry and process 1 tonne of limestone is 964 MJ/tonne. Water consumption for limestone quarrying and processing is 75,708 litres/tonne.

## 4.4.2 Calcination of Limestone in Kilns

The name calcination is derived from the Latin word *calcinare*, which means 'to burn lime'. Calcination of $CaCO_3$ is a highly endothermic reaction, requiring 755 MCal (3.16 GJ) of heat input to produce a tonne of lime. Calcination of limestone using charcoal fires to produce quicklime has been practised since antiquity by cultures all over the world. In modern lime plants, a calcining kiln is used, which is mainly of two types: (a) rotary kiln and (b) vertical kiln. Both the types of kilns can be designed to use any solid (wood or coal), liquid, or gaseous fuels.

The commonly used rotary or horizontal kiln is a long, cylindrical, refractory-lined furnace. The cylinder of the rotary kiln is oriented at about 3° to 4° from the horizontal and rotates at about 1–3 rpm. Limestone is fed into the upper or 'back end' of the kiln, while fuel and combustion air are fired into the lower or 'front end' of the kiln (Fig. 4.4). Typical size of limestone for this type of kiln is between 40 and 50 mm. Uniformity of limestone size for charging the kiln is of utmost importance for a uniform calcining process. The limestone is heated as it moves down the kiln towards the lower end. As the preheated limestone moves through the kiln, it is 'calcined' into lime. The lime is discharged from the kiln into a cooler where it is used to preheat the combustion air. Coal, oil, and natural gas may all be fired in rotary kilns. Rotary kilns often operate at higher temperatures, usually in the range of 1000–1100°C. Product coolers and kiln feed preheaters are commonly used to recover heat from the hot lime product and exhaust gases.

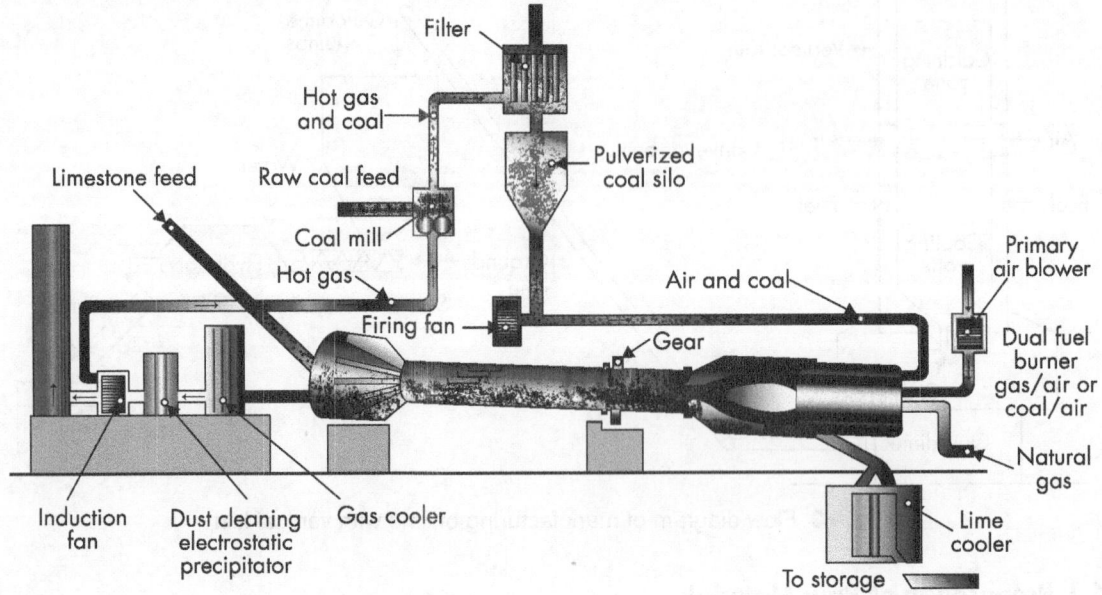

**Fig. 4.4** Typical rotary kiln
(*Source*: www.eula.eu/kiln-types, with permission from Industrial Minerals Association Europe)

The next most common type of kiln is the vertical, or shaft, kiln [IS 1861(Part 1):1990]. This kiln is as an upright heavy steel cylinder lined with refractory material. These kilns usually use limestone sizes between 130 and 200 mm. The limestone is charged at the top, calcined as it descends slowly, and is discharged at the bottom of the kiln (Fig. 4.3). Vertical limekilns typically have three zones: preheating, calcining, and cooling, and the lime is heated from 20 to 900°C, calcined up to 1200°C, and cooled down to 50°C. A primary advantage of vertical kilns over rotary kilns is their higher average fuel efficiency. Typically, vertical kilns use oil or natural gas as fuel. The primary disadvantages of vertical kilns are their relatively low production rates and the fact that coal cannot be used without degrading the quality of the lime produced. Most kilns employ a draft, whether natural or induced ventilation device, for removing the evolved $CO_2$, which increases and regulates the lime output.

Other kiln types include rotary hearth and fluidized bed kilns, which are not common. Both these kilns can achieve high production rates. The rotary hearth kiln is circular in shape and has a doughnut-shaped hearth, which revolves slowly. In fluidized bed kilns, finely divided limestone comes into contact with hot combustion air in a turbulent zone. These kilns have dust-collection equipment to collect the lime that is carried over with the exhaust gases.

In the United States and European Union, Parallel-flow Regenerative (PR) limekiln is used, which provides optimum heating conditions. More details about the principle and operation of limekilns, PR kilns, and energy efficiency of different types of kilns may be found in Adams (1996), Nimbalkar et al. (2014), and Canadian Lime Institute (2001).

Limestone calcination is very energy-intensive process and consumes a considerable amount of fuel [3.5 to 12.6 GJ/t, depending on the type of kiln used (UNIDO Report 2010)]. A number of processes have been tried in the past to improve the fuel-consumption efficiency, which include the following (IS 14860:2000; Hassibi, 2009, Canadian Lime Institute, 2001; and Nimbalkar et al., 2014):

1. Before the limestone enters the kiln, it could be preheated by the hot exhaust gasses. Using this process, the size of kiln can be reduced, as the residence time of limestone in the kiln is minimized. In addition, substantial heat from the exhaust gases could be recovered.
2. To recover the heat from the red hot limestone (when it is calcined and exits the kiln it has a temperature of about 1200°C), the combustion fresh air is used to cool the quicklime. The resulting heated air is then fed into the kiln. This process improves the fuel consumption efficiency.
3. Limestone could be calcined in a continuous process, thus reducing fuel consumption and minimizing the degradation of the refractory lining of the kiln.

## 4.4.3 Method of Storage of Lime

Once limestone is converted to quicklime (CaO), as it is very susceptible to moisture, it must be stored in airtight silos to avoid 'air slaking,' which will result in the deterioration of quality of quicklime. Air slaking is basically the process of converting the quicklime to $Ca(OH)_2$ at ambient temperature by moisture present in the air. Though air slaking takes several weeks for converting the lime to $Ca(OH)_2$, 'air slaked' lime is not very reactive and will have extremely large hydroxide particles. Exposure of quicklime to an atmosphere containing a high percentage of $CO_2$ also must be avoided, as the latter will be absorbed by lime, which will revert back to its original form of $CaCO_3$. This conversion is accelerated at elevated temperatures. Hence, quicklime lumps may be ground and processed further by hydrating them with just enough water to produce hydrated lime (see Sections 4.4.7 and 4.5). This hydrated lime can be bagged, stored, and handled for distribution.

## 4.4.4 Chemical Reaction that occurs during Manufacturing

As seen in Section 4.4.2, lime is manufactured in various kinds of kilns, and one of the following reactions will occur during its manufacture, depending on the type of limestone used:

1. Using high-calcium limestone (at about 900°C):

$$CaCO_3 + Heat \rightarrow CO_2\uparrow + CaO \text{ (calcium oxide)} \tag{4.1}$$

Using the atomic weight of the various elements in the reaction as (Hassibi, 2009):

Ca = 40 g, O = 16 g, C = 12 g, we get,
$(40 + 12 + 3 \times 16) \rightarrow (12 + 2 \times 16) + (40 + 16)$
i.e., 100 g $\rightarrow$ 44 g + 56 g

Or, in other words, 100 g of $CaCO_3$, when calcined, will result in 56 g of CaO and 44 g of $CO_2$. As already noted, CaO is unstable in the presence of moisture and $CO_2$. A more stable form of lime is $Ca(OH)_2$.

$$CaO + H_2O \rightarrow Ca(OH)_2 + Heat \tag{4.2}$$

Using the atomic weight of the various elements in the reaction as (Hassibi, 2009):

Ca = 40 g, O = 16 g, C =12 g, and H = 1 g, we get,

The reaction in terms of atomic weight is: $(40 + 16)$ g $+ (2 + 16)$ g $= 74$ g.

This shows that 56 units of CaO plus 18 units of $H_2O$ result in 74 units of $Ca(OH)_2$.

2. Using dolomitic limestone (at about 1100°C):

$$CaCO_3 + MgCO_3 + \text{Heat} \rightarrow 2CO_2 \uparrow + CaO + MgO \qquad (4.3)$$

Using the atomic weight of the various elements in the reaction as (Hassibi, 2009):

Ca = 40 g, O = 16 g, C =12 g, and Mg = 24 g, we get,

$(40 + 12 + 3 \times 16) + (24 + 12 + 3 \times 16) \rightarrow [2(12 + 2 \times 16)] + (40 + 16) + (24 + 16)$

i.e., 100 g + 84 g $\rightarrow$ 88 g + 56 g + 40 g.

This shows that 184 units of dolomitic limestone result in 96 units of dolomitic quicklime and 88 g of $CO_2$. Thus, with pure high-calcium limestone, 44% of the stone weight is lost as $CO_2$, while the loss with dolomitic limestone is 48%.

## 4.4.5 Estimating the Quantity of Lime

The quantity of quicklime produced by using high-calcium limestone in the kilns can be approximately estimated from the calcination equation as illustrated in Example 4.1 (Hassibi, 2009). It has to be noted that the actual quantity of quicklime produced will depend on the type of limestone and the type of fuel used in the kiln.

**Example 4.1**   Determine the theoretical weight of quicklime obtained and $CO_2$ emitted by burning 500 g of limestone.

**Solution**: The reaction takes place in the kiln is:

$$CaCO_3 + \text{Heat} \rightarrow CO_2 \uparrow + CaO$$

Using the atomic weights of C = 12 g, Ca = 40 g, and O = 16 g, we get,

$(40 + 12 + 3 \times 16) = (12 + 2 \times 16) + (40 + 16)$

That is 100 g = 44 g + 56 g

Thus, lime produced by 500 g of limestone = (56/100) 500 = 280 g

$CO_2$ emitted = (44/100) 500 = 220 g

## 4.4.6 Effect of Impurities in Limestone

In some locations, limestone with high content of $CaCO_3$ may not be available and dolomitic limestone has to be used. As indicated earlier, such limestone, having impurities, could also be used advantageously in the construction industry. When dolomitic limestone is calcined in kilns, it decomposes to MgO at about 775°C. When the temperature is further raised to 900°C [the required temperature to break down $CaCO_3$ to calcium oxide (CaO)], the already produced MgO will be over-burnt and hence will not readily convert to $Mg(OH)_2$, while hydrating. If hydration is not fully accomplished, it can lead to problems of expansion due to delayed hydration of MgO, in mortars or plasters. In plasters, it will result in blistering of plastered walls. To avoid this, it is better to use pit slaking over an extended period of time to form lime putty and then use it for plastering. While pit slaking, it may be better to add dolomitic lime to water instead of water to dolomitic lime for satisfactory results.

Limestone containing clays, such as *kankar* lime, will result in hydraulic lime, which can be used like Portland cement. They also gain greater strength at a faster rate than high-calcium limes. They will also set under water and have more durability than high-calcium lime (see Section 4.4.7).

### 4.4.7 Hydrators used in Lime Production

The last operation, before bagging lime for commercial distribution, is hydration (Fig. 4.3). As quicklime is not stable, it is usually converted to hydrated or slaked lime (Section 4.5.2). Generally, water sprays or wet scrubbers perform the hydrating process and prevent product loss. Following the hydration, the product may be milled and then conveyed to air separators for further drying and removal of coarse fractions.

## 4.5 Different Types of Limes

As we have learnt, when limestone are burnt in a kiln at 900°C to 1200°C, quicklime or hydraulic lime is produced, depending on whether the raw material is high-calcium limestone or dolomitic limestone, having impurities such as MgO. In addition, quicklime could be hydrated with sufficient water to produce slaked or hydrated lime. These different types of lime are discussed in this section (Fig. 4.5).

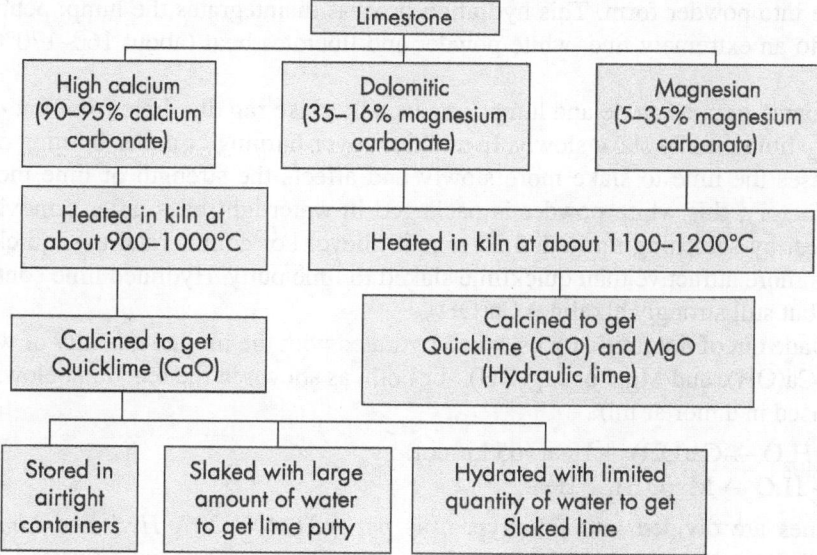

**Fig. 4.5** Production process of different types of lime

### 4.5.1 Quicklime or Lime Putty

The production of good quality lime depends upon the type of kiln, conditions of calcination, and the nature of the limestone used in the production. Use of correct temperature in the calcination process results in a relatively soft product, containing small lime crystals and having open porous structure. Such a lime has the optimum properties of high reactivity, high surface area, and low bulk density. In quicklime, the impurities are limited to less than 5%. If they exceed 10%, the product is termed as *hydraulic lime*.

As discussed earlier, the caustic quicklime is unstable and therefore has to be 'slaked' or hydrated with water to prevent carbonation. Ordinary (non-hydraulic) lime is produced by slaking quicklime with large amount of water (2 to 3 times the weight of quicklime), resulting in a chemical reaction involving sufficient heat (about 76 to 85°C) to boil the entire mass (The amount of water to be added is decided based on the composition, degree of burning and slaking methods, because a part of the water used is vaporized by the released heat.). The resulting product is called *quicklime putty* or *lime putty*. Lime putty is matured for several months in tanks/pits or under a thin film of water.

During this process, the lime crystals change shape, becoming smaller and flatter, thus aiding workability (Hansen et al., 2008). It is also known as 'air' limes, 'fat' limes, and 'high-calcium' limes. The conversion of quicklime on the job to quicklime putty is laborious, expensive, time consuming, and, in addition, the quality of the product varies considerably. More information on the methods of lime slaking and the factors that affect the process may be found in Boyton (1980) and Hassibi (2009).

## 4.5.2 Hydrated or Slaked Lime

Quicklime can also be 'dry slaked' or hydrated with limited quantity of water to produce what is known as *slaked lime* or *hydrated lime*. This process is called *slaking of lime* and is done using slakers (see Section 4.5.4). If it is done on a smaller scale at site, quicklime is heaped on a watertight platform and water is gradually sprinkled over it until quicklime is slaked and reduced to powder form. During the sprinkling of water, the heap should be turned over several times until sufficient water is added to make quicklime into powder form. This hydration process disintegrates the lump, pebble, or granules of quicklime into an extremely fine, white powder and liberates heat (about 160–170°C) as shown in Equation 4.2.

Lime from coarse-grained stone and lump-lime usually slake rapidly; lime from fine-grained stones, and dense lumpy lime usually slake slowly. In addition, over-burning or under-burning of the limestone in the kilns causes the lime to slake more slowly and affects the strength of lime mortar. If slaking is done in the factory, this white powder is packaged in watertight bags, after removing undesirable oversize particles, by screening through 3.35 mm IS Sieve. For commercial use, quicklime slaked to hydrated lime is more attractive than quicklime slaked to lime putty. Hydrated lime consists essentially of less caustic (but still strongly alkaline) $Ca(OH)_2$.

Quicklime made out of dolomitic rocks, when hydrated with the limited quantity of water, will result in a mixture of $Ca(OH)_2$ and $MgO$ or $Mg(OH)_2$, or both, as shown in the reaction below. Hydrated lime can be readily used in a mortar mix.

$$CaO + H_2O \rightarrow Ca(OH)_2 + Heat \ (65 \ kj/mol)$$
$$MgO + H_2O \rightarrow Mg(OH)_2 + Heat \tag{4.4}$$

Hydrated limes are divided into four types, as per ASTM C 207, *Hydrated Lime for Masonry Purposes*, as (a) Type N - Normal Hydrated Lime, (b) Type S - Special Hydrated Lime, (c) Type NA - Normal air-entraining hydrated lime, and (d) Type SA - Special air-entraining hydrated lime. Types S and SA are differentiated from Types N and NA by their ability to develop high early plasticity, higher water retentivity, and 8% limitation on unhydrated oxide content, and minimal coarse fraction. Type S limes are typically dolomitic. As these properties enhance the performance of mortar as well as plaster, Type S hydrated lime is used almost exclusively in the United States (Thomson, 2005). It has to be noted that in many countries in Western Europe use of lime in building construction is limited to the restoration of medieval castles and lime is not used for building new masonry buildings and their rendering (Thomson, 2005).

## 4.5.3 Equipment used for Slaking Process

Four different types of lime slakers are available in the market. They are (Hassibi, 2009):

1. Slurry detention slakers
2. Paste slakers
3. Ball mill slakers
4. Batch slakers

A slaker must mix the correct amount of quicklime (CaO) and water, hydrate the quicklime, and separate the impurities and grit from resultant $Ca(OH)_2$ slurry.

**Slurry slakers**   Typically, a *slurry slaker*, sometimes called a detention type slaker, consists of two chambers (Fig. 4.6). The first chamber where lime and water are mixed is called the slaking chamber. The second chamber is usually used as a grit remover chamber. The first chamber is at a higher level so that lime slurry can flow by gravity from it to the grit chamber. The first chamber is first filled with water to half depth. Quicklime is then added gradually until it fills half the depth of water (it has to be noted that lime is added to the water and not vice-versa). It is then stirred, taking care that the lime is not exposed to air above the water. Mixing is continued until the boiling of the liquid stops and the mixture thickens. More water is added, if required, and the contents are allowed to flow to the second lower tank.

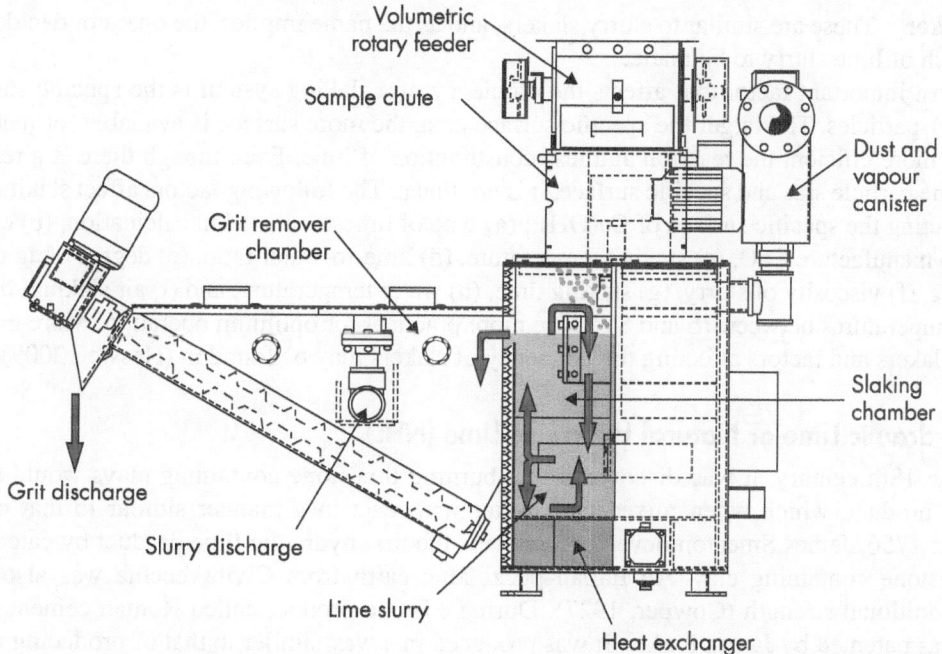

**Fig. 4.6** Typical slurry slaker (*Source*: Hassibi, 2009)

The mixture is allowed to stand in the lower tank for at least 72 h. The slurry viscosity is reduced in the second chamber by the addition of cold water to allow the heavier grit to settle to the bottom of the second chamber where the grit is elevated and discharged and used as slurry. If protected from drying out, the slurry can be stored for about 2 weeks. The slurry slakers are generally designed for a retention time of 10 min at a full-rated capacity. This means that the time quicklime enters the slaker and exits into the grit remover takes an average of 10 min. The slurry slakers are available in a variety of sizes ranging from 68 kg/h to 15 t/h. Slurry slakers are also available with external vibratory grit-separation screens. Slurry or detention-type slakers are the most common types in India, Europe, and the United States.

**Paste slakers**   As the name implies, these slake the lime into a paste form. The lime-to-water ratio is generally 1 to 2.5. Paste slakers are compact in size and designed for a retention time of 5 min in the slaking chamber. In a paste slaker, since the hydroxide paste is too heavy to flow by gravity, a pair of horizontal rotating paddles pushes the paste forward towards the discharge point. Once the paste exits the slaking chamber, it is diluted to approximately one part lime to four parts water. This dilution allows grit separation by gravity or by an external vibrating grit screen. The slurry consistency from a paste slaker and a slurry slaker is exactly the same after dilution and grit removal. Paste slakers are available in sizes ranging from 450 kg/h to 3600 kg/h. Paste slakers are mostly used in the United States.

**Ball mill slakers**   These are usually used for wet-and-dry grinding, and can be modified as ball mill slakers in order to slake lime. Such ball mills slakers can be classified as horizontal and vertical types. These ball mill slakers can be used when: (a) the capacity of other types of slakers is exceeded; (b) no grit discharge is allowed at sites; and (c) the available water at the site has high-sulphate content and hence regular slakers cannot be used. The available capacity of these slakers ranges from 450 kg/h to 50 t/h, but they are very expensive than paste or slurry slakers. They have an external classifier, which separates slurry from the oversized grit and impurities. The separated oversize grit is reground and recycled back into the mill.

**Batch slaker**   These are similar to slurry slakers, and as the name implies, the operator decides the size of the batch of lime slurry to be made.

The most important factor that affects the efficiency of a slaking system is the specific surface area of $Ca(OH)_2$ particles. The larger the specific surface area, the more surface is available for reaction, and hence the more efficient the reaction and less consumption of lime. Even though there is a relationship between the particle size and specific surface, it is not linear. The following factors affect slaking efficiency by affecting the specific surface of $Ca(OH)_2$: (a) type of limestone used in calcination, (b) calcination process to manufacture CaO, (c) slaking temperature, (d) lime-to-water ratio, (e) degree of agitation during slaking, (f) viscosity of slurry, (g) slaking time, (h) water temperature, and (i) air slaking. In practice, slaking temperatures between 76 and 85°C are more practical for optimum operation. More information on these slakers and factors affecting the efficiency of slakers may be found in (Hassibi, 2009).

## 4.5.4 Hydraulic Lime or Natural Hydraulic Lime (NHL)

During the 18th century, it was discovered that burning limestone containing clays would produce a hydraulic product, which when mixed with water will react in a manner similar to that of modern cement. In 1756, James Smeaton developed perhaps the first hydraulic lime product by calcining Blue Lias limestone containing clay. An Italian pozzolanic earth from CivitaVecchia was also added to provide additional strength (Cowper, 1927). During 1796, a product called Roman cement or natural cement was patented by James Parker. It was produced in a way similar to that of producing cement by burning a mixture of limestone and clay in kilns. The product so obtained from the kilns was ground and stored in waterproof metal vessels. As natural cements had higher clay content than hydraulic lime, it resulted in better strength development. Lime was used as the principal binder for mortars, especially in rural areas, until 1940s. Lime and the traditional techniques associated with it were slowly replaced by cement due to its speed of application and the employment of semi-skilled labour. The aggressive marketing by the cement companies also is a reason for the increased use of cement (Edwards, 2005).

---

### Natural Hydraulic Lime vs Portland Cement

The essential difference between modern hydraulic limes and Portland cement is that hydraulic lime contains lime instead of tricalcium silicate ($3CaO \cdot SiO_2$ designated as $C_3S$). Various types of hydraulic limes are available and they are produced in various grades. They are now produced by burning limestone containing clay and/or silica in a kiln at about 1200°C. This product can be hydrated with water to convert the calcium oxide to $Ca(OH)_2$, but not to hydrate the dicalcium silicate ($2CaO \cdot SiO_2$ designated as $C_2S$), which, in any case, is slow to hydrate. The setting process is a combination of the hydration of $C_2S$ and carbonation of the lime. In most hydraulic limes, a proportion of uncombined reactive silica and alumina is also present, and these will react with lime in the mortar to produce calcium silicate hydrates and calcium aluminate hydrates. Hydraulic limes are characterized by good workability, low shrinkage, salt and frost resistance, adequate compressive, and good flexural strength.

Two basic types of hydraulic lime may be available in the market: (a) natural hydraulic lime (NHL) and (b) artificial hydraulic lime (AHL). NHL is produced by calcining limestone that naturally contains clay and other impurities; no materials may be added to create the hydraulic property. As per EN-459, NHL is classified as feebly hydraulic (NHL 2), moderately hydraulic (NHL 3.5), and eminently hydraulic (NHL 5). AHL is similar to NHL, except that pozzolanic materials are added either before or after burning in a limekiln.

## 4.6 Cementing Action of Lime

The cementing action of lime is due to the following reaction of $Ca(OH)_2$ with the $CO_2$ present in the atmosphere to form $CaCO_3$.

$$Ca(OH)_2 + CO_2 \rightarrow CaCO_3 + H_2O \qquad (4.5)$$

Because of the nature of the aforementioned reaction, lime will not set without access to the atmospheric air, such as in conditions under water. Hydraulic lime or artificial hydraulic lime (containing pozzolanic materials) can combine chemically with lime and produce cementing compounds in the presence of moisture and such limes do not require atmospheric air containing $CO_2$ for the reaction. Thus, naturally occurring *kankar* lime is hydraulic and adding *surkhi* to fat or ordinary lime can also produce hydraulic lime, which will set under water. Thus, the main difference between fat lime and hydraulic lime is that the former sets only in the presence of atmospheric air and the latter acts like cement and can set even under water.

## 4.7 Classification of Lime

As per IS 712:1984, building limes are classified as follows:

*Class A*   Essentially hydraulic lime and used for structural works such as arches. It has the following characteristics: (a) grey colour, (b) calcium oxide and clay content are 60% and 15%, respectively, (c) slakes with difficulty, and (d) sets and hardens readily under water with initial setting time of 2 h and final setting time of 48 h.

*Class B*   Semi-hydraulic lime, containing both hydraulic lime and fat lime and is used for masonry mortars, lime concrete, and plaster undercoat. It has the following characteristics: (a) grey colour, (b) calcium oxide and clay content are 70% and 10%, respectively, and (c) slakes and sets at a slow rate and may take a week to set under water.

*Class C*   Basically fat lime used for finishing coat in plastering, whitewashing, composite mortars, etc., or could be used as masonry mortar with the addition of pozzolanic materials or *surkhi*. It has the following characteristics: (a) white colour, (b) calcium and MgO and clay content are 91% and 2%, respectively, and (c) slakes vigorously and increases to 2.0 to 2.5 times its original volume.

*Class D*   Magnesium/dolomitic lime used for finishing coat in plastering, white washing, etc. It has the following characteristics: (a) white colour, (b) calcium and MgO content is 85%, and (c) slakes readily.

*Class E*   *Kankar* lime used for masonry mortars. It has the following characteristics: (a) grey colour, (b) calcium oxide and MgO contents are 50% and 6%, respectively, and the remaining are impurities like clay (25%), silica and alumina (20%), and (c) slakes and hardens slowly.

*Class F*   Siliceous dolomitic lime used for undercoat and finishing coat of plaster. It has the following characteristics: (a) grey colour, (b) calcium oxide and MgO contents are 70% and 6–14%, respectively, and the remaining are impurities like clay (10%), silica and alumina (10%), and (c) slakes and hardens slowly.

The aforementioned classes of lime, except Classes A and E, are available either in hydrated or quick form. Classes A and E are supplied in hydrated form only.

Some of the physical properties of different classes of lime, as per IS 712:1984, are provided in Table 4.1. Free moisture content of up to 2% is allowed in hydrated lime only. A comparison of the compressive strength of lime and cement as per IS 712:1984 is given in Table 4.2. Other chemical and physical properties of the different classes of lime are provided in IS 712. IS 1760 (Part 1–6):1991 may be used for the determination loss on ignition, silica, iron oxide, alumina, calcium oxide and magnesia, and chlorides, through a chemical analysis.

**Table 4.1** Chemical requirements of different classes of lime

| Characteristics | Standard lime class | | | | | | | | | |
|---|---|---|---|---|---|---|---|---|---|---|
| | A | B | | C | | D | | E | F | |
| | Hydrated | Quick | Hyd. | Quick | Hyd. | Quick | Hyd. | Hydrated | Quick | Hyd. |
| Calcium and MgO, percent, Minimum | 60 | 70 | 70 | 85 | 85 | 85 | 85 | 50 | 70 | 70 |
| MgO, percent, Maximum | 6 | 6 | 6 | 6 | 6 | – | – | 6 | – | – |
| Minimum | – | – | – | – | – | 6 | 6 | – | 6 | 6 |
| Silica, alumina, and ferric oxide, percent, Minimum | 20 | 10 | 10 | – | – | – | – | 20 | 10 | 10 |
| Unhydrated MgO, percent, Maximum | – | – | – | – | – | 8 | 8 | – | 8 | 8 |
| Insoluble residue in dilute acid and alkali, percent, Maximum | 15 | 10 | 10 | 2 | 2 | 2 | 2 | 25 | 10 | 10 |
| $CO_2$, percent, Maximum (on oven dry basis) | 5 | 5 | 5 | 5 | 5 | 5 | 5 | 5 | 5 | 5 |
| Available lime as CaO, percent, Minimum | – | – | – | 75 (on dry basis) | 75 (on ignited basis) | – | – | – | – | – |

**Table 4.2** Compressive strength of lime and cement, minimum, and MPa

| Compressive strength, Minimum, MPa | Standard lime class | | | | | | Cement (33 grade) |
|---|---|---|---|---|---|---|---|
| | A | B | | E | F | | |
| | Hydrated | Quick | Hyd. | Hydrated | Quick | Hyd. | |
| At 14 days | 1.75 | 1.25 | 1.25 | 1.0 | 1.25 | 1.25 | 22.0 |
| At 28 days | 2.8 | 1.75 | 1.75 | 1.75 | 1.75 | 1.75 | 33.0 |

## 4.8 Storing of Lime

As discussed earlier, lime may react with the moisture present in the atmosphere or from the ground. Hence, the following precautions should be taken while storing it (IS 14401:1996).

1. Quicklime should be stored in airtight containers. If necessary, it may be stored in closed store rooms in compact heaps to avoid air slaking. But when quicklime is delivered to site for slaking, it should be used within a week.

2. Hydrated lime must be kept dry and should be stored under cover and off the ground, in weather-proof sheds with impervious floor and side walls.

3. Lime putty under water could be stored without any deterioration for several months, and its properties were found to improve with time. Cazalla et al. (2000) studied the influence of storing slaked lime under water for extended periods of time, using X-ray diffractometry, phenolphthalein tests, porosity measurements, electron microscopy, and ultrasonic wave propagation analyses and found that mortars prepared using traditional aged lime putties (up to 14 years storage under water) show rapid, extensive carbonation, resulting in porosity reduction and ultrasonic speed increase.

4. Hydraulic lime could be stored for 3 to 4 months in gunny bags lined with polythene lining or high-density polythene woven bags lined with polythene or craft paper bags and stored in weather-proof sheds with impervious floor and side walls. In the case of hydrated lime, which is used within 30 days, use of liner in the bags is not necessary, as per IS 712:1984.

## 4.9 Precautions to be Taken while Handling Lime

Workers handling lime must be trained and should wear proper protective dress/equipment, as described below (www.lime.org, IS 14401:1996):

**Skin burns**  Lime can cause irritation and burns to unprotected skin, especially in the presence of moisture. Quicklime is especially dangerous as it reacts with moisture and generates heat capable of causing thermal burns. Hence, prolonged contact with unprotected skin should be avoided. Workers should be advised to wear protective gloves and clothing that fully covers arms and legs. Workers should wear at least one long-sleeved shirt. Rolling up sleeves or short-sleeved shirts should not be permitted. In cool weather, a second long-sleeved shirt is advisable. They should also wear high top shoes or laced boots. Skin protecting creams may also be helpful. It is advisable to take bath or shower after a workday to cleanse the body entirely of lime and protective cream.

**Damage to eyes**  Lime can cause severe eye irritation or burning, including permanent damage. Hence, eye protection devices such as goggles, safety glasses, and/or face shield should be worn where there is a risk of lime exposure. In the event of contact, eyes should be rinsed immediately with plenty of water and medical attention should be sought. Contact lenses should not be worn when working with lime products.

**Damage to respiratory system**  Lime dust is irritating if inhaled. In most cases, dust masks provide adequate protection. Hat or cap should be worn to protect scalp from accumulated lime dust. Lime dust should also be kept out of reach of children.

**Fire accidents**  As quicklime gives out immense heat while slaking, suitable precautions are to be taken to avoid fire accidents.

## 4.10 Uses and Products of Lime in Construction

Lime is used in buildings for preparing mortar mixes, which are used for constructing masonry, pointing, plastering, rendering and also for limewashing. Lime may be available as dry hydrated quicklime powder, bagged lime, or as putty. The choice depends on the availability of space and water at construction sites and the available means of transport.

### 4.10.1 Lime Putty

Lime putty is used for the production of lime plasters, mortars, and limewash. To be used as mortar, lime putty is mixed with coarse sand. It is used to construct masonry and also in pointing and rendering of brickwork. Plasters are made with lime putty mixed with fine sand and is used for plastering brickwork or tone walls. As will be discussed in Section 4.10.3, limewash is made by diluting lime putty with sufficient water and small amount of glue. To colour wash, pigments of required colour are added to the solution of limewash.

### 4.10.2 Lime Mortar and Plaster

Lime mortar or plaster is a workable mixture of lime, sand, and water. This can be made using lime putty or natural hydraulic lime. Lime mortar is applied as a paste, which on drying, sets hard and binds masonry units or blocks together and fills the gaps between them. Lime mortar can also be used to fix or point masonry. Slow setting NHL (NHL 2) is used for internal plasterwork on soft stones and bricks. NHL 3.5 or NHL 5 is used on relatively strong blocks or granite walls and also for rendering an exposed wall. Artificial hydraulic lime can also be used for preparing mortar and plaster. There are many advantages of using lime mortar or plaster over cement mortar or plaster. Lime mortar and plaster are discussed in Chapter 7. IS 2451:1991 describes the methods of preparation and use of lime concrete.

### 4.10.3 Limewash

Hydrated lime or lime putty can be blended with water to produce *limewash* (also called *whitewash*). As per IS 6278:1971, the fat lime should be slaked at site and mixed and stirred with 5 L of water for 1 kg of unslaked lime to make a thin lime cream (This will cover approximately 10 m$^2$ of surface area, depending on the absorbency of substrate.). About 1 kg of gum/glue dissolved in hot water may be added to each m$^3$ of this lime cream. About 1.3 kg of sodium chloride (common salt) dissolved in hot water may also be added for every 10 kg of lime (The addition of sodium chloride to limewash hastens carbonation of Ca(OH)$_2$ and also makes the coating hard and rub-resistant.). Usually, a small quantity of ultra-marine blue (up to 3 g per kg of lime) is also added to the limewash solution, which is used for last two coats. This limewash solution should be allowed to stand for a period of 24 h and before using screened through a clean coarse cloth. The whole solution should be stirred thoroughly before use.

For exterior walls of stone or masonry, whitewash or colour wash should be made to adhere well to the surfaces. Such adhering quality to the white/colour wash can be achieved by using the following procedure:

1. Spread one part by weight of small lumps of tallow (rendered form of beef or mutton fat) over 12 parts of quicklime (Instead of tallow, linseed oil, or caster-oil, having 10% weight of dry lime, may also be used).
2. Slake it with sufficient quantity of water, and stir until it forms a thick paste.
3. Allow it to stand for a few hours so that it is cooled to the room temperature.
4. Add sufficient water to this paste to obtain a thin wash and filter it with a coarse cloth.

If the oil does not mix well with lime, heat the solution until the oil disappears. As the oil mixes with lime to form an insoluble soap, the white/colour wash made with it will not dissolve in heavy rain. Addition of skimmed milk will also help whitewash to bond with less porous materials. Limewash naturally looks white, but it can be coloured by adding natural/manufactured pigments using iron oxide, chromium oxide, or carbon black, which are not affected by lime. It should be noted that the use of linseed oil will give slight yellow tinge to the whitewash.

**Surface preparation**    The surface on which limewash is to be applied should be thoroughly cleaned of all dirt, dust, mortar drops, and other foreign matter. Very smooth surfaces, which don't have enough roughness for the lime to bond with, may require texturing with medium abrasives or a wire brush. If it an old surface, which has already been whitewashed or colour washed, it should be brushed using a wire brush to remove all dust and dirt. All loose scales of limewash and other foreign matter should also be removed. Where heavy scaling has taken place, the entire surface should be scraped clean. These preparations are necessary also when a colour wash has to be given on an already white-washed surface.

According to IS 6278:1971, any mould found on the surface should be removed by using a steel scraper. In addition, the surface should be applied with an ammoniacal copper solution consisting of 15 g of copper carbonate dissolved in 60 ml of liquor ammonia in 500 ml water or 2% sodium pentachlorophenate solution in water. After the surface dries thoroughly, limewash or colour wash may be applied.

Any unsound portions of the surface plaster should be removed to full depth of plaster in rectangular patches and plastered again after raking the masonry joints properly. Such portions should be cured properly with water and then allowed to dry. They should then be given one coat of limewash, before the actual limewashing.

**Required atmospheric conditions**    Proper temperature and humidity are required for the carbonation process to be successful (the most critical period for carbonation is the first 2 to 3 days).

1. Ambient temperature should not be below 5°C (7.5°C for dark colours) at any time during the application/carbonation process.
2. In temperatures above 30°C or in very low humidity, surface and subsequent coatings must be misted with water.
3. Exterior coatings should be protected from wind and sun to protect it from drying too quickly. An opaque covering such as damp canvas or burlap may be used.
4. Surfaces must not have a surface temperature below 5°C (7.5°C for dark colours) and must not be excessively hot or excessively wet.

The applicator should wear suitable clothing, dust mask and eye protection (see Section 4.9). Powders (lime, pigment, or additive) should be added to water, and not water into powder to minimize dust. Only the required amount of limewash to be used in one day should be mixed.

**Application techniques**    Brushes made specifically for limewash (such as those made out of grass plant *munj*), or a 4″ flexible paint brush (never use nylon) may be chosen. A sponge can also be helpful to apply patinas and soften brush marks, if desired. Lime is hard on tools – hence, it is necessary to rinse all brushes thoroughly after use and hung to dry, as the metal of the ferrule will be prone to rust.

Limewash should be applied on the surface using a brush in a cross-hatch motion (also known as figure eight or butterfly), always from top to bottom. This creates a more pleasing appearance than using a simple up and down motion with the brush. It is important to note that lime and/or pigment particles will not remain suspended in water for long and hence must be frequently agitated. Hence, during the application of whitewash or patina coats, it is better to transfer portions of freshly agitated mixture to a shallow container so that the brush bristles can touch the bottom and with each pass of the brush, the lime and pigment particles are swept back up into suspension. Each coat should be allowed to dry be-fore the next coat is applied. No portion of the surface should be left out initially to be patched up later on. The brush should be dipped in whitewash pressed lightly against the wall of the container, and then applied by lightly pressing against the surface with full swing of hand. The whitewashing on ceiling should be done prior to that on walls.

For new work, minimum two coats should be applied so that the surface presents a smooth and uniform finish through which the plaster is not seen. After moistening, the first diluted white coat can be applied to unify the substrate and after 24 h the second coat, white or pigmented, is applied. For old work, after the surface has been prepared, a coat of whitewash should be applied over the patches and repairs. Then one, two, or more coats of whitewash should be applied over the entire surface. For colour washing on new works, after the surface has been prepared, the first primary coat should be of whitewash and the subsequent coats (minimum two) should be of colour wash. While colour washing, a small area (about 0.1 m²) may be colour washed to the required coats over a first coat of whitewash and examined before the entire wall is applied with the colour wash. It should be noted that small areas of colour wash will appear lighter in shade than when the same shades are applied to large surfaces. For colour washing an old work, after the surface has been prepared, a coat of colour wash should be applied for the patches and repairs. Then, the specified number of coats of colour wash should be applied over the entire surface. No primary coat is needed for old surface bearing colour of the same shade.

Basically, limewash consists of hydrated lime and water. When applied on outside surfaces, hydrated lime absorbs $CO_2$ from the air and converts back as limestone, providing a protective layer on the limewashed surface. In addition, the pH of hydrated lime helps to sanitize surfaces. Hence, it is used in dairy barns, poultry houses, or similar buildings to reduce odours. The white colour of whitewash also minimizes the absorption of heat and hence may be used on roof tiles in hot climates. Limewash is highly porous and a breathable system, as opposed to modern emulsion paints, which lock-up the walls in a plastic film. In the United States, limewash has to meet the criteria of ASTM C206 for Finishing Hydrated Lime.

Some surfaces that are unsuitable for limewash are given below, with the reasons:

1. Asbestos cement (chemically incompatible)
2. Dry wall (porosity inconsistencies and chemical incompatibility)
3. Gypsum plasters and products (chemically incompatible)
4. Previously painted or sealed surfaces (lack of porosity, chemically incompatible)
5. Smooth surface or non-porous surfaces (no porosity, no room for lime crystals to penetrate for bonding).
6. Wood (too high in PH, will cause wood grain to swell)

### 4.10.4 FaL-G Technology

In India, fly ash bricks are made by using fly ash, sand, lime, and water – the mixing proportion is generally 40–50% fly ash, 50–40% sand, 10% lime, and 4% water (see also Section 3.10.1 of Chapter 3). Another method developed by Bhanumathidas and Kalidas, called FaL-G technology, is based on two principles:

1. Fly ash-lime pozzolanic reaction does not need external heat under tropical temperature condition.
2. The rheology and strength of fly ash-lime mixtures can be greatly augmented in the presence of gypsum.

This technology has dispensed the need for heavy-duty press and autoclave. It also made the process more energy efficient, and within the reach of tiny sector entrepreneurs (It has to be noted that FaL-G does not stand for any brand name; it only bears the first letters of the constituent materials.). In 2003, as against a handful of autoclaved plants in India, there were over 800 FaL-G brick plants, manufacturing more than one billion bricks or 2 million m³ of blocks annually (Bhanumathidas and Kalidas, 2003). More details about this eco-friendly technology could be had from www.fal-g.com.

### 4.10.5 Hempcrete and Limecrete

*Hempcrete* or hemplime is a composite material consisting of chopped hemp shiv and binder comprising of natural hydraulic lime and a small amount of cement. Hempcrete is used as an infill material in timber frames and acts as insulator and moisture regulator. It may not require expansion joints. It is considered a lightweight insulating material ideal for most climates as it combines insulation and thermal mass.

Hempcrete improves air quality, reduces energy consumption for heating and cooling, and provides a comfortable environment, which is cool in summer and warm in winter. As it is breathable, it improves the health and comfort of the occupants. Like other plant products, hemp absorbs $CO_2$ from the atmosphere as it grows. In addition, lime absorbs $CO_2$ during curing as lime turns to limestone. Owing to these reasons, hempcrete is considered carbon negative; it may be a better choice to achieve a low carbon footprint and sustainability. Being breathable, hempcrete is ideal for use in historic buildings and modern buildings using natural materials. More information and properties of hempcrete can be obtained from http://limecrete.co.uk.

*Limecrete* is a combination of natural hydraulic lime (instead of cement), sand, and lightweight aggregates, which can be used as an alternative to concrete. This creates a breathable floor slab with a certain amount of flex. Limecrete is widely specified by architects to protect ancient buildings, as it brings the advantages of lime to the concrete application.

## 4.11 Advantages and Disadvantages of using Lime

Recent research has revealed enormous benefits of natural lime over modern cement that dominates the current building industry. Some of the advantages are listed below. In addition, lime may also have disadvantages as compared to modern cement.

### 4.11.1 Advantages of Lime

Lime as a building material has the following advantages (Holmes and Wingate, 2002):

**Breathable** Lime works in harmony with seasonal and long term changes in the building, accommodating tiny movements without much cracking. It is breathable (vapour permeable), absorbing and evaporating moisture from surrounding masonry. This reduces the risk of trapped moisture and consequent damage to the building fabric. There is also less risk of salt and frost damage. Adjacent materials frequently affected this way include timber and iron as well as stone and brick masonry.

**Self-healing** If lime is used as mortar, when subjected to small movements because of climatic changes, many fine cracks are developed. This is in contrast to individual large cracks occurring in buildings in which stiffer cement is used. During rain, water may penetrate through these fine cracks, but at the same time will be able to dissolve the available unhydrated 'free' lime and transport it. As the water evaporates, this lime will 'self-heal' the cracks.

**Flexible** It has a low modulus of elasticity. This means it is extremely flexible and allows for movement and thermal expansion, thus expansion joints could be avoided.

**Provides a comfortable environment** As lime mortars have porous and open texture, they help to stabilize the internal humidity of a building by absorbing and releasing moisture. Insulation is improved and cold bridging reduced. This makes for a more comfortable environment and reduces surface condensation and mould growth. On the other hand, cement and gypsum being very hard and impervious materials, movements cause cracks in them. Since, moisture drawn into these cracks can't escape, condensation problems are created.

**Energy efficient and eco-friendly**   Lime has less embodied energy than cement (Although cement and lime are obtained by burning raw materials in the kiln, lime is burnt at a temperature of 900 to 1000°C as compared to 1300 to 1400°C required for cement). $CO_2$ emissions in the manufacture of lime are 20% less than ordinary cement. Furthermore, it reabsorbs $CO_2$ while setting, thus lowering its carbon footprint even further. It is also possible to produce lime on a small scale, thus reducing long distance transport. The gentle binding properties of lime enable full reuse of other materials. A very low proportion of quicklime will stabilize clay soils. Small quantities of lime can protect otherwise vulnerable, very low energy materials such as earth construction and straw bales (King, 1996).

**Provides better adhesion**   Owing to the fine particle size, far smaller than cement, lime mortar mixes often penetrate minute voids in the bricks deeply than other materials. They bind gently and the stickiness gives good adhesion to other surfaces.

**Lime mortars have good workability**   The workability provided by lime allows the inclusion of widely graded and sharp aggregates in the mix. These enhance both the performance and the aesthetics of the finished work.

**Lime binders are durable**   When used properly, lime is exceptionally durable. Many examples could be cited for its exceptional durability. An outstanding example is the Pantheon Temple in Rome, built circa 128 AD, having a 43.3 m diameter lime concrete dome. This has survived over 2000 years.

**Sympathetic**   It is softer than modern cement; a softer mortar will not wear away the surrounding masonry over time. Bricks and stones are also reclaimed more easily, when dismantled.

**Local limes enhance diversity and are aesthetic**   The diversity of limestone available in nature provides variety and local distinctiveness in colour, texture, and setting properties. It has attractive traditional appearance, pale in colour. Natural limes tend to reveal the colour and characteristics of the aggregate with which they're mixed.

**Indefinite shelf life**   Non-hydraulic limes have an indefinite shelf life when stored without access to air, usually as putty under water or in sealed containers. In fact, the quality of putty improves with time.

**Low-cost**   Lime-pozzolana mixes provide cheaper and structurally more suitable substitutes for cement mixes. Thus, they may help conserve more costly cement in important applications.

### 4.11.2 Disadvantages of Lime
Though lime has several advantages, it also has a few drawbacks.

1. The hydrated lime sets and hardens relatively slower compared to cement. The speed of setting may be increased by using limestone with impurities like clay or by adding pozzolanic material such as fly ash.
2. Quicklime will hydrate quickly and hence should be stored carefully in watertight sheds. To avoid this, quicklime may be hydrated as soon as it is unloaded from the kiln, as hydrated lime is much easier to store and transport.
3. Hydrated lime stored for long periods of time, gradually reacts with $CO_2$ in the air and reverts back to its original form. Hence, it should be stored in airtight bags.
4. Because quicklime can cause burns, and hydrated lime can irritate skin and eyes, precautions should be taken when handling these products.

5. Traditional small-scale lime producers may use fuels inefficiently and often produce low-quality lime (over or under burnt). These small-scale lime producers should be assisted by the governments to use more efficient kilns.
6. Plain limewash puts rigid demands to weather, temperature, and moisture-conditions during the application. It takes a long time to harden, and may be easily rubbed off.
7. Hydration of some lime may take place long after the component has dried, causing blisters, cracks, and unsightly surfaces. To avoid this, it is better to use lime putty, which not only hydrates all the particles of lime but also improves the quality of lime, as the time of storage increases.

## 4.12 Lime vs Cement

A comparison of the binding materials, lime and cement is presented in Table 4.3.

**Table 4.3** Comparison of lime and cement

| S. no. | Characteristic | Lime | Cement |
|---|---|---|---|
| 1 | Colour | White to greyish white | Grey (white cement is also available) |
| 2 | Slaking | Addition of water results in generation of immense heat within 5 to 15 min | Chemical reaction takes place when water is added leading to hardening |
| 3 | Workability of mortar | Very good | Good |
| 4 | Setting | Sets slowly by absorbing $CO_2$, leading to carbonation | Sets rapidly by reacting with water |
| 5 | Hardening | Slow | Hardens quickly and gains sufficient strength within 3 days |
| 6 | Mechanical properties (compressive strength, abrasion, etc.) | Less | High |
| 7 | Binding properties | Good | Best |
| 8 | Water retention | Eliminates water through vapour exchange and hence provides more durability to brickworks | Traps water and not highly suitable for brickworks, less durable than lime |
| 9 | Cost | Cheaper | Expensive |
| 10 | Uses | Suitable for ordinary construction works | Suitable for all construction works |
| 11 | Corrosion | May corrode iron and steel | No corrosion of iron and steel |
| 12 | Repair of older buildings | Owing to its flexibility, softness, and breathability, it is more suitable | Should not be used |
| 13 | Eco-friendly | Green material as it is more durable and absorbs $CO_2$ | Production emits considerable $CO_2$ and hence not eco-friendly |

## 4.13 Use of Lime as Green Building Material

Lime is a green as well as sustainable material. This may be assessed by considering the life cycle of lime by asking specific questions (Freed, 2008). The questions normally asked and the replies with respect to lime are as follows:

*Where does lime come from?* Lime is a low-cost material produced by heating limestone, which are abundant in nature. It can be produced at a small scale using relatively simple technology. The energy requirement of quicklime production depends on the type of kiln used and varies between 3.2 and 9.2 GJ/tonne (UNIDO Report, 2010).

***What is the carbon footprint?*** Manufacturing process of lime produces less $CO_2$ than ordinary cement. Furthermore, it re-absorbs $CO_2$ while setting, thus lowering its carbon footprint even further.

***How is lime delivered and installed?*** Lime production is widely distributed just as the availability of limestone. It is also possible to produce lime on a small scale, thus reducing long distance transport.

***How is lime maintained and operated?*** Buildings built using lime during the Roman period are still in good condition. It shows that the use of lime requires only occasional maintenance. In addition to its durability, it also has self-healing properties.

***How healthy is lime?*** Lime does not release any toxic or harmful chemicals. In fact, while hardening hydrated lime absorbs $CO_2$, and hence increases indoor air quality. As lime mortars have porous and open texture, they help to stabilize the internal humidity of a building by absorbing and releasing moisture. This makes for a more comfortable environment and reduces surface condensation and mould growth.

***What do we do with lime after we are done with it?*** It is softer than modern cement; a softer mortar will not wear away the surrounding masonry over time. Bricks and stones are also reclaimed more easily, when dismantled. Old lime mortar and plaster could be crushed and reused.

In addition, wall systems that use lime provide high thermal mass and improve energy efficiency. The white colour of lime, when utilized in mortar or limewash, will produce more reflective surfaces on the exterior of the building. These aspects of lime show that it is an excellent sustainable and green material.

## SUMMARY

- Lime, which is manufactured from limestone, has been used all over the world as a material of construction from very ancient days. Limestone is generally classified as high-calcium, dolomitic, and magnesian.
- Lime is derived from calcination of limestone, which is a process of heating pieces of limestone, in rotary/vertical/ PR kilns, to 900 to 1100°C.
- Dolomitic limestone and limestone containing clays (e.g., kankar) when heated to 1100°C result in hydraulic lime, which can be used like Portland cement.
- Different types of limes are produced such as quicklime/lime putty, hydrated/slaked lime, and hydraulic lime. Once limestone is converted to quicklime, it must be stored in airtight containers.
- As per IS 712:1984, building limes are classified into (a) Class A–Essentially hydraulic lime, (b) Class B–Semi-hydraulic lime, (c) Class C–Basically fat lime, (d) Class D–Magnesium/dolomitic lime, and (e) Class E–Kankar lime used for masonry mortars.
- If the lime is not slaked immediately using lime slakers, it tends to react with carbonic acid from the atmosphere and returns to its original form of $CaCO_3$. This process is called lime cycle.
- The cementing action of lime is due to the reaction of $Ca(OH)_2$ with the $CO_2$ present in the atmosphere to form $CaCO_3$.
- Precautions should be taken while storing lime, as it may absorb moisture. Workers must be trained to handle it and wear proper protective dress/equipment.
- Several field tests done as per IS 1624:1986, will reveal the quality of lime. The laboratory tests on lime are discussed in Section 25.6 of Chapter 25.
- Lime is used in mortar mixes for constructing masonry, and also in pointing, plastering, rendering, and for limewashing.
- Calcium-silicate (sand-lime) bricks may be used for masonry construction just like common burnt clay bricks (see Section 3.10.2 of Chapter 3). FaL-G technology, developed in India, may be used to manufacture eco-friendly bricks, made of fly ash, lime, and gypsum. Hempcrete is a composite material consisting of chopped hemp shiv and binder of natural hydraulic lime and a small amount of cement.
- Lime is a natural, low-cost, and sustainable material but under-utilized due to the preference of cement in modern buildings.

## EXERCISES

### Multiple-choice Questions

1. Which of the following is a false statement?
   (a) Limestone is the raw material for manufacturing lime.
   (b) White chalk is an impure limestone.
   (c) Lime varies in quality from place to place.
   (d) Lime is obtained by calcination of shells and corals.

2. Calcination of limestone occurs in kilns when they are heated to
   (a) 500–600°C
   (b) 800–1000°C
   (c) 900–1100°C
   (d) None of these

3. 100 g of $CaCO_3$, when calcinated, will result in
   (a) 44 g of CaO and 56 g of $CO_2$
   (b) 56 g of CaO and 44 g of $CO_2$
   (c) 60 g of CaO and 40 g of $CO_2$
   (d) None of these

4. Quicklime coming out of the kiln is also known as
   (a) hydrated lime
   (b) lump-lime
   (c) fat lime
   (d) hydraulic lime

5. When calcining dolomitic limestone in the kilns, it decomposes to MgO at
   (a) 775°C
   (b) 900°C
   (c) 1025°C
   (d) 1300°C

6. The hydration of Class C lime is accompanied by an increase in volume of
   (a) 1.5 to 2 times
   (b) 1 to 2 times
   (c) 1.7 to 2.5 times
   (d) 2 to 2.5 times

7. As per IS 712, Class A lime has calcium oxide and clay content, respectively, as
   (a) 40% and 5%
   (b) 50% and 10%
   (c) 60% and 15%
   (d) 70% and 10%

8. As per IS 712, Class B lime has calcium oxide and clay content, respectively, as
   (a) 40% and 5%
   (b) 50% and 10%
   (c) 60% and 15%
   (d) 70% and 10%

9. As per IS 712, Class D lime has calcium and MgO content as
   (a) 65%
   (b) 75%
   (c) 80%
   (d) 85%

10. As per IS 712, Class C lime is
    (a) hydraulic lime
    (b) semi-hydraulic lime
    (c) dolomitic lime
    (d) fat lime

11. As per IS 712, *kankar* lime can have clay and silica and alumina content, respectively, as
    (a) 20% and 10%
    (b) 25% and 15%
    (c) 25% and 20%
    (d) 30% and 25%

12. White colour of lime will indicate
    (a) Class A lime
    (b) Class B lime
    (c) Class C lime
    (d) Class E lime

13. Limewash is made by mixing
    (a) 1 kg of unslaked lime with 3.0 L of water
    (b) 1 kg of unslaked lime with 3.5 L of water
    (c) 1 kg of unslaked lime with 4.5 L of water
    (d) 1 kg of unslaked lime with 5.0 L of water

14. In India, fly ash bricks are made by using fly ash, sand, lime, and water in the proportion of
    (a) 20–30% fly ash, 70–60% sand, 10% lime, and 4% water
    (b) 30–40% fly ash, 60–50% sand, 10% lime, and 4% water
    (c) 40–50% fly ash, 50–40% sand, 10% lime, and 4% water
    (d) 50–60% fly ash, 40–30% sand, 10% lime, and 4% water

### Review Questions

1. What are the main two classifications of limestone?
2. What is *kankar* limestone? How is it different from high-calcium limestone?
3. What are the different steps in the manufacture of lime? How the manufacture of lime from dolomitic limestone differs from high-calcium limestone?
4. What are the two main types of kilns used in the manufacture of lime? Describe these two types of kilns briefly.
5. List three procedures that are usually adopted to increase fuel consumption efficiency of kilns.
6. Why is it necessary to store lime coming out of the kilns in airtight compartments?
7. What should be done to reduce the effect of over burnt $Mg(OH)_2$ in the case of dolomitic limes?
8. What are the two types of hydrators used?
9. What is meant by slaking of lime? Why is it necessary to slake quicklime immediately after burning?
10. Distinguish between quick, hydrated, and hydraulic limes. What is lime cycle? How does dolomitic lime cycle differ from the lime cycle of high-calcium lime?

11. Name two equipment used for slaking. Explain the working of a typical slurry slaker.
12. What is hydraulic lime? How is hydraulic lime classified?
13. Describe the cementing action of lime? How is hardening of hydrate lime different from that of hydraulic lime?
14. How is lime classified according IS specifications?
15. State the precautions to be taken while storing lime.
16. State the precautions to be taken while handling lime.
17. Write short notes on the following: (a) Use of lime as a building material (b) Quicklime (c) Hydrated/slaked lime (d) Hydraulic lime (e) Lime putty (f) Lime mortar and plaster.
18. What is limewash? How is plastered surface prepared for limewash? How is it applied on plastered surfaces?
19. Name three surfaces that are unsuitable for limewash.
20. Describe FaL-G technology. How is it advantageous over other brick technologies?
21. Differentiate between hempcrete and limecrete.
22. State any five advantages of using lime.
23. State any five disadvantages of using lime.
24. Compare lime and cement.
25. Why is lime considered a green and sustainable material?

## Exercises

1. Determine the theoretical weight of quicklime obtained and $CO_2$ emitted by burning 750 g of limestone.
2. Determine the theoretical weight of limestone required to be burnt in the kilns to produce 1000 g of quicklime. How much $CO_2$ will be emitted during the burning of limestone?

---

# ANSWERS

### Multiple-choice Questions

| | | | | | | |
|---|---|---|---|---|---|---|
| **1.** (b) | **2.** (c) | **3.** (a) | **4.** (a) | **5.** (a) | **6.** (d) | **7.** (c) |
| **8.** (d) | **9.** (d) | **10.** (d) | **11.** (c) | **12.** (c) | **13.** (d) | **14.** (c) |

### Exercises

1. As shown in Example 4.1, 100 g ($CaCO_3$) = 44 g ($CO_2\uparrow$) + 56 g (CaO)
   Thus, lime produced by 750 g of limestone = (56/100) 750 = 420 g
   $CO_2$ emitted = (44/100) 750 = 330 g
2. From Example 4.1, 100 g ($CaCO_3$) = 44($CO_2\uparrow$) + 56 g (CaO)
   Hence, to produce 1000 g of CaO, we need to burn 1000/56 ×100 =1785.7 g of limestone.
   The amount of $CO_2$ emitted = 17.857 × 44 = 785.7 g

# CHAPTER 5
# CEMENT AND CEMENTITIOUS MATERIALS

## 5.1 Introduction

Cement is the most important material of modern building construction. The cement currently being used had undergone a number of changes. From the beginning, civilizations sought a material that would bind stones. The Assyrians and Babylonians were perhaps the first to use clay for this purpose, and the Egyptians developed lime and gypsum mortar for building their Pyramids. The Greeks made further improvements and finally the Romans developed cement that built structures of remarkable durability. Volcanic ash (called *pozzuolana*, found near Pozzouli by the bay of Naples) is a key ingredient in the Roman cement that was popular during the days of the Roman Empire. Vitruvius, a Roman scientist, is considered to be the first to develop the chemistry of the cementitious lime (Vitruvius and Morgan, 1960). The great Roman baths, built around 27 BC, the Coliseum, and the huge Basilica of Constantine are examples of early Roman structures. Concrete was more extensively used again during the Renaissance and its manufacture was described in a work by De Lorme, published in 1568.

In the 18th century, with the advent of new technical innovations, a greater interest in concrete developed. In 1756, John Smeaton, a British Engineer, rediscovered hydraulic cement through the repeated testing of mortar in both fresh and salt water. Smeaton's work was followed by Joseph Aspdin, a bricklayer and mason in Leeds, England, who, in 1824, patented the first 'Portland' cement, so named since it resembled the stone quarried on the Isle of Portland off the British coast (Reed et al., 2008). Aspdin established a plant in Wakefield to manufacture Portland cement (PC), some of which was used in 1828 in the construction of the Thames River Tunnel. During 1959–1967, PC was used in the construction of the London sewer system. During 1845, Isaac Johnson made the first modern PC by firing a mixture of chalk and clay at much higher temperatures. At these temperatures (1400–1500°C), clinkering occurs and minerals form, which are very reactive and highly cementitious. It is interesting to note that cement is until made in this way. Thus, the PC used now is a predetermined and carefully proportioned chemical combination of calcium, silicon, iron, and aluminium.

From that time, gradual improvements in the properties and qualities of cement were made possible by several researchers in the USA, the UK, France, and Germany. From the turn of the 20th century, rotary cement kilns gradually replaced the original vertical shaft kilns, which were used originally for making lime. Rotary kilns (IS 8125:1976) heat the clinker more efficiently at higher temperatures, enabling higher burning temperatures to be achieved. Since the clinker is constantly moving within the kiln, a fairly uniform clinkering temperature is achieved in the hottest part of the kiln. Thomas A. Edison pioneered further development of the rotary kiln. Today, some kilns are more than 150 m long. The other two principal technical developments, the addition of gypsum in the kiln to control setting and the use of ball mills to grind the clinker, were also introduced at around the start of the 20th century.

Manufacturing of cement was started in India during 1904, but was fully established only in 1912. Until that time, cement was imported from England. In 2017, the world production of hydraulic cement was 4.1 BT (http://minerals.usgs.gov). The top three producers were China with 2.4 BT, India with 280 MT, and the USA with 86.3 MT. In general, the term cement refers to a powdery material manufactured from limestone and clay, which has adhesive properties and when mixed with water, undergoes chemical reactions to form a durable mass and also binds other materials together. For civil engineering works, its primary function is to bind together the particles of fine (sand) and coarse aggregates to form concrete. Cement is also used to bind stones or bricks to build walls; cement mixed with sand and water is used as plaster. The different types of tests conducted on cement are discussed in Chapter 25.

The manufacture of cement results in the depletion of precious raw materials [1 tonne of cement requires about 2 tonnes of raw materials (shale and limestone)] and also in the generation of greenhouse gases; it accounts for about 5% of global $CO_2$ emissions. Cement manufacturing is also highly energy intensive—producing a tonne of cement requires 5.16 GJ of energy, equivalent to about 200 kg of coal, and releases 0.87 tonne of $CO_2$, about 3 kg of nitrogen oxide ($NO_x$), which is an air contaminant that contributes to ground level smog and 0.4 kg of $PM_{10}$ (particulate matter of size 10 μm, which is harmful to respiratory tract when inhaled). Hence, attempts have been made to substitute cement with other industrial waste and by-products, which also have cementitious properties. These cement replacement products include fly ash, ground granulated blast furnace slag, and silica fume. These, along with novel cements, are also discussed.

Cements used in the construction industry may be classified as hydraulic and non-hydraulic.

*Non-hydraulic cement* (e.g., slaked lime) will not set in wet conditions; rather, it sets as it dries and reacts with carbon dioxide in the air. It can be attacked by some aggressive chemicals after setting. On the other hand, *hydraulic cements* (e.g., PC) set and harden when mixed with water due to a chemical reaction between the dry ingredients and water. We are concerned about hydraulic cements in this chapter.

## 5.2 Portland Cement

Also referred to as hydraulic cements, this cement not only hardens by reacting with water but also forms a water-resistant product. The raw materials used for the manufacture of PC include limestone, chalk, seashells, shale, clay, slate, silica sand, alumina, and iron ore; lime (calcium) and silica make up about 85% of the mass. About 3–5% of calcium sulphate (gypsum) is added in the final stages of manufacture.

PC (often referred to as Ordinary Portland Cement or OPC) is the most common type of cement in general use around the world. Cement production in India consists mainly of the following three types only (Fig. 5.1): OPC ~ 25%, Portland Pozzolana Cement (PPC) ~ 66%, and Portland Slag Cement (PSC) ~ 8%. All other varieties put together comprise only 1% of the total production (Mullick, 2017).

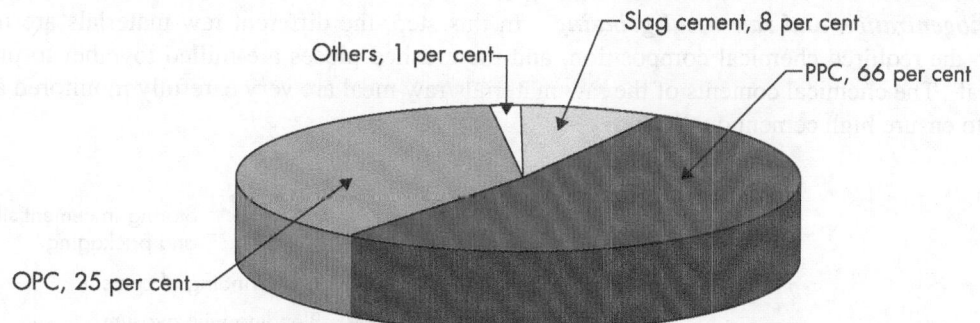

**Fig. 5.1** Production trend of different varieties of cement in India
(*Source*: Mullick, A.K., 'Sustainability of Concrete and Durability go Hand in Hand - Almost!', *The Bridge & Structural Engineer, Journal of the ING-IABSE* Vol. 47, No. 3 Sep. 2017, pp. 34–47.)

### 5.2.1 Grades of Cements in India

OPC is the most important cement and is often used, though the current trend is to use PPC (Fig. 5.1). Most of the discussions to follow in this chapter pertain to this type of cement. The Bureau of Indian Standards has classified OPC into the following three grades (for producing different grades of concrete) to meet the demands of the construction industry:

1. 33 Grade OPC (as per IS 269:2013)
2. 43 Grade OPC (as per IS 8112:2013)
3. 53 Grade OPC (as per IS 12269:2013)

The number in the grade indicates compressive strength of the cement in $N/mm^2$ at 28 days, tested using 1:3 (cement-sand ratio) mortar cubes of size 70.6 mm, as per IS 4031(Part 6):1988. Thus, Grade 33 cement (C-33) means cement with standard mortar cube strength of 33 $N/mm^2$ at 28 days. C-33 is suitable for producing concrete up to M 25. Both C-43 and C-53 cements are suitable for producing higher grades of concrete. It has to be noted that only C-53 is freely available now in the market.

## 5.3 Manufacture of Portland Cement

Two different processes, known as 'dry' and 'wet', are used in the manufacture of PC, depending on whether the mixing and grinding of raw materials are done in the wet or dry conditions. In addition, semi-dry process is also sometimes employed in which the raw materials are ground dry, mixed with water, and then burnt in the kilns. Most of the modern cement factories use either dry or semi-dry process. The schematic representation of dry process of cement manufacture is shown in Fig. 5.2. The dimensions and details of six sizes of standard rotary kilns, their components, and auxiliaries (dry process with suspension preheater) are given in IS 8125:1976.

The manufacture of PC consists of the following steps (see also, Fig. 5.2 [IEA-WBCSD Report, 2009]).

*Quarrying raw materials* Naturally occurring calcareous deposits such as limestone, marl, or chalk are extracted from quarries, often located close to the cement plant. They provide calcium carbonate ($CaCO_3$). Very small amounts of materials such as iron ore, bauxite, shale, clay, or sand may be needed to provide iron oxide ($Fe_2O_3$), alumina ($Al_2O_3$), and silica ($SiO_2$), and fulfil the chemical composition of the raw mix.

*Crushing* The quarried raw materials are transported to the primary/secondary crushers and broken into large pieces of about 100 mm size.

***Pre-homogenization and raw meal grinding***   In this step, the different raw materials are mixed to maintain the required chemical composition, and the crushed pieces are milled together to produce a 'raw meal'. The chemical contents of the raw materials/raw meal are very carefully monitored and controlled, to ensure high cement quality.

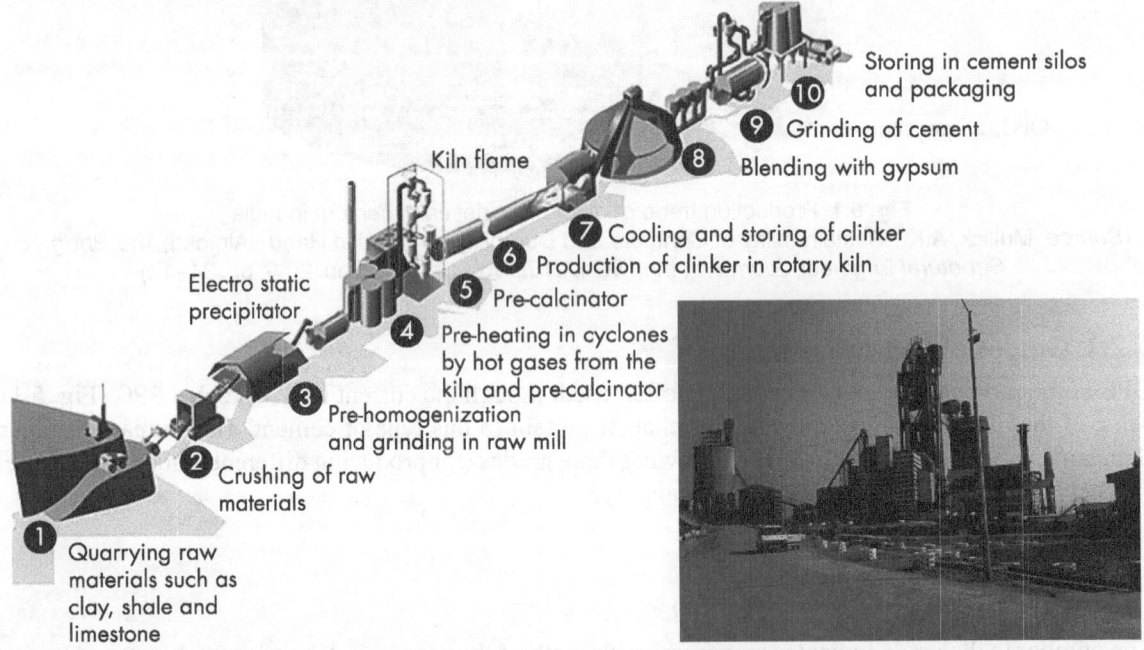

**Fig. 5.2** Schematic representation of the dry process cement manufacture and the view of MCL Cement plant, Thangskai, Meghalaya (adapted from www.cement.org/basics/images/flashtour.html and http://en.wikipedia.org/wiki/File:Cement_Plant_MCL.jpg)

***Preheating***   A preheater is a series of vertical cylinders through which the raw meal is passed, so that they come into contact with the swirling hot kiln exhaust gases moving in the opposite direction. In these cylinders, thermal energy is recovered from the hot flue gases, and the raw meal is preheated before it enters the kiln, in order that the chemical reactions are faster and efficient. Depending on the moisture content of the raw material, a kiln may have up to six cylinders with increasing heat recovery with each extra stage.

***Precalcinator***   Calcination is the process of decomposition of limestone into lime. Part of the reaction takes place in the precalcinator (a combustion chamber at the bottom of the preheater above the kiln), and the remaining part in the kiln.

***Clinker production in the rotary kiln***   The precalcined meal then enters the kiln. Fuel is fired directly into the kiln so that temperature of up to 1450°C is reached. As the kiln rotates, the material slides and tumbles down through progressively hotter zones towards the flame. This intense heat causes chemical and physical reactions to occur and partially melts the meal so that *clinker* is formed.

***Cooling and storing***   The hot clinker from the kiln is made to fall onto a grate-cooler where it is cooled by incoming combustion air, thereby minimizing energy loss in the system. The cooled clinker is usually stored.

***Blending***   At this step, the cooled clinker is mixed with other mineral components. Usually, around 3–5% gypsum will be added to control the setting time of the product. Significant amounts of slag, fly ash, limestone, or other materials can also be added at this stage to replace clinker and produce *blended cement*.

*Cement grinding*   OPC is produced when the cooled clinker and gypsum mixture is ground into a grey powder (when other mineral components are also added, blended cement is produced). Old plants use ball mills for grinding; modern plants use more efficient technologies such as roller presses and vertical mills.

*Storing in the cement silo*   The final product is homogenized and stored in cement silos and dispatched from there to either a packing station (for bagged cement) or a silo truck.

As discussed, the process of manufacture of cement consists of grinding the raw materials finely, mixing them thoroughly in certain proportions, and then feeding and heating to about 1450°C in huge cylindrical steel rotary kilns having diameter of 3.7–10 m, 50–150 m long, and lined with special fire-brick (The rotary kilns are inclined from the horizontal by about 3°, and rotate on its longitudinal axis at a slow and constant speed of about 3–5 revolutions/min). The heated materials sinter and partially fuse to form nodular shaped and marble-to-fistsized material called *clinker* (Note that at a temperature range of 600–900°C, calcination takes place, which results in the release of environmentally harmful $CO_2$). The clinker is cooled (the strength properties of cement are considerably influenced by the cooling rate of clinker) and ground into fine powder after mixing with 3–5% gypsum to form PC. (In modern plants, the heated air from the coolers is returned to the kilns, to save fuel and increase the burning efficiency). It is then loaded into bulk carriers or packaged into bags. In India, typically 50 kg bags are used, though 25 kg, 10 kg, 5 kg, 2 kg, or 1 kg bags are also allowed. As per Indian specifications, the cement should be packed in any of the following bags:

1. Jute sacking bag conforming to IS 2580
2. Multi-wall paper sacks conforming to IS 11761
3. Light weight jute conforming to IS 12154
4. HDPE/PP woven sacks conforming to IS 11652
5. Jute synthetic union bags conforming to IS 12174
6. Any other approved composite bag

Currently (2015), there are 209 large cement plants (producing 270.3 MT of cement) and 365 mini cement plants (producing 33.66 MT) in India (www.ibef.org).

The tolerance allowed is ±2.5% in weight per bag. If supplied in wagon loads of 20 to 25 tonnes, the overall tolerance is ±0.5% of wagon load. In the case of major projects such as bridges, dams, etc., the cement will be procured in bulk and stored in large bins at site.

Fuel used for cement manufacture amounts to 6–8% of the world's fuel consumption. Since fuel costs amount to about 40–60% of the manufacturing costs, different types of fuels are often selected to achieve economy: natural gas (2%), oil (7%), coal (71%), and trash such as wood chips, tires, rice husks, etc., (20%) (Madloola et al., 2011).

## 5.4 Chemical Composition of Raw Materials of Portland Cement

The three main chemical ingredients of PCs are lime (CaO), silica ($SiO_2$), and alumina ($AL_2O_3$). In addition, most cements contain small proportions of iron oxide ($Fe_2O_3$), magnesia (MgO), sulphur trioxide ($SO_3$), and alkalis ($Na_2O + K_2O$). As discussed earlier, gypsum (calcium sulphatedihydrate-$CaSO_4 \cdot 2H_2O$) is also added during the grinding of clinker to control the setting time of cement. There has been a change in the composition of PC over the years. Tricalcium silicate ($C_3S$), the compound primarily responsible for early strengths, has increased, while dicalcium silicate ($C_2S$), the compound responsible for later age strength, has decreased. The mean $SiO_2$ content is slightly lower for all the modern cements, by 0.3% to 2.2%, but on average about 1% lower. The sulphate content for modern cements is about 0.5% to 1.2%, higher than 1950s cements, which is consistent with changes in fineness.

Average alkali contents indicate some increases since the 1950 (PCA-1996). These composition changes, along with an increase in fineness (Blaine), resulted in modern cements having higher early strengths. This is mainly due to the demand in construction industry to increase early strengths to speed up construction processes, allowing wall forms to be stripped earlier and pavements and slabs to be opened to traffic soon. The lime content when exceeds a certain value makes it difficult to combine completely with other compounds. As a result of this, the clinker will contain free lime leading to unsound cement. Similarly, excess silica content, at the expense of alumina and ferric oxide, makes the cement difficult to fuse and form clinker. Typical chemical composition of PC is provided in Table 5.1.

**Table 5.1** Typical chemical ingredients of Portland cement

| Ingredient | Cement chemists notation | Function | Composition (%) |
|---|---|---|---|
| Lime (CaO) | C | Controls strength. Its deficiency reduces strength and setting time. | 61–67 |
| Silica ($SiO_2$) | S | Gives strength, prolongs setting time. | 18.6–23.4 |
| Alumina ($Al_2O_3$) | A | Responsible for quick setting. But in excess quantity lowers strength. | 4.7–6.3 |
| Iron Oxide ($Fe_2O_3$) | F | Gives colour and acts as a catalyst for the fusion of different ingredients. | 1.3–6.1 |
| Magnesia (MgO) | M | Affects colour and hardness. Excess quantities will result in unsoundness. | 0.6–4.8 |
| Sulphur trioxide ($SO_3$) | $\bar{S}$ | If in excess, increases setting time and causes unsoundness. | 1.7–4.6 |
| Alkalis ($Na_2O + K_2O$) | N + K | If in excess may react with aggregates, cause efflorescence and stain concrete or masonry. | 0.1–0.7 |

## 5.5 Chemical Compounds of Cement Clinker

During the burning operation in the manufacture of PC clinker, calcium combines with the other components of the raw mix to form four principal compounds that make up 90% of cement by mass. Gypsum (3–5%), or other calcium sulphate source, and grinding aids are also added during grinding. These *Bogue compounds* are given in Table 5.2 and Fig. 5.3.

**Table 5.2** Chemical composition of ordinary Portland cement (Bogue's compounds) (*Source*: Moir, 2003)

| S. no. | Compound | Cement chemist notation (CCN)[1] | Typical composition as % | Mineral phase | Function |
|---|---|---|---|---|---|
| 1 | Tricalcium silicate 3(CaO) · $SiO_2$ | $C_3S$ | 45–65 | Alite | Mainly responsible for early strength (1 to 7 days) |
| 2 | Dicalcium silicate 2(CaO) · $SiO_2$ | $C_2S$ | 20–30 | Belite | Mainly responsible for later strength (7 days and beyond) |
| 3 | Tricalcium aluminate 3(CaO) · $Al_2O_3$ | $C_3A$ | 5–10 | Aluminate | $C_3A$ increases the rate of hydration of $C_3S$. $C_3A$ gives flash set in absence of gypsum[2]. |

*(Contd)*

**Table 5.2** (*Contd*)

| S. no. | Compound | Cement chemist notation (CCN)[1] | Typical composition as % | Mineral phase | Function |
|---|---|---|---|---|---|
| 4 | Tetracalciumalumino ferrite 4(CaO) · $Al_2O_3$ · $Fe_2O_3$ | $C_4AF$ | 5–10 | Ferrite | It hydrates rapidly but its contribution to strength is uncertain and generally very low. |
| 5 | Gypsum (Calcium Sulphate) $CaSO_4$ · $2 H_2O$ | | 4–5 | – | It controls the setting of cement |

[1]Cement chemists use the shorthand notation as: $C = CaO$, $S = SiO_2$, $A = Al_2O_3$, $F = Fe_2O_3$, $M = MgO$, $H = H_2O$, $N = Na_2O$, $K = K_2O$, $\bar{S} = SO_3$.
[2]Flash set refers to the rapid development of rigidity in freshly mixed mix.

As seen in Table 5.2 and Fig. 5.3, there are four major compounds in cement and these are known as Tricalcium silicate ($C_3S$), Dicalcium silicate ($C_2S$), Tricalcium aluminate ($C_3A$), and Tetracalciumalumino ferrite ($C_4AF$). Their composition varies from cement to cement and plant to plant (The levels of the four clinker minerals can be estimated using a method of calculation first proposed by Bogue in 1929, or exactly by X-ray diffraction analysis.). In addition to the above, there are other minor compounds such as MgO, $Na_2O$, $K_2O$, $SO_3$, fluorine, chloride, and trace metals, which are present in small quantities (Moir, 2003). Of

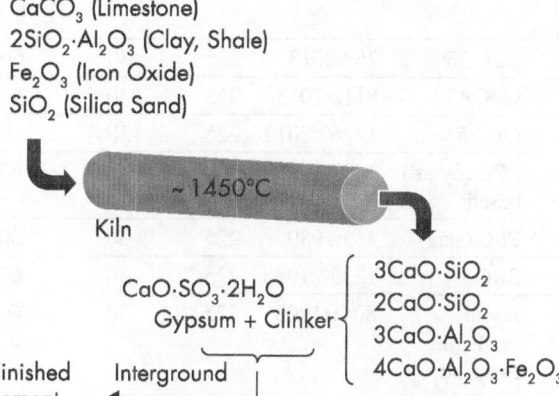

**Fig. 5.3** Chemical compounds of cement

these, $K_2O$ and $Na_2O$ are called as alkalis and found to react with some aggregates, resulting in alkali-silica reaction (ASR), which causes the disintegration of concrete at a later date.

The silicates $C_3S$ and $C_2S$ are the most important compounds and are mainly responsible for the strength of the cement paste. They constitute the bulk of the composition. $C_3A$ and $C_4AF$ do not contribute much to the strength, but they facilitate the combination of lime and silica, and act as a flux in the manufacturing process. The role of the different compounds on different properties of cement is shown in Table 5.3.

**Table 5.3** Role of different compounds on properties of cement

| Characteristic | Different compounds in cement | | | |
|---|---|---|---|---|
| | $C_3S$ | $C_2S$ | $C_3A$ | $C_4AF$ |
| Setting | Quick | Slow | Rapid | – |
| Hydration | Rapid | Slow | Rapid | – |
| Heat Liberation (Cal/gm) 7 days | Higher | Lower | Higher | Higher |
| Early Strength | High up to 14 days | Low up to 14 days | Not much beyond 1st day | Insignificant |
| Later Strength | Moderate at later stage | High at later stage (after 14 days) | – | – |

## 5.6 Physical Properties of Portland Cement

The important physical properties of the three grades of OPC and other types of cements, i.e., fineness, setting time, soundness, and compressive strength, are compared in Table 5.4. The specific gravity of PC is approximately 3.15.

**Table 5.4** Physical properties of the various types of cements

| S. no. | Type of cement | IS code | Fineness m²/kg (min) | Setting time in minutes | | Soundness | | Compressive strength in MPa (min.) | | |
|---|---|---|---|---|---|---|---|---|---|---|
| | | | | Initial (min.) | Final (max.) | Le Chatelier, Max. (mm) | Auto Clave, for MgO, Max. (%) | 3 days | 7 days | 28 days |
| 1 | OPC 33 | 269:2013 | 225 | 30 | 600 | 10 | 0.8 | 16 | 22 | 33 |
| 2 | OPC 43 | 8112:2013 | 225 | 30 | 600 | 10 | 0.8 | 23 | 33 | 43 |
| 3 | OPC 53 | 12269:2013 | 225 | 30 | 600 | 10 | 0.8 | 27 | 37 | 53 |
| 4 | PPC (fly ash based) | 1489:1991 (Part 1) | 300 | 30 | 600 | 10 | 0.8 | 16 | 22 | 33 |
| 5 | PSC (slag) | 455:1989 | 225 | 30 | 600 | 10 | 0.8 | 16 | 22 | 33 |
| 6 | SRC | 12330:1988 | 225 | 30 | 600 | 10 | 0.8 | 10 | 16 | 33 |
| 7 | Rapid hardening Portland cement | 8041:1990 | 325 | 30 | 600 | 10 | 0.8 | 27 | – | – |
| 8 | Low-heat Portland cement | 12600:1989 | 320 | 60 | 600 | 10 | 0.8 | 10 | 16 | 35 |

**Fineness**    It is a measure of the size of the particles of cement and is expressed in terms of *specific surface of cement,* and expressed in terms of m²/kg. It can be calculated using the sieving or one of the air permeability methods (see Section 25.7.2 of Chapter 25). It is an important factor in determining the rate of gain of strength. Approximately 95% of cement particles are smaller than 45 μm, with the average particle around 10 μm. For a given weight of cement, the surface area is more for finer cement than coarser cement. The finer the cement, the higher is the rate of hydration; greater fineness causes greater early strength (especially, during the first 7 days) and more rapid generation of heat. Increase in fineness is also found to increase the drying shrinkage of concrete. It has to be noted that the fineness of cement has increased over the years but has got stabilized recently. The disadvantage of fine grinding is that it is susceptible to air-set and early deterioration.

**Soundness**    It refers to the ability of the cement paste to retain its volume after setting. Cements which exhibit expansion in volume after setting are known as *unsound cements.* Unsound cements will result in serious difficulties for the durability of structures. Unsoundness in cement and changes in volume may take place due to excessive amounts of free lime, magnesia, and calcium sulphate. When the raw materials that are fed into the kiln contain more lime (CaO) than that can combine with the acidic oxides, the excess will remain in a free condition. This free lime will hydrate very slowly in a subsequent stage and the mortar/concrete using such cements will expand and crack after a few months or even a year. Excess magnesia (MgO) also reacts with water similar to quicklime (CaO). However, only the crystalline variety of magnesia is deleterious, and magnesia present in glass form is harmless. For this reason, magnesium content is usually restricted to 0.8% in cements (Table 5.4). As mentioned earlier, gypsum, which is a hydrate of calcium sulphate, is added to cement clinker to prevent flash setting. But when excess amount of gypsum is present, it reacts with $C_2A$ during setting, resulting in unsoundness in the form of a low expansion. Hence, the amount of gypsum in cement manufacture is also strictly restricted to about 3–5%. As unsound cement used in concrete will go unnoticed for a considerable amount of time, accelerated tests are required to test it. The main tests for predicting the soundness of cement are the Le Chatelier and Autoclave tests (see Section 25.7.3 of Chapter 25).

**Setting time**    When water is added to cement and mixed, it becomes a paste and will be in a plastic state. Owing to hydration, this paste will become less plastic and finally hardens to a rigid mass, which can resist some amount of pressure. This process is called *setting* and the time to reach this stage is termed as *setting time*. This time is calculated from the instant water is added to the cement. The setting time is normally considered to have two phases: *the initial setting time* and *the final setting time*. The initial setting time is the time between the adding water to cement and when the paste starts to lose its plasticity and may vary between 30 min (OPC/PPC) to 60 min (low heat cement). The cement is considered to have reached its final setting time when the paste completely loses its plasticity and attains sufficient strength and hardness. Usual final setting time for cement is 600 min. These two setting times are determined using the Vicat's apparatus (see Section 25.7.5 of Chapter 25).

**Compressive strength**    It is the most important property of cement. It has to be noted that the grades mentioned in the cement bags such as 43/53 grade OPC/PPC actually indicate the strength of cement. Thus, 43 grade OPC cement indicates that the 28th day compressive strength of cement-mortar cubes prepared using this cement will have a strength of 43 MPa. Tests to determine the compressive strength of cement are carried out on *cement-mortar cubes* and not on cement cubes. Generally, the strength of cement may be expressed in three ways: compressive, tensile, and flexural. However, only compressive strength test is carried out (see Section 25.7.6 of Chapter 25 for details). The compressive strength of cement-sand cubes is affected by a number of parameters, which include: water-cement ratio, cement-sand ratio, type and grading of sand, how the mix is prepared, size and shape of specimen, curing conditions, rate of loading in the testing machine, and the age of specimen. Since the strength of cement varies with time, it is important to specify the time at which the strength test is performed. Typically, compression tests are conducted on the 1st day (for high early strength cement), 3rd day, 7th day, 28th day, 56th, day or the 90th day (last two are usually done for PPC). The cement-mortar cube strength is used to check the quality of cement. It is important to realize that there is no relation between the cement-mortar cube strength and the concrete cube strength. It is because the strength of concrete cube is governed by several other parameters, especially the strength of coarse aggregate used.

**Consistency**    It may be defined as the amount of water needed to prepare a plastic mix. Consistency indicates the degree of density or stiffness of cement. It is necessary to find the consistency because the amount of water present in the cement paste may affect the setting time. Consistency test is done using Vicat's apparatus. The standard consistency of a cement paste, as per IS 4031(Part 4), is defined as that consistency at which the Vicat plunger (IS 5513:1976) penetrates to a point 5–7 mm from the bottom of the Vicat mould (see Section 25.7.4 of Chapter 25). The consistency is expressed as a percentage of the weight of dry cement, which usually varies from 26 to 33%.

## 5.7 Chemical Properties of Portland Cement

There are a number of chemical requirements of cement such as lime saturation factor, alumina-iron ratio, insoluble residue, loss of ignition, percentage of silica, ferric oxide, alumina, calcium oxide, magnesia, sulphuricanhydride, sodium and potassium oxide, water-soluble alkali, free lime, sulphurtrioxide, sulphur as sulphide, and slag/pozzolana content. Some of the important ones are shown in Table 5.5.

**Table 5.5** Chemical properties of various cements

| Characteristics | Ordinary Portland cement | | | Portland pozzolana cement (IS 1489-Part 1:2015) | Rapid hardening Portland cement (IS 8041:1990) | Low heat Portland cement (IS 12600: 1989) | Portland slag cement (IS 455–2015) |
|---|---|---|---|---|---|---|---|
| | 33 Grade (IS 269: 2013) | 43 Grade (IS 8112: 2013) | 53 Grade (IS 12269 :2013) | | | | |
| Lime saturation factor (%)[1] | 0.66–1.02 | 0.66–1.02 | 0.8–1.02 | – | 0.66–1.02 | – | – |
| Alumina-iron ratio (%) Min. | 0.66 | 0.66 | 0.66 | – | 0.66 | 0.66 | – |
| Insoluble residue (%) Max. | 5.0 | 4.0 | 4.0 | $P + \dfrac{4(100-P)}{100}$ | 2.0 | 2.0 | 4.0 |
| Magnesia (%) Max. | 6.0 | 6.0 | 6.0 | 6.0 | 6.0 | 6.0 | 10.0 |
| Sulphuric Anhydride-$SO_3$ (%) Max. | 3.5 | 3.5 | 3.5 | 3.5* | 3.0** | 2.5* | 3.5** |
| Loss on ignition (%) Max. | 5.0 | 5.0 | 4.0 | 5.0 | 5.0 | 5.0 | 5.0 |
| Permitted additives-other than gypsum (%) Max. | 1.0 | 1.0 | 1.0 | 1.0 | 1.0 | 1.0 | 1.0 |
| Performance improver (e.g., Fly ash, granulated slag, silica fume, limestone, or rice husk ash) (% addition by mass), Max. | 5 | 5 | 5 | Fly ash 15–35 GGBS 5 Limestone 5 | – | – | 25–70 (slag content) Fly ash 5 Limestone 5 |
| Total alkalies ($Na_2O$) (%) Max. | 0.6 | 0.6 | 0.6 | 0.6 | 0.6 | 0.6 | 0.6 |

[1] Lime saturation factor is calculated as $\dfrac{CaO - 0.7SO_3}{2.8SiO_2 + 1.2Al_2O_3 + 0.65Fe_2O_3}$

**when $C_3A$ is greater than 5%
The tricalcium aluminate content ($C_3A$) is calculated as $C_3A = 2.65(Al_2O_3) - 1.69(Fe_2O_3)$
P is the declared percentage of Pozzolana in cement

## 5.8 Hydration of Cement

When PC is mixed with water, a series of chemical reactions takes place, which results in the formation of new compounds and progressive setting, hardening of the cement paste, and development of strength. The overall process is referred to as cement hydration. Hydration involves many different reactions, often occurring at the same time. When the paste (cement and water) is added to aggregates (course and fine), it acts as an adhesive and binds the aggregates together to form concrete. Most of the hydration and about 90% strength development take place within 28 days; however, the hydration and strength development continue, though more slowly, for a long time with adequate moisture and temperature. See Section 5.8.1 for the heat liberated with respect to age. Hydration products formed in hardened cement pastes are more complicated, and the chemical equations are shown in Table 5.6. More details of these chemical reactions may be found in Johansen et al., 2002; Lea, 1971; Powers, 1961; Taylor, 1997.

**Table 5.6** Portland cement compound hydration reactions (Lea 1971, Tennis and Jennings, 2000)

| | | | | |
|---|---|---|---|---|
| $2\,(C_3S)$<br>Tricalcium silicate | $+11H$<br>Water | | $=C_3S_2H_8$<br>Calcium silicate hydrate (C-S-H) | $+3\,(CH)$<br>Calcium hydroxide |
| $2\,(C_2S)$<br>Dicalcium silicate | $+9H$<br>water | | $=C_3S_2H_8$<br>Calcium silicate hydrate (C-S-H) | $+CH$<br>Calcium hydroxide |
| $C_3A$<br>Tricalcium aluminate | $+3(C.\bar{S}.H_2)$<br>Gypsum | $+26H$<br>Water | $=C_6.A.\bar{S}_3.H_{32}$<br>Ettringite (AFt) | |
| $2(C_3A)$<br>Tricalcium aluminate | $+C_6.A.\bar{S}_3.H_{32}$<br>Ettringite (AFt) | $+4H$<br>Water | $=3(C_4.A.\bar{S}.H_{12})$<br>Calcium monosulfoaluminate (AFm) | |
| $C_3A$<br>Tricalcium aluminate | $+CH$<br>Calcium hydroxide | $+12H$<br>Water | $=C_4.A.13H$<br>Tetracalcium aluminate hydrate | |
| $C_4AF$<br>Tetracalcium alumino ferrite | $+10H$<br>Water | $+2(CH)$<br>Calcium hydroxide | $=6C.A.F.12H$<br>Calcium alumino ferrite hydrate | |

C-H-S is amorphous and fibrous, makes up 50–60% of solids in hardened paste and it forms a continuous binding matrix with a large surface area and it is the component responsible for the development of strength in the cement paste.

CH is a crystalline, thick hexagonal plate. Imbedded in the C-H-S matrix filling the pores, CH is not important in the strength of the hardened paste.

Ettringite is a long, slender, prismatic, crystalline, needle-like material that makes up for 10–20% of the solids in the paste. It is stable as long as long as gypsum is present. It plays a minor role in strength development but a major role in durability.

As shown in Fig. 5.4, tricalcium silicate ($C_3S$) hydrates and hardens rapidly and is mainly responsible for initial set and early strength of concrete. Thus, OPC containing increased percentage of $C_3S$ will have high early strength. The hydration of dicalcium silicate ($C_2S$) is the same as the hydration of $C_3S$, which produces the same hydrates with similar properties. However, the reaction of $C_2S$ is slower and generates less heat because it is less active. Hence, $C_2S$ contributes to strength increase only after about 7 days, as against the early strength development of $C_3S$.

Tricalcium aluminate ($C_3A$) reacts with water in a very violent and rapid way leading to flash setting-a rapid development of rigidity in freshly mixed PC paste, mortar, or concrete. $C_3A$ is responsible for the large amount of heat of hydration during the first few days of hydration and hardening. It also contributes slightly to the strength development in the first few days. $C_3A$ will react with gypsum and water (until all the gypsum is consumed) to produce *Ettringite*, a needle-like material (Fig. 5.5). Then, $C_3A$ will start to react with Ettringite to form Monosulfoaluminate, which will set the paste. Monosulfoaluminate is a stable hydration product and is seen in electron microscope images as a thin, irregular

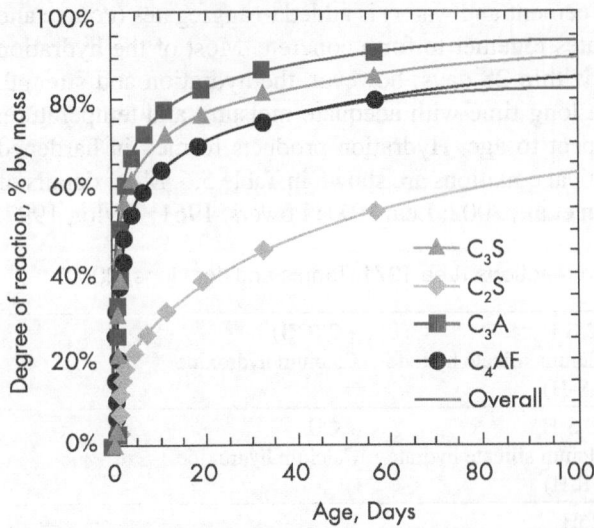

**Fig. 5.4** Relative reactivity of cement compounds

(*Source*: Reprinted from Tennis, P. D. and H.M. Jennings 2000, 'A Model for Two Types of Calcium Silicate Hydrate in Microstructure of Portland Cement Pastes', *Cement and Concrete Research*, Vol. 30, No. 6, pp. 855–63, permission from Elsevier).

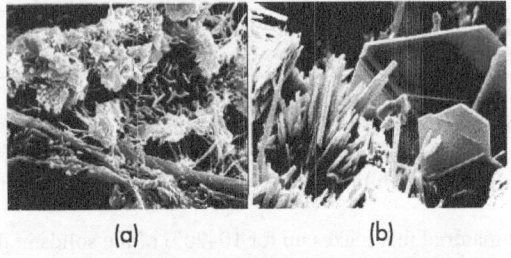

(a)  (b)

**Fig. 5.5** Scanning electron microscope (SEM) images. (a) Image showing platy or foil-like CSH, fine bundles of CSH fibres and platy CH. (b) Image showing platy CH and ettringite needles and also plate-like CH morphology (*Source*: www. fhwa.dot.gov)

plate, clustered-like flower petals and fairly crystalline. It fills the pores but can be reformed into ettringite, if sulphates ions are able to enter the concrete (Sulphate attack). This new formation causes volume to increase and leads to tensile cracking. This tendency is the basis for sulphate attack of PCs. Cements with low percentages of $C_3A$ are more resistant to soils and waters containing sulphates. Tetracalcium aluminoferrite ($C_4AF$) contributes little to strength. The grey colour of cement is due to $C_4AF$ and its hydrates.

As mentioned earlier gypsum (calcium sulphate dehydrate) is added to cement during final grinding to retard the rapid hydration of the calcium aluminates and to provide sulphate to react with $C_3A$ to form ettringite (calcium trisulfoaluminate or AFt). In addition to controlling setting and early strength gain, gypsum also helps to control drying shrinkage, and acts as a strength accelerator (Kosmatka et al., 2003, Bhanumathidas and Kalidas, 2004). If inadequate amounts of gypsum are added to the cement, the latter will react with $C_3A$ too vigorously and flash setting will occur. Further mixing will not dispel this rigidity, and a large amount of heat will be produced in the process. If more than required gypsum is added, an expansive reaction will take place. Figure 5.4 shows the relative reactivity of cement compounds. The 'Overall curve' has a composition of 55% $C_3S$, 18% $C_2S$, 10% $C_3A$, and 8% $C_4AF$.

### 5.8.1 Heat of Hydration

When PC is mixed with water, as a result of the exothermic chemical reaction, heat is liberated. This heat is called the *heat of hydration*. Usually, the greatest rate of heat of hydration occurs within the first 24 h (around 30% of the total heat), around 50% of heat is generated within the first 3 days; 75% in 7 days; and 83–91% at 180 days. Temperature rises of up to 55°C have been observed with mixes having high cement content. Such a temperature rise will result in cracking of the concrete. As a rule of thumb, the maximum temperature differential between the interior and exterior concrete should not exceed 20°C to avoid crack development. ACI 211.1 states that as a rough guide, hydration of cement will generate a concrete temperature rise of about 4.7–7.0°C per 50 kg of cement per $m^3$ of concrete.

Factors influencing heat development in concrete include the cement content (cements with higher contents of tricalcium silicate ($C_3S$) and tricalcium aluminate ($C_3A$), and higher fineness have higher rates of heat generation), water-cement ratio, placing and curing temperature, the presence of mineral and chemical admixtures, and the dimensions of the structural element. Higher temperatures greatly accelerate the rate of hydration and the rate of heat liberation at early ages (less than 7 days). Kulkarni (2012) observed that over the years there is a large increase in the $C_3S$ content and fineness of cement, both of which speed up the hydration reaction, and provide high early strength and accompanying side effect of higher heat of hydration (For example, in 1920s the cement in the USA contained 21% of $C_3S$ and 48% of $C_2S$. Now their proportion is completely reversed; $C_3S$-56% and $C_2S$-17%). In view of these changes in the cement characteristics, design strengths could be achieved with low cement content and low water-cement ratio using water-reducing chemical admixtures.

Mineral admixtures, such as fly ash, can significantly reduce the rate and amount of heat development. The methods to minimize the concrete temperature rise include cooling the mixing water, using ice as part of the mixing water, using a moderate-heat PC or moderate- or low-heat blended cement, and keeping cement contents to a minimum level. Fly ash, other pozzolan, or slag can be used with chemical admixtures (retarder, water-reducer/retarder), or the aggregate can be cooled. Also, curing with water helps to control temperature increases and is better than other curing methods.

## 5.8.2 Rate of Hydration

The rate at which hydration of cement takes place will occur in the following order: first, $C_3A$, followed by $C_3S$ and $C_4AF$, and finally $C_2S$. Ettringite forms first, followed by CH and C-S-H, and a little change in the total volume of the paste as a result of the hydration.

The actual process of cement hydration may be divided into five stages as shown in Fig. 5.6 (Gartner et al., 2002)

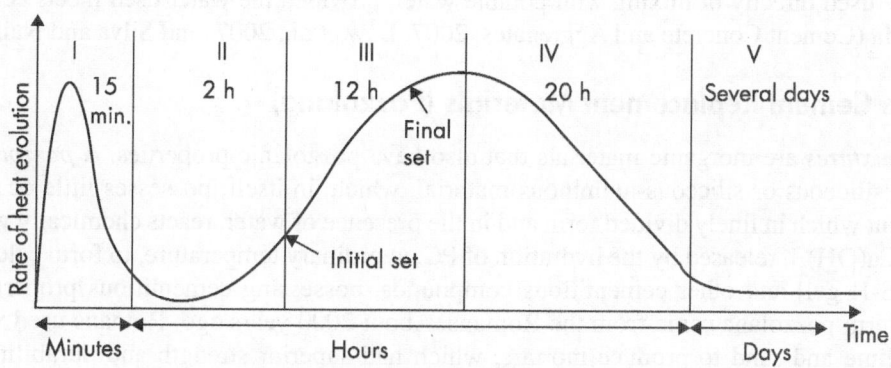

**Fig. 5.6** Rate of heat evolution

These stages are (http://people.ce.gatech.edu/~kk92/hyd07.pdf):

Stage I:  Initial stage (0–15 min) - dissolution of solids (increasing ionic concentration) and increase in heat of hydration

Stage II:  Dormancy stage or Induction period (15 min–4 h)

Stage III:  Acceleration stage (4–8 h)- heat of hydration increases causing the cement to set (initial set)

Stage IV:  Deceleration stage (8–24 h)

Stage V:  Steady stage- heat of hydration is very less

The rate of hydration can be increased in the following ways:

1. *Casting temperature:* For each 10°C increase in temperature, the rate of hydration doubles.
2. *Fineness of cement*: As the fineness of cement is increases, the heat of hydration will also increase, increasing the fineness from 300 $m^2/kg$ to 500 $m^2/kg$ will increase the heat of hydration by about 50%.

3. *Changing proportions of components:* By changing the proportion of cement components, rate of hydration can be increased; for example, increasing $C_3S$ and $C_3A$.

The final product of hydration is a hardened paste consisting of solids (C-H-S, CH, Ettringite, Monosulphate, and unhydrated cement), water, air voids, and capillary voids.

### 5.8.3 Water-Cement Ratio

Generally, water required to be added for cement hydration is very less as compared to water required for workability. For the complete hydration of PC, only about 38% water (This is represented by the water/cement or water/cementitious ratio, usually denoted by *w/c ratio* or *w/cm ratio*), i.e., w/c of 0.38, is needed. If a w/c ratio greater than about 0.38 is used, the excess water, which is not required for cement hydration, will remain in the capillary pores or may evaporate in due course. This process leads to *drying shrinkage* (drying shrinkage is destructive as it leads to micro-cracking and may eventually weaken concrete). Similarly, when a w/c ratio of less than about 0.38 is used, some cement will remain unhydrated. The space initially taken up by water in a cementitious mixture will be partially or completely replaced over time by the hydration products. If a w/c ratio of more than 0.38 is used, porosity in the hardened material will remain, even after complete hydration. This is called *capillary porosity*, and will lead to the corrosion of reinforcement.

Currently, all the batching plants are using a substantial amount of potable water to batch concrete and in the construction industry, a huge amount of wastewater is generated every day. Moreover in many parts of the world, water resources are either limited or the cost of potable water is escalating. Studies conducted recently for the sustainable use of water show that reclaimed water or slurry water from batching plants can be used directly or mixing with potable water, provided the water used meets certain performance criteria (Cement Concrete and Aggregates, 2007, Low et al., 2007, and Silva and Naik, 2010).

## 5.9 Green Cement Replacement Materials (Pozzolana)

*Mineral admixtures* are inorganic materials that also have pozzolanic properties. A *pozzolana* may be defined as a siliceous or siliceous-aluminous material, which, in itself, possesses little or no cementitious value but which in finely divided form and in the presence of water, reacts chemically with calcium hydroxide [$Ca(OH)_2$], released by the hydration of PC, at ordinary temperature, to form calcium silicate hydrate [C-S-H gel] and other cementitious compounds, possessing cementitious properties (Mehta, 1987). The term pozzolana came from the Romans. About 2000 years ago, Romans used volcanic ash along with lime and sand to produce mortars, which had superior strength and durability. In general, pozzolanic materials are classified as (a) Natural Pozzolana such as volcanic ash and (b) Artificial Pozzolana such as fly ash, blast furnace slag, and *surkhi*. India does not possess many deposits of natural Pozzolana. Hence, artificial Pozzolana is added to the concrete mix to improve its properties or as a replacement for PC leading to sustainable constructions (Subramanian, 2007).

The C-S-H gel produced due to the pozzolanic reaction can reduce the size of the pores of crystalline hydration products, make the microstructure of concrete more uniform, and improve the impermeability and durability of concrete [Hydrated cement paste contains large capillary pores with diameters between 50 and 10,000 nm, medium capillary pores with diameters of 10–50 nm, and gel pores with diameter of less than 10 nm (Mindess et al., 2003). Additional material resulting from reaction between liberated surplus lime and mineral admixtures block these capillary pores and reduce permeability of concrete]. These improvements can lead to an increase in strength and service life of concrete. Some of the mineral admixtures are briefly discussed in the following sections. These materials, such as fly ash, ground granulated blast furnace slag, silica fume, and calcined clay, are also called *supplemental cementitious materials (SCMs)*.

### 5.9.1 Surkhi

*Surkhi* is an artificial pozzolana made by grinding to powder burnt bricks, brick-bats, or burnt clay (under- or over-burnt bricks should not be used, nor bricks containing high proportion of sand). When clay balls are specially burnt for making *surkhi*, an addition of 10–20% of quicklime may improve its quality. As *surkhi* is not a standardized product, its properties vary widely. Replacement of 10–30% of cement by *surkhi* in concrete produces a more plastic, less bleeding, and less segregating concrete, compared to ordinary cement concrete (see IS 1344:1981 and IS 1727:1967). When PC was not popular, *surkhi* was used as an admixture in the construction of several dams in India (Shetty, 2005).

### 5.9.2 Fly Ash

*Fly ash* is a by-product of coal-fired thermal power plants. It may vary in colour from light grey to dark grey or even brown. In the UK, it is referred as *pulverized fuel ash* (PFA). Indian coal has low calorific value (3000–3500 Kcal.) and very high ash content (30–45%) resulting in huge quantity of fly ash. As of 2013–14, more than 225 MT of fly ash is produced every year in India and is expected to increase to 300–400 MT by 2020. The disposal of fly ash poses a serious environmental problem (more than 70,000 acres of land in India is presently occupied by ash ponds).There are several environmental benefits of recycling fly ash includes saving of landfill space (about 1.5 acre of land is needed for ash disposal per MW of power produced), and reduced demand for virgin materials (Senapati, 2011). Realizing that the utilization of fly ash is very low, the Indian Government set up the *Fly ash Mission*, under the Department of Science & Technology, New Delhi.

Any coal-based thermal power station may produce four kinds of ash: fly ash, bottom ash, pond ash, and mound ash. The ash disposed in ash ponds as slurry is known as *pond ash*, and that is disposed in the form of mounds is known *mound ash*. Fly ash constitutes about 75–85% of the total ash produced. The *bottom ash* and boiler slag, which are collected at the bottom of the furnace, are much coarser and are not pozzolanic in nature. Bottom ash is a coarse ash and is an excellent substitute for soil, for geotechnical applications. After processing for removal of carbon and grinding to the size of sand, bottom ash may be used as a part-substitute of sand in mortar and concrete. An important utilization of pond ash is in the manufacture of clay bricks. When 30–80% of pond ash is mixed with good clay, it improves the quality of clay bricks (called *fly ash brick*), reduces breakage at the kiln as well as during transit/use, and also reduces fuel consumption in the kiln (also see Section 3.10.1 of Chapter 3).

Physically, fly ash occurs as very fine spherical glassy particles, majority having sizes in the range from 1 µm to more than 100 µm, with typical particle size measuring under 20 µm (finer than PC). It also has low to medium bulk density, high surface area, and sandy-silt to silty-loam texture. The specific gravity of fly ash varies over a wide range of 1.6 to 2.6 (www.c-farm.org). Fly ash may contain about 35–40% silica and much lower calcium oxide than PC, resulting in better durability. In some cases, alumina and iron oxide content can be quite high, leading to problems of lower strength and unusual setting time. When fly ash is also included in the mix of concrete, it will react with lime to form additional C-S-H binder. The additional binder produced by the fly ash reaction allows fly ash concrete to continue to gain strength over time. The replacement of cement by fly ash reduces the water demand for a given slump. This decrease in water content combined with the production of additional cementitious compounds, reduces the inter-connectivity of pores leading to decreased permeability. The decrease in free lime, the increase in cementitious compounds, and the decreased permeability enhances the durability of concrete. Detailed information on the nature of fly ash and pozzolanic reactions in concrete can be found in the ACI Committee 232.2R-03 report on fly ash in concrete. A comparison of chemical properties of typical fly ash, slag, silica fume, calcined clay, and metakaolin is given in Table 5.7.

**Table 5.7** Comparison of chemical properties of typical fly ash, slag, silica fume, calcined clay, and metakaolin

| Ingredients | Cement (OPC) | Grade I fly ash | Grade II fly ash | GGBS | Silica fume | Calcined clay | Metakaolin |
|---|---|---|---|---|---|---|---|
| $SiO_2$, % | 20–25 | 45–52 | 35–40 | 35 | 85–90 | 58 | 53 |
| $Al_2O_3$, % | 4.5–6.3 | 23–28 | 18 | 12 | 0.4–0.7 | 29 | 43 |
| $Fe_2O_3$, % | 3.80 | 11 | 6 | 0.3–1.0 | 0.4–0.8 | 4 | 0.5 |
| CaO, % | 63.50 | 2.5–5.0 | 21 | 40 | 0.1–4.2 | 1 | 0.1 |
| $SO_3$, % | 1.5 | 0.2–0.8 | 4.1 | 2.0–9.0 | 0.05–0.4 | 0.5 | 0.1 |
| $Na_2O$, % | 0.15 | 0.04–1.0 | 5.8 | 0.34 | 0.2–0.5 | 0.2 | 0.05 |
| $K_2O$, % | 0.50 | 0.9–2.0 | 0.7 | 0.42 | 0.22–1.45 | 2 | 0.4 |
| Total Na eq. alkalis, % | | 2.2 | 6.3 | 0.6 | 1.9 | 1.5 | 0.3 |
| Loss of ignition, % | 1.15 | 2.8 | 0.5 | 1.30 | 3.0–4.5 | 1.5 | 0.7 |
| Blaine fineness, m²/kg | 225 | 420 | 420 | 400 | 20,000 | 990 | 19,000 |
| Relative density | 3.15 | 2.38 | 2.65 | 2.94 | 2.20 | 2.50 | 2.50 |

The quality of fly ash to be used in concrete is governed by IS 3812 (Parts 1 & 2):2013, which groups all these types of ash as PFA. IS 3812 categorizes fly ash into two types, namely, *siliceous fly ash* (SFA) or Grade I, having reactive calcium oxide content less than 10%, and *calcareous fly ash* (CFA) or Grade II, having reactive calcium oxide content generally between 10 and 25%. Grade I can be used in cement, mortar, and concrete and in lime pozzolana mixture, and for the manufacture of PPC and Grade II for cement, mortar, and concrete and in lime pozzolana mixture. Both these grades can be used as admixtures. The chief difference between these classes is the amount of calcium, silica, alumina, and iron content in the ash. Most of the particles of processed fly ash should pass through the 45 μm sieve. More than 40% of the particles, which are under 10 μm, contribute to the early age strength (7 and 28 days). Particles of sizes 10–45 μm react slowly and are responsible for gain in strength from 28 days to 1 year. Physical and chemical requirements of fly ash are given in Tables 5.8 and 5.9.

**Table 5.8** Physical requirements of fly ash, as per IS 3812-Parts 1 & 2

| Characteristics | Requirements | |
|---|---|---|
| | Grade I | Grade II |
| Fineness-specific surface in m²/kg, (Blaine's permeability method), Minimum | 320 | 200 |
| Particles retained on 45 μm IS 50 sieve (wet sieving) in%, Maximum | 34 | 50 |
| Lime reactivity – Average compressive strength in N/mm², Minimum | 4.5 | 3.0 |
| Compressive strength at 28 days in N/mm², Minimum | Not less than 80% of the strength of corresponding plain cement mortar cubes | |
| Soundness-expansion of specimen, (autoclave test)%, Maximum | 0.80 | 0.80 |

**Table 5.9** Chemical requirements of fly ash, as per IS 3812-Parts 1 & 2

| Characteristics | Requirements | |
|---|---|---|
| | Grade I (%) | Grade II (%) |
| Silicon dioxide ($SiO_2$) + Aluminium oxide ($Al_2O_3$)+ Iron oxide ($Fe_2O_3$), Minimum | 70 | 50 |
| Silicon dioxide ($SiO_2$), Minimum | 35 | 25 |
| Magnesium oxide (MgO), Maximum | 5 | 5 |
| Sulphur Trioxide($SO_3$), Maximum | 5 | 5 |
| Available alkalis such as Sodium Oxide ($Na_2O$), Maximum | 1.5 | 1.5 |
| Loss of ignition by mass, Maximum | 5 | 5 |

## 5.9.3 Ground Granulated Blast Furnace Slag (GGBS)

*Blast furnace slag,* or simply *slag,* is a by-product of steel industry, is non-metallic by-product, consisting essentially of silicates and alumino-silicates of calcium, developed in a molten condition simultaneously with iron in a blast furnace or electric pig iron furnace. Every year, about 9 million tonnes of blast furnace slag is produced in India. The *ground granulated blast furnace slag* (GGBFS or GGBS) is obtained by further processing the molten slag (which is at a temperature of about 1500°C), by rapidly chilling or quenching it with water or air to form a glassy sand-like granulated material. It has to be noted that the air-cooled slag does not have the hydraulic properties of water-cooled slag. GGBS has been used as a cementitious material since the 18th century (Granulated blast furnace slag was first developed in Germany in 1853). The granulated material, which is ground to less than 45 μm, has a surface area fineness of about 400–600 m²/kg. Its relative density is in the range of 2.85–2.95 and the bulk density varies from 1050 to 1375 kg/m³. The GGBS should conform to IS I2089:1987 (Table 5.10). The rough and angular-shaped ground slag, in the presence of water and an activator such as Sodium hydroxide (NaOH) or Calcium hydroxide (CaOH)-both supplied by PC, hydrates and sets in a manner similar to PC. Currently, it is inter-ground with PC to form blended cement, thus partially replacing PC. It reduces temperature (in mass concrete), permeability, expansion due to alkali-aggregate reaction, and improves sulphate resistance. See Section 5.10.2 for more details on PSC.

**Table 5.10** Specifications of GGBS (IS 12089)

| Constituent | Percentage |
|---|---|
| Manganese oxide, maximum | 5.5 |
| Magnesium oxide, maximum | 17.0 |
| Sulphide sulphur, maximum | 2.0 |
| Glass content, minimum | 85.0 |

Also, the following should be satisfied

$$\frac{CaO + MgO + AL_2O_3}{SiO_2} \geq 1.0 \; ; \quad \frac{CaO + MgO + \left(\frac{1}{3}\right)AL_2O_3}{SiO_2 + \left(\frac{2}{3}\right)AL_2O_3} \geq 1.0$$

When there is more than 2.5% of manganese oxide (MnO)

$$\frac{CaO + CaS + \left(\frac{1}{2}\right)MgO + AL_2O_3}{SiO_2 + MnO} \geq 1.5$$

## 5.9.4 Silica Fume

Silica fume, which is also referred to as *micro-silica* or *condensed silica fume*, is a light to dark grey cementitious material, and is a by-product of the production of silicon and ferrosilicon alloys. Silica fume rises as an oxidized vapour from these furnaces at about 2000°C. When it cools and condenses, it is collected in large cloth bags. This condensed silica fume is then processed to remove impurities and control particle size. More than 85% of condensed silica fume consists of silicon dioxide ($SiO_2$) in amorphous non-crystalline form. Silica fume is similar to fly ash with particles of spherical shape, and has a particle size in the range of 0.01–1 μm (hence, it is about 50–100 times smaller than the average cement particle). Condensed silica fume has a surface area of 13,000 to 30,000 $m^2$/kg as against the surface area of 225–400 $m^2$/kg of cement (The very high surface area can be estimated using the nitrogen adsorption method only.). This results in a higher surface to volume ratio and a much faster pozzolanic reaction. The relative density of silica fume is about 2.20, whereas cement has a relative density of 3.15. Silica fume can be obtained in powder form but usually available as liquid slurry. It has a bulk density of 130 to 430 kg/$m^3$ (as a slurry, it is 1320 to 1440 kg/$m^3$). The efficiency of silica fume depends upon its mineralogy and particle size distribution. The extremely fine particle size, large surface area, and high content of highly reactive amorphous silicon dioxide give silica fume superior pozzolanic properties. Silica fume used in concrete should conform to IS 15388:2003 (Table 5.11). As per clause 5.2.1.2 of IS 456, its usual proportion is 5–10% by mass of the total cementitious material. Silica fume appears to be the best-performing siliceous product among the pozzolanic materials, but it is not available in large quantities and also is the most expensive. It has to be noted that in comparison, fly ash is available in large amounts and is relatively cheap. However, the performance of fly ash is lower than that of silica fume because of the lower amount of amorphous silica (35–40%) and the larger size (0.1–40 μm) of its spherical grains.

**Table 5.11** Physical requirements for silica fume as per IS 15388:2003

| Characteristics | Requirement |
| --- | --- |
| Specific surface, $m^2$/kg, Minimum | 15,000 |
| Oversize percentage retained on 45 μm IS sieve, %, Maximum | 10 |
| Oversize percentage retained on 45 μm IS sieve, variation from average, %, Maximum | 5 |
| Compressive strength at 7 days, MPa, Minimum | 85 |

A new pozzolanic material produced synthetically, in the form of water emulsion of ultra-fine amorphous colloidal silica (UFACS), is available in the market and it appears to be potentially better than silica fume as it has higher content of amorphous silica (>99%) and the size of its spherical particles is much less (1–50 nm): due to this reduced particle size, UFACS is also called *'nano-silica'* in comparison with the term 'micro-silica', used with silica fume.

## 5.9.5 Metakaolin

*High reactivity Metakaolin* (HRM) or simply *Metakaolin* is obtained by the calcination of pure or refined 'kaolintic' clay at a temperature between 650°C and 850°C followed by grinding to achieve a fineness of 700 to 900 $m^2$/kg. Controlled blending of the kaolinite, processing, and calcinations provide a consistent, high-purity, quality-controlled product with excellent physical and chemical properties. It is a highly reactive pozzolan that reacts with calcium hydroxide produced from free lime during the hydration of PC, forming calcium silicate and alumina silicate hydrates. In many ways, the pozzolanic reaction of HRM is similar to that of fly ash; however, finer particle size and higher surface area of HRM enable it to react much faster and more frequently (www.flyash.com). HRM is available in two forms: (a) *HRM Clinker*: A metakaolin 'clinker' (20 mm to 25 mm in size) with good handling ability and resistance to deterioration. This product can be shipped in bulk and is ideal for inter-grinding with PC clinker for the production of blended cement (Type IP). (b) *HRM Powder*: This is produced by fine grinding the clinker and is sold in bags or in bulk.

This cementitious product supplements the binding action in concrete. It provides increased density, resulting in reduced porosity and permeability and increased chemical resistance. Use of HRM produces concrete with strength and durability similar to concrete made with silica fume. Other benefits of using HRM include: (a) increased compressive and flexural strength, (b) reduced degradation caused by sulphate attack or the ASR, and (c) while silica fume is usually dark grey or black in colour, high-reactivity metakaolin is usually bright white in colour, making it the preferred choice for architectural concrete where appearance is important.

### 5.9.6 Rice Husk Ash

Because rice husk ash (RHA) is available in many parts of the world, and in large quantities, it is an interesting by-product for use as supplementary cementing material in concrete. India produces about 125 million tonnes of paddy and 30 million tonnes of rice husk. RHA is produced by burning rice husk in controlled temperature, without causing environmental pollution. RHA is slightly blackish due to the presence of unburned carbon. Burned rice husk results in 14–20% RHA by weight. RHA consists of approximately 85–90% silica by weight and minor amounts of other elements with a relatively low loss on ignition value (2.1%). The high silica content is in the form of non-crystalline or amorphous silica. RHA exhibits high pozzolanic characteristics (It is classified as an artificial pozzolan.).The reactivity of RHA as pozzolanic material depends on the crystalline/amorphous ratio and varies with the source of RHA. The specific surface area of RHA is generally about 40,000–80,000 $m^2$/kg. The relative density of RHA is in the range of 2.0 to 2.05 and the mean grain size of RHA particles is about 3.8 µm (Genesan et al., 2008). The quality of RHA depends on various heating conditions such as temperature, rate of heating, and soaking time.

More details about mineral admixtures may be found in Bapat (2012).

## 5.10 Types of Cements Produced in India

The following are the various types of cements generally available in the market:

1. OPC - 33/43/53 Grade (IS 269/IS 8112/IS 12269)
2. PPC - fly ash based (as per IS: 1489 part-1- 1991- Reaffirmed 2005)
3. PPC - calcined clay based (as per IS: 1489 part-2 1991 - Reaffirmed 2005)
4. PSC (as per IS: 455 1989 - Reaffirmed 2005)

### 5.10.1 Portland Pozzolana Cement

As mentioned already, the Romans and Greeks were aware that the addition of volcanic ash results in better performance of concrete. Artificial pozzolans such as fly ash, ground granulated blast-furnace slag, silica fume, and natural pozzolans, such as calcined shale, calcined clay, or metakaolin, are used in conjunction with PC, to improve the properties of the hardened concrete. The generally used pozzolanic materials in India are fly ash (IS 1489-Part 1) or calcined clay (IS 1489-Part 2).

PPC (PPC-IS 1489 Part 1) based on fly ash is manufactured either by intimately inter-grinding PC clinker and fly ash [conforming to IS 3812 (Part 1)] with addition of gypsum or calcium sulphate or by intimately and uniformly blending with OPC (conforming to IS 269) and fine fly ash with addition of ground gypsum, if required. The latest amendment 3 to IS 1489 or the draft code released in 2014 requires that the fly ash constituent should not be less than 15% and not more than 35% by mass of Portland Pozzolana cement. Not more than a total of 1.0% of air-entraining agents or other agents including colouring agents is permitted to be added. The code limits the addition of performance improvers such as granulated slag or limestone to 5% only. For the chemical and physical requirements for PPC, IS 1489 (Part 1) should be consulted. It is particularly useful in marine and hydraulic construction and other mass construction structures. This cement is equivalent to OPC on the basis of the 3, 7, and 28 days compressive strength. Until now, in India, PPC is considered equivalent to C-33 OPC. UltraTech PPC, Suraksha, and Jaypee Cement (PPC) are some of the brand names of PPC in India.

PPC (PPC-IS 1489-Part 2) based on calcined clay is manufactured either by intimately inter-grinding PC clinker and calcined clay with the addition of gypsum. It can also be manufactured by intimately and uniformly blending with PC (conforming to IS 269) and fine calcined clay with the addition of ground gypsum, if required. Pozzolana used in the manufacture of calcined clay-based PPC should be either calcined clay conforming to IS 1344 or a mixture of calcined clay and fly ash conforming to IS 3812 (Part 1). The pozzolanic constituent should not be less than 10% and not more than 25% by mass of PPC. The code limits the addition of performance improvers such as granulated slag or limestone to 5% only. For the chemical and physical requirements for PPC, IS 1489 (Part 2) should be consulted. This type of PPC is not being manufactured currently in India by cement manufacturers because of non-availability of good quality calcined clay and huge availability of good quality of fly ash in thermal power stations.

PPC offers the following advantages:

1. It is economical than OPC, as the costly clinker is replaced by cheaper pozzolanic material.
2. It converts soluble calcium hydroxide into insoluble cementitious products, thus improving permeability, and durability.
3. It consumes calcium hydroxide and does not produce calcium hydroxide as much as that of OPC.
4. It improves the pore size distribution and reduces the micro cracks at the transition zone due to the finer particles than OPC.
5. It reduces the heat of hydration and thermal cracking.
6. It has a high degree of cohesion and workability in concrete and mortar.

The main disadvantage is that the rate of development of strength is initially slightly slower than OPC. In addition, its effect of reducing the alkalinity may reduce the resistance to corrosion of steel reinforcement. However, as the PPC significantly improves the permeability, the risk of corrosion is reduced. The setting time is slightly longer.

## 5.10.2 Portland Slag Cement

Portland slag cement (PSC) (as per IS 455:1989) is manufactured either by intimately inter-grinding a mixture of PC clinker and granulated slag with the addition of gypsum (natural mineral or chemical). It can also be manufactured by an intimate and uniform blending of PC (conforming to IS 269) and finely ground granulated slag (see Section 5.9.3). The granulated slag used in PSC should conform to IS 12089. Amendment No. 4 of IS 455 and the draft requires that the slag content should be neither less than 25% nor more than 70% of the PSC. Not more than a total of 1.0% of air-entraining agents or other agents (including colouring agents) are permitted to be added. The code limits the addition of performance improvers such as granulated slag or limestone to 5% only. For the chemical and physical requirements for PPC, IS 455:1989 should be consulted. PSC has physical properties similar to those of OPC.

PSC has the following advantages:

1. Utilization of slag cement in concrete not only lessens the burden on landfills, but also conserves a virgin manufactured product (OPC), and decreases the embodied energy required to produce the cementitious materials in concrete. Embodied energy can be reduced by 390 to 886 mJ with 50% slag cement substitution. This is a 30 to 48% reduction in embodied energy per cubic meter of concrete (http://www.slagcement.org).

2. Between 98 and 222 kg of $CO_2$ are saved per cubic meter of concrete by using a 50% slag cement substitution, a 42 to 46% reduction in greenhouse gas emissions (http://www.slagcement.org). Table 5.12 shows the environment benefits comparison of slag cement and fly ash in a typical M 20 concrete.

3. Using slag cement to replace a portion of PC in a concrete mixture is a useful method to make concrete better and more consistent. PSC has a lighter colour, better concrete workability, easier finishability, higher compressive and flexural strength, lower permeability, improved resistance to aggressive chemicals, and more consistent plastic and hardened consistency.

4. The lighter colour of slag cement concrete also helps reduce the heat island effect in large metropolitan areas. It has low heat of hydration and is relatively better resistant to soils and water containing excessive amounts of sulphates and hence used for marine works, retaining walls, foundations, and in mass concreting.

5. Both PPC and PSC will give more strength than that of OPC at the end of 12 months. UltraTech Premium, Super Steel (Madras Cement), and S 53 (L&T) are some of the brand names of PSC available in India.

**Table 5.12** Environmental benefits comparison of slag cement and fly ash in M20 concrete
(*Source*: http://www.slagcement.org)

| Environmental benefit (Substitution rate for OPC) | Slag cement (35%) | Slag cement (35%) | Fly ash (20%) |
| --- | --- | --- | --- |
| $CO_2$ emissions savings* | 30% | 43% | 17% |
| Energy savings* | 21% | 30% | 14% |
| Reduction in extracted material* | 5% | 7% | 3% |

*Percentages listed for savings are based on concrete with 100% OPC compared with concrete containing slag cement or fly ash substitution.

## 5.10.3 Blended Cements

Mixtures using three cementitious materials, called ternary mixtures, are becoming common, but no Indian specification has been developed yet. Blended cements are produced by blending ground PC and other SCMs or inter-grinding SCMs with cement clinker at the finish mills in tightly controlled proportions. They often have enhanced properties. As discussed earlier, the SCMs will chemically interact with the hydration products of PC, resulting in concrete with reduced permeability and increased durability.

Use of blended cements can mitigate ASR in concrete containing reactive aggregates. Blended cements can also be proportioned to increase the sulphate resistance in situations where the concrete is exposed to soils or water having high sulphate content. It is also possible to lower the heat of hydration of mass concrete, used in dams or large foundations, using a proper mix of blended cement. The gypsum content of blended cement could be optimized to produce high-performance cementitious materials.

Blended cements should comply with the provisions of ASTM C 595, Standard Specification for Blended Cements. Type IP(X) denotes pozzolana blended cement with fly ash or natural pozzolana such as calcined clay and Type IS(X) denotes slag blended cement, where (X) is the% of SCM in the blend. For example, Type IS(20) denotes a blended cement with 20% slag.

The different types of cements covered by Indian and American standards and their chemical compounds are shown in Table 5.13.

**Table 5.13** Types of Portland cements

| India/The UK | The USA (ASTM) | Typical compounds[3] |
|---|---|---|
| OPC (IS 269, IS 8112 and IS 12269) | Type I[1] | $C_3S$- 55%, $C_2S$-19%, $C_3A$- 10%, $C_4AF$- 7%, MgO-2.8%, $SO_3$-2.9%, Ignition loss- 1.0%, and free CaO- 1.0%. [$C_3A$ < 15%] |
| – | Type II[1] | $C_3S$- 51%, $C_2S$- 24%, $C_3A$- 6%, $C_4AF$- 11%, MgO-2.9%, $SO_3$-2.5%, Ignition loss- 1.0%, and free CaO- 1.0%. [$C_3A$ < 8%] |
| Rapid hardening Portland Cement (IS 8041:1990) | Type III[1] | $C_3S$- 57%, $C_2S$- 19%, $C_3A$- 10%, $C_4AF$- 7%, MgO- 3.0%, $SO_3$-3.1%, Ignition loss- 0.9%, and free CaO- 1.3%. Its 7-day compressive strength is almost equal to types I and II 28-day compressive strengths. |
| Low heat Portland Cement (IS 12600:1989) | Type IV | $C_3S$- 28%, $C_2S$- 49%, $C_3A$- 4%, $C_4AF$- 12%, MgO- 1.8%, $SO_3$-1.9%, Ignition loss- 0.9%, and free CaO- 0.8%. [$C_3A$ < 7 % and $C_3S$ < 35%] |
| Sulphate resisting Portland Cement (SRC) (IS 12330:1988) | Type V | $C_3S$- 38%, $C_2S$- 43%, $C_3A$- 4%, $C_4AF$- 9%, MgO- 1.9%, $SO_3$-1.8%, Ignition loss- 0.9%, and free CaO- 0.8%. [$C_3A$ < 5% and $(C_4AF) + 2(C_3A) < 25\%$] |
| PSC (IS 455:1989) | Type IS | Made by grinding granulated high-quality slag with Portland cement clinker. |
| PPC [IS 1489-Part I:1991(fly ash based), IS 1489-Part II:1991 (calcined clay based)] | Type IP | Blended cement made by inter-grinding Portland cement and pozzolana without burning. |
| Ternary blended cement | Type IT(SX) (P Y)[2] | Blended cement made by inter-grinding Portland cement, slag, and pozzolana without burning. |

**Note:**

1. Types Ia, IIa, and IIIa have the same composition as types I, II, and III, but have an air-entraining agent ground into the mix.
2. The letters X and Y stand for the percentage of supplementary cementitious material (SCM) included in the blended cement, and S and P are the Types of SCMs, S for slag or P for pozzolana. For example, Type IT(S25)(P15) contains 25% slag and 15% pozzolana.
3. See Table 5.1 for the explanation of these compounds.

## 5.10.4 Special-purpose Cements

There are other types, such as high alumina cement, supersulphated cement, hydrophobic PC, white cement, concrete sleeper grade cement (IRS-T 40:1985, now designated as OPC 53-grade S), expanding cements, masonry cement, and oil-well cement (IS 8229:1986), which are used only in some special situations (Mehta and Monteiro, 2006 and Shetty, 2005). Details of some of these cements are provided in Table 5.14. Geopolymer cements are inorganic hydraulic cements that are based on the polymerization of minerals (see Section 5.11.1).

**Table 5.14** Other special cements

| Types of cement | Composition | Comments |
|---|---|---|
| Rapid hardening cement also called high early strength cement (IS 8041:1990) | Increased $C_3S$ content (>55%) and fine grinding (specific surface > 325 m²/kg) | Attains high strength at an early age, with higher rate of heat development during hydration. Used in places where form work has to be removed quickly, that is, precast construction, cold weather concreting, road/bridge repairs. Should not be used in mass concreting or thick structural members. |
| Quick setting cement | Small percentage of aluminium sulphate as accelerator and reduced percentage of gypsum with fine grinding of OPC | Setting starts within 5 m of addition of water and setting completes in 30 min. Used in places where the work has to be completed in a very short period and concreting in static or running water. |
| Low heat Portland cement (IS 12600:1989) | Produced by increasing $C_2S$ content and reducing $C_3S$ and $C_3A$ content with fine grinding (specific surface > 320 m²/kg) | Slow development of strength but ultimate strength is similar to OPC. Initial setting time is 1 h and final setting time is 10 h. Used in mass concrete constructions like large foundations, retaining walls, bridge piers, and gravity dams. Heat generated at the end of 7 days-272 kJ/kg compared to OPC of 350 kJ/kg. |
| Sulphate resisting Portland cement (SRC) (IS 12330: 1988), e.g., Birla Coastal | Low $C_3A$ (<5%) and $C_4AF$ (6–12%) content, $2(C_3A) + C_4AF < 25\%$, and high $SiO_2$ content. | Used in places where structural elements may be exposed to sulphate attack in soils or water such as canal linings, foundations, piles, and culverts, retaining walls, sewage pipes. |
| White cement (IS 8042:1989), e.g., Birla White | Prepared from raw materials free from iron oxides and manganese | Expensive and is finer than grey cement. Has high strength and sets rapidly. Used for architectural purposes such as precast curtain walls, mosaic tiles, pointing work, and rendering of walls. |
| Coloured cement | Produced by mixing up to 10% mineral pigments with OPC/white cement | Used for decorative purposes mainly in floors |
| Air-entraining cement | It is produced by adding air-entraining agents such as natural wood resins, synthetic detergents, sulfonated lignins, petroleum acids, and animal/vegetable fats during the grinding of clinker of OPC or slag cements. | This type of cement is especially suited to improve the workability with smaller water-cement ratio and improve frost and sulphate resistance of concrete. |
| Hydrophobic cement (IS 8043:1991) | It is obtained by adding water repelling chemicals such as oleic acid, naphthenic acid, stearic acid, pentachlorophenol, or vegetable oils to OPC during grinding of cement clinker (Justnes, 2009). | This cement has high workability and strength. Useful in places having high humidity, marine concrete, tunnel works, foundations below water table, swimming pools, green roofs, parking structures, and rooftops. |
| Masonry cement (IS 3466:1988) | It is obtained by intimately grinding a mixture of OPC clinker and gypsum with Pozzolana and non-pozzolanic materials such as limestone, and air-entraining plasticizer in suitable proportions, to fineness greater than that of OPC. | Masonry cement is intended for use in masonry mortars for brick, stone, and concrete block masonry, and for rendering and plastering work. When mixed with fine aggregates (sand), it results in a smooth, plastic, cohesive and strong, yet workable, mortar. Masonry cement is considered superior to lime mortar, lime-cement mortar or cement mortar. Masonry cement should not be used in structural concrete, for flooring and foundation work or reinforced and prestressed concrete work. |

## 5.10.5 Portland Limestone Cement

In recent years, another type of blended cement is gaining popularity in North America, Type IL. It is cement containing ground limestone as a partial replacement for PC clinker. The amount of clinker replaced is typically more than 5% and may be as high as 35%. One of the advantages of Portland limestone cement (PLC) is that it has a lower carbon footprint than OPC due to its reduced clinker content and, thus, contributes to sustainable concrete construction. Cements blended with up to 35% ground limestone have been used for many years in other parts of the world. Contrary to the European standards for cement, the ASTM and CSA specifications both limit the quantity of limestone in PLC to 15% (Hawkins et al., 2003).

## 5.11 Novel, Resource-efficient Cements

A number of low-carbon or carbon-negative cements have been developed recently by a few companies. The mechanical properties of these cements appear to be similar to those of PC. However, they are currently neither proven to be economically viable nor tested for their long-term suitability. Nor have their products been accepted in the construction industry where strong material and building standards exist. However, they may have an impact on the future cement industry. These cements may contain geopolymers, activated glassy cements, hydraulic fly ash cements, activated slag cements, calcium aluminate cements, calcium sulfoaluminate cements, magnesia-based cements, or $CO_2$-cured cements. Although geopolymers, activated glassy cement, and hydraulic fly ash cements may contain fly ash, they have distinctive differences. Geopolymers are typically produced from low-calcium content fly ash and are activated by alkali hydroxides, sodium silicate, or both with water. Activated glassy cement is typically produced from high-calcium fly ash and is activated with a high-pH activator coupled with an organic acid and water. Finally, hydraulic fly ash cement is typically produced from high-calcium fly ash and activated with a pH-neutral activator coupled with a retarder and water.

Only geopolymer cement and self-cleaning cement are described below. Details of other cements may be found in the chapter references listed in the ORC.

### 5.11.1 Alkali-activated/Geopolymer Cement

Joseph Davidovits, a French materials scientist, in 1978, invented *geopolymer concrete*, which is completely different from the normal concrete in which PC glues together the other ingredients of concrete. Geopolymer concrete requires two sources to bind the ingredients: alumina silicates and alkaline activator. When the two are mixed together with other ingredients, a chemical reaction occurs and results in the creation of a strong concrete. According to the Geopolymer Institute, there are nine different classes of geopolymers. These geopolymers are produced with some natural materials (e.g., kaolinite clay) or industrial by-products (e.g., fly ash or slag), which are the sources for silicon (Si) and aluminium (Al). An additional source of silica (sodium silicate) is also most commonly used. These are then dissolved in an alkaline activator solution [sodium hydroxide (NaOH) and potassium hydroxide (KOH)] and subsequently polymerized, and are often referred to as *alkali-activated cements* or *inorganic polymer cements*. The performance of such a system is dependent on the chemical composition of the source materials, the concentration of alkaline activators, and the concentration of soluble silicates (Rangan, 2008).

Although the mechanism of polymerization is yet to be fully understood, it is important to realize that water is present only to facilitate workability and does not become a part of the resulting geopolymer structure. Thus, water is not involved in the chemical reaction and instead is expelled during curing and subsequent drying. This is in contrast to the hydration reactions that occur when PC is mixed with water, which produce the primary hydration products such as C-S-H and calcium hydroxide [$Ca(OH)_2$]. This difference has a significant impact on the mechanical and chemical properties of the resulting geopolymer concrete, making it more resistant to heat, water ingress, alkali-aggregate reactivity, and other types of chemical attack.

As mentioned earlier, in geopolymers based on alumino-silicate, the used materials, i.e., kaolinite clays, Class-F fly ash or ground granulated blast furnace slag, must be rich in amorphous forms of Si and Al. Otherwise, an additional source of silica (usually sodium silicate) should be used. When geopolymers are made from fly ash, the amount of calcium must be controlled carefully, because its presence can result in flash setting (Lloyd and Rangan 2009).

The curing temperature is very important, and depending upon the source materials and activating solution, heat often must be applied to facilitate polymerization, although a few systems have been developed that can be cured at room temperature. Although several patented geopolymer systems have been proposed, most of them require great care in their production. In addition, there is a safety risk associated with the high alkalinity of the activating solution; high alkalinity also requires more processing, resulting in increased energy consumption and generation of greenhouse gases. In addition, the polymerization reaction is very sensitive to temperature and usually requires that the geopolymer concrete be cured at elevated temperature (60 to 150°C, for the initial few days) under a strictly controlled temperature regime (Lloyd and Rangan, 2009). As a result, geopolymer cement concrete has had limited acceptance in the construction industry. During 2008, a company called Zeobond launched commercial production of geopolymer concrete in Melbourne, Australia, under the brand name E-Crete™. It is interesting to note that structures using similar geopolymer concretes were constructed in ancient Rome, as well as in the former Soviet Union in the 1950s and 1960s, and are still in service.

## 5.11.2 Self-cleaning Cement

In the early 1990s, scientists at the Italcementi Group in Bergamo, Italy, produced a self-cleaning concrete that keeps buildings clean from pollutants in the atmosphere. They made the concrete by adding nano-sized particles of white pigment titanium dioxide to the cement component. When light and heat strike the concrete's surface, titanium oxides use that energy to break down the dirt into molecules such as oxygen, water, carbon dioxide, nitrates, and sulphates. Gases float away, whereas liquids or solids are left on surface to be washed away by rain. The reactive material can kill bacteria and fungi as well as break down pollutants. Thus, buildings stay cleaner and do not require any chemical treatment that may be potentially harmful to the environment. Maintenance costs are also reduced.

Although the technology had been used for over a decade in other materials such as tiles and glasses, the first application of this technology in concrete was on the Dives in Misericordia Church in Rome – the architect wanted a building material that would stay bright. Italcementi scientists, in response to the challenge, developed this cement having photocatalysts. Photocatalytic concrete has received increased attention due to the current interest in sustainability and environmental issues. Buildings in highly polluted locations can benefit from this technology as is seen in the Air France headquarters at Roissy–Charles de Gaulle International Airport near Paris; this white concrete building has remained white for several years.

There are two major cement companies distributing this type of cement in the United States. Italcementi, the Italian parent company of Essroc Cement, originally brought this product to market. They have recently licensed Lehigh White Cement Company to produce and market the material in Europe and North America. Brand names for the white cement are TX Active® and Tiocem.

## 5.12 Storage and Handling of Cement

In any construction site, it may be necessary to keep a good stock of cement at site. The cement should always be stored in such a way that it is available for ready inspection. Cement is very finely ground and readily absorbs moisture; hence, care should be taken to ensure that the cement bags are not in contact with moisture. The cement storage shed should be provided with airtight doors and windows. The walls must be plastered and made damp proof and the roof must be given an appropriate water proofing treatment. The floor must be raised by at least 800 mm above the ground level to prevent any inflow of water. The flooring may consist of a solid cement concrete floor or 150 mm thick layer of dry bricks laid in two courses over a layer of earth consolidated to a thickness of 150 mm above the ground level. A newly constructed storage shed/godown should not be used for the storage of cement unless its interior is thoroughly dry. For working out the inside dimensions of a cement shed, it has to be noted that the average length, width, and thickness of cement bags are 700 mm, 350 mm, and 140 mm, respectively.

The cement bags should be kept on a raised wooden platform (at least 150 mm above ground level). Bags should be stacked close together to reduce air circulation but should never be stacked against outside walls. A space of 600 mm all-round should be left between the exterior walls and the stack of cement. In between individual piles, a passage of 600 mm may be provided for easy access. The number of bags in one pile should not usually be more than 10 to avoid the lumping of cement under pressure. In stacks more than 5 bags high, the cement bags should be arranged alternately length-wise and cross-wise, so as to tie the stacks together and minimize the danger of toppling over. They should be stored in such a way that the bags that come in first are the first to go out. For this reason, each consignment as it comes should be stacked separately and a placard indicating the date of arrival should be pinned into the pile. When cement bags are to be stored for a long period or during rainy season, the stack should be enclosed completely in 700 gauge polythene sheet, tarpaulin, or any other suitable water-proofing material. Different types of cement must be stacked and stored separately. Cement stored for a long time tends to lose its strength (loss of strength ranges from 10–20% in 3 months to 40–50% in 1 year). It is better to use the cement within 90 days of its production. In case it is used at a later date, it should be tested before use.

Bulk cement should be stored in weather-tight concrete or steel bins or silos. Vibration or dry, low-pressure aeration should be used to keep the cement free-flowing and prevent arching or bridging inside the silo, which can get clogged. In addition, silos should be drawn down every few months in order to prevent the caking of older cement around the perimeter of the silo.

PFA (IS 3812-Parts 1 & 2) should be stored in accordance with the recommendation given in IS 4082. Additionally, during bulk storage, the fly ash should be suitably covered to avoid getting airborne. Although moisture will not affect the performance of some SCMs, they should be protected from moisture nonetheless to reduce problems with handling. Cementitious materials can cause allergic reactions upon skin contact and may irritate skin, eyes, and lungs; thus, handlers should wear dust mask, goggles, and protective gloves.

## 5.13 Uses of Cement

Cement is used as an important binding material in construction. There are numerous applications of cement as a building material. Some of them are listed below:

**Used as a mortar and plaster** Cement, mixed with water and fine aggregate (sand), is used for (a) binding the various types of masonry units such as bricks, stones, and hollow cement blocks; (b) pointing in masonry joints; (c) plastering the surfaces of masonry for protecting it from weather and providing decorative finish; (d) as filler material in ferrocement units; (e) in damp-proof course below ground level, (f) water proofing of roofs (some chemicals are added to the mortar); (g) as topping in cement concrete floors; and (h) repairing cracks in concrete elements/structures.

**Making concrete components**    In concrete, cement binds the components of concrete such as coarse and fine aggregates into a solid mass. Concrete is used for (a) the construction of building components such as foundations, columns, beams, slabs, and stairs; (b) the construction of the various structures such as dams, bridges, chimneys, silos, tanks, tunnels, and pavements; (c) the manufacture of precast concrete products such as lintels, pipes, piles, fencing posts, and the various components of buildings; and (d) the manufacture of solid and hollow concrete blocks.

**Decorative applications**    White cement is used as a decorative material in mosaic floorings, floor tiles, cement paints, and pointing in stone masonry and floor tiles. In addition, concrete can be transformed into decorative concrete through the use of variety of materials that may be applied during the pouring process or after the concrete is cured. These materials and/or systems include stamped concrete, acid staining, decorative overlays, and polished concrete and concrete countertops. More information of *decorative concretes* can be gathered from http://www.decorativeconcreteinstitute.com/. One such application in the construction of a polymer cement overlay to change asphalt pavement to brick texture and colour to create decorative pedestrian area is shown in Fig. 5.7. It shows a construction worker using a propane torch to speed cure the coloured polymer cement slurry that has been applied over a brick stencil to create a decorative crosswalk. The steps of crosswalk construction include spraying a layer of grey polymer cement slurry to the crosswalk area, placing brick pattern stencil and spraying terra cotta tinted slurry (notice the grey and terra cotta colours that show brick appearance). Then, heat is applied to speed cure the surface.

**Fig. 5.7** Construction of a polymer cement overlay to change asphalt pavement to brick texture and colour to create decorative crosswalk (*Source*: Wikipedia)

## SUMMARY

- Cement is one of the most used binding materials in building and civil engineering construction. The cement that we use now had undergone a number of changes, from the early cements used by Romans.
- Modern cement is based on the development by John Smeaton (1756) and the patent of Joseph Aspdin (1824).
- Cement is a powdery material, with an average particle size of 10μm, manufactured from limestone and clay, and has adhesive properties.
- When mixed with water, it undergoes chemical reactions yielding gel-like materials with a high surface area, resulting in setting and hardening and binding other materials such as aggregates and bricks together. Because of its hydrating properties, cement is often called hydraulic cement.
- PC (referred to as OPC) is the most common type of hydraulic cement in use. It is used in mortar, plaster, and also in the production of concrete.
- Of the three grades (C-33, C-43, and C-53) of cement, only C-53 is freely available in the market. Although a higher percentage of cement in mortar, plaster, or concrete may give higher strength, it may lead to shrinkage and cracking.
- Most of the modern cement factories use either dry or semi-dry process. Cement clinker is produced by heating limestone, marl/chalk, and small quantities of iron ore, bauxite, shale, clay/sand, in rotating kilns for about 1,450°C. It is ground to powder, after adding with a small quantity of gypsum (3–5 %) to produce cement. In India, cement is usually available in 50 kg bags.
- The three main chemical ingredients of PC are lime, silica, and alumina. When heated in kilns, they fuse to form Bogue's compounds such as tricalciumsilicate ($C_3S$), dicalciumsilicate ($C_2S$), tricalciumaluminate ($C_3A$), and tetracalciumaluminoferrite ($C_4AF$). The silicates $C_3S$ and $C_2S$ are the most important compounds and are mainly responsible for the strength of the cement paste.

- The important physical properties of cements are fineness (expressed as specific surface), setting time, soundness, and compressive strength.
- There are a number of chemical requirements for cement. The reaction of cement with water is called hydration. OPC containing increased percentage of $C_3S$ will have high early strength; $C_2S$ contributes to strength increase only after about 7 days. $C_3A$, which reacts with gypsum, is responsible for the large amount of heat of hydration during the first few days of hydration. The greatest rate of heat of hydration occurs within the first 24 h, and hence curing of concrete is important during this time. For the complete hydration of PC, w/c ratio of 0.38 is needed.
- Several physical tests (such as for fineness, soundness, consistency, initial and final setting, compressive and tensile strength, heat of hydration, and specific gravity) and chemical tests may be conducted on cement to ascertain its quality, ingredients, and properties, based on IS 4031 (different parts), and IS 4032 codes (These tests are explained in Chapter 25).
- Use of SCMs, such as surkhi, fly ash, GGBS, silica fume, calcined clay, RHA, and metakaolin, as a replacement of PC, result in improved properties of concrete. As the production of cement contributes to about 5% of global green-house gas emissions, the use of SCMs will also reduce global warming potential.
- A variety of cements is available and the main types are three grades of OPC, PPC, and PSC. Blended cements can also be used advantageously. A number of low-carbon or carbon-negative cements, such as calcium sulfoaluminate cement, calcium aluminate cement, supersulphated cement, and geopolymer cement, have been developed recently.
- Cement should be stored in such a way that contact with moisture is avoided.

# EXERCISES

## Multiple-choice Questions

1. World's highest producer of cement in 2015 is
   - (a) the USA
   - (b) China
   - (c) India
   - (d) Brazil
2. The temperature range in a cement kiln is
   - (a) 500 to 1000°C
   - (b) 1000 to 1200°C
   - (c) 1300 to 1500°C
   - (d) 1600 to 2000°C
3. The three main chemical ingredients of Portland cements are
   - (a) Lime (CaO), silica ($SiO_2$), and alumina ($AL_2O_3$)
   - (b) Lime (CaO), sulphur trioxide ($SO_3$), and alumina ($AL_2O_3$)
   - (c) Lime (CaO), silica ($SiO_2$), and sulphur trioxide ($SO_3$)
   - (d) Sulphur trioxide ($SO_3$), magnesia (MgO), and alumina ($AL_2O_3$)
4. Composition of lime in cement is
   - (a) 41–47%
   - (b) 51–57%
   - (c) 61–67%
   - (d) 71–77%
5. Composition of tricalcium silicate ($C_3S$) in clinker is
   - (a) 25–45%
   - (b) 35–55%
   - (c) 45–65%
   - (d) 55–65%
6. Minimum fineness (m²/kg) of OPC should be
   - (a) 225
   - (b) 300
   - (c) 350
   - (d) 400
7. Minimum initial and final setting times (minutes) of OPC should be
   - (a) 30 and 600
   - (b) 60 and 600
   - (c) 30 and 500
   - (d) 60 and 500
8. The amount of gypsum in cement manufacture is strictly restricted to
   - (a) 2–3%
   - (b) 3–5%
   - (c) 4–6%
   - (d) 5–7%
9. Compressive strength of OPC is tested typically on
   - (a) 1st, 7th, and 28th day
   - (b) 1st, 7th, and 90th day
   - (c) 3rd, 7th, and 56th day
   - (d) 3rd, 7th, and 28th day
10. The water-cement ratio (by weight) % required for the complete hydration of cement is about
    - (a) 15
    - (b) 23
    - (c) 38
    - (d) 40
11. To produce low heat cements, it is necessary to reduce the compound
    - (a) $C_3S$
    - (b) $C_2S$
    - (c) $C_3A$
    - (d) $C_4AF$
12. Increase in the fineness of cement
    - (a) Reduces the rate of strength development and leads to higher shrinkage
    - (b) Increases the rate of strength development and reduces the rate of deterioration
    - (c) Decreases the rate of strength development and increases the rate of deterioration
    - (d) Increases the rate of strength development and leads to higher shrinkage
13. Minimum fineness of Grade I fly ash as per IS 3812 is (m²/kg)
    - (a) 220
    - (b) 320
    - (c) 350
    - (d) 400

14. Condensed silica fume has a surface area of (m²/kg)
    (a) 5,000–10,000
    (b) 8,000–15,000
    (c) 15,000–25,000
    (d) 13,000–30,000

15. For mass concrete works such as dams, the best suited cement is
    (a) low heat Portland cement
    (b) rapid hardening cement
    (c) ordinary Portland cement
    (d) blast furnace slag cement

16. Match the type of cement with characteristics

| (Type of cement) | (Characteristics) |
| --- | --- |
| (a) Ordinary Portland cements | (i) the amount of $C_3S$ is maximum (about 57%) |
| (b) Rapid hardening cements | (ii) the amount of $C_2S$ and $C_3S$ are the same (about 40%) |
| (c) Low heat cements | (iii) the amount of $C_3S$ is about 50–55% |
| (d) Sulphate resistant cements | (iv) the amount of $C_2S$ is maximum (about 49%) |

17. Match the type of cement with characteristics

| (Type of cement) | (Characteristics) |
| --- | --- |
| (a) High early strength cement | (i) should not be used with any Pozzolana |
| (b) Sulphate resisting Portland cement | (ii) is extremely resistant to chemical attack |
| (c) High alumina cement | (iii) gives higher rate of heat development during the hydration of cement |
| (d) Rapid hardening Portland cement | (iv) has a higher content of tricalcium silicate |

18. Match the type of cement with uses

| (Type of cement) | (Uses) |
| --- | --- |
| (a) Rapid hardening | (i) green roofs |
| (b) Quick setting | (ii) dams |
| (c) Hydrophobic cement | (iii) concrete under water |
| (d) Low heat | (iv) repair of bridges |

## Review Questions

1. What are hydraulic and non-hydraulic cements?
2. What are the ingredients of Portland cement?
3. What are the three different grades of cements used in India? How is the grade of cement fixed?
4. What are the processes by which modern cement is made? Explain the dry process of cement manufacture.
5. What are the three main chemical ingredients of Portland cement?
6. What is the purpose of adding gypsum while manufacturing cement?
7. Why do modern-day cements have high early strength?
8. State the function and limits of lime, silica, and alumina in cement.
9. What are the four principal compounds of cement clinker? How do they affect different properties of concrete?
10. Write short notes on the following:
    (a) Fineness of cement
    (b) Soundness
    (c) Setting times
    (d) Compressive strength of cement
11. How does the fineness of cement affect cement?
12. What are the initial and final setting times of OPC? What is their importance?
13. How is lime saturation factor calculated?
14. What tests would you specify to ensure if the cement supplied at the site is of good quality?
15. Describe the process of hydration of cement.
16. What are the factors influencing heat development in concrete during hydration?
17. Describe the five stages of rate of heat evolution during the hydration of cement.
18. How can we increase the rate of hydration of cement?
19. What is the water-cement ratio required for complete hydration? What are the drawbacks of using excess water-cement ratio?
20. What are the precautions to be taken while storing cement?
21. State the different uses of cement.
22. What are the supplemental cementitious materials used in cement?
23. What are the advantages of using mineral admixtures such as fly ash in cement?
24. Name any three mineral admixtures used in concrete.

25. Write short notes on the following:
    (a) *Surkhi*
    (b) Fly ash
    (c) Ground Granulated Blast furnace slag (GGBS)
    (d) Silica fume
    (e) Metakaolin
    (f) Rice Husk Ash
26. What are the major cements manufactured in India?
27. How is PPC manufactured? What are the advantages of the PPC?
28. How is PSC manufactured? What are the advantages of the PSC?
29. What is blended cement?
30. What is rapid hardening cement? What is responsible for its high early strength? How does it differ from ordinary Portland cement?
31. State the conditions under which you will recommend the following cements. Give also the reasons.
    (a) Pozzolana cement
    (b) Low heat Portland cement
    (c) Air-entraining Cement
    (d) Rapid hardening cement
    (e) Quick setting cement
32. What is the effect of mixing lime in cement to produce Portland limestone cement?
33. Write short notes on the following:
    (a) White cement
    (b) Sulphate nesisting cement
    (c) Hydrophobic cement
    (d) Masonry cement
    (e) Geopolymer cement
    (f) Self-cleaning cement

# ANSWERS

**Multiple-choice Questions**

| | | | | | | |
|---|---|---|---|---|---|---|
| **1.** (b) | **2.** (c) | **3.** (a) | **4.** (c) | **5.** (c) | **6.** (a) | **7.** (a) |
| **8.** (b) | **9.** (d) | **10.** (c) | **11.** (a) | **12.** (d) | **13.** (b) | **14.** (d) |
| **15.** (a) | **16.** (a)–(iii), (b)–(i), (c)–(iv), (d)–(ii) | | | **17.** (a)–(iv), (b)–(ii), (c)–(i), (d)–(iii) | | |
| **18.** (a)–(iv), (b)–(iii), (c)–(i), (d)–(ii) | | | | | | |

# CHAPTER 6
# AGGREGATES

## 6.1 Introduction

The term 'aggregates' represents a broad category of particulate material used in concrete, which includes sand, gravel, crushed stone, slag, recycled concrete, and geosynthetic aggregates. Aggregates may be classified based on geological origin (natural or artificial), size (fine and coarse), shape (rounded, irregular, angular, and flaky), and unit weight (normal, heavy, and light). The coarse aggregates form the main matrix of concrete and the fine aggregates (and also cement and cementitious materials) usually fill the voids created by coarse aggregate in the volume of concrete. Aggregates, especially the coarser ones, provide strength to the concrete. In addition, aggregates provide volume stability and durability to the hardened concrete. Aggregates are the most mined materials in the world. As per Satish Kumar et al. (2016), India consumed 3330 MT of total aggregates (coarse and fine) in 2015, which may escalate to 5075 MT by 2020. Natural gravel and sand are usually dug or dredged from a pit, river, lake, or seabed. Crushed aggregates are produced by crushing quarry rock, boulders, cobbles, or large-size gravel. After harvesting, aggregates are processed, i.e., crushed, screened, and washed to obtain proper cleanliness and gradation. After processing, the aggregates are handled and stored to minimize segregation and degradation and prevent contamination. The depletion of natural aggregates in several parts of India has resulted in the use of artificial aggregates such as bottom ash from thermal power plants, recycled concrete aggregates, manufactured sand (M-sand), and other industrial wastes, such as foundry wastes, blast furnace slag, copper slag, ceramic waste, waste glass powder, etc.

Aggregates typically occupy 60–75% of the volume (or, 70–85% of the weight/mass) of concrete and strongly influence the properties of fresh and hardened concrete, mixture proportions, and economy. Thus, though aggregates are largely inert, their large proportion and variation in their properties will significantly affect the properties or performance of concrete, such as strength, water demand for a given slump, and green concrete properties, such as cohesiveness, harshness, segregation, bleeding, ease of consolidation, finishability, and pumpability; each of which may not always correlate with slump.

## 6.2 Classification of Aggregates

Aggregates may be classified according to geological region, size, shape, and unit weight. These different classifications are discussed in this section.

## 6.2.1 According to Geological Origin

According to the geological origin, the aggregates may be classified as natural and artificial aggregates.

**Natural aggregates**    As mentioned already, natural aggregates are obtained from quarries, by crushing igneous (e.g., granite, basalt, pumice, gabbro, aplite, dolerite, and rhyolite), sedimentary (e.g., limestone and sandstone), and metamorphic (e.g., granulite, gneiss, schist, and marble) rocks (see Annex C of IS 383–16). Reduced-size gravels and sand produced due to the action of natural agents, such as water and wind, can also be classified under this category. The most widely used aggregates in concrete have igneous origin, as they are normally hard, tough, and dense. Aggregates obtained by crushing sedimentary rocks varies from soft to hard, porous to dense, and light to heavy, based upon their history of formation. They may also yield flaky aggregates. Limestone and some siliceous stones have proven to be good concrete aggregates. Metamorphic rocks exhibit foliated structure; hence, they are carefully examined before being used as aggregates. However, metamorphic rocks, such as quartzite and gneiss, have been found to be satisfactory concrete aggregates.

Sand obtained from river beds are common and are of good quality. Sand obtained from pits or dredged from other places may not be clean or well-graded, and hence may require sieving and washing before they are used in concrete.

**Artificial aggregates**    Examples of artificial aggregates include broken bricks, iron-making and steel-making slags, copper and ferro-nickel slags (collectively termed as *pyrometallurgical slags*), crushed rock-manufactured sand (M-sand or ROBO sand), bottom ash from thermal power stations, foundry wastes, ceramic waste, stone dust, marble powder, municipal solid waste ash, and recycled aggregates. Broken bricks, known as brick bats, are suitable only for low-strength/mass concrete applications and are not used for RCC. Iron and steel slags are air cooled, crushed, and screened. They do not have organic impurities and have uniform chemical composition. More information on artificial aggregates and sintered fly ash coarse aggregates may be found in IS 9142 (Parts 1 and 2):2018. Slag aggregates can be used both as coarse and fine aggregates in mortar and concrete (Satish Kumar et al., 2016). The specific gravity of these ranges from 2.0 to 2.8, and bulk density from 1120 to 1360 kg/m$^3$. The blast furnace slag aggregate has good fire-resisting properties. Although blast furnace slag contains 1–2% of sulphur, it does not present a corrosion risk to steel reinforcement in concrete made with blast furnace slag cement or aggregates (NSA 165–1). As per IS 383–16, manufactured aggregates are permitted to be used in plain and reinforced concrete only as a percent of total mass of fine or coarse aggregates, as shown in Table 6.1. Manufactured aggregates are not permitted in prestressed concrete.

**Table 6.1** Extent of utilization of manufactured aggregates as percent of total mass of fine or coarse aggregate, as per IS 383-2016

| Type of aggregate | Maximum utilization in % | | |
|---|---|---|---|
| | Plain concrete | Reinforced concrete | Lean concrete (<M15 grade) |
| **Course aggregate** | | | |
| Iron slag aggregate | 50 | 25 | 100 |
| Steel slag aggregate | 25 | Nil | 100 |
| Recycled concrete aggregate (RCA) | 25 | 20 (only up to M25 grade) | 100 |
| Recycled aggregate (RA) | Nil | Nil | 100 |

*(Contd)*

**Table 6.1** (*Contd*)

| Type of aggregate | Maximum utilization in % | | |
|---|---|---|---|
| | Plain concrete | Reinforced concrete | Lean concrete (<M15 grade) |
| **Fine aggregate** | | | |
| Iron slag aggregate | 50 | 25 | 100 |
| Steel slag aggregate | 25 | Nil | 100 |
| Copper slag aggregate | 40 | 35 | 50 |
| RCA | 25 | 20 (only up to M25 grade) | 100 |

## 6.2.2 According to Size

According to size, aggregates are classified as coarse aggregate, fine aggregate, all-in-aggregate, and graded aggregate.

**Coarse aggregates**    These are those having size between 10 and 40 mm and are usually retained on the 4.75 mm IS Sieve. As per IS 383–16, coarse aggregate may be one of the following types:

1. Uncrushed gravel or stone due to disintegration of natural rock;
2. Crushed gravel or stone due to crushing of natural stone or gravel;
3. A product obtained by blending type 1 and type 2 gravel or stone; and
4. Manufactured aggregate by using thermal processes or crushing, scrubbing, washing, and separation of construction and demolition (C&D) waste [Examples: RCA or RA (See Section 6.11.1)].

We normally specify graded course aggregate by its nominal size, i.e., 40 mm aggregate, 20 mm aggregate, and so on. Thus, 40 mm aggregate is one that passes through the 40 mm IS sieve. The maximum size of aggregate can be 80 mm; but for RC works, it is usually 20 mm or less. The adopted size is governed by the thickness of section, spacing of reinforcement, clear cover, mixing, handling and placing methods (see Section 6.9, for the maximum size of aggregates).

**Fine aggregates**    The aggregates passing through 4.75 mm sieve are considered as fine aggregates. As per IS 383–16, the fine aggregates used may be any one of the following:

*Natural sand*    They are naturally available due to the disintegration of rocks, made by rivers or glaciers by grinding the stones, and by oceans degrading seashells and corals. Sand in the beaches is normally not used due to its salt content.

*Crushed-stone sand*    They are manufactured by crushing quarried hard stone to a size that will pass completely through the 4.75 mm sieve.

*Crushed-gravel sand*    They are manufactured just like crushed-stone sand by crushing natural gravel.

*Mixed sand*    They are produced by blending natural sand and crushed-stone sand or crushed-gravel sand in suitable proportions.

*Manufactured sand (M-sand)*    They are manufactured from natural or other sources by using thermal or other processes, such as crushing, scrubbing, washing, and separation.

Nearly 70% of all sand grains on earth are quartz ($SiO_2$, also known as silica). The smallest size of fine aggregate (sand) is 0.075 mm. Depending upon the particle size, fine aggregates are described as fine, medium, and coarse sands. On the basis of particle size distribution (PSD), IS 383–16 has been classified fine aggregates into four zones – the grading zones becoming progressively finer from grading zone I to grading zone IV.

**All-in-aggregates** Sometimes, combined fine and coarse aggregate are available in nature, which are termed as *all-in-aggregates*. If such aggregates are used, they need not be separated into fine and coarse, but necessary adjustments should be made in the grading by the addition of single-sized aggregates (The grading of aggregates is discussed in Section 6.10.).

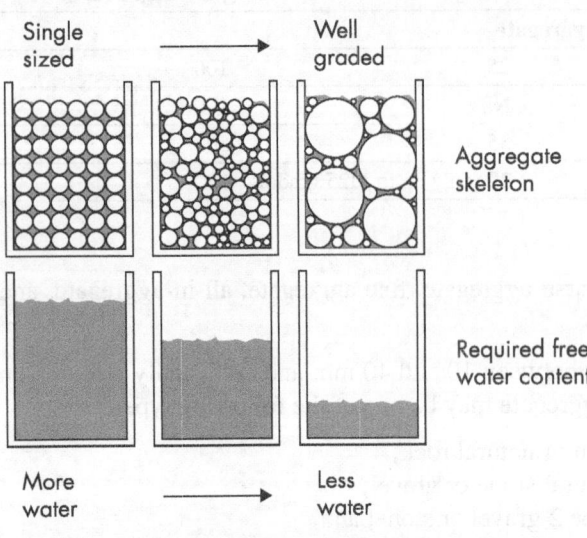

**Fig. 6.1** Different types of aggregates and water demand (*Source:* CCAA Tech Note 73-Reprinted with permission from CCAA, Australia)

**Single-sized aggregates** Aggregates consisting of particles within a narrow limit of size fractions are called *single-sized aggregate*. For example, a 20 mm single-sized aggregate means aggregates passing through a 20 mm IS sieve and the major portions of which are retained in a 10 mm IS sieve.

**Graded aggregates** These aggregates have particle size ranging from coarse aggregates to graded sand. They are characterized by the S-shaped gradation curve (see also, Section 6.10). Figure 6.1 shows the water demand of different types of aggregates. When well-graded aggregates are used in concrete, required free water content will be less resulting in greater strength and lower drying shrinkage.

### 6.2.3 According to Shape

According to the shape, aggregates are classified as rounded, irregular or partly rounded, angular, and flaky (see Figs. 6.2 and 6.3). In addition, based on the surface characteristics, IS 383–16 classifies aggregates into five groups as shown in Table 6.2. Particle shape, surface texture, and porosity of coarse aggregates influence the properties of freshly mixed concrete more than the properties of hardened concrete.

**Fig. 6.2** Angular and rounded aggregates

**Table 6.2** Surface characteristics of aggregates

| Group | Surface texture | Examples |
|---|---|---|
| 1 | Glassy | Black flint |
| 2 | Smooth | Chert, marble, and slate |
| 3 | Granular | Oolites and sandstone |
| 4 | Crystalline | *Fine*: Basalt, keratophyre, and trachyte<br>*Medium*: Dolerite, several dolomites, and a few limestone<br>*Coarse*: Gabbro, gneiss, and granite |
| 5 | Porous and honeycombed | Pumice, scoriae, and trass |

**Rounded aggregates** These are generally obtained from river or seashore, and are shaped by attrition (coastal or river erosion) or weathering and produce minimum voids (about 32–33%%) in the concrete. Round-shaped and smooth-textured aggregates are better for workability and will have lowest water demand and mortar paste requirement. Hence, they will result in most economical mixes for concrete grades up to M35. The only drawback of rounded aggregates is that interlocking between its particles is less and hence the development of bond is poor, making it unsuitable for high-strength concrete and pavements.

**Irregular or partly rounded aggregates** These aggregate are partly shaped by attrition (coastal or river erosion) or weathering and may have partly rounded edges. Owing to this, using them will result in a higher percentage of voids ranging from 35 to 37%. It gives lesser workability than rounded aggregate for the given water content. Water requirement is higher and hence more cement is needed for constant water cement ratio. The interlocking between aggregate particles is better than rounded aggregate but not adequate to be used for high-strength concrete and pavements. Large amounts of flaky and elongated particles can make concrete mixtures too harsh resulting in voids, honeycombing, or pump blockages.

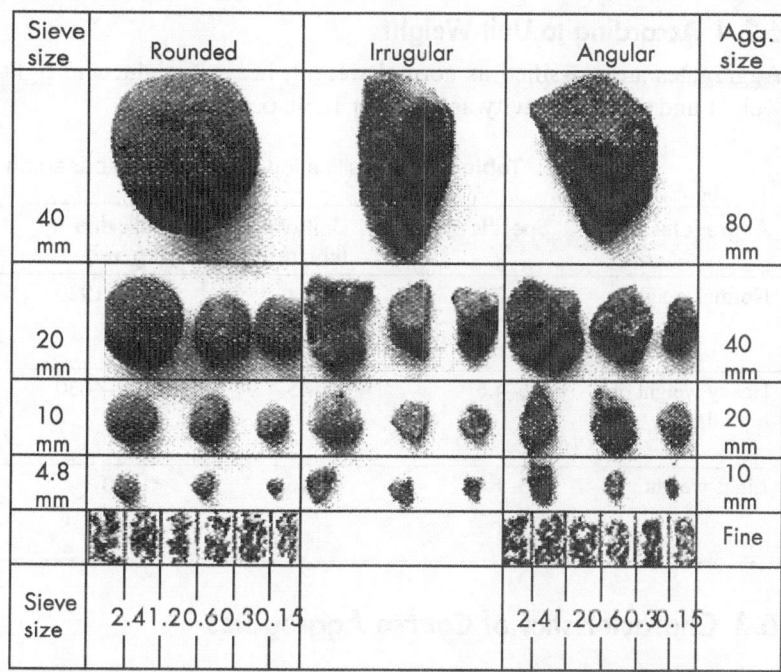

**Fig. 6.3** Different shapes and sizes of aggregates (*Source*: Ambuja technical booklet, 125:2007)

**Angular aggregates** These aggregates are obtained from crushing the rocks and have sharp, angular, and rough particles having maximum voids (about 38–45%%). Angular aggregate exhibit a better interlocking effect in concrete, which makes it suitable for high-strength concrete and for concrete used in roads and pavements. However, the requirement of cement paste is relatively more.

**Flaky aggregates** The angular aggregates obtained from laminated rocks are said to be flaky when its least lateral dimension (thickness) is less than 3/5th (or 60%) of its mean dimension. The mean dimension of the aggregate is the average of the sieve sizes through which these particles pass and is retained, respectively. These are sometimes wrongly called as elongated aggregate. Elongated aggregate is that aggregate whose length is 1.8 times greater than its mean dimension. Aggregate is said to be flaky and elongated when it satisfies both the above conditions. However, both of them influence the concrete properties adversely. Flaky aggregates tend to be oriented in one plane with water and air voids forming underneath them, thus affecting the durability of concrete. Generally, elongated or flaky particles in excess of 10 to 15% are not desirable.

Flakiness and elongation indices should be determined as per IS 2386 (Part 1). The *flakiness index* of an aggregate is the percentage by the weight of particles in it, whose least dimension (i.e., thickness) is less than three-fifths of their mean dimension. The *elongation index* of an aggregate is the percentage by the weight of particles, whose greatest dimension (length) is greater than nine-fifths times their mean dimension. As per IS 383:2016, the combined flakiness and elongation index should not exceed 40% for uncrushed or crushed aggregate.

## 6.2.4 According to Unit Weight

Aggregates are classified as normal-weight, heavy-weight, and light-weight aggregates depending on weight and specific gravity as given in Table 6.3.

**Table 6.3** Classification of aggregates based on unit weight

| Aggregates | Specific gravity | Unit weight (kN/m³) | Bulk density (kg/m³) | Examples |
|---|---|---|---|---|
| Normal weight | 2.4–2.9 | 23–28 | 1280–1920 | Sand, gravel, granite, sandstone, limestone, crushed stone, and air-cooled blast-furnace slag |
| Heavy weight or high-density | 3.6–4.6 | 35–45 | 2100–4650 | Limonite, goethite, barite, imenite, hematite, magnetite, and steel/iron pellets/punchings |
| Light-weight | 1.3–1.6 | 13–16 | <1120 | Expanded shale, clay, slate, slag, pumice, scoria, perlite, vermiculite, and diatomite |

## 6.3 Characteristics of Coarse Aggregates

It is significantly important to use right type and quality of aggregates in concrete. Aggregates used in concrete should be clean, strong and hard, impermeable, durable, properly shaped, as well as graded properly in size, to achieve economy and better performance. In addition, they should have chemical stability and resistance to abrasion, freezing, and thawing (in cold climates). They should not contain much deleterious material, which may cause physical or chemical changes to concrete, such as cracking, swelling, softening, or leaching. The amount of water in the concrete mixture must be adjusted to include the moisture conditions of the aggregate. These important characteristics of aggregates for concrete, their significance, and test methods are listed in Table 6.4 and most are discussed in the following section.

**Table 6.4** Summary of aggregate properties

| Characteristic | Significance | Test/Code | Requirement or limits |
|---|---|---|---|
| Particle shape and surface texture | Affects workability of fresh concrete | Visual inspection, flakiness, and elongation index test IS 2386 (Part 1):1963 | Maximum percentage of flaky or elongated particles. Flakiness index <30–40 desirable |
| Grading | Well-graded aggregates reduce cement requirements and improve workability | IS 2386 (Part 1):1963 | Grading limits for coarse, fine, and all-in aggregates are prescribed in IS 383–16. |
| Deleterious substances | Organic and other impurities may affect the strength or durability of concrete | Tests for the determination of clay lumps, coal, lignite, soft particles, fine silt/dust, and clay and estimation of organic impurities are given in IS 2386 (Part 2):1963. | Limits on impurities are prescribed in IS 383–16. |
| Relative density (specific gravity) | Required in mix design calculations, unit weight, and yield of concrete | Specific gravity determined as per IS 2386 (Part 3):1963 | – |

*(Contd)*

**Table 6.4** (*Contd*)

| Characteristic | Significance | Test/Code | Requirement or limits |
|---|---|---|---|
| Bulk density (unit weight) | Mix design calculations to convert proportions by weight to proportions by volume | Bulk density determined as per IS 2386 (Part 3):1963 | Compact weight and loose weight |
| Absorption and surface moisture | Control on the quality of concrete based on water-cement ratio | Test for absorption and surface moisture as per IS 2386 (Part 3):1963 | – |
| Resistance to crushing | The aggregate crushing value gives a relative measure of the resistance of an aggregate to crushing under a gradually applied compressive load. | Tests for the determination of aggregate crushing value, impact value, abrasion value, and crushing strength of rock are to be done as per IS 2386 (Part 4):1963. | As per IS 383–16, the aggregate impact value should not exceed 30% for wearing surfaces (runways, roads, and pavements) and 45% for other applications. The aggregate crushing value for wearing surfaces should not exceed 30% for wearing surfaces; for other surfaces, when it exceeds 30%, test for '10% fines' should be conducted. Aggregate abrasion value should not exceed 30% for wearing surfaces and 50% for other applications. |
| Chemical stability | Important for structures situated in chemical environments | Tests to determine resistance to disintegration of aggregates by saturated solutions of sodium sulphate or magnesium sulphate as per IS 2386 (Part 5):1963. | The average loss of mass after five cycles should not exceed 10% when tested with sodium sulphate ($Na_2SO_4$), and 15% with magnesium sulphate ($MgSO_4$) for fine aggregates. Similar percentages for coarse aggregates are 12% and 18%, respectively. |
| Resistance to freezing and thawing | Important in cold regions | For concrete exposed to the action of frost, the coarse and fine aggregates should pass a sodium or magnesium sulphate accelerated soundness test specified in IS 2386 (Part 5). | Same as above |

## 6.3.1 Particle Size, Shape, and Texture

As already mentioned, the aggregates may be classified according to shape as rounded, irregular or partly rounded, angular, and flaky. The particle shape and surface texture of an aggregate influence the properties of freshly mixed concrete more than the properties of hardened concrete. Rough-textured, angular, elongated particles require more water and cement paste to produce workable concrete than smooth, rounded, compact aggregates. Also, the mixing water and cement requirements increase as aggregate void content increases. Voids between aggregate particles increase with aggregate angularity. Hence, aggregates having angular shape require more cement to maintain the same water-cement ratio. However, by the proper gradation of crushed and non-crushed aggregates, we may achieve same strength for the same w/c ratio (Kosmatka et al., 2011). See Section 6.10 for a discussion on grading. In addition, angular or poorly graded aggregates pose difficulties while pumping concrete.

In general, the bond between cement paste and aggregates is more for rough and angular aggregates than smooth and rounded aggregates. Hence, when flexural strength or high compressive strength is required, it is better to use rough and angular aggregates. Owing to stronger aggregate bond, concretes having rough angular aggregates may have 10–15% higher compressive strength.

It is better to avoid flaky and elongated aggregates or at least limit their content to about 10–15% by mass of the total aggregate. Flakiness and elongation also reduce the flexural strength of concrete. Flaky and elongated coarse aggregate particles not only increase the water demand, but also increase the tendency of segregation. It has to be noted that fine aggregate made by crushing stone also may contain flaky and elongated particles. Such a fine aggregate also requires an increase in mixing water and thus may affect the strength of concrete.

## 6.3.2 Crushing or Impact Strength

The strength of aggregates is assessed by aggregates crushing test. The aggregates crushing value provides a relative measure of resistance to crushing under gradually applied compressive load. The strength of concrete cannot exceed that of coarse aggregate used to make that concrete. Rocks, commonly used as aggregates, have a compressive strength much higher than the usual range of concrete strength. In practice, the strength of concrete grade will be much less than the strength of the aggregate, due to stress concentrations generated at the interface of aggregates and cement paste. As M65 and higher grades of concrete require stronger aggregates, both crushing and impact values should be less than 22%. The aggregate impact value of Indian aggregates usually varies from 18 to 27%. The aggregate crushing value of Indian aggregates usually varies from 15 to 30%.

## 6.3.3 Relative Density

The relative density (specific gravity) of an aggregate is defined as the ratio of the mass of the aggregate in a given volume in air to the mass of an equal volume of water, at the same temperature. An aggregate with a relative density of 2.0 will thus be two times denser than an equal volume of water. Since the aggregates generally contain voids (which may hold water), their specific gravities may vary. Because the mass of aggregate will vary depending on its moisture content, specific gravity has to be determined at constant moisture content. Most natural aggregates have specific gravity ranging from 2.4 to 2.9 with corresponding mass densities of 2400 and 2900 kg/m³ (Kosmatka et al., 2011). The specific gravity is required in the calculation of mix design of concrete to convert proportions by weight to proportions by volume.

## 6.3.4 Bulk Density

The bulk density of aggregate is defined as the mass of the material in a given volume and expressed as kg/m³ [In IS 2386 (Part 3), it is expressed in kilograms per litre.]. It depends upon the packing of particles, particles shape and size, the grading, and the moisture content. For coarse aggregate, a higher bulk density is an indication of fewer voids to be filled by sand and cement.

Increasing moisture content of coarse aggregate will increase its the bulk density; however, for fine aggregate, increasing the moisture content, beyond the saturated surface-dry (SSD) condition, may decrease the bulk density. This is because thin films of water on the sand particles cause them to stick together, making them difficult to compact. The resultant increase in volume decreases the bulk density. This phenomenon is called *bulking of sand* and is discussed in Section 6.12.3. Bulking must be taken into account, when volumetric batching is used.

The bulk density of an aggregate may also be affected by grading, relative density, surface texture, shape, and angularity of particles. Well-graded aggregates will usually have a higher bulk density than those with one particle size. Higher relative density of the particles results in higher bulk density for a particular grading. Smooth rounded aggregates will normally have a higher bulk density than rough angular particles of the same mineral composition and grading.

### 6.3.5 Voids

When the aggregates are put in a container, the empty spaces between them are called voids. Determination of voids is discussed in Section 25.10.3 of Chapter 25.

### 6.3.6 Porosity

The entrapped air bubbles in the rocks during their formation lead to minute holes or cavities known as *pores*. The porosity of some of the commonly used rocks is generally less than 20%. Since aggregates constitute about 65–75% of the concrete, the porosity of aggregates will affect the porosity of concrete mass. In addition, the permeability and absorption of moisture will affect the bond between aggregate and cement paste, the resistance to freezing and thawing, and the resistance to abrasion. Porous aggregates may also absorb more moisture, resulting in the loss of workability of concrete at a much faster rate.

### 6.3.7 Moisture Content

The percentage of water absorbed by the aggregate when immersed in water is called the *absorption of aggregate*. The surface moisture expressed as a percentage of the weight of the SSD aggregate is known as *moisture content*. High moisture content in the aggregate will increase the effective water/cement ratio considerably and subsequently reduce the strength and durability of concrete.

Four moisture conditions are defined for aggregates depending on the amount of water held in the pores or on the surface of the particles. These conditions are shown in Fig. 6.4 and described as follows:

*Damp or wet*    Aggregates in which all the pores are completely filled with water and the surface of the aggregate contains free water. Aggregates in a stockpile will typically have this condition. This condition will contribute additional water to the concrete mix and hence proper adjustments are to be made in the mix design.

*Saturated surface-dry (SSD)*    Aggregate in which all pores are completely filled with water but no free water remains on the surface of the aggregate. Aggregates in this condition will not contribute free water nor absorb water from the concrete mix. This condition is achieved in the laboratory.

*Air-dry (AD)*    Aggregate that has a dry surface but contains some water in the pores. This condition may occur on a hot summer day in an arid region. These aggregates will absorb water from the concrete mix, which may affect the workability of concrete, unless proper adjustments are made in the mix design.

*Oven-dry (OD)*    Aggregate that contains no water in the pores or on the surface. This condition is achieved under laboratory conditions when the aggregate is heated to 105°C, for an extended period.

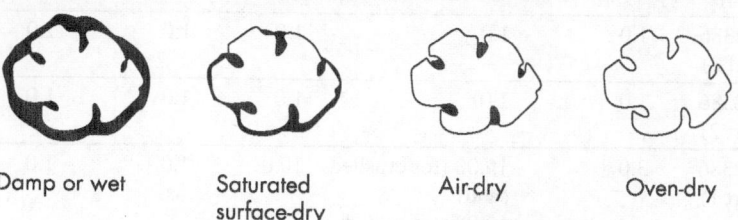

Damp or wet          Saturated          Air-dry          Oven-dry
                     surface-dry

**Fig. 6.4** Moisture condition of aggregates
(*Source*: ACI E1-16 , reprinted with permission from ACI)

Of these four states, only OD and SSD states correspond to specific moisture contents. Thus, for calculating moisture content, either of these states can be used as reference states. Stockpiled aggregates will have variable moisture content throughout the stockpile, with wetter aggregates found near the bottom of the pile and hence they may have AD and wet states. As self-consolidating concrete (SCC) is more sensitive to changes in aggregate moisture content, we need to calculate the aggregate moisture content at least once a day or even more frequently when producing SCC. Moisture meters can be used in batching systems of ready mixed concretes (RMCs) to read the moisture content of aggregates, when the aggregates are discharged from the hopper. As these meters are directly connected to the batch computer, the batch weights, for correct proportions and w/c ratio, can be automatically adjusted. However, in batching systems without moisture meters, manual computation of moisture content of aggregates is necessary.

## 6.4 Deleterious Substances and Organic Impurities

Substances such as organic matters, clay, silt, shale, coal, and iron pyrites may adversely affect the strength, workability, and long-term performance of concrete and are termed as deleterious substances. They are considered undesirable, as their intrinsic weakness, softness, fineness, or other physical or chemical characteristics may be harmful to the performance of concrete. They affect the properties of concrete in green as well as hardened state. They may be classified as those interfering with the process of hydration of cement (organic matters), coatings such as clay, affecting the development of bond between aggregate and the cement paste, and, unsound particles, which are weak or bring about a chemical reaction between aggregate and cement paste. The organic matter, such as decayed vegetable matter, can be easily washed (They are normally present in the fine aggregate rather than in the coarse aggregate.). The surface coated impurities in aggregate, such as clay, silt, and crusher dust, could be removed by adequate washing. Unsound particles such as shale, clay lumps, wood, and coal may lead to pitting and scaling. If their presence is more than 2–5%, they may affect the strength of concrete. Mica, if present in sand, may reduce the strength of concrete. Iron pyrites and sulphates, mostly found in natural aggregate deposits, produce surface staining and pop-outs, especially in warm and humid conditions. Hence, limits are specified for the maximum quantity of deleterious materials in IS 383–16 and are shown in Table 6.5 (Also see, Table 2 of IS 383–16). However, the code allows some relaxation of these limits, based on tests and the evidence of satisfactory performance of the aggregates.

**Table 6.5** Limits of deleterious substances (maximum percentage by mass)

| Deleterious substance | Method of test, ref. to | Fine aggregate | | | Coarse aggregate | | |
|---|---|---|---|---|---|---|---|
| | | Uncrushed | Crushed/Mixed | M-sand | Uncrushed | Crushed | Manufactured |
| Coal and lignite | IS 2386 (Part 2) | 1.0 | 1.0 | 1.0 | 1.0 | 1.0 | 1.0 |
| Clay lumps | IS 2386 (Part 2) | 1.0 | 1.0 | 1.0 | 1.0 | 1.0 | 1.0 |
| Materials finer than 75 μm IS Sieve | IS 2386 (Part 1) | 3.0 | 15.00 (for crushed sand) 12.00 (for mixed sand) | 10.0 | 1.0 | 1.0 | 1.0 |
| Soft fragments | IS 2386 (Part 1) | – | – | – | 3.0 | – | 3.0 |
| Shale[1] | | 1.0 | – | 1.0 | – | – | – |

*(Contd)*

**Table 6.5** (*Contd*)

| Deleterious substance | Method of test, ref. to | Fine aggregate | | | Coarse aggregate | | |
|---|---|---|---|---|---|---|---|
| | | Uncrushed | Crushed/Mixed | M-sand | Uncrushed | Crushed | Manufactured |
| Total of all deleterious substances (except mica)[2] | – | 5.0 | 2.0 | 2.0 | 5.0 | 2.0 | 2.0 |

**Note:**

1. When the clay stones are harder, platy, and fissile, they are known as shales. The presence and extent of shales are determined by petrography, at the time of selection and change of source.

2. Where no tests for strength and durability are conducted, the mica in the fine aggregate may be limited to 1.00% by mass (also see, Table 2 of IS 383). The total is including coal and clay lumps only for crushed sand and M-sand.

## 6.5 Soundness of Aggregate

Soundness is defined as the ability of aggregate to resist excessive changes in volume due to changes in environmental conditions, such as freezing and thawing, temperature changes, and alternate wetting and drying. The freeze-thaw resistance of aggregate is related to its porosity, absorption, and pore structure. Porous and weak aggregates containing undesirable matter undergo excessive volume changes under the aforementioned environmental conditions. This may lead to local scaling to extensive surface cracking or disintegration over a considerable depth, resulting in impaired appearance and sometimes even to structural failures. Aggregate with certain chemical constituents may react with alkalis in cement and can cause the abnormal expansion and cracking of concrete.

For concrete liable to be exposed to the action of frost, the coarse and fine aggregates should pass a sodium sulphate ($Na_2So_4$) or magnesium sulphate ($MgSO_4$) accelerated soundness test specified in IS 2386 (Part 5). As per IS 383–16, the average loss of mass after five cycles should not exceed 10 and 15% for fine aggregates when tested with sodium sulphate and magnesium sulphate, respectively. For coarse aggregate, they are prescribed as 12 and 18%.

For slag aggregates, additional tests, such as *iron unsoundness*, should not exceed 1% (When ferrous oxide content exceeds 3.0% and sulphur content exceeds 1.0%, this test should be conducted as per Annex D of IS 383–16), *volumetric expansion ratio* (as per Annex E of IS 383–16 and should not exceed 2.0%), and *unsoundness due to free lime* (using petrographic examination and the number of particles containing free lime should not exceed 1 in 20) should be conducted.

## 6.6 Alkali–Aggregate Reaction

Reaction between the alkali components ($Na_2O$ and $K_2O$) in Portland cement and certain active minerals in some rocks is well recognized as a potential cause of concrete deterioration, affecting the durability. The reactivity is harmful only when it produces significant expansion and subsequent cracking. Alkali–Aggregate reaction (AAR) has two forms—alkali-silica reaction (ASR) and alkali-carbonate reaction (ACR), sometimes called alkali-carbonate rock reaction (ACRR). ASR is more prevalent than ACR because of the common occurrence of aggregates containing reactive silica minerals. Essentially, ASR forms a gel, which has affinity for moisture. As it draws water from the surrounding cement paste, these gels swell and induce pressure, expansion, and cracking of the aggregate and surrounding paste, leading to deterioration of concrete. Such a classical ASR is more pronounced with high-alkali cement and potentially reactive siliceous aggregates containing minerals, such as opal, chalcedony, chert, etc., (see Table 6.6). It has to be noted that the presence of gels does not necessarily indicate destructive ASR. Some gels expand very little and may not show evidence of harmful reactivity for several years, say about 20 years. If a gel is low swelling, it may not create any problems.

The harmful effects of ASR in concrete were recognized as early as 1930s, whereas ACR in concrete was not documented until 1957. For ASR to occur, all the three following conditions must be present (Clause 5.6 of IS 383:2016):

1. Aggregate containing an alkali reactive constituent;
2. A cement with high-alkali content, or another source of alkali; and
3. A high moisture level within the concrete.

An aggregate that presents a large surface area for reaction – poorly crystalline, many lattice defects, amorphous, glassy, or micro-porous – is susceptible to ASR. As the pH or alkalinity of the pore solution increases, potential for the ASR increases. Structures in warmer exposures are more susceptible to ASR than those in colder exposures because ASR rate usually increases with increasing temperature. Any reduction in permeability, by using a low water-cement ratio or supplementary cementitious materials (SCMs), reduces the movement of moisture and alkalis into and within the concrete. It is better to avoid certain types of aggregates (dolomitic rocks) to prevent alkali-aggregate reaction. It has to be noted that the use of low-alkali cements alone may not be sufficient to control ASR when highly reactive aggregates have been used. Certain chemical admixtures or cement additives can also be used to control ASR.

**Table 6.6** Alkali-silica and alkali-carbonate reactive materials [based on IS 2386 (Part 8) and Farny and Kerkoff, 2007, ACI 221.1R-98]

| Alkali-silica reactive materials[1] | | Alkali-carbonate reactive materials[2] |
|---|---|---|
| Andesite | Opal | Calcitic dolomites |
| Argillites | Certain phyllites | Dolomitic limestone |
| Certain siliceous limestone and dolomites | Quartzites | Fine-grained dolomites |
| Chalcedonic cherts | Cherts | |
| Chert and flint containing Chalcedony | Glassy to cryptocrystalline rhyolites | |
| Cristobalite | Schists | |
| Dacites | Siliceous or opalineshales | |
| Glassy volcanic rocks, such as rhyolite, dacite, latite, and andesite, | Strained quartz and certain other forms of quartz | |
| Granites and granodiorites | Tridymite | |
| Granite gneiss | | |
| Graywackes | | |
| Heulandite | | |
| Siliceous carbonate rocks | | |
| Volcanic glasses and volcanic rocks | | |
| Zeolite | | |

1. Several of these rocks (e.g., granite gneiss) react very slowly and may not show the evidence of harmful reactivity for over 20 years.
2. Only certain sources of these materials have been reactive.

IS 2386 (Part 7):1963 describes the following two tests for determining the potential reactivity of aggregates:

1. Mortar bar method for determining the potential alkali reactivity of cement-aggregate combinations
2. Chemical method for determining the potential alkali reactivity of aggregates

Petrographic tests as per IS 2386(Part 8):1963 can also be conducted to confirm the presence of reaction products and verify whether the cause of deterioration is due to ASR.

## Case Study    The Sixth Street Viaduct, Los Angeles

The Sixth Street Viaduct, the architecturally significant, art-deco landmark of Los Angeles, was constructed in 1932 using then state-of-the-art concrete technology and an onsite mixing plant. However, just 20 years after it was constructed, the concrete began to deteriorate due to ASR. Over the years, various costly restorative methods were tried, but none resulted in overcoming the problem. Seismic vulnerability studies, completed in 2004, concluded that the viaduct, because of the ASR deterioration, has a high vulnerability to failure in the event of a major earthquake. Owing to the additional deficiencies in geometric design and safety, it was decided to replace the viaduct. (The 1067 m viaduct has appeared in numerous music videos, TV commercials and movies, including a couple of Arnold Schwarzenegger 'Terminator' films).

*Source*: https://upload.wikimedia.org/wikipedia/commons/5/5f/Sixth_Street_Bridge_over_Los_Angeles_River.jpg

The demolition of the viaduct began in Feb. 2016 (www.sixthstreetviaduct.org). In its place, a new bridge, known as 'The Ribbon of Light', will be constructed, which is scheduled to open in 2019.

It is interesting to note that the Rihand Dam in UP and the Hirakud dam in Odisha were also affected by ASR. More details on the ASR attack and the remedial measures taken at the Hirakud dam may be found in Ramachandran (1993).

## 6.7 Thermal Properties of Aggregates

The fire resistance and thermal properties of concrete, such as conductivity, diffusivity, and coefficient of thermal expansion, may depend on the mineral constituents of the aggregates, as aggregates occupy 70–75% volume of concrete. Concrete-containing calcareous coarse aggregate performs better under fire exposure than a concrete containing quartz or siliceous aggregate, such as granite or quartzite. The coefficient of thermal expansion of aggregates typically ranges from $4 \times 10^{-6}$ per °C to $13 \times 10^{-6}$ per °C (see Table 6.7). However, the coefficient of thermal expansion of cement paste is higher than that of aggregate (see Table 6.7). In hot weather, sprinkling or shading of stockpiles of aggregate reduces the temperature of concrete. Coarse aggregate may also be cooled by immersing them or spraying the stockpile with chilled water or by adding ice cubes (ACI 305R). In cold weather, aggregates may be heated to obtain desired concrete temperatures (ACI 306R). Frozen aggregates should not be used in concrete mixtures.

**Table 6.7** Coefficient of thermal expansion of aggregates and cement paste (*Source*: https://www.fhwa.dot. gov/publications/research/infrastructure/pavements/pccp/thermal.cfm)

| Material | Coefficient of thermal expansion $\times 10^{-6}/°C$ |
| --- | --- |
| **Aggregates** | |
| Granite | 7–9 |
| Basalt | 6–8 |
| Limestone | 6 |
| Dolomite | 7–10 |
| Sandstone | 11–12 |
| Quartzite | 11–13 |
| Marble | 4–7 |
| **Cement paste (saturated)** | 18–20 |
| **Concrete** | 7.4–13 |
| **Steel** | 11–12 |

All aggregates produce fire-resistant concrete and the degree of fire resistance is related to the type of aggregate used. Aggregates containing high levels of quartz offer relatively less thermal resistance, as quartz undergoes an expansive solid phase change at high temperature (at about 590°C). Limestone is low in silica and has a low coefficient, and hence may be specified for concrete exposed to early age thermal cracking. Manufactured and some naturally occurring lightweight aggregates are more fire resistant than normal-weight aggregates due to their insulating properties.

## 6.8 Fineness Modulus and Sieve Analysis

The *fineness modulus* is often computed using the *sieve analysis* results. The fineness modulus may be defined as an empirical factor obtained by adding the cumulative percentages of a sample of the aggregate retained on each of a specified series of sieves and dividing the sum by 100. The specified sieves are the 80.0, 40.0, 20.0, and 10 mm (3, 1–1/2, 3/4, and 3/8 in.), and 4.75 mm for coarse aggregates, and 2.36 mm, 1.18 mm, 600 µm, 300 µm, and 150 µm (Nos. 4, 8, 16, 30, 50, and 100) for fine aggregates. It has to be noted that the lower limit of the specified series of sieves is the 150 µm (No. 100) sieve and that the actual size of the openings in each larger sieve is twice that of the sieve in the following. The fineness modulus is a numerical index of fineness, giving some indication about the mean size of the particles contained in the sample of aggregates. Greater value of fineness modulus indicates that the aggregate is coarser and small value of fineness modulus indicates that the aggregate is finer. Determination of fineness modulus consists of shaking a representative sample of the aggregate, which has been properly prepared, through a series of sieves nested one above the other in order of size, with the sieve having the largest openings on top and the one having the smallest openings at the bottom (Fig. 6.5). These wire-cloth sieves have square openings.

A pan is used to catch material passing the smallest sieve. It has to be noted that the coarse and fine aggregates are sieved separately. Sieve sizes, commonly used for concrete aggregates, are given in Table 6.8. The various physical properties of normal-weight aggregates used in concrete are given in Table 6.9.

**Table 6.8** Sieves commonly used for concrete aggregate sieve analysis

| IS Sieve classification | ASTM E11 Sieve classification |
|---|---|
| **Coarse sieves** | |
| 80 mm | 3 in. (75 mm) |
| 63 mm | 2.5 in. |
| | 2 in. (50 mm) |
| 40 mm | 1.5 in. (37.5 mm) |
| 20 mm | 1.0 in. (25 mm) |
| 16 mm | ¾ in. (19 mm) |
| 12.5 mm | ½ in. |
| 10 mm | 3/8 in. (9.5 mm) |
| **Fine sieves** | |
| 4.75 mm | No. 4 |
| 2.36 mm | No. 8 |
| 1.18 mm | No. 16 |
| 600 μm* | No. 30 |
| 300 μm | No. 50 |
| 150 μm | No. 100 |
| | No. 200 (75 μm) |

**Note:** 1000 μm = 1 mm.

(a) Fine sieves

(b) Coarse sieves mounted in shaker

**Fig. 6.5** Series of sieves and small sieve shaker (*Source:* www.iowadot.gov)

**Table 6.9** Physical properties of normal-weight aggregates used in concrete (*Source:* ACI E1–16, reprinted with permission from ACI)

| Property | Typical range |
|---|---|
| Fineness modulus of fine aggregate | 2.0–3.1 |
| Nominal maximum size of aggregate | 40 to 10 mm |

*(Contd)*

**Table 6.9** (*Contd*)

| Property | | Typical range |
|---|---|---|
| Absorption of moisture | | 0 to 8% |
| Bulk relative density (bulk specific gravity) | | 2.30 to 2.90 |
| Dry-rodded bulk density of coarse aggregate | | 1280 to 1920 kg/m³ |
| Surface moisture content | Coarse aggregate | 0 to 2% |
| | Fine aggregate | 0 to 10% |

Figure 6.5(b) shows a small mechanical sieve shaker (designed for 8 in. round sieves), used mainly for fine aggregates. The mechanical shaker imparts lateral (or lateral and vertical) motion to the set of sieves, causing the particles thereon to bounce and turn so as to present different orientations to the sieving surface. This aspect is important, because the sieve openings are square; whereas, the particles are neither square nor round. Small shakers of this type require shaking times of 15 min to adequately grade the fine aggregate sample.

The large tray shaker with rectangular sieves is usually used for the coarse aggregate (see Fig. 6.6). Shakers of this make need to run 5 min for size 10 mm or larger and 10 min for sizes smaller than 10 mm. It is interesting to note that a number of automated test machines (e.g., VDG 40-videograder) have been developed as a faster alternative to the standard sieve analysis test, using which we can rapidly determine the PSD of aggregates (Kosmatka et al., 2011).

**Fig. 6.6** Large tray shaker (*Source*: www.in.gov/indot/div/mt/aashto/testmethods/aashto_t27.pdf)

The number and size of sieves selected for a sieve analysis depend on the particle sizes present in the sample and the specified grading requirements (see Section 6.10 for grading). Coarse and fine aggregates are generally sieved separately. After sieving, the weight of material retained on each sieve, and also in the bottom pan, is measured using a balance accurate to 0.1% of the weight of the test-sample. It must be ensured that all material entrapped within the openings of the sieves are cleaned out and included in the weight retained. This may be done using brushes to gently dislodge entrapped materials (Any sieve with an opening size smaller than the 300 μm sieve should be cleaned with a softer cloth hair brush.). Results are recorded in tabular form and typically reported as weight retained on each sieve, cumulative weight retained on each sieve, and cumulative percent of weight retained (see Examples 6.1 and 6.2). Individual percent retained is the percentage of material contained between successive sieves, recorded to the nearest whole percent. It is calculated by dividing the mass retained on each sieve by the sum of the masses retained on each sieve and the pan and multiplying by 100. Cumulative percent retained is calculated by successively summing the numbers in the individual percent retained column. The cumulative percent passing is calculated by subtracting the cumulative percent retained from 100. The fineness modulus is calculated as the sum of the cumulative percentages retained on the sieves divided by 100. A fineness modulus of 3.0 can be interpreted to mean that the third sieve, i.e., 600 μm is the average size.

The fineness modulus varies from 2.0 to 3.2 for fine aggregate, from 5.5 to 8.0 for coarse aggregate, and from 3.5 to 6.5 for all-in aggregate. In addition, for the better performance of any concrete, the following values of fineness modulus should be adopted for sand: fine sand- 2.2 to 2.6, medium sand- 2.6 to 2.9, and coarse sand- 2.9 to 3.2. The object of finding fineness modulus is to grade the given aggregate for the required strength and workability of concrete mix with minimum cement. Aggregates with higher fineness modulus result in harsh concrete mixes and with lower fineness modulus result in uneconomical concrete mixes. It is better to avoid adding too much fine aggregate to a concrete mixture and avoid using extremely fine sand.

**Example 6.1**  Determine the fineness modulus of sand from the sieve analysis result of weight of sand retained and as given in Table 6.10. Note that the sieves that are not specified in the definition (for example, 25.0 and 12.5 mm [l and 1/2 in.] sieves) are excluded and all of the specified finer sieves are included.

**Table 6.10** Example calculation of fineness modulus of fine aggregate

| Sieve size | Weight of sand retained (g) | Individual percent retained (%) | Cumulative percentage of sand retained (%) | Cumulative percentage of sand passing (%) |
|---|---|---|---|---|
| 10 mm | 0 | 0 | 0 | 100 |
| 4.75 mm | 10 | 2 | 2 | 98 |
| 2.36 mm | 50 | 10 | 12 | 88 |
| 1.18 mm | 50 | 10 | 22 | 78 |
| 600 μm | 105 | 21 | 43 | 57 |
| 300 μm | 160 | 32 | 75 | 25 |
| 150 μm | 90 | 18 | 93 | 7 |
| Pan | 35 | 7 | – | |
| Total | 500 | | 247 | |

Fineness modulus = 247/100 = 2.47

**Example 6.2**  Determine the fineness modulus of coarse aggregates from the sieve analysis results. The given data and the required calculations are given in Table 6.11. It has to be noted that though the 63, 16, and 12.5 mm sieves were used in the sieve analysis, they are not included in the calculations. Because the total percent retained on the 1.18 mm sieve was 100%, 100% will also be retained on the smaller sieves specified in the fineness modulus definition.

**Table 6.11** Example calculation of fineness modulus of coarse aggregate

| Sieve size | Weight of aggregate retained (g) | Individual percent retained (g) | Cumulative percentage of retained (%) | Cumulative percentage of passing (%) |
|---|---|---|---|---|
| 80 mm | – | – | – | – |
| 40 mm | 0 | 0 | 0 | 100 |
| 20 mm | 440 | 5 | 5 | 95 |
| 10 mm | 2800 | 34 | 39 | 61 |
| 4.75 mm | 2515 | 31 | 70 | 30 |
| 2.36 mm | 2065 | 25 | 95 | 5 |

*(Contd)*

**Table 6.11** (*Contd*)

| Sieve size | Weight of aggregate retained (g) | Individual percent retained (g) | Cumulative percentage of retained (%) | Cumulative percentage of passing (%) |
|---|---|---|---|---|
| 1.18 mm | 380 | 5 | 100 | 0 |
| 600 μm | 0 | 0 | 100 | 0 |
| 300 μm | 0 | 0 | 100 | 0 |
| 150 μm | 0 | 0 | 100 | 0 |
| Total | 8200 | – | 609 | |

Fineness modulus = 609/100 = 6.09

## 6.9 Maximum Size of Aggregates

In general, larger the size of aggregate, the smaller will be the requirement of cement for a particular w/cm ratio. In addition, as the size of aggregate increases, the workability of the concrete mix improves, due to the smaller surface area of large size aggregates. In other words, for a given workability the use of larger aggregates allows for a mix design with less cement. However, larger aggregates tend to form weaker *transition zones*, resulting in more micro-cracks. As the smaller size aggregates provide a larger surface area for bonding with the mortar matrix, the mix with smaller size aggregates may result in increased compressive strength. In general, aggregates of size up to 40 mm may be used in concretes of grade M20 or below and 20 mm aggregates may be used in concretes of grade M20 and above. In high-strength concretes, aggregates of size varying from 10 mm to 20 mm are preferable.

It is important to place concrete in any RC member in such a way that it fills the formwork uniformly without any segregation. This is possible only when the largest aggregate passes smoothly between the reinforcing bars and fills the narrowest corners of the formwork. To achieve this, the following two general rules are often specified: the maximum size of aggregates should not be more than 3/4th the size of the space between the reinforcing bars, and should not exceed 1/4th of the smallest form dimension. Clause 5.3.3 of IS 456 specifies that the maximum size of aggregates should not be greater than 1/4th the minimum thickness of the member and that the maximum size should not be more than 5 mm less than the minimum clear spacing of main bars and 5 mm less than minimum cover (see Tables 6.12 and 6.13). Moreover, as the maximum aggregate size increases, the size and robustness of the mixers, pumps, and other equipment needed to make and place the concrete also increase.

**Table 6.12** Maximum size of aggregate

| Situation | Maximum aggregate size |
|---|---|
| Reinforced concrete member | 1/4th of minimum thickness of member (IS 456) |
| Clear spacing between reinforcement or pre-stressing tendons | 3/4th of minimum clear space<br>5 mm less than minimum clear spacing of main bars (IS 456) |
| Clear space between reinforcement and form | 3/4th of minimum clear space<br>5 mm less than minimum cover (IS 456) |
| Unreinforced slab | 1/3rd of thickness |

**Table 6.13** Maximum size of aggregate based on type of construction as per IS 456

| Type of construction | Maximum aggregate size |
| --- | --- |
| Reinforced concrete members such as beams, columns, and slabs and prestressed concrete | 20 mm |
| Reinforced concrete in foundation | 40 mm |
| Reinforced concrete in thin members such as shells or domes | 10 mm |
| Unreinforced construction and mass concrete such as dams and pavements | 40 to 75 mm |

It has to be noted that in the mix design process, maximum aggregate size is used only at one place. Nominal maximum aggregate size is used in all other places. It has to be remembered that the maximum aggregate size is generally one sieve size larger than the nominal maximum aggregate size.

## 6.10 Grading of Aggregates

The determination of the particle size distribution (PSD) of the aggregates, using the results of sieve analysis, is termed as the *grading of aggregates*. Grading limits and maximum aggregate size are specified in codes because these properties affect the amount of aggregate used as well as cement and water requirements, workability, pumpability, and durability of concrete. In general, if the w/c ratio is chosen correctly, a wide range in grading can be used without a major effect on strength. If all the aggregate particles are of the same size, the compacted mass will contain more voids, whereas aggregate particles comprising different sizes will result in a compacted mass having lesser voids (see Fig. 6.7). Hence, it is important to have continuously graded aggregates having various sizes of particles such that the smaller size particles fill the gap between larger particles (see Fig. 6.7b). It has to be noted, however, that the production of satisfactory and economical concrete requires aggregates of low-void content, but not the lowest (ACI E1–16). Hence, it is important that the coarse and fine aggregates are well graded to produce a concrete that has better strength and greater durability.

Based on the grading, we can have the following three types of aggregates:

**Poorly graded aggregates (Fig. 6.7a)**   They have aggregate particles characterized by a small variation in size. It may contain aggregate particles that are almost of the same size. Thus, when the particles are packed together, they leave large voids in the concrete. Poorly graded concretes generally require excessive amounts of cement paste to fill the voids, making them uneconomical. It has to be noted that the amount of cement paste required in concrete is greater than the volume of voids between the aggregates (Kosmatka et al., 2011). They are also called uniform graded and characterized by a steep gradation curve (see Fig. 6.8).

**Well-graded aggregates (Fig. 6.7b)**   They have a gradation of particle size that is distributed evenly from the finest to the coarsest. Thus, when the particles are packed together, they leave minimum voids in the concrete. They are also termed as 'continuously graded' to distinguish from 'gap-graded'. Hence, they produce dense concrete, with the least amount of cement paste. It is characterized by the S-shaped gradation curve (see Fig. 6.9).

**Gap-graded aggregates (Fig. 6.7c)**   They consist of aggregate particles in which some intermediate particle sizes are omitted from the size continuum. It is characterized by a gradation curve with a horizontal line over the range of omitted sizes. Gap-graded aggregates fall in between well graded and poorly graded in terms of performance and economy. Close control of mix proportions is necessary to avoid segregation. Specific surface area of gap-graded aggregates is lower because of the presence of higher percentage of coarse aggregates. Thus, gap-graded aggregates require relatively lesser cement and lower w/c ratio. Owing to this, the concrete using gap-graded aggregates will have reduced drying shrinkage. Gap-graded aggregates are used in architectural concrete to obtain uniform textures in exposed-aggregate finishes.

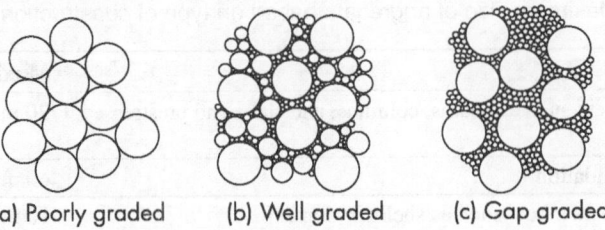

(a) Poorly graded    (b) Well graded    (c) Gap graded

**Fig. 6.7** Grading of aggregates

The grading of an aggregate is expressed in terms of percentages by weight retained (or passing through) through a series of sieves taken in order, as described in Section 6.8. Results of a sieve analysis are usually represented graphically in the form of *grading charts,* as shown in Fig. 6.8. In these charts, the percent passing of aggregates is plotted on the vertical axis and the sieve sizes are plotted on the horizontal axis. Upper and lower limits specified in codes for the allowable percentage of material passing each sieve may also be included on the grading chart. The following points are to be noted.

1. If the actual grading curve is below the code specified grading curve, the aggregates in the concrete mix will be coarser and there is a possibility of segregation.
2. If the actual grading curve is above the code specified grading curve, the aggregates in the concrete mix will be finer and hence more water will be required for proper workability. Thus, for a constant w/c ratio, the cement requirements will be higher.
3. If the actual grading curve is flatter than the code specified grading curve, it will indicate the absence of middle-sized particles (around the 10 mm size), resulting in a concrete with high shrinkage properties, high water demand, poor workability, poor pumpability, and poor placeability. Strength and durability may also be affected.
4. If the actual grading curve is steeper than the code specified grading curve, it will indicate excess of middle-sized particles, and the resulting mix will be harsher.

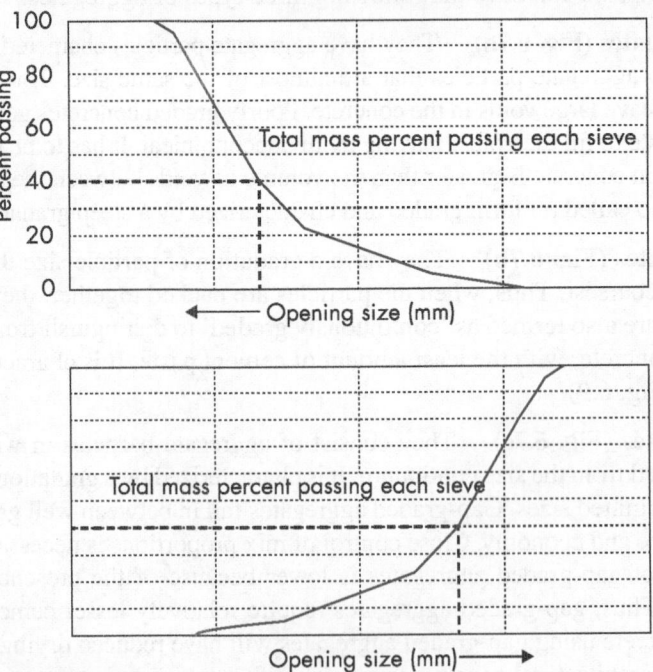

**Fig. 6.8** Gradation charts used for recording sieve analysis results

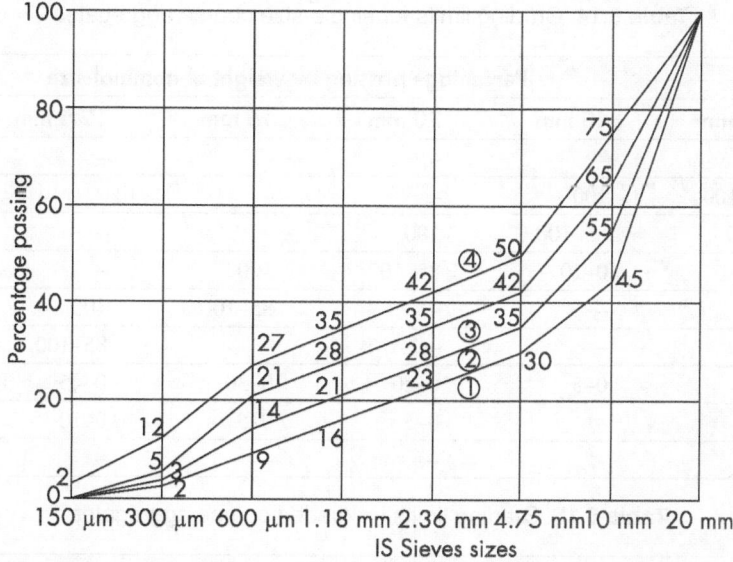

**Fig. 6.9** Grading curve for 20 mm nominal maximum size of aggregate

The Road Research Laboratory of the Department of Scientific and Industrial Research, London, has pre-pared a series of grading curves, which may be used in the design of concrete mixes. One such grading curve for the aggregate of maximum nominal size of 20 mm is shown in Fig. 6.9. It has to be noted that these are not the ideal or standard curves but represent grading used in the road research laboratory testing. Higher number of grading curve shows that there will be higher proportion of finer aggregates. The grading curve 1, shown in Fig. 6.9, will have more coarser aggregates and may be suitable for harsher mixes, which will be the most economical mix having highest permissible aggregate-cement ratio. The finest grading curve 4 shown in Fig. 6.9 is suitable for lean mixes, where a high workability is required. Thus, the outer curves 1 and 4 represent the limits for the normal continuous grading. The saving in cement by using a coarser grading can be significant. If locally available aggregate does not conform to the desired grading, the finer and coarser fractions of aggregates should be suitably combined to achieve the desired grading. Thus, for concrete mixture optimization, it is necessary to have combined gradation charts, which show the PSD of a mixture of fine and coarse aggregate. Combined gradation charts can identify gaps or deficiencies and an excess in a particular size range. However, it has to be remembered that the ideal proportions for concrete depend not only on the grading curve of fine and coarse aggregates, but also on the packing characteristics of the fine components such as cement, fly ash, and micro silica. The desired grading/packing can be achieved either by analytical calculations or graphically as shown in Fig. 6.9. Problems due to a poor gradation may be solved by considering alternative aggregates, blending, or special screening of existing aggregates.

## 6.10.1 Grading Limits

It has to be remembered that there is no universal or ideal grading curve. Hence, satisfactory performance of concretes can be obtained with various grading of aggregates. However, IS 383–16 has recommended certain limits within which the grading should lie in order to produce a concrete with satisfactory per-formance, fulfilling the properties of aggregates, such as shape, surface texture, type of aggregate, and the amount of flaky and elongated materials. The grading limits for single-size coarse aggregates and for graded coarse aggregates as per IS 383–16 are given in Tables 6.14 and 6.15, respectively. It has to be noted that the grading limits of coarse aggregates may be varied through wider limits than that of fine aggregates as they do not significantly affect the workability and finishing qualities of the concrete.

**166** Building Materials, Testing, and Sustainability

**Table 6.14** Grading limits for single-size coarse aggregates

| IS sieve designation | Percentage passing by weight of nominal size | | | | | |
|---|---|---|---|---|---|---|
| | 63 mm | 40 mm | 20 mm | 16 mm | 12.5 mm | 10 mm |
| 80 mm | 100 | – | – | – | – | – |
| 63 mm | 85–100 | 100 | – | – | – | – |
| 40 mm | 0–30 | 85–100 | 100 | – | – | – |
| 20 mm | 0–5 | 0–20 | 85–100 | 100 | – | – |
| 16 mm | – | – | – | 85–100 | 100 | – |
| 12.5 mm | – | – | – | – | 85–100 | 100 |
| 10 mm | 0–5 | 0–5 | 0–20 | 0–30 | 0–45 | 85–100 |
| 4.75 mm | – | – | 0–5 | 0–5 | 0–10 | 0–20 |
| 2.36 mm | – | – | – | – | – | 0–5 |

**Table 6.15** Grading limits for graded coarse aggregates

| IS sieve designation | Percentage passing by weight of nominal size | | | |
|---|---|---|---|---|
| | 40 mm | 20 mm | 16 mm | 12.5 mm |
| 80 mm | 100 | – | – | – |
| 63 mm | – | – | – | – |
| 40 mm | 90–100 | 100 | – | – |
| 20 mm | 30–70 | 90–100 | 100 | 100 |
| 16 mm | – | – | 90–100 | – |
| 12.5 mm | – | – | – | 90–100 |
| 10 mm | 10–35 | 25–55 | 30–70 | 40–85 |
| 4.75 mm | 0–5 | 0–10 | 0–10 | 0–10 |
| 2.36 mm | – | – | – | – |

Fine aggregates are generally classified into different zones based on the percentage passing through the IS 600 μm sieve. IS 383–16 classifies fine aggregates into four zones, namely Zone I, Zone II, Zone III, and Zone IV, in such a way that the range of percentage passing the 600 μm sieve in each zone does not overlap (see Table 6.16). Grading limits for fine aggregates as per this zoning is given in Table 6.16. It has to be noted that the fine aggregates become progressively finer as the grading zone increases from Zone I to Zone IV. It may be difficult to control the grading of fine aggregates. The grading may be controlled by combining fine aggregates from two or more different sources.

**Table 6.16** Grading limits for fine aggregates, as per IS 383–2016

| IS Sieve designation | Percentage passing by weight for grading | | | |
|---|---|---|---|---|
| | Zone I (Very coarse) | Zone II (Coarse) | Zone III (Fine) | Zone IV (Very fine) |
| 10 mm | 100 | 100 | 100 | 100 |
| 4.75 mm | 90–100 | 90–100 | 90–100 | 95–100 |
| 2.36 mm | 60–95 | 75–100 | 85–100 | 95–100 |
| 1.18 mm | 30–75 | 55–90 | 75–100 | 90–100 |

*(Contd)*

**Table 6.16** (*Contd*)

| IS Sieve designation | Percentage passing by weight for grading | | | |
|---|---|---|---|---|
| | Zone I (Very coarse) | Zone II (Coarse) | Zone III (Fine) | Zone IV (Very fine) |
| 600 μm | 15–34 | 35–59 | 60–79 | 80–100 |
| 300 μm | 5–20 | 8–30 | 12–40 | 15–50 |
| 150 μm | 0–10 | 0–10 | 0–10 | 0–15 |

Where the grading falls outside the limits of any particular grading zone of sieves other than 600 μm IS sieve, by a total amount not exceeding 5%, it should be regarded as falling within that grading zone. This tolerance should not be applied to percentage passing the 600 μm IS sieve or to percentage passing any other sieve size on the coarse limit of Grading Zone I or the finer limit of Grading Zone IV. For crushed stone sands, the permissible limit on 150 μm IS sieve may be increased by 20%. However, this does not affect the 5% allowance permitted above as applied to other sieve sizes. IS 383–16 also recommends that fine aggregate conforming to Grading Zone IV should not be used in reinforced concrete unless tests are made to ascertain the adequacy of the mix proportions.

In the case of all-in-aggregates, necessary adjustments may be made in the grading by the addition of single-sized aggregates without separating them into fine and coarse aggregates. As per IS 383–16, the grading of the all-in-aggregate should be as per Table 6.17.

**Table 6.17** All-in-aggregate grading

| IS Sieve designation | Percentage passing for all-in-aggregate of | |
|---|---|---|
| | 40 mm nominal size | 20 mm nominal size |
| 80 mm | 100 | – |
| 40 mm | 95 to 100 | 100 |
| 20 mm | 45 to 75 | 95 to 100 |
| 4.75 mm | 25 to 45 | 30 to 50 |
| 600 μm | 8 to 30 | 10 to 35 |
| 150 μm | 0 to 6 | 0 to 6 |

## 6.11 Green Substitutes for Coarse Aggregates

As already indicated, the demand for aggregates is continuously increasing, due to the increased construction activity throughout India. A recent estimate by the Freedonia Group, Inc., USA, shows that a total of 3330 MT of aggregates (sand-950 MT, crushed stone-1310 MT, gravel-950 MT, and others-120 MT) were consumed in India during 2015, which is likely to increase to 5075 MT by 2020 (www.freedoniagroup.com). Coarse aggregates are presently sourced from breaking large rocks and mountains, which will have long-term environmental impacts. These factors, in addition to the depletion of natural aggregates in several parts of India, have resulted in the use of alternative aggregates. Several foreign and Indian codes have been revised to include provisions on alternative aggregates. For example, IS 383:2016 includes provisions regarding quality requirements of iron slag, steel slag, copper slag, recycled aggregate (RA), and recycled concrete aggregates (RCA), along with necessary provisions relating to their utilization (see Table 6.1). It has to be noted that the manufactured aggregates should meet some additional requirements, as given in Table 3 of IS 383:2016. Some of these green alternatives are discussed briefly in the following.

### 6.11.1 Recycled Aggregates

The C&D waste is a major waste stream, the quantum of which is increasing as a result of increasing construction, maintenance, retrofitting, and demolition activities in several countries of the world, especially in China and India where there is a boom of construction activities. Demolition represented more than 90% of total C&D debris generation, whereas construction represented less than 10%. However, recycled aggregates, obtained from C&D debris, require necessary care while producing such that they are suitable for use in concrete. It is to be noted that recycling is not the best option as it also requires energy for processing and transportation and thus creates pollution. Recycling of concrete is a relatively simple process. It involves separation from C&D debris, breaking, removing, and crushing existing concrete into materials of specified size and quality. Some amount of washing may also be necessary. Significant contamination will largely be avoided if concrete is separated from other building materials early in the demolition process. The source concrete for recycled concrete aggregates should not be deteriorated concrete and it is desirable to source these from site being redeveloped for use in the same site. Similarly, if demolition concrete is from an unknown source, or multiple sources, it is more difficult to assess for hazardous contamination and hence such materials should be rejected. Even though there is high potential for recycling and reuse of this waste stream, the quantity of C&D waste being recycled in India is very low (*ICI Handbook*, 2016). The RCA have been successfully used in several countries in applications, such as bulk fills, bank protection, base or fill for drainage structures, road construction, noise barriers, and embankments.

Recycled aggregates may be of two types namely RA and RCA. RA is made from C&D waste, which may comprise concrete, brick, tiles, stone, etc., and RCA is derived from concrete after requisite processing. As per IS 383:2016, RA can be used as coarse aggregate and RCA can be used as coarse and fine aggregates. We will only consider RCA in further discussions. RCA contain not only the original aggregates, but also hydrated cement paste adhering to its surface. This paste reduces the specific gravity and increases the porosity compared to similar virgin aggregates. Figure 6.10 shows typical C&D waste and recycled concrete aggregate obtained from it. As with any new aggregate source, RCA should be tested for durability, gradation, and other properties. Table 6.18 shows typical properties of RCA.

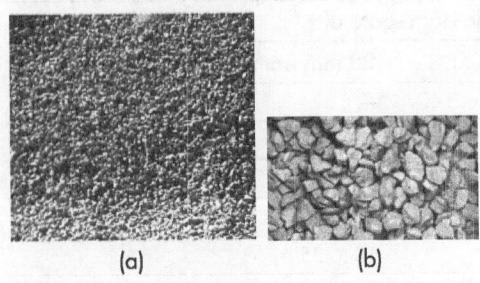

(a)                        (b)

**Fig. 6.10** Recycled concrete aggregate (a) Crushed RCA (b) Close-up view (*Courtesy*: Prof. Jake Hiller of Michigan Technological University)

If the old concrete contained alkali-reactive aggregate, petrographic examination and expansion tests should be conducted to check the suitability of aggregates. Concrete trial mixtures should be made to check the new concrete's quality and determine the proper mixture proportions.

**Table 6.18** Properties of RCA (*Courtesy*: Prof. M.C. Nataraja)

| Properties | Value |
|---|---|
| Shape and texture | Angular with rough surface (similar to crushed rock) |
| Water absorption,% | 4.5–8 |
| Specific gravity | 2.44–2.46 |
| L.A. abrasion,% | 20–45 |
| Sodium sulphate soundness,% | 18–59 |
| Magnesium sulphate soundness,% | 1–9 |
| Chloride content, kg/m³ | 0.6–7.1 |
| Fineness modulus | 3.47 |

The quality of concrete with recycled concrete aggregates is much dependent on the quality of the recycled material used. The workability of concrete made with recycled aggregates is lower than that of concrete made with natural aggregates because of the greater water absorption of the aggregate. This can be partially offset by pre-wetting the recycled aggregate. Several research studies have shown that RCA concrete has decreased compressive strength, decreased modulus of elasticity (15 to 50% lower), increased creep and shrinkage strains and higher rate of carbonation as compared to normal aggregate concrete. Some studies suggest that SCMs (fly ash and slag) may reduce these negative effects of RCA. However, new concrete made from RCA generally has good durability. In addition, carbonation, permeability, and resistance to freeze-thaw action have been found to be the same or even better than concrete with conventional aggregates (Kosmatka et al., 2011).

### 6.11.2 Iron and Steel Slags

*Iron and steel slags* are industrial by-products produced in steelmaking operations in integrated iron and steel plants under strict quality control. Steel slag is the by-product of steel making. It is produced in one of three furnace types: air-cooled Blast Furnace Slag (BFS)-Iron slag, Basic Oxygen/ Steel Furnace Slag (BOF or SFS), or Electric Arc Furnace (EAF)/ Ladle Metallurgical Furnace (LMF). The BOF is also referred to as 'LD converter slag' since it is generated from the Linz-Donawitz process. Some of these slag aggregates are shown in Fig. 6.11. Most of the available steel slag is from EAF. Steel slag consists principally of calcium silicates and calcium aluminoferrites and fused oxides of aluminium, iron, magnesium, and manganese. The compositions vary with the type of furnace, composition of furnace charges, grades of steel produced, and with individual furnace operating practices (www.nationalslag .org). These slags are made by pouring them from the furnace at a temperature of 1400–1700°C, into a cooling yard and cooling by air. After solidification, cooling may be accelerated by the application of water. Steelmaking slag contains about 10 to 20 metallic iron percent by mass, and is recovered by magnetic separation. The metal-free slag is then crushed and screened to meet the specified grading requirements for the particular application and stockpiled. The BFSs are also classified based on the processing technique used. Four main types of BFSs are: (a) air-cooled BFS, (b) expanded or foamed BFS, (c) pelletized BFS, and (d) granulated BFS. Steel slags obtained from the steel processing plants may also be classified into the following size categories: (a) below 16 mm - Fine, (b) 16–64 mm - Medium, and (c) 64–204 mm – Coarse.

Steel slags contain no impurities, such as clay, shells, and similar materials and have uniform chemical composition. Steel slags can be used as both coarse and fine aggregates in mortar and concrete. Steel slag aggregates generally exhibit a tendency to expand and become unstable because of the presence of free calcium oxide. The free lime and magnesium oxides that have not reacted with silicate structures can hydrate and expand in humid environments. To control this reaction, steel slag should be stockpiled outdoors for several months to expose the material to moisture from natural precipitation and/or application of water by spraying. Up to 18 months may be needed to hydrate the expansive oxides. These aggregates are hard, dense, angular, and have roughly cubical particles.

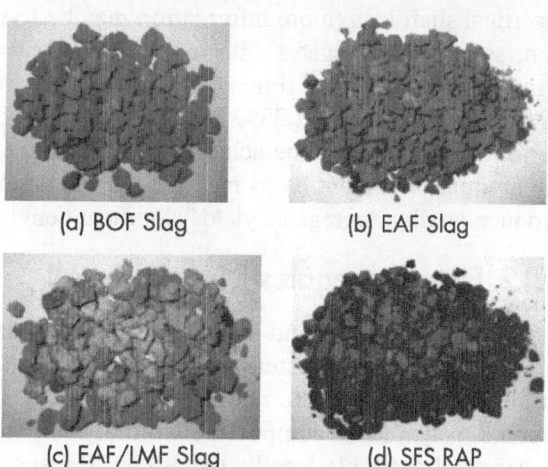

(a) BOF Slag      (b) EAF Slag

(c) EAF/LMF Slag      (d) SFS RAP

**Fig. 6.11** Types of steel slag (a) Basic oxygen furnace (BOF) slag (b) Electric arc furnace (EAF) slag (c) Ladle metallurgical furnace (LMF) slag (d) Steel furnace slag (SFS) (*Courtesy*: Prof. Jake Hiller of Michigan Technological University)

Common uses of these are as aggregates in ready-mix concrete, precast concrete, and hot mix asphalt aggregate. Processed steel slag has favourable mechanical properties for aggregate use, including good abrasion resistance and high bearing strength. Typical physical properties of steel slags are given in Table 6.19. They mainly consist of CaO (40–52%), $SiO_2$ (10–19%), and FeO (10–40%). The slags are highly resistant to weathering action such as freezing and thawing. Sulphate soundness losses are low for the steel slags. LA abrasion testing shows steel slag to be a hard aggregate.

**Table 6.19** Typical physical properties of steel slags (*Courtesy*: Prof. M.C. Nataraja)

| Property | Value |
|---|---|
| Unit weight, kg/m³ | 1600–1920 |
| Specific gravity | 3.1–3.6 |
| Maximum Dry Unit Weight (kN/m³) | 15.7–18.9 |
| Porosity,% | Up to 3 |
| Water absorption,% | 0.2–2 |
| California Bearing Ratio (CBR) | Up to 300 |
| Sodium Sulphate Soundness Losses,% | <12 |
| Total deleterious material | Nil |

Owing to their high heat capacity, steel slag aggregates retain heat considerably longer than conventional natural aggregates. This heat retention characteristic of steel slag aggregates can be used advantageously in hot mix asphalt repair work, especially in cold weather.

### 6.11.3 Sintered Fly Ash Lightweight Aggregates

Sintered fly ash lightweight aggregates (SFLA) have been commercially produced in several countries such as the UK, the USA, Japan, India, and North Europe. They can be produced by a two-step process. In the first step, the fly ash is pelletized and in the second, these pellets are sintered at 1100–1200°C in a vertical shaft kiln (more information may be found in IS 9142 (Part 2): 2018). This process produces light spherical aggregates with high voids. Concrete produced using these aggregates is around 22% lighter and at the same time 20% stronger than normal weight aggregate concrete. Drying shrinkage of this concrete is about 33% less than that of normal weight concrete (Kayali, 2008). In addition, superior properties could be achieved without increasing the cement content – it was possible to reduce the amount of cement by as much as 20% without affecting the required strength. Utilizing fly ash to produce quality aggregates yields significant environmental benefits.

### 6.12 Fine Aggregates (Natural Sand)

Traditionally, natural sand having particle size between 4.75 mm and 0.15 mm has been used as a fine aggregate in mortar, plaster, and concrete. Sand is essentially a granular form of silica (silicon dioxide or $SiO_2$), usually in the form of quartz. Typically, any sand, which lacks any impurities such as clay and organic matter and is composed of silicates, will not absorb any water. For economic reasons, it is better to use sand available locally. Otherwise, transportation costs will become a major portion of the cost of sand. Sand used for concrete is known as standard sand (IS 650:1991). As per IS 650, the standard sand should be obtained from Ennore, Tamil Nadu. However, it can also be obtained from local river beds or pits. It should be of quartz, light grey, or whitish variety and should be free from silt. The sand grains should be angular, preferably spherical. It should (100%) pass through 2 mm IS sieve and should be (100%) retained on 90 μm IS sieve with the PSD, as shown in Table 6.20. The fineness of the natural sand should be determined by sieve analysis and depending on its fineness it should be used in different items of construction. The standard sand should be free from organic impurities. As per IS 650:1991, the loss of mass on extraction, when tested with hot hydrochloric acid of relative density 1.21 (as per IS 265:1987), should not be more than 0.25%.

**Table 6.20** Particle size distribution of standard sand

| Particle size | Percent |
| --- | --- |
| Smaller than 2 mm and greater than 1 mm | 33.33 |
| Smaller than 1 mm and greater than 500 µm | 33.33 |
| Smaller than 500 micron but greater than 90 µm | 33.33 |

Owing to increased construction activity, natural sand has become rare and expensive. Also, due to environmental and ecological concerns of erosion of soil, change in the course of river, and stability of riverbanks, dredging of rivers is banned in several states of India. Hence, alternative materials such as crushed stone, slag, recycled concrete and geosynthetic aggregates (polymer based and made from recyclable plastics) are being used as a substitute for sand (see Section 6.13).

### 6.12.1 Sources of Sand

The various sources of sand, which are used in building construction, are described below:

*Pit sand*   This type of sand is obtained from old stream beds or by forming pits in soils. These sands consist of sharp angular grains, which are free from salts. These serve as excellent material for mortar or concrete. These may contain clay or other organic material and may have coating of oxide of iron. These impurities should be removed before using them in concrete or mortar.

*River sand*   It is obtained from the beds of rivers and is widely used for all purposes. Owing to attrition under the action of water current, this sand consists of more or less rounded and smooth grains. It has very low silt and clay contents and is white in colour. While using river sand, caution should be exercised because of the following: (a) since river sand is brought down by river water from upstream, it may have widely different mineralogy and may contain substances leading to alkali-aggregate reaction and (b) river sand dredged from river estuaries close to the sea have high chloride content.

*Stream sand*   It is obtained from small streams that may dry out during summer. It is generally coarse in nature.

*Crushed stone/crushed gravel sand*   This is a substitute for river sand. It is manufactured by crushing either granite or basalt rocks or gravel using a three-stage crushing process. Mortar made with crushed rock fine is very sticky and hence difficult to trowel. As this mortar has a higher water demand, this mortar will have higher tendency to drip downwards after trowelling; the mortar will also have higher shrinkage cracking after hardening. Hence, crushed stone/gravel sand should not be considered as the best substitute for mortar.

*Sea sand*   This sand is obtained from seashores. It is brown in colour and also has fine rounded particles. As it is obtained from the sea it contains salt and hence should not be used in reinforced concrete works. It is better to wash this sand in fresh water before use.

### 6.12.2 Classification of Sand

The classification of sand is done on the basis of source, mineralogical composition, size of the particles, or the PSD. Thus, based on source, sand is classified as natural sand, and crushed-stone sand, or crushed-gravel sand (see also, Section 6.2.2) Based on mineralogical composition, sand is classified as quartz, haematite, mica, feldspar, or shale (Dell, 1959). Based upon its size, sand is classified as coarse sand (having fineness modulus of 2.90 to 3.20 and with particle size of 2 to 4.5 mm), medium sand (having fineness modulus of 2.60 to 2.90 and with particle size of 0.425 to 2 mm), and fine sand (having fineness modulus of 2.20 to 2.60 and with particle size of 0.075 to 0.425 mm). Based on the PSD, it is classified in four grades: grading zone I (coarse) to grading zone IV (fine), as shown in Table 6.16.

### 6.12.3 Bulking of Sand

The presence of moisture in fine aggregates will have a tendency to increase the volume – this is termed as the *bulking of sand*. Owing to this moisture, each particle of fine aggregate will be coated with a thin film of moisture, which will then exert a force on each particle, which is known as *surface tension*. This surface tension will push each particle away from each other, reducing the direct contact among individual particles and resulting in increased volume. The extent of bulking is proportional to the percentage of moisture present in the sand and also its fineness. When it is absolutely dry or completely saturated, finer aggregates will not exhibit any bulking. Up to about 4–6% of moisture content, there will be a gradual increase in the bulking of sand. After reaching a maximum value of bulking, it begins to decrease in volume, because of the merging of moisture films, until the sand is inundated with water. At this stage, the bulking practically vanishes. The variation of percentage increase in the volume of typical sand with moisture content is shown in Fig. 6.12. It is seen from this figure that finer sand bulks considerably more than the coarse sand. With coarse and medium sands, the bulking is in the range of 15–30%. Extremely fine sand, especially the manufactured fine aggregate, may bulk as much as 40%, at a moisture content of 5–8%, and hence such sand is not suitable for making concrete. The percentage bulking may be obtained by using the test procedure described in Clause 4.3 of IS 2386 (Part 3):1963.

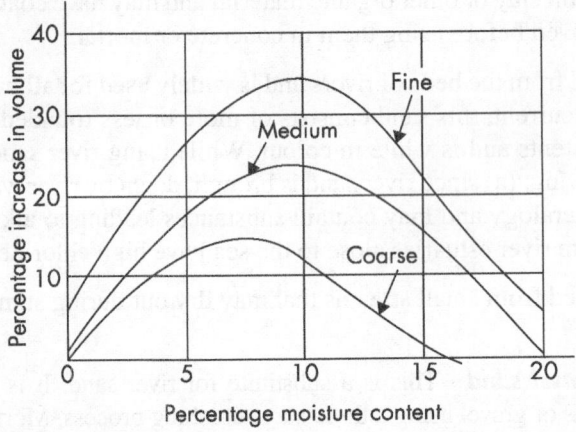

**Fig. 6.12** Effect of moisture content on the bulking of sand

When designing concrete mixes, if sand is measured by volume and no allowance is made for bulking, the mix will be harsher, and rich in cement as moist sand will occupy considerably larger volume than the same mass of dry sand. This results in a mix containing less sand. The *yield of concrete* will also be reduced (The yield of concrete is defined as the volume of freshly mixed concrete from a known quantity of ingredients.). Hence, in such cases, a correction factor called *bulking factor* must be applied to the volume of sand, in order that correct amount of sand is used in the mix. As per Neville (2011), the value of bulking factor ranges from 1.22 (medium sand) to 1.4 (crushed sand). For example, if the bulking of sand is 10% and if mix ratio is 1:1.5:3, the actual volume of sand used should be

1.1 × 1.5 = 1.65 instead of 1.5 per unit volume of cement. If this correction is not applied, the actual dry sand in the concrete will be 1/1.1 × 1.5 = 1.36, instead of 1.5 per unit volume of cement. Correspondingly, the mix proportion will be 1:1.36:3 instead of 1:1.5:3. Owing to this, there will be chances of segregation, honeycombing, and increase in concrete strength.

Because the volume of saturated sand is the same as that of dry sand, the most convenient way of determining bulking at sites is by measuring the decrease in volume of the given sand when inundated. Thus, a 250 cm³ cylindrical vessel is filled with loosely packed moist sand. The filled sand is consolidated by simple shaking so that the sand is levelled. The height of sand in the container is measured. Let this height be $h_1$. Then, the contained is emptied carefully, ensuring that no sand is lost during this transaction. The container is now partially filled with water and the sand is gradually fed back, with stirring and rodding to expel all air bubbles. The decreased volume of sand is noted by measuring the height of saturated sand as $h_2$. Now, the percentage of bulking is calculated as

$$Percentage\ bulking = \frac{h_1 - h_2}{h_2} \times 100 \tag{6.1}$$

It has to be noted that correction for bulking is not required when mix design is based on weigh batching. Coarse aggregate shows only a negligible increase in volume due to the presence of moisture, as the thickness of moisture films is very small compared to their particle size.

## 6.12.4 Uses of Sand

Sand is used as fine aggregate in concrete and also in mortars and plasters. It is also used for filling.

**Sand for concrete work** Very fine sands of Zone IV (see Table 6.16) are not recommended to be used for structural concrete. Very coarse sands pose difficulties while surface finishing of concrete but provide good strength. Use of fine sand results in more cohesion than coarse sand and hence results in the use of lesser quantity of sand. However, very fine sand will increase the water demand of the mix, whereas very coarse sand could compromise its workability. The amount of material passing the 300 μm and 150 μm sieves will affect workability, slab surface texture, and bleeding. Increased bleeding will occur as the portion passing the 300 μm sieve increases. Sand having fineness modulus of 2.3 to 3.1 may be suitable for concrete work.

**Sand for mortars and plasters** Locally available sand is usually used for making mortars or plasters. Though sharp and coarse sands give higher strength for mortars, they are unsuitable for brickwork as they lack the required workability or plasticity. Moreover, mortars made of coarse sands do not adhere easily to bricks. Soft-sand, on the other hand, is ideal for making mortar and plaster for brickwork (Sand which is not washed and containing some clay particles is called soft-sand because, it feels soft and smooth when held in the hand. Sand which is washed feels coarse in the hand, hence the term sharp.). As per IS 2116:1980, the maximum quantities of clay, fine silt, fine dust and organic impurities in the sand shall not exceed 5% sand by mass. If only coarse sand is available, it is better to sieve it and use only the finer sand for plastering.

Generally, the sand passing through 2.36 mm IS sieve is considered suitable for mortar in brickwork and the sand passing through 1.18 mm IS sieve is considered suitable for plastering and pointing. The particle size grading of sand for use in mortars and plasters, specified in IS 2116:1980 and IS 1542:1992, respectively, is given Table 6.21. A sand whose grading falls outside these specified limits due to excess or deficiency of coarse or fine particles may be modified to comply with the standard by screening through a suitably sized sieve and/or blending with required quantities of suitable sizes of sand particles.

**Table 6.21** Grading of sand for masonry mortars and plasters

| IS sieve designation | Percentage passing by weight | |
| --- | --- | --- |
| | Masonry mortar (IS 2116:1980) | Plaster (IS 1542:1992) |
| 4.75 mm | 100 | 95–100 |
| 2.36 mm | 90–100 | 95–100 |
| 1.18 mm | 70–100 | 90–100 |
| 600 μm | 40–100 | 80–100 |
| 300 μm | 5–70 | 20–65 |
| 150 μm | 0–15 | 0–15 |

It is seen from the table that the sand used for mortar is one in which the material passing through 600 μm sieve should be 40–100% and, for plaster, it is 90–100%. As per the grading given in Table 6.16, the fineness modulus of sand used for plasters could vary from 1.2 to 2.2. However, it is preferable to have minimum fineness modulus of 1.4 in case of crushed stone sands and crushed gravel sands and a fineness modulus of 1.5 in case of naturally occurring sands. Generally, the best sand for lime mortar is sharp sand with angular particles ranging in size from 4 mm to 150 μm, which will result in strong mixes, whereas soft sands with fine circular shaped particles make weak mixes. Sea sand should be avoided, even after washing, as any residual salt may be harmful to the masonry and the aesthetic appearance (due to efflorescence).

**Sand for filling** Sand is also used in building construction for filling below underground floors and behind retaining walls. When used as filling below underground floors, it reduces the action of capillary suction, through which water travels from foundation soil to the floor level. For this purpose, coarse sand with large voids between grains should be used. When used as filling behind the retaining walls, it drains off water, which will exert additional pressure on the retaining wall. Well-graded sand is suitable for this purpose. In addition, about 100 to 150 mm thick sand layer is provided below footings in clayey soils, to act as buffer layer between levelling concrete and soil. This sand layer, in addition to preventing dampness, will ensure a separation between soil and structure and protect concrete from the harmful action of sulphates and other acidic substances. A sand base is also used below brick walk ways, which will allow the bricks to maintain their position securely. Sand may be used for filling of trenches that will support structures.

## 6.13 Green Substitutes for Sand

Fine aggregate, in the form of sand, is mined from river beds. Excessive sand and gravel mining cause the degradation of rivers and lowers the stream bottom. Excessive sand mining is also a threat to bridges, riverbanks, and nearby structures. Sand mining is a direct cause of erosion, and also affects the local wildlife. River mining imposes severe damages to the physical and biological environments of the river systems. In addition, sand mining generates extra vehicle traffic, which negatively impairs the environment. Hence, the National Green Tribunal on 5 August 2013 banned sand mining of rivers across India without license and environmental clearance from the Union Ministry of Environment and Forests, which made the procurement of natural sand difficult and expensive. Hence, several green substitutes for sand have been experimented. The prominent is the manufactured sand, which can be used instead of sand, as per IS 383. It is also permitted in IS 383 to substitute iron slag (25% in RCC) and steel slag (not permitted in RCC but 25%, in plain concrete), copper slag (35%), and RCAs (20% only up to M20 concrete). These green materials are briefly discussed in the following sub-sections.

### 6.13.1 Manufactured Sand

Crushed stone fine aggregate, also called as *manufactured sand* or *M-sand*, in place of natural sand, either as full or part replacement, is becoming a common practice in several parts of India, especially in Kerala and Tamil Nadu (see Fig. 6.13). Manufactured sand (M-sand) is permitted by IS 383 and may be defined as a purpose-made crushed fine aggregate produced from suitable source materials. While producing M-sand, only source materials with suitable strength, durability, and shape characteristics have to be considered. The process of manufacturing of this sand involves three stages: (a) crushing of stones into aggregates by Vertical Shaft Impactor (VSI), which are then fed into Rotopactor for further crushing to get sand of required grain sizes (as fines); (b) screening to eliminate dust particles; and (c) washing of sand to eliminate very fine particles. Separation into discrete fractions, recombining, and blending may be necessary. It is usually tested for various quality aspects before distribution as a construction aggregate. The size specification for M-sand is that it should pass completely through a 10 mm sieve. The end product will satisfy all the requirements of IS 383 and can be used in concrete and other constructions. The VSI crushers are available with capacity up to 400 tonne per hour. Sand manufactured by the VSI crushers/Rotopactors is cubical and angular in shape and are better suited for concrete (Nataraja et al., 2016). On the other hand, manufactured sand from jaw crusher, cone crusher, and roll crusher often contain a higher percentage of dust and has flaky particles and is not recommended (Nataraja et al., 2016). A comparison of the M-sand with the natural river sand is provided in Table 6.22.

(a)                                                            (b)

**Fig. 6.13** (a) Manufactured sand (*Photo*: Er Rahul Leislie) (b) River sand

**Table 6.22** Comparison between manufactured sand and natural river sand (*Courtesy*: Prof. M.C. Nataraja)

| Property | River sand | M-sand | Remarks |
|---|---|---|---|
| Colour | Whitish grey | Grey | – |
| Shape of particles | Spherical | Cubical | Good |
| Gradation | Cannot be controlled | Can be controlled | – |
| Particle passing 75 µm | Presence of silt should be less than 3% as per IS 383 | Presence of silt should be less than 10–15% as per IS 383 | Greater limit in M-sand |
| Specific gravity | 2.3–2.7 | 2.1–3.2 | Many vary depending on source |
| Water absorption,% | 1.5–3 | <5 | – |
| Ability to hold surface moisture,% | Up to 7 | Up to 10 | – |
| Grading zone | Zones II and III FM 2.2–2.8 | Zone II FM 2.6–3.0 | Zone II is recommended for mass concrete |

*(Contd)*

**Table 6.22** (*Contd*)

| Property | River sand | M-sand | Remarks |
|---|---|---|---|
| Soundness (Sodium sulphate-ss and Magnesium sulphate-ms) (5 cycles) | Relatively less sound | Relatively sound | Limit 10% ss and 15% ms as per IS 383:2016 |
| Alkali-silica reactivity,% | Mortar bar expansion at 38°C < 0.05 (90 days) and 0.10 (180 days) | $Na_2O < 0.3$ (also see clause 5.7 of IS 383) | Limit 0.10% expansion at 180 days |

From Table 6.22, it is seen that there are some inherent differences between M-sand and natural river sand, mainly due to the geological process of shaping of the particles of natural sand. Thus, the individual particles of natural sand tend to be rounded to sub-rounded and have smooth surface texture. Moreover, natural sand may have high silt and clay content. It can be damaging for screed and concrete, if the sand is not sufficiently processed to bring down clay and other impurities to acceptable levels. By contrast, M-sand typically consists of graded and angular particles with a rough surface structure. Manufactured sand is free of silt and clay particles, and has denser particle packing than natural sand. It has to be noted that as per Table 9 of IS 383, the permissible limit of materials finer than 150 μm sieve for natural sand is 0–10%, whereas it is 0–20% for M-sand. The cohesiveness and workability of fresh concrete depend mainly on particles finer than 300 μm and 150 μm, respectively. Sands deficient in these sizes make concrete harsh and prone to segregation and excess amount of these sizes result in stickiness. Hence, IS 383 lays down a limit of 8–30%, for Zone II sand, and 12–40%, for Zone III sand (see Table 6.16). It has to be noted that these limits are the same for natural as well as manufactured sand. Similarly, the limit for clay lumps is 1.0% in either case.

---

**Case Study** | **Signature Bridge across river Yamuna in East Delhi**

It has to be noted that M-sand was used extensively in the construction of Signature Bridge across river Yamuna in East Delhi. More details about sieve analysis, mix design details of M35 and M40 grade concrete used in this bridge are provided by Mullick (2015). It was necessary to use higher dosage of superplasticizer, while using M-sand. The compressive strength of concrete using M-sand was slightly higher than with normal sand. Similarly, the bond strength of 12 mm high yield strength deformed (HYSD) bar was slightly higher.

## 6.13.2 Copper Slag

Copper slag is produced as a by-product from copper smelter, while producing copper through pyrometallurgical process. In the process of smelting, the iron present in the copper concentrate combines chemically at 1200°C with silica present in flux materials such as river sand/silica sand/quartz fines to form iron silicate, which is termed as copper slag (IS 383:2016). The copper slag is quenched with water to produce granulated copper slag. Copper slag is blackish granular, similar to medium-to-coarse sand, and has size ranging from 150 μm to 4.75 mm. Primary constituents of this slag are iron oxide $(Fe_2O_3)$ (55–60%) and silica $(SiO_2)$ (27–33%). The properties of copper slag are shown in Table 6.23. Addition of slag in concrete increases the density (up to 19%), thereby the self-weight. Most mixes with copper slag resulted in concrete with adequate properties, with optimal sand replacement in the range of 30–60%. Low water absorption and the coarser (than natural sand) and glassy surface of copper slag increase the workability of concrete.

**Table 6.23** Properties of copper slag (Satish Kumar et al., 2016, *Courtesy*: Prof. M.C. Nataraja)

| Properties | Value |
|---|---|
| Total deleterious material | Nil |
| Bulk density, g/cc | 1.7–1.9 |
| Specific gravity | 3.5–3.9 |
| Water absorption% | 0.14–0.17 |
| Chloride content, ppm | <15 |
| Free moisture,% | <1 |
| Angularity number | 49 |
| Fineness modulus | 3.47 |
| Alkali reactivity | Nil |

### 6.13.3 Granulated Blast Furnace Slag (GBFS)

*Iron slag* is a by-product of integrated iron and steel plants, while producing iron in blast furnaces or basic oxygen furnaces. Limestone is added during the production process of iron/steel to remove the silica, alumina ($Al_2O_3$), and other non-ferrous components contained in iron ore. The added limestone fuses with these components and lowers their melting point, making it easier to separate them from the iron and recover them. This recovered substance is called the iron/*blast furnace slag*. The molten slag at a temperature of approximately 1400–1600°C is taken out of the furnace and quenched rapidly using large volume of water. This results in vitrified (glassy) material with a sand-like appearance, with particles typically 1–5 mm size. Because of the rapid quenching process, its structure is more amorphous. Granulated slag is used primarily in the production of ground granulated blast furnace cement. It can also be used as normal weight aggregate, when further processing is done to improve the bulk density to more than 1.35 kg/L. Though not widely used, this type of slag is generally considered suitable as an aggregate in concrete. The compressive strengths of concrete made with blast furnace slag aggregate are typically comparable to equivalent conventional concretes (CCANZ Report 14, 2011, ACI E1–16). Several hundred cubic metres of concrete made with granulated slag sand have been used at Port Kembla, Australia (CC&AA Report, 2008).

### 6.13.4 Foundry Waste Sand

Metal foundries use large amounts of sand in the metal casting process. The sand is reused several times. When it cannot be reused, it is discarded as foundry waste sand (It is estimated that approximately 100 MT of sand is used and that 6–10 MT of foundry sand is discarded.). It consists primarily of silica sand ($SiO_2$, 88%), coated with thin film of burnt carbon, residual binder (bentonite and resins), and dust. In India, about 6–10 MT of foundry waste sand are generated every year. The specific gravity of foundry sand varies between 2.39 and 2.79. Waste foundry sand has low absorption capacity and is non-plastic. To satisfy concrete specifications, it is often necessary to remove fine material passing a mesh 75 μm sieve or blend used foundry sand with coarser sands. Foundry sand is black in colour. Up to 15% fine aggregate replacement with foundry sand may produce minimal colour change; above this percentage, it may give the finished concrete a greyish/black tint, which may not be desirable.

In addition to the above, bottom ash from thermal power plants, glass waste, quarry dust (IS 383 prohibits its use in concrete), fine fractions of recycled concrete aggregate (<4.75 mm), as part replacement of fine aggregates from natural resources have also been reported.

## 6.14 Storing and Handling of Aggregates

Aggregates should be stored at site on a hard, dry, and level ground. If such a surface is not available, a platform of wooden planks or iron sheets, or a floor of bricks, or a thin layer of lean concrete should be used. Contact with clay, dust, vegetable, and other foreign matters should be avoided. Storage piles are usually left uncovered, partially because of the need for frequent material transfer into or out of storage. At ready mixed concrete plant, aggregates are typically stored in a bin or a stockpile. Only one size aggregate should be stored in each storage bin or stockpile. Segregation of aggregate means separation of the course particles from the fine particles. Aggregates with a wide range of particle sizes tend to be most susceptible to segregation. Segregation of aggregate should be avoided. The following factors should be considered while handling and storing aggregates (http://precast.org/2010/05/raw-material-handling-and-storage/):

1. Aggregate should be stored in small piles or in thin horizontal layers of uniform thickness to minimize segregation. It has to be noted that some types of aggregates are prone to segregation-rounded aggregate are likely to segregate more than crushed aggregate; larger size aggregates segregate more than smaller size aggregates.

2. Aggregates should not be stored in large conical piles, since there is probability of larger size aggregates to roll out and separate from the smaller size aggregates.

3. Fine and coarse aggregates should either be stored separately or heaps be separated by dividing walls. Fine aggregate should be stored in such a way that loss due to the effect of wind is minimum, viz., in the leeward side behind a wall, or by covering with a polyethylene sheet. On large projects, dividing walls should be constructed to give each type of aggregates its own compartment.

4. Aggregates should be discharged into bins, directly above and onto the centre of the pile. Discharging aggregate against the side of a bin or a wall will tend to cause segregation.

5. Storage bins can be square or circular but their bottoms should slope not less than 50° from horizontal on all directions towards a central outlet. Flat bottomed square storage bins with angular corners, should be avoided. Instead, square storage bins with rounded corners, which ensure movement of all material towards the outlet, are to be used. It is preferable to keep the storage bins as full as practical to minimize possible changes in the gradation caused by withdrawal of the material and crushing of aggregates.

6. The piled aggregates can be reclaimed with a front-end loader, removing slices from the edges of the pile from bottom to top.

7. Since the surface moisture on aggregates can add a significant amount of water to a concrete mix, it is important to control the moisture in aggregate stock. The moisture content may be maintained by sprinkling the aggregates with water, keeping them in a constantly saturated condition. Moisture probes may be placed in the sand bins to automatically measure the amount of free surface moisture in the aggregates being batched.

8. Washed aggregates should be stockpiled for sufficient period before use, so that they can drain and have uniform moisture content.

9. In cold climates, it is preferable to cover overhead or underground bins. Plants may also use heaters to maintain a reasonable aggregate temperature or to prevent freezing of aggregates.

10. Trucks and loaders should not be driven on and stockpiling equipment should not be allowed over aggregate stockpiles, as their weight can crush the aggregates and introduce mud and dirt into the piles.

Dust emissions occur during the storage cycle especially when material is loaded onto or loaded out from the pile, or during strong wind currents. The movement of trucks and loading equipment in the storage pile area can also result in substantial dust. These dust emissions may be controlled by watering the piles or by using chemical wetting agents (such as *surfactants*). Covering inactive piles or storing aggregates within enclosed areas (to reduce wind erosion) can also reduce dust emissions. Note that continuous exposure to silica dust may lead to a fatal disease called silicosis.

## SUMMARY

- Aggregates typically occupy 60–75% of the volume (or 70–85% of the mass) of concrete and strongly influence the freshly mixed and hardened properties, mixture proportions, and economy of concrete.
- Aggregates may be classified according to geological region (natural and artificial), size (coarse and fine, single-sized, or graded), shape (rounded, angular, or flaky), and based on their unit weight (normal weight, heavy weight, and lightweight).
- Aggregates used in concrete should be clean, strong and hard, impermeable, properly shaped as well as properly graded. In addition, they should have chemical stability, resistance to abrasion, freezing, and thawing (in cold climates).
- It is better to avoid flaky and elongated aggregates. High moisture content in the aggregate will increase the effective water/cement ratio considerably and subsequently reduce the strength and durability of concrete.
- Deleterious substances in aggregates are considered undesirable, as their intrinsic weakness, softness, fineness, or other physical/chemical characteristics may be harmful to the performance of concrete. They affect the properties of concrete in green as well as hardened state.
- Porous and weak aggregates undergo excessive volume changes under adverse environmental conditions, and hence are to be avoided.
- Alkali-aggregate reaction (AAR) has two forms—ASR and ACR—of which ASR is more common. For ASR to occur, all the three following conditions must be present: (a) aggregate containing an alkali reactive constituent; (b) a cement with high-alkali content, or another source of alkali; and (c) a high moisture level within the concrete.
- The coefficient of thermal expansion of aggregates typically ranges from $4 \times 10^{-6}$ per °C to $13 \times 10^{-6}$ per °C.
- The fineness modulus is the sum of the cumulative percentages of aggregates retained on each of the specified series of sieves divided by 100. It varies between 2.0 and 3.5 for fine aggregate, between 5.5 and 8.0 for coarse aggregate, and from 3.5 to 6.5 for all in-aggregate. An efficient mix design should use maximum aggregate and minimum cement, and have trade-off between efficient grading and workability.
- Based on the grading, there can be three types of aggregates: poorly graded, well graded, and gap graded. Concrete with satisfactory performance is obtained by using the grading prescribed clause 6 of IS 383-16.
- Natural sand having particle size between 4.75 and 0.15 mm, is used as a fine aggregate. In the presence of moisture, natural sand bulks and the extent of bulking depend upon the percentage of moisture present in the sand and its fineness. Bulking has to be considered in the mix design. Sand is used for concrete, mortar, plastering, and filling in basements.
- Several tests are performed on aggregates, as per IS 2386-(Parts 1–8). These tests are described in Section 25.10 of Chapter 25.
- The depletion of natural aggregates and ban on aggregate mining in several parts of India has resulted in the use of alternative aggregates such as bottom ash from thermal power plants, recycled concrete aggregates, and M-sand, and other industrial and agro-waste products, such as foundry sand, blast furnace slag, iron and steel slag, copper slag, fly ash aggregates, waste glass powder, and oil palm/coconut shell aggregates.
- Aggregates should be properly stored and handled, without altering their original properties.

## EXERCISES

### Multiple-choice Questions

1. According to IS 383–16, the allowable percentage of recycled coarse aggregate in reinforced concrete is
   - (a) 20% up to M20
   - (b) 25% up to M20
   - (c) 20% up to M25
   - (d) 50% up to M25

2. According to IS 383–16, the allowable percentage of recycled fine aggregate in reinforced concrete is
   - (a) 20% up to M20
   - (b) 25% up to M20
   - (c) 20% up to M25
   - (d) 50% up to M25

3. According to IS 383–16, the allowable percentage of copper slag fine aggregate in reinforced concrete is
   - (a) 20%
   - (b) 25%
   - (c) 35%
   - (d) 40%

4. The maximum size of coarse aggregates allowed in RCC works is
   - (a) 12 mm
   - (b) 20 mm
   - (c) 25 mm
   - (d) 40 mm

5. Which is the size of particle below which the aggregate considered as fine aggregate?
   - (a) 2.36 mm
   - (b) 4.75 mm
   - (c) 10 mm
   - (d) All of these

6. An aggregate is considered flaky when its least lateral dimension (thickness) is
   - (a) less than 2/5th of its mean dimension
   - (b) less than 3/5th of its mean dimension
   - (c) less than 3/4th of its mean dimension
   - (d) less than 65% of its mean dimension

7. An aggregate is considered elongated when its length is
   - (a) 1.5 times greater than its mean dimension
   - (b) 1.6 times greater than its mean dimension
   - (c) 1.8 times greater than its mean dimension
   - (d) 2.0 times greater than its mean dimension

8. Aggregates are considered normal weight when
   - (a) specific gravity is 2.4–2.9 and bulk density is 1280–1920 kg/m$^3$
   - (b) specific gravity is 3.6–4.6 and bulk density is 2100–4650 kg/m$^3$
   - (c) specific gravity is 1.3–1.6 and bulk density is <1120 kg/m$^3$
   - (d) none of these

9. Aggregates are considered light weight when
   - (a) specific gravity is 2.4–2.9 and bulk density is 1280–1920 kg/m$^3$
   - (b) specific gravity is 3.6–4.6 and bulk density is 2100–4650 kg/m$^3$
   - (c) specific gravity is 1.3–1.6 and bulk density is < 1120 kg/m$^3$
   - (d) none of these

10. The aggregate impact value of coarse aggregates, used in concrete for purposes other than wearing surfaces, should not exceed
    - (a) 30%
    - (b) 40%
    - (c) 45%
    - (d) 50%

11. The aggregate crushing value of coarse aggregates used in concrete of wearing surfaces should not exceed
    - (a) 30%
    - (b) 40%
    - (c) 45%
    - (d) 50%

12. The abrasion value of coarse aggregates used in concrete of wearing surfaces should not exceed
    - (a) 30%
    - (b) 40%
    - (c) 45%
    - (d) 50%

13. As per IS 383–16, the maximum percentage by mass of materials finer than 75 μm IS Sieve allowed in natural sand is
    - (a) 3%
    - (b) 7%
    - (c) 12%
    - (d) 15%

14. As per IS 383–16, the maximum percentage by mass of materials finer than 75 μm IS Sieve allowed in M-sand is
    - (a) 3%
    - (b) 10%
    - (c) 12%
    - (d) 15%

15. The average loss of mass after five cycles for fine aggregates, when tested with sodium sulphate, should not exceed
    - (a) 10%
    - (b) 12%
    - (c) 15%
    - (d) 18%

16. The average loss of mass after five cycles for coarse aggregates, when tested with magnesium sulphate, should not exceed
    - (a) 10%
    - (b) 12%
    - (c) 15%
    - (d) 18%

17. The coefficient of thermal expansion of aggregates ranges from
    - (a) $3 \times 10^{-6}$ per °C to $12 \times 10^{-6}$ per °C
    - (b) $3 \times 10^{-6}$ per °C to $12 \times 10^{-6}$ per °C
    - (c) $4 \times 10^{-6}$ per °C to $13 \times 10^{-6}$ per °C
    - (d) $6 \times 10^{-6}$ per °C to $13 \times 10^{-6}$ per °C

18. The fineness modulus of fine aggregate varies from
    - (a) 2.2 to 2.6
    - (b) 2.6 to 2.9
    - (c) 2.9 to 3.2
    - (d) 2.0 to 3.2

19. The maximum size of coarse aggregates used in concrete elements should be less than
    - (a) 1/4th the minimum thickness of the member
    - (b) minimum clear spacing of main bars- 5 mm
    - (c) minimum cover - 5 mm
    - (d) all of these
    - (e) none of these

20. The maximum size of coarse aggregates in RCC work should be
    - (a) 16 mm
    - (b) 20 mm
    - (c) 25 mm
    - (d) 40 mm

21. The minimum thickness of a concrete structural member is 250 mm. The nominal maximum size of the aggregate can be
    - (a) 60 mm
    - (b) 40 mm
    - (c) 12 mm
    - (d) All of these

22. For crushed stone sands, the permissible limit on 150 μm IS sieve may be increased by
    - (a) 10%
    - (b) 15%
    - (c) 20%
    - (d) 25%

23. With coarse and medium sands, the bulking is in the range of
    (a) 10–20%   (c) 15–30%
    (b) 15–20%   (d) 20–30%
24. Which of the following can be considered as aggregate for mortar/concrete?
    (a) Iron slag
    (b) Copper slag
    (c) M-sand
    (d) All of these
25. Does the strength of aggregate influence the strength of concrete in conventional grades of concretes?
    (a) Yes
    (b) No
    (c) Depends on the mix proportions
    (d) None of these

## Review Questions

1. How are aggregates classified?
2. What are the two types of aggregates, according to the geological origin?
3. How are aggregates classified, according to size?
4. What are (a) all-in-aggregates, (b) single-sized aggregates, and (c) graded aggregates?
5. When is an aggregate considered flaky?
6. When is an aggregate considered elongated?
7. How are aggregates classified, according to unit weight?
8. Define the relative density of an aggregate.
9. Why should we not use porous aggregates in concrete?
10. What are the four moisture conditions of aggregates?
11. What are the deleterious substances in aggregates? What are their harmful effects?
12. What do you mean by the soundness of aggregate? What is alkali-aggregate reaction? What are the three factors, which affect this reaction? How can this reaction be controlled?
13. What is fineness modulus? How would you find the fineness modulus value of coarse and fine aggregates?
14. What are the IS Sieves used with (a) coarse aggregates (b) fine aggregates?
15. Determine the fineness modulus of sand from the sieve analysis result of weight of sand retained and as given in Table 6.24.

**Table 6.24** Sieve analysis results

| Sieve size | Weight of sand retained (g) | Individual percent retained (%) | Cumulative percentage of sand retained (%) | Cumulative percentage of sand passing (%) |
|---|---|---|---|---|
| 10 mm | 0 | | | |
| 4.75 mm | 10 | | | |
| 2.36 mm | 50 | | | |
| 1.18 mm | 70 | | | |
| 600 μm | 90 | | | |
| 300 μm | 160 | | | |
| 150 μm | 80 | | | |
| Pan | 40 | | | |
| Total | 500 | | | |
| Fineness modulus = | | | | |

16. Determine the fineness modulus of coarse aggregates from the following sieve analysis results given in Table 6.25.

**Table 6.25** Sieve analysis results

| Sieve size | Weight of aggregate retained (g) | Individual percent retained (g) | Cumulative percentage of retained (%) | Cumulative percentage of passing (%) |
|---|---|---|---|---|
| 80 mm | – | – | – | – |
| 40 mm | 0 | | | |
| 20 mm | 700 | | | |
| 10 mm | 2600 | | | |
| 4.75 mm | 1800 | | | |
| 2.36 mm | 1500 | | | |
| 1.18 mm | 420 | | | |
| 600 µm | 0 | | | |
| 300 µm | 0 | | | |
| 150 µm | 0 | | | |
| Total | 7020 | | | |
| Fineness modulus = | | | | |

17. How are aggregates classified based on the grading?
18. What is meant by all-in-aggregate?
19. What are the main factors governing the grading of aggregates?
20. How does IS 383:16 classify fine aggregates into four zones, based on 600 µm sieve? Which zone is coarser?
21. State the characteristics of good sand.
22. What is the particle size distribution of standard sand?
23. What are the sources of natural sand?
24. How is sand classified?
25. What is the bulking of sand? How does it affect concrete mix? How can it be determined in the field?
26. What are the different uses of sand?
27. What are sintered fly ash lightweight aggregates? How are they produced? List their advantages.
28. What is M-sand? How is it produced? What are its advantages?
29. Compare manufactured sand and natural river sand.
30. How is quarry dust differentiated from M-sand? Why is it not recommended to be used in concrete as fine aggregate?
31. Write short notes on the following:
    (a) Effect of moisture on aggregates    (d) Iron and steel slags    (g) Foundry waste sand
    (b) Properties of good coarse aggregate    (e) Copper slag
    (c) Recycled aggregates    (f) Granulated blast furnace slag (GBFS)
32. What are the alternate coarse aggregates specified in IS 383–16?
33. What are the precautions to be taken while storing and handling of aggregates?

# ANSWERS

## Multiple-choice Questions

| | | | | | | |
|---|---|---|---|---|---|---|
| **1.** (c) | **2.** (c) | **3.** (c) | **4.** (b) | **5.** (b) | **6.** (b) | **7.** (c) |
| **8.** (a) | **9.** (c) | **10.** (c) | **11.** (a) | **12.** (a) | **13.** (a) | **14.** (b) |
| **15.** (a) | **16.** (d) | **17.** (c) | **18.** (d) | **19.** (d) | **20.** (b) | **21.** (c) |
| **22.** (c) | **23.** (c) | **24.** (d) | **25.** (a) | | | |

# CHAPTER 7
# MORTARS AND PLASTERS

## 7.1 Introduction

Historically, mortar has been made from a variety of materials. The first mortars were made of mud and clay. The most ancient mortar discovered to date is in Galilee, Israel, near Yiftah'el and is reputed to be 10,000 years old (Cemex Guide, 2008). The earliest known mortar was made from gypsum and used by the ancient Egyptians, supposedly in the pyramids (Lucas, 2003). This was essentially a mixture of burned gypsum and sand and was quite soft (Regourd et al., 1988). The earliest known use of lime mortar dates back to about 4000 BC in ancient Egypt and Greece, when it largely replaced the clay and gypsum mortars common to ancient Egyptian construction. Lime mortars have been used throughout the world, notably in Roman Empire buildings throughout Europe and Africa. The vast majority of pre-1900 masonry buildings in Europe and Asia was built using lime mortar (www.sustainableconcrete.org.uk). Lime mortar has been used in India for monumental structures such as the Taj Mahal (completed in c. 1648) and several forts. After the introduction of Portland cement in 1824, it is considered to be the strongest binding material for making mortar; thus, it replaced lime mortar. Currently, the basic dry ingredients for mortar include some type of cement, hydrated lime, and sand. Each of these materials makes a definite contribution to mortar performance (BIA TN-8, 2008).

*Mortar* is as a workable, homogeneous, and cohesive paste (capable of setting and hardening) obtained by adding (a) water and any admissible admixture to a mixture of (b) fine aggregates such as sand (of particle size less than 5 mm) or *surkhi* (finely ground broken brick/burnt clay fine aggregate), and (c) binding material, such as cement, lime, gypsum, clay, or their combinations. Thus, it is a material that is plastic and can flow when fresh, but hardens (when cured for a few days) to a solid and hard mass over a period of hours to days. Building mortars are used for joining together bricks, stones, or concrete blocks, and for pointing bricks. Its purpose is to bind together different stones or bricks, such that the masonry acts as a unit and is stable.

*Plaster* is a building material similar to mortar, but contains finer sand than in mortar, in order to obtain a better finish. Plaster is used as a protective and/or decorative coating on walls and ceilings and for moulding and casting decorative elements. Plaster offers protection against moisture penetration. It has to be noted that the term mortar is often loosely used to refer to both plasters and mortars. Another term that is used to represent plaster is *stucco*, which is often used for plasterwork to produce relief decoration, rather than flat surfaces. The difference in nomenclature among stucco, plaster, and mortar is based more on use rather than composition. Until the latter part of the 19th century, plaster, which was used on the interior walls of a building, and stucco, which was used on the exterior face of walls, would consist of the same primary materials: lime and sand (which are also used in mortar). In those days, animal or plant fibres were often added to get additional strength. After Joseph Aspdin, a bricklayer and mason in Leeds, England, patented the first 'Portland' cement in 1824, Portland cement replaced lime, due to ease of preparation and to improve the durability of stucco. At the same time, traditional lime plasters were also replaced by gypsum plaster.

The composition and properties of building mortars are similar to that of concrete, but mortars do not contain any coarse aggregates. Sand in the mortar basically acts as filler and reduces shrinkage of the mortar. The mortar composition is designed by the volume or weight of material in 1 $m^3$ of mortar or by the relative amount of materials with the amount of binding material taken as unity. For simple mortars composed of only one binding material, the composition is designated as 1:$X$ (e.g., 1:4), that is, one part (by weight or volume) of binding material (say, cement) and $X$ parts of sand. Combined mortars, composed of two binding materials, are designated as 1:$X$:$Y$, (e.g., 1:1:6), that is, one part of cement, $X$ parts of lime, and $Y$ parts of sand.

A good mortar should possess adequate strength, workability, durability, and compatibility with the substrate (stones/bricks) and the proposed painting. The choice of mortar and its grade for a particular application is governed by several considerations, such as type of masonry, severity of the environment, bond and durability requirements, rate of setting and hardening, insulation, and fire resistance. As already mentioned, there are various types of binders, such as cement, lime, mud, and combined cement, lime, and fly ash, which can be used for the preparation of mortar. Mortars with lime and/or fly ash content are considered green products. In addition, some admixtures could be used to improve the properties of mortar. As per IS 2250, a number of tests can be performed on mortars to find their consistency, compressive and tensile strength, and water retention and adhesion properties. Plasters could also be based on cement, lime, or polymers. In addition to plasters and mortars, grouts are also used in certain places as a filler product. As a grout is intended to flow, it usually has higher water content than a mortar/plaster.

## 7.2 Characteristics of Good Mortar

As already mentioned, mortar is the bonding agent that integrates bricks into a masonry assembly. The mortar must be workable, sufficiently strong, durable, and capable of keeping the masonry intact, and provide a water-resistant barrier. The following are the important general requirements of a good mortar:

***Workability and water content***   Workability is the most important physical property of plastic mortar. A workable mortar can be spread with little effort using trowels, and readily adheres to vertical surfaces of masonry. Unlike concretes, mortars should contain the maximum amount of water, consistent with optimum workability. It is often subjectively determined by the mason by assessing the mortar's workability. Naik et al. (1992) found that the optimum water/cement (w/c) ratio of mortar ranges from 0.35 to 0.6. That length of time the mortar retains its workability is often termed as its *stiffening time* or *board life*. It has to be noted that lime mortars are more workable than cement mortars.

***Compressive strength***   The mortar must have adequate strength to match the strength of materials, such as bricks or concrete blocks, to be joined. It has to be noted that there is no advantage of using an over-strong mortar.

***Good bond and adhesion***   The mortar should provide adequate bonding of bricks or concrete blocks that it is joining.

***Durability***   The mortar should be durable. Properties of masonry mortar related to its durability include the following (Wright et al., 1993): (a) resistance to freeze thaw deterioration, (b) drying shrinkage characteristics, (c) resistance to sulphate attack, (d) water absorption characteristics, and (e) soundness. It has been found that air entrainment levels of at least 10 to 12% are needed to provide effective resistance to freeze-thaw deterioration in masonry mortars.

*Rate of stiffening*   The rate of stiffening of mortar should be adequate. The stiffening of mortar is due to the chemical reaction of the binder-hydration in the case of cement and carbonation in the case of lime. However, in the case of mud mortar, the stiffening is due to the loss of water through evaporation and suction. The rate of development of strength will be proportional to the stiffening. In addition, greater the rate of stiffening, greater will be the speed of construction.

*Compatibility*   The mortar should be compatible with the substrate (bricks/stones) and the proposed painting work. It has to be noted that lime plaster does not match well with many modern paints.

*Appearance*   Depending on the application, it may be desirable to have a mortar that complements or blends with the colour of masonry units. The desired colour may be obtained either by using coloured masonry cement, or by adding pigment to the mortar.

*Resistance to seepage of water*   It should have high resistance to rain water penetration. In addition, it should retain water for sufficient time for setting of the mortar. However, it should not trap water. Mineral and organic plasticizing agents may be added to enhance water retention.

*Volume stability*   A mortar should have minimum amount of volume changes after laying the units. It is because volume change may result in poor initial bond and subsequent destruction of bond between masonry and mortar.

*Resistance to efflorescence*   A mortar should be resistant to salt efflorescence.

*Reaction to other materials*   Mortar should not affect the durability of bricks or concrete blocks, which it joins.

Of all the requirements, the four main requirements are the workability, compressive strength, durability, and compatibility with the proposed painting. Even though the addition of lime in mortars and plasters may increase their plasticity, and even specified in codes, it is not practised in the field, due to the easy mixing of cement mortars.

## 7.2.1 Ingredients of Mortar

Mortars consist of a mixture of a *binder* (any one or combination of Portland cement, Pozzolana, clay, or masonry cement), *plasticizing materials* (such as limestone or hydrated lime), *fine aggregates* (any one or combination of sand, crushed stone, or manufactured sand), water, and *admixtures*.

The binders impart adhesive power and strength. *Surkhi* is used for economy and for furnishing hydraulic properties to lime mortar. Fly ash and cinders may be used in lime mortar as fine aggregate in place of *surkhi*. The details of various binders are discussed in Sections 7.4–7.8.

*Fine aggregates* (sand) act as a filler material in mortar, providing for an economical mix and controlling shrinkage. These sand particles are bonded together as the paste hardens to provide the required structural properties (PCA-IS 275, 1998). Sand used for masonry construction should be clean and well graded. M-sand can also be used (Mullick, 2015). ASTM C270 and IS 383:2016 provide gradation requirements for both natural and manufactured sand.

Potable water, free of contaminants, should be used for mixing mortar (PCA-IS 275, 1998). Since some mixing water may be lost due to absorptive units and evaporation, the amount of water required for optimum workability should be added to mortar (PCA-IS 275, 1998). The pH value of water used should not be less than 6.

*Admixtures* are sometimes used in mortar to obtain a specific mortar colour, and to enhance one or more properties, such as to increase workability, increase /decrease setting time, increase flexural bond strength or act as a water repellent (Beall, 1989). Admixtures to achieve a desired colour of the mortar are the most widely used. The admixtures used include the following: (a) air entraining agents, (b) bond enhancers, (c) accelerators, (d) retarders, (e) workability enhancers, and (f) colour pigments. Coloured mortar should not be re-tempered. More details on these admixtures may be obtained from Beall (1989) or BIA TN-8, 2008.

Though concrete and mortar contain the same principal ingredients (except coarse aggregate), it is important to realize that mortar differs from concrete in working consistencies, methods of placement, and structural performance. Mortar may have a high w/c ratio when mixed, but this ratio may change to a lower value when it comes in to contact with the absorbent masonry units.

## 7.3 Classification of Mortars

Mortars may be classified on the basis of the following: (a) type of binding material, (b) nature of applications, (c) bulk density, (d) physical and mechanical properties, and (e) type of additives used.

### On the Basis of Binding Material

The governing factors, while selecting a particular type of mortar for a specific application, depend upon the desired strength of masonry, resistance to penetration of rain water, immediate and long-term appearance, hardening temperature, expected working conditions of the building, and cost.

The most commonly used classes of mortar in building construction are:

*Cement mortar*   These mortars in different proportions are prepared from Portland cement or its varieties, sand, and water.

*Lime mortar*   These are mixtures of hydrated or hydraulic lime, sand, and water in different proportions.

*Gypsum mortar*   It is a mortar prepared from gypsums or anhydride binding materials, sand, and water (Gypsum wall plaster is discussed in Chapter 10).

*Mud mortar*   This mortar is prepared by simply mixing clayey soil with water until it becomes a plastic (workable) mix and is used in the construction of houses with adobe blocks and also in temporary structures.

*Composite mortar or gauged mortar*   This mortar may be made by mixing water with cement-lime-sand *or* cement-lime-*surkhi*-sand, *or* cement-lime-sand-air entraining agent in different proportions.

Moreover, pozzolana, such as fly ash, could also be added to the cement or lime mortar as per IS 2250:1981.

### On the Basis of Nature of Application

Based on the nature of application, the mortar is classified as follows:

*Masonry mortar*   This mortar is used for the construction of brick work. Masonry mortars may be either cement mortar, lime mortar, or lime-cement mortar.

*Finishing or decorative mortars*   These mortars are intended for architectural or ornamental jobs such as plastering, pointing, and ornamental finishing. Decorative finishes are obtained using coloured cements or pigments or using fine aggregates of appropriate colour and texture.

*Special mortars*   These mortars are used for obtaining acoustics, X-ray shielding, plugging concrete at oil fields, and other similar special applications.

## On the Basis of Physical and Mechanical Properties

Building mortars are subdivided into nine grades (MM 0.5 to MM 7.5), as per Table 1 of IS 2250:1981 on the basis of compressive strength from 0.5 to over 7.5 N/mm$^2$.

## On the Basis of Bulk Density

Based on the bulk density of the mix in dry state, they are classified into the following two types:

*Lightweight mortar*   These mortars are prepared using lightweight sands, such as pumice sand, tufa sand (a porous rock composed of calcium carbonate and formed by precipitation from water), slags, or cinders (partly burned coal or wood particles), and have bulk density in the range of 15 kN/m$^3$.

*Heavyweight mortar*   These mortars are obtained using dense or heavyweight sands, such as *quartz* sand, and have bulk density in the range of 15–22 kN/m$^3$. They are used in load bearing constructions.

## Type of Additives Used

Admixtures and additives could also be added to mortars to achieve the required properties. Based on the additives, they may be classified as follows:

*Aerated or air-entrained mortar*   These are cement mortars containing a small quantity of air-entraining agent and provide increased workability, and resistance to freeze-thaw deterioration. The entrained air, in the form of minute air bubbles, improves the flow characteristics and workability of the mix. The entrained air also makes the mortar lightweight and improves its insulating and acoustic properties.

*Surkhi mortar*   Sand may be replaced completely or partially by *surkhi* to obtain *surkhi* mortar.

## 7.4 Cement Mortar

Cement mortar is prepared by mixing cement, sand/M-sand, and water in desired proportions. It contributes to durability, high strength, and early setting of the mortar. Cement mortar is generally preferred over lime mortar in new buildings, due to its fast setting and hardening properties, which allow for faster construction and require semi-skilled labour. Cement mortar attains higher strength compared to other types of mortar and, hence, is more suitable for being used in structures carrying heavy loads. Portland cement and blast furnace slag cement mortars are usually used to construct walls built with bricks, stones, and large blocks. Pozzolana Portland cement and sulphate-resisting cement mortars are used when there is a likelihood of exposure to aggressive environments or soils/water contaminated with chemicals. Masonry cement is special cement containing Portland cement and pozzolanic or inert materials, and is discussed in Section 7.10.4.

### 7.4.1 Preparation

Cement mortar, required in small quantities, may be mixed manually; for large quantities, mechanical mixers are usually employed.

**Manual mixing**   This (volume mix) of cement motor consists of the following steps: sand is sieved, cleaned with water to remove dirt and dust, and dried. This dry sand is measured by boxes and the cement by weight of bags or assuming that one bag of cement has a volume of 0.035 m$^3$. The specified quantity of sand is first spread uniformly, on a watertight platform, over which the required quantity of cement is uniformly spread. The whole mass is then thoroughly dry mixed with a shovel or spade, several times, until the mixture is of uniform colour. The quantity of mix that can be used within 30 min is then taken apart, mixed separately, and formed into a heap. A small depression is then made on top of the heap. The required quantity of water is added in the depression and mixed well again until the water is completely absorbed by the mix. Clean and potable water, which is about 60% of the weight of cement, is required for 1:3 mortar. The wet mix is then worked with spades for 10 to 15 min to give a uniform consistency to the mortar.

**Machine mixing**   In this, the mortar is mixed using a concrete mixer. The calculated quantity of cement and sand is first fed into the cylindrical container of the pan mixer. After thorough dry mixing of the ingredients, the required quantity of water is added gradually. Wet mixing is continued for more than 1 min, until the mixture is brought to a plastic state with uniform consistency. The mixed mortar is then poured out for use. It is important to clean the drum of the mixer every time before suspending the work, to prevent any setting of mortar that will stick to the inside walls of the drum.

## 7.4.2 Precautions to be Taken

The following precautions are necessary while mixing and using the cement mortar:

1. Mixing mortar with the maximum amount of water consistent with workability will provide maximum bond strength. Hence, water content should be determined by the mason to produce the best workable mixture. The mortar should be used within 1–2 h after initial mixing, or as determined by the prevailing weather conditions (in hot weather, it should be used quickly).
2. Re-tempering, by adding additional water, can be permitted, but only to replace water lost by evaporation. But it is important to note that such re-tempering should be permitted only up to 60 min from the time of addition of water to cement (IS 2250:1981); some specifications may allow up to 2 h. Mortar unused for more than 2 h should be discarded.
3. The masonry units should be saturated with water, before laying them on a course of cement mortar. Otherwise, the masonry units will absorb the water from the mortar, making the mortar joint weak due to the lack of water required for proper hydration.
4. Completed masonry walls or plastered surfaces should be moist cured for at least 7 days to gain proper strength.
5. Cement mortar should not be used for the repair of older buildings constructed using lime mortar, which requires the flexibility, softness, and breathability of lime, for their proper functioning.
6. Higher grade cements, like 53 grade cement, and high cement content are generally not desirable for plasterwork as they result in more shrinkage cracking. Grade 33 cement is the best for this purpose. High workability mixes with low shrinkage may be obtained by the addition of lime to cement-sand mixes.
7. There is also no advantage of using over-strong mortar.

## 7.4.3 Properties

The setting and hardening characteristics of cement mortar depend on the setting and hardening properties of the cement-water paste, which binds the sand particles together, while binding the masonry units also. When the mortar hardens, it provides a strong joint to make the individual masonry units to act integrally. Strength of mortars to be used for joining bricks, stones, or concrete blocks should depend on the strength of materials to be joined. When the strength of mortar is higher than that of the masonry units it is joining, cracks due to shrinkage or settlement may appear and may be concentrated in a few places instead of being distributed. The use of mortars with lower cement content may result in lower workability and cohesion, leading to porous joints with low frost resistance. The approximate strength of various proportions of cement-sand mortars is given in Table 7.1 (see also IS 2250:1981, ACI 530–13, and ACI 530.1–13).

**Table 7.1** Approximate strength of cement-sand mortars (based on NBC, 2005)

| Grade of mortar | Cement-sand ratio | Minimum (28th day) strength of mortar, MPa | Optimum for masonry unit strength, MPa |
|---|---|---|---|
| H1 | 1:3 | 10.0 | 25 or above |
| H2 | 1:4 | 7.5 | 15 to 24.9 |
| M1 | 1:5 | 5.0 | 5 to 14.9 |
| M2 | 1:6 | 3 | 3.5 to 4.9 |
| M3 | 1:7 | 1.5 | 3.5 |
| L1 | 1:8 | 0.7 | 3.5 |

Strength of cement-sand mixture may be affected by a number of parameters including w/c ratio, cement-sand ratio, the type and grading of sand, manner of mixing, size and shape of specimen, curing conditions, rate of loading, and age of specimen. It has to be noted that cement mortar strength is not directly related to concrete strength. The compositions of cement-sand mortars that are used in different applications are given in Section 7.11.

### 7.4.4 Applications

Cement mortars may be used in the following applications:

1. For binding together various types of masonry units such as bricks, stones, hollow cement, or burnt clay blocks;
2. For pointing masonry joints;
3. For plastering the inner and outer surfaces of masonry, to protect it from environmental actions and also provide architectural renderings;
4. As damp-proof course in masonry below the ground level;
5. In concrete, to bind together the coarse and fine aggregates;
6. For repairing masonry or concrete structures;
7. For manufacturing hollow concrete blocks and precast elements;
8. For waterproofing of roofs; and
9. As filler in ferrocement elements.

## 7.5 Lime Mortar

*Lime mortar* is made by mixing lime (binder), sand/M-sand (fine aggregate), and water. It is one of the oldest known types of mortar, dating back to the 4th century BC, and was widely used in Ancient Rome and Greece, until it was replaced by clay and gypsum mortars in Ancient Egyptian constructions (Boynton, 1980; Lucas, 2003). The application of lime mortar in historic buildings is shown in Fig. 7.1.

Only recently, laboratory testing provided a scientific understanding of the remarkable durability of lime mortar. The durability of lime mortar is mainly due to its composition (calcium carbonate), resistance to salts and frost (no sulphate attack or alkali-silica reactions), flexibility (It has a low modulus

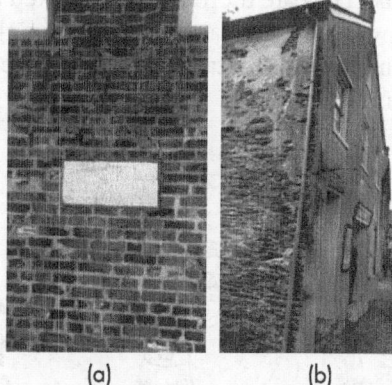

(a)  (b)

**Fig. 7.1** Use of lime mortar in buildings at Harpers Ferry National Park, W Virginia, USA (a) Wall built in 1910 with lime mortar (b) Wall with lime plaster

of elasticity and, hence, is extremely flexible and allows for movement and thermal expansion.), and breathability (It is vapour-permeable, absorbing and evaporating moisture from surrounding masonry.). Lime mortar is also softer than modern cement mortar and hence will not wear away the surrounding masonry over time. Bricks and stones can also be reclaimed more easily if dismantled. Lime mortar also has an attractive traditional appearance, which is pale in colour. Natural limes tend to reveal the colour and characteristics of the aggregate with which they are mixed.

Lime mortar is a green and eco-friendly material, as its manufacturing process produces less carbon dioxide ($CO_2$) than ordinary cement. Fat lime has high calcium oxide content. Quicklime or hydrated lime mortars harden and gain strength by the evaporation of water and the absorption of $CO_2$ from the atmosphere, a process called as *carbonation* (This re-absorption of $CO_2$ lowers lime mortar's carbon footprint even further.). The sand in the mortar assists the hardening of fat lime by facilitating air and $CO_2$ to penetrate resulting in carbonization of lime. Carbonation results in the gradual conversion of lime into its original calcium carbonate form, but strength gain in the mortar is very slow for practical buildings. Though the 28th day, mortar cube strength of 43 grade cement will be around 43 MPa, the strength of lime mortar in 28 days will be about 1.75 MPa only (Varghese, 2006). Unlike cement mortar, water is not required for the chemical reaction, setting, and hardening. For this reason, bricks need not be wetted while using lime mortar; whereas, bricks should be wetted while using cement mortars, so that the water in the mortar is not absorbed by the bricks.

Hydraulic lime contains silica, alumina, and iron oxide in small quantities. When mixed with water, it forms putty or mortar having the property of setting and hardening under water. Slaked fat lime is used to prepare mortar for plastering, while hydraulic lime is used for masonry construction.

### 7.5.1 Preparation

The method of manufacturing lime mortars and the manner in which they are used in construction work differ from one part of the country to another. For instance, in the South, lime mortar is generally prepared by grinding a mixture of slaked lime and sand in suitable proportions in a bullock mill, while, in Punjab, lime putty is mixed with sand and the mix used as mortar directly.

**Manual mixing**   Lime and sand in required quantities are placed on an impervious platform or in a tank. They are thoroughly dry mixed by turning them up and down with spades. The necessary quantity of water is added at stages and mixing is done again with spades until the mortar of uniform colour and consistency is obtained. The mixture can be ground to a plastic mix by pounding with heavy wooden hammers.

**Mill mixing**   When a large quantity of mortar is required, mills, such as *chakki or ghanni* run by bullocks (Fig. 7.2) or a pan mill (Fig. 7.3), are used. *Ghanni* consists of a trench along the circumference of a circle of radius 3–4.5 m, 0.3 m wide, and about 0.4 m deep. A wooden shaft, which is pivoted at the centre of the circle, carries a stone wheel of width 0.2–0.25 m that rotates in the trench, when the bullocks are driven round the trench (see Fig. 7.2). The required quantity of ingredients in the form of putty is placed in

Trench filled with mortar ingredients

**Fig. 7.2** Bullock-driven mortar mill (Ghanni)

the trench. Water is added gradually, while bullocks drive the wheel. A worker is employed to turn the mix, up and down, regularly. This method of mixing needs about 4–6 h to produce about 1.7–1.8 m³ of lime mortar; it has to be noted that pan mills produce much greater quantities.

In the power-driven pan mill mixing, a cast-iron circular pan is made to revolve using oil or electric power under a pair of stone or heavy cast-iron rollers filled with concrete (see Fig. 7.3). The ingredients of the mortar are thrown into the pan, while it is revolving. During mixing, the required quantity of water is added to produce a stiff mix. The mortar is raked up continuously during the process. These mills are made in different sizes, the pans varying in diameter from 1.2 to 1.8 m; the rollers from 0.8 to 1 m, as per IS 2438: 1963. The minimum weight for each roller differs from

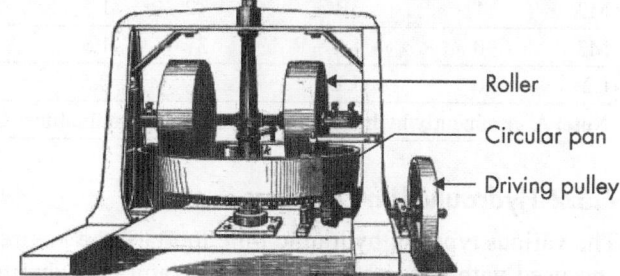

**Fig. 7.3** Power-driven mortar mill (pan mill)

100 kg for 1.2 m pan to 160 kg for 1.8 m pan. The speed of the rollers should not be less than 1.2 m/s. Grinding is done for at least 3 min, for mortars, and 4–5 min, for plasters. The duration of grinding thus depends on the required fineness of work.

The lime mortar made of hydraulic lime should be used as soon as possible after mixing preferably on the same day and that the other lime mortars (fat lime or hydrated) can be stored after mixing and used within 2–3 days. Fat lime mortar should be kept wet by covering with wet jute bags or by any other suitable means and should never be allowed to dry.

As discussed in Chapter 4, lime used for mortar may be fat lime (quick or hydrated lime) or hydraulic lime. Mixing hydraulic lime or hydrated lime, or both, with water and sand produces lime mortar and plaster. Hydrated lime is most commonly used, because of its hardening through the absorption of $CO_2$ in the atmosphere. However, hydrated lime reacts extremely slowly with $CO_2$ (www.buildinglime.org). Typically, only 6–10 mm depth of mortar in the joint will be carbonated over 100 years. However, hydrated lime may be combined with Portland cement and sand. In this system, lime functions as a plasticizer contributing to workability, board life, and water retention of the mortar. Air-entrainment of cement or hydrated lime may also be used to improve the workability and durability (PCA-IS 275, 1998). The maximum air content of cement-lime mortar made with Types NA and SA is 14% and with Types N or S is 7% (www.lime.org) - See Chapter 4 for the classification of lime (N, S, NA, and SA). The approximate strength of various cement-lime-sand mortars are provided in Table 7.2.

**Table 7.2** Approximate strength of cement-lime-sand mortars (based on NBC, 2005)

| Grade of mortar | Mix proportion by loose volume | | | Minimum (28th day) strength of mortar, as per IS 2250:1981, MPa | Optimum for masonry unit strength, MPa |
| --- | --- | --- | --- | --- | --- |
| | Cement | Lime[1] | Sand | | |
| H1 | 1 | 0.25C or B | 3 | 10 | above 25 |
| H2 | 1 | 0.25C or B | 4 | 7.5 | 15 to 24.9 |
| H2 | 1 | 0.5C or B | 4.5 | 6 | 15 to 24.9 |
| M1 | 1 | 1C or B | 6 | 3 | 5 to 14.9 |
| M2 | 1 | 2B | 9 | 2 | 3.5 to 4.9 |
| M2 | 0 | 1A | 2 | 2 | 3.5 to 4.9 |

*(Contd)*

**Table 7.2** (Contd)

| Grade of mortar | Mix proportion by loose volume | | | Minimum (28th day) strength of mortar, as per IS 2250:1981, MPa | Optimum for masonry unit strength, MPa |
|---|---|---|---|---|---|
| | Cement | Lime[1] | Sand | | |
| M3 | 1 | 3B | 12 | 1.5 | 3.5 |
| M3 | 0 | 1A | 3 | 1.5 | 3.5 |
| L2 | 0 | 1 B | 3 | 0.5 | 3.5 |

**Note:** A - Eminently hydraulic lime; B - Semi hydraulic lime; C - Fat lime

## 7.5.2 Hydraulic Lime Mortars

The various types of hydraulic lime mortars can be prepared and are to be treated like cement mortars, and used within 2–4 h after mixing. Natural Hydraulic Lime (NHL 3.5) is ideal for re-pointing and building walls. NHL 2 is weaker, generally more flexible, and sets slowly, and, hence, is ideal for internal applications or where conservation is a primary concern with soft or deteriorating stones and bricks. NHL 5 is stronger and has a faster set, which is primarily used for granite, engineered brick, basalt, flint, paving, roofing, and chimneys (see also Section 4.5.4 of Chapter 4).

Currently, NHLs are most frequently used as replacement mortars, when specific performance parameters are specified (www.naturalmortars.com). As they use lower energy during the production compared to modern cements, they are considered environment friendly and green. Recommended mortars for brick masonry as per ASTM C270–06 are given in Table 7.3. Hydraulic lime and lime-pozzolana mortars should be treated with moist curing during early stages, to improve the hydration reactions (Cizer, 2009).

**Table 7.3** Recommended mortars for brick masonry as per ASTM C270–06

| Mortar | Type | Cement-sand ratio/Cement-lime-sand ratio | Average compressive strength of mortar at 28 days, Minimum, (MPa) | Water retention, Minimum (%) | Air content, Maximum (%) | Brick strength, MPa |
|---|---|---|---|---|---|---|
| Cement or Ce-ment-lime | O (low strength) | 1:6 /1:2:9 | 2.4 | 75 | 14 | <5 |
| | N (normal brickwork) | 1:5/1:1:6 | 5.2 | 75 | 14 | 5 to 15 |
| | S (stronger brickwork) | 1:4/1:0.5:4.5 | 12.4 | 75 | 12 | 15 to 25 |
| | M (high strength) | 1:3/1:0.25:3 | 17.2 | 75 | 12 | >25 |
| Masonry cement | O | 1:6 | 2.4 | 75 | 20 | <5 |
| | N | 1:5 | 5.2 | 75 | 20 | 5 to 15 |
| | S | 1:4 | 12.4 | 75 | 18 | 15 to 25 |
| | M | 1:3 | 17.2 | 75 | 18 | >25 |

Aggregate ratio (measured in damp, loose conditions): Not less than 2.25 and not more than 3.5 times the sum of the separate volumes of the cementitious materials

## 7.5.3 Advantages and Disadvantages

Lime mortar has several advantages over cement mortar. Some of these advantages are described as follows:

1. Historic buildings were frequently constructed with relatively soft masonry units. Minor movements in such buildings will result in the breakage of masonry when cement mortar is used.

However, when lime mortar is used, the mortar will crack instead of the masonry. This results in much less damage, and is relatively simple to repair. Lime mortars are often used in the repair of historic buildings.

2. Lime mortars are exceptionally durable. Many buildings with lime mortars have sustained for over 2000 years of existence.

3. The size of cracks in cement mortar may be large, while lime mortar often produces numerous micro-cracks. These micro-cracks close themselves due to self-healing.

4. Lime mortar does not adhere as strongly to masonry as cement. The mortar will crack before the bricks. It is less expensive to replace cracked mortar than cracked bricks.

5. Lime mortar is more porous and, hence, absorbs any dampness, which is evaporated from the surface of the wall. This property is known as *breathability*. Thus, any salt content in the water crystallizes and damages the lime, thus saving the masonry. But the use of cement mortar causes salt formation and spalling, and the subsequent disintegration of bricks.

6. Lime mortars have good workability, which assists good workmanship.

7. Any dampness in the wall will cause the lime mortar to change colour, indicating the presence of moisture. As and when the moisture level changes, the shade of the limewash will also change. This effect will be more pronounced, when darker shade of limewash is used.

8. The mixed lime mortar may be allowed to sit as a lump for some time, without it drying out (it may get a thin crust). When ready to use, this lump may be remixed ('knocked up') again and then used. This process cannot be done with cement mortar.

9. The use of lime, instead of cement, in the mortar or plaster, may reduce $CO_2$ emissions by about 75–80%.

**Disadvantages**   The main disadvantage of lime mortar is its slow rate of setting. However, the rate of setting may be increased by adding pozzolanic materials, such as fly ash, to the mortar mix. Lime mortar should not be used below temperatures of 5°C, as it may take longer time to set; it should be protected from freezing.

### 7.5.4 Precautions to Be Taken

The precautions to be taken while preparing and handling lime, as discussed in Section 4.9 of Chapter 4, are applicable to the handling and mixing of lime mortars also. The mortar should be mixed thoroughly, avoiding 'balling'. The lime mortar should not be used, when the temperature is below 5°C. In addition, in mild winter weather, in the event of rain and/or frost, the walls with hydraulic lime mortars should be protected by covering them with hessian or tarpaulin.

## 7.6 Composite (Lime-Cement) Mortar

To improve the quality of lime mortar and achieve early strength, cement is sometimes added to it. This process is known as *gauging*. This type of *composite mortar* (also called as *gauged mortar* or *lime-cement mortar*) is economical, strong, and dense. It is used for bedding and in thick brick walls. The advantages of composite mortar are its increased water retention capability, bond, workability, and frost resistance. The mortar gives a good and smooth plaster finish.

**Preparation**   When lime is used in the combination mortar, it should be slaked well. If coarse sand is used, it should be ground well with lime to form an intimate mixture. For this purpose, first the sand and lime are ground dry and then ground again with the addition of necessary water (If fine sand is used, the first grinding of lime and sand may be omitted. Similarly, when factory made dry hydrated lime is used, grinding of lime and sand in the mortar is not necessary.). This lime mortar mix is transferred to the mechanical pan mill, to which required cement and additional water are added. The mixing is done in the pan mill for 3–5 min, until the required consistency is obtained. Composite mortar may also be obtained by making lime and cement mortars separately and then mixing them together. Composite mortar should be consumed within 30 min of adding water to the cement. If the mix is agitated well for every 10 to 15 min, it may be used up to 2 h. The mix proportions of lime-cement mortar are given in Table 7.4. The 1:2:9 (cement:lime:sand) mix can be used in the inside wall as well as the outside wall; whereas, the 1:3:12 mix should be used only in the inside wall.

**Table 7.4** Mix proportions for lime-cement mortar (based on IS 2250:1981)

| Mortar designation | Mix proportion | | | Compressive strength at 28 days, N/mm$^2$ |
| --- | --- | --- | --- | --- |
| | Cement | Lime | Sand | |
| MM 0.7 | 1 | 3 C or D | 12 | 0.7–1.5 |
| MM 2 | 1 | 2 C or D | 9 | 2–3 |
| MM 5 | 1 | 0.25 B, C, D or E | 4 | 5–7.5 |
| MM 7.5 | 1 | 0.25 C or D | 3 | > 7.5 |
| MM 7.5 | 1 | 0. 5 C or D | 4.5 | > 7.5 |

**Note:** A, B, C, D, and E denote the class of lime (see Section 4.7 of Chapter 4).

**Applications**  Lime-cement mortar can be used in a similar way as lime or cement mortar. However, now-a-days cement mortar is only used in many constructions, due to the easy availability and strength of cement and also due to the difficulty of making lime mortar.

## 7.7 Surkhi Mortar

*Surkhi* is used as a substitute for sand in mortar (and also in concrete). Though it has almost the same function as that of sand, it also improves the strength and hydraulic properties of mortar. *Surkhi* is made by grinding to powder burnt bricks, brick-bats, or burnt clay; under-burnt bricks should not be used, nor bricks containing a high proportion of sand (An easy test for under-burnt bricks is that they should not dissolve in water.). IS 2250 stipulates that the *surkhi* should conform to IS 1344.

**Preparation**  *Surkhi* mortar is prepared in the same way as lime mortar, with *surkhi* replacing sand or by replacing half of sand in case of fat lime mortar. The *surkhi* powder should be so ground such that it passes through a 4.75 mm sieve and the residue is not more than 10% by weight. As *surkhi* is not standardized, its properties vary widely. *Surkhi* must be mixed well with lime, preferably in a mortar-mill. The *surkhi* mortar can be used for ordinary masonry work of all kinds in foundation and superstructure.

**Applications**  However, it should not be used for plastering or pointing on surfaces exposed to weathering and humid conditions, as it may disintegrate after a period of time. *Surkhi* mortar gains strength if it is left immersed in water. *Surkhi* makes cement mortars and concretes more water proof and more resistant to alkalis and salt solutions. *Surkhi* for plaster may be made from slightly under-burnt bricks, and ground very fine; this will improve the hydraulic properties of fat lime. The mix proportions of *surkhi* mortar are given in Table 7.5.

**Table 7.5** Mix proportions for surkhi mortar (based on IS 2250:1981)

| Mortar designation | Mix proportion | | | Compressive strength at 28 days, N/mm² |
|---|---|---|---|---|
| | Lime | Surkhi | Sand | |
| MM 0.5 (Lime-*Surkhi*) | 1 C or D | 1 | 2 | 0.5–0.7 |
| MM 2 (*Surkhi*) | 1 C or D | 3 | – | 2–3 |
| MM 3 (*Surkhi*) | 1 C or D | 2 | | 3–5 |

**Note:** C and D denote the class of lime (see Section 4.7 of Chapter 4).

## 7.8 Mud Mortar

Mud mortar can be prepared by mixing soil with water, until it becomes workable. Once applied, it sets quite rapidly on drying without any curing. It is characterized by earthen colour, and it can be easily scratched from the wall using any sharp object, such as a knife. More details of buildings using mud mortar may be found from the book by Minke (2006). Guidance for the preparation and use of mud mortar in masonry is provided in IS 13077:1991. Mud mortar is used to construct temporary sheds such as watchman's shed, which will be dismantled later. Mud mortar is a low-strength mortar, and it is the weakest of all mortar types. Walls constructed with mud mortar or using mud plaster should be protected from rainwater, by providing necessary eves to the roof.

### 7.8.1 Preparation

The soil selected for mud mortar should be of tenacious nature. It should also pass through 2.36 mm IS sieve and not less than 75% through 600 μm IS sieve. The clay content should not exceed 10%. The plasticity index (PI) should be between 6 and 7. If this limit is exceeded, the suitable quantity of sand passing through 2.36 mm IS sieve should be added to the soil, to bring the PI value within the specified range. Loamy soil is preferable as it may have better adhesion to brick walls than granular soil.

A sufficient quantity of sand may be added to the selected soil, such that when the sand-soil mixture is mixed well with water and made into a ball, it should show no signs of cracking on drying. This mixture of sand and soil is well trodden and worked into the consistency of thick paste by adding sufficient amount of water. All the lumps and stones in the soil should be removed during this process, and the mixture is allowed to mature for about a week, with water standing on its top in a shallow pool. Then, it is kneaded well by treading on it to the suitable consistency to be used as mortar.

To improve the waterproofing qualities, hydrated lime (3% by weight) can be added first, followed by bitumen (2% by weight in liquid form, by using kerosene). Water is added finally to the above mixture and the whole mass is thoroughly mixed and trodden to the desired workable consistency.

### 7.8.2 Limitations

Mud mortars have a rather low tensile strength and are subject to shrinkage. Increased proportion of clay in a soil, or clays with high plasticity, will result in increase in the tensile strength but with an associated increase in shrinkage. Above certain clay content, shrinkage effects will cause large numbers of micro-cracks in the mortar, which will reduce its bond strength and, therefore, the strength of the whole structure, and possibly cause visible cracks to appear in the structure several days after construction (suitable clay content is usually around 10 to 15%). In humid or wet conditions, use of mud mortars result in increased erosion and loss of strength.

**Improving performance**   The performance of mud mortars can be improved with various additives such as (a) straw, which reduces the shrinkage without decreasing the bond strength; (b) sand, which reduces shrinkage and also the bond strength; and (c) cement, which increases bond strength and reduces the shrinkage slightly. A number of other additives such as vegetable fibres, vegetable oils, and fats have been used in the past. In addition, animal hair/fur/blood, cow dung, and glues have also been tried.

### 7.8.3 Rules for application

There are general rules to follow when applying all plasters. First, the wall surface has to be prepared well. This can be done by scrubbing off all the surface dust and loose material with a metallic brush. Then, the wall surface must be moistened to stop water being drawn out of the plaster layer into the wall. The 20 mm thick first layer is applied with force. When the first layer is still green, its surface is roughened by light grooving or scratching, to provide good bonding for the second coat. The second coat is applied after the first is completely dried. When cement or lime stabilizer is used in the plaster, the plaster has to be cured with water two or three times a day for about 3 days, especially during hot weather to reduce the development of cracks. In general, plaster work should be shaded from the sun, and it is better to avoid plastering on very hot or windy days.

### 7.8.4 Precautions to be taken

The use of mud mortars does not require any special precautions. However, they should not contain any particles larger than one third the thickness of the joints. To avoid possible shrinkage problems, the prepared mix should be workable enough for smooth and easy laying of bricks. In the case of stabilized mud mortar, to get a good homogenous distribution of the stabilizer in the mixture, it is necessary to sieve out or crush the lumps in the soil that are bigger than 5 to 8 mm. It is also better to mix the materials in small batches so that the mortar mix is used quickly. This practice will also avoid any significant setting of the stabilizer before the mortar is used.

## 7.9 Special Mortars

Certain applications may require special considerations for the selection of mortar. The following are some of these special mortars:

**Repointing mortars**   Repointing is the process of removing deteriorating mortar from the joints of old masonry walls and replacing it with a new mortar. If done properly, repointing restores the visual and physical integrity of the masonry. Compatibility between existing brick and mortar is the most important consideration while selecting a repointing mortar. In general, the compressive strength of a repointing mortar should not exceed the strength of existing mortar. In the USA, Type O mortar is used for repointing older brickwork, while Type N mortar is used repointing newer brickwork. It is better to pre-hydrate repointing mortars before use, such that when formed into a ball, it retains its shape (see BIA TN-46, 2005, for more information).

**Fire-resistant mortars/refractory mortars**   Typical fire-resistant mortars can be prepared by mixing one part of the calcium aluminate cement (CAC) with two parts of finely crushed powder of fire bricks (in place of sand), that is, the mix proportion is 1:2. A mix of one part of CAC, three parts of sand, one part of hydrated lime, and one part of fire clay has also been used. Commercial products such as Sikacrete® -223F, IR 4020-TUN, and Accoset 50F are also available. These are used for setting refractory bricks in furnace linings. They are used also as liners in RC tunnel walls, as tunnel fires will result in temperatures of over 400°C. It has to be noted that CAC mortars are used as a protective liner in sewers due to their higher resistance to biogenic sulphide and microbial corrosion.

**X-ray shielding mortars**   Following the discovery of X-rays in 1985, and the phenomena of radio-activity, it is necessary to develop shields to protect people from their harmful effects. Heavy mortars having bulk density greater than 22 kN/m$^3$ are used for plastering walls and ceilings of X-ray cabinets. The mortar may obtained by mixing water with binding materials such as cement, lime or slag cement, and fine aggregates derived from heavy rocks such as barite, magnetite, goethite or haematite. Suitable admixtures may be added to enhance the properties. Recycled glass derived from cathode ray tubes may be used as fine aggregate (Ling et al., 2012).

**Sound-absorbing mortars**   These are prepared by mixing water with binding materials such as cement, slag cement, lime, or gypsum, and lightweight porous fine aggregates made from pumice, cinders, and ceramsite. They have a bulk density of 6–12 kN/m$^3$.

**Damp-proofing mortars**   These mortars are prepared using high-grade sulphate-resisting cement or sulphate-resisting puzzolana cement as binding material and quartz sand or sand from crushed solid rock. An approximate composition of the mortar is 1:2.5 or 1:3.5. Water proof seams and joints are made from damp-proofing mortars prepared with expanding cement.

**Packing mortars**   These mortars are used for packing oil wells. They may consist of water mixed with either of the following: cement-sand, cement-loam, and cement-sand-loam. Slag cement, pozzolana, and sulphate-resisting cements are used when water pressure is expected. These mortars should have the following properties: (a) high homogeneity, (b) high water resistance, (c) predetermined setting time, and (d) resistance to sub-soil water pressure.

**Polymer modified mortar (PMM)**   These mortars are those in which cement is replaced partially with polymers. A number of thermoplastic or thermosetting polymers are used in modifying mortars and concrete (Ohama, 1995). These are used in various forms, such as liquid resins, latexes or emulsions, dispersible polymer powders, and water-soluble homopolymers, or copolymers (Aggarwal et al., 2007). The incorporation of polymers greatly improves strength, adhesion, resilience, impermeability, chemical resistance, and durability properties of mortars and concretes (ACI 548.1R-09 and ACI 548.3R-09). Dosage is in the range of 10–20% of the cement weight. Usually, 10 L of styrene-butadiene rubber (SBR) latex is added to one bag of cement. The mortar made using this cement with sand in the ratio 1:3 is extensively used as a concrete repair material. It has to be noted that before the application of the PMM, the surface to be repaired has to be treated with a 1:1 (Cement:SBR) cement slurry bond coat. This bond coat improves the adhesion of the new PMM to the old concrete surface.

**Chemical-resistant mortars**   Chemical-resistant masonry is often used in food processing plants, refineries, or breweries. Chemical-resistant mortars may include silicate mortars, sulphur mortars, and various resin mortars. They are used for jointing acid-proof bricks or tiles in construction of towers, stacks, tank-lining, sumps, drains, and chemical-resistant floors. Specifications for chemical-resistant mortars are prescribed in IS 4832(Parts 1 to 3):1969 and their use in IS 4441:1980, IS 4443:1980, and IS 4442:1980, respectively. Methods of test for chemical-resistant mortars of silicate and resin type are described in IS 4456(Part 1):1967 and of the sulphur type in IS 4456(Part 2):1967.

## 7.10 Green Mortars

As discussed earlier, lime mortar has been successfully used for many centuries and is an essential repair material for the conservation of heritage buildings. The beneficial technical performance of this material and its holistic compatibility with many traditional forms of construction are well known. Moreover, lime mortars have additional benefits of reducing $CO_2$, as compared to the modern cement mortars. The use of low-carbon materials is clearly an essential requirement for meeting $CO_2$ construction reduction targets. These dual characteristics have resulted in significant renewed interest in these materials. In addition, use of waste materials, such as fly ash, is also preferable in green constructions. Some of these green mortars are discussed in this section.

### 7.10.1 Lime Mortar

Lime mortars have been discussed in Section 7.5. Lime mortar is durable and long-lasting. Older mortar and plaster could be crushed and screened for use as aggregate in mortar mixes. In addition, lime mortar contributes to the high thermal mass of wall assemblies. On warm summer days, walls and floors with thermal mass will steadily absorb heat at their surface, conducting it inwardly, and storing it until exposed to cooler air of the evening/night. At this point, heat will begin to migrate back to the surface and is released. In this way, heat moves in a wave-like motion alternately being absorbed and released in response to the change in day and night-time conditions. This reduces peak energy loads and may reduce the size of required heating or cooling units. Moreover, there are many other advantages of lime over cement (as discussed in Section 4.12 of Chapter 4) and advantages of lime mortar over cement mortars, as discussed in Section 7.5.3 (also see www.hempcrete.com.au).

### 7.10.2 Lime-Pozzolana Mixture

Mortars using lime-pozzolana mixtures should be prepared in the same manner as described in Section 7.4.1, for cement mortars. The specification for pulverized fuel ash for use as pozzolana in cement mortar is provided in IS 3812(Part 1):2003 and for lime-pozzolana mixture applications in IS 15648:2006. The specifications for lime-pozzolana mixture are provided in IS 4098:1983 and that of quick setting lime-pozzolana mixture in IS 10772:1983. Lime-pozzolana mixtures are normally manufactured by inter-grinding the ingredients in a mill or by blending in the form of powder. The hardening of lime-pozzolana mixture depends upon the lime reactivity of the pozzolana and, in general, is slower than that of cement. However, it is satisfactory for most of the normal uses. Higher the lime reactivity of pozzolana, quicker will be the rate of setting and hardening of the lime-pozzolana mixture. Improved early strength may be obtained by inter-grinding hydrated class C lime, pozzolana, together with 6% type IV gypsum (as per, IS 1290:1973), and/or suitable hardening accelerators (less than 1%) in suitable proportions.

Lime-pozzolana mixture is supplied in three types, as shown in Table 7.6, and should conform to the requirements of IS 4098:1983.

**Table 7.6** Types of lime-pozzolana mixture (as per IS 2250 and IS 4098)

| Type | Mix proportion by loose volume | | Use |
|------|-------------------------------|------|-----|
| | Lime-pozzolana mixture | Sand | |
| LP 7 | 1 | 1.25 | For masonry mortars up to Grade MM 0.5, and for foundation concrete |
| LP 20 | 1 | 1.5 | For masonry mortars up to Grade MM 2 and for foundation concrete |
| LP 40 | 1 | 2.0 | For masonry mortars up to Grade MM 3 |

Mortars with lime-pozzolana mixture of type LP 20 and LP 40 as binder should be used within 4 h from the time of mixing of the mortar, whereas mortars which have hydraulic lime (Class B) or fat lime (Class C) and pozzolana or lime-pozzolana mixture of type LP 7 as ingredients, but do not have either Portland cement or eminently hydraulic lime (Class A), should be used within 12 to 24 h from the time of mixing of the mortar.

### 7.10.3 Cement-Fly Ash-Sand Mortar

Use of fly ash (by replacing up to 40 % of cement) will save cement content as well as reduce heat of hydration in a mortar mix. Thus, the use of cement-fly ash-sand mix provides an environmentally safe and economical option. Pulverized fuel ash (It is available in four forms, namely, fly ash, bottom ash, pond ash, and mound ash.) is a residue resulting from coal-based power plants [See for specifications in IS 15648:2006 and IS 3812(Part 1):2003]. Fly ash has high fineness, which decreases the porosity and pore size and increases the compressive strength. The use of fly ash saves precious raw materials and helps in disposing harmful waste materials. It can contribute towards LEED Materials and Resources points for use of recycled materials.

### 7.10.4 Masonry Cement Mortar

*Masonry cement* is obtained by intimately grinding a mixture of Portland cement clinker with pozzolanic materials or non-pozzolanic materials (such as limestone and granulated slag), waste materials (such as carbonated sludge and mine tailings), gypsum, and an air-entraining plasticizer in suitable proportions, as per IS 3466:1988. It is generally ground to a fineness greater than that of ordinary Portland cement. Masonry cement can be used as mortars in different types of masonry, and in plastering work. Masonry cements are also used to produce stucco. *Masonry cement mortar* may be considered superior to other types of mortars, because it produces a smooth, plastic, cohesive, strong, and, at the same time, workable mortar.

In masonry cements, Portland or blended cement contributes to compressive and bond strength of mortar. Plasticizers and other materials optimize workability, stiffening time, and water retention, contribute to improved durability, and reduce water absorption of mortar. Owing to these advantages, they are widely used in brick and block construction in the USA and a number of other countries. In the USA, masonry cements, conforming to ASTM C 91, are produced in Type N, Type S, and Type M strength levels for preparing ASTM C270 Type N, S, or M mortar, respectively, without the further addition of cements (PCA IS 181.04M, PCA IS 281, 2002, and PCA IS 282, 2002). Masonry cements are available in a variety of colours as well as natural grey. Ready-to-use Type N mortar mix bags available in the USA are shown in Fig. 7.4.

**Fig. 7.4** Type N mortar mix

Specification for sand for masonry mortars is stipulated in IS 2116:1980. Masonry cement, however, should not be used for structural concrete and prestressed concrete works. The physical requirements of masonry cement, as per IS 3466:1988 and ASTM C91 are given in Table 7.7. More information on masonry cement may be obtained from PCA IS 282: 2002 and PCA IS 181.04M (also see Table 5.14 of Chapter 5). Masonry cement is not widely used currently in India.

**Table 7.7** Physical requirements of masonry cement, as per IS 3466:1988 and ASTM C91

| S. no. | Characteristic | Requirement | | | |
|--------|----------------|-------------|---|---|---|
| | | IS 3466: 1988 | ASTM C91-Type N | ASTM C91-Type S | ASTM C91-Type M |
| 1 | Fineness: Residue on 45 μm IS Sieve, Maximum, (by wet sieving) | 15 % | 24% | 24% | 24% |

*(Contd)*

**Table 7.7** (Contd)

| S. no. | Characteristic | Requirement | | | |
|---|---|---|---|---|---|
| | | IS 3466: 1988 | ASTM C91-Type N | ASTM C91-Type S | ASTM C91-Type M |
| 2 | Setting Time (using Vicat Apparatus) | | | | |
| | (a) Initial, Minimum | 1.5 h | 2 h | 1.5 h | 1.5 h |
| | (b) Final, Maximum | 24 h | 24 h | 24 h | 24 h |
| 3 | Soundness | | | | |
| | (a) Le-Chatelier expansion, Maximum | 10 mm | – | – | – |
| | (b) Autoclave expansion, Maximum | 1% | 1% | 1% | 1% |
| 4 | Compressive strength: mortar* cubes 50 mm size, Minimum | | | | |
| | 7 days | 2.5 MPa | 3.4 MPa | 9.0 MPa | 12.4 MPa |
| | 28 days | 5 MPa | 6.2 MPa | 14.5 MPa | 20 MPa |
| 5 | Air Content: air content of mortar, | | | | |
| | Minimum | 6% | 8% | 8% | 8% |
| | Maximum | – | 21% | 19% | 19% |
| 6 | Water retention: flow after suction as percentage of original flow of mortar*, Minimum | 60% | 70% | 70% | 70% |

*mortar composed of one part of masonry cement and three parts of standard sand by volume

## 7.11 Selection of Mortar

Proper selection of mortars for various uses depends upon the following factors (IS 2250:1981):

1. Type of masonry (e.g., brick work, stone work, and concrete block work), and strength of individual masonry unit
2. Location of the masonry in the structure (whether masonry is in foundation or superstructure) and conditions of surrounding soil in the case of foundation work
3. Loading on the masonry
4. Environmental exposure or soil conditions in the case of masonry situated below ground level
5. Type and grading of fine aggregates to be used in the mortar, namely, whether sand, burnt-clay aggregate, or cinder aggregate
6. In the case of hydraulic structures, weathering conditions under water contact and underwater head action
7. In case of use in storage of acidic or alkaline substances such as fertilizers

The selection of masonry mortars from durability considerations should have to cover both the loading and exposure conditions of the masonry. If the masonry is exposed frequently to rains and when finishes are provided, not less than grade MM 0.7 should be used. When no protection is provided, minimum grade to be used is MM 2. In the case of load-bearing internal walls, grade MM 0.7 or more is to be used (see also Table 1 of IS 2250:1981). Where masonry is subject to the vibration of machinery, or in parapet walls (where the height is greater than thrice the thickness), grade MM 3 or more should be used.

Some recommended mix proportions for mortars/plasters (by volume), based on the practice followed in Tamil Nadu (where the strength of brick is low), are given in Table 7.8.

**Table 7.8** Recommended mix proportions for mortars/plasters by volume

| S. no. | Type of work | Mix proportions | | Grade of mortar as per IS 2250 |
|---|---|---|---|---|
| | | Cement-sand mortar | Cement-lime-sand mortar | |
| 1 | Masonry in foundation up to plinth | 1:3 to 1:6 | 1:0.25:3 to 1:2:9 | MM 0.7- MM 2.0, MM 2.0 when soil is damp, MM 3 when soil is saturated |
| 2 | 230 mm thick brick/stone masonry wall in superstructure | 1:6 to 1:8 (as per brick strength) | 1:2:9 to 1:3:12 | MM 0.5 to MM 0.7 |
| 3 | 115 mm brick partition walls and parapet walls | 1:5 | 1:1:6 | MM 0.5 |
| 4 | Wall with large concrete blocks | 1: 6 | – | MM 3 |
| 5 | Masonry in buildings subject to vibration of machinery | 1: 5 | – | MM 3 |
| 6 | Internal plaster | 1:5 to 1:6 | 1:1:6 to 1:2:9 | – |
| 7 | Ceiling plaster | 1:4 | 1:1:5 | – |
| 8 | External plaster | 1:5 | 1:1:6 | MM 0.7 to MM 2.0, MM 2.0 is preferable |
| 9 | Pointing | 1:2 to 1:3 | 1:1:3 to 1:0.25:3 | – |
| 10 | Arch work | 1:3 | 1:0.25:3 | – |
| 11 | Damp proof course | 1:2 | 1:1:3 | – |

## 7.12 Plasters

*Plaster* is defined as a material used for the protective and/or decorative coating of walls and ceilings. It provides a waterproof layer and protects the masonry from weathering effects of rain, temperature variations, and frost action. Normally, the word 'plaster' is used for the material used in the interiors of a building, while the word '*render*' is used for the external application. Another term used in connection with plastering is 'stucco', which is also applied on the outside walls to produce relief decoration, rather than flat surfaces. *Stucco* is usually a mix of sand, cement, lime, and water, but may also consist of a proprietary mix of additives including fibres and synthetic acrylics that add strength and flexibility. Similar to mortar, plaster is a homogeneous and cohesive material made by mixing inert aggregates, such as sand, a binder, such as cement or lime, and water. Though plaster is similar to mortar, in order to obtain better finish, we need to use more percentage of finer sand in plaster than in mortar. Crushed stone sand or crushed gravel sand or a combination of any of these is allowed by IS 1542:1992. However, crushed or manufactured sands are not generally suitable for plasters due to their angular particle shape. As per IS 1542:1992, the fineness modulus of sand should be not less than 1.4 in case of crushed stone sands and crushed gravel sands and not less than 1.5 in case of naturally occurring sands (100% should pass through 10 mm IS sieve and 0–15% through 150 μm IS sieve).

Plastering is the word used to describe the process of applying plaster on surfaces, such as walls and ceilings. Plastering can be done manually by masons or by machines, which can render up to a height of wall of 3.5–5 m, thickness of plaster of 5–30 mm, and at speeds of 45 m²/h. The surface on which the plaster is to be applied should be rough, absorbent (to a limited extent), strong, and clean, i.e., free of dust, oil, or paint that could impair the bond between the plaster and the surface. Usually, the surface is roughened by sharp tools or using a dash bond coat (a mixture of one part of cement to one-and-a-half parts of coarse sand). This cement-rich slurry is dashed against the base surface by hand with a brush, trowel, or by machine, such that the entire surface to be plastered is covered. The high-cement content provides a tenacious bond. This material is left unfinished so that a rough base is created. Plaster is then applied on this surface in a wet state in thin layers and hardens to a very dense solid when cured properly (moist curing for a minimum of 7 days is advisable). Recommended thicknesses of plaster are: first undercoat: 10–15 mm, second undercoat (if any): 3–8 mm, and finish coat: 3–5 mm. If plaster is applied in a single coat, thickness should be 10–15 mm (more details of thickness for different coats may be found in Table 2 of IS 1661:1972). A single coat should not be thicker than 15 mm. Plaster provides an economical hard surface that is often fire-resistant and can be coloured and finished in a wide range of textures, as per the architectural requirements.

Plastering should be protected from the sun and drying winds. The plaster should be used up within 2 h of being mixed and never be re-tempered by mixing with additional water. Plaster should not be continuous across the line of a damp-proof course as this will allow moisture to travel above the level of the damp-proof course.

### 7.12.1 Cement or Lime Plasters

Cement plaster is a mixture of cement, sand, and water, which is applied to interior and exterior surfaces of masonry, in two or three coats, to achieve a smooth surface (ACI 524R-08). It is applied directly to the surfaces of masonry or concrete walls. Interior surfaces may be applied with a final layer of gypsum plaster, for smooth and polished surface. Various cement-based fireproofing plasters are also available. These usually use vermiculite as lightweight aggregate. The advantages of cement plaster are its strength, hardness, quick setting time, and durability.

Lime plaster is a workable mixture of lime, sand, and water. It is preferable to use lime putty for making lime plaster. It is better to first grind the mortar and store it in damp condition for a few days, for thorough slaking. It should be ground again to a fine paste and used immediately. Plaster is applied as a paste in the form of thick coat on the surface. It absorbs $CO_2$ present in the atmosphere, and hardens to a durable surface.

Lime is considered a better alternative to cement in plasterwork. Lime mortar has higher degree of workability, which is desirable in mortars and plasters. As the setting of lime mortar takes considerable time, large quantities of the mix could be prepared at the site at a time, without having the fear of it going as a waste. More details on the application of cement and cement-lime plaster finishes may be found in IS 1661:1972. Guidance for the practice of lime plaster finish suitable to Indian conditions with respect to availability of materials and equipment and conditions of exposure is provided in IS 2394:1984.

### 7.12.2 Polymer-based Plaster Systems

Polymer-based (PB) spray-applied plasters have been widely used in Europe since the 1960s and is today one of the main forms of plastering used in modern building systems. PB systems are the most common in countries like the USA. Such exterior insulation finishing systems (EIFS) consist of a PB laminate that is wet applied, usually in two coats, to rigid insulation board that is fastened to the wall with adhesive, mechanical fasteners, or both (www.cement.org). There have been fewer problems with EIFS used over solid bases such as concrete or masonry because these substrates are very stable and are not subject to rot or corrosion. However, EIFS experienced performance problems in the 1990s, including water leakage and low-impact resistance. Though the PB skin repels water very effectively, when moisture gets behind the skin and trapped inside the wall, it may affect insulation, sheathing, wood/metal framing and metal attachments (www.cement.org).

Spray-plaster is a range of ready-mixed and polymer modified cementitious, spray-applied plasters. These are long-life products with guaranteed performance. Designed for thin coat application, spray-plasters significantly reduce the deadweight of buildings – a weight reduction of 45% compared to traditional plasters. In addition, their advantages include substantial cost and time savings, flexibility, reduced cracking, reduced curing time, perfect adhesion, high quality, and ease of application.

Polymer-modified cement plaster may be applied to surfaces that are clean and firm, free of dust, and other contaminants. The substrate should be made wet prior to the application of the mixture. The powder is mixed with clean, drinkable water in the ratio 5–7 L of water to 20 kg of powder. The mixture is allowed to stand for 10 min, and mixed again before applying. Best results are obtained with mechanical mixing, i.e., using mixing drill in a 20 L bucket. After sufficient plaster has been applied the surface may be levelled using a steel trowel. In one application, 2–5 mm thick coat is applied. When applying more than one layer the preceding layer must be completely dry before applying the next. Excess plaster can be remixed and used within its pot life. The surface will be dry in 24 h.

### 7.12.3 Stucco Plasters for Exterior Walls

*Stucco* plasters are usually applied, either by hand or machine, on exterior walls. They usually consist of a mix of sand, cement/lime and water, but may also contain proprietary mix of additives including fibres and synthetic acrylics that add strength and flexibility. Usually, three layers of stucco are applied, called the scratch coat, the brown coat and the finish coat. As in the case of cement mortar, a dash bond coat may be used to create a rough base for the scratch coat. As per ASTM C926, scratch coats contain a mix of one part cement and 2.25 to 4 parts sand, brown coats contain a mix of one part cement and one to one parts sand, and finish coats one part cement and 1.5 to 3 parts sand. The first layer of plaster consists of plastic cement and sand. As in the case of cement plaster, a trowel is used to scratch the surface in a regular or criss-cross pattern to provide a key for the second layer. The first coat is allowed to cure for 48 h before the second layer is applied. The next layer is called the levelling coat, and may consist of sand, cement, and lime. The second layer is levelled with tools, and scraped smooth to provide a smooth, even surface onto which the finish coat can be applied. It is then allowed to cure for about 7 days to allow shrinkage to take place. The final layer may be made with a mix of cement, lime, and sand (coloured sand may also be used) and typically applied for a thickness of about 3 mm. The final layer can also be made with an acrylic-based finish of about 1 to 4 mm thick; it can be applied in many ways and in different colours. It is possible to create a sandy finish or various styles of textured finishes using trowel or other tools.

Modern synthetic stucco can be applied as one base layer and a final layer, which is thinner and fast to apply, compared to the traditional application of three-coat stucco. Stucco finish having about 20 mm nominal thickness, may provide 1-h fire rating.

For more than 50 years, metal wire mesh or expanded diamond-mesh metal lath has been regarded as an integral component of stucco building exteriors, especially when cement or gypsum plaster is applied over wood or gypsum sheathing (see Fig. 7.5). It has to be noted

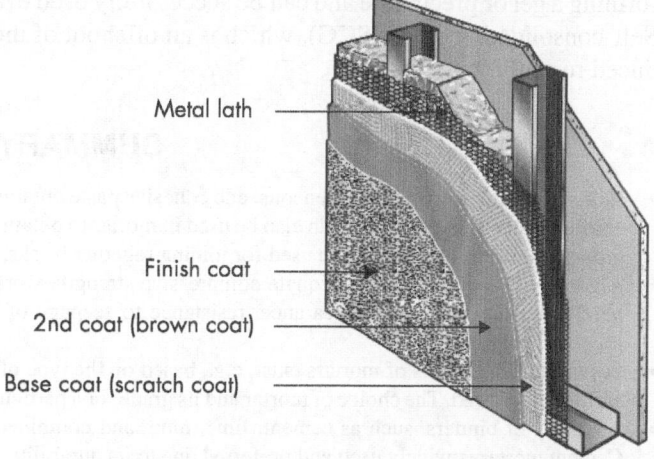

Metal lath

Finish coat

2nd coat (brown coat)

Base coat (scratch coat)

**Fig. 7.5** Conventional three-coat stucco plaster system
*Source*: United States Gypsum Company

that proper lath installation is of equal importance to the longevity of plaster-based building exteriors. Stucco has also been used as a sculptural and artistic material from the early 20th century.

## 7.13 Grout

Grout is a cementitious material, primarily composed of Portland cement, fine or coarse sand, possibly coarse aggregate, and in some cases a small amount of grouting admixture. The ingredients of grout are combined with a sufficient amount of water to produce a fluid mixture (IMI Technology Brief, 2009). Thus, grout is neither concrete nor mortar. Grout is a liquid material and poured, as against mortar, which is a workable mixture spread into place with a trowel. For example, the grout can be poured into the cores of masonry units – the initial high w/c ratio of the grout will be rapidly reduced on application, since the masonry absorbs much of the water.

Selection of grout materials depends on various desirable characteristics such as particle size, viscosity, gelification, gel strength, stability, and permanence coupled with economy. Grouts may also be classified into (a) particulate grouts and (b) chemical grouts. Details about these two types of grouts may be found in IS 14343:1996.

Grouts may be used to increase the bearing capacity of soil by injecting the grout under pressure (see IS 4999:1991 for the procedures, limitation of grouting techniques, and choice of grouting materials). Grout are also extensively used in dams to fill the cracks formed after the concrete sets and hardens [IS 11293 (Parts 1 & 2):1985]; spaces between tunnel walls and the surrounding earth – to spread the earth stresses uniformly over the structures; and in hollow concrete blocks – to develop bond between steel reinforcement and concrete. Specification for Bentonite used in grouting is provided in IS 12584:1989.

To increase the strength and at the same time reduce the shrinkage, it may be necessary to keep the w/c ratio of the grout as low as possible. This is achieved by the use of admixtures, such as accelerators, retarders, gas forming, and workability agents (IS 14343:1996). Fly ash can be added to a grout mixture as a partial substitute for Portland cement or lime. Also, fly ash helps grouts to maintain a given slump and improves grout pumpability. Note that the strength gain of the grout may be slower with fly ash than with Portland cement and this must be factored into construction sequencing. More details on masonry grouts may be had from IMI Technology Brief, 2009.

For finer cracks, chemicals grouts may be used. These consist of solution of two or more chemicals forming a gel or precipitate and can be successfully used even in the moist environment (IS 14343:1996). Self-consolidating grout (SCG), which is an offshoot of the self-consolidating concrete, has been introduced recently in the market.

## SUMMARY

- Mortar is as a workable, homogeneous, and cohesive paste obtained by adding water to a mixture of sand or *surkhi* and binding material. Admixtures can also be used in mortar to obtain a specific mortar colour, and to enhance one or more properties. Building mortars are used for joining together bricks, stones, or concrete blocks, and for pointing bricks.
- A good mortar should have adequate compressive strength, workability, durability, adequate rate of stiffening, good bond and adhesion, good appearance, resistance to seepage of water, volume stability, and compatibility with the proposed painting.
- Several classifications of mortars exist, e.g., based on the type of binding material, nature of applications, properties, and additives used. The choice of mortar and its grade for a particular application is governed by several considerations.
- A variety of binders, such as cement, lime, mud, and combined cement, lime and fly ash, can be used in mortars. Cement mortar is widely used and preferred due to its durability, high strength, fast setting, and hardening properties.
- Lime mortars improve the plasticity of mortars/plasters and are more durable and breathable than cement mortars. Some precautions should be taken while handling with cement or lime mortars. Mortars with lime and or fly ash content are considered green products.

- Quicklime or hydrated lime mortars harden and gain strength by the evaporation of water and the absorption of $CO_2$ from the atmosphere. However, hydraulic mortars, made with materials such as limestone or by adding pozzolanic materials behave like cement mortars. It is also possible to have composite mortar such as lime-cement mortar. *Surkhi* can be used as a substitute for sand in mortars.
- A number of special mortars such as fire-resistant mortars, and X-ray shielding mortars are used in certain special applications. Masonry cement is a product specifically formulated and manufactured for use in masonry mortars for brick, stone, and concrete block masonry.
- IS 2250:1981 stipulates a number of tests on mortars to find their consistency, compressive and tensile strength, water retention, and adhesion properties.
- Plaster is a building material similar to mortar, but will contain finer sand than in mortar, in order to obtain better finish. Plaster is used as a protective and/or decorative coating on walls and ceilings. Different types of plasters may be used such as cement plaster, lime plaster, PB systems, and stucco plaster.
- Grout is a cementitious material, primarily composed of Portland cement, fine or coarse sand, possibly coarse aggregate, and in some cases a small amount of grouting admixture. They are used in dams and other places to fill the cracks and cavities.

## EXERCISES

### Multiple-choice Questions

1. In the designation of mortar 1:X:Y, the numbers 1, X, and Y represent
   (a) One part of lime, X parts of cement, and Y parts of sand
   (b) One part of cement, X parts of lime, and Y parts of sand
   (c) One part of sand, X parts of lime, and Y parts of cement
   (d) None of these

2. Which of the following are the most important general requirements of a good mortar?
   (a) Tensile strength
   (b) Compatibility with the proposed painting
   (c) Appearance
   (d) All of these

3. Which of the following is an essential ingredient of mortar?
   (a) Admixtures
   (b) Manufactured sand
   (c) Lime or cement
   (d) All of these

4. One bag of cement has a volume of
   (a) $0.025 \text{ m}^3$   (c) $0.045 \text{ m}^3$
   (b) $0.035 \text{ m}^3$   (d) $0.05 \text{ m}^3$

5. In general, after initial mixing, cement mortar should be used within
   (a) 1–2 h   (c) 3–4 h
   (b) 2–3 h   (d) 4–5 h

6. As per IS 2250:1981, from the time of addition of water to cement, re-tempering of cement mortar is permitted only up to
   (a) 30 min   (c) 90 min
   (b) 60 min   (d) 120 min

7. Cement best suited for plaster work is
   (a) 33 Grade   (c) 53 Grade
   (b) 43 Grade   (d) All of these

8. As per IS 2250:1981, the strength of 1:2 lime mortar in 28 days will be about
   (a) 0.5–0.7 MPa   (c) 5–7.5 MPa
   (b) 2–3 MPa   (d) >7.5 MPa

9. One of the main demerits of using lime mortar is that it
   (a) does not adhere strongly
   (b) does not set quickly
   (c) absorbs dampness
   (d) is plastic

10. As per ASTM C270–06, the strength of 1:3 masonry cement (Type M) mortar in 28 days will be about
    (a) 17.2 MPa   (c) 5.2 MPa
    (b) 12.4 MPa   (d) 2.4 MPa

11. Which of the following mortars is most suitable for construction work in water-logged areas?
    (a) Lime mortar   (c) Cement mortar
    (b) Gauged mortar   (d) Mud mortar

12. A composite mortar is obtained by adding which of the following ingredients to cement?
    (a) Fly ash   (c) Sand and lime
    (b) Sand and *surkhi*   (d) *Surkhi* alone

13. After adding water to the mortar, the gauged mortar should be used without agitation within
    (a) 30 min   (c) 8–10 h
    (b) 1–2 h   (d) 24 h

14. The optimum fly ash content in terms of replacement of the weight of cement in cement-fly ash-sand mortar is
    (a) 20%   (c) 40%
    (b) 30%   (d) 50%

15. As per IS 2250:1981, the grade of mortar for durability should be more than
    (a) MM 3
    (b) MM 2
    (c) MM 0.7
    (d) MM 0.5

16. Preferred proportion of cement mortar for ceiling plastering is
    (a) 1:2
    (b) 1:3
    (c) 1:4
    (d) 1:5

17. Preferred proportion of cement mortar for external plastering is

18. If plaster is applied in a single coat, thickness should not be greater than
    (a) 10 mm
    (b) 15 mm
    (c) 20 mm
    (d) 25 mm

19. A weight reduction achieved using polymer-based systems is approximately
    (a) 30%
    (b) 35%
    (c) 40%
    (d) 45%

## Review Questions

1. What is mortar? How is it differentiated from plaster?
2. What are the requirements of a good mortar?
3. What are the necessary ingredients of a mortar?
4. Why are admixtures used in mortar? Give some examples.
5. List the various types of classifications of mortar with examples. Discuss briefly the classification based on binding materials.
6. Briefly describe the various types of mortars.
7. Describe the method of preparing cement mortar by (a) manual mixing and (b) machine mixing.
8. State the precautions to be taken while mixing and using cement mortar.
9. State a few applications of cement mortar.
10. (a) Describe the different procedures for making lime mortar.
    (b) State the functions of sand in lime mortar.
    (c) What machine would you suggest for preparing lime mortar in a large project?
    (d) How does hydraulic lime mortar differ from lime putty or hydrated lime mortars?
    (e) Why are lime mortars ground while cement mortars are not ground?
11. (a) State any five advantages of using lime mortar over cement mortar.
    (b) Under what conditions will you recommend cement mortar over lime mortar for masonry?
12. What are the precautions to be taken while mixing and using lime mortar? Why are they required?
13. Explain why lime mortar is suitable for repairs in historic buildings.
14. What is a gauged mortar? How is it prepared?
15. How is *surkhi* mortar prepared? What are the advantages of adding *surkhi* to lime mortar?
16. Describe briefly the preparation of mud mortar, its applications, and limitations.
17. How can the performance of mud mortar be improved?
18. What are special mortars? Describe briefly any four of them.
19. How does lime mortar qualify as a green material?
20. How are lime-pozzolana mixtures prepared? Why are they considered green materials?
21. What are the advantages of adding fly ash to cement mortar? How much percentage of cement will you recommend to be replaced by fly ash?
22. What is masonry cement mortar? How does it differ from lime or cement mortars?
23. What are the factors that influence the selection of a mortar?
24. What types of mortars will you recommend under the following situations? Give reasons for your selection and write the proportions of the ingredients.
    (a) Masonry in foundation
    (b) Brick work in superstructure
    (c) Wall with large concrete blocks
    (d) Internal plastering
    (e) External plastering
    (f) Ceiling plastering
25. Distinguish between mortar and plaster. What are the differences in the preparation of cement mortar and cement plaster?

26. Why is lime considered a better alternative to cement in plasterwork?
27. What are polymer-based plaster systems? How are they applied? How do they perform compared to regular cement plaster?
28. What is stucco and how does it differ from common plaster?
29. What are grouts? How grout materials are selected? How are they classified?
30. Give reasons for the following:
    (a) Why are bricks and stones soaked with water before they are laid in cement mortar?
    (b) Why is sand added to mortar?
    (c) Why is pozzolana added to mortar?
    (d) Why is it necessary to cure mortar?
31. Distinguish between the following:
    (a) Ordinary cement mortar and aerated cement mortar
    (b) Cement mortar and gauged mortar
    (c) Lime-*surkhi* mortar and lime-pozzolana mortar
    (d) Hydrated and hydraulic mortar
    (e) Bullock-driven mortar mill and pan mill
32. Write short notes on the following:
    (a) Lime-*surkhi* mortar
    (b) Fire-resistant mortar
    (c) Repointing mortar
    (d) X-ray shielding mortar
    (e) Polymer-modified mortar
    (f) Chemical-resistant mortar
33. Why is lime mortar not used extensively in India, though it offers many advantages over the cement mortar?

## ANSWERS

**Multiple-choice Questions**

|   |   |   |   |   |   |   |
|---|---|---|---|---|---|---|
| **1.** (b) | **2.** (d) | **3.** (d) | **4.** (b) | **5.** (a) | **6.** (b) | **7.** (a) |
| **8.** (b) | **9.** (b) | **10.** (a) | **11.** (c) | **12.** (c) | **13.** (a) | **14.** (c) |
| **15.** (c) | **16.** (c) | **17.** (d) | **18.** (b) | **19.** (d) | | |

# CHAPTER 8
# CONCRETE AND REINFORCED CONCRETE

## 8.1 Introduction

The word concrete comes from the Latin word 'concretus' (meaning condensed or 'grows together'). This name was chosen, perhaps, because concrete, due to the process of 'hydration', transforms from a visco-elastic, mouldable liquid into a hard, rigid, and solid rock-like substance. The Romans first invented what we call today hydraulic cement-based concrete or simply *concrete*. They built numerous concrete structures, including the 43.3-m diameter concrete dome, Pantheon, in Rome, now over 2000 years old but still in use and remains as the world's largest unreinforced concrete dome.

Second only to water, concrete is the most consumed material in the world. Concrete is used in nearly every type of construction. Traditionally, concrete was made primarily of cement, water, and aggregates (Aggregates can be of different materials such as gravel, limestone, and sand.). Thus, concrete is not a homogeneous material and its strength and structural properties may vary greatly depending upon how it is made (However, reinforced concrete is normally treated in design as a homogeneous material with different tensile and compressive properties.).

As of 2006, about 7.5 billion $m^3$ of concrete are produced each year – that equates to about 1 $m^3$ for every person each year on Earth (see Table 8.1). The National Ready Mixed Concrete Association estimates that ready mixed concrete production in the United States alone is about 262 million $m^3$ in 2016 and that there are about 6,000 ready mixed concrete plants in the United States. For a typical concrete mix, 1 tonne of cement (powder) will yield about 3.4 to 3.8 $m^3$ of concrete weighing about 7 to 9 tonne (i.e., the density is typically in the range of about 2.2 to 2.4 tonne/$m^3$). Although aggregates make up the bulk of the mix, it is the cement paste that binds the aggregates together and also contributes to the strength of concrete.

**Table 8.1** Annual consumption of major structural materials in the world (*Source*: US Geological Survey, International Iron and Steel Institute, NIST, US Department of Agriculture)

| Material | Unit weight (kg/m³) | Million Tonnes | Tonnes/person |
|---|---|---|---|
| Structural Steel | 7850 | 1244 | 0.18 |
| Cement | 1440 | 3400 | 0.48 |
| Concrete | 2400 | ~18000 | 2.4 (990 L) |
| Timber | 700 | 277 | 0.04 |
| Drinking water⁺ | 1000 | 5548 | 0.73 (730 L) |

**Notes:** Estimated world population as on July 2018 = 7.6 billion
⁺Assumed as 2 L/day/person

Concrete technology has come a long way since the Romans discovered the material. Now concrete is truly an engineered material, with a number of ingredients, which include a host of mineral and chemical admixtures. These ingredients should be precisely determined, properly mixed, carefully placed, vibrated (not required in self-compacting concrete!), and cured properly so that the desired properties are obtained; they should also be inspected at regular intervals and maintained adequately until their intended life. Even the cement we use has undergone a number of changes. We also use variety of concrete, some tailored for their intended use, and many with improved properties. Few specialized concrete have compressive strength and ductility matching steel. Owing to the extensive developments in concrete technology, it is not possible to describe about all types of concrete, their ingredients, their chemistry, and their properties in this chapter. Hence, only a brief description is given about them, and the interested reader should consult a book on concrete technology (many references are given in the ORC), for more details.

## 8.1.1 Brief History

Many researchers believe that the first use of a truly cementitious binding agent (as opposed to the ordinary lime commonly used in ancient mortars) occurred in southern Italy in about the 2nd century BC. Volcanic ash (called *pozzolana*, found near Pozzouli by the bay of Naples) is a key ingredient in the Roman cement that was popular during the days of the Roman Empire. Roman concrete bears little resemblance to modern Portland cement concrete. It was never put into a mould/formwork in a plastic state and made to harden, as we do today. Instead, Roman concrete was constructed in layers by packing mortar by hand in and around stones of various sizes. The Pantheon, constructed in AD 126, is one of the structural marvels of all times. It is a highly sophisticated unreinforced domed structure having many weight-reducing voids and small vaulted spaces, and uses lighter aggregates to reduce weight (Shaeffer, 1992).

**Case Study    Pantheon in Rome**

The oldest known concrete shell, the Pantheon in Rome, Italy, was completed about AD 126, and is still standing as the world's largest unreinforced concrete dome. It has a massive concrete dome, 43.3 m in diameter, with an open sky-light, called oculus, at its centre.

The downward thrust of the dome is carried by eight barrel vaults in the 6.4 m thick drum wall into eight piers. The thickness of the dome varies from 6.4 m at the base to 1.2 m around the oculus. The stresses in the dome were found to be substantially reduced by the use of successively less dense stone aggregates in the higher layers of the dome. The interior 'waffle'-like coffering not only was decorative, but also reduced the weight of the roof, as did the elimination of the apex by means of the oculus.

During the Middle Ages, the use of concrete declined, although isolated instances of its use have been documented and some examples have survived. Concrete was more extensively used again during the Renaissance and its manufacture was described in a work by de Lorme, published in 1568. At this time, mass or plain concrete was used in structures such as bridge piers. Pozzolanic materials were added to lime, as done by the Romans, to increase its hydraulic properties (Reed et al., 2008).

The small rowboats, built by Jean-Louis Lambot in the early 1850s, are cited as the first successful use of reinforcements in concrete. During 1850–1880 the French builder, Francois Coignet, built several large houses of concrete in England and France (Reed et al., 2008). Joseph Monier of France, who is considered the first builder of reinforced concrete, built RC reservoirs in 1872. In 1861, Monier published a small book, *Das System Monier*, and presented applications of reinforced concrete. During 1871–75, William E. Ward built the first landmark building in reinforced concrete in Port Chester, NY, USA. In 1892, François Hennebique of France patented a system of steel-reinforced beams, slabs, and columns, which was used in the construction of various structures built in England between 1897 and 1919. In Hennebique's system, steel reinforcement was placed correctly in the tension zone of the concrete; this was backed by a theoretical understanding of the tensile and compressive forces, which was developed by Cottançin in France in 1892 (Reed et al., 2008).

Earnest L. Ransome patented a reinforcing system using twisted rods in 1884; he also built the first RC framed building in Pennsylvania, USA, in 1903. In 1889, the first concrete reinforced bridge was built. In 1902, the rotary kiln was improved by Thomas Edison. The Ingalls building, first concrete sky-scraper, using the Ransome system, was built in 1904, which is still in use. In 1930, a Spanish engineer, Eduardo Torroja, designed the first thin shelled roof at Algeciras (Shaeffer, 1992). A few years later, another Spanish engineer, Felix Candela, built several outstanding shell structures. In 1935, an Italian architect-engineer, Pier Luigi Nervi, applied precast techniques to the construction of domed structures. The first major dam, the Hoover Dam, was built in 1936.

By the 1900s, concrete was generally used in conjunction with some form of reinforcement, and steel began to replace wrought iron as the predominant tensile material. A significant advance in the development of reinforced concrete was the pre-stressing of the steel reinforcing, which was developed by Eugene Freyssinet, in the 1920s, but the technique was not widely used until the 1940s (Walnut Lane Memorial Bridge in Philadelphia, built in 1951 was the first prestressed concrete bridge, in the USA). One of the first prestressed concrete bridges in India was built in 1954 over the Palar River in Tamil Nadu. Victoria in Montreal, constructed in 1964 with a height of 190 m and utilizing 41 MPa concrete in the columns, paved the way for high-strength concrete (Shaeffer, 1992).

### 8.1.2 Advantages and Disadvantages of Concrete

Concrete (usually, reinforced with steel bars) has been used in a variety of applications, such as buildings, bridges, roads and pavements, dams, retaining walls, tunnels, arches, domes, shells, tanks, pipes, chimneys, cooling towers, poles, foundations, piles and pile caps, and other structures and elements, due to the following advantages.

*Moulded to any shape*   Concrete can be poured and moulded into any shape varying from simple slabs, beams, columns to any complicated shells and domes, by using *formwork*. Thus, it allows the designer to develop a system satisfying the architectural and structural needs efficiently. Concrete also gives freedom to the designer to select any size or shape, unlike steel sections where the designer is constrained by the standard manufactured/available member sizes.

*Availability of materials*   The materials required for concrete (sand, gravel, and water) are often locally available and relatively inexpensive. Only small amounts of cement (about 14% by weight) and reinforcing steel (about 2–4% by volume) are required for the production of reinforced concrete, which may have to be shipped from other parts of the country. Moreover, reinforcing steel can be transported to most construction sites more easily than structural steel sections. Hence, reinforced concrete is the material of choice in remote areas.

*Low maintenance*   Concrete members require less maintenance compared to structural steel or timber members.

*Water and fire resistance*   Reinforced concrete offers great resistance to the actions of fire and water. A concrete member having sufficient cover can have 1 to 3 h of fire rating without any special fire-proofing material. Note that steel and wood need to be fireproofed to obtain similar fire rating – steel members are often enclosed by concrete for fire resistance. If constructed and cured properly, concrete surfaces could provide better resistance to water than steel sections, which requires expensive corrosion resistant coatings.

*Good rigidity*   Reinforced concrete members are very rigid. Owing to the greater stiffness and mass, vibrations are seldom a problem in concrete structures.

*Compressive strength*   Concrete has considerable compressive strength compared to most other materials.

*Economical*   It is economical, especially for footings, basement walls, and slabs.

*Low-skilled labour*   Comparatively lower grade of skilled labour only is required for the fabrication, erection, and construction of concrete structures.

In order to use concrete efficiently, one should also know the weakness of the material. The disadvantages of concrete include the following:

*Low tensile strength*   Concrete has a very low tensile strength, which is about 1/10$^{th}$ of its compressive strength, hence cracks when subjected to tensile stresses. Hence, reinforcements are often provided in the tension zones to carry tensile forces and limit crack widths. If proper care is not taken in design and detailing and also during construction, wide cracks may occur, which will subsequently lead to the corrosion of reinforcement bars (which are also termed as *rebars* in the USA) and even the failure of structures.

*Requires forms and shoring*   *Cast-in-situ* concrete construction involves the following three stages of construction, which are not required in steel or wooden structures: (a) construction of formwork over which concrete will be poured; the formwork holds the concrete in place until it hardens sufficiently, (b) removal of these forms, (c) propping or shoring of new concrete members until they gain sufficient strength to support themselves. Each of these stages involves labour and material and will add to the total cost of the structure. The formwork may be expensive and may be in the range of 1/3$^{rd}$ the total cost of reinforced concrete structure. Hence, it is important for the designer to make efforts to reduce the formwork cost, by reusing or reducing formwork.

*Relatively low strength*   Concrete has relatively low strength per unit of weight or volume (The compressive strength of normal concrete is about 5 to 10% of steel, and its unit density is about 31% of steel.). Owing to this, larger members may be required compared to structural steel. This aspect may be important for tall buildings or long-span structures.

*Time-depended volume changes*   Concrete undergoes drying shrinkage, and if it is restrained, it will result in cracking or deflection. Moreover, deflections will tend to increase with time due to creep of concrete under sustained compressive stress (the deflection may possibly double, especially in cantilevers). Note that both concrete and steel undergo approximately same amount of thermal expansion or contraction.

*Variable properties*   The properties of concrete may vary widely due to the variation in its proportioning, mixing, placing, curing, and due to its inherent cracking. Also, as *cast-in-situ* concrete is site controlled, its quality may not be uniform compared to materials such as structural steel and laminated wood, which are produced in the factory.

*CO₂ emission*   Cement, commonly composed of calcium silicates, is produced by heating limestone and other ingredients to about 1450°C by burning fossil fuels, and accounts for about 5–7% of $CO_2$ emissions globally. Production of 1 T of cement results in the emission of approximately 1 T of $CO_2$. Hence, the designer should specify cements containing cementitious and waste materials, such as *fly ash* and slags, wherever possible. Use of fly ash and other such materials, not only reduces $CO_2$ emissions, but also results in economy and improvement of properties, such as reduction in heat of hydration, enhancement of strength and/or workability, and durability of concrete (Neville, 2012; Subramanian 2007, 2012).

## 8.2 Ingredients of Concrete

As mentioned already, present-day concrete is made up of cement, coarse and fine aggregates, water, and a host of mineral and chemical admixtures. When mixed with water, the cement becomes adhesive, capable of bonding the aggregates into a hardened mass, called concrete. We have already discussed about cement and aggregates in Chapters 5 and 6, respectively. Hence, we will briefly discuss about water here. Chemical admixtures are discussed in Section 8.2.2. Mineral admixtures are inorganic materials that also have pozzolanic properties. These very fine-grained materials are added to the concrete mix to improve the properties of concrete, or as a replacement for Portland cement (blended cements). They include fly ash, ground granulated blast furnace slag (GGBFS or GGBS), silica fume, rice husk ash, and high-reactivity metakaolin. They have already been discussed in Section 5.9 of Chapter 5.

### 8.2.1 Water and Water-Cement Ratio

Water plays an important role in the workability, strength, and durability of concrete. Too much water reduces strength of concrete, whereas too little will make the concrete unworkable. Water used for mixing and curing should be clean and free from injurious amounts of oils, acids, alkalis, salts, sugars, or organic materials, which may affect concrete or embedded steel. As per Clause 5.4 of IS 456, potable water is considered satisfactory for mixing as well as curing concrete; otherwise, water to be used should be tested as per IS 3025- Parts 1–32 (1984 to 1988). Generally, sea water should not be used for mixing or curing of concrete. The pH value of water used for mixing should be greater than 6. The permissible limits of impurities as per Clause 5.4 of IS 456 are compared with ASTM C94 in Table 8.2.

**Table 8.2** Permissible limits for impurities in mixing water

| Impurity | Maximum permissible limit | |
| --- | --- | --- |
| | IS 456 (mg/L) | ASTM C 94 (ppm) |
| Organic | 200 | – |
| Inorganic | 3000 | – |
| Sulphates (such as SO₃ or SO₄) | 400 | 3000 |
| Chlorides (such as Cl) | 2000 - for plain concrete work 500 - for RCC | 1000[1] |
| Suspended matter | 2000 | 50,000 |
| Alkalis, as (Na₂O + 0.658 K₂O) | – | 600 |

[1]Pre-stressed concrete or concrete in bridge decks 500 ppm (ppm and mg/L are approximately equal).

It has to be noted that there is a noticeable difference in the suspended matter of these specifications. The current water shortage in many parts of the world has forced authorities to look for alternative sources of water for use in concrete production. ASTM C 1602M-04 and ASTM C 1603M-04 provide performance basis for the qualification of water for use in concrete. They distinguish between sources of water as potable, non-potable, and water from concrete production operations (CPO), and establish qualification requirements by requiring a certain frequency of tests depending on what the producer plans to use. *Combined water* is the one in which more of the sources are used in combination. The CCAA report (2007) provides a review of the current information on the quality of concrete mixing water in terms of mandatory limits on impurities, permissible performance variations stipulated in various standards, and the impact of combined wash-water and slurries from the CPO.

As mentioned in Section 5.8.3 of Chapter 5, cement reacts with mixing water and produces new compounds leading to progressive setting, hardening of the concrete, and development of strength. This overall process is called *hydration*. Generally, water required to be added for cement hydration is very less as compared to water required for workability. For complete hydration of Portland cement, only about 38% water, i.e., *w/c* of 0.38, is needed. See Section 5.8.3 of Chapter 5 for further discussions on *w/c* ratio and *w/cm* ratio and their effect on concrete.

## 8.2.2 Concrete Chemicals

It is interesting to note that Romans were the first to use admixtures in concrete in the form of blood, milk, and lard (pig fat). Present-day admixtures may be classified as chemical and mineral admixtures. As mineral admixtures have already been discussed in Section 5.9 of Chapter 5, only chemical admixtures are discussed here.

*Chemical admixtures* are materials in the form of powder or fluids that are added to the concrete immediately before or during mixing in order to improve the properties of concrete. They should comply with the requirements of IS 9103:1999. Admixtures are used for several purposes, which include: to increase flowability or pumpability of fresh concrete, obtain high strength through lowering of w/c ratio, retard or accelerate time of initial setting, increase freeze-thaw resistance, and for corrosion inhibition (Krishnamurthy, 1997). The effectiveness of an admixture depends upon factors such as type, brand, and amount of cementing materials; water content; aggregate shape, gradation, and proportions; mixing time; slump; and temperature of the concrete (Kosmatka et al., 2011).

The common types of *chemical admixtures* are as follows (Kosmatka et al., 2011; Krishnamurthy, 1997, and IS 9103:1999).

**Accelerators** Accelerators interact with $C_3S$ (tricalcium silicate) component of the cement and enhance the rate of hydration of concrete, resulting in high early strength of concrete, and early removal of formwork (Myrdal, 2007). They are useful for modifying the properties of concrete in cold weather. Typical materials used are calcium chloride, triethenolamine, sodium thiocyanate, calcium formate, calcium nitrite, and calcium thiosulfate. Typical commercial products are Mc-Schnell OC and Mc-Schnell SDS. Typical dosage is 2 to 3 % by weight of cement. As the use of chlorides causes corrosion in steel rebars, they are not used now. The effect of accelerators on concrete setting time is shown in Fig. 8.1(a).

**Retarders** They slowdown the initial rate of hydration of cement, thus delaying the initial set of the concrete, and are used more frequently than accelerators. Most retarders also function as water reducers and may entrain some air in concrete. Retarders keep concrete workable during placement and delay the initial set of concrete. Retarders neither affect the final setting time of cement, nor do they influence the 28-day strength of concrete. They assist in hot-weather concreting and in transporting fresh concrete over long distances; hence, they are used extensively in ready mixed concrete technology. They also prevent the formation of cold joints while casting and consolidating concrete pours. Most common retarder is calcium sulphate. Others include sugars, hydroxides of zinc and lead, calcium, and tartaric acid. Typical dosage is 0.4 to 0.5% by weight of cement. Chemicals such as lingosulphonic acids and hydroxylated carboxylic acids act as water reducing and retarding admixtures. The effect of retarders on concrete setting time is shown in Fig. 8.1(b).

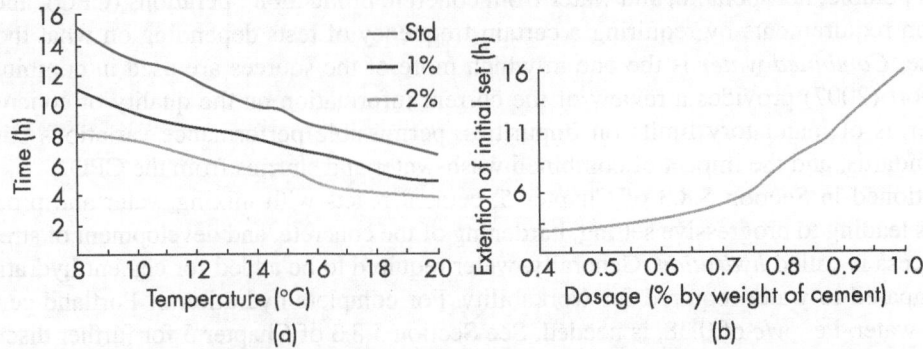

**Fig. 8.1** Effect of accelerators and retarders on concrete setting time
(a) Accelerators (b) Retarders (Reprinted with permission from Allied concrete, New Zealand)

**Water-reducing admixtures**  These are used to reduce the quantity of mixing water required to produce concrete with a certain slump, reduced w/c ratio, reduced cement content, or increased slump. Water-reducing admixtures (WRA) are available as ordinary WRAs and *high-range water-reducing admixtures* (HWRA). WRAs enable up to 15% water reduction, whereas HWRAs up to 30%. Popularly, the former are called *plasticizers* and the latter *super-plasticizers*. In modern-day concreting, the distinction seems to be disappearing. WRAs are generally supplied as liquid formulations, with the active solids content in the range of 30–40%. Common plasticizers are hydroxylated carboxylic acids, carbohydrates (corn syrup and dextrin), calcium, sodium, and ammonium lignosulphonates. Dosage ranging from 0.15 to 0.5% by weight of cement (different dosages for different products) are recommended by the manufacturers. At higher dosages, there is a danger of excessive retardation, bleeding, and air entrainment. Commercial plasticizers include Conplast 211, Conplast P509, Roff Plast 330, Plastiment BV 40, and Emceplast BV.

With super-plasticizers, it is possible to use w/cm ratio as low as 0.25. Compounds used in India as super-plasticizers include sulphonated naphthalene formaldehyde condensates, and sulphonated melamine formaldehyde condensates (SMF), and modified lignosulphonates. The optimum dosage of super-plasticizers to be used with concrete can be determined by using Marsh cone test, mini slump test, and flow table test. Some new generation super-plasticizers include: acrylic polymer based, copolymer of carboxylic acid with acrylic ether, polycarboxylate ether (PCE), and multi-polycarboxylate ethers. Currently, SMFs are more widely used and there is gradual shift towards PCEs, due to their better performance. The naphthalene and melamine types of super-plasticizer or HWRAs are typically used in the range 0.7 to 2.0% by weight of cement and give water reductions of 16 to 30%. The polycarboxylate ethers are more powerful and used at 0.3 to 8.0% by weight of cement to typically give 20 to over 40% water reduction (Clause 10.3.3 of IS 456 restricts the dosage to 1.0%.). The effect of super-plasticizers may last only 30 to 60 min, depending on the brand and dosage rate, and is followed by a rapid loss in workability. As a result of the slump loss, super-plasticizers are usually added to concrete at the jobsite. Use of super-plasticizers with reduced water content and w/c ratio can produce concrete with (a) high workability of fresh concrete, with increased slump, allowing it be placed more easily, with less consolidating effort, (b) high compressive strengths, (c) increased early strength gain, (d) reduced chloride-ion penetration, and (e) high durability. Note that it is important to consider the compatibility of super-plasticizers with certain cements (www.theconcreteportal.com; Aïtcin & Mindess, 2015; and Jayasree et al., 2011). Properties such as tensile strength, modulus of rupture, modulus of elasticity, drying shrinkage, and creep are not affected to any significant extent due to the addition of WRA or HWRA. Commercial super-plasticizers include Zentrament Super BV, Muraplast FK 61, Roff Super Plast 321, Roff Super Plast 820 & 840, Sikament FF & 600, and Conplast 337 & 430.

**Air-entraining admixtures** These are used to entrain tiny air bubbles in the concrete, which will reduce damage during freeze-thaw cycles, thereby increasing the concrete's durability. Furthermore, the workability of fresh concrete is improved significantly, and segregation and bleeding are reduced or eliminated. However, entrained air entails a trade off with strength, as each 1% of air may result in 5% decrease in compressive strength. Materials used as air-entraining admixtures include salts of wood resins, some synthetic detergents, salts of petroleum acids, fatty and resinous acids and their salts, and salts of sulphonated hydrocarbons. Some commercial air-entraining agents are MC-Mischoel AEA, Complast AE 215, and Roff AEA 330. Air-entrained concrete has been used in the construction of Hirakud dam, Koyna dam, and Rihand dam (Shetty, 2005).

**Corrosion inhibitors** These are used to minimize the corrosion of steel reinforcements in concrete. A corrosion inhibitor can affect steel reinforcement in concrete in two ways: (a) By delaying the time of depassivation by strengthening the passive film, or (b) by reducing the corrosion rate after depassivation. They may be classified into three broad electrochemical classes: (a) anodic, (b) cathodic, and (c) mixed, depending on whether it affects the anodic reaction, the cathodic reaction, or both.

Inhibitors for concrete are either mixed into fresh concrete (mixed-in inhibitors-MIIs) or applied onto the surface of hardened concrete (migrating inhibitors -MCIs) to penetrate the concrete cover. Sodium nitrite, an anodic inhibitor, was the first to be used in concrete about 50 years ago. Since then, a number of inhibitors have been tested, of which nitrates, phosphates, amines, alkanolamines, and carboxylic acids are the most investigated.

Twenty years ago, amines and alkanolamines became the most common MIIs. These are mixed inhibitors suppressing both the anodic and cathodic reactions. According to Myrdal (2010), blends of nitrite/nitrate salts (preferably calcium) are the most used common commercial MIIs, today. Nearly all MCIs in the market are blends of amines/alkanolamines and carboxylic acids, i.e., aminocarboxylates. Calcium nitrite is widely used, with a dosage of $18–30$ L/m$^3$ of concrete, depending on the chloride levels of concrete (Shetty, 2005). Examples of commercial products are MC-Corrodur, Sika® FerroGard®-901(IN), Sika® CNI, and KemCorpro710. It has to be noted that there is still a debate on whether these products really act as anticipated. Examples of commercial MCIs are Emceecolour Flex and KemCorpro 810 M. They have been used as a protective and decorative coating on several flyovers in Mumbai.

The other chemical admixtures include foaming agents (to produce lightweight foamed concrete with low density), alkali-aggregate reactivity inhibitors, bonding admixtures (to increase bond strength), colouring admixtures, shrinkage reducers, and pumping aids. Normally, admixture dosage of about 2 to 5% by mass of cement is recommended by different manufactures for different products. However, exact dosage rate should be determined based on trial mixes at site. It is important to test all chemical admixtures adequately for their desired performance (see also Clause 5.5 of IS 456). They should not be used in excess of the prescribed dosages, as they may become detrimental to the concrete.

## 8.3 Classification of Concrete

Concrete may be classified in different ways based on the binder used, proportion of ingredients, strength and density, and the place of production.

***Based on the binder*** Based on this, concrete may be classified as lime concrete, gypsum concrete, and cement concrete.

***Based on proportion of ingredients*** Concrete may be specified by the proportions of its ingredients, such as 1 (cement): 2 (fine aggregate): 3 (coarse aggregate). This type of specifying the concrete mix is known as *nominal mix*. Nominal mixes will not result in targeted strength as the quantity of fine aggregate is fixed irrespective of the cement content, w/c ratio, and the maximum size of aggregate to be used. When concrete proportions are fixed based on some mix design and trial mixes, it is designated as designed-mix concrete. IS 456 restricts the use of nominal mix only up to M20 grade, where M denotes the mix.

***Based on grade of concrete*** Based on the 28th day compressive strength of concrete cubes (150 mm side), expressed in $N/mm^2$, concrete may be classified as M10, M20, or M30 (see Table 2 of IS 456:2000). It may further be classified as ordinary concrete ($<20 N/mm^2$), standard concrete ($25–60 N/mm^2$), and high-strength concrete ($65–100 N/mm^2$).

***Based on density*** On its basis, concrete may be classified as super heavy (over 2500 $kg/m^3$), dense ($1800–2500 kg/m^3$), or lightweight ($500–1800 kg/m^3$).

***Based on place of casting*** Concrete made and placed in position at site is called *in-situ concrete*, whereas the concrete made in the factory for making prefabricated units is called *precast concrete*.

## 8.4 Mixing the Materials for Concrete

The measurement of materials for making concrete is called as *batching* (See also Clause 10.2 of IS 456). Though volume batching is used in small works, it is not a good method and weigh batching should always be attempted—water and liquid admixture may be weighed or measured by volume (fully automatic weigh batching equipment are used in RMC plants—see Section 8.4.1). The grading of aggregates should be controlled by obtaining them in right proportions, the different sizes being stocked in separate stockpiles. The grading of coarse and fine aggregates should be checked as frequently as possible. The accuracy of measuring equipment should be within ±2% of the quantity of cement being measured and within ±3% for other materials such as aggregate, chemical admixtures and water. These can be realized only by proper batching and mixing plants.

It is important to maintain the w/c ratio constant at its intended value. In order to maintain the w/c ratio, the moisture contents of both fine and coarse aggregates should be determined as frequently as possible and the amount of water added should be adjusted to compensate for any observed variations in the moisture contents of aggregates. The moisture contents of aggregates may be determined as per IS 2386(Part 3). To allow for the variation in mass of aggregates due to variation in their moisture content, suitable adjustments in the masses of aggregates should also be made (see also Section 8.9). Where batching plants are used, moisture content may be determined by moisture probes fitted to the batching plants. It is of interest to note that majority of batching plants in India are not fitted with moisture probes (Reddi, 2012). Fitted at the entry point of fine aggregates into batch hopper, the probe monitors moisture content, quantifies the value, and via the *Program Logic Control* automatically adjusts water content in each batch, ensuring the designed water content in each mix.

The mixing of materials should ensure that the mass becomes homogenous, uniform in colour and consistency. Again hand mixing is not desirable for obvious reasons and machine mixing is to be adopted for better quality. Several types of mixtures are available; pan mixtures with revolving star blades are more efficient (Shetty, 2005; IS 1791:1985; IS 12119:1987). It should be ensured that stationary or central mixers and truck mixers comply with the performance criteria of mixing efficiency as per IS 4634. Mixing efficiency test should be performed at least once in a year. Clause 10.3 of IS 456 stipulates that if there is segregation after unloading from the mixer, the concrete should be remixed. It also suggests that when using conventional tilting type drum mixtures, the mixing time should be at least 2 min, and the mixture should be operated at a speed recommended by the manufacturer (normal speeds are 15–20 revolutions/min). Clause 10.3.3 of IS 456 restricts the dosage of retarders, plasticizers, super-plasticizers, and polycarboxylate based admixtures to 0.5, 1.0, 2.0, and 1.0%, respectively, by mass of cementitious materials. IS 456 allows a higher percentage of the aforementioned admixtures based on performance tests relating to workability, setting time, and early age strength.

**Consistency**  It may be defined as the relative ability of freshly mixed concrete to flow; it is also a measure of the ease with which concrete can be consolidated. It is related to but not synonymous with workability. Once we select the materials and proportions of a particular mix, the required workability and also the consistency of the mix may be achieved through minor variations in the water content. The slump test (see Section 25.12.1 of Chapter 25) is often used to find the consistency of mix – the higher the slump, the more mobile the mixture. The Vebe test is generally recommended for stiffer mixtures. Values of slump, compacting factor, and VB time for the entire range of consistencies used in construction are given in Table 8.3 (also see Secs. 25.12.1 to 25.12.3 of Chapter 25).

**Table 8.3**  Consistencies of concrete mix (Based on American Concrete Institute (ACI) 309R-05 and IS 1199)

| Consistency description/use | Slump, mm | VB time, sec | Average compacting factor |
|---|---|---|---|
| Extremely dry | – | 32–18 | – |
| Very stiff | – | 18–10 | 0.70 |
| Stiff – To be used with vibration | 0–25 | 10–5 | 0.75 |
| Stiff plastic – Mass concrete | 25–75 | 5–3 | 0.85 |
| Plastic – Normal RCC | 50–100 | 3–0 | 0.90 |
| Pumped concrete, slipform work | 75–100 | – | 0.90 |
| Highly Plastic – *cast-in-situ* piles | 100–150 | – | 0.95 |
| Flowing – RCC with congested reinforcement, not suitable for vibration | >190 | – | 0.95 |

**Note:** Maximum size of aggregate is 38 mm.

## 8.4.1 Batching Plants

As mentioned earlier, batching is a process whereby cement is mixed with graded aggregates (coarse and fine aggregates), water, as well as chemical and mineral admixtures (see Fig. 8.2). Concrete batching plants are plants used to mix the various ingredients to form concrete. There are two types of concrete plants:

(a)  (b)

**Fig. 8.2** (a) Concrete batching plant (b) Concrete mixer truck

*Ready mix plants*  In this plant, all ingredients, except water, are weighed in weigh batchers, which may be computer controlled at the concrete plant. Then, they are mixed and the mixture is then discharged into a ready mix truck. Water is weighed or volumetrically metered and the contents are then mixed for a minimum of 10 min during transportation to the site.

*Central mix plants*  Such a plant combines some or all of the above ingredients (including water) at a central location. The mixture is then transported to the job site. Central mix plants differ from ready mix plants in that they offer the end user a much more consistent product, since all the ingredient mixing is done in a central location and is computer-assisted to ensure uniformity of product. It has to be noted that temporary batching plants are usually set up at the job site for larger projects.

Concrete batching plants are widely used to produce the various kinds of concrete suitable for large- or medium-sized building projects, roads, bridge projects, and precast concrete plants.

More recently, *mobile concrete batch plants,* patented in 1975 by Vincent Hagan, are also made available. Mobile batch plants are typically single units, made up of a concrete silo, an aggregate bin and batcher, an aggregate conveyor and the cement batcher, pulled by a tractor. Sometimes, a water batch is also included in the unit. Aggregate bins may have three to four compartments for storing sand and various aggregate sizes. Conveyors are between 0.6 to 1.2 m wide and carry aggregate from the batcher. The mobile batching plant is easy to transport. It can be fixed-mounted on a truck, mounted on a truck with tipping box or mounted on an interchangeable cradle.

## 8.5 Transporting and Placing Concrete

Concrete can be transported from the mixer to the formwork by a variety of methods and equipment: mortar pans, wheel barrows, belt conveyors, truck-mixer-mounted conveyor belts, buckets used with cranes and cable ways, truck mixer and dumpers, chutes/drop chutes, skip and hoist, transit mixer [in the case of ready mixed concrete (RMC)], tremies (for placing concrete underwater), or by pumping through steel pipes. As the great danger during transportation is segregation, care should be taken to avoid it. More details about the methods of transportation may be found in Panarese (1987), Kosmatka et al. (2011), and Shetty (2005).

It is also important that the concrete is placed in the formwork properly to yield optimum results. Prior to placing, reinforcements must be checked for their correctness, cover requirements, and any loose rust must be removed. The formwork must be cleaned, adequately braced, joints between planks or sheets effectively plugged, and the inside of formwork applied with mould releasing agents for easy stripping (Details of different kinds formwork and their design may be found from Hurd [2005] and IS 14687:1999). It is necessary to clean the surface of previous lifts thoroughly with water jet and treat it properly. Concrete should be deposited continuously as near as possible to its final position without any segregation. In general, concrete should be placed in thicker members in horizontal layers of uniform thickness [about 150 mm thick for reinforced concrete (RC) members] and each layer is thoroughly consolidated before the next is placed (see Section 8.6). Chutes and drop chutes may be used when the concrete is poured from a height, to avoid segregation. Though Clause 13.2 of IS 456 suggests a permissible free fall of 1.5 m, it has been found that a free fall of even a high slump concrete of up to 46 m, directly over reinforcing steel, does not result in segregation or reduction of compressive strength (Suprenant, 2001).

Concreting during hot or cold weather should conform to the requirements of IS 7861(Part 1):1975 and IS 7861(Part 2):1988. More guidance on hot weather concreting is given in ACI 305R:10 and Venugopal and Subramanian (1977). Guidance for underwater concreting is provided in Clause 14 of IS 456.

## 8.6 Consolidation of Concrete

Right after placement, concrete may contain 5 to 20% entrapped air. The amount varies with mix type and slump, form size and shape, the amount of reinforcing steel, and the concrete placement method (Suprenant, 1988). At a constant w/c ratio, each percent of air decreases compressive strength by about 3 to 5%. The concrete should be deposited and compacted before initial setting of concrete commences and should not be disturbed subsequently. *Consolidating* the concrete, usually by vibration, increases concrete strength by driving out entrapped air; it also improves bond strength and decreases concrete permeability. It also improves the appearance of hardened concrete by minimizing surface blemishes, such as honeycombing and bug-holes. When there is a congestion of reinforcement, vibration alone is not enough to consolidate concrete adequately. In such cases, to ensure adequate consolidation it may be necessary to use super-plasticizers, reduce aggregate maximum size, or adjust rebar spacing.

### 8.6.1 Hand Consolidation

*Hand compaction* was used earlier when large w/c ratio was adopted (>120 mm slump). This method may be useful to compact concrete in thin elements such as slabs, in members with congested reinforcements, and in sharp corners and edges. Hand consolidation is laborious, costly, and suitable only for workable and flowing concrete mixtures. It is done by rodding the concrete with a 16 mm diameter steel rod, continuously during concreting, to pack the concrete.

### 8.6.2 Compaction by Vibration

High-frequency power driven internal/external vibrators (as per IS 2505, IS 2506, IS 2514, and IS 4656) permit the easy consolidation of stiff mixes having low w/cm ratio. *Compaction by vibration* is the most common and widely used method of compacting concrete for any structural member. The vibration sets the different components of fresh concrete in motion, reducing the friction between them and liberating the entrapped air to rise to the surface. This results in a temporary liquefaction of concrete and subsequent settlement, which produces a dense and compact mass. Vibrations produce higher strength when a stiff mix is used with low w/c ratio (with a slump of about 40 mm). It is important to recognize the two stages in the compaction process:

**Initial consolidation of the concrete**   This is often achieved relatively quickly using vibration; the concrete liquefies and a levelled surface is achieved, giving the impression that the concrete is compacted.

**Release of entrapped air**   This air takes a little longer to rise to the surface. Hence, compaction has to be prolonged until air bubbles no longer appear on the surface (see Fig. 8.3). It is usually not practical to remove all the entrapped air with standard vibrating equipment; hence, concrete with 1–2% entrapped air is considered to have good field consolidation.

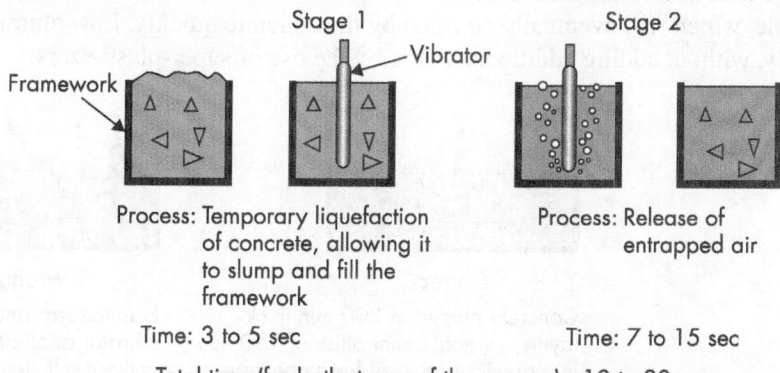

Fig. 8.3 The two-stage process of compaction (CCAA Data Sheet, 2006b, reprinted with permission from Cement Concrete & Aggregates Australia)

It must be noted that the level of compaction has a great effect on the compressive strength; for example, the strength of concrete containing 10% of entrapped air is only 50% of that concrete when fully compacted (see Fig. 8.4). Plastic mixes need less time of vibration than stiff/dry mixes. The various types of mechanically operated vibratory equipment, such as: (a) internal vibrator (needle vibrator), (b) formwork vibrator

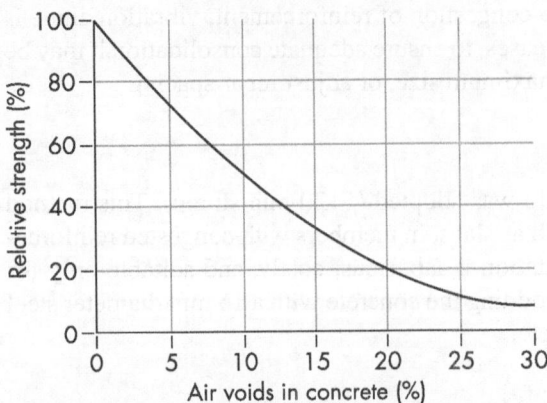

**Fig. 8.4** Loss of strength due to incomplete compaction (CCAA Data sheet, 2006b, reprinted with permission from Cement Concrete & Aggregates Australia)

(external vibrator), (c) table vibrator, (d) platform vibrator, (e) surface vibrator (screed vibrator), and (f) vibratory roller may be used.

**Internal vibrator** Of all the vibrators, this is most commonly used. These are also known as immersion, needle, or *poker vibrator*. A typical needle vibrator is shown in Fig. 8.5(a). Immersion vibrator can be used in any type of concrete work. The most common immersion vibrator is the flexible-shaft type. It has a steel tube, called poker (with 20–150 mm diameter and length of 250–900 mm), connected to an electric motor or diesel engine through a flexible tube (see Table 8.4). The frequency of vibration used ranges from 5500 to 15,000 vibrations per minute (see Table 8.4). As shown in Fig. 8.5(b), the vibrator should penetrate about 150 mm into the previous layer of fresh concrete to meld the two layers together (While consolidating the first layer, the vibrator should be kept 100 to 150 mm above the bottom of the form.). As shown in Fig. 8.6, the vibrator should be immersed into concrete in a definite pattern so that the radius of action overlaps and covers the whole area of the concrete (see Table 8.4, for the value of radius of action – it is approximately four times the diameter of vibrator.). Usually, a spacing of 1.5 times the radius of action or six to eight times the diameter of the poker (ranging from 120 to 900 mm) is adopted to achieve compaction of all the poured concrete. The thumb rule is to allow the vibrator to sink under its own weight and then remove it at the rate of 3 s per 300 mm. The concrete should not be placed in layers greater than 300 mm height. The vibrator should be allowed to penetrate the concrete vertically (with an inclination of less than 10°) under its own weight. The consolidation is considered adequate, when the surface becomes level, no larger air bubbles escape, and a thin film of glistening mortar appears on the surface of concrete. After adequate consolidation, the vibrator is slowly withdrawn from the concrete using up-and-down motions. After withdrawal, the vibrator will leave a hole, which will eventually be filled by the concrete quickly. Low-slump concrete can be consolidated easily, without adding additional water, by the use of super-plasticizers.

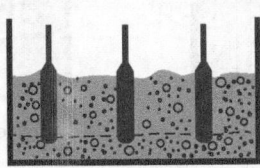

**Correct**

Concrete placed in 300 mm thick layers, vertical penetration of 150 mm into previous layer of fresh concrete to meld the two layers together and avoid 'cold-pour' lines on the finished surface; insertion at systematic regular intervals.

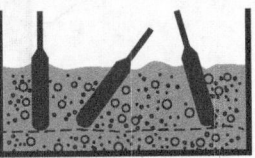

**Wrong**

Haphazard random penetration of vibrator at all angles and spacings without sufficient depth will not assure melding of the two layers

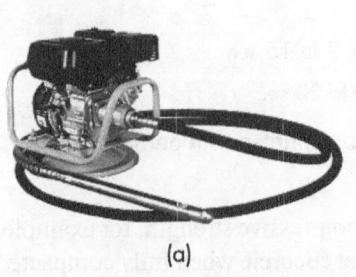

(a)

(b)

**Fig. 8.5** Needle vibrator and systematic vibration (a) Needle vibrator (b) Right and wrong methods of compacting

**Table 8.4** Characteristics and applications of immersion vibrators (Based on ACI 309R-05, reprinted with permission from the American Concrete Institute)

| Diameter of head | Recommended frequency,[1] vibrations per minute (Hz) | Approximate radius of action, e(mm) [2,4] | Rate of concrete placement [3,4] (m³/h per vibrator) | Application |
|---|---|---|---|---|
| 20–40 | 9000–15,000 (150–200) | 80–150 | 0.8–4 | High-slump concrete in very thin members and lab test specimens; also used to supplement larger vibrators where reinforcement or ducts cause congestion in forms. |
| 30–60 | 8500–12,500 (140–210) | 125–250 | 2.3–8 | Plastic concrete of 75–125 mm slump in thin walls, columns, beams, precast piles, thin slabs, and along construction joints. |
| 50–90 | 8000–12,000 (130–200) | 175–360 | 4.6–15 | Stiff plastic concrete (less than 75 mm slump), in general construction such as walls, floors, beams, and columns. |
| 80–150 | 7000–10,500 (120–1800) | 300–500 | 11–31 | Mass and structural concrete (0 to 50 mm slump) deposited in quantities up to 3 m³ in relatively open forms of heavy construction. |
| 130–150 | 5500–8500 (90–140) | 400–600 | 19–38 | Mass concrete in gravity dams, large piers, massive walls, etc. |

**Notes:**
[1] While vibrator is operating in concrete.
[2] Distance over which concrete is fully consolidated.
[3] Assumes insertion spacing of 1½ times the radius of action, and that vibrator operates two-thirds of time concrete is being placed.
[4] Reflects not only the capability of the vibrator, but also differences in workability of the mix, degree of de-aeration desired, and other conditions experienced in construction.

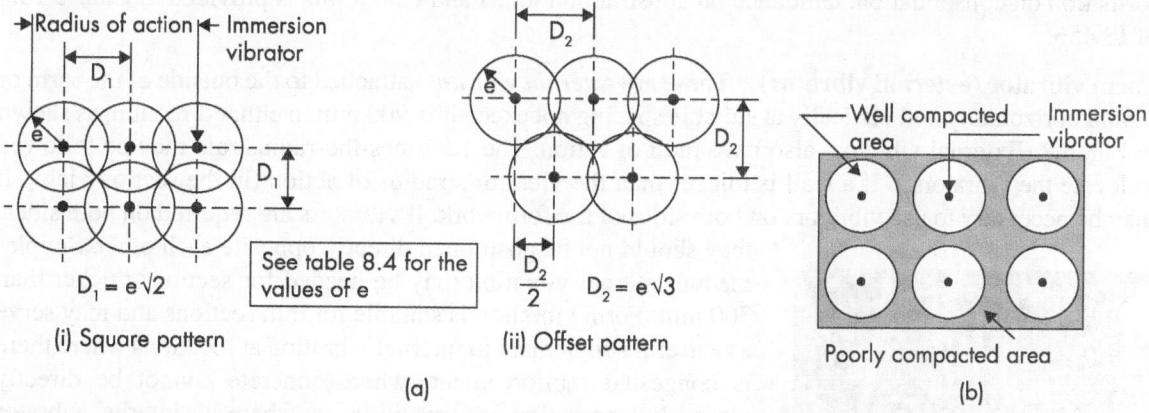

**Fig. 8.6** Vibrator insertions for proper compaction (a) Correct vibrator locations for full compaction (b) Wrong vibrator insertion locations

Generally, the vibrator should be kept about 50 mm away from the face of the formwork. Similarly, the vibrator should not be held against the reinforcement as this may cause its displacement. Concrete should not be spread sideways with an immersion vibrator as this may cause segregation of the mix. The vibration of very wet mixes should be avoided (CCAA, 2006).

When concrete is poured at site using pumps, concrete may be heaped up due to fast rate of dumping; in such situations, the position of vibrator should be changed gradually from the side of the heap and moved away from the heap in all directions, as shown in Fig. 8.7(a) for proper compaction. Poker should not be introduced in the centre of the heap, as shown in Fig. 8.7(b).

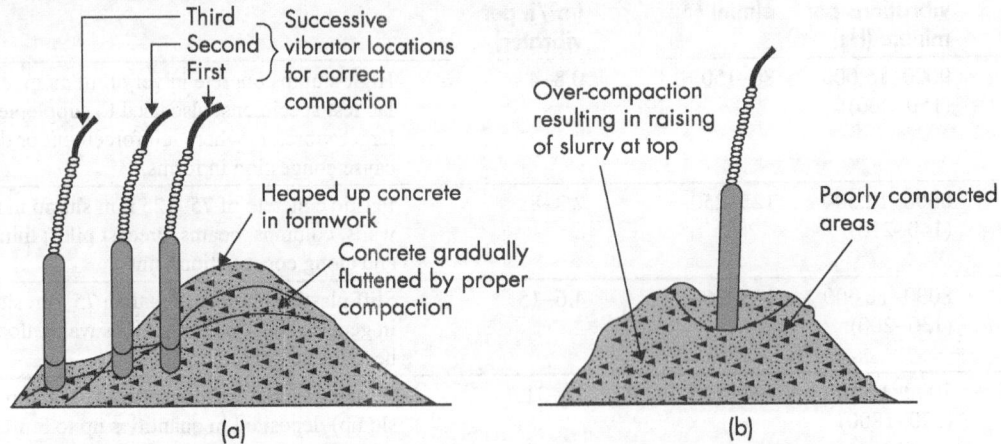

**Fig. 8.7** Compacting heaped-up concrete (a) Correct (b) Wrong (*Courtesy*: Dr N.V. Nayak, Advisor, Gammon India Ltd)

It has to be noted that *over-vibration* as well as *under-vibration* are harmful to concrete and should be avoided (Under-vibration of concrete is more common than over-vibration.). Inadequate compaction results in honey-combing, pour lines, sand streaking, cold joints, subsidence cracking, and excessive entrapped air (ACI 309R:05 may be consulted for discussions on these defects).

In walls and beams, it is better to use two vibrators, one for levelling the concrete immediately after placement and the other for further consolidation. ACI 309R:05 and CCAA [2006(b)] provide more information on consolidation. Guidance on construction joints and cold joints is provided in Clause 13.4 of IS 456.

**Form vibrator (external vibrator)**　These are *external vibrators* attached to the outside of the form or mould horizontally and vertically at suitable spacing not exceeding 900 mm in either direction, as shown in Fig. 8.8 (External vibrators also have radii of action. The 1.5 times-the-radius rule may be followed to locate the vibrators.). If a wall is thicker than the vibrator's radius of action (in the section view), it may be necessary to use vibrators on both sides of the formwork. If vibrators are required on both sides, they should not be positioned directly opposite each other. Supplemental internal vibration may be needed for sections thicker than 300 mm. Form vibration is suitable for thin sections and may serve as a useful supplement to internal vibration at locations where there is congested reinforcement, where concrete cannot be directly placed but must flow into position, or where an internal vibrator cannot be inserted. Formwork vibrators are used for concreting columns, thin walls or in the casting of precast units. When form vibrators are used, the design of formwork and the disposition of vibrators should be considered carefully to ensure efficient compaction and avoid surface blemishes. The formwork has to be rigid, strong, and watertight. Forms must also withstand the lateral pressure of the vibrating liquefied concrete.

**Fig. 8.8** Formwork external vibrator AR 36 attached to the formwork (*Courtesy*: www.wackerneuson.us)

Different types of form vibrators are available: rotary, reciprocating, electromagnetic, and pneumatic (ACI 309R:2005). External vibrators are rigidly attached to steel/aluminium channels or I-beams, running along and attached to the formwork. These vibrators operate typically at a frequency of 3600 to 6000 vibrations per minute (www.wackerneuson.us). They consume more power and are less efficient than needle vibrators. Concrete compacted by form vibration should be deposited in layers 250 to 400 mm thick, with each layer being vibrated separately. Vibration time is considerably longer than for internal vibration, and is generally between 1 to 2 min.

**Table vibrator**   It is a special case of formwork vibrator, used in laboratories, in the form of a vibrating table, consisting of a rigidly built steel platform mounted on flexible springs and driven by an electric motor. Typical frequency of vibration is 4000 vibrations per minute. They are commonly used for vibrating concrete cubes.

**Platform vibrator**   It is nothing but a table vibrator, but is larger in size. This is used in the manufacture of large prefabricated concrete elements such as electric poles, railway sleepers, prefabricated roofing elements, etc.

**Surface vibrator**   These are also called *screed board vibrators*. Surface vibrators are applied to the top surface and consolidate the concrete from the top down. Their effect also assists in the levelling and finishing operations of the surface. They are used mainly in floor/roof slab and industrial floors, road pavements, and similar flat surfaces. There are three principal types of surface vibrators: vibrating-beam screed (typical frequencies of 3000 to 6000 vibrations per min), plate or grid vibratory tampers (work well on stiffer concrete), and vibratory roller screed (ACI 309R:05). The most common type is the single or double vibrating-beam screed. These are effective only up to a thickness of 100 to 200 mm.

## 8.6.3 Bleeding of Concrete

When concrete is being compacted, the heavier aggregate and cement particles settle at the bottom due to the force of gravity, forcing excess and lighter mixing water upward, which collects at the top of the surface. This upward migration of water is termed as *bleeding* (When water collects at the bottom of the aggregates, it is called as *internal bleeding*.). Bleeding ceases either when the solid particles cannot settle any more, or when the concrete stiffens due to cement hydration and prevents further movement.

Mixes which bleed excessively are those which are harsh and not sufficiently cohesive. The amount of bleeding depends on mix properties, primarily water content and amount of fines (cement, fly ash, and fine sand). Increasing water content increases bleeding, and increasing the amount of fines reduces bleeding. Bleeding may also occur due to excessive vibration imparted to concrete to achieve full compaction.

All concrete bleed. Bleeding may be observed during the first hours of concreting of thin slabs that are cast in sunny weather-using mixtures with more flaky aggregates. In general, bleeding may not be bad for the concrete, as it may lower the w/c ratio and produce a dense concrete. However, concrete that bleeds too fast or too long can cause a number of problems: sand streaks in walls, weak horizontal construction joints, and reduced bond between aggregate, cement, and reinforcement. Finishing the concrete surface before the evaporation of bleed water can cause problems such as dusting, craze cracking, scaling, and low-wear resistance. Working bleed-water into the surface also increases permeability; water, deicing salts (in the case of bridges), and other harmful chemicals can enter the concrete more easily. Highly permeable concrete may increase the possibility of rebar corrosion. If the surface is sealed before the end of bleeding, bleed water will be trapped beneath the surface of concrete, resulting in blistering leading to possibility of peeling off the whole surface.

Excessive bleeding may be avoided by not adding excess water to the concrete. Instead of adding water for increasing workability, super-plasticizer may be used. Increasing the fines of concrete will also reduce bleeding; this can be achieved by:

1. Using more cement. Rich mixes with same water content, bleed less than lean mixes.
2. Using more finely ground cement, such as high early strength cement.
3. If the sand used does not contain much material passing through the No. 50 and 100 sieves, fine sand may be blended at the batch plant.
4. Using fly ash or other pozzolana in the concrete, which may break continuous water channels.
5. By using an air-entraining agent- entrained air bubbles act as additional fines. Air entrainment also lowers the amount of water needed to achieve the desired slump.

Proper finishing should not start until bleeding has ceased and the bleed water has disappeared or has been removed (ACI 308R:01).

**Laitance**   It is a condition in which cement slurry and water raise to the top and get settled on the surface, due to the presence of too much mixing water, excessive tamping or vibration of concrete. This is a very dangerous condition, since the top surface will weather out fast with larger shrinkage cracks. Laitance is a major cause of failure in flooring installations. If laitance is formed in a lift of concrete, it should be removed before next lift is placed.

**Finishing**   The time period between the initial and final set of concrete is suitable for finishing the concrete. Concrete may be assumed to have reached initial set at about the time when standing on the concrete leaves a boot-print of about 6 mm deep [about 30 min (minimum) as per IS 269:1989]. This is about the time when the first machine-float pass can begin. After the final set of concrete [about 600 min (maximum) as per IS 269:1989], it is not possible to finish the surface of concrete. In recent past, efforts have been made to develop surface finishes that will give a better appearance to concrete surfaces. A few of these are discussed as follows:

*Formwork finish*   Concrete is usually cast using formwork to get the required shape. By careful preparation of the formwork, proper mix design and good workmanship smooth surfaces can be achieved. The dimensions, rigidity, joint tightness, and texture of formwork are important for the required finish. For example, timber formwork can be sand blasted to raise the grain, rough cut; bevelled or champhered for fluting; and made clapboard fashion for bold texture.

*Surface treatment*   The type of surface treatment depends on the purpose for which the concrete surface is used. For example, a pavement surface should not only be plane, but also have sufficient roughness to provide skid resistance. After levelling, the required surface finish may be obtained by screed, float or trowel action, and the required texture provided by additional measures such as brooming, raking, grinding or scabbling (*Scabbling* is a mechanical process of removing a thin layer of concrete surface using compressed air powered machines.). An acid, detergent, or high-pressure water-wash may be used to clean the surface of the concrete prior to the application of subsequent surface treatments. Such processes add considerable water to the concrete and unless sufficient drying time is allowed, alternate means of cleaning such as abrasive blasting should be considered. Brush blasting produces a surface, which has a light sandpaper-like texture, which is suitable for most surface finishes.

*Applied finishes*   Before applying any external finish, concrete surfaces have to be first roughened, cleaned, and wetted. Over this, a cement mortar of ratio 1:3 may be applied to provide a number of surface finishes such as sand facing, rough cast finish, and pebble dash stucco. A range of test methods, such as hygrometer test, electrical resistance test, calcium chloride test, and gravimetric moisture content test are available to determine whether the moisture in concrete is at an acceptable level for the application of various finishes. Moisture sensitive finishes such as timber, vinyl, carpet on a rubber underlay, epoxy-based terrazzo tiles, and surface coatings should not be applied before the concrete has dried up (A rule of thumb for drying time is one month for every 25 mm of concrete thickness, from the time curing is completed.). More details may be found from CCAA datasheet (2007c).

### 8.6.4 Revibration

It is the intentional systematic vibration of concrete, which has already been compacted. It should not be confused with unintentional revibration, which happens when placing successive layers of concrete, and the vibrator extends down into the underlying layer (which was previously vibrated). Though revibration of concrete can be beneficial to its strength and may improve the bond of top reinforcement and its surface finish, the practice is not widely used, because engineers are not sure of the time up to which it can be applied. A good rule of thumb is that revibration may be used as long as the vibrator is capable of liquefying the concrete and sinking into it under its own weight or just prior to the time of initial setting of concrete for mixtures with slumps of 75 mm or more. Revibration is found to be more beneficial in the top 0.5 to 1 m depth of a placement, where air and water voids are most prevalent (ACI 309R:05). Studies have shown that revibration is detrimental to the bond strength of bottom reinforcing steel (Altowaiji et al., 1986).

### 8.7 Curing of Concrete

All newly placed and finished concrete slabs should be cured and protected from drying, from extreme changes in temperature. For concrete to achieve its potential strength and durability, it has to be properly cured. *Curing* is the process of maintaining an adequate moisture content and temperature in concrete at early ages so that it can develop properties the mixture was designed to achieve. Curing begins immediately after placement and finishing so that the concrete may develop the desired strength and durability. Prevention of moisture loss is particularly important when the adopted w/cm is low, the cement used has high rate of strength development (Grade 43 and higher cements) or when supersulphated cement is used in concrete (it requires moist curing for at least 7 days). Curing affects primarily the concrete in the cover to the reinforcement, and basically the cover protects the reinforcement from corrosion by the ingress of aggressive agents. Curing is often neglected in practice and it is the main cause of deterioration of concrete structures in India and abroad.

The duration of curing depends on the required properties of concrete, the purpose for which the concrete is used, and the ambient conditions, i.e., the temperature and relative humidity of the surrounding atmosphere. Wet curing should start as soon as the final set occurs and should be continued for a minimum period of 7 to 10 days (longer duration of curing in the case of concrete with fly ash). Note that in concrete without the use of retarders/accelerators, final set of cement takes place at about 5 h. As per ACI 308R:01, curing should be started when the concrete surface begins to dry, i.e., when the accumulated bleed water evaporates faster than it can rise to the concrete surface. Concreting in hot-weather conditions requires special precautions against rapid evaporation and drying due to high temperatures.

Curing may be applied in a number of ways and the most appropriate method of curing is usually dictated by the site environmental conditions and/or by the selected construction method. Methods of curing may be divided broadly into the following three categories (some of these curing methods are shown in Fig. 8.9):

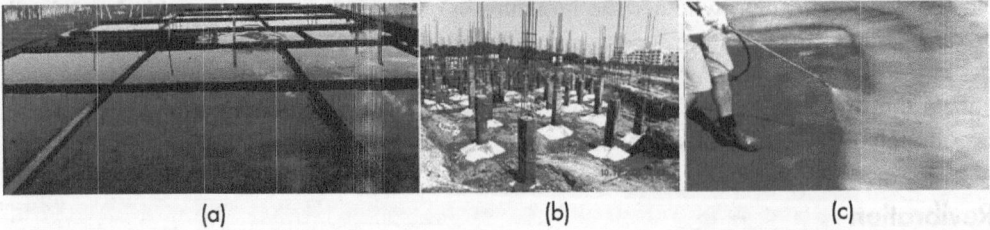

(a)                                    (b)                                    (c)

**Fig. 8.9** Methods of curing (a) Ponding (b) Moist curing of columns using wet hessian (called burlap in the USA) sacks (c) Spraying water-based curing compound

***Curing that prevents loss of moisture from the concrete*** This can be achieved by either one of the following methods: (a) leaving the concrete formwork in place (done in the case of sides of beams; in very hot dry weather, it may be necessary to moisten timber formwork during the curing period); (b) covering the concrete with an impermeable membrane (polyethylene sheets or water proof paper kept securely in place), with adequate lapping at the junctions, as soon as concreting is completed to prevent evaporation of moisture from the surface (also called *Membrane curing*); (c) using membrane-forming curing compounds, which can be applied by hand spray, power spray, brush, or roller on all exposed surfaces (water/paraffin wax/synthetic and natural resin-based emulsions, acrylic blended compounds, and chlorinated rubbers are available – approximate coverage rate: 4 m²/L for untextured surface and 6 m²/L for textured surface); they also should be applied as soon as concreting is completed; and (d) by a combination of the above methods.

***Water curing*** Preventing the loss of moisture from the exposed surface by (a) immersing in water (used in the case of concrete cubes/cylinders), (b) *ponding* of water on the surface of concrete slabs, (c) spraying or fogging, and (d) *moist curing* using wet hessian (called burlap in the USA) sacks, canvas, or straw on concrete columns.

***Steam curing*** It is also called *accelerated curing* (high temperature in the range of 40–100°C in the presence of moisture accelerates the hydration process resulting in faster development of strength). This method is useful in cold weather or when early strength gain is important. It is often used in the precast concrete industry as it allows increased production by a more rapid turnover of moulds and formwork. Techniques used for steam curing include: high-pressure steam using autoclaves, insulating blankets, electrical, and oil and infrared curing. With careful temperature selection, the strength of concrete after 3 days can exceed the 28-day strength of normally cured concrete. In addition to the early strength gain, steam-cured concrete has reduced drying shrinkage and creep as compared to normally cured concrete.

Water curing is the best method of curing as it satisfies all the requirements of curing, namely, promotion of hydration, elimination of shrinkage, and absorption of the heat of hydration. Temperature is an important factor in proper curing. The temperature of concrete should be maintained above 10°C for an adequate rate of strength development. The curing water should not be more than about 11°C cooler than the concrete in order to prevent cracking due to thermal stresses. Keeping the formwork intact and sealing the joints with any sealing compound may prove to be adequate for the curing of beams. It has to be noted that curing compounds might prevent bonding between hardened concrete and any anticipated freshly placed concrete overlay. Most curing compounds must be removed before the application of any applied floor finishes such as carpets and vinyl, epoxy, or polyurethane coatings and ceramic tile adhesives (CCAA, 2006b). Acrylic and resin-based curing compounds are expensive and hence are not used widely, whereas bitumen-based curing compounds are predominantly used for large surfaces that are constantly exposed to sunlight and wear-and-tear, such as in pavements. It is worth noting that ACI 308R-01 recommends the use of membrane forming curing compounds in lieu of moist curing only in cold weather concrete, after initial prevention of freezing. In a recent study on water, resin, wax, and acrylic-based curing compounds, it was found that the efficiency of a curing compound at 28 days, in terms of compressive strength, is only about 72 to 86%, for the recommended dosage (varying from 0.6–0.95 L/m² for water based, 0.25–0.75 L/m² for resin based, 0.40–0.75 L/m² for wax based, and 0.40–0.75 L/m² for acrylic based curing compounds, in ambient conditions) (Vandana and Gettu, 2016).

As shown in Fig. 8.10(a) and (b), there is a correlation between the strength of concrete and the duration for which it is cured. Concrete that is allowed to dry out immediately achieves about 40% of the strength of the same concrete water cured for a period of 10 days. Even 3 days of water curing achieves 60% strength, while 28 days water curing achieves 95% strength. Curing concrete for about 28 days is, therefore, the most effective strategy of increasing its strength. It has been found that the concrete that is allowed to dry out quickly also undergoes considerable early age drying shrinkage. Over time, water curing causes hydration products to fill (either partially or completely) the pores and capillaries present; this reduces the porosity of the paste and reduces the permeability and absorptivity of the concrete considerably. Higher curing temperatures promote an early strength gain in concrete but may decrease its 28-day strength.

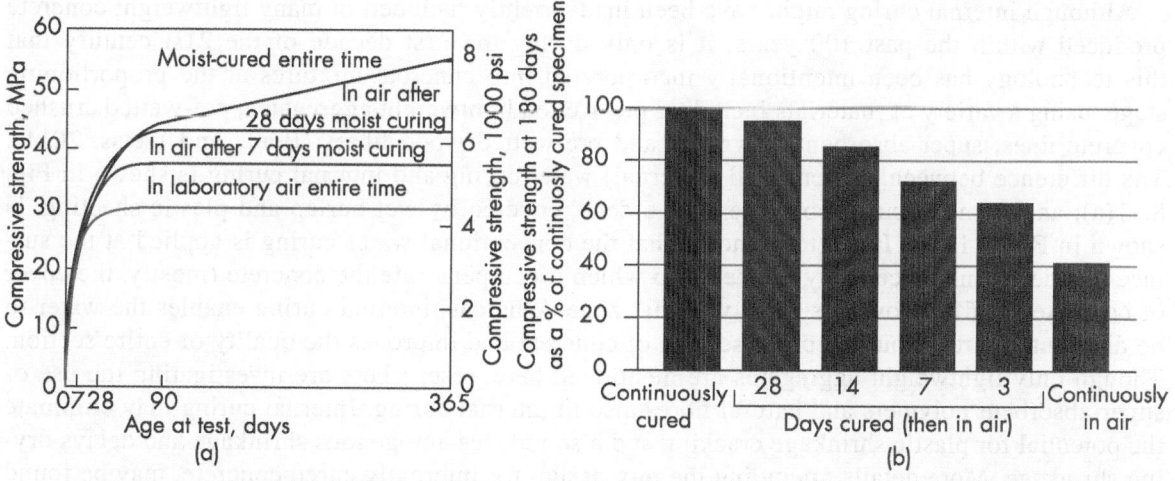

**Fig. 8.10** (a) & (b) Duration of water curing on strength of concrete (Gonnerman and Shuman, 1928)

In India, several builders adopt the wrong practice of commencing curing only on the next day of concreting. Even on the next day, curing is started after making arrangements to build bunds with mud or lean mortar to retain water [see Fig. 8.9(a)]. This further delays the curing. The time of commencement of curing depends on several parameters such as, prevailing temperature, humidity, wind velocity, type of cement, fineness of cement, w/cm used, and size of member. However, the main objective is to keep the top surface of concrete wet. Enough moisture must be present to promote hydration. Curing compound should be applied or wet curing started immediately after bleeding water, if any, dries up. In general, concrete must be cured until it attains about 70% of specified strength. IS 456 Clause 13.5.1 suggests curing for a period of 7 days (with temperature maintained above 10°C) in case of ordinary Portland cement concrete, and 10 days (with a recommendation to extend it for 14 days) when mineral admixtures or blended cements are used or when concrete is exposed to dry and hot weather conditions. At lower temperatures, curing period must be increased. Mass concrete, heavy footings, large piers, and abutments should be cured for at least 2 weeks. Further precautions to be undertaken during hot or cold weather concreting may be found in IS 7861 (Parts 1 and 2) or Venugopal and Subramanian (1977). High-temperature, high-pressure, or steam curing are beyond the scope of this book. More details on curing, including curing in hot and cold weather and accelerated curing, may be found in ACI 308R (2001); Subramanian (2002), Kosmatka et al., (2011), and CCAA Datasheet, (2006a).

## 8.7.1 Internal Curing of Concrete

When curing is done only to prevent moisture loss (as in the case of covering the surface of concrete with impermeable membrane or by using membrane-forming curing compounds), *self-desiccation* (loss of water in the concrete due to hydration, which is similar to the effect of drying) takes place resulting in autogenous shrinkage, especially when the concrete has lower w/cm ratio, as in the case of high-performance concrete (see Chapter 9). In such cases, internal curing may be necessary. As per the ACI, *internal curing* may be defined as 'supplying water throughout a freshly placed cementitious mixture using reservoirs, via pre-wetted lightweight aggregates, that readily release water as needed for hydration or to replace moisture lost through evaporation or self-desiccation'. This definition concisely identifies the two major objectives of internal curing: maximizing hydration and minimizing self-desiccation (and its accompanying stresses that may produce early age cracking).

Although internal curing might have been inadvertently included in many lightweight concrete produced within the past 100 years, it is only during the first decade of the 21st century that this technology has been intentionally incorporated into concrete mixtures at the proportioning stage, using a variety of materials including pre-wetted lightweight aggregates, pre-wetted crushed concrete fines, super-absorbent polymers, and pre-wetted wood fibres (Bentz and Weiss, 2011). The difference between conventional (external) water curing and internal curing is shown in Fig. 8.11(a); an internally cured concrete bridge deck, covered by wet burlap and plastic sheeting, is shown in Fig. 8.11(b). It has to be noted that the conventional water curing is applied at the surface and hence influences only the depth to which it can penetrate the concrete (mostly, the cover of concrete), and improves its quality in that zone. Whereas, internal curing enables the water to be distributed throughout the cross-section of concrete and improves the quality of entire section. Though only lightweight aggregates are mentioned here, researchers are investigating the use of super-absorbent polymers and natural fibres also in internal curing. Internal curing may eliminate the potential for plastic shrinkage cracking and also reduces autogenous shrinkage and delays drying shrinkage. More details, including the mix design for internally cured concrete, may be found in Bentz and Weiss (2011) and Bentz et al. (2005).

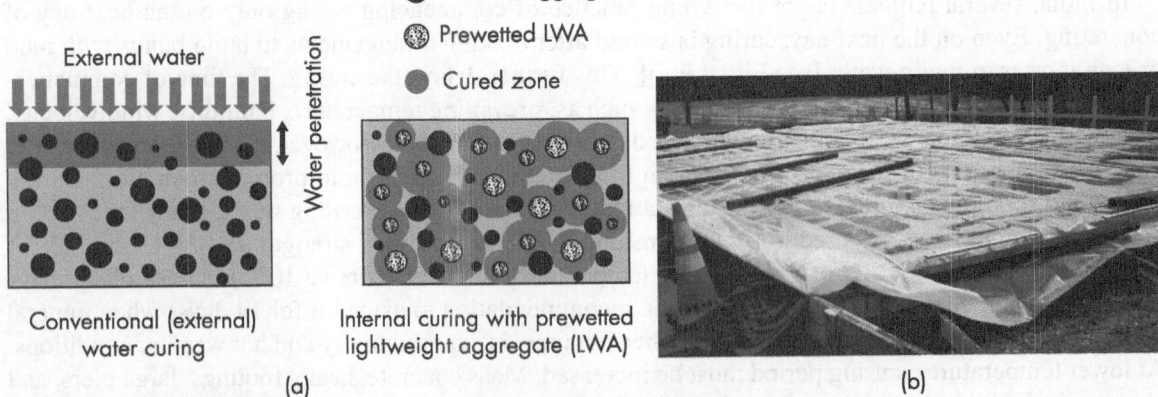

**Fig. 8.11** Internal curing (a) Comparison of conventional water curing and internal curing using pre-wetted lightweight aggregates (b) Plastic sheets covering the wet burlap on an internally cured concrete bridge deck (*Source*: Bentz and Weiss, 2011)

### 8.7.2 Removal of Forms

It is advantageous to leave forms in place as long as possible to continue the curing period. The vertical supporting members of formwork (shoring) should not be removed until the concrete is strong enough to carry at least twice the stresses to which the concrete may be subjected to at the time of removal of formwork as per Clause 11.3 of IS 456. When the ambient temperature is above 15°C and where Portland cement is used and adequate curing is done, the vertical formwork to columns, walls, and beams can be removed at 16 to 24 h after concreting. Beam and floor slab forms and supports (props) may be removed between 3 and 21 days, depending on the size of the member and the strength gain of the concrete (see Clause 11.3.1 of IS 456). If high early-strength concrete is used, these periods can be reduced. Because the minimum stripping time is a function of concrete strength, the preferred method of determining stripping time should be based on tests of job-cured cubes or concrete in place. More details including shoring and reshoring of multi-storey structures may be found in ACI 347–04.

## 8.8 Properties of Fresh and Hardened Concrete

A thorough knowledge of the properties of concrete is required for the designer while designing reinforced concrete structures. As seen in the previous sections, today's concrete is much sophisticated with the use of several different types of materials, which affect the overall strength and other properties considerably. More knowledge of these materials, their use, and effects on concrete can be had from Gambhir (2004), Mehta and Monteiro (2006), Mindess et al. (2003), Neville (2012), Neville and Brooks (2010), Santhakumar (2006), and Shetty (2005). An introduction to some of the properties, which are important for the construction professionals, is presented in this section.

### 8.8.1 Workability

Water is added to the concrete mix not only for hydration purposes, but also for workability. *Workability* may be defined as that property of the freshly mixed concrete, which determines the ease and homogeneity with which it can be mixed, placed, compacted, and finished. The desired degree of workability of concrete is provided in Table 8.5.

**Table 8.5** Workability of concrete (Based on IS 456:2000)

| Placing conditions | Degree of workability | Slump, mm |
|---|---|---|
| Mud mat, shallow section, and pavement/road using pavers | Very low | 0.70–0.80 (compacting factor) |
| Mass concrete, lightly reinforced slabs, beams, walls, columns, and strip footings | Low | 25–75 |
| Heavily reinforced slabs, beams, walls, and columns | Medium | 50–100 |
| Slip-form work and pumped concrete | Medium | 75–100 |
| In-situ piling and trench fill | High | 100–150 |
| Tremie concrete | Very high | 150–200 (flow test as per IS 9103:1999) |

**Note:** Internal (needle) vibrators are suitable for most of the placing conditions. The diameter of the needle should be determined based on the density and spacing of reinforcements, and the thickness of sections. Vibrators are not required for Tremie concrete.

The main factor that affects workability is the water content (in the absence of admixtures). The other interacting factors that affect workability are: aggregate type and grading, aggregate/cement ratio, presence of admixtures, fineness of cement, and temperature. It has to be noted that finer particles require more water to wet their large specific surface, and irregular shape and rough texture of angular aggregate demand more water. Slump test (see Section 25.12.1 of Chapter 25) is used to measure the workability of concrete.

**Segregation**    It usually implies separation of (a) coarse aggregate from fine aggregate, (b) paste from coarse aggregate, or water from the mix and results in the ingredients of fresh concrete not being uniformly distributed. If the mix is too wet, segregation may occur with resulting honeycombing, excessive bleeding, and sand streaking of the formed surface. On the other hand, if the mix is too dry, it may be difficult to place and compact and segregation may occur due to the lack of cohesiveness and plasticity of the mix. Segregation occurs due to dropping concrete from greater heights, badly designed mixes, concrete carried over long distances using pumping or belt conveyor systems, over vibrations, and extra floating and tamping during concrete finishing. Segregation mainly occurs in dry non-sticky concrete mixes. Segregation can be reduced by increasing small size coarse aggregates, air entrainment, using dispersing agents and by increasing the pozzolanic content of the concrete mix.

## 8.8.2 Compressive Strength

Compressive strength at a specified age, usually 28 days, measured on standard cube/cylinder specimens, has traditionally been used as the criterion for the acceptance of concrete. It is very important for the designer because concrete properties such as stress-strain relationship, modulus of elasticity, tensile strength, shear strength, and bond strength are frequently expressed in terms of the uniaxial compressive strength. The compressive strengths used in structural applications vary from 20 N/mm² to as high as 100 N/mm² (In One, World Trade Center, New York, USA, a concrete with a compressive strength of 96.5 MPa was used with a modulus of elasticity of 48,265 MPa). As per Table 5 of IS 456, a minimum grade of concrete of M20 should be used in RCC, in mild exposure conditions (see Table 8.16). It has to be noted that when cementitious materials are added to concrete, compressive strength may be specified at 56 days.

### Factors Affecting Compressive Strength

The compressive strength of concrete is affected by the following important factors: w/c or w/cm ratio, type of cement, use of supplementary cementitious materials, type of aggregates, mixing water, moisture condition during curing, temperature conditions during curing, age of concrete, rate of loading during cube/cylinder test (the measured compressive strength of concrete increases with increasing rate of loading), and the size of specimen.

The w/c ratio is inversely related to concrete strength: the lower the ratio, the greater the strength. It is also directly linked to the spacing between cement particles in the cement paste. The smaller the spacing, the faster the cement hydrates fill in the gaps between cement particles, the stronger the links created by the hydrates, and more importantly the stronger the concrete (Bentz and Aïtcin, 2008). Various mathematical models have been developed to link strength to the porosity of the hydrates. In 1918, Abrams presented his classic *Abrams's law* in the following form (Shetty, 2005):

$$f_{c,28} = \frac{k_1}{k_2^{wc}}$$

(8.1a)

where, $f_{c,28}$ is the 28-day compressive strength, $k_1$ and $k_2$ are empirical constants, and $wc$ = water/cement ratio by volume. For 28 day strength of concrete, ACI 211.1–91 recommends the constants $k_1$ and $k_2$ as 124.45 and 14.36 MPa, respectively. Popovics (1998) observed that these values are conservative and suggested the values 187 and 23.07 MPa, respectively. Abrams's w/c ratio law states that the strength of concrete is dependent only upon w/c ratio, provided the mix is workable. *Abrams's law* is a special case of the following *Feret formula* developed in 1897 (Shetty, 2005):

$$f_{c,28} = k\left(\frac{V_c}{V_c + V_w + V_a}\right) \tag{8.1b}$$

where, $V_c$, $V_w$, and $V_a$ are the absolute volumes of cement, water, and entrained air, respectively, and $k$ is a constant. In essence, strength is related to the total volume of voids and the most significant factor in this is the w/c ratio. The graph showing the relationship between the strength and w/c ratio is approximately hyperbolic in shape (see Fig. 8.12).

At a more fundamental level, this relation can be expressed as a function of the gel/space ratio ($x$), which is the ratio of the volume of the hydrated cement paste to the sum of the volumes of the hydrated cement and the capillary voids. The data from Powers (1961) give the following relationship:

$$f_{c,28} = 234x^3 \text{ MN/m}^2 \tag{8.1c}$$

where $x$ is the gel-space ratio and 234 is the intrinsic strength of the gel in MPa for the type of cement and specimen used by Powers. It has to be noted that this relation is independent of the age of the concrete and the mix proportions. This equation is valid for many types of cement, but the values of the numerical coefficients vary a little depending on the intrinsic strength of the gel. Such models that focus only on the cement paste ignore the effects of the aggregate characteristics on strength which can be significant. A comparison of these mathematical models is provided by Popovics (1988). Based on the strength versus w/c ratio curves provided in the earlier version of IS 10262:2009, Rajamane, 2005, derived the following equation.

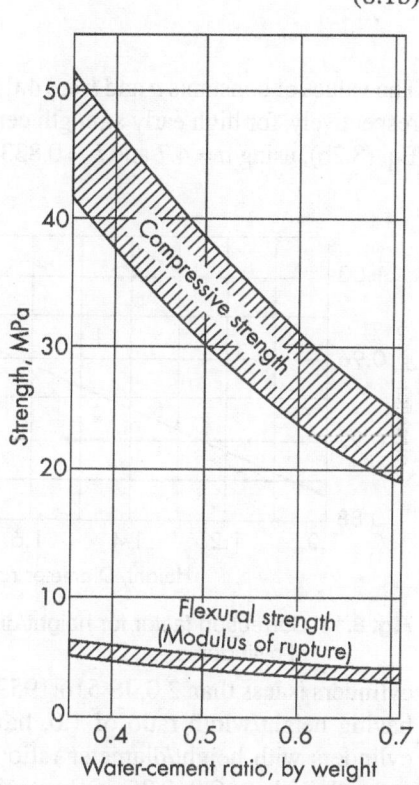

**Fig. 8.12** Relation between strength and w/c ratio of normal concrete

$$f_{c,28} = 0.39 f_{cem}\left[(1/wc) - 0.50\right] \tag{8.1d}$$

where $f_{cem}$ is the 28-day compressive strength of cement tested, as per IS 4031 (MPa), and $wc$ = water/cement ratio by weight.

Many researchers have also attempted to estimate the strength of concrete at 1, 3, or 7 days and correlate it to the 28-day strength. This relationship is useful for formwork removal and monitoring early strength gain; however, it depends on many factors such as chemical composition of cement, fineness of grinding, and temperature of curing. The 7-day strength is often estimated to be about 75% of the 28-day strength (Neville, 2012). Neville, however, suggests that if the 28-day strength is to be estimated using the strength at 7 days, a relationship between the 28-day and 7-day strengths has to be established experimentally for the given concrete. For concrete specimens cured at 20°C, Clause 3.1.2(6) of Eurocode 2(EN 1992–1-1:2004) provides the following relationship.

$$f_{cm}(t) = \exp\left[s\left(1 - \left(\frac{28}{t}\right)^{0.5}\right)\right] f_{cm} \tag{8.2a}$$

where $f_{cm}(t)$ is the mean compressive strength at age $t$ days, $f_{cm}$ is the mean 28-day compressive strength, and $s$ is a coefficient depending on the type of cement; $s = 0.2, 0.25$, and $0.38$, for high early strength, normal early strength, and slow early strength cement, respectively. ACI committee 209.2R-08 recommends the relationship for moist-cured concrete made with normal Portland concrete as

$$f_{cm}(t) = \left(\frac{t}{a + bt}\right) f_{c28} \qquad (8.2b)$$

The values of constants $a$ and $b$ are 4.0 and 0.85, respectively, for normal Portland cement, and 2.3 and 0.92, respectively, for high early strength cement. The 1978 version of IS 456 specified an 'age factor', based on Eq. (8.2b), using $a = 4.7$ and $b = 0.833$, but that provision has been deleted in the 2000 version of the code.

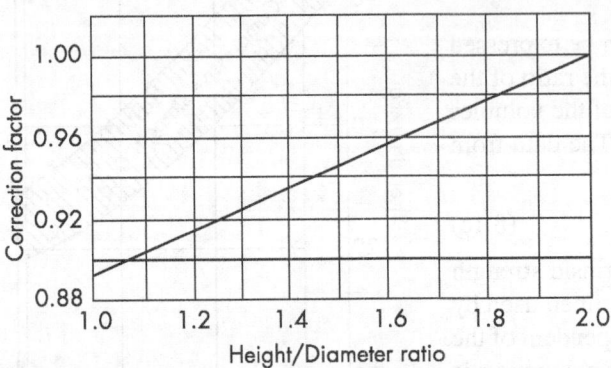

**Fig. 8.13** Correction factor for height/diameter ratio of cylinder

### Influence of Size of Specimen

The pronounced effect of the height/width ratio and the cross-sectional dimension of test specimen on the compressive strength have been observed by several researchers. The difference in compressive strength of different sizes of specimens may be due to several factors such as St Venant's effect, size effect, or lateral restraint effect due to the testing machine's platen (Pillai and Menon, 2003). In addition, the preparation of the end conditions (capping) of the concrete cylinder can significantly affect the measured compressive strength. When the height/diameter ratio of cylinders is less than 2.0, IS 516:1959 suggests a correction factor as shown in Fig. 8.13. Standard cubes, having height/width ratio of 1.0, have been found to have higher compressive strength than standard cylinders with height/diameter ratio of 2.0. The ratio of standard cylinder strength and standard cube strength is about 0.8–0.95; higher ratio is applicable for HSC. Similarly, $100 \times 200$ mm cylinders exhibit 2–10% higher strengths than $150 \times 300$ mm cylinders; the difference is less for higher strength concrete (Graybeal and Davis, 2008). It has to be noted that the ACI code formulae, which are based on standard cylinder strength, $f_c'$, have been converted to standard cube strength, $f_{ck}$, for easy comparison, by using

the relation $f_c' = 0.8 f_{ck}$, in this book. A more precise coefficient $R$ to convert cylinder strength to cube strength is $R = 0.76 + 0.2 \log \left(\frac{f_c'}{20}\right)$.

In the case of cubes, the specimens are placed in the testing machine in such a way that the load is applied on opposite sides of the cube as cast, i.e., not to the top and bottom. On the other hand, cylinders are loaded in the direction in which they are cast. Owing to this reason and also because the standard cylinders have height/width ratio of two, the compressive strengths predicted by cylinders are more reliable than cubes.

### 8.8.3 Stress-Strain Characteristics

Typical stress strain curves of normal-weight concrete of various grades, obtained from uniaxial compression tests, are shown in Fig. 8.14(a), and a comparison of normal-weight and light weight concrete is shown in Fig. 8.14(b). It has to be noted that, for design, the value of maximum compressive strength of concrete in structural elements is taken as 0.85 times the cylinder strength, $f_c'$, which is approximately equal to $0.67f_{ck}$.

It is seen from Fig. 8.14 that the curves are linear initially and becomes nonlinear when the stress level exceeds about 40% of the maximum stress. The maximum stress is reached when the strain is approximately 0.002; beyond this point, the stress-strain curve descends. IS 456 limits the maximum failure strain in concrete under direct compression to 0.002 [Clause 39.1(a)] and under flexure to 0.0035 [Clause 38.1(b)]. The shape of the curve is due to the formation of micro-cracks within the structure of concrete. The descending branch of the curve can be fully traced only with rigid testing machines. In axially flexible testing machines, the test cube/cylinder will fail explosively when the maximum stress is reached.

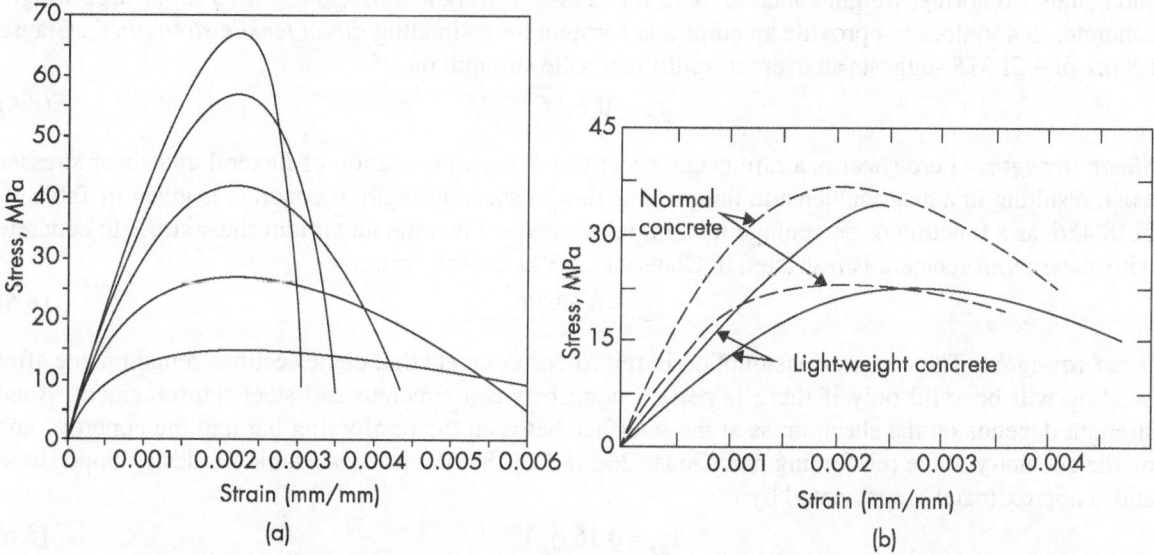

**Fig. 8.14** Typical stress-strain curves of concrete in compression (a) Concrete with normal-weight aggregates (b) Normal-weight vs light-weight aggregates concrete

Numerical approximations of stress-strain curves of concrete have been provided by various researchers, and a comparison of these formulae is provided by Popovics (1988). HSCs exhibit more brittle behaviour, which is reflected by the shorter horizontal branch of stress-strain curves.

### 8.8.4 Tensile Strength and Shear Strength

As mentioned earlier, concrete is very week in tension, and direct tensile strengths is only about 8–11% of compressive strength for concrete of grade M 25 and above (Shetty, 2005). The use of pozzolanic admixtures increases the tensile strength of concrete. Although the tensile strength of concrete increases with an increase in compressive strength, the rate of increase in tensile strength is of the decreasing order (Shetty, 2005). The tensile strength of concrete is generally not taken into account in the design of concrete elements. Knowledge of its value is required for the design of concrete structural elements subject to transverse shear, torsion, shrinkage, and temperature effects. Its value is also used in the design of prestressed concrete structures, liquid retaining structures, roadways, and runway slabs. Direct tensile strength of concrete is difficult to determine. The *splitting (cylinder) tensile test* on 150 mm × 300 mm cylinders, as per IS 5816:1999, or *the third-point flexural loading test* on 150 mm × 150 mm × 700 mm concrete beams, as per IS 516:1959, is often used to find the tensile strength. The splitting tensile test is easier to perform and gives more reliable results than other tension tests; though splitting strength may give 5 to 12% higher value than direct tensile strength (Shetty, 2005). According to Mehta and Monterio (2010), the third-point flexural loading test tends to overestimate the tensile strength of concrete by 50 to 100%.

The theoretical maximum flexural tensile stress occurring in the extreme fibres of RC beams, which causes cracking, is referred to as the *modulus of rupture*, $f_{cr}$. Clause 6.2.2 of IS 456 gives the modulus of rupture or flexural tensile strength as

$$f_{cr} = 0.7\sqrt{f_{ck}} \tag{8.3}$$

It should be noted that Clause 9.5.2.3 of ACI 318 code suggests a lower, conservative value for modulus of rupture, which equals $\lambda 0.55\sqrt{f_{ck}}$, where $\lambda$ is the modification factor for lightweight concrete and equals 1.0 normal weight concrete; 0.85 for sand-lightweight concrete and 0.75 for all-lightweight concrete. IS 456 does not provide an empirical formula for estimating *direct tensile strength*, $f_{ct}$. Clause R8.6.1 of ACI 318 suggests an average splitting tensile strength of

$$f_{ct} = 0.5\sqrt{f_{ck}} \tag{8.4}$$

***Shear strength*** Pure shear is a rare occurrence; usually, a combination of flexural and shear stresses exist, resulting in a diagonal tension failure. The design shear strength of concrete is given in Table 19 of IS 456, as a function of percentage flexural reinforcement and the maximum shear stress in concrete with shear reinforcement is restricted in Clause 40.2.3 to the following:

$$\tau_{c,\max} = 0.63\sqrt{f_{ck}} \tag{8.5}$$

***Bond strength*** The common assumption in reinforced concrete that plane sections remain plane after bending will be valid only if there is perfect bond between concrete and steel reinforcement. Bond strength depends on the shear stress at the interface between the reinforcing bar and the concrete, and on the geometry of the reinforcing bar. Clause 26.2.1.1 of IS 456 provides a table for design bond stress and is approximately represented by

$$\tau_{bd} = 0.16\,(f_{ck})^{2/3} \tag{8.6}$$

More discussions on bond strength of concrete is provided in Subramanian (2013).

## 8.8.5 Bearing Strength

The compressive stresses at supports, e.g., at the base of column, must be transferred by bearing (Niyogi, 1974). Clause 34.4 of IS 456 stipulates that the permissible bearing stress on full area of concrete in the working stress method can be taken as $0.25\,f_{ck}$ and for limit state method it may be taken as $0.45\,f_{ck}$. According to Clause 10.4.1 of ACI 318, the design-bearing strength of concrete should not exceed $\phi\,0.85\,f_c{}'$, where $\phi$ is the *strength reduction factor*, which is taken as 0.65. Thus, it is approximately equal to $0.442\,f_{ck}$.

## 8.8.6 Modulus of Elasticity and Poisson's Ratio

Concrete is not an elastic material, i.e., it will not recover its original shape on unloading. In addition, it is nonlinear and exhibits a nonlinear stress-strain curve. Hence, the elastic constants, such as *modulus of elasticity* and *Poisson's ratio,* are not strictly applicable. However, they are used in the analysis and design of concrete structures, assuming elastic behaviour. The modulus of elasticity of concrete is a key factor for estimating the deformation of buildings and members as well as a fundamental factor for determining *modular ratio, m.* The use of HSC will result in a higher modulus of elasticity, reduced deflection, and increased tensile strengths. The modulus of elasticity is primarily influenced by the elastic properties of the aggregates and, to a lesser extent, by the curing conditions, age of the concrete, mix proportions, porosity of concrete, and the type of cement. It is normally related to the compressive strength of concrete and may be determined by means of an extensometer attached to the compression test specimen as described in IS 516:1959.

The Young's modulus of elasticity may be defined as the ratio of axial stress to axial strain, within the elastic range. When the loading is of low intensity and of short duration, the initial portion of the stress-strain curve of concrete in compression is linear, justifying the use of modulus of elasticity. However, when there is sustained load, inelastic creep occurs even at relatively low stresses, making the stress-strain curve nonlinear. Moreover, the effects of creep and shrinkage will make the concrete to behave in a nonlinear manner. Hence, the initial tangent modulus is considered to be a measure of *dynamic modulus of elasticity* (Neville and Brooks, 2010).

When linear elastic analysis is used, one should use the *static modulus of elasticity*. The various definitions of modulus of elasticity are available: initial tangent modulus, tangent modulus (at a specified stress level), and *secant modulus* (at a specified stress level), as shown in Fig. 8.15. Among these, the *secant modulus*, which is the slope of a line drawn from the origin to the point on the stress-strain curve corresponding to 40% of the failure stress, is found to represent the average value of $E_c$ under service load conditions (Neville and Brooks, 2010). Clause 6.2.3.1 of IS 456 suggests that the *short-term static modulus of elasticity* of concrete, $E_c$, may be taken as:

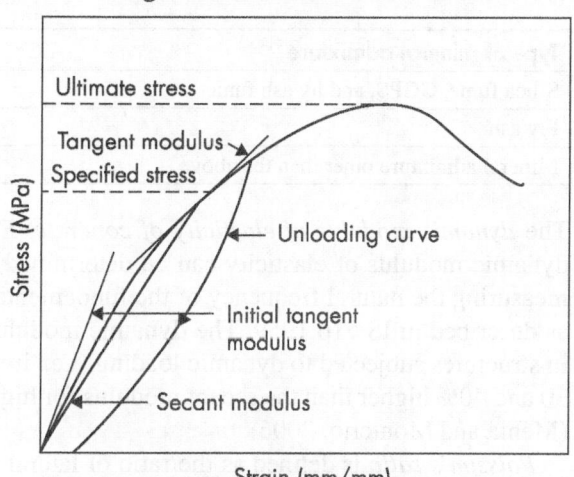

**Fig. 8.15** Various definitions of modulus of elasticity of concrete

$$E_c = 5000 \sqrt{f_{ck}} \text{ N/mm}^2 \tag{8.6a}$$

As per Clause 8.5.1 of ACI 318, the modulus of elasticity for concrete may be taken as

$$E_c = \rho_c^{1.5} 0.038 \sqrt{f_{ck}} \text{ N/mm}^2 \tag{8.6b}$$

where $\rho_c$, the unit weight of concrete (varies between 1440 and 2560 kg/m³). For normal weight concrete, ACI code allows it to be taken as (assuming $\rho_c$ = 2300 N/mm²)

$$E_c = 4200 \sqrt{f_{ck}} \text{ N/mm}^2 \tag{8.6c}$$

Both IS 456 and ACI 318 caution that the actual measured values may differ by ±20% from the values obtained from Eq. (8.6). Moreover, the US code value is 16% less than the value specified by the Indian Code. It has to be noted that the use of lower value of $E_c$ will result in conservative (higher) estimate of the short-term elastic deflection.

The ACI committee report on HSC (ACI 363R:92) suggests the following equation, which has been adopted by NZS 3101-Part 1:2006 and CSA A23.3–04.

$$E_c = (2970 \sqrt{f_{ck}} + 6900) (\rho_c/2300)^{1.5} \text{N/mm}^2 \qquad \text{for 26 MPa} < f_{ck} < 104 \text{ MPa} \tag{8.6d}$$

Noguchi et al. (2009) proposed the following equation, which is applicable to a wide range of aggregates and mineral admixtures used in concrete.

$$E_c = k_1 k_2 \times 3.36 \times 10^4 (\rho_c / 2400)^2 (f_{ck} / 75)^{1/3} \text{ N/mm}^2 \tag{8.6e}$$

Where the correction factors $k_1$ and $k_2$ are given in Tables 8.6 and 8.7.

**Table 8.6** Values of correction factor $k_1$

| Type of coarse aggregate | Value of $k_1$ |
|---|---|
| Crushed limestone and calcined bauxite | 1.20 |
| Crushed quartzite aggregate, crushed andesite, crushed basalt, crushed clay slate, and crushed cobblestone | 0.95 |
| Coarse aggregate other than above | 1.0 |

**Table 8.7** Values of correction factor $k_2$

| Type of mineral admixture | Value of $k_2$ |
|---|---|
| Silica fume, GGPS, and fly ash fume | 0.95 |
| Fly ash | 1.10 |
| Mineral admixture other than the above | 1.0 |

The *dynamic modulus of elasticity of concrete, $E_{cd}$,* corresponds to a small instantaneous strain. The dynamic modulus of elasticity can be determined by the non-destructive electro-dynamic method, by measuring the natural frequency of the fundamental mode of longitudinal vibration of concrete prisms, as described in IS 516:1959. The dynamic modulus of elasticity has to be used when concrete is used in structures subjected to dynamic loading (i.e., impact or earthquake). The value of $E_{cd}$ is generally 20, 30 and 40% higher than the secant modulus for high-, medium-, and low-strength concrete, respectively (Mehta and Monterio, 2006).

*Poisson's ratio* is defined as the ratio of lateral strain to the longitudinal strain, under uniform axial stress. Experimental studies have predicted the widely varying values of Poisson's ratio – in the range of 0.15 to 0.25. A value of 0.2 is usually suggested for design for both normal-strength and high-strength concrete. For lightweight concrete, the Poisson's ratio has to be determined from tests.

### 8.8.7 Shrinkage and Temperature Effects

As *shrinkage and temperature effects* are similar; they are considered in this section.

### Shrinkage Effects

*Shrinkage* and *creep* are not independent phenomena. For convenience, their effects are treated as separate, independent, and additive. The total shrinkage strain in concrete is composed of the following:

1. *Autogenous shrinkage* which occurs during the hardening of concrete (Holt, 2001),
2. *Drying shrinkage* which is a function of the migration of water through hardened concrete.

Drying shrinkage, often referred to simply as *shrinkage*, is caused by the evaporation of water from the concrete. Both shrinkage and creep introduce time-dependent strains in concrete. However, shrinkage strains are independent of the stress conditions of concrete. Shrinkage can occur before and after the hydration of the cement is complete. It is most important, however, to minimize it during the early stages of hydration in order to prevent cracking and improve the durability of the concrete. Shrinkage cracks in RC are due to the differential shrinkage between the cement paste, the aggregate and the reinforcement. Its effect can be reduced by the prolonged curing, which allows the tensile strength of the concrete to develop before evaporation occurs. The most important factors which influence shrinkage in concrete are: (a) type and content of aggregates, (b) w/cm ratio, (c) effective age at transfer of stress, (d) degree of compaction, (e) effective section thickness, (f) ambient relative humidity, and (f) presence of reinforcement (ACI 209R:92).

Shrinkage strain is expressed as a linear strain (mm/mm). In the absence of reliable data, Clause 6.2.4.1 of IS 456 recommends an approximate value for the total shrinkage strain for design as 0.0003 (ACI 209R:92 suggests an average value of $780 \times 10^{-6}$ mm/mm for the ultimate shrinkage strain, $\varepsilon_{sh}$). Different models for the prediction of creep under compression and shrinkage induced strains in hardened concrete are presented and compared in ACI 209.2R-08.

## Temperature Effects

Concrete expands with rise in temperature and contracts with fall in temperature. The effects of thermal contraction are similar to the effects of shrinkage. Concrete placed during hot mid-day temperatures will contract as it cools during the night. A 22°C drop in temperature between day and night (common in semi-arid zones) would cause about 0.7 mm of contraction in a 3 m length of concrete, sufficient to cause cracking if the concrete is restrained (PCA R&D Serial No. 2155, 2001). To limit the development of temperature stresses, expansion joints are to be provided, especially when there are marked changes in plan dimensions. In addition, when the length of the building exceeds 45 m, expansion joints are to be provided, as per Clause 27 of IS 456. Temperature stresses may be critical in the design of concrete chimneys and cooling towers. Roof slabs may also be subjected to thermal gradient due to solar radiation. In large exposed surfaces of concrete such as slabs, nominal reinforcements are usually placed near the exposed surface, to take care of temperature and shrinkage stresses. The coefficient of thermal expansion depends on type of cement and aggregate, cement content, relative humidity, and the size of section. Clause 6.2.6 of IS 456 provides a table to choose the value of *coefficient of thermal expansion* based on the aggregate used. However, SP 24:1983 recommends a value of $11 \times 10^{-6}$ mm/mm per °C for the design of liquid storage structures, bins, and chimneys, which is close to the thermal coefficient of steel (about $11 \times 10^{-6}$ mm/mm per °C). More information on thermal and shrinkage effects and reinforcements to be provided for the same may be found in Section 3.9.2 of the book by Subramanian (2013), Suprenant (2002), and Gilbert (1992).

## 8.8.8 Creep of Concrete

It is the gradual increase in deformation (strain) with time in a member subjected to sustained loads. The creep strain is much larger than the elastic strain on loading (Creep strain is typically two to four times the elastic strain.). If the specimen is unloaded, there is an immediate elastic recovery and a slower recovery in the strain due to creep (see Fig. 8.16). Both amounts of recovery are much less than the original strains under load. If the concrete is reloaded at a later date, instantaneous and creep strains develop again. Creep occurs both under compressive and tensile stresses and always increases with temperature. HSCs creep less than NSCs. When stress in concrete does not exceed one-third of its characteristic compressive strength, creep may be assumed proportional to the stress (Clause 6.2.5 of IS 456). It has to be noted that steel will creep only above 700°F.

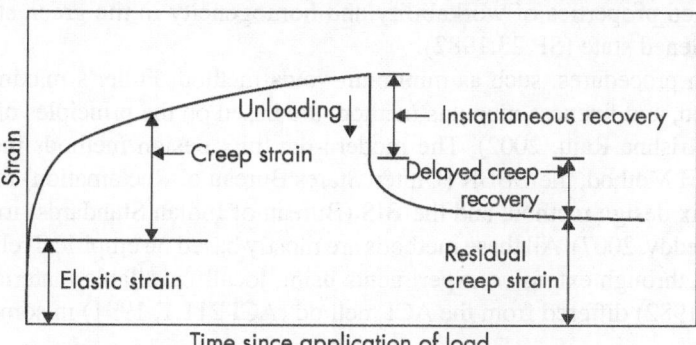

**Fig. 8.16** Typical creep curve of concrete

The main factors affecting creep strain are the concrete mix and strength, the type of aggregate used, curing, ambient relative humidity, and the magnitude and duration of sustained loading. As per IS 456, ultimate creep strain, $\varepsilon_{cp}$, is to be calculated from the creep coefficient $C_t$ ($\theta$ in IS nomenclature) given in Clause 6.2.5.1. Calculation of long-term deflection due to creep is provided by Subramanian (2013). More information on creep, shrinkage, and temperature effects may be obtained from Bamforth et al. (2008).

---

**Case Study** | **Collapse of Koror–Babeldaob Bridge, Republic of Palau, Micronesia**

Koror–Babeldaob Bridge—Before and after collapse
(*Source*: Bazant et al. 2010, reprinted with permission from Concrete International, ACI)

The Koror-Babeldaob Bridge was built in 1977 in the Republic of Palau. It is a reinforced concrete, balanced cantilever prestressed concrete box girder bridge with a total length of 385.6 m. The main span had a length 241 m. Its two-lane single cell box girder superstructure was built using cast-in-place segments and a permanent mid-span hinge. After 18 years, the deflection in the main span was found to be excessive (the total deflection was 1.61 m compared to the calculated final sag of 0.46 m to 0.58 m, measured from the design camber of –0.3 m), and the prestress loss was measured as 50%. The bridge was considered safe by two independent studies and, hence, it was repaired and reopened. But it collapsed suddenly, 3 months after reopening, on 26th September 1996, with two fatalities. A new cable-stayed bridge was built in its place by the Japanese Kajima Corporation in December, 2001. The main span of the new bridge is still a prestressed concrete box girder, but has a depth of only 7 m at the main pier and 3.5 m at the centre; it is now supported by stay cables. Further studies showed that the current models for creep and shrinkage grossly underestimate the deflections and prestressing loss (Bazant et al., 2011).

## 8.9 Mix Design of Ordinary Grade Concrete

*Concrete mix design* is the process of proportioning various ingredients such as cement, cementitious materials, aggregates, water, and admixtures optimally in order to produce a concrete at minimal cost, and will have specified properties of workability and homogeneity in the green state and strength and durability in the hardened state (SP 23:1982).

Earlier mix design procedures, such as minimum voids method, Fuller's maximum density method, Talbot-Richart method, and fineness modulus method, are based on the principles of minimum voids and maximum density (Krishna Raju, 2002). The modern-day mix design methods include the Road Note No. 4 method, the ACI Method, the USBR (United States Bureau of Reclamation) method, the Bolomeya model, the British mix design method, and the BIS (Bureau of Indian Standards) method (Krishna Raju, 2002; Nataraja and Reddy, 2007). All these methods are mostly based on empirical relations, charts, graphs, and tables developed through extensive experiments using locally available materials. Though the older BIS code (IS 10262:1982) differed from the ACI method (ACI 211.1, 1991) in some aspects, the present the BIS code (IS 10262:2009) is in line with the ACI code method (Nataraja and Das, 2010). In all these mix-proportioning methods, the ingredients are proportioned by the weight per unit volume of concrete.

The main objective of any concrete mix-proportioning method is to make a concrete that:

1. satisfies workability requirements in terms of slump for the easy placing and consolidating;
2. meets the strength requirements as measured by the compressive strength;
3. can be mixed, transported, placed, and compacted as efficiently as possible;
4. will be economical to produce
5. fulfils durability requirements to resist the environment in which the structure is expected to serve.

### Changes in Procedure for Mix Proportioning in IS 10262:2019

As per Clause 9.1.1 of IS 456, the minimum grade of concrete to be used in RCC should not be less than M 20. Moreover, all concrete above M20 grade for RCC work must be *design mixes*. Concrete grades above M 60 fall under the category of HSC.

The 2019 version of the code contains the graph of w/c ratio versus 28-day compressive strength. It is also preferable to establish the relationship between w/c ratio and the compressive strength of concrete for the materials actually used based on experiments. Maximum w/c ratio given in IS 456:2000 for various environmental conditions may be used as a starting point. The water content per cubic metre of concrete in the earlier version of the standard was a constant value for various nominal maximum sizes of aggregates. However, in the revised versions, maximum water content per cubic metre of concrete is suggested. Another major inclusion in the revised standards is the estimation of volume of coarse aggregate per unit volume of total aggregate for different zones of fine aggregate. Approximate air content in normal (non-air entrained) concrete is provided in Table 3 of IS 10262:2019; it is not considered in IS 456–2000. Note that the 2019 version of IS 10262 incorporates the design provisions for high strength concrete, self-compacting concrete, and mass concrete in addition to normal concrete, along with worked out examples of mix design for these cases.

### Data for Mix Proportioning

The following basic data are required for concrete the mix proportioning of a particular grade of concrete:

1. Exposure condition of the structure under consideration [see Table 3 of IS 456:2000 (Table 8.8) for guidance]
2. Grade designation- The minimum grade of concrete to be designed for the type of exposure condition under consideration (see Tables 3 and 5 of IS 456:2000- Tables 8.8 and 8.9 of this book for guidance).
3. Type of cement [Ordinary Portland cement (OPC), Portland pozzolana cement (PPC), Portland slag cement (PSC), etc.] and grade of cement (if applicable).
4. Minimum and maximum cement content [see Tables 3, 4, 5, and 6 of IS 456:2000 (Tables 8.8 and 8.10) for guidance].
5. Type of aggregate (basalt, granite, natural river sand, crushed stone sand, etc.).
6. Maximum nominal size of aggregate to be used (40, 20, or 12.5 mm).
7. Maximum w/c ratio [see Tables 3 and 5 of IS 456–2000 (Tables 8.8 and 8.10) for guidance].
8. The desired degree of workability (see Table 8.5, which is based on Clause 7 of IS 456–2000).
9. Use of admixture, its type, and conditions of its use.
10. Maximum temperature of concrete at the time of placing.
11. Method of transporting and placing.
12. Early age strength requirements, if required.

**Table 8.8** Different exposure conditions for concrete

| S. no. | Environment | Exposure conditions Table 3 of code IS 456 | Allowable maximum crack width as per Clause 35.3.2 (mm) |
|--------|-------------|---------------------------------------------|---------------------------------------------------------|
| 1. | Mild | Protected concrete surfaces, except those situated in coastal area | 0.3 |
| 2. | Moderate | Concrete surfaces sheltered from rain, continuously under water or in contact with non-aggressive soil/ground water | 0.25* |
| 3. | Severe | Concrete surfaces exposed to severe rain, coastal environment, alternate wetting and drying, or completely immersed in sea water | 0.20 |
| 4. | Very severe | Concrete surfaces exposed to sea water spray, corrosive fumes, severe freezing conditions while wet or in contact with aggressive sub-soil or ground water | 0.10 |
| 5. | Extreme | Concrete surface of members in tidal zone or in direct contact with aggressive chemicals | <0.10 |

*Assumed to be in between severe and mild

**Table 8.9** Grades of concrete

| Group | Grade designation | Specified characteristic 28-day compressive strength of 150 mm cube, N/mm² |
|-------|-------------------|-----------------------------------------------------------------------------|
| Ordinary concrete | M 10–M 20 | 10–20 |
| Standard concrete | M 25–M 60 | 25–60 |
| High strength concrete | M 65–M 100 | 65–100 |

**Table 8.10** Prescriptive durability requirements of cement content, water/cement ratio, and grade of concrete for different exposures

| S. no. | Exposure | Reinforced concrete | | |
|--------|----------|---------------------|---|---|
| | | Minimum cement content (kg/m³) | Maximum free w/c ratio | Minimum grade of concrete |
| 1 | Mild | 300 | 0.55 | M 20 |
| 2 | Moderate | 300 | 0.50 | M 25 |
| 3 | Severe | 320 | 0.45 | M 30 |
| 4 | Very severe | 340 | 0.45 | M 35 |
| 5 | Extreme | 360 | 0.40 | M 40 |

**Notes:** Cement content prescribed in this table is irrespective of the grades and types of cement and grade of concrete and it is inclusive of mineral admixtures. The additions such as fly ash or GGBFS may be taken into account in the concrete composition with respect to the cement content and w/c ratio, not exceeding the limit specified in IS 1489 (Part 1) and IS 455, respectively.

The step-by-step mix proportioning procedure as per IS 10269 is as follows (IS 10262:2009; Nagendra, 2010):

**Step 1: Calculate the target mean compressive strength for mix proportioning** The 28-day target mean compressive strength as per Clause 4.2 of IS 10262 is

$$f_{ck}' = f_{ck} + 1.65 \times s \quad \text{(or)} \quad f_{ck}' = f_{ck} + X \tag{8.7}$$

whichever is higher, where $f_c'$ is the target mean compressive strength at 28 days (N/mm²), $f_{ck}$ is the characteristic compressive strength at 28 days (N/mm²), $s$ is the standard deviation (SD) (N/mm²), and X is a factor based on the grade of concrete as per Table 8.11.

**Table 8.11** Assumed standard deviation

| S. no. | Grade of concrete | Assumed SD, N/mm² | Value of X, N/mm² |
|--------|-------------------|-------------------|-------------------|
| 1. | M 10 | 3.5 | 5.0 |
| 2. | M 15 | | |
| 3. | M 20 | 4.0 | 5.5 |
| 4. | M 25 | | |
| 5. | M 30 | 5.0 | 6.5 |
| 6. | M 35 | | |
| 7. | M 40 | | |
| 8. | M 45 | | |
| 9. | M 50 | | |
| 10. | M 55 | | |
| 11. | M 60 | | |
| 12. | M 65–M 80 | 6.0 | 8.0 |

**Note:** These values correspond to strict site control of storage of cement, weigh batching of materials, controlled addition of water, and so on. The values given in this table should be increased by 1 N/mm², when the aforementioned are not practiced.

SD should be calculated for each grade of concrete using at least 30 test strength of samples (taken from site), when a mix is used for the first time. In case sufficient test results are not available, the values of SD, as given in Table 8.11, may be assumed for proportioning the mix in the first instance. As soon as sufficient test results are available, actual SD should be calculated and used to proportion the mix properly.

**Step 2: Select the w/c ratio** The concrete made today has more than four basic ingredients. We now use both chemical and mineral admixtures, to obtain concrete with improved properties both in fresh and hardened states. Even the qualities of both coarse and fine aggregates in terms of grading, shape, size, and texture have improved due to the improvement in crushing technologies. As all these variables will play a role, concrete produced with same w/c ratio may have different compressive strengths. Therefore, for a given set of materials, it is preferable to establish relationship between the compressive strength and free w/c ratio. If such a relationship is not available, maximum w/c ratio for various environmental exposure conditions as given in Table 5 of IS 456–2000 (Table 8.10) may be taken as a starting point. Any w/c ratio assumed based on the previous experience for a particular grade of concrete should be checked against the maximum values permitted from the point of view of durability, and lower of the two values should be adopted.

**Step 3: Select the water content** The quality of water considered per cubic metre of concrete decides the workability of the mix. The use of water-reducing chemical admixtures in the mix helps to achieve increased workability at lower water contents. Water content given in Table 8.12 (Table 4 of IS 10262) is the maximum value for a particular nominal maximum size of

**Table 8.12** Maximum water content per cubic metre of concrete for the nominal maximum size of (angular) aggregate

| S. no. | Nominal maximum size of aggregate, mm | Maximum water content*, Kg |
|--------|---------------------------------------|---------------------------|
| 1 | 10 | 208 |
| 2 | 20 | 186 |
| 3 | 40 | 165 |

**Note:** These quantities of mixing water are for use in computing cementitious material contents for trial batches.
*Water content corresponding to saturated surface dry aggregate.

(angular) aggregate, which will achieve a slump in the range of 25 to 50 mm. Water content per unit volume of concrete can be reduced when increased size of aggregate or rounded aggregates are used. On the other hand, the water content per unit volume of concrete has to be increased when there is increased temperature, cement content, and fine aggregate content.

In the following cases, a reduction in water content is suggested by IS 10262:

1. For sub-angular aggregates, a reduction of 10 kg
2. For gravel with crushed particles, a reduction of 20 kg
3. For rounded gravel, a reduction of 25 kg

For higher workability (greater than 50 mm slump), the required water content may be established by trial, or an increase by about 3% for every additional 25 mm slump, or alternatively by the use of chemical admixtures conforming to IS 9103–1999.

**Use of water-reducing admixture**   Depending on the performance of the admixture (conforming to IS 9103–1999) that is proposed to be used in the mix, a reduction in the assumed water content can be made. As mentioned earlier, water-reducing admixtures (plasticizers) may decrease water content by 5 to 10% and super-plasticizers by 20–30% at appropriate dosages. Polyether-polycarboxylate (PCE)-based super-plasticizers may reduce water usage up to 30–40%.

**Step 4: Calculate the content of cementitious material**   The cement and supplementary cementitious material content per unit volume can be calculated from the free w/c ratio of Step 2. The total cementitious content so calculated should be checked against the minimum content for the requirements of durability and the greater of the two values should be used. The maximum cement content alone (excluding mineral admixtures such as fly ash and the GGBS) should not exceed 450 kg/m$^3$ as per Clause 8.2.4.2 of IS 456.

**Step 5: Estimate the proportion of coarse aggregate**   Table 8.13 (Table 5 of IS 12062) gives the volume of coarse aggregate for the unit volume of total aggregate for different zones of fine aggregate (as per IS 383–2016) for a w/c ratio of 0.5, which requires to be suitably adjusted for other w/c ratios. This table is based on ACI 211.1–1991. Aggregates of essentially the same nominal maximum size, type, and grading will produce the concrete of satisfactory workability when a given volume of coarse aggregate per unit volume of total aggregate is used. It can be seen that for equal workability, the volume of coarse aggregate in a unit volume of concrete is dependent only on nominal maximum size, and grading zone of fine aggregate.

**Table 8.13** Volume of course aggregate per unit volume of total aggregate for different zones of fine aggregate

| Nominal maximum size of aggregate, mm | Volume of coarse aggregate* per unit volume of total aggregate for different zones of fine aggregate (for w/c ratio = 0.5) | | | |
|---|---|---|---|---|
| | Zone IV | Zone III | Zone II | Zone I |
| 10 | 0.54 | 0.52 | 0.50 | 0.48 |
| 20 | 0.66 | 0.64 | 0.62 | 0.60 |
| 40 | 0.73 | 0.72 | 0.71 | 0.69 |

**Note:** The volume of coarse aggregate per unit volume of total aggregate needs to be changed at the rate of ±0.01 for every ±0.05 change in w/c ratio.

*Volumes are based on aggregate in saturated surface dry condition.

**Step 6: Identify the combination of different sizes of coarse aggregate fractions**   Coarse aggregate from stone crushes are normally available in two sizes, namely 20 and 12.5 mm. Coarse aggregates of different sizes can be suitably combined to satisfy the gradation requirements (cumulative percent passing) of Table 7 in IS 383:2016, for the given nominal maximum size of aggregate.

**Step 7: Estimate the proportion of fine aggregate**   The absolute volume of cementitious material, water, and the chemical admixture is found by dividing their mass by their respective specific gravity, and multiplying by 1/1000. The volume of all aggregates is obtained by subtracting the summation of the volumes of these materials from the unit volume. From this, the total volume of aggregates, and the weight of coarse and fine aggregate, is obtained by multiplying their fraction of volumes (already obtained in Step 5) with the respective specific gravities and then multiplying by 1000.

**Step 8: Perform trial mixes**   The calculated mix proportions should always be checked by means of trial batches. The concrete for trial mixes may be produced by means of actual materials and production methods. The trial mixes may be made by varying the free w/c ratio by ±10% of the preselected value, and a suitable mix selected based on the workability and target compressive strength obtained. Ribbon type mixer or pan mixer is to be used to simulate the site conditions where automatic batching and pan mixers are used for the production of concrete. After successful laboratory trials, confirmatory field trials are also necessary.

The guidelines for mix proportioning for HSC are provided by IS 10262:2019 and ACI 211.4R:93, for concrete with quarry dust by Nataraja et al. (2001) and for concrete with internal curing by Bentz et al. (2005). Rajamane (2004) explains a procedure of mix proportioning using the provisions of IS 456:2000. Optimal mixture proportioning for concrete may also be performed using online tools such as COST (Concrete Optimization Software Tool) developed by NIST, USA (http://ciks.cbt.nist.gov/cost/).

### Example 8.1   (Mix proportioning for M25 concrete)

Calculate the mix proportioning for M25 concrete, if the following are the stipulations for proportioning:

1. Grade designation: M25
2. Type of cement: OPC 43 grade conforming to IS 8112
3. Maximum nominal size of aggregate: 20 mm
4. Exposure condition: Moderate
5. Minimum cement content (Table 5 of IS 456): 300 kg/m$^3$
6. Workability: Slump -75 mm
7. Method of concrete placing: Pumping
8. Degree of supervision: Good
9. Type of aggregate: Crushed angular aggregate
10. Maximum cement content: 450 kg/m$^3$
11. Chemical admixture type: Super-plasticizer

The test data for materials are as follows:

1. Cement used: OPC 43 grade conforming to IS 8112
2. Specific gravity of cement: 3.15
3. Chemical admixture: Super-plasticizer conforming to IS 9103
4. Specific gravity of materials used is as follows:
    (a) Coarse aggregate: 2.68
    (b) Fine aggregate: 2.65
    (c) Chemical admixture: 1.145
5. Water absorption is as follows:
    (a) Coarse aggregate: 0.6%
    (b) Fine aggregate: 1.0%
6. Free (surface) moisture data are as follows:
    (a) Coarse aggregate: Nil (absorbed moisture also nil)
    (b) Fine aggregate: Nil

7. Sieve analysis
   (a) Coarse aggregate: Conforming to Table 7 of IS 383:2016
   (b) Fine aggregate: Conforming to grading Zone 1 of Table 9 of IS 383:2016

## Solution:

### Step 1 Calculate the target strength for mix proportioning
From Eq. (8.7)

$f_{ck}' = f_{ck} + 1.65 \times s$  (or)  $f_{ck}' = f_{ck} + X$

From Table 8 of IS 456 (see Table 8.11), SD for M25, $s = 4$ N/mm$^2$
Therefore, target strength = 25 + 1.65 × 4 = 31.6 N/mm$^2$ (or) 25 + 5.5 = 30.5 N/mm$^2$
Hence, take it as 31.6 N/mm$^2$

### Step 2 Selection of w/c ratio
From Table 5 of IS 456 (Table 8.10), maximum w/c ratio for moderate exposure = 0.50. Adopt w/c ratio as 0.45 < 0.50.

### Step 3 Selection of water content
From Table 4 of IS 10262, maximum water content = 186 kg (for 25–50 mm slump and for 20 mm aggregate).

Estimated water content for 75 mm slump = 186 + 3/100 × 186 = 191.58 kg.

As super-plasticizer is used, the water content can be reduced to more than 20%. Based on trials with super-plasticizer, water content reduction of 20% has been achieved. Hence, the assumed water content = 191.58 × 0.80 = 153.2 kg.

### Step 4 Calculation of cement content
w/c ratio = 0.45

Cement content = 153.2/0.45 = 340.4 kg/m$^3$

From Table 5 of IS 456 (Table 8.10), minimum cement content for moderate exposure condition = 300 kg/m$^3$. Since 340.4 > 300 kg/m$^3$, it is acceptable.

### Step 5 Determine the proportion of volume of coarse aggregate and fine aggregate content
From Table 5 of IS 10262 (Table 8.13), the volume of coarse aggregate corresponding to 20 mm size aggregate and fine aggregate (Zone 1) for w/c ratio of 0.50 = 0.60. We now have w/c ratio as 0.45. Therefore, the volume of coarse aggregate has to be increased to decrease the fine aggregate content. As the w/c ratio is lower by 0.05, the proportion of volume of coarse aggregate is increased by 0.01 (at the rate of ±0.01 for every +0.05 change in the w/c ratio). Therefore, the corrected proportion of volume of coarse aggregate for the w/c ratio of 0.45 = 0.61.
*Note*: Even if the selected coarse aggregate is not angular, the volume of coarse aggregate has to be increased suitably, based on experience.

For pumpable concrete, these values should be reduced by 10%.
Therefore, the volume of coarse aggregate = 0.61 × 0.09 = 0.55
The volume of fine aggregate content = 1 – 0.55 = 0.45

### Step 6 Perform the mix calculations
The mix calculations for the unit volume of concrete are as follows:

1. Volume of concrete = 1 m$^3$
2. Volume of cement = Mass of cement / Specific gravity of cement × (1/1000)

   a = 340.4 / 3.15 × (1/1000) = 0.108 m$^3$

3. Volume of water = Mass of water / Specific gravity of water × (1/1000)

   b = 153.2 /1 × (1/1000) = 0.153 m³

4. Volume of chemical admixture (super-plasticizer) (at 1.0% by the mass of cementitious material)

   c = Mass of chemical admixtures / Specific gravity of admixture × (1/1000)

   = 3.4 / 1.145 × (1/1000) = 0.00297 m³

5. Total volume of aggregate (coarse + fine)

   d = [1 − (a + b + c)] = 1 − (0.108 + 0.153 + 0.00297) = 0.736 m³

6. Mass of coarse aggregate = d × Volume of coarse aggregate × Specific gravity of coarse aggregate × 1000 = 0.736 × 0.55 × 2.68 × 1000 = 1084.86 kg

7. Mass of fine aggregate = d × Volume of fine aggregate × Specific gravity of fine aggregate ×1000

   = 0.736 × 0.45 × 2.65 × 1000 = 877.68 kg

**Step 7 Determine the mix proportions for trial number 1**

Cement = 340.40 kg/m³

Water = 153.2 kg/m³

Fine aggregate = 878 kg/m³

Coarse aggregate = 1085 kg/m³

Chemical admixture = 3.4 kg/m³

w/c ratio = 0.45

The following are the adjustments for moisture in aggregates and water absorption of aggregates correction for aggregates:

Free (surface) moisture is nil in both fine and coarse aggregates.

Corrected water content = 153.2 + 878 (0.01) + 1085 (0.006) = 168.49 kg

The estimated batch masses (after corrections) are as follows:

Cement = 340.4 kg/m³

Water = 168.5 kg/m³

Fine aggregate = 878.0 kg/m³

Coarse aggregate = 1085 kg/m³

Super-plasticizer = 3.4 kg/m³

Two more trial mixes with variation of ±10% of w/c ratio should be carried out, to achieve the required slump and dosage of admixtures. A graph between three w/c ratios and their corresponding strengths should be plotted to correctly determine the mix proportions for the given target strength.

**Example 8.2** (Mix proportioning for M 25 concrete, using fly ash as part replacement of OPC):

Calculate the mix proportioning for M25 concrete with the same stipulations for proportioning and the same test data for materials given in Example 8.1, except that fly ash is used as part replacement of OPC.

**Solution**: Considering the same data as in Example 8.1 for M 25 concrete, mix proportioning steps from 1 to 3 will remain the same.

The procedure of using fly ash as a partial replacement to OPC has been explained in Step 4.

**Step 4 Calculate the cement content**

From Example 8.1, cement content = 340.4 kg/m³

Now, to proportion a mix containing fly ash, the following steps are suggested:

1. Decide percentage of fly ash to be used based on project requirement and quality of materials.
2. In certain situations, increase in cementitious material content may be warranted.

The decision to increase cementitious material content and its percentage may be based on experience and trial. Let us consider an increase of 10% in the cementitious material content.

Cementitious material content = 340.4 × 1.1 = 374.4 kg/m³

Water content = 153.2 kg/m³ (from Example 8.1)

Hence, w/c ratio = 153.2/374.4 = 0.41

Fly ash at 35% of total cementitious material content = 374.4 × 35% = 131 kg/m³

Cement (OPC) content = 374.4 − 131 = 243.4 kg/m³

Saving of cement while using fly ash = 374.4 − 243.4 = 97 kg/m³, and

Fly ash being utilized = 131 kg/m³

**Step 5 Determine the proportion of volume of coarse aggregate and fine aggregate content**

From Table 5 of IS 10262 (Table 8.13), the volume of coarse aggregate corresponding to 20 mm size aggregate and fine aggregate (Zone I) for w/c ratio of 0.50 = 0.60. In this example, w/c ratio is 0.41. Therefore, the volume of coarse aggregate is required to be increased to decrease the fine aggregate content. As the w/c ratio is lower by approximately 0.10, the proportion of volume of coarse aggregate is increased by 0.02 (at the rate of ±0.01 for every +0.05 change in w/c ratio). Therefore, the corrected proportion of volume of coarse aggregate for the w/c ratio of 0.41 is 0.62.

*Note:* Even if the selected coarse aggregate is not angular, the volume of coarse aggregate has to be increased suitably, based on experience.

For the pumpable concrete, these values should be reduced by 10%.

Therefore, the volume of coarse aggregate = 0.62 × 0.09 = 0.56.

The volume of fine aggregate content = 1 − 0.56 = 0.44.

**Step 6 Perform the mix calculations**

The mix calculations per unit volume of concrete shall be as follows:

1. Volume of concrete = 1 m³
2. Volume of cement = Mass of cement / Specific gravity of cement × (1/1000)

    a = 243.4 / 3.15 × (1/1000) = 0.0773 m³
3. Volume of fly ash = Mass of fly ash / Specific gravity of fly ash × (1/1000)

    b = 131 / 2.0 × (1/1000) = 0.0655 m³
4. Volume of water = Mass of water / Specific gravity of water × (1/1000)

    c = 153.2 /1 × (1/1000) = 0.153 m³
5. Volume of chemical admixture (super-plasticizer) (at 0.8% by mass of cementitious material)

    d = Mass of chemical admixture / Specific gravity of admixture × (1/1000)

    = 3 / 1.145 × (1/1000) = 0.0026 m³
6. Total volume of aggregate (coarse + fine)

    e = [1 − (a + b + c + d)]

    = 1 − (0.0773 + 0.0655 + 0.153 + 0.0026) = 0.7016 m³
7. Mass of coarse aggregate

    = e × Volume of coarse aggregate × Specific gravity of coarse aggregate × 1000

    = 0.7016 × 0.56 × 2.68 × 1000 = 1053 kg
8. Mass of fine aggregate

    = e × Volume of fine aggregate × Specific gravity of fine aggregate × 1000

    = 0.7016 × 0.44 × 2.65 × 1000 = 818 kg

**Step 7 Determine the mix proportions for trail number 1**

Cement = 243.4 kg/m$^3$
Fly ash = 131 kg/m$^3$
Water = 153 kg/m$^3$
Fine aggregate = 818 kg/m$^3$
Coarse aggregate = 1053 kg/m$^3$
Chemical admixture = 3 kg/m$^3$
w/c ratio = 0.41

*Note*: The aggregate should be used in saturated surface dry condition. As mentioned in Example 8.1, three trial mixes with slightly varying w/cm ratio has to be made to determine experimentally the exact mix proportions that will result in the required workability, strength, and durability.

## 8.10 Reinforced Cement Concrete (RCC)

Plain concrete is weak in tension and hence used only in limited applications. To increase the tensile strength, reinforcements in the form of steel or other rebars are often provided in the tension face of structural elements, such as beams, columns, foundations, walls, and slabs (see Fig. 8.17). Such a

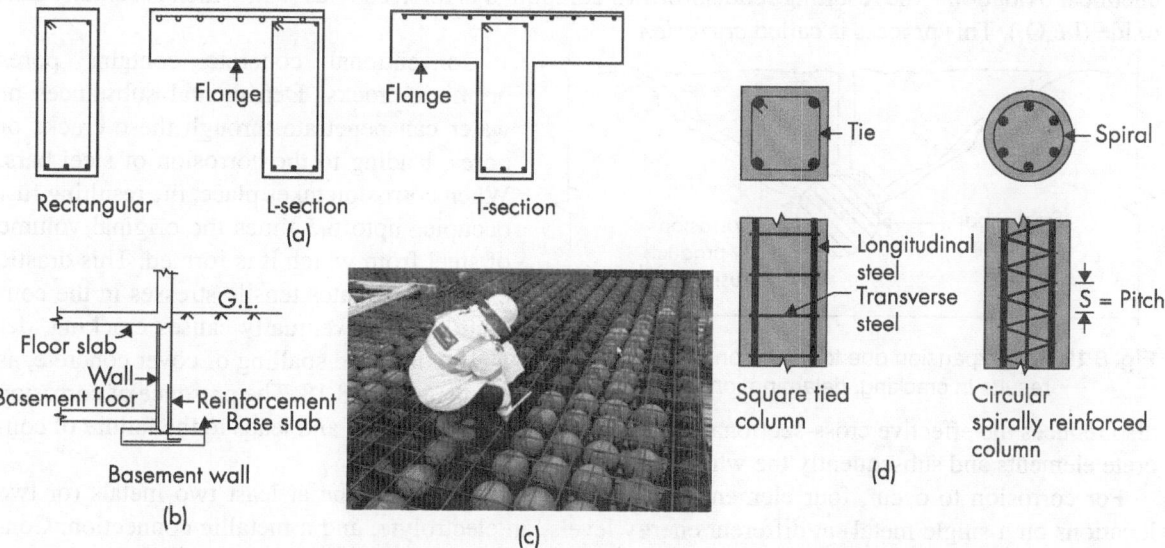

**Fig. 8.17** Reinforced concrete elements (a) Beams (b) Walls (c) Slabs (d) Columns

concrete is often termed as *reinforced cement concrete (RCC)* or simply RCC/RC. Though traditionally plain mild steel rebars were used as reinforcement, a number of different types of rebars are now available and include hot-rolled high-strength deformed bars, hard-drawn wire fabric, thermo-mechanically treated rebars (TMT bars), and TMT-corrosion resistant bars (see Section 13.13 of Chapter 13, for the details of these steel bars). The reinforcements that mitigate corrosion are discussed in Section 13.13.6 of Chapter 13. Complete details, including the design of various RC elements, such as foundations, columns, beams, and slabs, may be found in the book by Subramanian (2013).

## 8.10.1 Corrosion Problem of Reinforced Concrete

Some of the Roman structures, built approximately 2000 years ago, are still in good condition. Two factors have contributed to the success of these structures, some of which are functional even today. These are the excellent quality of the mortar in the concrete mixture and the careful selection and grading of the aggregate material. Ancient concrete mixtures were generally characterized by low-cementitious material content, low water content (consolidation was achieved by tamping), a very slow rate of strength development, and almost no shrinkage strains from cooling and drying. A mono-lith foundation of concrete has recently been designed for the first all-stone Hindu Temple in the island of Kauai, USA, to last 1000 years (Mehta and Langley, 2000). Most of the aforementioned structures do not have steel reinforcement. However, several reinforced concrete structures built in the past few years have shown severe deterioration. This is mainly due to the corrosion of high-strength steel re-inforcement and the poor quality of concrete cover. According to the World Corrosion Organization (www.corrosion.org), the annual cost of corrosion worldwide is about US $2.2 trillion, which is over 3% of the world's GDP.

Steel, just like most metals except gold and platinum, is thermodynamically unstable under normal atmospheric conditions and, in the presence of moisture, reverts back to its natural state due to electro-chemical oxidation – the resulting reddish-brown compound is referred to as *rust*, which is actually iron oxide ($Fe_2O_3$). This process is called *corrosion*.

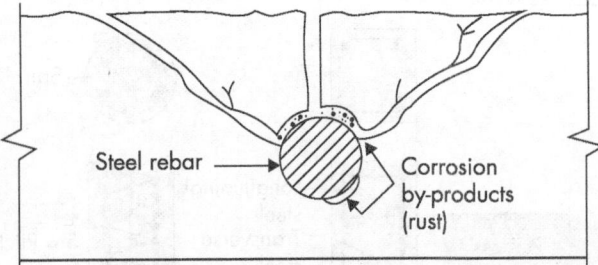

Conventional concrete contains pores or micro-cracks. Detrimental substances or water can penetrate through these cracks or pores, leading to the corrosion of steel bars. When corrosion takes place, the resulting rust occupies upto 6.5 times the original volume of steel from which it is formed. This drastic expansion creates tensile stresses in the con-crete, which eventually causes cracking, de-lamination, and spalling of cover concrete, as shown in Fig. 8.18. The presence of corrosion

**Fig. 8.18** The expansion due to corrosion of steel rebars results in cracking, delamination, and spalling

also reduces the effective cross-sectional area of the steel reinforcement, and leads to the failure of con-crete elements and subsequently the whole structure.

For corrosion to occur, four elements must be present: there must be at least two metals (or two locations on a single metal) at different energy levels, an electrolyte, and a metallic connection. Con-crete acts as the electrolyte, and the metallic connection is provided by wire ties, chair supports, or the rebar itself. In fact, the alkaline environment of concrete (pH of 12–13) provides a thin oxide passive film over the surface of steel rebars, and reduces the corrosion rate considerably. For steel bars sur-rounded by sound concrete, the passive corrosion rate is typically 0.1 μm per year. Without the passive film, the steel would corrode at rates at least 1000 times higher (ACI 222R-01). The destruction of the passive layer occurs when the alkalinity of the concrete is reduced (e.g., due to carbonation) or when the chloride concentration in concrete is increased to a certain level (due to the application of deicing salts on bridges during snow seasons or due to the presence of seawater). In many cases, exposure of reinforced concrete to chloride ions is the primary cause of premature corrosion of steel reinforcement. Although chlorides are directly responsible for the initiation of corrosion, they appear to play only an indirect role in the rate of corrosion after initiation.

The primary factors controlling the corrosion rate are the availability of oxygen, electrical resistivity, relative humidity, pH of the concrete, and prevailing temperature. *Carbonation* is another cause for corrosion. Carbonation-induced corrosion often occurs in building façades that are exposed to rainfall, and shaded from sunlight, and have low concrete cover over the reinforcing steel. Carbonation occurs when carbon dioxide from the air penetrates the concrete and reacts with calcium hydroxide to form calcium carbonate. This reaction reduces the pH of the pore solution to as low as 8.5, destroying the passive film on steel rebars. It has to be noted that carbonation is generally a slow process. The rate at which carbonation occurs is a function of concrete quality; in particular, the cement content, the w/c ratio, and the compaction. The depth of carbonation may be calculated as

$$\text{Depth of carbonation in mm} = (\text{Age in years})^{0.5} \quad\quad (8.8)$$

The highest rate of carbonation occurs at a relative humidity of 50 to 75%. The amount of carbonation is significantly increased in concrete with a high w/c ratio, low cement content, short curing period, low strength, and highly permeable paste. The depth of carbonation can be determined by applying phenolphthalein solution to a fractured concrete surface. The outermost sections, where concrete is carbonated, will not show any colour change and the inner-sections will display a shade of fuchsia (pink-purple).

Mitigation measures to reduce the occurrence of corrosion include (a) decreasing the w/c or w/cm ratio of concrete and using pozzolana and slag to make the concrete less permeable (pozzolana and slag also increase the resistivity of concrete, thus reducing the corrosion rate, even after it is initiated); (b) providing dense *concrete cover*, as per Table 16 of IS 456, and using *controlled permeability formwork* (CPF), thus protecting the embedded steel rebars from corrosive materials (see Section 8.10.2 for the details of cover requirements as per IS 456 and Section 8.10.3 for the details of CPF); (c) including the use of corrosion-inhibiting admixtures; (d) providing protective coating to reinforcement; and (e) using of sealers and membranes on the concrete surface (ACI 5152R-13). It should be noted that sealers and membranes, if used, have to be reapplied periodically (Kerkhoff, 2007).

## 8.10.2 Cover for Steel Reinforcement

*Cover* is the shortest distance between the surface of a concrete member and the nearest surface of the reinforcing steel. Concrete cover protects steel reinforcement against corrosion in two ways: providing a barrier against the ingress of moisture and other harmful substances, and by forming a passive protective (calcium hydroxide) film on the steel surface. Cover provides corrosion resistance, fire resistance, and a wearing surface, and is required to develop a bond between reinforcement and concrete. Cover should exclude plaster and any other decorative finish. Too large a cover reduces the effective depth of a RC member and is prone to cracking, whereas too less cover may lead to corrosion due to the carbonation of concrete.

Nominal cover required to meet durability requirements is given in Table 8.14. These values should be increased when lightweight or porous aggregates are used. Nominal cover is the design depth of

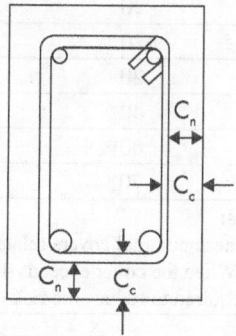

**Fig. 8.19** Clear ($c_c$) and nominal ($c_n$) covers to reinforcements

cover to all steel reinforcements including links (see Fig. 8.19). Moreover, according to Clause 26.4 of IS 456, the nominal cover for longitudinal reinforcement in columns should not be less than 40 mm, and it should not be less than 50 mm for footings. In addition to providing the nominal cover, it should be ensured that the cover concrete is well compacted, dense, and impermeable. Otherwise, heavy corrosion of reinforcement will take place as shown in Fig. 8.20.

**Table 8.14** Required cover (mm) for durability

| Exposure condition | Concrete grade with aggregate size = 20 mm | | | | |
|---|---|---|---|---|---|
| | M20 | M25 | M30 | M35 | M40 |
| Mild | 20 | 20 | 20 | 20 | 20 |
| Moderate | – | 30 | 30 | 30 | 30 |
| Severe | – | – | 45 | 40* | 40* |
| Very Severe | – | – | – | 50 | 45* |
| Extreme | – | – | – | – | 75 |

**Notes:**
1. For main reinforcement up to 12 mm diameter bar in mild exposure, the nominal cover may be reduced by 5 mm.
2. A tolerance in nominal cover of +10 and –0 mm is permissible as per IS 456.
3. To develop proper bond, a cover of at least one bar diameter is required.
4. Cover should allow sufficient space so that concrete can be placed or consolidated around bars. For this reason, the cover should be 5 mm more than the size of aggregates.
5. Cover at the end of bars should be ≥25 mm and ≥2.0 $d_b$, where $d_b$ is the diameter of the bar.
*For severe and very severe conditions, 5 mm reduction in cover is permissible, if M35 and above concrete is used.

Adequate cover, in thickness and quality, is necessary for other purposes, too, to transfer the forces in the reinforcement by bond action, provide fire resistance to steel, and provide alkaline environment at the surface of steel. Nominal cover requirement for different hours of fire resistance is given in Table 8.15.

**Table 8.15** Nominal cover (mm) for fire resistance

| Fire resistance (hours) | Beams | | Slabs | | Ribs | | Columns |
|---|---|---|---|---|---|---|---|
| | Simply supported | Continuous | Simply supported | Continuous | Simply supported | Continuous | |
| 0.5 | 20 | 20 | 20 | 20 | 20 | 20 | 40 |
| 1.0 | 20 | 20 | 20 | 20 | 20 | 20 | 40 |
| 1.5 | 20 | 20 | 25 | 20 | 35 | 20 | 40 |
| 2.0 | 40* | 30 | 35 | 25 | 45* | 35 | 40 |
| 3.0 | 60* | 40* | 45* | 35* | 55* | 45* | 40 |
| 4.0 | 70* | 50* | 55* | 45* | 65* | 55* | 40 |

**Notes:**
1. These nominal covers relate to the minimum member dimensions given in Fig. 1 of IS 456.
2. *When the cover exceeds 40 mm in flexural members, additional measures, such as sacrificial steel in tensile zone, are required to reduce the risk of spalling (see Clause 21.3.1 of the code).

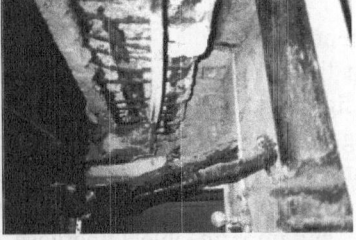

**Fig. 8.20** Heavy corrosion of rebars in a 4-star hotel in Chennai due to permeable or less than nominal cover (*Photo*: Dr N. Subramanian)

It has been found that thick cover leads to increased crack widths in flexural reinforced concrete members, defeating the very purpose for which it is provided. Hence, the engineer should adopt a judicious balance between cover depth and crack width requirements. The German code, DIN 1045, stipulates that concrete cover greater than 35 mm should be provided with wire mesh within 10 mm of surface to prevent spalling due to shrinkage or creep. A novel method called *supercover concrete* has been developed by researchers at South Bank University, UK, for preventing reinforcement corrosion in concrete structures with thick covers using Glass-fibre reinforced Plastic (GFRP) rebars (see Fig. 8.21). This method involves using conventional steel reinforcement together with concrete covers in excess of 100 mm, with a limited amount of GFRP rebars in cover zones. This method is found to be cheaper than cathodic protection. (Arya and Pirathapan, 1996; Subramanian and Geetha, 1997)

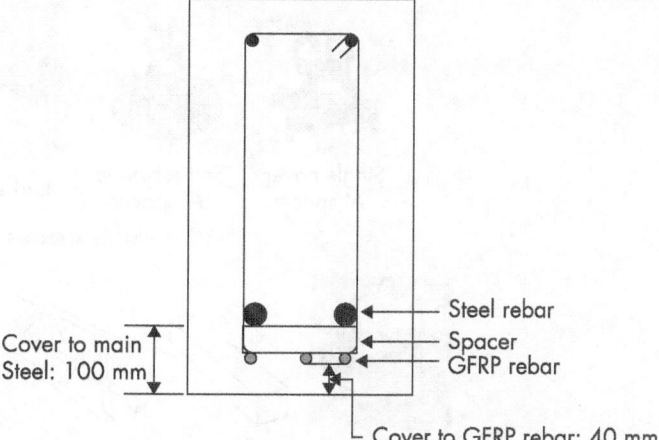

**Fig. 8.21** Schematic diagram of supercover concrete system (*Source*: Arya and Pirathapan, 1996)

A holistic approach to the durability of concrete structures must consider the following: component materials, mixture proportions, placement, consolidation and curing, and structural design and detailing. Air-entraining admixture has to be used under conditions of freezing and thawing.

The philosophies to tackle corrosion in concrete and their representative costs (given as a percentage of the first cost of the concrete structure) include (Mehta, 1997):

1. Use of fly ash or slag as a partial replacement of the concrete mixture (0%).
2. Pre-cooling of the concrete mixture (3%)-pre-cooling will mitigate the effects of heat of hydration and may reduce the extent of cracking.
3. Use of silica fume and a super-plasticizer (5%).
4. Increasing cover by 15 mm (4%).
5. Addition of corrosion-inhibiting admixture (8%).
6. Using epoxy-coated or galvanized reinforcing bars (8%).
7. External coatings (20%).
8. Cathodic protection (30%).

Where thermal cracking is of concern, the most cost-effective solution would be to use as low Portland cement content as possible with large amounts of cementitious or pozzolanic admixture (Mehta, 1997).

Plastic and cementitious spacers and steel wire chairs should be used to maintain the specified nominal cover to reinforcement (see Figs. 8.22 and 8.23). *Spacers* go between the formwork and the reinforcement, and *chairs* go between layers of reinforcement (e.g., top reinforcement supported off bottom reinforcement). *Spacers* and *chairs* should be fixed at centres not exceeding 50d in two directions at right angles for reinforcing bars and 500 mm in two directions at right angles for welded steel fabric, where d is the size of the reinforcement to which the spacers are fixed. The material used for spacers should be durable, and it should neither lead to corrosion of the reinforcement nor cause spalling of the concrete cover. Cementitious spacers must be factory-made and should be comparable in strength, durability, porosity, and appearance of the surrounding concrete. It is important to check the cover before and during concreting. The position of reinforcement in the hardened concrete may be checked using a *cover meter*.

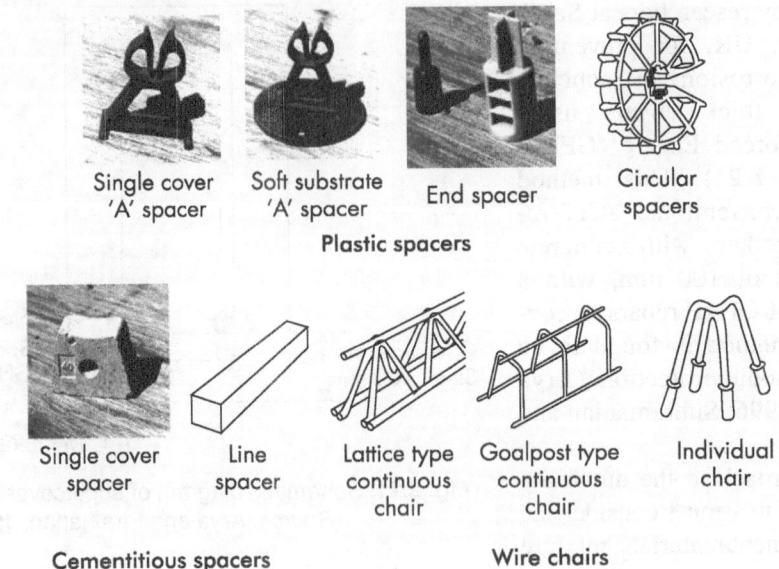

Single cover 'A' spacer    Soft substrate 'A' spacer    End spacer    Circular spacers

**Plastic spacers**

Single cover spacer    Line spacer    Lattice type continuous chair    Goalpost type continuous chair    Individual chair

**Cementitious spacers**        **Wire chairs**

**Fig. 8.22** Spacers and chairs to ensure good and uniform concrete cover (*Source*: Shaw, C., *Durability of Reinforced Concrete*, http://www.localsurveyorsdirect.co.uk/sites/default/files/attachments/reinforced concrete.pdf)

The reinforcements need to be tied together to prevent displacement of the bars before or during concreting. The six types of ties used in practice are shown in Fig. 8.24. Slash ties are used in slabs, ring slash and crown ties in walls, and crown or hairpin ties in beams and columns. British Standard (BS) 7973–1 contains complete details of the product requirements for the spacers and chairs, and BS 7973–2 specifies how they are to be used, including the tying of the reinforcement. More discussions on cover, spacers, and chairs may be found in Prakash Rao (1995) and Subramanian and Geetha (1997).

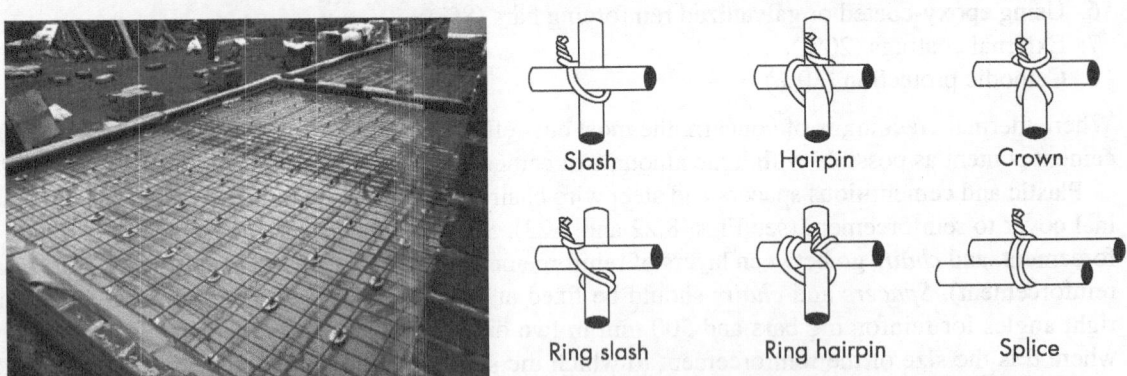

Slash      Hairpin      Crown

Ring slash      Ring hairpin      Splice

**Fig. 8.23** Spacers for welded steel fabric with new soft substrate spacers (*Photo copyright*: Mr. C. B. Shaw, UK)

**Fig. 8.24** Six types of ties used in binding wires (*Source*: Shaw, C., *Durability of Reinforced Concrete*, http://www.localsurveyorsdirect.co.uk/sites/default/files/attachments/reinforced concrete.pdf)

To assist designers to choose concrete mix, minimum cover, and minimum thickness of slab based on Tables 3, 5, 16, and 16A of IS 456:2000, Varyani (2001) developed Table 8.16. Reinforcements to mitigate corrosion are discussed in Section 13.3.6 of Chapter 13.

**Table 8.16** IS 456:2000 requirements for durability and fire resistance (Varyani, 2001)

| Exposure zone | Where applicable | Minimum concrete mix | Nominal cover for members, mm | | Minimum thickness of slabs, mm | Remarks |
|---|---|---|---|---|---|---|
| Mild | Concrete surface protected against weather or aggressive conditions in non-coastal regions | M 20 | Slab | 20 | 110 | For buildings in mid- land areas like Delhi |
| | | | Beam | 25 | | |
| | | | Column | 40 | | |
| | | | Footing | 50 | | |
| Moderate | Concrete surface sheltered from saturated salt air in coastal areas | M 25 | Slab | 30 | 110 | For buildings in coastal areas such as Mumbai, Chennai, and Kolkata |
| | | | Beam | 30 | | |
| | | | Column | 40 | | |
| | | | Footing | 50 | | |
| Severe | Concrete exposed to coastal environment or completely immersed in sea water | M30 | Slab | 45 | 140 | For structures immersed in sea water |
| | | | Beam | 45 | | |
| | | | Column | 45 | | |
| | | | Footing | 50 | | |

**Notes:**
1. The other two exposure zones, very severe and extreme, being applicable to special situations, have not been given here.
2. Slab covers have been reduced in moderate and severe exposure zones in order to restrict crack width and also to reduce dead load of buildings.

## 8.10.3 Controlled Permeability Formwork (CPF) Systems

It is well known that the use of conventional impermeable formworks (wood or steel) results in cover zones having reduced cement content and increased w/cm ratio. As a result, the presence of blowholes and other water-related blemishes is often observed upon removal of formwork. The concept of using permeable formwork (PF) to produce better quality cover concrete was first originated by John J. Earley in the 1930s. The US Bureau of Reclamation developed the first type of PF, known as *absorptive form liner*, in 1938. This technology was revived in Japan during 1985, and a number of Japanese companies have developed the CPF systems, using textile and silk form. The company, DuPoint, has also developed a less-expensive CPF liner system known as *Zemdrain*. CPF systems have been used in a number of projects in Europe and Australia (Basheer et al., 1993). CPF systems have proven, both in the laboratory and in the field, to increase the cement content of the cover region, while, at the same time, reducing the w/cm ratio, porosity, and permeability (Basheer et al., 1993).

Typically, CPFs are thermally bonded permeable liners that consist of a polyester filter and polyethylene drain elements, attached in tension to the internal face of a structural support, as shown in Fig. 8.25 (Reddi 1992; Annie Peter and Chitharanjan, 1995). During concreting, owing to the action of vibrators, the entrapped air and excess mix water, which would otherwise become trapped at the surface causing blemishes, pass through the liner, as shown in Fig. 8.25. The pore structure of the liners is so chosen that they will retain majority of cement and other smaller fines. A proportion of water is held within the liner and under capillary action, imbibes back into the concrete to assist curing. The forms can be removed with the normal level of care and cleaned with high-pressure water and reused. Release agents are not required as CPF liners easily debond from the concrete during formwork striking. The main advantages of the CPF are surface finish with very few blow holes, aesthetically pleasing textured surfaces giving good bond for plaster or tiles, and improved initial surface strength, allowing earlier formwork striking. Recently, the influence of self-compacting concrete (SCC), which does not require any vibration effort for its compaction, on the CPF was studied by Barbhuiya et al. (2011). They found that the degree of improvement in the cover region is significantly lower in the case of SCC compared to conventional concrete.

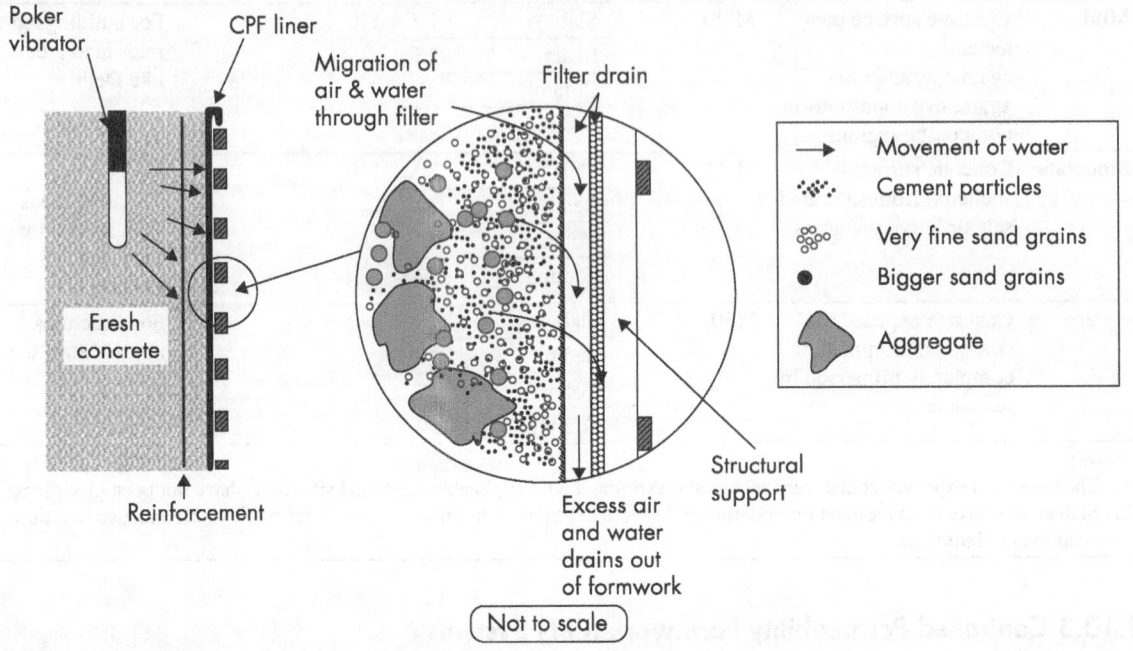

**Fig. 8.25** Controlled permeability formwork

## Case Study | Partial collapse of congress hall, Berlin, Germany

The Benjamin Franklin Hall, Berlin, Germany, was designed by American architect Hugh A. Stubbins, Jr., in collaboration with two Berlin architects Werner Duettman and Franz Mocken in 1957. This one-third curved and cantilevered roof (see picture) partly collapsed on 21 May 1980, killing one and injuring numerous people. It started rumbling and vibrating in the morning and hence most people inside had time to leave the building before it collapsed. The report of the failure of this 76 mm thick RC shell roof reasoned that the collapse was mainly due to the cracks, which lead to corrosion and finally to the failure of tension-

*Source*: http://commons.wikimedia.org/wiki/File:Berlin_Kongresshalle_BW_1.jpg

ing elements. The prestressing steel used was placed in an oval reinforcing metal tendon duct with 7–10 wires made of ribbed high-strength steel Sigma St 145/160, (area of 30 mm²). This was tempered steel with yield strength of 1450 N/mm² and ultimate strength of 1600 N/mm², and is especially sensitive to stress corrosion cracking. The hall was rebuilt in its original style and reopened again in 1987 at the 750 year's jubilee of Berlin. More details of the failure may be found in Schlaich et al. (1980), Subramanian (1982), and Helmerich and Zunkel (2014).

### 8.10.4 Vacuum Dewatering of Concrete

It is a process using which the excess water in the concrete, after placement and compaction, is sucked out from the surface of slabs using surface mats connected to vacuum pumps. Thus, the w/c ratio is reduced to the optimum level, which increases the compressive and flexural strength (by about 25 to 50%) and improves the abrasion resistance and impact strength of concrete. This technique is often used in industrial floors, parking lots, and deck slabs of bridges. The magnitude of applied vacuum is usually about 0.08 MPa, which may reduce the water content by up to 20–25%. The reduction is effective up to a depth of about 100 to 150 mm of the slab.

## 8.11 Durability of Concrete

As discussed already, several unreinforced concrete structures, which are more than 2000 years old, such as the Pantheon in Rome and several aqueducts in Europe, made of slow-hardening, lime-pozzolana cements are still in excellent condition, whereas many reinforced concrete structures built in the 20th century, constructed with Portland cement, have deteriorated within 10–20 years (Subramanian, 1979; Mehta and Burrows, 2001). In most countries in the European Union and other countries such as the USA, approximately 40–50% of the expenditure in the construction industry is spent on repair, maintenance, and remediation of existing structures. The growing number of deteriorating concrete structures not only affects the productivity of the society, but also has a great impact on our resources, environment, and human safety. It has now been realized that the deterioration of concrete structures is due to the main emphasis given to the mechanical properties and structural capacity, while neglecting construction quality and life-cycle management (ACI 202.2R:2008). Strength and durability are two separate aspects of concrete; neither guarantees the other. Hence, clauses on durability were included for the first time in the fourth revision of IS 456, published in 2000 (see Clause 8 of the code).

A *durable concrete* is one that will continue to perform its intended functions, i.e., maintain its required strength and serviceability, in the working environment during the specified or traditionally expected service life. The durability of concrete may be affected by a number of parameters, which include: the environment (see Table 8.8), temperature or humidity gradients, abrasion and chemical attack, permeability of concrete to the ingress of water, oxygen, carbon dioxide, chloride, sulphate and other deleterious substances, alkali-aggregate reaction (chemical attack within the concrete), chemical decomposition of hydrated cement, corrosion of reinforcement, concrete cover to the embedded steel, the quality and type of constituent materials, cement content and w/c ratio, the degree of compaction and curing of concrete, shape and size of member, and presence of cracks. The prescriptive requirements given in the code relate to the use of specified maximum w/c ratios, minimum cement content, and minimum grade of concrete for various exposure conditions, and minimum cover (see Table 8.14).

Low w/cm ratio produces dense and impermeable concrete, which is less sensitive to carbonation. Well-graded aggregates also reduce w/cm ratio. The coefficient of permeability increases more than 100 times from w/cm ratio of 0.4 to 0.7. It is now possible to make concrete with w/cm ratio as low as 0.25 using *super-plasticizers*, also called *high-range water-reducing admixtures*. As mentioned earlier, the super-plasticizer used must be compatible with the other ingredients such as Portland cement (Aïtcin and Mindess, 2015; Jayasree et al. 2011). Micro-cracks that are produced in the interface between the cement paste and aggregates (called the transition zone) are also responsible for increased permeability. Use of pozzolanic material, especially silica fume, reduces permeability of the transition zone as well as permeability of the bulk cement paste. When silica fume is included, use of super-plasticizers is mandatory. The SCC, in which the ingredients are proportioned in such a way that the concrete is compacted by its own weight without the use of vibrators and assures complete filling of the formwork even when access is hindered by congested reinforcement detailing, may be adopted for severe and extreme environmental conditions.

Currently available cements are more finely ground and are hardened rapidly at an earlier age. Moreover, they may contain more tricalcium silicate ($C_3S$) and less dicalcium silicate ($C_2S$), resulting in the rapid development of strength. Compared to old concrete mixtures, modern concrete tends to crack more easily due to higher thermal shrinkage, drying shrinkage, and elastic modulus (Mehta and Burrows, 2001). There is a close relationship between the cracking and deterioration of concrete structures exposed to severe exposure conditions.

## 8.12 Defects in Concrete

The hardened concrete may contain some defects. Some of these defects are obvious only to a trained eye, others, such as cracking, are obvious to anyone. Some common defects, their causes, and how to prevent and repair them are explained below. An expert should be consulted for better repairs.

**Blistering**    *Blisters* are hollow low-profile bumps on the concrete surface filled with either air or bleed water. Finishing the concrete when it is still spongy (premature troweling), insufficient vibration, or even overuse of vibration during compaction may cause blisters. This is more likely to occur in thick slabs or on hot, windy days when the surface is prone to drying out. Reducing the amount of sand in the mix (A reduction of 60 to 120 kg of sand per cubic meter of concrete may be enough.) and replacing it by adding a similar amount of the available smallest size coarse aggregate may eliminate blisters. Use of cement contents in the range of 305 to 335 kg/m³ and not using concrete with high slump, excessively high-air content, or excess fines may also help. If blisters are forming, delay troweling as long as possible and take steps to reduce evaporation by using an evaporative retarder. When blisters have formed, grind off the weakened layer to an even finish, or remove blisters and apply a repair mortar or epoxy coating.

**Colour variation**    Variations in colour across the surface of concrete may appear as patches of light and dark. It may be caused by uneven or variable compaction and curing conditions, addition of excess water, or segregation of materials and variable colour dosage (in coloured concrete). Colour variation may be prevented by the use of uniform concrete mix and the use of consistent placing, compacting, finishing, and curing procedures. Many colour variations from workmanship will be permanent. To hide the variation, a surface coating may be applied. Rectification of colour variation from stains is a very difficult operation and may need repeated gentle treatments with a weak acid.

**Cracking**    These are frequently encountered in concrete and can be the result of one or a combination of the following factors: drying shrinkage, thermal expansion or contraction, restraint (external or internal) to shortening, subgrade settlement, and applied loads. Cracking can be significantly reduced when their causes are taken into account in the design or/and preventive steps are taken at the appropriate time of concreting. For example, the provision of expansion joints during design and properly executed during construction will force cracks to occur at such joints. *Settlement cracks* are due to insufficient consolidation (vibration), high slumps (overly wet concrete), or a lack of adequate cover over steel bars. Relatively short *plastic-shrinkage cracks* may occur before final finishing on windy days with low humidity and high temperature. These cracks have varying lengths and are spaced at few centimetres to even 3 m, and may penetrate up to mid-depth of slabs. Fog spraying the concrete and providing wind-shades above concrete lessen the risk of plastic-shrinkage cracking. Cracks that occur after hardening usually are the result of drying shrinkage, thermal contraction, or subgrade settlement. While drying, hardened concrete will shrink about 0.5 mm per m length. To accommodate this shrinkage and control the location of cracks, joints have to be placed at regular intervals [at about 3 m intervals in each direction in a 100 mm thick unreinforced concrete slabs on grade and at about 6 m intervals in 200 mm thick RC slabs (PCA R&D Serial No. 2155, 2001)]. The major factor influencing the drying-shrinkage properties of concrete is the total water content of the concrete. Increased shrinkage will result by the use of high-shrinkage aggregates or calcium chloride admixtures. See also Section 8.8.7 for more information on the shrinkage of concrete and thermal expansion and contraction. Cracks can also be caused by freezing and thawing of saturated concrete, alkali-aggregate reactivity (see Section 6.6 of Chapter 6), sulphate attack, or corrosion of reinforcing steel (see Section 8.10.1). It has to be noted that cracks from these sources may appear after many years of service. Proper mix design and selection of suitable concrete materials can significantly reduce or eliminate the formation of cracks and deterioration related to freezing and thawing, alkali-aggregate reactivity, sulphate attack, or steel corrosion. Cracking in concrete can be reduced significantly or eliminated by the following practices (PCA R&D Serial No. 2155, 2001).

1. Using proper sub-grade preparation, including uniform support and proper sub-base material with adequate moisture content;
2. Minimizing the water content of concrete mix by maximizing the size and amount of coarse aggregate and the use of low-shrinkage aggregate;
3. Using the optimum amount of water required for workability;
4. Avoiding calcium chloride admixtures;
5. Preventing the rapid loss of surface moisture while the concrete is still plastic through proper curing or by covering the surface using plastic sheets to avoid plastic-shrinkage cracks;
6. Providing contraction joints at reasonable intervals, approximately at 30 times the slab thickness;
7. Providing isolation joints to prevent restraint from adjoining elements of a structure;
8. Avoiding extreme changes in temperature;
9. Properly placing, consolidating, finishing, and curing the concrete;
10. Avoiding the excessive amount of cementitious materials;
11. Using shrinkage-reducing admixtures to reduce drying shrinkage, which may reduce shrinkage cracking; and
12. Using synthetic fibres to help control plastic shrinkage cracks.

**Crazing**   It is a network of fine cracks across the surface of concrete. The cracks rarely exceed 3 mm deep, and are more noticeable on steel-towelled surfaces. These cracks are not serious, but crazed surfaces may be unsightly. Crazing is caused by minor surface shrinkage in rapid drying conditions (i.e., low humidity and high temperatures, or alternate wetting and drying). They develop at an early age and are seen within 5–6 days after placement. The causes include poor or inadequate curing, too wet a mix and excessive floating, finishing operations performed while bleed water remains on the surface, and sprinkling cement on the surface to dry up the bleed water. Crazing may be prevented by using moderate slump (75–125 mm) concrete with reduced bleeding characteristics, using an evaporative retarder, starting curing operations immediately after initial set of concrete, and by avoiding excessive manipulation of the concrete surface. Troweling should be avoided as it will result in crazing. Repair may not be necessary because crazing will not weaken concrete. If the appearance is unacceptable, a surface coating of paint or other overlay sealer can be applied to hide these cracks.

**Delamination**   It is similar to blisters in that delaminated areas of surface mortar result from bleed water and bleed air being trapped below the prematurely closed (densified) mortar surface.

**Dusting**   It is the result of a thin, weak layer, called *laitance*, composed of water, cement, and fine particles. It results in fine powdery substance on the concrete surface, which comes off on fingers. Finishing before the bleed water has dried, finishing during rain, improper curing, or the surface drying too quickly, and using low-grade concrete in locations of severe abrasion are the causes of dusting. It may be prevented by troweling after the bleed water dries up, proper curing, protecting concrete from drying out quickly in hot or windy conditions, and using stronger concrete when there is a likelihood of abrasion. Where surface dusting is minimal, the application of a surface hardener can be beneficial. If the surface is showing significant wear, it is essential to remove all loose material by grinding or scraping the surface to a sound base and then applying a suitable topping, if required.

**Efflorescence**   The appearance of fluffy white crystalline deposit sometimes found on the surface of concrete is called *efflorescence*. It is caused by poorly washed aggregate and salty water used in making concrete. Excess bleeding may also result in efflorescence. As the water evaporates, white patches appear on the surface. It can be prevented by using clean, salt-free water and washed sands and avoiding excessive bleeding. Efflorescence can be repaired by removing it by dry brushing and washing with clean water; wire brush should not be used. If this fails to remove the deposit, the surface may be washed with a dilute solution of hydrochloric acid.

**Honeycombing**   With honeycombing, the concrete surface looks like a coarse, stony surface with air voids. The causes are stiff or unworkable concrete, segregation during placing or paste leakage from forms, poor compaction, congested rebar, and improper placing practices. Preventing honeycombing starts with proper concrete mix proportioning – with more fine content, increased slump, and reduced aggregate size (Concrete not containing enough cementitious material and fine aggregate will be prone to segregation and will not flow freely. Similarly, slow placement rates and high ambient and concrete temperatures can cause concrete to stiffen, reducing its flowability.). Providing good watertight formwork, better rebar and concrete placement, and better compaction with vibrators will also prevent honeycombing. If honeycombing occurs only in a thin surface layer, it can be rectified by applying a layer of mortar over the surface. If honeycombing occurs to a greater depth, the concrete may have to be removed and replaced.

More details about defects in concrete structures, their causes, and remedial measures may be found in Subramanian (1979).

## 8.13 Formwork

It provides a mould, into which concrete is placed. When concrete hardens and attains sufficient strength, the formwork is removed. Formwork must be accurate, strong, and well made. Formwork should be designed and constructed so as to remain sufficiently rigid during placing and compaction of concrete (should not sag, bulge, or move), and should prevent loss of slurry from the concrete. Especially in large constructions, it should be designed to carry the loads applied to it safely – several failures in the past are due to the failure of formwork. Tolerances on the shapes and dimensions of formwork are given in Clause 11.1 of IS 456. Design and detailing of formwork must be carried out as per IS 14687:1999. The design of formwork has to be given more importance in the case of self-compacting concrete.

Formwork is normally made from steel or timber. Timber formwork is easier to make whereas steel formwork will allow a greater number of re-uses. Timber formwork could be made at site while steel formwork is usually bought from formwork suppliers. Special forms made from materials, such as GFRP, can be purchased for forming waffle slabs, circular columns, and other special profiles. The surface of the forms in contact with concrete affects how concrete will look. If the final appearance of the concrete is important, choose a material that will give the required surface texture. The CPF may be used to provide better quality cover to RC structures (see also Section 8.10.3).

All rubbish shall be removed from the interior of the formwork before using it. Form release agent should be applied to the inside of the formwork to stop it sticking to the concrete and thus make removal easier. Coating the formwork with form release agent should be done before the reinforcement is put in place. If formwork is placed in awkward positions or tight corners, it may be difficult to remove the formwork when the concrete has hardened. It is better if formwork is made simple to build, easy to handle, and re-usable. Formwork sections should be of simple design, not too big and of standard sizes if they are to be re-used.

Formwork may be left in place to help curing. Removal time will vary according to the weather – in cold weather, concrete will take longer time to gain strength than in warm weather; hence, removal times will therefore be longer (see also Section 8.7.2). In mild conditions (around 20°C), 7 days is long enough a time period to leave the forms in place (see Clause 11.3 of IS 456, for the stripping time of formwork). More details of different kinds of formwork and their design may be found from Hurd (2005) and IS 14687:1999. It is also possible to have the formwork left in place, acting as permanent formwork and contributing to the load carrying capacity of the structure.

## SUMMARY

- Concrete technology has advanced considerably since the Romans discovered the material more than 2000 years ago. Present-day concrete is made up of cement, coarse and fine aggregates, water, and a host of mineral and chemical admixtures.
- Mixing water plays an important role in the workability, strength, and durability of concrete. Potable water is considered satisfactory for mixing as well as curing concrete. If the w/c ratio greater than about 0.38, the excess water, which is not required for cement hydration, will result in the shrinkage of concrete.
- Mineral admixtures are added to the concrete mix to improve the properties of concrete, or as a replacement for Portland cement (blended cements).
- Chemical admixtures (such as accelerators, retarders, water-reducing admixtures, air-entraining admixtures, and corrosion inhibitors) are added to the concrete immediately before or during mixing in order to improve the properties of concrete. Chemical interaction of these admixtures with the ingredients of cement, may affect the performance of concrete.
- In order to obtain good quality, proper proportioning, mixing, transporting, placing, consolidation, and curing of concrete is important.
- Curing is very important to the strength and durability of concrete but is often neglected at site. The different types of curing include preventing loss of moisture from the concrete, water curing, accelerated (steam) curing, of which water curing is the best method. Curing should be done at least for a period of seven days for OPC concrete and 10 days for PPC concrete.
- Internal curing which uses pre-wetted lightweight aggregates, is being researched and adopted in a few applications.
- Forms should be removed only after concrete has gained sufficient strength to carry at least twice the stress it may be subjected at the time of removal of forms. Forms (made of ferrocement) could also be left in place.
- The important properties of fresh and hardened concrete are the workability of concrete (measured by slump test), compressive strength (measured on the 28th day by compressive tests on cubes/cylinders), stress – strain characteristics, tensile and bearing strength, modulus of elasticity, modulus of rupture, and the Poisson's ratio.
- The mix design of concrete, for the required strength should be done as per IS 10262:2019.
- Tests on fresh concrete are explained in Section 25.12 of Chapter 25. Tests on hardened concrete are described in Section 25.13 of Chapter 25. Various non-destructive tests performed on concrete are discussed in Section 25.14 of Chapter 25.
- Quality assurance of concrete has to be ensured by proper design of the concrete mix, use of proper materials, and proper workmanship while production at site/factory. Proper maintenance and repair of the completed concrete structure throughout its intended life period is also important. The sampling and acceptance criteria of concrete are explained in Section 25.16 of Chapter 25.
- Reinforced concrete is one in which reinforcing steel is used in tension zone. Different types of rebars could be used as reinforcement and include mild steel rebars, hot-rolled high-strength deformed bars, hard-drawn wire fabric, TMT bars, and TMT-corrosion resistant bars (see Section 13.13 of Chapter 13).
- These steel bars are the cause for corrosion and deterioration of concrete structures. Such corrosion may be prevented by providing good quality concrete in the cover region or by the use of CPF. Corrosion may also be mitigated by the use of fusion bonded epoxy coated rebars, galvanized rebars, GFRP bars, basalt bars, or TMT-corrosion resistant bars.
- Durability of concrete is enhanced when the defects in concrete are eliminated.

# EXERCISES

## Multiple-choice Questions

1. To be used for mixing concrete, water should have pH greater than
   (a) 4
   (b) 5
   (c) 6
   (d) 8

2. For the complete hydration of Portland cement, the minimum required w/c ratio is
   (a) 0.26
   (b) 0.38
   (c) 0.40
   (d) 0.42

3. Typical percentage dosage of accelerators by weight of cement is
   (a) 1 to 2%
   (b) 2 to 3%
   (c) 1 to 3%
   (d) 2 to 4%

4. Using water-reducing admixtures, the achievable percentage of water reduction is
   (a) 10%
   (b) 15%
   (c) 20%
   (d) 25%

5. Using high-range water-reducing admixtures, the achievable percentage of water reduction is
   (a) 20%
   (b) 25%
   (c) 30%
   (d) 35%

6. As per IS 456:2000, the following is considered standard concrete (NSC)
   (a) M25-M40
   (b) M30-M50
   (c) M25-M60
   (d) M30-M60

7. As per IS 456:2000, the following is considered high strength concrete (HSC):
   (a) Above M40
   (b) Above M50
   (c) Above M55
   (d) Above M60

8. The dosage of retarders, plasticizers, super-plasticizers are restricted by Clause 10.3.3 of IS 456 as
   (a) 0.5, 1.0, and 2.0%
   (b) 0.4, 0.6, and 2.5%
   (c) 0.5, 1.5, and 2.5%
   (d) 0.6, 0.5, and 2.0%

9. Normal RCC will have a slump of
   (a) 25–50 mm
   (b) 50–100 mm
   (c) 75–100
   (d) 100–150 mm

10. Right after placement, entrapped air contained in concrete ranges from
    (a) 5 to 10%
    (b) 5 to 15%
    (c) 5 to 20%
    (d) 10 to 25%

11. The strength of concrete containing 10% of entrapped air will be the following percentage of fully compacted concrete
    (a) 30%
    (b) 40%
    (c) 45%
    (d) 50%

12. Poker vibrator diameter varies from
    (a) 10 to 50 mm
    (b) 15 to 75 mm
    (c) 20 to 100 mm
    (d) 20 to 150 mm

13. The approximate diameter of radius of action poker vibrators is
    (a) 3 times the diameter of poker
    (b) 4 times the diameter of poker
    (c) 5 times the diameter of poker
    (d) 6 times the diameter of poker

14. While vibrating concrete using poker vibrators, they should be poked at a spacing of
    (a) 1.2 times the radius of action
    (b) 1.5 times the radius of action
    (c) 1.75 times the radius of action
    (d) 2 times the radius of action

15. External vibrators should be located at a spacing of
    (a) 1.2 times the radius of action
    (b) 1.5 times the radius of action
    (c) 1.75 times the radius of action
    (d) 2.0 times the radius of action

16. The best method of curing concrete is by
    (a) water curing
    (b) leaving formwork in place
    (c) membrane curing
    (d) using curing compounds

17. Wet curing should start as soon as the following takes place:
    (a) initial setting of concrete
    (b) final setting of concrete
    (c) formwork is stripped
    (d) bleeding water dries up on the surface

18. Curing of fly ash-containing concrete should be done for at least
    (a) 7 days
    (b) 10 days
    (c) 15 days
    (d) 28 days

19. Internal curing may be achieved using
    (a) pre-wetted lightweight aggregates
    (b) superabsorbent polymers
    (c) pre-wetted wood fibres
    (d) all of these

20. Vertical formwork to columns, walls and beams can be removed after concreting at
    (a) after 1 day
    (b) after 3 days
    (c) after 7 days
    (d) after 10 days

21. Compressive strength of ordinary Portland cement concrete is specified at
    (a) 1 week
    (b) 2 weeks
    (c) 4 weeks
    (d) 56 days

22. As per IS 456, minimum grade of concrete to be used in RCC is
    (a) M15
    (b) M22.5
    (c) M20
    (d) M27.5

23. The following factor affects the compressive strength of concrete:
    (a) water-cementitious ratio (w/c or w/cm ratio)
    (b) type of cement
    (c) type of aggregates
    (d) all of these

24. The ratio of standard cylinder strength and standard cube strength is
    (a) 0.5 to 0.6
    (b) 0.6 to 0.75
    (c) 0.8 to 0.95
    (d) >1.0

25. IS 456 limits the maximum failure strain in concrete under direct compression and under flexure as
    (a) 0.002 and 0.003
    (b) 0.0025 and 0.003
    (c) 0.002 and 0.0035
    (d) 0.0025 and 0.0035

26. Tensile strength of concrete, as per IS 456, is
    (a) $0.5 \sqrt{f_{ck}}$
    (b) $0.7 \sqrt[3]{f_{ck}}$
    (c) $0.7 \sqrt{(f_{ck})}$
    (d) $\sqrt{(f_{ck})}$

27. Bearing strength of concrete as per limit states method of IS 456 is
    (a) $0.25 f_{ck}$
    (b) $0.30 f_{ck}$
    (c) $0.40 f_{ck}$
    (d) $0.45 f_{ck}$

28. The short-term static modulus of elasticity of concrete, as per IS 456 in N/mm², is
    (a) $4200\sqrt{f_{ck}}$
    (b) $4500\sqrt{f_{ck}}$
    (c) $5000 \sqrt{f_{ck}}$
    (d) $5500 \sqrt{f_{ck}}$

29. The coefficient of thermal expansion of concrete as per SP 24 is about
    (a) $10 \times 10^{-6}$ mm/mm per °C
    (b) $11 \times 10^{-6}$ mm/mm per °C
    (c) $11 \times 10^{-5}$ mm/mm per °C
    (d) $12 \times 10^{-6}$ mm/mm per °C

30. The allowable maximum crack width for mild exposure, as per IS 456, is
    (a) 0.1 mm
    (b) 0.2 mm
    (c) 0.3 mm
    (d) 0.4 mm

31. As per IS 456, minimum cement content in concrete in mild exposure should be
    (a) 200 kg/m³
    (b) 300 kg/m³
    (c) 360 kg/m³
    (d) 400 kg/m³

32. Nominal cover for longitudinal reinforcement in columns should not be less than
    (a) 25 mm
    (b) 30 mm
    (c) 35 mm
    (d) 40 mm

33. Nominal cover for reinforcement in footings should not be less than
    (a) 35 mm
    (b) 40 mm
    (c) 45 mm
    (d) 50 mm

34. Crazing may be prevented by using concrete with a slump of
    (a) 50–100 mm
    (b) 75–125 mm
    (c) 100–125 mm
    (d) 100–150 mm

## Review Questions

1. Write a short history of concrete, beginning with the Roman concrete.
2. What are the advantages and drawbacks of concrete?
3. What are the five components that can be used in concrete?
4. Name any three mineral admixtures used in concrete.
5. What kind of water is used for mixing or curing concrete?
6. When more water than necessary for hydration is added to concrete, what happens?
7. Can sea water be used for the mixing or curing of concrete? State the reason.
8. Name any three chemical admixtures used in concrete.
9. Name any two compounds used as super-plasticizers in India.
10. What are corrosion inhibitors? Why are they used?
11. State the different classification of concrete.
12. What are the main objectives of concrete mix proportioning?
13. How is the consistency of mix related to the slump of concrete?
14. What are the two types of concrete plants?
15. How can concrete be transported from the place of mixing to the formwork?
16. What are the precautions to be taken before concrete is placed in the formwork?
17. What are the two stages of concrete compaction using vibration?
18. Name three types of mechanically operated vibratory equipment used for consolidating concrete.
19. Write short notes on internal vibrator.
20. What are the precautions to be taken when using vibrators?
21. What are the problems of inadequate compaction?
22. How does form vibrator differs from internal vibrator?
23. When are table vibrators used?

24. What are surface vibrators? When are they used?
25. What is meant by the bleeding of concrete? How can it be controlled?
26. What is laitance?
27. Can we revibrate already vibrated concrete? What are its advantages and drawbacks?
28. Why is curing considered important? What are the different methods of curing? Which curing method is more efficient?
29. What is internal curing? How can it be achieved?
30. Write a short note on removal of forms.
31. What is meant by workability? When does segregation occur?
32. At what age is compressive strength specified for concrete?
33. What are the factors that affect the compressive strength of concrete?
34. What is Abrams's law? What is Feret's formula?
35. How does the size of cubes/cylinders affect the compressive strength of concrete?
36. Sketch the typical stress-strain of M20 and M50 concrete. What are your observations on the shape of these two curves?
37. Write short notes on (a) tensile strength, (b) modulus of elasticity, (c) Poisson's ratio, (d) shrinkage, (e) temperature effects, and (f) creep of concrete.
38. Write the expressions of modulus of elasticity, tensile, shear and bearing strength of concrete, as per IS 456.
39. How is target mean compressive strength fixed for mix proportioning?
40. How is initial w/c ratio assumed in mix proportioning?
41. Why is it necessary to produce trial mixes at the laboratory and at site, even after doing mix design?
42. When is concrete denoted as reinforced concrete?
43. When does the corrosion of rebars take place? What are the different methods adopted to mitigate corrosion?
44. Why is cover provided to reinforcements in concrete?
45. What is super cover?
46. What are the requirements of spacers used in concrete?
47. What is controlled permeability formwork? How does it aid to provide better quality cover?
48. Name three types of rebars that are used in corrosive environments.
49. What are the parameters that affect the durability of concrete?
50. What are the prescriptive requirements given in the code for achieving durable concrete?
51. Name any four defects in concrete.
52. State the reasons for concrete to crack. How can we accommodate shrinkage cracks?
53. Name at least five practices that should be followed to reduce or eliminate cracking of concrete.
54. How can honeycombing be prevented in concrete?
55. Write a short note on formwork required for casting concrete elements.

## Exercises

1. Do mix proportioning for M30 concrete for the data given in Example 8.1.
2. Do mix proportioning for M30 concrete for the data given in Example 8.1, with fly ash as part replacement of the OPC.

# ANSWERS

## Multiple-choice Questions

| | | | | | | |
|---|---|---|---|---|---|---|
| 1. (c) | 2. (b) | 3. (b) | 4. (b) | 5. (c) | 6. (c) | 7. (d) |
| 8. (a) | 9. (b) | 10. (c) | 11. (d) | 12. (d) | 13. (d) | 14. (b) |
| 15. (b) | 16. (a) | 17. (d) | 18. (b) | 19. (d) | 20. (a) | 21. (c) |
| 22. (c) | 23. (d) | 24. (c) | 25. (c) | 26. (c) | 27. (d) | 28. (c) |
| 29. (b) | 30. (c) | 31. (b) | 32. (d) | 33. (d) | 34. (b) | |

# CHAPTER 9
# SPECIAL STRUCTURAL CONCRETE

## 9.1 Introduction

Depending on where it is mixed, concrete may be classified as site mixed or factory (ready) mixed concrete (RMC). Site mixing is not always recommended as the mixing may not be thorough and the control on water-cement or water-cementitious ratio (w/c or w/cm ratio) cannot be strictly maintained. Hence, it is used only in locations when RMC is not readily available. Depending on the strength it may attain in 28 days, concrete may be designated as normal/standard strength concrete (NSC) and high strength concrete (HSC). The grades of concrete are designated as per Table 8.9 of Chapter 8. As per Clause 6.1.1 of IS 456, the characteristic strength is defined as the strength of the concrete below which not more than 5% of the test results will fall. The minimum grade for RC as per IS 456 is M20; note that other international codes specify M25 as the minimum grade. In general, the usual concrete fall in the M20 to M50 range. In normal buildings, M20 to M35 concrete are used; whereas, in bridges and prestressed construction, strengths in the range of M40 to M60 are common. Very HSC in the range of M65 to M100 are used in columns of tall buildings and are normally supplied by RMC companies.

Concrete using fly ash is called *fly ash concrete*. Fly ash is generally used to replace cement up to 35%. However, recent research has resulted in *high volume fly ash (HVFA) concrete*, which has more than 50% replacement of cement by fly ash. Similarly, concrete with 5–10% of silica fume content is called *silica fume concrete*. The use of fly ash and/or silica fumes in concrete results in concrete having improved properties and increased durability.

Concrete with enhanced performance characteristics is called *high-performance concrete* (HPC). *Self-compacting concrete* (SCC) is a type of HPC in which the maximum compaction is achieved using special admixtures, and without using vibrators. The structural engineer should aim to achieve HPC through suitable mix proportioning and the use of chemical and mineral admixtures.

When fibres are used in concrete, it is called *fibre reinforced concrete (FRC)* (Fibres are usually used in concrete to control cracking due to plastic shrinkage and drying shrinkage). High performance FRC are called *Ultra*-HPC. Owing to the non-availability of standard aggregates or to reduce the self-weight, we may use lightweight aggregates; such concrete are called *Structural Lightweight Concrete* or *Autoclaved Aerated Concrete* (AAC). A brief description of these concrete is given in this chapter.

**Case Study    Chicago's 311 South Wacker Drive**

311 South Wacker Drive, Chicago, completed in 1989, became the world's tallest concrete building at that time. The structural system used is modified tube with a RC peripheral frame, with the shear walls interacting with the frames. It had interior steel columns and a composite steel and concrete slab. It was designed in such a way that the relative stiffness of both internal and external elements remained the same throughout the entire height. HSC with a compressive strength of 68.9 MPa and 82.7 MPa were used in the construction. For concreting the top elements, a self-climbing pump with a separate, mounted, placing boom was used. The concrete quantity was optimized due to the use of high strength thinner elements. In addition, the post-tensioned floor slabs reduced steel consumption. The construction involved the cycling of two sets of flying forms every 5 days.

Instead of the Sears Tower's black glass squares, 311 South Wacker uses an octagonal main shaft clad in pink granite from Texas (*Photo*: Er M. Karthik)

## 9.2 Ready Mixed Concrete

Ready mixed concrete (RMC) is a type of concrete that is manufactured in a factory or batching plant, based on standardized mix designs, and then delivered to the work site, by truck mounted transit mixers. This type of concrete results in more precise mixtures, under strict quality control, which are difficult to adopt on construction sites. Although the concept of RMC was known in the 1930s, the industry expanded only during the 1960s (It is interesting to note that as early as 1909, concrete was delivered by a horse-drawn mixer that used paddles turned by the cart's wheels to mix concrete en route to the jobsite.). The first Indian RMC plant operated in the City of Pune, in 1987, but the growth of RMC picked up only after 1997. Most of the RMC plants are located in seven large cities of India, where they contribute to 30–60% of total concrete used in these cities (Even today, a substantial proportion of concrete produced in the country is volumetrically batched and site-mixed, involving a large number of unskilled labourers in various operations.). RMC is being used for bridges, flyovers, and large commercial and residential buildings (Alimchandani, 2007).

RMC plants should be equipped with up-to-date equipment, such as concrete batching plant, transit mixer, and concrete pump. RMC is manufactured under computer-controlled weight-batched operations using sophisticated equipment and methods, and mixed at the plant itself. As per IS 4926:2003, RMC should have a slump: +25 mm or +1/3$^{th}$ of the specified value, whichever is less, and compacting factor: ±0.03, where the specified value is 0.90 or greater, ±0.04, where the specified value is less than 0.90 but more than 0.80, and ±0.05, where the specified value is 0.80 or less. The transport trucks are equipped with revolving drum mixtures, as shown in Fig. 9.1, which continuously mix the concrete until it is delivered at construction site, in ready-to-use condition. The major disadvantage of RMC is that since the materials are batched and mixed at a central plant, travelling time from the plant to the site is critical over longer distances. It is better to have the ready-mix be placed within 2 h of batching at the plant (The average transit time at Mumbai is 4 h during the daytime!). However, a longer period may be

**Fig. 9.1** Special transport truck with a revolving drum

permitted, if retarding admixtures are used or in cool humid weather or when chilled concrete is produced (It has to be noted that retarding admixture, added to the mix, may affect the properties of concrete.).

Frequently, the concrete is partially mixed in transit and mixing is completed at the jobsite. Materials, such as water and some kinds of admixtures, are often added to the concrete at the jobsite, after it has been batched, to ensure that the specified properties are attained before placement (www.nrmca.org).

Most of the RMC is placed using *concrete pumps*, which are specially designed to place concrete to heights up to 400 m and also to horizontal distances of about 200 m (German engineers, Max Giese and Fritz Hull, conceived the idea of pumping concrete through pipes in 1927, and the concrete pump was patented in Holland in 1932 by Jacob Cornelius Kweimn.). Hence, it is important to design such pumpable concrete, which can flow through pipes without segregation. The RMC plants usually provide the necessary concrete pumps for hire. Concrete pumps have revolutionized the construction of large high-rise buildings. Before the invention of concrete pumps, cranes were employed to lift large buckets of concrete (one bucket at a time), of capacity 1.53 m$^3$. (Putzmeister's specially designed BSA 14000 SHP-D pump reached a world record vertical concrete pumping height of 606 m topping out Burj Khalifa, and has the maximum output of 71 m$^3$/h and at a pressure of 220 bar. Concrete has been pumped as far as 1250 m with a Putzmeister BSA 14000 trailer pump and a 125 mm pipeline.).

Usually, the RMC producer, who is an expert in the field, accepts the responsibility of the design of the mixture and its performance (As per IS 4926, RMC plants should have the minimum testing frequency of one sample for every 50 m$^3$ of production, with three test specimens for each sample for testing at the 28$^{th}$ day.). Thus, the specifications of the mix are agreed between the user and the producer, before the delivery of RMC at site. Usually, the producer adopts performance-based specifications. The information to be provided by the RMC plant to the purchaser are, the precautionary steps to be conducted on materials, during mixing and on machines, and also the quality control measures (including testing) to be taken at the plant, which are spelt out in IS 4926. RMC is ordered and supplied in terms of volume of concrete (cubic meters in one-quarter increments). When placing order to an RMC plant, it is advisable to order 5–10% extra concrete for larger projects and 15% for smaller jobs. This extra is required to compensate for spillage or wastage, some loss of entrained air, and settlement of wet mixture. Concrete suppliers usually demand that a minimum order of 1 m$^3$ is placed as smaller batches may not have consistent quality. The user should remember to place order for the correct quantity of concrete, as disposal of extra concrete may be difficult and prove to be expensive.

RMC is useful in congested sites, where there is no space for storage or handling of concrete materials. Advantages of RMC include the following:

*Assured quality of concrete*   is produced under controlled conditions using consistent quality of raw material and weight-batching.

*High speed of construction*   Use of RMC can speed up construction, as it is ready to use.

*Reduction in cement consumption*   Reduction of 10–12% cement is possible due to better handling and proper mixing. Further reduction is possible if mineral admixtures are used.

*Versatility in uses and methods of placing*   The mix design of concrete can be tailor made to suit the placing methods used by the contractor at site.

*Reduced pollution*   Since it is produced at the RMC plant, noise and air pollution are reduced at site.

*Better quality of concrete*   Owing to the better quality of concrete, durability of structures and their overall service life is increased.

***Elimination of errors*** Results in elimination or minimization of human error while producing concrete and reduced dependency on labour.

***Timely schedules*** Timely deliveries in large as well as small pours of concrete.

***Other benefits*** No botheration of procurement, storage, and handling of materials such as coarse and fine aggregate, cement, water, and admixtures. No delay due to site-based batching plant erection/dismantling; hiring of equipment such as mixers, and no depreciation of costs.

However, use of RMC may have the following limitations:

1. As RMC has to be transported to considerable distances before placement at site, loss of workability may occur. Plasticizers and super-plasticizers are added to RMC at the plant to increase workability, as these concrete will have optimum w/c ratio. In addition, to prevent the setting of concrete during the transportation, admixtures such as retarders are used. Such addition of retarders may delay the setting time substantially and may result in placement problems. Retarders may also affect the early strength (first few days and not the 28$^{th}$ day) of concrete and result in an increased rate of slump loss. Hence, it is important that the admixtures (plasticizers, super-plasticizers, or retarders) used in RMC are properly tested for their suitability with the cement and concrete. De Weerdt et al. (2006) found that a system of sodium gluconate (set retarder, water reducer, and plasticizer) and calcium nitrate (accelerator, corrosion inhibitor, and antifreeze admixture) might prove to be a good system for long transport of concrete. The system will, however, have to be combined with a plasticizer in order to obtain sufficient workability.
2. Since a large quantity of concrete is made available in short duration, it is necessary to make proper arrangements for quick placing of concrete and to design form work for the extra loads.

The requirement for the production and supply of RMC are specified in IS 4926:2003.

**Example 9.1** Calculate the quantity of concrete to be ordered from RMC for a slab measuring 6 m long, 4 m wide, and 125 mm thick.

**Solution:** Volume of concrete in proposed slab = $6 \times 4 \times 0.125 = 3$ m$^3$
Quantity of concrete to be ordered = $3 \times 1.15 = 3.45$ m$^3$

## 9.3 Green Concrete (with Supplementary Cementitious Materials)

Climate change resulting from high concentration of greenhouse gases (GHGs), such as carbon dioxide ($CO_2$), methane ($CH_4$), nitrous oxide ($N_2O$), and fluorinated gases, in the atmosphere is threatening the World's environment. According to the National Oceanic and Atmospheric Administration, USA, the concentration of $CO_2$, one of the primary GHGs, has risen from 316 ppm in 1959 to about 412.14 ppm in March 2019 (www.esrl.noaa.gov/gmd/ccgg/trends/);. It is increasing by about 2 ppm per year, though this varies somewhat from year to year. To avoid a global warming threshold of an increase in average temperature of 2.1°C, it is estimated that a $CO_2$ concentration of less than 450 ppm needs to be maintained. In order to stabilize $CO_2$ concentrations at about 450 ppm by 2050, global emissions should have to decline by about 60% and the GHG emissions of industrialized countries by about 80% by 2050.

Construction industry consumes 40% of the total energy and about one half of world's major resources. Hence, it is imperative to regulate the use of materials and energy in this industry. $CO_2$ is a major by-product in the manufacturing of two most important materials of construction: Portland cement and steel. The concrete industry is responsible for creating up to 5% of worldwide manmade emissions of $CO_2$, of which 50% is from the chemical process and 40% from burning fuel. Thus, while selecting the material and system for the structure, the designer has to consider the long-term environmental effects. The long-term environmental effects to be considered include maintenance, repair and retrofit, recyclability, environmental effects of demolished structure, adoptability of fast-track construction, demountability, and dismantling of the structure at a future date. In concrete structures, in addition to performance, the concrete mixture has to be considered in terms of waste or by-product material content, embodied energy, and carbon footprint. The $CO_2$ produced due to the manufacture of structural concrete (using ~14% cement) is estimated at 410 kg/m$^3$ (~180 kg/tonne at density of 2.3 g/cm$^3$), which can be reduced to 290 kg/m$^3$ with 30% fly ash replacement of cement.

Mehta (2009) has shown that by simultaneously using the following three tools, major reductions in concrete consumption and carbon emissions can be achieved.

*Consuming less concrete*   This can be done by increasing the service life of structures from the present 50 years to 100–150 years, enhancing the long-term durability (by careful selection of constituents of concrete), and rehabilitating old buildings. Use of demountable precast products, that can be reused, is also an efficient solution.

*Using less cement in concrete mixtures*   This can be done by the following: (a) using high-range water-reducing admixtures, which will reduce 20 to 25% of water and in turn reduce cement content; (b) using optimizing aggregate size and grading; and (c) using 56–90 days compressive strength instead of traditional 28th day strength (especially in the Portland Pozzolana Cement (PPC), which will have higher strength on 56th day) in the design, which may result in 15–20% cement savings. Note that the 28th day strength was adopted as the standard when concrete were made only with Portland cement.

*Substituting cement with waste industrial by-products in a concrete mix*   The use of industrial by-products such as fly ash, blast furnace slag, silica fume, reactive rice-husk ash (RHA), etc., can lead to significant reductions in the amount of cement needed to make concrete, and hence reduces the emissions of $CO_2$ and consumption of energy and raw materials, as well as reduced landfill/disposal burdens. (India produces over 270 million tonnes of fly ash per year, which is harmful and difficult to dispose.). Fly ash can be readily substituted for over 30% of cement volume; blast furnace slag for more than 35%. HVFA concrete with 50 to 70% of cementitious content has been studied extensively and found to be feasible in certain situations and found to have better properties than concrete produced with Portland cement (Malhotra, 2002).

Combined use of tools 1 and 2 will reduce cement consumption by 30% (2.80 billion tonnes in 2010 to 1.96 billion tonnes in 2030). Clinker factor is reduced by 20–30% by the use of alternate cementitious materials. Carbon emission factor is decreased by 10–20% by the use of waste material as fuel.

It is interesting to note that the use of Portland cement containing limestone filler (which does not have pozzolanic properties) is a common practice in European countries, especially in France. Bentz et al. (2009) carried out a study using the Power's model and suggested that for low w/cm ratios in the range of 0.30 to 0.35, it is possible to replace cement with limestone powder to the extent of 15%. Such incorporation of coarse limestone powder, with median particle diameter of about 100 µm could significantly increase durability also, by reducing autogenous deformation and inclination for related early age cracking.

The use of RMC can also help in obtaining quality concrete, which will increase the durability and life of concrete structures. Modern concrete, such as the FRC, geopolymer concrete, HPC, reactive powder concrete, SCC, self-curing concrete, etc., not only enhance the properties of concrete, but also increase the life of structures built with them.

## 9.3.1 Fly Ash Concrete

Fly ash is a waste material produced in coal-based thermal power plants (see Section 5.9.2 of Chapter 5). As per IS 1489 (Parts 1 & 2):1991, fly ash in Portland pozzolana cement should not be less than 15% and not more than 35% by mass. As per the note of Table 5 of IS 456, the addition of fly ash quantity can be taken into account in the concrete composition with respect to the cement content and w/c ratio (see also Section 8.9 on mix design).

The potential for using fly ash in concrete was known for several years, but only during the mid-1900s the use of fly ash as a supplementary cementitious material in concrete began following the pioneering research conducted at the University of California, Berkeley. In recent years, the use of fly ash in concrete has grown dramatically.

The main objective of using fly ash in the cement concrete is to obtain durable concrete at reduced cost and this can be achieved by adopting one the following two methods:

1. Using fly ash-based PPC (which is manufactured as discussed in Section 5.10.1 of Chapter 5)
2. Using fly ash as an ingredient in cement concrete

When PPC is made in the factory, it does not require any additional quality check for fly ash during the production of concrete. However, the proportion of fly ash and cement is fixed, limiting the fly ash content in concrete mixes.

The addition of fly ash as an additional ingredient at concrete mixing stage as part replacement of the ordinary Portland cement (OPC) and fine aggregates is a more flexible method. It allows for the maximum utilization of fly ash as an important component (cementitious and as fine aggregates) of concrete. There are three basic approaches for selecting the quantity of fly ash in cement concrete, at site:

1. Partial replacement of the OPC—the simple replacement method
2. Addition of fly ash as fine aggregates—the addition method
3. Partial replacement of the OPC, fine aggregate, and water—the modified replacement method

These three basic approaches are briefly discussed here.

**Simple replacement method**   In this method, a portion of the OPC is replaced by fly ash (up to 30%) on a one-to-one basis by mass of cement. The early (i.e., 7–28 days) strength of concrete produced will be lower but higher strength will be achieved after 56–90 days. It is because, fly ash exhibits very little cementing value at early ages. At later ages, liberated lime resulting from the hydration of cement reacts with fly ash and contributes considerable strength to the concrete. The modulus of elasticity is also lower at early ages and reaches a higher value at later ages. The simple replacement method is adopted for mass concrete works where the initial strength of concrete is of little importance compared to the reduction of temperature rise.

**Addition method**   In this method, fly ash is added to the concrete as fine aggregate, without corresponding reduction in the quantity of the OPC. This increases the effective cementitious content and results in increased strength of concrete at all ages. This method is useful when there is a minimum cement-content criterion due to some design considerations.

**Modified replacement method**   This method is useful to make strength of fly ash concrete equivalent to the strength of control mix (without fly ash) at early ages, i.e., between 3 and 28 days. This method is most commonly used, and the mass of fly ash used is relatively more than the cement replaced (often, around the replacement ratios of 1:1:1 or 1:1:2) with additional adjustments in water and fine aggregate to obtain the desired performance and offset the reduction in early strength.

Fly ash particles are generally spherical in shape and help to reduce friction between aggregates and between concrete and pump line. The important characteristics, which affect the performance of fly ash in concrete are: (a) loss on ignition, (b) fineness, and (c) calcium (CaO) content. The calcium content of the fly ash is perhaps the best indicator of how the fly ash will behave in concrete. Pozzolanic reactivity of fly ash is more in high-calcium fly ash than low-calcium fly ash. If the calcium content of the fly ash is high enough (>20% CaO), it is even possible to make concrete with moderate strength using the fly ash as the sole cementing material (Thomas, 2007). The advantages of using fly ash in cement concrete include the following (see also Table 9.1):

1. Reduction of heat of hydration resulting in reduced thermal cracks and improvement in the soundness of concrete mass. A substitution of 30% fly ash may result in a reduction of 30–60% heat of hydration. However, it has to be noted that fly ash high in calcium may produce little or no decrease in the heat of hydration (compared to plain Portland cement) when used at normal replacement levels (Thomas, 2007). Inclusion of fly ash also reduces drying shrinkage.

2. Improvement of workability/pumpability of concrete. The use of good quality fly ash with a high fineness and low-carbon content reduces the water demand of concrete and, consequently, the use of fly ash should permit the concrete to be produced at lower water content when compared to a Portland cement concrete (PCC) of the same workability. An approximate thumb rule is that each 10% of fly ash should allow a water reduction of at least 3% (Thomas, 2007).

3. Increase in fines volume and decrease in water content and thus resulting in reduced bleeding and segregation of concrete.

4. As the level of replacement of cement increases the early-age strength of concrete decreases. However, finer the fly ash and lower the carbon content, the greater will be the pozzolanic activity and greater will be its contribution to the long-term strength in concrete of the same workability.

5. Reaction between fly ash and liberated lime results in pore and grain refinement resulting in improved impermeability, provided the concrete is cured properly.

6. Improved impermeability increases resistance against the ingress of moisture, sulphate resistance, carbonation, penetration of chlorides, and any other harmful gases resulting in increased durability.

7. If the sand is coarse, the addition of fly ash produces beneficial results; for fine sands, its addition may increase the water requirement for a given workability.

8. If loaded at an early age, fly ash concrete may exhibit higher amounts of creep than PCC because it has a lower compressive strength. However, at ages beyond 56 days, it may exhibit less creep because of its continued strength gain.

9. The amount of drying shrinkage of well-cured and properly proportioned fly ash concrete should be equal to or less than an equivalent PCC mix.

10. It is well established that low-calcium fly ash is capable of controlling damaging alkali-silica reaction (ASR) in concrete at moderate levels of replacement (20 to 30%) and this effect may be due to the reduced concentration of alkali hydroxides in the pore solution when fly ash is present. However, high-calcium Grade II fly ashes may not be very effective in controlling ASR.

11. Reduces requirement of cement for same strength thus reduces cost of concrete.

**Table 9.1** Effect of fly ash on the properties of concrete (adapted from Thomas, 2007, printed with permission from Portland Cement Association)

| Property | Effect of fly ash* | Comments |
| --- | --- | --- |
| Fresh concrete | Workability is improved and water demand is reduced. Concrete is more cohesive and segregates less-improved pumpability. Bleeding is reduced especially at high replacement levels. | Reduces water content by approximately 3% for each 10% addition. Precautions to be taken to protect concrete during hot weather, as it may accelerate the rate of moisture loss. Final finishing operations should be started after bleeding has stopped. |

(Contd)

**Table 9.1** (*Contd*)

| Property | Effect of fly ash* | Comments |
|---|---|---|
| Setting time | Significant delay in initial and final setting times, especially in cold weather (Naik and Singh, 1997). | Reduce the level of fly ash during cold weather. Test fly ash-cement-admixture compatibility. |
| Heat of hydration | Grade I fly ash at normal levels of replacement, reduces heat of hydration. If Grade II fly ash is used, higher levels of replacement (>50%) is necessary. | Use Grade I fly ash, if temperature control is critical. Otherwise, use high levels of Grade II fly ash and/or take other measures to reduce temperature, such as: reduce cement content, use low-heat or moderate heat cement, or lower concrete placing temperature (use crushed ice or liquid-nitrogen cooling). |
| Early-age strength | Strength reduced, especially at 1–3 days. Reduction is greater for Grade I fly ashes and for higher replacement levels. | Consider reducing fly ash content if early-age strength is critical. Use accelerating admixtures, high-early strength cement, or silica fume to compensate for reduced early-age strength. |
| Long-term strength | It increases with the level of fly ash content and extended curing. | Better to use 56th day strength in design calculations. |
| Permeability and chloride resistance | Permeability reduced significantly and improvement of chloride resistance, especially at later ages. | Adequate curing is essential if these benefits are to be achieved in the concrete close to the surface (cover concrete). |
| Expansion due to ASR | ASR is reduced. Grade I fly ash (with up to 20% CaO) of 20 to 30% and Grade II fly ash of greater than 40% replacement may completely suppress the reaction. | If reactive aggregate are used, use Grade I fly ash. If Grade I fly ash is not available, consider using combinations of Grade II fly ash with silica fume or slag. The level of fly ash usage must be determined using appropriate testing. |
| Sulphate resistance | Sulphate resistance increased by Grade I fly ash. Use of Grade I fly ash of about 20 to 30% will provide performance equivalent to that of sulphate resistance Portland cement. | Do not use Grade II fly ash. Consider using Grade I fly ash with sulphate-resisting Portland cement. |
| Resistance to carbonation | Similar or slightly greater degree of carbonation for concrete with up to 30% fly ash as compared to the OPC concrete. Concrete containing 50% fly ash carbonated at significantly greater rates (Thomas and Matthews, 1992). | Concrete containing fly ash should be cured properly and should have adequate cover for rebars. |

*Unless indicated otherwise, a minimum amount of 15% fly ash is needed to achieve the desired properties. Optimum dosage levels are dependent on the composition of the fly ash, mix design, exposure conditions, and required service.

The properties of fresh concrete and the mechanical properties and durability of hardened concrete are strongly influenced by the incorporation of the fly ash into the mixture. The extent to which fly ash affects these properties is dependent not only on the extent and the composition of the fly ash, but also on other parameters, such as the proportion of other ingredients in the concrete mixture, the type and size of these ingredients, the exposure conditions during and after placement, and the construction practices. From these, it is clear that no one replacement level is best suited for all applications.

It is fairly well established that low-calcium fly ash extend both the initial and final set of concrete, especially in cold weather. It may be better to reduce fly ash content of concrete produced during cold weather. Testing may be required to study the fly ash-cement-admixture compatibility. Because most fly ash react at a much slower rate than Portland cement, it is better to extend the curing period to 14 days, where possible, or to place curing membrane after 7 days of moist curing (Malhotra and Mehta, 2005). If the specified strength is required at 28 days or earlier, lower values of w/cm ratio should be used. A lower w/cm ratio can be achieved by a combination of (a) reducing the water content by either taking advantage of the lower demand in the presence of fly ash, or by using a water-reducing admixture, or both; and (b) increasing the total cementitious content of the mix.

The relationships between the tensile strength, flexural strength, elastic modulus, and the compressive strength of concrete are not significantly affected by the presence of fly ash at low and moderate levels of replacement. Fly ash concrete has been used in practically all types of concrete applications from residential foundations to HPC for highway and marine structures, and high-strength concrete for high-rise construction. Fly ash has also been used in readymixed and precast concrete, pumped concrete, slip-formed concrete, roller-compacted concrete, and shotcrete (Thomas, 2007).

**Case Study** | **Lower Notch dam, Ontario**

The Lower Notch dam, Ontario, Canada, which was completed in 1969, is situated on the Montreal River at Lake Timiskaming in the Canadian province of Ontario. Although petrographic examination and expansion testing of mortar bars (ASTM C227) did not identify the aggregates used as reactive (expansion <0.10% at 6 months), similar rock types from the Montreal River area had been implicated in 1965 as the cause of ASR in the nearby Lady Evelyn Dam (This structure was eventually replaced.). Testing indicated that expansion due to ASR could be prevented by using either a low-alkali cement or fly ash. The final decision was to use a combination of Class F fly ash combined with high-alkali cement. A replacement level of 20% fly ash was used for the structural concrete in the powerhouse and 30% fly ash in the massive concrete structures (Thomas, 1996). A recent inspection of the structure (Sept. 2010) confirmed that there were no visible signs of ASR (i.e., no signs of expansion or cracking), even after 41 years after construction (Thomas et al., 2012).

## 9.3.2 Silica Fume and RHA Concrete

*Silica fume* is a by-product of the production of silicon and ferrosilicon alloys (see Section 5.9.4). Silica fume used in concrete should conform to IS 15388:2003. Typically, 4–15% weight of the OPC can be replaced with silica fume. Concrete with added silica fume content is called *silica fume concrete*. Because of its extreme fineness (with specific surface about 40–60 times higher and particle size 100 times smaller than that of cement) and high-silica content, silica fume is a very effective pozzolan. The higher surface area and amorphous nature of silica fume make it highly reactive. The hydration of $C_3S$, $C_2S$, and $C_4AF$ are accelerated in the presence of silica fume.

As already mentioned in Section 5.9.4, silica fume for use in concrete is available in wet or dry forms. It is usually added during concrete production at a concrete plant. The fineness of silica fume greatly increases the water demand of a concrete mixture. Hence, it has to be used with high-range water-reducing admixture to maintain low w/c materials ratios and have proper workability. The mix with silica fume remains highly cohesive. Silica fume reduces bleeding and segregation of fresh concrete significantly because the free water is consumed in wetting of the large surface area of the silica fume and thus reducing the free water left in the mix for bleeding. Silica fume also blocks the pores in the fresh concrete so water within the concrete is not allowed to come to the surface. Hence, it is essential to finish the silica fume concrete as soon as possible, after it has been placed and compacted, to protect the surface from drying out. It, therefore, requires early membrane curing. Also, there is greater tendency for plastic shrinkage cracks to occur, to avoid them proper and prolonged curing of concrete is necessary.

Though silica fume can be used to replace cement in a given concrete mixture while maintaining the same compressive strength, it is not economical. If cost saving is the goal, fly ash or blast furnace slag is a better replacement material for cement; such blending also helps to develop strength at later ages. By using silica fume along with super-plasticizers, it is relatively easier to obtain compressive strengths of the order of 100–150 MPa in laboratory. The first person who realized the beneficial use of silica fume in concrete was Hans Bache of Aalborg Portland in Denmark (Bache, 1987a).

The use of silica fume increases the performance of concrete in two distinct ways, due to its very small particle size and its high silicon dioxide content:

1. As the Portland cement in concrete begins to react chemically, it releases calcium hydroxide $[Ca(OH)_2]$. The silica fume reacts with this calcium hydroxide to form additional binder material called calcium silicate hydrate (C-S-H gel), which is very similar to the calcium silicate hydrate formed from Portland cement. It is an additional binder that gives silica fume concrete its improved properties, such as compressive strength, bond strength, and abrasion resistance.
2. The presence of very smaller particles, smaller than that of cement, improves the properties of concrete. This effect is termed as '*micro filling effect*'. The extreme fine silica fume particles fill the microscopic voids between cement particles, especially in the voids at the surface of the aggregates. This so-called *interface zone* influences the properties of the concrete considerably.

The addition of silica fume to concrete improves the durability of concrete through reduction in the permeability, refined pore structure, leading to a reduction in the diffusion of harmful chloride ions and calcium hydroxide content, which results in a higher resistance to sulphate attack. Improvement in durability will also improve the ability of silica fume concrete in protecting the embedded steel from corrosion. Improvements in paste to aggregate bond, and the ultimate strength of concrete are also achieved.

Hence, silica fume is often used in high-strength and high-performance concrete and in applications where a high degree of impermeability is needed. Examples of structures using silica fume concrete are: (a) Norway's Gullfaks Offshore platform built in 1981, it has cement content of 400 kg/m³, water content of 165 L/m³, 6 L of admixtures, and 10 kg of silica fume to get a concrete strength of 79 MPa with a slump of 240 mm; and (b) The Key Tower in Cleveland, USA (Designed by architect César Pelli, and structural engineers Skilling Ward and Magnusson Barkshire), completed in 1990, it used 65% Portland cement, 27% slag cement, and 7% silica fume to achieve a design strength of 85 MPa. More details of silica fume concrete and its use in structures may be found in www.silicafume.org.

If silica fume is added at site, it must always be added first with coarse aggregate and some water. Silica fume alone should not be batched first into the mixer. These above should be mixed for 1.5 min. Then, Portland cement and any other cementitious material, such as fly ash or slag cement, may be added and mixed for an additional 1.5 min. Finally, fine aggregate and the remaining water should be added and mixed for 5 min; after a rest of 3 min, it should be mixed again for a minimum of 5 min (www.silicafume.org).

## RHA Concrete

As seen in Section 5.9.6, RHA consists of approximately 85–90% silica by weight and minor amounts of other elements. Up to 30% by weight of the OPC can be replaced with RHA without any adverse effect on strength and permeability properties (Ganesan et al., 2008). Use of RHA in concrete has been found to improve the properties of concrete and results in high strength and impermeability of concrete. Water demand and drying shrinkage should be studied before using rice husk. Coutinho (2003) studied the behaviour of concrete with 10, 15, and 20% replacement of cement with Portuguese RHA. It was found that using RHA as a partial cement replacement enables higher compressive strength than control concrete (0% RHA) and also higher than 10% silica fume concrete (after 80 days of curing). Cement replacement of up to 40% with RHA was found to provide sulphate resistance.

Other researchers have found that the compressive strength of concrete with RHA dosage of 10 to 30% increases in the range 34–80% compared to the control concrete with 0% RHA. The strength of concrete with RHA increased with time for all dosages of RHA. Ganesan et al. (2008) found that the replacement of cement with 30% of RHA leads to substantial improvement in the permeability properties of blended concrete, when compared to that of unblended OPC concrete – about 35% reduction in water permeability, about 28% reduction in chloride diffusion, and about 75% reduction in chloride permeation. This observation shows that the use of RHA will enhance the durability and design life of RC constructions.

### 9.3.3 Slag Cement Concrete

As discussed in Section 5.9.3 of Chapter 5, ground granulated blast furnace slag (GGBS) is a molten, non-metallic by-product having 90% glass and silicates and aluminosilicates of lime obtained in the blast furnace from the iron industry. The use of ground granulated iron blast-furnace slag cement, usually called as *Portland slag cement* (PSC) or simply as *slag cement*, as a cementitious material, dates back to 1774, when Loriot made a mortar using slag cement in combination with slaked lime (ACI 233R-03). Until the 1950s, slag cement was used in two basic ways: (a) as a raw material for the manufacture of Portland cement and (b) as a cementitious material combined with Portland cement, hydrated lime, or gypsum. After 1950s, slag cement was used as a separate cementitious material in the concrete mixer along with Portland cement (ACI 233R-03). The manufacture of PSC is discussed in Section 5.10.2 of Chapter 5. In India, several cement companies, such as ACC and JFK Cement, manufacturer the PSC as per IS 12089:87, by combining up to 45–50% slag, 45–50% clinker, and 3–5% gypsum.

The proportion of slag cement in a concrete mixture will depend on the purposes for which the concrete is used, the curing temperature, the grade (activity) of the slag cement, and the Portland cement. Slag cement, up to 25 to 70% of the mass of the total cementitious material, have been successfully used. The optimum amount of slag in cement, that produces greatest strength at 28 days, is usually 40 to 50% by mass of the total cementitious material, although this percentage varies depending on the grade of slag cement. Apart from being more environment-friendly, the use of slag cement in concrete offers the following advantages:

1. Increased compressive strength (GGBS is a low-performance cementitious material, but can achieve high compressive strength when an alkaline activator is used)
2. Excellent resistance to chloride and sulphate attacks
3. Reduced risk of cracking
4. Improved workability
5. Better compatibility with all types of admixtures
6. Ease of pumping
7. Superior finish
8. Better resistance against ASR (ACI 233R-03)
9. Increased sulphate resistance
10. Reduced drying shrinkage, penetration of chloride ions, and permeability

As the percentage of slag cement increases, a slower rate of strength gain should be expected, particularly at early ages, unless the w/cm ratio is substantially reduced, chemical accelerators are used, or accelerated curing is provided. The proportioning techniques for concrete incorporating slag cement are similar to those used in proportioning concrete made with Portland cement or blended cement. Slag cement is usually substituted for Portland cement on a one-to-one basis by mass, and should be included in the determination of the w/cm ratio. The effects of chemical admixtures on concrete-containing slag cement are similar to those for concrete made with Portland cement. When more than 25% slag cement is used as a replacement for Portland cement, there may be delays in the setting time. Bleeding can be reduced by using finer slag. Longer curing period may be necessary for concrete-containing slag cement. Slag cement can be used as an effective means of controlling temperature rise in mass concrete.

The PSC has been successfully used in concrete pavements, mass concrete applications, high performance or HSC, structures and foundations, precast concrete, concrete exposed to sea water and marine application. A 50% substitution of slag cement in a concrete mix can save 44% of embodied energy and 60% of embodied $CO_2$ emissions per cubic meter. According to the Slag Cement Association (SCA), Georgia, slag cement can contribute to achieving 13–17 LEED-NC points (Sustainable sites: 2 points, Material and resources: 12 points, and Innovation and design: 3 points). In the 225.8 m tall, 7 World Trade Center of New York, slag cement replaced 40% of the Portland cement in the concrete mix, and achieved strengths of over 69 MPa. This LEED™ Gold Certified structure also received SCA's 2006 award for 'Best Use of Slag Cement for Strength'. For more information on the use of slag cement in concrete, refer ACI 233 R-03 and www.slagcement.org.

## 9.3.4 Concrete with Recycled Aggregates

According to the Construction & Demolition Recycling Association, the amount of Construction and Demolition (C&D) waste generated in the United States (which is made up primarily of concrete, asphalt, wood, gypsum, demolition metals, and asphalt shingles), in 2012, was estimated at 480 MT (www.cdrecycling.org). (Construction waste is generated at the rate of about 0.5 T/person each year in the United States). This waste has to be transported to the disposal site, thus consuming more energy and pollution. A recent study by IIT Madras has shown that the quantity of C&D debris generated in Chennai city in 2013 was around 1.25 MT and C&D debris forms around 32% of the total solid waste generated in the city.

In many developed countries, more than 70% of the total C&D waste being generated is recovered and put to beneficial use by the C&D recycling industry (corresponding to a 35% recycling rate for mixed C&D, an 85% recycling rate for bulk aggregate, and an over 99% recycling rate for RAP). In the USA, the area of landfill avoided by recycling this amount of C&D is equivalent to over 1740 ha (at a waste depth of 15 m). Methods of recycling concrete and concrete components are discussed in Swamy (2003) and Sakai and Sordyl (2009).

---

**Case Study**    **Four Times Square, New York, NY**

Four Times Square, New York, is a 48-storey, 148,645 m² office tower, completed in July 1999. The building was designed to address environmental building issues, such as energy efficiency and indoor air quality. Before construction could begin 42,968 m² of built-up area was demolished. Prior to demolition, private groups salvaged more than 110 T of wood beams and architectural features. Primary materials diverted during demolition were scrap metal, brick, concrete and dirt, and during construction, scrap metal, cardboard, wood, dirt, and rock. Materials salvaged prior to demolition included wood timbers, ornate stone work, office doors, copper, and facial corners. The environmental consultant helped contractors anticipate and reduce packaging waste during construction. Savings from recycling and reuse efforts were calculated at $895,000, which were estimated to exceed the added costs involved with planning and instituting these recovery practices.

---

Recycling of concrete involves breaking, removing, and crushing existing concrete into materials of specified size and quality. Figure 9.2 shows the typical flow of concrete recycling system. It has to be noted that the Netherlands, with a population of about 17 million, has about 120 C&D recycling facilities. However, India with a population of over a billion relatively has very few C&D recycling facilities (*ICI Handbook*, 2016). This statistic illustrates the magnitude of the problem with respect to C&D waste management in India.

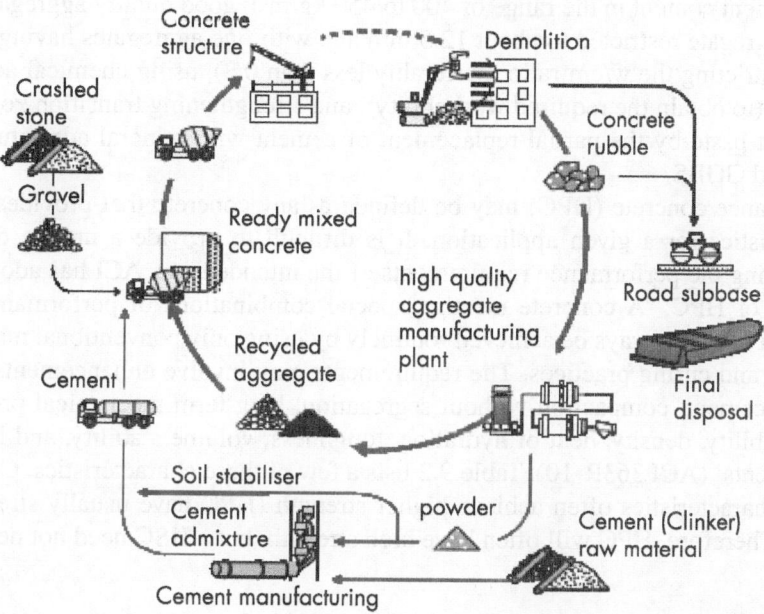

**Fig. 9.2** Schematic flow of concrete recycling system (from Shima et al., 2005, by JCI, reprinted with permission)

The quality of concrete with recycled concrete aggregates (RCA) is much dependent on the quality of the recycled material used. RCA have been successfully used in applications such as bulk fills, bank protection, base or fill for drainage structures, road construction, noise barriers and embankments. Often, recycled aggregate is combined with virgin aggregate when used in new concrete. RCA usually present greater porosity and absorption, and lower density and strength than natural aggregates. Micro-structural studies on RCA indicated differences in the characteristics of the interfacial transition zones between the cement paste and the aggregates. It was also found that the reduction in concrete stiffness is higher than the strength. They result in concrete with higher drying shrinkage and creep (Kerkhoff, and Siebel, 2001). More details about recycling of concrete may be found in the study by Noguchi (2012).

We can conserve our natural resources and also save landfill space by using recycled aggregates in concrete. River sand can also be substituted with environment friendly options of manufactured sand, iron/steel/copper slag aggregates, dredged sand, and mining wastes. Other options tried with concrete include used foundry sand and cupola slag from metal-casting industries, post-consumer glass, wood ash from pulp/ saw mills, de-inking solids from paper-recycling companies, and sludge from various industrial units (Naik, 2002). It is important to note that recycling also requires energy for processing, transportation, etc.

Many cement plants have transformed old abandoned mines into huge reservoirs of water and maintain a norm of zero-water discharge. ACC Ltd. in India treats its waste-water and recycles it and hence self-reliant with respect to its water requirements.

## 9.4 High Strength/High Performance Concrete

As per Table 2 of IS 456:2000, any concrete that has a characteristic (28th day) compressive strength of 65 to 100 MPa (Grade M65-M100) is considered high strength concrete (HSC). In HSC, the aggregate plays an important role on the strength of concrete. HSC is often produced with high-quality Portland cement (with cement content in the range of 400 to 450 kg/m³), good quality aggregates (the maximum size of coarse aggregate restricted to about 12.5 mm and with fine aggregates having fineness modulus of about 3.0), restricting the w/cm ratio (preferably less than 0.3), using chemical admixtures, such as super-plasticizer (to obtain the required workability), and strengthening transition zone between aggregates and cement paste by the partial replacement of cement with mineral admixtures, such as silica fume, fly ash, and GGBS.

High performance concrete (HPC) may be defined as any concrete that provides enhanced performance characteristics for a given application. It is difficult to provide a unique definition of HPC, without considering the performance requirements of the intended use. ACI has adopted the following broad definition of HPC: 'A concrete meeting special combinations of performance and uniformity requirements that cannot always be achieved routinely by using only conventional materials and normal mixing, placing, and curing practices. The requirements may involve enhancements of characteristics such as easy placement, compaction without segregation, long-term mechanical properties, early-age strength, permeability, density, heat of hydration, toughness, volume stability, and long service life in severe environments' (ACI 363R-10). Table 9.2 lists a few of these characteristics. Concrete possessing many of these characteristics often achieve higher strength (HPC have usually strengths greater than 50 to 60 MPa). Therefore, HPC will often have high strength, but a HSC need not necessarily be called as HPC.

**Table 9.2** Desired characteristics of HPC

| Property | Criteria that may be specified |
|---|---|
| High strength | 70 to 140 MPa at 28 to 91 days |
| High-early compressive strength | 20 to 28 MPa at 3 to 12 h or 1 to 3 days |
| High-early flexural strength | 2 to 4 MPa at 3 to 12 h or 1 to 3 days |
| High modulus of elasticity | More than 40 GPa |
| Abrasion resistance | 0 to 1 mm depth of wear |
| Low permeability | 500 to 2000 coulombs |
| Chloride penetration | Less than 0.07% Cl at 6 months |
| Sulphate attack | 0.10% /0.5 % max. expansion at 6 months for moderate/severe sulphate exposures |
| Low absorption | 2 to 5% |
| Low diffusion coefficient | $1000 \times 10^{-14}$ m/s |
| Resistance to chemical attack | No deterioration after 1 year |
| Low shrinkage | Shrinkage strain less than 0.04% in 90 days |
| Low creep | Less than normal concrete |

HPCs are made with carefully selected high-quality ingredients and optimized mixture designs (see Table 9.3). These ingredients are to be batched, mixed, placed, compacted, and cured, with superior quality control, to get the desired characteristics. To achieve the desired characteristics, HPC may contain materials such as Portland cement of high early strength cement (with high-cement contents ranging from 400 to 550 kg/m³), blended cement, fly ash, slag, silica fume (5–15%), metakaolin, super-plasticizers (5–15 L/m³), high-range water reducers, retarders, accelerators, corrosion inhibitors, shrinkage reducers, ASR inhibitors, polymer/latex modifiers, fibres, and optimally graded aggregates in various combinations. Typically, such concrete will have a low w/cm materials ratio of 0.22 to 0.40.

**Table 9.3** Typical HPC mixtures used in some structures

| Ingredients | Structure | | | | |
|---|---|---|---|---|---|
| | Two Union Square, Seattle, 1988 | Great Belt Link, East Bridge, Denmark, 1996 | Kaiga Atomic Project Unit 2, India, 1998 | Petronas Tower, Malaysia, 1999 | Urban Viaduct, Mumbai, India, 2002 |
| Water kg/m³ | 130 | 130 | 136 | 152 | 148 |
| Portland Cement, kg/m³ | 513 | 315 | 400 | 186 | 500 |
| Fly ash, kg/m³ | – | 40 | – | 345* | – |
| Slag, kg/m³ | – | – | – | – | – |
| Silica fume, kg/m³ | 43 | 23 | 25 | 35 | 50 |
| Coarse aggregates, kg/m³ | 1080 | 1140 | 1069 | 1000 | 762 (20 mm) + 384 (10 mm) |
| Fine aggregates, kg/m³ | 685 | 710 | 827 | 725 | 682 |
| Water reducer, L/m³ | – | 1.5 | – | – | – |
| Retarder, L/m³ | – | – | – | – | – |
| Air content % | – | 5.5 | 2 | – | 1.5 |
| Super-plasticizer, L/m³ | 15.7 | 5.0 | 5.82 | 9.29 | 8.25 |
| W/cm ratio | 0.25 | 0.34 | 0.32 | 0.25–0.27 | 0.269 |
| Slump, mm | – | – | 175+25 | 180–220 | 130–180 (at plant) 80–120 (at site) |
| Strength at 28 days, MPa | 119 | – | 75.9 | 80 | 79.6–81.3 |
| Strength at 91 days, MPa | 145 | – | 81.4 (180 days) | 100 (56 days) | 87.2–87.4 |

*Mascrete- A cement-fly ash compound in the ratio 20:80.

Super-plasticizers are usually used to make these concrete fluid and workable. Note that without super-plasticizers, the w/cm cannot be reduced below a value of about 0.40. Typically, 5 to 15 L/m³ of super-plasticizers can effectively replace 45 to 75 L/m³ of water (Aïtcin and Neville, 1993). This drastic reduction in mixing water reduces the distance between cement particles, resulting in much denser cement matrix than NSC. With the currently available cement and super-plasticizers, and the present-day mixing, placing, and curing practices, the optimum value of w/cm is about 0.22; values lower than this are harmful, because an adequately high density of the cement matrix cannot be achieved (Aïtcin, 1998). Since mixing water reacts chemically with cement and is lost by self-desiccation, the resulting cement paste in HPC has a very low porosity. This high-density matrix, in addition to the chemical bonds created by the hydrates (which also exist in NSC), results in high compressive strengths (Aïtcin and Neville, 1993). Studies have shown that 35% of cement paste by volume in HPC represents an optimum solution in balancing the conflicting requirements of strength, workability, and dimensional stability. Note that despite the use of low w/cm, HPC may require air entrainment for protection against repeated cycles of freezing and thawing. Though the cement mixes in Table 9.3 show the presence of silica fume, which is about 10 times costlier than cement, it is not an essential ingredient of HPC, though it makes it easier to attain high strengths above 60 MPa (Neville, 2012). The quantity of silica fume necessary to achieve very high strength is about 10% of the volume of Portland cement. The optimum particle-packing mixture design approach may be used to develop a workable, and highly durable design mixture (with cement content less than 300 kg/m$^3$), having compressive strength of 70–80 MPa (Kumar and Santhanam, 2004).

In the design of HPC mixes, one needs to consider the cement-super-plasticizer compatibility, as their physical and chemical interaction is very complex and may result in rapid slump loss. The controlling factor is the solubility of $SO_3$ (which has a maximum content of 3.0 to 3.5% depending on the content of $C_3A$) in the given cement. Ideal cement for HPC from rheological point of view may be the one which is not too fine, has a low $C_3A$ content, rich in belite content, moderate alite content, low alkali content (<0.6%), and with as little interstitial phase content as possible (i.e., $C_3A + C_4AF$ less than 10%, with $C_3A < 3\%$ and $C_4AF < 7\%$), and should have high strength – OPC 53 grade is commonly chosen (*Rheology* is the science of the deformation and flow of materials). The behaviour of super-plasticizer is a function of its structure and the degree of polymerization. Polycarboxylates and acrylic copolymers are the most effective of all the chemicals used in concrete. An ideal super-plasticizer should consist of rather long molecular chains. In the absence of guidelines, the final selection of the cement and the super-plasticizer should be made based on trial concrete mixes. More information on cement- super-plasticizer compatibility may be found in Ramachandran (1995), Rixom and Mailvaganam (1999), Mullick (2008), and Jayasree et al. (2011). The cement substitutes, such as fly ash, slag, and silica fume, refine the pore structure and block the capillaries due to 'secondary hydration' process (Mullick, 2005).

As the crushing process takes place along any potential zones of weakness within the parent rock, and thus removes them, smaller particles of coarse aggregates are likely to be stronger than the large ones. Hence, for strengths in excess of 100 MPa, the maximum size of aggregates should be limited to 10 to 12 mm; for lesser strengths, 20 mm aggregates can be used (Aïtcin and Neville, 1993). Strong and clean crushed aggregates from fine-grained rocks, mostly cubic in shape, with a minimum of flaky and elongated shapes, are suitable for HPC. In order to have good packing of the fine particles in the mixture, as the cement content increases, the fine aggregates should be coarsely graded and have fineness modulus of 2.7 to 3.0.

As the HPC has very low water content, there will not be any bleeding, leading to plastic shrinkage cracking (However, the drying shrinkage will be minimum due to the low w/cm ratio.). Note that hydration of cement in HPC is very rapid. The silica fume, if used, also reacts very early, rapidly using up the available water and contributing to self-desiccation. This causes autogenous shrinkage, resulting in micro-cracking though out the concrete mass. Hence, it is important to effectively cure HPC, as early as possible. Membrane curing is not suitable for HPC, and hence fogging or wet curing should be adopted to control plastic and autogenous shrinkage cracking.

High-performance concrete has been primarily used in tunnels, bridges, pipes carrying sewage, offshore structures, tall buildings, chimneys, and foundations, and piles in aggressive environments for its strength, durability, and high modulus of elasticity. It has also been used in shotcrete repair, poles, parking garages, and agricultural applications. Note that in severe fires, HPC results in bursting of the cement paste and spalling of concrete. More information on HPC may be had from ACI 363R-10, Zia et al. (1991, 1993), Aïtcin and Neville (1993), Aïtcin (1998), and IS 9103:1999.

HPC, with strength exceeding 100 MPa, has been used in the 451.9 m tall, 88-storey Petronas twin towers in Kuala Lumpur, Malaysia. In the Rs. 16 billion ($19.2 million) prestressed concrete Bandra–Worli Sea link project, M60 concrete was used.

## Case Study    Burj Khalifa, Dubai

Burj Khalifa, located near down-town Dubai, United Arab Emirates, is the world's tallest tower. The construction of this 828 m tall, RC tower structure, broke several records during its construction. The basic structure of Burj Khalifa is a central hexagonal core with three lobes (wings) clustered around it. As the tower rises, one wing at each tier sets back in an upward spiralling pattern, decreasing the cross section of the tower as it reaches towards the sky in 26 helical levels. The Y-shaped floor plan along with the upward spiralling pattern of setbacks in the wings, was determined based on extensive wind tunnel tests, which also helped to reduce the wind forces on the tower. The superstructure is supported by a large RC raft, which is in turn supported by bored RC piles, is made of M50 grade SCC. The piles were made high density, low permeability M60 grade SCC concrete, placed by tremie method utilizing polymer slurry.

High-performance SCC concrete, with a mix designed to provide a low-permeability and high-durability, was used in the walls and columns of Burj Khalifa tower. The C80 to C60 cube strength concrete used Portland cement, fly ash, and local aggregates. The C80 concrete had a specified Young's Modulus of 43,800 N/mm² at 90 days. Two of the largest concrete pumps in the world were used to deliver concrete to heights over 600 m in a single stage. To reduce the cracks due to the high temperatures of Dubai (about 50°C), the concrete was poured at night, when the air is cooler and the humidity is higher, with ice added to the mix.

## 9.5 Self-compacting Concrete

*Self-compacting concrete (SCC)* also known as *high-workability concrete, self-consolidating concrete*, or *self-levelling concrete*, is a HPC, developed by Prof. Okamura and associates at the University of Tokyo (now, Kochi Institute of Technology), Japan, during 1988 (Okamura et al., 2000). The SCC is highly workable concrete that can flow under its own weight and fill all voids within the formwork without segregation, excessive bleeding, excessive air migration, and without any vibration or other mechanical consolidation. The highly flowable nature of the SCC is due to very careful mix proportioning, usually replacing much of the coarse aggregate with fines and cement, and adding chemical admixtures (EFNARC, 2005). The SCC may be manufactured at a site batching plant or in a RMC plant and delivered to site by truck mixer. It may then be placed either by pumping or pouring into horizontal or vertical forms.

Advantages of SCC include the following:

1. Placed at a faster rate without any mechanical vibration and less screeding, resulting in savings in placement costs
2. Reduced vibration effort and noise during placing, potentially increasing construction hours in urban areas
3. Ability to fill complex forms, which have limited accessibility
4. More uniform distribution of concrete in regions of congested reinforcement
5. Improved consolidation around reinforcement and better bond with reinforcement
6. Improved pumpability of concrete
7. Quicker RMC trucks turn-around time, enabling efficient service by ready-mix producer
8. Shorter construction time, reduced labour, and improved labour safety thus resulting in cost savings
9. Uniformly compacted surface
10. Less surface voids and need for patching
11. Improved and more uniform aesthetics of flatwork with less effort

The following are some of the disadvantages of SCC:

1. Production of SCC places stringent selection of materials as compared to conventional vibrated concrete.
2. The plant/site personnel should be trained to handle SCC and need experience to successfully produce SCC.
3. SCC requires a larger number of trial batches, both in the laboratory/plant and at site.
4. The mixing time of SCC may be longer than that of conventional concrete. SCC is more sensitive to the total water content in the mix. It is necessary to take into account the moisture/water content in the aggregates and the admixtures before adding the remaining water in the mix. The mixer must be clean and moist, and should not contain any free water.
5. The truck drivers must check the concrete drum before filling with SCC to make sure that the drum is clean and moist, but with no free water. Extra care must be taken for long deliveries.
6. It may require design of formwork for hydrostatic pressure.
7. SCC requires a higher level of quality control than conventional slump concrete. In addition to the normal concrete testing, it is desirable to test the slump flow, T50, and L-box tests at the jobsite before placement.
8. It is better to limit the distance of horizontal flow to 10 m and the vertical free fall to 5 m.
9. SCC tends to dry faster than conventional vibrated concrete, as there is little or no bleeding water at the surface. SCC should be cured as soon as practicable after placement to prevent surface shrinkage cracking.
10. SCC is costlier than ordinary concrete, though the cost may be offset by improvement in productivity, reduction in vibration cost, and other advantages. Life cycle cost may prove to be in favour of SCC.

The original methodology of producing SCC was based on the following:

1. limiting aggregate content,
2. using low water/powder ratio, and
3. using conventional super-plasticizers.

The key for creating SCC is producing a mixture that will flow like a liquid and at the same time will be stable, without any segregation. Note that these two are opposing requirements. To achieve fluidity, a new generation super-plasticizers based on polycarboxylic esters (PCE) is used nowadays, as it gives better water reduction and slower slump loss than traditional super-plasticizers. The stability of a fluid mix may be achieved either by using high fines content or using viscosity-modifying agents (VMA). The developments undertaken in Europe, during 1990s, resulted in adding powders (cement, supplementary cementitious materials, and inert materials, such as limestone, dolomite, and granite dust) passing through the 150 μm sieve to increase plastic viscosity.

Although the mix proportion of SCC varies with the particular application, some typical mixes are given in Table 9.4.

**Table 9.4** Typical SCC mixes used in Europe [Ouchi et al. (2003) and Kumar and Kaushik (2003)]

| Ingredients | Mix 1 | Mix 2 | Mix 3 | Mix 4 |
| --- | --- | --- | --- | --- |
| Cement, kg | 280 | 330 | 310 | 343 |
| Fly ash, kg | 0 | 0 | 190 | 244 |
| Limestone powder (filler), kg | 245 | 0 | 0 | 0 |
| Ground granulated blast furnace slag, kg | 0 | 200 | 0 | 0 |

*(Contd)*

**Table 9.4** (Contd)

| Ingredients | Mix 1 | Mix 2 | Mix 3 | Mix 4 |
|---|---|---|---|---|
| Silica fume, kg | 0 | 0 | 0 | 17.2 |
| Crusher dust, kg | – | – | – | 858 |
| Coarse aggregate, kg | 750 | 750 | 750 | – |
| Fine aggregate, kg | 865 | 870 | 700 | 686 (2–6 mm) |
| *HRWR, kg | 4.2 | 5.3 | 6.5 | 5.8 (PCE-based) |
| **VMA, kg | 0 | 0 | 7.5 | 0 |
| Water, kg | 190 | 192 | 200 | 213 |
| Slump flow test-diameter of spread, mm | 600–750 | 600–750 | 600–750 | 700 |

*HRWR = High-range water reducing admixture.
**VMA = Viscosity-modifying admixture.

The tests to be conducted for evaluating the SCC are explained in Section 25.17 of Chapter 25 on testing.

## 9.6 Structural Lightweight Concrete

Conventional cement concrete is usually heavy, with a density of 2400 kg/m³. For structures such as multi-storey buildings, it is desirable to reduce the dead loads; reduction of weight is also necessary for buildings in earthquake zones. In such cases, air-entrained concrete/lightweight concrete could be used advantageously. Structural lightweight concrete (SLWC) is particularly suitable when low density, good thermal insulation, or fire protection is required, but not many aggregates are suitable for such applications. It has to be noted that weight reduction could also be done by entraining air in the concrete by using a chemical agent—such a concrete is termed as *aerated air-entrained concrete,* is discussed in Sec 9.6.1.

The first known use of lightweight concrete is over 2250 years old. Some of the early structures from the Roman Empire, including Port of Cosa (built around 273 BC using lightweight concrete made from natural volcanic materials), the Pantheon Dome (This 43.3 m diameter dome finished in 27 BC is still in use), and the Coliseum (It was built in 75 to 80 AD, and has foundations cast with lightweight concrete using crushed volcanic lava, and walls made using porous, crushed brick aggregate), have elements that were constructed with lightweight concrete. The use of lightweight concrete in modern times started when Steven J. Hayde, a brick-maker from Kansas City, MO, developed a rotary kiln method during 1908, for expanding and vitrifying select clays, shale, and slates, at temperatures of about 1093°C, and patented it in 1918 as Haydite®. This process produced a high-quality ceramic aggregate that is structurally strong, durable, environmentally inert, low density, and with high insulation properties. It is a natural, nontoxic, absorptive aggregate that is dimensionally stable and will not degrade over time. Expanded shale, clay, and slate are collectively termed as *ESCS lightweight aggregates*. The method by Hayde has been perfect-

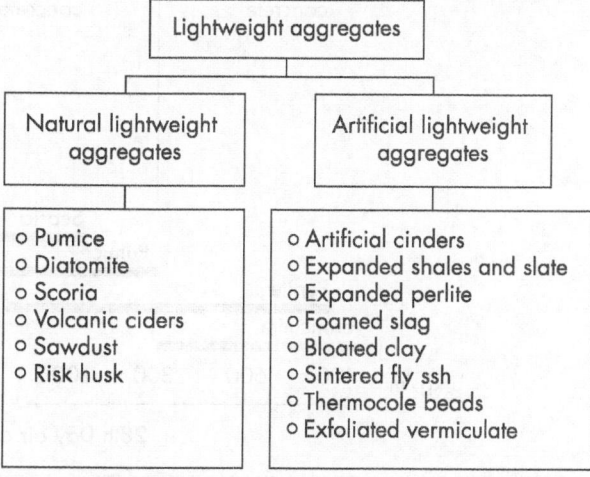

**Fig. 9.3** Natural and artificial lightweight aggregates

ed by The Expanded Shale, Clay and Slate Institute (ESCSI), which was founded in 1952. Companies like Lafarge produce varieties of industrial lightweight aggregates; examples include Aglite™, Haydite™, Leca™, Litex™, Lytag™, True Lite™, and Vitrex™ (www.escsi.org).

As per ACI E1–07 and ACI 213R-14, SLWC can be made with lightweight (a) natural aggregates such as pumice, scoria, volcanic cinders, tuff, and diatomite; (b) rotary-kiln expanded sintered clay, slate, shale, perlite, vermiculite or slag; and (c) sintered fly ash or industrial cinders (see Figs. 9.3 and 9.4). Structural lightweight aggregates have densities ranging from 560 to 1120 kg/m³ compared to 1200 to 1760 kg/m³ for normal-weight aggregates (Kosmatka et al., 2016). The density (unit weight) of such structural lightweight concrete will be of the order of 1680 to 1920 kg/m³, compared to the density of normal weight concrete of 2240 to 2400 kg/m³; this reduction in density allows for a concrete weight savings of over 20%. Normal strengths of SLWC will be in the range of 21 to 35 MPa. HSC can also be made with structural lightweight aggregates. SLWC mixtures can be designed to achieve similar strength and other mechanical and durability performance requirements of similar normal-weight concrete. As with normal-weight concrete, entrained air in SLWC (of about 5 to 8%) provides resistance to freezing and thawing. It also improves workability, reduces bleeding and segregation, and may compensate for minor grading deficiencies in the aggregate (Kosmatka et al., 2016).

The use of SLWC allows us to reduce the dead weight of concrete elements, thus resulting in overall economy. In most cases, the slightly higher cost of SLWC is offset by reductions in volume of concrete and steel used. Seismic performance is also improved because the lateral and horizontal forces acting on a structure during an earthquake are directly proportional to the inertia or the mass of a structure. Other advantages of SLWC include higher fire rating, and improved insulation properties of walls due to higher R-values. SLWC is more fire resistant than normal-weight concrete because of its lower thermal conductivity, lower coefficient of thermal expansion, and the inherent fire stability of aggregates, which are obtained by heating to about 1090°C (Wolfe, 2008). In countries where there is a shortage of natural dense graded aggregates, use of locally produced lightweight aggregates may offer an economical solution. Owing to these advantages, SLWC has been used in a variety of applications in the past 80 years. The reduced strength of SLWC is considered in the design by ACI code by the factor λ. Prestressed lightweight concrete has also been widely used for more than 50 years in North America, similar to prestressed normal-weight concrete. Summary of the properties of prestressed lightweight concrete may be found in ACI 213R-14. Fatigue properties of lightweight concrete were found to be similar to the fatigue properties of normal-weight concrete.

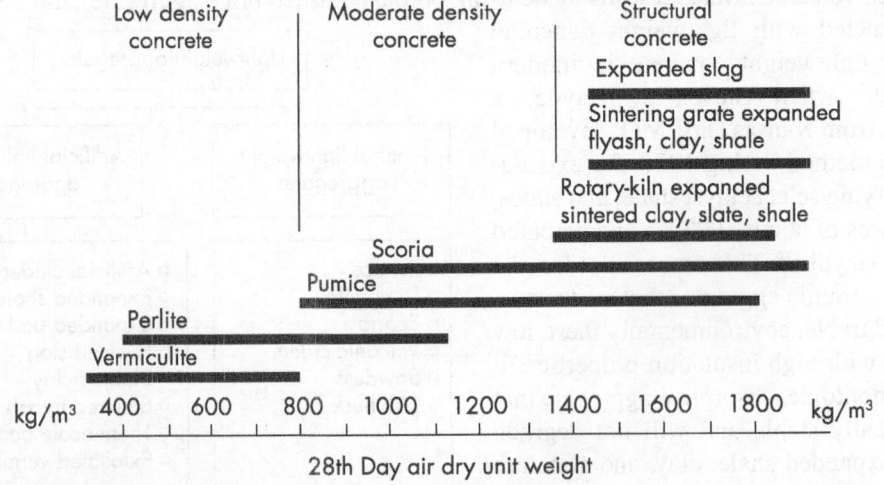

**Fig. 9.4** Range of lightweight concrete (Based on ACI 213R-87, reprinted with permission from ACI)

Note that lightweight aggregates usually absorb water when placed in a concrete mixture, and the resulting rate of absorption is important in proportioning lightweight concrete (They absorb water from 5 to 20% by mass of dry aggregate, depending on the aggregate pore system.). It has to be noted that the internally absorbed water in the lightweight aggregate will not be available to the cement during mixing, and hence should not be counted as mixing water. Usually, the dry lightweight aggregates are pre-wetted before mixing to fully saturate them. The technique of thermal and vacuum saturation could be adopted to efficiently saturate lightweight aggregates and overcome the difficulty of lightweight aggregates absorbing mix water during batching (Neville, 2012, ACI 211.2:98, and ACI 213R-14). As per ACI 211.2:98, the compressive strength of SLWC is related to the cement content at a given slump and air content, rather than the w/cm ratio. It is mainly due the difficulty of determining the amount of mixing water absorbed into the aggregates, and not available for reaction with cement. Some guidelines on the relationship between cement content and compressive strength is provided by ACI 211.2:98. The E-value is not directly related to strength and density, and ranges between 0.5 to 0.75 of the E-value of NWC. A useful empirical relationship to determine the E-value is given below (Newman and Choo, 2003)

$$E = D^2 \sqrt{f_{ck}} \times 10^{-6} \, \text{kN/mm}^2 \tag{9.1}$$

where $D$, $E$, and $f_{ck}$ are nominal density, Young's modulus, and cube strength of concrete, respectively.

Grading requirements for lightweight aggregates deviate from those of normal-weight aggregates. As per ASTM C 33, a larger mass of the lightweight aggregates have to pass through the finer sieve sizes. This modification in grading is necessary due to the increase in density with decreasing particle size of lightweight expanded aggregates. It also yields the same volumetric distribution of aggregates retained on a series of sieves for both lightweight and normal-weight aggregates. Air entrainment in lightweight concrete, as in normal-weight concrete, in the range of 4.5 to 7.5% by volume for 19 mm aggregates, is required for resistance to freezing and thawing. Admixtures are used to entrain air, reduce mixing water requirements, and modify the setting time or other properties of the concrete. As air entrainment also improves workability, it is adopted even when the concrete is not subjected to freezing and thawing. As with normal-weight concrete, water-reducing plasticizing and mineral admixtures are frequently used with lightweight concrete mixtures to increase workability and facilitate placing and finishing. Lightweight concrete, with compressive strengths in the range of 48 to 69 MPa, have been achieved by incorporating various pozzolans (fly ash, silica fume, metakaolin, calcined clays, and shale) combined with mid-range or high-range water-reducing admixtures, or both (ACI 213R-14). The placing, handling, and compaction of lightweight concrete are similar to those used in normal-weight concrete. Care should be exercised while finishing SLWC. More details about the mix design, production techniques, properties, etc., may be found in Newman and Choo (2003), Clarke (1993), and Chandra and Berntsson (2002).

SLWC has contributed to sustainable development by conserving energy, lowering transportation requirements, lowering dead weight, better fire resistance resulting in reduced concrete thickness, reduced structural members such as columns and beams, less reinforcement and concrete required in foundations, and increasing service life of concrete structures. In addition, ESCS aggregates may also help to reduce heat island effects. Use of ESCS aggregates could earn 33 or more points in the LEED-NC Rating system: Sustainable Sites (7), Water Efficiency (6), Energy & Atmosphere (1+), Materials & Resources (9), Indoor Environmental Quality (1), Innovation & Design (up to 5), and Regional priority (up to 4). More details may be found in Chapter 13 of the ESCSI reference manual (ESCSI, 2007).

SLWC has many and varied applications, including multi-storey building frames and floors, curtain walls, shell roofs, folded plates, bridges, prestressed or precast elements of all types, marine structures, offshore platforms, and also in ships. SLWC has a proven performance in marine and bridge construction for more than 80 years. Some of examples of bridges include the Martinez-Benicia Bridge in

California built in 1962, the Silver Creek Overpass Bridge, Utah, USA, built in 1968, and Boknasundet Bridge, Rogaland County, Norway, built in 1990.

High-performance lightweight concrete has been used in the precast prestressed members of the 1994 Wabash River Bridge, Lafayette, Indiana, The 1991 North Pier Apartment Tower, Chicago, The 1994 Bank of America building, Charlotte, N.C., and in numerous bridges (e.g., Chesapeake Bay Bridges, Annapolis, MD) and offshore concrete structures (e.g., Heidron floating concrete platform, North Sea, Norway). Details of these and also more structures built using SLWC may be found in ACI 213R-14, Holm and Bremner (1990, 2000), and ESCSI (2007).

## 9.6.1 Autoclaved Aerated Concrete

The aerated concrete is a one type of lightweight concrete. It can be divided into two main types according to the method of production (Neville and Brooks, 2010): (a) autoclaved aerated concrete (AAC) and (b) foamed concrete (see Fig. 9.5). Foamed concrete is discussed in Section 9.6.2. AAC, also known as *autoclaved cellular concrete* (ACC), *autoclaved lightweight concrete* (ALC), autoclaved concrete, cellular concrete, porous concrete, and with commercial names such as Aircrete, Hebel Block, Siporex, e-crete, AERCON, and Ytong, is a lightweight, precast, concrete building material. AAC was invented in the mid-1920s by Swedish architect and inventor Dr Johan Axel Eriksson, working with Prof. Henrik Kreüger at the Royal Institute of Technology, Sweden. Dr Eriksson patented his 'gas concrete', known locally as 'porenbetong', in 1924. It went into production in Sweden in 1929, with the name Y-Tong, and quickly became very popular. During 1934, another company started to manufacture AAC blocks under the brand name Siporit (renamed in 1937 as the well-known Siporex). Siporex also introduced AAC reinforced elements during 1935, such as roof/floor panels and lintels. The German Hebel and the Dutch Durox were the other two companies that provided AAC factories and plant. Whereas the tilt-cake system of manufacturing is used by Ytong to produce AAC blocks, Durox, Siporex, and later Hebel used the flat-cake system of manufacturing for producing both AAC blocks and reinforced elements. Cellular concrete blocks were first produced in India by the Hindustan Prefab Limited (HPL), New Delhi, during 1960s. It was called as *Vayutan* and had a size of 500 × 250 × 100 mm. In addition to HPL, it was manufactured by the Tamil Nadu Housing Board (TNHB), under the joint Indo-Polish collaboration at its plant at Ennore near Chennai (called *Celcrete*; it had a size of 600 × 200 × 125 or 200 mm) and the Siporex (India) Ltd., Pune. It is estimated that the World now has more than 3000 AAC production facilities with a capacity to produce 450 million m³ of non-reinforced blocks per year. Also, AAC accounts for more than 40% of all construction in the United Kingdom and 60% of new constructions in Germany. In India, the demand for AAC blocks has grown by 10 fold in the past few years.

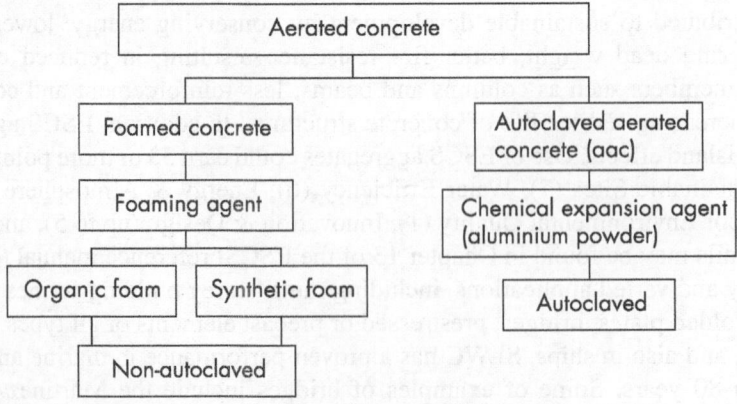

**Fig. 9.5** Classification of aerated lightweight concrete

AAC products include blocks, lightweight wall partitions, load-bearing wall panels, large reinforced panels for floors and roofs, cladding (façade) panels, and lintels. Large blocks are also available with hand holds, which can be quickly and easily placed with thin bed mortar. Blocks and panels are available in a verity of block sizes and shapes, with tongue and groove joints for easy and better connection. Unlike conventional concrete blocks, AAC blocks are not hollow, but solid, light, and porous. AAC blocks are used for both load-bearing and non-load bearing internal walls, partition and panel walls, inner leaf of cavity walls or as backing to brick masonry. When used for external load-bearing walls as well as panel walls in steel or RC framed construction, they should be protected from weather by waterproof cladding or stucco, as they are more porous.

AAC is a precast product manufactured by mixing gas-forming chemicals with lime or cement mortar made using silica (either in the form of quartz/silica sand, or pulverized fly ash), calcined gypsum, cement or lime, and water. The sand and other siliceous material are generally ground in ball-mills to their required degree of fineness, which is about the fineness of ordinary Portland cement [IS 2185(Part 3):1984]. They are mixed during the liquid or plastic stage, resulting in a mass of increased volume and when the gas escapes, leaves a porous structure (in the same way as $CO_2$ is formed and used in aerating bread and baked products). The gas-forming chemicals such as aluminium powder, hydrogen peroxide/bleaching powder and calcium carbide, which liberate hydrogen, oxygen, and acetylene are used (Narayanan and Ramamurthy, 2000). Among these, aluminium powder is the most commonly used aerating agent. Aluminium powder should be finer than about 50 μm (Aircrete Europe uses average grain size of 20–45 μm) and added at a rate of 0.05 to 0.08% by volume. Because aluminium powder is highly flammable, it needs to be added to the aerated concrete slurry through a special 'aluminium dosing system'. The chemical reactions that take place are as follows (www.aircrete-europe.com):

$$CaO + H_2O = Ca(OH)_2 + 65.2 \text{ KJ/mol}$$

$$3Ca(OH)_2 + 2Al + 6H_2O = 3CaO \cdot Al_2O_3 \cdot 6H_2O + 3H_2 \uparrow$$

$$6SiO_2 + 5Ca(OH)_2 = 5CaO \cdot 6SiO_2 \cdot 5H_2O$$

The efficiency of this process is influenced by fineness of the aluminium powder used, purity and alkalinity of cement, and also the procedures taken to prevent the escape of gas before hardening of mortar. It has to be noted that no coarse aggregates are used in AAC.

With respect to structurally reinforced AAC products such as lintels or roof panels, steel rebar or mesh is also placed in the mould. When AAC is mixed into slurry and poured into greased moulds, several chemical reactions take place that give AAC its lightweight (20% of the weight of concrete) and thermal properties. Aluminium powder reacts with calcium hydroxide and water to form numerous hydrogen bubbles. The hydrogen bubbles cause the concrete to expand to roughly two to five times its original volume. The hydrogen subsequently evaporates, creating micro-pores, with diameters ranging from 0.5 to 1.5 mm, to form. When the forms are removed from the material, it is solid but still soft. It is then cut into either blocks or panels, and placed in an autoclave chamber for 12 h. When the temperature reaches 190°C and the pressure reaches 1.2 N/mm², quartz sand reacts with calcium hydroxide to form calcium silicate hydrate. After the autoclaving process, the material is ready for immediate use. Depending on its density, up to 80% of the volume of an AAC block is air. AAC blocks are available in lengths of 400, 500, and 600 mm (half-length blocks may also be available), heights of 200, 250, or 300 mm, and widths of 100, 150, 200, or 250 mm. They are stacked one over the other using thin-set mortar, as opposed to the traditional concrete masonry unit (CMU) construction.

AAC blocks can be produced with a variety of surface textures ranging from a very fine texture to a coarse open texture by selecting, grading, and proportioning aggregates during manufacture. In addition, when the blocks are still green, different textures can be produced by using wire brushes on their faces. It is also possible to get some textures by slightly eroding the surfaces using a fine spray of water. By using non-fading mineral pigments in the mix, it is possible to get coloured blocks. Colour can also be achieved, after the blocks are removed from the moulds, by applying coloured cement grout or paint to the surface of the units.

AAC's low-density accounts for its low structural compressive strength (about 50% of normal concrete). AAC has thermal properties that allow a building to consume less energy and, therefore, increase its energy efficiency. The AAC blocks are classified into two grades, as per IS 2185 (Part 3):1984, according to their compressive strengths, as shown in Table 9.5.

**Table 9.5** Physical properties of AAC blocks [IS 2185 (Part 3):1984; Newman and Choo, 2003]

| Density in oven-dry condition, kg/m³ | Compressive strength (min.), MPa | | E-value, GPa | Thermal conductivity in air dry condition, W/(m.K) |
|---|---|---|---|---|
| | Grade 1 | Grade 2 | | |
| 451–550 | 2.0 | 1.5 | 1.6 | 0.21 |
| 551–650 | 4.0 | 3.0 | 2.4 | 0.24 |
| 651–750 | 5.0 | 4.0 | 2.5 | 0.30 |
| 751–850 | 6.0 | 5.0 | 2.7 | 0.37 |
| 851–1000 | 7.0 | 6.0 | 3.0 | 0.42 |

Specifications for AAC blocks are provided in IS 2185 (Part 3):1984 and IS 6598:1972 provides specification for cellular concrete for thermal insulation. Guidance for low-density cellular concrete is available in ACI 523.1R-06 and for high density (800 to 1920 kg/m³) cellular concrete in ACI 23.3 R-93. ACI 523.4R-09 provides guidance for the design and construction of AAC panels.

Compared to conventional concrete, AAC has the following advantages (also see www.aerconaac.com):

***Economy*** As explained already, the weight of AAC is considerably less than conventional concrete. Concrete blocks that are made out of ACC weigh about one-fifth of typical concrete. This leads to sufficient economy, particularly in foundations and basements, and may be in the order of 10%. Lightweight also saves cost and energy in transportation, labour expenses, and increases chances of survival during earthquakes.

***Environment-friendly*** Environmental friendly and energy-conserving AAC consumes approximately 50 and 20% less energy than that needed to produce concrete and CMUs, respectively, and may use waste materials such as copper mine tailings and fly ash in its production. Despite the energy-intensive autoclaving process, manufacturers say it takes about 50% less energy to make, because of the lower Portland cement content by volume. Absolutely no pollutants or hazardous wastes are generated in the manufacturing process and there is no waste of precious raw materials. Even the production method conserves energy since steam curing is carried out at high pressures and thermal energy is recovered and reused for maximum efficiency. Production trimmings can also be fully recycled. Additionally, the final product is completely recyclable and able to store and release thermal energy over time; hence, it qualifies as a green and sustainable material.

***Transportation*** Since it is lighter, it cuts down on transportation costs and fuel use.

*Thermal insulation* These properties of AAC provide solid insulation without thermal bridging or cold spots. As a result, buildings using AAC tend to be cooler in the summer and warmer in the winter and often lead to lower utility bills because of the insulation benefits. When it is used, there is no need for any supplementary insulation.

*Acoustic insulation* AAC has superior acoustic properties as compared to other dense-weight concrete.

*Fire safety* It is non-combustible and does not emit any toxins or gases when exposed to fire. AAC is a virtually fire-resistant masonry material that can withstand a 1093 F fire for 4 h.

*Aesthetics* Owing to the possibility of producing the blocks with a variety of surface textures and colours, it has been used extensively by architects.

*Speed of construction* Owing to its lightweight and large size of blocks and panels (as compared to bricks), it can be laid at high speeds. In addition, it is easy to cut them to desired sizes, nail or drill holes for electric wiring, sockets, and pipes.

*Quality and accuracy* Because AAC is produced in factories with a higher level of quality control than that associated with concrete poured in the field, it has uniform properties. The panels and blocks can be produced to the exact required sizes in the factory. Hence, there is less need for onsite trimming. Since the blocks and panels fit so well together, there is a reduced use of finishing materials such as mortar.

*Provides ventilation* This material is very airy and allows for the diffusion of water. This will reduce the humidity within the building. ACC will absorb moisture and release humidity, which helps to prevent condensation and other problems that are related to mildew.

*Long lasting* The life of this material is extended because it is not affected by harsh climates or extreme changes in weather conditions. It will also not degrade under normal climate changes. It does not attract rodents or other pests, nor can it be damaged by them.

The following are the disadvantages of AAC blocks:

*Reduced strength* Compressive and flexural strengths reduce with its density.

*Fixings* The cellular nature of the blocks requires specialized fasteners for the attachment of both structural framing members and non-structural fittings.

*Require special drilling bits* All drilled holes require the use of high-speed twist drills suitable for wood or wood; masonry drill bits and hammers cannot be used.

*Brittle nature* They need to be handled more carefully than clay bricks to avoid breakages.

*Thicker walls* Insulation requirements in newer building codes of northern European countries would require very thick walls when using AAC alone.

IS 6041:1985 suggests that the walls using AAC blocks should be constructed with a mortar, which has a strength that is relatively lower than that of the mix used for making blocks in order to avoid the formation of cracks. A 1:2:9 cement-lime-sand mortar may be used; however, when either the intensity of load is high or wall is exposed to severe condition, 1:1:6 mortar is suggested. If good quality lime is not available, 1:6 cement-sand mortar may be used. ACC blocks should not be used in foundations and for masonry below damp-proof course. The design and construction of cellular concrete masonry walls should conform to the requirements of IS 1905:1987. More details of construction using AAC blocks are provided in IS 6041:1985 and the various tests conducted on AAC blocks for determining their properties are explained in IS 6441(Parts 1–9):1972.

## 9.6.2 Lightweight Foamed Concrete

Lightweight foamed concrete (LFC), also referred to as *lightweight cellular concrete* (LCC) or non-autoclaved aerated concrete (NAAC), has structure similar to that of AAC, but does not involve expensive autoclaving process. Foamed concrete is produced by adding a special air-entraining admixture (called foaming agent) or injecting preformed stable foam (such as those used in fire-fighting) to the slurry of Portland cement, fly ash, water, and sand. In both methods, the foam concentrate should be of such chemical composition that is capable of producing stable foam cells in concrete, which can resist the physical and chemical forces imposed during mixing, transporting, pumping, placing, and setting of concrete. The foaming agent should meet the requirements of IS 9103:1999 and the foam produced shall be stable for duration beyond the final setting time of Portland cement. Such foaming agents should be completely harmless to concrete and embedded steel reinforcement and be nontoxic, non-flammable, and biodegradable. The various foaming agents used are detergents, resin soap, glue resins, saponin, hydrolyzed proteins such as keratin, with the addition of foam stabilizers, metal salts, highly surface-active fluorotensides and compensating agents [IS 2185(Part 4):2008; Narayanan and Ramamurthy, 2003]. The properties of foamed concrete are mainly dependent upon the quality of the foam. The resulting bubbles in the hardened concrete should be discrete and the usual bubble size is between 0.1 and 1 mm (Newman and Choo, 2003).

Owing to the absence of coarse aggregate and the ball bearing effect of minute foam bubbles, the fluid mass of foamed concrete fills up and levels into the moulds by itself without requiring any external vibration or compaction. Depending on the ambient temperature and quality of cement used, the building blocks may be de-moulded after 24 h from pouring of foamed concrete unless accelerated curing processes are used, and shifted to the curing/stacking yard. After curing as per the requirements of IS 456, the blocks should be allowed to dry under shade for a period of 2 to 3 weeks, so as to complete their initial shrinkage before being used in the work. Steam curing of the blocks, after 5 h of casing at temperatures of 70°C, is also possible. Similar to the AAC blocks, the foamed concrete blocks may be given a variety of surface finishes on the exposed face by casting them against textured surface plates. Colour may be introduced by incorporating non-fading mineral pigments in the facing concrete or applying a coloured Portland cement grout or paint to the face of the units soon after they are removed from the moulds.

### Typical Types

The foamed concrete is generally classified according to its density. Some typical types are given here (www.litebuilt.com):

*Foamed concrete with density 300–600 kg/m³*  This is made with cement and foam only. It is flowable and pumpable. It is used in roof and floor as insulation against heat and sound and applied on rigid floors. It is also used as insulation in hollow blocks and any other filling situation where high insulating properties are required.

*Foamed concrete with density 600–900 kg/m³*  It is made with sand, cement, and foam. It is used for the manufacture of precast blocks and panels for curtain and partition walls, slabs for false ceilings, thermal insulation, and soundproofing screeds in multi-storey residential and commercial buildings. This density range is also ideal for bulk-fill applications.

*Foamed concrete with density 900–1200 kg/m³*  It is made with sand, cement, and foam. It is used in concrete blocks and panels for outer leaves of buildings, architectural ornamentation as well as partition walls, concrete slabs for roofing and floor screeds.

*Foamed concrete with density 1200–1600 kg/m³* It is made with sand, cement, and foam. It is used in precast panels of any dimension for commercial and industrial use, *in-situ* casting of walls, garden ornaments, and other uses where structural concrete of lightweight is recommended.

Specifications for preformed foam cellular concrete blocks are provided in IS 2185(Part 4):2008. Physical properties of preformed foam cellular concrete blocks are provided in Table 9.6. Drying shrinkage should be a maximum of 0.05% for the load bearing class of blocks and a maximum of 0.08% for the non-load-bearing class of blocks.

**Table 9.6** Physical properties of foamed concrete [IS 2185(Part 4): 2008; Newman and Choo, 2003]

| Bulk density in oven-dry condition, kg/m³ | Compressive strength, MPa | | E-Value, GPa | Water absorption (oven dry density) | Thermal conductivity in air dry condition, W/(m.K) |
| --- | --- | --- | --- | --- | --- |
| | Average, Min. | Individual, Min | | | |
| 800 | 2.5 | 2.0 | 2.0–2.5 | 12.5 | 0.372 |
| 1000 | 3.5 | 2.8 | 2.5–3.0 | 12.5 | 0.418 |
| 1200 | 6.5 | 5.2 | 3.5–4.0 | 10.0 | 0.442 |
| 1400 | 12.0 | 9.0 | 5.0–6.0 | 10.0 | 0.523 |
| 1600 | 17.0 | 14.5 | 10.0–12.0 | 7.5 | 0.581 |

Preformed foam cellular concrete blocks are available in lengths of 400, 500, and 600 mm (half-length blocks may also be available), heights of 250 or 300 mm, and widths of 100, 150, 200, or 250 mm. The blocks in the density range of 600 to 900 kg/m³ are used in non-load-bearing walls and those in density range of 1200 to 1600 kg/m³ are used in load-bearing walls.

Foamed concrete blocks have the same advantages of AAC blocks such as economy, durability, providing thermal and acoustic insulation, fire safety, aesthetics, resulting in speedy construction, high quality because of factory production, reduction in fuel and transportation costs due to reduced weight, and high environmental friendliness. In addition, as they do not require the autoclaving process of AAC blocks, their energy requirements are also reduced. The disadvantages are also similar to the AAC blocks. More information on foamed concrete and its applications may be found in the works of Wee et al. (2006), Nambiar and Ramamurthy (2007), Hamad (2014), and www.litebuilt.com.

## 9.7 Fibre Reinforced Concrete

Fibres are added to concrete to control cracking due to plastic shrinkage and drying shrinkage. The addition of small closely spaced and uniformly dispersed fibres will act as crack arresters and enhance the tensile, fatigue, impact, or abrasion resistance of concrete. They also reduce the permeability of concrete. Though the flexural strength may increase marginally, fibres cannot totally replace flexural steel reinforcement (the concept of using fibres as reinforcement is not new; fibres have been used as reinforcement since ancient times, for example horsehair in mortar, and asbestos fibres in concrete).

Clause 5.7 (Amendment No. 3) of IS 456:2000 permits the use fibres in concrete for special applications to enhance its properties. Steel, glass, polypropylene, carbon, and basalt fibres have been used successfully; steel fibres are the most common (see Fig. 9.6). Steel fibres are manufactured by at least three processes: (a) cutting cold drawn wire, (b) slitting steel sheet, and (c) extracting them from a pool of molten steel. Steel fibres may be crimped, hooked or flat. Wire fibres with bent or deformed ends have a higher pull-out resistance than straight fibres and may be used in smaller quantities to achieve similar properties. The ultimate tensile strength of fibres varies from 345 to 2070 MPa. This type of concrete, using the various types of fibres in the mix, is known as *fibre reinforced concrete* (FRC).

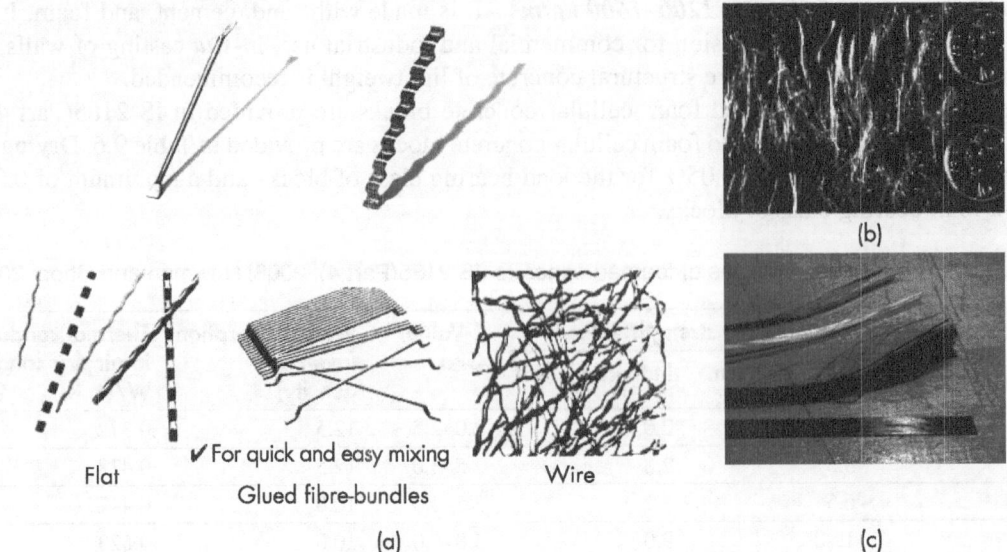

Flat

✔ For quick and easy mixing

Glued fibre-bundles

Wire

(a)

(b)

(c)

**Fig. 9.6** Fibres used in concrete (a) Different types and shapes of steel fibres (b) Fine fibrillated polypropylene fibres (c) Glass fibres (*Courtesy*: Dr V.S. Parameswaran)

The amount of fibres added to a concrete mix is expressed as a percentage of the total volume of the composite (concrete and fibres), and termed volume fraction. Volume fraction (denoted by $V_f$) typically ranges from 0.25 to 2.5% (0.75 to 1.0% is the most common). The aspect ratio of a fibre is the ratio of its length to its diameter. Typical aspect ratio ranges from 30 to 150. The fibre sizes range from $13 \times 0.25$ mm to $64 \times 0.75$ mm. Increasing the aspect ratio of the fibre usually increases the flexural strength and toughness of the matrix. However, fibres that are too long tend to 'ball' in the mix and create workability problems (Subramanian, 1976a). To obtain adequate workability, it is necessary to use super-plasticizers.

Another interesting and useful development in FRC construction has been to provide non-metallic fibres in small, cylindrical bundles, approximately 50 mm in length and 55 mm in diameter, wrapped in a water-soluble compound. This permits the easy addition of the fibres, by hand, into the mixing drum of a truck mixer, either during charging or at the job site. The wrapper disintegrates, allowing the fibres to disperse into the concrete mixture with little balling or segregation, thus improving quality control. Successful field application of this method is described in Ramakrishnan and Kakodkar (1995). More information on FRC may be had from Parameswaran and Balasubramanian (1993), Bentur and Mindess (2007), and ACI 544.1R-96.

### 9.7.1 Ultra-high Performance Concrete

*Ultra-high performance concrete* (UHPC) is a high-strength, high-stiffness, self-consolidating, and ductile material formulated by combining Portland cement, silica fume, quartz flour, fine silica sand, high-range water reducer, water, and steel or organic fibres. Originally, it was developed by the Laboratoire Central des Pontset Chaussées (LCPC), France, containing a mixture of short and long metal fibres, and known as Multi-scale FRC (Rossi, 2001). Typical composition of UPHC is shown in Table 9.7 (note that there are no course aggregates). Note that a low w/cm ratio of about 0.2 is used in UHPC compared to about 0.4 to 0.5 in NSC. The material provides compressive strengths of 120 to 240 MPa, flexural strengths of 15 to 50 MPa, post-cracking tensile strength of 7.0 to 10.3 MPa, and has a modulus of elasticity from 45 to 59 GPa [*Ductal®* (Lafarge, France), CoreTUFF® (US Army Corps of Engineers),

BSI®, Densit® (Denmark), Ceracem® (France and Switzerland), are some of the examples of commercial products]. The enhanced strength and durability properties of UHPC are mainly due to optimized particle gradation that produces a very tightly packed mix, use of steel fibres, and extremely low water-to-powder ratio.

Some of the potential applications of UHPC are in prestressed girders and pre-cast deck panels in bridges, columns, piles, claddings, overlays, and noise barriers in highways. The 60 m span Sherbrooke pe-destrian Bridge, constructed in 1997 at Quebec, Canada, is the World's first UHPC bridge, without any bar reinforcement (see Fig. 9.7). This bridge uses a kind of UHPC containing a maximum of 2.5% metal fibres, 13 mm long and 0.16 mm in diameter, developed originally in France during 1990s by Bouygues in cooperation with other industrial partners like Lafarge and marketed as *Reactive Powder Concrete* (RPC). It consists of three-dimensional open web prestressed space truss, whose diagonals are made with RPC 200 and confined in thin-walled stainless steel tubes. The walkway deck, which also serves as the top chord of the truss, is only 30 mm thick. The RPC 200 used in this bridge consists of the following: cement 710 kg/m³, silica fume 230 kg/m³, ground quartz 210 kg/m³, silica sand 1010 kg/m³, super-plastisizer 19 L/m³, steel fibres 190 kg/m³, and water 200 L/m³ with a water-powder ratio of 0.21. The RPC made from a mix of small particles (with maximum particle size of 0.6 mm compared to a NSC with 20 mm maximum course aggregates) provided a dense mixture, minimized void spaces in the concrete and greatly enhanced durability. Its properties were further enhanced by heat treatment under pressure to a temperature of 90°C (More details of this bridge may be had from Blais and Couture, 1999; Subramanian, 1999). The 15 m span Shepherds Creek Road Bridge, NSW, Australia, built in 2005, is the world's first UHPC Bridge for nor-mal highway traffic. Since then, a number of bridges, and other structures have been built utilizing UHPC all over the world (FHWA Report FHWA-HRT-13–060, 2013 provides details of many such structures.). Other potential applications of UHPC include sewer pipes, precast spun columns and poles, barrier walls, field-cast thin-bonded overlays, cable-stayed bridge superstructure, bridge bearings, precast tunnel seg-ments, and seismic retrofit of bridge columns (FHWA-HRT-13–060, 2013).

The materials for UHPC are usually supplied by the manufacturers in a three-component premix: powders (Portland cement, silica fume, quartz flour, and fine silica sand) pre-blended in bulk-bags; super-plasticizers; and organic fibres. Care should be exercised during mixing, placing, and curing. The ductile nature of this material makes concrete to deform and support flexural and tensile loads, even after initial cracking. The superior durability characteristics are due to the combination of fine powders with a small grain size (maximum 600 µm) and chemical reactivity. The net effect is maximum compactness and a

**Table 9.7** Typical composition of UHPC (Extracted from Graybeal and Davis, 2008)

| Material | Amount kg/m³ | % by weight |
|---|---|---|
| Portland cement | 711 | 28.3 |
| Silica fume | 232 | 9.2 |
| Ground Quartz | 210 | 8.3 |
| Fine sand | 1019 | 40.5 |
| Steel fibres | 155 | 6.2 |
| Super-plasticizer (High range water reducing admixture) | 30 | 1.2 |
| Accelerator | 26 | 1.0 |
| Water | 134 | 5.3 |

**Fig. 9.7** The Sherbrooke Footbridge in Sherbrooke, Quebec, Canada, across the Magog River, with a precast truss is the world's first bridge made of reactive pow-der concrete (*Source*: www.fhwa.dot.gov)

disconnected pore structure. The use of this material for construction is simplified by the elimination of reinforcing steel and its ability to be virtually self-placing. More details about UHPC may be found in Schmidt et al. (2004), Fehling et al. (2008), and Schmidt et al. (2012). A comparison of stress-strain curves concrete is provided in Fig. 9.8 and, another comparison, based on strength, is provided in Table 9.8, which was suggested by Prof. J. Francis Young of the University of Illinois at Champaign-Urbana.

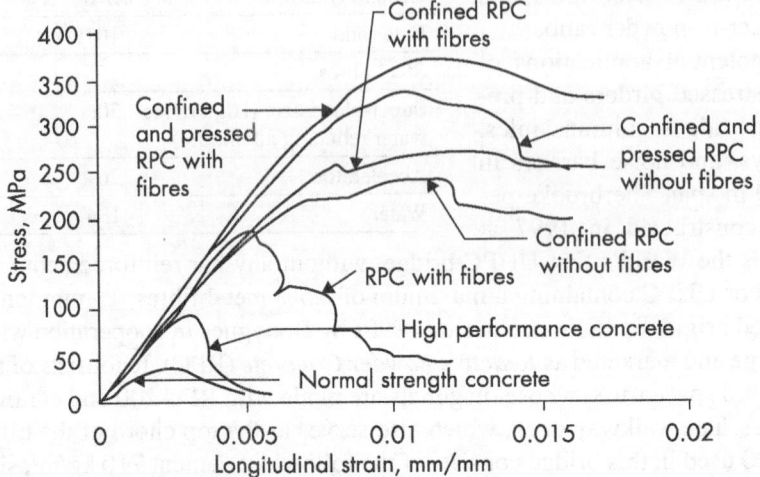

**Fig. 9.8** Comparison of stress-strain curves of NSC, HPC, and reactive powder concrete
(*Source*: Blais and Couture, 1999)

**Table 9.8** Comparison of concrete based on strength (*Source*: Farny and Panarese, 1994;
reprinted with permission from Portland Cement Association)

| Parameter | Conventional concrete | High-strength concrete | Very-high strength concrete | Ultra-high strength concrete |
|---|---|---|---|---|
| Strength, MPa | <60 | 60–100 | 100–150 | >150 |
| Water-cement ratio | >0.45 | 0.45–0.30 | 0.30–0.25 | ~0.25 |
| Chemical admixtures | Not-necessary | WRA/HRWR[1] | HRWR | HRWR |
| Mineral admixtures | Not-necessary | Fly ash | Silica fume[2] | Silica fume[2] |
| Permeability coefficient | $>10^{-10}$ | $10^{-11}$ | $10^{-12}$ | $10^{-13}$ |
| Freeze-thaw protection | Needs air entrainment | Needs air entrainment | Needs air entrainment | No freezable water |

**Notes:**
[1]WRA = Water reducing admixture, HRWR- High range water reducer
[2]May also contain fly ash

**Sustainability of UHPC**   The UHPC is a very durable product; hence, the structures that use it will have a longer service life and require less maintenance than structures built with conventional concrete. In addition, UHPC will have greater frost and deicing salt resistance, a lower rate of carbonation, and greater chloride resistance than conventional concrete. Although the UHPC may have higher initial costs, life cycle cost analysis conducted by researchers in Germany, on two replacement methods for the Eder Bridge in Felsberg, found that the life cycle cost over 100 years would be less for the UHPC bridge (FHWA-HRT-13–060, 2013). Another study in Germany concluded that the environmental impact of structures made with state-of-the-art UHPC may be up to 2.5 times greater than with conventional concrete. However, the environmental impact is decreased by reducing the amount of Portland cement, steel fibres, and high-range water-reducing admixtures in the UHPC.

## 9.8 Polymer Concrete

*Polymer concrete* (PC) was introduced in the late 1950s and became well known in the 1970s for its use in repair, thin overlays, in floors, and precast components. Polymer concrete is obtained by impregnating ordinary concrete with a monomer material and then polymerizing it by radiation, or by heat and catalytic ingredients, or by a combination of these two techniques. Depending on the process by which the polymeric materials are incorporated, they are classified as (a) *polymer impregnated concrete* (PIC), (b) *polymer concrete* (PC), (c) *polymer-Portland-cement concrete* (PPCC) now called *polymer-modified concrete* (PMC). Polymers essentially occupy the micro-pores, which are present in ordinary concrete; hence, the porosity of concrete is considerably reduced. In addition, the polymerization also enhances strength and other properties. Such concrete have also been used in repairing damaged concrete structural members (Subramanian and GnanaSambanthan, 1979). However, it has to be noted that polymer concrete is more expansive than ordinary concrete. Cost studies show that using unit material cost/strength as an index, polymer concrete is superior to conventional concrete, plastic, steel, and aluminium in respect to compressive strength. In cost comparisons dealing with tension, polymer concrete is superior to conventional concrete and plastic, but does not equal steel or aluminium.

PIC is a hydrated PCC that has been impregnated with a monomer that is subsequently polymerized in situ. For the monomer to fill the voids, free water in the pores of concrete should be removed by drying the concrete by some means. Vacuum drying followed by elevated temperature drying could be used to dry the concrete to a temperature of 150°C at a rate that should be less than 35°C/h. The monomer is introduced into the concrete by soaking at atmospheric pressure or above. When 85% of the void space is filled with monomer, it is considered as full impregnation. Once the desired amount of monomer is impregnated, the monomer has to be converted into a polymer. The two most common methods used for polymerization are called thermal-catalytic (best for full impregnation) and promoted-catalytic. Another method, involving radiation, is not used commonly due its cost, requirement of biological shielding, and the low polymerization rates. In practice, impregnation is usually done using vinyl monomers that contain a polymerization initiator that can be activated by heat (examples include methyl methacrylate, acrylonitrile, and other acrylic monomers, styrene, and vinyl acetate). In addition to these monomers, various co-monomers in the form of plasticizers and extenders [e.g., vinyl stearate or butyl acrylate (BA)], cross-linking agents [e.g., trimethylolpropane trimethacrylate], initiators (e.g., isobutyronitrile), promoters (e.g., cobalt napthenate), and silane coupling agents are also used to improve the properties of PIC. Increases in strength in compression, tension, and flexure up to four to five times those of unimpregnated concrete strengths could be achieved. Increases in modulus of elasticity (up to 80%), modulus of rupture (up to 250%), freezing/thawing resistance (up to 300%), hardness, or impact (up to 70%) have been reported. In addition, water permeability was virtually eliminated, water absorption decreased up to 95%, and corrosion resistance greatly improved. Examples of fully impregnated concrete elements include precast tunnel lining and support systems, beams, pipes, curbstones, plumbing and electrical fixtures, prestressed piling, fender piling, wall panels, and trench covers.

PIC (sometimes called surface impregnated concrete) is usually accomplished by impregnating conventional PCC to a less-than full depth using a simple soaking technique. The main aim of developing PIC was to protect concrete bridge decks and spillways from damage caused by deicing salts and freeze-thaw deterioration. All the methods used for full impregnation are applicable to this partial impregnation also. The potential applications include treatment of precast concrete elements and existing concrete structures to improve durability, reduce maintenance requirements, and restore deteriorated structures.

PC is a material in which the aggregate is bound together with a polymer instead of cement. The composites do not contain a hydrated cement phase, although Portland cement can be used as an aggregate or filler. It is manufactured in a manner similar to that of cement concrete. Monomers and co-monomers are added to the aggregates and the mixture is thoroughly mixed and the PC mixture is then placed in formwork. The formwork used must be extremely durable, solvent resistant, have a low coefficient of thermal expansion, have smooth surfaces, and preferably should be a good heat conductor. Conventional concrete vibrators may be used for consolidation and entrapped air removal. After the forms are removed, the product is polymerized at room temperature or at elevated temperatures. PC has been used as a patching material suitable for use in the repair of PCC structures, skid-resistant protective overlays and wearing surfaces on concrete, linings in carbon-steel pipes for geothermal applications, swimming pool and patio decking, in sewer pipes, equipment vaults, and drainage channels. Many PC products can be precast at plants more efficiently and economically than cast at site. Most of the procedures used for precast PCC can be applied to making PC. An obvious advantage of PC for precasting is its extremely short hardening time. Depending on monomer used, forms could be removed in about 40 sec after form filling. Such a rapid removal allows for the efficient use of forms and production facilities.

PMC, also called as *latex-modified concrete* (LMC), is Portland cement and aggregate combined at the time of mixing with organic polymers that are dispersed or redispersed in water. As the cement hydrates, coalescence of the polymer occurs, resulting in a comatrix of hydrated cement and polymer film throughout the concrete. The addition of polymers to Portland cement results primarily in improvements in adhesion, resistance to permeation of water, durability, and some strength properties. In general, polymer levels of 10 to 25% by mass of the hydraulic cement are required for optimum performance. W/cm material ratios of PMC are typically 0.30 to 0.40 by mass. Essentially, the mixing and handling of PMC is similar to conventional PCC. The majority of PMC placed today uses a polymer in latex form. *Latex* has been defined as a dispersion of organic polymer particles in water. These latexes include vinyl acetate copolymers (VAC), acrylic esters, and copolymers (PAE and S-A), styrene-butadiene copolymers (S-B), epoxy resins, and vinyl acetatecopolymers in the form of redispersible powders; the use of polyvinyl acetate powders is declining because of their tendency to hydrolyze. Industrial floors in wineries, citrus juice plants, dairies, and chemical plants have been constructed or repaired with PMC because of its resistance to dilute aqueous solutions of salts and acids. Exterior and interior walls are often treated with a polymer-modified, and sometimes a fibre-reinforced, mortar. PMC, often with no coarse aggregate, is used for corrosion-resistant coatings of metallic pipes and beams. PMC is used extensively in the application of ceramic tiles, both in the thin-set bedding mortar and in the grout between the tiles. In addition, LMC has been widely used in the field of patching, resurfacing or overlaying works of damaged bridge decks for the past 40 years, because of its ease of execution, excellent adhesion to the base concrete, high freeze-thaw durability, and resistance to chloride penetration.

While using polymers in concrete construction, careful consideration should be given to the safety of workers, who are handling them. Although the chemical systems used create the most obvious hazards, the use of high-temperature drying and curing equipment and their energy sources also pose potential dangers, and hence they should be tackled properly. More information on different types of polymer concrete can be obtained from ACI 548.1R-09, ACI 548.3R-09, ACI 548.4–11, and ACI 548.5R-16.

## 9.9 Precast Concrete

Unlike concrete that is poured at site (referred to as *in-situ* concrete), *precast concrete* is produced in a controlled factory environment using better manufacturing techniques and hence will have high quality. Reinforcements are placed into reusable and adjustable moulds, and concrete is poured, vibrated, and cured in the factory. Architectural finishes, if specified, may also be applied on components, while they are still in the factory. The finished precast elements are transported to site and placed into position using cranes. Precast concrete may be produced in a vast range of different sizes, shapes, and finishes. Precast products may include walls, floors, columns, beams, stairs, bridges, pipes, culverts, tunnels, and even furniture. Precast products are used in a variety of applications, including buildings, civil construction, and landscaping. Precast concrete offers the following advantages:

1. As precast products are manufactured in factory, all process of making concrete can be controlled properly.
2. Better quality of the product is obtained.
3. As the products are mass produced and moulds are reused, they result in lower costs.
4. Problems due to weather during casting are eliminated.
5. As soon as the product is delivered, it can be installed.
6. The modularity of precast products makes installation quicker.
7. As all the processes, such as mixing, consolidation, and curing, are controlled, and extremely durable concrete is obtained.

## 9.10 Decorative Concrete

Decorative concrete may be obtained by integrally mixing colour pigments with concrete made with white cement rather than the conventional grey cement. It can also be made by using chemical stains, or by exposing colourful aggregates at the surface. Various textured finishes can also be made on the concrete using various tools. The polymer concrete, discussed in Section 9.8, is a good material for decorative concrete, because colour dyes may be incorporated in the monomer. Polymer concrete also has the ability to retain its attractive appearance despite weathering, rendering it promising for decorative concrete work. The decorative concrete may be classified as: (a) stained/coloured concrete, (b) stamped concrete, (c) polished concrete, and (d) exposed aggregates concrete.

More information on decorative or architectural concrete can be obtained from ACI 303R-12, ACI 303.1–97, and www.concretenetwork.com.

## 9.11 Ferrocement

*Ferrocement* also known as *ferrocrete*, invented by Jean Louis Lambot of France, in 1848, is a composite material like RCC. In RCC, reinforcement consists of steel bars placed in the tension zone, whereas ferrocement is a thin RC made of rich cement mortar (cement to sand ratio of 1:3) based matrix reinforced with closely spaced layers of relatively small diameter wire mesh/welded mesh/chicken mesh (The diameter of wires ranges from 4.20 mm to 9.5 mm and spaced up to 300 mm apart.). The mesh may be metallic or synthetic (Naaman, 2000). The mortar matrix should have excellent flow characteristics and high durability. The use of pozzolanic mineral admixtures like fly ash (50% cement replacement with fly ash is recommended) and use of super-plasticizers will not only permit the use of water-binder ratio of 0.40 to 0.45 by mass, but will also enhance the durability of the matrix. A mortar compressive strength of 40 to 50 MPa is recommended.

During the 1940s, Pier Luigi Nervi, an Italian engineer, architect, and contractor, had used ferrocement for the construction of aircraft hangars, boats, and buildings. More information about the design and construction of ferrocement may be had from Naaman (2000) and ACI 549.1R-93.

## 9.12 Geopolymer Concrete

The *geopolymer concrete* can be used as a greener alternative to PCC. The term 'geopolymer' was first introduced by Davidovits in 1978, who proposed that an alkaline liquid could be used to react with the silicon and aluminium in by-product materials such as fly ash, GGBS, silica fume, or RHA to produce binders. Because the chemical reaction that takes place in this case is a polymerization process, he coined the term geopolymer to represent these binders. In this product, water is used only to facilitate workability and is not involved in the chemical reaction and actually expelled during curing and subsequent drying. This is in direct contrast to the hydration reactions that occur in PCC. Owing to this, the mechanical and chemical properties of geopolymer concrete are also affected; hence, the geopolymer concrete is more resistant to heat, water ingress, alkali-aggregate reactivity, and other types of chemical attack (Davidovits, 2008).

Geopolymer concrete mixes are often made using fly ash, GGBS, fine aggregates, coarse aggregates, and alkaline liquids. Geopolymer concrete requires great care in its production. Geopolymer concrete may be made to harden at room temperatures with compressive strengths of 25–35 MPa after 24 h and up to 70–100 MPa after 28 days. It generally offers better protection to embedded steel from corrosion as compared to cement concrete.

## 9.13 Prestressed Concrete

Prestressed concrete (PSC) is a HSC (with 28th day compressive strength not less than 40 MPa), where external loads are applied through prestressing steel reinforcement, which introduces compressive stresses in the concrete. These compressive stresses will counteract the desired degree of tensile stresses caused by the applied loads. High tensile steel wires, or strands (which are obtained by twisting wires together) with ultimate strength in the range of 1000 to 2100 MPa, are used as prestressing steel.

In the *pre-tensioning system*, wires/strands are stretched between bulk heads placed several meters apart on prestressing bed, using hydraulic jacks. HSC is then cast in the moulds around the stressed tendons. After the concrete has attained sufficient strength, the jacking pressure is released, and the tendons are cut off at the ends of the beams called anchorages. In the *post-tensioning system*, steel strands are inserted into a metal or plastic duct that is embedded in the concrete element. The strands are then stressed with hydraulic jacks and anchored at the ends of the concrete element. The duct is then filled with a cementitious grout that provides corrosion protection to the strand and bonds the tendon to the concrete surrounding the duct. Some of the popular, patented post-tensioning systems are (a) Freyssinet, (b) Magnel, and (c) Leonhardt. More details of PSC may be found in IS 1343:2012.

## 9.14 Shotcrete

Shotcrete (called in Europe as sprayed concrete) is basically a construction technique, using which concrete or mortar is deposited through a hose and pneumatically projected at high velocity onto a surface. Shotcrete can be applied by two distinct application techniques, the dry-mix process and the wet-mix process. In the *dry-mix process*, the dry mixture of cementitious material and aggregate is fed to a pneumatically operated gun using compressed air through the delivery hose. The required amount of water is added at the nozzle and jetted on to the surface. In the *wet-mix process*, the ingredients are thoroughly mixed and fed into a concrete pump, which sprays the mixture on to the required surface through the nozzle with the help of compressed air.

## 9.15 Other Types of Concrete

Some of the other types of concrete, which are used in some special applications, are discussed very briefly below:

*Shrinkage compensating concrete*   This concrete contains an expansive cement or a mixture of normal cement and expansive cement (Types K, M, or S). The initial expansion, if properly restrained by reinforcement or by other means, offsets strains caused by drying shrinkage. This concrete is used to minimize cracking and structural movement caused by drying shrinkage in concrete. More details of this concrete can be found in ACI 223R-10.

*Heavyweight concrete*   It is a concrete in which heavy natural aggregates such as barites or magnetite or manufactured aggregates such as iron or lead shots are used. Its main use is in radiation shielding (medical or nuclear applications), and also for ballasting of pipelines and similar structures in offshore uses. The cement content and w/c ratio of this concrete are similar to that of normal concrete, but the aggregate/cement ratio will be higher, due to the higher density of aggregates. The density of this concrete depends on the aggregate used (When barites are used- 3500 kg/m$^3$, if magnetite is used- 3900 kg/m$^3$, and when iron or lead shot are used as aggregate, it will be 5900 kg/m$^3$ and 8900 kg/m$^3$, respectively).

*Roller-compacted concrete (RCC)*   This concrete is a lean, dry (no slump) concrete, delivered to the site by dump trucks, spread by small bulldozers and consolidated using vibratory rollers. It offers a rapid and economical method of construction of dams and pavements. More information on RCC may be found in ACI 327R-14.

*Fal-G concrete*   The FaL-G binder, innovated by the N Bhanumathidas and N Kalidas of INSWAREB in 1989, is a cementitious blend of fly ash, lime, and gypsum (www.fal-g.com). This environment friendly binder can be used instead of cement to produce building products such as bricks, hollow bricks, and structural concrete.

*Bacterial concrete*   It is a type of *self-healing concrete*, in which bacteria (*Bacillus psuedofirmus* or *Sporosarcina pasteurii*, which have a life of about 200 years) are added and mixed with concrete. When the concrete cracks, rainwater, or atmospheric moisture seeps into the cracks, activating the bacteria, which consume the calcium lactate, causing a chemical reaction that creates limestone, which then fills the gap of the cracks. For more information, visit http://selfhealingconcrete.blogspot.com.

*Light Transmitting Concrete (Litracon™)*   It is a concrete consisting of fine aggregates and 4–5% optical glass fibres of thickness 2 μm to 2 mm. It is used to make prefabricated blocks or panels. It was first developed in 2001 by Hungarian architect Aronlosonzi. The optical glass fibres can transmit light even at an incident angle greater than 60° and allows the light to pass through in a diffused manner. Persons standing inside will appear as silhouettes through the material.

*Photo-catalytic concrete*   This concrete has in addition to Portland cement, a proprietary formulation of photo-catalytic titanium dioxide particles. This concrete decomposes dirt, soot, mould, bacteria, and chemicals that cause odours, which are washed away by subsequent rains. As it reflects much of the Sun's heat, it results in reduced heat gain, which can make cities cooler, reducing air-conditioning loads and smog.

# SUMMARY

- There are a number of special types of concrete. RMC is manufactured in a factory under computer-controlled weight-batched operations based on standardized mix designs with precise mixtures, under strict quality control. The concrete from RMC plants are transported to the site by trucks equipped with revolving drum mixtures and concrete pumps. RMC offers better quality, reduced cement consumption and pollution at site, and high speed of construction.
- Concrete is responsible for about 5% of $CO_2$ in the atmosphere. Such GHG emissions can be reduced by consuming less concrete/cement, and substituting some percentage of cement with waste by-products, such as fly ash, blast furnace slag, silica fume, and reactive RHA.
- Fly ash is a waste product of coal-based thermal power plants. PPC is produced with 15 to 35% by mass of fly ash. Fly ash may also be added at site by the simple replacement method, the addition method, and the modified replacement method. Use of fly ash in cement concrete reduces heat of hydration, improves workability/pumpability, reduces bleeding, segregation and creep, improves impermeability, chloride and sulphate resistance, and provides better strength at ages beyond 56 days. Use of fly ash can control expansion due to ASR. Fly ash concrete should be cured for at least 14 days.
- Silica fume is a by-product of silicon and ferrosilicon industry, and has particle size 100 times smaller than that of cement. Available in wet or dry forms, it is added to concrete at RMC plant, by replacing 4–15% weight of OPC. Silica fume needs high-range water-reducing admixture for better workability. Silica fume reacts with calcium hydroxide to form additional binder material. It also improves the compressive and bond strength, and abrasion resistance of concrete. Silica fume concrete also requires careful curing.
- RHA has 85–90% silica by weight. Up to 30% by weight of OPC can be replaced with RHA. The addition of RHA increases strength, permeability, and durability of concrete.
- GGBS is a waste of iron and steel industry. The optimum amount of slag cement is about 40 to 50% by mass of the total cementitious material. GGBS improves strength, workability, ease of pumping, resistance to ASR, and reduces shrinkage. However, it may delay the setting time of concrete.
- RCA can be combined with virgin aggregates when used in new concrete. Concrete with RCA may have lower density and strength. Recycling concrete not only conserves resources, but also saves landfill space.
- HPC is made with high-quality ingredients and optimized mixture design (with 10–12 mm coarse aggregates and coarsely graded sand), and cast at site with superior quality control. In addition to high strength, HPC has high-early compressive strength and flexural strength, high modulus of elasticity, low permeability, resistance to chemical attack, and low shrinkage and creep. HPC may contain high-early strength cement, mineral and chemical admixtures, and low w/cm ratio of 0.22 to 0.40. HPC mixes should consider compatibility of cement with super-plasticizer.
- SCC is highly flowable concrete that can fill the formwork without segregation or bleeding. The advantages of SSC include no vibration to compact concrete, reduced noise during placing, ability to fill complex forms, and improved pumpability. However it requires stringent selection of materials, proper mix design, trained workers at site, and high quality control. Additional tests for SCC at the job site include slump flow, T50, and L-box tests.
- SLWC can be made with lightweight natural aggregates such as pumice or with manufactured aggregates such as expanded sintered clay, shale, or slag. SLWC has a density of 1680 to 1920 kg/m$^3$. SLWC has advantages of reduced dead weight leading to economy and improved seismic performance, higher fire rating, and improved insulation properties of walls.
- Aerated lightweight concrete can be AAC or foamed concrete. AAC is manufactured by mixing gas-forming chemicals with lime or cement mortar (without any coarse aggregates). Aluminium powder is used commonly as aerating agent, which expands the material to 2–5 times its original size, with numerous micro-pores. The cut blocks are heated in an autoclave chamber for 12 h at 190°C. AAC blocks offer several advantages due its light weight and factory production and provide thermal and acoustic insulation, and fire safety.
- Foamed concrete is produced by adding a special foaming agent (e.g., detergents or resin soap) or injecting preformed stable foam to the slurry of Portland cement, fly ash, water, and sand. Foamed concrete blocks need not be autoclaved and have the same advantages of AAC blocks.
- FRC is produced by adding steel, glass, polypropylene, carbon, and basalt fibres to concrete. The fibres act as crack arresters and enhance the tensile, fatigue, impact, or abrasion resistance of concrete.
- Ultra-high Performance Concrete (UHPC) is a high-strength, high stiffness, self-consolidating, ductile material, formulated by combining Portland cement, silica fume, quartz flour, fine silica sand, high-range water reducer, water, and steel or organic fibres.
- Polymer concrete is obtained by impregnating ordinary concrete with a monomer material and then polymerizing it. The three types are PIC, PC, and PMC. PIC is a hydrated Portland cement concrete that has been impregnated with a monomer that is subsequently polymerized in situ. Polymerization can be done by thermal-catalytic and promoted-catalytic methods. In PC, monomers and co-monomers are added to the aggregates and the mixture is

placed in formwork; after the forms are removed, the product is polymerized at room or elevated temperatures. In PMC, cement and aggregates are mixed with organic polymers. As the cement hydrates, coalescence of the polymer occurs, resulting in a comatrix of hydrated cement and polymer film throughout the concrete.

- Precast concrete is produced in a controlled factory environment using better manufacturing techniques and, hence, results in products of high quality and durability at a lower cost.
- Decorative concrete may be obtained by integrally mixing colour pigments with concrete made with white cement. It can also be made by using chemical stains, or by exposing colourful aggregates at the surface.

## EXERCISES

## Multiple-choice Questions

1. To achieve the desired properties, the minimum amount of fly ash to be added to concrete is
   (a) 5%          (c) 15%
   (b) 10%         (d) 20%
2. Substitution of 30% fly ash results in the reduction of heat of hydration to an extent of
   (a) 20–30%      (c) 30–60%
   (b) 30–40%      (d) 40–60%
3. Each 10% addition of fly ash reduces the water content of concrete by
   (a) 3%          (c) 5%
   (b) 4%          (d) 6%
4. When fly ash or silica fume is used as a replacement of Portland cement, curing should be done for a minimum of
   (a) 7 days      (c) 14 days
   (b) 10 days     (d) 21 days
5. When fly ash is used, the early age strength is reduced, but higher strength is achieved after
   (a) 28 days     (c) 56 days
   (b) 4 weeks     (d) 5 months
6. Concrete is called high volume fly ash (HVFA) concrete, if it contains cementitious content of
   (a) 20–30%      (c) 35–45%
   (b) 30–40%      (d) 50–70%
7. Using silica fume along with super-plasticizers, the compressive strength obtained is in the range
   (a) 40–60 MPa   (c) 80–90 MPa
   (b) 60–80 MPa   (d) 100–150 MPa
8. The amount of OPC that can be replaced with RHA without any adverse effect on strength is
   (a) 10%         (c) 30%
   (b) 20%         (d) 40%
9. The amount of slag that is used in the manufacture of Portland slag cement as per IS 12089 is
   (a) 5–10%       (c) 30–35%
   (b) 15–20%      (d) 45–50%
10. High-performance concrete is supposed to have the following strength at 28–91 days
    (a) 50 to 70 MPa    (c) 70 to 120 MPa
    (b) 50 to 100 MPa   (d) 70 to 140 MPa
11. High-performance concrete is supposed to have the following strength at 1–3 days
    (a) 5 to 7 MPa      (c) 20 to 28 MPa
    (b) 10 to 18 MPa    (d) 40 to 58 MPa
12. The optimum value of w/cm ratio in HPC is about
    (a) 0.22       (c) 0.28
    (b) 0.25       (d) 0.30
13. The quantity of silica fume necessary to achieve very high strength in HPC is
    (a) 5%         (c) 15%
    (b) 10%        (d) 20%
14. For strengths of HPC in excess of 100 MPa, the maximum size of aggregates should be limited to
    (a) 8 to 10 mm     (c) 12 to 16 mm
    (b) 10 to 12 mm    (d) 16 to 20 mm
15. It is better to limit the horizontal flow and vertical fall of SCC to
    (a) 10 and 5 m     (c) 15 and 7.5 m
    (b) 12.5 and 7.5 m (d) 20 and 10 m
16. The density (unit weight) of structural lightweight concrete is in the range of
    (a) 1200 to 1520 kg/m³
    (b) 1680 to 1920 kg/m³
    (c) 2000 to 2200 kg/m³
    (d) 2240 to 2400 kg/m³
17. AAC blocks should be constructed with a cement-sand mortar of mix
    (a) 1:3        (c) 1:5
    (b) 1:4        (d) 1:6
18. Foam cellular concrete blocks with the following density are used in load-bearing walls
    (a) 300 to 600 kg/m³
    (b) 600 to 900 kg/m³
    (c) 900 to 1200 kg/m³
    (d) 1200 to 1600 kg/m³
19. Typical aspect ratio of fibres in FRC is in the range of
    (a) 20 to 100      (c) 50 to 200
    (b) 30 to 150      (d) 60 to 250
20. Typical compressive strengths of UHPC is in the range of
    (a) 60 to 100 MPa   (c) 120 to 240 MPa
    (b) 80 to 160 MPa   (d) 200 to 300 MPa

# Review Questions

1. What is ready mixed concrete? How is it transported? How is RMC poured at site?
2. What are the advantages and drawbacks of RMC as compared to site-mixed concrete?
3. What are the three tools suggested by Mehta (2009) to reduce concrete consumption and carbon emissions?
4. How does the addition of 20 to 30% of fly ash control the damaging alkali-silica reaction (ASR) in concrete?
5. What is silica fume? State the two distinct ways in which silica fume increases the performance of concrete.
6. What is RHA? What are the advantages of using it in concrete? How much weight of OPC can be replaced with RHA without compromising on strength?
7. What is GGBS? How Portland slag cement is manufactured. List at least five advantages of using PSC. What is the main disadvantage of using PSC?
8. What is RCA? In what areas of construction has RCA already been used? What are the problems of using RCA in new concrete construction? What are the advantages of using RCA?
9. What is the definition of high-performance concrete adopted by the ACI? Give at least five desired characteristics of HPC.
10. What is the difference between high-strength concrete and high-performance concrete? How is HPC made?
11. What are the considerations of aggregates and super-plasticizer in the design of HPC?
12. Why is curing very important for HPC?
13. What is SCC? What are the advantages and disadvantages of SCC?
14. What is structural lightweight concrete? What aggregates are used in it? What are its advantages?
15. How is aerated lightweight concrete classified? How is AAC blocks produced? Where are AAC blocks used?
16. What are the advantages and drawbacks of using AAC blocks?
17. How is the production of lightweight foam concrete different from that of AAC? What are the ingredients used in it? What are its advantages?
18. How is 'foam concrete' classified according to its density?
19. What is FRC? What kinds of fibres are used in FRC? What is the typical volume fraction and aspect ratio of fibres?
20. What is ultra-high performance concrete (UHPC)? Where is it used?
21. What is polymer concrete? How is it produced?
22. Differentiate between polymer impregnated concrete (PIC), polymer concrete (PC), and polymer-modified concrete (PMC).
23. What is precast concrete and what are its advantages?
24. How are precast concrete products produced? What are the advantages of precast products?
25. How is decorative concrete produced?

# ANSWERS

## Multiple-choice Questions

| | | | | | | |
|---|---|---|---|---|---|---|
| **1.** (c) | **2.** (c) | **3.** (a) | **4.** (c) | **5.** (c) | **6.** (d) | **7.** (d) |
| **8.** (c) | **9.** (d) | **10.** (d) | **11.** (c) | **12.** (a) | **13.** (b) | **14.** (b) |
| **15.** (a) | **16.** (b) | **17.** (d) | **18.** (d) | **19.** (b) | **20.** (c) | |

# CHAPTER 10
# GYPSUM

## 10.1 Introduction

Chemically known as 'calcium sulphate dihydrate', gypsum ($CaSO_4 \cdot 2H_2O$) contains calcium, sulphur bound to oxygen, and water (79% calcium sulphate and 21% water). Gypsum is a soft non-metallic mineral commonly found within layered sedimentary rock deposits, with its mineral structure, containing some water. Gypsum almost exclusively occurs in settings where seawater is evaporating or where groundwater containing dissolved ions from rocks reforms these ions as gypsum. As a consequence, gypsum is typically found in beds or bands in sedimentary formations, such as limestone, sandstone, and shale, and known to occur in several varieties viz., selenite (crystalline), satinspar (fibrous), alabaster (pure, compact, and fine grained), gypsite (earthy), and anhydrite (anhydrous). The content of gypsum in sedimentary rock varies from 75 to 95%, the rest being clay and chalk. It has hardness equal to 2 on Mohs' scale and has a specific gravity of 2.3. Gypsum deposits are found in about 85 countries, including the USA, Canada, Mexico, Germany, the UK, France, Spain, Italy, Turkey, Poland, India, China, Iran, and Thailand. The USA, Canada, and Mexico have some of the largest reserves of high-quality gypsum. The White Sands National Monument in New Mexico is the world's largest gypsum dune-field. India has huge reserves of natural gypsum, in Rajasthan, Tamil Nadu, Jammu and Kashmir, Uttar Pradesh, and Gujarat, of the order of 1120 million tonnes, of which recoverable reserves are estimated at 237 million tonnes (http://tifac.org.in).

When heated to about 190°C, pure gypsum loses its luster and its specific gravity is increased from 2.3 to 2.95 due to the loss of water of crystallization. Heating gypsum will drive off the water in its crystalline structure to produce a dry powder that can later be combined with water to again produce gypsum. Although gypsum can form clear transparent crystals, it usually occurs as massive white chalky deposits. Usually white, colourless or grey in massive form, crystals are clear, transparent to translucent (see Fig. 10.1). If impurities are present, gypsum may also appear to be red, pink, green, or blue. Gypsum is the most common sulphate mineral and one of the most widely used non-metallic minerals in the world. It has been extracted, processed, and used by humans in construction or decoration since 9000 BC. Gypsum has been extensively used since the days of ancient Egypt, where it was used in building the Pyramids. Some of this construction is still visible over 5000 years later, a tribute to gypsum's durability

**Fig. 10.1** Gypsum

as a building material. Alabaster is a form of gypsum used for carvings and ornamental purposes. The use of gypsum was probably developed by the Greeks who called it 'Gupsos', which means 'to cook' or 'burned', a reference to how gypsum is commonly prepared for use as plaster. In the 18th century, French chemist Lavoisier began modern research on gypsum by studying its chemical properties. Large deposits of gypsum were discovered near Paris, and 'Plaster of Paris' became a popular building material. Plaster of Paris is raw gypsum that is chemically altered by heat to remove much of the water naturally occurring in gypsum (www.gypsum.org).

Gypsum has many applications in art and pottery, but is also used in medicine as casts for broken bones or as dental moulds for making artificial teeth. Its many other uses include the 'paste' component of toothpaste, modern chalk used in classrooms, and as a filler for paper and paints. It is also an important component added to cement, to slow the rate at which concrete hardens. A large amount of gypsum is used as a conditioner for soil, displacing sodium in the soil and allowing the soil to hold more moisture and improve crop yields. Benjamin Franklin brought this product to the USA and the use of gypsum in agriculture expanded dramatically when gypsum beds were discovered in New York State, and later across the USA, notably near Ft. Dodge, Iowa.

The modern use of gypsum, as a building material, was discovered in 1888 when America-born Augustine Sackett invented a machine for producing plasterboards (also known as wallboards and dry walls) composed of several layers of paper with gypsum in between. The first plasterboard plant was built in the USA in 1901. In 1908, the plasterboard technique was improved by America-born Stephen Kelly, who patented plasterboard with a gypsum core and one layer of paper on the front and back side (www.eurogypsum.org).

Products of gypsum have a number of desirable properties such as relatively small bulk density, incombustibility, good fire resistance, good sound absorbing capacity, rapid drying and hardening with negligible shrinkage, superior surface finish, and resistance to insects and rodents. Moreover, gypsum requires only low energy input during burning to produce gypsum plaster. Gypsum, due to its very low thermal conductivity, is considered an excellent, low-cost insulating material for buildings. The water within the crystalline structure of gypsum also helps to minimize fire damage. During a fire breakout, heat drives the water out of gypsum walls to cool and protect the wood or steel supporting the walls. The major shortcomings of gypsum are its poor strength in wet state and high creep under load. Gypsum-based items should be used only in dry state and where relative air humidity is not more than 60%. Currently, gypsum-based products are the most widely used materials in the USA for walls and ceilings in both residential and commercial constructions. The most important applications of gypsum in the construction industry are in the production of plaster, plasterboards, gypsum fibreboard, and gypsum blocks.

## 10.2 Effect of Heat and Moisture

As stated earlier, the mineral structure of gypsum contains some water ($CaSO_4 \cdot 2H_2O$), which is not held firmly by the mineral. When calcined (using pan, kettle, or rotary kilns), i.e., heated to about 130–150°C, three-quarters of its combined water is removed producing a fine white powder [($CaSO_4) \cdot 0.5H_2O$], as shown below. Based on the experiments conducted by Le Chatelier in Paris during 1887, this white powder is termed as plaster of Paris, and contains about 6% of crystalline water.

$$CaSO_4 \cdot 2H_2O + heat \rightarrow (CaSO_4) \cdot 0.5H_2O + 1.5H_2O \text{ (released as steam)}$$

This calcined gypsum, or calcium sulphate hemihydrate (commonly called *stucco*), becomes the base for gypsum plaster, gypsum board, and other gypsum products.

When calcination is carried out at a temperature of about 180–200°C, all the crystalline water is removed to produce gypsum anhydrite ($CaSO_4$), as shown below.

$$CaSO_4 \cdot 2H_2O \rightarrow CaSO_4 \text{ (Gypsum anhydrite)} + 2H_2O$$

This is called the high burning process or complete calcination.

Under damp or moist conditions, both of the above products revert back to form gypsum rock by combining with water as shown below.

$$(CaSO_4) \cdot 0.5H_2O + 1.5H_2O \rightarrow CaSO_4 \cdot 2H_2O$$
$$CaSO_4 + 2H_2O \rightarrow CaSO_4 \cdot 2H_2O$$

These processes, known as dehydration or rehydration, form the basis of gypsum technology. This process can be repeated almost indefinitely, with important implications for recycling.

## 10.3 Classification

Gypsum is usually classified as natural gypsum and synthetic gypsum. As already mentioned, *natural gypsum* that occurs in sedimentary rock formations is mined or quarried, crushed, and screened to about 50 mm diameter. If necessary, it is dried and ground to the extent that 90% of it is less than 149 μm. It is then heated in kettle or flash calciners at 120 to 150°C to remove three-quarters of the chemically bound water to form stucco (1 tonne of gypsum calcines to about 0.85 tonne of stucco). The calcined gypsum is the base for gypsum plaster, gypsum board, and other gypsum products. In the manufacture of plasters, stucco is ground further in a tube or ball mill and then batch-mixed with retarders and stabilizers to produce plasters with specific setting rates. The thoroughly mixed plaster is fed continuously from intermediate storage bins to a bagging operation.

*Synthetic gypsum* (also called as *by-product gypsum* or *chemical gypsum*), commonly known as the Flue Gas Desulphurization (FGD) gypsum, is primarily derived from coal-fired electrical utilities which are provided with equipment in the smoke stacks to remove sulphur dioxide from flue gases. These systems capture the sulphur dioxide by passing the gasses through scrubbers that contain limestone (calcium carbonate), which absorbs and chemically combines with the sulphur dioxide to form pure calcium sulphate or gypsum. The synthetic gypsum is then transported to the gypsum board manufacturer; the production process for calcining synthetic gypsum is largely the same as with mined gypsum except that primary crushing is not necessary. It has been in use for more than 30 years and presently, almost half of all gypsum used in the USA is the FGD gypsum. Some of the other important by-product gypsums are phosphor-gypsum, fluorogypsum or anhydrite, and marine gypsum. It has to be noted that some types of the FGD gypsum, for example, phosphor-gypsum that may contain radon and radio nuclides, are generally considered unsuitable for use in gypsum board due to their potential environmental hazards. For the chemical requirements of by-product gypsum, reference should be made to IS 12679:1989. The rise in consumption of the FGD gypsum in Europe and North America is slowing the rate at which natural gypsum reserves are exploited. A further factor that may reduce the demands on natural gypsum in future is the increase in the amount of recycled plasterboards.

Gypsum may also be classified as low-strength gypsum [obtained by heating natural gypsum rock at normal pressure (β modification)- it has an amorphous structure and requires more mixing water, and has lower strength; used for plastering applications and filling compounds], and extra-strong gypsum [obtained by heating gypsum at pressure of 2–3 atm followed by drying at 100–150°C (α modification)], which is less porous and has particles of uniform size. The extra-strong gypsum is used for moulding plasters, floor troweling, and levelling compounds.

## 10.4 Plaster of Paris

Plaster of Paris is a quick-setting gypsum plaster consisting of a fine white powder (calcium sulphate hemihydrates), which hardens when moistened and allowed to dry. Known since ancient times, plaster of Paris is so called because of its preparation from the abundant gypsum found near Paris. Plaster of Paris does not generally shrink or crack when dry, making it an excellent material for casting moulds. It is commonly used to precast and hold parts of ornamental plasterwork placed on ceilings and cornices. As mentioned earlier, plaster of Paris is obtained by heating gypsum, to about 130°C. This low-burning process is called the incomplete calcination. When the dry plaster powder is mixed with water, it re-forms into gypsum. The setting of unmodified plaster starts in about 10 min after mixing and is fully set in 72 h. When an additive is added to retard the set, it is called wall or hard wall plaster, which can provide passive fire protection for interior surfaces (the retarders used are tartaric acid, malic acid; polyphosphate is used as an additional retarder).

## 10.5 Gypsum Wall Plasters

Gypsum building plasters are used extensively in many countries of the world including Australia, Canada, the UK, the USA, and Russia, for general building operations and for the manufacture of preformed gypsum building products, which have the specific advantages of lightness and high-fire resistance. Wet gypsum plaster consists of gypsum powder with other ingredients to which water is added at the site to make it a paste-like consistency. This paste can be spread on the wall with trowels, in two or three successive coats, the last coat carefully smoothened to produce the required finish for the wall. The plasters used for the first (*scratch*) coat and second (*brown*) coat are called the *base coat plasters*, and that used for the final coat are called *finish coat plasters*. Typical thicknesses of these coats are 12.5 mm, 16 mm, and 20 mm, respectively. These plasters are replaced by drywall construction, and hence are useful in certain applications only.

Wet plaster has to be applied on a suitable base, usually some type of *lath*, which helps the plaster to develop mechanical and adhesive bond with the wall (see also Sec. 7.12.3 and Fig. 7.5 of Chapter 7). Gypsum lath (gyp lath) was used for interior gypsum plaster work before drywall was introduced in the latter part of the 1960s. It is simply a gypsum board, which has a highly absorptive and fibrous surface, which will quickly absorb the moisture of wet gypsum plaster. It may also be perforated, which allows the plaster to squeeze through the holes and lock it to the lath, as shown in Fig. 10.2. In addition, due to this locking of plaster and lath, the perforations generally produced better fire resistance than plain gyp lath. Metal lath (see Fig. 7.5 of Chapter 7) consists of sheet metal, which is perforated or expanded, creating openings through which the plaster base coat interlocks. It has to be noted that the wet plaster may be applied directly over masonry or concrete surfaces, over a base coat of bond plaster.

The various types of gypsum plasters are used to simulate the appearance of wood, stone, or metal surfaces. In India, gypsum plaster should conform to IS 2547 (Part 1). By-product gypsum conforming to the requirements of IS 12679 can also be used for the preparation of plaster. The various types of plasters include plaster of Paris, hard wall plaster, wood fibre plaster, bond plaster, gauging plaster, Keene's Cement, veneer plaster, and acoustic plaster. These are discussed briefly in the following:

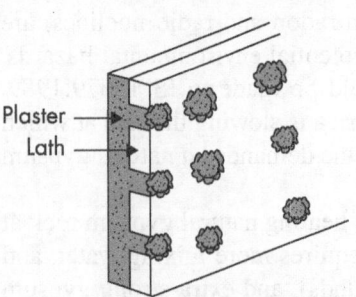

Plaster
Lath

**Fig. 10.2** Wet plaster squeezes through holes of perforated lath to form self-locking plugs when made to dry

**Plaster of Paris**    Calcined gypsum, to which no retarders are added, is called plaster of Paris. It sets in about 10 to 15 min and hence not suitable for use as wall or ceiling plaster. It is used for sculpturing, ornamental work, and small repair works. The hydration of plaster of Paris relies on the reaction of water with the dehydrated or partially hydrated calcium sulphate present in the plaster.

**Hard wall plaster**    It has a setting time of about 1 h and has a compressive strength of 7 MPa. Admixtures are used to increase the plasticity and setting time. Sand, vermiculate, or perlite may be added to this gypsum plaster and used in the base coat. A mix of two parts of sand to one part of plaster is usually adopted. Often, hair or cellulose fibre is added to increase the bulk and reduce the excessive squeezing through perforated laths.

**Wood fibre plaster**    It is a high-strength basecoat plaster. It provides excellent fire-resistance and sound dampening in plastered wall and ceiling assemblies. It is manufactured with fine particles of selected wood fibre, so it needs the addition of only water. It may also be used with sand aggregate for providing strong and hard base coats.

**Bond plaster**    Non-porous surfaces like concrete may require a plaster with strong bonding characteristics. Bond plaster may be obtained by adding a small amount of lime, as well as chemicals to improve bond, to the gypsum powder. Bond plasters may be applied directly on concrete surfaces, without requiring any laths. However, the surfaces must be prepared properly before the application of bond plaster.

**Gauging plaster**    This is the most common and economical finish coat mortar. It is mixed with two parts of lime, to control the setting time. It is a grey to white plaster (depending on gypsum rock source) that provides smooth-trowel and sand-float lime-putty finishes for normal interior walls and ceilings. Quick-set and slow-set formulations are un-aggregated for use over sanded basecoats. Perlite aggregate formulation includes mill-mixed fine aggregates for use over lightweight basecoats. It produces hard, crack- and abrasion-resistant surfaces over gypsum basecoats. Quick-set formulation sets in 30–40 min; slow-set formulation sets in 50–75 min. Smooth finish plasters should be applied at a thickness of not more than 1.5 mm. Texture finishes should be applied at a thickness of not more than 3 mm.

**Keene's cement**    It was invented by R.W. Keene of England, in 1841. It is a white cementitious material manufactured from gypsum that has been burned at a high temperature of about 190°C and ground to a fine powder. Alum is added to accelerate the set. The resulting plaster is hard, strong, and has a high polish; hence, it is used as finishing plaster. It produces durable, highly crack-resistant smooth and sand-float finishes when mixed with lime. It sets in 3 to 6 h and is retemperable; thus, it can be floated for an extended time period to provide a densified finish. Rapid setting types are also available, which have a setting time of 20 min. Keene's cement requires a high-strength gypsum base coat. It is to be noted that it is not resistant to cold weather and should never be used outside.

**Veneer plaster**    Introduced in the 1960s, these systems are thin coat, high-strength plaster systems that can be quickly applied over a variety of substrates. They provide hard, dense surfaces that are resistant to scuffing, denting, cracking, and abrasion. It is possible to provide veneer plaster finishes ranging from a mirror smooth surface to any type of floated, swirled, or light texture. Veneer plaster can be integrally coloured with pigments or finished with any number of decorating products to achieve a broad array of looks. Gypsum veneer plasters are manufactured to comply with ASTM C 587 and are complete when bagged and require only the addition of potable water. Similarly, the special gypsum board product designed as a substrate for veneer plaster, known as 'gypsum base', is specifically manufactured, as per ASTM 1396, to work with a corresponding plaster product. Gypsum base has blue tinted face paper that distinguishes it from regular gypsum board. Veneer plaster may be applied in one (with a thickness of 1.5 mm to 2.5 mm) or two coats. Although one-component systems are designed to work best over gypsum base, they can also be applied over other bases such as monolithic concrete or masonry block with good results. Coarser float or deeper texture finishes can be made by the addition of sand. Two component systems consist of two different products packaged in separate containers. Gypsum basecoat plaster, many times stronger than conventional plaster, is applied over the base. The finish coat is applied approximately 2 h after the base coat is set (If the second coat is applied more than 12 h after the first coat, a bonding agent is recommended.). Two-component veneer plaster systems are usually applied on masonry block walls: some surface preparation is often required and for better performance, a bonding coat should be applied on the masonry surface.

**Acoustic plaster**   It is a porous and sound-absorbing plaster incorporating fibrous or porous aggregates, such as wood, mineral wool, cork, or asbestos. Chemical ingredients, which will result in the development of air bubbles thus increasing porosity, can also be used (A method called Hushkote, which incorporated yeast in the plaster mixture to generate bubbles has also been developed.). Such a plaster is applied in thicknesses of up to 37 mm. Compared to other sound insulation, it is easy to apply and is fireproof but it can be more fragile, being affected by physical stress and humidity. Acoustic plaster is used in the construction of rooms, which require good acoustic qualities such as auditoria and libraries. Several proprietary acoustic plasters were developed in the 1920s, which include Macoustic Plaster, Sabinite (developed by Paul Sabine at the Riverbank Acoustical Laboratory), Kalite, Wyodak, Old Newark, and Sprayo-Flake, produced by companies such as US Gypsum. Spray-applied acoustical coatings were developed during 1945, which incorporated mineral wool or asbestos in a fireproof binder. These superseded felts and quilts as a common preference of architects but were difficult to apply and, hence, superseded in turn by acoustic tiles (see Sec. 21.6 of Chapter 21).

**Pre-mix gypsum plaster**   This is a gypsum plaster premixed and ready to use, with only water to be added before use. It avoids the messy, wasteful operation of mixing dry powdered gypsum at the site. This may or may not contain aggregate. This material is applied as one undercoat to a finished thickness of 8 to 11 mm. In addition to the standard lightweight aggregate undercoat, other gypsum undercoat plasters are also available.

The specifications for gypsum building plaster, excluding premixed lightweight plaster is provided in IS 2547(Part 1):1976 and Part 2 of this code deals with premixed lightweight plasters. The users may refer IS 2547 (Part 1) for the chemical composition of various plasters. The physical requirements of these plasters, based on IS 2547, are given in Table 10.1.

**Table 10.1** Physical requirements of gypsum plasters

| S. no. | Characteristics | Requirement | | | |
|---|---|---|---|---|---|
| | | Plaster of Paris | | Anhydrous gypsum plaster | Keene's plaster |
| | | Type A (short-time setting) | Type B (long-time setting) | | |
| 1 | Setting time, minutes (a) Plaster-sand mixture | 45–120 | 120–900 | – | – |
| | (b) Neat plaster | 20–40 | 60–180 | 20–360 | 20–360 |
| 2 | Transverse strength N/mm$^2$ | 0.5 | 0.4 | – | – |
| 3 | Soundness | Set plaster pats should not show any sign of disintegration, popping, or pitting | Set plaster pats should not show any sign of disintegration, popping, or pitting | Set plaster pats should not show any sign of disintegration, popping, or pitting | Set plaster pats should not show any sign of disintegration, popping, or pitting |

*(Contd)*

**Table 10.1** (*Contd*)

| S. no. | Characteristics | Requirement | | | |
|---|---|---|---|---|---|
| | | Plaster of Paris | | Anhydrous gypsum plaster | Keene's plaster |
| | | Type A (short-time setting) | Type B (long-time setting) | | |
| 4 | Mechanical resistance of set neat plaster | – | Diameter of the indentation >3 mm and <4.5 mm | Diameter of the indentation <4 mm | Diameter of the indentation <3.5 mm |
| 5 | Residue on 90 µm sieve, Max. (%) | 5.0 | 5.0 (under coat) 1.0 (final coat) | 2.0 | 2.0 |
| 6 | Expansion on setting, Max. (%) | – | 0.2 at 24 h (board finish plasters only) | – | 0.5 at 96 h |

IS 9498:1980 provides the specifications for inorganic aggregates that can be used in gypsum plaster. IS 13001:1991 provides guidelines for the manufacture of gypsum plaster in mechanized pan system.

### 10.5.1 Characteristics of Wet-plastered Walls

Gypsum plaster is comparatively easy to spread and level. Wet plaster provides uniform monolithic surface over an entire wall. Gypsum allows the surface to be finished smooth or with the various degrees of texture from a sand-textured finish to a sprayed-on acoustical finish. The advantage of gypsum is that it does not shrink on drying out and it forms a sufficiently dense surface. The ability of the surface to absorb water, called *suction*, will affect the drying rate of plaster. When the suction of the background is too high, the plaster will fail to set properly, resulting in the loss of adhesion of the plaster. Whereas, when the suction of the surface is too low, the plaster will retain excessive moisture, and drying shrinkage cracks will occur in the plaster.

Plastered walls are able to resist abrasion and indentation due to normal wear. When painted, the walls are also moisture-resistant. When gypsum plaster is supplied pre-mixed for undercoats, it is less messy to use than fine, dry lime, and cement powder. The main drawback is that gypsum plaster is more expensive than lime or cement plaster. Severe moisture exposure conditions will damage gypsum plastered walls and, hence, they should be covered with either tile or other waterproof material, in locations such as bath rooms.

## 10.6 Gypsum Plaster Boards

In order to reduce the demand of site labour involved in plastering, the use of building board, such as gypsum plaster board, fibre hard board, cement coir board, and asbestos cement building board, as covering for walls and ceiling is recommended (note that asbestos is banned in several countries). *Gypsum board* is the generic name for a family of panel products that consist of a non-combustible core, composed primarily of gypsum, and a paper surfacing on the face, back, and long edges (see Fig. 10.3). In the manufacture of gypsum wallboard, stucco from storage is first mixed with dry additives such as perlite, starch, fibreglass, dolomite, or vermiculite to improve core cohesion at high temperature. This dry mix is combined with water, soap foam, accelerators, and shredded paper, or pulpwood in a pin mixer at the head of a board

**Fig. 10.3** Typical gypsum board

forming line. The slurry is then fed between two paper sheets that serve as a mould. As the wet board travels the length of a conveying line, the calcium sulphate hemihydrate combines with the water in the slurry and sets to form solid calcium sulphate dihydrate, or gypsum, resulting in rigid board. Panels emerge with the face, back, and log edges covered with paper (see Fig. 10.3). The board is rough-cut to length, and it enters a multi-deck kiln dryer, where it is dried by direct contact with hot combustion gases or indirect steam heating. The dried board is conveyed to the board end sawing area and trimmed and bundled for shipment.

All gypsum panel products contain gypsum cores; however, they can be faced with a variety of different materials, including paper and fibreglass mats. Gypsum board is often called *drywall*, *wallboard*, or *plasterboard*. The paper surfaces may vary according to the use of the particular type of board, and the core may contain additive to impart additional properties. The longitudinal edges are paper covered and profiled to suit the application. The standard edge profiles are shown in Fig. 10.4.

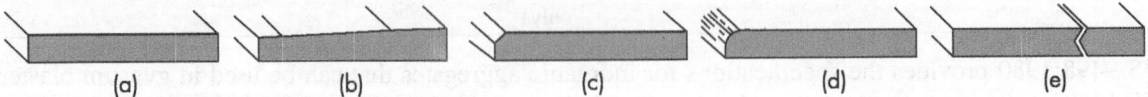

(a)       (b)       (c)       (d)       (e)

**Fig. 10.4** Standard edges of gypsum board (a) Square (b) Tapered (c) Beveled (d) Rounded (e) Tongue and groove

It differs from other panel-type building products, such as plywood, hardboard, and fibreboard, because of its non-combustible core and paper facers. When joints and fastener heads are covered with a joint compound system, gypsum wall board creates a continuous surface suitable for most types of interior decoration. The boards are available in thicknesses of 6, 8, 9.5, 12.5, 16, and 25 mm thick, width of 1200 mm, and lengths of 2400, 3000, 3600, and 4200 mm. A 12.5 mm thickness board is generally used in a single-layer wall and ceiling construction, and in double-layer systems for greater sound and fire ratings. A typical board application is shown in Fig. 10.5. Gypsum boards have the specific advantage of being lighter than the boards of similar nature, such as fibre hard boards and asbestos cement building boards (asbestos cement is banned in several countries). Gypsum boards also possess better fire-resisting, thermal, and sound insulating properties. Tall buildings of the world such as the 100-storey John Hancock tower and the 110-storey Sears Tower in Chicago, USA, have used gypsum board in their construction.

### 10.6.1 Types of Gypsum Boards

The various types of gypsum boards that are available in the market and their uses are given below (www.gypsum.org):

*Regular gypsum wallboard* This is the most commonly used gypsum board, and is used as a surface layer on finished walls and ceilings (see Fig. 10.7). Backing paper is grey, whereas face paper is ivory (Fig. 10.7). Edges are tapered to allow joint finishing. It is manufactured as per ASTM C 36 or IS 2095. The dimensions of gypsum wall boards as per IS 2095(Part 1) are given in Table 10.2.

*Type X gypsum board* This is similar to regular gypsum wallboard, except that the core has an improved fire resistance, due to special glass fibres that are intermixed with the gypsum to reinforce the core of the panel. It is available in 12.5 and 16 mm thicknesses. It may also be available with a pre-decorated finish. Type X gypsum board is used in most fire-rated assemblies (Fig. 10.6). As per ASTM C 36, 12.5 mm thick board should provide a fire rating of 45 min and 16 mm thick board should provide a fire rating of 1 h (see Fig. 10.6).

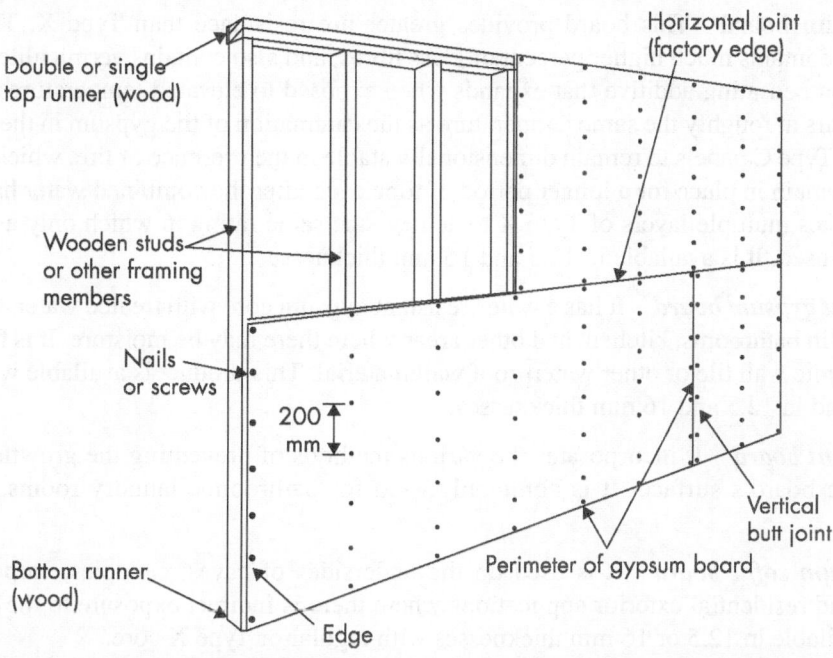

**Fig. 10.5** Horizontally applied gypsum wallboard, with paper bound edges at right angles to the framing members

**Fig. 10.6** Type X gypsum board – 16 mm thick

**Fig. 10.7** Lightweight dry wall – 12 mm thick

**Table 10.2** Dimensions of gypsum wallboards

| Dimensions of wallboard | Size, mm | Tolerance, mm |
|---|---|---|
| Width | 1800 to 3600 in steps of 100 mm | 0 to 6 |
| Length | 600, 900, 1200 | 0 to 5 |
| Thickness | 9.5, 12.5, 16, 19, 23, 25 | ±0.5 to ±0.6 |

**Type C gypsum board** This board provides greater fire resistance than Type X. The core of the Type C panels contains much higher percent of glass fibres, and also contains vermiculite, which acts as a shrinkage-compensating additive that expands when exposed to elevated temperatures of a fire. This expansion occurs at roughly the same temperature as the calcination of the gypsum in the core. It allows the core of the Type C panels to remain dimensionally stable in the presence of fire, which in turn allows the panels to remain in place for a longer period of time even after the combined water has been driven. Typically, it takes multiple layers of Type X to achieve the same rating in which only a single layer of Type C can be used. It is available in 12.5 and 16 mm thicknesses.

*Water-resistant gypsum board* It has a water-resistant gypsum core with treated water-repellent paper. It may be used in bathrooms, kitchen, and other areas where there may be moisture. It is frequently used a base for ceramic wall tile or other waterproof wall material. This product is available with a regular or Type X core and in 12.5 and 16 mm thicknesses.

*Mould-resistant board* It incorporates the various methods of preventing the growth of mould and mildew on the board's surface. It is commonly used for bathrooms, laundry rooms, kitchens, and basements.

*Exterior gypsum soffit board* It is used on the undersides of eaves, canopies, carports, and other commercial and residential exterior applications where there is indirect exposure to the weather. Soffit boards are available in 12.5 or 16 mm thicknesses with regular or Type X core.

*Impact-resistant gypsum panels* It consists of a mould-, mildew-, moisture- and fire-resistant Type X gypsum core with a specially designed Purple® paper. The purple face paper is a heavy paper that is 100% recycled and offers superior abrasion, mould, mildew, and moisture resistance. The 100% recycled grey back paper is also mould, mildew, and moisture resistant. Additionally, it has a fibreglass mesh embedded into the core, providing more impact and penetration resistance.

*Sag-resistant board* It is a ceiling panel that offers greater resistance to sagging than regular gypsum boards used for ceilings, where framing is typically spaced at 600 mm.

*Gypsum base board* This board serves as a base for thin coats of hard, high-strength gypsum veneer plaster. The dimensions of gypsum base boards, as per IS 2095(Part 1), are given in Table 10.3.

**Table 10.3** Dimensions of gypsum baseboards

| Dimensions of baseboard | Size, mm | Tolerance, mm | |
|---|---|---|---|
| | | Non-perforated | Perforated |
| Width | 400 and 900 | 0 to 8 | 0 to 8 |
| Length | 1200, 1500, and 1800 | 0 to 6 | 0 to 16 |
| Thickness | 9.5 and 12.5 | ±0.6 | ±0.6 |

*Gypsum liner board* It has a non-combustible gypsum core that is encased in a water-resistant, 100% recycled green face and back paper. They are used as a liner panel in shaft walls, stairwells, area separation fire walls, and corridor ceilings. Liner board is available in 19 to 25 mm thickness, width of 600 mm, and lengths of 2400–4200 mm, and feature double-bevelled edges for easy installation. Gypsum liner board is also available with fibreglass mat facing that increases its weather and mould resistance.

*Exterior Gypsum sheathing* It is a water-resistant product designed for attachment to exterior side-wall framing as an under-layer for various exterior siding materials, such as wood, masonry veneer, stucco, and shingles. The panel is manufactured with a wax-treated, water-resistant core faced with water-repellent paper on both face and back surfaces and long edges. It may also add structural rigidity to the framing system. It is available in 1200 mm width and 12.5 and 16 mm thickness and feature square edge. It is also available with Type X core, for use in fire-rated assemblies.

***Non-paper-faced gypsum board*** It is unfaced or has a facing other than paper.

***Abuse-resistant gypsum panel*** It offers greater resistance to surface indentation, abrasion, and penetration than standard gypsum panels.

***Eased edge gypsum board*** It has a tapered and slightly rounded or bevelled factory edge (see Fig. 10.4), felicitating much stronger and concealed joint than a tapered square edge.

Breaking load for gypsum plaster boards, as per IS 2095 (Part 1) should be in accordance with Table 10.4.

**Table 10.4** Breaking load for gypsum plaster boards

| Type of board | Thickness, mm | Breaking load, Minimum, N | |
|---|---|---|---|
| | | Transverse direction | Longitudinal direction |
| Wall board | 9.5 | 140 | 360 |
| | 12.5 | 180 | 500 |
| | 15.0 | 220 | 650 |
| | 19.0 | 250 | 750 |
| | 23.0 | 300 | 850 |
| | 25.0 | 380 | 1000 |
| Baseboard | 9.5 | 125 | 180 |
| | 12.5 | 165 | 235 |

Specifications for gypsum plaster boards are provided in IS 2095 (Part 1): 2011, for coated laminated gypsum plaster boards in IS 2095 (Part 2), and for reinforced gypsum plaster boards in IS 2095 (Part 3). Methods of test for gypsum plaster and concrete products are stipulated in IS 2542 (Part 1/Sec. 1–12): 1978, and IS 2542 (Part 2/Sec 1–8): 1978. Under the Environment (Protection) Act, 1986, and the Rules made thereunder to BIS, in order to obtain the ECO Mark, the gypsum boards used for partitioning, panelling, cladding, and false ceiling should be made from industrial wastes, such as by-products of gypsum. IS 12654:1989 deals with the use of low-grade gypsum in building industry. More details about the application and finishing of gypsum panel products may be found in GA-216–2013.

## 10.6.2 Characteristics of Drywall Construction

Gypsum boards are available in the market as plain, laminated, and reinforced boards. The boards may be reinforced with glass, paper, and vegetable fibres. Plain or laminated gypsum boards are used as finishes to masonry walls, ceilings, steel, or timber-framed partitions. They also can be used as claddings to structural steel columns and beams, and manufactured as prefabricated partition panels. Gypsum may be mixed with cement and used for interior plastering. Glass fibre reinforced gypsum boards (GFRG) may have considerable flexural and impact strengths. GFRG can be used similar to timber and can be sawn, drilled, screwed, or nailed. The advantages of GFRG boards are that they are resistant to white ants and termites, and are non-combustible. Thin GFRG panels have isotropic properties and are cost effective than thick timber panels. Hence, they can be used advantageously for panel doors, wall panelling, partitions, false ceiling, and furniture components (IS 2095:2011).

The gypsum boards may be fixed by nailing, screwing, or sticking with gypsum based or other adhesives. They may also be inserted in lay-in grids and/or secured by clips (DOC.CED 4(7782):2011). Finished walls constructed using gypsum board have uniform smooth surfaces suitable for painting, papering, or by any other means of decorating. It is also possible to provide a variety of surface textures, by using paints, or machines like drywall texture sprayer, or by using joint taping compound (also known as drywall mud). Drywall panels could be installed quickly and since only a thin layer of joint compound is used, the wall can be decorated sooner than a wet-plastered wall. The gypsum core as well as the facing paper could give the panels a fire-resistance rating similar to wet plaster. It has to be noted that just like regular plaster, gypsum board is not waterproof and, hence, special precautions have to be taken to protect it from moisture. Similarly, gypsum boards should not be exposed to temperatures exceeding 52°C for extended periods of time. More information on fire and sound insulation may be found in GA-600–2012.

## Case Study    Demonstration house built by IITM students using GFRG panels

Gypsum panels are not widely used in India. During 2013, a team of students at the IIT Madras demonstrated the use of pre-fabricated GFRG panels, by building a house inside the IIT Madras campus (the photo at the top shows four storeyed GFRG hostel buildings at IIT Tirupati). This low-cost, eco-friendly house comprised four flats, with two on each floor. It took only less than a month for them to construct it. The panels used were developed by RBS Australia and intended as RapidWall panels, ideal for rapid erection of walls in buildings. The use of prefabricated lightweight GFRG panels not only resulted in faster overall construction time, but also provided a safer working environment. The building has a total built-up area of about 184 m², with two flats having a carpet area of 25 m² each and two more with 46 m² each – mainly intended for the low-income group. The cost of construction, with all amenities, was only Rs. 1,250 per square foot.

Calcined gypsum plaster combined with glass fibres and special additives were used to produce the final GFRG panels. Each panel was about 12 m long, 3 m tall, and about 124 mm thick. The IIT Madras team extended the use of these panels to construct the entire building – right from floors, roofs, and staircases. The method also allowed for drastically reducing the overall consumption of reinforced cement concrete (RCC). They also developed a waterproofing material to protect the GFRG panels used in the roofs and toilets. Another team at IIT Madras, which is working on decentralized solar photovoltaic systems with DC appliances, will use the newly constructed building to demonstrate the overall savings in electricity consumption and, thus, reducing the electricity bills.

According to Prof. Devdas Menon and Prof. A. Meher Prasad who developed the building concept, filling the cavities in the wall with concrete increases the vertical load-carrying capacity almost tenfold, and inserting vertical steel bars in these cavities, contributes to their earthquake resistance. In a multi-storeyed building, the number of concrete-filled cavities and steel bars can be reduced at the higher floor levels. When used as floor slabs, reinforced concrete beams can be embedded and hidden in some of the cavities, as per the design. The overall weight of the structure and consumption of concrete is reduced significantly, and conventional plastering is eliminated.

## 10.7 Non-load Bearing Gypsum Partition Blocks

Another gypsum product used for interior walls is *partition block*, which consist of set gypsum plaster complying with IS 2547 (Part 1):1976, with or without aggregates (In the USA, the standard is ASTM C52 and Europe it is EN 12989.). As per IS 2849:1983, the mass of combustible materials should not exceed 15% of the mass of the dry block. The partition block may be solid type or hollow type and shall be truly rectangular in shape with straight and square edges and true surfaces. As per IS 2849:1983, in hollow blocks, the hollow spaces should be symmetrically spaced. The sum of the thicknesses of the two side shells, in addition to the thickness of the central vertical web of 150 mm wide blocks, should be greater than 50 mm for blocks having circular holes and greater than 70 mm for blocks having elliptical or rectangular holes, as shown in Fig. 10.8. There can be only one row of holes lengthwise in the case of 75, 100, or 125 mm wide blocks. The dimensions of the block as recommended by IS 2547 (Part 1) is given in Table 10.5.

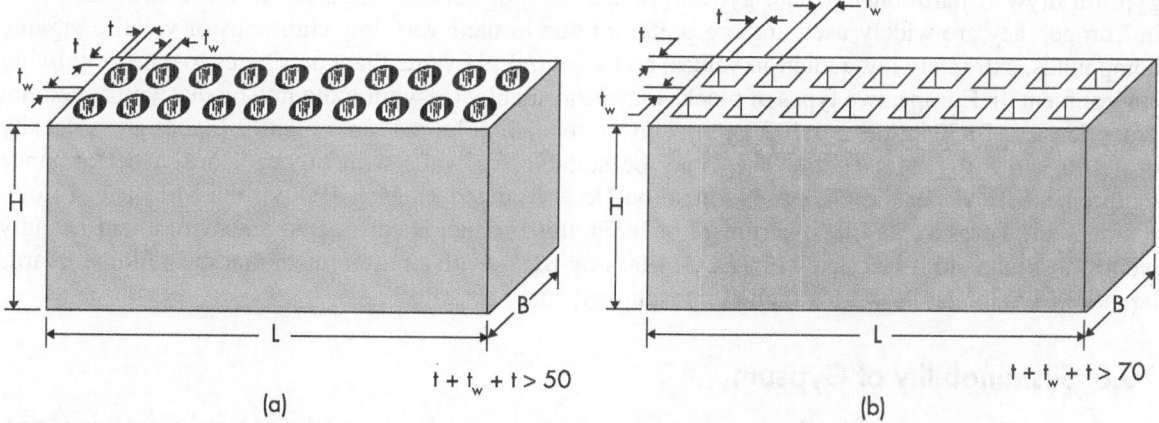

$$t + t_w + t > 50$$

(a)

$$t + t_w + t > 70$$

(b)

**Fig. 10.8** Gypsum hollow partition block (a) Block with circular holes (b) Block with elliptical or rectangular holes

**Table 10.5** Sizes of gypsum partition blocks (Solid and hollow)

| Length, mm | Height, mm | Breadth, mm | Hollow blocks side and edge thickness, Min, mm | |
|---|---|---|---|---|
| | | | Circular holes | Elliptical or rectangular holes |
| 700 maximum in multiples of 100 | 700 maximum in multiples of 100 | 60 | – | – |
| | | 75 | 15 | 20 |
| | | 80 | – | – |
| | | 100 | 20 | 20 |
| | | 125 | 25 | 30 |
| | | 150 | 15 | 20 |

When tested in accordance with IS 2542 (Part 2/Secs. 1 to 8), the compressive strength of the block should be greater than 0·5 N/mm$^2$ based on gross area. This code also states that when tested no block should: (a) cause the temperature readings of the furnace thermocouple to rise by more than 50°C above the initial furnace temperature, (b) cause the temperature readings of the specimen thermocouple to rise by more than 50°C above the initial furnace temperature, or (c) flame for more than 10 sec.

These partition blocks are laid up with gypsum mortar, and then covered with a coat of finish plaster. To distinguish this construction from conventional drywalls (with wall studs and gypsum panels or gypsum fibre boards), the construction with gypsum blocks is referred to as *massive drywall construction*. The width of the partition walls is thin (about 60 to 100 mm) compared to brick walls and, hence, results in an increase of effective floor space. These blocks enable quick construction and offer very flexible floorplanning, as walls can be added or removed quickly at any time. To improve the sound proof qualities of gypsum block partition walls, insulation strips may be used on all sides to connect the partition walls to adjacent walls, ceilings, and floors. The acoustic decoupling of the walls in this way reduces the acoustic transmissions of these lightweight partition walls significantly. Gypsum blocks have a high level of passive fire protection: 60 mm thickness offers 30 min, 80 mm thickness offers 2 h, and 100 mm thickness offers 3 h of fire resistance (see also GA-600–2012).

Gypsum partition blocks were commonly used in the USA during the early 20th century (1900–1926); after 1926, they were largely replaced by gypsum wallboards, concrete masonry units, and framed gypsum drywall partitions. Hence, gypsum blocks are not manufactured in the USA now. However, in Europe, they are widely used and are preferred due to their very low emissions of volatile organic compounds, extremely low radiation values, and a neutral pH value that contributes to a healthy living environment. In Europe, two types of blocks may be available: (a) white-coloured blocks with a medium gross density of 850 kg/m$^3$ to 1,100 kg/m$^3$, and (b) reddish-coloured blocks with a higher gross density of 1,100 kg/m$^3$ to 1,500 kg/m$^3$ (They may be suitable for walls with higher acoustic performance requirements.). As per the Gazette Notification No. 170 dated 18 May 1996 of the Ministry of Environment and Forests (MEF), Government of India, this product is considered an environment friendly wood substitute and is labelled with (ECO) Mark by BIS – with a requirement that the building boards be made from industrial wastes such as phospho-gypsum.

## 10.8 Sustainability of Gypsum

Gypsum products are amongst the very few construction materials where 'closed loop' recycling is possible. It has to be noted that closed loop recycling is different from 'down-cycling' of some other construction materials such as waste concrete and bricks, where the materials are merely recovered for reuse in low-strength applications, such as aggregates in road construction. In the closed-loop recycling, the waste can be used directly to make the same product again. Furthermore, a major advantage of gypsum is that it is eternally recyclable. The closed-loop recycling of gypsum products involves the collection and processing of the gypsum waste, and the delivery of the obtained recycled gypsum to the manufacturer of gypsum products. It is, therefore, essential that the recycled gypsum achieves a predetermined quality suitable for the manufacturing of new gypsum products. Hence, the contaminants are removed from the gypsum waste and the paper facing of the plasterboard is separated from the gypsum core through mechanical processes including grinding and sieving in specialized equipment. Gypsum waste, such as gypsum blocks and plaster, do not require such a removal of paper, as paper is not included in their production. Gypsum recyclers typically accept up to 3% of contamination from other materials. Professional recyclers are capable of handling gypsum waste with nails and screws, and wall coverings. Once collected, the plasterboards are broken down into a fine powder, which is then re-introduced, in a controlled blend, into the manufacturing process. It has to be noted that the tonnages derived from this source are increasing (www.eurogypsum.org).

By choosing the closed-loop recycling, the need for manufacturers to acquire virgin gypsum is reduced, thus resulting in sustainable manufacturing process. In the most advanced plants, such as those in the Nordic countries of Europe, up to 30% of virgin gypsum raw materials are substituted with recycled gypsum.

# SUMMARY

- Gypsum is a soft non-metallic mineral commonly found within layered sedimentary rock deposits, such as limestone, sandstone, and shale, and occurs in several varieties.
- Products of gypsum have a number of desirable properties such as relatively small bulk density, incombustibility, good fire resistance, good sound-absorbing capacity, rapid drying and hardening with negligible shrinkage, superior surface finish, and resistance to insects and rodents. Gypsum requires only low-energy to produce gypsum plaster.
- When heated in rotary kilns to about 130–150°C, three-quarters of gypsum's water content is removed producing a fine white powder, called plaster of Paris, which acts as a base for gypsum products.
- Gypsum is usually classified as natural gypsum (produced by heating mined sedimentary rock pieces) and synthetic gypsum (produced by capturing sulphur dioxide in chimneys of coal-fired power plants using scrubbers).
- Wet gypsum plaster consists of gypsum powder with other ingredients to which water is added at the site to make it a paste-like consistency. It may be applied in two to three coats on walls to get the required finish.
- The various types of plasters include plaster of Paris, hard wall plaster, wood fibre plaster, bond plaster, gauging plaster, Keene's Cement, veneer plaster, and acoustic plaster. Gypsum-plastered walls should be protected from moisture.
- The family of panel products that consist of a non-combustible core of gypsum (with or without fibres or aggregates) sandwiched between strong waterproof papers is called gypsum board or drywall. They are lighter than similar boards made of other materials. They also possess better fire resistance, thermal, and sound insulating properties.
- The partition block is a thicker product, without paper on the faces, and may be available as solid or hollow types.
- Gypsum products can be recycled in a 'closed loop' and, hence, result in sustainable manufacturing.

# EXERCISES

## Multiple-choice Questions

1. In general, gypsum is made up of
   (a) 59% calcium sulphate and 41% water
   (b) 69% calcium sulphate and 31% water
   (c) 79% calcium sulphate and 21% water
   (d) 89% calcium sulphate and 11% water

2. Pure gypsum
   (a) Is transparent to translucent crystalline mineral
   (b) Was used in building the Pyramids
   (c) When heated up to about 190°C, its specific gravity increases from 2.3 to 2.95.
   (d) All of these

3. Gypsum has
   (a) High bulk density
   (b) Negligible shrinkage
   (c) Damp proofing property
   (d) Low creep

4. Hardened gypsum
   (a) Is fire resistant
   (b) Can be used in places of high relative humidity
   (c) Has high shrinkage
   (d) Has high bulk density

5. Gypsum has
   (a) Hardness of 3 on Mohs' scale and the specific gravity of 2.3.
   (b) Hardness of 2 on Mohs' scale and the specific gravity of 2.7.
   (c) Hardness of 2 on Mohs' scale and the specific gravity of 2.3.
   (d) Hardness of 3 on Mohs' scale and the specific gravity of 2.7.

6. Plaster of Paris is formed when pieces of gypsum rock are heated at
   (a) 100–120°C     (c) 160–180°C
   (b) 130–150°C     (d) above 200°C

7. Plaster of Paris contains about
   (a) 3% of water     (c) 8% of water
   (b) 6% of water     (d) 11% of water

8. Gypsum loses all its water of crystallization at about
   (a) 130–150°C     (c) 180–200°C
   (b) 150–170°C     (d) above 300°C

9. Synthetic gypsum is also called as
   (a) By-product gypsum
   (b) Chemical gypsum
   (c) Flue gas desulphurization (FGD) gypsum
   (d) All of these

10. The setting time of plaster of Paris is delayed by adding a retarder such as
    (a) Tartaric acid     (c) Polyphosphate
    (b) Malic acid        (d) All of these

11. Keene's cement is made from
    (a) Lime     (c) Gypsum
    (b) Slag     (d) Pozzolana

12. As per IS 2547, the setting time of neat plaster made of plaster of Paris is
    (a) 20–40 min
    (b) 60–180 min
    (c) 20–360 min
    (d) >360 min
13. As per IS 2547, the setting time of neat plaster made of Keene's cement is
    (a) 20–40 min
    (b) 60–180 min
    (c) 20–360 min
    (d) >360 min
14. The usual thickness of gypsum board in single-layer wall and ceiling construction is
    (a) 8 mm
    (b) 9.5 mm
    (c) 12.5 mm
    (d) 16 mm
15. As per ASTM C 36, Type X and 16 mm thick board provide a fire rating of
    (a) 30 min
    (b) 45 min
    (c) 1 h
    (d) 2 h
16. Gypsum partition blocks of 60 mm thickness offer a fire rating of
    (a) 30 min
    (b) 1 h
    (c) 2 h
    (d) 3 h

## Review Questions

1. (a) What is gypsum?
   (b) What are the desirable properties of gypsum?
   (c) What are the shortcomings of gypsum?
2. Briefly describe the effect of heat and moisture on gypsum.
3. (a) How is gypsum classified?
   (b) What type of by-product gypsum is not used in gypsum boards, due to environmental concerns?
4. (a) What is plaster of Paris?
   (b) How is it manufactured?
   (c) What are its uses?
5. (a) How many layers of gypsum plaster are applied?
   (b) What are the usual thicknesses of these layers?
6. (a) What is the function of lath, when using wet plaster? (b) List two types of lath that are used with wet plasters.
7. What are the various types of gypsum wall plasters?
8. (a) What is gauging plaster? (b) What is Keene's cement?
9. Write notes on the following:
   (a) Bond plaster
   (b) Veneer plaster
   (c) Acoustic plaster
10. (a) How is gypsum plaster board manufactured?
    (b) What are the advantages of using gypsum plaster board?
11. (a) List at least five types of gypsum plaster boards. (b) How is Type C different from Type X gypsum board?
12. What are the advantages of glass fibre reinforced gypsum boards?
13. (a) What is non-load bearing gypsum partition block?
    (b) How does it differ from gypsum plaster board?
    (c) What are the advantages of using them in partition walls?
14. Why gypsum is considered an excellent green material?
15. What is 'closed loop' recycling? How is it different from down-cycling?

## ANSWERS

### Multiple-choice Questions

| | | | | | | |
|---|---|---|---|---|---|---|
| 1. (c) | 2. (d) | 3. (b) | 4. (a) | 5. (c) | 6. (b) | 7. (b) |
| 8. (c) | 9. (d) | 10. (d) | 11. (c) | 12. (a) | 13. (c) | 14. (c) |
| 15. (c) | 16. (a) | | | | | |

# CHAPTER 11
# WOOD AND WOOD PRODUCTS

## 11.1 Introduction

*Wood* or *timber* has been used as a building material for thousands of years and is still one of the most widely used. An excellent example is the Europe's Neolithic long house – a long, narrow timber dwelling built in 6000 BC. In the Middle Ages, carpenters were considered one of the most skilled craftsmen and were in high demand, as the construction of every building required wood. Certain properties of wood are complex; but, despite this, civil engineers have successfully harnessed wood to build a variety of structures ranging from houses to boats, and used it successfully in doors, windows, furniture, and décor. Timber framing was first developed by the Romans in 50 AD. Some of the impressive buildings built with timber include the Westminster Hall, London, built in 1045–1050 AD, and the pagoda at the Horyu-Ji Temple, built in 607 AD (which has withstood 7.0 magnitude or greater earthquakes 46 times). Timber is a complex material available from a large variety of trees, which should be felled soon after they reach maturity, during mid-summer or mid-winter (when the movement of sap in the wood is minimal). Timber used in buildings is usually from exogenous trees. The wood used in south India differs from that used in north/east/central/western India, as the trees grown in these regions differ from one another (see IS 399:1963, for the type of trees available in each zone). The Indian construction industry generally uses the word 'timber' to describe structural products of wood; whereas, in North America, the word 'lumber' is used.

Wood, as a building material, offers several advantages; the main is that it is a natural resource, and, hence, readily available and economically feasible. If harvested from a sustainably grown forest, wood can be an environment-friendly alternative to concrete, metal, and other building materials. Wood has long life and tends to decay only when used in damp locations. However, it can be easily treated with preservatives to allow it to be used in the most severe moisture conditions. Wood can be worked with simple carpenter tools to fabricate all kinds of shapes and sizes to fit practically any construction need. It is light and easy to handle, and, at the same time, very strong and has very good strength-to-weight ratio. Wood also can provide good insulation. It has been used in several countries for structural load-bearing members, such as piles and trusses, and light framing members, such as studs and rafters. It is extensively used in doors, windows, and several other decorative items. However, in India, due to indiscriminate deforestation, availability of quality wood has become scarce and even wooden doors and windows are being made with alternate materials such as aluminium and plastics.

Another type of wood, commonly used in construction, is the engineered wood. As its name implies, it is manufactured by binding together various wood strands, particles, fibres, veneers, or boards of wood, using adhesives, or other methods of fixation to form a type of composite material. Common examples of engineered wood include plywood, glued laminated timber (glulam), oriented strand board (OSB), fibreboard, and particle board. Engineered wood products are used in a wide variety of residential, commercial, and industrial construction projects.

## 11.2 Classification of Trees

There are numerous species of trees, each with a different kind of timber. In India alone, over 150 species are available. For engineering purpose, they are classified as: endogenous and exogenous, and are briefly described below.

**Endogenous trees** The stem of these trees grows by the addition of tissues inwards – the new wood-strand are interspersed among the old – and, hence, causes the cross-sections to appear dotted (examples: palm, bamboo, yuccas, and cane). Timber from these trees has very limited application in buildings.

**Exogenous trees** The stem of these trees grows outwards from the centre, in approximate concentric rings across the longitudinal section of the stem. They are subdivided into *deciduous* and *conifers*.

*Deciduous* These broad-leaved trees shed their foliage in the autumn of each year, and yield hardwood (examples: oak, dogwood, mahogany, maple, poplar, *sal*, *sheesham*, and teak). The annual rings are indistinct with the exception of poplar and bass wood. They are non-resinous, close-grained, dark-coloured, heavy, strong, and durable. Owing to their long, straight trunks and better quality of wood, deciduous trees are very frequently used in engineering applications.

*Conifers* These trees, with needle-shaped leaves and cone-shaped fruits, are almost without exception, evergreen and mostly yield *softwood* (examples: cedar, deodar, chirr, Douglas fir, pine, and larch). They show distinct annual rings, are light in colour, resinous, and lightweight.

## 11.3 Parts of the Stem of a Tree

The basic parts of a tree are the roots, trunk or stem, branches, twigs and leaves. The trunk, or stem, of a tree supports the crown and gives the tree its shape and strength. The cross-section of the stem of a typical exogenous tree is given in Figs. 11.1 and 11.2. The hard outer *bark* is akin to our skin and protects the tree from the extremes of weather (storms and temperatures), insects, any mechanical injury, and, in some cases, even from fire. The inner bark or the *phloem*, which is found between the cambium and the outer bark, is softer and moist than the outer bark; it carries sap produced by the leaves to the rest of the tree.

A thin layer of growing tissues, next to inner bark, is called the cambium, and is responsible for making the trunk, branches, and roots of the tree grow larger in diameter. The new tissues that form during the growing season will become sapwood (xylem) or phloem. Thus, this process adds a new layer of xylem to the trunk of the tree during every growing season, which will be seen as a growth ring in most of the trees. The *sapwood* (xylem) is the youngest soft and moist layer of wood that transports water and minerals to the various parts of the tree. The starch in sapwood may attract fungal and insect attack. As a tree grows, older sapwood cells in the centre of the tree become inactive and die, forming the *heartwood*. Heartwood is essentially the inner core of dead wood that provides strength to the tree. It is darker than sapwood. At the centre of the stem is the *pith*, which is the first formed part of the stem and decays when the tree reaches maturity. Pith may vary in size and shape, depending on the type of tree. In a felled tree, the pith may easily crumble and rot.

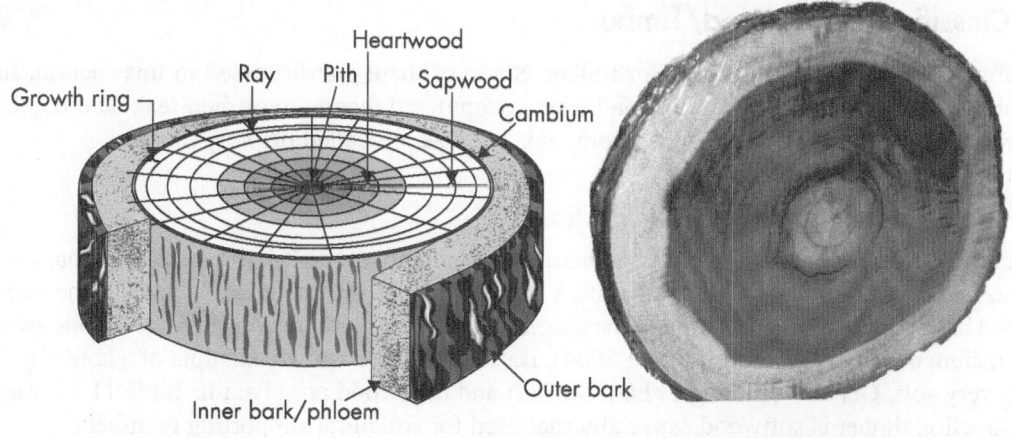

**Fig. 11.1** Cross-section of the stem of a typical exogenous tree

**Fig. 11.2** Vertical and horizontal cut section of an exogenous tree from Alpenzoo, Innsbruck, Austria

The nature of growth of exogenous trees produces a new layer of tissues outwards every season, and this produces a concentric layer of tissues called the *growth ring*. The total number of growth rings can be used to estimate the age of the tree. The thin radial fibres, which grow in the horizontal direction from the pith to the cambium, are known as *rays*. Their function is to carry sap from the outermost layer to the innermost. These are more easily visible in hardwoods than in softwoods.

## 11.4 Classification of Wood/Timber

There are a number of classifications for timber. Some of them are discussed in this section. In addition to these classifications, timber obtained from coconut and mango trees, ben-teak, and bamboo, are called *secondary timber,* whereas timber from teak, *sal*, oak are called *primary timber*.

### 11.4.1 Commercial/Hardness Classification

Commercial timber is usually classified as hardwood and softwood. It is important to note that these terms do not have any relation to the softness or hardness of the material. In fact, some species of hardwood have softer wood than some of the species of softwood, whereas certain softwoods are as hard as the medium density hardwoods (AWC, 2004). Balsa is the best-known example of a hardwood that is actually very soft. General differences between soft and hardwood are given in Table 11.1. Practically, all construction timber is softwood especially that used for structural supporting members.

**Table 11.1** Difference between softwood and hardwood

| Property | Softwood | Hardwood |
|---|---|---|
| Growth | Fast | Slow |
| Colour | Light | Dark |
| Annual rings | Seen distinctly | Indistinct and narrow |
| Structure | Open cell; resinous | Close celled |
| Heartwood and sapwood | Not easy to identify | Can be identified |
| Conversion | Easy | Difficult |
| Weight | Light | Heavy |
| Density | Low | High |
| Strength | Lesser strength in compression and shear; strong only along grains. | Higher strength; strong along and across grains. |
| Fire resistance | Poor | Moderate |
| Examples | Coniferous trees such as cedar, cypress, fir, pine, spruce, deodar, and redwood. | Deciduous trees such as ash, beech, birch, mahogany, maple, oak, teak, and walnut, *sal* and *babool*. |

### 11.4.2 On the Basis of Grading

The different grade classifications of timber as per IS 6354:1971 are as below:

1. **A Grade** (used in Kerala and Mysore): It is based purely, and sometimes arbitrarily, on dimensions and general appearance. Under these classifications, teak is placed in four grades with two sub-classes in each grade.
2. **B Grade** (used in Andhra Pradesh and Tamil Nadu): The logs are classified into grades on the best use possible as for beams, planks, scantlings, etc., and each grade is further divided into 'A', 'B', and 'C' classes to indicate occurrence of defects. Sometimes, another letter is also added to indicate the type of species, for example, 'T' for teak.
3. **C Grade** (used in Madhya Pradesh): It is based purely on defects and rough estimate of out-turn of utilizable material. Under this, there are two to four classes of teak.
4. **D Grade** (used in Mumbai, by the BIS, and also internationally): Two to three grades of sawn timber, based purely on defects (see also Table 11.2).

**Table 11.2** Permissible defects for the various grades of timber (as per IS 4021:1995)

| Defects | First grade | Second Grade |
|---|---|---|
| Cross-grain (not steeper than) | 1/15 | 1/10 |
| Knots and live knots<br>(a) Size (maximum)<br>(b) Number per meter | <br>20 mm<br>1 | <br>35 mm<br>2 |
| Decayed knots, dead knots, and knot holes. | They should be less than 10 mm in size and not more than 1 knot/m. They should be properly plugged with seasoned timber of the same quality so that its grains run in the direction of main pieces. | They should be less than 10 mm in size and not more than 2 knots/m. They should be properly plugged with seasoned timber so that its grains run in the direction of main pieces. |
| Pitch pockets or streaks | None | Permitted but must be filled with filler/putty. |
| Sapwood | Should not exceed 5 mm in width and 150 mm long/m. | Should not exceed 10 mm in width and 300 mm long/m. |
| Pinholes | Permitted if not in clusters | Permitted |
| Worm holes | None | Permitted provided if less than 10 mm in diameter and not more than 1/m and treated. |
| Checks, depth (maximum) | 3 mm, provided it is fully stopped | One-fourth the total thickness or 6 mm whichever is less, provided it is fully stopped. |

Besides such classifications, there is a system in Madhya Pradesh of grading of *sal* into three grades, according to its fitness for disposal.

### 11.4.3 According to Availability

Based on their availability, depending upon the figures supplied by the forest departments, timber is classified as below (IS 399:1963):

1. X: Most common, 1415 m³ (1,000 tonnes) and more per year
2. Y: Common, 355 m³ (250 tonnes) to 1415 m³ (1,000 tonnes) per year
3. Z: Less common, below 355 m³ (250 tonnes) per year

### 11.4.4 According to Durability

Based on its durability, timber may be classified, according to their average life as below (IS 399:1963):

1. High: Timbers having an average life of 120 months and over (examples: *anjan*, deodar, *sal*, teak, and rosewood)
2. Moderate: Timbers having an average life of 60 to 120 months (examples: Indian oaks, *salai*, and *kanju*)
3. Low: Timbers having an average life of less than 60 months (examples: fir, spruce, babul, and elm)

The durability is determined based on test specimens of size 600 × 50 × 50 mm, which are buried in the ground to half their lengths. The condition of the specimen at the various intervals of time is noted and, from these observations, their average life is calculated.

### 11.4.5 Based on Modulus of Elasticity

Based on the value of Young's modulus of elasticity ($E$) and extreme fibre stress in bending and tension ($f_b$), determined by conducting bending test on specimens, the species of timber are classified for structural purposes as below (IS 883:1994):

1. Group A: The value of $E$ above 12.6 GPa; $f_b$ above 18 MPa
2. Group B: The value of $E$ between 9.8 and 12.6 GPa; $f_b$ between 12 and 18 MPa
3. Group C: The value of $E$ between 5.6 and 9.8 GPa; $f_b$ between 8.5 and 12 MPa

### 11.4.6 Based on Treatability

Based on treatability with preservatives, timber may be classified as below (IS 399:1963):

1. Heartwood easily treatable
2. Heartwood treatable, but complete penetration of preservative not always obtained
3. Heartwood only partially treatable
4. Heartwood refractory to treatment
5. Heartwood very refractory to treatment, penetration of preservative practically nil from side or end

This classification represents the approximate degree of resistance offered by the heartwood of a tree to the penetration of preservative fluids under a working pressure of 10.5 kg/cm².

### 11.4.7 Based on Refractoriness to Air-Seasoning

Depending upon their behaviour with respect to cracking and splitting, and drying rate during normal air-seasoning, timber may be classified as (IS 399:1963, IS 1141:1993):

1. Class A: *Highly refractory* (slow drying and difficult to season and free from cracking and splitting. Examples are heavy structural timbers, such as Indian oaks, *sal*, *anjan*, and laurel).
2. Class B: *Moderately refractory* (may be seasoned free from surface defects and end cracking within reasonable short periods, given a little protection against rapid drying conditions. Examples are moderately heavy furniture class of timbers, such as *sissoo* and teak, babul, and rosewood).
3. Class C: *Non-refractory* (may be rapidly seasoned free from defects even in the open air and sun. If not rapidly dried, they develop blue stain and mould on the surface. Examples are light-broad leaved [hardwood] species such as *semul* and *salai* and all coniferous species).

## 11.5 Defects in Wood/Timber

There may be several defects in timber. A defect in timber is an abnormality or irregularity in wood, which lowers its technical quality or commercial value by decreasing its strength and affecting adversely its use or its appearance or in further conversion [IS 3364(Parts 1 & 2):1976]. Most of these defects may weaken the timber or make it visually not acceptable. However, some defects may be advantageously used; for example, twisted wood is used for making timber bowls. It has to be noted that the timber with defects may be cheaper than good timber. Before selecting timber for a particular application, it is advisable to look for the following defects (Eatonand & Hale, 1993).

The five main types of defects in timber may be due to: (a) natural forces, (b) attack by insects, (c) fungi, (d) defective seasoning, and (e) defective conversion. Out of these, only the defects due to natural forces and defects that develop after felling trees are described here. Other defects are discussed in subsequent sections.

## 11.5.1 Defects due to Natural Forces

The defects in timber due to natural forces may be present in the form of knots, shakes, upsets, twisted fibres, cracks, fissures, and pitch pockets.

**Knots**   These are the most common defects caused due to natural forces. When a tree grows, many of its branches fall and the stump of these branches remain in the trunk. In the sawn pieces of timber, the stump of fallen branches appear as knots [see Fig. 11.3(a)]. Knots appear as dark and hard pieces. Grains are distorted in this portion. Knots are more common and dark coloured in softwoods like pine. In many hardwoods, they are of the same colour as that of the parent wood; they are noticed only when a stain and clear finish is applied. Knots lead to a localized slope of grain in the timber adjacent to the knot as shown in Fig. 11.3(a), and may become a source of weakness. Knots cause serious defects when the load is applied perpendicular to the grains. The degree of weakness is dependent on the position, size, and the extent of grain distortion. Large knots are those having diameters greater than 40 mm, and those with less than 6 mm are called the *nail knots*. There are two different types of knots: (a) live knot and (b) dead knot. A *live knot* is usually lighter in colour and intact with surrounding wood (see Fig. 11.4). Whereas, a *dead knot* is darker and can fall out leaving a hole and eventually reduce the strength of timber. A *knot cluster* is created when two or more knots are joined together. Timber with large dead knots should be avoided.

**Twisted fibres**   The main reason of this defect is twisting of trees by strong winds. Owing to this, the fibres of the timber are twisted in one direction, as shown in Fig. 11.3(b), making the timber hard to saw. Such wood is used in unsawn condition, such as those used in poles.

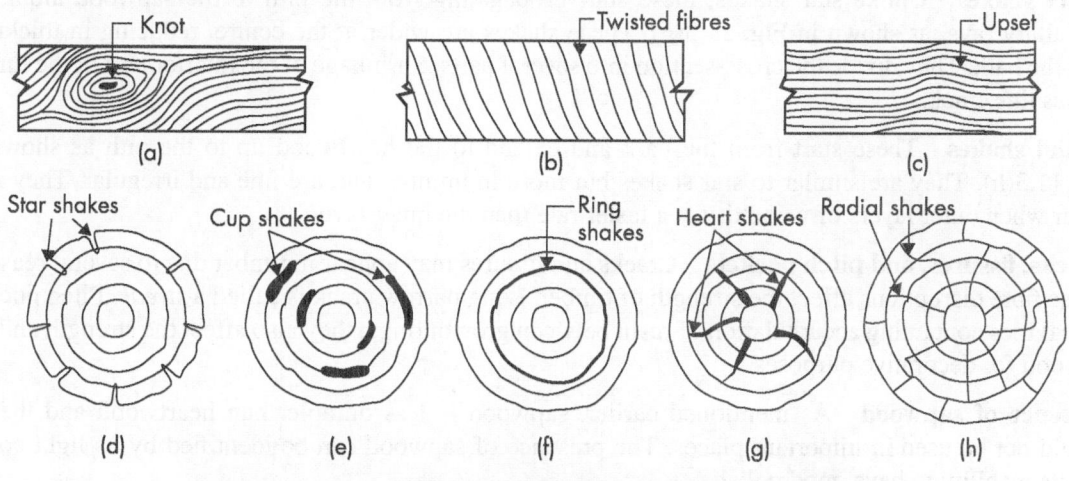

**Fig. 11.3** Defects in timber

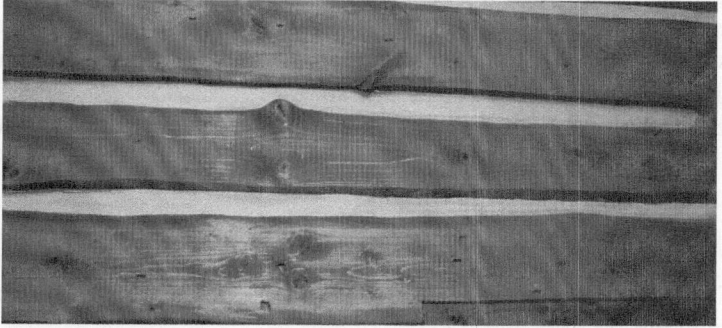

**Fig. 11.4** Knots in timber (note the live knots)

**Upsets or ruptures** Upsets or ruptures may be due to improper felling of trees, injury during the growth of the tree, or heavy winds blowing during the young age of the tree. These actions result in the change in direction of wood fibres, as shown in Fig. 11.3(c).

**Shakes** These are cracks or splits that occur around the annual ring or growth ring of a timber. These occur due to extreme climatic conditions, such as high temperature or severe frost, which may occur during the growth period of the tree or rapid or uneven seasoning after cutting off the timber. Extreme heat or frost causes temperature differences, causing shrinkage, which eventually leads to cracks. Shakes may or may not cause any structural defect depending upon their depth and usage. However, when the appearance is important, shakes are undesirable.

The different types of shakes include: star shakes, cup and/or ring shakes, and heart shakes.

**Star shakes** Star shakes start propagating from the bark towards the sapwood and sometimes even towards the heartwood along the lines of medullary rays, as shown in Fig. 11.3(d). Cracks are wider on the outer edge and narrower on the inside, and usually confined to the sapwood.

**Cup and/or ring shakes** It is a curved split as shown in Fig. 11.3(e), along the annual growth rings. They may separate the growth ring partially or completely. When the crack separates the annual ring completely, it is called *ring shakes* [see Fig. 11.3(f)]. These shakes are formed due to excessive frost action or non-uniform growth. Ring shakes are normally not visible in greenwood and detected only when the wood dries. Ring shakes are considered more detrimental than cup shakes.

**Heart shakes** Unlike star shakes, these start propagating from the pith to the sapwood along the medullary rays, as shown in Fig. 11.3(g). These shakes are wider at the centre, reducing in thickness near the bark. They divide the cross-section into several parts. Shrinkage of the interior part of the timber causes this crack.

**Radial shakes** These start from the bark and extend to the heartwood up to the pith as shown in Fig. 11.3(h). They are similar to star shakes but more in number and are fine and irregular. They may occur when outer layers of wood dry at a faster rate than the inner layers.

**Cracks, fissures, and pitch pockets** Cracks and fissures may appear in timber due to several reasons. These fibre disruptions affect the strength of timber. Long narrow crack is called a *streak*. Pitch pockets are cavities containing accumulation of resin between growth rings; they also affect the strength and use of wood for decorative purposes.

**Presence of sapwood** As mentioned earlier, sapwood is less durable than heartwood and, hence, should not be used in important places. The presence of sapwood can be identified by its light colour and its inability to have good polish.

**Sloping or spiral grains** Generally, the cells in trees do not grow perfectly vertical or straight and parallel to the entire length of trunk. They usually taper from bottom to the top, causing sloping of grains when the timber is sawn parallel to the pitch. Only when the cells grow in a spiral pattern around the axis of the tree, the sloping grains are considered undesirable.

Permissible defects for the various grades of timber, as stipulated in IS 4021:1995, are given in Table 11.2.

### 11.5.2 Defects that Develop after Felling Trees

The following are the defects that may occur after the felling of trees:

*Bow* This defect is caused due to shrinkage, resulting in a curvature in the direction of the length of the timber, as shown in Fig. 11.5(a).

*Cup*   When curvature is formed in the transverse direction of the timber, as shown in Fig. 11.5(b), it is called cup/cupping.

*Twist*   It occurs when the ends of timber twist in opposite directions [see Fig. 11.5(c)].

*Check*   These are small cracks that separate fibres, running lengthwise, but do not extend from one end to another. Fig. 11.5(d) shows surface checks and Fig. 11.5(e) shows internal checks. They may be due to the drying or seasoning of wood and may not penetrate the full depth of swan timber. As they do not normally affect the structural properties, they may be filled with appropriate coloured fillers.

*Honeycombing*   During the drying process or seasoning, stresses may be developed in the heart-wood, resulting in various radial and circular cracks; this kind of honeycombed texture renders the timber week.

*Split*   It is as a separation of fibres as seen both on the cross-section as well as on the longitudinal surface of the log, near the ends [see Fig. 11.5(f)]. This may result due to handling or drying stresses. Splits may reduce the shear strength of timber and make it unsuitable for decorative applications.

*Wane*   It is the lack of wood on any face or edge of sawn timber, due to the presence of the original rounded surface of a log, with or without bark, as shown in Fig. 11.5(g). Wane reduces the mechanical properties of wood and the volume of useful timber.

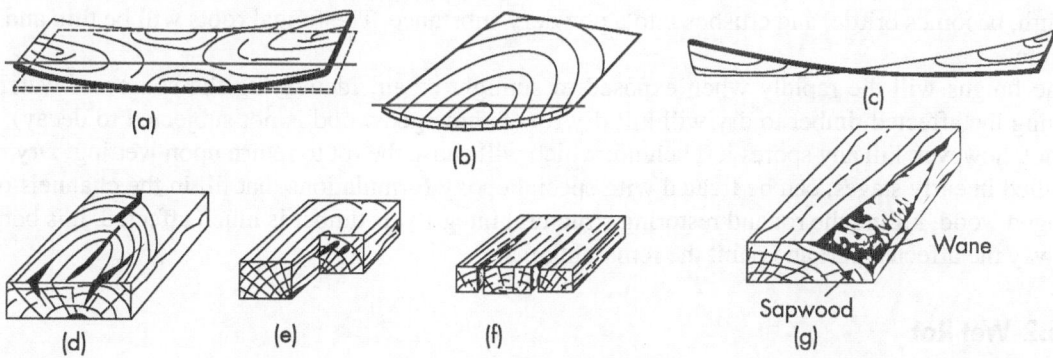

**Fig. 11.5** Defects in timber that develop after felling trees

## 11.6 Rot and Decay of Timber

Timber can have a long service life, provided it is maintained properly and protected from weathering actions (such as warm and moist conditions, alternate wetting and drying, freezing and thawing, and lack of ventilation), and attack from bio-organisms, such as bacteria and fungi, and insects, such as termites, beetles, carpenter ants, etc. Wood is essentially an organic substance, made up of a skeleton of cellulose impregnated with *lignin*. The organic substances are susceptible to attack by both bacteria and fungi. Bacteria do not cause any serious damage to timber, except some discolourations.

Moulds, most sapwood stains, and decay are caused by *fungi*, which are microscopic, thread-like microorganisms that must have organic material to live. Fungi attack timber causing stains, rot, and decay. Stain-producing fungi does not affect the strength of wood but affects the appearance – by a blue stain in softwoods. In order for the fungi to grow, four conditions are necessary: it must have a source of food (wood), a suitable temperature (10° to 32°C), sufficient air (oxygen), and sufficient moisture in wood (usually, 30% or more). There are many types of wood rot, but two of the most common types are dry and wet rots. Rot is the decomposition of felled timber due to the action of various fungi. The names dry and wet rots refer to the appearance of the affected timber rather than the moisture content of the timber. Wood-preservation methods attempt to poison the wood, which is the food source for fungi (see Section 11.9).

## 11.6.1 Dry Rot

Dry rot is caused by certain species of fungi that digest parts of the wood. It occurs in imperfectly seasoned timber that has at least 20% moisture content; it cannot survive in very wet wood. Dry rot, which is more common than wet rot, occurs in areas, which lack ventilation and in dark, humid areas; hence, usually, it will not be visible externally. Dry rot occurs in wood that is in contact with soil, bathroom door frames, joints of beams, and in window frames that are built against the sill. Dry rot can remain dormant when the wood is dry and can be active when it becomes wet. Dry rot is more serious than wet rot because, once established, the spores of dry rot can spread to other timbers (that have lower moisture content of about 14%) through air; therefore, it may be very difficult to eradicate. In fact, the fungal roots can travel through mortar and brickwork to affect wood in other areas. The typical growth rate is about 1 m per year, but can reach up to 4 m per year. The decayed wood will be darkened with large cracks or checks with yellow or red coloured crumbly appearance. Eventually, the timber loses strength, becomes brittle, and crushes into a powdery substance. The fungal roots will be thin and grey or white.

The fungus will die rapidly when exposed to sunlight or air. Identifying the source of water and allowing the affected timber to dry will kill dry rot (Submerged wood is not subjected to decay). This will not, however, kill any spores left behind, which will cause the rot to return upon wetting. Dry rot, if identified in early stages, can be treated with special epoxy formulations that fill in the channels of the damaged wood, killing the rot and restoring structural integrity. If timber is much affected, it is better to cut away the affected part and paint the remaining part.

## 11.6.2 Wet Rot

Wet rot occurs in timber that is subjected to alternate wet and dry conditions with moisture content above 40%, leading to the decomposition of tissues of timber. This may also occur due to a damp wall, poor or damaged paint work, water leakage on the timber, or when timber is seasoned by exposure to moisture. Wet rot is produced by a different type of fungus, the spores of which do not spread through air. It is less dangerous as it will not spread beyond the wet area and simply drying out the area will kill it. Timber with wet rot will feel spongy through the paint when poked. Affected wood will be darkened with small cracks and there usually is a skin of sound timber on the face of the affected wood. The fruiting body (a multicellular structure on which spore-producing structures are borne) will usually be a greenish brown. The fungal roots, if present, will be black or dark brown. Wet rot can be treated by fixing the source of the dampness and drying out the affected timber. The two commonly available inexpensive materials that will kill rot in wood and prevent its recurrence are:

1. Borate (borax-boric acid mixture), which has an established record of preventing rot in new wood and in killing rot organisms and wood-destroying insects in infested wood
2. Propylene glycol (used in antifreeze and deicing solutions)

Both borate and glycol solutions penetrate dry and wet wood well because they are water soluble. Safety measures have to be taken while applying them as they are toxic. To avoid wet rot, well-seasoned timber should be used with preservatives and paints. In addition, any timber used in a structure should not contain sapwood as it is more susceptible to fungal attack.

### 11.6.3 Damage due to Insects

The insects that attack wood include termites, beetles, borers, and carpenter ants (Eatonand Hale, 1993).

**Termites**   These are the most destructive of all insects. Although they are also called *white ants*, they resemble to cockroaches more than ants. There are over 2300 different species of termites, of which 220 are found in India [IS 6313(Part 1)]. The two major types of termites are: (a) subterranean or ground-nesting termites (they require moisture to sustain their life), and (b) *non-subterranean* (also called as *dry-wood termites*) or *wood-nesting termites* having no contact with soil (they are confined to coastal areas and interior of eastern India). The *subterranean termites* are most destructive and are mainly responsible for the damage caused in woodwork of buildings. Termite control in buildings is very important as the damage caused by these termites is huge. Termites get nutrients from cellulose, the organic fibre found in wood and plant matter. They also damage materials of organic origin having a cellulosic base, i.e., household articles such as furniture, furnishings, clothing, stationery, etc. Termites are also known to damage non-cellulosic substances in their search for food such as rubber, leather, plastics, neoprene as well as lead coating used for covering of underground cables [IS 6313(Part 3)]. All attacks by subterranean termites originate from their nest near or below soil level. They can reach to the timber, either lying on or buried in the ground, through underground galleries from which they spread well above the ground level, either inside the wood or by way of mud-walled shelter tubes, built by them, on the outside. More than one million termites can live in a colony. When a termite colony has matured, winged and swarming termites can be seen around windows and doors. Winged termites are highly attracted to sources of light and are most active in springtime. The sight of discarded wings of these termites and the mud-walled shelter tubes indicate the presence of subterranean termites.

Attack by termites, especially where there is no human habitation in the building, is of common occurrence in the tropics. Damage to wood by subterranean termite will not be evident from the exterior, as they do not leave powdered particles of wood as wood borers (The presence of dry-wood termites on the contrary may be identified by their pellets of excreta). Subterranean termites eat the wood at the centre completely, leaving an intact outer shell of wood. This type of interior damage may not become apparent until infestations are full-blown. Termite damage sometimes appears similar to water damage. Outward signs of termite damage include buckling wood, swollen floors and ceilings, areas that appear to be suffering from slight water damage, and visible mazes within walls or furniture. Termite infestations also can exude a scent similar to mildew or mould (www.orkin. com). Softwood is eaten more quickly than hardwood. Seasoned timber that is naturally durable in heartwood and treated to withstand the attack of-subterranean termites should be used in buildings. Periodic and thorough inspection (the presence of termites is indicated by a hollow sound heard while tapping wooden members), and taking control measures are the most important steps in checking termite damage to buildings. Some of the preventive methods adopted to protect buildings from termites are:

1. Chemical barriers which prevent the termites from reaching the superstructure
2. Physical barriers such as concrete aprons, with anti-termite grove at floor level around the building, as shown in Fig. 11.6
3. Using preservative-treated wood

4. Providing a continuous layer of sheet metal around the periphery of a building (similar to metal flashing) – they are usually employed in grain storage godowns and warehouses

5. Installation of termite caps at plinth level around wooden members and in down-water pipes

6. Installation of termite frames using metal sheets to cover all the sides of openings [see also IS 6313(Parts 1 to 3)]

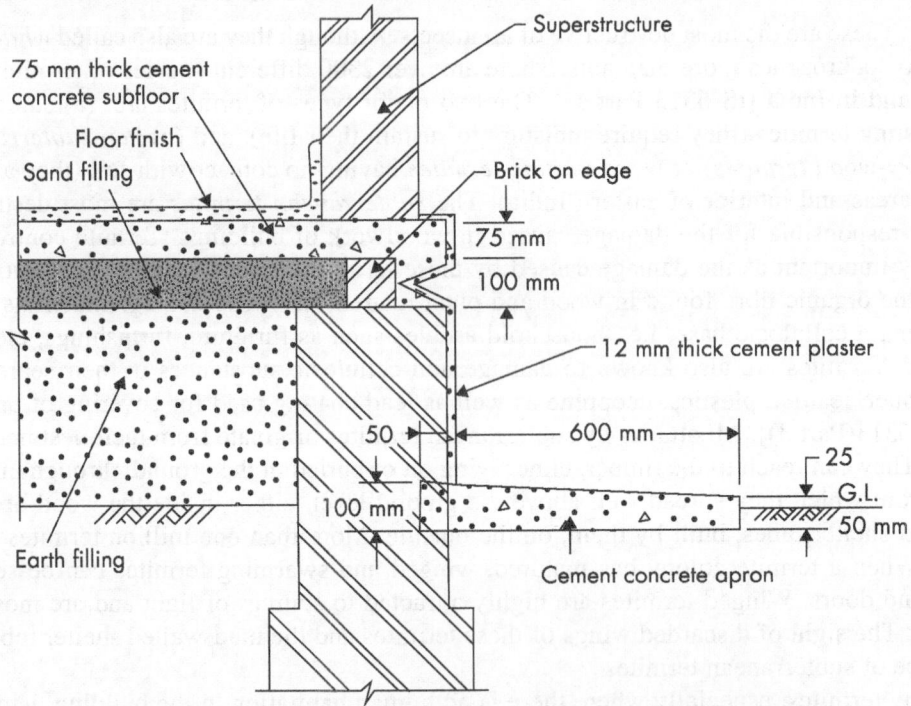

**Fig. 11.6** Anti-termite measures in the construction of buildings

Some termite prevention and treatment methods require the application of termiticides, which are pesticides that can control or kill termites. The exact treatment approach depends on the termite species and the location of the termite infestation. Termiticides can be used in liquid, foam, granule, or gaseous forms to control subterranean and dry-wood termite infestations in and around buildings. Termite control needs a knowledgeable, trained professional to identify signs of infestation, recommend the nature of treatment, and apply the treatment, as the chemicals involved are toxic in nature. Water emulsion consisting of Chlorpyrifos 20 EC (conforming to IS 8944) and Lindane 20 EC (conforming to IS 8944), each at 1% concentration, is suggested by IS 6313(Part 3), for use in the soil treatment in order to protect a building from termite attack. Oil or kerosene-based solution of Chlorpyrifos 20 EC or Lindane 20 EC, with a concentration 1.0% (by weight), is suggested for the treatment of wood.

Other liquid termiticides applied directly to the soil include Fipronil (Termidor®), chlorfenapyr (Phantom®), and imidacloprid (Premise®). These chemicals affect the nervous system of termites which come in contact with them and can cause their death within a few days. Sulfuryl fluoride is the most common fumigant used to control dry-wood termites in the US. Borate formulations are often used, as they are effective on many species of wood-destroying insects and are low in toxicity to humans. Hexaflumuron is specifically designed for termite baiting systems and exploits the social behaviour of termite colonies. Hexaflumuron is considered a reduced-risk pesticide and is much less toxic to the surrounding environment than the other liquid pesticides (www.hunker.com). If there are signs of reinfestation after treatment, it is necessary to repeat appropriate treatment, depending upon the termite infestation.

The soil in contact with the external wall of the building should be treated with chemical emulsion at the rate of 7.5 L/m² of the vertical surface of the substructure to a depth of 300 mm. This may be done by excavating a shallow channel along and close to the wall, and applying 1.75 L/m to 300 mm depth from the ground level. If there is a concrete or masonry apron around the building, 12 mm diameter holes are drilled as close as possible to the plinth wall, at 300 mm apart so that the soil below is reached. Then, the chemical emulsion is pumped into these holes to soak the soil below at a rate of 2.25 L/m. In walls, holes should be drilled at a downwards angle of about 45°, preferably from both sides of the walls, at 300 mm intervals and the emulsion is applied through these holes to soak the masonry using a hand-operated pressure pump [IS 6313 (Part 3)]. Similar treatment has to be done at the base of wooden window and door frames. An oil-based termiticide should be sprayed on all the woodwork infested by the termites. Visible shelter mud tubes are first sprayed, then removed and infested area is also given similar treatment. The safety precautions to be considered during the application are described in IS 6313 (Part 3), which also provides guidance for termite detection.

**Beetles and borers**  These are small insects that bore tunnels through the timber and live in them, causing a rapid decay of timber by converting them into fine powder. Usually, the outer shell of timber remains intact and, hence, the timber looks sound from outside until it fails completely. The most effective treatment is to apply insecticide and preservatives. Marine borers can attack susceptible wood rapidly in salt water harbours, where they are the principal cause of damage to piles and other wood marine structures.

**Carpenter ants**  These are usually black in colour and vary in size within the same nest. Unlike termites, they do not eat wood but merely tunnel it out for habitation. They normally attack slightly rotted or water-softened wood but may continue into wood, which appears perfectly sound. One common material used for the treatment of beetles, borers, and carpenter ants is turpentine mixed with a small quantity of orthodichlorobenzene. The vapour is found to be deadly for these insects, but is not much poisonous to human beings.

In any case, timber used in buildings should be preserved by using oils, paints, or varnish at regular intervals. There should be proper ventilation around woodwork, to keep it in dry condition. They should also be not placed in corrosive environments such as in lime or cement mortar or subject to damp conditions or alternate wetting and drying.

## 11.7 Moisture Content of Timber

The moisture content of freshly cut logs and newly sawn lumber may range from 45 to 200%. As the wood dries out, water and water vapour in the cell cavities (free water) first evaporates leaving only the water in the cell walls (bound water). At this stage, the timber is said to have reached its *fibre saturation point* (FSP). The FSPs for most wood species are in the range of 25 to 30% moisture content. Wood shrinks as the moisture content of most cells fall below the FSP (timber shrinks more across the grain than along it). If moisture is added, wood swells until the moisture content of most cells reaches the FSP.

As timber is *hygroscopic*, its moisture content will fluctuate based on the relative humidity (RH) of the surrounding air (RH is the ratio of the amount of moisture in the air to the amount that the air could hold at the same temperature). Hence, when the relative humidity increases, the moisture content of timber also increases, and the timber expands. Similarly, as the relative humidity decreases, the moisture content of timber also decreases, resulting in shrinkage of timber. Of course, the moisture content is also dependent upon the species of wood and cross-section of member. In the same way, the moisture content of seasoned wood also changes from season to season, depending chiefly upon the fluctuations in atmospheric humidity, and also on the nature of the species, the cross-sectional area, and the surface treatment applied to it. The moisture content at which timber neither gains nor loses moisture when subjected to a given constant condition of temperature and humidity is known as the *equilibrium moisture content* (EMC). The value of the EMC depends on the material and the relative humidity and temperature of the air with which it is in contact (see Table 11.3). As an example, consider a piece of timber dried to moisture content of 12% and then introduced into an environment with relative humidity of about 30%. The timber eventually will reach the moisture content of about 6%. This means the timber will lose moisture content from 12% to 6%. Since the wood is already below its FSP, it will shrink.

**Table 11.3** Relative humidity vs equilibrium moisture content for 0 to 32°C

| RH (%) | 10 | 20 | 30 | 40 | 50 | 60 | 70 | 80 | 90 |
|---|---|---|---|---|---|---|---|---|---|
| EMC (%) | 2.3–2.6 | 4.3–4.6 | 5.9–6.3 | 7.4–7.9 | 8.9–9.5 | 10.5–11.3 | 12.6–13.5 | 15.4–16.5 | 20–21 |

The excess moisture content is responsible for over 90% of wood flooring problems. The application of a paint or varnish to the surface of timber reduces changes in moisture content due to changes in humidity of the surrounding air. Dipping or pressure treatment with oil or creosote-oil mixture will also have a similar effect (IS 287:1993).

Based on the data collected by the Forest Research Institute, Dehradun, India, our country has been broadly divided India into four zones, depending on the seasonal changes in the moisture content of timber, as shown in Table 11.4 (see the map of India showing these zones in IS 287:1993). The permissible moisture content or the EMC of timber for various uses in these zones, as per IS 287:1993, is given in Table 11.5. At a constant relative humidity of air, the EMC will drop by about 0.5% for every increase of 10°C air temperature.

**Table 11.4** Zones of India based on annual relative humidity

| Zone | Typical city/region | Average annual relative humidity | Nature of location |
|---|---|---|---|
| Zone I | Rajasthan and Ahmadabad | Less than 40% | Dry zone |
| Zone II | Delhi, Srinagar, and Lucknow | Between 40 and 50% | Moderate dry zone |
| Zone III | Chennai, Kolkata, and Mumbai | Between 50 and 67% | Moist zone |
| Zone IV | Assam, Cochin, Mangalore, and Vizag | Greater than 67% | Very moist zone |

**Table 11.5** Permissible moisture content of timber for different uses

| Use | Moisture content (%), Maximum | | | |
|---|---|---|---|---|
| | Zone I | Zone II | Zone III | Zone IV |
| Beams and rafters | 12 | 14 | 17 | 20 |
| Doors and windows: (a) 50 mm and above in thickness | 10 | 12 | 14 | 16 |
| (b) Less than 50 mm thick | 8 | 10 | 12 | 14 |
| Flooring strips for general purposes | 8 | 10 | 10 | 12 |
| Timber meant for further conversion, post, and poles. | 20% in all zones, moisture content being determined within a depth of 20 mm from the surface and excluding 300 mm from each end. | | | |

Hot-pressed plywood and other board products, such as particle board and hardboard, usually will not have the same moisture content as timber. The high temperatures used in the manufacture of these products will result in lower moisture content for a given relative humidity. As this lower EMC varies widely, such products may be conditioned at 30 to 40% relative humidity for interior use and 65% for exterior use (Simpson, 2010).

## 11.7.1 Shrinkage of Timber

As wood dries, it shrinks differently in different directions. The greatest shrinkage occurs parallel to the growth rings. This is called the tangential surface. Wood shrinks about 8% along this surface as it dries from the FSP to 0% moisture content. Thus, a 250 mm wide quarter-sawn board dried from the FSP (assume 30% moisture content) to 15% moisture content would shrink about 10 mm in width. This same board would shrink almost 20 mm, if dried to 0% moisture content. Wood shrinks about 4% along the radial surface (perpendicular to the growth rings) as it dries from the FSP to 0% moisture content. The least shrinkage occurs along the longitudinal axis of the tree; about 0.2–0.4%, as it dries from the FSP to 0% moisture content (see Fig. 11.7). As the timber shrinks on drying, the shrinkage in fabricated parts leads to unsightly gaps between planks and loosening of joints besides deformation, cracks, etc. The timber used should, therefore, be properly seasoned before use. It is essential that timber used should have attained moisture content as near as possible to the value that will be attained in equilibrium with the average atmospheric conditions in service.

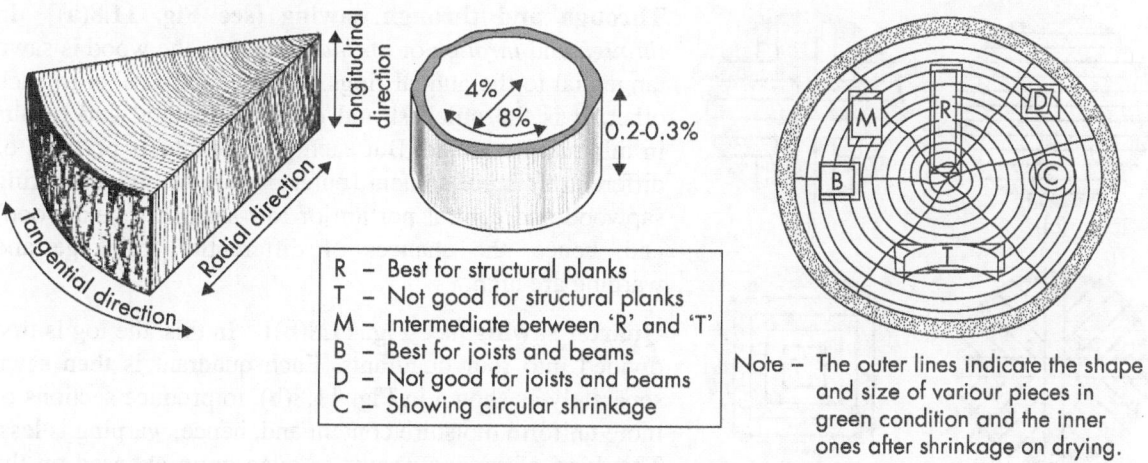

R – Best for structural planks
T – Not good for structural planks
M – Intermediate between 'R' and 'T'
B – Best for joists and beams
D – Not good for joists and beams
C – Showing circular shrinkage

Note – The outer lines indicate the shape and size of variour pieces in green condition and the inner ones after shrinkage on drying.

**Fig. 11.7** Shrinkage of timber

As wood dries, its strength properties improve. For example, the compression and bending strength of wood increase by about two-fold as wood dries from fresh to 12–15%. The tensile strength of wood is at its greatest in the 6–12% moisture content range.

## 11.8 Conversion and Seasoning of Timber

After felling the trees, the timber is cut and sawn into suitable sizes. This process is called *conversion*. Felling and conversion of trees into logs are discussed in IS 9561:1985 and IS 4895:1985. Seasoning is the process of reducing the moisture content from wood after it is converted, and is covered in IS 1141:1993. Dried wood, although lighter than greenwood, will still contain moisture, but will be stronger, less likely to warp or mould and is easier to finish with paint or varnish. The length of the process depends on the type of wood used along with relative humidity in the area where the wood is seasoned.

## 11.8.1 Conversion and Market Forms of Timber

After felling the tree, the branches are removed from the main stem to obtain a clean trunk. The trunk is then cross cut to the appropriate required lengths depending on the use of wood. These logs are then sawn into suitable sections of timber. This kind of conversion of logs should be done as soon as possible after felling the tree; otherwise, the outer rings will shrink disproportionately with respect to the central portion. Conversion exposes a greater area of timber for drying, thus accelerating seasoning. The conversion should be done in such a way that there is minimum wastage of timber. The sawing, though done manually using saws in olden days, is usually carried out by using power machines and skilled workers (guidance for hand sawing given in IS 4423:1999). The method of sawing depends upon factors such as: (a) the size of the log, (b) the condition of the log, (c) timber species, (d) size of sawing machine, (e) end use of converted timber, and (f) minimizing wastage. Guidance for mill sawing of timber is provided in IS 9576:1980, specification for cut sizes of timber in IS 1331:1971 and IS 4891:1988, and specification for coniferous and non-coniferous sawn timber in IS 190:1991, IS 1326:1992, and IS 5966:1993, respectively.

The sawing of timber can be done in several different ways, such as (a) through and through sawing, (b) quarter sawing, (c) radial sawing, and (d) tangential sawing (see Fig. 11.8).

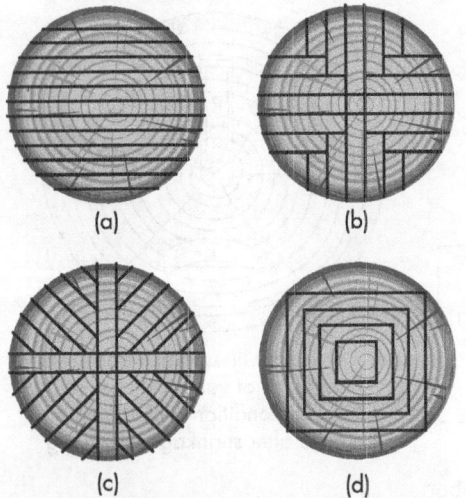

(a)    (b)

(c)    (d)

**Fig. 11.8** Sawing of timber (a) Through and through sawing (b) Quarter sawing (c) Radial sawing (d) Tangential sawing

**Through and through sawing [see Fig. 11.8(a)]** In *through* and *through* or *ordinary sawing*, the wood is sawn tangential to the annual rings. All cuts are parallel to each other. It is the quickest and cheapest method and results in minimum wastage. But each plank contains portions of different moisture content (outer portion of high-shrinking sapwood and central portion of low-shrinking heartwood) and, hence, the chances of differential shrinkage and warping are high.

**Quarter sawing [see Fig. 11.8(b)]** In this, the log is first divided into four quadrants. Each quadrant is then sawn separately as shown in Fig. 11.8(b), to produce sections of more uniform moisture content and, hence, warping is less. This kind of sawing results in edge grain exposed on the surface of timber. Hence, it is more decorative and also less prone to cupping or distorting. However, the timber has a tendency to bend in the transverse direction.

**Radial sawing [see Fig. 11.8(c)]** In this, the cuts are radial and parallel to medullary rays. In this case, there is minimal shrinkage and warping. However, there is more wastage and it also requires more labour. This method is generally used for hardwood.

**Tangential sawing [see Fig. 11.8(d)]** This is done tangentially to the annual rings. This method is adopted when medullary rays are not distinct and annular rings are very distinct. As medullary rays, which impart strength to the timber are cut, the planks obtained warp significantly.

The market names of converted timber are battens, plank, pole, scantling, etc. The preferred length is 0.5 m and downwards in 0.1 m steps. The nominal size varies from 10, 15, 20, 25, 30, 40, 50 and 60 mm and then onwards increasing by 20 mm, and going up to 200 mm.

The width of cut sizes is as follows:

1. 10 mm thick: 40, 50, 60, 80, 100, and 120 mm wide
2. 15 mm thick: 140, 160, and 180 mm wide
3. 20 mm thick: 200, 220, and 240 mm wide
4. More than 20 mm thick: 260, 280, and 300 mm wide

Some of the market forms of timber are as follows (IS 707:2011):

*Batten*   A piece of sawn timber whose cross-sectional dimensions do not exceed 50 mm in either direction.

*Baulk*   A piece of sawn timber whose cross-sectional dimensions exceed 50 mm in one direction and 200 mm in other direction.

*Planks*   A piece of sawn timber whose thickness does not exceed 50 mm but the width exceeds 50 mm. Thin planks are sometimes called as *boards*.

*Pole*   Round log used as support for communication lines, power lines, whose diameter does not exceed 200 mm.

*Post*   A general term for timber used in an upright position in building, fencing, or other structural work.

*Log*   A log is the trunk of a tree that is felled and prepared for conversion.

*Lumber*   Converted timber.

*Scantling*   A piece of timber whose cross-sectional dimension is between 50 and 200 mm in both directions.

**Loss in conversion**   The losses in conversion from logs to timber may be as follows:

*Planks*   From round logs: up to 40%; from square logs: up to 30%.

*Scantlings*   From round logs: up to 50%; from square logs: up to 40%.

## 11.8.2 Seasoning of Timber

For the efficient utilization of timber, seasoning should be done first. This is important especially in India where the climatic conditions are harsh for wood to grow. In addition, the commercial species of wood grown in India are hard, heavy, and liable to develop defects during drying and in use, if not seasoned properly. As proper seasoning will result in satisfactory service life, it should be regarded as an integral part of timber utilization (IS 1141:1993).

*Seasoning* is a process of drying timber under controlled conditions, to reduce its moisture content. The main objective of seasoning is to eliminate shrinkage and the associated drying defects. Air drying is the traditional method but seasoning is mostly done now using industrial kilns. Drying, if carried out as soon as the trees are felled, protects timber against primary decay, fungal stain, and attack by certain kinds of insects. Seasoned timber is also lighter to transport, stronger than green timber, has improved strength, electrical and thermal insulation properties, and will have better wood-working qualities such as gluing, painting, and polishing. Seasoning will also make the timber amenable for impregnation with preservatives that will prevent attack by insects and fungi.

When outer layers are dried out more rapidly than the interior layers, the timber may shrink, split, and crack. Such defects could be avoided by preventing rapid drying conditions, proper stacking and top-weighting of the stacked timber. The external drying conditions that control the rate of internal moisture movement and drying rate are the temperature, relative humidity, and air velocity. They can be better controlled in kilns rather than by air drying. However, wood kilns produce harmful air emissions and require significant heat source. Seasoned timber should be stored under cover, protected from rain, and other forms of precipitation.

A well-seasoned wood has an average moisture content of 10 to 12%. Such moisture content will be in equilibrium with the atmospheric humidity present in most parts of India. Thus, such seasoned wood will retain the shape and size of components of cabinet work, railway carriages, panelling, etc. For rough work and outdoor uses, it is enough to reduce the moisture content to 15–25%. For the purpose of seasoning, timbers are classified into three classes (see Section 11.4.7).

Converted timber can be seasoned either by (a) natural or air-seasoning or (b) artificial seasoning.

## Natural or Air-Seasoning

It is a traditional method for drying wood; it is the cheapest and the longest, taking 6 to 12 months, depending on the type of timber. To air season wood, stacks of sawn timber products are placed horizontally in layers in a covered shed, on raised brick/concrete pillars, 300 mm above ground. The raised brick pillars keep timber away from any damp ground. Timber is stacked in such a manner that air can freely circulate vertically and horizontally through the timbers (see Fig. 11.9). Crossers of uniform thickness and cross-section, say 40 mm × 25 mm, are used for stacking planks up to 50 mm in thickness. The distance between the successive crossers in the layer is kept at 600 mm for 25 mm thick planks. For thicker planks, the spacing is increased to 750 mm. The crossers should be in vertical alignment in a stack. In the case of mixed lengths, the longest planks should be at the bottom and the shortest at the top (see Fig. 11.9). A gap of 25 mm is left between adjoining planks in all the layers to allow free vertical movement of air. The width and height of stacks is restricted to 1.5 and 3 m, respectively. The ends of timber pieces dry out quickly than the central portion, resulting in long, wide cracks. To prevent this, the ends of planks or logs are wrapped or painted with suitable moisture-proof coatings. Weights are also placed at the top of the stacks (a minimum of 380 kg/m² is recommended by IS 1141), acting through a top layer of crossers and not directly over the top layer of planks, for controlling warping and cupping.

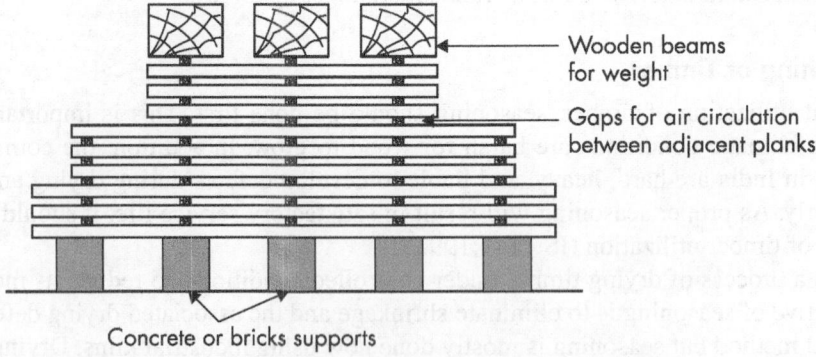

Fig. 11.9 Horizontal stacking of planks for air-seasoning

Air-seasoning is suitable for thick timber sections, which dry slowly. As there is no control over the rate of drying, drying 'degrade' cannot be controlled; forced air drying using fans in closed sheds can fairly control the rate of drying. In addition, the seasoning may not be uniform and moisture content of less than 18% cannot be attained. Such a process also requires considerable land area and may present fire hazards. If not regulated properly, the long periods of this seasoning may result in the timber being attacked by fungi and insects. For protecting timber from insect attack and decay during storage and seasoning a suitable prophylactic treatment, as discussed in Section 11.9.1, should be adopted.

**Artificial seasoning** Seasoning of timber may be accelerated by using *artificial seasoning*. Several types of artificial and special seasoning methods are available, which include (a) kiln seasoning, (b) high temperature and superheated steam drying, (c) solar drying, (d) dehumidification drying, (e) pre-drying or forced air drying, and (f) baking over open fire. Some of these methods are described briefly in the following. More details about the same may be found in IS 1141:1993.

**Kiln seasoning** It is the most common, effective, and quick commercial process of seasoning timber to the desired moisture content, using external energy. Seasoning kilns are also used for the sterilization of timber against insects and decay. The timber is stacked in kilns similar to that of air drying, and is placed inside an airtight chamber (called *batch chamber*) in which the conditions can be varied to suit the type of timber used and obtain best results. The seasoning of timber is started at a comparatively lower temperature and higher humidity. These conditions are gradually altered as the timber dries, and at the end of seasoning the temperature is kept fairly high, whereas the humidity is low. Air is circulated around the stacked timber. In modern kilns, fans are also used for a rapid circulation of air. Each species has different cell characteristics and, therefore, requires different drying schedules. For details of seasoning of Indian timbers, as per Schedules I to VII, reference should be made to IS 1141. For example, Schedule I timbers (examples: alder, fir, spruce, *maharukh*, and *gokul*) will take about 4 to 5 days to season. Schedule VI timbers, used for structural purposes and heavy planking, will take about 16 to 20 days to season. Initial steaming of the charge for about 2 h, at 55°C and 100% relative humidity, is carried out to kill mould growth. To prevent deterioration by checking, the stack is allowed to cool inside the kiln to within 15 to 20°C of the outside temperature, before removal.

Two methods, progressive and compartmental, are used for kiln seasoning. In progressive kilns, timber enters at one end and travels on a trolley through chambers, with different air conditions, to progressively dry the timber. This method produces a constant flow of seasoned timber. Wood seasoned via the compartmental process remains in a single building, where it is subjected to a programme of varying conditions, until the moisture content is reduced to the desired level. This process is used for hard-to-dry or expensive wood.

**High temperature and superheated steam drying** In this process, timber is dried in kiln at temperatures in excess of the boiling point of water. Drying times are reduced several times compared to kiln drying under medium temperature schedules. Freshly swan green hardwoods may suffer severe warping, crook, and collapse by this process. However, the process can be continuously used for quick drying without excessive degrade of hardwood timber after partial air drying or low temperature pre-drying to about 30% moisture content (IS 1141).

**Solar drying** It offers a compromise between the slow, low energy air-seasoning, and the fast and expensive steam-heated kilns. Solar kilns have single-thickness windows on the south side of the structure that work as collectors to trap the energy of the Sun. Heat collectors, made from black metal, are attached near the top of the window sashes. Various methods force the heated air to circulate through the kiln to dry the wood. Some solar kilns have insulation to retain heat at night. Two types of solar kilns are available: (a) forced air circulation design and (2) thermal circulation design (IS 1141). The forced air circulation design is suitable for rapid and uniform drying of timber without any 'degrade'. In this design, a water spray is also provided for additional humidification required while drying thick and refractory timber. Solar kilns of capacity up to 50 m³ of sawn timber per charge have been used in India (see also IS 15890:2010). As the entire energy input required in the form of heat for the evaporation of water is derived from the solar energy, it is energy efficient and pollution free. Electric energy is required, only for driving the kiln fans. This process takes approximately twice as long as traditional kiln seasoning; however, it is much less than that of air-sasoning.

**Pre-drying or forced air-drying** In this method, unheated atmospheric air or air heated to about 42°C is forced through the timber stack by means of fans in a single pass or by a recirculating system. Low temperature pre-drying is carried out inside sheds with lightly insulated walling. A solar kiln may also be very effectively used for partial pre-drying.

**Chemical seasoning** In this type of seasoning, green sawn timber is first soaked in a solution of hygroscopic chemical for an appropriate period, depending upon the species and thickness of section, so as to treat the outer layers of the section by diffusion. The treated timber is then air or kiln seasoned. Any of the following chemicals may be used: polyethylene glycol, commercial fertilizer grade urea, or urea-sorbitol solution (IS 1141).

**Electrical seasoning** In this method, the logs are placed such that their two ends touch the electrodes. Current is passed through the setup. As timber is a bad conductor, it resists the flow of current, generating heat in the process, which results in its drying. The drawback is that the wood may split.

**Microwave seasoning** Microwave seasoning, used in Canada, consists of directing pulsed microwave energy into layers of timber, such that the moisture in the timber is driven out, at rates that will not cause seasoning 'degrade'. It is possible to deliver energy that can be varied to suit the moisture content of the timber and the ambient conditions.

After seasoning, the timber undergoes *planning* (also called *dressing*). The planning not only removes the rough sawn surfaces, but also reduces the timber to uniform standard thicknesses and widths.

### 11.8.3 Seasoning Degrade

Most seasoning degrades that develop in timber can be classified as fracture or distortion, warp, or discolouration. Wood shrinkage is mainly responsible for ruptures and distortion of shape (twisting, bow, or cupping). Cell structure and chemical substances in wood contribute to defects associated with uneven moisture content, undesirable colour, and undesirable surface texture. Drying temperature is the most important factor because it is the main cause for all these defects.

## 11.9 Preservation of Timber

There are many treatments done to timber to improve its dimensional stability, resistance to biological degradation, thermal stability, or fire resistance, UV resistance, and mechanical properties. Wood treatments normally use one of three strategies: modification of the cell wall, impregnation, and coating. Modification of the cell wall can be further divided into thermal modification and chemical modification, both of which are active strategies. Some of these strategies are explained in this section.

In India, the supply of timber from traditional naturally durable species, such as teak and *sal,* from the forest is decreasing. However, timber from plantation species such as rubber, eucalyptus, and poplar, is becoming increasingly available. Such species are normally non-durable, but when suitably treated, would give adequate life under service conditions. Hence, IS 401:2001 suggests that timber with more sapwood content (easily treatable) or timber having good treatability can also be substituted and used appropriately. Preservative treatment of timber conserves wood and greatly increases the life of wooden structures, thus reduces replacement costs, and allows more efficient use of forest resources.

The efficacy of preservative treatment depends on the proper choice of preservative and the treatment process, which ensure the required absorption and penetration of the preservative. *Preservatives* protect wood by poisoning the fibres to create conditions that will kill fungi and other insects. Preservative treatment, however, will not improve the density or hardness or mechanical, electrical, or chemical properties of wood. Preservative treatment and seasoning make the wood stable and durable. Preservative treated and seasoned wood is distinct from densified wood of the same species and should not be classified as densified wood.

Preservatives used to poison the food supply of fungus are of the following three types (IS 401: 2001): (a) oil type, (b) organic solvent type, and (c) water-soluble type (leachable and fixed).

**Oil type**  It is mainly coal/lignite tar creosote (conforming to IS 218), with or without the admixture of coal tar petroleum oil, fuel oil, or other suitable oil having a high boiling range. *Creosote* is one of the oldest and still used widely (In the USA, during 1986, creosote became a restricted-use pesticide, and its use is currently restricted to pressure-treatment facilities). It is a viscous liquid that is normally heated and forced into wood under pressure. The advantages of using creosote are that it has high toxicity and relatively high permanence and is non-corrosive. It offers good protection against termites. However, it is difficult to handle (leaves the surface of wood rather dirty), has an unpleasant odour, and is not suitable when timber is to be painted. It is extensively used in the piles, fence and utility poles, and railway sleepers.

**Organic solvent type**  These preservatives are insoluble in water and consist of toxic chemical compounds, such as copper naphthenate orabietate, pentachlorophenol, and gamma-BHC. These are dissolved in suitable organic solvents like naphtha, or in petroleum products such as kerosene or spirit. These preservatives are permanent and the treated material is clean to handle when light organic solvents are used. As some of the solvents are inflammable, care is necessary in handling the solutions, and, in such cases, the preservative are applied cold. In most cases, treated timber can be painted, waxed, or polished.

**Water-soluble type**  They are two types: leachable and fixed type. Water-soluble (leachable) salt types include zinc chloride, boric acid, borax, gamma-benzene hexachloride (water dispersible powder), sodium fluoride, and synthetic pyrethroids. These are adopted for inside locations only. If applied over outside surfaces, the salts maybe leached by rainwater. Water-soluble (fixed) types include copper-chrome-arsenate (CCA), copper-chrome boron, ammoniacal copper arsenate (ACA), and ammoniacal copper zinc arsenate (ACZA) [see also IS 10013(Parts 1 to 3)]. The role of chromium is to fix the toxic element, arsenic, copper, or boron, in the timber, so that it becomes difficult for the toxic salts to leach out by the action of water. The CCA was invented in 1933 by Dr Sonti Kamesam, an Indian scientist, and patented in 1934. The CCA is mainly recommended in heavy termite and marine borer infested areas and in cooling tower timbers where soft rot is the chief deteriorating factor. Though toxic to insects and fungi, the CCA [conforming to IS 10013 (Part 2)] treated timber is non-toxic to humans under normal conditions of use. However, the USA, Canada, the European Union (EU), and Australia have banned the use of the CCA, due to the fear that arsenic may leach out of the treated timber. The ACA is recommended for treating refractory wood species like eucalyptus, and can be used at elevated temperatures to reduce the treating period. The water-soluble types are popular because water is a cheap solvent, and the surface of wood is left clean. The treated material should be allowed to dry for 2 to 3 weeks to complete the fixation process and get the appropriate moisture content. These preservatives should be applied cold as these are liable to get precipitated when heated. Timber treated with these preservatives may be used outdoors and can also be painted.

Wood preservatives may be applied by brushing, spraying, dipping, or by other non-pressure processes and pressure processes. For the oil-type preservatives, the moisture content in timber should not be more than 20% (IS 401). Whereas for water-soluble preservatives, moisture content of 20 to 30% is permissible. At least two coats should be applied. The second and subsequent coats should not be applied after the first has dried. Where possible, the treatment is done hot. Surface treatment is used mostly for treating timber at site and for the retreatment of cut surfaces. The only method that forces the preservative to penetrate deep into the timber and provides protection under severe conditions is the one using pressure processes. Several pressure processes such as the Full Cell or Bethel Process, Empty Cell Process, Lowry process, Fast Fluctuating Pressure Process, and Alternating Pressure Method, are available (see IS 401, for the details). The Bethel Process involves placing the timber in large cylinders, removing the air, introducing the preservative, and then applying a pressure of about 35 to 125 MPa, so that the preservative is forced deep into the timber. In the case of oil-type preservative, a temperature of 80 to 90°C should be maintained during the pressure period. The degree of penetration of the preservatives into the timber depends on its permeability and treatment method. Naturally, sapwoods are more permeable than heartwood.

Most of the preservatives are poisonous or skin irritant and some are inflammable or explosive or yield poisonous vapour, which may catch fire. Hence, both workers in the field and factories and actual users of treated material should be careful while handling preservatives, treating solutions, as well as freshly treated wood. It is also important that such preservative treatment be carried out by well qualified and experienced technicians, who have access to laboratory and field equipment to identify and evaluate the preservatives in respect of their purity, correctness of the composition, and their presence quantitatively in the treated material. More details of the different preservatives, the methods of their application, and tests to evaluate new wood preservatives may be found in IS 401:2001; Lebow, 2010; Richardson, 1993; and Archer & Lebow, 2006.

## 11.9.1 Prophylactic Treatment

Freshly felled timber is liable to be attacked immediately by fungi and insects and may also develop splits and cracks. Splits and cracks develop because of the wide difference in moisture content in the fresh logs and the relative humidity of the surroundings; this condition is pronounced in the hot and dry seasons. This may be controlled by applying coating to the ends of logs and also by ponding the logs in fresh water. In species such as teak, rosewood, *padauk*, and walnut, the bark is removed completely and an artificial bituminous bark containing preservatives is provided to protect both mechanical and biological damage. This artificial bark is removed after warming at the processing centres, before further processing. When the logs are stored for long periods, they are ponded in fresh water, or properly stacked high above the ground and continuously sprayed with water containing insecticides and fungicides. When there is slight insect or fungal attack, the logs are sterilized by boiling in hot water or steaming (at 15 MPa pressure, for a period depending on the size and moisture content of the material). Prophylactic treatments are then applied to prevent fresh infection.

## 11.9.2 Precautions to be taken while Handling Chemically Treated Wood

When sawing, sanding, and machining treated wood, a dust mask should be used. Whenever possible, these operations should be performed outdoors to avoid indoor accumulations of airborne sawdust from treated wood. When power-sawing and machining, goggles should be worn to protect eyes from flying particles. Prolonged skin contact with creosote – or pentachlorophenol-treated wood – should be avoided. Vinyl coated gloves, long sleeved shirts, and long pants should be worn when handling such harmful preservatives. Only treated wood that is visibly clean and free of surface residue should be used for patios, decks, and walkways. Chemically treated wood should not be used in the interiors of farm buildings, or in places where the timber will be in frequent contact with bare skin (for example, chairs), or with drinking water. In building components that are in contact with ground and are subject to decay or insect attack, two coats of an appropriate sealer should be applied at the installation site. Urethane, shellac, latex epoxy enamel, and varnish may be used for pentachlorophenol-treated wood. Sealers such as urethane, epoxy, and shellac are acceptable sealers for all creosote-treated wood. The CCA-treated wood should be disposed carefully – disposed treated wood should not be used as firewood or in stoves or fireplaces, because toxic chemicals may be produced as the part of smoke and ashes.

### 11.9.3 Thermal or Heat Treatment

Thermal or heat treatment decreases the hygroscopicity of timber and improves the dimensional stability and durability against biodegradation. However, thermally modified wood becomes more brittle, with lower bending and tensile strength, abrasion resistance, and toughness; thus, it is not suitable for load-bearing applications. However, poly-condensation reactions of lignin result in higher strength in the longitudinal direction, along with an increase in compressive strength and stiffness. Wood can be heated in various ways: (a) heating dry wood, (b) heating wood in the presence of moisture, (c) heating dry wood followed by compression, and (d) heating wood in the presence of moisture followed by compression. Commercial heat treatment is usually performed in the absence of air at temperatures ranging from 180 to 260°C, for a few minutes to several hours. Such heat-treated wood products are called by various names in different parts of the world (examples: Staypak and Staybwood in the USA, ThermoWood in Finland, Lingostone and Lignofol in Germany, and Jablo in the UK). All heat-treated wood can be glued and painted and may be used for furniture, flooring, decking, door, and window components. More details of heat treated wood may be found in Ross (2010).

### 11.9.4 Wood-Polymer Composites

In this treatment, wood is vacuum impregnated with certain liquid vinyl monomers that do not swell wood and are later polymerized *in situ* by gamma radiation or chemical-catalyst-heat systems. This method does not alter the hygroscopic characteristics of wood but the high void volume is greatly decreased, improving the water-repellent effectiveness of wood. Wood-polymer composites offer better aesthetic appearance, have high compressive strength, abrasion and wear resistance, and increased hardness. These products are used mainly in hardwood flooring. More details about wood polymer composites and impregnation may be found in Ross (2010).

### 11.9.5 Protective Coating

A *protective coating* or sacrificial layer can be painted on the surface of wood products, providing a physical barrier against weathering and degradation, and increasing the aesthetics of the product. As a surface treatment, coating usually is the final operation of wood processing.

Protective coatings can retard dimensional changes in wood but do not prevent them. Moisture-excluding effectiveness (MEE) of coatings depends on a number of variables: coating film thickness, defects and voids in the film, type and amount of pigment, chemical composition and amount of resin, vapour-pressure gradient across the film, and length of exposure (Ross, 2010). Maleic-alkyds, two-part polyurethane, and paraffin wax have high MEE. Coatings that retard water vapour diffusion also repel liquid water. Porous paints, such as latex and low-lustre (flat) paints, afford little protection against water vapour transmission.

Paints are highly pigmented film-forming coatings and give the most protection against UV radiation. Paints protect wood surfaces from weathering, conceal some surface defects, provide a cleanable surface, offer many colours, and give high gloss. However, paint is not a preservative. It will not prevent decay if conditions are favourable for fungal growth. Paint is available in two general types: solvent-borne oil-alkyds and waterborne latexes (usually, acrylic or vinyl acrylic polymers). Oil-alkyd primers block water absorption into end grain and, to a limited extent, can penetrate wood cell walls, thus modifying the surface and improving its dimensional stability. Latex primers do not penetrate cell walls but merely flow into cut cells and vessels. Latex primers do not seal the end grain as well as oil-alkyd primers do. Latex primers are porous and thus permeable to water and water vapour; oil-alkyd paints are less permeable to water and water vapour. Latex top-coats can be applied over oil-alkyd primers. A primer and two top-coats should be applied to achieve a dry film thickness of 0.10–0.13 mm (4–5 mil). More information on paints and the methods of their application may be found in Ross (2010), and in Chapter 18.

## 11.10 Storage of Timber

IS 3629:1986 suggests that timber, after conversion and preservation, should be stored in stacks, similar to those used in air-seasoning, as described in Section 11.8.2 and Fig. 11.9.

## 11.11 Qualities of Good Timber

In general, a good timber should have good strength, high durability, and a good finished appearance. The physical properties of the wood are also depend on the following:

1. The type of soil in which the tree grew
2. The type of tree
3. The age and maturity of tree
4. Time of felling the tree
5. Method of seasoning
6. The type and process of preservation

The required characteristics are as follows:

**Strength**   It should have sufficient strength to resist heavy structural loads.

**Density**   *Wood density* has long been considered the most important wood quality. To a large extent, the density determines the suitability of wood for a specific end use. High-density wood is usually associated with high lumber strength and stiffness.

**Hardness**   It should be hard enough to resist deterioration. It is important consideration for flooring. Timbers having hardness of 4.5 to 9.1 in Janka scale are found to be optimal in flooring.

**Durability**   It should be durable and able to resist attacks of fungi and worms and also atmospheric effects for a longer period of time.

**Appearance**   Freshly cut surface of timber should give sweet smell and have smooth and shining surface. It should have uniform dark colour, as light coloured timbers are generally considered weak. It is to be noted that sapwood is weak and has white to light brown colour.

**Dimensional stability**   It should retain its shape and size during the processes of conversion and seasoning, and also during service. It should not warp, bow, twist, split, or expand during its service life.

**Moisture content**   It affects operations such as conversion and seasoning. Moisture content affects shrinkage, warping, strength, stiffness, adhesive bonding, and the formation of mould and decay. It should be limited to 10–20% depending on the use of wood.

**Toughness**   It should have enough toughness to resist shocks due to vibrations. It should not break in bending and should resist splitting.

*Structure*   The structure of timber should be uniform with narrow annual rings. Timber with close, narrow angular rings and compact medullar rays are generally the strongest.

*Defects*   Timber should be prepared from the heartwood of a sound tree and be free from sap, dead knots, twits, shakes, and other similar defects.

*Elasticity*   It should have the property of elasticity so as to regain its original shape after removal of loads.

*Weight*   Heavy timbers are always stronger than lightweight timbers.

*Workability*   It should be well seasoned and easily workable. Teeth of saw should not get clogged during the process of sawing. It should provide smoothened surface easily.

*Fire resistance*   A dense timber is a bad conductor of heat and, hence, offers good resistance to fire and requires sufficient heat to cause flame. Fire resistance of any wood can be enhanced by various treatments such as pressure-impregnated fire retardants or intumescent surface coatings.

*Sound*   When two pieces of the same timber are struck against each other, it should emit a clear ringing sound.

*Polishing*   Timber used in doors, windows, balusters, handrails, etc., should be able to take good polish.

*Others*   Other wood characteristics include acoustic, thermal, electrical, and other properties that may be of importance to specific end uses. Lightweight timber is a better insulator than dense timber.

## 11.12 Selection of Timber

As there is a wide variety of timber available in India (about 150 species), the type of timber chosen depends on the chosen project, and its use. Thus, timber suitable for doors and windows is different from that used for beams and columns. In India, teak wood is considered ideal for all types of uses. Few other wood species that are commonly used in India are given in Table 11.6. Generally, timber is selected based on (a) local availability, (b) grade (see Section 11.4.2), (c) closeness of grains, (d) hardness, (e) durability (see Section 11.4.4), (f) ease of working, (g) colour, and (h) the ease of polishing.

Table 11.6 Types of timber used in India (see also www.wood-database.com)

| Common name | Colour | Density, kN/m³ | Location | Characteristics and uses |
|---|---|---|---|---|
| Ben-teak | Brown | 6.75 | Kerala, Tamil Nadu, Maharashtra, and Karnataka. | Strong and takes up a smooth surface; used for building constructions and furniture. |
| Deodar | Yellowish brown | 5.60 | Jammu and Kashmir, Himachal Pradesh and Uttarakhand. | This softwood with distinct annual rings can be easily worked and is moderately strong; used for making furniture, railway sleepers, packing boxes, and structural work. |
| Indian Elm | Red | 9.60 | Throughout India | Moderately hard and strong; used for door and window frames. |
| Mahogany | Reddish brown | 7.20 | Kerala, Tamil Nadu, Karnataka, Andhra, and West Bengal. | Takes a good polish and is easily worked; most commonly used for furniture and cabinet work. |

*(Contd)*

**Table 11.6** (*Contd*)

| Common name | Colour | Density, kN/m³ | Location | Characteristics and uses |
|---|---|---|---|---|
| Rosewood | Dark brown | 8.50 | Kerala, Karnataka, Maharashtra, M.P., Tamil Nadu, and Odisha. | Strong, tough, and close-grained; provides aesthetic appearance, takes high polish, and has good dimensional stability; used for quality furniture, cabinet work and ornamental carvings. |
| Sal | Brown | 8.80–10.50 | Karnataka, Andhra Pradesh, Maharashtra, U.P., Bihar, M.P., and Odisha. | Hard, fibrous, and close-grained and requires slow and careful seasoning; durable underground and in water; does not take up good polish; used for railway sleepers and bridges. |
| Sissoo | Yellowish white | 7.70 | Mysore, Maharashtra, Assam, West Bengal, U.P., and Odisha. | Strong and tough; it is durable, attractive, and has dimensional stability; easily seasoned; difficult to work but takes good polish; used for high-quality furniture, plywood, bridge piles, railway sleepers, and in decorative works and carvings. |
| Teak (Balharshah, Malabar, Dandeli, C.P. teak, Burma teak) | Deep yellow to dark brown | 6.40 | Central India and southern India. | Moderately hard, durable, and fire-resistant; easily seasoned and worked; takes good polish and not attacked by insects and dry rot. |

Timber used in building is usually not tested in the laboratory before using it. It is chosen by its species and visual examination for any aforementioned defects.

## 11.13 Fire Proofing of Timber

Timber is combustible but its behaviour in fire is fundamentally different to that of steel and reinforced concrete. The strength and stiffness of timber are both reduced at higher temperatures. Thus, its strength is reduced by more than 50% at 100°C, compared to that at 20°C. In the case of fires, the charred outer layer of timber can insulate the inner material. When steel is used to connect timber elements, heat is quickly conducted through the connectors, degrading the strength and stiffness of the wood around them.

One method of fire-proofing consists of pressure impregnating the wood with waterborne or organic solvent-borne fire retardant chemicals. The second method consists of applying fire-retardant chemical coatings to the wood surface. The pressure impregnation method is usually more effective and longer lasting; however, this technique is standardized only for plywood (Ross, 2010). For wood in existing constructions, surface application of fire-retardant paints offers a possible method to reduce flame spread. Fire-retardant finishes have low surface flammability, and when exposed to fire, they 'intumesce' to form an expanded low-density film. The expanded film insulates the wood from heat and retards combustion. The finishes have additives, which promote the decomposition of wood to charcoal and water rather than flammable vapours.

## 11.14 Structural Properties of Timber

Unlike other building materials like steel and concrete, timber is as an orthotropic material and has different mechanical properties in three mutually perpendicular axes (longitudinal, radial, and tangential). Timber has high compressive strength when it is loaded parallel to grains. In addition, hardwood will be slightly stronger than softwood and the amount of seasoning has considerable effect on strength.

Hence, well-seasoned timber only should be used in structural applications. The strength of timber even for the same species varies with its location and its position on the same tree. Timber has a low density compared to concrete or steel. As knots affect the strength, such knots occurring in tension members or the tension side of beams, and in short columns should be carefully considered. 'Close grained timber' is stronger than 'loose grained timber'. In structural timber, a minimum of 12 rings per 50 mm is required for strength. Larger slopes of grain result in reduction of strength. Limiting slopes are given as: select grade (with minimum or no defects): one in 20, Grade-I (having defects not larger than specified in IS 1331): one in 15, and Grade-II (poorer in quality than Grade-I): one in 12.

Since moisture content affects the strength of timber, three locations are specified for strength in IS 883:1994: (a) *inside location-* where timber remains dry and protected from weather, (b) *outside location-* where timber is subjected to occasional wetting and drying, (c) *wet location-* where timber is continuously damp or wet, being in contact with moist earth or water. Structural properties and safe working stresses for some selected Indian timbers are given in Table 11.7. These values for other types of timber are provided in IS 883:1994. Different factors of safety have been used for different locations, while arriving at these permissible stresses. These factors of safety are given in Table 11.8 (Table 2 of IS 3629:1986). A factor of safety of 1.25 has been used for the modulus of elasticity. The permissible values given in Table 11.7 have to be multiplied with modification factors to take into account for change in the slope of grain and creep due to sustained load as specified in IS 883.

**Table 11.7** Properties of a few selected species of timber (see also IS 883:1994)

| Trade name | Average mass at 12% moisture content, kN/m³ | Modulus of elasticity, MPa | Permissible stress for Grade-I, MPa | | | |
|---|---|---|---|---|---|---|
| | | | Bending and tension along grain | Horizontal shear | Compression parallel to grain | Compression perpendicular to grain |
| Ben-teak | 6.17–6.75 | 107.6 | 10.6 | 0.84 | 7.3 | 2.6 |
| Coconut | 7.61 | 73.4 | 7.7 | 0.74 | 8.4 | 3.0 |
| Deodar | 5.60 | 94.8 | 8.7 | 0.7 | 6.9 | 2.1 |
| Oak | 8.7–10.08 | 108.2–124.4 | 12.1–13.1 | 1.15–1.22 | 7.1–7.8 | 2.9–3.9 |
| Rosewood | 8.50–8.84 | 83.9 | 10.8 | 1.08 | 7.1 | 3.3 |
| Sal | 8.80–10.50 | 126.7 | 14.0 | 0.94 | 9.4 | 3.5 |
| Sissoo | 7.70–8.0 | 71.4 | 10.7 | 1.25 | 7.3 | 3.3 |
| Teak | 6.40–6.60 | 99.7 | 12.9 | 1.15 | 8.3 | 3.5 |

**Note:** Only outside location values for bending, and compression are given. IS 883 gives three values: inside, outside, and wet locations.

**Table 11.8** Factors of safety applied to obtain safe permissible stress

| Type of stress | Grade-I (standard) | | |
|---|---|---|---|
| | Inside location | Outside location | Wet location |
| Extreme fibre stress in beams for broad leaved species, Min | 5 | 6 | 7.5 |
| Extreme fibre stress for beams in conifers, Min | 6 | 7 | 8.5 |
| Horizontal shear in beams | 10 | 10 | 10 |
| Shear along grain | 7 | 7 | 7 |
| Compressive stress parallel to grain, Min | 4 | 4.5 | 5.5 |
| Compressive stress perpendicular to grain | 1.75 | 2.25 | 2.75 |

Timber will be structurally efficient when high proportion of the load to be resisted is the self-weight of the structure. During an earthquake, the forces imposed on the structure depend strongly on the mass of members, with heavier structures experiencing larger seismic forces. Light timber residential buildings will therefore perform well in seismic events, such as that happened during the 2011 earthquake in Christchurch, New Zealand (Ramage et al., 2017). More details about properties of the various species of timber may be found in Ross (2010).

## 11.15 Advantages and Problems of Timber

The use of timber in building construction has the following advantages:

*Sustainable*   If reforestation is properly planned and implemented, timber is the ideal natural material for construction, as it is renewable (see also Section 11.25).

*Versatile*   Timber can be used for both load-bearing and ornamental applications. It can be handled by traditional labour with simple tools.

*High strength-to-weight ratio*   Most species have very high strength-to-weight ratio, making them ideal construction materials, especially in earthquake zones due to its lightweight.

*Durable*   Wood has long life (greater than 100–125 years), and subject to decay only when used in damp locations. If properly seasoned and preserved, timber is very durable.

*Easy to fabricate*   Timber can be sawn, cut, and joined and formed into any shape and size with simple carpentry tools.

*Easy and fast to build*   A timber construction can be built very easily and fast in contrast to the common brick or concrete construction.

*Aesthetic appeal*   Timber, with its beautiful grains, can provide aesthetic appearance for both internal and external applications.

*Excellent insulator*   Houses built out of timber have outstanding insulating properties, and better sound absorption, providing comfortable and healthy living. Timber can be used in all climatic zones.

*Better fire resistance*   Thicker timber members have better fire resistance than steel, as the charred outer surfaces protect the inner fibres.

*Easily repaired*   Alterations and repairs can be easily carried out on timber buildings.

*Can be easily recycled*   Demolished timber structural or other members can be easily recycled or even burnt as fuelwood – the ash can even serve as a fertilizer.

*Use of non-traditional timber*   These, like coconut timber (coco-wood), are now being utilized as an alternative to scarce timber resources. Structural applications of laminated bamboo have been demonstrated in full-scale construction such as short-span bridges and two-storey housing (Sharma et al., 2017; Kaminski et al., 2016, 2017).

Although, timber is considered the healthiest and most sustainable building material, the use of timber may have the following problems:

*Misuse of forest produce*   India is facing a severe shortage of timber supplies from domestic sources, due to unsustainable extraction and illegal trade resulting in the loss of forests and biodiversity (World's forests are depleting at the rate of 16 million ha per annum). Although demand for timber in India is estimated as 153 million cubic meters in 2020, supply will be around 60 million cubic meters only.

To address this problem, conservation measures (including comprehensive long term re-forestation programmes), which regulate extraction and use of forest produce, should be effectively enforced.

*Attack by insects*   The secondary timber species are generally susceptible to fungal decay, rot, and attack by insects. These problems are minimized by harvesting timber in mid-summer or mid-winter season.

*Problems due to seasoning*   Seasoning of timber, to less than 20% moisture content, results in defects such as cracking or splitting. These defects can be minimized by controlled slow drying.

*Health hazards of preservatives*   Most effective chemical preservatives are highly toxic and pose serious health hazards. To avoid this, correct application methods and precautionary measures should be undertaken.

*Shrinking and swelling*   Timber has a natural ability to absorb water, resulting in shrinkage and swelling. Proper seasoning, coating, etc., will minimize this problem.

*Fire resistance*   Timber is highly combustible. To prevent this, sections that are slightly bigger than required should be chosen and fire-resistant treatment, which ensures specified fire rating, should be provided.

*Discolouration*   Untreated timber exposed to weather results in the discolouration of timber. Hence, contact of timber with ground should be avoided and timber exposed to exterior should be protected from rain, sunlight, and wind.

## 11.16 Industrial Timber Products

Solid timber of good quality is in short supply, especially when wide pieces are required. To rectify this, timber is made into industrial products by special processes. For the manufacture of such products, timber is converted into particles, strands, or laminates, which are then combined using glues to produce *timber composite products*. These composites can be broadly divided into the following four categories:

1. Layered timber composites [examples: glulam, plywood, and laminated veneer lumber (LVL)]
2. Particle composites [examples: particle boards and parallel strand lumber (PSL)]
3. Fibre composites [examples: hardboards and medium density fibreboard (MDF)]
4. Timber-concrete composites (TCC)

Some of these products are shown in Fig. 11.10 and described briefly in the following sections.

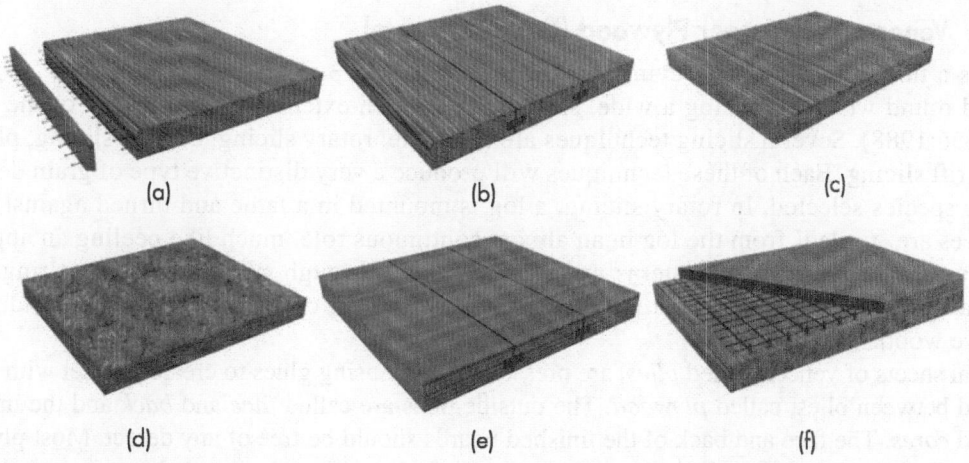

**Fig. 11.10** Mass timber panellized products (a) Nail-laminated timber (NLT) (b) Glued-laminated timber (GLT) (c) Cross-laminated timber (CLT) (d) Laminated strand lumber (LSL) (e) Laminated veneer lumber (LVL) (f) Timber-concrete composites (TCC) (*Courtesy*: Jackson et al., 2017)

*Structural composite lumber* (SCL) products are characterized by smaller pieces of wood glued together into sizes common for solid-sawn lumber. Examples of SCL include LVL, laminated strand lumber (LSL), PSL, and oriented strand lumber (OSL). All types of SCL products can be substituted for sawn timber products. The various types of SCL are also used in a number of non-structural applications, such as the manufacture of windows and doors (Ross, 2010).

Wood-based composites are used for a number of structural and non-structural applications including interior panelling, sheathing, furniture, and support structures in many different types of buildings. Several structural assemblies can be built-up using these composites such as I-beams, T-beams, box-beams, and stressed-skin panels. Synthetic resin thermosetting adhesives used in the production of these composites are formaldehyde-based: urea-formaldehyde (UF), melamine-urea-formaldehyde (MUF), phenol-formaldehyde (PF), phenol-resorcinol-formaldehyde (PRF), and isocyanate (see also IS 848:2006; IS 15684:2006). Extender should not be added to the adhesive by the plywood manufacturer and fillers, if used, should not exceed 10% by mass of solid content of the glue. The resin acts as the enabler for these composites, providing strength, durability, performance, and enhanced wood resource use and efficiency. Preservation of plywood and other panel products should be done as per IS 12120:1987. Wood-based composites represent more than 40% of the total materials used in residential construction in North America, making them the largest single material type used in residential construction (www.fpl.fs.fed.us). Wood-based composites have become so widely used because they are readily available, light, strong, easily worked, and cost-effective.

## 11.17 Layered Timber Composites

Layered composites can be used to produce sheets as well as structural sections. Layered composites can also be reinforced or interleaved with other materials to further increase their strength and dimensional stability. Layered timber composites may be sub-classified as:

1. Cross laminates: Plywood, cross-laminated timber (CLT), nail-laminated timber (NLT);
2. Parallel laminates: glued-laminated timber (GLT or Glulam) and LVL
3. Block boards

All these categories consist of multiple layers of timber sections glued together to act as a single unit.

### 11.17.1 Veneers and Veneer Plywood (Cross-laminate)

*Veneer* is a thin sheet of wood having thickness of less than 3 mm sliced from 1.5 to 2.5 m long, streamed round wood logs using a wide, sharp knife, which extends the full length of the log (see also IS 656:1988). Several slicing techniques are available: rotary slicing, quarter slicing, plain slicing, and rift slicing. Each of these techniques will produce a very distinctive type of grain depending upon the species selected. In rotary slicing, a log is mounted in a lathe and turned against a blade. Thin slices are 'peeled' from the log in an almost continuous role, much like peeling an apple. This method produces a variety of patterns as the blade slices through successive growth rings. Using veneer, an attractive wooden finish is obtained at much lower cost than by using a solid piece of expensive wood.

Several sheets of veneer, called *plies*, are pressed together using glues to create a panel with an adhesive bond between plies, called *plywood*. The outside plies are called *face* and *back* and the inner plies are called *cores*. The face and back of the finished boards should be free of any defect. Most plywood is always made from an odd number of layers, (typically 3, 5, or 7), with an equal number of plies sandwiched on each side of the centre ply. In this manner, the surface plies (face and back) are always parallel, and the grain of the surface plies usually follows the longest side of the sheet. Commonly, an outer layer of expensive and attractive wood is used and a core of cheaper materials.

Plywood can be made from either softwoods or hardwoods – typical species that are used are walnut, teak, sissoo, and rosewood (also see Annex B of IS 303:1989; IS 12077:1987). Hardwood plywood is used mainly for panelling, furniture and cabinets, whereas softwood plywood is used in construction purposes. Most softwood plywood has a veneer core, whereas hardwood plywood is made with several different core materials. Hardwood and softwood species can be mixed in both types of plywood. Usually, hardwood plywood has a hardwood face. Veneers from non-durable species and sapwood of all species should be treated as given in IS 303:1989.

There are many types of classifications of plywood. Based on quality, it is classified into three grades: (a) ordinary grade, (b) exterior grade, and (c) marine grade (IS 710:2010). In India, plywood is commonly called as *kitply*, after a leading brand that pioneered waterproof plywood in 1970s (www.kitply.com). Ordinary grade plywood is used for packing. Exterior grade is bonded with waterproof glue and marine grade has core plies as well as face veneers of good quality (also see IS 9188:1979). As per IS 303, plywood for general purposes has the following grades: (a) boiling water resistant or BWR Grade, and (b) moisture resistant or MR Grade. Plywood may also be classified into three types, namely, AA, AB, and BB, based on the quality of the two surfaces (The first letter denoting the quality of 'face' and the second by the quality of 'back'. Thus, Type AA will denote both surfaces of better quality A). Specifications for veneered decorative plywood, which may have ornamental veneers on one or both faces, are provided in IS 1328:1996. This type of plywood is used for decorative purposes, such as in furniture, panelling of all kinds, interior lining of railway coaches, buses and ships, and general interior decoration. Such plywood may be having a chosen 'face' of teak or rosewood to match the solid wood used in a building. Standard dimensions of plywood boards are: (a) 2400 mm × 1200 mm, (b) 2100 mm × 1200 mm, (c) 1800 mm × 1200 mm, (d) 2100 mm × 900 mm, and (e) 1800 mm × 900 mm (also see IS 12049:1987). The following thicknesses of plywood are specified in IS 303:

1. 3 ply: 3, 4, 5, 6 mm
2. 5 ply: 5, 6, 8, 9 mm
3. 7 ply: 9, 12, 15, 16 mm
4. 9 ply: 12, 15, 16, 19 mm
5. 11 ply: 19, 22, 25 mm

In the case of 3-ply boards, the thickness of the centre veneer (core) should be at least equal to that of one of the face veneers, but should not exceed the combined thickness of the two face veneers. In the case of multi-ply boards, the thickness of any veneer should not exceed twice the thickness of any other veneer in the same board. Other quality requirements of plywood for general purposes are given in IS 303:1989. The board should be given treatment by pressure impregnation with fixed type either water-soluble or oil-based preservatives, as per IS 5539:1969 (also see IS 12120:1987).

The properties of plywood depend on the quality of the different layers of veneer, order of layer placement, adhesive used, and control of bonding conditions. Specifications for structural plywood are available in IS 10701:2012, fire retardant plywood in IS 5509:2000, and plywood for concrete shuttering work in IS 4990:2011.

**Advantages of using plywood** The cross-graining of plies used in plywood provides dimensional stability, warp resistance, and increased strength. The high surface dimensional stability makes it well suited to formwork applications (see also IS 4990:2011). It also makes the strength and stiffness properties of the plywood panel consistent across both directions. The laminated construction distributes defects, markedly reduces splitting when the plywood is penetrated by fasteners (compared with the splitting of solid wood), and improves resistance to checking. The use of plywood may also result in better utilization of wood. Plywood can cover large areas with a minimum amount of wood fibre because plywood that is thinner than sawn timber can be used in some applications. Nails and screws can be placed close together and close to the edges of panels. Panel shear strength is approximately double that of solid timber of the same thickness. High strength and stiffness-to-weight ratios make the handling and installation of plywood cost effective. Plywood is easily pressure treated with waterborne preservatives and fire retardants (see IS 12120:1987; IS 5509:2000).

**Decorative laminates**   Also called *laminated plastics* (popularly known as *sun-mica* in India), it was invented in 1913 by Herbert A. Faber and Daniel J. O'Conor of the USA. *High pressure decorative laminates* (HPL) may be used on plywood sheets, to provide decorative surfaces and have relatively hard and resistant surfaces for wear, scratching, impact, boiling water, domestic stains, and moderate heat. These sheets are intended for interior applications and are ready for use. The back surface of these sheets is suitable for adhesive bonding to a substrate. As described in Chapter 17 on plastics, these plastic laminates are made using melamine resin, paper, and phenolic resin, by hot pressing under high pressure. *Formica* is one of the popular brands. Thin, single-faced decorative laminates are usually less than 2 mm thick (see IS 2046:1995, for more details). Standard size of HPL sheets is 1.22 m × 2.44 m. These sheets are bonded to plywood or wood surfaces with suitable glue (example: Fevicol SH Synthetic resin glue) and pressed down with weight for a few hours to obtain uniform bonding. Single- or double-faced laminates, between 2 and 5 mm thick and double faced, self-supporting compact laminates (thicker than 5 mm) are also available. The thickness of these laminates is selected according to the application. The details of the use and specification of thin metal plates to protect the top face of plywood is given in IS 13957:1994.

## 11.17.2 Cross-laminated Timber (CLT)

CLT panels were developed in Europe in the early 1990s and are now considered the most versatile and robust product for use in mass timber buildings. The CLT can be used for floor, roof, and wall panels. The CLT is made of a number of layers of timber, glued together with the grain of alternate layers laid at right angles to one another, much like the veneers of plywood. The CLT combines the advantages of large sized load-bearing panels, stability, fire resistance, long-term carbon storage, and renewability. A related product, made by nailing wood components together, is marketed as the NLT. Typical species used in the CLT are Spruce-Pine-Fir, Douglas fir, or Black Spruce. Owing to the CLT's cross laminations, the panels afford significant in-plane shear capacity and can be used as diaphragms or shear walls in building lateral systems. In these cases, panel-to-panel joints need to be carefully detailed for in-plane shear transfer. However, current US building codes do not recognize the use of the CLT in a lateral capacity (Jackson et al., 2017).

Although, only a few fabrication plants produce CLT panels in North America, many more plants exist in Europe. In-plane panel dimensions are quite stable due to the cross laminations, but the thickness of the panels remains susceptible to swelling and shrinkage. For buildings that load the CLT perpendicular to grain over multiple stories, it is important to consider the shrinkage and compression that can occur through the depth of the panel as they accumulate over the height of the building. The CLT is a standardized product that is often used in a one-way decking capacity. However, it can also be used as a two-way bending member. Typical panel dimensions are as given below (Jackson et al., 2017):

1. Thicknesses: 100 mm (3-ply) to 300 mm (9-ply): max lengths: 9 to 18 m depending on supplier
2. Typical spans: 3 to 10 m
3. Max widths: 2.4 and 3 m depending on supplier

More information on the CLT may be found in Yeh et al., 2012.

### 11.17.3 Nail-laminated Timber (NLT)

As mentioned already, the NLT is similar to the CLT, and has been used, since the early 1900s, for floor, roof, and wall panels. These panels are typically comprised of Spruce-Pine-Fir or Douglas fir 2X timber (35 mm in net thickness after surfacing prior to gluing), placed on edge and nailed together side-by-side, as shown in Fig. 11.11(a). However, any other wood species could also be used. Plywood is used to sheathe the panels, providing in-plane stiffness and shear resistance for lateral diaphragm loads. The panels can be fabricated in a shop environment or can be nailed together on-site by any reputable carpenter.

Robust moisture protection during fabrication and erection is a must for NLT panels, as they are susceptible to swelling perpendicular to the grain. To mitigate these potential swelling issues, a 2X lamination should be left out every 6 m and installed back in after the panels have acclimatized. The NLT is a non-standardized one-way panel system. Typical panel dimensions are as follows (Jackson et al., 2017):

1. Thicknesses: 62.5 to 287.5 mm
2. Lengths: 2.4, 3, 3.6, 4.8 m
3. Max width (prefabricated): 1.2 m

Where spans exceed 4.8 m, interlocking finger-jointed lumber may be used or a staggered butt-jointed pattern may be specified. However, these longer panels tend to be less cost-effective when compared to simple span panels less than 4.8 m. NLT panels used in the floor and roof systems is the Mountain Equipment Co-op Head Office in Vancouver, BC, Canada, are shown in Fig. 11.11. Panels were prefabricated in 1.2 m width and 12 m length using butt-jointed laminations.

**Fig. 11.11** NLT panels used in Mountain Equipment Co-Op Head office, Vancouver, Canada (Jackson et al., 2017)

## 11.18 Parallel Laminates

Glulam, LVL, and block board/lamin board/sandwich panels are examples of parallel laminates.

### 11.18.1 Glued Laminated Timber (GLT or Glulam)

The GLT or Glulam is not made of veneers but with solid wood. It is manufactured by bonding together a number of wood laminations, or *'lams'*, using durable, moisture-resistant structural adhesives. The grain of all laminations runs parallel with the length of the member. Typical species used in glulam are Spruce-Pine-Fir, Douglas-Fir-Larch, Southern Pine, yellow-cedar, and Hem-Fir. Individual lams typically are 35 mm thick for pine, although other thicknesses may also be used. The laminating process allows timber to be used for much longer spans, heavier loads, and complex shapes. Glulam is available in both stock and custom sizes. Common custom shapes include straight beams, curved beams, pitched and curved beams, and radial arches. Typical stock beam widths used in residential construction include 80, 89, 130, 140, and 170 mm. Glulam is available in a range of different appearances (such as framing, industrial, architectural, and premium), but having the same structural characteristics for given strength grades. Properties of glulam products may be found in Ross (2010). Although used in several countries, glulam products are not yet popular in India.

Floor and roof panels of the GLT are similar to glulam beams laid on their sides, with the lamination lines running vertically. Similar to the NLT, plywood is used to stitch the panels together and to act as a diaphragm. Some glulam suppliers are able to provide a fluted soffit, which can help with acoustics and give a unique visual appearance. GLT panels also require robust moisture protection during erection, as they are susceptible to swelling perpendicular to the grain.

Typical panel dimensions of the GLT are as follows:

1. Thicknesses: 80 to 215 mm
2. Max lengths: 12 to 18 m depending on supplier
3. Typical spans: 4.5 to 9 m
4. Max widths: 0.6 or increments of 38 mm

The Kin Centre Complex in Prince George, British Columbia, Canada, in which GLT panels are used as the secondary roof framing, is shown in Fig. 11.12. The gently arched roof contains 122,000 m board decking. Recesses were provided for piping and conduits to showcase the warmth and beauty of wood. The roof consists of glulam panels 600 mm wide by 130 mm deep and constructed from locally sourced spruce, pine, and fir. The wood panels became a solid diaphragm enabling the roof to be designed without additional steel braces in the roof plane (www.naturallywood.com).

**Fig. 11.12** Kin Centre Complex, Canada, uses the GLT panels (Jackson et al., 2017)

## 11.18.2 Laminated Veneer Lumber (LVL)

The LVL was invented by Arthur Troutner of Idaho, USA, in 1971. The LVL is made by laminating veneers, 2.5 to 4.2 mm thick, with suitable adhesive and with the grain of veneers in successive layers aligned along the longitudinal (length) dimension of the composite. Veneers used to make the LVL are given suitable preservative treatment before lamination, with a preservative that is compatible with the adhesive. IS 14616:1999 suggests the use of fixed type of water soluble preservatives, the CCA or the CCB, or non-leachable, solvent soluble preservatives as per IS 401 (also see IS 16171:2014). After preservative treatment, the veneers are dried and conditioned so that the moisture content is below 10% before bonding. After gluing, the assembly is hot pressed under controlled conditions of temperature, pressure and pressing time, depending upon the thickness of the composite. The completed board is called *billet* in the USA.

Some of the minimum requirements of the LVL, as per IS 14616, are: (a) modulus of rupture 50 MPa, (b) modulus of elasticity 7500 MPa, (c) compressive strength: parallel to grain 35 MPa, and perpendicular to grain: 35 MPa, (d) tensile strength parallel to grain 55 MPa, and (e) horizontal shear, parallel to grain 6 MPa and perpendicular to grain 8 MPa.

In the USA, Douglas fir is used as the veneer in LVL panels. Plywood is typically used to sheathe the panels, providing in-plane stiffness for lateral diaphragm loads. It is not recommended to use LVL panels in diaphragm applications, as the product is not cross-laminated. The in-plane panel dimensions are stable; however, the thickness of the panels remains susceptible to swelling and shrinkage.

Typical LVL panel dimensions are as follows (Jackson et al., 2017):

1. Thicknesses: 44 and 88 mm
2. Max length: 20 m
3. Typical spans: 3 to 6 m
4. Max width: 1.2 m

The LVL may not be suitable for outdoor load-bearing use. In addition to panel application, the LVL is used as flanges in timber I-beams, beams, rafters, and ply-webs in box beams. It is possible to cut holes in the webs of LVL beams at selected locations, through which services could be installed. The properties of the LVL show much less variation than sawn timber or even glulam.

### 11.18.3 Block Board, Lamin Board, and Sandwich Panels

*Block board*, also called *batten board* or *solid-core plywood*, is similar to plywood in composition, but is made up of a core of blocks of wood. It has a core made up of softwood strips of wood, not exceeding 30 mm in width (IS 1659:2004). The core strips are placed edge-to-edge (which may or may not be glued together) and glued to two or more hardwood veneers (each with a thickness of 0.5 to 1.5 mm), on either side, with grain direction of core and veneers running at right angles to one another [see Fig. 11.13(a)]. Commonly used species of timber for face veneers for both these types are to be as per IS 1659. Adhesives used for bonding should be as per IS 848. All timber used must be treated as per IS 401. Trimmed and cut ends of a finished block board may be given a protective treatment. Two grades of block boards are available: BWP grade – used in situations where the members are exposed to high humidity and external use – and MR (moisture resistant) grade – used in interiors. Each of these grades may be of decorative or commercial type. Block boards are being used in the construction of railway carriages, bus bodies, marine and river crafts, and for furniture making, partitions, panelling, and even in prefabricated houses. When using block board, it is important to ensure that the core blocks run along

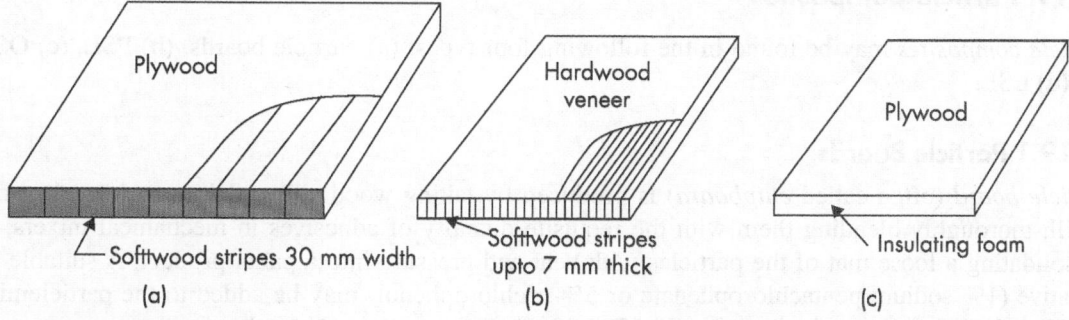

**Fig. 11.13** Block board, lamin board, and sandwich panels (a) Blockboard (b) Laminboard (c) Sandwich panels

the length in order to achieve maximum strength. Block board is sold in sheets of 2440 mm × 1220 mm, 2140 mm × 1220 mm, 2140 mm × 920 mm, 1830 mm × 1220 mm, and 1830 mm × 920 mm, and in thickness of 12, 15, 19, 25, 30, 35, 40, 45, or 50 mm. Permissible defects and tolerances on thickness should conform to IS 303 and IS 1328, for commercial and decorative veneers, respectively.

*Lamin board* looks very similar to block board but it is made up of softwood strips, 5–7 mm in width as shown in Fig. 11.13(b). In both cases, the strips are sandwiched between two outer veneers with the grain running at right angles to the core strips. The facing veneers may be of birch. *Sandwich panels* may have cores of insulating foam used in façade panels as shown in Fig. 11.13(c), or timber spacers, which are used for stressed skin panels.

### 11.18.4 Flush Doors

The plywood and block board industry has resulted in the manufacture of flush doors, which are very popular and are used in interior doors. *Flush doors* are typically entirely flat on both sides, and, hence, the name 'flush' within the construction industry. However, it is possible to manufacture flush doors that look like panelled doors with timber veneers. Flush doors are usually cheaper than panelled doors made of wood, and are more modern in appearance. They have high levels of fire resistance; sound attenuation and impact resistance (laminate faced doors). They are available in three types (a) hollow core, (b) solid core, and (c) cellular core, as shown in Fig. 11.14.

The solid core type, which is popular, has solid block board, particle board, or medium density fibre board, which are faced with sheets of commercial or decorative plywood [IS 2202 (Part 1):1999] or particle board/hardboard [IS 2202 (Part 2):1983] to provide a flat surface that can be painted, veneered, or faced in laminate. The edges are usually lipped with hardwood. These doors are available in modular sizes, with heights of 1905 and 2005 mm, widths of 700, 800, 900, and 1100 mm, and thicknesses of 25, 30, and 35 mm. Plywood sheet is glued under pressure to the assembly of core fixed in the frame on the both faces. Instead of plywood sheets, separate cross bands and face veneers are also used. The solid core type doors are strong and heavy and may be used as exterior doors, as they provide better strength and sound insulation.

Hollow core doors are hollow inside with only a supporting honeycomb structure, with voids not exceeding 500 cm$^2$ in area [IS 2191 (Parts 1 and 2):1983]. The cellular core is formed by fixing wooden or plywood battens. These battens are not less than 25 mm in width and uniformly distributed with the voids and these voids do not exceed 25 cm$^2$ in area. Hollow core and cellular core types are lighter than solid core type, but are not as strong as solid core door.

## 11.19 Particle Composites

*Particle composites* may be found in the following four types: (a) particle boards, (b) PSL, (c) OSBs, and (d) LSL.

### 11.19.1 Particle Boards

*Particle board* (often called *chipboard*) is produced by taking wood chips, flakes, and sawdust from a mill, thoroughly blending them with the requisite quantity of adhesives in mechanical mixers, and consolidating a loose mat of the particles with heat and pressure into a panel product. A suitable preservative (1% sodium pentachlorophenate or 5% trichlorophenol) may be added to the particlemix at the time of mixing the synthetic resins (IS 3087:2005). The hot-pressed boards are subsequently cooled, conditioned to attain equilibrium moisture content and sanded on both sides to attain uniform thickness and finally trimmed and cut to standard sizes.

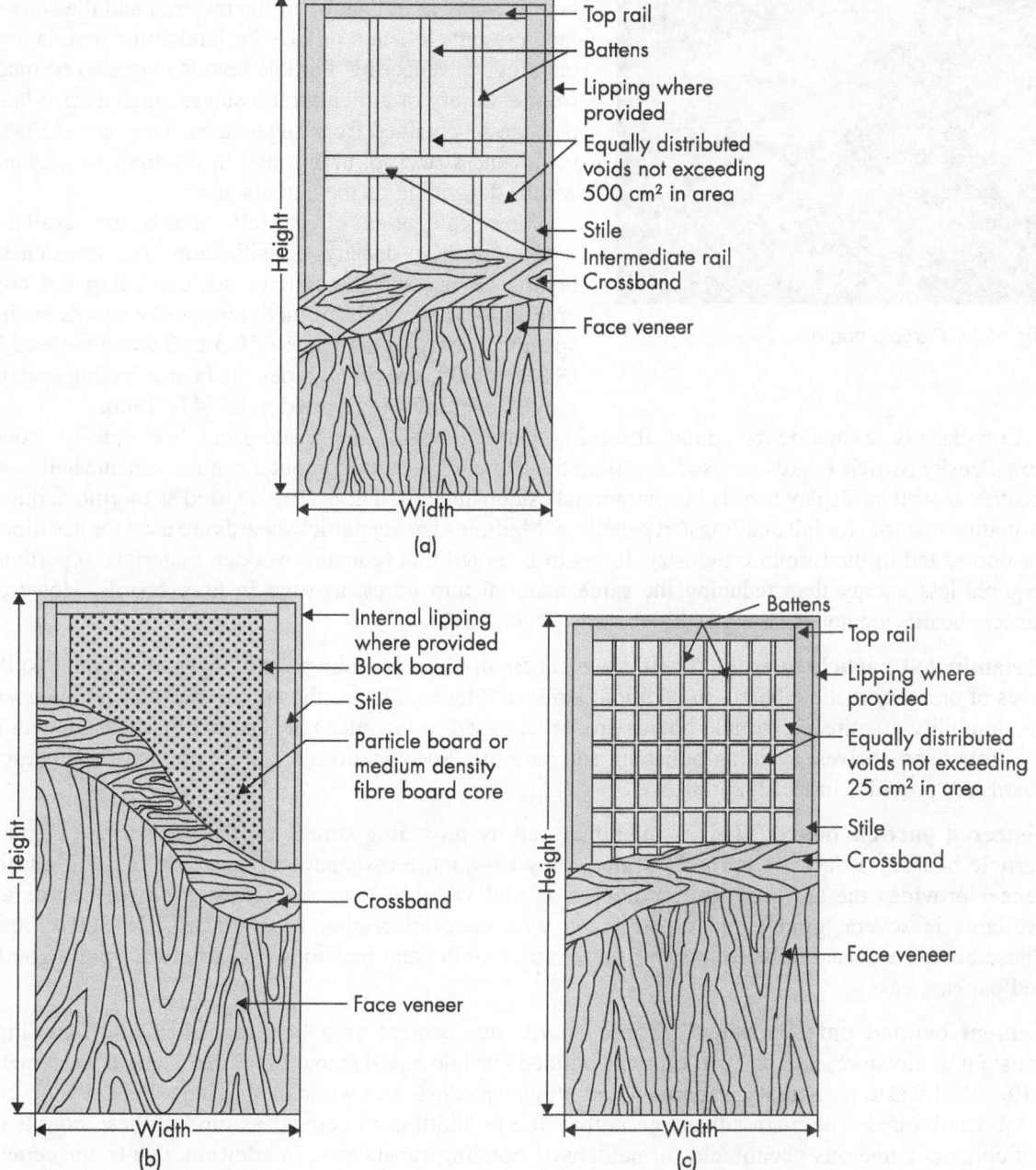

**Fig. 11.14** Types of flush doors (a) Hollow core (b) Solid core (c) Cellular core

All particle boards are currently made using a dry process, where air or mechanical formers are used to distribute the particles prior to pressing (Ross, 2010). Since most applications are interior, particle boards may be bonded with a UF resin, although PF and MUF resins are sometimes used in applications where more moisture resistance is required. Particle board is typically made in three layers (see Fig. 11.15). The faces of the board consist of fine wood particles, and the core is made of coarser material. Such particle

**Fig. 11.15** Particle boards

boards improve utilization of the material and the smooth face presents a better surface for laminating, overlaying, painting, or veneering. Particle boards may also be made from a variety of agricultural residues, such as rice husk or bagasse obtained from sugarcanes. They are available in thickness of 5 to 50 mm and in different lengths and widths depending on the manufacturer.

Three categories of particle boards are available based on their density classification: (a) low-density boards having specific gravity not exceeding 0.4 covered in IS 3129:1985, (b) medium-density boards having specific gravity in the range of 0.5 to 0.9 and covered in IS 3087:2005, and (c) high-density boards having specific gravity over 0.9 and covered in IS 3478:1966.

Low-density insulating or sound absorbing particle boards are made from low density wood. Low-density particle boards are used as ceiling tiles for auditoriums, computer centres, cinema halls, and theatres as well as display boards in commercial establishments. These are also used as thermo-acoustic insulation material for false ceiling and panelling. Medium-density particle boards are used for partitions, for doors, and in the furniture industry. It has to be noted that reducing wooden materials to particles requires less energy than reducing the same material into fibres, as used in fibre boards. However, particle boards are not as strong as fibreboards (Ross, 2010).

**Prelaminated particle boards**  These are manufactured by providing a thin layer of laminate to the faces of ordinary particle board. In addition, laminated faces enhance the appearance and also increase the durability. Laminated particle boards are widely used in modular kitchen, and to make shutters of wardrobes and shelves, avoiding polishing and painting. Specifications for these prelaminated particle boards are provided in IS 12823:2015.

**Veneered particle board**  This is manufactured by providing veneer on the top surface of plain particle board. As they are properly sealed, they have more resistance to warping. In addition, the veneer provides the aesthetic appearance of natural wood at low cost. These types of boards are available in several grades and types, faced with decorative/other type veneers (IS 3097:2006). These boards are used in buildings, furniture, automobile, and bus body construction, sports goods, and packing cases.

**Cement bonded particle board**  These boards use cement as a bonding agent and have high resistant to moisture, fire, and rot. Cement-bonded particle board is manufactured using 65% cement, 24% wood waste particulate, such as wood chips, sawdust, and wooden shavings, 8.5% water, and 2.5% admixtures. The main advantage is that due to addition of cement as binder, these boards do not contain dangerous phenol and formaldehyde bonding substances. In addition, due to the cement content, these boards are more durable, fire-resistant, and termite-resistant. The cement bonded particle board has high expansion and shrinkage properties in the presence of moisture. They are used to produce fire-resistant furniture, false ceiling, internal and external walls, and permanent shuttering for concrete floors and walls. Specifications for cement bonded particle boards are provided in IS 14276:2016, and for prelaminated cement-bonded particle boards in IS 15786:2008. Cement bonded particle boards use treated wood flakes, whereas *cement fibre boards* use cellulose fibre, which is extracted from plants.

## 11.19.2 Parallel Strand Lumber (PSL)

The PSL is made from parallel wood strands bonded together with adhesive. It is used for beams, headers, and columns [see Fig. 11.16(a)]. These clipped veneer strands in the PSL have least dimension less than 6.4 mm and average length more than 300 times this least dimension. As the randomly oriented strands are glued together under high pressure, the PSL is mush denser and has design strength greater than that of sawn lumber. In addition, as the knots and other imperfections are randomly dispersed through the product and filled up and fortified by the glue, the PSL has more uniform properties than solid-sawn wooden beams. Thus, the PSL may possess higher strengths in bending, tension parallel to grain, and compression parallel to grain. The PSL can be made from any wood species, but Douglas-fir, southern pine, western hemlock, and yellow-poplar are commonly chosen due to their superior strength.

The product is manufactured as a 300 mm × 300 mm or 300 mm × 450 mm billet in a rectangular cross-section, which is then typically sawn and trimmed to smaller cross-sectional sizes. Typical widths are 88, 130, or 175 mm; typical depths are 240, 300, 350, 400, and 450 mm. Typically, the beams are made to a maximum length of 18 m. Similar to the LVL, two to three PSLs can be connected together to create a larger beam.

## 11.19.3 Oriented Strand Board (OSB)

The OSB was invented by Armin Elmendorf in California in 1963. It is also known as *flakeboard* or *sterling board*. It is similar to particle board and manufactured in wide mats from cross-oriented layers of thin, rectangular wooden strips, which are compressed and bonded together with synthetic resin adhesives [see Fig. 11.16(b)]. Wood strips on the surface layers are arranged in the long panel direction to give strength, whereas the strips in the internal layers are perpendicular to them. The number of layers used is dependent on the thickness of the panel. The OSB comes in a variety of types and thicknesses. It is used extensively for roof, wall, and floor sheathing in residential and commercial construction. Typical properties of the OSB may be found in Ross (2010).

(a)  (b)

**Fig. 11.16** Parallel strand lumber and oriented strand board (a) Parallel strand lumber (PSL) (b) Orientation of wood strands in oriented strand board (OSB)

## 11.19.4 Laminated Strand Lumber (LSL)

The LSL is the latest engineered wood product to come onto the market. The LSL resembles the OSB in appearance because like the latter, the former is made from long strands coming from fast-growing aspen or poplar. However, unlike the OSB, the strands are arranged parallel to the longitudinal axis of the member (www.cwc.ca). The LSL panels made from flaked wood strands having length-to-thickness ratio of about 150, are one-way systems. These strands of the LSL are oriented randomly, glued with adhesive, and pressed together to form large mats or billets. These billets can be reduced in size to be used as smaller beams and rim boards for lightwood frame construction, but can also be used as larger panels (Jackson et al., 2017).

These panels can be sheathed with plywood to provide in-plane stiffness and shear resistance to resist lateral diaphragm loads. The semi-random orientation of the flakes provides some in-plane member stiffness and stable in-plane panel dimensions. However, similar to the CLT, the thickness of the panels are susceptible to swelling and shrinkage (Jackson et al., 2017). Typical panel dimensions of the LSL are (Jackson et al., 2017):

1. Thicknesses: 38 and 89 mm
2. Max length: 18–19.5 m
3. Typical spans: 3 to 6 m
4. Max width: 1.2 m

Although mass timber products are often used for floor and roof panels, the larger scale of LSL billets offers unique opportunities to machine large structural components out of a single piece, minimizing connections. The Guildford Recreation Centre in Surrey, British Columbia, Canada, using the computer numerical control (CNC) cut timber strand LSL billets, to create the webs of 27.5 m span and 3 m deep prefabricated three-dimensional roof trusses, is shown in Fig. 11.17. The top and bottom chords consist of glulam members. Full threaded screws acting in tension/compression connect the web members to the top and bottom chords. The wood trusses were painted with highly reflective paint to increase the light levels.

**Fig. 11.17** Guildford Recreation Centre, using machined LSL billets (Jackson et al., 2017)

## 11.20 Fibre Composites

Fibre composites are available as fibreboards (soft and medium density) and hardboards.

### 11.20.1 Fibreboards

The MDFs are made from very fine particles of wood [see Fig. 11.18(a)]. The logs are debarked and, after removing the cambium layer, chipped into fine pieces, using chipping machines. These chips are then steamed and defibrated in suitable defibrating machine. The fibres thus produced are dried in flash dryers and blended with resin and wax. A preservative, such as 2% sodium pentachlorophenate or 5% trichlorophenol, may be added to the fibre mix at the mixing stage of adhesive. Then, they are formed into mats by air felting and pressed into panels under controlled heat and pressure conditions. These boards are then dried, may be laminated, trimmed, and packed for distribution (IS 12406:2003; IS 14587:1998). They have a standard width of 1.22 m and available in lengths

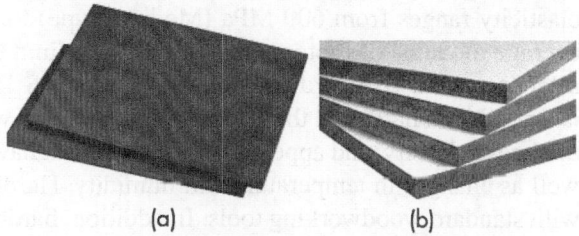

**Fig. 11.18** Fibreboard (a) Medium density fibreboard (MDF) (b) Prelaminated MDF

ranging from 1.22 to 5.49 m and may have thickness of 6 to 40 mm (IS 12406:2003). They are available in two grades (Solid board Grade I and Solid board Grade II) with a density of 6–9 kN/m³ and moisture content of 5–10%. They are used for the mass production of furniture, cabinets, and flush doors.

The MDF may be laminated on both surfaces by synthetic resin impregnated base papers with or without impregnated overlay by using heat and pressure [see Fig. 11.18(b)]. Common surface finishes of the base papers are glossy and matt textured (soft, swede, wood pore, and leather). *Prelaminated MDF* boards are also available in two grades, with four types under each grade, depending upon the abrasion resistance of the prelaminated surface. Type I is used in flooring applications, Type II in horizontal applications such as cash counters and restaurant table tops, Type III in normal horizontal applications such as office table tops and domestic furniture tops, and Type IV in vertical applications such as panelling, partitioning, and false ceilings (see IS 14587). Other physical and mechanical properties of the MDF may be found in IS 12406:2003. All MDF boards are recommended for use in interior dry locations only. Guidance for the selection and use of the various types of the MDF are provided in IS 15512: 2004.

*Soft boards*, having density less than 500 kN/m³, are used as thermal insulation boards for interior walls and ceilings and as acoustical ceiling material. They are normally 12 mm thick, 1.2 m wide, and either 2.4 or 3.6 m long. They are economical, light, and easy-to-work-with. An insulation board has low structural strength and may burn easily, although flame resistant coating and fire retardants may be added during manufacture.

## 11.20.2 Hardboard

Hardboards are made from wood or any other lignocellulosic material, which are cut into chips, cooked in a preheater at 170–200°C to yield pulp, and then pressed into panels under controlled heat and pressure (higher than that are used for fibre board) (IS 1658:2006). The pressed panels of uniform thickness and density are then cut to the required sizes using trimming machines. The primary bond is derived from the felting of fibres and their inherent adhesive properties. Bonding material and/or additives may also be used. As per IS 1658:2006, these homogeneous fibre hardboards are classified into the following three types according to their method of manufacture, thickness, density, specific properties, and application purposes:

1. Medium hardboard - (density 350–800 kg/m³) available in 6, 8, 10, and 12 mm nominal thickness;
2. Standard hardboard - (density 800–1025 kg/m³), available in 2.5, 3, 4, 5, 6, and 7 mm nominal thickness; and
3. Tempered hardboard - standard hardboard, which has been further treated to improve or modify one or more properties of the final board and available in 2.5, 3, 4, 5, and 6 mm nominal thickness.

All these boards have a standard width of 1.22 m. Medium and standard hardboards are available in lengths of 1.2, 1.83, 2.44, 3, and 3.66 m, and tempered hardboard in lengths of 4.26, 4.85, and 5 m. They may have moisture content of 1–15% and the maximum swelling of 20–25%. The average modulus of elasticity ranges from 600 MPa (Medium type) to 3000 MPa (Standard and Tempered type), and the average modulus of rupture from 6 MPa (Medium type) to 50 MPa (Tempered type). Guidance for the selection and use of hardboards is provided in IS 15932 (Part 2):2017. The face surface of hardboards is smooth and hard, and the opposite side is rough with pattern of cross-lines. Hardboard has a uniform thickness, density, and appearance, and does not have grain. It resists marring, scuffing, and abrasion, as well as changes in temperature and humidity. Hardboard sheets can be cut, routed, shaped, and drilled with standard woodworking tools. In addition, hardboard can be securely glued or fastened with screws, staples, or nails. Hardboard panels can be laminated with paper overlays, plastic laminates, and veneers. Hardboard is used in a variety of applications including furniture components, wall panelling, doors, underlayment, and perforated boards.

## 11.21 Timber-concrete Composites (TCC)

*Timber-concrete composite* (TCC) panels consist of a thicker layer of concrete on the top side and a mass timber panel on the bottom side. The concrete acts as a compression element, whereas the timber acts as the tension element, giving flexural stiffness. There are several ways to engineer the connection between the concrete and timber, for generating the required shear flow. They range from glued-in perforated steel plates to fully threaded screws installed at an angle (Jackson et al., 2017).

TCC floor systems are very efficient and can achieve high span-to-depth ratios. The depth of the concrete topping also allows hiding of electrical conduits and in-floor heating lines inside the floor system. Any of the aforementioned mass timber panels, as shown in Fig. 11.10(f), can work as the tension lamination in a TCC floor system.

Typically, the concrete is cast on-site directly on top of the timber panels, connecting them together. However, precast versions of the concrete compression element are also possible by installing proprietary screw sleeves within the concrete. Additionally, timber-concrete composite T-beams are possible with wood beams rather than a wood panel (Jackson et al., 2017).

In addition to the aforementioned wooden products, coir and bamboo based products are also available (see IS 707:2011; IS 1902:2006; IS 10388:1982; IS 6874:2008; IS 9096:2006, IS 14842:2000; IS 15476:2004; IS 15877:2010; IS 15878:2010; Kaminski et al., 2016–2017). Bamboo is covered in Section 23.4 of Chapter 23. Rubber is covered in Section 17.14 of Chapter 17.

## 11.22 Advantages and Drawbacks of Wooden Composites

The structural wooden composites, reconstituted from plantation grown wood, are eco-friendly products. They effectively overcome all the problems associated with direct use of plantation timber. Through a redistribution of natural defects such as knots and cross grain, these manufactured wooden products are engineered to give more uniform strength properties, greater reliability, and lower factors of safety than are possible with natural solid wood. They also do not have the dimensional limitations of timber and can utilize low-grade timber, minimize waste, and maximize the use of timber. The following are some of the advantages of wooden composites:

*Possibility of using unusable wood*   These composite products help to conserve timber, and are also able to utilize lower-grade timber, branches of trees, and even waste wood, which cannot otherwise be advantageously used in construction. Even low-grade materials are transformed into quality homogenous products.

*Lower costs*   By converting expensive wood into thin veneers, these are able to cover larger surface area, and also the use of lower-grade timber results in economy and lower costs.

*Dimensional stability and strength* Marine plywood and highly compressed boards made with moisture-resistant resins and do not absorb moisture from air in wet weather. Hence, they have dimensional stability. In addition, stiffness and strength of wooden products are greatly improved than sawn wood. However, particle board may absorb moisture.

*Dimensional limitations* They do not have dimensional limitations of sawn wood.

*Durability* The durability of these products depends on the grade of adhesive used. With better adhesives, these products (marine plywood and block board) can be very durable even in wet locations.

*Less energy for production* Their production and processing require less energy than concrete or steel.

*Workability* They can be easily sawn, cut, and joined to any form, shape, and size using ordinary carpentry tools.

*Speed of construction* The lightweight of wooden products facilitate faster and cost-effective handling for transportation and installation at site. On mass timber projects, using panellized products, the superstructure can be completed 25% faster than a steel, concrete, or lightwood frame counterpart. Such reduction of time at sites reduces construction costs. Fully coordinated shop and erection drawings can create a smooth and efficient flow on site, where small and large prefabricated elements can be installed within a matter of days.

*Less transportation cost* Mass timber project sites typically require only 10% of the number of trucks to service them compared to a concrete alternative. As most of these projects use large, prefabricated timber panels for the decking system, the labour required on the active deck can be reduced to 25% of that of a concrete alternative. With all of these factors working together, site noise is significantly reduced thereby reducing the impact on the local community.

*Thermal insulation and sound absorption* Like natural timber, they are suitable for construction in all climatic zones and provide good thermal insulation and sound absorption, resulting in healthy and comfortable living. In addition, their very high strength-to-weight ratio makes them ideal materials for construction in earthquake zones.

*Easily repaired* Alterations and repairs to wooden products are relatively easy and faster.

A few disadvantages of wooden products include the following:

*Swelling and shrinkage* Swelling can be a problem with plywood, particularly when fir plywood is used for sheathing and flooring in the presence of moisture. Hence, while using them in exterior applications or situations such as bathroom doors, care must be taken to use moisture-resistant grades.

*Toxic glues and preservatives* They may contain formaldehyde based glues or preservatives such as chromated copper arsenate (CCA), creosote, and pentachlorophenol (PCP), which are toxic. Particle boards are known to give off formaldehyde and can be a serious health risk. In the USA, CCA, creosote, and PCP are banned in residential constructions and are only used for commercial purposes such as utility poles, railroad ties, and wharf pilings. Examples of less toxic preservatives include: alkaline copper quaternary (ACQ), copper naphthenate, disodium octaborate tetrahydrate (DOT), copper azole, copper napthenate, copper-HDO, and polymeric betaine (www.epa.gov).

## 11.23 Applications of Wood and Wood Products

Wooden components used in construction, furniture, packaging, and other industrial and domestic requirements fall into three categories: (a) panels for sheathing and diaphrams, siding, casing, partitions, panelling, shelving, topping, and decking; (b) load-bearing structural members such as beams, posts, members in trusses, structural frame work in construction, joinery; and (c) furniture. As already mentioned, reconstituted lignocellulosic products, such as plywood, particle board, fibre hardboard, and the MDF, are used as substitutes of wood for panel material, which is well established (IS 14616:1999).

Softwood is commonly used for timber structures as it is readily available, easily worked, and relatively low cost. In addition, the fast rate of growth of softwood provides a continuous supply from regenerated forest areas. As softwoods often come from very tall, straight trees, they are better suited for construction work (in the form of planks and poles). Hardwoods are typically used for exposed structures and claddings where durability and particular aesthetic characteristics, such as colour or grain pattern, are required. Owing to their beautiful grain pattern, hardwoods are used in furniture and decorative woodwork.

The different colours and structures of types of wood and the different interior products made of wood such as panels, veneers, plywood, and glued laminated boards, enable the use of wood in countless different ways. The use of wood is also attractive because of its ease of use. Wood products are handy in terms of their size and lightness to transport, simple to put up, and no special expertise or tools needed to work with them. If desired, wood can be carved, milled, or lathed into very intricate interior elements. Wood is used in interiors as a surface material for walls, ceilings and floors, furniture, stairs, windows and doors (frames and shutters), and in decorative products (examples: statues, sculptures, and relief carvings). It is also used for making tools, music instruments such as piano, violin, drums, flute, and guitar, sports equipment and toys, packaging material, dishes and utensils, and paper. In prefabricated wood building systems, floors, walls, and ceilings are made of timber studs (typically of size 50 mm × 100 mm or 50 mm × 150 mm, and spaced at 300 to 400 mm) and plywood sheeting with a certain nailing pattern (for the maximum strength and rigidity) is attached to the wood studs (see Figs. 11.19 and 11.20). Wood is also used in poles and posts, piles, members of Fink, Pratt, and Warren trusses, (see Fig. 11.20) bowstring girders, lattice girders, arches timber shuttering and scaffoldings. Uses of engineered timber products are shown in Table 11.9.

(a)                          (b)                          (c)

**Fig. 11.19** Application of wood in house building (a) Timber building in europe (b) Typical wooden studs (c) Typical beams in wooden houses

(a)                                              (b)

**Fig. 11.20** Wooden trusses and external appearance of a wooden house in the USA (a) Wooden trusses (b) External appearance of wooden house

**Table 11.9** Uses of engineered timber products

| Engineered timber product | Parallel stand lumber (PSL) | Laminated veneer lumber (LVL) | I-Joist | Glulam | Structural insulating panel (SIP) | Cross-laminated timber (CLT) |
|---|---|---|---|---|---|---|
| Application | Beams and columns | Beams, columns, and cord | Joist and beams | Long span beams | Roof, wall, and floor | Roof, wall, and floor |
| Usage | Interior | Interior | Interior | Interior/exterior | Interior/exterior | Interior |

## 11.23.1 Tall Buildings in Wood

A few timber buildings up to six storeys have been constructed recently, and the possibility of building much taller buildings with timber is being considered. In low-rise buildings, where the forces to be resisted are relatively low, it is possible to resist lateral loads by bending stresses in walls which form a vertical cantilever. This is the approach widely used in some of the buildings with the CLT construction. A common system for very tall buildings in concrete is a central core coupled with shear walls near the outer edges of the building by stiff link beams. The 10-storey, 45-m tall office tower at 5 King Street in Brisbane, Australia, opened in Nov. 2018. Constructed using a hybrid of Glulam and CLT load-bearing members resulted in a 74% reduction in embodied carbon, 46% reduction in energy, 20% weight saving compared to concrete, and a construction period of just 15 months aided by offsite prefabrication. In such wooden buildings, speed is achieved by the use of prefabricated materials, including the CLT slab panels, glulam columns, steel connectors, and facade elements. It is interesting to note that the world's tallest timber building is the 18-story, 80-m tall building in Brumunddal, Norway known as the Mjøsa Tower. Designed by Voll Arkitekter, the Mjøsa Tower is taller than the next tall timber building, the 53 m tall Brock Commons Student Housing building at the University of British Columbia, Canada.

## 11.24 Cork and Linoleum

Compressed cork strips and granules, derived from the outer bark of the cork oak trees of southern Europe, are used to make sheets, wall and floor tiles, and cork wall coverings. It is a natural, healthy material with good thermal- and noise-insulating properties. Cork is also resistant to rot and moulds. It has several desirable properties such as impermeability, elasticity, hard wearing, and fire retardant. Hence, it is used as an acoustic and thermal insulation in house walls, floors, ceilings, and façades. Its natural colour and grainy surface require little or no treatment except when used on floors. Corkboard is considered a non-allergenic, easy-to-handle, and safe alternative to petrochemical-based insulation products. The composite material made by mixing cork granules and cement has lower thermal conductivity, lower density, and good energy absorption. Portugal's pavilion at Expo 2000 demonstrated that cork can be used to make bricks for the outer walls. Cork has been used as a core material in sandwich composite construction, resulting in lightweight, high-performance, and low-maintenance structures (Gil, 2015). Cork production is highly sustainable because the cork tree is not cut down to obtain cork; only the bark is stripped to harvest the cork and used without waste. The tree continues to live and grow and the bark regenerates itself after stripping.

*Linoleum,* invented by Englishman Frederick Walton in 1855, is made from powdered cork, linseed oil, wood resin, and wood flour mixed with chalk and pressed on a burlap or canvas backing. Pigments are often added to the material to create the desired colour finish. Linoleum is strong and flexible, and is available as sheets or tiles, but needs a firm, damp-proof surface, since moisture can rot the fabric backing. It is better to use wood lignin paste as an adhesive instead of petrochemical adhesives. Linoleum has all the advantages and none of the disadvantages of PVC floor coverings.

## 11.25 Timber: The Ultimate Green Building Material

Wood has many positive characteristics, including low-embodied energy, low-carbon impact, and sustainability. Embodied energy refers to the quantity of energy required to harvest, mine, manufacture, and transport to the point of using a material or product. Wood, a material that requires a minimal amount of energy-based processing, has the lowest level of embodied energy – wood products that require more processing steps (for example, plywood, engineered wood products, flake-based products) may require more energy to produce but still require significantly less energy than other materials used in construction (such as steel, concrete, aluminium, or plastic). A substantial amount of carbon can be sequestered in forest trees, forest litter, and forest soils. If tree is used to produce a wood or paper product, these products store carbon while in use. For example, solid wood lumber, used in building construction, sequesters carbon for the life of the building (Ross, 2010).

Wood products have a low level of embodied energy as compared to other building products. In addition, wood is one-half carbon by weight. Owing to this, wood products are actually carbon negative, as shown in Table 11.10 (Ross, 2010). Unlike any other construction material, trees are renewable and with proper management a flow of wood products can be maintained indefinitely. Unfortunately, sustainable practices have not always been applied in the past, nor are they followed in Asia and Africa today. Forest certification programmes not only ensure that the forest resource is harvested in a sustainable fashion, but also that issues of biodiversity, habitat protection, and the rights of indigenous people are included in land management plans (Ross, 2010). North America has certified more than one-third of its forests and Europe more than 50% of its forests; however, Africa and Asia have certified less than 0.1% (www.metafore.org). More information on the sustainability of wood and forest certification programmes may be found in Calkins (2009) and Ross (2010).

**Table 11.10** Net carbon emissions in producing a tonne of various materials (adapted from Ross, 2010)

| Material | Material net carbon emissions (kg C/t) | Near-term net carbon emissions including carbon storage within material (kg C/t) |
|---|---|---|
| Lumber | 33 | −457 |
| MDF | 60 | −382 |
| Brick | 88 | 88 |
| Concrete | 265–291 | 265–291 |
| Steel(virgin) | 694 | 694 |
| Recycled steel (100% from scrap) | 220 | 220 |
| Plastic | 2502 | 2502 |
| Aluminium | 4532 | 4532 |

**Note:** The carbon stored within wood will eventually be emitted back to the atmosphere at the end of the useful life of the wood product.

Source: Forest Products Laboratory, 2010, Wood handbook—Wood as an Engineering Material, General Technical Report FPL-GTR-190, Madison, WI: U.S. Department of Agriculture, Forest Service, Forest Products Laboratory, p. 508.

If reforestation is properly planned and implemented, timber is the ideal material for construction, as it is renewable; several secondary species are at least available in most arid zones. Use of fast-growing species and properly preserved softwoods instead of the slow growing primary species can reduce the effects of excessive timber harvesting and deforestation.

A scheme of labelling environment friendly products, known as ECO Mark, has been introduced at the instance of the Ministry of Environment and Forests, Government of India. For the ECO Mark, only species of wood from sources other than natural forests, such as wood from rubber, coconut, cashew, industrial and social forestry plantations, etc., and shade trees from tea and coffee estates, should be used.

## SUMMARY

- Wood or timber has been used as a building material for several years. On the basis of mode of growth, trees are classified as exogenous and endogenous. Exogenous trees can be either deciduous (yielding hardwood) or conifers (mostly softwood).
- Commercial timber is often classified as hardwood and softwood. Other classifications are based on grading, availability, durability, modulus of elasticity, treatability, and refractoriness to air-seasoning.
- The five main types of defects in timber are due to: (a) natural forces (knots, shakes, upsets, twisted fibres, cracks, and fissures), (b) attack by insects, (c) fungi, (d) defective seasoning, and (e) defective conversion.
- Any piece of wood will absorb moisture until its moisture content is in equilibrium with the existing atmospheric conditions. When the wood dries, it shrinks differently in different directions. The moisture content of timber may be determined using moisture meters or by the methods described in IS 1708.
- After a log is sawn and converted into commercial sizes such as planks, battens, posts, and beams, it is known as converted timber. Sawing of timber can be done by (a) through and through sawing, (b) quarter sawing, (c) radial sawing, and (d) tangential sawing.
- Seasoning is the process of reducing moisture content of timber to the desired level. Air-seasoning, by stacking timber in open air requires several months. The various methods of artificial seasoning are: boiling, kiln seasoning, solar drying, and chemical and electrical seasoning.
- Several treatments are given to timber to improve its dimensional stability, resistance to biological degradation, fire resistance, and to improve the properties.
- Preservatives used are of three types: (a) oil type, (b) organic solvent type, and (c) water-soluble type. As most of the preservatives are toxic, care must be taken while handling them. Protective coatings can retard dimensional changes in wood but do not prevent them.
- Timber should have good strength, high durability, and a good finished appearance. Generally, timber is selected based on (a) local availability, (b) grade, (c) closeness of grains, (d) hardness, (e) durability, (f) ease of working, (g) colour, and (h) the way it can take polish.
- A number of timber composite products are available and may be divided into the following four categories
  - Layered timber composites [examples: glulam, plywood, and LVM]
  - Particle composites [ examples: particle boards and PSL],
  - Fibre composites [examples: hardboards, and MDF]
  - Timber-concrete composites
- Timber composites have several advantages over timber: they do not have the dimensional limitations but have dimensional stability, and even low-grade materials are transformed into quality homogenous products.
- Timber and its composites can be used in several structural as well as non-structural applications. Practically, all construction timber is softwood especially that used for structural supporting members.
- Wood has many positive characteristics, including low-embodied energy, low-carbon impact, and sustainability. It is being used increasingly even in tall buildings.

## EXERCISES

### Multiple-choice Questions

1. Identify the false statement regarding conifer trees.
   - (a) They are endogenous trees.
   - (b) They grow outwards.
   - (c) They have needle-like leaves and cone-shaped fruits.
   - (d) They mostly yield softwood.

2. The age of trees can be predicted by
   (a) measuring the diameter of pith
   (b) counting number of rings
   (c) the length of medullary rays
   (d) the thickness of bark
3. Sapwood is
   (a) thin growing tissue near bark
   (b) youngest soft and moist layer of trunk
   (c) the inner core of dead wood .
   (d) none of these
4. Identify the wrong statement in the following:
   (a) Growth of softwood is fast.
   (b) Heartwood and sapwood are not easily identified in softwood.
   (c) Conversion of hardwood is difficult.
   (d) Fire resistance of hardwood is moderate.
5. Match the list on right with that on the left.
   (a) Sapwood    (i) Cracks or splits that occur around the annual ring
   (b) Knot       (ii) Disintegration caused by fungi
   (c) Shake      (iii) Stump of fallen branches of tree
   (d) Rot        (iv) Outer layers of a log of wood
6. Which of the following in timber is caused by fungus?
   (a) Upsets          (c) Dry rot
   (b) Shake           (d) Wet rot
7. Which of the following statements is not correct?
   (a) Upsets are due to improper felling or heavy winds blowing during the young age of the tree.
   (b) Cup shakes are longitudinal cracks.
   (c) Star shakes are radial cracks widest at the circumference and diminishing towards the centre.
   (d) Heart shakes are cracks widest at the centre and diminishing towards the outer circumference.
8. Match the list on right with the one at left.
   (a) Wane    (i) Caused by stump of fallen branches of tree.
   (b) Bowing  (ii) Small cracks due to the drying or seasoning of wood.
   (c) Checks  (iii) Due to shrinkage, resulting in a curvature in the length direction of timber.
   (d) Knots   (iv) Lack of wood on any face or edge of sawn timber.
9. Subterranean termites
   (a) originate from their nest near or below soil level
   (b) may be identified by their mud-walled shelter tubes.
   (c) eat the wood at the centre completely, leaving an intact outer shell of wood.
   (d) all of these
10. Fibre saturation point (FSP) for most wood species falls in the range

(a) 5–10%          (c) 25–30%
(b) 10–20%         (d) 25–35%
11. At Chennai, the average relative humidity is about 70%. What will be the equilibrium moisture content of wood?
    (a) 8.7–9.5%       (c) 12–13%
    (b) 10–11%         (d) 15–16%
12. The moisture content recommended for doors less than 50-mm thick is
    (a) 4–8%           (c) 12–20%
    (b) 8–14%          (d) 2–4%
13. The ratio of tangential shrinkage to the radial shrinkage of wood due to reduction in moisture content is
    (a) about 2        (c) greater than 5
    (b) about 3        (d) less than 1
14. By reducing moisture content of timber, which one of the following is not correct?
    (a) Tensile strength is increased.
    (b) Compressive strength is increased.
    (c) Bending strength is increased.
    (d) Shear strength is increased.
15. Which of the following is the quickest and produces timber pieces without much wastage?
    (a) Ordinary sawing
    (b) Tangential sawing
    (c) Quarter sawing
    (d) Radial sawing
16. Which of the following is not an objective of seasoning timber?
    (a) Reduction in shrinkage
    (b) Reduction of weight
    (c) Increase in strength and durability
    (d) Reduction of natural defects in timber
17. Average moisture content of well-seasoned wood is about
    (a) 6–8%           (c) 10–12%
    (b) 8–10%          (d) 12–14%
18. How many months a timber may require for natural seasoning?
    (a) 4 months       (c) 1.5–2 years
    (b) 6–12 months    (d) More than 2 years
19. How many days a timber, used for structural purposes, will require kiln seasoning?
    (a) 1 week         (c) 16 to 20 days
    (b) 15 days        (d) 20 to 25 days
20. The drawback of electric seasoning of timber is
    (a) checks         (c) cracks
    (b) splitting      (d) reduced strength
21. The wood preservative 'Creosote' is derived from
    (a) coal/lignite tar
    (b) zinc chloride
    (c) copper-chrome-arsenate
    (d) pentachlorophenol

22. Which of the following methods of the preservation of timber is the most effective?
    (a) Dipping
    (b) Brushing
    (c) Spraying
    (d) Pressure impregnation
23. Thermal/heat treatment of timber results in
    (a) lower bending and tensile strength
    (b) lower toughness
    (c) increased strength in the longitudinal direction
    (d) all of these
24. Which of the following woods is suitable to build bridges?
    (a) Babool
    (b) Bamboo
    (c) *Sal*
    (d) Mahogany
25. Fire proofing of timber with fire-retardant finishes
    (a) will not allow timber to catch fire
    (b) insulates the wood from heat and retards combustion
    (c) does not allow the fire to come closer to wood
    (d) none of these
26. The following factor of safety is used in timber to arrive at allowable stresses in bending
    (a) 2–4
    (b) 4.5–5.5
    (c) 5–8.5
    (d) above 10
27. Plywood is specified by
    (a) weight
    (b) volume
    (c) thickness
    (d) number of plies/layers
28. Plywood has great stiffness and strength
    (a) across the grains
    (b) along the grains
    (c) both (a) and (b)
    (d) tangential to the grain
29. The expansion and shrinkage of plywood are comparatively very low as
    (a) they are held in position by adhesives
    (b) they are glued under pressure
    (c) plies are places at right angles to each other
    (d) they are prepared from veneers

30. As a construction material, plywood is preferred to thin planks of timber because
    (a) it has good dimensional stability in both directions, warp resistance, and increased strengths
    (b) it looks better than wood
    (c) it has better fire resistance
    (d) nails and screws can be placed close to the edge of panels
31. Which of the following about glulam is incorrect?
    (a) All laminations runs parallel with the length of the member.
    (b) It is made by gluing several veneers.
    (c) It is used in longer spans, heavier loads, and complex shapes.
    (d) Individual lams typically are 35-mm thick for pine.
32. The core of block board is made up of softwood strips of wood, with widths not exceeding
    (a) 20 mm
    (b) 25 mm
    (c) 30 mm
    (d) 40 mm
33. The core of lamin board is made up of softwood strips of wood, with widths not exceeding
    (a) 5–7 mm
    (b) 8–10 mm
    (c) 10–12 mm
    (d) Greater than 20 mm
34. Which one of the following statements is correct with respect to particle board?
    (a) The slices of superior quality of wood are glued and pressed on the surface of inferior wood.
    (b) Wood chips, flakes, and sawdust are mixed with adhesives and pressed hard into a panel.
    (c) Thin and narrow strips of wood are soaked in a refractory binder material and then hard pressed.
    (d) It is obtained as wood veneer backed by a fabric mat.
35. Oriented strand board is similar to
    (a) plywood
    (b) fibre board
    (c) particle board
    (d) block board

## Review Questions

1. Distinguish between the following:
    (a) Endogenous and exogenous trees
    (b) Deciduous and conifer trees
    (c) Hardwood and softwood
    (d) Sapwood and heartwood
    (e) Shakes and knots
2. What are the various parts of the trunk of a tree? With a neat sketch of cross-section of stem, explain their functions briefly.
3. What is the best season for felling a tree?

4. List the various types of classification of timber. Discuss briefly the commercial or hardness classification.
5. What are the five main types of defects in timber?
6. Differentiate between nail knot, live knot, dead knot, and knot cluster.
7. Explain the following defects of timber.
   (a) Upsets
   (b) Different types of shakes
   (c) Sloping or spiral grains
8. List the defects that may develop after felling trees. Describe briefly with neat sketches the wane and checks in timber.
9. State the principal causes of decay of timber. Describe dry and wet rot. What are the differences between them? How are they caused and how can they be prevented?
10. What are the two types of termites? Describe the type of damage on timber due to these two types of termites. How will you identify that timber has been affected by termites? List four preventive methods adopted to protect buildings from termites.
11. After discovering active infestation of termites, what procedure should be adopted to eradicate them?
12. How is the soil around the building treated for termites?
13. How beetles, marine borers and carpenter ants affect timber? How can we prevent their attack?
14. What is the moisture content of freshly sawn lumber? What will be the moisture content at fibre saturation point of timber? How does the change in moisture content affect the properties of timber?
15. How does timber shrink in tangential, radial, and longitudinal directions? How can one solve this shrinkage problem?
16. What is meant by the conversion of timber? Describe the four methods of sawing used to convert wood into timber products. Which is the quickest and cheapest sawing method? Which sawing method results in less warping?
17. List at least five market forms of timber.
18. What is seasoning of timbers and why is it done? What is the main objective of seasoning? List the two methods of seasoning.
19. How is converted timber natural/air-seasoned? What are the drawbacks of this method of seasoning?
20. List at least four methods of artificial seasoning. Describe briefly kiln seasoning and solar seasoning.
21. What are seasoning degrades? Which is the main cause of such degrades?
22. What is meant by preservation of timber? Discuss the methods of preserving timbers.
23. What precautions are to be taken while handling chemically treated wood?
24. Describe the heat treatment of wood. What are the problems of heat-treated wood?
25. How is wood treated with polymers?
26. How is wood protected by coatings? Will painting preserve wood? Describe the two types of paints used with timber.
27. State the characteristics of good timber.
28. State the qualities you will consider while selecting timber for construction purposes?
29. How is timber fire-proofed?
30. What are the factors that affect the strength of timber? What are the three locations in timber at which strength has to be known as per IS 883? Why is it necessary?
31. State at least five advantages and disadvantages of timber used as a building material.
32. What are timber composites? What are the four broad categories of timber composites?
33. What synthetic resin thermosetting adhesives are used in the production of timber composites?
34. What are layered timber composites? What are their sub-classifications?
35. What are veneers? What are the four slicing techniques used to produce veneers?
36. How plywood is made? Why is plywood made with an odd number of layers? What are the three grades of plywood available in the market? What is veneered decorative plywood? What are the advantages of using plywood?
37. What are the uses of decorative laminates?
38. How are cross-laminated timber made? What are its uses?
39. What is the difference between cross-laminated timber and nail-laminated timber?
40. What are the examples of parallel laminates?
41. What is glulam? What are its advantages and where are they used?
42. Describe briefly laminated veneer lumber and what are its minimum requirements as per IS 14616? What are its applications?
43. Differentiate between block board, lamin board, and sandwich panels. Where is block board used?
44. What are the four types of particle composites? Describe briefly how particle boards are made and its uses.

**45.** Write short notes on
   (a) Veneered particle board
   (b) Prelaminated particle boards
   (c) Parallel strand lumber
   (d) Oriented strand board
   (e) Laminated strand lumber
**46.** What are cement bonded particle boards? How are they made?
**47.** How are medium density fibreboards made? Where are they used?
**48.** How are hardboard made? What are the three types of classifications of hardboard?
**49.** Briefly describe timber-concrete composites.
**50.** Give at least five advantages and two drawbacks of wooden composites.
**51.** Describe briefly the applications of wood and wood products.
**52.** Why is timber considered the ultimate green building material?
**53.** Suggest suitable timber for the following purposes. Give also the reasons for your choice.
   (a) Doors
   (b) Railway sleepers
   (c) Piles
   (d) Shuttering
   (e) Furniture
   (f) Packing boxes
   (g) Scaffolding

# ANSWERS

## Multiple-choice Questions

| | | |
|---|---|---|
| **1.** (a) | **2.** (b) | **3.** (b) |
| **4.** (d) | **5.** (a)–(iv) (b)–(iii) (c)–(i) (d)–(ii) | **6.** (c) |
| **7.** (b) | **8.** (a)–(iv) (b)–(iii) (c)–(ii) (d)–(i) | **9.** (d) |
| **10.** (c) | **11.** (c) | **12.** (b) |
| **13.** (a) | **14.** (d) | **15.** (a) |
| **16.** (d) | **17.** (c) | **18.** (b) |
| **19.** (c) | **20.** (b) | **21.** (a) |
| **22.** (d) | **23.** (d) | **24.** (c) |
| **25.** (b) | **26.** (c) | **27.** (d) |
| **28.** (c) | **29.** (c) | **30.** (a) |
| **31.** (b) | **32.** (c) | **33.** (a) |
| **34.** (b) | **35.** (c) | |

# CHAPTER 12
# FERROUS METALS OTHER THAN STEEL

## 12.1 Introduction

Iron is the second most abundant metal (after aluminium) in the earth's crust. However, pure iron is seldom found in nature. The common ores containing iron are hematite ($Fe_2O_3$) (which contains about 70% of iron), magnetite ($Fe_3O_4$) (which has about 70–75% iron), and goethite ($HFeO_2$). Other ores that contain about 40–50% of iron are limonite ($2Fe_2O_3 \cdot 3H_2O$), pyrite ($FeS_3$), and siderite ($FeCO_3$). These compounds, both containing iron and oxygen, are called iron oxides.

Iron is an important metal that has been used by mankind as early as 5000 years ago. Different kinds of iron have been used in the past, such as pig iron, wrought iron, cast iron, and steel (see Fig. 12.1). The Chinese were making pig iron during 1122–256 BC. The 7-m tall Ashokan iron pillar in the Qutub Complex, Mehrauli, Delhi, as shown in Fig. 12.2, which has withstood corrosion for the past 1600 years, is testimony to the high level of skill achieved by ancient Indian ironsmiths (Balasubramanian, 2005). Similarly, iron beams have been used at the Konark Sun Temple, which dates back to the early 9th Century (Chakrabarti, 1992). They demonstrate that this expertise was available even before the use of modern blast furnace (BF) technology, which was developed only in AD 1350 (Gupta, 1998).

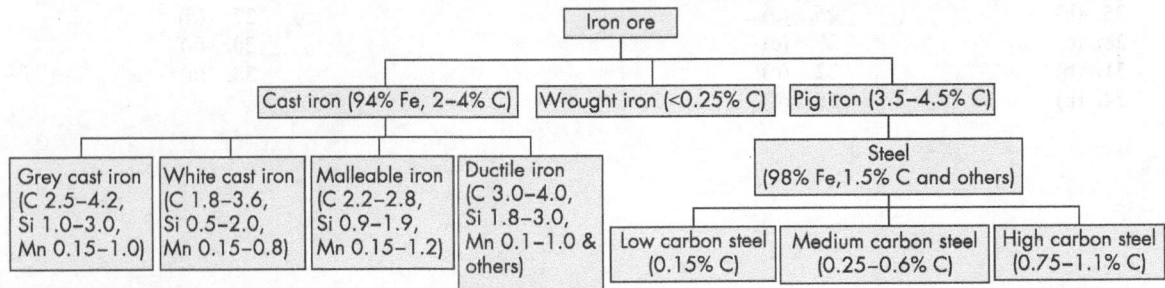

**Fig. 12.1** Types of iron and iron alloys

All the ironwork in very old buildings is wrought iron. It was continued to be used in railway tracks until it was replaced by steel. The first major application of cast iron was in the 30.4-m span Coalbrookdale arch bridge constructed by Abraham Darby III in England, in 1779, over the River Severn. Cast iron varies according to its silicon and carbon content – grey cast iron is the most common. The use of cast iron (which is weak in tension) as primary construction material was continued up to about 1840.

Modern extraction techniques use BFs, which are character-ized by their tall stacks (chimney-like structures). During 1826, wrought iron chains were used in Menai Straits suspension bridge designed by Thomas Telford (which has since been re-placed by steel chains). Robert Stephenson's Britannia Bridge was the first box girder wrought iron bridge. It was in use until around the 19th Century.

Excavations in Middle Ganges Valley show that iron working in India might have begun as early as 1800 BC. In fact, the prac-tice of manufacturing metals first began in India (Radhakrishna, 2007). Archaeological sites in India show iron implements in the period between 1800–1200 BC. There are archaeological evi-dences of use of iron in India from the Indus valley civilization. The two excellent examples are the iron pillar near Qutub Minar erected in the 5th Century (Fig. 12.2) and the 14-m tall iron post in Kodachadri village in Karnataka.

## 12.2 Pig Iron

Iron making involves the separation of iron from iron ore. Iron making is not only the first step in steelmaking, but also the most capital- and energy-intensive process in the production of steel. Iron produced in BF is called *pig iron*.

**Fig. 12.2** 1600-year old, 7-m tall iron pillar in the Qutub Complex, Mehrauli, Delhi

### 12.2.1 Production of Pig Iron

There are three basic methods of producing iron: the BF method, direct reduction, and iron smelting. Here, only the BF method is described.

The BF method of production is a continuous process. It consists of chemically reducing iron ore (iron ores are compounds of iron with non-metallic elements and contain impurities such as carbon, manganese, phosphorus, silicon, and sulphur) in a BF using coke (carbon-rich coal, as fuel) and crushed limestone (calcium carbonate, as flux) at temperatures of about 1900°C [Flux is a substance used to low-er the melting point of the ore and remove impurities such as silica ($SiO_2$), alumina ($Al_2O_3$), sulphur (S), and phosphorus (P). It combines with these impurities to form a slag that consists of low melting point complex compounds such as calcium silicate and calcium aluminate]. The principle of working of a BF is shown in Fig. 12.3. The BF is a huge, steel stack lined with refractory brick, where a carefully controlled mixture of iron ore, coke, and limestone are continuously charged from the top (Each tonne of iron ore is approximately mixed with about three-quarters of a tonne of coke and a quarter of a tonne of limestone). Air (often enriched with oxygen) that has been preheated to temperatures of about 900–1300°C, together with injected fuel such as oil or natural gas, is blown into the furnace through multiple tuyeres (nozzles) located around the circumference at the bottom of the furnace.

The raw materials require 6 to 8 hours to descend to the bottom of the furnace, where they become the final product of liquid slag and liquid iron. These liquid products are drained from the furnace at regular intervals. The hot air that was blown into the bottom of the furnace ascends to the top in about 6 to 8 sec after going through numerous chemical reactions. Once a BF is started, it is continuously run for 4 to 10 years, with only short stops to perform planned maintenance (www.steel.org).

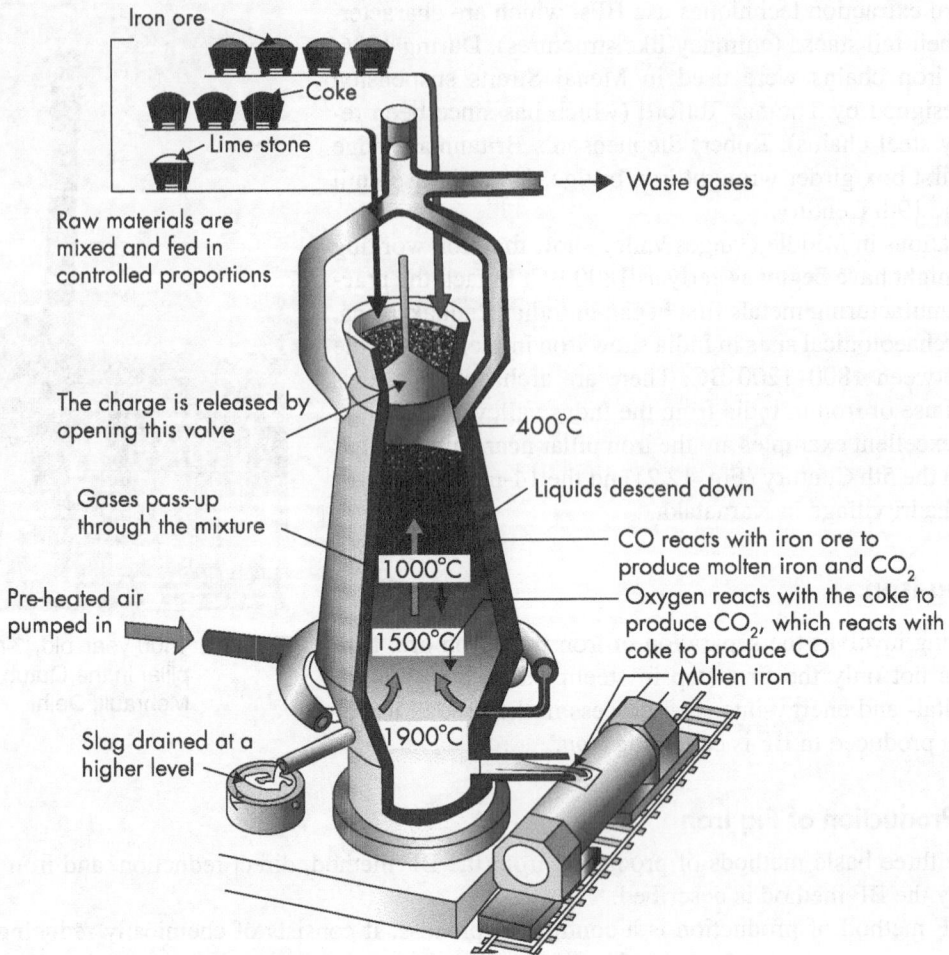

**Fig. 12.3** Smelting of iron in blast furnace (BF)

The raw ore is broken into pieces that range in size from 10 to 25 mm. Iron ore used can be in the form of iron-rich raw ore, pellets, or sinter. When the iron ore has lower iron content, it must be processed to increase its iron content. Pellets produced from this lower iron content ore are then used. Sinter is produced from the appropriate mixture of fine raw ore, small coke, sand-sized limestone, and other steel plant waste materials that contain some iron. The iron ore, pellets, and sinter are reduced to obtain liquid iron (i.e., the oxygen in the iron oxides is removed by a series of chemical reactions). These chemical reactions that occur are as follows (http://metallics.org.uk):

$$3Fe_2O_3 + CO \rightarrow CO_2 + 2Fe_3O_4 \qquad \text{Beginning at } 455°C \qquad (12.1)$$

$$Fe_3O_4 + CO \rightarrow CO_2 + 3FeO \qquad \text{Beginning at } 595°C \qquad (12.2)$$

$$FeO + CO \rightarrow CO_2 + Fe \text{ or } FeO + C = CO + Fe \qquad \text{Beginning at } 700°C \qquad (12.3)$$

While these reactions take place, they also soften and melt the iron, which trickles down as liquid through the coke to the bottom of the furnace and taken out for cooling.

The coke also descends to the bottom of the furnace as shown in Fig. 12.3. When it meets the blast of preheated air that is pumped from the bottom of BF, the coke gets ignited and reacts to generate heat and $CO_2$ as follows (www.steel.org):

$$C + O_2 \rightarrow CO_2 + Heat \tag{12.4}$$

Owing to the presence of excess carbon at a high temperature, the carbon dioxide thus generated is reduced to carbon monoxide as shown follows:

$$CO_2 + C \rightarrow 2CO \tag{12.5}$$

The carbon monoxide produced by the aforementioned reaction is used to reduce the iron ore as shown in Eqs. (12.1) to (12.3).

The limestone, which is fed from the top of the furnace (see Fig. 12.3), also descends in the BF and remains as solid until the first reaction takes place at about 870°C as follows:

$$CaCO_3 \rightarrow CaO + CO_2 \tag{12.6}$$

The sulphur present in the iron ore is removed by using the CaO produced from the aforementioned reaction as shown follows (http://metallics.org.uk):

$$FeS + CaO + C \rightarrow CaS + FeO + CO \tag{12.7}$$

$$CaO + SiO_2 \rightarrow CaSiO_3 \tag{12.8}$$

Owing to the aforementioned reactions, a liquid slag is produced which may contain CaS, silica ($SiO_2$), alumina ($Al_2O_3$), magnesia (MgO), and calcia (CaO), which might have been present the iron ore, coke, or limestone. This liquid slag settles at the bottom of the furnace, after trickling through the coke bed (which is formed from the fallen coke). As it is less dense, it floats on top of the liquid iron and is drained at a higher level than liquid iron, as shown in Fig. 12.3.

The iron-making process also results in hot dirty gases. These gases escape through vents provided the top of the BF, passing through gas-cleaning equipment, which removes particulate matter from the gases, after which the gas is cooled. As these gases has a considerable energy value, they are collected and cooled and burnt as a fuel in the stoves that are used to preheat the air entering the BF.

The resulting material, called *pig* iron, contains carbon, sulphur, and phosphorus. Even though the design of BF has improved and higher production rates are achieved, the processes inside the BF still remain the same. A review of the recent developments in iron and steel-making industry is provided by Dash and Das (2009) and Naito (2006). Modern BFs have heights in the range of 20 to 35 m, have diameters of 6 to 14 m, and can produce from 1000 to almost 10,000 tonnes of pig iron daily.

## 12.2.2 Properties and Uses of Pig Iron

The term 'pig iron' comes from the old method of casting BF iron into moulds arranged in sand beds such that they could be fed from a common runner; with the side channel looked like a 'sow' and the individual ingots looked like 'piglets' (pigs) lined-up to feed. Pig iron made from hematite ore has a very high carbon content, typically 3.5–4.5%, along with less than 1.5–3.5% silicon, 0.5–1.0% manganese, 0.02–0.05% sulphur, and 0.03–0.12% phosphorus. Owing to the high carbon content, it is very brittle, hard, and wear resistant. Hence, it is rarely used without further chemical treatment, except for a few limited applications. It has a low melting point (around 1200°C) as compared to steel. Typically, its hardness varies between 250 and 450 BHN (Brinell hardness number), and the percentage elongation of 0–0.5% (www.ispatguru.com). Pig iron does not rust and cannot be riveted or welded (see also IS 13502:1992; IS 2084:1991).

The liquid iron from a BF, which is not sent for steel making, is cast into ingots for use in steel making later as cold charge or is sold to foundries or mini steel plants having induction furnaces as merchant pig iron. *Merchant pig iron* is basically of the following two types:

1. Basic grade (with silicon and manganese, each having less than 1%)
2. Foundry grade (with 1.5 to 3.5% silicon)

Pig iron is the foundation material for other kinds of iron and steel. Until recently, molten pig iron (referred to as *hot metal*), collected from the bottom of BF, is poured into a ladle car and transferred to the steel mill to produce steel, typically with an electric arc furnace (EAF), induction furnace (IF) or basic-oxygen furnace (BOF), by burning off the excess carbon in a controlled fashion and adjusting the alloy composition (see Chapter 13).Pig iron can also be used to produce grey cast iron. This is achieved by re-melting pig iron, often along with substantial quantities of scrap iron and steel, removing undesirable contaminants, adding alloys and adjusting the carbon content (www.ispatguru.com).

In India, the National Mineral Development Corporation (NMDC), and Sesa Goa (Sesa) are the major merchant producers of iron ore (The Kudremukh mine, one of the largest iron ore mines in the world, was closed in 2006). SAIL and Tata Steel have their captive iron ore mines. Pig iron is mainly produced by Sesa Goa and Usha Ispat. In addition, there are many mini BF pig iron producers. Even integrated steel plants like SAIL and RINL produce significant amount of pig iron. In 2015, India was the third largest producer of raw steel in the world. In the year 2017–18, India produced 9.39 MT of pig iron, and 104.98 MT of finished steel.

## 12.3 Wrought Iron

The term *wrought iron* is used specifically for finished iron goods, as manufactured by a blacksmith. The term wrought means 'work by hand'. Wrought iron is considered to be pure iron, with a soft fibrous structure. It is produced by removing the impurities of cast iron. The total impurities are limited to 0.5% with a maximum percentage of carbon as 0.15% (see Table 12.1).

**Table 12.1** Comparison of chemical composition of pig iron, wrought iron, and steel

| Material | Iron (%) | Carbon (%) | Silicon (%) | Manganese (%) | Sulphur (%) | Phosphorus (%) |
|---|---|---|---|---|---|---|
| Pig iron | 91–94 | 3.5–4.5 | 0.25–3.5 | 0.5–2.5 | 0.018–0.1 | 0.03–0.1 |
| Wrought iron | 99–99.8 | 0.05–0.25 | 0.02–0.2 | 0.01–0.1 | 0.02–0.1 | 0.05–0.2 |
| Cast iron | 91.9–96.4 | 2.0–4.0 | 1.0–3.0 | 0.4–1.0 | 0.06–0.12 | 0.10–1.0 |
| Low carbon Steel | 98.1–99.5 | 0.10–0.25 | 0.005–0.4 | 0.3–1.5 | 0.02–0.05 | 0.002–0.05 |

A comparison of the physical and mechanical properties of wrought iron, cast iron, mild steel, and stainless steel are provided in Tables 12.2 and 12.3.

**Table 12.2** Physical properties of ferrous metals

| Base metal or alloy | Specific gravity | Melting point, °C | Boiling point, °C | Coefficient of linear expansion × $10^{-6}$ per °C | Specific heat (cal/g per °C) | Electrical conductivity (%) (Copper = 100%) | Relative thermal conductivity (copper = 1) |
|---|---|---|---|---|---|---|---|
| Iron, Cast | 7.50 | 1260 | NA | 10.8 | 0.119 | 2.9 | 0.12 |
| Iron, wrought | 7.70 | 1540 | 3000 | 12.1 | 0.115 | 15.0 | 0.16 |

*(Contd)*

**Table 12.2** (*Contd*)

| Base metal or alloy | Specific gravity | Melting point, °C | Boiling point, °C | Coefficient of linear expansion × 10–6 per °C | Specific heat (cal/g per °C) | Electrical conductivity (%) (Copper = 100%) | Relative thermal conductivity (copper = 1) |
|---|---|---|---|---|---|---|---|
| Steel, low carbon | 7.94 | 1483 | NA | 12.1 | 0.118 | 14.5 | 0.17 |
| Steel, stainless (austenitic) | 7.90 | 1395 | NA | 17.3 | 0.117 | 3.0 | 0.12 |

**Table 12.3** Mechanical properties of ferrous metals

| Base metal or alloy | Yield strength, MPa | Tensile strength, MPa | Elongation,% in 50 mm | Hardness (BHN) | World resources |
|---|---|---|---|---|---|
| Iron, Cast | – | 172.4 | 0.5 | 180 | 230 BT |
| Iron, wrought | 159–221 | 234–372 | 25 | 100 | – |
| Mild steel | 250 | 415 | 35 | 110 | – |
| Steel, stainless (austenitic) | 275 | 620 | 23 | 160 | – |

## 12.3.1 Manufacture

Wrought iron may be manufactured by the *puddling process* (invented by Henry Cort in 1784) in a reverberatory furnace (Other processes include the Bloomery process, developed in Europe, the Lancashire process, used in Sweden, and the Aston process, developed in USA).In the reverberatory furnace, as shown in Fig. 12.4, the metal does not come into contact with the fuel, and so is not contaminated by its impurities. The heat of the combustion products pass over the surface of the puddle and the roof of the furnace reverberates (reflects) the heat onto the metal puddle on the fire bridge of the furnace.

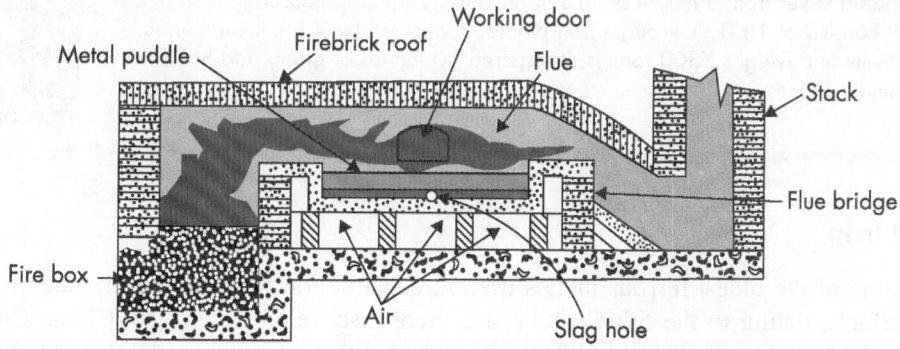

**Fig. 12.4** The reverberatory furnace using puddling process

The molten metal is subjected to a strong current of air and stirred with long bars, called puddling bars through working doors. The air and the stirring action help impurities to oxidize and drive the carbon out of the pig iron. The molten slag that is formed is drained while the iron solidifies into spongy wrought iron and floats on the top of the puddle and is fished out as puddle balls using puddle bars.

### 12.3.2 Properties

Wrought iron is an iron alloy with very low carbon content with respect to cast iron. It is soft, ductile, magnetic, malleable, and has high elasticity and tensile strength. It can be heated and reheated and worked into various shapes. Its yield strength is in the range of 159–221 MPa and ultimate tensile strength is 234–372 MPa (see also Tables 12.2 and 12.3). The modulus of elasticity of wrought iron is 193 GPa. The melting point of wrought iron is 1540°C and specific gravity is about 7.70 (www.azom.com). Wrought iron effectively resists corrosion. It is tough and withstands shocks and can neither be hardened nor tempered. It is generally fatigue resistant. Wrought iron may be welded in the same manner as mild steel (Cary, 2002). Two pieces of wrought iron can also be joined by forge welding - by heating them white hot (to about 900°C), and then fusing the elements into a single unit under the blows of a hammer.

### 12.3.3 Uses of Wrought Iron

Wrought iron replaced bronze in ancient civilizations and led to the Iron-age. It is a relatively soft material that is easily worked with hammers (it is said to be easily forged). Wrought iron is suitable for members in tension or compression (whereas cast iron is suitable for members in compression only). It was used to make decorative items such as outdoor stairs, grills, fences and gates, railway tracks, nuts and bolts, straps for timber roof trusses, nails, wire, chains, water and steam pipes, and handrails. Wrought iron furniture has a long history, dating back to Roman times. Wrought iron gates of the 13th-Century are still used in Westminster Abbey in London, and wrought iron furniture was very popular (especially in Britain) in the 17th Century (demand for it reached its peak in the 1860s). Wrought iron is no longer produced on a commercial scale. Many products described as wrought iron, such as guard rails and gates, are now made of mild steel.

---

**Case Study | The Eiffel Tower, Paris**

The Eiffel Tower, the iconic landmark of Paris, was designed by French civil engineer, Gustave Eiffel, as the entrance to the 1889 World's Fair. This 324-m tall wrought iron tower is composed of four arched legs, set on masonry piers that curve inward and are joined as a single, tapered tower from a height of 115 m. Its base is square, measuring 125 m on each side. It consists of 18,000 wrought iron pieces, connected by 2.5 million thermally assembled rivets and weighs 7300 tonnes. It required 60 tonnes of paint, and has since been repainted 18 times.

---

## 12.4 Cast Iron

Cast iron is one of the oldest ferrous metals used in construction and outdoor ornament. The earliest cast-iron artefacts, dating to the 5th Century BC, were discovered in Jiangsu, China. Cast iron was used in ancient China for warfare, agriculture, and architecture. It is primarily composed of iron, carbon, and silicon, but may also contain traces of sulphur, manganese and phosphorus (see Table 12.1). It has a relatively high carbon content of 2% to 4%. It is hard, brittle, non-malleable (i.e. it cannot be bent, stretched or hammered into shape) and more fusible than steel (www.gsa.gov). Cast iron, unlike wrought iron, is brittle and cannot be worked either hot or cold. Cast iron may break if struck with a hammer. Its structure is crystalline and it fractures under excessive tensile loading with little prior distortion. Cast iron is, however, very good in compression and hence commonly found in columns, but not in structural beam. The composition of cast iron and the method of manufacture are critical in determining its characteristics. Its usefulness derives from its relatively low melting temperature of 1260°C.

## 12.4.1 Manufacture

*Cast iron* is made by re-melting pig iron, often along with substantial quantities of scrap iron, scrap steel, limestone, coke (enriched carbon) and taking steps to remove undesirable contaminants. Re-melting is sometimes done using a special type of BF known as a *cupola*, but more often melted in electric induction furnaces or EAFs. After it is melted, the molten iron is poured into a ladle or moulds of desired size and shape (*Casting* is defined as the forming of shapes by pouring molten metal into a mould).

Similar to the BF, the *cupola furnace* is a vertical, 6 to 11 m high and 450 to 2000 mm diameter steel shell, which is lined with refractory bricks (see Fig. 12.5). The charge is introduced into the furnace body by means of an opening approximately half way up the vertical shaft. The charge consists of alternate layers of the metal to be melted, coke fuel, and limestone flux. The fuel is burnt in air, which is introduced through tuyeres positioned above the hearth. The hot gases generated in the lower part of the shaft ascend and heat the descending charge. Most cupolas have hinged doors under the hearth, which allows the bottom to drop down, so that cleaning and repairs can be done. In the front, at the bottom, is a tap-hole for the molten iron and at the rear, positioned above the tap-hole is a slag-hole. The top of the stack is capped with a spark/fume arrester hood.

The first cupola furnace was built by René-Antoine Ferchault de Réaumur (1683–1757), in France, during 1720. Cupola melting is still recognized as the most economical melting process; most grey iron is melted by this method.

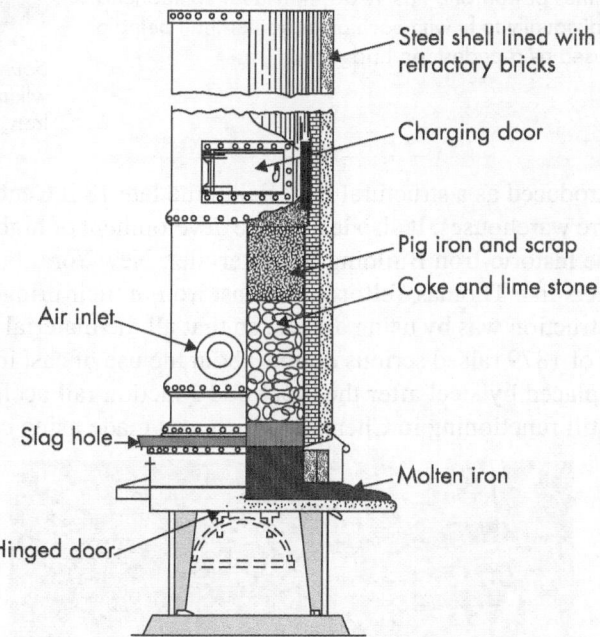

**Fig. 12.5** Schematic sketch of a cupola furnace

The word cast iron is a misnomer as steel with carbon content less than 2% can also be cast. The striking difference between cast steel and cast iron is that the former is plastic and forgeable while the latter is not. However, some of the modern cast iron can develop a fair degree or plasticity and toughness.

Depending on the application, carbon and silicon content are adjusted to the desired levels, which may be anywhere from 2 to 3.5% and 1 to 3%, respectively. Other elements are then added to the melt before the final form is produced by casting.

The basic strength and hardness of all iron alloys are derived from the metallic structures of *graphite*, which is pure and soft carbon. The various types of steel ranging from soft, low-carbon steel (with ultimate tensile strength 415 MPa) to hardened, high-carbon steel (with ultimate tensile strength 2000 MPa), can be manufactured. Owing to the free graphite and high silicon content of cast iron, an insoluble graphitic protective film/scale is formed on the surface, which forms a barrier against further corrosion (Santner and Goodrich, 2006). Thus, iron castings are used in applications where this resistance is required to provide long life. Alloyed irons have enhanced resistance to heat, oxidation, and corrosion. Properties of cast iron can be modified and enhanced by heat treatment. Annealing produces soft machineable ferrite. This engineered material can be hardened and tempered using conventional heat treating or surface hardening (Santner and Goodrich, 2006). In building construction, cast iron is used in soil, waste, ventilating, and rainwater pipes, cast iron spigot and socket, fittings, and accessories, manhole covers and frames (IS 7181:1986; IS 3989:2009; IS 1729:2002; IS 1536:2001; IS 1537:1976; IS 1538:1993; IS 1726:1991; IS 6629:1972).

### Case Study  Coalbrookdale Cast Iron Bridge, England

The Coalbrookdale Bridge crosses the River Severn in Shropshire, England. Opened in 1781, it was the first major bridge in the world to be made of cast iron. Designed by architect Thomas Farnolls Pritchard, this 30.63-m span bridge used 378 tonnes of iron and has 1700 individual components. Components were cast individually to fit with each other, rather than being of standard sizes. It is now used as a pedestrian bridge.

*Source*: https://commons.wikimedia.org/wiki/File:The_Iron_Bridge_(8542).jpg

Cast iron was first introduced as a structural material in the late 18th Century. The first buildings to use this new material were warehouses. It also lead to the development of high-rise framed construction in the 20th Century. The historic Iron Building in Watervliet, New York, built in 1859, is a cast-iron building. Several engineers like Thomas Telford used cast iron in their bridges. The best way of using cast iron for bridge construction was by using arches, so that all the material will be compression. The Tay Rail Bridge disaster of 1879 raised serious doubts about the use of cast iron. Cast-iron beam structures were eventually replaced by steel after the Norwood Junction rail accident of 1891. Figure 12.6 shows an old building, still functioning in Chennai, which was made using cast iron columns.

**Fig. 12.6** Cast iron columns of Government Press, Anna Salai, Chennai

## 12.4.2 Classification

The properties of cast iron are changed by adding various alloying elements. The various types of cast iron can be classified by their microstructure. This classification is based on the form and shape in which the major portion of carbon is present in the iron. Based on these, cast iron is classified as grey, white, malleable, compacted graphite, ductile, and alloyed. A comparison of these various types of cast iron, their important properties, and uses are given in Table 12.4.

**Table 12.4** Types of cast iron, important properties, and uses

| Name | Nominal composition (% by weight) | Form and condition | Yield strength (MPa(0.2% offset)) | Tensile strength (MPa) | Percentage elongation (%) | Hardness (Brinell scale) | Uses |
|---|---|---|---|---|---|---|---|
| **Grey cast iron** (IS 210:2009, ASTM A48) | C 2.5–4.2, Si 1.0–3.0, Mn 0.15–1.0, S 0.02–0.25, P 0.02–1.0 | Cast | – | 150–400 (Grades FG150-FG 400) | 0.5 | 130–270 | Pipes, fittings, flywheels, Automobile engine blocks & heads, gearbox cases, machine-tool bases, soil pipes |
| **White cast iron** ASTM A532 | C 1.8–3.6, Si 0.5–2.0, Mn 0.15–0.8, S 0.02–0.2, P 0.02–0.2 | Cast (as cast) | – | 170 | 0 | 450 | Crushing & grinding applications; rolling mill rolls, crushers, pulverisers and ball mill liners. |
| **Malleable iron** (IS 14329:1995, ASTM A47) | C 2.2–2.8, Si 0.9–1.9, Mn 0.15–1.2, S 0.02–0.2, P 0.02–0.2 | Cast (annealed) | 190–530 | 300–700 Grades WM 350–400, BM 300–350 & PM 450–700 | 2–12 | 150–290 | Chains; sprockets; tool parts & hardware; connecting rods; drive train & axle components; and spring suspensions. |
| **Ductile or nodular iron** (IS 1865: 1991, ASTM A395) | C 3.0–4.0, Si-1.8–3.0, Mn 0.1–1.0, S 0.01–0.03, P 0.01–0.1, Ni 1.0, Mg 0.02–0.1 | Cast | 210–400 | 350–700 (Grades SG 350/22 to SG 700/2) | 2–22 | 130–360 | Gears; automotive & truck suspension components; brake components; valves; pumps; linkages; hydraulic components; and wind turbine housings |

*(Contd)*

**Table 12.4** (*Contd*)

| Name | Nominal composition (% by weight) | Form and condition | Yield strength (MPa(0.2% offset)) | Tensile strength (MPa) | Percentage elongation (% ) | Hardness (Brinell scale) | Uses |
|---|---|---|---|---|---|---|---|
| Ni-resist type 2 ASTM A518 | C 3.0, Si 2.0, Mn 1.0, Ni 20.0, Cr 2.5 | Cast | – | 185 | 2 | 140 | Parts for chemical processing plants; petroleum refining; food handling & marine service; control of corrosive fluids; and pressure valves. |

**Designation system for cast iron**  The designation of ferrous castings is given in IS 4843:1968. The nomenclature white and grey signify the colour of the fractured surface of a casting. The classification for alloyed irons has a wide range of carbon, silicon, and manganese compositions with minor additions of other elements, such as nickel, chromium, molybdenum, and copper. Thus, the designation system for cast iron as per IS 4843:1968 consists of two letters designating the type of iron (Grey iron: FG; Malleable iron: BM for black-heart, PM for Pearlitic, and WM for white-heart; Nodular graphite iron: SG: Austenitic flake graphite iron: AFG; Austenitic nodular graphite iron: ASG; Abrasion-resistant iron: ABR; Steel castings; CS, Heat-resistant steel castings: CSH; and Corrosion-resistant steel castings: CSC). It is followed by a space and then two digits, which show the minimum tensile strength (in MPa, divided by 10) – examples: FG 40, BM 35, PM 70, WM 35, SG 42/12, and CS 125. In the case of grey iron/alloy iron castings, where chemical composition is more important than the tensile strength, the aforementioned designation is followed by the chemical symbol, as per IS 1762:1961 – examples: FG 35 Si 15, AFG Ni 16 Cu 7 Cr 2, ASG Ni 20 Cr 2, ABR 33 Ni 4 Cr 2, CS 50 Cr 1 V 20, and CSH 130 Ni 6 Cr 28.

## 12.4.3 Grey Cast Iron

The most common traditional form of cast iron is *grey cast iron*. Grey cast iron is easily cast but it cannot be forged or worked mechanically, either hot or cold. Grey cast iron is characterized by its graphitic microstructure, and when it is fractured has a grey appearance. The carbon content is in the form of graphite flakes distributed throughout the metal. Since the melting temperature of grey cast iron is about 1260°C (which is about 250°C lower than the melting point of wrought iron), it is extensively used for castings. Most grey cast irons have a chemical composition of 2.5 to 4.2% carbon, 1 to 3% silicon, and the remainder is iron (see Table 12.4). Grey cast iron has less tensile strength and shock resistance than steel, but has high compressive strength (600 to 1200 MPa for grades FG 150 to FG 400). Its compressive strength is typically three to four times more than its tensile strength. As per IS 210:2009, there are seven grades of grey iron castings, namely, grades FG 150, FG 200, FG 220, FG 260, FG 300, FG 350, and FG 400. The unique combination of high hardness, high damping capacity, and excellent machinability of grey cast iron makes it ideally suitable for machine bases and supports, engine cylinder blocks, and brake components. The damping capacity of grey cast iron is about 20–25 times higher than steel [see also IS 210:2009; IS 3005(Parts 1–4):1979].

## 12.4.4 White Cast iron

White cast iron shows a white surface when fractured. White cast iron is hard and essentially free of graphite. The metal solidifies with a compound called cementite or iron carbide ($Fe_3C$), which dominates the microstructure and properties of white iron. The carbides are in a matrix that may be pearlitic, ferritic, austenitic, martensitic, or a combination of these. The white cast iron is unique in that it is the only member of the cast iron family in which carbon is present only as carbide. The presence of different carbides, which depends upon the alloy content, makes white cast iron hard, and resistant to wear and abrasion. However, white cast iron is very brittle, but has a high compressive strength. It is impossible to machine white cast iron due to its brittleness, and it is difficult to cut it due to its high hardness. White cast irons are limited in application because of the lack of impact resistance and the difficulty in maintaining the structure in thicker sections.

## 12.4.5 Malleable Cast Iron

Malleable cast iron is essentially white cast iron, which is modified by heat treatment. It is obtained by heating white cast iron to around 920°C for over 12h and then leaving it to cool in air at a very slow, controlled rate. It solidifies to a microcrystalline structure and the carbon content is present in the form of cementite ($Fe_3C$). The microstructure consists of ferrite with irregularly shaped nodules of tempered carbon, rather than the graphite flakes of white cast iron. The chemical composition of malleable cast iron is generally in the ranges as given in Table 12.4. Malleable cast iron may also contain small amounts of chromium (0.01 to 0.03%), boron (0.002%), copper (up to 1.0%), nickel (0.5 to 0.8%), and molybdenum (0.35 to 0.5%). Malleable cast iron has superior mechanical properties compared to grey cast iron with increased strength and wear resistance and considerable ductility and toughness (Properties of all types of malleable cast iron are provided in IS 14329:1995). Malleable cast iron castings are produced in section thicknesses ranging from about 1.5 to 100 mm. Thin section castings are used in parts that are to be pierced, coined, or cold formed, which require maximum machinability, good impact resistance at low temperatures, and wear resistance.

There are three groups of malleable cast iron: *white heart malleable cast iron* (WM cast iron), *black-heart malleable cast iron* (WM cast iron), and *Pearlitic malleable cast iron* (PM cast iron). These three groups are differentiated by chemical composition, temperature, and time cycles of the annealing process, the annealing atmosphere and properties, and the resulting microstructure. All their microstructure does not contain flake graphite.

The WM cast iron is obtained after annealing in a decarburizing atmosphere and has a silvery-grey fracture surface. Its microstructure depends on the size of the section – smaller sections have a ferrite, pearlite, and temper carbon microstructure, whereas larger sections have an outer ferrite layer and a tougher ferrite, pearlite, and temper carbon core.

The BM cast iron is obtained after annealing in an inert atmosphere and has black fracture. It has a ferrite and temper carbon microstructure and is softer and more ductile. The PM cast iron is obtained after heat treatment and has a homogeneous matrix microstructure, according to the specified grade, of pearlite or other transformation products of austenite. Graphite is present in the form of temper carbon nodules. The PM cast iron is harder and less ductile (http://ispatguru.com).

Malleable cast iron can be specified either by its tensile properties or hardness of the casting. Unless the relationship between the test bar properties and the specific casting hardness is established, both strength and hardness should not be specified together. IS 14329 specifies two grades of the WM cast iron (WM 350 and WM 400), three grades of the BM cast iron (BM 300, BM 320, and BM 350), and five grades of the PM cast iron (PM 450, PM 500, PM 550, PM 600, and PM 700).

### 12.4.6 Spheroidal Graphite Iron (Ductile Iron)

Developed in 1948, spheroidal graphite or nodular ferritic iron is obtained by adding tiny amounts of magnesium(0.02 to 0.1%). The added magnesium reacts with the sulphur and oxygen in the molten iron and changes the way the graphite is formed. The graphite is now formed in concentric layers resulting in nodules (spheroids) rather than as individual flakes as in grey iron. As a result, the ductile cast iron has properties similar to that of malleable iron, without the stress-concentration effects that flakes of graphite would produce. Eight grades of spheroidal cast iron are specified in IS 1865: 1991 – SG 350/22A, SG 400/15A, SG 400/18A, SG 500/7A, SG 600/3A, SG 700/2A, SG 800/2, and SG 900/2. Ductile iron, like malleable iron, exhibits a linear stress-strain relation, a considerable range of yield strengths and ductility.

Ductile irons have excellent machinability and wear resistance. Castings are made in a wide range of sizes with sections that can be either very thin or very thick. Ductile iron has the ability to be used as cast and without heat treatments or other further refining. It has a tensile strength comparable to many steel alloys and a modulus of elasticity between that of grey iron and steel. It has found wide acceptability and even competes favourably with steel.

Ductile iron pipe is unique when it comes to corrosion control. An oxide layer is formed on the inside and the outside of all ductile iron pipes during the manufacturing process. This oxide layer provides adequate corrosion protection in many circumstances. Additional corrosion protection may be provided in the form of polyethylene encasement, which has proven to control corrosion in aggressive soils for more than 50 years. Ductile iron pipes are also very cost effective compared to steel pipes with bonded coating (see also IS 1865:1991; IS 1729:2002; IS 9523:2000) Malleable cast iron and ductile iron are used for some of the applications in which ductility and toughness are important (IS 1879:2010).

### 12.4.7 Alloyed Iron

*Alloyed iron* includes grey irons, ductile irons, and white irons that have more than 3% alloying elements (nickel, chromium, molybdenum, silicon, or copper). Malleable irons are not heavily alloyed because many of the alloying elements interfere with the graphite-forming process that occurs during heat treatment. These irons are classified into two types: corrosion-resistant and elevated-temperature resistant. Corrosion-resistant alloyed cast iron is used to produce parts for engineering applications that operate in environments such as sea water, organic and inorganic acids, and alkalis (IS 7520:1974). Silicon and chromium increase resistance to heavy scaling by forming a light surface oxide that is impervious to oxidizing atmospheres. Both elements reduce the toughness and thermal shock resistance.

Elevated-temperature resistant alloyed iron may have nickel, molybdenum, and aluminium. Nickel increases strength and toughness at elevated temperatures above 540°C by promoting an austenitic structure that is significantly stronger than ferritic structures. Molybdenum increases high-temperature strength in both ferritic and austenitic iron alloys. Aluminium, however, may adversely affect mechanical properties at room temperature. Elevated-temperature resistant alloyed iron resists fracture under service loads, oxidation at ambient atmosphere, and instability at temperatures up to 600°C (Santner and Goodrich, 2006). Abrasion-resistant cast irons may be of the following grades (IS 4771:1985; IS 7925:1976):

1. Nickel-chromium martensite white iron (examples: NiLCr 30/500 and NiHCr 34/600)
2. Chromium-molybdenum martensite white iron (example: CrMoHC 34/500)
3. High-chromium white iron (HCrNi 27/400)

In these specifications, the first two digits show the tensile strength (in MPa, divided by 10) and the last three digits the Brinell hardness.

## 12.4.8 Corrosion Resistance of Cast Iron

Cast iron is extremely strong and durable when used appropriately and protected from adverse exposure. Alloying elements play a greater role in providing corrosion resistance to cast iron. The alloying elements normally used to improve the corrosion resistance of cast irons include silicon, nickel, chromium, molybdenum, and copper. The addition of silicon in range of 3 to 14% results in increases in the corrosion resistance of cast iron. However, increase in silicon content beyond 14%, although increases corrosion resistance considerably, results in the loss of strength and ductility. As mentioned earlier, alloying iron with silicon results in the formation of strongly adherent surface film, which provides resistance to corrosion. Owing to the corrosion resistance of such cast iron, it has been used in drain and waste water pipes, which can be made at low cost, and have good strength and long service life. The addition of small amount of chromium alone or in combination with silicon and/or nickel increases the corrosion resistance of cast iron by forming protective oxides on the surface of the metal. Protective coatings may also be used on iron, which include bituminous coatings, paints, and metallic coatings. Corrosion-resistant high silicon iron castings are also available (see IS 7520:1974).

## 12.5 Steel

Steel is an alloy of iron and various other elements, which are used to enhance the properties, such as strength, resistance to corrosion, and tolerance of heat, of iron. Changing the type and amount of the elements alloyed with iron can produce different types of steel (IS 2062:2011). Steel is made using *Basic Oxygen Steel making* (BOS) process or the *electric arc method*. As steel is an important building material compared to iron, it is considered separately in Chapter 13. A comparison of the properties of cast iron, wrought iron, and steel is provided in Table 12.5.

**Table 12.5** Comparison of cast iron, wrought iron, and mild steel

| S. no. | Property | Cast iron | Wrought iron | Steel |
|---|---|---|---|---|
| 1. | Composition | It is a crude form of iron containing 2.5–4.5% carbon. | It is the purest form of iron containing only up to 0.20% carbon. | It is midway between cast iron and wrought iron, containing 0.1 to 1.1% carbon. |
| 2. | Structure | It has a crystalline structure. | It has fibrous structure with a silky lustre. | It has a granular structure. |
| 3. | Specific gravity | Its specific gravity varies from 7 to 7.5. | Its specific gravity is 7.80. | Its specific gravity is 7.85. |
| 4. | Melting point | Its melting point is about 1250°C. Its contracts on melting. | It melts at about 1500°C | Its melting point is between 1300 and 1400°C. |
| 5. | Hardness | It is quite hard and can be hardened by heating and sudden cooling. | It cannot be hardened or tempered. | It can be hardened and tempered. |

*(Contd)*

**Table 12.5** (*Contd*)

| S. no. | Property | Cast iron | Wrought iron | Steel |
|--------|----------|-----------|--------------|-------|
| 6. | Strength | Its ultimate compressive strength is 600–700 MPa and ultimate tensile strength 120–150 MPa. | Its ultimate compressive strength is 200 MPa and ultimate tensile strength is about 400 MPa. | Its ultimate compressive strength is 180–350 MPa and ultimate tensile strength is 310–700 MPa. |
| 7. | Reaction to sudden shock | It does not absorb shocks. | It cannot stand sudden heavy shocks. | It absorbs shocks. |
| 8. | Magnetization | It cannot be magnetized. | It does not form permanent magnets but can be temporarily magnetized. | It can form permanent magnets. |
| 9. | Rusting | It does not rust easily. | It rusts more than cast Iron. | It rusts easily. |
| 10. | Malleability and ductility | It is neither malleable nor ductile. | It is tough, malleable, ductile, and moderately elastic. | It is tough, malleable, and ductile. |
| 11. | Forging and welding | It is brittle and cannot be welded or rolled into sheets. | It can be easily forged or welded. | It can be rapidly forged or welded. |
| 12. | Uses | Because of its non-rusting property, it is used in the manufacture of parts most likely to rust such as water pipes, sewers, and drain pipes. It is used for making such parts of machines which are not likely to be subjected to shocks or tension. Lamp posts, carriage wheels and rail chairs are usually made of cast iron. | As it can withstand sudden shocks without permanent injury, it is used for chains, crane hooks, railway couplings, etc. | It is used as reinforcement in RCC and as structural members, bolts, rivets, and sheets (plain and corrugated). High-carbon steel is used for those parts of machinery where hard, tough, elastic, and durable material is required. It is used for making cutlery, files, and machine tools and rails. |

## 12.6 Environmental Effects of Iron/Steel Production

Every year, in addition to the 300 MT of recycled iron, around 500 MT of iron is produced. Iron and steel production results in about 6.5% of $CO_2$ emissions. Most iron ore is extracted through opencast mines. The environmental effects due to opencast mining include the generation of dust through blasting, raw-ore haulage and processing activities, and contamination of ground water (Mining has high-water demands for extraction, processing, and waste disposal – these processes can pollute water sources nearby and deplete freshwater supplies in the surrounding region). These will have serious impacts on the air quality at opencast mines and on health, and the ambient air quality surrounding opencast mines. In addition, the flux limestone is also quarried, crushed, and transported to steel mills.

Production of steel is a high energy-intensive process and the average energy required for production is 18.68 GJ/t (which include 50% for coking coal, 35% for electricity, and 5% for natural gas). Average $CO_2$ emitted is 1.77 t/tonne of steel. Steel production has a number of effects on the environment, including air emissions (CO, $SO_x$, $NO_x$, and particulate matter), wastewater contaminants, hazardous wastes, and solid wastes. The major environmental impacts from integrated steel mills are from coking and iron-making (www.greenspec.co.uk).

Coke production is one of the major pollution sources from iron/steel production: air emissions from coke ovens include benzene, benzene-soluble organics, naphthalene, ammonium compounds, crude light oil, sulphur $SO_x$, volatile organic compounds, and coke dust. Coke plant operators have developed efforts to control/extract and use these emissions. Water emissions come from the water used to cool coke after it has finished baking. Quenching water becomes contaminated with coke breezes and other compounds. The typical volume of process wastewater generated at a well-controlled coke plant is approximately 570 L/tonne of coke. However, quenching water is fairly easy to reuse. Most pollutants can be removed by filtration.

Slag and the limestone and iron ore impurities collected at the top of the molten iron are the by-products of iron production (www.greenspec.co.uk). Air-cooled BF slag/steelmaking slag is used in road base course material in the same way as gravel. Granulated BF slag is used to produce Portland BF slag cement (www.slg.jp). Sulphur dioxide and hydrogen sulphide are volatized and captured in air emissions control equipment.

Iron and steel are the world's most recycled materials, and among the easiest materials to reprocess, as they can be separated magnetically from the waste stream. Globally, around 85% of construction steel is currently recovered from demolition.

## SUMMARY

- Iron is the second most abundant metal found on earth. Important iron ores are hematite, magnetite, and goethite. Different kinds of iron have been used in the past such as pig iron, wrought iron, cast iron, and steel.
- There are three basic methods of producing iron: the BF method, direct reduction, and iron smelting. In the BF, iron ores are chemically reduced to pure iron using coke and crushed limestone at temperatures of about 1900°C.
- Owing to the high carbon content, pig iron is very brittle, hard, and wear resistant. It does not rust and cannot be riveted or welded. Pig iron is the foundation material for other kinds of iron and steel.
- Wrought iron is pure iron, with a soft fibrous structure. Wrought iron may be manufactured by the puddling process. Wrought iron is soft, ductile, magnetic, malleable, and has high elasticity and tensile strength. It is tough and withstands shocks and can neither be hardened nor tempered. It was used to make decorative items.
- Cast iron is primarily composed of iron, carbon (2 to 4%), and silicon. Cast iron, unlike wrought iron, is brittle and cannot be worked either hot or cold. Cast iron is made by re-melting pig iron in cupola furnace. By adding various alloying elements, the properties of cast iron can be changed. Based on this, cast iron is classified as grey, white, malleable, compacted graphite, ductile, and alloyed. The designation of ferrous castings is given in IS 4843:1968.
- Grey cast irons have 2.5 to 4.2% carbon, and 1 to 3% silicon. It has graphitic microstructure, and when fractured shows a grey appearance. They have combination of high hardness, compressive strength, high damping capacity, and excellent machinability. White cast iron shows a white surface when fractured. White cast iron is hard and essentially free of graphite. It has cementite microstructure and is brittle and has high compressive strength. Malleable cast iron is obtained by heating white cast iron to around 920°C and cooling it very slowly. Its microstructure consists of ferrite with irregularly shaped nodules of temper carbon. There are three groups of malleable cast iron.
- Spheroidal graphite or ductile iron is obtained by adding tiny amounts of magnesium. Ductile cast iron has properties similar to that of malleable iron. It has a tensile strength comparable to many steel alloys. Cast iron pipes are relatively cheap, have good strength, and a long life.
- Alloyed iron is made with alloying elements such as nickel, chromium, molybdenum, silicon, or copper. These irons are classified into two types: corrosion-resistant and elevated-temperature resistant.

# EXERCISES

## Multiple-choice Questions

1. The crudest form of iron is
   (a) mild steel
   (b) pig iron
   (c) wrought iron
   (d) cast iron
2. The purest form of iron is
   (a) mild steel
   (b) pig iron
   (c) wrought iron
   (d) cast iron
3. Carbon content in pig iron
   (a) <0.25%
   (b) 2–4%
   (c) 1.5%
   (d) 3.5–4.5%
4. Carbon content in wrought iron
   (a) <0.25%
   (b) 2–4%
   (c) 1.5%
   (d) 3.5–4.5%
5. Carbon content in cast iron
   (a) <0.25%
   (b) 2–4%
   (c) 1.5%
   (d) 3.5–4.5%
6. Carbon content in steel
   (a) <0.25%
   (b) 2–4%
   (c) 1.5%
   (d) 3.5–4.5%
7. Match the metal on the left with the melting temperature on the right
   (a) Pig iron
   (b) Wrought iron
   (c) Cast iron
   (d) Steel
   (i) 1260°C
   (ii) 1483°C
   (iii) 1540°C
   (iv) 1200°C
8. Match the metal on the left with the tensile strength on the right
   (a) Stainless steel
   (b) Wrought iron
   (c) Cast iron
   (d) Mild steel
   (i) 172 MPa
   (ii) 620 MPa
   (iii) 415 MPa
   (iv) 234–372 MPa
9. Match the metal on the left with the BHN hardness on the right
   (a) Stainless steel
   (b) Wrought iron
   (c) Cast iron
   (d) Mild steel
   (i) 100
   (ii) 160
   (iii) 110
   (iv) 180
10. Match the metal on the left with the manufacturing processes on the right
    (a) Pig iron
    (b) Wrought iron
    (c) Cast iron
    (d) Mild Steel
    (i) Puddling process
    (ii) Blast furnace
    (iii) BOS Process
    (iv) Cupola
11. Match the following tensile strengths:
    (a) Grey cast iron
    (b) White cast iron
    (c) Malleable iron
    (d) Ductile iron
    (i) 300–700 MPa
    (ii) 350–700 MPa
    (iii) 170 MPa
    (iv) 150–400 MPa
12. Which of the following types of steel is used in the manufacture of present day rails?
    (a) Mild steel
    (b) Pig iron
    (c) Cast steel
    (d) Malleable iron
13. Cast iron is found to resist the following effectively
    (a) tension
    (b) compression
    (c) shear force
    (d) none of these
14. Which of the following types of steel is used for sewer pipes?
    (a) Mild steel
    (b) Pig iron
    (c) Cast steel
    (d) Malleable iron
15. Iron and steel production accounts for the global $CO_2$ emission of
    (a) 3.5%
    (b) 4.0%
    (c) 5.5%
    (d) 6.5%

## Review Questions

1. Name the ores required for making steel.
2. What is pig iron? Describe the method of its manufacture from iron ore.
3. List some of the properties and uses of pig iron.
4. What is wrought iron? How is it manufactured? How do its properties differ from that of cast iron or pig iron?
5. List a few uses of wrought iron. Why is it replaced by steel?
6. Describe the differences between pig iron, wrought iron, and cast iron.
7. What is cast iron? How is it manufactured?
8. Can cast iron be used as a structural element? Why is its use replaced by steel?
9. What are the different types of cast iron? State their advantages and disadvantages.
10. How is cast iron designated as per IS 4843?
11. Write short notes on (a) grey cast iron, (b) white cast iron, (c) malleable cast iron, (d) ductile cast iron, and (e) alloyed cast iron.
12. What are three groups of malleable cast iron? How are they differentiated?
13. What are alloyed irons? What are their main uses?

14. Why cast iron pipes are preferred over steel pipes?
15. State the carbon contents and uses of the following metals: (a) pig iron, (b) cast iron, (c) wrought iron, and (d) low-carbon steel.
16. Compare the properties of cast iron, wrought iron, and mild steel.
17. Write a short note on the environmental effects of iron/steel production.

## ANSWERS

### Multiple-choice Questions

1. (b)      2. (c)      3. (d)      4. (a)      5. (b)      6. (c)
7. (a)-(iv), (b)-(iii), (c)-(i), (d)-(ii)        8. (a)-(ii), (b)-(iv), (c)-(i), (d)-(iii)
9. (a)-(ii), (b)-(i), (c)-(iv), (d)-(iii)     10. (a)-(ii), (b)-(i), (c)-(iv), (d)-(iii)
11. (a)-(iv), (b)-(iii), (c)-(i), (d)-(ii)     12. (a)     13. (b)     14. (c)     15. (d)

# CHAPTER 13
# STEEL

## 13.1 Introduction

Steel is the most extensively used of all metals. As per the World Steel Association (www.worldsteel.org), the global steel production of crude steel during 2018 is estimated as 1808 MT, and consumption was 1658 MT (China and India held the first and second ranks with consumption of 928.3 MT and 106.5 MT, respectively). Thus, the consumption of steel is exceeded only by concrete. It is also one of the sustainable building materials in the world. Steel was first introduced in 1740, but was not available in large quantities until Sir Henry Bessemer of England invented and patented the process of making steel in 1855. In 1865, Siemens and Martin invented the open-hearth process and this was used extensively for structural steel production. The carbon content of steel varies from 0.25 to 1.5%. (Carbon is the most important alloy of steel. Increasing carbon content increases hardness and strength and improves hardenability; at the same time, carbon also increases brittleness and reduces weldability because of its tendency to form martensite.) The first major structure to use steel exclusively is the Railway Bridge at the Firth of Forth, UK, built in 1890. Designed by Sir John Fowler and Sir Benjamin Baker and built by Sir William Arrol & Co, this bridge still has the world's second-longest cantilever spans, and 190 to 200 trains still run over it every day. Companies such as Dorman Long started rolling steel 'I' sections by 1880. Components or members made of steel can be joined easily by using riveting, bolting, or welding. Riveting was used as a fastening method until around 1950, when it was superseded by welding. Bessemer steel production in Britain ended in 1974 and the last open-hearth furnace closed in 1980. Basic oxygen steel (BOS) making process using CD converter was invented in Austria in 1953. In the later part of the 19th century and early 20th century, newer technologies resulted in better and new grades of steel. Today, we have several varieties of steel made with alloying elements such as carbon, manganese, silicon, chromium, nickel, and molybdenum (see Secs. 13.8, 13.15, and 13.16). The electric arc furnace is used to make special steels such as stainless steel.

In this chapter, the making, metallurgy, and heat treatment of structural steel, which are important for the selection of the material for a particular situation, are briefly discussed. The various types of steels, such as hot and cold rolled (CR) steel, high-performance steel, weathering steel, and stainless steel, and their properties are also explained. The various types of reinforcing steel used in RC structures are also provided. Corrosion of steel and the methods to overcome the same are also included. Other steel products used in construction, are also briefly described.

## 13.2 Steel Making

Three main processes exist for the production of steel. The oldest of this is the *open-hearth process*. Since it was slow and uneconomical, it has been replaced largely by the *BOS making* process and the *electric arc method*. (The electric arc furnace is used mainly to make special steels such as stainless steel.) Steel production is basically a batch process and involves reducing the carbon, sulphur, and phosphorus levels and adding, when necessary, manganese, chromium, nickel, or vanadium.

### 13.2.1 Integrated Steel Plants

Today, most structural steel is made in *integrated steel plants* using the BOS process shown in Fig. 13.1. Iron ore lumps, scrap steel (up to 30%), pellets, coke (made from cooking coal), and fluxes, such as limestone and dolomite, are used as the major raw materials. The main steps involved in the manufacturing process are as follows.

**Melting**  Raw materials are charged in a blast furnace, where hot air is pumped to melt iron and fluxes at 1600°C. The molten metal, when cooled and solidified, is called *pig iron*. Alternatively, it can be further refined to make steel. The excess carbon and other unwanted impurities are floated off as slag (this slag is blended with clinker to make blast furnace cement, which is used in high performance concretes).

**Refining**  Molten metal from the blast furnace is taken to steel melting shop where the impurities are further reduced in a basic oxygen furnace (LD converter) or open-hearth furnace (see Fig. 13.2). (This process was invented in Austria in 1953 and first adopted in two towns –Linz and Donawitz–and hence called LD converter). The working of the BOS process is as follows:

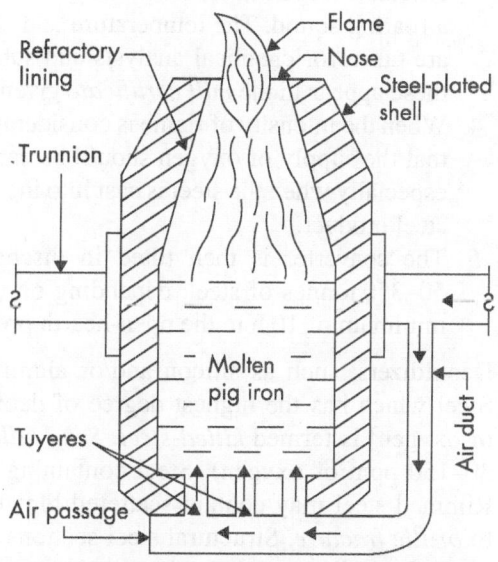

**Fig. 13.1** Basic oxygen steel (BOS) making process

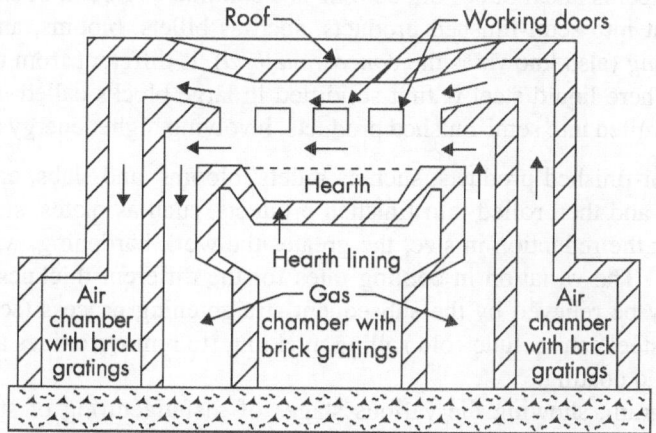

**Fig. 13.2** Open-hearth furnace (Subramanian, 2016)

1. The converter is tilted and charged with molten pig iron from cupola furnace or sometimes directly from the blast furnace. (The converter is mounted on two horizontal trunnions as shown in Fig. 13.1), so that it can be tilted or rotated at any suitable angle.
2. The converter is brought in an upright position and a jet of pure oxygen is blown at extraordinary speed through the tuyeres (There exist several variations: top blowing, bottom blowing, and a combination of both).
3. The oxygen passes through the molten pig iron. A high temperature is developed and the excess elements present in pig iron, such as, carbon, silicon, manganese, sulphur, and phosphorus, are oxidized. At this time, a reddish-yellow flame is seen at the nose of the converter, accompanied by a roaring sound. The temperature and chemical composition are carefully monitored and samples are taken for chemical analysis and subsequent examination of physical properties; the results of these appear in the *mill certificate* given to the purchaser of the steel.
4. When the intensity of flame is considerably reduced, the oxygen supply is shut off. It has to be noted that the supply of oxygen should be carefully controlled to avoid trapping of gas pockets in steel, especially when the steel is cast into ingots. These gas pockets may lead to defects in the final rolled steel product.
5. The converter is then tilted in discharge position and this batch process typically produces 50–350 tonnes of steel, depending on the size of furnace, every one hour to 8 h (compared to a minimum of 10 h in the open-hearth process).

Deoxidizers, such as silicon and/or aluminium, are used to control the dissolved oxygen content. Steel which has the highest degree of deoxidation [containing less than 30 parts per million (ppm) of oxygen] is termed *killed-steel*. *Semi-killed steel* has an intermediate degree of deoxidation (about 30–150 ppm of oxygen). Steel containing the lowest degree of deoxidation is called *rimmed steel*. Rimmed steel may contain scattered blowholes throughout its structure. Such a steel is most prone to *brittle fracture*. Structural steel sections are often produced using either killed steel or semi-killed steel, depending upon the intended use and the thickness. During continuous casting, only killed steel is used. Generally, structural steel contains carbon (in the range of 0.10–0.25%) manganese (0.4–0.12%), sulphur (0.025–0.05%), and phosphorus (0.025–0.050%) depending on end use and specifications (see also Section 13.7). The crude steel in liquid form is taken in a ladle for further refining/addition of ferroalloys.

**Casting**  The liquid steel is taken out of the bottom as a continuous ribbon of steel. When this is sufficiently cooled, it is cut into semi-finished products, such as billets, blooms, and slabs. This process, called *continuous casting* (also known as the *concast method*), is different from the old method (still in use in older plants), where liquid steel is first solidified in large blocks called *ingots* (weighing about 5–40 tonnes) and then rolled into semi-finished products, involving higher energy and waste in reheating.

**Hot rolling**  The semi-finished products, such as billets, blooms, and slabs, are heated at 1200°C to make metal malleable and then rolled into finished products, such as plates, structural sections, bars, and strips. The greater the reduction in size, the greater the work hardening, which produces varying properties in a section. The variation in cooling rates for the different thicknesses introduces *residual stresses*, which may be relieved by the subsequent straightening process (see also Section 13.10). Further processing of steel can include cold rolling, pickling (to remove oxides and mill scale from the surface of the steel), and coating.

The schematic diagram, showing the various stages of manufacturing of structural steel sections from the iron ore, is shown in Fig. 13.3. Figure 13.4 shows the relative proportion of the semi-finished products.

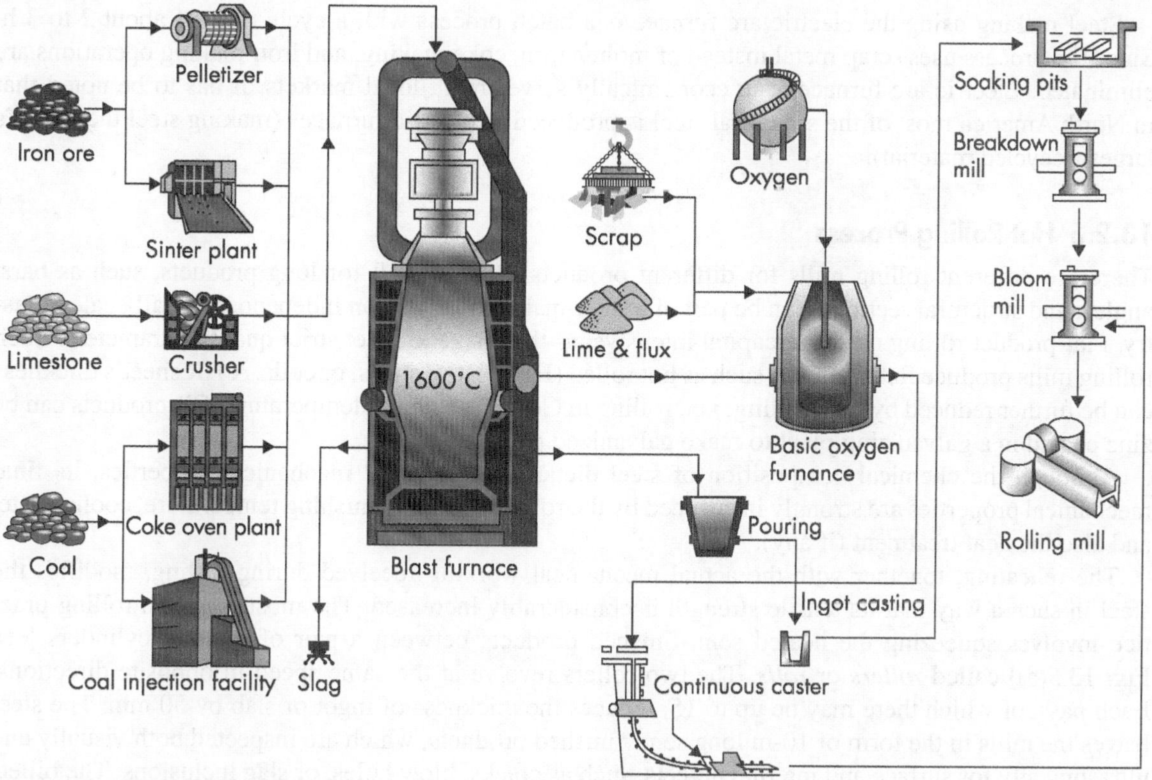

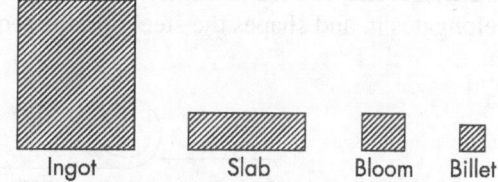

**Fig. 13.3** Schematic diagram showing the various stages of manufacturing structural steel sections from iron ore

## 13.2.2 Mini Steel Plants

Mini steel plants use raw materials such as scrap, fluxes, and ferroalloys. Sponge iron can also be used to substitute scrap up to 50%. The main steps in the manufacturing process in electric arc furnace are as follows:

**Melting**  Scrap/sponge iron, fluxes (minerals such as limestone, which are used to collect impurities), and ferroalloys are melted in an electric arc furnace, wherein

**Fig. 13.4** Relative proportions of semi-finished products

electric current is passed through three large graphite electrodes, thus creating an arc between the electrodes and scrap steel. The heat so produced is sufficient to melt the scrap steel. An alternate method involves the use of induction furnaces, which are very small and being phased out.

**Refining**  The molten metal from electric arc furnace is taken in a ladle, for refining. The metallurgy of steel in terms of carbon, phosphorus content, alloy elements, etc., is controlled at this stage. Manufacture of stainless steel requires the addition of nickel (7–8%) and chromium (15–18%).

**Casting**  The liquid steel is cast into semi-finished products such as billets, blooms, and slabs.

**Hot Rolling**  The semis, such as billets, blooms, and slabs, are heated at 1200°C to make metal malleable and then rolled into finished products.

Steel making using the electric arc furnace is a batch process with a cycle time of about 2 to 3 h. Since the process uses scrap metal instead of molten iron, coke making, and iron-making operations are eliminated. Electric arc furnaces can economically serve small, local markets. It has to be noted that in North America most of the structural steel is produced in electric furnaces (making steel the world's largest recycled material).

### 13.2.3 Hot Rolling Process

There are different rolling mills for different products. Rolling mill for long products, such as bars, angles, and structural sections, can be part of a steel-making plant, or an independent small-scale industry. Flat-product rolling mills are capital intensive, as they have to meet strict quality parameters. Such rolling mills produce flat products, such as hot rolled (HR) plates, strips, or coils. A HR sheet's thickness can be further reduced by cold rolling, i.e., rolling in CR mills at room temperature. CR products can be zinc coated in a galvanizing plant to make galvanized plates or coils.

Although the chemical composition of steel dictates its potential mechanical properties, its final mechanical properties are strongly influenced by the rolling process, finishing temperature, cooling rate, and also the heat treatment (if any).

The reheating, together with the actual mechanical working received during rolling, modifies the steel in such a way that its tensile strength is considerably increased. The most common rolling practice involves squeezing the heated semi-finished products between a pair of rotating cylinders [see Fig. 13.5(a)] called *rollers* or *rolls*. The two rollers revolve at the same speed in opposite directions. Each pass, of which there may be up to 15, reduces the thickness of ingot or slab by 50 mm. The steel leaves the mills in the form of 10-m long semi-finished products, which are inspected both visually and ultrasonically for surface and internal defects, such as cracks, blow holes, or slag inclusions. The billets coming out of bloom mills are then reheated by using a series of furnaces and then passed through grooved or profiled rolls [Fig. 13.5(b)], which operate on all four edges to turn the flat products into structural shapes such as angles, I sections, and channels. The rolling process reduces it in cross-section, elongates it, and shapes the steel into the required shape by refining the grain sizes of the material.

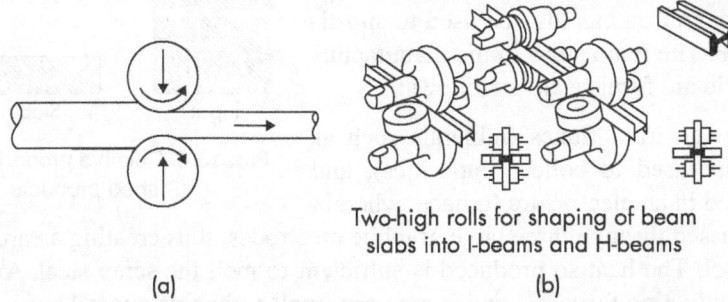

Two-high rolls for shaping of beam slabs into I-beams and H-beams

(a)                                          (b)

**Fig. 13.5** Rolls used for rolling plates and shaping I- and H-beams (Subramanian, 2016)

### 13.3 Production of Steel in India

As early as in 1907, Jamsetji Nusserwanji Tata set up the first integrated steel manufacturing plant at Jamshedpur in Bihar, which started production in 1913 (Note that India possesses enormous deposits of very rich iron ores, but there is non-availability of coal mines in their vicinity). First major steel plants were set up in Bhilai, Rourkela, and Durgapur. Sir M. Visweswarayya established the Bhadravathi steel plant in Karnataka. The steel sector now consists of seven integrated plants and about 180 mini steel plants and rerollers. The Howrah Bridge and second Hooghly cable-stayed bridge in Calcutta are examples of steel intensive bridge construction. Steel has been extensively used in numerous bridges (built by railways) and in industrial buildings.

At present, India is the fourth largest producer of steel in the world, with 101.4 MT in 2017. However, the per capita consumption of steel in India is low and is about 60 kg/person/year as compared to 515 kg in China and 225 kg for the world. SAIL, RINL, Tata Steel, and Jindal Vijayanagar (JVSL) are the largest primary steel producers. Essar steel, Ispat Industries, and Lloyds steel are the largest secondary steel produces in India. To produce 1 tonne of crude steel, nearly 1.75 tonnes of iron ore, 1.35 tonnes of coal, 0.5 tonne of limestone, and 0.5 tonne of other materials (dolomite, manganese, and other alloy materials) are required. The primary energy requirement to produce 1 tonne of finished steel in India is about 430–540 Kwh.

## 13.4 Metallurgy of Steel

The metallurgy of steel is outside the scope of this book. However, a brief discussion about the metallurgical composition of steel is included in this section.

### 13.4.1 Crystalline Structure of Metals

The regular repeating pattern in which the atoms are arranged in a metal is called the *crystalline lattice*.

It has been found that 14 different types of crystal structures exist in nature. The simplest of these crystalline unit cells is the cubic, where the atoms are arranged in a square, 3D grid fashion. The unit cell can be assumed as a box with an atom at each corner. However, such simple cubic crystals are relatively rare, due to the fact that they can easily distort. But, many crystals have face-centred-cubic (FCC), body-centred-cubic (BCC), or hexagonal close-packed (HCP) structure. The FCC structure will have 14 atoms – eight at each corner and another six in the centre of each face of the cube, as shown in Fig. 13.6(a). The BCC structure will have nine atoms - eight at the corners and an additional one right at the centre of the cube, as shown in Fig. 13.6(b). The HCP structure is hexagonally shaped as shown in Fig. 13.6(c) (www.nde-ed.org).

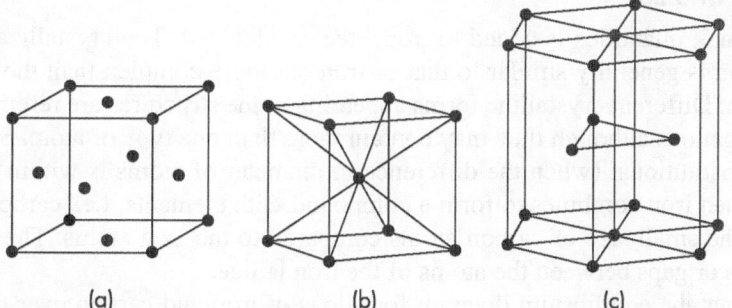

(a)         (b)         (c)

**Fig. 13.6** The three most common crystal structures in metals and alloys (a) Face-centred cubic (FCC) (b) Body-centred cubic (BCC) (c) Hexagonal close-packed (HCP)

The HCP structure has three layers of atoms. In both the top and bottom layers, there are seven atoms, six arranged at the corners of hexagonal shape and the seventh in the middle of the hexagon, as shown in Fig. 13.6(c). The middle layer has three atoms forming a triangular "pattern (www.nde-ed.org).These BCC, FCC, and HCP structures can change depending on the temperature. The FCC crystal structure will exhibit more ductility (deform more readily under load before breaking) than a BCC structure.

The BCC lattice, although cubic, is not closely packed and forms strong metals. HCP lattices are closely packed, but not cubic.

The metals that have a BCC structure include alpha-iron, chromium, vanadium, and tungsten. Metals which have a BCC structure are usually harder and less malleable than close-packed metals such as gold. The metals that have the FCC structure include aluminium, copper, gold, lead, nickel, and silver. The HCP structure is found in metals, such as magnesium, titanium, zinc, and zirconium (www.nde-ed.org).

Pure iron when heated to its melting point undergoes several crystalline transformations. Iron up to a temperature of 910°C has 'BCC' crystalline structure and remains as 'ferrite' or '$\alpha$-iron'. (The iron that exists between 768 and 910°C and having 'BCC' structure is called the '$\beta$-iron'). Between 910 and 1400°C, it transforms to 'austenite' or '$\gamma$-iron' with 'FCC' structure (see also Fig. 13.7). From 1400°C up to its melting temperature of about 1539°C, iron reverts back to the 'BCC' structure and is called '$\delta$-ferrite'. On cooling the molten iron back to room temperature, the transformations are reversed almost at the same temperature when heated. These different phases of iron are summarized in Table 13.1.

**Table 13.1** Different phases of iron

| Stable temp. range °C | Form of matter | Phase | Identification symbol |
|---|---|---|---|
| >2740 | Gaseous | Gas | Gas |
| 1539–2740 | Liquid | Liquid | Liquid |
| 1400–1539 | Solid | BCC | $\delta$-ferrite |
| 910–1400 | Solid | FCC | $\gamma$-austenite |
| <910 | Solid | BCC | $\alpha$-ferrite |

Many structural metals undergo some special treatment to modify their properties so that they will perform better for their intended use. This treatment can include mechanical working, such as rolling or forging, alloying, and/or thermal treatments.

## 13.4.2 Structure of Steel

When carbon in small quantities is added to iron, steel is obtained. The crystallization of such alloys during solidification is generally similar to that of iron but more complex than the crystallization of a pure metal like iron. Different crystalline forms appearing in the structures are referred to as *phases* and appear to be homogenous although they may contain more than one type of atom. Solid structures may be described as substitutional (when the difference in diameter of atoms is within 14%) or interstitial solid solutions. When iron combines to form a compound with elements, i.e., carbon, the compound is interstitial due to the small size of carbon atoms compared to the iron atoms. These small atoms will enter the interstices or gaps between the atoms of the iron lattice.

Figure 13.7 shows the equilibrium diagram for alloys of iron and carbon over the range 0.2 to 5% carbon. Ferrite or $\alpha$-iron, up to 910°C, dissolves carbon in interstitial solid solution to a maximum of 0.025%. $\gamma$-iron is capable of dissolving carbon to a maximum of 2%. Cementite or iron carbide is an interstitial compound of iron and carbon, containing 6.7% carbon with the chemical formula $Fe_3C$. It is extremely hard and brittle. At E (0.8% carbon), the austenite solid solution reacts to form a fine laminated mixture of ferrite and iron carbide. This eutectoid mixture, known as *pearlite,* is hard and has low ductility (Ductility is the ability of a substance to undergo large plastic deformation).

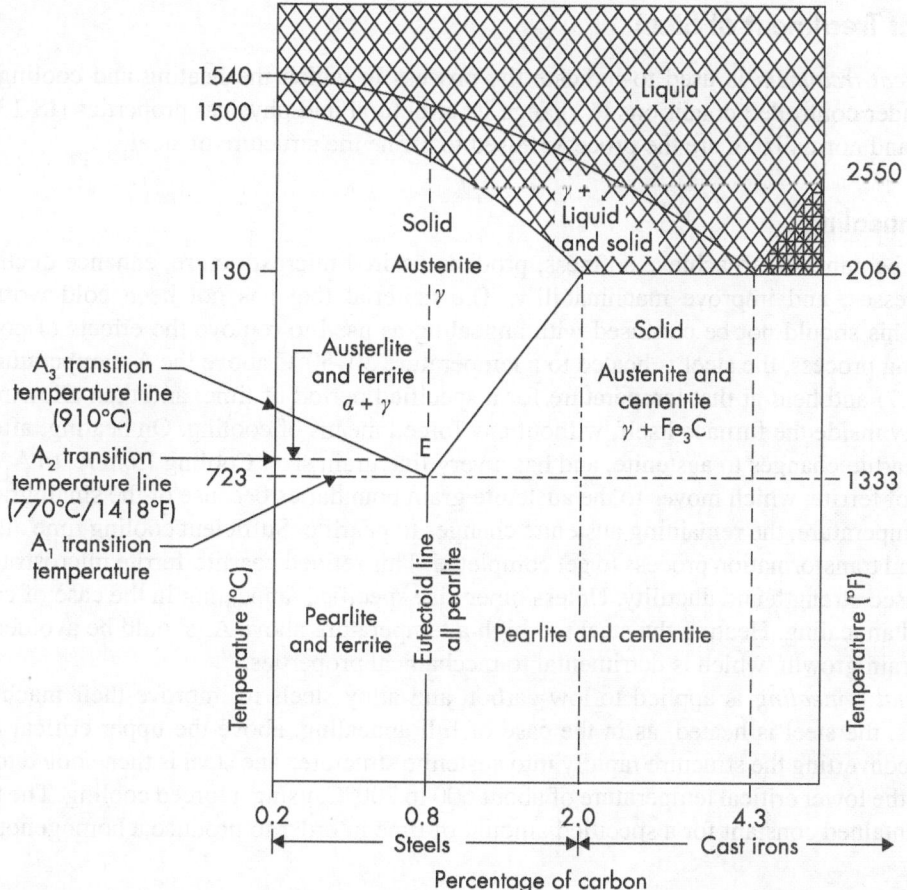

**Fig. 13.7** Equilibrium diagram for alloys of iron and carbon

It has to be noted that as carbon dissolves in the interstices, it distorts the original crystal lattice of iron and provides increased mechanical strength. However, the addition of more carbon reduces ductility and causes problems during the welding process.

Structural engineers are interested in the portion of the phase diagram shown in Fig. 13.7 that has up to 2.2% carbon. Iron containing very low-carbon (up to 0.025%) is called very *low-carbon steel* and is very soft, ductile, and has low mechanical strength.

During manufacture, when the steel is cooled very slowly from a higher temperature, ordinary structural steel with a pearlite-ferrite microstructure is formed. As mentioned previously, ferrite is soft and ductile, whereas pearlite is hard and hence imparts mechanical strength to steel. The higher the carbon content, the higher will be the pearlite content and hence the higher will be the mechanical strength. The amount of pearlite for a given carbon content is calculated according to the following formula.

Volume fraction of pearlite = (% of carbon) / (0.8% of carbon)  (13.1)

However, as the pearlite content increases, the ductility is reduced. Steel with more than 0.85% carbon is of no great significance in civil engineering, though it is used in the manufacture of cutting tools, where high hardness is required. More details about microstructure of steel may be found in Llewellyn and Hudd (1998) and Theling (1984).

## 13.5 Heat Treatment of Steel

The term *heat treatment* is used to indicate the process in which the heating and cooling of steel is involved under controlled conditions to change its structural and physical properties (IS 13417:1992). Annealing and normalizing are the processes used to refine the structure of steel.

### 13.5.1 Annealing

Steels may be annealed to achieve softness, produce desired microstructure, enhance ductility, relieve internal stresses, and improve machinability. The material that has not been cold worked can be annealed. This should not be confused with annealing as used to remove the effects of cold work. In the annealing process, the steel is heated to a temperature 30–50°C above the $A_3$ temperature (>910°C, see Fig. 13.7) and held at that temperature for a specified period of time, and then allowing it to cool down slowly inside the furnace itself, without any forced means of cooling. On heating, after $A_1$ line is reached, pearlite changes to austenite, and has a very fine grain size. Cooling from $A_3$ to $A_1$ permits the separation of ferrite, which moves to the austenite grain boundaries because of the small grain size, and at the $A_1$ temperature, the remaining austenite changes to pearlite. Sufficient cooling time allows carbon diffusion and transformation process to get completed. This refined pearlite-ferrite microstructure shows both increased strength and ductility. Unless otherwise specified, annealing in the case of carbon steels implies full annealing. Heating the steel too high a temperature above $A_3$ should be avoided to prevent austenite grain growth, which is detrimental to mechanical properties.

*Isothermal annealing* is applied to low-carbon and alloy steels to improve their machinability. In this process, the steel is heated, as in the case of full annealing, above the upper critical temperature (>910°C), converting the structure rapidly into austenite structure. The steel is then cooled to a temperature below the lower critical temperature of about 600 to 700°C, using a forced cooling. The temperature is then maintained constant for a specified amount of time in order to produce a homogenous structure.

### 13.5.2 Normalizing

Steel is normalized to refine grain size, ensure structural homogeneity, modify residual stresses, and enhance machinability. The process of normalizing is similar to annealing, except that in normalizing the steel is removed from the furnace and allowed to cool in still air. The changes occurring are the same as that of annealing but less time at high temperature and the faster cooling rate give a slightly finer grain structure and finer laminations in the pearlite. These finer structures result in slightly improved properties compared to those obtained as a result of annealing. Though the heating rate in normalising is not critical, the cooling rate is of significant importance as it influences the amount of pearlite, its size and spacing of the pearlite lamellae. Higher cooling rate may be employed to increase the strength and hardness of the parts. Normalizing is cheaper than annealing since the steel is kept in the furnace for less time. However, it can only be used for fairly uniform sections, where air cooling is unlikely to cause distortion due to differential cooling and contraction.

Mild steel plates and structural sections show very good properties of strength with ductility in the normalized condition. Heat treatment is costly and hence for many purposes the exposure of normalizing can be avoided, provided the finishing temperature during the hot rolling of steel is appropriate. A finishing temperature for hot rolling which is only slightly above the $A_3$ temperature (see Fig. 13.7) gives a very fine austenite grain size and, on air cooling, a microstructure very similar to that obtained by normalizing. It may be noted that in India the heat treated steels amount hardly to about 5% of steel production (Rangwala, 1997).

### 13.5.3 Quenching and Tampering

*Quenching* involves rapid cooling of the steel from the austenizing temperature by immersion in water, oil, molten salts, brine, or polymer solution to develop a suitable microstructure (IS 13417:1992). When small sections of steel are water quenched from the γ-region, the cooling rate is too great to allow the separation of ferrite and formation of pearlite by the nucleation and growth process. The FCC austenite is unstable, however, and the change to a BCC structure similar to ferrite cannot be prevented. This structure, known as *martensite*, is extremely hard and brittle, owing to the distortion produced in the lattice by the carbon retained in the super-saturated solution (see also Section 13.13.3 and Fig. 13.30). Oil quenching, where cooling rate is slightly slower, results in a mixture of martensite and pearlite. Martensite is not used in structural steel construction due to its brittle nature and because it is difficult to weld. However, it is used in high strength bolts.

*Tempering* is a process to achieve greater ductility and toughness by decreasing the hardness of the alloy. The reduction in hardness is usually accompanied by an increase in ductility, thereby decreasing the brittleness of the metal. Tempering is usually performed after quenching. Tempering is accomplished by controlled heating of the quenched steel to a temperature below its lower critical temperature, $A_1$ (about 600°C) holding and subsequent cooling at a specified rate. Heating above lower critical temperature is avoided, so as not to destroy the very-hard, quenched martensite microstructure. Tempering also helps in relieving quenching stresses and ensures dimensional stability.

### 13.5.4 Stress Relieving

Stress relieving is carried out to reduce internal stresses that remain within the members made of steel due to the manufacturing processes in order to stabilize its properties. It consists of heating the material slowly and uniformly to a suitable temperature below the lower critical temperature (600 to 675°C), holding at this temperature for a predetermined time (1 h per 25 mm of thickness), followed by slow and uniform cooling to minimize the development of new internal stresses. The relief of residual stresses is related to both temperature and time and similar relief can be achieved by prolonged holding at a lower temperature. More details about heat treatment of steels, required equipment, and inspection, and testing may be found in IS 13417:1992.

### 13.5.5 Case Hardening

It could be applied to harden only the surface of the steel product, obtain a wear resistant skin rich in carbon, and, at the same time, get a relatively tough and ductile interior (IS 4432:1988). Such a treatment is required for components such as forgings and fabrication of case-hardened machine or automobile parts, which require a hardened bearing surface for carrying higher stresses. It has to be noted that mild steels are not very hardnen-able, but alloy steels with high-carbon content always have adequate hardenability. The five common surface-hardening processes are:

1. Carburizing
2. Carbonitriding
3. Nitriding
4. Induction hardening
5. Flame hardening

In its earliest application of carburizing, low-carbon steel products were placed in a suitable container, covered with a thick layer of carbon powder (pack carburizing), and heated to a temperature of about 900–950°C. The products were kept at this temperature for about 6 to 8 h and then cooled slowly to room temperature. In this process, carbon diffuses into the surface of the product, enhancing its surface toughness. Although effective, this method was exceedingly slow, and as the production demands grew, several new methods, such as gas carburizing, vacuum carburizing, or low-pressure carburizing, plasma carburizing, and salt bath carburizing, have been developed. All of these methods have limitations and advantages, but gas carburizing is used most often for high-volume production because of its accurate controllability and the absence of any special handling. More details about case hardening process may be found in the work of Schneider and Chatterjee, 2013.

## 13.6 Alloying Elements in Steel

The physical properties of steel, such as ductility, elasticity, strength, toughness, etc., are greatly influenced by the following factors.

1. Carbon content
2. Heat-treatment process
3. Alloying elements

We have already discussed about the heat treatment processes. Non-alloy steel has no alloying component in it except the presence of carbon. In India, non-alloying steels constitute about 95% of the total finished steel production, and a major quantity of it is mild steel, having a carbon content of less than 0.10%. Depending upon the carbon content, the steel is designated as *low-carbon steel* (carbon content 0.10–0.25%), *medium-carbon steel* (carbon content 0.25–0.60%), and *high-carbon steel* (carbon content 0.60–1.10%). All types of steel other than mild steel are called special steel. It is mainly because special care has to be taken to maintain the level of chemical composition in such steel. Table 13.2 shows the various uses of steel of each category. Structural steels normally have carbon content less than 0.6%. As already discussed, increasing the carbon content increases the hardness, yield, and tensile strength of steel. However, it decreases the ductility and toughness. Carbon also has greater influence on weldability. Mild steel is widely used for structural work.

**Table 13.2** Uses of steel

| Name of steel | Carbon content | Uses |
|---|---|---|
| Mild steel | Up to 0.10% | Steel sections used in buildings, bridges, trusses, and space frames |
| Medium-carbon steel | Up to 0.25% | Boiler plates, structural steel, etc. |
| | Up to 0.45% | Rails, tires, etc. |
| | Up to 0.60% | Hammers, large stamping and pressing dies, etc. |
| High-carbon steel or hard steel | Up to 0.75% | Sledgehammers, springs, stamping dies, etc. |
| | Up to 0.90% | Miner's drills, smith's tools, stone mason's tools, etc. |
| | Up to 1.00% | Chisels, hammers, saws, wood working tools, etc. |
| | Up to 1.10% | Axes, cutlery, drills, knives, punches, etc. |

Alloying elements, such as manganese, silicon, sulphur, phosphorus, copper, vanadium, nickel, chromium, columbium, molybdenum, and aluminium, may be added to structural steel (see Section 13.7). The alloy steels are designated as *low-alloy steels* (total alloy content <5%), *medium-alloy steels* (total alloy content 5–10%), and *high-alloy steels* (total alloy content >10.0%). Based on manganese content, steels are also classified as carbon manganese steels (Mn > 1%) and carbon steels (Mn < 1%). (It has to be noted that the atomic diameter of manganese is larger than that of iron. Hence, manganese exists as substitutional solid solution in the ferrite crystal, by displacing the smaller iron atoms.).

If silicon content is less than 0.2%, it has no appreciable effect on the physical properties of steel. If it is raised to about 0.30 to 0.40%, the elasticity and strength of steel are considerably increased without serious reduction in its ductility. More than 2% of silicon causes brittleness.

If the sulphur content is between 0.02 and 0.10%, it has no appreciable effect on the ductility or strength of steel. The sulphur content, however, decreases malleability (*malleability* is the ability of materials, both in cold and hot states, to be bent or pressed to form different shapes without fracture) and weldability of hot metal. A sulphur content of more than 0.10% decreases the strength and ductility of steel.

It is desirable to keep the phosphorus content of steel below 0.12%. It reduces the shock resistance, ductility, and strength of steel. If present in quantities between 0.30 and 1.00%, manganese helps to improve the strength and hardness of mild steel in more or less the same way as carbon; it also increases the toughness of steel. However, when its content exceeds 1.50%, it increases the formation of martensite and hence decreases ductility and toughness.

## 13.6.1 Weldability of Steel

In most cases, members of steel are welded during fabrication. Hence, steels must not only possess high strength, but must also be suitable for welding. For good weldability, steel should not show high hardness in welded parts, but should have adequate elongation and notch toughness even in the *heat-affected zone* adjacent to a weld. Since weldability is affected by the kinds and amounts of alloying elements present in the steel, it is important to restrict both to the extent possible.

A major factor in weldability is the carbon equivalent, $C_{eq}$, of the chemical components in steel. The smaller this value, the better is the weldability. The carbon equivalent may be calculated by an equation such as that shown below, in which each symbol refers to the proportion of weight of that particular element in percentage (IS 2062:2011).

$$C_{eq} = C + \frac{Mn}{6} + \frac{(Cr + Mo + V)}{5} + \frac{(Ni + Cu)}{15} \tag{13.2a}$$

Where $C$ is carbon, $Mn$ is manganese, $Cr$ is chromium, $Mo$ is molybdenum, $V$ is vanadium, $Ni$ is nickel, and $Cu$ is copper.

When micro-alloy/low-alloys are used, the value of $C_{eq}$ should not be more than 0.53%. When micro-alloys/low-alloys are not used, carbon equivalent is calculated using the formula:

$$C_{eq} = C + \frac{Mn}{6} \tag{13.2b}$$

In this case, $C_{eq}$ value should not be more than 0.42%.

High-strength steels tend to have high-carbon equivalent. When the carbon equivalent exceeds a certain limit ($C_{eq} = 0.30$–$0.42$), the loss of weldability is compensated by reheating or post-heating of the weld zone. However, if carbon content is less than 0.12%, $C_{eq}$ can be tolerated up to 0.45%.

Several varieties of steel are produced in India. The Bureau of Indian Standards (BIS) classifies structural steels into different categories based on ultimate yield strength of basic material and their use (see IS 7598). They are listed along with the appropriate codes of practice issued by the BIS in Table 13.3.

**Table 13.3** Types of steel and their relevant IS standards

| Type of steel | Relevant IS standards |
|---|---|
| Structural Steel | 2062, 5517, 11587, 12145, 15103, 15911, 15962 |
| Steel for rivets | 1148, 1149, 1929, 2155, 2998, 7557 |
| Steel for tubes and pipes | 806, 1161, 1239, 1914, 4923, 10748 |
| Steel for sheets and strips | 277, 513, 1079, 3502, 10748, 12313, 12367, 14246 |
| Steel for rerolling | 2830, 2831 |
| Steel for reinforced concrete | 432, 1786, 2090 |
| Steel for bolts, nuts and washers | 730, 1363, 1364, 1367, 3640, 3757, 4000, 5624, 6623, 6639, 6649, 8412, 10238, 12427 |
| Welding electrodes | 814, 1395, 816, 819, 1024, 1261, 1323 |
| Steel for filer rods / wires | 1278, 1387, 2879, 4972, 6419, 6560, 7280 |
| Steel casting | 276, 1030, 2708, 2644 |

## 13.7 Chemical Composition of Steel

The chemical compositions of some typical steels specified by the BIS are listed in Table 13.4. For details of chemical composition of other steels, refer IS 15911 [structural ordinary (low tensile) quality] and IS 2062.

**Table 13.4** Chemical compositions of some typical structural steels (Extracted from IS 2062:2011)

| Designation | IS code | C (max.) | Mn (max.) | S (max.) | P (max.) | Si (max.) | | Carbon equivalent |
|---|---|---|---|---|---|---|---|---|
| E250 A[a] | | 0.23 | 1.5 | 0.045 | 0.045 | 0.40 | SK[b]/K | 0.42 |
| E250 BR/B0 | | 0.22 | 1.50 | 0.045 | 0.045 | 0.40 | SK/K | 0.41 |
| E250 C | | 0.20 | 1.5 | 0.040 | 0.040 | 0.40 | K | 0.39 |
| E300 A/BR/B0 | | 0.20 | 1.5 | 0.045 | 0.045 | 0.45 | SK/K | 0.44 |
| E350 A/BR/B0 | | 0.20 | 1.55 | 0.045 | 0.045 | 0.45 | SK/K | 0.47 |
| E650 A/BR | | 0.22 | 1.70 | 0.015 | 0.025 | 0.50 | SK/K | 0.55 |

[a]E Stands for steel and the number after E is the tensile strength in $N/mm^2$ or MPa
[b]K–killed steel, SK–Semi killed steel (Explained in Section 13.2.1)
C = Carbon, Mn = Manganese, S = Sulphur, P = Phosphorus, Si = Silicon

## 13.8 Types of Structural Steel

Civil engineers are now in a position to select structural steel for a particular application from the following general categories:

***Carbon steel (IS 2062)*** Carbon and manganese are the main strengthening elements. The specified minimum ultimate tensile strength for these steels varies from about 410 to 440 MPa and their specified minimum yield strength from about 230 to 300 MPa (see Table 13.5 and Table 1 of IS 800:2007).

***High-strength carbon steel (IS 2062)*** As discussed already, such a steel has high-carbon content and hence shows reduced ductility, toughness, and weldability. This steel is specified for structures such as transmission line and microwave towers, where relatively light members are joined by bolting. Such steels have a specified ultimate tensile strength ranging from about 480 to 550 MPa and a minimum yield strength of about 350–400 MPa.

*Medium- and high-strength microalloyed steel (IS 2062)*   Such steel has a low-carbon content but achieves high strength due to the addition of alloys, such as niobium, vanadium, titanium, or boron (total microalloying elements restricted to less than 0.25%). Such a steel has a specified ultimate tensile strength ranging from 440 to 590 MPa and minimum yield strength of about 300–450 MPa.

*High-strength low-alloy (HSLA) steels*   In recent years, high-strength low-alloy (HSLA) steels have been developed (see also Section 13.16.1). They are basically low-carbon-manganese steels to which small amounts of chromium, nickel, molybdenum, copper, nitrogen, vanadium, niobium, titanium, and zirconium are added in various combinations.

*High-strength quenched and tempered steels (IS 12145:1987)*   These steels are heat treated to develop high strength. Though they are tough and weldable, they require special welding techniques. They have a specified ultimate tensile strength between 700 and 950 MPa and minimum yield strength between 550 and 700 MPa.

*Weathering steels (IS 11587)*   These are low-alloy atmospheric corrosion- resistant steels, which are often left unpainted (see Section 13.14.1, for the details of these steels). They have an ultimate tensile strength of about 480 MPa and yield strength of about 350 MPa.

*Stainless steels*   These are essentially low-carbon steels to which a minimum of 10.5% (maximum 20%) chromium and 0.50–10% of nickel is added. More details about stainless steel are given in Section 13.15.

*Fire-resistant steels (IS 15103)*   Also called as *thermo–mechanically treated steels*, they perform better than ordinary steel under fire. They can be used in structural work up to a maximum temperature of 600°C for a maximum duration of 3 h. Two grades, namely, FR-Fe410, FR-Fe490, are available as per IS 15103. More details about their chemical composition may be found in IS 15103:2002.

*Steels with improved seismic resistance (IS 15962)*   These are steels with microalloying elements, such as niobium, boron, vanadium, and titanium added singly or in combination to obtain higher strength-to-weight ratio combined with better toughness, formability, and weldability as compared to unalloyed steel of similar strength level. Four grades (E250S, E300S, E350S, and E450S), conforming to IS 15962 may be available.

## 13.9 Mechanical Properties of Steel

Mechanical properties of steels depend upon the following factors:

1. Chemical composition
2. Rolling methods
3. Rolling thickness
4. Heat treatment
5. Stress history

Important mechanical properties of steel are ultimate strength (also called tensile strength), yield stress (also called proof stress), ductility, weldability, toughness, corrosive resistance, and machinability.

The last four properties are often associated with fabrication of steel structures and important for the durability of the material.

### 13.9.1 Tensile Strength/Yield Strength

The tensile strength, yield strength, modulus of elasticity, etc., are determined by conducting tensile test on specimen called *coupons*. The tensile test is described in Section 25.20 of Chapter 25. The steel is specified by the guaranteed yield stress, which is designated in the code as *characteristic yield stress*, $f_y$ (which is defined as the minimum value of stress below which not more than a specified percentage [usually, 5%] of corresponding stresses of samples tested are expected to occur). See Section 13.9.2 for the discussion on characteristic strength. Note that in the earlier versions of IS 2062, the steel was specified by the (characteristic) ultimate tensile strength.

The steel is designated in India as E250, E275, E300, E350, E410, E450, E550, E600 and E650, where $E$ stands for steel and the number after $E$ is the characteristic yield stress in MPa. For grades E250 to E410 there are four sub-grades $A$, $BR$, $B0$, and $C$, whereas grades E450 to E650 have only two sub-grades ($A$ and $BR$) [see Tables 13.4 and 13.5]. The sub-grade $A$ denotes that impact test is not required. For sub-grade $BR$, impact test is optional at room temperature. For sub-grade $B0$, impact test is mandatory at 0°C. For sub-grade $C$, impact test is mandatory at −20°C. Only sub-grade $C$ is made as killed and all other sub-grades are made as semi-killed or killed. For semi-killed steel, the silicon content must be less than 0.10%. For killed steel, when they are used alone, aluminium content should be greater than 0.02% and in the case of silicon it should be greater than 0.10%. Sometimes, the letter $W$ is used to denote that the steel is weldable (copper-bearing quality is designated with a suffix $Cu$, e.g., *E250 Cu-WA*. Copper may be present between 0.20 to 0.35%). Table 13.5 indicates the minimum ultimate tensile stress (UTS) and other important mechanical properties of steel produced in India. Grade $A$ steel, specified by IS 2062, is intended for use in structures subject to normal conditions and for non-critical applications (for parts not prone to *brittle fracture*, see Section 13.9.4, for discussions on brittle fracture). Grade $B$ is intended for use in structures subject to critical loading applications, where service temperature does not fall below 0°C. Grade $BR/B0$ steel is generally specified for those structural parts which are prone to brittle fracture and/or are subjected to severe fluctuations of stresses (for example, members in bridges). Grade $C$ steel has guaranteed low temperature (up to 40°C) and impact properties. Grade $C$ steel is used in members or structures where the risk of brittle fracture requires consideration due to their design, size, and/or service conditions. Steel with designation A, BR, B0, and C are generally suitable for welding. The weldability increases from designation A to C for grades E250 to E275. Welding, however, is not permitted in grades E250 C, E275 C, E300 to E650 (See IS 2062:2011).

**Table 13.5** Mechanical properties of some typical structural steels
(a) Ultimate tensile strength, yield strength, and percentage elongation

| Designation | UTS (MPa) | Yield strength (MPa) Thickness (mm) | | | Min. percentage elongation (gauge length = 5.65 $\sqrt{A_o}$) | Charpy V–notch impact energy (min.) |
|---|---|---|---|---|---|---|
| | | <20 | 20–40 | >40 | | |
| E250 A | 410 | 250 | 240 | 230 | 23 | – |
| E250 B0/BR | 410 | 250 | 240 | 230 | 23 | 27 |
| E250 C | 410 | 250 | 240 | 230 | 23 | 27 |
| E300 A/BR/B0 | 440 | 300 | 290 | 280 | 22 | 27 |
| E350 A/BR/B0 | 490 | 350 | 330 | 320 | 22 | 27 |
| E410 A/BR/B0 | 540 | 410 | 390 | 380 | 20 | 25 |
| E650 A/BR | 780 | 650 | 630 | 620 | 12 | 15 |

*(Source*: Extracted from IS 2062:2011)

**Table 13.5** (b) Other mechanical properties as per IS 800:2007

| Property | Value |
|---|---|
| Modulus of elasticity (E) | $2 \times 10^5$ MPa |
| Shear modulus (G) | $E/[2(1 + \mu)] = 0.769 \times 10^5$ MPa for $\mu = 0.3$ |
| Poisson's ratio ($\mu$) | |
|    (i) Elastic range | 0.3 |
|    (ii) Plastic range | 0.5 |
| Unit mass of steel, $\rho$ | 7850 kg/m³ |
| Coefficient of thermal expansion, $\alpha_t$ | $12 \times 10^{-6}$ / °C |
| Brinell hardness number | 150–190 |
| Vickers hardness number | 157–190 |
| Approximate melting point | 1530°C |
| Thermal conductivity | 0.14 cal / cm² s / 1°C / cm |

Mild steel has a linear stress-strain curve whose slope is the *Young's modulus of elasticity*, E. Thus, the modulus of elasticity is defined as,

$$E = f/\varepsilon \qquad (13.3)$$

where E is the modulus of elasticity also called Young's modulus, $f$ is the uniaxial stress below the proportional limit, and $\varepsilon$ is the strain corresponding to stress $f$. The values of E vary in the range 200,000–210,000 MPa and the approximate value of 200,000 MPa is assumed in the code.

The yield strain for mild steel is of the order of 0.00125 or 0.125%. Depending on the steel used, $\varepsilon_{sh}$ generally varies between $5\varepsilon_y$ and 15 $\varepsilon_y$. The average value of 10 $\varepsilon_y$ is taken as yield plateau of structural steels. The value of $\varepsilon_u$ is taken as 100 $\varepsilon_y$ and the value of $\varepsilon_{br}$ is taken as 0.23 mm/mm. The initial slope of the strain hardening part of the curve is termed the *strain hardening modulus*, $E_{sh}$. It is much less steep than the elastic part, with $E_{sh}/E$ being typically between 1/30 and 1/100 (Alpsten, 1973). The strain-hardening range is not consciously used in design, but some of the buckling limitations are conservatively derived to preclude buckling even at strains well beyond onset of strain hardening.

Yielding is sometimes accompanied by an abrupt decrease in load, as shown in Fig. 13.8, which results in upper yield ($f_{yu}$) and lower yield ($f_{yl}$) points. Typical values of the ratio $f_{yu}/f_{yl}$ for normal structural steel range from about 1.05 to 1.10. The term *yield stress* is commonly used to mean either yield point or yield strength when it is not necessary to make the distinction. Steel in compression has the same modulus of elasticity as in tension. The lower yield stress is also the same and there is about the same length of level yielding (contraction).

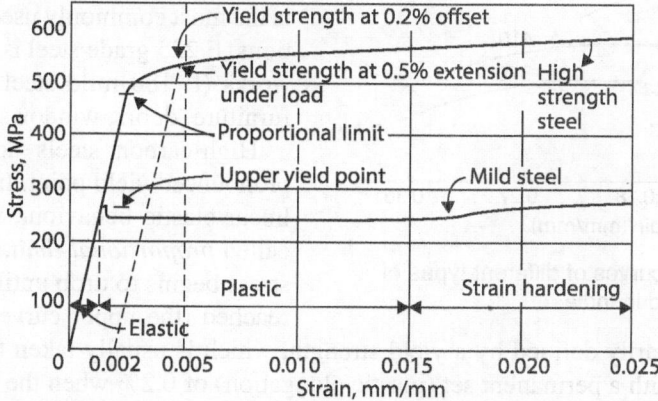

**Fig. 13.8** Stress-strain curves of different types of steel produced in India

## Parameters that Influence Yield Stress

The strain rates used in tests to determine the yield stress of a particular steel type are significantly higher than the nearly static rates often encountered in actual structures (Alpsten, 1973; McGuire, 1968). However, since the majority of the mill test results are not more than 10% higher or lower than the static rate values, the net effect, when averaged over a complete design, may not be significant (Nethercot, 2001). At higher temperatures, the reverse takes place (i.e., at higher strain rates, there is reduction in yield strength). This fact need to be considered only in blast resistant design. This is of less practical importance in earthquake engineering applications, since the strain rate will be well within the range. Mild steel and medium-strength steels have clear yield points and should not be stressed beyond the yield point as the deformation will be large and uncontrollable beyond yield. At strain rates characteristic of seismic response (0.01 to 0.10/s), steel exhibits a significant increase in yield strength (10–20%) above static test values. However, under cyclic straining, i.e., straining under cyclic loads, the effective strain rate decreases, minimizing this effect.

Yield stress may also be influenced by the position from which the test coupons are taken. For example, the webs of I-section will be thinner than those of the flanges and hence tend to possess a slightly finer grain structure as a result of faster cooling after rolling. Owing to this, the yield stress will be higher than that at the flange (Alpsten, 1973). It has to be noted that in most situations, the flanges of I-sections contribute most to their load carrying capacity, since most of the area is concentrated in the flanges. Hence, structural engineers must be careful in selecting the appropriate value for material strength for use in their calculations. In order to use plastic design or in earthquake resistance structures, the steel should satisfy the following criteria.

1. The yield plateau extends for at least six times the strain at first yield.
2. The ultimate/yield stress ratio must be greater than 1.25 (To develop inelastic rotation capacity, a structural member needs adequate length of yield region along the axis of the member. The larger the ultimate to yield ratio, the longer is the yield region).
3. The minimum elongation must be 15% on a gauge length of $5.65 \sqrt{A_o}$, where $A_o$ is the area of tensile test specimen.

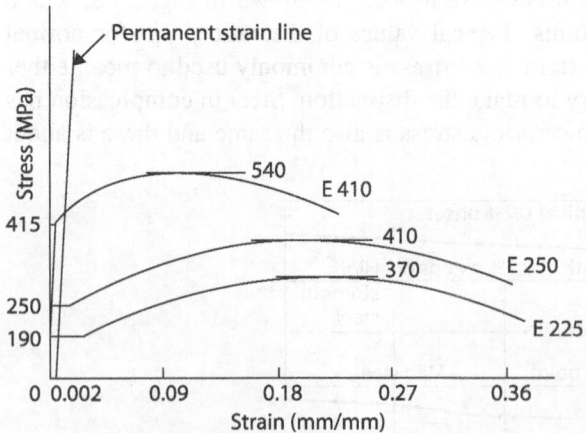

**Fig. 13.9** Stress-strain curves of different types of steel produced in India

It is also preferable that the actual yield strength based on tensile test of steel does not exceed the specified yield strength by more than 120 MPa. Figure 13.9 shows the stress-strain curves of different types of steels produced in India and the permanent strain line. E 250 grade mild steel is the most commonly used in structural applications. E 225 grade steel is used in less important works (E 180 mild steel is used primarily for furniture, doors, windows, etc.).

High-carbon steels do not usually have a pronounced yield point. Instead, after a range of linear elastic behaviour, which ends at a point called *proportional limit*, the rate of increase in stress begins to drop until the tensile strength is reached (the upper curve of Fig. 13.9). In this case, yielding is arbitrarily defined by a yield strength, which is usually taken to be that stress which leaves the specimen with a permanent set (plastic elongation) of 0.2% when the specimen is unloaded.

It is obtained by drawing a line parallel to the elastic portion at 0.2% strain, which intercepts the stress-strain curve as shown in Fig. 13.9. However, some standards (e.g., ASTM specification, A370) define the yield stress as the stress corresponding to a 0.5% elongation under load.

The yield stress $f_y$ also varies significantly with the chemical constituents of the steel (e.g., carbon and manganese), the heat treatment used, and with the amount of working, which occurs during the rolling process. Thus, thinner plates which are more worked have higher yield stresses than thicker plates. The yield stress is also increased by cold working.

### 13.9.2 Characteristic Strength

Variations in material properties (due to the non-uniform molecular structure of the material, and variations and inconsistencies in the manufacturing process, which depend on the degree of control, etc.) should be recognized and taken into consideration during the design process. The material properties that are of greatest importance in the design of structures using steel are yield strength, maximum percentage elongation, and the Young's modulus. Other properties that are of less importance are hardness, impact resistance, and melting point.

If a number of samples are tested for a particular property (e.g., yield strength) and the number of specimen with the same strength (frequency) is plotted against the strength, the results approximately fit a normal distribution curve as shown in Fig. 13.10.

This curve can be mathematically expressed by the equation shown in Fig. 13.10, which can be used to define 'safe' values for design purposes. When defined, this yield strength, is called *characteristic strength*. If the characteristic strength is defined as the mean strength,

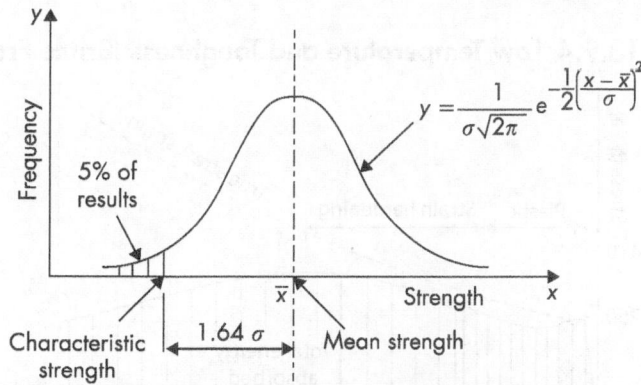

**Fig. 13.10** The normal-distribution curve

then from Fig. 13.10, 5% of the material has a characteristic strength below this value and hence not acceptable. Hence, a characteristic value which has a particular chance (often 95%) of being exceeded in any standard tension test is chosen.

Thus, the characteristic strength is calculated from the equation

$$f_k = f_{mean} - 1.64\,\sigma \qquad (13.4)$$

where $\sigma$ is the standard derivation for $n$ samples, and is given by

$$\sigma = \left[ \sum (f_{mean} - f)^2 / (n-1) \right]^{0.5} \qquad (13.5)$$

The characteristic strength of steel is the value obtained from tests at the rolling mills, but by the time the steel becomes part of the finished structure, its strength might have been reduced (e.g., by corrosion or accidental damage). The strength to be used in design calculations is therefore the characteristic strength divided by a partial safety factor. The value of partial safety factor adopted for steel, as per IS800:2007, is 1.10 for yielding resistance.

### 13.9.3 Ductility

It may be described as the ability of a material to change its shape without fracture. In order words, ductility of a structure or its members is the capacity to undergo large inelastic deformations without a significant loss of strength or stiffness. The stress-strain curve also indicates the ductility. It is the amount of permanent strain, i.e., strain exceeding proportional limit up to the point of fracture.

Values of 20% ductility can be obtained for mild steel but are less for high-strength steel. By improper testing, one may get percentage elongation values of 20–60% and hence the test houses should be careful with the testing process. A high value is advantageous because it allows the redistribution of stresses at ultimate load and the formation of plastic hinges. Design procedures based on inelastic behaviour require large ductility, particularly for the treatment of stresses near holes or abrupt changes in member shapes as well as for the design of connections. The minimum required percentage elongation of steel produced in India as per the code is given in Table 13.5(a). Designers normally use specified properties, only rarely calling for their own material tests in the case of steel. For most standard mild steels, the values are greater than the required minimum. However, rerolled steel with impurities or improperly controlled steel, which may absorb more carbon, has higher strength but less percentage elongation.

### 13.9.4 Low Temperature and Toughness (Brittle Fracture)

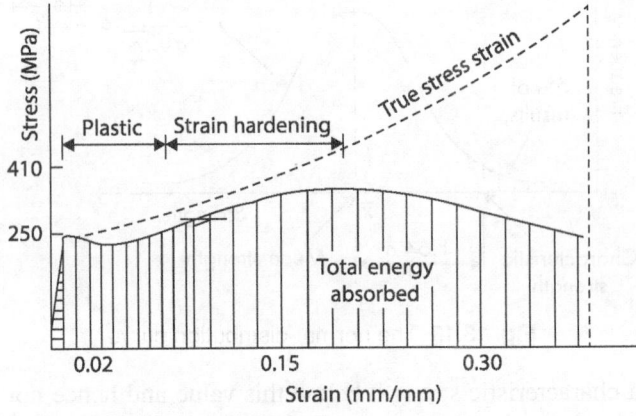

**Fig. 13.11** Toughness of steel

In structural steel design, toughness is a measure of the ability of steel to resist fracture under impact loading, i.e., the capacity to absorb large amounts of energy. *Toughness* can be an important design criterion, particularly for structures subject to impact loads (e.g., bridges) and for those subject to earthquake loads. The area under the stress-strain curve is a measure of toughness (see Fig. 13.11). The triangular area shown in Fig. 13.11 is elastic strain energy. Hence, both strength and ductility contribute to toughness. Thus, because its modulus of elasticity is only one-third that

of steel, an aluminium structural member can absorb three times the energy for the same stress, as a steel member of the same dimensions, provided the stress does not exceed the proportional limit.

At the room temperature, common structural steel is very tough and fails in a ductile manner. At lower temperatures below 0°C, the yield strength of steel as well as ultimate tensile strength are affected, and there is a substantial reduction in ductility and toughness (The ductility of a material is sometimes indicated by the ratio of total energy absorbed by the material to the elastic energy.). Owing to this phenomenon, at low temperature, steel structures sometimes fail suddenly and without warning. As we are aware, when the material elongates, there is a contraction in the lateral dimension owing to the *Poisson's ratio* effect. If the lateral dimensions are fully (or even partially) restrained, the steel will pull apart without fully developing the yield potential. This type of failure is termed as *brittle fracture*. Stress-concentration effect occurs at a notch (a *notch* may be any abrupt change in cross-sectional area) or a hole.

A right combination of low temperature, an abrupt change in section size (notch effect) or an imperfection, and the presence of tensile stress can initiate a brittle fracture. This may begin as a crack, which may propagate and cause the member to fail. Most brittle fractures occur under static load at stress levels which are not excessive, but they may also be due to the dynamic application of load or some over load.

The first failure due to brittle fracture was identified in 1886 (McGuire, 1968). Similar sudden failures of steel water tanks, oil tanks, transmission lines, ships, plate girders, and bridges have occurred in the past (McGuire, 1968). Most of these failures have occurred under normal service conditions, in welded structures and at low temperatures. During the cold winter of 1977, several spectacular failures occurred in bridge structures in Illinois, Minnesota, and Pennsylvania in the USA. On January 22, 1988, several brittle failures occurred in a bridge in Providence, Rhode Island, the USA (Gaylord et al., 1992).

Extensive research on brittle fracture using the concepts of fracture mechanics has shown that material toughness, crack size, and stress are the primary factors which determine the susceptibility to brittle fracture. The *Charpy V-notch test* is commonly used to evaluate the behaviour of metals as they are affected by an abrupt change in cross-section (IS 1757). See Section 25.20.2 of Chapter 25, for the details of this test.

Structural steels differ greatly in toughness. Highly killed, fine grain steel with a suitable chemical composition or specially heat-treated steel will exhibit considerable toughness. IS 2026 and IS 1757 codes allow the use of only those steels that exhibit a minimum energy absorption capacity at a predetermined temperature (e.g., 20 J at 23 ± 5°C). The best combination of properties, a yield strength of 400–450 MPa, and Charpy V–notch (CVN) impact value of 27 J down to –50°C is produced in steels by treating them with a combination of niobium/vanadium and aluminium. By lowering the carbon content to about 0.18–0.16%, we may obtain CVN impact values of 27 J at 30°C. In addition to the chemistry of steel, size of plates, residual stress, and cold work also affect the toughness. (Thick plates, large residual stress, and cold work are detrimental.)

From the designer's point of view, the most satisfactory way of dealing with brittle fracture is to reduce the likelihood of its occurrence by a sensible choice of material. In other words, the service temperature should be above the specified low temperature of the steel and the CVN impact value more than the one specified, to prevent brittle fracture. Using thick plates with mutually perpendicular welds stressed in the 'through thickness' direction should be avoided.

Clause 2.4.4 of BS 5950–1: 2000 and Tables 3.7 and 3.8 of Hong Kong steel code provide recommendations for the selection of thickness of each element such that brittle fracture could be eliminated.

### 13.9.5 Lamellar Tearing

It is a form of brittle fracture that may occur in certain welded joints. For example, a tear can occur if a large weld (or welds from both sides) is placed on a thick plate, since the shrinkage strains from the welding operation will be large and restrained (see Fig. 13.12). The restraint may be developed due to the weld on the far side, because of the member thickness or due to a combination of both the factors.

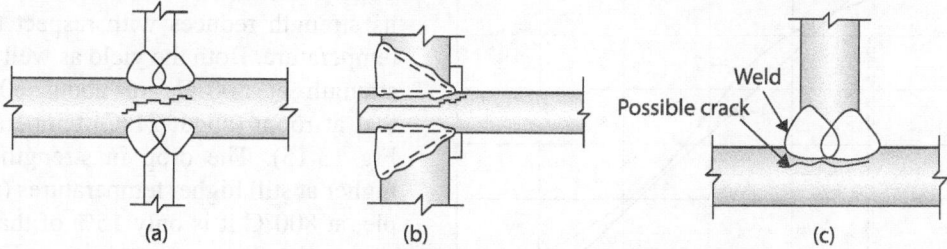

(a)  (b)  (c)

**Fig. 13.12** Lamellar tearing at welded joints

Generally, I-sections are adequately ductile when loaded either parallel or transverse to the rolling direction (see Fig. 13.13). However, when the strain is localized in the 'through-thickness' direction (at one thick flange of the section), a restrained situation exists because the strain cannot redistribute

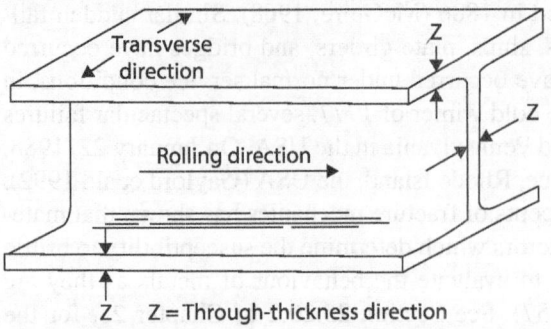

**Fig. 13.13** Definition of through-thickness direction

from the flange through the web to the opposite flange. The large localized 'through-thickness' strain may exceed the yield strain and lead to lamellar tear (Salmon & Johnson, 1996). A thin, stiffened column is also susceptible to lamellar tearing, since the flange stiffeners that are welded to the column flange will produce a restraint. A large overmatch of electrode and base metal in full penetration butt weld also tends to increase the possibility of tearing.

The use of fillet welds, a joint design that allows weld shrinkage to occur in the rolling direction (so that the shrinkage pulls on the fibres longitudinally in their strongest orientation) and the order of welding to minimize shrinkage strains are some of the practical methods used to avoid lamellar tearing (ASCE paper, 1982; Thornton, 1973). A number of joint alternatives (see Fig. 13.14) that may be used to reduce lamellar tearing are given in AISC, Engineering Journal, 1973.

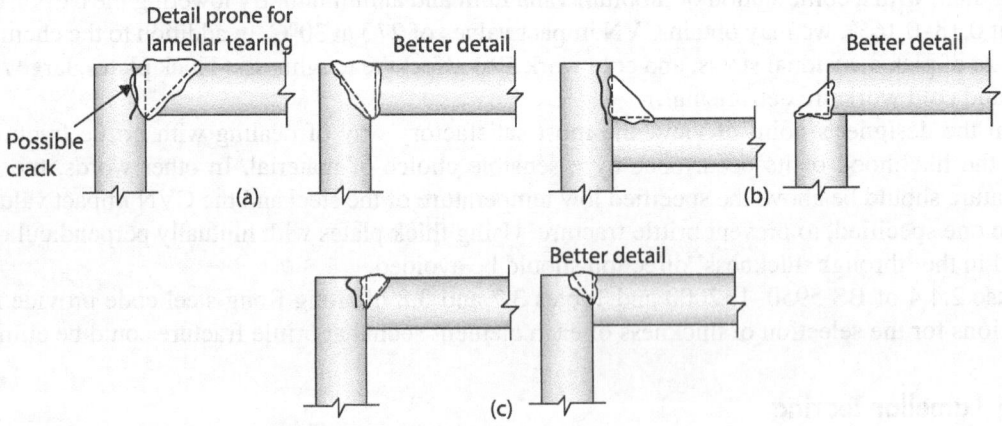

**Fig. 13.14** Joint alternatives to reduce lamellar tearing

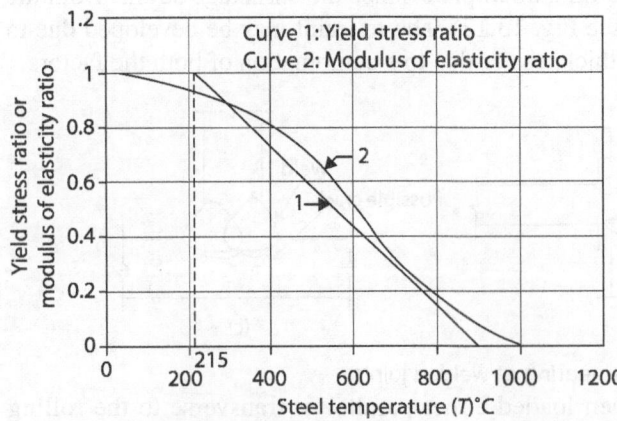

**Fig. 13.15** Variation of mechanical properties of steel with temperature
(*Source*: Based on IS 800)

### 13.9.6 High Temperature Effects

Steel is not a flammable material. However, its strength reduces with respect to raising temperature. Both the yield as well as tensile strength at 500°C are about 60–70% of that at room (about 21°C) temperature (see Fig. 13.15). The drop in strength is much higher at still higher temperatures (for example, at 800°C it is only 15% of that at room temperature).

Hence, steel frames enclosing materials that are flammable require fire protection, to control the temperature of steel members for a sufficient time for the occupants to seek safety or for the fire be extinguished before the building collapses. In many cases, the building does not collapse, even at high temperatures. But the deformation of the members will be more than the acceptable limits, and hence have to be replaced.

### 13.9.7 Hardness

It is a measure of the resistance of the material to indentations and scratching. Several methods are available to determine the hardness of steel and other metals. These methods are described in Section 25.20.3 of Chapter 25. Typical values of hardness for some metals and alloys are given in Table 13.6.

**Table 13.6** Typical values of hardness for some metals and alloys

| Metal | Brinell hardness (BHN) | Vickers hardness (VHN) | Rockwell hardness (RHN) |
|---|---|---|---|
| E 250 structure steel | 122 | 126 | 70.5 |
| E 320 Structural steel | 149 | 157 | 80.8 |
| Aluminium | 20 | – | – |
| Grey iron castings (FG 150)-IS 210:1993 | 130–180 | 139–190 | 75–89 |

### 13.9.8 Fatigue Resistance

It is the term used in connection with the initiation and propagation of microscopic cracks into macrocracks by the repeated application of alternating stresses. The damage and failure of materials under cyclic loads is called *fatigue damage*. Fatigue need not be considered unless numerous significant fluctuations (usually, taken as $2 \times 10^6$ to $5 \times 10^6$ cycles) of stress are anticipated. As per IS 800:2007, ordinary reversals due to wind may be ignored. Wind-induced oscillations have to be taken into account in special cases. For instance, oscillations of $2 \times 10^6$ cycles can be easily reached in lighting masts. The IS 800:2007 code states that the designer should check the following members for fatigue.

1. Members supporting lifting or moving loads
2. Members subjected to wind-induced oscillation of a large number of cycles
3. Members subjected to repeated stress cycles from vibrating machinery
4. Members subjected to crowd-induced oscillations

Thus, fatigue effects are more likely to occur in bridges and cranes, due to the cyclic nature of loading, which causes reversal of stresses. Welds are susceptible to a reduction in strength due to fatigue because of the presence of small cracks, local stress concentrations, and abrupt changes of geometry. It has to be noted that the incidence of fatigue and fatigue crack growth is independent of steel grade. Also, fatigue cracks are far more common than brittle fracture.

### 13.10 Residual Stresses

Higher temperatures in the range of 600–700°C are involved during the rolling of steel sections. Steel members are also subjected to high temperatures during fabrication by welding and also when material or members are cut by flame-cut (in these cases, heat is applied locally to selected parts of the cross-sections). Cooling of these members or materials always takes place unevenly (even for a hot-rolled member placed on a cooling bed after rolling, some parts of the section, for example, the flange tips of an I-section, cool faster than the flange-to-web junctions. Similarly, the central portion of the web tends to cool faster than the junctions). Owing to this uneven heating and cooling, structural members normally contain *residual stresses*. Residual stresses may also result from cold straightening of bent members (Gaylord et al., 1992). Although it is possible to remove these stresses by subsequent reheating and slow cooling, this process is very expensive and hence is not attempted in normal structural engineering applications (such a removal of residual stress is mainly used in pressure vessels).

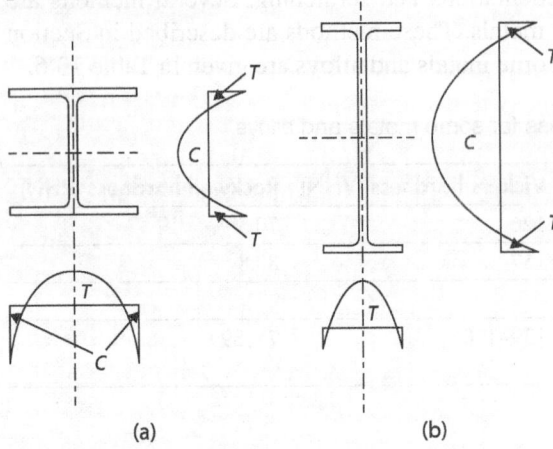

**Fig. 13.16** Residual stress distribution in standard I-section

As discussed earlier, in an I-section, the flange tips and web interiors cool faster than other parts of the cross-section, resulting in residual tensile stresses in the region of junctions (which cool more slowly) and compressive stresses in the reminder of the cross-section (which cool first). The typical distribution of residual stresses in a standard I-section is shown in Fig. 13.16(a). The stresses also vary across the thickness, and the pattern shown in Fig. 13.16(a) represents average of the values across the thickness. Residual stresses tend to increase in magnitude with increase in size of the element. Both magnitude and distribution of thermal residual stress are influenced to a considerable degree by the geometry of the cross-section as shown in Fig. 13.16(b) (Gaylord et al., 1992). The magnitude of tensile residual stresses may reach up to $0.3 f_y$ and the compression residual stress up to $0.5 f_y$ is rolled I-sections (Trahair et al., 2001).

Owing to the high concentration of heat, tensile residual stresses at the weld in welded members usually equal the yield strength of the weld metal itself; the compensating longitudinal compressive stresses also will be quite high. Residual stresses in welded shapes are determined by the section geometry and the method of preparation of the components. Thus, a welded I-section fabricated from rolled plates has a different residual stress distribution from that of an I-section welded from plates flame-cut to width.

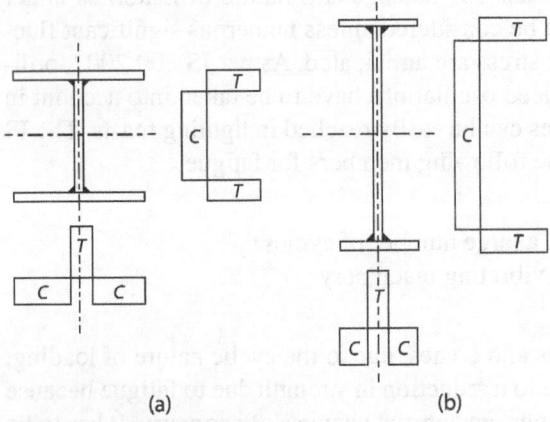

**Fig. 13.17** Residual stress distribution in welded I-beams

The typical distribution of residual stresses in welded sections made of plates with rolled edges is shown in Fig. 13.17. Large residual tensions develop at the corners of the welded box (Fig. 13.18). On the other hand, residual stresses, in the hot-rolled square box are very low.

In steel sections which are made of quenched and tempered steels, the residual stresses are small. This is because such steels are partially stress-relieved. Thermal residual stresses extend almost the full length of the member.

Fabricating operations such as cambering and straightening by cold bending also induce residual stresses. These residual stresses have the same magnitude as those of thermal residual stresses but differ in distribution. They are super-imposed on the thermal residual stresses. If a member is straightened by a continuous straightening process (rotorizing), the residual stress pattern is changed throughout the length of the member. On the other hand, if it is straightened at discrete points by gagging, thermal residual stresses are not altered over much of the length (Gaylord et al., 1992).

Because residual stresses must themselves be in equilibrium, their effect on structural behaviour is limited. Although such stresses do not affect the yield strength of the member, they do lower the proportional limit and increase the strain at initiation of overall yielding. The most important consequence of this in statically loaded structure is to cause the member to behave as if it has a non-uniform distribution of yield stress over its cross-section. This factor is important in the design of compression members (columns), since under compressive

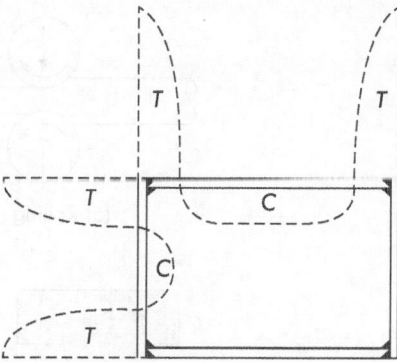

**Fig. 13.18** Residual stresses in a welded box section

stress, the regions that contain residual compressive stress yield earlier (at loads which produce an applied stress less than $f_y$). Similarly, members in bending (beams) also yield early and hence tend to deflect more (Tall, 1974). Residual stresses have to be considered in the design of members subject to fluctuating loads (fatigue), since they reduce the member's resistance to the growth of stable cracks. Residual stresses also reduce the member's resistance to the growth of cracks in an unstable manner due to brittle fracture (see Section 13.9.4).

## 13.11 Mechanical Working of Metals

Metals are crystalline in nature and consist of irregularly shaped grains of various sizes. Each grain is made up of atoms in an orderly arrangement, known as a *lattice*. The orientation of the atoms in a grain is uniform but differs in adjacent grains. When a force is applied to deform it or change its shape, a lot of changes occur in the grain structure. These include grain fragmentation, movement of atoms, and lattice distortion. The properties of metals can be modified as well as desired shapes and thicknesses can be obtained by processes like hot working and cold working. The difference between these processes is based on the amount of heat applied to the metal before applying any mechanical force. Under the action of heat and force, the atoms of the metal reach a certain higher energy level, leading to the formation of new crystals. This is called *recrystallization*. When this happens, the old grain structure deformed by any previously carried out mechanical working is lost, and strain-free new crystals are formed.

### 13.11.1 Hot Working

When metals are plastically deformed above their recrystallization temperature of about 60% its melting temperature (for steel, it is approximately 1090°C), it is termed as *hot working* (it has to be noted that for lead and tin, the recrystallization temperature is below the room temperature of about 20°C, and hence working of these metals at room temperature is always considered as hot working). Working below the recrystallization temperature is called *cold working*. Different hot working processes are: rolling, drawing, forging, extruding, and pressing (see Fig. 13.19). Of all these operations, rolling and drawing are extensively used.

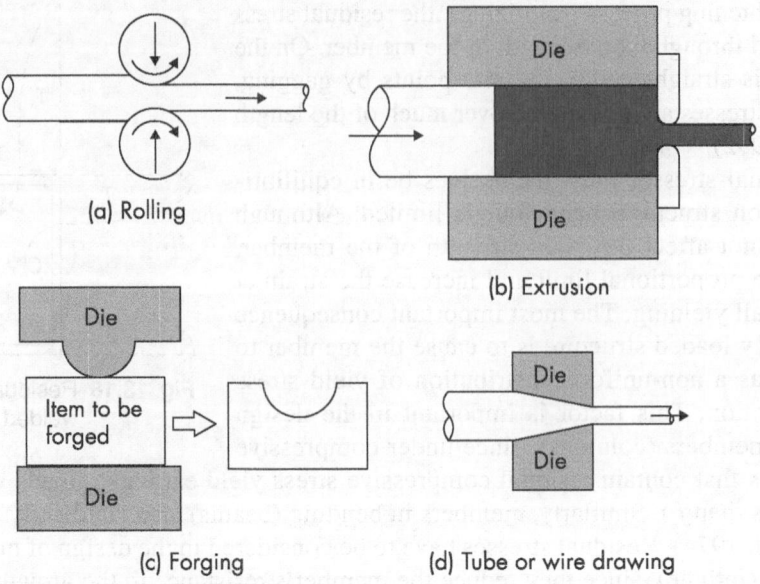

**Fig. 13.19** Hot working processes

As discussed in Section 13.2.3, in hot rolling, the red-hot ingot is forced between two rollers, which have a narrow gap between them, to obtain desired shapes of steel sections such as I, T, channel, or angle. In drawing, the hot metal is drawn through different dies and specially shaped tools; drawing is often used for the manufacture of steel reinforcements and tubes. In forging, the metal is pounded by hammers or squeezed between a pair of shaped dies.

As compared to cold working, the advantages of hot working are as follows:

1. There is no strain hardening
2. Lesser forces are required for deformation
3. Possibility of more deformation in ductile metals
4. Better mechanical properties of material due to favourable grain size
5. Use of equipment which require lesser power
6. Elimination of residual stresses in the material

Some disadvantages associated with hot working are as follows:

1. Requirement of heat energy
2. Poor surface finish of material due to scaling of surface
3. Poor accuracy and dimensional control of parts
4. Troublesome handling of hot metal
5. Lower life of tooling and equipment

## 13.11.2 Cold Working

Cold working or cold-forming consists of shaping the metal at a temperature substantially below the recrystallization temperature. Suppose that a tensile specimen is loaded beyond the yield point, such as point A, as shown in Fig. 13.20. When it is unloaded, it does not return to point O, but to point B. This means that a permanent set OB has occurred and the ductility has been reduced from OF to BF. When reloaded, the specimen will behave as if the stress-strain origin is shifted to point B. Thus, the plastic zone prior to strain hardening is also reduced.

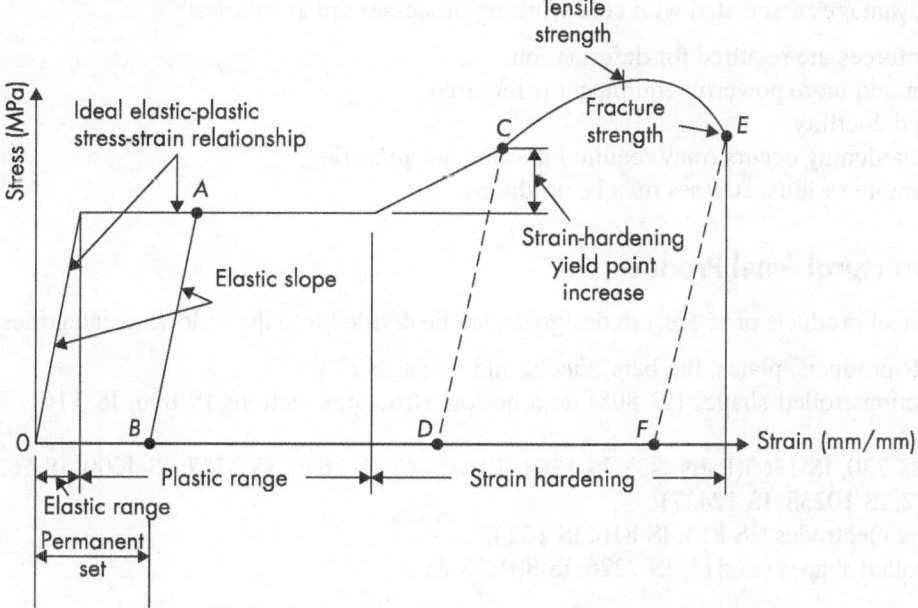

**Fig. 13.20** Effect of cold work and strain hardening

If the specimen is loaded up to point C in the stress-strain curve, the unloading follows the dashed line CD. In this case, the origin is shifted to point D; but the yield point has increased. This increased yield point is known as *strain hardening effect*. Now, the available ductility is much reduced, from OF in the original state to only DF as shown in Fig. 13.20. This kind of loading beyond the elastic range to cause reduction in ductility, when done in atmospheric temperature is referred to as *cold work*. (It may be noted that in real structures the state of stress will not be in uniaxial tension or compression, and hence the cold work effect is more complex.)

Inelastic deformation occurs at the bends of plates when they are made by cold forming (*cold forming* is a process by which the sheets [HR or CR] are folded into the desired section profile by a series of forming rolls in a continuous train of roller sets or brakes). Cold working into the strain hardening range at the bend locations increases the yield strength and this increased strength is permitted in design by IS 801. Upon unloading and after a period of time, due to *strain ageing*, there will be an increase in the yield point beyond point C in Fig. 13.20 and the original shape of stress-strain diagram will be restored (with a plastic zone and strain hardening zone) with reduced ductility (Salmon & Johnson, 1996). Cold twisted deformed bars are manufactured by this principle.

The process of *annealing* [heating to a temperature beyond transformation range (750–780°C) and allowing the material to cool slowly] results in recrystallization and elimination of the effects of cold work.

As compared to hot working, the advantages of cold working are as follows:

1. No heating is required
2. Better surface finish is obtained
3. Better dimensional control and hence no secondary machining needed
4. Better strength, fatigue, and wear properties of material
5. Easier handling of cold metal

Some disadvantages associated with cold working processes are as follows:

1. Higher forces are required for deformation
2. Heavier and more powerful equipment is required
3. Reduced ductility
4. Strain hardening occurs (may require intermediate annealing)
5. Undesirable residual stresses may be produced

## 13.12 Structural Steel Products

Structural steel products of interest to designers can be divided into the following categories.

1. Flat HR products: plates, flat bars, sheets, and strips(IS 1730)
2. HR sections:rolled shapes (IS 808), and hollow structural sections IS 806, IS 1161, IS 4923, IS 10748)
3. Bolts[IS 730, IS 1363(Parts 1-3), IS 1364 (Parts 1-6), IS 3640, IS 3757, IS 4000, IS 5624, IS 6639 IS 8412, IS 10238, IS 12427]
4. Welding electrodes (IS 814, IS 816, IS 1024)
5. Cold-rolled shapes (IS 513, IS 7226, IS 801, IS 811)

### 13.12.1 Hot-rolled Steel Sections (IS 808)

The HR sections and products consist of the following (see Fig. 13.21).

1. Rolled beams

    (a) Junior beams (ISJB meaning Indian Standard Junior Beams)
    (b) Lightweight beams (ISLB)
    (c) Medium-weight beams (ISMB)
    (d) Wide-flange beams (ISWB)
    (e) Heavyweight beams/columns (ISHB)
    (f) Column sections (ISSC)

2. Channels: junior, light, and medium and parallel flange (ISJC, ISLC, ISMC, and ISMCP)
3. Equal angles (ISEA or ISA)
4. Unequal angles (ISA)
5. T sections (ISDT, ISJT, ISLT, ISMT, ISNT, and ISHT)
6. Rolled bars

    (a) Round (ISRO)
    (b) Squares (ISSQ)

7. Tubular sections (ISLT, ISMT, and ISHT)
8. Plates (ISPL)
9. Strips (ISST)
10. Flats (ISFl)

IS Handbook No. 1 and IS 808, published by the BIS, provide the dimensions, weights, and geometrical properties of steel beam, column, channel, and angle sections. Steel tubes are designated by their nominal bore in mm and self-weight. Rolled steel circular or square rods are designated by diameter or side, respectively (e.g., ISRO 10 mm or ISSQ 10 mm).

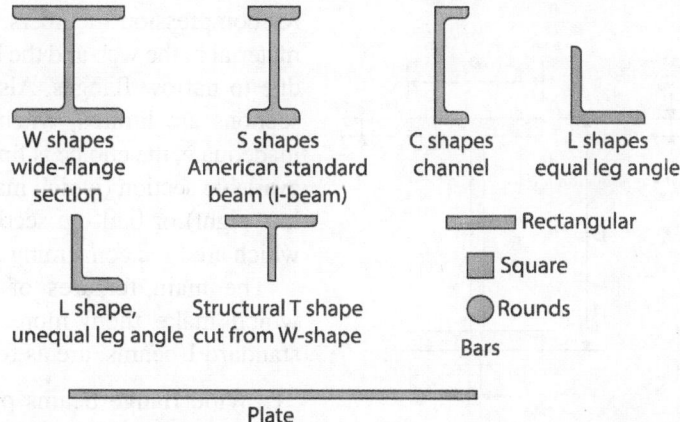

**Fig. 13.21** Types of hot-rolled sections produced by steel producers

Angle sections are mainly used in trusses, microwave and transmission line towers, and open web girders (https://steeljoist.org). Channel sections are used as purlins or small joists, stair stringers, wind girts, framing around openings, and as pipe supports. Paired channel built-up sections are also used as columns. I- and H-sections are used as beams and also as columns in multi-storey buildings and bridges. I- and H-sections (and also heavy steel pipes) can be used as end-bearing piles. They are advantageously used in situations where bending stresses are the maximum. The materials in I- and H-sections are distributed in such a way that the maximum material exists where maximum bending stresses occur. I-sections are also used in grillage foundations and railway tracks. T-sections are widely used as top and bottom chords of trusses. Steel bars, mats, and wire meshes are used as reinforcements in concrete and ferro-cement. Steel plates are used as base plates, gusset plates, chequered plates (IS 3502:2009), batten plates in built-up columns, bracings, brackets, to build steel tanks (IS 804:1967; IS 805:1968), and for creating built-up or welded sections. Plates are also used to fabricate welded I- and compound/built-up sections.

Steel sheets, usually with corrugated or trapezoidal shape for stability, are produced out of steel zinc coated sheets (with thicknesses ranging from 0.50 to 1.50 mm), and used as roofing sheets. Most modern flooring systems in buildings use concrete slab with cold-formed profiled steel sheeting having a thickness of about 1 mm. Pretreated galvanized steel sidings with polyvinylidene fluoride (PVDF) coatings are available in a wide variety of shapes, patterns and colours, and are used in the USA in commercial constructions. Most of them include ribbing (to give stiffness to thin material) and an interlocking joint at the edges to secure adjacent pieces together and give weather-tight seal. The wall, utilizing steel siding, may be completed by adding an interior lining panel with insulation in the wall cavity.

More than half of the steel produced worldwide is used in the construction of steel buildings and infrastructure. Steel is used for building industrial buildings, multi-storey buildings, bridges, tunnels, rail tracks, train stations, ports, and airports. It is also used in underground pipelines to distribute water, gas, or oil. Steel is also used in many non-structural applications in buildings, such as heating and cooling equipment and interior ducting, internal fixtures and fittings such as rails, shelving and stairs, door and window frames, collapsible gates and grills, etc.

### 13.12.2 Wide Flange Sections (IS 12778:2004)

As mentioned in Section 13.12.1, ISMB sections are the only I-sections that are normally produced in India on account of calibre rolling method. These sections have relatively narrow and sloping flanges and a thick web compared to wide flange sections (see Fig. 13.22). ISMB beams are not economical, especially

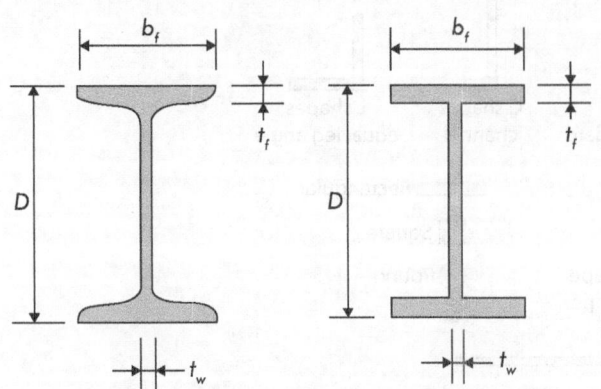

**Fig. 13.22** Standard and wide-flange sections

for compression members, because of excessive material in the web and the lack of lateral stiffness due to narrow flanges. Also, since the available sections are limited, when a section is slightly inadequate, the choice is limited to either the next available section (which may be 25–45% heavier in weight) or built-up sections through welding, which are time consuming and expensive.

The main features of wide flange beams, which make them more popular than Indian standard I-beams, are as follows:

1. Wide flange beams provide excellent sectional performance, with high bending and buckling resistance due to the H-shaped arrangement of flanges and the web.
2. The use of such beams reduces fabrication difficulties – since there is no taper in the H-beam flange, no tapered washer is necessary for bolting, and the gussets can be welded to the inner surface of the beam flange. Unlike tapered-flange beams, H-beams can be readily butt welded, and a sound welding is assured.
3. Since H-beams have higher section modules for the same weight, using them is economical (Savings of the order of 10–24% can be achieved.).

Using a new manufacturing technology, it is now possible to have beams with same depth but with different flange and web thickness and also flange width. This facilitates simple design and improves fabrication efficiency (Subramanian, 1992). These wide parallel flange beams and columns are manufactured in India by M/s Jindal Steel and Power Limited (JSPL), at Raigarth, Chhattisgarh.

### 13.12.3 Welded and Hybrid Sections

Hot-rolled plates or flame-cut plates can be welded together to form I-sections or box girders as shown in Fig. 13.23 (such built-up sections can also be made by using riveting or bolting). *Welded I-beams* with top and bottom plates and welded stiffeners are often used as plate girders. Welding makes it possible to combine any structural shape to get the desired properties.

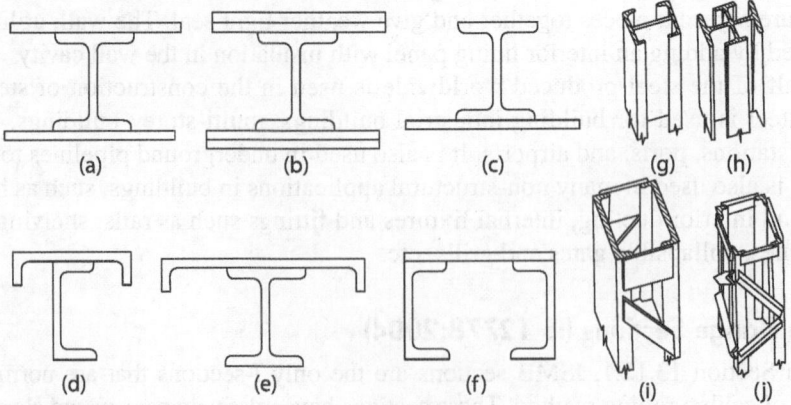

**Fig. 13.23** Built-up sections

*Tapered girders* are fabricated either by welding two flange plates to a tapered web plate or by cutting a rolled I-beam lengthwise along its web at an angle, turning one half end for end, and then welding the two halves back together again along the web as shown in Fig. 13.24(a). Tapered girders are widely used in the framing of roofs over large areas, where it is desirable to minimize the number of interior columns or eliminate them altogether.

Similarly, *castellated beams* can be made economically by flame-cutting a rolled I-beam web in a zigzag pattern along its centreline [see Fig. 13.24(b)]. One of the two equal halves are turned end for end and is welded to the other half. The result is a deeper beam, stronger and stiffer than the original. Castellated beams with circular holes can also be fabricated as shown in Fig. 13.24(c). Castellated beams have more section modulus and moment of inertia and result in greater economy. Figure 13.25 shows a structure with castellated beams that led to significant savings in construction costs.

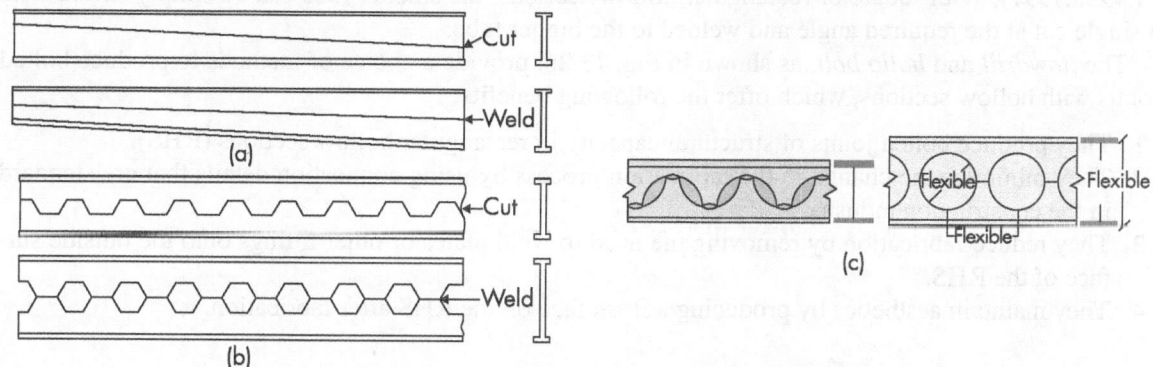

**Fig. 13.24** Tapered girders and castellated beams

Since the web of a beam or a plate girder contributes only a little to the bending resistance and because its strength in shear depends on its slenderness ratio $h/t_w$, it is economical to have the web of a lower-strength steel than the flange. Beams with stronger steel in the flanges than in the web are called *hybrid beams*. Such beams are often fabricated by welding plates of different steel strengths.

With the development of high-performance steels (see also Section 13.16), it is now considered advantageous to replace the flat webs of conventional built-up *I-beams with corrugated webs*. These corrugated plates could be produced by cold-forming long, flat plates. The corrugations can have a trapezoidal shape or a

**Fig. 13.25** Castellated beams over a car park in Connecticut, USA

sine-wave cross-section. A Japanese company has developed welding equipment that automatically senses and adjusts the equipment position as it welds corrugated web plate to a flange plate. Corrugated webs allow deeper girders that are not susceptible to web stability problems, resulting in thinner webs and smaller flanges. A plate girder with corrugated web, therefore, would weigh less than a conventional plate girder (Kulicki, 2000).

## 13.12.4 Hollow Steel Sections (HSS)

Tubular members are being used extensively in plane and space trusses as tubes are more efficient in resisting compressive forces (IS 806-1968, IS 1161, IS 10748). The advent of welding has made the connection between tubular members, using gusset plates, possible and resulted in the widespread use of tubular sections. Welded connections without gusset plates require proper planning for profiling and welding the ends, and have to be executed carefully. A tube is an efficient section, having the same radius of gyration in all directions. Care should, however, be taken to see that the *d/t* (diameter to thickness) ratio of the section is small enough to prevent premature failure by local buckling before the failure of the whole member. Recently, *square and rectangular tubes* have been introduced in India and these, of course, are much easier to connect because of their flat surfaces (see http://www.tatastructura.com and IS 4923:1997). With square or rectangular hollow sections, the smaller tube can be simply sawed with a single cut at the required angle and welded to the bigger tube.

The *flowdrill* and *hollo bolt,* as shown in Fig. 13.26, provide a choice of methods to produce bolted joints with hollow sections, which offer the following benefits:

1. They produce bolted joints of structural capacity in rectangular hollow sections (RHS).
2. They minimize the change in the fabrication process by using connection details that are standard in the construction industry.
3. They reduce fabrication by removing the need to weld plates or other fittings onto the outside surface of the RHS.
4. They maintain aesthetics by producing a flush face on the RHS after fabrication.

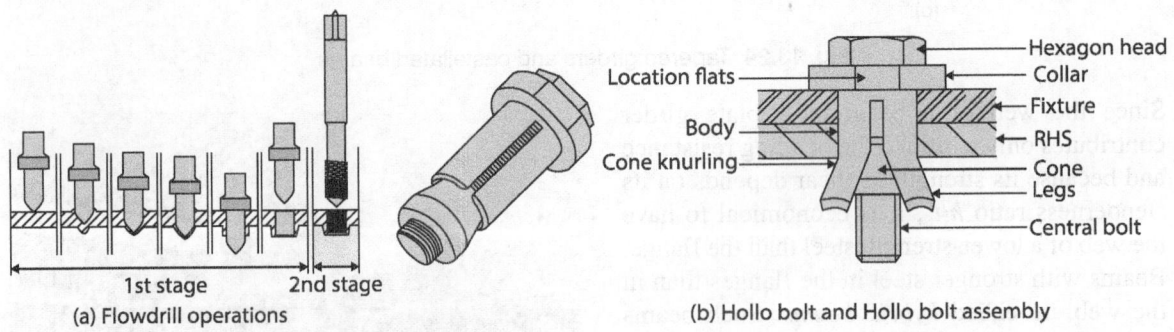

(a) Flowdrill operations                    (b) Hollo bolt and Hollo bolt assembly

**Fig. 13.26** Flowdrill and hollo bolt assembly

In India, until recently, only two grades of RHS, i.e., $f_y$ = 210 MPa ($f_u$ = 330 MPa) and $f_y$ = 240 MPa ($f_u$ = 450 MPa) were produced. Now, HSS as per the specifications given in IS 4923 with $f_y$ = 310 MPa ($f_u$ = 450 MPa) are also available. RHSs are available in sizes $50 \times 25 \times 2$ mm to $300 \times 200 \times 10$ mm and square hollow sections in sizes from $25 \times 25 \times 2$ mm to $250 \times 250 \times 10$ mm.

Basically, there are two major classifications of steel hollow sections: welded and seamless tubes. Hot-rolled or cold-rolled steel sheets are rolled into tubes and welded along the length, resulting in a longitudinal welded joint. Since they are easy to manufacture, they are made by even small-scale industries. Hence, the thickness and welding quality may vary and the designer has to be careful in choosing them. The tubes may also be classified as

1. hot-finished welded (HFW),
2. hot-finished seamless (HFS),
3. cold-drawn seamless (CDS),
4. electric-resistance welded (ERW),
5. cold-drawn electric-resistance welded (CEW), and
6. oxy-acetylene welded (OAW).

As per IS 1161, tubes for structural purposes are available in three grades of steel (see Table 13.7) in light, medium, and heavy categories. The yield stress of $Y_{st}$ 240 is the same as that of mild steel and this is often the grade (medium-section) available in the market. $Y_{st}$ 310 Grade tubes are also now available ranging in diameter from 15 to 300 mm (NB) with thicknesses of 2 to 10 mm. Engineers are advised to check the tubes used at site, as tubes confirming to IS 1239, which are less expensive and easily available, should not be used for structural purposes – they are intended for use for water, and non-hazardous gas, air and steam pipes.

**Table 13.7** Strength of steel used for circular tubes as per IS 1161

| Grade | Ultimate tensile stress, $f_u$ MPa (min.) | Yield stress, $f_y$, MPa (min.) |
|---|---|---|
| $Y_{st}$ 210 | 330 | 210 |
| $Y_{st}$ 240 | 410 | 240 |
| $Y_{st}$ 310 | 450 | 310 |

It has to be noted that tubes having less than 3.25 mm have to be carefully welded. If the tubes and hollow sections are not plugged at the ends properly, they retain the moisture inside and corrosion starts from the inside, which will not be noticed at the early stages.

## 13.12.5 Light-gauge Cold-formed Steel Sections

*Cold-formed steel (CFS) sections* are made from light-gauge steel strips, 2 to 4 mm thick (20–8 B.G.) and occasionally as thick as 5 mm, cold formed to shape in a rolling mill or press-brake, and are known as cold-rolled and pressed-steel sections (IS 513, IS 7226, IS 801, IS 811).

Considerable development has taken place in the production of floor and roof units and metal trim, and also in the field of structural members. Pressed-steel sections are largely used for flooring and roofing units and wall panels, and for metal trim such as skirting and sub-frames, the lengths of which are limited by the maximum width of the press-brake. The press consists of a die and a punch. The die and punch are suitably shaped to get the sections of desired shape. The steel strip is placed on the die and the punch is then lowered under a very heavy pressure. The steel strip is thus pressed between the die and the punch, and the section of desired shape is obtained without involving any shock. Using this process, sections up to 3-m long can be produced; although a few machines can produce lengths up to 6 m.

For structural members of greater length, cold-rolled sections are used. These are formed into the required shape by passing metal strip between six to 15 progressive sets of forming spindles or rollers, each pair of which adds successively to the shaping of the strip, the final pair producing the final section. As per IS 801, cold-formed steel sections are available in four grades of steel (see Table 13.8); the grade St 42 is similar to mild steel and is the only grade often available in the market. The basic sections rolled are plain angles and channels, lipped angles and channels, and zeds (see Fig. 13.27). Outwardly lipped channels are commonly called as hat sections. The length is limited only by considerations of transport. There is virtually no limit to the shape that can be rolled – the designer can choose a shape best suited to any particular purpose.

**Table 13.8** Strength of steel used for cold formed steel sections

| Grade | Ultimate tensile stress, $f_u$, MPa (min.) | Yield stress, $f_y$, MPa (min.) |
|---|---|---|
| St 34 | 340 | 210 |
| St 42 | 420 | 240 |
| St 50 | 500 | 300 |
| St 52 | 520 | 360 |

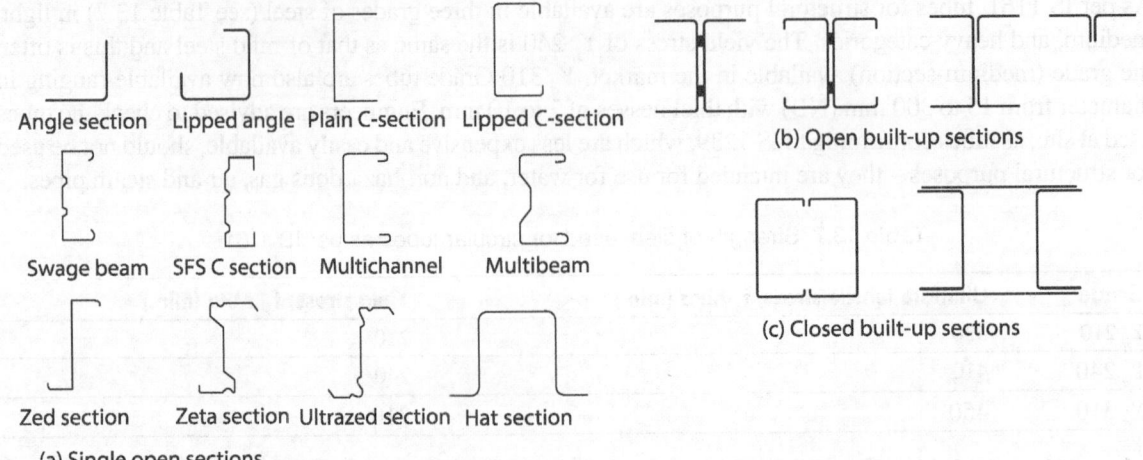

Angle section   Lipped angle   Plain C-section   Lipped C-section

(b) Open built-up sections

Swage beam   SFS C section   Multichannel   Multibeam

(c) Closed built-up sections

Zed section   Zeta section   Ultrazed section   Hat section

(a) Single open sections

**Fig. 13.27** Various shapes of cold-formed steel sections

Structures of moderate loads and span can be built more efficiently with cold-rolled sections; in such cases, they are cheaper than using hot-rolled members. Erection of the structure is often cheaper and easier because of its light weight and rigidity. Cold-rolled sections are used in India for purlins, girts, wind bracings and roof trusses. In countries like the USA, they have been used in columns (with channels placed back to back), I-section beams formed by channels (for light loads over short spans), lattice beams, and rigid frames.

Connections in CFS can be made by using welds, self-tapping screws, bolts, cold rivets, and hot rivets. It is also possible to make the sections to push-fit into each other, avoiding the use of gusset plates. For example, beams made of top-hat flange sections into which the bracing members can be made to fit, providing a direct node connection. Owing to the possibility of local bucking, the shape and design of CFS sections should be properly made. Often, lips to the edges of flanges or web stiffeners are provided, as seen in Fig. 13.27(a), to increase the local buckling strength, similar to those of aluminium sections (Yu, 2000). The need for protection from corrosion is important, because of the following factors:

1. Such structures are made up of very thin plates, less than 6-mm thickness (mostly, 2.0–4.0 mm thick).
2. It is difficult to reach inside faces for painting.
3. The re-entrant corners may hold dust and moisture.

The structural steel or strip steel should confirm with IS 11513 and the design is done as per IS 801.

## 13.12.6 Steel Wires and Ropes

Mild steel wires are available in diameters ranging from 0.125 to 12.5 mm. These wires are manufactured by cold drawing from the steel rods conforming to IS 7887. In this process, the steel bars are cold drawn through a series of successive smaller dies. Cold drawing increases the strength of the metal in each drawing stage. Hence, smaller the wire, the higher will be its yield and ultimate tensile strength (As per IS 280:2006, non-galvanized annealed wires have a maximum tensile strength of 500 MPa and galvanized annealed wires 300–550 MPa). However, these wires will have less plasticity and ductility, with every pass. Wire smaller than 5-mm diameter should be subjected to wrapping test as per IS 1755. As per this test, the wire should withstand wrapping eight times round its own diameter and subsequently straightened, without breaking or splitting. Wires having diameters of 5 mm or more are subjected to bend test. According to this test, the wire of diameter $d$ should withstand bending through an angle of 90° round a shaft or spindle of diameter equal to $2d$, without breaking or splitting. As per IS 280, the wire may be (a) annealed, (b) galvanized, (c) coppered,

(d) tinned, or (e) coated, and drawn (coating may-be of tin, copper, or zinc). The galvanized coating should conform to IS 4826 and the electro-galvanized wire should conform to the requirements of IS 12753. Wires of various types and diameters are manufactured for various applications, e.g., galvanized steel wire for tying steel reinforcements, and thick galvanized steel used as lightening conductors.

*Steel wire ropes* are made from round steel wires as per IS 1835. Steel wire ropes should conform to IS 6594 and can be of strands (a *strand* is an element of rope consisting of an assembly of several wires of appropriate shape and dimensions, spun helically in one or more layer), or strand ropes (A *strand rope* is an assembly of several strands spun helically in one or more layers around a core) (see Fig. 13.28). The strands can be of round strand, flattened strand, and multi-strand rotation resistant types, in ordinary or Lang's lay construction (see IS 2363:1981, for the different types of lay construction). Ropes can also be locked coil wire ropes [see Fig. 13.28(d)]. Ropes of round strand and flattened strand types are used for hoisting purposes in mines (winding and man riding haulages). Specifications and different arrangement of strands with fibre or steel core for steel wire ropes are provided in IS 2266. Some typical constructions of wire ropes are illustrated in IS 2363. Locked coil winding ropes (IS 3626) are essentially smooth surface ropes, composed of shaped wires laid in concentric layers, around a centre of round wires. These ropes consist of an outer layer of shaped wires with inner layers of either a combination of shaped and round wires, or round wires over a core. Locked coil ropes are being increasingly used in mining because of their inherent superiority over the usual stranded wire ropes. Specification for wire ropes and strands for suspension bridges are provided in IS 9282:2002. More information on wire ropes and their construction may be found in the Wire Rope Handbook (Martin, 2014).

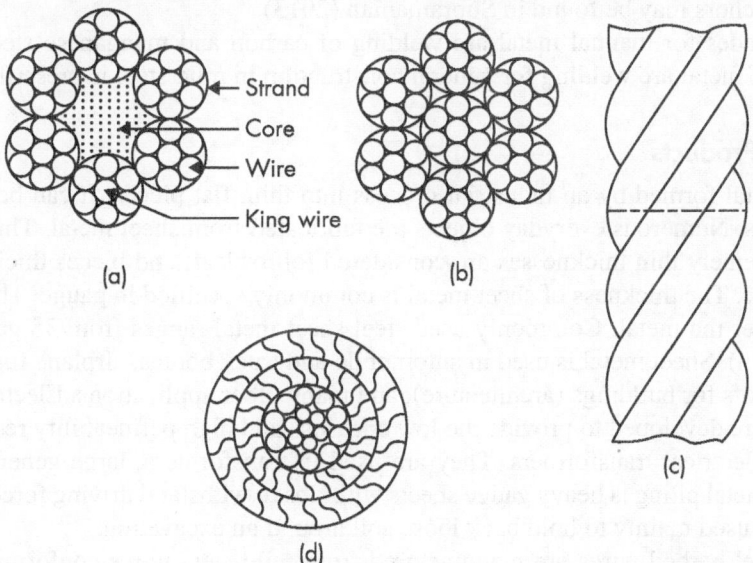

**Fig. 13.28** Steel wire ropes (a) Wire rope with fibre core (CF) [with round strand 6/7 (6/1)] (b) Wire steel core (CWR) (c) Longitudinal view of wire rope (d) Fully locked coil ropes

### 13.12.7 Steel Fasteners

There are several types of bolts used to connect structural members. Some of them are listed as follows:

1. Unfinished bolts or black bolts or C Grade bolts (IS 1363),
2. Turned bolts
   (a) Precision bolts or A Grade bolts (IS 1364),
   (b) Semi-precision bolts or B Grade bolts (IS 1364; IS 3640),

3. Ribbed bolts
4. High strength bolts (IS 3757; IS 4000).

Figure 13.29 shows typical hexagonal head black bolt and nut.

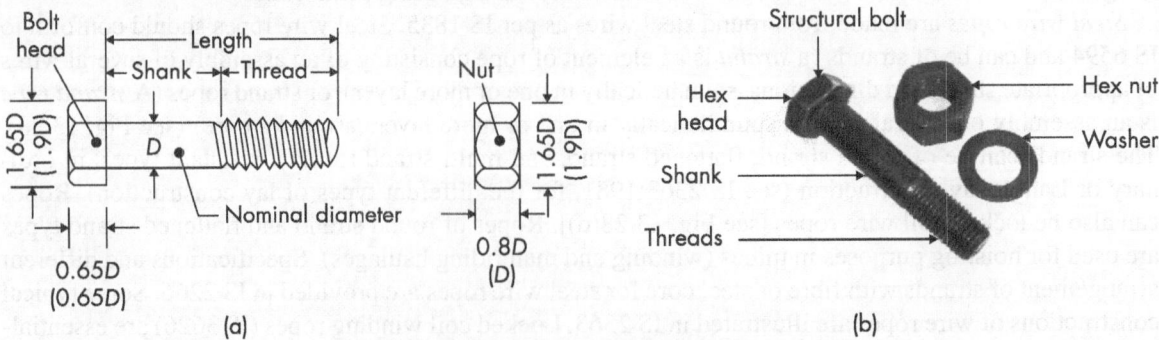

Specification for *foundation bolts* are provided in IS 5624. *Anchor bolts* are used to connect steel base plates of steel columns, light poles, highway sign structures, bridge rails, and equipment with concrete foundations. They may be broadly classified as cast-in-place anchors and post-installed anchors. More information on anchors may be found in Subramanian (2013).

Welding electrodes for manual metal arc welding of carbon and manganese steel are specified in IS 814. The use of metal arc welding for general construction in mild steel is prescribed in IS 816.

## 13.12.8 Other Products

*Sheet metal* is metal formed by an industrial process into thin, flat pieces. It can be cut and bent into a variety of shapes. Numerous everyday objects are fabricated from sheet metal. Thicknesses can vary significantly; extremely thin thicknesses are considered foil or leaf, and pieces thicker than 5 mm are considered as plate. The thickness of sheet metal is commonly specified in gauge. The larger the gauge number, the thinner the metal. Commonly used steel sheet metal ranges from 38 gauge (0.10 mm) to 6 gauge (4.115 mm). Sheet metal is used in automobile and truck bodies, airplane fuselages and wings, medical tables, roofs for buildings (architecture), and many other applications. Electrical steels are silicon alloys that were developed to provide the low core loss and high permeability required for efficient and economical electrical transformers. They are used in transformers, large generators, and electric machines. Sheet metal piling is heavy gauge sheets with ribs to withstand driving forces and lateral earth pressure; they are used mainly to hold back loose soil around an excavation.

Galvanized steel barbed wires are manufactured from mild steel wires conforming to IS 280, and formed by twisting together two-line wires, one or both containing the barbs (IS 278). Gabions, revet mattresses (wire mesh container uniformly partitioned into internal cells of relatively small height and having smaller mesh openings than those used for gabions), and rock-fall netting produced using double-twisted galvanized wire mesh (PVC coated after galvanization) are covered by IS 16014. Galvanized steel wire meshes (both in woven form or welded form), available in wide variety of opening sizes and wire diameters are used as base for plaster, fencing, caging, and enclosures, window and safety guards, and in a number of other general industrial uses.

In addition to the above, a large variety of steel products are used in building construction, which include galvanized steel nails, screws, bolts, nuts and washers. Steel nails may have smooth, spiral, or ring shanks. Framing nailer is a tool that makes nailing faster – they may use clip-headed and round-headed nails. Galvanized roofing nail is a shorter style nail with a ringed shaft and a large round flat head. All these products are required for connecting different kinds of construction components. Steel fibres having diameters varying from 0.25 to 0.75 mm and with aspect ratios (the ratio of length to diameter) of 30 to 150 are used in fibre reinforced concrete. These steel fibres may be flat, hooked, or crimped.

## 13.13 Reinforcing Steel Bars

Steel reinforcements are provided in RCC to resist tensile stresses. The quality of steel used in RCC work is as important as that of concrete. Steel reinforcement used in concrete may be of the following types (see Table 1.1 of SP 34(S&T): 1987, for the physical and mechanical properties of these different types of bars).

1. Mild steel and medium tensile steel bars (MS bars) conforming to IS 432 (Part 1): 1982
2. High strength deformed steel bars (HYSD bars) conforming to IS 1786: 2008
3. Hard drawn steel wire fabric conforming to IS 1566:1982
4. Structural steel conforming to Grade A of IS 2062:2006

It should be noted that different types of rebars, such as plain and deformed bars of various grades, say grade Fe 415 and Fc 500, should not be used side by side, as this may lead to confusion and error at site. Mild steel bars, which are produced by hot rolling, are not generally used in RCC as they have smooth surface and hence their bond strength is less compared to deformed bars (when they are used they should be hooked at their ends). They are used only as ties in columns or stirrups in beams. Mild steel bars have characteristic yield strength ranging from 240 N/mm² (grade I) to 350 N/mm² (medium tensile steel) and percentage elongation of 20 to 23% over a gauge length of 5.65√area.

The nominal size (in mm) of the available bars, as per IS 1786:2008, are 4, 5, 6, 8, 10, 12, 16, 20, 25, 28, 32, 36, 40, 45, and 50 (It is of interest to note that the cross-sectional area of these bars is approximately equal to the sum of the cross-sectional areas of two preceding lower size bars, for example, $10^2 = 6^2 + 8^2$). Generally, 6 to 16 mm bars are used in slabs, whereas 12 to 36 mm bars are used in beams and columns. The nominal size of deformed bar is taken as the equivalent diameter of smooth bar having the same weight per unit length. Allowable tolerance in the weight of bars are: ±7% up to 10 mm bars, ±5% over 10 and up to 16 mm bars, and ±5% over 16 mm bars. Usually, the bars are supplied in lengths of 9–12 m. A density of 7850 kg/m³ may be taken for calculating the nominal mass.

Hot-rolled mild steel ribbed bars are obtained by providing ribs or ridges on plain mild steel bars. Such ribbed bars, even though not recommended in the code, may be available in the market. They are not recommended for use in RC structures.

### 13.13.1 Hot-rolled Bars

HR HYSD bars were first introduced in India in 1967 by Tata steel as Tistrong bars and later as Tiscon/ Torsteel bars. Mild steel bars, used until then, were completely replaced by HYSD bars, except in situations where acute bending was required, especially in bars of 30 mm or more in diameter.

### 13.13.2 Cold Twisted Deformed (CTD) Bars (Torsteel Bars)

*Cold twisted deformed bars (CTD bars* or *Torsteel bars)* are made in two stages. First, the high-strength mild steel bars, with two or three parallel straight ribs is made by hot rolling. Second, the bars are cooled to room temperature and then cold twisted to strain the steel beyond the elastic limit and released. This steel is not recommended to be welded. The original straight ribs after cold twisting will form a helix around the CTD bars. Over-twisting of the bars will result in the pitch of the helixes to be too close. As cold twisting introduces residual stresses in the CTD bars, the latter were found to corrode much faster than other bars and hence are not used now.

### 13.13.3 Thermo-mechanically Treated (TMT) Bars

*Thermo-mechanically treated reinforcement bars (TMT Bars)* are also HR HYSD bars. When they are leaving the last stand of the rolling mill at a temperature of about 950°C, they are rapidly cooled by water by a controlled quenching process, to about 450°C. This sudden quenching and final cooling transforms the surface austenite layer of the bars to tempered martensite and having a fine grained ferrite-pearlite core as shown in Fig. 13.30. TMT bars can be welded using ordinary electrodes without any extra precautions, as per IS 9417. The advantages of TMT bars include strength, weldability, ductility, and economy. To distinguish TMT bars from mild steel deformed bars, IS 1786 suggests the following procedure: cut a small 12-mm long piece of bar and apply progressively finer emery papers of size '0' on the transverse face. When this face is macro-etched with nital liquid (5% nitric acid in alcohol) at ambient temperature, the TMT bars will reveal the darker edge ring of martensite/bainite microstructure and a lighter core in a few seconds as shown in Fig. 13.30.

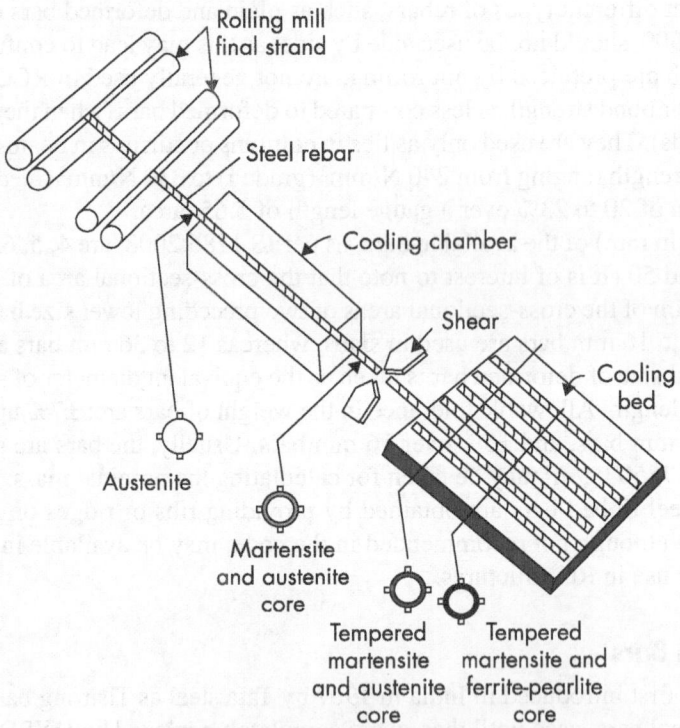

**Fig. 13.30** Manufacturing process of TMT bars

*Thermo-mechanically treated corrosion resistant steel bars* (TMTCRS bars) are produced by micro-alloying steel with copper, phosphorous, and chromium. TMTCRS bars have better corrosion resistance than ordinary TMT bars. Even these bars are not 100% corrosion-resistant and hence should be used with certain amount of caution. Six grades, designated as Fe 415, Fe 500, Fe 550, Fe 600, Fe 650, and Fe 700 and an additional three grades with a suffix D (Fe 415D, FE500D, and Fe 550D), to denote that they are ductile, are specified in IS 1786:2008 (Note that the numbers after Fe denote the 0.2% proof or yield stress, in N/mm$^2$). There are also two more grades, Fe 415S and Fe 500S (where the suffix S denotes special, which have the tensile strength to the 0.2% proof stress ratio of 1.25. However, there is only limited availability of Fe 550, Fe 600, Fe 415D, Fe 500D, Fe 550D, Fe 415S, and Fe 500S grades in the market.

The most important property of the reinforcing bar is its characteristic yield strength or 0.2% proof stress as the case may be and the important characteristic is its stress-strain curve (see Fig. 13.31 and Table 13.9). Clause 5.6.3 of IS 456 suggests that the modulus of elasticity $E_s$ may be taken as 200 kN/mm$^2$. [Note that for HYSD bars the yield stress is not easily defined. Hence, an offset yield point at 0.2% of the strain is considered. This point on the stress-strain curve is located by drawing a line parallel to the elastic portion of the stress-strain curve from the 0.2% strain as shown in Fig. 13.31(b)]. The design stress-strain curves for both mild steel and HYSD bars are given in Fig. 23 of IS 456. The chemical composition of various grades of steel is given in IS 1786:2008.

As per Clause 5.3.1 of IS 13920:2016, bars of grade Fe 415 or less should only be used in structures situated in earthquake zones. However, TMT bars of grades Fe 500 and Fe 550, having elongation more than 14.5% are also allowed. In addition, as per Clauses 5.3.2 and 5.3.3 of IS 13920, the actual 0.2% proof strength of steel bars determined based on tensile tests should not exceed the characteristic 0.2% proof strength by more than 20% and the ratio of the actual ultimate strength to the actual 0.2% proof strength should be at least 1.15. In order to have sufficient bond between the bars and the concrete, the area, height, and pitch of the ribs should satisfy Clause 5 of IS 1786 (see Fig. 13.32).

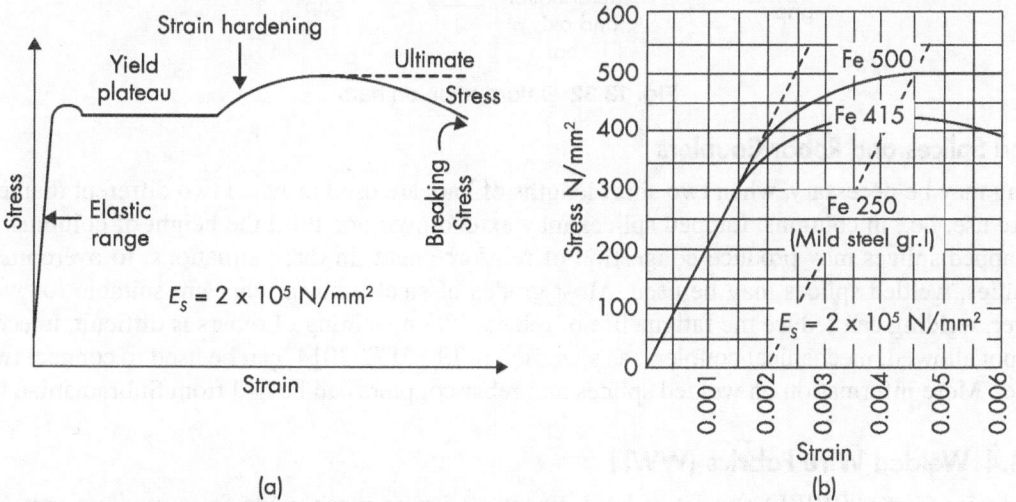

**Fig. 13.31** Stress-strain curve (a) Mild steel (b) HYSD bars (only initial portions of the curves are shown)

**Table 13.9** Mechanical properties of high strength deformed bars (see also IS 1786:2008)

| S. no. | Property | Fe 415 | Fe 415D | Fe 500 | Fe 500D | Fe 550 | Fe 550D | Fe 600 |
|---|---|---|---|---|---|---|---|---|
| 1 | 0.2% proof stress or yield stress, Min. (N/mm$^2$) | 415.0 | 415.0 | 500.0 | 500.0 | 550.0 | 550.0 | 600.0 |
| 2 | Elongation, percent, on gauge length 5.65 $\sqrt{A}$*, (Min.) | 14.5 | 18.0 | 12.0 | 16.0 | 10.0 | 14.5 | 10.0 |

*(Contd)*

**Table 13.9** (Contd)

| S. no. | Property | Fe 415 | Fe 415D | Fe 500 | Fe 500D | Fe 550 | Fe 550D | Fe 600 |
|---|---|---|---|---|---|---|---|---|
| 3 | Ratio of ultimate tensile strength to the 0.2% proof strength (TS/YS ratio) | ≥1.10, but TS ≥485.0 N/mm² | ≥1.12, but TS ≥500.0 N/mm² | ≥1.08, but TS ≥545.0 N/mm² | ≥1.10, but TS ≥565.0 N/mm² | ≥1.06, but TS ≥585.0 N/mm² | ≥1.08, but TS ≥600.0 N/mm² | ≥1.06, but TS ≥660.0 N/mm² |
| 4 | *Total elongation at maximum force percentage, on gauge length 5.65 √A*, (Min.) | – | 5% | – | 5% | – | 5% | – |

*A is the cross-sectional area of the test specimen

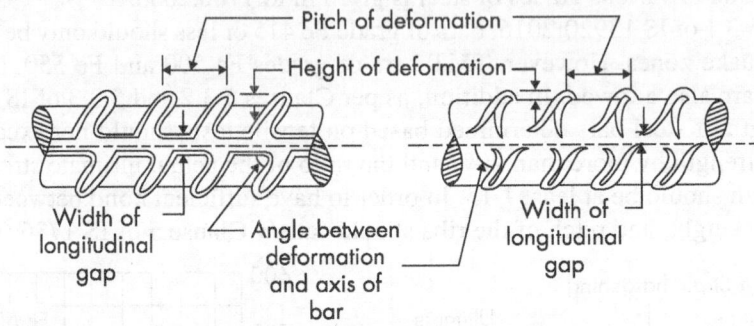

**Fig. 13.32** Deformation on bars

## Welded Splices and Rebar Couplers

Lapping may be necessary, when two short lengths of bars are used or when two different diameters of bars are use, i.e., in columns. Lapped splices may extend over one-third the height of columns. Moreover, lapped splices may produce congestion of reinforcement. In these situations, to overcome these difficulties, welded splices may be used. Most grades of steel used in rebar are suitable for welding; however, welding can reduce the fatigue life of rebars. When welding of rebars is difficult, uneconomical or not allowed, mechanical couplers, as specified in IS 16172:2014, can be used to connect two bars together. More information on welded splices and rebar couplers can be had from Subramanian (2013).

### 13.13.4 Welded Wire Fabrics (WWF)

*Welded wire fabrics* (WWF) consist of hard drawn steel wire mesh made from medium tensile steel, drawn out from higher diameter steel bars. As they undergo cold working, their strength is higher than that of mild steel. WWF consists of longitudinal and transverse wires (at right angles to one another) joined by resistant spot welding using machines. They are available in different widths and rolls and as square or oblong meshes; see Table C-1 of SP 34 (S &T): 1987 and IS 1566:1982. Their use in India is limited to small size slabs.

### 13.13.5 Steel for Prestressed Concrete

High-tensile steel wires, strands (an assembly of several wires of appropriate shape and dimensions spun helically in one or more layer, as shown in Fig. 13.28), and bars are used in prestressed concrete construction. Where large prestressing forces are involved, it is advantageous to specify strands instead of single wires. Strands consisting of a seven-wire bundle, in which six wires are helically wrapped around a single straight wire, are normally used. The high tensile strength obtained in cold-drawn, high-carbon steel wire is due to the following three factors:

1. Chemical content
2. Thermal treatment
3. Cold drawing

To achieve the minimum required mechanical properties, the prestressing steel is alloyed with high-carbon and manganese, and also with small amounts of (generally, less than 0.20%) chromium, vanadium, or both. Two different thermal treatments have been used in the past to produce a very fine lamellar pearlitic microstructure with proper tensile strength and ductility for wire drawing. Before the 1980s, the most widely used thermal treatment for rods was *lead patenting*. In the current *Stelmore Process*, hot-rolled rod emerging from the last stand on the mill is rapidly cooled with water to about 815°C and laid in a continuous spiral on a moving roller bed. This spiral of rod immediately passes over large fans that blow air against the exposed hot strands (ACI 222.2 R-01). The rapid cooling of the rod by the air transforms the steel into the fine pearlite microstructure required for drawing high-tensile-strength wire. The tensile strength of the finished, thermally treated rod will be 1100 MPa or higher. Cold drawing is carried out by drawing the wire through a number of consecutive wire drawing dies of decreasing diameter in a continuous operation. This process increases the ultimate tensile strength (by about 700 MPa) as well as yield strength, in each successive drawing, but decreases the ductility of wires [As per IS 1785 (Part 1), the wire should withstand three 90° reverse bends without fracture]. The wires are subjected to straightening and stress-relieving processes after cold drawing. Stress relieving is performed by heating the wire to a temperature of 150–420°C and subsequent cooling to room temperature. Prestressing strand is also manufactured in a stabilized, stress-relieved condition. Stabilization is performed to reduce relaxation characteristics of the strand by thermal-mechanical treatment-simultaneously stretching and heating the strand (ACI 222.2-R01).

Some prestressing systems, such as the German Dywidag System, use high tensile rods instead of wires or strands. The diameters of these rods range from 20 to 40 mm, with characteristic strength of about 1000 N/mm². According to Clause 5.6.1.1 of IS 1343:2012, any one of the following materials can be used as prestressing steel:

1. Plain hard-drawn steel wire of diameter 2.5–8 mm (cold drawn stress relieved wire) conforming to IS 1785(Part 1) – with tensile strength varying from 1375 N/mm² (8-mm diameter wire) to 2010 N/mm² (2.5-mm diameter wire)
2. Indented wire conforming to IS 6003:2010
3. High tensile steel bar (diameter 10 to 32 mm) with the tensile strength of 980 N/mm² and proof stress not less than 80% of maximum specified tensile strength, conforming to IS 2090
4. Uncoated stress relieved strand conforming to IS 6006
5. Uncoated stress relieved low relaxation seven ply strand conforming to IS 14268

The modulus of elasticity of high tensile steel may be taken as 200 GPa. In most of the high tensile wires, creep is negligible up to $0.45\,f_p$ and it is about 3% at $0.55\,f_p$ stress level, where $f_p$ is the ultimate tensile strength of pretension steel. Welding is not permitted in prestressing steel.

## 13.13.6 Corrosion of Rebars and Mitigation Methods

Corrosion of steel rebars is considered the main cause of deterioration of numerous RCC structures throughout the world (see also Section 8.10.1 of Chapter 8). Detrimental substances like chlorides or water can penetrate through cracks or pores in concrete, leading to corrosion of steel bars. Corrosion can also occur due to carbonation of cover concrete or when two different metals are in contact within concrete. For example, *dissimilar metal corrosion* can occur in balconies where embedded aluminium railings are in contact with the reinforcing steel.

Corrosion of rebars can be mitigated in concrete by using the following reinforcements:

***Fusion-bonded epoxy-coated reinforcing bars (IS 13620:1993)*** Typical coating thickness of these bars is about 130 to 300 μm. Damaged coating on the bars, resulting from handling and fabrication and the cut ends, must be properly repaired with patching material prior to placing them in the structure. These bars have been used in RC bridges from the 1970s and their performance is found to be satisfactory (Smith & Virmani, 1996). They may have reduced bond strength.

***Galvanized reinforcing bars (IS 12594:1988)*** The precautions mentioned for epoxy coated bars are applicable to these bars as well. The protective zinc layer in galvanized rebars does not break easily and results in better bond.

***Stainless steel bars (IS 16651:2017)*** Stainless steel is an alloy of nickel and chromium. Four stength grades (SS 500, SS 550, SS 600, and SS 650) and seven designation numbers (A to G) are specified in IS 16651. Thus, a bar with a nominal diameter of 20 mm, steel designation number 'B' and strength grade 550 N/mm$^2$ shall be designated as 20-B-SS550 (See IS 16651 for more information on these different grades). Though the initial cost of these bars is high, life cycle cost is lower and they may provide 80–125 years of maintenance-free service. The Progreso Pier in Yucatan, Mexico, was built during 1937–1941 using stainless rebars and has not required maintenance until now.

***Fibre-reinforced polymer bars (FRP bars)*** These are aramid fibre (AFRP), carbon fibre (CFRP), or glass fibre (GFRP) reinforced polymer rods. They are non-metallic and hence non-corrosive. Although their ultimate tensile strength is about 1500 MPa, their stress-strain curve is linear up to failure. In addition, they have one-fourth the weight of steel reinforcement and are expensive. The modulus of elasticity of CFRP is about 65% of steel bars and the bond strength is almost the same. As the Canadian Highway Bridge Design Code, CSA-S6–06, has provisions for the use of GFRP rebars, a number of bridges in Canada are built using them. More details about them may be obtained from the work of Ganga Rao, et al. (2006) and ACI 440R-07 report.

***Basalt bars*** These are manufactured from continuous basalt filaments, epoxy, and polyester resins using a pultrusion process. It is a low-cost, high-strength, high-modulus, and corrosion-resistant alternative to steel reinforcement. More information about these bars may be found in the study of Subramanian (2010).

In addition, *Zbar*, a pretreated high-strength bar with both galvanizing and epoxy coating, has been recently introduced in the USA. High-strength ChromX® (formerly MMFX) *steel bars,* conforming to ASTM A1035, with yield strength of 827 MPa and having low-carbon and 8–10% chromium have been introduced in the USA recently, which are also corrosion-resistant, similar to TMT-CRS bars (www.mmfx.com).

Viswanatha et al. (2004), based on their extensive experience of testing rebars, caution about the availability of sub-standard rebars in India, including rerolled bars and inadequately quenched or low-carbon content of TMT bars. Hence, it is important for the engineer to accept the rebars only after testing them in accordance with IS 1608:2005 and IS 1786:2008 (Rai et al., 2012). Basu et al. (2004) also provides an overview of important characteristics of rebars, and a comparison of specifications of different countries.

## Corrosion Mitigation of Prestressed Steel

It is a common belief that pre-tensioned and post-tensioned structures, when properly detailed and constructed, will be corrosion free throughout their life, as prestressed concrete is considered to be crack-free. The corrosion of prestressing steel (called tendons) can be much more serious than in conventional RC structures, because the prestressing tendons have a relatively small cross-sectional area and are under very high constant stress of about 55 to 65% of its ultimate tensile strength. Moreover, high-strength steel is more susceptible to brittle fracture due to stress corrosion or hydrogen embrittlement. In addition, stress-corrosion cracking, fretting fatigue, and corrosion fatigue may also significantly affect the performance of the prestressing steel. It has to be noted that unlike in RC structures, there are no visible warning, such as corrosion product, before failure. The failures also tend to be brittle, involving little elongation before fracture.

In pre-tensioned members corrosion of tendons is prevented by having adequate cover to the tendons and also by using concrete with a sufficiently low w/cm ratio, which is usually the case for the high-strength concretes used for prestressed concrete. The ducts in post-tensioned members were, for many years, grouted/pressure grouted after tensioning. However, as several failures have shown corrosion of partly grouted tendons, it may be preferable to leave the post-tensioned tendons unbonded. In order to avoid the partial grouting of ducts, pressure-testing of ducts may be done, before grouting. A combination of greasing and coating with plastic has been used successfully as an alternative protection to the tendons (ACI 423.4R-98). Unbonded tendons are generally protected by a plastic sheath and anticorrosive grease placed in the annular space around the strand. Epoxy coatings and zinc galvanizing have also been successfully applied to prestressing steel. Epoxy-coated metal ducts, spirally corrugated, and hoop-corrugated polyethylene ducts have been introduced in an effort to provide additional corrosion protection. For bridges, it is essential to inspect the tendons at regular intervals during the design life of the structure. More information on corrosion of different types of prestressing steel and their mitigation may be found in ACI 222.2 R-01, ACI 423.4R-98, ACI 423.8R-10, and Podolny (1992).

## 13.14 Protection of Structural Steel from Corrosion

Steel readily corrodes in moist air. The rate of corrosion depends primarily on the type of metal or alloy, rainfall, humidity, temperature, degree of atmospheric pollution, and the extent of exposure to the prevailing wind and rain. Aggressive environments such as sea water, acid, or alkaline vapours will hasten the process. The loss of thickness of steel members in industrial areas with high humidity and aggressive atmosphere or in coastal areas with high salinity may be in the range of 80–200 µm/year (BS EN ISO 12944–2:2017). Hence, structural steel members should be protected effectively against corrosion. Corrosion prevention methods may be classified into four main groups (IS 9172:1979):

1. Treatment of the environment to render it non-corrosive
2. Protective coatings
3. Cathodic protection
4. Use of corrosion resistant structural steels

Corrosion may be reduced by correct planning of the structural layout and the arrangement of details, particularly of those members/details most liable to corrosive attack. Some typical examples of details that have been found to initiate local corrosion and alternative arrangements to avoid them are provided in IS 9172. Free circulation of air around and through the structure should be arranged. Arrangements should be made for shedding dripping water and condensation. Wherever necessary, short length of pipes should be attached to the drain hole to carry the water clear and prevent water being blown back on to the structure. Water traps should be avoided, especially where steel stanchions are embedded in a concrete base. Figure 13.33 shows some bad and better detailing to avoid water entrapment (also see IS 9172:1979). It has to be noted that the corrosion process requires the simultaneous presence of water and oxygen. In the absence of either, corrosion does not occur.

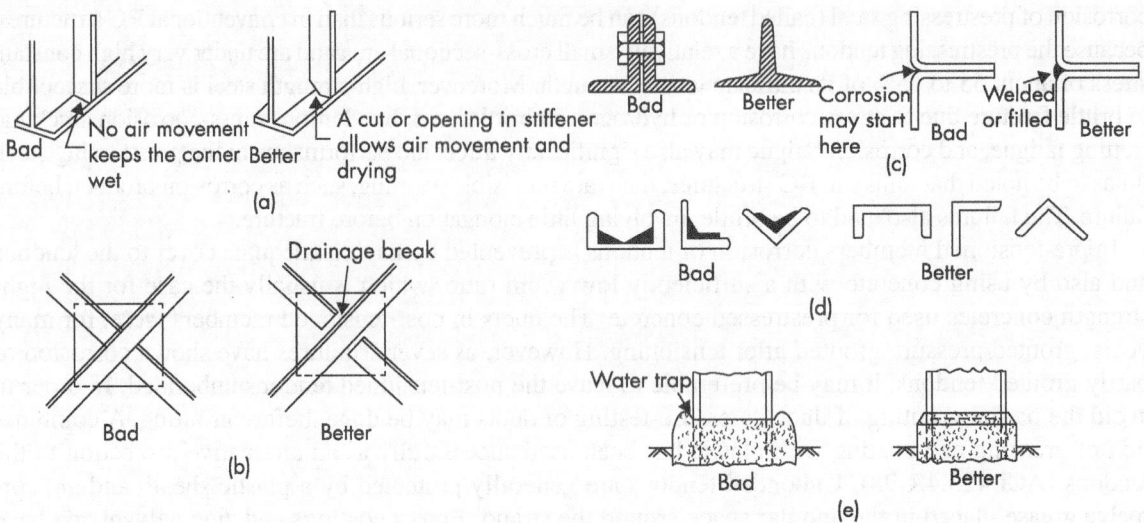

**Fig. 13.33** Detailing to prevent water entrapment (a) Encourage air movement (b) Prevent retention of water/dirt at junction plates by means of 'breaks' (c) Avoid open crevices (d) Avoid entrapped dust/water by proper orientation of members (e) Avoid water trap at the foundation level

The most common form of protecting a steel member involves the use of paints or metallic coating or plastic coat in the case of metallic sheeting. Steels with copper content of 0.2 to 0.5% have improved resistance to atmospheric corrosion but still need to be protected. The selection of a suitable primer and paint for metals depends on many factors including the type of metal to be coated, the type of surface preparation to be used, environmental and surface conditions, desired appearance, performance requirements, method of application, and type and level of exposure. Paint systems consist of zinc- or aluminium-based primary coat on which two or three layers of finishing coats (zinc phosphate modified alkyd or two pack zinc-rich epoxy) are applied. Metallic coatings include galvanizing and sherardizing (both of which use zinc), electroplating (usually applied to fasteners), and metal spraying using either zinc or aluminium. For the coating or painting to be effective, the surfaces of steel members have to be cleaned effectively before treatment. Several methods are available and the usual one consists of blast cleaning the surface using small abrasive particles, such as iron, which are directed to the surface of steel members by using compressed air or an impeller. In order that coatings are applied with ease and efficiency, details such as back-to-back angles, recesses, deep corners, and behind holes should be avoided. The coating should also be reapplied in intervals of 8–20 years, depending on the environmental condition and the coating system employed. Section 15 of IS 800:2007 provides some guidelines on different coating systems for different environmental conditions (also see Chapter 18).

Instead of protective treatment, special corrosion resistant steel could be used, which on exposure to weather forms a protective surface layer of oxide film. Such *weathering steels* contain an increased amount of phosphorus, chromium, and copper, compared to the normal steel (see Section 13.14.1, for more details). Stainless steel, as discussed in Section 13.15, could also be used. They cost 20% more than normal steel but this initial cost is offset by savings in weight, protective treatment, and maintenance. Where *cathodic protection* is to be applied, the design should ensure good electrical conductivity throughout and, if necessary, insulation from neighbouring structures (more details on cathodic protection may be gathered from IS 8062 (Parts 1–4): 2006).

### 13.14.1 Weathering Steel

These are intended for applications where weight saving, along with improved atmospheric corrosion resistance, is important. The atmospheric corrosion resistance of these steels is approximately four times to that of carbon structural steels (IS 11587:1986). This steel is also manufactured by the open-hearth, electric, basic-oxygen, or a combination of these processes, and supplied in killed condition. Weathering steel is a steel alloy containing small amounts of copper, chromium, manganese, nickel, phosphorus, and other alloying elements to enhance corrosion resistance. As per IS 11587, it is available in three grades: (a) WR-Fe 480A ($f_y$ = 345 MPa), (b) WR-Fe 480B ($f_y$ = 345 MPa), and (c) WR-Fe 500 ($f_y$ = 355 MPa). Note that the number after WR-Fe refers to the ultimate tensile strength. The $f_y$ values given here are for members with thickness up to 12 mm. For other thicknesses, refer IS 11587. The chemical composition of weather-resistant steels is given in Table 13.10. All these steel grades are of weldable quality, and are supplied with a carbon equivalent value of 0.54 (maximum) based upon the ladle analysis.

**Table 13.10** Chemical composition of weathering steel

| Grade | C | Si | Mn | P | S | Cr | Ni | Cu | V |
|---|---|---|---|---|---|---|---|---|---|
| WR-Fe 480A | 0.12 Max. | 0.25–0.75 | 0.60 Max. | 0.07–0.15 | 0.050 Max. | 0.30–1.25 | 0.65 Max. | 0.25–0.55 | – |
| WR-Fe 480B | 0.10–0.19 | 0.15-0.50 | 0.90–1.25 | 0.04 Max. | 0.050 Max. | 0.40–0.70 | – | 0.25–0.40 | 0.02–0.10 |
| WR-Fe 480A | 0.17 Max. | 0.40 Max. | 1.00 Max. | 0.07–0.10 | 0.050 Max. | 0.70–1.00 | – | 0.25–0.55 | 0.10 Max. |

C = Carbon, Cr = Chromium, Cu = Copper, Mn = Manganese, Ni = Nickel, P = Phosphorus, S = Sulphur, Si = Silicon, and V = Vanadium

**Case Study   Richard J. Daley Center, Chicago**

When the weathering steel rusts under normal atmospheric exposure, it forms a rusty-orange-brown protective impervious layer that prevents further corrosion. Minor damage to the coating is self-healing. Over time, as the thickness of coating increases, the texture roughens and the oxide coating changes from rusty-orange-brown to dark purple-brown patina. Thus, weathering steel does not require any painting to protect it and provides significant life-cycle cost savings.

Richard J. Daley Center and Pablo Picasso Sculpture (*Photo*: Dr. N. Subramanian)

The Richard J. Daley Center, Chicago, built in the year 1965, was the first building to be constructed entirely with a type of weathering steel, called Cor-Ten. The Pablo Picasso sculpture outside the centre is also made of weathering steel. As the picture shows, even after more than 50 years of existence, they look great due to proper maintenance of the building and the sculpture.

Of course, weathering steel is not a material of choice in environments that are constantly wet or humid or for coastal structures. It has been used in several short span bridges such as the award-winning Box Elder Creek Bridge in Colorado. More details about weathering steel are available online at: http://www.aisc.org/content.aspx?id=17892.

## 13.15 Stainless Steel

It is not advisable to use structural steel in coastal areas due to its potential for corrosion. Hence, stainless steel was developed to rectify this problem at the beginning of the 20th century. However, it was not used extensively except in military supplies and the chemical industry, due to its prohibitive cost.

The advancement in refining processes during 1950s made it possible to mass produce stainless steel at low cost. Stainless steel is now being used in the construction industry in roofing, interiors and exteriors of buildings, and also in structural applications.

Stainless steel is a low-carbon steel to which chromium is added in amounts greater than 10.5% by weight, thus giving it the unique 'stainless' corrosion resistant properties. The various grades of stainless steel is produced in India are designated as SS 301, SS 304, SS 304L, SS 310S, SS 316, SS 316L, SS 321, SS 409, SS 409M, SS 410S, SS 420, and SS 430. Its production in India has increased from 20,000 tonnes in 1978, to 3.6 MT in 2017.

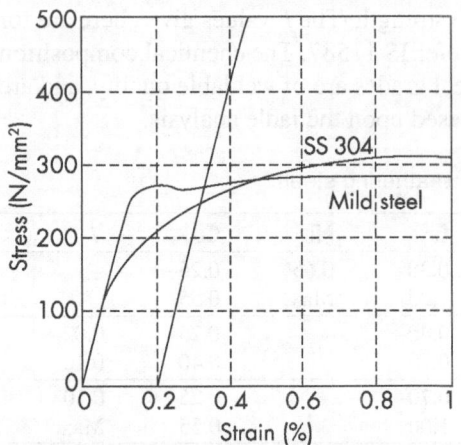

**Fig. 13.34** Stress-strain curves of stainless steel and carbon steel

Stainless steel also has a high strength to weight ratio, is weldable and environment friendly. Its main advantages over the other steels are its aesthetically pleasing appearance, corrosion resistance, high tensile strength, high toughness, and impact and heat resistance. It has a nonlinear stress-strain curve, similar to that of high-strength steels, as shown in Fig. 13.34. Hence, an offset yield point at 0.2% of the strain is considered as shown in Fig. 13.34. The initial modulus is similar to that of structural steel, but the proportional limit is usually low. The stress-strain curve of certain stainless steels may be different in tension and compression and may also be different in the longitudinal and transverse direction, due to their method of production. Hence, separate guidelines have been developed for the design of stainless steel structural members (e.g., Euro Inox, 1994, ASCE/ANSI-8–90; ENV 1993–1–4). Indian code provisions are not yet available, which is one of the reasons why stainless steel members have not been used in engineering structures. Figure 13.35(a) shows the Cloud Gate or the Bean at Millennium Park, Chicago, Illinois, which is a stainless steel, $10 \times 13 \times 20$ m³ sculpture, designed by Indian-born British artist Anish Kapoor, and opened in 2006. The outer shell comprises 168 stainless steel panels, each 10-mm thick. It received the American Welding Society's Extraordinary Welding Award. Figure 13.35(b) shows stainless steel railing and a sculpture installed in Lake Forest Mall, Gaithersburg, built during the year 2000.

(a)

(b)

**Fig. 13.35** Use of stainless steel (a) Cloud Gate or the Bean at Millennium Park, Chicago, Illinois (b) Railing and sculpture in Lake Forest mall, Gaithersburg

Apart from being more resistant to corrosion, stainless steel is superior to carbon steel in such qualities as elongation and fire resistance. While the proof strength of ordinary steel begins to decline at a steel temperature of 300 to 500°C, that of stainless steel has a small rate of decrease up to about 700°C. This means that stainless steel has far superior fire resistance, making possible the construction of buildings using stainless steel without fire-proof insulation.

Of the various available grades, SS 304, SS 304L, SS 306, SS 409, and SS 430 are suitable for structural applications. The chemical, mechanical, and physical properties of these grades are given in Tables 13.11 and 13.12 (The properties for mild steel are also given for comparison). According to IS 6603 and Salem Steel (1998), stainless steel has a density of 75.5–77.5 kg/m³, the modulus of elasticity of 193–200 GPa, and the coefficient of thermal expansion of $11.34–19.8 \times 10^{-6}/°C$ (also refer IS 16651).

**Table 13.11** Chemical composition (in percentage) of stainless steels (Salem Steel 1998)

| Grade | C (max.) | Si (max.) | Mn (max.) | Cr | Ni | P (max.) | S (max.) |
|---|---|---|---|---|---|---|---|
| SS 316 | 0.08 | 0.75 | 2.00 | 16–18 | 10–14 | 0.045 | 0.03 |
| SS 304 | 0.08 | 0.75 | 2.00 | 18–20 | 8–10.5 | 0.045 | 0.03 |
| SS 304L | 0.03 | 0.75 | 2.00 | 18–20 | 8–12 | 0.045 | 0.03 |
| SS 409 | 0.08 | 1.00 | 1.00 | 10.50–11.75 | 0.50 (max.) | 0.045 | 0.03 |
| SS 430 | 0.12 | 1.00 | 1.00 | 16–18 | 0.75 (max.) | 0.040 | 0.03 |
| Mild steel (E 250) | 0.22 | 0.40 | 1.50 | – | – | 0.045 | 0.045 |

**Table 13.12** Mechanical properties (annealed condition) of cold-rolled stainless steel (IS 6603; Salem Steel, 1998)

| Grade | UTS (N/mm²) (min.) | 0.2% Proof Stress (N/mm²) (min.) | Min. percentage Elongation on 50 mm GL | Hardness RH (max.) |
|---|---|---|---|---|
| SS 316 | 515 | 205 | 40 | 95 |
| SS 304 | 515 | 205 | 40 | 92 |
| SS 304 L | 485 | 170 | 40 | 92 |
| SS 409 | 380 | 205 | 20 | 88 |
| SS 430 | 450 | 205 | 22 | 89 |
| Mild steel E 250 | 410 min | 250 min | 23 | 80–105 |

Most stainless steels fall into one of three main classes – martensitic, ferritic, and austenitic, according to their metallurgical structures. Martensitic types are rarely used in buildings, only when high strength or hardness is required. Ferritic steels, which are magnetic, contain chromium to the extent of 12–30%. They are identified in the AISI 400 Series (SS 409 and SS 430 grade). They are low-priced stainless steel useful in less aggressive environment, where environmental condition is of less importance (Mathur, 2000).

Austenitic steels contain 16–26% chromium and 6–22% nickel. They are in annealed condition, non-magnetic, have excellent corrosion resistance, and are weldable. Grades SS 304, SS 304L, and SS 316 (best grade for all external applications) are used in structures. Grade 304 should not be used for external elements near coastal areas. Grade SS 316 is basically a 304 grade with the addition of 2–3% molybdenum.

Owing to its plastic deformation capacity, stainless steel is used for space trusses, bracings, column bases, and base-isolated structures to control earthquake-response. Based on life-cycle cost analysis, it is found that stainless steel structures will be economical after 20 years. Stainless steel will be ideal for structures where more than 100 year life is required.

### 13.5.1 Surface Finish

Cold-rolled (thin gauge up to 3 mm) stainless steels are available with the following finishes (Mathur, 2000):

1. No. BA: A bright, smooth mirror finish
2. No. 2B: A bright, smooth, silvery grey, moderately reflective finish(most common finish for sheet material)
3. No. 2D: A matt, dull, silvery grey, non reflective finish (gives better paint adhesion)
4. No. 3: A directional, uniform, polished finish, using 100 to 120 grit abrasive.
5. No. 4: A bright, polished finish with a visible directional grain from a 120 to 180 grit abrasive.

The aforementioned finishes are mill finishes. Other finishes such as various polished finishes, coloured BA finishes, regidized patterns, and decorative steel are also commercially available in stainless steel. More information on various stainless steels and their uses may be found from www.nickelinstitute.org and www.stainless-steel-world.net

## 13.16 High-performance Steel

The Federal Highway Administration, the American Iron and Steel Institute, and the Department of the Navy jointly developed the *high-performance steel*, in 1994. Three grades of HPS are available: HPS 345W, HPS 485W, and HPS 690W, out of which HPS 485W (yield strength = 485 MPa) is widely used in the USA. The benefits of HPS include enhancements in weldability, toughness, corrosion resistance, ductility, fatigue and fire resistance, formability, and strength. These factors result in higher economic efficiency, ease of maintenance, and longer service life.

HPS 485W has the same strength as AASHTO N270 Grade 485W steel. But, its unique chemical and physical properties result in economical fabrication practices. The main difference between N270 and HPS steel are that the latter contains only 50% of carbon, and one-tenth of sulphur. Toughness values of over 200 J are obtained, as compared to values of 30 to 50 J for the N270 grade steel.

The low-carbon levels require minimum or no preheat allowing an increased productivity of fabrication at reduced cost and eliminates post weld treatment and hydrogen induced cracking. As it is possible to recycle 100% of HPS steel bridges, it is also environment friendly.

HPS steels have the ability to perform without painting under normal atmospheric conditions; they have slightly better atmospheric corrosion resistance than the conventional grade 345W or 485W steels. HPS 485 W is produced by quenching and tempering or thermal-mechanical controlled processing (TMCP). It has been found that HPS 485W remains fully ductile even at lower temperatures, whereas conventional 345W steel begins to show brittle behaviour.

The high-performance steel bridge in Tennessee, USA, is a two-span continuous structure and provides 8.5-m wide roadway over two spans, both 72-m long. This bridge consists of three continuous welded plate girders, fabricated from HPS-485W, 2-m deep, and spaced at 3.2-m centres. These plate girders act compositely with a cast-in-place concrete dock slab of thickness 212.5 mm. The bridge is jointless, having integral, pile supported abutments. The design was fully optimized for the 485 MPa steel, using the AASHTO Load and Resistance Factor (LRFD) Bridge Design Specifications.

*Source*: www.fhwa.dot.gov

The weight of steel is about 25% less compared to the original grade 345W design. Because HPS-485W costs slightly more than grade 345W steel, this resulted in a 16% reduction in the total cost to fabricate and erect steel for this bridge.

Nebraska Department of Transportation was the first to use HPS 485W in the design and construction of the 45-m simple span Snyder bridge, a welded plate girder (1.37 m) bridge, which opened to traffic in October 1997. HPS is currently produced by Bethlehem–Lukens plate, Oregon Steel Mills, and U.S. Steel. Since the late 1990s more than 250 HPS steel bridges have been opened to traffic and 150 more bridges are under various stages of construction in the USA alone.

Though HPS is considered superior to normal mild steel, the following barriers have also been identified.

1. High yield-to-tensile stress ratios (less ductile behaviour, which is not favourable in earthquake zones),
2. Cost premium for base material, and
3. Currently, the AASHTO Manual for Design of Steel Bridges includes certain limitations that prevent full utilization of the higher yield strength of HPS (Jamshidi et al., 1997; Lwin, 2002).

## 13.16.1 High-strength Low-alloy (HSLA) Steel

HSLA steel is a type of alloy steel that provides better mechanical properties and greater resistance to corrosion than carbon steel. HSLA steels are made to meet specific mechanical properties and to have some specific chemical composition. They have carbon content between 0.05 and 0.25% to obtain formability and weldability, and manganese content up to 2.0%. Small quantities of chromium, copper, molybdenum, nickel, niobium, nitrogen, titanium, vanadium, and zirconium are also added. These elements are intended to alter the ferrite-pearlite microstructure of carbon steels. Special processing techniques such as controlled rolling and accelerated cooling methods are also employed in the manufacture. The yield strength of HSLA steel is in the range of 275–590 MPa, depending on its constituents. Owing to their higher strength and toughness, HSLA steels usually require 25 to 30% more power to form, as compared to carbon steels.

Depending upon the alloying elements used, HSLA steels are divided into the following six categories (www.asminternational.org):

1. Weathering steels (see Section 13.14.1)
2. Micro-alloyed ferrite-pearlite steels
3. As-rolled pearlitic steels
4. Acicular ferrite (low-carbon bainite) steels
5. Dual-phase steels
6. Inclusion-shape-controlled steels

They are covered by several ASTM specifications, for example, ASTM A242, A572, A 588, and A 656.

They are used in oil and gas pipelines, cars, heavy-duty highway vehicles such as trucks, construction and farm machinery such as cranes, industrial equipment, storage tanks, bridges, offshore structures, power transmission towers, light poles, roller coasters, and other structures where high stresses occur or higher strength-to-weight ratio is required. Use of HSLA steel may result in 20 to 30% lighter sections than mild steel with the same strength.

## 13.17 Storage and Handling of Steel

Structural steel of different classifications, sizes, and lengths should be stored separately. It should be stored above ground level by at least 150 mm on platforms or any other suitable supports to avoid distortion of sections. In coastal areas or in the case of considerable storage periods, suitable protective coating of primer (red oxide) paint should be given to prevent scaling and rusting.

As per IS 4082 :1996, steel reinforcement should be stored properly to avoid corrosion and distortion, by stacking them above ground level by at least 150 mm. If they are to be stored for a longer time or in coastal areas, some roof covering should be provided to keep off the rain. It is not a good practice to cover steel with a cement wash, as the cement grout dries quickly and gets scaled off as powder. Slight rusting on the surface of steel bars is not detrimental, but scales of rust must be cleaned properly before using them in reinforced concrete. It is preferable to store bars of different classification, sizes, and lengths separately to facilitate later issues in such sizes and lengths, and also to minimize wastage in cutting from standard lengths. For each classification of steel, separate areas should be earmarked. It is desirable to paint the ends of bars and sections of each class by distinct, separate colours.

All reinforcements used in reinforced concrete construction should be free from paint, oil, grease, loose rust, and loose mild scale, as they will reduce the bond between concrete and steel. Oil may be removed by thoroughly washing with petrol; before using such treated steel, it should be allowed to dry and brushed with a wire brush. The rods when bent into hooks should not crack or split, as they indicate a brittle material.

Any doubt in the quality of steel being used should be checked with the test certificate from the manufacturer or from any approved laboratory. The report should contain data regarding yield strength, ultimate tensile strength, and percentage elongation at failure.

Results of bend test, as per IS 1599:2012, may also be useful.

## 13.18 Advantages and Disadvantages of Steel

As mentioned in the introduction, structural steel offers several advantages over other competing materials. These advantages are listed below:

*High strength* The superior strength-to-weight ratio of steel results in structures bearing higher loads using less material. This also results in smaller foundations. Less material use and less transportation can also lower building costs. The stiffness of steel allows it to span greater distances and hence provides more design freedom than other materials. If bucking is prevented, steel has equal strength in compression and tension.

*High ductility*   Properly designed and detailed steel, when highly overloaded (for example, due to earthquakes), will not fail but will deflect abnormally to provide ample warning due to the ductile nature of steel. This aspect is very important for the safety of occupants.

*Uniformity*   The quality of steel-intensive construction is invariably superior, when compared to other materials. Moreover, the properties of steel do not change appreciably with time as do those of reinforced concrete.

*Environment friendly*   Steel is amenable for prefabrication and steel structures can be demounted. Steel components can be reused/recycled many numbers of times; even steel waste can be recycled (see Section 13.19).

*Versatile*   Structural steel offers the ability to fasten different members together by simple connection techniques such as welding, bolting, and riveting. Steel members can also be rolled into a wide variety of sizes and shapes, as described in Section 13.12.

*Prefabrication*   Often, steel components are manufactured at the factory (which means that they are produced using strict supervision and quality control), transported to site, and erected using bolting and minimum amount of welding. Prefabrication of steel structures results in the proper planning of construction, saving in time and money, speedy erection, and better quality of finished structures. Lighter steel members facilitate easy handling and erection.

*Permanence*   Steel frames that are properly maintained will last indefinitely. Several structures are available to testify the long-term performance of steel structures (e.g., Eiffel Tower and the Railway Bridge across Firth of Forth both built in 1890). The use of weathering steels under certain conditions does not require any painting or maintenance.

*Additions to existing structures*   The repair and retrofit of steel members and their strengthening at a future date (for example, to take into account enhanced loading) is simpler than in concrete members. Thus, new bays or even entire new wings, can be added to existing steel-frame buildings, and steel bridges may often be widened. Of course, special precautions have to be taken while welding on a member already carrying loads.

*Elasticity*   Steel behaves closer to design assumptions than most materials because it follows Hooke's law up to fairly high stresses. The moments of inertia of steel members can be calculated accurately while the values obtained for a reinforced-concrete members are not accurate and depend on the extent of cracking (Subramanian, 2013).

Though steel has several advantages, it also has the following disadvantages:

*Maintenance costs*   Most steels are susceptible to corrosion when freely exposed to air and water and must therefore be periodically painted. However, weathering steels do not need painting.

*Fireproofing costs*   Members of steel frames may have to be protected by other materials, with certain insulating characteristics, for better fireproofing.

*Susceptibility to buckling*   The longer and more slender the compression members, the greater is the danger of buckling. Though steel has a high strength per unit of weight, steel columns have to be stiffened against buckling, hence resulting in uneconomical solutions sometimes.

***Fatigue*** Steel members may fail if they are subjected to a large number of stress reversals or several variations of tensile stress.

## 13.19 Steel and Sustainability

Structural steel is 100% recyclable but accounts for 8% of $CO_2$ emissions worldwide (About 1.8 tonne of $CO_2$ is emitted per tonne of steel produced). The metallurgical properties of steel allow it to be recycled continually with no degradation in performance and from one product to another. Over 500 MT of steel is recycled annually worldwide, which represents 50% of all steel produced. In North America alone, each year more steel is recycled than aluminium, paper, glass, and plastic combined (www.recycle-steel.org). Recycling 1 tonne of steel saves about 1130 kg of iron ore, 635 kg of coal, and 55 kg of limestone. Steel structures can be readily disassembled at the end of their useful lives. This has many environmental and economic advantages. It means that the steel components can be reused in future structures without the need for recycling; resulting in saving of energy and avoidance of $CO_2$ emitted from the steel production processes. The R&D efforts by the steel industry has resulted in sustainable manufacturing processes. It decreased energy use by 40%, carbon footprint by 35%, and increased productivity by a factor of 24, since 1990 (www.recycle-steel.org). Using advanced technologies, steel plants in areas of water scarcity are able to recycle and reuse about 98% of their water. While making steel, more than 400 MT of iron and steel slags are produced every year – 80% of steel-making slag and almost 100% of iron-making slag are recovered and used as construction aggregates, admixture in cement, and in roofing and mineral wool production.

Use of epoxy coated steel bars will result in the following LEED Credits: recycled materials MR 4.1 and MR 4.2; regional materials MR 5.1 and MR 5.2 (Subramanian & Mota, 2016). It has been shown that by using steel as a material, it is possible to get 10 LEED points, under the categories of building reuse, construction waste management, resource reuse, and local/regional materials (MSC, 2003). By using innovation in the design process, four more points could be achieved. Examples of this are as follows:

1. The Greater London Authority building, designed by architect Foster and Partners, which used HSS structural members in the atrium and allowed hot water to run through these members, creating a giant radiator.
2. German engineer Werner Sobek, who designed his residence at Stuttgart in such a way that the entire steel structure can be disassembled and reused.
3. The Utah Olympic Oval used an innovative cable suspension system to support a very shallow steel truss roof, reducing the total arena volume by 73,624 m³ (making it easier to heat and cool) and weight by 860 tonnes, against competing designs.

## SUMMARY

- Basic oxygen making process is used in the steel making.
- Different heat treatment process of steel, such as annealing, normalization, quenching and tempering, stress relieving, and case hardening are used to improve the properties of steel.
- Steel can be alloyed with other materials to improve its properties and weldability. Depending upon the carbon content, the steel is designated as low-carbon steel, medium-carbon steel, and high-carbon steel.
- Several types of steels are produced in India, by changing its chemical composition and alloying elements. Special steels such as stainless steel, weathering steel, high-performance steel, and cold formed steel are used in special applications.
- The important properties of structural steel are its ultimate tensile and yield strength, ductility, toughness, hardness, and weldability. The yield stress, as measured by the tension coupons, is affected by several factors such as the rate of loading and the position from which the test coupons are taken. Ductility and toughness are important when a steel structure is subjected to earthquake or impact loads.

- Variation in material properties can be incorporated in the design by the concept of characteristic strength.
- The CVN impact toughness, the quality of design, detailing, and fabrication are important to prevent brittle fracture. Severe stress raisers should be avoided to achieve smooth stress flow.
- In crane supporting structures, bridges, and structures supporting rotating machinery, fatigue may be an important consideration.
- A variety of structural steel products are available which include rolled steel sections, welded and hybrid sections, hollow steel sections, light-gauge cold-formed sections, steel wires and ropes, steel fasteners, etc. Similarly, the various types of steel rebars are used in RC structures such as HR mild steel bars, Tor Steel bars, TMT bars, and high-tensile wires/bars (used in prestressed concrete).
- Weathering steel or stainless steel, containing 10.5% or more of chromium and nickel, could be used to mitigate corrosion. Proper detailing should be adopted to avoid water entrapment. Structural steel sections may also be prevented from corrosion by a zinc- or aluminium-based primary coat and two or three layers of finishing coats of zinc phosphate modified alkyd or zinc-rich epoxy paints. It is necessary to blast clean the surface, before applying paints.
- Though the cost of high-strength steel may be 10–20% higher, it may result in reduction in steel weight and subsequent foundation cost.

# EXERCISES

## Multiple-choice Questions

1. Unit mass of steel is
   (a) 2400 kg/m³
   (b) 6850 kg/m³
   (c) 7500 kg/m³
   (d) 7850 kg/m³
2. Young's modulus of steel is
   (a) 2 2000 MPa
   (b) $2 \times 10^5$ MPa
   (c) $2 \times 10^4$ MPa
   (d) $2 \times 10^6$ MPa
3. Poisson's ratio of steel in the elastic range is
   (a) 0.25
   (b) 0.22
   (c) 0.30
   (d) none of these
4. The oldest steel-making process is
   (a) basic oxygen steel making
   (b) electric arc method
   (c) open-hearth process
   (d) none of these
5. Structural steel (low-carbon steel) normally contains carbon in the range of
   (a) 0.4 to 1.2%
   (b) 0.02 to 0.08%
   (c) 0.10 to 0.25%
   (d) 0.5 to 1.0%
6. To produce one tonne of steel, the following materials are required
   (a) 1.5 tonne of iron ore, 1.25 tonne of coal, 1.5 tonne of limestone, and 1.5 tonne of other materials
   (b) 1.25 tonne of iron ore, 1.50 tonne of coal, 1.0 tonne of limestone, and 1.0 tonne of other materials
   (c) 1.75 tonne of iron ore, 1.25 tonne of coal, 0.75 tonne of limestone, and 0.5 tonne of other materials
   (d) 1.75 tonne of iron ore, 1.35 tonne of coal, 0.5 tonne of limestone, and 0.5 tonne of other materials

7. Up to the following temperature iron has a bcc crystalline structure and remains as ferrite
   (a) 820°C
   (b) 910°C
   (c) 950°C
   (d) 1000°C
8. Carbon equivalent is calculated using
   (a) $C + \dfrac{Mn}{5} + \dfrac{Cr + Mo + V}{10} + \dfrac{Ni + Cu}{15}$
   (b) $C + \dfrac{Mn}{6} + \dfrac{Cr + Mo + V}{5} + \dfrac{Ni + Cu}{15}$
   (c) $C + \dfrac{Mn}{10} + \dfrac{Cr + Mo + V}{6} + \dfrac{Ni + Cu}{15}$
   (d) None of these
9. Coefficient of thermal expansion of steel is
   (a) $11 \times 10^{-6}/°C$
   (b) $12 \times 10^{-6}/°C$
   (c) $13 \times 10^{-6}/°C$
   (d) None of these
10. The magnitude of tensile and compressive residual stresses in rolled sections may reach
    (a) $0.2 f_y$ and $0.8 f_y$
    (b) $0.4 f_y$ and $0.5 f_y$
    (c) $0.3 f_y$ and $0.5 f_y$
    (d) $0.3 f_y$ and $0.6 f_y$
11. Which of the following is a false statement?
    (a) Use of CR sections results in saving of weight than HR section.
    (b) CR sections are prone for local buckling.
    (c) CR sections offer a wide variety of shapes.
    (d) The maximum thickness of CR section may be 20 mm.
12. As per IS 13920:2016, which of the following should not be used in earthquake zones?
    (a) Bars of grade Fe 500 and below
    (b) Bars of grade Fe 550 and below
    (c) Bars of grade Fe 600 and above
    (d) None of these

13. Which of the following is more corrosion resistant?
    (a) Medium tensile steel bars
    (b) HYSD bars
    (c) TMT bars
    (d) TMT CRS Bars
14. Which of the following is not made of steel?
    (a) Fusion-bonded epoxy coated bar
    (b) Galvanized bars
    (c) FRP bars
    (d) Basalt bars
15. Stainless steel has chromium to the extent of
    (a) less than 5.75%
    (b) greater than 10.5%

(c) about 8%
(d) greater than 15%
16. State which of the following is a false statement.
    (a) Steel is an environment friendly material.
    (b) Steel is amenable for prefabrication.
    (c) Steel offers high ductility.
    (d) Steel is highly resistant to fire.
17. Match the code on right with the material on left.
    (a) Cold formed steel        (i) IS 280
    (b) Bolts                    (ii) IS 806
    (c) Tubes                    (iii) IS 801
    (d) Mild steel wires         (iv) IS 1367

## Review Questions

1. Describe the production of structural steel in integrated steel plants.
2. What is the difference between killed and semi-killed steels?
3. What are the differences in the production of steel in integrated steel plants and in mini steel plants?
4. What is the difference between hot rolling and cold rolling?
5. What are the three different types of crystal structures or lattices found in metals?
6. Describe the various crystalline transformations that take place when pure iron is heated to its melting point.
7. Describe the phase diagram of steel.
8. What are the heat treatments that are employed to improve the properties of steel?
9. How does isothermal annealing differ from ordinary annealing?
10. How is normalizing different for annealing?
11. Describe quenching and tempering processes.
12. What is stress relieving and why is it necessary?
13. What is case hardening? On what components is it required?
14. How does carbon content affect the properties of steel?
15. Name at least five alloying elements added to steel to modify its properties.
16. What is the relation between weldability and carbon equivalent? Up to what value of carbon equivalent, steel is generally weldable?
17. List at least four types of structural steel.
18. List the important mechanical properties of steel along with the factors that influence them.
19. Give the values of yield strength, young's modulus, coefficient of thermal expansion, and ultimate tensile strength for mild steel as per IS 800:2007.
20. What are the parameters that influence the yield stress of steel?
21. How is yield strength determined in high-strength steel?
22. What is meant by characteristic strength?
23. What is meant by ductility? Why and where is it considered important?
24. What circumstances lead to brittle failure of steel?
25. List the methods by which brittle fracture may be controlled.
26. Write short notes on
    (a) Lameller testing
    (b) High temperature effects on steel
    (c) Hardness
    (d) Fatigue resistance
27. Sketch the residual stress in a typical rolled I-beam, welded I-beam, and welded box section. How are residual stresses induced in steel sections?
28. How does residual stress affect the design of compression members?
29. What is meant by recrystallization in metals?
30. What are the different hot working processes?

31. Compared to cold working, what are the advantages of hot working?
32. Compared to hot working, what are the advantages of cold working?
33. List at least five HR steel sections used in practice.
34. Give three types of structures where HR angle sections are used.
35. What are the uses of I- or H-sections?
36. Where are steel sheets used?
37. State the advantages of using wide flange beams over narrow ISMB beams.
38. How is tapered beam produced? What is its advantage?
39. What are castellated beams and how are they produced? What is their advantage?
40. Give three types of structures where hollow sections are used. What is the main advantage of hollow sections over rolled I or channel sections?
41. Write short notes on steel hollow sections and cold-formed steel sections.
42. How are steel wires produced? What are the advantages and drawbacks of cold drawing of wires? How is bend test performed on wires?
43. Describe the possible ways in which wire ropes are made.
44. What are the different types of bolts that are used in steel structures?
45. What are the different types of steel reinforcement bars that can be used in RC construction?
46. What is the different between cold twisted deformed bars (CTD bars) and thermo-mechanically treated reinforcement bars (TMT Bars)?
47. How are TMT bars manufactured?
48. What is the difference between Fe 415, Fe 415D, and Fe 415S bars?
49. Draw the stress-strain curve for mild steel bars and HYSD bars.
50. When are welded splices and rebar couplers used?
51. What are welded wire fabrics (WWF)? How are they made?
52. What type of steel is used in prestressed concrete? Can the steel used in prestressed concrete be welded?
53. When does corrosion of rebars take place? What are the different methods adopted to mitigate corrosion of rebars?
54. What are the corrosion mitigation methods of prestressing steel?
55. What are the methods of preventing corrosion in structural steel?
56. Sketch some method of detailing to prevent water entrapment.
57. What technique has to be adopted, for the painting to be effective on steel structures?
58. How does weathering steel differ from mild steel? Is it necessary to paint weathering steel?
59. What are the advantages of stainless steel? What is the main difference in chemical composition between stainless steel and normal mild steel? Is stainless steel weldable?
60. What are the characteristics of high-performance steel?
61. What is high-strength low-alloy steel (HSLA)?
62. How should we store (a) reinforcing bars and (b) structural steel?
63. State the main advantages and a few drawbacks of using steel as a structural material.
64. Why is steel considered a sustainable material?

## ANSWERS

### Multiple-choice Questions

| | | | | | |
|---|---|---|---|---|---|
| **1.** (d) | **2.** (b) | **3.** (c) | **4.** (c) | **5.** (c) | **6.** (d) |
| **7.** (b) | **8.** (b) | **9.** (b) | **10.** (c) | **11.** (d) | **12.** (c) |
| **13.** (d) | **14.** (c) & (d) | **15.** (b) | **16.** (d) | **17.** (a)-(iii), (b)-(iv), (c)-(ii), (d)-(i) | |

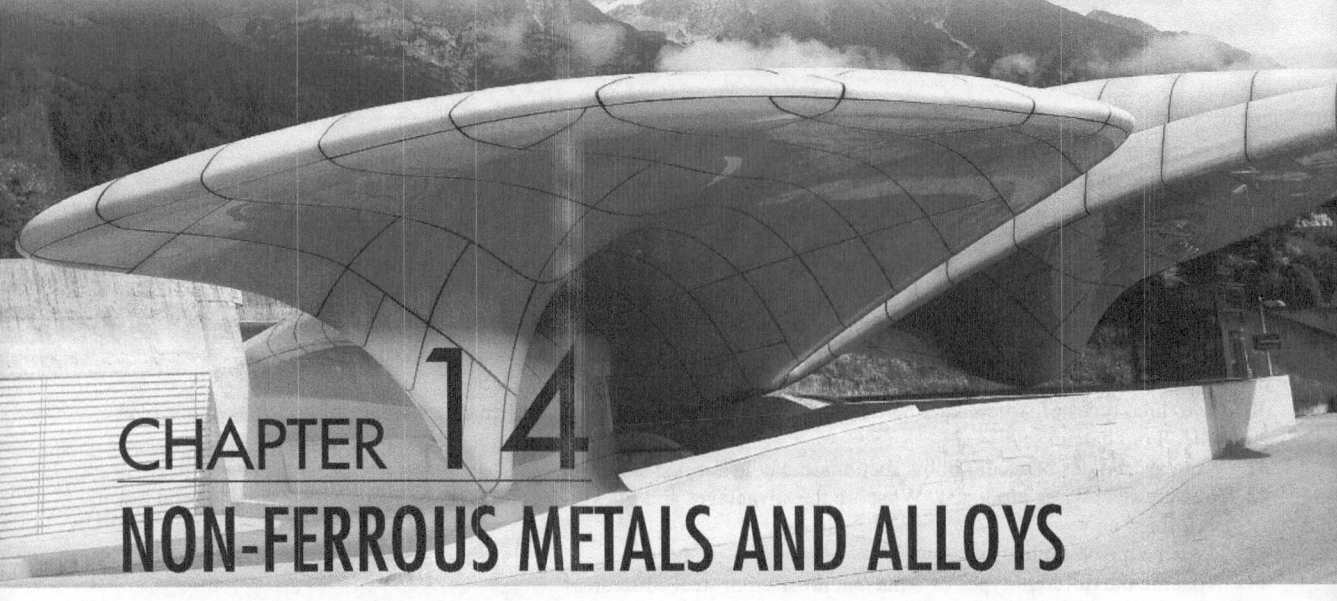

# CHAPTER 14
# NON-FERROUS METALS AND ALLOYS

## 14.1 Introduction

No metal is used as widely as steel for construction purposes, but there are many other non-ferrous metals, such as aluminium, copper, zinc, lead, chromium, nickel, magnesium, and their alloys, which are used in a variety of applications. Aluminium is the most abundant metal in the earth's crust. Owing to its unique combination of being lightweight and having relatively high strength, aluminium is used extensively. However, it is highly energy intensive and relatively expensive. Copper has long been popular in architectural works due to its ease in working, aesthetic appearance, ductility, and permanency. Alloys of copper, such as brass and bronze, are also used as fittings and ornaments. Lead, which is injurious to heath, is used for its properties of corrosion resistance and ornamental purposes. Zinc, chromium, and nickel are mainly used as alloying elements for various purposes, or as coatings to improve the durability of metals. Magnesium alloys have strengths equal to that of mild steels and can be cast, forged, rolled into plate, strips, and structural shapes. These non-ferrous metals can be made into sheets.

## 14.2 Aluminium and its Alloys

*Aluminium* (Al) is a shiny, silvery, and light metal (called aluminum in the USA). By mass, aluminium makes up about 8% of the Earth's crust. It is the third most abundant element after oxygen (47%) and silicon (28%). Owing to its versatility, aluminium is the second widely used metal after steel. It conducts heat and electricity and reflects heat and light. It is strong but easily workable, and retains its strength even under extreme cold without becoming brittle. The surface of aluminium quickly oxidizes to form an invisible barrier to corrosion. Furthermore, aluminium can be easily and economically recycled into new products. It has been discovered that the strong vessels made by Persian potters around 5000 B.C. from clay contained aluminium oxide. Ancient Egyptians and Babylonians used aluminium compounds in fabric dyes, cosmetics, and medicines. Global resources of bauxite ore are estimated to be between 55 and 75 BT. Primary aluminium production in 2016 was about 57.6 MT, with China producing a major share of 31.6 MT; about 31% of this production is from scrap metal (www.world-aluminium.org).

### 14.2.1 Manufacture of Aluminium

Unlike metals like silver and gold, aluminium never occurs alone in nature. Instead, it is available as compounds of other elements. To get the pure metal, these compounds must be broken down by an industrial process called refining, which requires a huge amount of energy. In 1886, two scientists independently developed a smelting process that made the mass production of aluminium economically feasible. Known as the *Hall-Héroult process* after its American and French inventors, this process is still used for the production of aluminium. An Austrian chemist, Carl Josef Bayer, also developed the *Bayer process* in 1887, for refining aluminium ore.

**The Bayer process**  Pure aluminium is obtained from the ore bauxite ($Al_2O_3 \cdot 2H_2O$), which consists of 40–60% aluminium oxide, along with other compounds, such as silica ($SiO_2$) and titanium oxide ($TiO_2$). Aluminium manufacture is accomplished in two phases: (1) the *Bayer process* of refining the bauxite ore to obtain aluminium oxide, and (2) the Hall-Héroult process of smelting the aluminium oxide to obtain pure aluminium. The *Bayer process* is carried out in four steps, as shown in Fig. 14.1(a) (www.aluminum.org).

1. First, the bauxite ore is crushed and mixed with sodium hydroxide (caustic soda, NaOH) and water ($H_2O$). This slurry is heated to about 110 to 270°C under a pressure of 340 kPa in pressure vessels, to get a boiling hot solution of sodium aluminate [$NaAl(OH)_4$].
2. This solution is pumped into settling tanks, where the impurities (called 'red mud' and do not dissolve in caustic soda) settle at the bottom of the tanks are removed and properly discarded. The finer impurities are also filtered during this step.
3. The filtered solution is made to cool in about 18-m tall precipitation tanks, where very small quantities of aluminium hydroxide are added as seeds. These stimulate the precipitation of solid aluminium hydroxide crystals ($Al.OH_3$), at the bottom of the tank, and are removed.
4. These crystals are washed to get rid of any caustic soda and heated in rotating cylindrical kilns to 960–1000°C to remove water molecules and to obtain fine white powder of alumina (aluminium oxide).

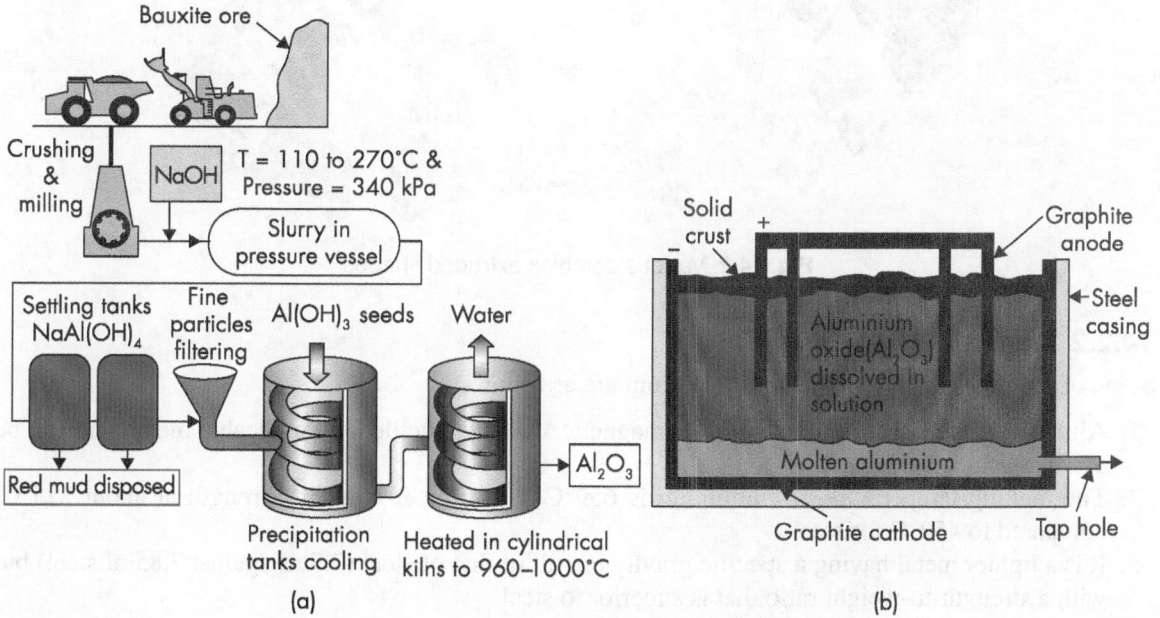

**Fig. 14.1** The Bayer and the Hall-Héroult refining processes
(a) The Bayer process (b) The Hall-Héroult refining process

**The Hall-Héroult process**  It is used to refine the alumina ($Al_2O_3$) to aluminium, using electrolysis. The difficulty of separating aluminium from oxygen in the oxide ore is overcome by the use of cryolite

(AlNa$_3$F$_6$) as a solvent (or flux) to dissolve the oxide minerals (The naturally occurring cryolite, which was mined from Greenland, became extinct in 1987; thus, synthetic sodium aluminium fluoride is used now.). Thus, the alumina dissolved in cryolite and heated to a temperature of 940–980°C is used as the electrolyte (current-conducting medium) in the smelting operation. The lining of the bath consists of the graphite (soft carbon) cathode [see Fig. 14.1(b)]. When electricity is passed into the electrolyte, the positively charged aluminium particles (called *ions*) move towards the negatively charged lining (cathode) of the bath. A solid crust forms at the top of the bath, and the pure molten aluminium, accumulated at the bottom, is removed and cast (poured into moulds) as ingots. Both the smelting and casting processes are continuous.

The world's largest smelters of alumina are located in China, Russia, and Canada (Quebec and British Columbia). Aluminium production involves large amounts of electrical energy. Production of 1 tonne of aluminium requires 14,000 kWh of energy. However, only 5% of this energy is required to recycle 1 tonne of aluminium. The cost of electricity is about one-third of the cost of smelting aluminium.

Since aluminium is soft, ductile, and malleable, its sections are manufactured by extrusion (*Extrusions* are produced by pushing solid material through a 'die' opening to form members with complex cross sections.). Even complicated sections are extruded with very little tolerance (see Fig. 14.2). These sections are used for sliding windows, doors, and façades of tall buildings.

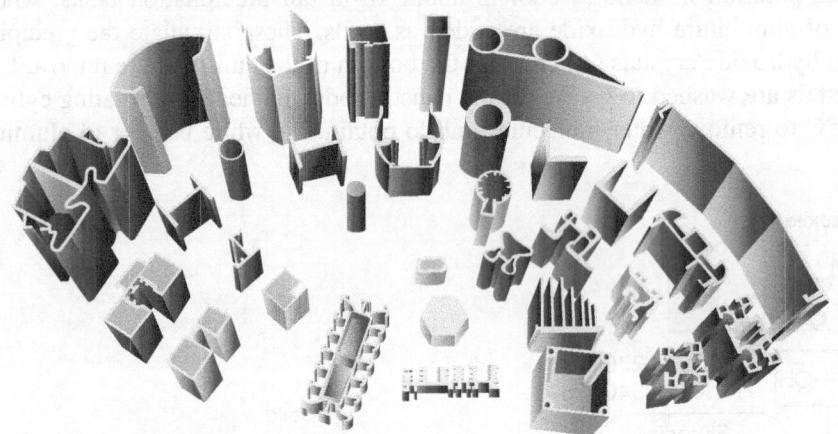

**Fig. 14.2** Various complex extruded shapes

## 14.2.2 Properties of Aluminium

Some of the important properties of aluminium are as follows:

1. Aluminium is a silvery-white, soft, nonmagnetic, durable, ductile, and malleable metal, and can be easily machined and cast.
2. The melting temperature of aluminium is 659°C. However, it loses its strength at about 225°C, compared to 450°C for steel.
3. It is a lighter metal having a specific gravity one-third that of steel (2.7 as against 7.85 of steel) but with a strength-to-weight ratio that is superior to steel.
4. Its modulus of elasticity is also about one-third that of steel (62–73 GPa). Owing to this, the deflection an aluminium structure will be thrice that of a similar structure made of steel.
5. The coefficient of thermal expansion of aluminium is double that of steel (about $23 \times 10^{-6}$/°C).
6. It has good heat and light reflectivity.

7. It is a good thermal and electrical conductor.
8. It has remarkable corrosion resistance. When aluminium sections are exposed to the atmosphere, a thin layer of white aluminium oxide ($Al_2O_3$) is formed on the surface, effectively preventing further oxidation. This phenomenon is called *passivation*. However, some stronger aluminium-copper alloys have less corrosion resistance than pure aluminium.
9. It does not have a well-defined yield point and hence it is taken as the 0.2% proof stress. The yield strength is in the range of 35 MPa, whereas aluminium alloys may have yield strengths in the range of 115 to 505 MPa.
10. One major disadvantage of pure aluminium is its softness. While this makes it easy to shape, it can be a disadvantage where strength is required.
11. It has a tensile strength of around 89.6 MPa, which limits its usefulness in structural applications. However, by cold-working aluminium, its strength can be approximately doubled (Kissell and Ferry, 2002). Also, its strength can be increased by adding alloying metals, such as manganese, silicon, copper, magnesium, and zinc. Further, many aluminium alloys are strengthened by heat treatment. Some heat-treated aluminium alloys may have tensile strengths exceeding 680 MPa.

Because of the greater strength-to-unit weight ratio compared to steel, aluminium alloys are being increasingly used for structural purposes.

It has to be noted that hydrochloric acid will dissolve aluminium and a strong alkali like sodium hydroxide also reacts with aluminium resulting in sodium aluminate and hydrogen. Hence, acidic fruit juices should not be stored in aluminium containers and alkaline detergents should not be used to clean aluminium surfaces.

The physical and mechanical properties of most of the important non-ferrous metals are compared in Tables 14.1 and 14.2, respectively (For comparison, the properties of steel are also given.).

**Table 14.1** Physical properties of metals

| Base metal or alloy | Specific gravity | Melting point, °C | Boiling point, °C | Coefficient of linear expansion × $10^{-6}$ per °C | Specific heat (cal/g per °C) | Relative electrical conductivity (%) (Copper = 100%) | Relative thermal conductivity (Copper = 1) |
|---|---|---|---|---|---|---|---|
| Aluminium | 2.70 | 659 | 2470 | 23.6 | 0.22 | 59 | 0.52 |
| Brass, navy | 8.60 | 900 | NA | 21.2 | 0.09 | 28.0 | 0.28 |
| Bronze (90 Cu-10 Zn) | 8.78 | 950–1000 | NA | 18.4 | 0.09 | 11.0 | 0.12 |
| Copper (de-oxidized) | 8.89 | 1081 | 2600 | 17.6 | 0.095 | 100 | 1.00 |
| Lead | 11.34 | 328 | 1740 | 29.5 | 0.03 | 8.0 | 0.08 |
| Magnesium | 1.74 | 650 | 1100 | 25.7 | 0.246 | 37.0 | 0.40 |
| Nickel | 8.80 | 1452 | 3000 | 13.3 | 0.105 | 23.0 | 0.16 |
| Tin | 7.29 | 232 | 2270 | 23.0 | 0.125 | 13.5 | 0.15 |
| Zinc | 7.13 | 419 | 907 | 39.8 | 0.093 | 30 | 0.27 |
| Steel, low carbon | 7.85 | 1530 | 2470 | 12.1 | 0.118 | 14.5 | 0.17 |

**Table 14.2** Mechanical properties of metals

| Base metal or alloy | Yield strength, MPa | Tensile strength, MPa | Elongation,% in 50 mm | Hardness (BHN) | World resources |
|---|---|---|---|---|---|
| Aluminium and alloys | 85–505 | 100–570 | 6–42 | 21–140 | 55 to 75 BT |
| Brass, navy | 206.8 | 427.4 | 47 | 89 | – |
| Bronze (90 Cu-10 Sn) | 103.4 | 275.8 | 52 | 119 | – |
| Copper (deoxidized) | 68.9 | 227.5 | 40 | 30 | 2.1–3.5 BT |
| Lead | 5.5 | 13 | 45 | 6 | 1.5 BT |
| Magnesium | 89.6 | 172.4 | 4 | 40 | 8.6 BT |
| Nickel | 58.6 | 317.1 | 40 | 85 | 130 MT |
| Tin | 11.8 | 21.6 | 50 | 5.3 | 11.7 MT |
| Zinc | 124.1 | 172.3 | 20 | 38 | 1.9 BT |
| Steel, low carbon | 250 | 415 | 20–35 | 150–190 | – |

## 14.2.3 Aluminium Alloys

Aluminium is most commonly alloyed with copper, zinc, magnesium, silicon, manganese, and lithium. Small additions of chromium, titanium, zirconium, lead, bismuth, and nickel are also made and iron is invariably present in small quantities.

**Designation of aluminium and its alloys**   There are over 300 wrought alloys with 50 in common use (see IS 733 to IS 740, IS 1284, and IS 1285). They are normally identified by a four- or five-digit system that originated in the USA and is now universally accepted. Table 14.3 describes the five-digit system adopted by IS 6051:1970 for wrought alloys (also see IS 5052:1993). Wrought alloys have less than 4% of alloying elements, whereas cast alloys have more than 4%. Cast alloys have similar designations and use a four digit system. Primary Ingots (IS 2590, IS 4026), and Castings (IS 617) also use a four-digit system.

**Table 14.3** Designation of wrought aluminium alloys

| Alloying element | Wrought alloy | Strength | Corrosion resistance |
|---|---|---|---|
| Un-alloyed Aluminium | 1xxxx | Fair | Excellent |
| Copper (Cu) | 2xxxx | High | Fair |
| Manganese (Mn) | 3xxxx | Fair | Good |
| Silicon (Si) | 4xxxx | Good | Good |
| Magnesium (Mg) | 5xxxx | Good | Good |
| Magnesium + Silicon | 6xxxx | Good | Good |
| Zinc (Zn) | 7xxxx | High | Fair |
| Other Elements [such as, Iron (Fe), Nickel (Ni), Titanium (Ti), Chromium (Cr), Lead (Ld), Bismuth (Bi)] | 8xxxx | – | – |
| Unassigned | 9xxxx | – | – |

The first digit identifies the major alloying element. The second digit indicates rounded off mean value of the percentage of the major alloying element, except for group 4 containing silicon, when the digit refers to the mean percentage halved and rounded off; and for group 6 containing magnesium and silicon, the second digit refers to five times the mean magnesium percentage rounded off. The third, fourth, and fifth digits identify the minor alloying elements in the descending order of their percentage or in case of same alloy percentage in the serial order [Except for group 6, in which case, the third digit refers to either magnesium or silicon which is in the excess of that required for magnesium silicate (i.e., 1.7, which is the ratio of Mg to Si)]. For balanced compositions, the third digit will be zero. In the case of high purity aluminium base alloy, the fifth digit is 1. Examples of this designation are shown Table 14.4. More examples are provided in IS 6051.

**Table 14.4** Examples of the designation system of IS 6051

| Alloying elements and their mean percentages | Designation as per IS 6051 |
| --- | --- |
| Cu 1.5, Si 1.0, Mg 0.85 | 22450 |
| Mn 1.2 | 31000 |
| Si 5.2 | 43000 |
| Mg 4.4, Mn 0.75 | 54300 |
| Mg 0.65, Si 0.50 | 63400 |
| Mg 0.65, Si 0.50 (EC grade) | 63401 |
| Mg 0.95, Si 0.95 | 65400 |

The 1xxxx series is pure aluminium that has excellent corrosion resistance, excellent workability, as well as high thermal and electrical conductivity, but has less strength. Hence, the 1xxxx series is commonly used in electrical power grid lines and chemical industries. Only 2xxxx, 6xxxx, and 7xxxx alloys are strengthened by heat-treatment and then by quenching (rapid cooling). Non-heat treated alloys, such as 3xxxx, 4xxxx, and 5xxxx, are strengthened through cold-working.

In the 2xxxx series, copper is the principal alloy and provides high strength when properly heat-treated, but has poor corrosion resistance. They are used in the aircraft industry. In the 3xxxx series, manganese is the major alloying element and limited to about 1.5%. These alloys have moderate strength and easily worked and used in applications, such as gutters, sidings, heat exchangers, aluminium beverage cans, and cooking utensils. Silicon is the major alloying element in the 4xxxx series and used for *brazing* alloys and welding electrodes (*Brazing* is a technique employed while joining two dissimilar metals- by heating and melting filler [alloy] that bonds the two pieces of metal and joins them). Magnesium is the primary alloying agent in the 5xxxx series and has medium strength, good welding characteristics, and corrosion resistance. It is one of the most effective and widely used alloying elements for aluminium and used in boats and docks. The 6xxxx series, containing silicon and magnesium, has medium strength and good corrosion resistance. Extrusion products from the 6xxxx series are used for architectural and structural applications. Zinc is the major alloying element in the 7xxxx series and is widely used in the aircraft industry.

**Temper designation system**   Tempering of aluminium increases its strength but reduces ductility. To be able to select a suitable aluminium alloy, it is essential to be familiar with the designation system used for their characteristic stages or tempers. As per IS 5052:1993, they are designated as follows:

*F—Fabricated*   This applies to wrought products without any special treatment.

*O—Annealed*   This applies to wrought products, which are fully annealed to obtain the lowest strength condition and cast products, which are annealed to improve ductility and dimensional stability.

*H—Strain hardened* This applies to wrought products whose strength is increased by strain hardening with or without supplementary treatment. The letter H is always followed by one or more digits. The first digit indicates the specific combination of basic operations as follows:

**H-l** Strain-hardened only

**H-2** Strain-hardened and then partially annealed

**H-3** Strain-hardened and then stabilized (generally applies to aluminium-magnesium alloys)

The digit following H-1, H-2, and H-3 indicates the final degree of strain hardening. Tempers between 0 (annealed) and 8 (full hardened) are designated by numbers 1 through 7 (see IS 5052, for more discussions).

*M—Manufactured* This applies to the cast products which acquired some temper from hot shaping processes for which mechanical property limits apply.

*T—Thermally treated to produce tempers other than M, F, 0 or H* This designation applies to products, which have their strength increased by thermal treatment, with or without supplementary strain hardening. The letter **T** is always followed by one or more digits indicating the specific sequence of treatments. For example, **T1**—cooled from hot working and naturally aged; **T2**—cooled from hot working, work-hardened, and naturally aged; and **T3**—solution heat-treated, work-hardened, and naturally aged (see IS 5052, for the description of **T4** to **T10**). The temper designations are important from a welding point of view, as welding, which is a thermal process, can change the characteristics of the metal in the heat-affected zone (they are assumed to extend 25 mm from the actual weld location). Some 2xxxx and 7xxxx alloys are not weldable. In addition to alloying, the strength of aluminium may be due to strain hardening. The heat from welding erases these effects. Moreover, the aluminium oxide firm that forms on the surface should be removed prior to welding. More information on welding is provided by Cary (2002).

**Important alloys of aluminium** The most common aluminium alloys contain copper, magnesium, chromium, silicon, iron, nickel, and zinc. Each of these adds its own unique properties to the final alloy. A few most important alloys are briefly described as follows.

**Duralumin** Duralumin, introduced in 1909, was developed by metallurgist Alfred Wilm at Dürener Metallwerke AG, Germany. It is composed of 94% aluminium, 4% copper, 0.5 to 1.5% magnesium, 0.5 to 1.0% manganese, and 0.5% silicon. Duralumin, a 2xxxx series alloy, is a strong, hard, but a lightweight alloy of aluminium and has typical yield strength of 450 MPa. The alloy has been used in the construction of aircraft and internal-combustion engines and automobile parts, where the strength as equivalent to steel is required. The American Buckminster Fuller used duralumin for his projects, including the Dymaxion House.

**Aluminium-bronze** It is an alloy C95400 of 10.00–11.50% aluminium, 83% copper, 3–5% iron, 1.5% nickel, and 0.5% manganese, and has a tensile strength of 586 MPa and yield strength of 221 MPa. It is used in heavy-duty, high-load, and high-wear applications that require tensile strength, good ductility, weldability, or exceptional resistance to fatigue and deformation in overload situations. Typical applications include gears, valves, bearings, and machine parts.

**Y-alloy** It was developed by the National Physical Laboratory during World War I, in an attempt to find an aluminium alloy that would retain its strength at high temperatures. Like duralumin, it is an alloy of 92.5% aluminium, 4% copper, 2% nickel, and 1.5% magnesium. The alloy can be used in cast and wrought forms. It is used in the pistons of aircraft engines. In the late 1920s, Rolls-Royce developed a similar alloy called Hiduminium or 'R.R. alloy', with 93.7% aluminium, 2% copper, 1.3% nickel, 1.4% iron, 0.8% magnesium, 0.7% silicon, and 0.1% titanium.

The following four principal aluminium alloys are used in general and structural engineering applications.

*Alloy 64430*    It is available in the following forms: sheet, plate, extrusion, tube, wire, and forgings. It is used in structural applications of all kinds, such as road and rail transport vehicles, bridges, cranes, roof trusses, rivets, cargo containers, and flooring.

*Alloy 65032*    It is available in the following forms: extrusion, wire, tube, forgings, sheet, and plates. It is used in structural applications of all kinds, similar to alloy 64430.

*Alloy 63400*    It is available as extrusion, wire, tube, rolled rod, and forgings. It is used in architectural and other similar applications where surface finish is important and medium strength would suffice. It is also used in electrical conduits, tubes for waveguide, gas, and oil transmission pipelines.

*Alloy 54300*    It is available as sheet, plate, extrusion and forgings and is used in welded structures, cryogenic applications, structural marine applications, rail and road tank cars, rivets, and missile components.

Aluminium alloys, such as 24345, 31000, 52000, and 53000, are considered secondary alloys and not used in building construction. For more information on the different alloys and their uses, refer to IS 1285:2002 and IS 617:1994.

## 14.2.4 Anodizing of Aluminium

*Anodizing* is a process that is used to obtain shiny protective oxide coating on the surface of aluminium, producing a thicker oxide layer than that would naturally occur on aluminium surfaces due to oxidation. An anodized coating is actually part of the metal and hence will not peel off and provides excellent durability. It converts the surface of the metal into a decorative, durable, corrosion-resistant, and anodic oxide finish. Aluminium is ideally suited to anodizing, although other non-ferrous metals, such as magnesium and titanium, also can be anodized.

The object to be anodized is immersed into a dilute sulphuric acid ($H_2SO_4$) bath and an electric current is passed through the medium. A cathode is mounted to the inside of the anodizing tank and the aluminium acts as an anode. In this process, the object is coated with a thick layer of oxide, which prevents aluminium from the attacks of atmospheric agents. More details of anodizing may be got from IS 1868. One of the primary advantages of anodizing is its resistance to corrosion and abrasion.

Anodizing may be coloured or clear. Clear anodizing retains the grey aluminium colour and results in a more even appearance than mill finish. The availability of other colours depends on the anodizing process used. The anodizing thickness ranges from 2.5 to 25 μm, depending on its use.

Aluminium can also be powder coated to different shades. Aluminium coatings on doors and windows can also be made to look like wood.

## 14.2.5 Available Forms of Aluminium

The products of aluminium are available in the following forms:

*Casting-based*    Sections such as baluster heads, railings, fittings, and grills (IS 617, IS 6751, IS 1981)

*Extrusion-based*    Sections for doors, windows, and other fabricated items (IS 738, IS 1285, IS 6477)

*Fasteners*   Bolts and rivets (IS 740, IS 1284, IS 3577)

*Foils*   Thin foils for various applications and powder for paints (IS 15392)

*Sheet-based*   Sheets of various thicknesses, used for roofing and utensils (IS 736, IS 737, IS 1254, IS 2676, IS 2677, IS 14712)

*Wires*   For transmission and distribution of power (IS 739, IS 4026); aluminium conductor steel-reinforced cables are used in over-head power lines and typically have a central strand of steel surrounded by concentric layers of aluminium wires.

## 14.2.6 Uses of Aluminium

By utilizing combinations of its advantageous properties such as strength, lightness, corrosion resistance, recyclability, and formability, aluminium has been used in a variety of applications. It is used in transportation (automobiles, airplanes, trucks, railcars, and marine vessels), packaging (cans and foils), construction (windows, doors, and siding), consumer durables (appliances and cooking utensils), electrical transmission lines, machinery, and many other applications. Some of these applications have already been mentioned while discussing the different alloys. Here, only the use of aluminium in building and construction is briefly discussed.

**Fig. 14.3** Aluminium frames with glass infills in the Gaithersburg Library, MD, USA

Since 1950s, aluminium has been used extensively in the construction of curtain walls. These walls act like large curtains hung from the building frame, serving as a weathertight envelope, and at the same time resisting wind loads and transmitting them to the frame (Kissell & Ferry, 2002). Curtain walls are typically designed with extruded aluminium, although earlier curtain walls were made of steel. The aluminium frame consisting of vertical and horizontal extruded aluminium mullions is typically infilled with glass, which provided architecturally pleasing appearance, as well as benefits such as day lighting (see Fig. 14.3). Aluminium extruded members are used extensively as frames for doors and windows and in storefronts (IS 1948:1961).

Standing seam aluminium roof sheeting, spanning between roof purlins, has been used in industrial buildings. Owing to its corrosive resistance and architectural appeal, aluminium sheet is employed for roofing and siding, as well as routine use for flashing, gutters, soffit, fascia, and downspouts in buildings (IS 736, IS 737, IS 1254). The use of aluminium siding in a building is shown in Fig. 14.4. These horizontal aluminium siding have thickness of 0.61 mm and a width of 200 mm, and are fastened with aluminium nails at 410 mm centres. Pool enclosures, canopies, and awnings are also frequently constructed using aluminium due to its ease of fabrication and corrosion resistance (Kissell & Ferry, 2002).

**Fig. 14.4** Aluminium sidings in buildings

The *aluminium composite material* (ACM) was developed in the 1970s. It consist of a plastic core sandwiched between adhesively bonded aluminium face sheets. These sandwich panels are used in buildings and cladding, walls, and sign boards. The thickness of aluminium face sheets is usually 0.51 mm. ACM panels are available in a wide variety of painted or anodized colours, with either polyethylene or proprietary fire-resistant plastic core, consisting of 70 to 90% mineral wool. These panels can be routed, punched, drilled, bent, and curved using common fabrication equipment. Concealed fastener panel attachment systems are also provided by the manufacturers (Kissell & Ferry, 2002). One of the earliest uses of it as cladding is for the EPCOT sphere in Disney World. An example of building in Chennai clad with such sandwich panels is shown in Fig. 14.5.

**Fig. 14.5** Aluminium cladding for a building in Chennai

Welded aluminium tanks are used for industrial storage and process vessels and pipe for corrosive liquids and are well suited to cryogenic applications due to aluminium's good low-temperature properties. Large aluminium roofs with bolted members, some clad with integral aluminium sheeting have been used to cover water storage, wastewater treatment, and petrochemical tanks (Mazzolani, 1994, 2006). Aluminium signs and sign structures, light poles, and guard rails are used for highways and railroads and aluminium has been used for a number of bridge decks and bridge structures in North America and Europe. Details of aluminium domes and double layer grids may be had from Subramanian (2008).

Culverts made of large diameter corrugated aluminium pipes are used for bridges, liners, and retaining walls. Aluminium is also used for portable bridges for military vehicles. Aluminium's corrosion resistance lends itself to marine applications, including gangways and floating docks, and decks in offshore floating platforms. The renovation in 1999 of the Washington Monument, which is capped by a 225-mm tall aluminium pyramid, used scaffolding made of aluminium pipes.

Because of the lower modulus of elasticity, aluminium members are more susceptible to buckling and hence the advantage of reduced weight is often offset by the use of thicker members. However, the use of aluminium results in economic structures due to the additional advantage of reduced maintenance. The design concepts of aluminium structures are similar to that of steel. More details about the design aspects may be found in IS 8147 and Kissell and Ferry (2002).

**Case Study** | **Use of aluminium composite panels in EPCOT'S Spaceship Earth**

Spaceship Earth is the iconic and symbolic structure of Epcot Center, built at the Walt Disney World Resort, Florida, USA. The structure is similar to the United States pavilion of Expo 67 in Montreal, but unlike that structure, Spaceship Earth is a complete sphere, supported by three pairs of legs. The architects were Wallace Floyd Design Group and the structural designers were Simpson Gumpertz & Heger Inc. of Boston. The geodesic sphere is composed of 11,324 ACM panels.

**Case Study** | **Fire at Grenfell Tower, London**

A huge fire at the 24-storey residential Grenfell Tower, built in 1974, in North Kensington, West London, started at 23:54 GMT on 14 June 2017, and was brought under control only after about 24 h. This fire destroyed 151 homes, both in the tower and surrounding areas, and killed 79 people. Investigations showed that the Reynobond PE, a composite aluminium panel with a polyethylene core, which was used in the tower during the renovation in 2016, was the main cause of the fast spreading of fire. This composite aluminium panel should not be used as cladding on buildings over 18-m high. Based on this fire, the government carried out tests on 600 buildings, and found that at least 190 high-rise buildings around London do not have fire-retardant cladding.

### 14.2.7 Advantages of using Aluminium

Physically, chemically, and mechanically aluminium is a metal like steel, brass, copper, zinc, lead, or titanium. It can be melted, cast, formed, and machined much like these metals and it conducts electric current. In fact, often the same equipment and fabrication methods are used as for steel. The following are the advantages of using aluminium in building construction:

*High strength-to-weight ratio* Structural components made from aluminium and its alloys are vital to aerospace industry. The use of aluminium in automobiles, aircraft, trucks, bicycles, and roof sheeting reduces dead-weight and energy consumption while increasing load capacity.

*High corrosion resistance* Owing to the naturally generated protective oxide coating, aluminium is highly corrosion resistant and durable. Surface treatment such as anodizing, painting, or lacquering can further improve this property. This is an important factor for buildings near seashores and also for tall buildings.

*Resistance to insect attack* Unlike wood, aluminium is not attacked by insects like white ants and borers. Hence, it can be advantageously used in making door and window frames and shutters.

*Low maintenance* It requires no paining and very little maintenance.

*Aesthetic appearance* It has a pleasing appearance. Aluminium can also be anodized and it is possible to give numerous textures, colours, and surface finishes.

*Capacity to withstand low temperatures* It is highly suitable for sub-zero temperatures as low as −270°C, whereas steel becomes brittle and does not perform adequately. Below 0°C, most aluminium alloys show little change in properties; yield and tensile strengths may increase; elongation may decrease slightly; impact strength remains approximately constant.

*Ease of fabrication and assembly* Even complicated sections can be made using the extrusion technique, producing sections to specific requirements, thus saving costs. As these sections are lightweight, they are easy to handle and assemble at site.

*Air tightness* As the extrusion technique allows sections to be made with very low tolerances, components can be made with high precision to be airtight. This is important for windows, doors, and facades of tall and air-conditioned buildings.

*Good noise control* Owing to its excellent reflectivity of sound, aluminium parts provide good noise control. It also reflects electromagnetic waves.

*Electrical and thermal conductivity*    Aluminium is an excellent heat and electricity conductor and, in relation to its weight, is almost twice as good as copper. This has made aluminium the most commonly used material in major power transmission lines.

*High reflectivity*    Aluminium is a good reflector of visible light as well as heat, and these properties, together with its low weight, make it an ideal material for reflectors in, for example, light fittings or rescue blankets. Aluminium roofs absorb less radiant heat. Aluminium paints and films are used for the thermal insulation of roofs.

*Easy to transport*    Owing to its lightweight, aluminium is easy to transport and thus saves fuels.

*Absorbing crash energy*    Aluminium absorbs twice the crash energy of steel and performs as well in an accident. Aluminium crash rails fold up like an accordion, which dissipates and directs energy away from the vehicle's occupants.

*Recyclability*    Aluminium is 100% recyclable with no downgrading of its qualities. The re-melting of aluminium requires only about 5% of the energy required to produce the primary metal in the recycling process.

## 14.3 Copper and its Alloys

*Copper (Cu)* is the oldest metal used by man. Its use dates back to prehistoric times. Copper has been mined for more than 10,000 years with a copper pendant found in current day Iraq being dated to 8700 B.C. But, much of all copper was mined and smelted only since 1900, and over 50% in the past 25 years. Copper is available in its pure form and frequently alloyed with zinc, tin, nickel, and aluminium to produce brass and bronze. Available worldwide, copper resources are estimated at about 2.10 BT (https://minerals.usgs.gov) and its production in 2016 was 19.1 MT. Copper has a very high thermal and electrical conductivity and is relatively corrosion resistant. Copper is used in diverse sectors, such as electrical engineering, automobiles, construction, plumbing, machinery, ship building, aircraft, and precision instruments.

### 14.3.1 Manufacture of Copper

The most abundant copper ores are chalcopyrite ($CuFeS_2$) and bornite ($Cu_5FeS_4$), which contain both copper and iron sulphides. These account for about 80% of the world's known ores (nine-tenths of world reserves are found in the Great Basin of the western USA, central Canada, the Andes region of Peru and Chile, and Zambia). Over 90% of copper ore is mined as copper sulphide using the open-pit method. Most ore contains only 0.5–1% metal and hence the ore must be concentrated before it is sent for smelting and refining. The manufacture of the copper has three stages as shown in Fig. 14.6. They are: (a) enriching the ore, (b) conversion of the sulphides/other copper compounds to copper, and (c) the purification of copper.

Thus, the ore is first enriched using a process called *froth flotation* (invented in the early 1900s in Australia by C. V. Potter and D. Delprat). In this process, the ore is first crushed and then ground to a fine grit. The grit is introduced to a bath of water containing a foaming agent, which produces a kind of bubble bath combined with a special oil-based chemical that makes the copper particles water-repellent. When jets of air are forced up through the bath, the water-repellent and lightweight particles of copper sulphide are carried to the top and float on the froth (see Fig. 14.7). The froth is skimmed off the surface and the enriched ore is taken for further refining. Unwanted material (called gangue) sinks to the bottom and is removed (Knapp, 1996).

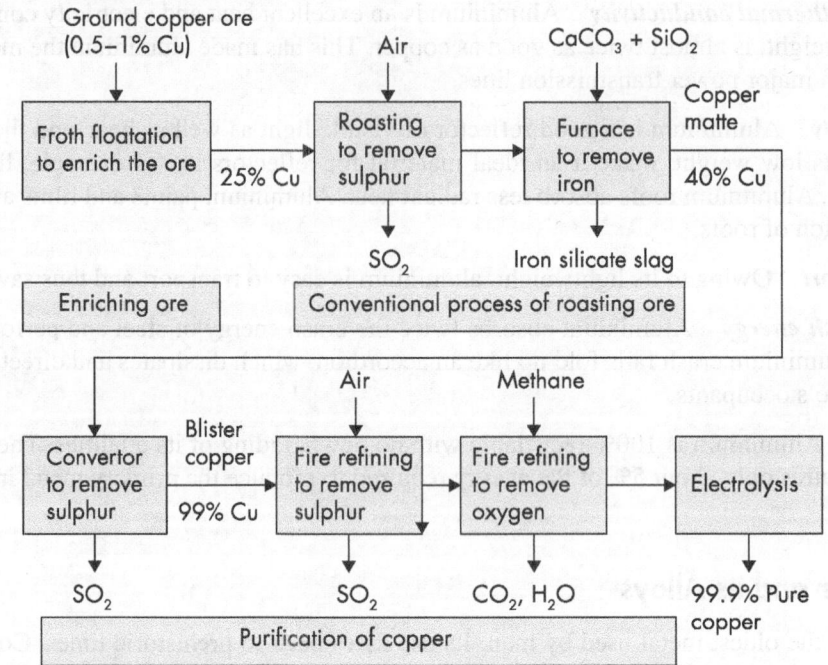

**Fig. 14.6** Three stages in the manufacture of copper

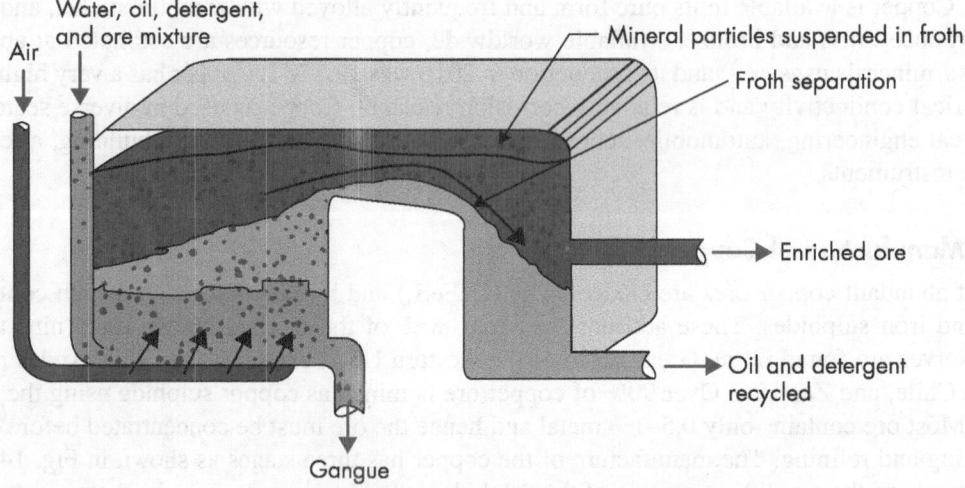

**Fig. 14.7** Principles of the froth-flotation process

The ore containing 25% copper may be further refined by any one of the following methods:

1. By roasting copper sulphide ores
2. Using the leaching process
3. By the bacterial method

Most copper ores are difficult to refine, although smelting removes most of the impurities from the enriched ore. In the conventional process of roasting the ore, the powdered, enriched ore is heated in air between 500 and 700°C to remove some sulphur (as $SO_2$) and dry the ore, to obtain a solid material and called calcine. Calcine is then mixed with limestone ($CaCO_3$) and sand ($SiO_2$) and smelted at about 1200°C in blast furnaces to remove iron (as iron silicate slag) and to produce '*copper matte*' (a mixture of liquid copper and iron sulphide), which may contain 40% of copper. The recent ISASMELT™ process, launched in 2009 by Glencore Technology, is an innovative, high intensity, low-cost submerged-lance smelting process (www.isasmelt.com).

In the next purification process, air is blown into the liquid matte paced in a convertor to remove further sulphur (as $SO_2$) and result in *blister copper*. Blister copper is again heated until it is molten and air is injected into it to remove unwanted sulphur. This is followed by the injection of methane to remove oxygen. These operations are known as fire-refining. Pure copper is obtained from blister copper by electrolysis (in which the anodes cast from processed blister copper are placed into an aqueous solution of 3–4% copper sulphate and 10–16% sulphuric acid with rolled sheets of highly pure copper acting as cathodes). Another important source of copper is recycled scrap, called as secondary copper, which in 2014 accounted for 34% of total copper production (www.thebalance.com). Copper is also obtained as a by-product of silver production. Copper is alloyed and further processed to produce rods, profiles, wires, sheets, strips, and tubes.

## 14.3.2 Properties of Copper

Copper is a metal with an orange-red colour that turns into reddish tarnish when exposed to air. Copper does not react with water, but it does slowly react with atmospheric oxygen to form a layer of brown-black copper oxide, which, unlike the rust that forms on iron in moist air, protects the underlying metal from further deterioration. When it is exposed to oxygen and moisture for a long time, a green deposit, called patina, is formed on the surface of copper. Copper is malleable and ductile and can be worked in hot and cold conditions (IS 191:2007). These properties make copper extremely suitable for tube and wire drawing. Of all common metals, copper possesses the highest rating for both electrical and thermal conductivity (see Table 14.1).

Copper is inherently antimicrobial, i.e., it will rapidly kill bacteria, viruses, and fungi that settle on its surface. Owing to this, surfaces made from copper and copper alloys are used in hospitals and other areas where hygiene is a key concern.

### Case Study  Statue of Liberty, USA

The Statue of Liberty is an iconic sculpture on Liberty Island in New York Harbor in New York City, USA. The copper statue, a gift from the people of France, was designed by French sculptor Frédéric Auguste Bartholdi and built by Gustave Eiffel. The statue was dedicated on 28 October 1886. Originally, the statue had a dull copper colour, but by 1906, it was entirely covered with a green patina, caused by the oxidation of copper skin. Believing that the patina was the evidence of corrosion, in 1906, US Congress authorized US$62,800 (equivalent to $1.7 million in 2017) for various repairs, and to paint the statue both inside and out. Fortunately, this plan failed and we still have the famous green look. During the renovation in 1984, with $1.7 million, larger holes in the copper skin of the statue were replaced with a skin that was taken from a copper rooftop at Bell Labs, because its patina closely resembled that of the statue.

## 14.3.3 Uses of Copper

Because of its properties (singularly or as alloys) of high ductility, malleability, and thermal and electrical conductivity, and its resistance to corrosion, copper is the third major industrial metal, after iron and aluminium, in terms of quantities consumed. About 50 to 65% of copper is used in the electrical industry as bus bars, conductors that distribute power, transformers, motor windings, building wiring, telecommunication wiring, and in electrical and electronic products (By weight, aluminium has higher conductivity than copper, but it has properties that cause problems when used for building wiring). Copper is mainly used to make semi-finished products (called semis), which are made from the refined metal, either as pure copper or as copper alloys. They can be in the form of wire, rod, bar, plate, sheet, strip, foil, or tube. Most of the copper produced is sold in the form of cables, wires, and tubes. Much of the rest is made into alloys. Copper and copper alloys can be used in an extraordinary range of applications (IS 725, IS 778, IS 781, IS 1703, IS 2501, IS 2963, IS 8931, IS 9861, IS 14810, and IS 14811).

**Fig. 14.8** Copper tubes and accessories

In addition to being used in buildings as copper piping (see Fig. 14.8) and wiring, copper is used as cladding, which results in a very attractive colour. There are many examples of historical buildings, palaces, and churches all over Europe with their green copper roofs and gutters proving copper to be both beautiful and durable. It is also used in heat exchangers like refrigerators and air conditioning units for its ease of fabrication and thermal properties. Countries like England and the United States have long used copper for their coins, such as the penny. Euro coins have copper as their base or copper coatings.

### 14.3.4 Copper Alloys (Brass and Bronze)

Copper can be alloyed easily with other metals. The first alloy produced was copper alloyed with tin to form *bronze* – this period in history is considered important and is called The Bronze Age (c. 3300–1200 B.C.). The discovery of *brass* (an alloy of copper and zinc) came much later. There are more than 400 copper alloys, each with a unique combination of properties, to suit many applications, manufacturing processes, and environments. The addition of alloying elements to copper increases the tensile strength, yield strength, and the rate of work hardening. In the following, only brass and bronze are briefly described.

**Brass**    It is the generic term for copper when it is alloyed with 5 to 40% zinc. Romans discovered how to make brass in about 500 B.C. It has higher malleability than zinc or copper and flows when melted. It has a muted yellow colour, somewhat similar to gold, but is duller. There are many types of brass, with properties tailored for specific applications by adjusting the percentage of zinc. Minor amounts of other alloying elements, such as aluminium, lead, tin, iron, silicon, manganese, and nickel, may also be added. For example, the addition of 3% of lead improves the machinability of brass considerably. Brass is used for its strength, corrosion resistance, appearance and colour, and also the ease with which it can be worked and joined. It is not as hard as steel (see also Tables 14.1 and 14.2). A few of the different types of brass are described as follows.

*Alpha brass*    Also known as common brass, it contains less than 36% zinc and is named for its formation of a homogenous (alpha) crystal structure. It is very ductile and easy to cold work, weld, and braze.

*Alpha-beta brass*    Also known as 'duplex brass', it contains about 37–45% zinc and is made up of both the alpha and a beta grain structure. The inclusion of alloy elements such as aluminium, silicon, or tin can also increase the amount of beta phase brass present in the alloy. More common than alpha brass, alpha-beta brass is both harder and stronger, and has a lower cold ductility than alpha brass. Alpha-beta brass is cheaper due to the higher zinc content and is usually hot worked by extrusion, stamping or die-casting (www.copper.org).

***Beta brass***   Although much more rarely used than alpha or alpha-beta brasses, beta brass makes up a third group of the alloy that contains about 45–50% zinc content. This brass has beta crystal structure and is harder and stronger than both alpha and alpha-beta brasses. As such, they can only be hot worked or cast (www.copper.org).

***Red brass***   Called by this name in the USA, it is an alloy having 85% copper, 15% zinc, and small amounts of tin and lead, and considered both a brass and a bronze. It is also known as *gunmetal*, and is used to make steam and hydraulic castings, valves, gears, and statues. *Yellow brass*, on the other hand, is composed of about 60–70% copper, 30–40% zinc, and trace amounts of tin and zinc, and has a golden yellow colour. In addition to its use in household plumbing, yellow brass in used in the automobile industry, and also to make keys.

***Applications of brass***   It is easy to fabricate by drawing, and has high cold-worked strength and corrosion resistance. Brass is blanked, coined, drawn, and used in many applications (see Table 14.5 and Fig. 14.9). Brass has excellent castability. Cast brass is used as plumbing fixtures, decorative hardware, architectural trim, low-pressure valves, gears, and bearings (www.copper.org). Cast brass can be chromium or nickel plated, for polished appearance. Plain brass fittings have to be polished frequently with polish called Brasso.

**Table 14.5** Important alloys of copper and their uses

| Copper alloy | Alloying element | Uses |
| --- | --- | --- |
| Brass | Zinc | Decorative, screws, wires, plumbing/electronics, musical instruments, low-friction applications (locks, hinges, tower bolts, gears, door-knobs, ammunition, valves), and ornaments |
| Bronze | Tin | Boat and ship fittings, propellers and submerged bearings because of resistance to salt water corrosion; widely used for cast bronze sculpture; bearings, clips, electrical connectors, and springs; for top-quality bells and cymbals |
| Phosphor bronze | Tin and phosphorus | Precision bearings, springs, cymbals, instrument strings, and coins |
| Aluminium bronze | Aluminium, tin (iron, nickel, and silicon) | Tools, high-temperature aircraft and automobile engine components |

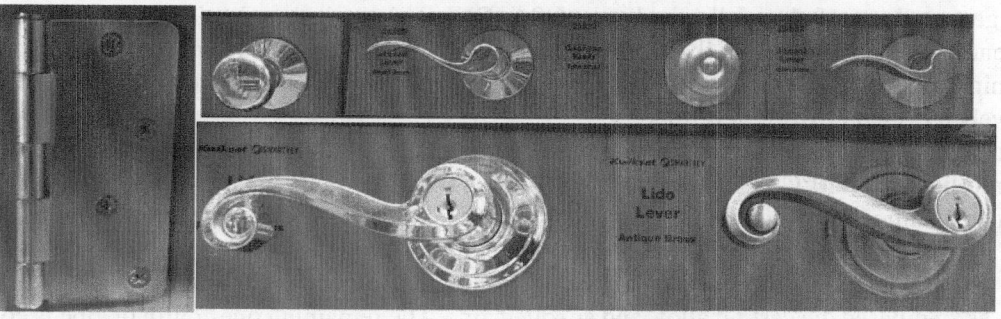

**Fig. 14.9** Typical applications of brass hardware

***Bronze*** It is an alloy of 90% copper and about 10–12% tin, and has a reddish brown colour. It dates back to about 3500 B.C. It is hard and brittle (less brittle than cast iron). Bronze resists corrosion (especially, sea water corrosion) and metal fatigue more than steel. It also conducts heat and electricity better than most steels. It melts at 950°C but this value depends on the amount of tin present (see Tables 14.1 and 14.2). Similar to brass, there could be different types of bronze depending on the percentage of copper and other alloys. *Phosphor bronze* (with less than 8% tin and 0.01 to 0.35% phosphorous) has a yellowish red colour, and has superb spring qualities, high fatigue resistance, excellent formability, solderability, and high corrosion resistance. It is used commonly for bearings in machines and engines where shafts commonly rotate, and also as coins. *Aluminium bronze* typically contains 9–12% aluminium and up to 6% iron and nickel. Aluminium bronze has high strength, and excellent corrosion and wear resistance. It is used in castings that have exceptional corrosion resistance, high strength, toughness, and wear resistance, and are weldable. The uses of different types of copper alloys are given in Table 14.5. Typical bronze hardware is shown in Fig. 14.10.

**Fig. 14.10** Typical bronze hardware

## 14.4 Zinc and its Alloys

*Zinc* (Zn) is a bluish-white, lustrous, and diamagnetic metal. The oldest evidence of pure zinc comes from Zawar, in Rajasthan, as early as the 9th century AD. The most common zinc ores are sphalerite, a form of zinc sulphide ($ZnS$), containing 60–62% zinc and smithsonite ($ZnCO_3$). Sphalerite is first concentrated by froth flotation. Then, it is roasted in air, and finally extracted either by distillation or electrolysis. In electrolysis, pure zinc rod is used as cathode whereas a block of impure zinc is used as anode. A mixture of zinc sulphate ($ZnSO_4$) and sulphuric acid ($H_2SO_4$) is used as electrolyte. Australia is the biggest producer of zinc, followed by the USA. Global zinc mine production in 2017 was 13.2 MT, and identified zinc resources of the world are about 1.9 BT. Zinc can be recycled many times without weakening the metal. About 25% of zinc produced is from recycled resources – over 95% of zinc is effectively recycled in Western Europe.

Zinc is hard and brittle at most temperatures but becomes malleable between 100 and 150°C, and turns into liquid at 420°C (also see Tables 14.1 and 14.2). The most important property of zinc is its resistance to atmospheric corrosion. Zinc is a fair conductor of electricity. Zinc stands fourth among all metals in world production – being exceeded only by iron, aluminium, and copper. About three-fourths of the zinc produced is consumed as a coating to protect iron and steel from corrosion (galvanized metal), as zinc-based die casting alloy, and as rolled zinc. The remaining one-fourth of zinc is consumed as zinc compounds mainly by the rubber, chemical, paint, and agricultural industries. Other applications are in electrical batteries, small non-structural castings, and alloys such as brass. A variety of zinc compounds are commonly used, for example, zinc sulphide and zinc oxide are used in luminescent paints.

Zinc is used to make many useful alloys. Brass, an alloy of zinc that contains between 55 and 95% copper, is among the best known alloys. Zinc is alloyed with lead and tin to make solder, a metal with a relatively low melting point and used to join electrical components, pipes, and other metallic items. Other zinc alloys include nickel silver and German silver. Alloys of zinc with small amounts of copper, aluminium, and magnesium are useful in die casting as well as spin casting, especially in the automotive, electrical, and hardware industries. These alloys are marketed under the name Zamak. Another alloy, marketed under the brand name Prestal, contains 78% zinc and 22% aluminium, and is reported to be nearly as strong as steel but as malleable as plastic. In addition, zinc is used as a lead replacement.

### 14.4.1 Zinc Galvanizing

The word *galvanized* comes from the Italian scientist Luigi Galvani (1737–1798), who discovered electric currents. Galvanized or zinc-coated steel sheet is used in many applications like roof sheeting, where they will come into direct contact with the atmosphere and is difficult to monitor. Manufacturers of truck bodies, buses, and automobiles are increasingly concerned about corrosion, particularly when chemicals are used on roads for ice control. Galvanized steel is also used in washing machines and dryers and in many industrial products such as air conditioning and processing tanks. Other uses include truss members, electrical transmission and communication towers, and highway sign masts.

During the *hot-dip galvanizing of steel,* the steel components are immersed in a molten zinc providing zinc coating of thickness 0.035 to 0.10 mm. Although zinc melts at 419°C, for the proper implementation of the galvanizing process, it has to be heated to over 455°C. Two common methods of hot-dip galvanizing exist. In the first method, called the *continuous hot-dip galvanizing*, zinc is applied to the surface of steel sheet as it passes as a continuous ribbon through a bath of molten zinc at high speeds (typically, 200 m/min). As the steel exits in the molten zinc bath, 'air knives' blow off the excess coating from the steel sheet and control the coating thickness to the required specification. When one coil is being processed through the coating bath, another coil is welded to it at the trailing end. Hence, the process is made continuous, and can be operated without interruption for a few days. The second method is called the batch or general galvanizing. In this process, after steel items are fabricated, a zinc coating is applied onto the surfaces, through molten zinc baths. Thus, in this batch process, discrete steel items are dipped into a zinc bath, containing 99.8% zinc and 0.2% aluminium. From large bridge girders to small items, such as steel fasteners, could be galvanized using this method. In the discrete process, before the part is immersed in the zinc bath, it is cleaned to remove oils, dirt, etc., pickled to remove the thin oxide coating present on the steel surface, and then rinsed. Sometimes, before dipping into the zinc bath, the steel items are passed through a flex of zinc ammonium chloride, to dissolve any remaining oxide film present on the steel surface, after pickling. The longer the immersion in the zinc bath, the ticker will be the zinc layer. Typical minimum required coating weight for both surfaces is 275 $g/m^2$ (nominal thickness per side of 19.4 μm). It has to be remembered that the life of zinc coating is directly proportional to the thickness of the coating.

The main mechanism by which galvanized coatings protect steel is by providing an impervious barrier that does not allow moisture to contact the steel, since without moisture (electrolyte) there is no corrosion. When the zinc coating is exposed to air, it reacts with oxygen to form a thin dull grey layer of zinc oxide (ZnO). When moist air is present, zinc reacts with the moisture to form zinc hydroxide. This zinc hydroxide reacts with carbon dioxide ($CO_2$) present in the atmosphere to form a thin and insoluble layer of zinc carbonate, and protects the underlying zinc. This is the reason for the low corrosion rate of galvanized steel in most environments. The second shielding mechanism is provided by the ability of zinc to galvanically protect steel. Thus, when base steel is exposed due to some cut or scratch, the steel is cathodically protected by the sacrificial corrosion of a zinc-bearing coating (www.astm.org).

When the object is too large to dip into the bath of zinc, such as hulls of ship, after abrasive cleaning of the steel surface, melted zinc is sprayed using compressed air onto the surface using a heated gun. The quality of this process, called *metalizing*, depends on the skill of the applicator and also the weather.

## 14.5 Lead

*Lead* (Pb) is a dense, soft, heavy, ductile, corrosion resistant, and blue-grey metal. It is a naturally bright, shiny metal like silver. But as soon as it is exposed to the oxygen in air, it develops a dull, bluish-grey covering of lead oxide. It also reacts with certain acids such as nitric acid and hydrochloric acid, and is a poor conductor of electricity. Lead mining predated the Bronze or the Iron Ages, with the earliest lead mine in Turkey dated about 6500 B.C. The ancient Romans used lead for water and drain pipes and some of these 2000-year-old pipes are still functional. Lead is not used for pipes now because of its toxicity. In addition to its abundance, the properties which make it commercially attractive include: easy workability, low melting point (328°C), ability to form carbon metal compounds, hold pigments well, easy recyclability, resistance to atmospheric elements, the high degree of corrosion resistance, and its inexpensiveness.

Lead is extracted from the main ore galena (lead sulphide, PbS), which are obtained by digging through underground mines. The ore is ground and impurities are separated by the flotation process. It is then filtered, and roasted with coal as fuel, on a moving grate with hot air at 1400°C blowing through the grate. The fused material called sinter obtained in this process is smelted in a blast furnace at 1200°C. The lead obtained is refined in a reverberatory furnace. Identified world lead resources total more than 2 BT and its production in 2017 was about 4.87 MT (https://minerals.usgs.gov).

Lead is used in storage batteries, cable coverings, bullets, as a sound absorber, a radiation shield for nuclear reactors and X-ray equipment, in making fine crystal glass and flint glass, as liners in containers for corrosive liquids such as sulphuric acid, roofing (as gutters, flashings, and sheets), alloys, such as solder, and in insecticides. About 75% of the world's lead production is consumed by the battery industry (The first lead battery is credited to a French physicist, Gaston Plante, who invented it in 1859.). Lead was used in lead-oxide paints and gasoline; however, due to lead's toxicity these uses are being phased out since 1990s (Recent studies on the hair of German composer Beethoven, who died in 1827, suggest that he might have died of lead poisoning). An alloy of tin and lead was used in the letters (called type) of earlier printing machines. Lead oxide has also been used in pottery glazes and paints, since ancient times to produce orange colour. The Golden Gate Bridge was originally painted International Orange, with a lead primer and a lead-based topcoat. In 1968, the original lead based paint (primer and topcoat) was removed and replaced with an inorganic zinc silicate primer and vinyl topcoats of the same colour. Lead, from lead-acid batteries, lead pipes, and sheets, is also easily recycled.

**Case Study** | **Lead poisoning in flint, Michigan, USA**

In April 2014, the city of Flint, Michigan, USA, opted out of Detroit's water supply and started using water from the Flint River. The corrosive river water allowed dangerously high levels of lead to leach out of outdated lead pipes and into the water supply, causing a public health crisis of lead poisoning in children and adults. A federal state of emergency was declared in January 2016. Although the water quality returned to acceptable levels in early 2017, residents are instructed to continue the use of bottled or filtered water until all the lead pipes are replaced, which is expected to be completed only by 2020.

## 14.6 Chromium

*Chromium* (Cr) is a hard, malleable, and bluish metallic element. The only ore of chromium is the mineral chromite and much of it is found in South Africa, Kazakhstan, India, Russia, and Turkey. World resources of chromium are greater than 12 BT and the world production in 2017 was about 31.0 MT (https://minerals.usgs.gov).

More than 95% of high purity chromium is produced using the aluminothermic or silicothermic process. In the aluminothermic process, the chromite ore is first roasted along with soda and lime (fluxing agent) in air at about 1000°C, to produce sodium chromate. This product is then leached away from the waste material, refined further to reduce other impurities and precipitated as chromic oxide ($Cr_2O_3$). Powdered aluminium is then mixed with chromic oxide and the mixture is put into a large container and barium peroxide and magnesium powder are spread over it. The container is then insulated by sand, and the mixture is ignited. Now, the oxygen in chromic oxide reacts with the aluminium to produce aluminium oxide and liberates 97–99% pure molten chromium metal.

Chromium is mostly used for making stainless steel and other metal alloys. Chromium is also used to make heat-resisting steel. Chromium plating can be used to give a polished mirror finish to steel (see Fig. 14.11). Chromium compounds are used as industrial catalysts and pigments (in bright green, yellow, red, and orange colours). Rubies get their red colour from chromium, and glass treated with chromium has an emerald green colour.

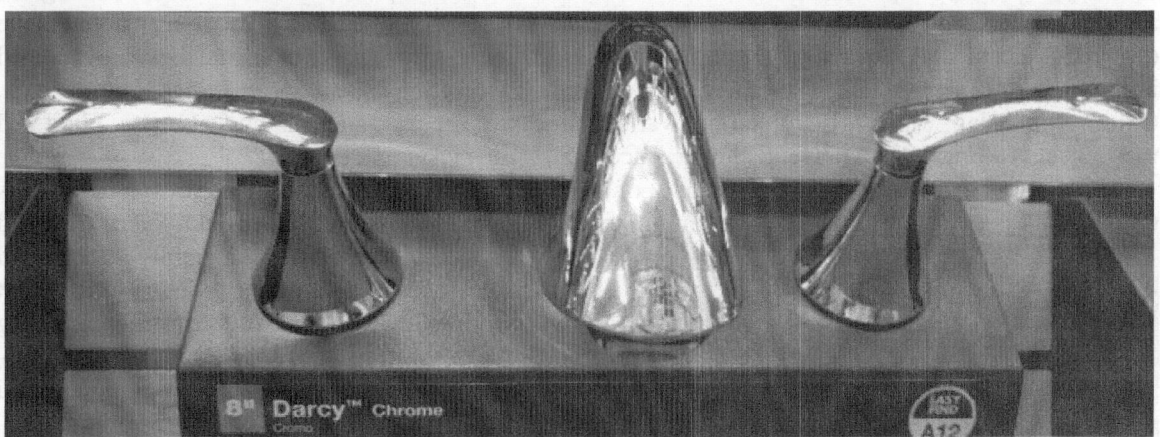

**Fig. 14.11** Chrome plated hardware

## 14.7 Nickel

*Nickel* (Ni) is a naturally occurring, lustrous, silvery-white metallic element. It is the fifth most common element on earth; however, it is more common deep inside the Earth, where it is completely out of reach (Many meteorites consist of a nickel-iron alloy.). Identified land-based resources averaging 1% nickel or greater contain at least 130 MT of nickel, with about 60% in laterites and 40% in sulphide deposits (https://minerals.usgs.gov). Although nickel was identified as a metal by Swedish scientist Axel Fredrik Cronstedt in 1751, an alloy of nickel, copper and zinc (called pait'ung) was found to have been used in China's Yunan Province in the 1500 B.C. Nickel has a high melting point (1452°C), resists corrosion and oxidation, very ductile, readily alloys with other metals, magnetic at room temperature, and can be deposited by electroplating (www.nickelinstitute.org).

## The Process of Electroplating

*Electroplating* is a process of depositing a thin layer of metal onto the surface of another substance using electrolysis. The item to be coated functions as the cathode and is submerged in a bath containing the metal, which acts as the anode. Anodes can be made from a variety of different metals or alloys, depending on the desired result, including aluminium, zinc, magnesium, titanium, tin, lead, copper, bronze, brass, copper-tin, and copper-zinc. The bath usually is a solution that contains dissolved metal salts and other ions that encourage the flow of electricity. Because the item (the cathode) is negative, when voltage is applied to the anode, metal ions in the solution are attracted to the cathode and are deposited onto the cathode's surface as pure metal. The most common form of electroplating is used for creating coins such as American pennies, which are small zinc plates covered in a layer of copper. Chromium plating is done on many objects such as car parts, bath taps, kitchen gas burners, and many others. Electroplating increases the life of the metal and prevents corrosion.

Primary nickel is produced from two types of ores: laterites, such as saprolite, limonite, and smectite, and sulphides such as pentlandite and pyrrhotite. The world's largest source of nickel is at Sudbury, Ontario, Canada. The ores are dug using open-pit mining or deep mining tunnels. The separation process for nickel is highly dependent upon the type of ore. For example, the sulphide ores are concentrated, smelted, and converted to metal-rich matte. The concentration using froth flotation is similar to that explained in Section 14.3.1, and shown in Fig. 14.7. The froth is filtered and dried to create a powder of nickel-sulphide. To extract pure nickel, the powder is smelted in a furnace. There the sulphides and the iron react with gases and produce large amounts of heat. The heat is adequate to smelt the concentrate, producing a liquid matte (up to 45% nickel) and a fluid slag. The furnace matte still contains iron and sulphur, and these are oxidized in the converting step to sulphur dioxide and iron oxide by injecting air or oxygen into the molten bath. Oxides forma slag, which is skimmed off. Various processes are used to refine nickel matte. Fluid bed roasting and chlorine-hydrogen reduction produce high-grade nickel oxides.

As nickel is corrosion resistant, about 65% of the nickel is used in the manufacture of stainless steels (most stainless steels are about 18% chromium and 8% nickel). Another 20% is used in other steel and non-ferrous alloys – often, for highly specialized industrial, aerospace, and military applications. About 9% is used in plating and 6% in other uses, including coins, electronics, in nickel-cadmium and nickel-metal hydride batteries for portable equipment and hybrid cars, as a catalyst for certain chemical reactions, and, as a colorant-nickel is added to glass to give it a green colour. Figure 14.12 shows the use of nickel in hardware. The 987-m long Québec bridge, the longest steel cantilever bridge in the world, built over the St. Lawrence River in 1917, near Québec

**Fig. 14.12** Brushed nickel hardware

City, Canada, is the first bridge to use 3% nickel steel in its construction. The first nickel-based coins were introduced in the 1860s. In the USA, nickels, dimes, quarters, and half-dollars are still made of a copper-nickel alloy. The nickel coin, quite aptly, contains the most nickel at 25%, while the other coins are 8.33% nickel. Nickel-coated plastic is used as trims in cars, bathroom fittings, and electrical connectors. Nickel is also readily recycled from many of its applications, and large tonnages of secondary or 'scrap' nickel are used to supplement newly mined metal.

Nickel-silver or German silver is an alloy with 60% copper, 20% nickel, and 20% zinc (note that it does not contain any silver). It is used as base metal for silver-plated cutlery and other silverware, and also in zippers, coins, better-quality keys, costume jewellery, for making musical instruments such as flutes and clarinets. The nickel-silver cutlery can also be covered with a thin layer of real silver using electroplating.

*Monel* is a group of expensive nickel alloys, primarily composed of nickel (up to 67%) and copper, with small amounts of iron, manganese, carbon, and silicon. Monel's corrosion resistance makes it ideal in applications such as piping systems, pump shafts, seawater valves, kitchen sinks, and in the frames of eyeglasses.

## 14.8 Magnesium

*Magnesium* (Mg) is a grey-white, lightweight, and strong metal. It tarnishes slightly in air, and is thus protected against further oxidation by a thin impermeable layer of oxide. However, it has poor corrosion resistance. There are over 60 different minerals known to have a magnesium content of 20% or greater, making it the eighth most abundant element in the Earth's crust. If sea water is also accounted, magnesium is the most abundant element on the surface of the earth, as sea water has about 1290 parts per million of magnesium. Although Sir Humphrey Davy discovered magnesium in 1808, it was not produced until 1831 when Antoine Bussy developed a method to produce it. Commercial production of electrolytic magnesium began in Germany in 1886. Currently, China is the dominant producer and uses the process of 'silico-thermal reduction' of dolomite (MgO.CaO); however, in the USA, it is produced by the electrolysis of hydrated $MgCl_2$.

Traditionally, magnesium is produced from dolomite and magnesite ore, as well as magnesium chloride containing salt brines (naturally occurring salt deposits). Its characteristics are similar to its sister metal, aluminium. It is the lightest structural metal and is approximately 33% lighter than aluminium, 60% lighter than titanium, and 75% lighter than steel (see Tables 14.1 and 14.2). Magnesium is highly flammable, and burns with intense heat; this is greatly reduced by alloying it with calcium (owing to this property, it is employed in flash bulbs, and in fireworks to produce brighter sparks). Magnesium alloys contain small amounts of aluminium, zinc, manganese, or zirconium, and have strengths equalling that of mild steels. Magnesium can be cast, forged, rolled into plate, strips, and structural shapes, fabricated, and machined. Because of their low modulus of elasticity, magnesium alloys can absorb energy elastically – it absorbs 16 times more shock and vibrations than aluminium. Magnesium's tendency to creep at high temperatures is eliminated by the addition of scandium and gadolinium.

As a structural metal, it is used in aircrafts. It is used by the materials-moving industry for parts of machinery and hand-powered tools due to its strength-to-weight ratio. Magnesium-aluminium alloys are used in various auto parts in the automobile industry, beverage cans, sports equipment, such as golf clubs, fishing reels, and archery bows and arrows. Owing to its shock-absorbing property, it is used in motorcycles and many types of sports equipment. Use of magnesium in laptops, cameras, and cell phones makes them lighter and thinner and also durable to withstand the daily wear and tear. Products from advanced medical equipment to vacuum cleaners are also using magnesium. Magnesium alloys have also been used as a replacement to some engineering plastics due to their higher stiffness, high recycling capabilities, and lower cost of production (www.intlmag.org). Magnesium could be welded by many of the arc and resistance welding processes as well as by the oxyfuel gas welding process, and it can be brazed (Cary, 2002). Magnesium compounds, primarily magnesium oxide (MgO), are used as a refractory material in furnace linings for producing iron, steel, non-ferrous metals, glass, and cement. They are also used in the agricultural, chemical, and construction industries. Magnesium is also used to remove sulphur from iron and steel.

## 14.9 Vanadium

Vanadium (V) was discovered twice. Andrés Manuel del Rio, Professor of Mineralogy, Mexico City, found it in a specimen of vanadite in 1801 and sent it to Paris. Unfortunately, French chemists declared it as chromium. It was again discovered by Swedish chemist Nil Gabriel Selfström at Stockholm in 1831. Pure vanadium was produced in 1869 by Henry Roscoe at Manchester, who showed that previous samples were really vanadium nitride (VN).

Vanadium is the 20th most abundant element in the earth's crust and its compounds occur naturally in about 65 different minerals. World resources of vanadium exceed 63 MT and its production in 2017 was 80,000 tonnes, with China's share being 50% (https://minerals.usgs.gov/minerals). Vanadium is a medium-hard, silvery-grey, ductile, and malleable metal. It has good corrosion resistance and is stable against alkalis and sulphuric and hydrochloric acids.

About 85% of the vanadium produced is used as a steel additive. Vanadium-steel alloys are very tough and used for armour plate, axles, tools, piston rods, and crankshafts. Less than 1% of vanadium, with little chromium, will make steel shock and vibration resistant. Owing to the low neutron-absorbing properties of vanadium alloys, they are used in nuclear reactors. Vanadium also has some applications in cathodes for lithium-ion batteries. More details of vanadium can be had from Kustin et al. (2007).

About nine-tenths of world's vanadium is used as an additive in high strength rebar, as it offers the best combination of high strength, good ductility, bendability, weldability, and reduced sensitivity to strain aging. China has tightened its standard on rebars, GB/T 1499.2–2018, by eliminating low strength Grade 2 (335 MPa) rebar and introduced three different high-strength bars: Grade 3 (400 MPa), Grade 4 (500 MPa), and Grade 5 (600 MPa). As per this standard, the vanadium content will be at 0.03% in Grade 3, 0.06% in Grade 4, and more than 0.1% in Grade 5 rebars, in order to make buildings in China more earthquake resistant. It has to be noted that as per Amendment no. 1 of IS 1786:2008 (Nov. 2012) for high strength deformed rebars, when Nb, V, B, and Ti are used individually or in combination, the total contents should not exceed 0.30%. The use of vanadium pentoxide as electrolyte in vanadium redox flow batteries (VRBS) for storing large amounts of wind and solar energy has resulted in more interest in this metal.

## 14.10 Problems with Non-ferrous Metals

Some of the problems associated with metal roofs and some possible solutions are briefly discussed as follows.

**Inferior products**   High costs and limited availability of good quality metal products in India have resulted in the supply of inferior quality products such as roofing sheets, which are thinner than standard sheets, and components with less than minimum galvanization thickness. Such thin roofing sheets tend to tear off at bolted points, with or without necessary washers, and may cause injury and damage.

**Heat and condensation**   Metal roofs are vulnerable to condensation whenever warm, moist interior air rising up through a building, comes into contact with a cold roof assembly, potentially causing water damage to the inside of the building. Probable causes of condensation are high interior humidity, reflective roofs (that are designed to reduce heat gain by using both solar reflectivity and thermal emissivity), and heating of interior during winter. Some of the solutions to this problem are to provide a vinyl-backed fiberglass insulation (which will prevent humid air from coming into contact with the cooler metal roof) or provide proper ventilation. Cool roofs and thermal insulation also reduce intolerable indoor temperatures, and significantly reduce the energy bills.

**Noise**   During rain, the metal sheets produce noise. Installing an underlayment (mats made from nylon filaments and foam insulating panels) between the metal roofing and the roof sheathing will reduce such noise. Flat or standing seam metal roof will vibrate less than corrugated roofing. Replacing or tightening loose fasteners, as well as adding additional fasteners, can also help manage the noise.

**Uplift**   The uplift resistance of the roof sheeting may be increased by using thicker gauge sheets and providing better fastening with washers.

**Fire resistance** Though metals are non-combustible, members made of metals may lose strength at high temperatures and may buckle or even collapse. Various passive protection methods have been tried in practice and include: cladding the members with insulating materials (using spray protection, board protection, and intumescent coatings), use of composite slabs, double skin walls, and fire-resistant steels. More details about these systems may be found in Subramanian (2008).

**Corrosion** Most of the metals corrode – ferrous metals corrode due the presence of moisture; aluminium in alkaline environment such as those present in wet concrete (it is better to provide a bitumen coating or a paint that tolerates alkaline environments – anodizing will not help); copper may corrode due to elevated concentrations of sulphate, chloride, ammonia compounds, or sulphide; and organic or inorganic acid. *Galvanic corrosion* may occur when two dissimilar metals (e.g., iron and aluminium) are in contact with each other in a corrosive electrolyte. A good example of this is the corrosion of steel bolts in aluminium members – a notable example of galvanic corrosion is that occurred in the Statue of Liberty, USA, when iron armature came in contact with the copper skin. Corrosion may be avoided by minimizing the contact with moisture and periodic coating of the surfaces with appropriate paints. Galvanic corrosion can be avoided by preventing the contact of dissimilar metals. More details about corrosion and its protection may be found in Subramanian (2008).

**Toxicity** As already mentioned, lead is toxic and hence it should not be used in paints and water pipes. Welding metals such as copper, chromium, lead, nickel, and zinc may emit toxic fumes during welding (www.ccohs.ca). Hence, while welding them, good ventilation should be provided. Other precautions to be taken while welding non-ferrous metals may be found in Cary (2002).

## 14.11 Sustainability of Non-ferrous Materials

The environmental impact of mining for metals includes erosion, formation of sinkholes, loss of biodiversity, and contamination of soil, groundwater and surface water by chemicals from mining processes. Besides creating environmental damage, the contamination resulting from the leakage of chemicals also affects the health of the local population. The huge holes created due to open-pit mining can damage the environment, unless it is carefully restored after the mining has ended. Other environmental effects of mining are discussed in web.mit.edu. Though the current end-of-life recycling rate for the metals discussed in this chapter is over 50%, this should reach higher percentages. In 2007, iron and steel production accounted for 30% and aluminium for 2% of global industrial $CO_2$ emissions. Improvements in production methods should reduce these emissions considerably.

Aluminium is theoretically 100% recyclable without any loss of its natural qualities. Recycling involves melting the scrap, a process that requires only 5% of the energy used to produce aluminium from the ore, though a significant part (up to 15% of the input material) is lost as dross (ash-like oxide). Europe has achieved high rates of aluminium recycling ranging from 42% of beverage cans, 85% of construction materials, and 95% of transport vehicles. Recycled aluminium is known as secondary aluminium, but maintains the same physical properties as primary aluminium. Electric power represents about 20 to 40% of the cost of producing aluminium, depending on the location of the smelter. Aluminium production consumes roughly 5% of electricity generated in the USA. Worldwide, 1 tonne of alumina is produced from around 2.9 tonnes of bauxite; the resulting bauxite residue, called the 'red mud' is toxic, and consists of insoluble particles. Bauxite residue from the Bayer process has long posed a problem for storage and management. The industry generates some 120 MT of residue per year from some 100 processors. Most of it is stored in holding ponds. Very little is reused. Research continues to look for suitable applications. However, aluminium-intensive LEED-certified buildings have won Platinum and Gold category awards.

Historically, the major environmental problem associated with the production of copper/zinc from primary sources was the emission of sulphur dioxide into air from roasting and smelting of sulphide concentrates. This problem has been fixed to produce sulphuric acid and liquid sulphur dioxide. In addition, both the primary and secondary copper production may result in the un-captured emission of harmful gases. Careful plant design is needed to capture process gases. About 30% of all copper production is recovered from secondary and scrap materials, which are recycled. Copper tailings, the waste products of copper mines, contain high concentrations of copper, which is poisonous to plants. Care has to be taken to make sure that contaminated water does not reach nearby rivers.

Zinc and zinc-containing products could be recycled considerably and a recovery rate of 80% has already been achieved. Zinc in rivers flowing through industrial and mining areas can be as high as 20 ppm. Using effective sewage treatment, this amount could be greatly reduced. For example, treatment along the Rhine, Germany, has decreased zinc levels to 50 ppb. Concentrations of zinc as low as 2 ppm adversely affect the amount of oxygen that fish can carry in their blood. Levels of zinc in excess of 500 ppm in soil interfere with the ability of plants to absorb other essential metals, such as iron and manganese.

Similar to zinc and copper, the production of primary lead also resulted in the emission of $SO_2$, and this problem has been fixed. Secondary lead production may also give off dangerous gases, which are often cleaned in fabric filters. Lead batteries are recycled effectively. The use of lead in gasoline, paints, solders, and water pipes is reduced and even eliminated.

Chromite ore mining and enrichment produces dust, overburden, waste rock, tailings, and tailings water. Ferrochrome production creates air pollution, dust, slag (waste produced during ferrochrome separation from ore), and process water. These waste materials have the potential to be contaminated with chromium and a number of other heavy metals and chemicals of concern. Chromium may cause health problems. The film Erin Brockovich highlighted the now well-known example of long-term Cr-VI contamination of groundwater in Hinkley, California.

Nickel, in certain forms and under particular circumstances, may generate detrimental environmental effects (including health and safety), notwithstanding the fact that it is considered to be a vital element for public health by some scientists.

## SUMMARY

- There are several non-ferrous metals that are used extensively in building construction, which include aluminium, copper, zinc, lead, chromium, nickel, vanadium, and their alloys.
- Aluminium is a silvery-white, soft, non-magnetic, ductile, and malleable metal, and can be easily machined and cast. It has excellent corrosion resistance and durability. Aluminium can be alloyed with metals such as copper, zinc, magnesium, and silicon. These alloys are identified by a five-digit system, as per IS 6051. Duralumin, aluminium-bronze, and Y-alloy are some of the alloys of aluminium. Aluminium is available as castings, extruded sections, fasteners, foils, sheets, and wires. Owing to its lightweight and non-corrosive properties, it is used in numerous applications.
- Copper is a soft, orange-red metal, having very high thermal and electrical conductivity and corrosion resistance. Copper is alloyed with zinc, tin, aluminium, nickel, or silicon, to produce a range of brasses and bronzes. Used copper can be recycled without any loss of quality.
- Zinc is a bluish-white, lustrous, diamagnetic, durable, and recyclable metal. It is mainly used in galvanized coating to protect iron and steel from corrosion. In the galvanizing process, steel/iron products are immersed in hot molten zinc to get the required thickness of zinc coating.
- Lead is a dense, soft, ductile, corrosion resistant, and blue-grey metal. Though it was used in water and drain pipes, due to its toxicity its use is now being curtailed.
- Chromium is a hard, malleable, and bluish metallic element. It is mostly used for making stainless steel and other metal alloys.

- Nickel is a lustrous, silvery-white metallic element. It is also mainly used to manufacture stainless steels.
- Vanadium is a medium-hard, silvery-grey, ductile, and malleable metal. It has good corrosion resistance and is stable against alkalis and sulphuric and hydrochloric acids. It is mainly used to harden steel.
- All metals and metal compounds have a certain level of toxicity and may cause adverse effects on living organisms. To be sustainable, these non-ferrous metals should be recycled to the maximum so that the energy for producing them is minimized. In addition, the toxic leftovers during their ore processing should be properly disposed, so that they do not pollute water, soil, and air.

## EXERCISES

### Multiple-choice Questions

1. Match the melting point on the left with the metal on the left
   - (a) Aluminium      (i) 650
   - (b) Copper      (ii) 1452
   - (c) Magnesium      (iii) 419
   - (d) Nickel      (iv) 1081
   - (e) Zinc      (v) 659

2. Match the specific gravity on the left with the metal on the left
   - (a) Aluminium      (i) 1.74
   - (b) Copper      (ii) 7.13
   - (c) Magnesium      (iii) 8.8
   - (d) Nickel      (iv) 8.89
   - (e) Zinc      (v) 2.7

3. Pure aluminium is obtained by using
   - (a) the Bayer Process
   - (b) the Hall-Héroult process
   - (c) both the Bayer and the Hall-Héroult processes
   - (d) none of these

4. Normally, members of aluminium are
   - (a) rolled      (c) forged
   - (b) extruded      (d) none of these

5. IS 6051:1970 uses the following designation system for aluminium.
   - (a) 3-digit system
   - (b) 4-digit system
   - (c) 5-digit system
   - (d) None of these

6. The designation 6xxxx shows the following wrought aluminium alloy.
   - (a) Un-alloyed
   - (b) Copper
   - (c) Magnesium + silicon
   - (d) Magnesium

7. The letter H in the temper designation system of aluminium shows that it is
   - (a) annealed      (c) manufactured
   - (b) strain hardened      (d) none of these

8. Anodizing of aluminium is done in order to
   - (a) increase fire resistance
   - (b) increase corrosion resistance

   - (c) increase corrosion and abrasion resistance
   - (d) none of these

9. Y-alloy is an alloy of aluminium and
   - (a) iron and nickel
   - (b) copper and nickel
   - (c) zinc
   - (d) tin

10. Which of the following is not an advantage of using aluminium?
    - (a) High strength-to-weight ratio
    - (b) High corrosion resistance
    - (c) Has modulus of elasticity one-third of steel
    - (d) Good electrical and thermal conductivity

11. Manufacture of copper uses
    - (a) froth-flotation process
    - (b) the Bayer process
    - (c) the Hall-Héroult process
    - (d) none of these

12. Brass is an alloy of
    - (a) copper and zinc
    - (b) copper and nickel
    - (c) copper and tin
    - (d) copper and aluminium

13. Bronze is an alloy of
    - (a) copper and zinc
    - (b) copper and nickel
    - (c) copper and tin
    - (d) copper and aluminium

14. Addition of which of the following metal improves the machinability of brass?
    - (a) Nickel      (c) Lead
    - (b) Aluminium      (d) Chromium

15. From which of the following ore zinc is extracted?
    - (a) Sphalerite      (c) Galena
    - (b) Bauxite      (d) Smectite

16. The main use of zinc is for
    - (a) electrical batteries
    - (b) galvanization
    - (c) to produce brass
    - (d) paints

17. Which of the following pairs are matched correctly?
    (a) Aluminium           (i) Electrolysis
    (b) Copper              (ii) Bayer's process
    (c) Zinc                (iii) Reverberatory furnace
    (d) Lead                (iv) Blast furnace
18. Nickel is mainly used to
    (a) manufacture non-ferrous alloys
    (b) manufacture stainless steels
    (c) make coins
    (d) none of these

19. Match the list in the right with that on the left.
    (a) Aluminium alloy     (i) Radiation shielding
    (b) Copper              (ii) Solder
    (c) Zinc alloy          (iii) Beverage cans
    (d) Lead                (iv) Ideal for hospital use
20. Which of the following is the densest and heavy metal?
    (a) Aluminium
    (b) Copper
    (c) Lead
    (d) Magnesium

## Review Questions

1. List the five important properties of aluminium.
2. How are aluminium alloys designated as per IS 6051? Give three examples of alloys using this system.
3. Explain the temper designation system of aluminium as per IS 5052.
4. Explain some of the problems associated with the welding of aluminium alloys.
5. Write short notes on (a) duralumin, (b) aluminium-bronze, and (c) Y-alloy.
6. Compare structural steel with structural aluminium.
7. What are the problems associated with aluminium alloys in structural applications?
8. What is meant by anodized aluminium sections? What are the advantages of anodized aluminium sections?
9. What are the available forms of aluminium?
10. What are the uses of aluminium products in building construction? How are extruded sections made?
11. What is aluminium composite material? Where is it used?
12. List 10 advantages of using aluminium.
13. Describe the manufacture, properties and some of the uses of:
    (a) Aluminium
    (b) Copper
    (c) Zinc
14. Discuss the important properties of copper and explain its uses.
15. Describe the properties and uses of brass and bronze.
16. Discuss a few important properties of zinc.
17. What is meant by hot-dip galvanization? How is continuous hot-dip galvanizing performed?
18. How do galvanized coatings protect steel members?
19. List a few applications of lead. Why is the use of lead diminished in the recent past?
20. Write short notes on the following:
    (a) Chromium
    (b) Nickel
    (c) Magnesium
    (d) Nickel-silver
    (e) Electroplating , and
    (f) Vanadium
21. What are the problems associated with the use of non-ferrous metals?
22. What are the problems that occur with metal roof sheets? How can they be remedied?
23. Write a brief note on the sustainability of non-ferrous metals.

## ANSWERS

### Multiple-choice Questions

**1.** (a)-(v), (b)-(iv), (c)-(i), (d)-(ii), (e)-(iii)    **2.** (a)-(v), (b)-(iv), (c)-(i), (d)-(iii), (e)-(ii)
**3.** (c)          **4.** (b)          **5.** (c)          **6.** (c)          **7.** (b)          **8.** (c)
**9.** (b)          **10.** (c)         **11.** (a)         **12.** (a)         **13.** (c)         **14.** (c)
**15.** (a)         **16.** (b)         **17.** (a)-(ii), (b)-(iv), (c)-(i), (d)-(iii)          **18.** (b)
**19.** 17. (a)-(iii), (b)-(iv), (c)-(ii), (d)-(i)       **20.** (c)

# CHAPTER 15
# GLASS

## 15.1 Introduction

Archaeological findings in Egypt and Eastern Mesopotamia indicate that the first manufactured glass dates back to 7000 B.C. The first glassmaking manual from the library of the Assyrian king Ashurbanipal (669–626 B.C.) dates back to around 650 B.C. The discovery of 'glass blowing', by the Syrian craftsmen around the end of the 1st century was revolutionary, as it made glass production easier, faster, and cheaper, and made it available to even ordinary citizens. The art of glass making flourished in the Roman Empire and spread across Western Europe and the Mediterranean. The Romans were the first to use glass for architectural purposes, when clear glass was discovered in Alexandria around A.D. 100. A flourishing glass industry developed in Europe at the end of 13th century, when the glass industry was established in Venice. The properties of glass, which is neither a true solid nor a true liquid, make it one of the most versatile materials manufactured by mankind. Further discussions of glass in this chapter are confined mostly to its use in the building industry.

Several types of glasses are used in buildings, which include normal (annealed), laminated, tempered, heat strengthened, reflective, insulating, and wire glass. Clear glass is used for windows and doors and they may be used in double/triple layers to get the required thermal insulation. Safety glass, in the form of laminated clear glass, is used in locations where breakage may be injurious; traditionally, embedded wire mesh was used to resist fire. A wide range of colours, textures, finishes, and shapes of glasses are used for ornamental purposes. The global market for flat glass (covering sheet glass, float glass, etc., in which the shape of the glass is flat) in 2018 is estimated to be 87.4 MT and may reach 112.4 MT in 2022 (www.statista.com). Nippon Sheet Glass (NSG), Pilkington (based at Japan/ United Kingdom), Saint-Gobain (based at France), Asahi (base at Japan), and Guardian (based at USA), produce 67% of the total high quality float glass in the world. China produces and consumes more than 50% of the global output of flat glass.

Scientifically, glass may be considered an inorganic product of fusion, which has cooled to a right condition without crystallizing. It is an amorphous mixture of silicates of sodium, potassium, and calcium. It is typically hard and brittle, and chemically inert. It may be colourless or tinted and transparent to opaque. Glass differs from other ceramics because it is shaped at high temperatures and made to cool, whereas most ceramics are shaped when they are cold and then fired to get the product. In addition, glass can be reheated and reshaped. Hence, it may be considered a thermoplastic material. It is the most common material glazed into frames for doors, windows, and curtain walls. Glass is considered a green product and almost 100% of it can be recycled.

## 15.2 Composition and Classification

Glass is composed primarily of silica sand ($SiO_2$) obtained from sandstone (represents 70–74% weight of modern glass). Sandstone may contain pure silica, whereas river sand and beach sand may have impurities and hence may not be suitable for glass making (see also IS 488:1980). Ancient glass making consisted of 90% of pure silica, but had a melting point of about 2000°C. Glass made of pure silica had the ability to block UV radiation. Soda (sodium carbonate, $Na_2CO_3$) and lime (calcium oxide, CaO) are also used in the glass manufacture, along with aluminium or magnesium oxides.

Other ingredients such as lead oxide, barium and lanthanum oxide are added, if necessary, to increase the refractive index of glass, making it more reflecting and suitable for optical purposes (eyeglasses and lenses). Thorium oxide was used in the past for this purpose, but is not used now due to its radioactivity. Lead oxide also gives colour to the glass. Addition of sodium sulphate and sodium chloride prevents air bubbles forming in the glass mixture. Iron can strengthen the ability of glass to absorb infrared energy and heat. Boron oxide has the ability to strengthen the structure of glass and protect it from thermal expansion, cracking, and thermal shock.

Based on the ingredients of glass, it may be classified as (a) soda-lime-silicate glass, (b) borosilicate glass or pyrex glass, (c) lead-oxide glass, and (d) aluminosilicate glass. The chemical compositions of these types of glass are given in Table 15.1. Other aspects are briefly discussed in the following sections.

**Table 15.1** Chemical composition of common types of glass

| Element (%) | Type of glass | | | |
| --- | --- | --- | --- | --- |
| | Soda-lime-silicate | Borosilicate | Lead glass | Aluminosilicate glass |
| Silica sand, $SiO_2$ | 70–74 | 70–81 | 54–65 | 62 |
| Lime, CaO | 5–12 | – | – | 8 |
| Soda, $Na_2O$ | 12–18 | 4–8 | 13–15 | 1 |
| Alumina, $AL_2O_3$ | 1 | 2–7 | 2 | 17 |
| Magnesia, MgO | 4 | – | – | 7 |
| Boron oxide, $B_2O_3$ | – | 7–13 | – | 5 |
| Lead oxide, PbO | – | – | 18–38 | – |

**Soda-lime-silicate glass**  It is called *soda-ash glass*, *soda glass,* or *window glass*, and is obtained from the fusion of silica, soda, lime, and alumina. It accounts for about 90% of all the manufactured glass. Though the soda acts as a flux and lowers the melting point of silica to about 1200°C, it has the undesirable effect of making the glass water-soluble. Lime (obtained from limestone/dolomite as per IS 997:1973) makes the glass more workable and improves the weather resistance of the finished glass. Magnesium oxide (MgO) and aluminium oxide ($Al_2O_3$) are added to provide for a better chemical durability. Broken glass, called *cullet*, is added as a recycled material; it acts as a flux, and results in savings of raw materials and energy cost. Soda-lime glass is relatively inexpensive to manufacture; its ingredients are common and are melted at a lower temperature than other glass compositions. It is suitable for the glazing of windows and doors. It has a high thermal expansion and poor resistance to heat.

**Borosilicate glass**  It is also called as *pyrex glass* (commercial name given by Corning, Inc.). The main constituents of this glass are silica and boron trioxide ($B_2O_3$) and some soda ($Na_2O$) and alumina ($Al_2O_3$) are also used. It has a low coefficient of thermal expansion of $3.25 \times 10^{-6}$/°C as compared to about $9 \times 10^{-6}$/°C for a typical soda-lime glass. This property makes them less susceptible to stress caused by thermal expansion and hence less vulnerable to cracking due to thermal shock. Borosilicate glass has the ability to resist both moisture and chemical attack for long periods of time. It is commonly used for chemical and laboratory glassware, industrial equipment, transformer bushings, optical components, household cookware, and aircraft.

**Lead-oxide glass**   It is also called as *crystal glass* or *lead glass,* and is obtained from the fusion of a mixture of silica, lead oxide (PbO) – typically, 18–38% –, soda, and alumina. Historically, it was also known as *flint glass*. Owing to the health risks of lead oxide, it is now replaced by barium oxide, zinc oxide, or potassium oxide. Because of its high density, lead glass has a high refractive index making it sparkle brightly with a relatively soft surface, which can be easily decorated by grinding, cutting, and engraving. Lead-oxide glass has a refractive index in the range of 1.7–1.8, as against 1.5 of ordinary glass. Owing to its low viscosity, it is more workable in the factory, but does not resist heat. It is easier to cut but more fragile than other glasses. Lead glass is used to make a wide variety of decorative glass objects. Glass with higher lead oxide content (typically, 65%) may also be used as radiation shielding because of the ability of lead to absorb gamma rays and other forms of harmful radiation.

**Aluminosilicate glass**   Aluminosilicate glass is made with silica, alumina, lime, magnesia, barium oxide (BaO), and boric oxide ($B_2O_3$). This type of glass is resistant to weathering and water erosion. It is extensively used to make fibreglass, which is used to make glass-reinforced plastics (boats, fishing rods, etc.) and halogen bulb glass.

## 15.3 Manufacture of Glass

In 1832, the British Crown Glass Company was the first to adopt the 'cylinder method' to produce sheet glass, by blowing long cylinders of glass, which were then cut along the length, flattened on a cast-iron table, and then annealed. An automated glass manufacture was patented in 1848 by engineer Henry Bessemer. His system produced a continuous ribbon of flat glass by forming the ribbon between rollers. Bessemer also introduced an early form of 'Float Glass' in 1843, which involved pouring glass onto liquid tin. Several manufacturing methods were developed from 1902 to 1959. But the revolutionary float glass process, developed by Sir Alastair Pilkington and Kenneth Bickerstaff in 1959 at the Pilkington Brothers Company in the UK, made glass to be mass produced easily and at the desired quality.

### 15.3.1 Fourcault System of Sheet Glass Manufacture

The method of manufacture of sheet glass was first developed by Belgian Émile Fourcault during 1902, and this process was used globally. It is a vertical draw process as shown in Fig. 15.1. In this process, the glass is first produced in a molten state and filled in a tank. A ceramic die, known as 'Debiteuse', floats on the molten glass, and acts as a 'bait' to draw part of the molten glass slightly above the top surface of the die. As the bait is slowly lifted, it draws a sheet of glass vertically in the form of a ribbon, between jets of flame. The drawn sheet of glass is then drawn through rollers. Glass rollers hold the ribbon throughout the process, supporting its weight and aiding the drawing process. As the molten glass is drawn upwards into a chimney-like structure, called *lehr*, (about 10-m tall), it is rapidly cooled and solidified at about 650°C. It is further cooled slowly in the *lehr*, until it can be cut into discrete sheets of flat glass. The glass surface undergoes rolling and annealing, and cooled gradually in a chamber. The thickness of the glass was depended on the speed of glass being pulled between the rollers. The glass produced by this process was called the *sheet glass*. This process resulted in some distortions, irregularities and in homogeneities, mainly as a result of small differences in viscosity due to chemical or even thermal

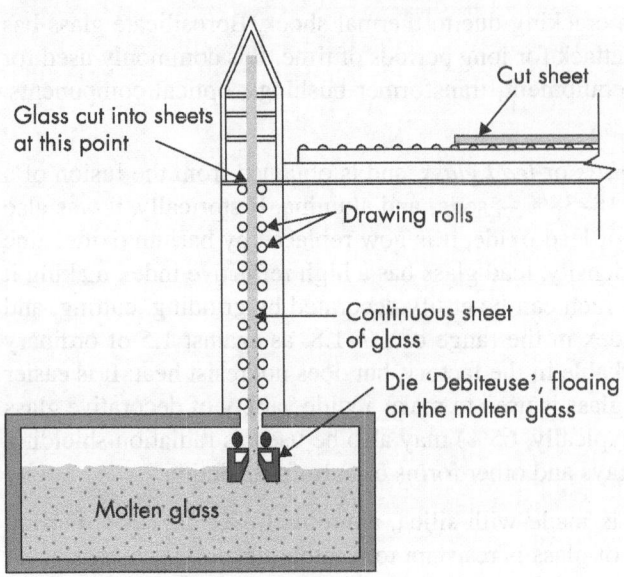

Glass cut into sheets at this point

Cut sheet

Drawing rolls

Continuous sheet of glass

Die 'Debiteuse', floaing on the molten glass

Molten glass

**Fig. 15.1** Schematic of the Fourcault system of sheet glass manufacture

variations. In addition, the rollers left marks on both sides of the glass, which needed grounding and polishing. Until the advent of the float process in the 1950s, perfect results were rare (wastages were about 20%) and it was very expensive (Nascimento, 2014).

### 15.3.2 The Float-glass Process

In 1959, British glassmaker Sir Alastair Pilkington (1920–1995) found a way to produce sheets of flat glass by floating molten glass on a bath of molten tin. Today, about 90% of glass sheet is made using Pilkington's method. It is called the *float-glass process*. In this process, the constituents of glass such as sand, limestone (calcium carbonate, $CaCO_3$), and soda (sodium bicarbonate, $NaHCO_3$) are mixed and heated to 1500°C in a furnace, where they melt to form molten glass. The molten glass then enters a 10-m wide and 50-m long float tank of molten tin; at a temperature of 1100°C (it leaves the float bath as a solid ribbon at 600°C). Molten glass is very viscous (thick), but the molten tin is very fluid. Hence, the glass flows over the tin surface, forming a flat layer, due to surface tension. As the bath cools, the glass solidifies, which is then transported to a cooling chamber. The resulting glass is flat and smooth on both sides and hence does not need to be ground or mechanically polished. This process is used to produce clear, tinted and coated glass for buildings and vehicles.

The principle of float-glass manufacture has not much changed from the 1950s. But the finished product has changed dramatically – from a single thickness of 6.8 mm to a range from 0.4 to 25 mm and in widths up to 3 m; from a glass ribbon frequently containing inclusions, bubbles, and striations to almost optical perfection. Present-day float glass has a 'fire finish' with the lustre of new chinaware. There are about 260 float plants worldwide with a combined output of about 800,000 tonnes of glass per week. A float plant, which operates non-stop for 11–15 years, makes around 6000 km of glass a year (www.britglass.org.uk).

### 15.3.3 Rolled Glass Process to make Plate Glass

Rolling was introduced to steel production in 1783 by Henry Cort, who successfully improved on earlier primitive attempts to use this technique. Rolling is a very rapid way to produce glass, but the surface of the produced glass is not even due to its contact with the metal rollers. The rolling process is used for the manufacture of solar panel glass, patterned flat glass, and wired glass. In this process, a continuous stream of semi-molten glass is squeezed between pairs of metal rollers to produce a glass ribbon with predefined thicknesses and patterned surfaces. The pairs of rollers revolve at the same speed but in opposite directions and spaced so that the distance between them is slightly less than the thickness of the glass. The hot glass cools to about 815°C and then goes into a metal *lehr* about 120-m long, where it is annealed (cooled). It is then taken out of the *lehr* and cut to the required lengths.

The production of *patterned glass* is a single-pass process, in which the molten glass passes through two cast iron or stainless steel rollers, kept at a temperature of about 1050°C. The top roller is made plain, whereas the bottom roller is engraved with the negative of the required pattern. The gap between the rollers are adjusted to get the required thickness of glass. After it attains the required thickness, the glass ribbon leaves the rollers at about 850°C. This glass ribbon is supported over a series of water-cooled steel rollers, until it reaches the annealing *lehr*, where it is annealed and cut to the required size.

The production of *wired glass* requires double pass process. In this process, molten glass from a common melting furnace is fed into two independently driven pairs of water-cooled forming rollers. A continuous ribbon of glass is produced through the first set of rollers, but it will have only half the thickness of the end product. A wire mesh with the required pattern is then laid over it, and a second feed of glass, with the same thickness as the first, is kept on top of the wire mesh. Now, this 'sandwiched' wire mesh between the two glass ribbons is rolled again through the second pair of forming rollers, to obtain the final product. After annealing, the finished wire glass ribbon is cut to size using special cutting tools.

## 15.4 Treatments given to Glass

Annealing and tempering are the two main treatments that are normally given to glass to improve its properties and appearance.

**Annealing** The manufactured glass ribbon is usually cooled in a controlled atmosphere. This process is called *annealing* and ensures slow and homogenous cooling of the produced glass. Without annealing, the surface of the glass will cool faster than the interior, resulting in the development of considerable internal stresses. During annealing, temperatures are closely controlled both along and across the ribbon.

**Heat treatment or tempering** Heat treatment, also called *tempering*, is used to improve the strength of sheet, plate, float, or patterned glass. In this process, the glass is heat treated to a uniform temperature of about 620°C (heated to its softening temperature) and then rapidly cooled. This produces compressive stresses in the surface of the glass, whereas the centre core remains in tension. The compressive stresses induced are in the range of 75 to 143 MPa on the surfaces. These stresses give the glass strength of four-to-five times their original values in order to resist impact forces and temperature variances. The heat treatment process does not change the light transmission and solar radiant heat properties of the glass. It has to be noted that tempering is not automatically done in the manufacturing process. It is a separate process. In 1874, Francois Barthelemy Alfred Royer de la Bastie invented tempered glass and hence it may sometimes be referred as Bastie glass. In 1877, German Frederick Siemens invented another method, but the first patent for the whole process of manufacturing was obtained years later by Austrian-American chemist Rudolph A. Seiden.

**Other treatments** In addition to the aforementioned treatments, the produced glass may be cut to the required size by either diamond cutter or a wheel made of hardened steel. The edges of cut glass may be ground and rounded by using grinding stones.
*Bent glass* is obtained by heating flat glass until it becomes red hot, carefully bending it to the desired shape, and then cooling it slowly. The resulting bent tempered glass has up to a fourfold strength increase compared to annealed glass of the same thickness.

*Frosted (opaque) glass* is produced by the sandblasting or acid etching of clear sheet glass. The frosted glass effect may also be achieved by the application of vinyl film. Silvering is the chemical process of coating glass with a reflective substance. During the 16th century, most glass was silvered with an amalgam of tin and mercury or by a layer of silver coating. Today, aluminium compounds are more often used for this purpose.

## 15.5 Properties of Glass

The physical and engineering or mechanical properties of glass are discussed in this section.

### 15.5.1 Physical Properties

The main physical properties of glass are given here.

1. It is a solid, hard, tough but brittle material, and does not deform plastically.
2. It has a disordered and amorphous structure. It is capable of being worked in many ways. It can be blown, drawn, or pressed. But it is difficult to cast in large pieces. It can be welded by fusion.
3. It absorbs, refracts, or transmits light. It can be made transparent or translucent, and polished. The transparency is the most used characteristic of glass.
4. It is fragile and easily breakable into large sharp pieces. In wired glass, the broken pieces stay in place due to the wires. Laminated glass, though subjected to cracking, will remain intact due to the plastic interlayer. Tempered glass will break into smaller pieces.
5. It is resistant to weather and chemicals except alkalis.
6. It is inert and biologically inactive material.
7. It is an excellent electrical insulator. However, molten glass is a conductor.
8. It is possible to obtain glass with diversified properties. It is possible to produce it as clear, colourless, diffused, and stained. It is available in a variety of colours. It can be frosted by sand blasting. It can be cleaned easily.
9. When there is fire, plain, heat-treated, or tempered glass will disintegrate and fall out of its frame, thus creating an opening, which will allow the fire to spread. Though wired glass may crack under heat, the pieces will be held for some time by the wires, preventing the creation of opening and the risk of the fire spreading to other areas.
10. Glass is 100% recyclable. It is also considered as one of the safest packaging materials due to its composition and properties.

These properties can be modified and changed by adding other compounds or heat treatment.

### 15.5.2 Engineering and Mechanical Properties

The mechanical properties of glass determine the amount of stress a glass can withstand. By varying the quantities of the basic constituents such as sand, soda, and lime, and by using certain special additives, the properties and characteristics of glass can be changed and a large variety obtained. The important properties as well as available sizes of some types of glass are provided in Table 15.2 [see also IS 16231 (part 1)]. It is seen from this table that the tensile strength of glass can be improved by heat-treatment. Full tempering increases the strength by 4–5 folds. Density of glass ranges from 2420 to 2530 kg/m$^3$. Modulus of elasticity of glass is around 70 GPa and the coefficient of thermal expansion is about $9 \times 10^{-6}$/°C (see IS 5623:1999, for the methods of determination).

**Table 15.2** Typical properties and available sizes of some types of glass

| Properties | Sheet glass | Float glass | Laminated glass | Tempered/toughened glass |
|---|---|---|---|---|
| Tensile strength (MPa) | 40 | 40 | 32 | 120–200 |
| Compressive strength (MPa) | 800–1000 | 1000 | 1000 | 1000 |
| U value (W/m². K) | 5.7 (12-mm thick) to 6.4 (19-mm thick) | 5.7 (12-mm thick) to 6.4 (19-mm thick) | 2.84 (3-mm thick) to 5.96 (6-mm thick) | 5.7 (12-mm thick) to 6.4 (20-mm thick) |
| Normally available sizes (mm) | 2440 × 3660 | 2440 × 3660 | 2000 × 3210 | 2440 × 3660 |
| Available thickness (mm) | (1–32) See IS 2835 | As per IS 14900 (normally, 2–19) | 4.38–20.76 | 3–19 |

## 15.6 Uses of Glass in Buildings

Glass is used for architecture applications (windows doors, ventilators, and display cabinets), illumination, electrical transmission, instruments for scientific research, optical instruments, domestic tools, and even textiles. Glass does not deteriorate, corrode, stain, or fade and therefore is one of the safest packaging materials. Figure 15.2 shows the glass staircase used in a jewellery shop in Austria.

**Fig. 15.2** Glass staircase and glass chandelier in Innsbruck, Austria

### Case Study | Louvre Pyramid

The Louvre Pyramid, designed by American architect I. M. Pei in 1989, consists of a large glass and metal pyramid surrounded by three smaller pyramids, in the main courtyard of the Louvre Palace in Paris. The large pyramid serves as the main entrance to the Louvre Museum. It has a height of 21.6 m and square base with side of 34 m. It consists of 603 rhombus-shaped and 70 triangular glass segments. The pyramid structure was engineered by Nicolet Chartrand Knoll Ltd. of Montreal and Rice Francis Ritchie of Paris.

(*Source*: M. Karthik)

## 15.7 Common Types

A number of manufactured glasses are available in the market with different names such as sheet glass, plate glass, and float glass. Some of the most commonly used types of glass are discussed in this section.

### 15.7.1 Sheet Glass

It is also called as *annealed glass,* and is the most extensively used in small panels of doors and windows, and partitions in all types of buildings. As per IS 2835:1987, it is a transparent, flat glass having glossy, fire-finished, apparently plane and smooth surfaces, but having a characteristic waviness of surface. It may have a greenish colour tint, due to the presence of some impurities such as iron. Sheet glass has very high-energy transmission when exposed to sunlight. It provides a clear view of the objects across it. It is also used for further processing to other glass types. It is normally available in thicknesses of 2, 2.5, 3, 4, 5, 5.5, and 6 mm. Additional glass thicknesses, up to a nominal thickness of 32 mm, are available (see IS 2835:1987). Usually, glass is specified in weight per square feet or square meter (see Table 15.4).

As per IS 2835, sheet glass is classified into four grades. These grades and their uses are given in Table 15.3. Of these, B Quality or Ordinary Quality (OQ) is mostly used in buildings.

**Table 15.3** Grades of sheet glass and common uses

| Grade of sheet glass | Uses |
| --- | --- |
| AA Quality or Special Selected Quality (SSQ) | Used where superior quality of safety glass is required. |
| A Quality or Selected Quality (SQ) | Used for selected glazing, in mirrors, and as safety glass. |
| B Quality or Ordinary Quality (OQ) | Used for glazing in buildings |
| C Quality or Greenhouse Quality (GQ) | Used for green house glazing, frosted glass, and strips for flooring |

**Table 15.4** Approximate weight of architectural flat glass

| Glass thickness (mm) | Approximate weight (kg/m$^2$) |
| --- | --- |
| 2.0 | 5.1 |
| 3.0 | 7.6 |
| 4.0 | 9.9 |
| 5.0 | 11.9 |
| 6.0 | 14.6 |

If the exact thickness is known, the exact weight may be calculated using a glass density of 2530 kg/m$^3$.

### 15.7.2 Plate Glass

Plate glass has the same composition as sheet glass. After the plate glass is manufactured and cut to size, it is run through huge grinders that grind the surface to a flat finish. After the grinding, it is polished to get a glossy surface. Plate glass can be twin ground or ground on one side then turned over and ground on the other side. Twin ground glass is ground on both surfaces at the same time, to produce glass of very even thickness. They are stronger, more transparent, and have less waviness than sheet glass. Though the term, plate glass, denotes the polished variety, it may be available as rough cast, rolled (patterned), and polished. Depending on the manufacturer they are produced in thickness ranging from 3 to 32 mm, widths from 1.8 to 3.6 m, and lengths from 2 to 7 m. They are available in two grades: (a) silvering quality, suitable for mirrors and (b) glazing quality intended for quality glazing.

### 15.7.3 Float Glass

The manufacturing process of float glass produces a fire polished, nearly optically flat surface that does not have to be ground or mechanically polished. It is of uniform thickness. It is further annealed to relieve all the stresses. Showcase windows, exposed windows in commercial buildings, and façades of tall buildings are glazed with this type of glass (see Fig. 15.3). Specifications for float glass may be found in IS 14900:2000.

### 15.7.4 Extra Clear Glass

It is a high-value glass, free from impurities such as iron. It has high light transmission of more than 92% and is free from interference with the true colour and sparkling of objects across it. Extra clear glass can be fabricated, laminated, painted, cut, and tempered just like ordinary float glass. It is used for a sparkling display of expensive materials like jewellery and watches, and for solar applications.

**Fig. 15.3** Use of float glass in a commercial building; note also the skylight

## 15.8 Special Types

Several types of glasses have been developed for special purposes or to overcome a particular weakness of glass. These special types include safety glass, translucent glass, etched glass, tinted glass, coated or reflective glass, patterned glass, stained glass, glass blocks, insulated glass/double-glazed window units, bullet proof glass, etc. They are briefly discussed in this section.

### 15.8.1 Safety Glass

It is of three types – tempered, laminated, and wired. These three types of glass have been developed to give glass more resistance to breakage and keep glass from flying and injuring persons, when it is accidentally broken.

**Tempered or toughened glass**   As per IS 2553(Part 1):1990, there are four types of safety glass: (a) toughened safety (tempered) (TS) glass, (b) toughened-float (TF) safety glass, (c) laminated safety (LS) glass, and (d) laminated float (LF) safety glass.

Although not unbreakable, *toughened glass* resists bending stress better than plate glass. Tempered glass cannot be cut or drilled and hence manufactured according to specification. Any holes or surface textures should be made before the glass is tempered. Any attempt to cut, drill, grind, or sand blast the toughened glass will result in the glass being broken into relatively small pieces. As there will not be any sharp corners or jagged edges like normal glass, these pieces will not cause any serious injury to people. Tempered glass provides increased resistance to both sudden temperature changes and temperature differentials up to about 150°C compared to normal glass, which can withstand up to 40°C only. It is available as flat, curved and as self-tinted/coloured or with coloured inter-layers. Heat soaking, in a chamber to about 290°C, may minimize the risk of nickel-sulphide-induced spontaneous fracture in toughened glass [IS 16231 (part 1):2016].

Tempered glass is used in commercial applications where wind, snow, or thermal loads exceed the strength capabilities of normal (annealed) glass. Tempered glass is used extensively in places where human safety is an issue such as in entrance doors, shower doors, storm doors, sliding/swing doors, skylights, escalator side panels, staircase handrails, curtain walls of high-rise buildings, showroom and lobby façades, microwave oven, side and rear windows in vehicles (the windscreen is made of laminated glass), and in viewing partitions of sports complexes, resorts, and airports.

**Heat strengthened glass**  It is a type of tempered glass in which a surface compression of 41 to 64 MPa is induced in the glass as compared to 75 to 143 MPa in the case of fully tempered glass. It is used for its increased mechanical strength, which is twice that of normal annealed glass, though it is only half of fully tempered glass. Except for strength and thermal resistance, the properties of this type of glass are similar to the annealed glass. Heat-strengthened glass provides adequate resistance to thermal stresses that are developed when using high performance glazing materials such as tinted glass and reflective glass. It also provides necessary resistance to heat building up when used as spandrel glass (An opaque glass used in windows and curtain walls to conceal internal construction). Heat-strengthened glass with its flatter surface also results in facades having less optical distortions [IS 16231(part 1)].

In addition to its use in facades, it is also used in applications such as double-glazing.

**Laminated glass**  Invented by Frenchman Édouard Bénédictus in 1903, *laminated glass* is composed of two or more layers of glass, with one or more layers of transparent/pigmented and specially treated plastic Polyvinyl Butyral (PVB), of thickness 0.40 to 0.60 mm, sandwiched between the glass layers. The glass layers of laminated glass can be float, plate, sheet, or tempered glass; heat bent or curved glass, wired glass, or any combination of these types. The PVB is quite strong and extremely flexible/stretchable. It makes the laminated glass a crystal clear safety glass. The PVB interlayer bonds to the glass so that when it breaks, the broken pieces are held in place and kept from flying, thereby reducing the risk of injury. Laminated glass is capable of stopping flying debris and limit or avoid splintering on the opposite side of the impact. If a hole is made on it, the edges are likely to be less jagged than that of ordinary glass. Laminated glass should conform to the requirements of IS 2553 (Part 1). Laminated glass of multiple layers, with thickness ranging from 20–75 mm is also used as *bullet-proof glass*. This glass can withstand high-power firearms at very close range, and will not allow bullet to pierce through it.

In addition to increased strength, laminated glass is an excellent barrier to noise. The damping performance of the plastic interlayer makes laminated glass an effective sound insulation product. Owing to this property, it is considered ideal for airports, hotels, data-processing centres, recording studios, and any building near airports, highways, or train lines. Ultraviolet light (UV) is the leading cause of deterioration and fading of several items such as furnishings, pictures, and fabrics. Laminated glass can screen out most of the UV radiation of the Sun, thus protecting interior furnishing, displays, etc., from fading. When exposed to heat, laminated glass may break, but stays in place longer. The risk of thermal breakage is avoided only when heat strengthened/tempered laminated glass is used. Where privacy is required, laminated glass can be made opaque and also coloured.

Laminated glass provides more safety because even if the glass is broken, the interlayer will be intact and will resist intrusion of any persons inside the building. Moreover, cutting the glass from one-side alone is very difficult and hence ordinary glasscutters cannot be used as break-in tools. In addition, laminated glass also resist impact. Thus, in multi-ply laminated glass, it is not possible to penetrate the glass, even if several blows are given at the same location. It is also possible to resist the impact of heavy objects, bullets, or even small explosions. Thus, laminated glass will be an ideal choice where terrorism and urban crime rates are high.

During earthquakes, normal glass in windows/doors is easily broken and shards of broken glass may injure or even kill persons. In addition, conventional glass could be shattered during heavy winds, tornados, and hurricanes, and the flying debris can cause injuries. Moreover, as the exterior glass is damaged, the interior spaces may also be exposed to the devastating outside weather. This situation could be avoided by using laminated glass. It is because, even when broken, it remains in the frame preventing glass shards from flying, maintaining a protective envelope, thus not allowing the harsh weather inside the building.

Furthermore, laminated glass retains its strength and also its colour throughout its life.

Because of the aforementioned characteristics, laminated glass is used in several applications. These includes residential and commercial buildings, hospitals, libraries, airport terminals, embassies, computer centres, aquariums, and also in high-security places such as banks, airport terminals, embassies, drive-through windows, armoured vehicles, jewellery shops, and burglar-resistant showcases. Other areas where laminated glass is used include curtain wall glazing, skylights, animal observatory windows, acoustic glazing and noise control, and in earthquake/high velocity winds/fire-resistance applications.

**Wired glass**    It is a product in which a wire mesh has been inserted in between two glasses during production (see Section 15.3.3). It has an impact resistance similar to that of normal glass. Wired glass was commonly used in the early 1970s in doors or windows to comply with fire safety provisions, to retard the spread of smoke or flame for a specific period of time. The sole purpose of the wire was to hold the glass shards together when the heat of a fire broke the glass. Keeping the broken shards from falling to the ground helped slow or stop the spread of smoke and flame from one specific area of a building to another. Specification for wired glass may be found in IS 5437:1969.

Contrary to popular belief, it has been found that the strength of wired glass is only about half of annealed glass. This is due to the fact that internal stresses develop during the manufacturing process, due to the difference in the cooling rates of glass and wire. The wires in the wired glass may even corrode due to imperfections. It has also been found that the wired glass is also not acting as 'safety glass' at all, as it actually breaks with much less force than it takes to break ordinary glass (laminated or tempered glass are more impact-resistant than wired glass). In addition, when the wire glass breaks due to human impact, the wire in the glass grabs onto human flesh like teeth, causing severe damage to the victim. Many students in the USA were injured due to this type of glass. Owing to this, the 2006 International Building code (IBC), banned traditional wired glass in all impact-prone locations of any new building.

## 15.8.2 Translucent Glass

It is also called as *frosted glass*, *obscured glass*, or *ground glass*. It is a decorative glass having texture or pattern on one or both faces. The pattern can be impressed during the rolling process or sandblasted or etched later (see Section 15.8.3). In addition to diffusing light and obstructing visibility from the outside, the texture or patterns soften the interior lighting. Usually, one side is textured (*frosted*) and the other side is flat. If installed in windows, the textured side is placed inside the room and the plane glass side is made to face the outside, to prevent dust collection and facilitate easy drainage of water. Some of the patterns available on this type of glass are floral, hammered, granular, pebbled, ribbed, corrugated, squares, and strips. This type of glass is usually more fragile and may be used in doors and windows of bathrooms, and toilets, partitions, and office doors (see Fig. 15.4).

**Fig. 15.4** Use of translucent glass in bathroom

## 15.8.3 Etched Glass

It is similar to patterned glass but is produced by spraying the surface of float glass with hydrofluoric acid or by dipping the glass in it. The amount of etching depends on the concentration of the acid and the length of exposure. The finished surface may be rough, frosted, or almost opaque. Thus, etched glass admits light while providing softening and vision control. If the glass is fixed by putty, the glass must be installed with the smooth side facing the putty side. Its use is similar to translucent glass.

### 15.8.4 Tinted Glass

It is produced by adding small amounts of metal oxides to the float or rolled glass composition. These small additions colour the glass bronze (selenium oxide), green (iron oxide), blue or grey (cobalt oxide) but do not affect the basic properties of the glass except for changes in the solar energy transmission. The colour is homogeneous throughout the thickness. Tinted glass has filtering properties that help reduce eyestrain due to dazzle. It absorbs 30–40% of solar radiation, depending on the tint and thickness of glass. Tints like green allow more visible light and cut out infrared radiation. Although darker shades reduce the amount of heat being transmitted to the interiors, they also reduce the amount of transmitted daylight. This type of glass is used in doors, windows, and partitions.

**Stained glass** It consists of pieces of glass of different colours, which are fitted together to make decorative windows. Stained glass windows are commonly used in churches, mosques, and other significant buildings, for decorative and informative purposes. Stained glass windows in churches and monasteries in Britain were used as early as the 7th century. Stained glass has also been used as a decorative element in public buildings.

**Fig. 15.5** Reflective glass used in the façade of a building in Gaithersburg, MD

### 15.8.5 Coated/Reflective Glass

The *coated glass*, which is also called as *energy efficient glass*, *low-emissivity glass (Low E-glass)*, or *heat-reflecting glass*, is obtained by coating one side of ordinary float glass with microscopically thin, virtually invisible, metal, or metallic oxide layer. Visibility is possible through coated glass from the dark side to the lighter side. These coatings have high reflective properties to control transmission of light, heat, and solar radiation. Thus, they provide energy conservation, comfortable living, and environmental protection from UV rays. The metallic coating can be applied to clear or tinted glass. An application of this type of glass is shown in Fig. 15.5.

**Low emissivity coating** Low-emissivity (low-e) coatings on glass control heat transfer through windows with insulated glazing. Windows manufactured with low-e coatings typically cost about 10–15% more than regular windows, but they reduce energy loss by as much as 30–50% (https://energy.gov). A low-e coating is a microscopically thin, virtually invisible, metal, or metallic oxide layer deposited directly on the surface of one or more of the panes of glass. This coating is applied under vacuum chamber conditions and precisely controlled. The low-e coating reduces the infrared radiation from a warm pane of glass to a cooler pane, thereby lowering the U-factor of the window. Use of low-emissivity glass results in more efficient windows because radiant heat originating from indoors in winter is reflected back inside, whereas infrared heat radiation from the sun during summer is reflected away, keeping it cooler inside. Different types of low-e coatings have been designed to allow for high solar gain, moderate solar gain, or low solar gain. A low-e coating can also reduce a window's visible transmittance (VT) unless a spectrally selective coating is used. This type of glass is used in school and office buildings to control sun glare and heat transmission.

Although low-e coatings are usually applied during manufacturing, some films are available which can be pasted on the surface of clear glass. These films are inexpensive compared to total window replacements, last 10–15 years without peeling, save energy, reduce fabric fading, and increase comfort.

**Spectrally selective coatings** A special type of low-e coating is spectrally selective, filtering out 40–70% of the heat normally transmitted through insulated window glass or glazing while allowing the full amount of light transmission. Spectrally selective coatings are optically designed to reflect particular wavelengths, but remain transparent to others. Such coatings are commonly used to reflect the infrared (heat) portion of the solar spectrum while admitting more visible light. They help create windows with a low U-factor and solar heat gain coefficient but a high visible transmittance (https://energy.gov).

Spectrally selective coatings can be applied on the various types of tinted glass to produce customized glazing systems capable of either increasing or decreasing solar gains according to the desired aesthetic and climatic effects. Computer simulations have shown that window glazing with such coatings can reduce the cooling requirements of new homes in hot climates by more than 40% (https://energy.gov).

**Reflective coatings** Reflective coatings on window glass reduce the transmission of solar radiation, blocking more light than heat. Therefore, they greatly reduce a window's visible transmittance and glare, but they also reduce a window's solar heat gain coefficient (SHGC). The reflective coating imparts an enhanced appearance to the exterior of the building. The special coatings also produce a mirror effect, preventing people from seeing through the glass. Reflective coatings usually consist of thin, metallic layers, and come in a variety of colours, including silver, gold, and bronze. Reflective window glazing is commonly used in hot climates to control solar heat gain. It has to be noted that heat-reflective glass by itself does not provide the needed insulation; it has to be a part of double or triple glazed window. If designed/used properly, reflective glass can reduce the air-conditioning bill. The reduced cooling energy demands can be offset by the need for additional electrical lighting, so reflective glass is used mostly for special applications (https://energy.gov). Figure 15.6 shows the use of reflective glass in the curtain wall of an office building.

**Fig. 15.6** The use of reflecting glass in the curtain wall of an office building

## 15.8.6 Insulated Glass/Double-glazed Window Units

Insulated window glazing refers to window units with two or more panes of glass. To insulate the window, the glass panes are spaced apart (ideally, at 12 mm, but can range from 10 to 16 mm) and hermetically sealed (completely sealed even for air), leaving an insulating air space (see Fig. 15.7). The units can be made up of the various types of glass including low-e, laminated, toughened, tinted, or reflective glass. Insulated window glazing primarily lowers the U-factor, but it also lowers the SHGC. The edge seal not only binds the individual sheets of glass together, but also protects the cavity between the glasses from outside influences. The edge seal may be a bonded metal seal, rubber or synthetic seal, lead seal, or fused glass edge. The spacer ensures the precise distance between the glass panes and can be made of aluminium, composites, or plastics. The space between the sheets of glass is normally filled with dehydrated air, but can be also be filled with desiccant to prevent moisture condensation inside the glass. Gases such as Argon and Krpton can also be used for better thermal performance and hydrogen fluro-oxide for better acoustic performance. The low-heat conductivity of the enclosed dry gas between the glass panes drastically reduces the thermal heat transmission through the glass to 2.8 W/m²K (with 12-mm thick spacer between two 6-mm glasses) as compared to 5.73 W/m²K for normal 6-mm glass [IS 16231 (part 1):2016]. The U-values for double/triple glazed windows and doors filled with air/argon gas with wooden or plastic frames are given in Table 15.5 (www.gov.scot). Insulated window glazing also helps is reducing the direct solar energy, when the outer pane is a solar control glass (see also Section 20.11 of Chapter 20).

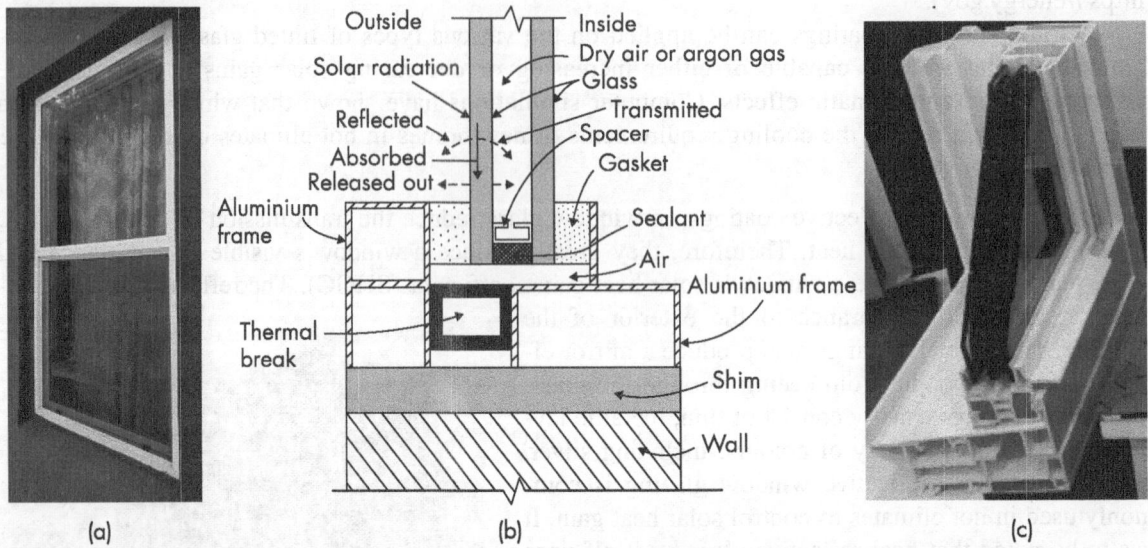

**Fig. 15.7** (a) Double-glazed insulating window (b) Schematic drawing of insulating glass unit
(c) 3-D view of the cut frame

**Table 15.5** U-values (W/m²K) for windows and doors with wood or plastic frames

| Type of window/door | Gap between panes | | |
|---|---|---|---|
| | 6 mm | 12 mm | 16 mm or greater |
| Double-glazing (air filled) | 3.1 | 2.8 | 2.7 |
| Double-glazing (low-e, $\varepsilon_n = 0.2$, air filled) [See Note 2] | 2.7 | 2.3 | 2.1 |
| Double-glazing (low-e, $\varepsilon_n = 0.1$, air filled) | 2.6 | 2.1 | 1.9 |
| Double-glazing (low-e, $\varepsilon_n = 0.05$, air filled) | 2.6 | 2.0 | 1.8 |
| Double-glazing (argon filled)[See Note 3] | 2.9 | 2.7 | 2.6 |
| Double-glazing (low-e, $\varepsilon_n = 0.2$, argon filled) | 2.5 | 2.1 | 2.0 |
| Double-glazing (low-e, $\varepsilon_n = 0.1$, argon filled) | 2.3 | 1.8 | 1.7 |
| Double-glazing (low-e, $\varepsilon_n = 0.05$, argon filled) | 2.3 | 1.8 | 1.7 |
| Triple-glazing (air filled) | 2.4 | 2.1 | 2.0 |
| Triple-glazing (low-e, $\varepsilon_n = 0.2$, air filled) | 2.1 | 1.7 | 1.6 |
| Triple-glazing (low-e, $\varepsilon_n = 0.1$, air filled) | 2.0 | 1.6 | 1.5 |
| Triple-glazing (low-e, $\varepsilon_n = 0.05$, air filled) | 1.9 | 1.5 | 1.4 |
| Triple-glazing (argon filled) | 2.2 | 2.0 | 1.9 |
| Triple-glazing (low-e, $\varepsilon_n = 0.2$, argon filled) | 1.9 | 1.6 | 1.5 |
| Triple-glazing (low-e, $\varepsilon_n = 0.1$, argon filled) | 1.8 | 1.5 | 1.4 |

*(Contd)*

**Table 15.5** (*Contd*)

| Type of window/door | Gap between panes | | |
|---|---|---|---|
| | 6 mm | 12 mm | 16 mm or greater |
| Triple-glazing (low-e, $\varepsilon_n$ = 0.05, argon filled) | 1.7 | 1.4 | 1.3 |
| Solid wooden door 3.0 | | | |

1. These U-values are for frames composed of 30% of the total window area.
2. $\varepsilon_n$ is the normal emissivity of the low-e coating (Corrected emissivity is used in the calculation of glazing U-values). Uncoated glass is assumed to have a normal emissivity of 0.89.
3. The gas mixture is assumed to consist of 90% argon and 10% air.
4. For doors which are half-glazed, the U-value of the door is the average of the appropriate window U-value and that of the non-glazed part of the door (e.g., 3.0 W/m²K for a wooden door).

(*Source*: https://www2.gov.scot/resource/doc/217736/0088291.pdf, reprinted with permission)

The following are some of the points to be noted while using double-glazed insulated windows.

1. Compared to normal glass, insulating glass reduces the heat transferred by conduction and convection, due to inside and outside temperature difference, to nearly half. This effect is very useful in winters as it may preserve inner heat and save energy costs.
2. Heat-absorbing or heat-reflective glass, when used on the outside walls, reduces the load on the cooling system, during summer.
3. In humid climates, the use of monolithic glass may lead to condensation, due the difference in temperature between inside and outside. However, the use of double-glazed window will prevent dew condensation and the glass becoming cold, due to the insulating effect of the air/argon gas layer between the glasses.
4. If glass panes of asymmetrical thickness are used in an insulating glass, the exterior noise pollution could be significantly reduced inside the room. The combination of insulating glass used has a bearing on the sound reduction. When one or both panes of the window glass are laminated or acoustic laminated, the reduction in sound will be considerable.
5. Double-glazed windows help in saving energy costs, offer increased personal comfort, and increased strength to resist wind load.

### 15.8.7 Glass Blocks

These are glass units, transparent and translucent, produced by a pressing process, in which two hollow half blocks are first formed in moulds and then fused together to form a hollow, hermetically sealed block. These blocks can be hollow with a dividing membrane, or can be even solid. The membrane, if used, is usually of glass fibre and may be coloured. Corner and radial blocks are also available to have the desired architectural finish. They have high degree of thermal insulation and noise reduction. Dividing the core by a membrane improves the insulating qualities of the glass block by decreasing the U-value by about 0.3 for a given size of block. The U-value without membrane is about 1.7 W/m²·K. A recent innovation is the inclusion of argon gas within the hollow of glass blocks, which decreases the U-value to 1.5 W/m²·K. Hollow glass blocks may be smooth faced, light-diffusing, sculptured, or even ceramic glazed.

Glass blocks are usually manufactured on a module of 100 mm and range in size up to 300 × 300 mm. They should not be used to construct load-bearing walls and used in the construction of partition walls (see Fig. 15.8). They are joined by cement or cement-lime mortar. The mortar

**Fig. 15.8** Use of glass blocks in a community room in USA

has to be stiff, like peanut butter, as the blocks will not absorb the moisture like concrete blocks. When used for walls, the height is limited to 6 m, and the width to 5 m, with total area less than about 13.5 m². Glass walls, in addition to low thermal conductivity, are sound proof, hygienic, ornamental, and provide excellent light transmission.

---

**Case Study | Crown Fountain, Chicago**

Crown Fountain is an interactive work of public art and video sculpture featured in Chicago's Millennium Park. The fountain is composed of a black granite reflecting pool placed between a pair of glass brick towers. The towers are 15.2-m tall and use light-emitting diodes (LEDs) to display digital videos on their inward faces. Designed by Catalan artist Jaume Plensa and executed by Krueck and Sexton Architects, it was constructed at a cost of $17 million. The Crown Fountain was opened to public on July 2004.

*Photo:* Er M. Karthik

---

### 15.8.8 Heat-absorbing and Glare-reducing Glass

Controlled quantities of ferrous iron admixture are added to *heat-absorbing glass* to absorb much of the energy of the Sun. Although it dissipates much of the absorbed heat, some of the heat is retained. Hence, heat-absorbing glass may become hotter than ordinary plate glass. It is available as plate and sheet glass and as patterned, tempered, wired, and laminated types. As heat-absorbing glass has a higher rate of expansion, proper care should be exercised while cutting, handling, and glazing. Edge stresses may develop, if it is subjected to sudden heating or cooling. Similarly, the heat-absorbing glass may break, if the edges are improperly cut or damaged.

*Glare-reducing glass* is available in two types: (a) a transparent glass with a neutral grey tint–this type preserves true colour vision but lowers light transmission, and (b) a translucent glass, usually in white colour – this type reduces glare in addition to providing wide light diffusion. Both these types have heat-absorbing qualities and absorb some radiant energy of the Sun. However, no special precautions are required as that of heat-absorbing glass. Glare-reducing glass has physical characteristics similar to that of plate glass.

### 15.8.9 Self-cleaning Glass

It is a type of glass with a surface that keeps itself free of dirt and grime through natural processes. The first self-cleaning glass, invented in 2001 by Pilkington Glass, was based on a thin film titanium dioxide coating. The glass cleans itself due to two mechanisms: (a) the photo-catalytic effect, in which the organic dirt on the glass is broken down by the UV rays even on overcast days; (b) the hydrophilic effect in which rain water is attracted to the surface of the glass, which washes away the broken-down organic compounds.

### 15.8.10 Associated Glazing Materials

Different types of sealers are required to install, hold fast, and seal a window in its setting; they include: sealants [IS 12118(Part 1)], wood-sash putty (IS 419), metal-sash putty, butyl and acrylic compounds, synthetic rubber compounds based on a polysulphide polymer [IS 11433 (Part 1)], preformed butyl and foam tapes, and rubber materials.

### 15.8.11 Types of Window Frames

The overall energy efficiency of a window could be enhanced by improving the thermal resistance of the frame, particularly its U-factor. There are some advantages and disadvantages to all types of frame materials. In general wood, fibreglass, vinyl, and some composite frame materials provide greater thermal resistance than metal.

**Aluminium or metal frames**  They are very strong, lightweight, and require little maintenance. However, as metals are poor insulators, the conduct heat very rapidly. To reduce heat flow and the U-factor, the windows should be provided with a continuous, 35 to 60 mm wide, insulating plastic strip (called *thermal break*) between the inner and outer frames. The wider the thermal break the better will be the window's insulation.

**Wooden frames**  The insulation provided by wooden window frames is relatively better. However, wooden frames also expand and contract in response to weather conditions, and require regular maintenance. By providing aluminium or vinyl cladding, the maintenance requirements can be reduced.

**Vinyl frames**  They are made of polyvinyl chloride (PVC) with UV stabilizers, thus preventing the deterioration of the material due to sunlight. Vinyl window frames have good moisture resistance and require no painting. To make these frames thermally superior to standard wooden frames, their hollow cavities should be filled with insulation.

## 15.9 Fibreglass

Another glass product that is used widely today to conserve energy is fibreglass. To produce fibreglass, the glass fibre is first extruded through small orifices (typically 5–25 $\mu m$ in diameter) in bushing plates made of platinum alloy, from the bottom of the molten glass furnace. Water jets cool the filaments as they exit the bushing at about 1204°C. A chemical coating (called size) is then applied. Finally, the drawn, sized filaments are collected together into a bundle, called *roving*. These rovings are then either used directly in a composite application such as pultrusion or in an intermediary step, to manufacture fabrics such as chopped strand mat. This is used as reinforcing agent for many polymer products. The resulting composite material, is called as fibre-reinforced polymer (FRP) or glass-reinforced plastic (GRP), and popularly known as *fibreglass*. GRP is used for many purposes including waterproofing of roofs and manufacture of water tanks. Glass fibres may also be used with concrete. GRP polymer rebars are used where there is a possibility of steel corrosion (also see Section 13.3.6 of Chapter 13).

## 15.10 Plastic Sheets for Glazing

Sheets made of thermoplastic acrylic resin (*Plexiglas*®, *Perspex* and *Lucite*, are trade names), and polycarbonate (trade names are *Lexan* and *Makrolon*), which are transparent like glass, are also available in flat and corrugated sheets. These materials are readily formed into curved shapes and hence often used in place of glass. Acrylic sheeting is shinier and less expensive, whereas polycarbonate sheet is much stronger and 35% more expensive. Acrylic sheeting cracks more easily than polycarbonate, and not recommended in areas where flames may be present. Polycarbonate sheets can handle temperatures up to 115°C. Both types have higher level of impact resistance than glass (acrylic sheets have 17 times and polycarbonate sheets 250 times). However polycarbonate sheets are easier to scratch and cannot be polished. Acrylic and polycarbonate both weigh less than half of comparable piece of glass; yet they are both much stronger than glass. Both materials are also very easy to clean. These sheets can be obtained in transparent, translucent, or opaque sheets and in a wide variety of colours. They are also commonly used as roofing sheets.

Recently, Dr Lars Berglund and associates of the Wallenberg Wood Science Center, KTH Royal Institute of Technology, Sweden, have developed an optically transparent wood veneer by removing lignin in the wood chemically and incorporating polymethyl methacrylate.

## 15.11 Characteristics and Performance of Glass

In residential and commercial buildings, glass is used primarily in windows and sometimes in doors. The main aim of using glass is to obtain as much daylight as possible (see IS 2440:1975 and IS 6060:1971, for more guidelines on day lighting of buildings). Direct sunlight (causing direct glare), bright sky (sky glare), or light reflected from a bright surface (reflected glare) may cause glare. In such situations, slightly tainted glass may be used in the windows to reduce the glare. The following three factors should be given due consideration, while selecting a glass for any building: (a) solar control, (b) wind loading, and (c) sound insulation. These three factors are discussed briefly here.

### 15.11.1 Solar Control

When solar radiation strikes glass, it is partly reflected, partly absorbed in the thickness of the glass and partly transmitted (see Fig. 15.7). The ratio of each of these three parts to the incident solar radiation defines the reflectance factor, the absorptance factor, and the transmittance factor of the glazing. Factors which will affect these ratios for a given incidence are the tint of the glass, its thickness, and in the case of a coated glass, the nature of the coating. The solar radiation that reaches the earth consists of 3% UV, 55% infra-red radiation (IR), and 42% visible light. UV radiation has a shorter wavelength and a higher energy level than visible light, whereas IR radiation has a longer wavelength and a weaker energy level.

If clear glass is used in a building, it transmits incident short wave radiation, which is absorbed by interior objects and surfaces, which then retransmit the energy as thermal radiation. In an air-conditioned room, the solar energy entering through the glass is thus trapped inside the room, tends to heat up, raising the temperature inside the room. This is called the *greenhouse effect*. The efficiency of the air-conditioning units is affected by this effect. Such solar gain can be reduced by using external/internal shading devices such as awning, sun shade, screens, and blinds. Obviously, external shading is more effective than internal shading. The location of windows in the building with respect to the sun in summer, and a high canopy of trees planted nearby (especially in the western side) are also important factors in the design of the air-conditioning system.

UV radiation can be virtually eliminated by the use of PVB laminated glass. Alternatively, a body-tinted glass may be used, which will filter light selectively; for example, yellow glass will absorb mainly violet and blue light. Glass with a low solar factor could also be used to reduce the thermal effect of the radiation. For improved heat and sound insulation, double or multiple glazing should be used. Double/triple glazing windows with clear glass on the inner side and solar reflecting type on the outer side, will result in improved performance. The heat insulation mainly depends on the spacing (increases only up to a spacing of 10 mm) and not on the thickness of glass. More information on these aspects may be found from IS 16231 (Part 2).

### 15.11.2 Wind Loading

In coastal regions and in tall buildings, the glass will be subjected to heavy wind loads. The design of these glasses for such loads must be carried out as per IS 16231 (Part 1):2019, which also contains programme for determining the thickness of glass plate for windowpanes. IS 16231 (Part 1) also contains dimensional and framing requirements, devises, and associated glazing materials for fixing the glass, and the methods of cleaning and maintenance (see also IS 1081:1960 and IS 1003(Parts 1 and 2):1994).

## Case Study   Glass window failures at John Hancock Tower at Boston

Designed by architect I.M. Pei, the 60-storey, 240-m tall John Hancock Tower at Boston, Massachusetts, USA, was built during 1969–1976 [see Figure (a)]. The beauty of the building was marred, when several of its 1.2 × 3.4 m windowpanes (each weighing 227 kg) detached from the building and crashed to the sidewalk hundreds of feet below during 1972–1973. Owing to this, the openings were covered with plywood before the repair work could be done [see Fig. (b)], causing this building to be nick named *Plywood Palace*. Police closed off surrounding streets whenever winds reached 72 km/h (Subramanian, 2014).

(a)   (b)

(a) John Hancock Tower, Boston designed by Architect I.M. Pei (b) Near ground level, most of the glass curtain wall replaced with plywood for repair [*Photo*: Michael Shellenbarger/University of Oregon Library, reprinted with permission]

### 15.11.3 Sound Insulation

For effective sound insulation, spacing of glass in double/multiple glazed windows should be of the order of 100 mm or more. Thicker glass also provides better insulation. Further improvements may be obtained by lining the surrounds of the glasses with a sound absorbing material. For further guidance, IS 1950:1962 and IS 3483:1965 should be consulted.

### 15.12 Advantages and Disadvantages

Use of glass in buildings has the following advantages (Freed, 2008).

1. Glass is a very green material, as it is made from naturally available abundant material of sand, especially silica.
2. Unlike most of the construction materials, glass is often produced locally and is available everywhere.
3. Except for some rare vandalism, glass is durable and easily maintained. As it has a smooth glossy surface, it is dust proof and requires only simple cleaning. It is weather resistant and hence can withstand the effects of wind, rain, or the Sun effectively and can retain its appearance and integrity. It is waterproof.
4. Glass is an inorganic material which can be blown, drawn, and pressed to any shape.
5. Glass can absorb, reflect, or transmit natural light, without any yellowing, clouding or weathering. It can be made transparent or translucent, or with different colours, to provide extraordinary beauty to any building.
6. Glass is completely inert; neither does it rust nor degrade gradually by chemical and surrounding environmental effects. In addition, it does not release any chemicals, while in use. It is also mould resistant, unlike other exterior materials like wood.

7. Glass allows natural light to enter the house even when doors/windows are closed, thus saving energy and lowering electricity bills. It can brighten up any room and boost the mood of the occupants.
8. Glass can be 100% recycled and does not degrade during the recycling process. Hence, it can be recycled endlessly without any loss of quality or purity. Glass is considered a green material because it is healthy and also due to the aforementioned characteristics. Recycling glass lessens its minimal environmental impact further. Producing recycled glass requires less energy because crushed recycled glass melts at a lower temperature.
9. As glass is stable over a wide range of temperatures, it is used for fireplace glass, high-temperature light lenses, wood-burning stoves, cooking tops, and high-temperature areas where low expansion is needed.
10. Glass may also provide sound insulation. In cold seasons, solar energy can be utilized by trapping the heat within the building to provide indoor comfort and save energy bills.
11. Glass is an ideal material to showcase a product.

Use of glass in buildings may have the some of the following disadvantages.

1. The production of glass requires heat and molten tin, which requires energy and produces some green-gas emission in the process.
2. Glass is a brittle material and requires careful handling and transport. Incorrect installation, impact or thermal stresses can break any glass. Broken glass may be sharp and hence can cause serious injuries.
3. Glass is affected by hydrofluoric and phosphoric acids, which can etch the surface. Alkalis such as caustic soda, alkaline paint removers, and cement solutions can dissolve the glass surface. Hence, water run-off from fresh concrete or mortar on any glass surface must be cleaned immediately to prevent any deterioration (Alkali-resistant glass fibres are used in glass fibre reinforced concrete).
4. Although glass is non-combustible, in fires, it can break and melt.
5. Small-scale manufacturers of glass may still use energy-inefficient and polluting kilns.
6. Glass may be an expensive product and may also enhance the cost of security. Unless double or multi-layered glazing is used, ordinary glass can absorb heat, resulting in greenhouse effect, leading to higher operating cost of air-conditioners; hence, they are not suitable in warm and hot climates.
7. Buildings may be painted once in every 5 years but the outside of glass may have to be cleaned every year, which may prove to be expensive in the case of façades of high-rise buildings (use of self-cleaning glass may reduce this recurring expenditure).
8. Ordinary glass is unsafe for severe earthquake prone areas. Annealed or heat strengthened laminated glass may help protect building occupants and pedestrians from falling glass during a severe earthquake, and also help maintain building envelope integrity (Behr, 2001). ASCE 7–16 contains specific requirements for earthquake resistant architectural glass.
9. Glare may be a major problem in glass façade buildings.
10. A study by the Dutch scientific institute TNO found that up to 90 MT of $CO_2$ emissions could be saved annually by 2020 if all Europe's buildings (existing and new residential and non-residential buildings) are fitted with double-glazed low-e insulating glass units. An additional seven million tonnes of $CO_2$ emissions could be cut through a greater use of triple-glazed low-e insulating glass units for new buildings (www.nsg.com).

## SUMMARY

- The types of glass include sheet glass, plate glass, and float glass. Other special types are safety glass, coated glass, patterned glass, stained glass, etched glass, glass blocks, and double/multi-glazed window units.
- The right type of glass should be selected for a particular application. While selecting, the factors that should be considered are the solar control, wind loading, and sound insulation.

- Glass has several advantages and only a few disadvantages. It is architecturally pleasing and can be made to reduce energy costs of heating.
- Glass is a green material due to its natural ingredients which are plentiful. In addition, it can be recycled completely without any degradation to its quality and strength.

## EXERCISES

### Multiple-choice Questions

1. Which of the following is not a constituent of glass?
   (a) Sand          (c) Lime
   (b) Clay          (d) Soda
2. The tensile strength of float glass is about
   (a) 10 N/mm$^2$          (c) 80 N/mm$^2$
   (b) 40 N/mm$^2$          (d) 100 N/mm$^2$
3. The flux added for the fusion of glass is
   (a) soda          (c) lime
   (b) silica          (d) nickel
4. Pyrex glass is an example of
   (a) soda-lime-silicate glass
   (b) lead-oxide glass
   (c) borosilicate glass
   (d) aluminosilicate glass
5. Most of the sheet glass is currently manufactured by
   (a) float-glass process
   (b) Fourcault system
   (c) rolled glass process
   (d) none of these
6. Heat treatment increases the strength of glass by
   (a) 2 to 3 times its original value
   (b) 3 to 4 times its original value
   (c) 4 to 5 times its original value
   (d) none of these
7. Density of glass is in the range of
   (a) 1920 to 2030 kg/m$^3$
   (b) 2100 to 2220 kg/m$^3$
   (c) 2420 to 2530 kg/m$^3$
   (d) None of these
8. The coefficient of thermal expansion is about
   (a) $8 \times 10^{-6}/°C$          (c) $10 \times 10^{-6}/°C$
   (b) $9 \times 10^{-6}/°C$          (d) $12 \times 10^{-6}/°C$
9. The modulus of elasticity of glass is about
   (a) 60 GPa          (c) 70 GPa
   (b) 65 GPa          (d) 75 GPa
10. Which of the following glass is most suitable to withstand high temperatures?
    (a) Soda-lime-silicate glass
    (b) Lead-oxide glass
    (c) Borosilicate glass
    (d) Tempered glass

11. Tempered glass can withstand temperature differential of about
    (a) 100°C          (c) 200°C
    (b) 150°C          (d) 290°C
12. In tempered glass, the induced surface compression is about
    (a) 41 to 64 MPa          (c) 61 to 94 MPa
    (b) 51 to 74 MPa          (d) 75 to 143 MPa
13. In heat strengthened glass, the induced surface compression is about
    (a) 41 to 64 MPa          (c) 61 to 94 MPa
    (b) 51 to 74 MPa          (d) 75 to 143 MPa
14. Bullet-proof glass is made with multiple layers of
    (a) laminated glass          (c) wired glass
    (b) tempered glass          (d) none of these
15. Translucent glass can be obtained by
    (a) sand blasting
    (b) etching with acids
    (c) impressed during rolling
    (d) all of these
16. Tinted glass can absorb solar radiation up to
    (a) 20 to 30%          (c) 45 to 50%
    (b) 30 to 40%          (d) 50 to 60%
17. Low emissivity coating can reduce energy loss by
    (a) 20 to 40%          (c) 50 to 60%
    (b) 30 to 50%          (d) 60%
18. To insulate a window the two glass panes are spaced apart ideally by
    (a) 10 mm          (c) 22 mm
    (b) 12 mm          (d) 32 mm
19. Thermal performance of windows is improved when the gap between two panes of glass is filled with
    (a) dry air
    (b) hydrogen fluro-oxide
    (c) oxygen
    (d) argon
20. Which of the following plastic sheet may be substituted for glass?
    (a) PVC
    (b) Nylon
    (c) Polystyrene
    (d) Thermoplastic acrylic resin

## Review Questions

1. List the various constituents of glass used for glazing. What are the functions of each of these constituents?
2. How is glass classified, based on the ingredients?
3. Write short notes on (a) soda-lime-silicate glass, (b) borosilicate glass, (c) lead-oxide glass, and (d) aluminosilicate glass.
4. Explain the Fourcault system of sheet glass manufacture.
5. How is float glass manufactured? What are the advantages of this kind of manufacturing?
6. How are patterned glass and wire glass manufactured?
7. What are the different types of treatments given to glass? Give a brief description of these treatments.
8. What is annealing? Why is it necessary to anneal the glass?
9. What is tempering? What are the advantages of tempering?
10. How is frosted glass made?
11. Enumerate the physical and mechanical properties of glass.
12. State the various applications of glass in buildings.
13. Name the three commonly used types of glass and briefly explain them.
14. What are the three types of safety glasses? Explain them.
15. Differentiate between tempered glass and heat strengthened glass.
16. What is laminated glass? What are its advantages?
17. How is bullet proof glass made?
18. Why is wired glass not recommended in schools?
19. Differentiate between low emissivity coating, spectrally selective coatings, and reflective coating.
20. Write short notes on the following:
    (a) Tempered glass        (b) Laminated glass        (c) Translucent glass
    (d) Insulated glass        (e) Coated glass        (f) Block glass
    (g) Wired glass        (h) Tinted glass        (i) Self-cleaning glass
    (j) Multi-glazed window units    (k) Heat-absorbing and glare-reducing glass
21. What is fibreglass? How is it produced?
22. What plastics are used for glazing? Where are they used?
23. Explain the factors to be taken into account while glazing rooms that will be air-conditioned.
24. Give any five advantages of using glass in buildings. State any four disadvantages also.
25. Why is glass considered a green material?

## ANSWERS

**Multiple-choice Questions**

| | | | | | |
|---|---|---|---|---|---|
| **1.** (b) | **2.** (b) | **3.** (a) | **4.** (c) | **5.** (a) | **6.** (c) |
| **7.** (c) | **8.** (b) | **9.** (c) | **10.** (d) | **11.** (b) | **12.** (d) |
| **13.** (a) | **14.** (a) | **15.** (d) | **16.** (b) | **17.** (b) | **18.** (b) |
| **19.** (d) | **20.** (d) | | | | |

# CHAPTER 16
# CLAY AND CERAMIC PRODUCTS

## 16.1 Introduction

Ceramics are defined as a class of inorganic, non-metallic, and inert solids that are subject to high temperature in manufacture and/or use. The word ceramic is derived from the Greek word 'Keramos' meaning 'potter' or 'pottery' (Keramos in turn was originated from a Sanskrit word, which meant 'to burn'). Clay and refractory products, and even glass, are categorized as ceramics. This is a vast subject taught in universities as Ceramic engineering. Hence, in this chapter, only the clay and ceramic products that are used in building construction are briefly discussed.

Ceramics have been found to be used by humans since ancient times. Archaeologists have unearthed man-made ceramics that date back to at least 26,000 B.C. Ancient Greeks used ceramic firing as early as 6,000 B.C. Porcelain, the first ceramic composite, was created by Chinese in 600 AD. Refractory materials, which can withstand extreme temperature, were introduced in 1870s, during the industrial revolution. The construction industry depends on the use of several ceramic materials such as bricks, cement, tiles, and glass.

The basic ingredient for ceramics is clay, which is defined as very fine soil particles, resulting from the chemical weathering of different types of rocks. Depending on the minerals present in the clay, the clay deposits on earth are differentiated. Each of the ceramic products requires its own type of special clay. Thus, ceramics can be defined as inorganic, non-metallic materials that are typically produced by using clays/other minerals from the earth or by using chemically processed powders. Ceramics are typically crystalline in nature and compounds formed between metallic and non-metallic elements such as aluminium and oxygen (alumina – $Al_2O_3$), silicon and nitrogen (silicon nitride – $Si_3N_4$), and silicon and carbon (silicon carbide – SiC).

## 16.2 Clay Tiles

Tiles, derived from word 'thacktyle' and mentioned in London building codes as early as 1212, were used as an alternative to thatch as early as 1189 in London. They are usually thin slabs of burnt clay or ceramics used for various purposes in building construction, especially for covering roofs and floors. But, the term 'tile' is now used to represent thin slabs of any material. Hence, we have terrazzo tiles made of concrete, rubber tiles made of rubber, and even glass, stone, and metal tiles. Clay tiles are expensive and labour intensive but have very pleasing appearance, long life, and good service properties. Based on their application, they are called roofing tiles, flooring tiles, wall tiles, glazed tiles for roof and walls, and vitrified tiles. Bricks and tiles are generally manufactured together (the manufacturing steps of bricks has already been discussed in Chapter 3). As tiles are available in different colours, before firing at temperatures of 900 to 1200°C, they are treated with engobe or slip (i.e., slurry of finely ground clay coloured with oxides, carbonates, and stains). Engobe or slip is applied by dipping, pouring over or spraying on the unfired clay tile. Engobed clay tiles can have matt, matt-glossy, or glossy surfaces. Each individual tile was made by hand until the 19th Century. Tile-making machines were then introduced for making pan tiles or other shaped tiles.

## 16.3 Clay Roofing Tiles

Clay roof tiles are durable, natural, and sustainable products that improve with age and weathering. Their appeal adds value to buildings and enhances the built environment. They require little maintenance and are designed to protect the interior of the building from rain. Roofing tiles can be of clay or concrete. Transportation cost is minimal, as locally available materials are used. They offer high levels of insulation and are fire-resistant. Choice between clay and concrete roofing tiles is made based on functional, aesthetic, and durability considerations. Tiles are available in large number of sizes, colours (about 50 different colours, ranging from deep reds, browns, warm oranges to the muted blues may be available), and profiles.

Roof tiles are hung from battens by fixing them with nails. Battens are thin strips of material that run horizontally along the length of the roof. The tiles usually run in parallel rows, with each row overlapping the row below to prevent rainwater from entering and cover the nails that hold the row below. In addition to the standard tiles, special roof tiles are required in ridges, eaves, valleys and hips. Where ridges and hip starters are laid, mortar mix is poured to seal the joints. Once the purlins are laid, roof tiles are laid from bottom up. More details of laying the roofing tiles may be found in Boral roof tiles technical information guide, 2017 or in textbooks on building construction.

### 16.3.1 Flat Plain Tiles

The plain flat tile is a small rectangular flat tile measuring 150 to 250 mm × 75 to 200 mm [IS 2690 (Parts 1 and 2)]. They are made in the same way as floor tiles but are not usually glazed.

Machine-pressed/machine extruded or hand-made burnt clay flat terracing tiles are used for flat roof finishing over lime concrete or cement concrete base. Depending on the degree of protection necessary, they are used in two or more courses. Hand-moulded tiles are identified by the letter H and machine-moulded tiles with the letter M, marked on them. The characteristics of these tiles are extracted from IS 2690 and provided in Table 16.1.

**Table 16.1** Characteristics of burnt clay flat terracing tiles

| Characteristics | Hand moulded | Machine moulded |
|---|---|---|
| Length, mm | 150 to 250 in steps of 25 mm | 150 to 250 in steps of 25 mm |
| Width, mm | 75 to 200 in steps of 25 mm | 100 to 200 in steps of 25 mm |
| Thickness, mm | 25 to 50 in steps of 5 mm | 15 and 20 |
| Tolerances, percent | ±3 | ±2 (Pressed) <br> ±3 (Extruded) |
| Water absorption | <20 | <15 |
| Warpage, percent of dimension | <2.0 | <1.0 |
| Flexural strength, MPa | <1.5 | <2.0 |

[*Source*: Extracted from IS 2960 (Parts 1 and 2)]

A minimum of 25° slope is necessary for any tiled roof, though in the past, much steeper, i.e., 33° or 45°, slopes have been provided. If the roof has a steep slope, battens may be required to hold the tiles in place (see Fig. 16.1). *Batten* or Purlins are 25-mm thick and 50-mm wide wood strips nailed to the roof or rafters, upon which the tiles are attached. Many tile varieties possess a lip or hook that will hang on these battens. Clips may also be used to attach the tiles to the battens. The side lap between tiles in subsequent courses must not be less than one-third of the width of a standard tile. The head-lap between tiles in the course below but one must not be less than 65 mm. Owing to their small size, plain flat tiles are labour intensive and heavier.

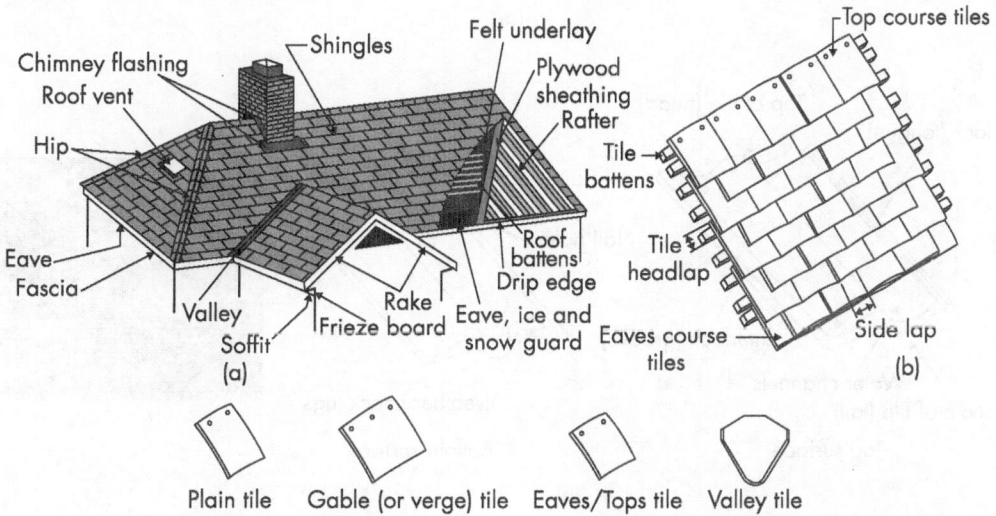

**Fig. 16.1** Roofing with plain tiles (a) Terms used in roofing (b) Flat plain tiles
(*Source*: Adapted from http://rosewellroofing.co.uk)

A number of *interlocking plain tiles* are now available, which will result in reduced labour costs, use of less material and energy, and require fewer tiles per m². These tiles reduce roof costs by around 40%. Interlocking plain tiles can also be laid at lower pitches than traditional plain tiles, allowing them to be used for extensions and additions.

### 16.3.2 Pan Tiles

These are used mainly in the UK and Scotland and were used in Holland from the early 17th century. Pan tiles were originally made from clay until the 1950s when more cost-effective concrete pan tiles appeared. They have a flattened S-shaped cross-section, creating a distinctive flowing appearance. They are laid *in such a way that* the downward curve *of one tile* overlaps *the* upward *curve of the* adjoining *tile* [see Fig. 16.2(a)]. A pan tile-covered roof is considerably lighter than a flat-tiled equivalent and can be laid to a lower pitch.

The concrete pan tile has a thicker and rougher front edge of 30 mm, as compared to 15 mm for a traditional clay pan tile and around 20 mm for an interlocking pan tile. The other key difference is the colour – the colours of clay pan tile have a more natural appearance, as they are fired in a kiln, and do not fade over time like concrete pan tiles. Interlocking pan tiles have interlocks and weather lugs, which enable them to be attached to lower pitches [see Fig. 16.2(b)]. They have a thicker leading edge and are slightly cheaper than the traditional pan tiles. Clay double unit pan tile, introduced in 2008, results in a significant reduction in the cost of a natural clay roof.

*Roman pan tiles* are flat in the middle, with a concave curve at one end and a convex curve at the other end, to allow interlocking. Imbrex and Tegula tiles have curved and flat patterns in order to create rain channels on a roof [see Fig. 16.2(b)].

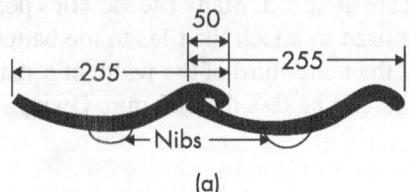

(a)

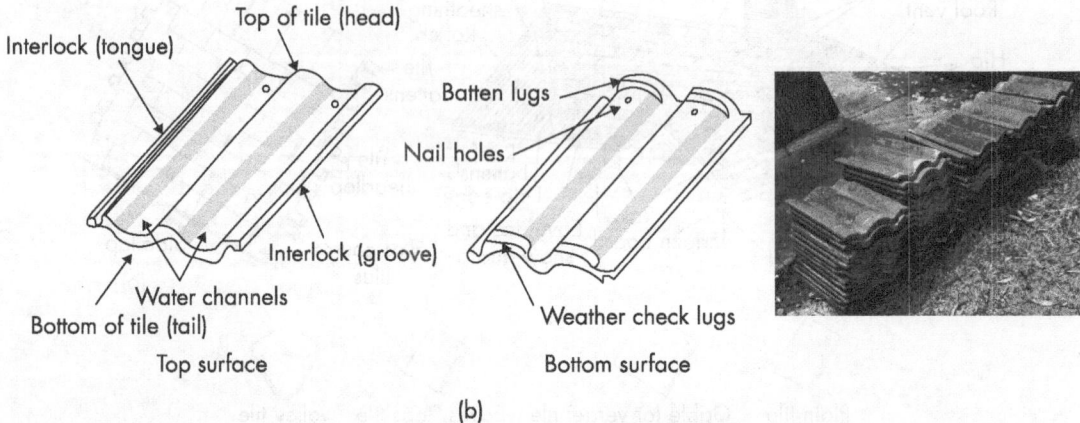

(b)

**Fig. 16.2** Interlocking roof tiles (a) Section through two pan tiles (b) Interlocking tiles

### 16.3.3 Mangalore Pattern Roofing Tiles

Clay roofing tiles of interlocking type, with a particular pattern known as *Mangalore roof tiles*, are being manufactured in large scale in southern parts of India. Glazed Mangalore tiles are also available in different sizes. The detailed classifications are given in IS 654:1992 (see Tables 16.2 and 16.3). The minimum overlap [see Fig. 16.3(a)] should be 60-mm lengthwise and 25-mm widthwise, for each type of tiles.

**Table 16.2** Classification of Mangalore pattern roofing tiles, as per IS 654:1992

| Characteristics | Requirements | |
|---|---|---|
| | Class AA | Class A |
| Water absorption (%) | 18 | 20 |
| Breaking strength (kN) Average | 1.0 (for 410 × 235 mm) 1.10 (for 420 × 250 mm and 425 × 260 mm) | 0.80 (for 410 × 235 mm) 0.90 (for 420 × 250 mm and 425 × 260 mm) |
| Individual | 0.9 (for 410 × 235 mm) 1.0 (for 420 × 250 mm and 425 × 260 mm) | 0.68 (for 410 × 235 mm) 0.78 (for 420 × 250 mm and 425 × 260 mm) |

**Table 16.3** Dimensions of Mangalore pattern roofing tiles, as per IS 654:1992

| Overall dimensions | |
|---|---|
| Length (mm) | Width (mm) |
| 410 | 235 |
| 420 | 250 |
| 425 | 260 |

The tiles have two batten lugs provided at the bottom to lock in with the purlin. Similarly, there are two eaves lugs [see Fig. 16.3(a)]. Historically, Mangalore roof tiles were laid with wooden purlins. As wood has become scarce, the tiles are now laid over steel frameworks. The lugs have a projection from the tile of 7 to 12 mm. There is no need of nailing them down. This may reduce labour costs considerably. These tiles should be tested for shape, dimensions, weight, water absorption, permeability, and breaking load. The sample size (selected at random) and criterion for conformability should be as per Table 16.4.

**Table 16.4** Sample size and criterion for conformability, as per IS 654:1992

| Lot size | Sample size | Permissible number of defective tiles |
|---|---|---|
| Up to 3000 | 32 | 3 |
| 3001 to 10,000 | 50 | 5 |
| 10,001 to 35,000 | 80 | 7 |
| Above 35,000 | 125 | 10 |

The average weight of the six tiles, when dried at 105 to 110°C, should be greater than 2 kg but not more than 3 kg. The tile-to-tile linking [see Fig. 16.3(d) and (e)] of double groove Mangalore roof tiles is better than the single groove Mangalore roof tiles. Because of the better interlocking, the chances of leakage are lesser compared to the single groove roof tiles. Single groove Mangalore tiles are used as decorative tiles over RCC sloped roofs (see Fig. 16.4). The cost of double groove tiles will be about 40% more than the single groove roof tiles. In cyclonic zones, at least one hole is provided in one of the cross ribs near the eves end, and the tile is secured to the batten with a wire, so that they are not lifted off during high winds.

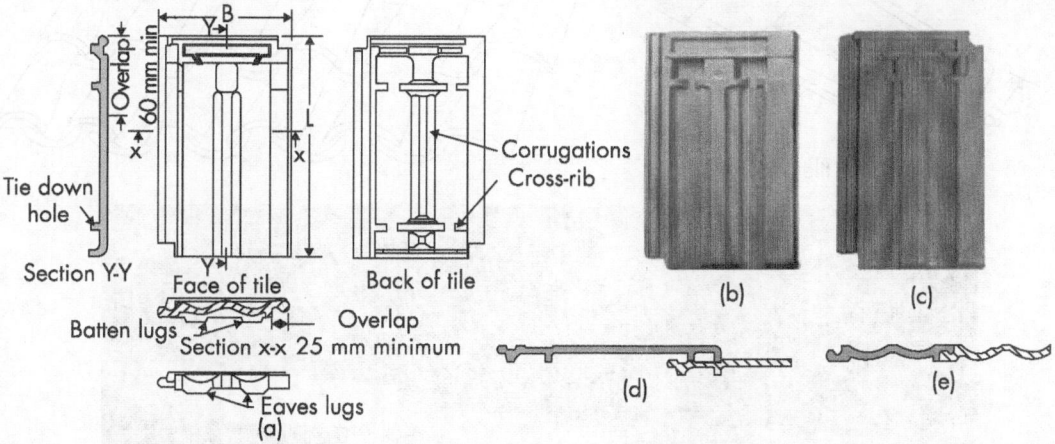

**Fig. 16.3** Single and double-grooved Mangalore tiles (Adapted from IS 654:1992 & *Courtesy*: www.btiles.com) (a) Details of mangalore tile (b) Single groove roofing tile (c) Double groove roofing tile (d) Arrangement of tiles lengthwise in tight position (e) Arrangement of tiles breadthwise in tight position

**Fig. 16.4** Mangalore roof tiles – over RCC sloped roof (*Courtesy*: www.btiles.com)

## 16.3.4 Clay Half-Round Country Tiles (Spanish Tiles)

Half-round country tiles are commonly used in villages. They are called as *Spanish tiles* and *tapered mission style clay roof tiles* (less-rounded tiles are called as *barrel-mission style clay roof tiles*). They are laid in pairs as under-tiles and over-tiles, as shown in Fig. 16.5. The under-tiles are usually laid with their narrow end towards the eves, whereas the over-tiles are laid with their wide end towards the eves.

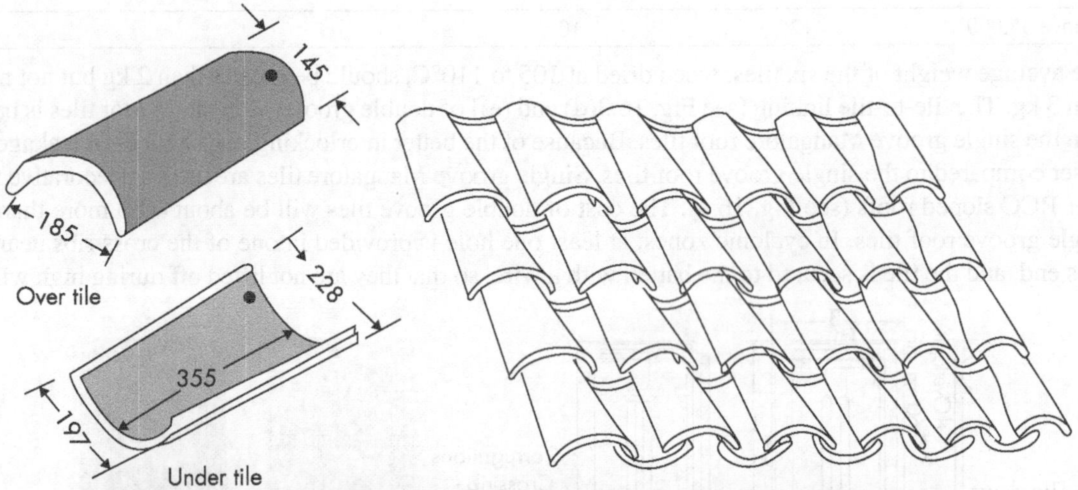

Over tile

Under tile

**Fig. 16.5** Spanish tiles (*Photo*: Dr N. Subramanian)

## 16.3.5 Allahabad Tiles (Italian Tiles)

These tiles consists of two types of tiles–over-tile and under-tiles. They are also called *Italian tiles*. The bottom tiles are flat, with tapering upturned flanges at the sides. The over-tiles are half-round and tapered as shown in Fig. 16.6. The trough tiles are 235 × 395 mm in size and the top tiles are 165 × 395 mm in size. Both these tiles are tapering towards one end. The taper in the over-tile allow the tile in next course to fit in. The under-tiles are laid side-by-side and the joints are covered by the half-round tiles. Over-tiles are fixed to vertical battens with 75-mm nails. More information about these tiles is given in the Central Public Works Department (CPWD) specifications.

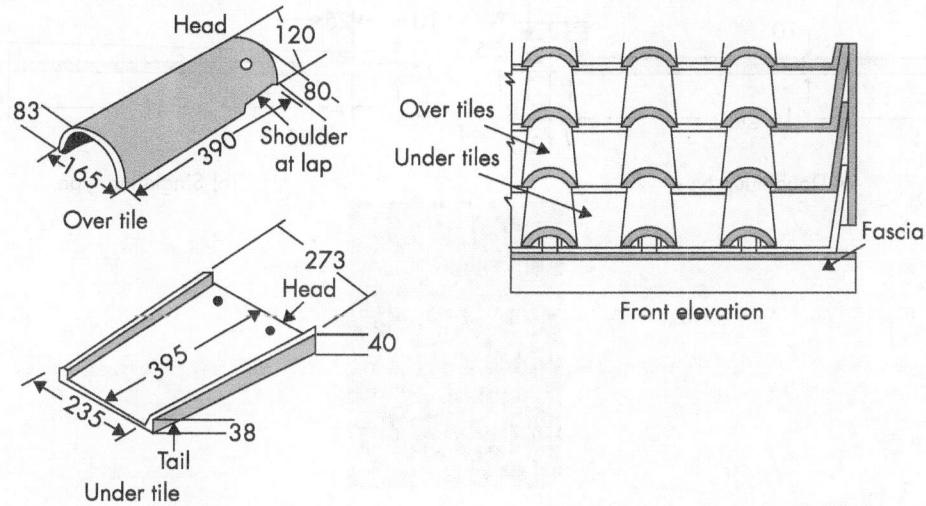

**Fig. 16.6** Allahabad tiles (Italian tiles)

## 16.3.6 Ridge and Ceiling Tiles

The apex of a roof can be given a sharp or soft outline by using different ridge profiles. The half-round ridge tiles are most commonly used. Angle ridge tiles (see Fig. 16.7) provide a sharp apex. The clay ridge and ceiling tiles, as shown in Figs 16.7 and 16.8 respectively, are classified as Class AA and Class A as per IS 1464:1992. The strength and water absorption requirements of these tiles are as per Table 16.5.

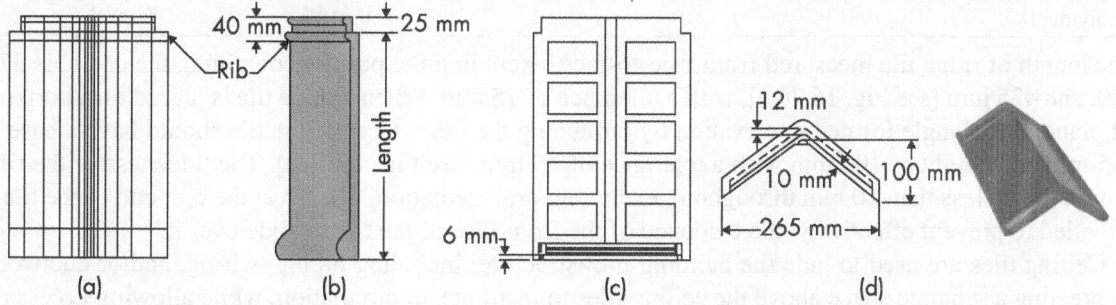

**Fig. 16.7** Ridge tiles as per IS 1464:1992 (a) Top view (b) Side view (c) Bottom view (d) Section

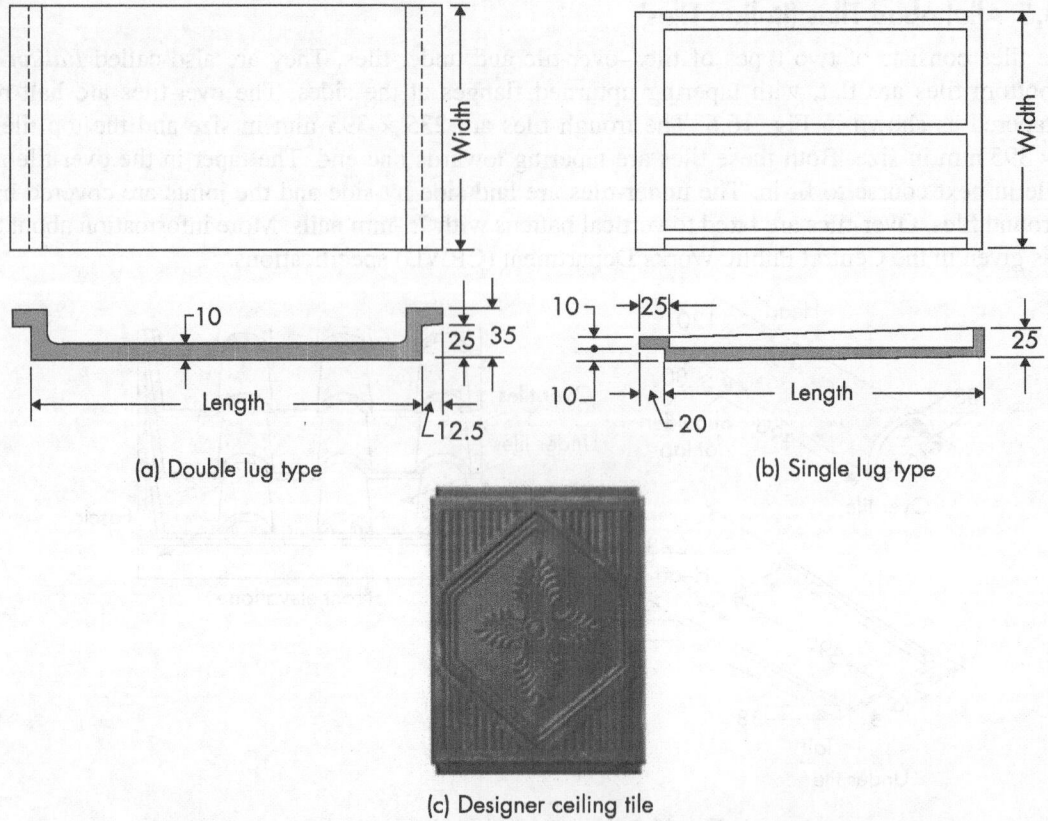

(a) Double lug type

(b) Single lug type

(c) Designer ceiling tile

**Fig. 16.8** Ceiling Tiles, as per IS 1464:1992 (*Photo courtesy*: www.btiles.com)

**Table 16.5** Classification of ridge and ceiling tiles as per IS 1464:1992

| Characteristics | Requirement for | |
|---|---|---|
| | Class AA | Class A |
| Maximum water absorption (for ridge and ceiling tiles) (%) | 18 | 20 |
| Minimum breaking strength (for ridge tile only), kN Average | 0.015 | 0.011 |
| Individual | 0.0125 | 0.0095 |

The length of ridge tile measured from face-to-face, excluding the portion containing the catch, is 375, 400, and 435 mm [see Fig. 16.7(b)], with a tolerance of ±5 mm. When a ridge tile is placed on a horizontal plane, the triangle formed in elevation by producing the inner faces of the tile should have a base of 265 mm and height of 100 mm with a tolerance of ±5 mm [see Fig. 16.7(d)]. The thickness of the tiles should not be less than 10 mm throughout, excluding ornamentation. The rib at the rear end of the tile is provided to prevent effectively, the tendency of the front face of the tile to slide over it.

Ceiling tiles are used to hide the building infrastructure; including piping, wiring, and/or ductwork, by creating a separate space above the ceiling tiles to facilitate air circulation, while allowing access for repairs and inspections. Ceiling tiles may also provide acoustic control (achieved by adding insulation known as *Sound Attenuation Batts*, above the panels), lighting and thermal system control, thermal comfort, and also aesthetic appearance. The ceiling tiles may have double leg or a single leg as shown in

Fig 16.8. The length of the double lug ceiling tile at the bottom should be such that when a tile is placed between two battens, the space between the face of the batten and that of end of tile is about 3 to 6 mm. The length of the single lug ceiling tile at the bottom should be 30 mm less than the face-to-face spacing of battens. The thickness of the tile or lug should not be less than 10 mm and the length of lug should be less than 20 mm (IS 1464:1992).

### 16.3.7 Solar Roof Tiles

Dow Chemical Company began producing *solar roof tiles* in 2005, and now several others are also manufacturing them. They are similar in design to conventional roof tiles but have photovoltaic cell within to generate renewable electricity. Tesla and SolarCity also developed a solar roof system using a high-efficiency solar cell manufactured by Panasonic and covered with a 'colour louver film', which allows cells to blend into the roof while exposing them to the sun above, and finally a tempered glass on top for durability. Solar roof tiles are more than three times stronger than standard roofing tiles. Tesla estimates that its solar roof will be cheaper than regular tile roofs or virtually pay for itself through electricity savings. The product is offered in four different styles – textured glass tile, slate glass tile, tuscan glass tile, and smooth glass tile (www.tesla.com/solarroof).

## 16.4 Clay Flooring Tiles

As per IS 1478:1992, the flooring tiles are classified as class 1, class 2, and class 3, depending on water absorption, flexural strength, and impact resistance as shown in Table 16.6. Class 1 bricks should be especially hard-burnt as they are used in industrial flooring where heavy wear may occur.

**Table 16.6** Classification of flooring tiles as per IS 1478:1992

| Characteristic | Requirement for | | |
|---|---|---|---|
| | Class 1 | Class 2 | Class 3 |
| Maximum water absorption (%) | 10 | 19 | 24 |
| Minimum flexural strength, N/mm width | | | |
| (a) Average | 6.0 | 3.5 | 2.5 |
| (b) Individual | 5.0 | 3.0 | 2.0 |
| Impact resistance, maximum height of drop of steel ball, mm | | | |
| (a) 15-mm thick | 25 | 20 | 15 |
| (b) 20-mm thick | 60 | 50 | 40 |
| (c) 25-mm thick | 75 | 65 | 50 |
| (d) 30-mm thick | 80 | 70 | 60 |

The flooring tiles are made from good soils of even texture and are well burnt, to enable thin tiles having smooth surface. To make the tiles hard and impervious, a mixture of ground glass and pottery ware may be added during manufacture. Generally, the tiles are extruded, pressed, and fired at about 1300°C. They may be salt glazed for better appearance. The faces of tiles may be plain, grooved, fluted, or figured as specified and the edges should be square. The backs of the tile may have engraved or embossed design. The depth of the grooves on the underside of flooring tiles should not exceed 3 mm. The warpage shall not exceed 2% along the edges and 1.5% along the diagonals. The dimensions of square tiles, as per IS 1478:1992, are given in Table 16.7. Half-tiles, both rectangular and triangular in shape, are also available.

**Table 16.7** Dimensions of sides, thickness, and tolerance of flooring tiles (adapted from IS 1478:1992)

| Length (mm) × Breadth (mm) | Minimum thickness, mm | Tolerance, mm | |
|---|---|---|---|
| | | Sides (average) | Thickness (average) |
| 150 × 150 | 15 | ±5[a] | ±2[b] |
| 150 × 150 | 20 | ±5 | ±2 |
| 200 × 200 | 20 | ±5 | ±2 |
| 200 × 200 | 25 | ±5 | ±2 |
| 250 × 250 | 30 | ±5 | ±2 |

[a]For a given area and space, the dimensions of individual tile should not vary by more than ±2%.
[b]For individual tile, ±1 mm

## 16.4.1 Clay vs Concrete Roof Tiles

A comparison of the clay and concrete roof tiles is as follows:

**Manufacture**   Clay tiles are manufactured from locally available earth and hence are environment friendly, and can be easily recycled. Whereas concrete is made from Portland cement/lime, aggregates, and water. It may contain pigments to get the desired colour.

**Durability**   Both materials are durable and strong. But the lifespan of clay is longer than concrete, which may have a life span of about 50 years. Clay tiles also resist wind, rain, or moisture effectively.

**Colour**   The natural colour of clay tiles is red or terracotta colours. Glazing may be used to change the colours. These colours will not fade with time regardless of weather. Although concrete can be finished with any desired colour it tends to fade with time.

**Cold weather**   In cold weather, clay tiles may crack or shatter due to freezing and thawing. Hence, they are preferable in warm climates. But concrete can be used in any climatic zone.

**Weight**   Although clay and concrete tiles are heavier than other roofing material, clay is a bit heavier than concrete. Hence, they should be used in existing buildings, only after checking the load carrying capacity of supporting members.

**Cost**   Concrete tiles cost less than clay tiles but life cycle cost of clay tiles may be less, as they have longer lifespan.

For the testing and evaluation of the properties of clay tiles, refer to Section 25.22 of Chapter 25.

## 16.5 Ceramics

There are many different types of *engineered ceramics*. They fall into three general categories and are based on nitrides, carbides, and oxides. The two most widely used nitrides are silicon nitride, and aluminium nitride. The most common oxides are alumina and zirconia. The most popular carbides are silicon carbide and boron carbide. Ceramic processing generally involves high temperatures, and the resulting materials are heat resistant or refractory. Based on their constituents and glaze, ceramic products exhibit resistance to corrosion and chemical substances, wear, and water. Engineered ceramic products are made from high-purity raw materials and hence their properties will be consistent. Ceramic products are composed of diverse materials, which make them suitable for varied applications including construction.

Ceramic products that are made from highly refined natural or synthetic compositions and designed to have special properties, are referred to as *advanced ceramics*. Advanced ceramics can be classified according to their application in electrical/electronics, magnetic, optical, chemical, thermal, mechanical, aerospace, biological, and nuclear fields (Freitag & Richerson, 1998). Ceramic materials are used for the manufacture of products such as construction materials, electrical equipment, glass products, grinding materials, dinner-ware, and heat-resistant materials. The applications of advanced ceramics include space shuttle tile, engine components, artificial bones and teeth, computers and other electronic components, and cutting tools.

The demand for ceramic products has grown significantly over the past couple of years. Ceramic products, such as facing bricks, roofing, flooring, and wall tiles, and sewer pipes, are increasingly being used in the construction industry. Ceramics ensure a longer life and low maintenance.

## 16.5.1 Manufacturing of Ceramic Products

The production process of any ceramic product is shown in Fig. 16.9. The basic steps are similar to that of brick manufacturing, and include raw material procurement, beneficiation, mixing, forming, green machining, drying, presinter thermal processing, glazing, firing, final processing, and packaging. These operations are briefly described in this section.

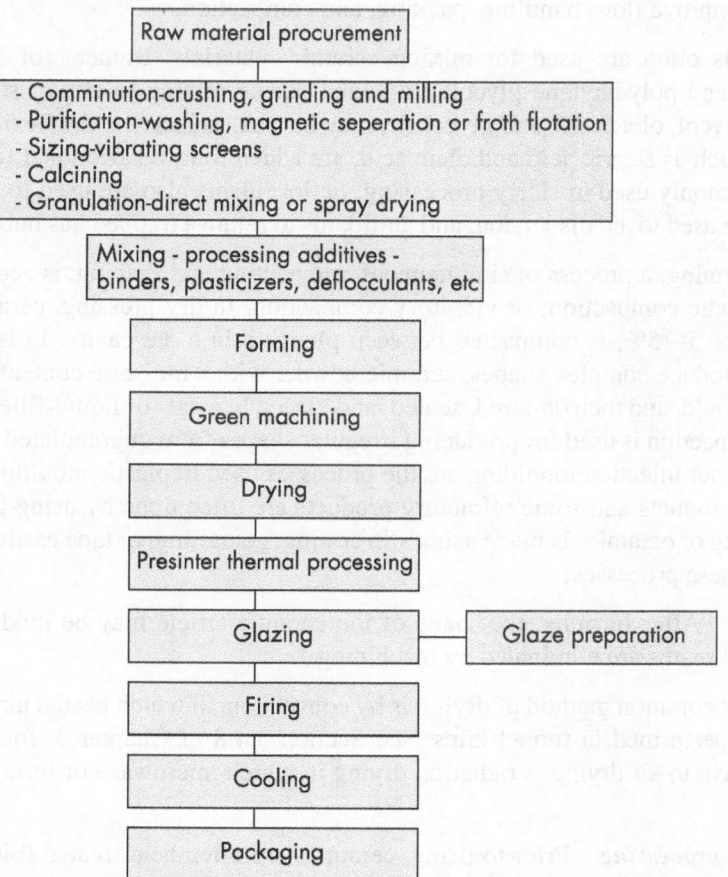

**Fig. 16.9** Process flow diagram of the manufacture of any ceramic product (Adapted from AP-42, 1996)

**Raw material procurement** The raw materials used in the manufacture of ceramics range from relatively impure clay materials mined from natural clay deposits to ultrahigh purity powders prepared by chemical synthesis. Naturally occurring raw materials used to manufacture ceramics include silica, sand, quartz, flint, silicates, and alumino-silicates (e.g., clays, feldspar, and pyrophyllite-mineral composed of aluminium silicate hydroxide, i.e., $Al_2Si_4O_{10}(OH)_2$). These raw materials, after extraction, are transported and stored at the manufacturing facility.

**Beneficiation** Naturally occurring raw materials often undergo some beneficiation process at the mining site/processing facility prior to transporting to the manufacturing facility. Some of the beneficiation processes include comminution, purification, sizing, classification, calcining, liquid dispersion, and granulation. *Comminution* is the process of reducing the particle size of the raw material to about 1.0 μm in diameter by crushing, grinding, and milling, or fine grinding. Ball mills are most commonly used for milling. Crushing and grinding typically are dry processes; milling may be a wet or dry process. Several procedures are used to purify the ceramic material. For example, water-soluble impurities are removed by washing and filtering. Acid leaching may be used to remove metal contaminants. Magnetic separation is used to extract magnetic impurities. Sizing and classification, using fixed or vibrating screens, separate the material into different size ranges. Calcining is done in rotary calciners by heating a ceramic material to a temperature well below its melting point to liberate undesirable gases. Dry powders often are granulated to improve flow, handling, packing, and compaction.

*Mixing* Pug mills often are used for mixing ceramic materials. Binders (of about 3%), such as polyvinyl alcohol and polyethylene glycol, are added during mixing to impart strength. Plasticizers, such as ethylene glycol, oleic acid, and glycerine, may be used to increase the flexibility of the ceramic mix. Lubricants, such as stearic acid and oleic acid, are added to lower frictional forces between particles. Water is commonly used in slurry processing; deflocculants also are used to improve dispersion. Surfactants may be used to aid dispersion, and antifoams to remove trapped gas bubbles from the slurry.

*Forming* Dry forming, a process of simultaneous compacting and shaping, is accomplished by using dry pressing, isostatic compaction, or vibratory compaction. In dry pressing, ceramic powder, with a moisture content of 5–15%, is compacted between plungers in a die cavity. In isostatic compaction, which is used to produce complex shapes, ceramic powder with a moisture content of 0–15% is placed inside a flexible mould, and then de-aired, sealed, and placed in a gas- or liquid-filled pressurized chamber. Vibratory compaction is used for producing irregular shapes from ungranulated powders. Extrusion, jiggering, and powder injection moulding are the processes used in plastic moulding. The manufacture of structural clay products and some refractory products are often done by using the process of extrusion. Slurry forming of ceramics is made using slip casting, gelcasting, or tape casting (see AP-42, 1996, for the details of these processes).

*Green machining* After forming, the shape of the ceramic article may be modified slightly, or the rough surfaces and seams are eliminated by machining.

*Drying* The most common method of drying is by convection, in which heated air is circulated around the ceramics and performed in tunnel kilns (see Section 3.4.8 of Chapter 3, for operation of tunnel kilns). An alternative to air drying is radiation drying in which microwave or infrared radiation is used to enhance drying.

*Presinter thermal processing* Prior to firing, ceramics are often heat-treated (bisque firing) at about 700 to 1000°C, to provide additional drying, vaporize or decompose impurities, and remove residual, crystalline, and chemically bound water.

*Glazing*   The surface of bisque-fired ceramic-ware is often applied with glaze coatings, prior to firing. These glazes, which resemble glass in structure and texture, provide a smooth and shiny surface to the ceramic articles, sealing the pores. Glazes are primarily oxides and may be classified as raw glazes or frit glazes. A frit glaze is a pre-reacted glass and the raw glazes often used are quartz, feldspars, carbonates, borates, and zircon ($SiO_2$, $Al_2O_3$, $B_2O_3$, $ZrO_2$, $Na_2O$, $K_2O$, MgO, CaO, and ZnO). Glazes are applied by spraying or dipping. Addition of copper oxide gives a green colour and addition of iron oxide gives a red colour to the product.

*Firing and cooling*   Firing also is referred to as *sintering* or densification. Ceramics generally are fired at 50–75% of the absolute melting temperature of the material (900–1300°C). The outcome of firing is affected by firing temperature, time, pressure, and atmosphere. Materials that are difficult to fire using conventional methods can be fired by applying pressure, which also decreases firing time. Electric resistance-heated furnaces with controlled atmosphere are used to fire advanced ceramics. After firing, some ceramic products require further processing to enhance their characteristics or meet dimensional tolerances. These processes include abrasive grinding, chemical polishing, annealing, oxidizing, or coating. The fired products are cooled to room temperature.

*Packaging*   The fired and cooled tiles are carefully packed and sent for storage and delivery.

## 16.5.2 Properties of Ceramics

The requirements for the three grades of China clay used in the ceramic industry, and the tests to be conducted on the clay for determining the physical and chemical properties are stipulated in IS 2840:2002. Ceramics are brittle, hard, and strong in compression, weak in shearing and tension. They possess excellent wear resistance, good frictional behaviour, low thermal conductivity, corrosion resistance in acids and alkalis, excellent surface finish ($R_a$ = 0.006 μm), modulus of elasticity similar to steel, and thermal expansion coefficient similar to cast iron. In addition, they are anti-static and non-magnetic.

## 16.5.3 Ceramic Tiles

Ceramic tiles are divided into groups according to their method of manufacture and their water absorption, as shown in Table 16.8 as per IS 13712:2006.

**Table 16.8** Classification of ceramic tiles according to their group

| Shaping | Water absorption (E) | | | |
|---|---|---|---|---|
| | **Group I(a)** | **Group I(b)** | **Group II** | **Group III** |
| | **E ≤ 0.08%** | **0.08% < E ≤ 3%** | **3%< E ≤ 6%** | **E >10%** |
| A | Group AI(a) | Group AI(b) | Group AII | Group AIII |
| B | Group BI(a) | Group BI(b) | Group BII | Group BIII |
| C | Group CI(a) | Group CI(b) | Group CII | Group CIII |

Thus, according to their method of manufacture, they are classified as follows:

1. Extruded tiles (shaping A) - These are the tiles whose body is shaped in the plastic state in an extruder, the column obtained being cut into tiles of predetermined length.
2. Pressed tiles (Shaping B) - These tiles are formed from a body reduced to powder or small grains and shaped in moulds at high pressure; they may be glazed or unglazed.
3. Cast tiles (Shaping C) - The body is cast into a mould or on to a porous refractory batt that absorbs water.

Both pressed and cast tiles can be glazed or unglazed. There can also be the following tiles:

1. Split tiles (Split Pattern) - These are formed as double tiles that are separated after firing to obtain single tiles. They can be glazed or unglazed and have parallel ridges on the back.
2. Quary tiles - These are the tiles that are cut in succession from a single extruded column, are either pressed or not pressed, and are sometimes glazed.

According to the water absorption (percent by mass), which is measured in accordance with IS 13630 (Part 2), the ceramic tiles are classified as (IS 15622:2006) follows:

1. Tiles of very low and low water absorption (Group I)
   (a) E < 0.08% [Group BI(a) ]- Ceramic fully vitrified tiles
   (b) 0.08% <E ≤ 3% [Group BI(b)]- Ceramic porcelain tiles

2. Tiles of medium water absorption (Group II)
   3%< E <6% (Group BII) - Ceramic floor tiles

3. Tiles of high water absorption (Group III)
   E >10% (Group BIII) - Ceramic wall tiles

Spacer lugs, with projections of about 0.6 mm, may be provided along certain edges of tiles so that when two tiles are placed together in line, the lugs on adjacent edges separate the tiles by a distance not less than the specified width of joints. Lugs are positioned so that the joint between the tiles are filled with grout without the lugs remaining exposed. Several sizes of the each category of tiles are available: modular tiles ranging in size from M100 × 100 mm to M300 ×300 mm, and non-modular tiles ranging from 100 × 200 mm to 400 × 400 mm (see IS 15622:2006, for more details). Typical dry pressed tiles may have thickness from 3 to 11 mm, whereas extruded products are thicker at 10–20 mm and more (Bechthold et al., 2015).

Ceramic tiles are less expensive than natural stone. It is possible to mix and match tiles of different colours, patterns, and sizes to create custom designs, intricate borders, and one-of-a-kind floors. Glazed ceramic tiles such as *Spartek* tiles, which are made from special clay and given coloured glazing, were originally used only for walls, where there is no traffic. Improved techniques of glazing have made them possible to be used in houses and even in high traffic areas like airports.

## 16.5.4 Vitrified Tiles

These are made with dust compression method and hence they are harder, denser, and less porous than normal ceramic tiles. They are fired at high temperatures exceeding 1250°C to obtain extremely hard, dense, homogenous tile. Vitrified tiles are then polished to a very high finish using equipment that are usually used to polish granite and marble. They bridge the gap between ordinary ceramic tiles and marble/granite flooring. Vitrified tiles may be available in the following sizes: 600 × 600 mm, 600 × 1200 mm, 800 × 800 mm, and 800 × 1200 mm, and thicknesses ranging from 7.5 to 10 mm.

Four types of vitrified tiles are available in the market are as follows:

1. Soluble salt
2. Double charge
3. Full body
4. Glazed (www.morbiceramicindustry.com)

*Soluble salt vitrified tiles* are printed with screen printing technology and then polished. They are the cheapest type of vitrified tiles in the Indian market. However, there have limited design options and the designs may fade away after few years. It is now possible to nano-polish these tiles making them more shiny and smooth and with less pores.

In *double charge (DC) vitrified tiles*, during the pressing process, two types of glazes are applied on the tile's bisque, providing a thicker design layer (about 3 to 4 mm) on top of them. This makes these tiles ideal choice for use in heavy traffic areas (see Fig. 16.10). DC tiles have very long life compared to traditional soluble salt tiles. However, the design patterns are limited.

*Full body vitrified tiles* have the same colour throughout the thickness of the tile, and hence scratches are less noticeable and the designs on them will not fade. These tiles can be distinguished from glazed tiles by examining the broken section of the tile. They are formed with a paint mixture pre-added while making the tiles. These are ideal for heavy traffic areas such as shopping malls and airports. However, they are expensive compared to other type of tiles and only limited designs are available.

*Glazed vitrified tile (GVT)* is the only tile in which it is possible to have any type of design/art work and several types of textures (such as wooden, bamboo, slate, or stone). The use of digital printing technology and the glazy layer makes it possible to have more design options possible in GVT (see Fig. 16.11). GVTs are the best for any kind of requirement. However, due to digital technology, the cost is higher.

Less water absorption property (less than 0.5%) makes vitrified tiles acid/alkali/chemical resistant, and gives it a greater strength. Since vitrified tiles are less porous than ceramic tiles, they are also easy to clean. These tiles are also available in different colours and shades. The edges of these tiles are usually ground again after manufacture and hence they can be laid with very thin joints. Vitrified tiles are used in kitchen floors as ordinary ceramic tiles are brittle and hence may chip when heavy objects fall on them. They are also often used outdoors due to their water and frost resistance. The specification for ceramic unglazed vitreous acid-resisting tiles, having thickness less than 20 mm, used as floor covering in chemical and allied industries, lining of tanks, etc., and used in conjunction with chemical resistant mortars, are provided in IS 4457:2007.

**Fig. 16.10** Vitrified tiles used in Charlotte Douglas International Airport, NC, USA

**Fig. 16.11** Numerous designs are possible with vitrified tiles

## 16.5.5 Fixing of Floor Tiles

Floor tiles are typically set on the floor by using cement mortar with added latex for better adhesion or using adhesives. As per IS 15477:2004, Type 1 Adhesive (polymer-modified adhesive) is used for fixing tiles of standard body composition with apparent porosity greater than 3%. Type I adhesive will be suitable for most ceramic (that is non-vitrified) tiles and the majority of porous stones and backgrounds. Type 2 Adhesive (highly polymer-modified adhesives) is used for fixing tiles of standard body composition with an apparent porosity less than or equal to 3%. Type 2 adhesive will be suitable for vitrified/fully vitrified tiles, dense and large dimension tiles (slabs), and where background and location is especially demanding. The spaces between tiles were traditionally filled with mortar, but nowadays sanded/un-sanded floor grout is used. A penetrating sealer may be used on unglazed tiles or grouted joints, which may provide as stain resistant shield.

## 16.6 Porcelain

*Porcelain* is a white vitrified translucent ceramic, and baked at higher temperatures in the kiln than most ceramic. High grade ceramic-ware originated in China during the T'ang dynasty (618–907 AD), and evolved during the Yuan dynasty (1279–1368). The Chinese made porcelain of kao-lin (white clay) and pe-tun-tse, a stone composed of feldspar, mica, and quartz. The name *Porcelain* was given by Marco Polo, from the Italian 'porcellana' (cowrie shell), to describe the pottery he saw in China, because it resembled the translucent surface of the shell. As porcelain is white in colour, it is also called white-ware.

Porcelain can be of soft or hard types. Hard paste type is usually composed of kaolin clay, quartz, and feldspar, or a combination of white clay, flint, and alabaster. The composition of porcelain is highly variable, but the kaolin clay (hydrous aluminium silicate) is often a raw material (porcelain often contains more kaolin and fewer impurities than ceramic). Alabaster, feldspar, quartz, and pe-tun-tse act as fluxes. A typical composition is – China clay 50–60%, ordinary clay up to 5%, whiting ($CaCO_3$) < 1%, and feldspar 20%. Hard paste-types are made by firing them at high temperatures of about 1450°C, when the feldspar particles fuse and bind the other constituents into a hard, dense, and vitreous object. Painted decoration on porcelain is usually executed over the fired glaze.

Soft paste porcelain, developed in Europe (also referred to as *China*) as a substitute for the original porcelain, although not much softer than the hard type, but can be cut with a file. It is manufactured by adding ground glass to clay and firing it at a lower temperature of 1200°C. In 1707, Germans Ehrenfried Walter von Tschimhaus and Johann Friedrich Bottger developed a type of porcelain by combining clay with ground feldspar instead of the ground glass previously used. Later in the 18th century, *bone china* was invented in the UK, by adding ash from cattle bones to clay, feldspar, and quartz. As some of these soft type porcelains may be porous, they may require a glaze. After the object has been fired, the glaze, consisting of oxides and salts or lead, is applied on the object and fired again to vitrify it. The glaze may be applied by means of painting, pouring, dipping, or spraying. Different types of glazes can be obtained by varying the proportions of the glazing ingredients. Porcelain has the following properties – low permeability and elasticity, considerable strength, hardness, toughness, whiteness, and translucency. It also has a high resistance to chemical attack and thermal shock.

### 16.6.1 Porcelain Tiles

Porcelain is usually used in the form of tiles, sanitary wares, fine vases, dishes, figurines, and other decorative items, as linings for tunnels and subway stations, and electric insulators. Porcelain tiles are often extruded and rectified (mechanically cut or grounded to an exact size, so that the edges will have perfect 90° angle). With rectified tiles, it is possible to have thin grout lines (about 3 mm) and unrectified tiles will have thicker grout lines (up to 12 mm). Porcelain tiles are available in many forms as plain, coloured, and also with decorative patterns and sizes. As they are waterproof and hygienic, they are ideal for use in bathrooms, kitchens, and laundry areas (see Fig. 16.12); unglazed tiles are well suited for outdoor installations. However, they are expensive than other tiles and are brittle.

**Fig. 16.12** Porcelain tiles

It has to be noted that bath tubs are not made of porcelain, but of porcelain enamel on a metal base, usually of cast iron. Porcelain enamel is a marketing term used in the USA, and is not porcelain but vitreous enamel.

## 16.7 Choosing Tiles

There are several types of tiles for floors and we must choose the type of tile, based on the following criteria.

*Shape and texture*   They should be uniform in size and shape and free from irregularities, flaws, laminations, and imperfections, which may affect the appearance or serviceability.

*Water absorption*   Ordinary clay tiles have an absorption capacity of more than 10% of its weight. However, by using ceramic fully vitrified tiles, water absorption of less than 0.08% could be obtained. In general, porcelain tile is harder and more impervious to moisture than ceramic tile (has less than 0.5% water absorption rate).

*Colour*   Unglazed tiles have terracotta red colour. Using the glaze, any desired colour could be obtained. Dark tiles add warmth, but should be used only when plenty of sun light is available through open spaces or windows. Chocolate brown, navy, or burnt-coloured floor tiles could be used in large kitchen space. To make a room look larger, lighter hues should be chosen.

*Glazing*   The quality and thickness of glaze used are important factors for choosing glazed floor tiles. They should be chosen keeping in mind the type of traffic. If a poor quality and thickness of glazing is chosen in heavy traffic areas, the flooring will deteriorate in short time. When we chip the ceramic tile, a different colour will be found underneath the top glaze. Whereas, in porcelain tiles the colour is uniform – it is because porcelain is fired at higher temperatures for a longer time than ceramic. In addition, porcelain has higher feldspar content, making it more durable.

*Durability*   Natural clay tiles are highly durable and last longer. Many tiles are available with 50 year warranty. Porcelain tiles are stronger than ceramic tiles because they are denser. Hence, it is more durable and better suited for heavy usage than ceramic tiles. Porcelain tiles are also suitable for outdoor use as they will not soak up water, which can freeze and crack the tile. For outside use, stone tiles are a much better option.

*Layout*   The size of floor tile determines how many grout lines are seen on the flooring, and fewer grout lines can create an illusion of more space. Large tiles make small areas seem more expansive and also take less time and effort to install than smaller tiles. Small tiles make a room feel more intimate and have more grout lines, which can add a complex look to the design.

*Aesthetics*   In addition to ceramic tiles, there are many alternatives for floor tiles such as marble. The material and pattern may be chosen depending on the aesthetic requirements.

*Function*   For high-traffic areas, it is better to use durable porcelain tiles that resist scratching and chipping. Use ceramic tile flooring for indoors only, as ceramic tiles are more porous than porcelain and hence will not last long against the weather. Textured tile should be used in bath room floors instead of tiles with smooth surface, as they become slippery when wet. However, in the entryway, textured tiles are tough to clean, whereas smooth tile may be easier to clean.

*Density and ease of working*   Since ceramic is less dense and more porous than porcelain, it is easier to cut by hand. Porcelain is more brittle and may require experienced person to cut it properly.

*Cost*  The floor material should be consistent with the type of construction envisaged. Porcelain tiles are usually expensive than ceramic tiles, which again are costlier than clay tiles.

## 16.8 Terracotta

Derived from the Italian, *terracotta* means burnt earth, and is a clay-based unglazed or glazed ceramic, where the fired body is porous. This refractory clay product is used to make sculptures and also vessels (flower pots), water and waste water pipes, roofing tiles, and brick ornamental parts of buildings. The clay used for its manufacture should be of superior quality and have sufficient iron and alkaline content. By varying the quantity of iron oxide in clay, various shades of yellow, orange, buff, red, pink, grey, or brown are obtained. It is usually left unglazed, and is cheap, versatile, and durable.

**Manufacture**  The clay is mixed thoroughly with water in a tub. Powdered glass (8–10%), crushed pottery (18–20%) and clean powdered white sand (10–20%) are added to it and the mixture is mixed thoroughly. The excess water is drained and pugged several times in pug mills, until the mix becomes soft and uniform for moulding. The pugged clay is then pressed into special porous moulds made of plaster of Paris or zinc. This shaped clay is taken out of the moulds after few days and then dried slowly. The dried shaped clay is burnt to get uniform colour in special kilns, called muffle furnace, at 1100–1200°C. If burnt properly, terracotta products are impervious and hard, weathers well and can be cleaned easily. The numerous figures at Qin tomb, near Xi'an, Shaanxi province, China, are famous examples of terracotta.

Terracotta is generally available in two types: the porous and the polished (Faience).

**Porous terracotta**  It is manufactured by mixing sawdust or finely fragmented cork to clay and may have the following characteristics.

1. Light weight, sound proof, and fire resistant
2. Resists weathering action
3. Can be sawn to various shapes and may be nailed
4. Has poor strength and hence used only for ornamental works

**Polished terracotta**  It is made by mixing quartz sand and chalk (which acts as a fusing agent), to refractory clays. The polished terracotta is burnt twice. First burning, called *bisque-firing*, is done at 650°C. Then, it is coated with a glazed solution to get the desired colour and texture and dried. It is again fired at 1200°C to obtain highly glazed architectural terracotta with relatively coarse body. The finished polished terracotta may have the following characteristics.

1. It is hard, strong, and durable.
2. It is also fire proof and can be given any desired shape and colour.
3. It has low water absorption (<12%) and can be cleaned easily.
4. It is resistant to weathering and chemical actions.

Applications of terracotta are as follows.

1. It is commonly used in ornamental works such as cornices, arches, and statuettes.
2. As it is fire proof, it can be installed around steel beams for fire-proofing. They have also been used as fillers in concrete floors, to reduce the weight, and in the construction of arches (http://historicbldgs.com).
3. Porous terracotta can be used for sound insulation.
4. Hollow terracotta tile blocks are often used as a fill between steel structural elements and for fire resistant wall construction, including interior partitions.

## 16.9 Earthenware

It is glazed or unglazed non-vitreous pottery, which is normally bisque fired at 1000–1150°C, and hence is slightly porous and coarser than stoneware or porcelain. Pit-fired earthenware dates back to as early as 29,000–25,000 BC (Richerson & Lee, 2018). Typical composition for earthenware is 25% kaolin, 25% ball clay, 35% quartz, and 15% feldspar. Owing to its porosity, it is often glazed to make it watertight (by covering the fired object with finely ground glass powder, and then firing it again at 950 to 1050°C). After firing, most earthenware objects will be coloured white, buff or red. Earthenware has lower mechanical strength than bone china, porcelain, or stoneware, and hence they are commonly made in thicker cross-section, although they are more easily chipped. There are two main types of glazed earthenware. One is covered with a transparent lead glaze (dating back to the Han period of 25–220 AD) – when a cream coloured glaze is applied and called cream-ware. The second type, covered with an opaque white tin glaze (originated in Iraq in the 9th century), and called faience, majolica, or Delftware. Earthenware is used to make cooking pots, pans, and decorative items. Earthenware and stoneware together is called *pottery*.

## 16.10 Stoneware Pipes and Fittings

*Stoneware*, a hard ceramic material resembling porcelain, is pottery that is fired at a high temperature (1100–1200°C) until vitrified. It is opaque and can be of any colour (usually, grey or brownish). It was found to have been used in China and the Indus Valley Civilization as early as 1400–1900 B.C. The key raw material is fine-grained plastic clay or non-refractory fire clay. Typical composition is 0–100% plastic fire clays, 0–15% ball clays, 0–30% quartz, and 0–15% feldspar. The glaze may be obtained by using limes of common salt during the process of burning or using glazing material, applied prior to firing (IS 651:2007). When the interior and exterior surfaces of the pipes and fittings are exposed after jointing, they should be glazed. When they are covered after jointing, they need not be glazed.

Based on the value of crushing strength, the stoneware pipes are classified into three classes SP1, SP2, and SP3 (IS 651:2007). Different tests that may be conducted on stoneware pipes and their sizes are specified in IS 651:2007. These pipes may be laid as per IS 4127:1983. Modern commercial tableware and kitchenware use stoneware rather than porcelain or bone china. Stoneware is also common in craft and studio pottery. Some of its characteristics are follows:

1. Hard, strong, and durable material
2. Resistant to chemical and weathering action when glazed
3. Good appearance and finish
4. Gives ringing sound when struck

It is used in light sanitary-ware such as wash basins and water closets, drain pipes and fittings, and as flooring tiles and wall tiles in toilets and kitchens. Although stoneware pipes are much cheaper than plastic pipes, they are being replaced by plastic pipes due to their ease of joining.

## 16.11 Ceramic Sanitary Appliances

*Sanitary appliances* are also made of ceramics (see Fig. 16.13). They consist of a strong high-grade ceramics made from a mixture of suitable clay and finely ground minerals, such as quartz and feldspar. After firing at a

**Fig. 16.13** Sanitary appliances

high temperature, the ware should not, even when unglazed, have an average value of water absorption greater than 0–5% of the dry weight of the ware [IS 2556 (Parts 1–17)]. It should be coated on all exposed surfaces with an impervious non-glazing vitreous glaze with a white or coloured finish. The lead content, if any in the glaze, should not exceed 5% of the weight of the glaze. The thickness at any place in an appliance should not be less than 6 mm. These appliances may be of the following two types:

1. Soil appliances for the collection and discharge of excretory matter such as water closets and urinals
2. Waste appliances for collection and discharge of waste water such as bidets, washbasins, and sinks

More details about these appliances may be found in IS 2556 (Parts 1–17), and IS 9140:1996.

**Fig. 16.14** Burnt clay *jalley* used in Indian restaurant in Washington D.C.

## 16.12 Burnt Clay Jallies

*Burnt clay jallies* (perforated screen blocks) may be used to provide a screen on *verandah* and also for decorative purposes (see Fig. 16.14). These are generally hand moulded but superior qualities can be produced by machines. The standard sizes of burnt clay *jallies* as per IS 7556:1988 are as follows: 190 × 190 × 100 mm, 190 × 190 × 50 mm, 190 × 140 × 100 mm, 190 × 140 × 50 mm, 140 × 140 × 100 mm, 140 × 140 × 50 mm, 140 × 90 × 50 mm, and 90 × 90 × 50 mm. The total void area of the *jallies* should not exceed 40%. The thickness of any shell should not be less than 10 mm and that of the web not less than 8 mm. Keys for bonding with mortar should be 10 mm wide and 3 mm deep. For more details, IS 7556:1988 should be consulted.

## 16.13 Environmental Effects of Ceramics

As clay, which is used in clay products and ceramics, is an abundantly available natural material, the environmental effect of its extraction is minimal. Moreover, clay can also be easily recycled. However, the production of these products may involve pollutants and particulate matter. The primary pollutants associated with raw material beneficiation are particulate matter (PM) and PM less than 10 μm in aerodynamic diameter. In addition, calciners emit products of combustion such as nitrogen oxides ($NO_x$), sulphur oxides ($SO_x$), carbon monoxide (CO), carbon dioxide ($CO_2$), and volatile organic compounds (VOC). For ceramic bodies that are dry-formed, PM is likely to be emitted from green machining activities and when the glaze is applied by spraying. Particulate matter emissions consisting of metal and mineral oxides also arise from glaze preparation. Heat-treating processes result in combustion products and filterable and condensable PM. Emissions of fluorine compounds also are associated with firing. Fabric filters, wet scrubbers, and electrostatic precipitators are used to control PM emissions. Lead emissions have been reduced in recent times, due to the use of low-lead or lead-free glazes. However, these newer glazes require the use of boron, which may result in arsenic emissions. However, ceramic tiles do not contain the chemicals such as formaldehyde, VOCs or PVC (polyvinyl chloride), which may be present in other floor coverings and cause adverse health effects. More information may be had from *https://ehs. princeton.edu.*

# SUMMARY

- Ceramics are inorganic, non-metallic, crystalline materials that are typically produced using clays/other minerals from the earth or chemically processed powders.
- Several types of ceramic and clay products are used in buildings. Clay tiles are used as roofing and flooring tiles.
- The various clay roofing tiles include plain tiles, pan tiles, Mangalore pattern tiles, Spanish tiles, Allahabad tiles, and ridge and ceiling tiles.
- Clay tiles may be tested for water absorption, permeability, flexural strength, breaking strength, and impact resistance as explained in Section 25.22 of Chapter 25.
- Ceramic processing generally involves high temperatures, and the resulting materials are heat resistant or refractory in nature.
- The constituents and glaze used in ceramic products make them resistant to corrosion and chemical substances, wear, and water. Manufacture of ceramics involves various steps, which are similar to that of brick making. Ceramics is used in sanitary appliances as well as in decorating buildings.
- Different types of vitrified tiles are made by firing tiles at high temperatures and applying glazes and are fixed using mortar or adhesives.
- Tiles should be chosen carefully by considering shape, texture, water absorption, colour, glazing, durability, and cost.
- Terracotta, porcelain, earthenware, and stoneware are used in certain applications.

# EXERCISES

## Multiple-choice Questions

1. Ceramics dates back to
   (a) 26,000 B.C.
   (b) 6,000 BC
   (c) 600 AD
   (d) none of these
2. Use of interlocking plain tiles instead of plain tiles may reduce roof cost by
   (a) 20%
   (b) 30%
   (c) 35%
   (d) 40%
3. The minimum overlap for Mangalore pattern roofing tiles is
   (a) 40-mm lengthwise and 30-mm widthwise
   (b) 50-mm lengthwise and 30-mm widthwise
   (c) 60 mm lengthwise and 25-mm widthwise
   (d) 60-mm lengthwise and 30-mm widthwise
4. Water absorption of Class AA and Class A Mangalore pattern roofing tiles are
   (a) 10 and 15%, respectively
   (b) 15 and 18%, respectively
   (c) 18 and 20%, respectively
   (d) 20 and 18%, respectively
5. Ceiling tiles are used to provide
   (a) aesthetic appearance
   (b) thermal comfort
   (c) acoustic control
   (d) all of these
6. Maximum water absorption of Classes 1 and 2 flooring tiles as per IS 1478 are
   (a) 19 and 24%, respectively
   (b) 10 and 19%, respectively
   (c) 10 and 24%, respectively
   (d) 20 and 18%, respectively

7. Average minimum flexural strength of Class 1 flooring tile is
   (a) 2.5 N/mm width
   (b) 3.5 N/mm width
   (c) 5 N/mm width
   (d) 6 N/mm width
8. In dry pressing, ceramic powder with a moisture content of the following percent is compacted between plungers in a die cavity
   (a) 5 to 10%
   (b) 10 to 15%
   (c) 0 to 15%
   (d) 5 to 15%
9. Ceramics generally are fired at
   (a) 700–1100°C
   (b) 800–1200°C
   (c) 900–1300°C
   (d) 1000–1400°C
10. The cheapest type of vitrified tile available in the Indian market is
    (a) soluble salt vitrified tile
    (b) double charge vitrified tile
    (c) full body vitrified tile
    (d) glazed vitrified tile
11. One of the following properties of finished polished terracotta is wrong. Identify it.
    (a) It is fireproof.
    (b) It is strong and durable.
    (c) It has water absorption of 20%.
    (d) It is resistant to chemicals.
12. Bisque firing of ceramics is done at
    (a) 600 to 800°C
    (b) 700 to 1000°C
    (c) 900 to 1200°C
    (d) >1200°C

## Review Questions

1. How are tiles manufactured? Briefly explain the various steps.
2. Write short notes on the following:
   - (a) Flat roof plain tiles
   - (b) Pan tiles
   - (c) Mangalore pattern roofing tiles
   - (d) Clay half-round country tiles
   - (e) Allahabad tiles
   - (f) Ridge and ceiling tiles
3. How are roof tiles attached to the roofs?
4. How do pan tiles differ from plain tiles? What are interlocking pan tiles?
5. What are Mangalore pattern roofing tiles? How are they classified as per IS 654?
6. How are Spanish tiles laid? Explain with a neat sketch.
7. How do Allahabad tiles differ from Spanish tiles? Explain their laying with a neat sketch.
8. What are ridge tiles? Explain with neat sketch. How are ridge and ceiling tiles classified as per IS 1464?
9. Why are ceiling tiles provided? Explain the single and double-lug type ceiling tiles with neat sketches.
10. What is the main advantage of using solar roof tiles?
11. How are clay flooring tiles classified as per IS 1478?
12. Compare clay and concrete roof tiles.
13. Explain the manufacturing process of ceramic products.
14. What are the typical properties of ceramics?
15. How are ceramic tiles classified as per IS 13712?
16. What are the four types of vitrified tiles available in the market? Explain them briefly.
17. How are floor tiles fixed to the floor?
18. List the criteria based on which we can choose tiles.
19. Write short notes on the following:
   - (a) Terracotta
   - (b) Earthenware
   - (c) Stoneware
   - (d) Porcelain
   - (e) Ceramic sanitary appliances
20. Write a brief note on the environmental effects of clay and ceramic products.

## ANSWERS

### Multiple-choice Questions

| | | | | | |
|---|---|---|---|---|---|
| **1.** (a) | **2.** (d) | **3.** (c) | **4.** (c) | **5.** (d) | **6.** (b) |
| **7.** (d) | **8.** (d) | **9.** (c) | **10.** (a) | **11.** (c) | **12.** (b) |

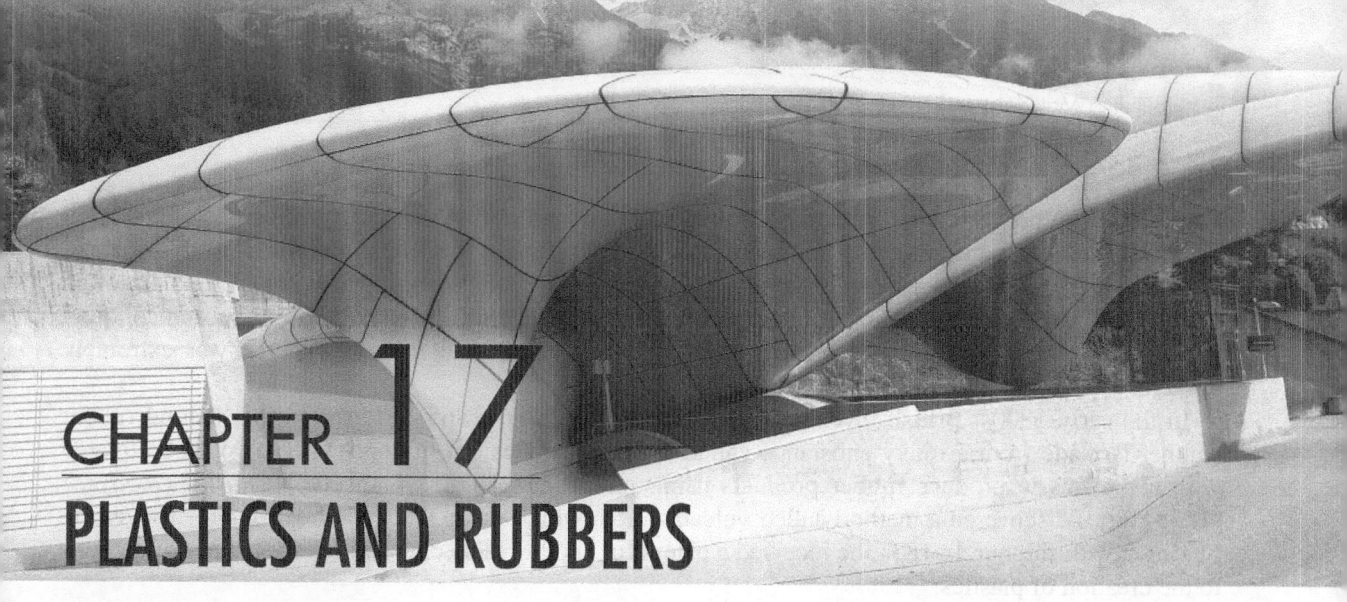

# CHAPTER 17
# PLASTICS AND RUBBERS

## 17.1 Introduction

We now live in the plastic age (from 1930 to present) [Historians and archaeologists have defined periods of human history approximately by the technologies/materials that made the greatest impact on society, such as stone age (2.6 M BC to 3300 BC), Bronze age (3000 BC to 1200 BC), Iron age (1200 BC to 568 AD), Industrial age (1760–1830 AD), and Plastic age]. The term *'plastics'* was derived in 1632 from the Greek word *'plastikos'* meaning fit for moulding, and *'plastos'* meaning moulded. It represents the malleability or plasticity of this material during manufacture, which allows it to be cast, pressed, or extruded into a variety of shapes – such as films, fibres, plates, tubes, bottles, and boxes. Plastics, unlike other building materials such as brick, stone, timber, or concrete, are of very recent origin. Plastics, also called as *polymers* or *synthetic resins*, are unique man-made materials of the 20th century. Most plastics are organic polymers derived from petrochemicals, although some come from coal, natural gas, or even cotton. Hence, mostly they are non-renewable materials.

The wide functionality of plastics provides a distinct advantage over many other traditional building materials due to their resistance to weathering, flexibility for many uses, and lower costs. Owing to their relatively low cost, ease of manufacture, versatility, and imperviousness to water, plastics offer a wide variety of properties and design flexibility. They are used in a range of products that include electric wiring, flooring, window and door parts, wall covering, waterproofing, pipes, valves and fittings, hinges, glues, insulation, and decorative items. They have already displaced many traditional materials, such as wood, stone, metal, glass, and ceramic, in most of their former uses. Industries such as those for fibres, rubbers, plastics, adhesives, sealants, and caulking compounds are based on polymers. Rubber is a natural polymer.

The building and construction industry consumes about 20–25% of plastics (amounts to over 5 MT annually in Europe alone), which is surpassed only by the packaging industry. Polymers, despite being the fourth major class of building material used after steel, wood, and cement, still represent only about 1% of the total volume of building materials used. According to World-watch institute, though about 348 MT of plastics were produced in 2017, recovery and recycling is insufficient resulting in about 22–43% ending up in landfills and oceans each year (The nine countries in Europe such as Switzerland, Austria, the Netherlands, Germany, Sweden, Luxembourg, Denmark, Belgium, and Norway reached a recovery ratio of 95% of the post-consumer plastic waste, as they have a landfill ban). Study of rubber and plastics is a special field in chemical engineering called polymer technology. Only some basic information is presented in this chapter about plastics and rubbers.

## 17.2 Brief History of Plastics and Polymers

Plastics are synthetic materials, meaning that they are not natural but created by humans. Some natural materials such as Japanese lacquer, amber (from sap of plants used by Greeks and Romans to make jewellery), tortoise shell, and shellac (a substance secreted by insects) are similar to plastic, and are often hard to find. The first synthetic plastics were derived during the 1800s from cellulose, a substance found in plants and trees – cellulose was heated with chemicals and resultant new material was extremely durable.

In the early 1800s, British inventor *Thomas Hancock* (1786–1865) discovered that natural rubber (rubber is made from a milky substance called latex that oozes out of rubber trees) could be treated with chemicals to produce rubber products using moulds. In 1839, US inventor *Charles Goodyear* (1800–1860) discovered a method called vulcanization, which made natural rubber much stronger, by mixing it with sulphur. In 1851, he received a patent for ebonite (also called vulcanite), thus paving way to the creation of plastics.

British inventor *Alexander Parkes* (1813–1890) developed a nitrocellulose-based product called Parkesine in 1866, as a substitute for rubber. In 1869, *John Wesley Hyatt* of USA (1837–1920) added camphor to nitrocellulose and invented celluloid, which was used by George Eastman (1854–1932) as the main ingredient in photographic films. *Parkes* and *Hyatt* share the credit as the first inventors of plastics. Belgian inventor *Leo Baekeland* (1863–1944) successfully produced the first thermosetting plastic in 1907 and named it as *Bakelite*. Bakelite was created by heating phenol (comes from coal tar) and mixing it with formaldehyde, and was an excellent insulator. During 1884, French chemist *Hilaire Chardonnet* (1839–1924) invented the artificial fibre called rayon, by dissolving nitrocellulose in alcohol and ether, and squeezing the substance through holes under pressure.

After this period, the development of plastics took place in the laboratories of giant chemical and oil companies. A team of researchers headed by *Wallace Carothers* (1896–1937) at the DuPont Chemical Corporation created nylon (polyamide) in 1935, which was used for making tents and parachutes during World War II (1939–1945). US chemist *Roy Plunkett* (1910–1994) accidentally discovered teflon (polytetrafluoroethylene or PTFE) in 1938, which was first used in the military before being used as a coating in cookware. Vinyl (polyvinyl chloride, or PVC) was created by US chemist *Waldo Semon* (1898–1999), and patented in 1933. Chemists *Eric Fawcett* and *Reginald Gibson*, while working in the British chemical company ICI, unexpectedly discovered polyethylene [known today as low-density polyethylene (LDPE)] in 1933, while heating a mixture of ethylene, benzaldehyde, and oxygen at 170°C, and under a pressure of 192 MPa. German professor *Karl Ziegler* found a way to polymerize ethylene at low pressure using catalysts in 1953; this is known today as high-density polyethylene (HDPE). Many industrial companies like ICI have developed special plastics, which are used in several applications in construction as paints, laminates, and bathroom fittings.

## 17.3 Composition of Plastics

Until the 1920s, no one understood the chemistry of polymers. German scientist *Hermann Staudinger* (1881–1965) showed in 1920 that *polymers* are long chains of single molecules known as *monomers* that are joined together (by covalent bonds) but are still one molecule (Staudinger, 1920). For his discovery that natural and synthetic polymers have 'macromolecules', Staudinger was awarded the Nobel Prize for Chemistry in 1953.

The formation of most plastics is a part of the field of *organic chemistry*, which is a chemistry based on carbon atoms. When atoms join together, they are called *molecules*. A molecule is the smallest and simplest unit of a substance. Plastics are made up of macromolecules (very large molecules composed of hundreds or thousands of atoms) linked together as huge chains called polymers (from the Greek *poly*, meaning 'many' and *meros*, meaning 'units'). A polymer chain, when magnified, may look like hundreds of paperclips hooked together.

## 17.4 Classification of Plastics

They are divided into two broad categories depending on their reaction to heat during and after manufacture as (a) thermoplastics and (b) thermosetting plastics.

### 17.4.1 Thermoplastics

About 92% of plastics are thermoplastics. Thermoplastics are polymers which soften on heating without undergoing any chemical change. They are shaped during this soft stage and then harden again as they cool. This process can be repeated several times, as there is no chemical change. Thermoplastic molecules are chain-like with very few cross-links between molecules – this gives them a high degree of elasticity and ability to return to the original shape, if not overstressed. Thus, thermoplastics can be recycled. Examples of thermoplastics are: polyethylene terephthalate (PET), polypropylene, acrylic, PVCs, polystyrene, and nylon.

### 17.4.2 Thermosetting Plastics

Thermosetting plastics are polymers which are liquid or soft at low temperatures, but which undergo irreversible chemical change to solidify and become hard when heated to about 125 to 175°C. Hence, the process cannot be repeated. Thermosetting plastics are generally stronger than thermoplastic materials due to the large number of cross-links between their molecules. They are better suited to high-temperature applications – up to the decomposition temperature of about 340°C. Hence, they cannot be recycled. Mechanical strength and hardness also improve with crosslink density, although at the expense of brittleness. Examples of thermosetting plastics are: melamine formaldehyde (Formica), phenol formaldehyde (Bakelite), urea formaldehyde, epoxies, polyurethane, and polyesters.

## 17.5 Polymerization

The chemical process of combining a large number of small molecular units ('*mers*') to form a long-chain of molecules is called *polymerization*. Some polymers, such as polypropylene and polystyrene, contain only hydrogen and carbon atoms. Others are made by adding other atoms such as oxygen, nitrogen, chlorine, sulphur, or hydrogen to the carbon atom – for example, nylon contains nitrogen, Teflon contains fluorine, and vinyl contains chlorine. The groups of atoms that are used to make unit cells are called *monomers*. For some plastics, such as polyethylene, the repeat unit can be just one carbon atom and two hydrogen atoms and is known by the chemical formula $(C_2H_4)_n$. For other plastics, such as nylons, the repeat unit can involve 38 or more atoms. Polymerization involves heat, pressure, mixing, and chemical action.

There are three basic types of polymerization, chain-reaction (or addition), step-reaction (or condensation) polymerization, and co-polymerization. They are described briefly.

**Chain-reaction (or addition) polymerization** In this process, identical molecules are combined to form an exact multiple of the original monomeric molecule. Polymers formed by addition polymerization are often thermoplastic in nature. Addition polymerization involves three basic steps: initiation, propagation, and termination. For example, during the initiation phase of the polymerization of polyethylene, the double bonds in the ethylene 'mers' break and begin to bond together (see Fig. 17.1). A catalyst may be necessary to start or speed up the reaction. The second phase of propagation involves the continued addition of monomers together into chains. During the final termination phase, the reaction is ceased, by quenching using water. Polymers formed by addition polymerization include acrylic, polyethylene, and polystyrene.

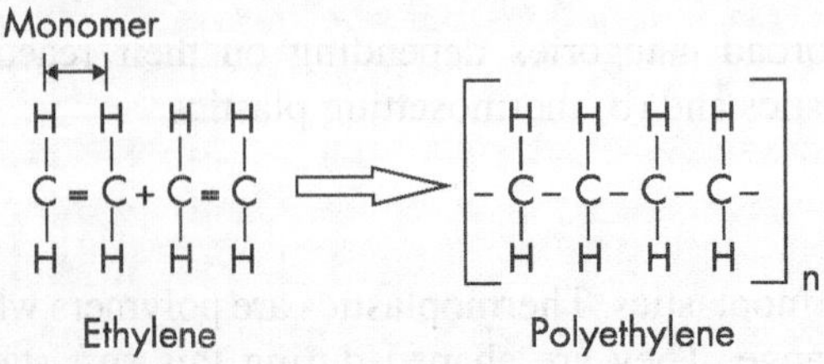

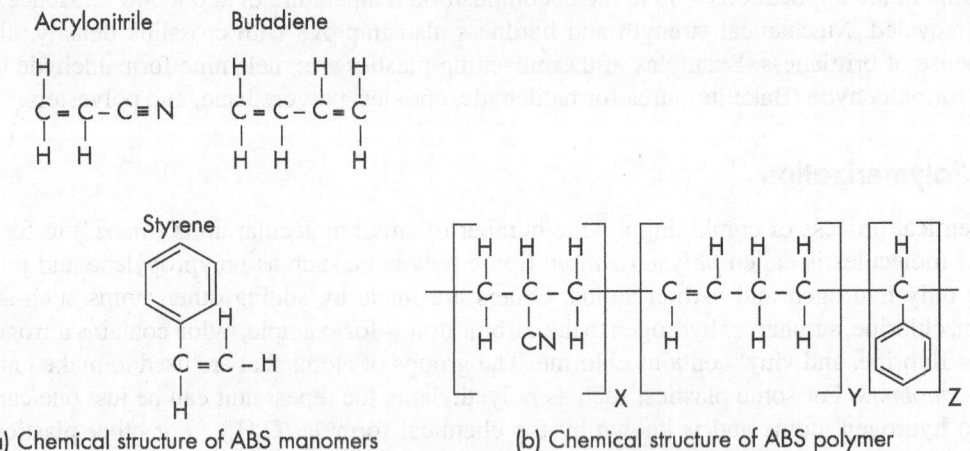

**Fig. 17.1** Addition polymerization of polyethylene

**Step-reaction (or condensation) polymerization**   In this process, a large number of identical or different molecules are combined, often eliminating low molecular substances such as water and methanol, which are the by-products of the chemical reaction. This polymerization method typically produces polymers of lower molecular weight than chain reactions and requires higher temperatures. Condensation polymers include nylons, some polyesters, urea formaldehyde, and urethanes. These polymers can be thermoplastic or thermosetting in nature. Both addition and condensation polymers can be linear, branched, or cross-linked. Condensation polymers, unlike addition polymers, may be biodegradable.

**Co-polymerization**   In this process, two or more different monomers are added together to form a different polymer (see Fig. 17.2). Examples of copolymers include acrylonitrile butadiene styrene (ABS) rubber, styrene butadiene rubber (SBR), butyl rubber and saran.

(a) Chemical structure of ABS manomers          (b) Chemical structure of ABS polymer

**Fig. 17.2** Co-polymerization of ABS polymer

# 17.6 Fabrication of Plastic Products

Polymers are initially produced as powders or granules. However, they are seldom used in this form but are changed by adding other additives to improve properties, reduce cost, or achieve other results. Thus, adding additives will make plastics resistant to sunlight, very flexible, or even less expensive.

## 17.6.1 Additives/Moulding Compounds

Additives/moulding compounds are available as fillers, reinforcements, catalysts, hardeners, plasticizers, lubricants, stabilizers, flame retardants, foaming agents, and colourants.

*Fillers or extenders*    Inert fibrous or powder fillers are used in relatively large quantity and help reduce cost. They may add strength, toughness, or thermal stability to the plastic. For example, wood flour is used as filler for PF (phenol formaldehyde). Addition of metallic flake to PVC results in electrically conductive plastics.

*Reinforcements*    They are fibres or additives that improve physical properties, especially the tensile and impact strength of the material. Mats or yarns of polymer or metal are examples of reinforcements. Glass fibres can be used to increase the strength of resins. Laminate structures are used in aircraft and car bodies, foam-backed carpet, and Formica® counter tops.

*Plasticizers*    These, such as dioctyl phthalate, are materials added to a resin to improve the plasticity and make plastics flexible at ordinary temperatures.

*Lubricants*    These are usually waxy materials applied to the surface of moulds for easy release of the product.

*Flame retardants*    They lower the flammability of the material; antimony trioxide or compounds of chlorine or bromide are used as flame retardants. Carpets are often stabilized with flame retardants.

*Stabilizers*    They prolong the lifetime of the polymer by suppressing degradation that results from UV-light, oxidation, or biological agents. Carbon black is used to absorb UV radiation. A special form of PVC, called UPVC, is used in the manufacture of drainpipes so that the pipes will not degrade when exposed to the Sun.

*Catalysts*    They assist and accelerate the hardening of resins.

*Hardeners*    They are used to increase the hardness of the resin.

*Foaming agents*    These are additives that give polymers a specific form. Cellulose sponges (open-cell) and polystyrene (closed- cell) are examples of the two different foams that are produced. They are used in expanded polystyrene (EPS) cups, building boards, and polyurethane carpet underlayment.

*Colourants*    These are either dyes or pigments, added in small quantities to obtain coloured plastic parts.

## 17.6.2 Solvents

Thermoplastics are used in paints, varnish, enamel, and lacquer. In order to bring the hard resin to liquid state, appropriate organic solvent has to be used. Examples of solvents include acetone, benzene (it is carcinogenic and hence replaced by toluene) and formic acid (before using the solvent, it is important to consider its compatibility with the resin). Once the paints are applied on the surface, the solvent evaporates and the resin returns to the hard state.

## 17.6.3 Fabrication Methods

Because of the properties of polymers, it is possible to mould them and change their shape using several manufacturing processes: compression moulding, extrusion, injection moulding, transfer moulding, jet and blow moulding, calendaring, and foaming. Compression moulding is the most common method of moulding for thermosetting plastics, whereas injection moulding is the preferred method for thermoplastics.

**Compression moulding** It was originally developed by Baekeland in 1909. In the simplest version, weighed pellets containing thermosetting plastics in powder form are placed inside a pre-heated steel lower mould. The mould is then closed by placing the upper half and subjected to further heat and pressure, provided by a press (see Fig. 17.3). The pressure and heat causes polymerization and flow of the plastic material in the mould. This process is suitable for thermosetting plastics. Curing is done by heating. After the curing is completed, the mould is opened and formed plastic part is taken out.

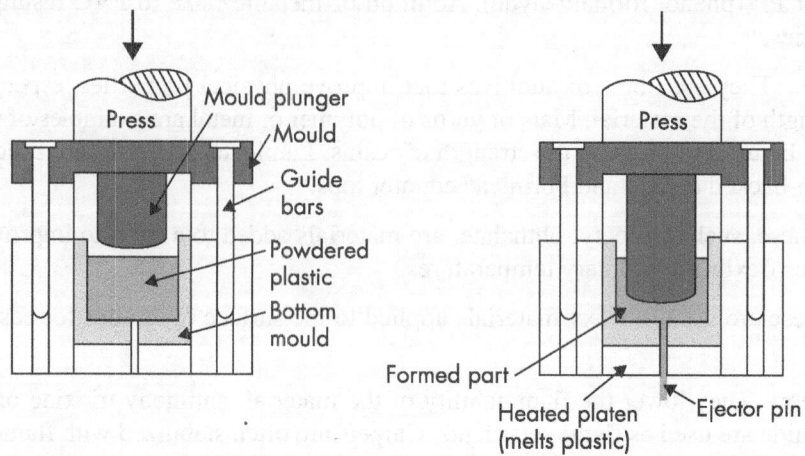

**Fig. 17.3** Compression moulding process

**Extrusion** It is a process that is used for continuous moulding of thermoplastic materials with uniform cross-section such as tubes, rods, and electric cables. As shown in Fig. 17.4, the thermoplastic pellets are fed into a machine through a hopper, and heated to a plastic state. Then, they are pushed into a die by a rotating screw to get the profile of the extruded component.

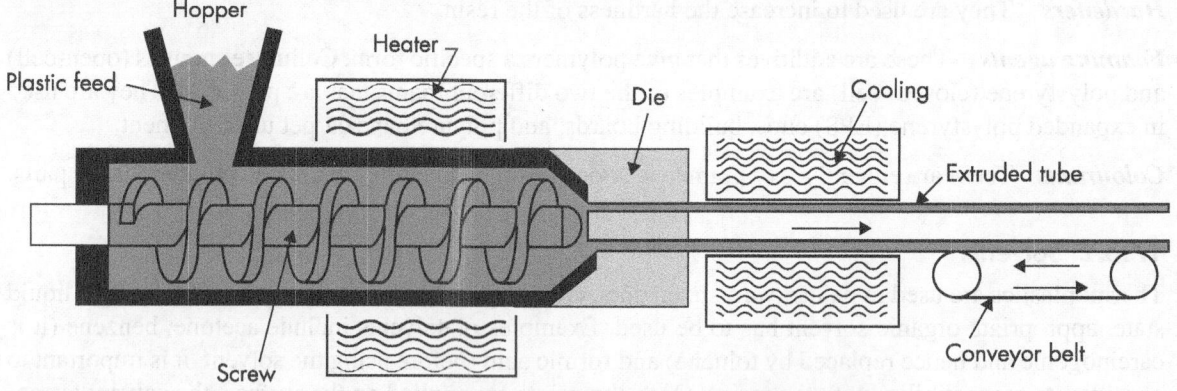

**Fig. 17.4** Moulding of tubes by horizontal extrusion

**Injection moulding** This is the most common of all plastics manufacturing processes and is adopted for both thermoplastic and thermosetting plastics. The polymer, in granule form, is fed through a hopper and heated inside a cylinder until fused. A screw pushes the melted material into the mould cavity through a nozzle (Fig. 17.5). Because of the viscous (thick) nature of the fused polymer, very high pressure is needed to make it flow. Thus, the machine and the mould have to be very strong to withstand the forces involved. The mould is kept cold to allow the hot plastic to cure and acquire the shape. Half of the mould is opened to obtain the finished article. Advantages of injection moulding include rapid processing, little waste, and easy automation.

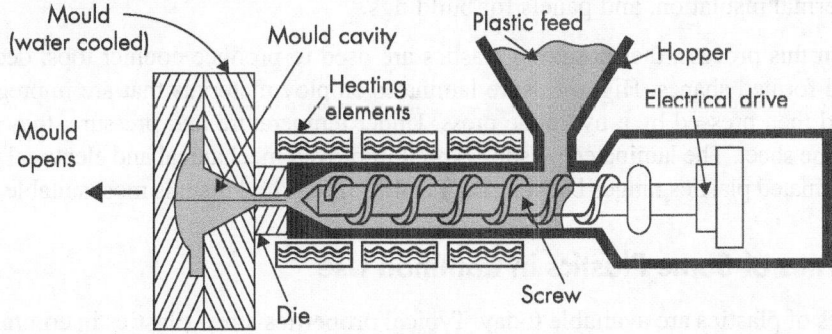

**Fig. 17.5** Injection moulding process

**Transfer moulding**  This process used for thermosetting plastics is a modification of compression moulding and is used to produce complex parts of varying thickness. The steps in this process are as follows:

1. A partially polymerized material is placed in a heated chamber.
2. A plunger forces the flowing material into moulds.
3. The material flows through an orifice.
4. The temperature and pressure inside the mould are higher than in the heated chamber and hence induces cross-linking.
5. The plastic is cured and hardened; then the mould is opened to remove the product.

**Jet moulding**  It is similar to extrusion moulding. The nozzles can also be heated in this method to obtain higher temperatures. It is suitable for both thermoplastics and thermosetting plastics.

**Blow moulding**  It is a simple process, which is rapid and relatively inexpensive. In this process, used for thermoplastics, compressed air is introduced underneath a warmed sheet material forcing the material into a mould cavity. Blow moulding is used to produce hollow plastic products such as tubes, bottles, tubs, and drums.

**Other processes**  The other processes of fabrication include calendaring, casting, foaming, and laminating. They are discussed briefly below:

*Calendaring*  In this process, fused thermoplastic is passed through a set of heated-rotating rollers that squeeze the material into a continuous sheet or film of desired thickness. The film is cooled by jets of air or water, before being cut to suitable lengths or loaded onto rolls. Film textures can also be produced by rollers with the desired pattern embossed on them.

*Casting*  This process is similar to that of metal casting.

*Foaming*  In this process, synthetic resin is converted into a sponge-like material with a closed-cell or open-cell structure, either of which may be flexible or rigid. Any thermosetting or thermoplastic resin can be converted into foams. Plastics that are commonly foamed include vinyls, polystyrene, polyethylene, phenolics, silicones, cellulose acetate, and urethanes.
Foams with a closed-cell structure are produced by incorporating a blowing agent that releases gas when heated to form bubbles in the plastic. Foams with an open-cell structure are produced by incorporating

an inert gas into the resin under pressure, releasing the mixture to the atmosphere, and then curing the resulting foam. They are used for a variety of products including cushioning materials, packaging, furniture, toys, thermal insulation, and panels for buildings.

*Laminating* In this process, thermosetting plastics are used to produce counter tops, decorative sheets, tubes, rods, and formed shapes. High-pressure laminates employ materials that are impregnated with the plastic resin and then pressed by a hydraulic press. Under temperature and pressure, they are bonded together to form one sheet. The laminated plastics exhibit improved mechanical and electrical properties. The thickness of laminated plastics ranges between 0.13 and 15 mm. Vinyl resin is most suitable for lamination.

## 17.7 Properties of Some Plastics in Common Use

Numerous types of plastics are available today. Typical properties a few plastics in common use are listed in Table 17.1. It has to be noted that as the density of a plastic within the family group is increased, the softening temperature, strength, hardness, stiffness, creep resistance, and impermeability to gas and liquid, will also increase. When the average molecular weight of the plastic is increased, all the aforementioned quantities will increase. At the same time, elongation and toughness will decrease. The additives used with the resins will cause greater changes in properties.

**Table 17.1** Typical properties of plastics

| Material | Specific gravity | Ultimate tensile strength, MPa | Modulus of elasticity, GPa | Coefficient of thermal expansion × $10^{-6}$/°C | Service temperature, °C |
|---|---|---|---|---|---|
| **Thermoplastics** | | | | | |
| Polyethylene (PE) | 0.91–0.96 | 20–44 | 0.62–0.89 | 120 | 80–130 |
| Polypropylene (PP) | 0.9–1.36 | 28–41 | 0.9–1.55 | 72–90 | 115–120 |
| Polystyrene (PS) | 1.04–1.07 | 35–56 | 2.3–3.3 | 70 | 52–96 |
| Polyvinyl chloride (PVC) | 1.37 | 40–65 | 2.4–4.1 | 50 | 66–104 |
| Acrylonitrile butadiene styrene (ABS) | 1.05 | 27–55 | 1.1–2.82 | 60–130 | 87–105 |
| Nylon | 1.13–1.15 | 90–165 | 2.6–3.2 | 80–150 | 120–150 |
| **Thermosetting plastics** | | | | | |
| Polyester | 1.28 | 45–90 | 2.5–4.0 | 100–110 | 120–200 |
| Epoxy resin | 1.03 | 45–90 | 3.5–7.0 | 48–85 | 80–260 |
| Phenolic | 1.28–1.9 | 33–59 | 2.0–4.8 | 60–80 | 120–200 |
| Alkyd resin | 2.02–2.24 | 20–62 | 5.0–6.0 | – | 150 |

## 17.8 Advantages and Disadvantages of Plastics

The main desirable properties and advantages of plastics are as follows:

*Mouldability* Plastics can be moulded to any desired shape or size. It can also be made in any desired colour and texture (it is not necessary to paint or polish the surface), thus making it an attractive architectural material.

*Cheap*    As its manufacturing cost is low, it is cheap. Hence, it is extensively used to replace scarce and/or most expensive materials like wood.

*Chemical resistance*    They have good resistance to almost all chemicals. In addition there is no corrosion.

*Resistance to biological hazards*    They are not attacked by insects and fungi and have high resistance to weathering conditions.

*Light in weight*    They are light in weight and hence plastic products can be easily handled and transported. A few varieties have a glossy appearance like glass.

*Electrical insulation*    They are good electrical insulators and hence extensively used in electrical applications such as plugs and switches.

*Low thermal conductivity*    They have low thermal conductivity, like wood.

*Easy to maintain*    They require little maintenance, as they are able to withstand moisture, oil, and grease well.

*Easiness in fixing*    They can be sawn, drilled, or punched and fused together easily.

*Durability*    Plastics are extremely durable, which means that they last longer. This is an advantage as well as a disadvantage. It is because plastics in the landfill will last for hundreds of years.

*Good strength*    They have high tensile (10–95 MPa) and compressive strength (20–100 MPa), and have high strength-to-weight ratio. Although they are relatively strong, their stiffness is too low to be used in most structural applications. However, they can be combined with fibres of high stiffness and strength, to form composites with improved structural properties.

*Absorbs shock*    They can act as shock absorbing material.

*Highly reflective*    They may have high refractive index.
Plastics, however, may have the following disadvantages:

*High thermal expansion*    Plastics have a high coefficient of thermal expansion, which is about 10 times as much as steel.

*Lack of fire resistance*    Most of the plastics cannot withstand high temperatures and are highly flammable. In addition, they emit toxic fumes and dense smoke, in case of fire in buildings.

*Lack of durability when exposed to the sun*    Most plastics deteriorate when exposed to the ultraviolet (UV) rays of the Sun for prolonged periods.

*High creep*    Plastics exhibit high creep and hence not suitable for long-term structural use.

*Lack of ductility*    Most plastics are not ductile.

*Cost of recycling*    Though plastics can be recycled, recycling of plastics is generally expensive.

*Occupies landfills*    In the United States alone, 20% of the landfill is made up of plastics.

## 17.9 Common Plastics and their Use

Plastics are produced with a wide variety of material properties that allow them to be adapted to many different applications. Because of their lightweight, low cost, and desirable properties, their use has rapidly increased; they are substituted for other materials such as wood, metals, and glass. They are used in hundreds of items, including pipes, doors, windows, formwork, cars, bulletproof vests, toys, hospital equipment, food containers, clothing, etc. The characteristics and typical applications of some of the widely used plastics are listed in Table 17.2 (see also Table 17.5).

**Table 17.2** Commonly used plastic materials, their characteristics, and uses

| Material | Characteristics | Typical uses |
|---|---|---|
| **Thermoplastics** | | |
| Polyethylene terephthalate (PETE or PET) | Low water absorption and permeability, high heat resistance, clear, hard, tough, chemically resistant, and shatterproof | Most widely produced plastic in the world. Polyester fibres for clothing and carpets, bottles and packaging, buckets, sheeting for waterproofing and as vapour barriers, water tanks. |
| High-density Polyethylene (HDPE or PE) | There are three versions: LDPE, HDPE, and UHMW. LDPE has high ductility but low tensile strength. HDPE has excellent moisture barrier properties and chemical resistance, hard to semi-flexible and strong, soft waxy surface, permeable to gas. UHMW: Extremely strong that can even exceed steel in strength | LDPE: Plastic bags and squeezable bottles. HDPE: Robust plastic packaging like milk and water jugs, laundry and shampoo bottles, recycling bins, agricultural pipe, wire insulation, pipes, playground equipment (see Fig. 17.6), toys, and water tanks UHMW: Medical devices (e.g., artificial hips) |
| Polypropylene (PP) Developed in Germany in 1957 | Relatively inert and semi-transparent, light but strong and can withstand higher temperatures, vulnerable to UV radiation, and can degrade in direct sunlight, not as impact-resistant as HDPE or LDPE; Somewhat permeable to highly volatile gases and liquids, has good electrical resistance | Second widely produced plastic because of its adaptability to a variety of manufacturing techniques. Packaging for consumer products, plastic parts for the automotive industry, plastic mouldings, valves, special devices like living hinges and textiles. |
| Polyvinyl Chloride (PVC) also known as Poly vinyl or Vinyl | There are three versions: Rigid PVC (RPVC or uPVC) (1872 by German Eugen Baumann), flexible or plasticized PVC (1926 by Waldo Semon of the USA), and chlorinated PVC (CPVC). Rigid PVC: Hard, rigid and dense (sp. gr. 1.4), good tensile strength, non-flammable, very resistant to chemicals and alkalis, good resistance to weathering. However, limited temperature range, subject to attack by several solvents. Flexible PVC: less brittle, more temperature resistant. | Third widely used. One of the cheapest. Rigid PVC: pipes (both water, sanitary, and sewer applications), doors and windows, flooring, gutters, and wall siding. Flexible PVC: plumbing, electrical cable insulation, signage, in applications replacing rubber, as foam, film and sheeting, coatings, paints and adhesives. CPVC: hot-water pipes up to 120°C |
| Polycarbonate (PC) | High strength, excellent transparency, high impact resistance, good resistance to weathering, and shatter resistant | Used as safety glazing in high-rise buildings, shop windows, railway station platform coverings, and green houses, cover for costly paintings, bulletproof protection in vehicles and VIP podiums |
| Polystyrene (PS) | Hard, brittle, clear to opaque, glassy surface, rigid or foamed, affected by fats and solvents | Kitchen items, food containers, toys, electronic goods and furniture, foamed insulation boards or blocks, widely used in packaging (trade name: Styrofoam), adhesives, and coatings |

*(Contd)*

**Table 17.2** (*Contd*)

| Material | Characteristics | Typical uses |
|---|---|---|
| Acrylic-Polymethyl Methacrylate: (PMMA, Perspex, Plexiglas) | *PMMA*: Transparent and scratch resistant<br>*Perspex*: Hard, stiff, transparent, lightweight, high breaking resistance, resistant to detergents, insulator, reduces UV rays considerably, can be formed into different shapes like domes (see also Section 15.10 of Chapter 15). | *PMMA*: Used in optical devices<br>*Perspex:* Replacement of glass in buildings (when large size is used provision should be made for thermal expansion or contraction), roof lighting, storm glazing, engineering laboratory models |
| Polyamide (Nylon) | Tough, smooth, lightweight, high tensile strength and lustrous fibre; has good elasticity, excellent abrasion resistance and resilience but low absorbency. | Carpets, brush bristles, and bearings; nylon yarn is woven into garments, ropes, and stockings. |
| Acrylonitrile Butadiene Styrene (ABS): Cycolac, Lustran, Novodur, and Ronfalin | Tough, hard, high tensile and flexural strength, chemically resistant, impact resistant, opaque, amorphous and inexpensive, easy to machine and has a low melting temperature | Numerous applications. Examples: pipe fittings, safety helmets, computer keyboard, power-tool housing, plastic face-guard on wall sockets (oftentimes a PC/ABS blend), LEGO toys, and 3D printing. |
| Polyvinyl Butyral | Strong binding, optical clarity, adhesion to many surfaces, toughness and flexibility | Major application in laminated safety glass, and automobile windshields (see also Section 15.8.1) |

**Thermosetting plastics**

| Material | Characteristics | Typical uses |
|---|---|---|
| Formaldehydes (Methanol): PF (phenol-formaldehyde or *Bakelite*), UF (urea-formaldehyde), and MF (melamine-formaldehyde or *Formica*) | *PF:* oldest plastic, hard and rigid, strong, light, insoluble in a wide range of solvents and unreactive, high electrical and heat resistance, low water absorption<br>UF: similar to PF; but colourless, less water and heat resistant than PF.<br>*Formica:* hard, resilient, wear and heat resistant, wipe-clean laminate of paper or textile impregnated with melamine resin, and topped with a decorative layer protected by melamine, then compressed and cured with heat to make a hard, durable surface. | *PF:* electrical insulator (switches) and for the manufacture of brake and clutch linings of vehicles, laminated wood parts, used as adhesives in plywood and hardboard. Resins used in paints.<br>*UF:* most used of the three types. Moulded into electrical plugs and sockets, toilet seats and some table wear. Binder in chipboard.<br>*Formica:* originally served as a substitute for mica in electrical applications. Now used in furniture, cabinetry, wall boards and other solid surfaces (see also: www.formica.com) |
| Epoxy resins (Polyepoxides) Example: *Araldite* | Excellent adhesion, chemical and heat resistance, good mechanical and electrical insulating properties. Should be cured by mixing with hardener to obtain better properties. | Coatings, adhesives, and composite materials with carbon fibre and fiberglass reinforcements. Sealing cracks in concrete and masonry, bonding coat for plaster and epoxy mortar for repairs.<br>*Araldite* is used in two parts: GY and HY, which are mixed together and the mix hardens to a solid material. |
| Polyester (Terylene) | Developed in 1942. Resistant to most solvents, acids, and salts. Resistant to stretching, shrinking, wrinkling, and abrasion. | Interior partitions, skylights, fabrics in apparel and home furnishings. Industrial polyester used in car tire reinforcements, fabrics for conveyor belts, safety belts, and coated fabrics. |
| Polyurethane | Wide range of hardness, high load-bearing capacity, flexibility, resistant to water, oil, grease, mould, mildew, and fungus. High abrasion, impact, and tear resistance | High-resilience foam seating, rigid foam insulation panels, durable elastomeric wheels and tires, high-performance adhesives, surface coatings and surface sealants, synthetic fibres (e.g., Spandex), carpet underlay, hard-plastic parts and hoses. |

**Fig. 17.6** Use of plastics in children playground

## 17.10 Use of Plastics In Buildings

As discussed in Section 17.9, plastics can be used in in-numerous applications, to either substitute other building materials, or improve the comfort conditions. However, because of relatively low stiffness they are not used as primary load-bearing materials. Only those applications in buildings are briefly explained in this section.

### 17.10.1 PVC Pipes in Buildings

Many types of plastic pipes are used in building construction. These are costlier than AC pipes but cheaper than GI or copper pipes. The available types of plastic pipes are as follows:

1. *Unplasticized PVC (UPVC)* or rigid PVC white or cream coloured pipes for use with cold water
2. *Plasticized PVC pipes* which are plasticized with the addition of rubber
3. Chlorinated PVC (CPVC) pipes which can withstand higher temperatures up to 120°C. Purple coloured CPVC pipes are also available in the USA for non-potable water distribution inside buildings, with clear marking 'Warning: Non-Potable Water. Do Not Drink'. It is commonly used in rainwater harvesting systems and grey water systems.
4. *PEX (cross-linked polyethylene)* is the newest pipe for residential use. It is resistant to freezing, meets most codes, can handle any normal domestic water temperature (up to 93°C). It is pressure resistant, adaptable, flexible (bent up to a radius of 90° change of direction without installing a fitting), shares many fittings with copper and so may be mixed into copper systems with considerable ease. PEX is easy to install because it can be cut easily, is flexible, and uses compression fittings. However, more permanent connections require a special crimping tool. PEX is three to four times more expensive than copper or plastic. Although most PEX pipes have some UV resistance, PEX pipes should not be stored outdoor where they are exposed to the Sun.
5. *ABS black pipes* and fittings are used for drain, waste, and vent purposes only. They are used in non-pressure systems where temperatures will not exceed 60°C. They are lightweight, have a PVC core, and require no special tools for cutting and easily installed by solvent cementing (no primer required). Limitations are UV resistance in some forms (and some forms are treated for UV).
6. *Polypropylene random (PPR) pipe* was developed in Europe in the early 1990s, and used in hospitals, and laboratories. PPR pipe can deliver hot water at 95°C with a pressure of 1.2 MPa, making it ideal for both hot and cold water use. PPR pipe system has a life expectancy of more

than 50 years, and is fully recyclable. Installations, additions or repairs of PPR pipes systems can only be done by making use of a *fusion-welding tool* and must be carried out by skilled and trained workers. PPR is joined by heating both the socket and the pipe. When melted, both parts are joined together to become one. It is impossible to predict whether fusion was done properly at the pipe joints visually, until blockages occur. PPR pipes are available in white, green, and grey colours. They are chemical resistant and may be subject to biological hazards and meet the various safety requirements.

Since copper pipes resist corrosion, they are used in water supply lines. Flexible copper pipe, which is easy to bend, is often used for dishwashers, and refrigerators. But they are expensive and require wheel/tube cutter or a hacksaw to cut and soldered fittings for their connections, and hence need more skill. In addition, pin-hole leaks occur in copper due to chemicals in water. Copper, however, is UV, cold, and heat resistant, and may be buried underground if proper grade is used.

In India, IS 13952, IS 4984, and IS 4985 give specifications respectively for the following:

1. *Unplasticized PVC-U*: Pipes for soil and waste discharge system for inside and outside buildings including ventilation and rain water system
2. *High density polyethylene*: Pipes for potable water supplies
3. *Unplasticized PVC*: Pipes for potable water supplies

**Advantages of using PVC pipes**  The following are the advantages of using PVC pipes over metal pipes:

1. PVC pipes are cost effective as compared to metal pipes.
2. PVC pipes do not corrode due to chlorides in brackish waters, whereas galvanized iron (GI) pipes tend to corrode.
3. They are not affected by atmospheric pollution.
4. As PVC pipes are light in weight, they are easy to transport, handle, and install.
5. Since the inside of PVC pipes is smooth, they have good flow characteristics to convey liquids. Thus, pipes smaller in size than metal pipes can be used for a given flow. CPVC pipes can be used to carry hot water also.
6. PVC pipes are inexpensive and simple to install using solvent (glue) fittings and do not require any special tools or equipment.
7. PVC pipes are extensively used for bore-well/tube-well applications, as large diameter pipes are easy to handle and are not affected by corrosive soils.
8. PVC is a good insulator. Owing to this, PVC pipes are used extensively in concealed electrical conduits. They can also be buried in brickwork or concrete.

**Disadvantages**  Some of the disadvantages of PVC pipes are as follows:

1. Some of the PVC pipes are brittle and may be broken easily in compression.
2. They creep under load much more than metals.
3. Some codes do not allow them to be buried under ground.
4. As they are thermoplastics, ordinary PVC pipes should not be used to carry hot waters. They should be used in limited pressure and temperature ratings (less than 80°C).
5. They have high coefficient of expansion, which is 10 times of steel. Hence, sufficient care should be taken to provide for their expansion.
6. Ordinary PVC pipes should not be directly exposed to the Sun.

### 17.10.2 Polyethylene Water Tanks

Tanks made of HDPE or LDPE are used as overhead water tanks. As these tanks are kept exposed to the Sun, about 2.5% carbon black is added as a stabilizer while manufacturing, which absorbs or screens out damaging UV rays. The UV rays are transformed into heat. The heat is harmlessly dissipated throughout the tank itself. Addition of carbon black makes it black in colour. These tanks are made as single piece and have square or circular shape. Tanks are provided with integral manhole at the top and have openings for inlet, overflow, outlet, and drain pipes. Tanks of capacity from 200 L to 50,000 L are available in the market. Single- or double-compartment plastic septic tanks constructed from polyethylene resins, as an alternative to concrete, has advantages such as lightweight, easy to install and clean, lower cost, and resistant to chemicals.

### 17.10.3 Doors and Windows

The doors and windows made of plastic sheets provide aesthetic appearance, preserve the texture and feel of fine wood grain, and at the same time eliminate the various problems associated with wooden products. Doors are available in attractive colours, as smooth flat 40–45-mm thick PVC panel with solid PVC frame with a weather strip sealed edge, available as single and double doors, in standard or custom sizes.

The use of PVC is becoming popular for doors and windows due to the following reasons:

1. PVC provides an alternate to timber of good quality, which has become scarce in India.
2. These doors and windows are lightweight, durable, and impact resistant.
3. These are insect, termite, and mildew resistant and also are not affected by water, saline air, or atmospheric pollution (will not rust), and hence require only low maintenance.
4. They can be made to tight tolerances resulting in airtight doors and windows that are required in air-conditioned rooms.
5. As they are water impermeable, they can be used as external doors and also in bathrooms and toilets, where wooden doors rot or warp. They also do not require painting.
6. They are energy efficient and economical.

Though the initial cost of PVC windows may be 20 to 30% costlier than aluminium windows, a PVC double-pane, low emissivity, argon gas-filled glass window may result in long-term energy savings.

### 17.10.4 Polycarbonate or PVC Roofing Sheets

Polycarbonate and PVC roofing sheets are used increasingly as an alternate to traditional roofing sheets such as asbestos sheets. Polycarbonate is a thermoplastic that is aesthetically pleasing and at the same time is tough, rigid, non-corrosive, and resistant to UV rays. Available in clear and opaque forms, they are almost unbreakable, fire retardant, and have high impact strength. Hence, they are used in bulletproof windows and police shields. Because they are light, it is easy to transport and install them, resulting in lower labour costs. Owing to their good optical, heat, and sound insulation properties, they are preferred in greenhouses and plant nurseries. They also last long without any fading, yellowing, or discolouration.

However, polycarbonate panel has the following disadvantages:

1. It is much expensive.
2. It is not resistant to scratching and abrasive surfaces.
3. It is sensitive to abrasive and alkaline cleaning products and solvents.
4. Its manufacture is not environment friendly, requiring very high temperatures, and often synthesized from phosgene, which is harmful to human health.

**Case Study | Munich Olympic Stadium**

The Olympiastadion, Munich, Germany, was built as the main venue for the 1972 Summer Olympics. Designed by German architect Günther Behnisch and engineer Frei Otto, with the assistance of John Argyris, the lightweight tent construction of the Olympiastadion was considered revolutionary for its time. The 105 × 68 m roof of the main stadium is a 4-mm thick polyester fabric coated with PVC suspended independently in each of the cells formed by a network of pre-stressing cables. The masts support the main cables.

### 17.10.5 Vinyl Flooring Sheets/Tiles/Planks

Vinyl flooring material can be used in office and residential building. Vinyl sheet works best in larger rooms, whereas tiles are easier for smaller rooms. Modern vinyl designs can mimic stone, ceramic, and even metallic elements. Vinyl sheets/tiles give a resilient, decorative, and non-porous surface, which can be easily cleaned. However, the surface should be protected from any burning items such as cigarette butts. Direct sunlight may damage or fade the vinyl floor. The material should conform to IS 3462:1986.

**Vinyl sheet**   It is available in 1.8–3.6 m wide and 10 m long rolls, respectively. There are two different kinds of sheet vinyl – inlaid and rotogravure. Inlaid vinyl sheets, which are durable and expensive, imbed colour granules into the vinyl sheet for a richer finish; thus, colours are an integral part of the floor. Rotogravure is more economical and consists of a foam base that is printed with ink just like newspapers and covered with a *wear layer*. Different backings are available. Felt-backed sheet, which is more common, features a felt layer that adds strength and cushion and requires the entire floor to be covered with adhesive. Vinyl-backed sheets are glued at the edges and installed by professionals. Modified loose-lay sheets feature a fiberglass backing that adds strength and stability, and do not require adhesive and is the easiest to install. Rubber based adhesives may also be used.

**Vinyl tile**   It is available in 300 mm², with 1.5–5.0 mm thickness, and feature adhesive backing. The thicker the tile, the longer will be its life. It often features peel-and-stick backing, making it easy to install.

**Vinyl plank**   Vinyl plank is commonly manufactured 100–150 mm wide by 0.9–1.2 m long. Like vinyl tile, the vinyl plank product is waterproof. Gaps need to be sealed in order to prevent moisture from seeping through the floor. Over the years, this flooring method has improved to realistically mimic different varieties of colours and types of wood.

**Wear-layer**   Tiles are covered with one of three surface coatings, called wear-layers, to resist dirt, stains, and scuff marks. Choosing the right finish for the wear-layer will help ensure high-quality and long-lasting performance. Available wear layers are no-wax vinyl (suitable for light traffic and minimal exposure to dirt), urethane (suitable for normal to heavy traffic), and enhanced urethane (provides highest quality surface coating). Thicker layers (thickness is measured in mils; 1 mil = 0.0254 mm) provide better protection.

### 17.10.6 Plastic Insulation

Up to 60% of the energy used in buildings is due to heating and or cooling. The key to minimize the environmental impact is to make buildings more energy efficient. Insulation is one of the easiest and most cost-effective ways to achieve this. As plastic insulation materials require only minimal thickness to achieve maximum energy efficiency, they are considered as the most thermally efficient insulation materials. They are simple to install both in existing and in new buildings, are very durable, and perform at the same high level over the whole life of the building.

There are four types of insulation: rolls and batts, loose-fill, rigid foam, and foam-in-place. Rolls and batts are typically flexible insulators that are available as fibres (e.g., fiberglass). Loose-fill insulation comes in loose fibres or pellets and should be blown into a space, using a suitable machine. Rigid foam is more expensive than fibre, but generally has a higher R-value per unit of thickness. Foam-in-place insulation can be blown into small areas to control air leaks, like those around windows, or can be used to insulate an entire building. As rigid foam insulation is not flexible, it may not fill all the gaps that may exist in the walls.

Rigid foam materials include EPS, extruded polystyrene (XPS, also called *Styrofoam*), and polyisocyanurate (POLYISO). EPS is the cheapest and POLYISO is the costliest. Unfaced EPS should be installed with felt paper. XPS may be available as unfaced and in blue, pink, or green colour. POLYISO is the best insulator and is faced with foils. The open cell urea-formaldehyde foam, which was used in the late 1970s, was banned in 1982. It has to be noted that several cities have completely or partially banned Styrofoam.

There are two types of spray foam: open-cell spray foam (density 4 kg/m$^3$) and closed-cell spray foam (density 32 kg/m$^3$). The higher the density of the foam, the greater will be the R-value. The ingredients of spray foam are usually mixed at site using special equipment mounted on a truck. It is then conveyed through heated hoses to a mixing gun and sprayed on the surfaces to be insulated. As soon as the chemicals are mixed, an exothermic chemical reaction begins and the liquid mixture foams, expands, and eventually hardens. It is advisable to avoid foams with hydrofluorocarbons as blowing agents, due to their high global-warming potential.

Fiberglass insulation is made by effectively weaving fine strands of glass. Fiberglass minimizes heat transfer, is cheaper, and commonly used. Installation of fiberglass requires safety equipment such as masks and gloves to protect eyes, lungs, and skin. The R-values of these materials are compared in Table 17.3. More information on plastic insulation may be found from Giannini (2011) and www.dow.com.

**Table 17.3** R-values of insulation materials

| Material | RSI value per m | R-value per inch |
|---|---|---|
| Expanded polystyrene (EPS) | 27 | 4 |
| Extruded polystyrene (XPS) | 34 | 5 |
| Polyisocyanurate (POLYISO) | 45 | 6.5 |
| Fiberglass loose-fill | 17–27 | 2.5–4 |
| Fiberglass rigid panel | 17 | 2.5 |
| Closed cell spraypolyurethane (SPF) foam (e.g., Corbond®) | 38–55 | 5.5–8 |
| Brick | 0.2 | 1.3–1.8 |
| Concrete | 0.08 | 0.43–0.87 |

## 17.10.7 Structural Insulated Panel

Structural insulated panels (SIPs) may be used for roofs, walls, and even floors. They are made of foam, usually *EPS foam*, sandwiched between layers of plywood, oriented strand board (a type of engineered lumber), or fibre cement. The foam may also be made out of *XPS foam, POLYISO foam, polyurethane foam*, or composite honeycomb. The panels may be from 1.2–7.2 m wide, up to 2.7-m high, and with thickness ranging from 112.5 to 162.5 mm for walls. The roof panels are as thick as 300 mm for added insulation. SIP walls may be erected on conventional foundations or a slab. SIP floors provide excellent insulation and may be installed over floor joists. Panels are lightweight and hence two to three persons can easily move them into position. The sheathing on most panels have a space at the top and bottom for the 'sole plate' and top plate. Each panel is placed over the sole plate and connected to the panel next to it, usually with glued 'splines or panel connectors' and screws, providing excellent air seal, as shown in Fig. 17.7. Splines were often made of wood resulting in thermal bridging (A *thermal bridge* is an area which has a significantly higher heat transfer than the surrounding materials), which reduces the overall thermal insulation of the building. Hence, present splines are often built from composite material or insulated materials. Some advanced systems have locking cams that snug up each panel to the one next to it.

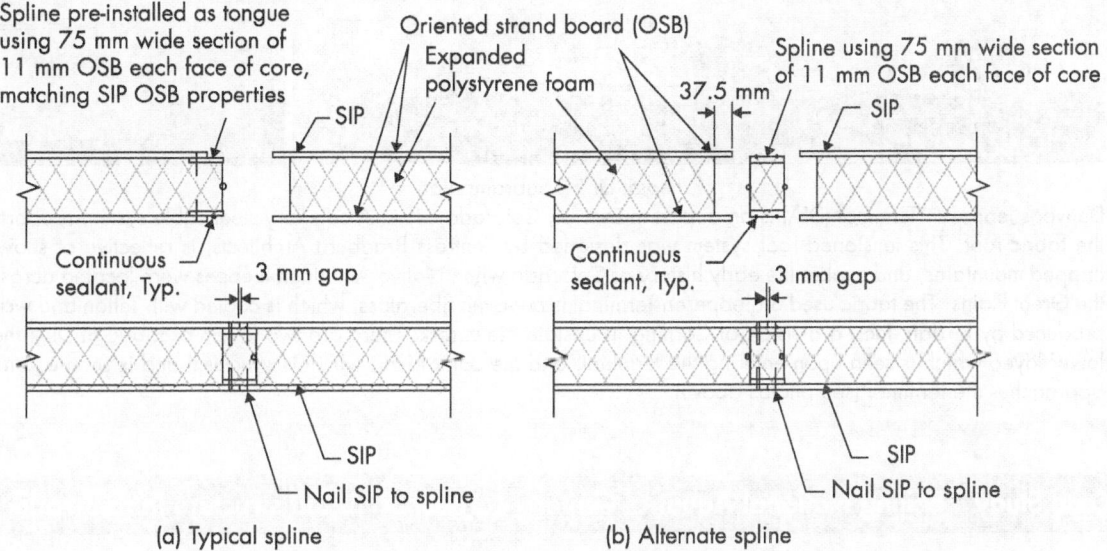

**Fig. 17.7** Connecting SIPs using splines

An experienced crew can assemble the shell of an average house in just a few days, thus providing savings in time and labour costs. Their insulation rating is from R-16 for 112.5-mm thickness to R-40 for 300-mm thickness. Openings for doors and windows can be cut before the panels are installed; it is also possible to request the manufacturer to cut the openings at the factory. Any conventional finish may be used on both the exterior and interior. Homes built with SIPs are close to airtight, reducing infiltration of air and heat loss. According to the Structural Insulated Panel Association (www.sips.org), operational energy costs may be reduced by as much as 50% compared to conventional homes. SIPs have the potential to contribute up to 25 points for LEED-NC (energy and atmosphere:1–19 points, materials and resources (waste management): 5 points, and indoor environmental quality: 1 point), and up to 48.5 points for LEED for homes. More information on SIPs may be obtained from Morely (2000) and www.sips.org.

## 17.10.8 Use of Plastic Fabrics in Tensile Structures

Pneumatic/tensile structures are mainly fabricated from woven synthetic fabrics such as the polyamide fabric, nylon, and the polyester fabrics, terylene, and dacron, which are coated on one or both sides with vinyl, butyl, neoprene, hypalon, or any other plasticized elastomers. Cloth membranes made of neoprene/cotton, polythene, or fibre glass reinforced vinyl are also used. The coating materials widely used are PVC, neoprene, PTFE, and hypalon [chlorosulfonated polyethylene (CSPE)]. PVC-coated cloths have most of the advantageous properties for use in temperate climate. Coated cloths of total weight 0.5–1.0 kg/m² are widely used. The thickness of these cloths is about 0.5 to 1 mm. The hypalon coating is superior to both in its resistance to sunlight and weather. The life of neoprene and hypalon coated fabrics can be increased by respraying or repainting. The life of these coated fabrics varies from

7–10 years. Fibreglass coated with Teflon Fluorocarbon resin has a life of 25 years. It is translucent and also has self-cleaning properties. More details may be found in Subramanian (2007).

**Case Study** | **Denver International Airport - Jeppesen Terminal PTFE fabric roof**

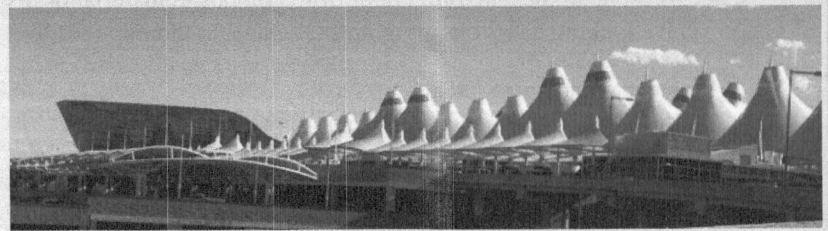 

*Photo*: Dr N. Subramanian

Denver's Jeppesen Terminal (DIA) is an airport in Denver, Colorado, USA. A catenary steel cable system supports the fabric roof. This tensioned roof system was designed by Fentress Bradburn Architects, is reflective of snow-capped mountains, and evokes the early history of Colorado when Native American teepees were located across the Great Plains. The fabric used on Jeppesen Terminal is a woven fiberglass, which is coated with Teflon and was produced by Birdair, Inc., a division of Corning in upstate New York. There are two layers of fabric at DIA: the lower layer, which is seen upon entering the terminal; and the outer layer, which is structural and is seen as one approaches the terminal (see photos above).

**Case Study** | **Beijing National Aquatics Center—Largest ETFE clad structure in the world**

(a) View from outside        (b) View from inside

*Source*: https://en.wikipedia.org/wiki/Beijing_National_Aquatics_Center

The Beijing National Aquatics Centre, nicknamed the Water Cube was built for the 2008 Summer Olympics, held at Beijing, China. Designed by the consortium of Ove Arup, Australian architecture firm PTW, the CSCEC (China State Construction and Engineering Corporation), and the CSCEC Shenzhen Design Institute (CSCEC+DESIGN), this aquatics centre measures 177 × 177 m in plan and has a height of 30.5 m. It houses five swimming pools and covers an area of 7.8 acres. The steel building is covered with ETFE (ethyl tetra fluoro ethylene) pillows. These are provided by two layers of durable and recyclable plastics (having a thickness of about 0.2 mm), inflated by air. This is the largest ETFE clad structure in the world with over 100,000 m² of ETFE pillows. These pillows allow internal noise to pass directly outside without reverberating the inner space. The ETFE cladding also allows better light and heat penetration than traditional glass, and resulted in 30% decrease in energy costs. For more details, see Subramanian (2008).

### 17.10.9 Geosynthetic Materials

*Geosynthetics* is the term used to refer all the polymeric materials used in a soil environment. Geosynthetics, all of which are thermoplastic polymers, can be of five broad categories, which are as follows:

**Geotextiles**   It is any permeable textile material used to increase soil stability, provide erosion control, or aid in drainage. Modern geotextiles are made from a synthetic polymer such as polypropylene, polyester, polyethylene, and polyamides. Geotextiles can be woven, knitted, or non-woven. By varying polymers and the manufacturing processes, an array of geotextiles, suitable for a variety of construction applications can be made. Although they have been used in Egypt thousands of years ago, the first modern application of geotextiles was in Florida, USA, for a waterfront structure in 1958, with woven fabrics. The first nonwoven geotextile was developed in 1968 by the Rhone Poulence company in France.

**Geogrids**   These, as shown in Fig. 17.8, are used to reinforce retaining walls, as well as sub-bases or sub-soils below roads or structures. As geogrids are strong in tension compared to soils, they allow the forces to be transferred to a larger area of soil. Geogrids are made polyester, polyvinyl alcohol, polyethylene, or polypropylene. They may be woven or knitted from yarns, heat-welded from strips of material, or produced by punching a regular pattern of holes in sheets of material, then stretched into a grid (www.geogrid.com).

**Fig. 17.8** Geogrids (*Source*: https://en.wikipedia.org/wiki/Geogrid)

**Geomembranes**   These are made from relatively thin continuous polymeric sheets. They can also be made from the impregnation of geotextiles with asphalt, elastomer or polymer sprays, or as multilayered bitumen geocomposites (www.geomembrane.com). They are impermeable sheets made of HDPE, LDPE, PVC, flexible polypropylene (fPP), CSPE, or ethylene propylene diene terpolymer. Geomembranes dominate the sales of geosynthetic products and are used in numerous applications such as liners for portable water or waste liquids, secondary containment of underground storage tanks, liners for the agriculture industry and landfills, covers for solid-waste landfills, within cofferdams for seepage control, etc.

**Linear strips**   These are long, slender strips made with polymer fibres, which are used as reinforcing tendons in reinforced earth-retaining walls. Glass-reinforced plastics can also be used for this purpose.

**Geocomposites**   These are polymeric materials used in a soil environment not covered by the above four categories.

Ingold and Miller (1988) provide more information about different geosynthetic materials and their properties.

Renzo Piano, working at the Polytechnic University of Milan, designed, manufactured, and constructed a number of amazing lightweight spatial structures in plastics, using various innovative techniques (Piano, 1967). More information on the use of plastics in buildings may be found in Koehler (1955), Subramanian & Ganapathy (1976), Rosato, et al. (1991), and Humphrey (2003).

## 17.11 Bioplastics

Bioplastics are a new group of plastics, which are either (a) made up of renewable biomass sources, (b) break down completely via a natural process (biodegradable), or (c) are both bio-based and biodegradable. Some natural materials that can be made into bio-based plastics are corn starch, soy protein, sugar, cellulose, vegetable fats and oils, and collagen (a protein found in mammals); one good example is polylactic acid, which is made from corn. Biodegradable plastics are those that degrade into carbon dioxide ($CO_2$), methane ($CH_4$), and water ($H_2O$), through biological action in a defined environment and in a defined timescale. These environments include composting, anaerobic digestion, and marine and soil environments. It has to be noted that 'bio-based' and 'bio-degradable' are not related. A bio-plastic that is bio-based may not necessarily biodegradable, and a biodegradable bioplastic may not be bio-based. Researchers have also developed photodegradable plastics that breakdown when they are exposed to the Sunlight. Bioplastics may be used in the same ways as other plastics-in packaging, agriculture, medical, automotive parts, 3D printing, etc. Typical environmental benefits of using bioplastics include: (a) reduction of fossil fuel usage, (b) reduction of carbon footprint, (c) reduction of global warming potential (GWP), and (d) any combination of these.

## 17.12 Fibre Reinforced Plastics

Fibre-reinforced plastic (FRP) is a composite material made of a polymer matrix reinforced with fibres. The fibres used are usually glass (in fiberglass), carbon, aramid, or basalt. Other fibres such as paper, wood, or asbestos have also been used. The polymer is usually an epoxy, vinylester, or polyester thermosetting plastic. FRPs are commonly used in the aerospace, automotive, marine, and construction industries. The applications in buildings include: (a) waterproofing and dampproofing of roofs and tanks, (b) use of translucent FRP sheets in roofs, (c) in water storage tanks, (d) in building components such as doors, window frames, wall panels, etc., and (f) strengthening of RC sections. The repair using FRP sheets is explained in this section.

FRP can be used to strengthen beams, columns, and slabs of buildings and bridges. It is also possible to increase the strength of structural members, even after they have been damaged by severe loading conditions such as an earthquake. FRP strengthening can be done in different ways, such as (a) side bonding, (b) U-wraps (U-jackets), and (c) closed wraps. Thus, FRP sheets may be attached to the tension face of the beam/slab (with fibres oriented in the longitudinal axis) or the web of a beam with fibres oriented transverse to the beam's longitudinal axis. Columns are usually wrapped completely with FRP around their perimeter. This not only results in higher shear resistance, but also in increased compressive strength. The FRP wrap works by restraining the lateral expansion of the column, which results in the confinement of concrete core similar to that provided by spiral reinforcement (Palanivel et al., 2014).

### Case Study  FRP bridges in Rotterdam, the Netherlands

Rotterdam, the second-largest city in the Netherlands, which has 850 pedestrian bridges, is replacing them with plastic instead of steel, concrete, or wood. These pedestrian bridges, made of lightweight FRP, are brought to the site by trucks and installed in an hour, compared with wooden/steel bridges, which might take more than three weeks to construct. A recent such bridge, 20-m long and 1.8-m wide, is only 250-mm thick. Thus, these bridges are more than three times lighter than a comparable RC bridge, and less than half the weight of a similar steel bridge. Using plastics also has environmental advantages, because steel and concrete may consume more than twice the energy. In addition, the FRP bridges will last for 100 years, while a comparable RC bridge may last only 50 to 60 years. FRP bridges will not rust resulting in lowest possible maintenance costs, and are simple to repair. The first such bridge was installed in the city in 2009, and it now has around 90, more than any other city in the world.

## 17.13 Thermocol/Geofoam

*Thermocol* is a trade-name for polystyrene foam created by BASF for EPS. Thermocol is a light cellular plastic material, which is used for thermal and sound insulation. It is also used in formwork of concretes to create special patterns.

*Geofoam* is EPS or XPS manufactured into large lightweight blocks. The blocks are made in different sizes but are often $2 \times 0.75 \times 0.75$ m. Both EPS and XPS are resistant to moisture; however, XPS is more common for below-grade waterproofing and roof systems where insulation is placed over the roof membrane. Geofoam is used as a lightweight void fill below a highway, bridge approach, embankment or parking lot. EPS Geofoam minimizes settlement on underground utilities. Geofoam may also be used in green roof fill, compressible inclusions, thermal insulation, and drainage (when appropriately formed). EPS and XPS geofoams are covered in ASTM D6817, ASTM D7180, and ASTM D7557 and some of these properties are shown in Table 17.4.

**Table 17.4** Physical property requirements of GPS geofoam as per ASTM D6817 and ASTM C578 (also see www.expol.co.nz)

| Property | Grade | | | | | | | | |
|---|---|---|---|---|---|---|---|---|---|
| | EPS12 | EPS15 | EPS19 | EPS22 | EPS29 | EPS39 | XPS20 | XPS26 | XPS36 |
| Density, min. $(Kg/m^3)$[1] | 11.2 | 14.4 | 18.4 | 21.6 | 28.8 | 38.4 | 19.2 | 25.6 | 35.2 |
| Elastic modulus, min. $(KN/m^2)$ | 1,500 | 2,500 | 4,000 | 5,000 | 7,500 | 10,300 | NA | NA | NA |
| Compressive resistance at 1% deformation, min. $(KN/m^2)$ | 15 | 25 | 40 | 50 | 75 | 103 | 20 | 75 | 160 |
| Flexural strength, min. $(KN/m^2)$ | 69 | 172 | 207 | 276 | 345 | 414 | 276 | 345 | 517 |
| Water absorption by total immersion, max. Volume (%) | 4 | 4 | 3 | 3 | 2 | 2 | 0.3 | 0.3 | 0.3 |
| Oxygen Index, min. volume (%) | 24 | 24 | 24 | 24 | 24 | 24 | 24 | 24 | 24 |

[1] See ASTM D6817 Standard for test methods and complete information about geofoam.

The first use of EPS Geofoam was in the embankments around the Flom Bridge in Oslo, Norway, in 1972, to reduce settlements. Prior to the geofoam use, the annual settlement was 200–300 mm, which caused extreme damage to the roadway. During 1985–1987, Japan used over 1,300,000 m$^3$ of geofoam in 2,000 projects. Geofoam was first used in the United States in 1989 on Highway 160 and I-70 in Colorado, with 51-mm thick XPS geofoam having nominal compressive strength of 276 kPa. The use of geofoam resulted in an 84% saving in the total cost and reduced repetitive repairs of frost-heave damage of the pavement (Negussey, 1997). The largest geofoam project in the United States, during 1997–2001, is on the Interstate 15 in Salt Lake City, Utah, involving 100,000 m$^3$ of geofoam.

Advantages of using geofoam include the following:

*Low density/high strength*    Geofoam, although having only 1–2% of the density of soil, but has equal strength

*Predictable behaviour*    Geofoam has uniform composition and properties, compared to other lightweight fillers, such as soil, that can have very variable composition.

*Inert*    As geofoam will not break down, it will not spread into surrounding soils and hence not pollute the surrounding soil. Geofoam can also be dug up and reused.

*Requires less labour*   As geofoam is lightweight, it can be installed by hand using simple hand tools, eliminating investment and operating cost of heavy machinery.

*Reduced construction time*   Geofoam can be installed quickly during any type of weather, and during day or night.

However, geofoam may have the following disadvantages:

*Fire hazards*   If geofoam is not treated it may become a fire hazard.

*Vulnerable to petroleum solvents*   Geofoam looses strength and turns into a glue-like substance, whenever it comes into contact with any petroleum solvent; hence, it will not support any load.

*Buoyancy*   Dangerous uplift forces may result due to buoyancy and hence structures supported by geofoam should be guarded against these forces. For example, in Crayford, UK, on 9 October 2016, floodwaters raised polystyrene below the floor of a car park and crushed sixteen cars against the ceiling.

*Susceptible to insect damage*   When geofoam is used in wooden buildings as insulation, it is better to treat it to resist insect infestation. However, in the often used application as lightweight fill for road construction, no insect damage has been observed till now.

More information about geofoams and their applications may be found in Horvath (1995) and Negussey (1997).

## 17.14 Environmental Effects of Plastics

A main disadvantage of plastics is that they are not renewable. It is because they are made of petrochemicals, a non-renewable source of energy. Other connected issues are: destruction of habitat, extraction of crude oil, security issues from the volatile countries where oil is produced, processing of petroleum, and chemical manipulation into the various types of plastics. The manufacture involves several chemicals, many of which have not been sufficiently tested for their toxicological impact on humans or animals. Although some common plastics can be recycled (#1 and #2 plastics used in common soda and milk bottles), the vast majority cannot. Thus, they take up considerable space in landfills and create air pollution when incinerated.

Although the use of plastics is widespread and advantageous in some aspects, plastic bags, and other plastic waste are found to cause considerable environmental damage. At present, plastics constitute between 14 and 22% of the volume of solid waste, and this percentage is increasing. It takes anywhere between 20 and 1000 years for a plastic bag to break up into smaller pieces, which may again contain harmful polymers and toxic chemicals. Thus, the plastics stay longer in the environment (much more than degradable materials like paper). In other words, the more plastics we use, the greater will be the chances of environmental damage.

Since the mass production of plastics began in the 1940s and 1950s, the amount of plastic debris entering marine and freshwater ecosystems has also increased exponentially. It is estimated that the plastic waste that entered the global marine environment in 2010 is about 4.8 to 12.7 MT (Beaman & Bergeron, 2016). According to www.nationalgeographic.com, 5.25 trillion pieces of plastic debris are in the ocean and out of that 269,000 tonnes float on the surface, while some four billion plastic microfibers per square kilometre litter the deep sea. As many as 100,000 whales, turtles, and birds are reported to die every year, mainly because of plastics in the ocean environment. Thus, plastic bags not only have adverse effects on our natural habitats, but also are responsible for the death of many animals on land, and even children mainly on account of the suffocation. The floods in Bangladesh in 1988 and 1998 were made more severe because of plastic bags clogging the drains.

Plastics often leech component chemicals, including hazardous chemicals, during common temperature changes. It is for this reason that toxicologists do not recommend storing very cold foods in plastics or heating foods in plastics (especially microwaving).

Plastics are durable materials. Thus, they are hard to eliminate once used and create tremendous waste. Recycling plastics is more complicated because there are several different kinds of plastics. Each type requires different processing methods due to the additives such as dyes and fillers that must be removed before it can be reused. To help with initial curb side sorting, the Society of the Plastics Industry (SPI) created the resin identification code, a set of numbers and symbols stamped on plastic consumer goods indicating the type of plastic and its potential for recycling (see Table 17.5). Plastics with code 1 (PET) and code 2 (HDPE) are easily recycled (It costs US$4000 to recycle 1 tonne of plastic bags). Polystyrene is seldom recycled due to its low density and hence hauled to landfills. Several countries, states, and large cities have banned disposable plastic bags to protect the environment. These include Bangladesh, Sri Lanka, Kenya, Taiwan, Zimbabwe, Canada, France, New Zealand, Israel, South Africa, The Netherlands, China, Seattle, New Delhi, and Tamil Nadu. The UK and Australia are experimenting with voluntary phase outs of plastic bags.

**Table 17.5** Society of the Plastics Industry (SPI) Code, uses, and recycling of plastics

| Plastics code | Plastic name | Common uses and health issues | Examples of recycled products |
|---|---|---|---|
| 1 PETE | polyethylene terephthalate (PET or PETE) | Polyester fibres, soft drink, water and beverage bottles, tote bags, furniture, carpet, panelling – no known health issues. | Most easily recycled into bottles, Tote bags, picnic tables, polyester for carpet and fabrics, fencing, paint brushes furniture, fibre fill. |
| 2 HDPE or PE-HD | High-density polyethylene (HDPE) | Milk and water jugs, recycling bins, agricultural pipe, base cups, car stops, laundry and shampoo bottles, playground equipment, and plastic lumber- no known health issues. | Clear HDPE containers are easily recycled back into new containers, Recycling bins, benches, drain pipes, and floor tiles. Coloured HDPE are converted into plastic lumber, lawn and garden edging, pipes, rope, and toys. |
| 3 PVC or V | Polyvinyl chloride (PVC) | Vinyl pipes, shower curtains, flooring, home siding, and window and door frames, fencing, and lawn chairs | PVC is one of the least recyclable plastic due to additives. Potentially harmful substances are also are produced in the manufacturing, disposal, or destruction of PVC. Recycled products include floor mats, pipes, hoses air bubble cushioning, decking, film, panelling, recycling containers, roadway gutters, and playground equipment. |
| 4 LDPE or PE-LD | Low-density polyethylene (LDPE) | Plastic bags and wraps, dispensing bottles, wash bottles, tubing, and various moulded laboratory equipment. While there is no known health effects, organic pollutants are formed during manufacturing | Garbage can liners, floor tile, furniture, compost bins, panelling, trash cans, landscape timber, and mud flaps (LDPE is not usually recycled). |
| 5 PP | Polypropylene (PP) | Auto parts, industrial fibres, diapers, food containers, and dishware (Tupperware)- No known health issues | PP is not easily recycled. Differences in the varieties of type and grade, results in difficulty of achieving consistent quality during recycling. Items include paint buckets, videocassette cases, signal lights, brooms and brushes, bicycle racks. |

*(Contd)*

**Table 17.5** (*Contd*)

| Plastics code | Plastic name | Common uses and health issues | Examples of recycled products |
|---|---|---|---|
| 6 PS | Polystyrene (PS), Formed Polystyrene (Styrofoam) | Desk accessories, cafeteria trays, plastic utensils, cups, toys, video cassette cases, and insulation board and other EPS products-Styrene can leach from polystyrene. Over the long term, this can act as a neurotoxin. | Recycling PS is possible, but not normally economically viable. Septic tank drainage systems, light switch plates, insulation, egg cartons, vents, desk trays, license plate frames, and fibres in concrete. Styrofoam packing and polystyrene cutlery can be reused. |
| 7 OTHER | Other plastics, such as acrylic, nylon, polycarbonate, and polylactic acid (a bioplastic), and multilayer combinations of different plastics | Bottles, plastic lumber applications, headlight lenses, and safety shields/glasses. Health effects vary depending on the resin and plasticizers used. Polycarbonate plastic leaches bisphenol A (BPA) a known endocrine disruptor. | Very difficult to recycle. |

**Tips for use of plastic containers with food:**

1. Avoid heating food in plastic containers, as they can release chemicals; drink only cold liquids from plastic containers.
2. Wash plastic containers in mild detergents. Using harsh detergents liberate chemicals, leaching them into food.
3. Avoid using plastic packaging. For example, use reusable bags when grocery shopping.
4. Only use plastic containers with the recycling #1, #2, #4 and #5 for food storage. Better to use glass storage containers since plastic containers can leach chemicals as they age.

## 17.15 Rubber

It is an *elastomer* type natural polymer, where the polymer has the ability to return to its original shape after being stretched or deformed (a variety of synthetic polymers have mechanical behaviour similar to that of an elastomer). It is harvested mainly in the form of latex from *Hevea Brasiliensis* trees (also called the Para rubber tree), which grow in tropical regions. Mesoamericans used rubber for balls and other objects as early as 1600 BC Thailand, Malaysia, Indonesia, India, and China are the largest producers of natural rubber in the world. World natural rubber production in 2018 was about 13.96 MT (India produced 700,000 tonnes). Natural rubber is used extensively in many applications and products, either alone or in combination with other materials. In most of its useful forms, it has a large stretch ratio and high resilience, and is extremely waterproof. It has to be noted that a majority of rubber products used today are made not from natural latex but from synthetic rubber. Natural rubber is a truly sustainable product, with a long life cycle and low maintenance requirement.

### 17.15.1 Types of Rubber

There are two types of rubber: (a) natural rubber and (b) synthetic rubber. They are discussed briefly in this section.

**Natural rubber** It is the sticky, milky latex which is drawn from rubber trees by making incisions in the bark, using a process called 'tapping'. It is coagulated by using dilute acids such as acetic or formic acid. The coagulated rubber is then rolled to remove excess water. Then, a final rolling is performed using a textured roller and the resultant rubber sheet is dried. This 'crepe' rubber is then processed to get commercial rubber products. The economic life period of rubber trees in plantations is around 32 years, out of which about 25 years are productive. Natural rubber belongs to the monomer isoprene. Since isoprene has two double bonds, it still retains one of them after the polymerization reaction.

**Synthetic rubber**   Most synthetic rubbers are polymers made by polymerizing a mixture of two or more monomers. There are two types of synthetic rubber, which are as follows:

*General purpose synthetic rubber*   An example of this type is SBR – which is a copolymer of butadiene and styrene, mixed in the ratio of 3 to 1, respectively. Both styrene and butadiene are currently obtained from petroleum. SBR rubber was developed during World War II when important supplies of natural rubber were cut off. SBR is more resistant to abrasion and oxidation than natural rubber and can also be vulcanized. More than 40% of the synthetic rubber production is SBR and about 50% of car tires are made from the various types of SBR.

*Special purpose synthetic rubber*   These synthetic rubbers are made from special materials for chemical and temperature resistant applications. An example of this is *Neoprene*, which is produced by the polymerization of chloroprene and used in the bearings of bridges. Neoprene was invented in 1930 by DuPont scientists.

*Recycled rubber* is obtained by grinding used and discarded tires. This tire crumb is produced in a variety of mesh sizes to be used in various applications. Tire crumb can also be mixed with coal, wood, or chemical wastes (called as *tire-derived fuel*, or TDF), and used in concrete kilns, power plants, or paper mills. However, studies have shown that the use of TDF results in drastic increase in atmospheric contamination. Recycled rubber has advantages such as the elimination of waste in landfills, and savings in energy, and cost of manufacture.

Although *reclaimed rubber* also uses discarded tires in its production, it undergoes de-vulcanization and vulcanization once again in order to create products similar to petroleum based goods. While these processes may involve more time, money, and energy to create, they result in durable and better products.

## 17.15.2 Vulcanization of Rubber

US inventor *Charles Goodyear* accidentally discovered in 1839 that by mixing sulphur and rubber, properties of rubber could be improved. The resultant rubber had higher tensile strength and resistance to swelling and abrasion, and was elastic over a greater range of temperatures. This process was later called *vulcanization* after the Roman god of fire. Soft rubber is obtained by the addition of one to 5% of sulphur and hard rubber by adding about 30% sulphur. The vulcanization process is, however, slow and the product is porous, and deteriorates on prolonged exposure to atmosphere. To overcome these shortcomings, some additives are added prior to vulcanization. These additives are classified as accelerators, softeners, reinforcing pigments, and antioxidants according to the functions they perform. Fillers like carbon black are also added to improve its rigidity. Tires are generally reinforced with nylon threads or steel wires to make them stronger and to resist impact loads and shocks.

## 17.15.3 Properties of Rubber

Natural rubber has a strong, unpleasant odour, and melts at about 180°C. Turpentine and naphtha (petroleum) are used as solvents of rubber. Rubber has excellent tensile, elongation, and tear resistance and resilience. It has good abrasion resistance and excellent low temperature flexibility. Without special additives, it has poor resistance to ozone, oxygen, sunlight, and heat. It has poor resistance to solvents and petroleum products.

SBR also has properties similar to natural rubber. Its resistance to solvents and petroleum products is about the same as natural rubber. Water resistance is better. Without special additives, it is vulnerable to ozone, oxygen, and sunlight.

### 17.15.4 Uses of Rubber in Buildings

Rubber is a very important industrial product. It has been used in a number of applications, which include automobile tires, rubber sheets, adhesives and coatings, airbags, clothing, balls, erasers, gaskets, variety of ducting products, gloves, lining, and flotation products.

In building construction, it has been used in the following applications:

**Flooring** Natural or synthetic rubber flooring has been used in gyms, commercial kitchens, animal shelters, and even playgrounds. Rubber provides a surface which prevents fatigue, provides padding, and is both slip-resistant and waterproof. It is easy to maintain and long-lasting, making it an ideal option for flooring in computer rooms and libraries where we need resilient floors. These floors are noise proof, shock absorbing, and provide protection against a variety of potential hazards. Use of rubber flooring tiles and mats are eco-friendly and offer two key advantages: (a) they help reduce non-biodegradable waste in the ecosystem by using old vehicle tires; (b) they tend to be more affordable than their non-natural synthetic rubber products. Rubber tiles can be laid on any rigid floor and are available in an array of colours, patterns, and textures.

**Cement and concrete** Cement mortars containing synthetic or polymer latex are called polymer-modified mortars. Such mortars have (a) improved adhesion, (b) higher tensile, compressive and flexural strengths, (c) excellent waterproofing qualities, and (d) higher resistance to chemicals. Addition of synthetic rubber will have better properties than natural rubber. Rubber bonding agents are added to cement when used for repairing concrete and plastered surfaces. SBR is often used in concrete to produce polymer modified concrete. Latex-modified concrete has been widely used in the field of patching, resurfacing, or overlaying works for damaged bridge decks for the past 30 years, because of its ease of execution, excellent adhesion to the base concrete, high freeze-thaw durability, and resistance to chloride penetration. Particularly in the USA, hundreds of bridge decks have been restored with the latex-modified concrete since 1957 (See also Section 9.8 of Chapter 9).

**To reduce sound and vibration** Rubber pads are used to mount machinery to ensure that vibration does not affect its performance. It can also be used to produce sound-proofing materials. Neoprene pads are used in bearings of bridges.

**Paints and foams** Latex paints are essentially a solution of coloured pigment and rubber latex. Latex foam is made by pumping air into the latex before coagulating it.

**Paving** Ground and crumb rubber, also known as size-reduced rubber, can be used in both paving type projects and in mouldable products. The paving are called *rubber modified asphalt* (RMA), and Rubber Modified Concrete. Some examples of rubber-moulded products include carpet padding or underlay, flooring materials, patio decks, sidewalks, and rubber tiles, and bricks. Athletic and recreational areas can also be paved with the shock absorbing rubber-moulded material. Rubber crumb can also be used as an infill, alone, or blended with coarse sand, and as infill for grass-like synthetic turf products.

**Back-fills** Shredded tires, known as *tire derived aggregate* (TDA), have many civil engineering applications. TDA can be used as a back-fill for retaining walls, back-fill for roadway landslide repair projects, and as a vibration damping material for railway lines. Recycled rubber can also be used to produce rubber floorings, playground safety products, rubber mulch, shingles, shoes, highway safety barriers, and even moulded recycled rubber sheets that compete with traditional vulcanized virgin sheets. Discarded whole tires can be used in the construction of barrier reefs.

## SUMMARY

- Most plastics are organic polymers derived from petrochemicals. They can be classified as thermoplastics or thermosetting. Only thermoplastics will regain its original properties after subjected to high temperature and pressure.
- The chemical process of combining a large number of small molecular units ('mers') to form a long-chain of molecules is called polymerization.
- Several manufacturing processes such as compression moulding, extrusion, injection moulding, transfer moulding, blow moulding, calendaring, foaming, and lamination are used.
- Advantages of plastics include mouldability, low cost, chemical and biological resistance, durability, and easy maintenance. However, plastics lack fire resistance, and have high creep and thermal expansion.
- Plastics are used in numerous building applications such as PVC pipes, HDPE, or LDPE water tanks, doors and windows, polycarbonate roofing sheets, vinyl flooring, insulation, woven synthetic fabrics, and as geosynthetic materials.
- FRP is produced using fibres of glass, carbon, aramid, or basalt, and used in many applications.
- Rubber is an elastomer derived from rubber trees. Most synthetic rubbers are polymers made by polymerizing a mixture of two or more monomers.
- Recycled rubber is produced by grinding old tires and can be substituted for virgin material. Reclaimed rubber however undergoes de-vulcanization and vulcanization.
- Natural rubber offers good elasticity, whereas synthetic rubber tends to offer better resistance to environmental factors such as oils, temperature, chemicals, and UV light. Both natural and synthetic rubbers are used in various applications of the building industry.
- Plastics are not green materials, though their use in thermal insulation may reduce the energy used to cool or heat the buildings. Recycling of plastics is generally expensive and plastics have polluted our lakes, rivers, and even oceans. In contrast, natural rubber is a truly sustainable product, with long life cycle and requiring low maintenance.

## EXERCISES

### Multiple-choice Questions

1. Thermoplastics are plastics which
   - (a) undergo no chemical change when heated
   - (b) undergo chemical changes when heated
   - (c) moulded twice only
   - (d) none of these
2. Thermoplastics molecules are
   - (a) chain like
   - (b) cross-linked
   - (c) complex
   - (d) none of these
3. Thermosetting plastics are materials which
   - (a) undergo no chemical change when heated
   - (b) undergo chemical change when heated
   - (c) are moulded several times
   - (d) none of these
4. To increase the strength of the plastics, which of the following methods is used?
   - (a) Linear linking of their molecules
   - (b) Adding metal powder
   - (c) Cross linking of their molecules
   - (d) None of these

5. The process of forming a long-chain of molecules is called
   - (a) diffusion
   - (b) polymerization
   - (c) recrystallization
   - (d) none of these
6. Which of the following is not an example of thermoplastic?
   - (a) Polyethylene
   - (b) Nylon
   - (c) PVC
   - (d) Bakelite
7. Compression moulding is used for
   - (a) thermoplastics
   - (b) thermosetting plastics
   - (c) both (a) and (b)
   - (d) none of these
8. Extrusion is used for
   - (a) thermoplastics
   - (b) thermosetting plastics
   - (c) both (a) and (b)
   - (d) none of these

9. Bottles are made by using
   (a) compression moulding
   (b) injection moulding
   (c) blow moulding
   (d) casting

10. Plastic sheets or films are made by using
    (a) compression moulding
    (b) injection moulding
    (c) calendaring
    (d) extrusion

11. The sheets making procedure in which the thermo-plastic material is squeezed between two rolls is called
    (a) laminating          (c) calendaring
    (b) blow moulding        (d) casting

12. Specific gravity of plastics varies from
    (a) 0.9 to 2.24          (c) 5.0 to 7.5
    (b) 3.0 to 5.0           (d) 7.5 to 9.0

13. Plexiglas is the trade name of
    (a) urea formaldehyde
    (b) PVC
    (c) cellulose acetate
    (d) acrylic

14. Bakelite is the trade name of
    (a) urea formaldehyde
    (b) phenol-formaldehyde
    (c) polyesters
    (d) PVC

15. Which of the following is used for making electrical switches?
    (a) PVC
    (b) Polypropylene
    (c) Bakelite
    (d) Polyvinyl acetate

16. Araldite is the trade name of
    (a) phenolics            (c) vinyls
    (b) epoxies              (d) PVC

17. For hot water of up to 100°C, we should use
    (a) PVC pipes
    (b) chlorinated PVC pipe
    (c) ABS black pipes
    (d) unplasticized PVC pipe

18. Check which one of the following is an incorrect statement.
    (a) PVC pipes are cost effective as compared to metal pipes.
    (b) PVC pipes do not corrode.
    (c) PVC pipes are not affected by UV radiation.
    (d) PVC pipes are inexpensive and simple to install.

19. Neoprene is suitable for use in
    (a) bearings of bridges
    (b) plastic insulation
    (c) water tanks
    (d) pipes

20. Resin identification code developed by SPI is used for
    (a) insulation
    (b) recycling
    (c) compatibility with food
    (d) none of these

21. The vulcanizing agent used in rubber is
    (a) nickel               (c) zinc
    (b) copper               (d) sulphur

22. To improve the rigidity of rubber tires, the reinforcing filler used is
    (a) carbon black         (c) nylon wires
    (b) sulphur              (d) steel wires

## Review Questions

1. What are the two major classifications of plastics? Distinguish between them.
2. What is meant by polymerization? Explain the mechanism of polymerization. Differentiate between addition and condensation polymerization.
3. How does co-polymerization differ from addition and condensation polymerization?
4. Why are fillers usually added to moulding plastics?
5. Name any five moulding compounds and explain their use in plastics.
6. What are solvents? What is their use?
7. List the different processes used in the fabrication of different types of plastic products.
8. Describe the compression moulding process.
9. How can moulding be performed through the extrusion process?
10. How is injection moulding different from the extrusion process?
11. Explain the process of transfer moulding.
12. When is blow moulding used? Explain the process.
13. Explain calendaring and foaming.
14. How is the process of laminating done?
15. What are the advantages of plastics? List a few disadvantages also.

16. Give two popular examples of thermosetting plastics and thermoplastics and indicate their uses.
17. Differentiate between LDPE, HDPE, and UHMW, including their uses.
18. What are the following plastics and where are they used in building constructions?
    (a) Polyvinyl chloride      (b) Polycarbonate      (c) Polystyrene      (d) Perspex
    (e) Nylon                    (f) ABS               (g) Urea-formaldehyde  (h) Formica
    (i) Araldite
19. What are the uses of PVC pipes in building construction? Differentiate between unplasticized PVC pipes, plasticized PVC pipes, CPVC, PEX, and PPR pipes.
20. Discuss the advantages and disadvantages of using PVC pipes instead of GI/Copper pipes in buildings.
21. Write short notes on the use of plastics in (a) water tanks, (b) doors and windows, (c) polycarbonate roofing sheets, (d) vinyl flooring, (e) plastic insulation, (f) geosynthetic materials, and (g) structural insulated panel.
22. What are bioplastics? What are the environmental benefits of using bioplastics?
23. What are fibre-reinforced plastics? What are the applications of FRP?
24. Why plastics are considered bad for the environment?
25. Explain the identification code developed by the Society of the Plastics Industry. How is it useful?
26. What is geofoam? Where is it used? What are its advantages and drawbacks while used in pavement applications?
27. What are the different types of rubber? Distinguish between natural rubber and synthetic rubber
28. Differentiate between recycled rubber and reclaimed rubber.
29. What is vulcanization of rubber and how is it made more efficient?
30. What are the important properties of natural and synthetic rubbers?
31. Explain the uses of rubber in (a) flooring, (b) cement and concrete, (c) paving, and (d) backfills.

# ANSWERS

**Multiple-choice Questions**

| | | | | | |
|---|---|---|---|---|---|
| 1. (a) | 2. (a) | 3. (b) | 4. (c) | 5. (b) | 6. (d) |
| 7. (b) | 8. (a) | 9. (c) | 10. (c) | 11. (c) | 12. (a) |
| 13. (d) | 14. (b) | 15. (c) | 16. (b) | 17. (b) | 18. (c) |
| 19. (a) | 20. (b) | 21. (d) | 22. (a) | | |

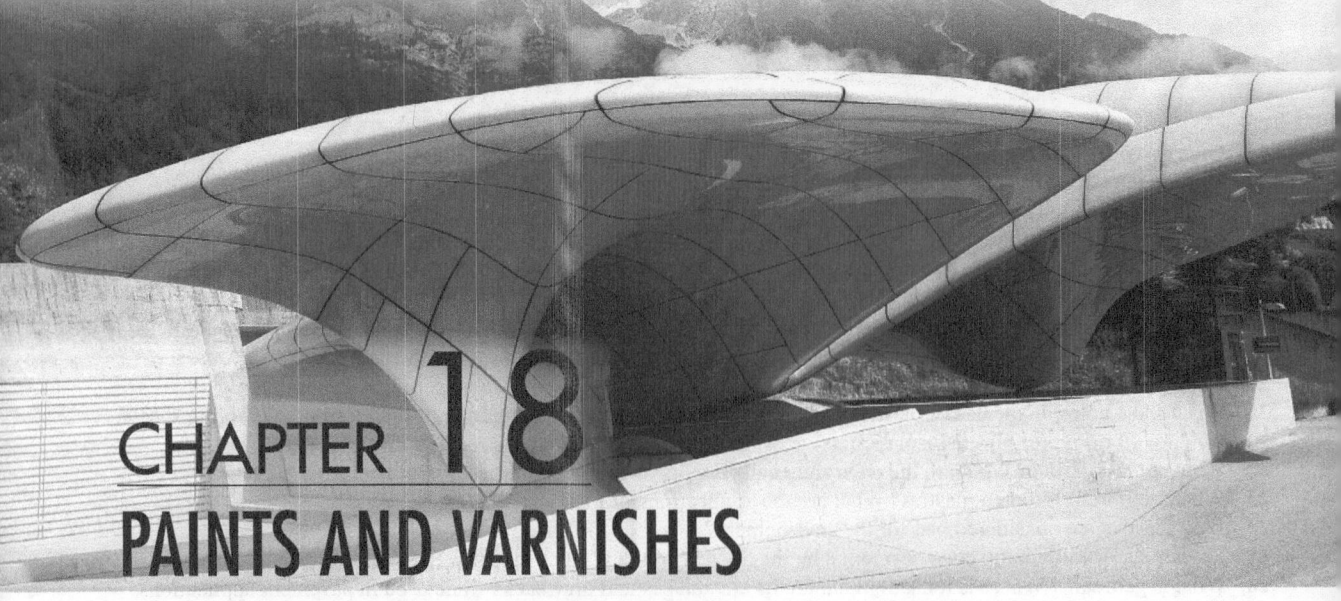

# CHAPTER 18
# PAINTS AND VARNISHES

## 18.1 Introduction

Paints are coatings applied over concrete, masonry, metals, plasters, and wood, to give these base materials a decorative finish as well as protect them from weathering, corrosion, or other chemical or biological attacks. In addition, painting may impart sanitation and improve illumination. There are several kinds of paints available in the market; however, it is important to identify and use them properly in a particular job. In addition to paints, there are different types of varnishes used on wooden surfaces, which provide glossy and transparent coatings. Paints and varnishes are usually applied in several layers depending on the environmental conditions, as primary coat, first, and subsequent coats, on surfaces which are properly prepared. Usually, they are available as liquids and applied by using brushes/rollers or spraying, which when dried form a protective coating on the surface.

The painting industry consists of two separate divisions: (a) *decorative paint industry* (also called *architectural coatings*) and (b) *industrial paint industry* (see Fig. 18.1). At present, the decorative paints account for about 75% of the overall paint market in India. Top organized players include Asian Paints, Kansai Nerolac, Berger Paints, and ICI. Major segments in decorative paints include interior wall and ceiling paints, exterior wall paints, wood finishes and enamels for wood and metal surfaces, and ancillary products such as primers and putties. Asian Paints is the market leader of decorative paint segment. Three main segments of the industrial paint sector include automotive coatings, powder coatings, and protective coatings. Kansai Nerolac is the market leader in this segment. The per capita paint consumption in India is about 4 kg, which is very low compared to consumption in developed western nations.

In this chapter, we will confine our discussions to the category of decorative paints used in building construction.

## 18.2 Components of Paints

Paint is essentially a mixture of (a) a *binder* or *vehicle*, which makes the paint to adhere to a surface, (b) *pigments*, which give colour, make it opaque, and occasionally make the paint corrosion resistant, and (c) *solvents,* which make the paint spreadable (IS 1303). In addition, several *additives* may be added to achieve the required characteristics. Modern paints can have 10 to 20 components, each enhancing a specific quality of the paint. The aforementioned four basic ingredients are explained briefly here.

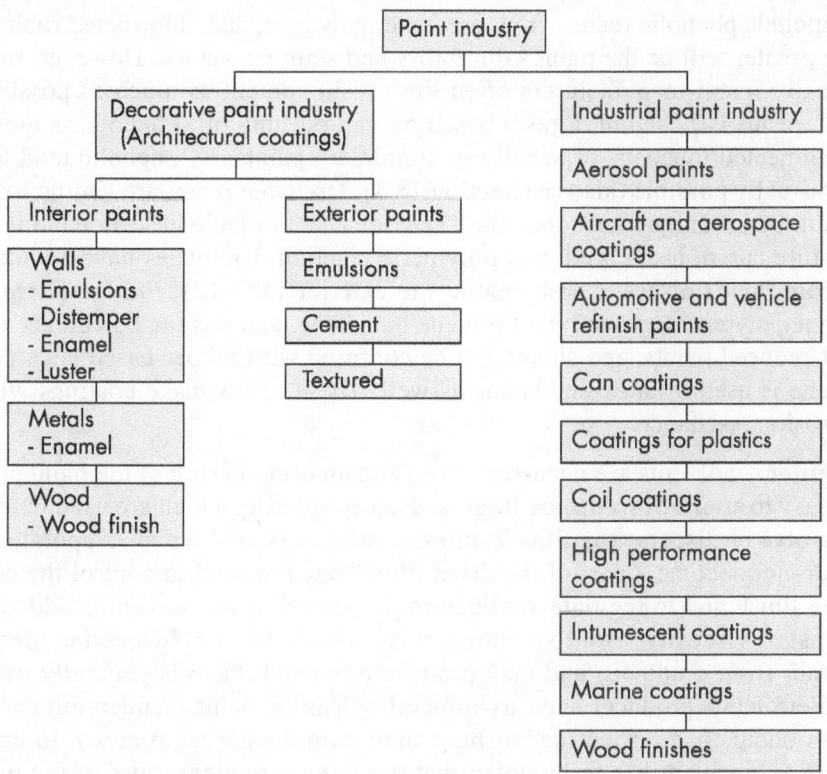

**Fig. 18.1** Different types of paints in the paint industry

**Base pigments**   These are finely ground insoluble solid particles that are suspended in a water-based or oil-/mineral-based liquid. When the liquid is applied to the surface, it dries to form a decorative and protective layer. Thus, base pigments serve three main functions: (a) the optical function of providing colour, opacity, and gloss; (b) a protective function of protecting the surface underneath the paint and the binder from being destroyed by ultraviolet (UV) light; and (c) a reinforcing function of helping the binder to stick to the surface. Some special purpose pigments may enable paint to resist heat, control corrosion, or reflect light. They provide durability and protection to the painted surface. In addition to the colour, pigments give paint *hiding power* (also called *opacity* or *coverage*), i.e., the ability to form an opaque film and cover the surface completely. Pigments are composed of tiny solid particles less than 1 μm in diameter, a size that enables them to refract light (light has wavelengths between 0.4 and 0.7 μm). Pigments are broadly classified as either organic or inorganic. Some examples of inorganic pigments are: carbon black (black), titanium dioxide and lithopone (white), iron oxides (yellow, red, brown, black), zinc chromates (yellow), azurite (blue), chromium oxides (green, blue), aluminium powder, and zinc phosphate (corrosion-protection). Titanium dioxide is probably the most important prime pigment used. The use of red lead and white lead is prohibited in most of the countries.

**Vehicle or binder**   The *vehicle* or *binder* in the form of oil and resins is responsible for the paint to adhere to a surface by forming a thin film and binding the pigments into the film. The binder is the one that is left behind after the paint has dried and all the diluents have evaporated. The first binders used in paints were the natural drying oils such as linseed, tung, and soybean, which have been replaced to a large extent by polymers, often referred to as *resins*. The three most important binders (resins) used in modern paints are: (a) acrylic polymers, (b) alkyd polymers, and (c) epoxy polymers.

Other binders include phenolic resin, vinyl, urethane, polyester, and chlorinated rubber. Higher their proportion, the greater will be the paint's durability and stain resistance. However, since the resin is the most expensive, paint manufacturers often limit resin content as much as possible. One of the oldest synthetic resins is *phenolic*, a resin based on natural tung oil. Phenolic is used in varnishes, primers, and pigmented topcoats, especially in aluminium paints. As phenolic tend to darken, light colours are not usually possible (also see Section 18.4). *Urethane resins* are similar to alkyds, but are available in a number of unique formulations. There are also available in several cuing modes such as air-cured, moisture cured, heat cured, and polymeric reaction. Urethanes have a wide range of uses in buildings, from floor finishes to wall coatings to exterior varnishes. *Rubber-based resins* include chlorinated rubber, styrene acrylate, vinyl toluene butadiene, and styrene butadiene, all of which are used in solvent-reduced paints, and should not be confused with rubber-based latex paints, although styrene butadiene is used in latex emulsions as well. These resins make coatings with good water, alkali, and corrosion resistance.

**Solvent or thinner**    Solvents are necessary to ensure an even mixing of the paint components and to make them easy to apply. *Solvents* are liquids, usually volatile, which are used in the manufacture of paint, to dissolve or disperse the film-forming constituents, and which evaporate during drying, and therefore do not become a part of the dried film. They are used to control the consistency and character of the finish and to regulate application properties. If the solvent is added to the coating material to adjust its viscosity prior to painting, it is called a *thinner*. *Turpentine* (derived from pine oil but sometimes from crude oil, and may produce noxious fumes) is generally used as a thinner in oil paints. Petroleum products such as mineral spirits or paint lacquer thinner are also used (They have less odour than turpentine but have their own dangerous fumes.). In emulsion paints, water is used as a thinner. It has to be noted that the thinner reduces gloss of the paint. A number of other solvents can be used to remove paint, such as the very caustic methyl ethyl ketone (MEK), acetone (the least toxic), turpentine, naptha, and xylene. While using thinners, the area should be well ventilated, safety goggles and gloves should be worn, and the remaining thinners should be disposed safely.

**Additives and extenders**    These are substances added during the manufacture and/or application of the paints, usually in small quantities, to enhance certain properties or selectively improve the film properties. Additives may provide specific performance characteristics, such as stain-blocking or mould-killing properties. Additives vary in volatility, and may remain either in the coalescing film or escape from it more or less completely.

*Colouring pigments* (also called *fillers*) are added to give different colours to the paint, which may also act as partially cheap substitutes for the expensive base pigment. The anti-corrosive pigment which has been used for several years in red lead, $Pb_3O_4$. As red lead is found to be toxic, it has been replaced by zinc phosphate, zinc chromate, zinc molybdate, and barium metaborate (alternates to chromates are being sought now as they are also toxic and environmentally hazardous).

*Extenders*, which are inorganic material in powder form and have a low refractive index, can be used to improve the application characteristics, provide flat or semi-gloss finishes, prevent the settlement of pigments, or provide better sticking properties for subsequent coatings. Examples of extender pigments are: whiting (calcium carbonate), talc (magnesium silicate), barytes, kaolin (aluminium silicate), silica, and mica (www.nzic.org.nz). Metallic pigments are also used. Modern cars have metallic finishes and this is possible due to the inclusion of finely divided aluminium. Bronze has also been used as a decorative pigment. Zinc and lead powders, when used as pigments, have anti-corrosive properties. A few of these additives are briefly discussed here.

**Drier**  These are small quantities of compounds that are added to accelerate the process of drying. They can be oxidizers, [octoates of metals such as cobalt (widely used), manganese, vanadium, cerium, iron, and lead], polymerizers [such as zirconium, aluminium, bismuth, and barium (substitute for lead)], and auxiliaries [such as calcium, potassium, lithium]. Lead compounds are rarely used in modern paints due to their high toxic nature. The drying process of paints having synthetic resins is a complex one of polymerization.

**Anti-foaming agents**  Additives such as mineral spirits, octyl alcohol, pine oil, etc., are used to reduce the surface tension of the paint and minimize foaming (formation of bubbles during manufacture or application).

**Preservatives**  To guard against bacteria which may ruin the performance of paints, preservatives like phenyl mercury salts were first used. They were replaced by formaldehyde condensates. At present, derivatives of isothiazolinone are used as both anti-bacterial, in-can preservatives and as anti-fungal, dry-film preservatives. Examples include methylchloroisothiazolinone (MIT), chloromethylisothiazolinone (CMIT), octylisothiazolinone (OIT), butyl-benzisothiazolinone (BBIT), and dichlorooctyl-isothiazolone (DCOIT).

In addition to these, we can have *anti-flooding agents* (which bind white hiding pigments to colour pigments), *anti-skinning agents* (which are volatile and prevent solvent-reduced paints skinning over in partly emptied cans), *thixotropic agents* (which keep paints in creamy consistency and drip-resistant), and *emulsifiers and coalescing agents* (emulsifier keeps particles separated until the paint is spread on a surface and then the coalescing agent draws and coalesces the particles together to form a thin film).

## 18.3 Manufacture of Paints

The manufacture of paint is fairly simple. The pigment, binder, and thinner are blended in the correct proportions such that, when the paint is applied, the finished film is continuous, smooth and attractive. The *paint technologist* (who formulates the paint to meet certain predetermined standards) and the *paint technician* (who manufactures the paint to achieve these properties) work together to obtain the required properties of paints. Usually, chemical engineers, who design the machinery and equipment for the manufacture of paints, are also responsible for the testing and quality control of paints. Details of these equipment are outside the scope of this book, but in general, the ball or pebble mills, horizontal and vertical bead mills, and high-speed intensive stirrers are used (www.nzic.org.nz). For high-quality finishes, triple or single roll mills are sometimes used, and there are many other varieties of machine available to the paint manufacturer. In the laboratory, the small-scale ball mill, bead mill, and high-speed mixer are commonly used. More information on the manufacturing of paints may be found in Lambourne and Strivens (1999) and Talbert (2008).

### 18.3.1 Packaging of Paints

After the manufacture, paints are packaged in metallic or plastic containers of different sizes—250 mL, 500 mL, 1 L, 4 L, 20 L, etc.,—as shown in Fig. 18.2. It is important that the cover of the packaged product should be kept airtight. Anti-skinning agents are necessary for paints packaged in metallic containers.

**Fig. 18.2** Paint available in containers of different sizes

## 18.4 Classification and Types of Paints

Standardizing the classification of paints is difficult due to the large number of variations in each of the constituents. However, in building industry, paints are classified based on the media or binder, and on the basis of its ultimate use and performance as follows:

**Based on solvent and binder**   (a) water-based (or emulsion): using acrylic, vinyl acrylic, and poly vinyl acetate (PVA) binders and are generally known as *Latex* or *Acrylic paints*; (b) oil based: alkyd (or enamel): using synthetic binders called alkyds; most alkyds are made from triglycerides derived from plant and vegetable oils (e.g., linseed oil); (c) *epoxy paints*, which have an epoxy resin as binder for increased resistance to corrosion, abrasion, and chemicals.

**Based on their function**   (a) primers or undercoat (applied before painting for better adhesion), (b) finishing coat (to obtain smooth and decorative surface), (c) floor paint (applied on concrete floors), (c) iron primer (water- or oil-based anti-corrosive coat for metal surfaces), (d) heat-resistant paint, (e) fire-retardant paint, (f) rust inhibiting paint, and (g) spray paint (applied with spray gun).

**Based on pigment used**   (a) zinc rich (used to withstand temperature up to 550°C, and to protect the surface against weathering and corrosion), (b) white lead paints (now banned due to its harmful effects), (c) graphite paint (has powdered graphite and oil and is used to coat metallic structures), (d) red lead paints (earlier used to provide anti-corrosive layer, but banned now), and (e) zinc chromate (anti-corrosive).

**Based on sheen of paint**   Sheen is the amount of light reflected by the painted surface: (a) flat, (b) satin, (c) semi-gloss, and (d) gloss.

**Based on appearance**   (a) eggshell, (b) matt, (c) satin finish, (d) semi-gloss, (e) gloss, (f) flat, (g) fluorescent, etc.

The first classification, which is important, is explained here.

### 18.4.1 Water-based Latex Paints

Water-based paint uses water as a solvent, but confusingly it is commonly called *latex paint,* although it does not contain any latex. Many water-based paints are made with acrylic and may be called *acrylic latex paint.* Water-based latex paints are more popular because of their environment friendliness, easy clean up with plain soap and water, and due to their significant performance (about 75% of all paints sold today are latex).

Top quality exterior latex paints have greater durability than oil-based paints and have better colour retention and chalk resistance. In addition, they have better resistance to cracking, as they do not become brittle with time like oil-based paints. Moreover, latex paints dry much faster than oil-based paints (through the evaporation of water), allowing the second coat to be applied in about 1 to 6 h. However, they are not effective in blocking stains from showing through. They are also more sensitive to temperature.

Latex paints of better quality, which contain '100% acrylic' binders, are more durable and highly flexible. In addition, they adhere to a variety of exterior surfaces nicely and hence are not susceptible to paint failures such as blistering, flaking, and peeling. Latex paint is the most common type of paint used in residences due to the advantages shown in Table 18.1.

### 18.4.2 Oil-based Alkyd Paints

These may be dissolved using natural resins such as linseed oil or alkyd resin dissolved in petroleum distillates or an organic solvent. Alkyd paint is not the paint of choice for home painting applications. They were normally used for bathrooms and other wet areas since these paints were easy to scrub, had a higher sheen, and were more durable than the early latex paints. But many present-day latex paints can match or even outperform alkyd paints. Top quality oil-based paints have excellent adhesion characteristics, which are essential for durability. However, oil-based coatings may oxidize and get brittle over time, leading to cracking problems in exterior applications, and yellowing and chipping problems in interior applications. Oil-based or alkyd paints should never be directly applied to fresh masonry or galvanized iron. Oil-based paints are also not recommended for surfaces submerged in water.

Oil-based coatings are still the best choice in the following two circumstances:

1. When repainting exterior surfaces with heavy 'chalking' (chalk is the powdery substance that comes off when we run hands across the surface)
2. When repainting any exterior or interior surface that has four or more layers of old oil-based paint (the number of layers can often be determined by removing some paint chips and examining them)

A comparison of water-based latex paints and oil-based alkyd paints is provided in Table 18.1 (Jenkins, 2014)

**Table 18.1** Comparison of latex and alkyd paints (www.californiapaints.com)

| Latex paints | Alkyd paints |
|---|---|
| Water-based; have fewer volatile organic compounds, hence environment friendly | Oil-based; contain more volatile organic compounds (VOCs); harmful to the environment |
| Easy to apply and flows smoothly and evenly, with less brush drag | Thick and sticky, requiring a little more effort to apply. Has better one-coat hiding and coverage. |
| Quick drying through evaporation(1 to 6 h); hence, second coat can be applied quickly | Longer time to dry through chemical means- 8 to 24 h; hence, second coat cannot be applied quickly. |
| Low odour | Strong-odours when drying-can last even weeks |
| Shows brush marks if used in low temperatures | Suitable for use in low temperatures |
| Can be cleaned with soap and water | Requires toxic chemical solvents (mineral spirits, or paint thinner like turpentine) to clean hands, spills, tools and brushes; extra care is required in handling and disposing of used rags |
| It resists mildew better and 100% acrylic paints resist abrasion | Vegetable oil base can provide nutrients for mildew growth; additives may minimize their growth. |
| Durable and excellent adhesion to most substrates, with proper surface preparation; flexible and expands and contracts with the surface. Its porosity allows moisture to escape from the painted surface. Current formulations may even exceed the durability of oil-paints | High durability; usually better adhesion than latex paints; especially on heavily chalked surfaces. Dries to a hard smooth finish and withstands abrasion well. It is so hard that it should not be painted over latex, because the softer coat beneath tends to flex and may crack the oil-coating. |
| Available in numerous colours and sheens | Usually, (but not always) a high-gloss finish |
| Usually, cheaper than alkyd paints. 100% acrylic paint costs more, but lasts longer. | More expensive than latex paint |
| Versatile; can be used on wood, concrete, stucco, brick, galvanized metal, vinyl, and aluminium siding. | Can be used on most materials, except on new concrete, stucco, and other masonry, where a sealer or pre-treatment is required; best suited to woodwork, doors or window trim, and metals; should not be applied directly on galvanized metal. |
| Good sheen and colour retention; More resistant to fading, yellowing, cracking and chalking than oil paint, especially when exposed to UV light | Not as good as latex; can be sensitive to UV, light colour may fade due to sunny exposure |
| No big restrictions | Their release in waste streams is regulated or even banned by some municipalities; old paint should be disposed only through an authorized recycling centre |

### 18.4.3 Application Methods

Numerous methods can be used to apply paints on different surfaces, such as brush, roller, dipping, flow-coating, spraying, hot spraying, electrostatic spraying, airless spraying, electro-deposition, powder coating, vacuum impregnation, and immersion. A few are shown in Fig. 18.3. Brushes made of hog or ox bristles should be used only with alkyd-based paints. All-purpose brushes blend polyester, nylon, and sometimes animal bristles. For painting walls and ceilings, a 75-mm (3 inch) straight-edged brush, and for woodwork, a 50-mm trim brush, and for painting corners, a tapered sash brush may be used. It has to be noted that proper thinning and substrate temperature allow the paint to flow-out and eliminate the brush marks. Paint contains solid particles that may sink to the bottom of the can. Hence, before using, it is better to stir the paint to uniform consistency using a clean, flat batten, or stick. More details about these methods of applying may be found in Talbert (2008), Lambourne and Strivens (1999), and Wicks, Jr., et al. (2007).

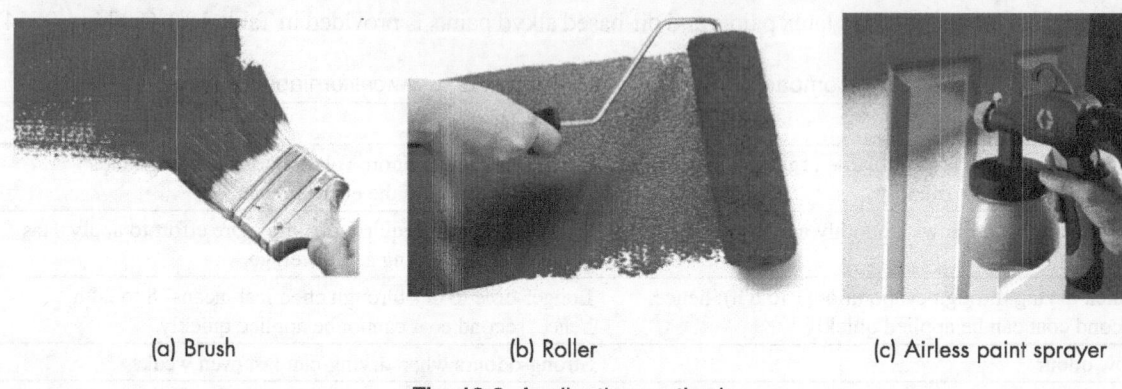

(a) Brush        (b) Roller        (c) Airless paint sprayer

**Fig. 18.3** Application methods

### 18.4.4 Characteristics of Good Quality Paint

The requirements of good quality paint are uniform spread as a thin film, high coverage, good workability and durability, sufficient elasticity to remain unaffected by expansion or contraction of the surface to be painted or by weathering action of atmosphere. Other desirable attributes include the ease of application, quick drying, high opacity, good flow-out without application marks (e.g., brush-marks), forming a continuous protective film, water resistance, colour stability (i.e., against visible and UV radiation), heat resistance, abrasion and scratch resistance, corrosion resistance, affordable cost, and ease of clean-up. The performance comparison of good quality latex and oil-based paints is provided in Table 18.1.

The key to any good paint job is always surface preparation. The surface should be washed to remove loose dirt and dust. Surfaces should be scraped or sanded to remove loose, peeling, and flaking paint. It is important to note that such scraping of older buildings (especially pre-1978) may release dust containing lead or asbestos. Once the surface is cleaned, any exposed substrates may be applied with appropriate primer before painting.

## 18.5 Considerations in Choosing Paints

Many factors determine the type of paint that should be used in a paint job, including the nature of the surface to be painted and its condition. The 100%-acrylic paint systems offer the best performance on a variety of surfaces; however, not all acrylic paint resins have the same performance. It is better to check for paint performance properties (e.g., adhesion, flexibility, colour fade, and gloss loss). Some of these considerations are briefly discussed here.

**Nature of the surface to be painted**   Different types of paints have to be used for different surfaces such as plaster, wood, or steel. The paint should be compatible with the surface. For example, wood trim and siding expand and contract depending on the season. Good elastic paint such as 100% acrylic paint may be chosen in this situation.

**Moisture present on the surface**   The paint that is used on a moist wall (e.g., newly constructed plastered brick wall) should have the capacity to resist the effects of moisture. Cement paints allow the moisture to pass through them (allow them to 'breathe') and hence are suitable for this purpose. As oil-based paints form an impermeable film, they should not be used. For the protection of exteriors of tall buildings, antifungal paints that last for a long time should be used (e.g., Weather-Coat-Biowash by Berger Paints), as repainting of tall buildings is difficult and expensive.

**Relative humidity**   Many paints will not stick to the surface when the relative humidity is high. Drying of paint will also be slow. In such cases, it is better to postpone painting to drier days or use paints such as cement paints, which are suitable for such conditions. Oil-based paints should not be used in such conditions.

**Colour**   Painting the interior with cool colours such as blues, greens, and purples makes small rooms appear larger and more airy while colours such as reds, yellows, and oranges will give a room a more vibrant appearance. As a general rule of thumb, light exterior colours will make any house appear larger and will be more affected by different lighting conditions, whereas dark shades can make the house appear smaller and help it blend with the surroundings. Darker colours absorb more heat and energy, which cause substrate movement that can shatter the bond between previous coats and the substrate – especially wood. In addition, bright exterior colours fade faster. Using lighter colours is beneficial, as they reflect more light and do not absorb as much heat and energy; thus, improving energy efficiency and film durability. A good indicator of how well a colour reflects light is to look at its *light reflectance value* or LRV (in the range 0–100). Higher values (closer to 100) indicate colours that reflect more light.

**Paint sheen**   It refers to how shiny the dried paint surface becomes. Selecting the ideal sheen or gloss-level for an interior or exterior paint job involves both aesthetic and practical considerations. The following sheens in increasing order of gloss are normally recognized as per IS 1303: (a) flat (matt) – practically free from sheen even when viewed from oblique angles; gloss value varying from 0 to 5 units (all values at 60°); (b) eggshell matt - gloss value from 6 to 15; (c) eggshell-gloss - gloss value from 16 to 30; (d) semi-gloss - gloss value from 31 to 50; and (e) full gloss - smooth and almost mirror like surface when viewed from all angles, gloss values 51 and above.

Flat paints have the least amount of sheen. Flat paint hides imperfections well and the painted surface creates very little glare. Flatter paint makes touch-ups easier and more seamless. However, the surface is not very washable and stain removal may be more difficult than other sheens. Owing to these, flat paint is typically used on the ceilings.

Sometimes called satin finish, eggshell paint has a moderate amount of sheen and hence hides some amount of imperfections and produces relatively little glare. It is also fairly washable. Owing to these characteristics, it is the standard choice for living/bed rooms and halls. Semi-gloss is used in kitchens, bathrooms, laundry rooms that require high washability and moisture-resistance of a glossier paint. The glossier the finish the greater will be the durability, and more washable and scrubbable is the surface. Gloss surfaces reflect lots of light and also offer more mildew resistance because they are less porous. Gloss paint is used on trims, doors, and cabinets as it is tough and highly washable. Eggshell is probably the most popular sheen as it can hide imperfections like a flat, easier to wash, durable, and provides a smooth surface.

**Alkalinity of the surface**   Lime-plastered surfaces should be coated with alkali-resistant paints. Otherwise, the paint may react with the lime and subsequently peeled off. Silicate paint will not peel or flake like the usual limewash, and will remain moisture-permeable, enabling water vapour to pass easily through the substrate thus allowing the wall to breathe.

**Cost of the paint** Often, the choice of paint is decided based on the cost. However, cheap options do not always produce the desired results. Flat paints may be cheaper than high-gloss paints, but makes the finish less durable. In the same way, expensive price tag does not always translate into premium performance. Eco-friendly, low-VOC options, although a bit expensive, but are better for health and the environment.

**Pigment volume concentration** (see also Section 18.5.1) This is an indicator of the amount of resin and pigment in the paint. A lower percentage means less resin and pigment, which can result in lower performance.

### 18.5.1 Pigment Volume Concentration

Pigment volume concentration (PVC) gives us an idea of how much pigment is there in the paint compared to the amount of binder. PVC is defined as the ratio of the volume of the pigment divided by the volume of both pigment and binder together, as follows:

$$PVC = \frac{V_p + V_f}{(V_p + V_f) + V_b} \tag{18.1}$$

where $V_p$ is the volume of pigments, $V_f$ is the volume of fillers (extenders), and $V_b$ is the volume of all binders.

The PVC controls factors such as gloss, washability, adhesion, hiding power, permeability, and durability. As PVC increases, hiding power and density increase, whereas gloss, scrubbability, adhesion, and exterior durability decrease. If paint has no pigments, it will usually be very glossy and have PVC of zero (see Table 18.2). Generally, the darker the colour of the glossy paint the lower will be the PVC. Note that the PVC of paint is known only to the paint chemists who formulate them and hence cannot be easily worked out.

**Table 18.2** Selected paints and their PVC

| Paint | PVC (%) | Paint | PVC (%) |
|---|---|---|---|
| Flat/latex paints | 55–80 | Exterior house paints | 28–36 |
| Semi-gloss and satin paints | 30–45 | Metal primers | 25–40 |
| Gloss paints | 25–35 | Wood primers/under coats | 35–40 |
| Glossy topcoat | 3–20 | Transparent varnishes | <10 |

The critical pigment volume concentration (CPVC) is the PVC at which the binder concentration is just barely sufficient to completely wet pigments and fillers. Above the CPVC, the porosity of the coating increases rapidly because of the voids in the coating and the coating become discontinuous. Below the CPVC, there is sufficient polymer for pigment wetting. The CPVC will be generally in the range of 30–60%. Latex paints may have a CPVC of about 60%. At CPVC, many physical and optical properties of paint change abruptly.

## 18.6 Paint Primers

A *primer* or *undercoat* is a preparatory coating given to the surfaces of porous materials (e.g., concrete and wood) before painting. Priming is mandatory if the surface is not water resistant and will be exposed to weathering. A good primer serves several functions. Whether used on interior or exterior surfaces, primers ensure that the painting surface has an ideal, uniform texture (slightly coarse), so that paint adheres effectively. In addition, primers seal up porous surfaces and prevent stains and previous colours from showing up underneath the paint job. Primer before painting often reduces the number of coats required to achieve adequate coverage (Jenkins, 2014). Using a primer also increases paint durability and provides additional protection for the surface being painted. Different primers are used for different materials such as plaster, wood, metal, and plastics, and for different types of paints. A primer consists of 20 to 30% synthetic resin, 60 to 80% solvent, and 2 to 5% additive agent. Exterior primers minimize cracking and mildew growth, and protect masonry surfaces from alkalinity and efflorescence. Exterior primers come in specific formulas for use on wood, masonry, or metal. Metal primers provide a tight bond between the surface and topcoat and inhibit corrosion. Tinted primers improve the quality of painting. There are also *all purpose primers*, which eliminate the need to buy several different primers, resulting in a material savings. *Paint and primer in one* has actually no primer in the mix; it is a thicker paint, giving a sturdier coat of paint (www.thespruce.com). When working on painted surfaces, priming is necessary if paint types are switched. For example, while changing from oil-based paints to latex-based paints and vice versa, or changing colours drastically.

The often used three types of primers are described here.

**Oil-based primers** They may be used with both oil paints, latex paints, on wood and metals, and also on surfaces with existing paint. They prevent stains from showing through new coats of paint, and slow down paint peeling, cracking, and blistering. Oil-based primers penetrate wood more thoroughly than latex primers, making them better at preparing weathered wood for paint. They are slow-drying and release high amounts of VOCs, and use harsh thinners and solvents to clean brushes and applicators. They should not be used on masonry.

**Latex-based primers** They are used with water-based paints, are more flexible and fast drying, and less brittle than oil-based primers. They are used for priming wood, plaster, masonry, concrete, and galvanized metals. Although they are not as effective at covering stains as the other two types of primers, they are easier to clean up and come in low- or no-VOC formulations, making them healthier alternatives.

**Pigmented shellac primer** They are the best stain-blocking primers, working well on severe water and smoke damaged surfaces. They work well on wood, metal, plaster, and even plastic, and are highly adhesive. They are the only primers that can be applied in freezing temperatures. They do not penetrate wood deeply, but dry very fast; a topcoat can be applied after about 45 min. They also can be used with both oil-based and latex paints. But they give off more fumes and require denatured alcohol or ammoniated detergent for clean-up.

## 18.7 Cement Paints

Cement paints are so called because the vehicle or binder used is white cement. Cement paint is a mixture of 50–60% Portland or white cement and 10–15% lime, with additions of mineral extenders (10–15% calcite/titanium dioxide, 3–5% alkali-resistance pigment, and 10% China clay), accelerator (2–5% calcium chloride), water repellents (5–10% calcium/aluminium stearate), and fungicides. The ingredients in powdered form are weighed and charged in the ball mill for grinding and mixing property. After obtaining the desired fineness, the products are packed in polyethylene bags in suitable metal/plastic containers. The product should conform to the requirements for quality, safety, and performance prescribed in IS 5410:1992. *Snowcem* and *Supercem* are examples of this type of paint. Cement paints readily mix with water, which is added to the powdered paint just before use. It is used on porous surfaces such as masonry, concrete, rough plaster, but should not be used on gypsum plaster, wood, and metal surfaces. Cement paint allows the painted surfaces to breathe. These paints have relatively better performance than distemper paints for outdoor protection and decoration of buildings. Normal cement paints require curing of the painted surface. However, cement paints that require no curing, and special waterproof cement paints are also available now.

## 18.7.1 Procedure for Applying Cement Paint

Cement paint is water-based paint and may be applied to either exterior or interior walls made of brick work and concrete block masonry. It is used for painting exterior wall surface mainly for preventing water penetration and reductions of dirt collection. They are not well suited for interior surfaces requiring frequent and thorough cleaning, as coatings of these paints are not easily washed and tend to erode with vigorous scrubbing. It is available in the market in different colours.

Some guidelines for the application of cement paint are given here [IS 2395(Parts 1 and 2):1992].

1. Cement paint should not be applied on surfaces already treated with whitewash, colour wash, dry distemper, or oil-bound distemper (OBD), unless the surface is thoroughly scraped and cleaned properly. Growth of vegetable material, which cannot be removed by brushing, should be destroyed by applying a wash of ammoniacal copper solution. Any patches should be repaired by plastering followed by wetting and application of one coat of cement paint.
2. Cement painting of new surface should be deferred until the cement concrete of stucco has aged at least three weeks. Then, the surface should be thoroughly brushed to remove all dirt and remains of loose or powdered materials.
3. Before applying cement paint, the cement or concrete surface should be wetted thoroughly to provide moisture to aid in the proper curing of the paint. The surface should be moist not wet when the paint is applied.
4. The paint is prepared by adding paint powder to water and stirring to obtain a thick paste, which is then diluted to a brushable consistency. Generally, equal volumes of paint powder and water make a satisfactory paint.
5. No painting should be done when the paint is likely to be exposed to a temperature below 7°C within 48 h after application.
6. Cement paint should be stirred well during the application and used within an hour after mixing with water; otherwise, the mixture would be set and thicken affecting the flow of paint and finishing.
7. The solution should be applied on the cleaned and wetted surface with brushes or spraying machine. It is better to apply the paint on shady surface of the buildings so that fast drying due to direct heat of the Sun is avoided. Cement paint should not be applied when it is raining.
8. Painted surfaces should be cured by sprinkling with water using a fog spray two or three times a day, for at least 24 h.
9. The second coat should be applied after the first coat has become sufficiently hard (i.e., at least after 24 h).
10. In special cases, a coat of cement primer is applied followed by two or more coats of waterproof cement paint.
11. All tools and application equipment should be cleaned with water immediately after use.
12. The covering capacity of cement paint for two coat brush work on plastered surfaces will be about 3.0 to 4.0 m²/kg.

## 18.7.2 Advantages and Disadvantages of Cement Paint

The following are some of the advantages of cement paints:

1. They can be applied on damp walls, whereas oil-based paints and OBD cannot be used on damp surfaces.
2. Painting with cement paint requires lesser skills.

3. Walls, once limewashed, can be painted with cement paint after a delay of few months, because cement paint is not affected by alkalinity.
4. Cement paint gives very good protection from UV rays present in sunrays.
5. It gives very good protection from severe climatic conditions such as rain, heat, water, humidity, salt atmosphere near seashores, to all types of cemented walls and surfaces.
6. It is possible to give fungicidal wash to the surfaces before applying cement paint.
7. It is economical and has good weather resistance and durability. It resists the penetration of water on external walls and at the same time provides a breathable film.
8. It is possible to paint with other types of paints on the surfaces already painted with cement paint. Generally, the walls are first painted with cement paint, and any other desired paint is painted over it after 1 or 2 years.

Some of the disadvantages of cement paints are as follows:

1. The walls are to be wetted before the cement paint is applied to give the paint a good bond.
2. Cement paint will not adhere properly on surfaces that are too smooth.
3. Most cement paints require water curing, except the recently developed verities. Water curing the inside walls of rooms is difficult when the building is occupied.
4. Cement paint should not be applied on surfaces, which already have oil-paints or have gypsum plaster or limewash, without thoroughly removing the old paint.
5. As compared to acrylic paints, it is not durable and may last up to 2 to 3 years only.

## 18.8 Distempers

These are water-based paints consisting of whiting (powdered chalk) or lime as the base in a glue called *size* (a gelatinous substance), with resins or acrylics (synthetic) as the binder, and water as carrier or thinner (toxic substances like white lead may be used as an alternative to chalk; hence, it is better to check the paint's product datasheet before use). Colouring pigments are also added to give various colours. Distemper is the most economical type of painting available in the Indian market today. Distempers should conform to IS 427:2013 and IS 428:2013, and are easy to apply. They are more decorative and durable than whitewash or colour wash. They are commonly used on plastered interior walls of buildings which are not exposed to weather. Distempers are available in two forms: (a) ordinary (soft) distemper and (b) OBD.

### 18.8.1 Soft Distempers

The requirements for *soft* (dry) *distemper* are as per IS 427:2013. These are available in powder form. New plastered walls must be allowed to dry completely for at least 28 days (3–6 months in high humidity areas), and the surface cleaned with wire brush. In addition, soft-distempered surfaces must be thoroughly removed by rubbing with sand paper and washing down with warm water. A priming coat of approved primer should be applied over the prepared surface. No whitewashing coat is used as priming coat. The distemper powder is mixed with suitable quantity of water/hot water, according to the direction given by the manufacturer for application by brushing (usually, 0.6 L of water per kg of distemper), and kept for 4 h to allow complete dissolution of glue. The mixture is stirred well before application (drying time is 3 h). Two coats of distemper are applied on the dry surface. Soft distemper has a velvety, matt finish and is used in internal walls due to its water solubility.

### 18.8.2 Oil-bound Distemper

OBD is made exclusively from natural plant resins, minerals, and pigments and bound with plant oils, such as linseed oil and casein. OBD is available in thick paste form, and can be thinned with water, if desired (1 L of water to every 1.5 to 2 kg of distemper). OBD comes in a variety of shades and has a smooth matt finish. As with oil-based paints, it should not be used on damp walls. Finished OBD surfaces are washable. The film dries in 30 min. Two coats are necessary, and the second coat is applied after 6–8 h. The advantages are as follows:

1. It is easy to use and has a variety of shades.
2. It is durable and affordable, and provides a sleek matt finish
3. It dries rapidly, i.e., touch dry within 4 h.
4. It is non-flammable and eco-friendly since it is lead free and includes no petrochemical dependent components.

## 18.9 Plastic Emulsion Paints

Water-based paints, which are based on acrylic and/or vinyl emulsions, are the most extensively used paints in the decorative paint market, and account for 50–70% of the volume. An emulsion is a liquid with fine suspended particles. Emulsion paints are so-called because liquid monomers to be polymerized are dispersed in water, as an emulsion (see also IS 15489). Thus, it consists of pigment and solid or semi-solid polymeric particles dispersed in a continuous aqueous medium in which they are insoluble (It has to be noted that usual alkyd and water-based paints, on the other hand, consist of a pigment, solvent, and binder which are all mutually soluble). On drying, the water evaporates leaving behind a smooth film.

*Plastic emulsion* paints should conform to IS 15489:2004. As per IS 15489, plastic emulsion paint is of two types: (a) type 1- for interior use and (b) type 2- for exterior use. These paints can be used get the finishes such as (a) matt, (b) egg shell/satin, (c) semi-glossy, and (d) glossy. When suitably thinned with water, the material will mix readily with the minimum amount of foaming to a smooth and homogeneous state. The foaming, if any, will dissipate rapidly. After thinning 50/50 with water, it can be applied by brush, spray, or roller. The resulting film will not show pigment flocculation, coarseness, or other undesirable characteristics.

The binder in many modern emulsion paints is based on resins such as vinyl acetate and acrylic ester. As the polymers used in these paints are carried in water, these paints are much better for the environment than paints in which the binders are in organic solvents. The addition of acrylic/vinyl resins makes them more hard-wearing than traditional emulsions. This results in varying degrees of sheen in the finish; as the shine increases, the paint tends to be more hard wearing.

Emulsion paints are very popular because of their ease of application, quick drying properties, good washability, and non-objectionable odour. They are available in a range of colours and decorative effects. Although normally thought as suitable for internal walls and ceilings, some water-based emulsion paints are specially produced for woodwork. These are easy to apply but do not give the same hard-wearing qualities as oil-based paints. Plastic emulsion is recommended for interior decorations of surfaces such as concrete, plaster, gypsum, and wood. Textured emulsions are also available for exterior painting.

### 18.9.1 Latex (Acrylic) Emulsion Paint

The name 'latex' paint is a misnomer because it does not contain any latex (*Latex* is a natural product that is derived from the Brazilian rubber tree). Instead, all latex paints consist of synthetic polymers that look like natural latex but have different chemical and other properties. It has to be noted that the paint which is called as 'latex paint' in the USA is called as 'emulsion paint' in India and the UK. Latex paint is a general term that includes all paints that use synthetic polymers such as acrylic, vinyl acrylic (PVA), styrene acrylic, etc., as binders. The term 'latex' is applied to most water-based paints, regardless of whether they are 100% acrylic, vinyl, or styrene. A typical latex paint has 25% resins (binder), 12.5% titanium dioxide (pigment), 12.5% limestone (extender), and 50% water (thinner), which are mixed together until they form an emulsion. As latex paints with 100% acrylic resins cost twice as much as vinyl, paint companies try to balance the ingredients to keep the cost down. For example, 20% acrylic and 80% vinyl resins may be found in typical common interior house paint, whereas paints that have more acrylic in the mixture will be of higher quality and cost more. PVA is even cheaper and is the main ingredient for white glues and most cheap paints.

Acrylic resins are better for interior and exterior applications because they offer: (a) better stain protection (washability), (b) water resistance, (c) better adhesion, (d) better hiding, (e) better resistance to cracking and blistering, and (f) resistance to alkali cleaners. The paint with a high acrylic content will have much better water and stain resistance. To have better effects of these paints, the surface should be prepared carefully with *putty* so that it is made smooth (Putty for concrete/cement plastered surfaces may be made of white cement based fine powder mixed with water. Oil-based paints may use putty consisting of whiting powder and boiled or raw linseed oil). The paint is thinned as per the manufacturer's recommendations. It is desirable to apply the paint on dry surfaces. Usually, two coats are applied. As per IS 15489, the paint film becomes surface dry in 45 (Class A and B) to 90 (Class C and D) min, and hard dry in 4 to 8 h. The second coat is applied after the first coat is fully dried. The dried surfaces can be washed with water.

### 18.9.2 Vinyl Emulsion Paint

These are basically same as acrylic emulsions but are cheaper as they contain less acrylic and more vinyl resins. Vinyl emulsion breaks more easily than acrylic emulsion when exposed to UV light. Three main types of vinyl emulsion are available, each giving a different finish. These are described here.

1. *Vinyl matt emulsion* provides a matt, non-shiny finish, hiding small imperfections on the wall/ceiling. The shinier finishes may reflect back more light highlighting imperfections.
2. *Vinyl satin emulsion* provides a subtle soft-sheen finish and is more durable than vinyl matt. It is suitable for walls that are occasionally washed.
3. *Vinyl silk emulsion* provides high sheen finish and is durable than the aforementioned two types. Hence, they are ideal for kitchen and bathrooms, where there will be more moisture.

## 18.10 Solvent or Oil-based Paints

Lustre paints, enamel paints and oil paints all come under the category of solvent/oil-based paints. They cannot be pre-mixed with water. Such oil-based paints take a longer time to dry and often produce strong odours, which are irritating and even toxic in nature. The advantages of these paints are that they last long and produce rich and desired effects on surfaces.

### 18.10.1 Enamel Paints

The term *enamel paint* is used to represent oil-based paints, and has significant amount of gloss. (It has to be noted, however, that many latex or water-based paints are also available, which match qualities of enamel paints.). Most enamel paints are alkyd resin based. Some enamel paints have been made by adding varnish to oil-based paint. Enamel paints should conform to IS 5:2007, IS 110:2017, and IS 133:2013. They dry quickly and result in non-breathing membrane, hence, should not be used in damp weather or on damp surfaces. Enamel floor coatings of a high gloss finish, high-heat enamel, rust-resistant enamel, and spray paints are also available. Enamels can be used for internal as well as external applications and can be used on all surfaces, especially on wood and metal. The waterproof film is acid resistant, not affected by alkalis/ gases, and more resistant to abrasion and moisture penetration. The surface preparation consists of rubbing wood with sand paper. After cleaning, a primer coat of titanium white in pale linseed oil, followed by two to three coats of enamel paint can be applied. Enamels are available in matt, semi-gloss, and gloss finishes. The extent of pigmentation in the enamel determines its gloss. Generally, gloss is reduced by adding extenders (lower-cost pigments). The level of gloss depends on the ratio of pigment to binder.

### 18.10.2 Synthetic Enamel Paints

Synthetic enamel paint is made from synthetic resins and drying oil with titanium dioxide and other selected pigments. They can be applied to all exterior and interior surfaces. They are high-gloss paints, resulting in smooth finish. They have superior brushability, excellent hardness, durability, and good stain resistance. They are available in various colours; white and pastel shades will resist yellowing and darkening with aging. This paint should conform to IS 2932:2003 and IS 2933:2013.

## 18.11 Exterior Paints

*Exterior paint industry* is a subset of the decorative paint industry. In addition to providing aesthetic appearance, exterior paints should be able to withstand fungus, moss, dirt, etc., which may accumulate on outer surfaces of walls. The most common chemical processes leading to degradation of exterior coatings are photo-initiated oxidation and hydrolysis resulting from exposure to sunlight, air, and water. Exterior paints should resist such chemical degradation, waterproof, and long lasting (longevity is important in tall buildings, where painting of external walls may not be easy). Exterior paints are expensive and are usually acrylic paints. Special waterproof cement paints for exterior use is also available, which are comparatively less expensive. Several brands of antifungal, anti-algae, exterior emulsion paints are available in the market, in wide price ranges and hence have to be chosen with care. Stain-blocking and 100% acrylic

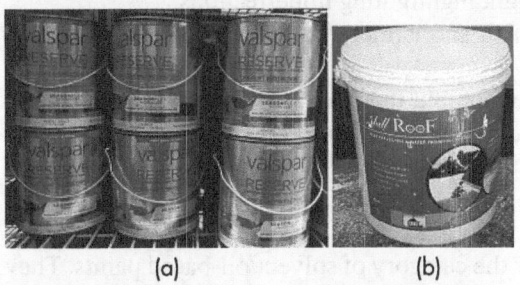

latex paint plus primer is also available that provides the perfect defence against dirt, moisture, and extreme weather, in just one coat [i.e., One-coat perfection™! Valspar® Reserve® (see Fig. 18.4(a))]. It also provides superior adhesion to any substrate (wood, stucco, masonry, or brick), gives mould, mildew, and algae resistant and scrubbable finish, and with HydroChroma® Technology, the colour stays true longer, even in the harshest climate. In order for the paints to function effectively, the crevices in the walls have to be filled, damaged places should be repaired, and a quality primer should be applied. Very bright colours are not normally used for exteriors, as they may absorb heat and also may fade sooner than light coloured paints.

(a)        (b)

**Fig. 18.4** Exterior paints (a) Extreme weather paint + primer (b) Reflective paint

High-performance acrylic-based heat reflective and waterproof paints significantly lower the surface temperature of the roofs and exteriors [Fig. 18.4(b)].

### 18.12 Whitewash and Colour Wash

Hydrated lime or lime putty can be blended with water to produce *limewash* (also called *whitewash*). Mineral colours, such as oxide of iron, red, and yellow colours, based on chromium oxide and carbon black and not affected by lime, may be added to produce colour wash. These are explained in Section 4.10.3

of Chapter 4. The approximate coverage of whitewash and colour wash, based on experience, is given in Table 18.3; actual value may vary depending on factors such as method and condition of application, and surface roughness.

**Table 18.3** Coverage of whitewash and colour wash (Adapted from CAMTECH Guide, 2007)

| Number of the coat | Approximate coverage | Remark |
|---|---|---|
| **Whitewash** | | |
| First | 5.0–8.0 m²/kg | Primer coat |
| Second | 3.5–5.0 m²/kg | Finishing coat |
| Third | 2.2–3.3 m²/kg | Finishing coat |
| **Colour wash** | | |
| First | 4.5–5.5 m²/kg | Primer coat |
| Second/third | 3.0–3.5 m²/kg | Finishing coat |

## 18.13 Painting on Different Substrates

As mentioned earlier, before painting the surfaces, the surfaces should be cleaned and properly prepared. Then, the primer is applied (base coat) and the required number of final coats (at least two coats) is applied by using brush/rollers; sprayer is also a good option for large surfaces. It has to be noted that oil-based or alkyd paint can be applied over existing water-based latex or acrylic paint, but the reverse should never be used. A primer must be used before applying latex paint over oil-based paint.

**Surface preparation all substrates**  Paint cannot take care of construction defects. Hence, before applying any paint, it is important to rectify surface defects and clean the surface, such that it is free from dust, dirt, loose matter, or grease. Then, it should be rubbed with emery paper, to provide a mechanical key between surface and paint for satisfactory adhesion. Recommended type and grade of sandpaper should be used to avoid scratching or gouging of surfaces.

**New woodwork**  For good results, woodwork should be well seasoned, and should not contain more than 15% moisture content. Soiled/greasy surfaces should be cleaned with turpentine, sanded smooth, and dusted off thoroughly (pressure washing may be done on exterior surfaces). Nails, if any, should be driven down the surface by at least 3 mm. A light wood-filler and a putty knife may be used to fill any notches or cracks. When it becomes dry, the surface should be sanded smooth. The knots in the wood may excrete resins that may cause defects such as cracking, peeling, and brown discolouration. To alleviate this, the surface should be primed with an alcohol-based shellac tinted primer. A good primer provides a surface for the paint to adhere, seals the pores of the wood, and will stop colours and saps from the knots from bleeding through the surface coat. On drying, the selected paint is applied with brushes. After painting the surface in one direction, the brush is worked in the perpendicular direction to eliminate brush marks. All the successive coats are applied after drying and slight rubbing of previous coats for proper bond. Refer to IS 2338 (Parts 1 and 2):1967.

**Old woodwork**  The old paint is removed with sand paper, or paint remover. Any smoky or greasy substance is washed with lime and subsequently rubbed with pumice stone. The surface is then washed with soap and water and dried completely. Then, two coats of paints are applied in a way similar to that described in painting new woodwork.

**New iron/steel work**   The surface should be free from scales, rust, and grease. Scales and rust are cleaned by hard wire brush. Grease is removed by using petrol, benzene, or lime water. Surface preparation using with power-driven tools, flame cleaning, or sand/grit/shot blasting are more efficient than cleaning with wire brush, which removes only 30% of the rust and scale (also see Subramanian, 2008; IS 1477:2000). A priming coat of high build zinc phosphate modified alkyd of thickness 60 μm is then applied over the prepared surface. After drying of the priming coat, intermediate coats are applied to 'build' total film thickness. Generally, the thicker the coating, the longer will be the life. The second coat is given only after the first coat has dried. The finish coat provides the required appearance and surface resistance of the system [see also Tables 29(a) and (b) of IS 800:2007]. Bituminous paints may be used to paint inside of pipes, on iron under water, piles, ships, and boats; they are unsatisfactory when exposed to sunlight.

## Case Study   Painting of Golden Gate Bridge

The Golden Gate Bridge, San Francisco, USA was painted international orange, with a lead primer and a lead-based top coat, as soon as it was constructed. Until 1965, only touch up was required. In 1968, due to advancing corrosion, the primer and the top coat were removed carefully and were replaced with an inorganic zinc silicate primer and acrylic emulsion top coat. The bridge requires routine touch up painting on an on-going basis.

*Photo*: Dr N. Subramanian

**Old iron/steel work**   The surface is prepared by scraping properly all the scales and rust with emery paper. The greasy substances are removed with lime water. The old paint may be removed by flame cleaning or using suitable solvents. After this, the surface is scrapped with wire brush, washed with a solution of caustic soda, and painted as for new iron/steel work.

**Masonry, concrete, or plaster**   Newly plastered surface may contain considerable moisture. Hence, it will be better to wait for at least 28 days (3–6 months in high humidity areas) before painting. Concrete, lime, and cement plasters are likely to cause alkali attack on paint. Hence, it is essential to apply alkali resistant primer (see also Table 18.4, which shows the pre-treatment required for different types of paints). Minor cracks, holes, and fissures are filled with cement grout and the surface is made smooth. Bonding agent is included in cement grout mix. Dents and cracks, if any, are filled with putty using a knife applicator. If the surface is very rough, putty may be applied in two coats. Putty should not be applied thick (maximum 1.5 mm). After the putty has dried, the whole surface should be rubbed well and cleaned. Since putty is absorbent, the coat of paint applied over it will be absorbed unevenly, leading to patches. Thus, a primer layer on top of the putty is a must. Two or three finish coats may then be applied. The surface after each coat must be dry, before the application of next coat.

**Table 18.4** Types of pre-treatment required on plastered surfaces [Adapted from IS 2395 (Part 1)]

| S. no. | Type of paint | Required pre-treatment |
|--------|---------------|------------------------|
| 1 | Size bound distemper | |
| | i.  One coat application | i.  A coat of clearcole |
| | ii. Two coat application | ii. A coat of size alone may be sufficient |

(Contd)

**Table 18.4** (*Contd*)

| S. no. | Type of paint | Required pre-treatment |
|--------|---------------|------------------------|
| 2 | Dry distemper | A coat of the same distemper thinned with water or Interior wall primer (water thinned) |
| 3 | Oil paint | Interior wall primer (solvent thinned), or primer-sealer recommended by manufacturer of paint |
| 4 | Emulsion paint | A coat of same paint thinned with water or primer/sealer recommended by manufacturer of paint |
| 5 | Cement paint or limewash | Wet the surface before applying paint |

**Painting old plastered surfaces**   The procedure depends on the condition of the existing coating. Dry distempers and limewash should be totally removed prior to repainting. Where solvent-based paint is used, all patches and repairs should be primed and applied with undercoating. Where oil-bound water paint or emulsion paint is used, a thin coat should be applied prior to the general application [IS 2395 (Part 1)]. If any of the defects discussed here is very much pronounced, it should be completely removed and the surface should be painted as a new surface.

*Chalking*   The surface is cleaned, rubbed with emery paper to remove the chalk. Then, one or two finish coats are applied.

*Efflorescence, blistering, flaking, etc.*   Old paint from the affected areas is scrapped off. These areas are touched up with primer and one or two finish coats are applied on the affected areas. Then, the entire surface is rubbed off, cleaned, and the finish coats are applied.

*Glossy surface*   All gloss is removed by rubbing with emery paper, surface cleaned, and then the finish coats are applied.

## 18.14 Varnishes

The term *varnish* has come to be used more generally for transparent coatings on wood, even though few of them today are varnishes in the original meaning of the word (Wicks Jr., et al., 2007). Varnishes were widely used in the 19th and early 20th centuries, but have been mostly replaced by a variety of other products, especially alkyds, epoxy esters, and uralkyds. Varnish, one of the toughest of the finishes, is superior to the other traditional finishes. It enhances and gives warmth to the grain of the wood and is resistant to impact, heat, abrasion, water, and alcohol. It can be used as a topcoat over worn finishes. Varnish provides a clear finish, but it darkens the wood slightly. It is available in high-gloss, semi-gloss or satin, and matte or flat surface finishes. There are several types of varnishes (see Section 18.14.2).

### 18.14.1 Ingredients of Varnish

Nearly all modern varnish contains the following three basic components:

1. Oils - Linseed oil, tung oil, or mixtures of the two
2. Resins - Alkyd, phenolic, or polyurethane
3. Solvents - Mineral spirits, naptha, or paint thinner

Varnish may also contain driers to accelerate drying. By modifying the types and amounts of these components, different types of mixtures that vary in price and suited for either indoor or outdoor use can be made (see also IS 340:1978). It has to be noted that they do not contain any pigment and hence provide a transparent finish.

Varnish is manufactured by cooking the drying oil and the resin together to high temperature to obtain a homogeneous solution of proper viscosity(linseed and tung oil are referred to as *drying oils*, because they will readily dry when exposed to oxygen). The traditional varnish, called *oil varnish*, was based on natural resins derived from animal (shellac), or from plants sources (*balsam, copal, damar*, and *kauri*) and oils, and can be thinned with mineral spirits or turpentine to the required application viscosity (see also IS 338:1952). When a varnish is made, the ratio of oil to resin can have a significant effect on the way the varnish will behave. A formulation that contains a greater percentage of oil (known as *long-oil varnish*) results in a softer, more flexible finish, that will not crack when the wood expands and contracts. The most common oil used to make varnish is linseed oil. Its lower cost makes it the most practical choice for both indoor and outdoor applications (IS 337:1995 and IS 338:1952). But tung oil is actually better for outdoor use.

*Synthetic varnishes*, based on synthetic resins such as alkyd, phenolic, or polyurethane, require special petroleum-based thinners (IS 339:1952). The best of the synthetic varnishes is the polyurethane types, which are clear, non-yellowing, and very tough. Polyurethane varnish made with both alkyd and polyurethane resins and called 'uralkyd'. Other synthetic varnishes are the phenolics, used for exterior and some marine work, and the alkyds, often used in coloured preparations. Phenolic and alkyd varnishes yellow with age and are not recommended for refinishing. A higher quality finish like *Epifanes* clear varnish contains phenolic-modified alkyd resins. Several outdoor formulations also use alkyd and urethane resins.

Water-based varnishes, consisting of acrylic resins dissolved in water, offer similar results without the clean-up hassle and toxicity. Modern regulations restrict the VOC of uralkyd varnishes to 275–300 g/L and water-based varnishes to 130 g/L.

## 18.14.2 Types of Varnishes

There are different types of varnishes and some of the frequently used types are briefly described as follows:

**Acrylic varnish** It is a water-borne varnish providing high transparency, good UV-resistance, and lowest refractive index. It resists yellowing. Acrylics have the advantage of water clean-up and lack of solvent fumes. Compared to solvent-based varnishes, they do not penetrate much into wood and lack brushability and self-levelling properties. They tend to dry quicker (most dry in 15 to 30 min) than oil-based varnishes, but they are not as durable as oil-based varnishes. They do not crack, chip, or bubble, and are water and alcohol-resistant.

**Alkyd varnish** Modern varnishes employ some form of alkyds (synthetic polyester resins). Various other resins may also be combined with alkyds. They are suitable in a wide range of conditions and can be made to dry and harden faster. Expensive exterior varnishes use alkyds made from high-performance oils and contain UV-absorbers, thus improving gloss and durability.

**Lacquer** It is a quick-drying, solvent-based varnish, which is similar to French polish. But, lacquer is different from shellac and cannot be dissolved in alcohol. Lacquer is dissolved in lacquer thinner, which is a highly flammable solvent. Lacquer is usually sprayed onto wooden furniture, in a separate spray booth that evacuates overspray and minimizes the risk of combustion. Lacquers of different shades can be formed by adding colouring pigments. Once applied, lacquer can be removed using lacquer thinner.

**Polyurethane varnish** It is a hard, tougher, abrasion-resistant, and durable coating. The temperature and humidity of the air has a huge influence on how fast a varnish dries and cures. They are mostly used indoors, especially on hardwood floors but may not be preferred for finishing furniture. A thick film of polyurethane may de-laminate when subjected to heat or shock, resulting in fractured film and white patches. Polyurethane is not UV resistant and when used outside UV-resistant modified polyurethane should be used. A good primer (epoxy or shellac) is helpful with polyurethane finishes. Polyurethane can be applied on top of properly cured oil-based finishes.

**Shellac/French polish**   Shellac (IS 347:1975), also called *French polish* in India (French polish is actually the name of the process using shellac, although there are several other shellac-based finishes), is a resin varnish that is soluble in alcohol. (*Shellac* is actually a resin secreted by the female Kerria laccainsect, on trees in the forests of countries like India and Thailand. It is processed and sold as dry flakes and dissolved in ethanol to make liquid shellac). As per IS 348:1968, French polish is prepared by dissolving 25 ± 1% by weight of shellac in denatured spirit or isopropyl alcohol or a mixture of the two or with other alcohol-soluble ingredients, at room temperature. The surface on which it is applied is made smooth by rubbing. Then, the filler (2 kg of whiting in 1.5 L of isopropyl alcohol) of the desired colour is prepared as a paste and applied to fill cracks and pores. The surface is rubbed after drying and dusted off. Two coats of polish are then applied. It dries in about 2 h to a smooth and glossy finish. French polish was used on expensive timbers like mahogany and on instruments like pianos and guitars. It should not be used in outdoor surfaces or at places where there is constant moisture. French polish is not used now because of its very labour-intensive process and its tendency to melt under low heat; for example, hot cups can leave marks on it. However, unlike lacquers, it can be easily repaired.

**Spar varnish**   In modern times, *spar varnish,* also called *marine varnish,* has become a generalized term for any outdoor varnish. It not only protects the wood, but also gives it the flexibility and UV protection (the term *spar* represents long wooden poles that support the sails of ships). Modified tung oil and phenolic resins are often used in the formulations. Many polyurethane finishes contain a lower concentration of oil and are consequently more brittle and are more prone to crack. It is used on boats, doors, furniture, bar tops, or any other marine application where a durable clear gloss finish is desired. If applied inside, proper air circulation must be ensured; otherwise, it may become tacky (sticky) even after 3 days.

**Stains**   These are colouring agents that are used to change the colour or shade of the wood. A simple stain requires a coat (or coats) of varnish or another finish on top to protect the wood. Stains can be used to highlight the grain, lighten, or darken the natural tones. Some combination products are sold in which a stain and a sealer are applied to the wood at one time. There are various kinds of stains that are distinguished by the vehicle or solvent in which the colour is suspended [e.g., linseed-oil based (wiped on or sprayed), turpentine-based, alcohol-based (available in aerosol can), and water-based stains (acrylic latex based, powdered, and need mixing)]. Most latex stains are opaque. Exterior stains are chemically similar to oil-based paint but with creosote added to the mix. Exterior varnish usage has declined steadily in favour of wood-stains, owing to a high incidence of flaking and discolouration, with consequent increased maintenance costs.

## 18.14.3 Application and Recoating

A sealer should be applied to wood surfaces before they are varnished, so that varnish is absorbed evenly; if wood is not sealed, the varnish may dry to mottled leather like finish (recall that primers are used to seal surfaces that are painted).

Natural varnish can be used with any stain or filler. The sealer for natural varnish may be thinned using shellac or a mixture of one part varnish and one part turpentine or mineral spirits. Polyurethane varnish is not compatible with all stains and fillers. Some polyurethanes can be thinned for use as a sealer; some do not require sealers. Water-based varnish can be used over stain and filler. Varnish is usually applied by special 'fine haired' brush but can also be sprayed (see IS 384:2002). It can also be done with a polishing pad of absorbent (woollen cloth) covered by a fine muslin cloth. Varnish should be applied first along the grain of the wood, and then across the grain to level and even the surface. Drying times for natural varnish average about 24 h, but water-base varnish and polyurethanes often dry more quickly. Dampness slows drying; hence, the drying time may be longer in humid or wet weather. It is better to let it dry for a couple of days more.

Many varnishes require two or even three coats for a smooth finish. The varnish should flow onto the surface of the wood, with no drag. The dust and lint should be picked from the wet finish with a rosin lint picker and not after it has dried. When the first coat is completely dry, the varnished wood should be lightly sanded in the direction of the grain, using grade 7/0 (very fine) sandpaper. All sanding residue should be cleaned with a tack cloth, and the second coat of varnish may be applied.

A comparison of the appearance, protection, durability, and safety of different clear varnish finishes is provided at https://en.wikipedia.org/wiki/Wood_finishing.

## 18.15 Industrial Paints

*Industrial coatings* cover a large range of diverse uses with many differing requirements and functions. They are paints or coatings defined by its protective, rather than its aesthetic properties, although they can provide both. The most common use of industrial coatings is for corrosion control of steel or concrete. Some of these paints are briefly discussed here (www.coatings.org.uk, Wicks, Jr., et al., 2007).

**Aircraft and aerospace coatings**   High solids chromated epoxy primer, a base coat of clear-coat polyurethane/high solids polyester urethane or acrylic urethane and a top coat of urethane paints (e.g., Jet Glo Express™) are applied to both civil and military aircraft and helicopters.

**Aluminium paints**   These consist of finely ground aluminium powder held in suspension by varnish (IS 13183:1991). They are highly heat reflective and resistant to acid fumes. Aluminium paints are used for painting metal roofs, silos, machinery, poles, towers, and storage tanks, and provide attractive appearance.

**Anti-corrosive/Bituminous paints**   Anti-corrosive paints and powder coatings have linseed oil as vehicle with red lead or zinc/iron oxide or zinc chromate as pigments. Anti-corrosive bitumen coating (as per IRS P-30:1968), made of asphalt bitumen dissolved in mineral spirit or naphtha, is used on railway carriage tracks, coaches, and wagons. Bituminous black paint, conforming to IS 158, is used as acid, alkali, water, and heat-resistant paint for boilers, pipes, structural steelwork, and corrugated iron sheets. Bituminous black paint, conforming to IS 9862, is used as acid, alkali, water, and chlorine-resistant paint for apparatus, equipment, and drinking water tanks. Where sunlight (UV) exposure is expected, an over coating of aluminium paint is used – it also provides a bleed through barrier when painting over bitumen-based coatings.

**Automotive and vehicle refinish**   *Automotive coatings* are applied at the factory during the manufacture of cars and other vehicles, and *vehicle refinish coatings* are applied in body-shops during repairs. They are used for both protection and decoration purposes. Earlier, lead, chromium, and other heavy metals were used in automotive paint. Enamel paint is currently the most widely used due to reasons of reducing environmental impact. The coating is applied by dipping the vehicle body into the electro-coat paint operation, and applying high voltage (www.electrocoat.org). After the primer coat, three categories of base coat, having solid, metallic, and pearlescent pigments, may be used. A glossy and transparent clear-coat is then sprayed on top of the coloured base coat.

**Can coatings**  These are a range of very special coatings that are applied at low thickness but provide a high degree of protection. These coatings are used for both internal and external coating on food and drink cans.

**Cellulose paints**  These are also known as *lacquers* and contain colloidal dispersion of cellulose derivative (e.g., nitrocellulose), resin, and plasticizers in petroleum solvents. Nitrocellulose was used in lacquers for automobile top coats before acrylic lacquers, which have better exterior durability, were developed. Nitrocellulose lacquer primers are still used in refinishing automobiles and wood finishing. Because of their high VOC, their use in the USA and Europe has declined substantially (Wicks, Jr., et al., 2007).

**Coatings for plastics**  These are special coatings designed for use on plastic substrates, such as car bumpers, dashboards and grills, plastic toys, and electronics.

**Coil coatings**  These are coatings applied to metal sheets – often, starting as large coils of steel, which are then formed into a wide range of uses, e.g., domestic appliances such as washing machines, automotive parts, and composite panels used in exterior wall cladding.

**Epoxy paints**  The *epoxies* require the use of a curing agent such as polyamide resins; the resin and catalyst components are usually packed separately, and mixed at the job site prior to use. Epoxy paints offer superior toughness, abrasion, moisture and corrosion resistance, and good cleaning properties. Although they are expensive, they are cost effective in the long run.

**Fire-retardant intumescent coatings**  *Intumescent coatings* can be soft char or hard char. Soft char paints have binder based on vinyl acetate copolymers or styrene acrylates, and significant amount of hydrates. Hard char products contain sodium silicates and graphite. These paints are applied to structural steelwork, and during fire produce a char on the surface, which is a poor conductor of heat.

**Marine coatings**  *Marine coatings* are applied to all sizes of ships and boats. The coatings cover topsides, above the waterline, and antifouling coatings which are applied to minimize marine fouling and reduce ships drag through the water. Marine fouling was prevented by using biocide-containing paints, but environmental concerns and legislation have resulted in non-biocidal solutions. Non-biocidal, fouling-resistant coatings are based on polymers designed to minimize molecular adhesive forces between the adhesives used by marine organisms and the coating.

**Rubber-based paints**  These contain rubber treated with chlorine gas and dissolved in solvent and desired pigment. These paints are resistant to acid, alkalis, and dampness. Latex paint is rubber-based, but is made from a polyvinyl material rather than natural rubber. Most products referred to as rubber paint are also made from synthetic rubber, but unlike latex paint, they form a protective coating of about 75–125 μm thick. Examples are synthetic rubber and chlorinated rubber paints. Rubber paints form a flexible coating that provides durability and protection from moisture, extreme temperatures, chemicals, sunlight, rust, and impact. They are used on concrete, plaster, and masonry surfaces, waterproofing decks, roof tiles and gutters, lining swimming pools, concrete water tanks, and steel works.

## 18.15.1 Innovations in Industrial Paint Industry

Some of the innovations in industrial paint industry are briefly listed here.

1. *Electrophoretic deposition* (EPD) is a process in which colloidal particles suspended in a liquid medium migrate under the influence of an electric field (electrophoresis) and are deposited onto an electrode. The colloidal particles that can be used to form stable suspensions include materials such as polymers, pigments, dyes, ceramics, and metals.

2. In Canada, Rona Inc. has introduced 'ECO *Recycled Latex Paint*', certified by EcoLogo, which is made from 90% recycled materials, and may generate only a quarter of the greenhouse gases emitted by other paints over its life cycle.

3. *Powder coatings*: In the area of powder coating, which was invented 40 years ago, several innovations are being made. Powder coatings are generally used for items such as bicycles, refrigerators, and washing machines. The powder is made up of a resin (often, an epoxy resin), pigments, additives, and a catalyst to promote cross-linking when the powder is heated. The powder is sprayed on to the desired surface using an *electrostatic spray gun* and is then heat cured to produce a hard coating. Recently, acrylic powder coatings have been introduced as clear-coats on car bodies. The powder coating application method provides a number of environmental and economic benefits. Because the process does not use solvents, little or no VOCs are emitted into the atmosphere. One of the benefits of this powder coating system is that the over-sprayed powder can be reclaimed and reused. Innovative new materials, such as BLOX adhesive resins, can replace the currently used thermoplastic and thermoset resins in the powder coating industry, which may result in the development of tougher, more durable coatings.

4. *Intelligent coatings*: Naturally occurring functional additives are added to develop 'intelligent coatings', which result in self-healing, detoxifying, and self-degreasing surfaces. Another additive, made using carbon-nano-tube technology, could reduce needed film thicknesses/costs in epoxies and urethanes, and strengthen integrity of the film. It will result in five–seven times more flexible film and increase corrosion protection more than four times (www.coatingsworld.com).

## 18.16 Covering Capacity of Paints

The covering capacity of paint is the area in square meters that can be covered by 1 L of paint, when the paint is bought in liquid form or 1 kg of paint when the paint is bought in powder form, per single coat. It is to be noted that the coverage of paint for new buildings will be less and it will be more when applied as second coat or when applied on old walls with the same colour. The approximate covering capacity of different paints is provided in Table 18.5.

**Table 18.5** Approximate covering capacity of different paints (Adapted from CAMTECH Guide, 2007)

| Type of paint | Covering capacity | Drying time | Recoating time for final finish (at 25°C and 50% RH) | Remark |
|---|---|---|---|---|
| *Primers on interior masonry surface* | | | | |
| Water-based wall primers | 18 to 22 m²/L/coat | 4–6 h | 6–10 h | Depending on surface and surface preparation |
| Solvent-based primer | 16 to 18 m²/L/one coat 12 to 14 m²/L/second coat | 12 h | 12–14 h | On smooth plastered surface by brushing. May also be used on wooden surface |
| *Interior wall paints* | | | | |
| Cement primer | 12 to 13 m²/L | 30 min-1 h | 3 h | On smooth plastered surface by brushing |
| Cement paint like Snowcem | 2 to 4 m²/kg (per coat) | 30 min | 24 h of water curing | On smooth primed plastered surface by brushing or spraying |

*(Contd)*

**Table 18.5** (*Contd*)

| Type of paint | Covering capacity | Drying time | Recoating time for final finish (at 25°C and 50% RH) | Remark |
|---|---|---|---|---|
| Dry distemper (water based) | 15 to 18 m²/kg/first coat 7 to 8 m²/kg/second coat | 30 min | 4–6 h | On smooth plastered surface by brushing |
| Oil-bound distemper (water based) | 15 to 18 m²/L/first coat 9 to 10 m²/L/second coat | 15–30 min | 6 for synthetic 3–4 for acrylic | On smooth primed plastered surface by brushing or spraying |
| Ordinary Emulsion (water based) | 24 to 26 m²/L/first coat 14 to 15 m²/L/second coat | 10–15 min | 4 h | On smooth plastered surface by brushing |
| Plastic acrylic emulsion (water based) | 22 to 25 m²/L/first coat 15 to 18 m²/L/second coat | 15–25 min | 3–4 h | On smooth primed plastered surface by brushing |
| Lustre (solvent based) paint | 18 to 22 m²/L/first coat 11 to 13 m²/L/second coat | 2–3 h | 8 h | On smooth primed plastered surface by brushing / rolling |
| Matt (solvent based) | 18 to 22 m²/L/first coat 9 to 12 m²/L/second coat | 45 min | 8 h | On smooth primed plastered surface by brushing followed by rolling |
| Enamel (water based)-semi glossy finish | 14 to 16 m²/L/first coat 9 to 10 m²/L/second coat | 20–30 min | 4–6 h | On smooth primed plastered surface by brushing; may be used on wooden/metal interior surfaces |
| **Exterior wall paints** | | | | |
| Emulsion (water based) | 10 to 12 m²/L/first coat 5 to 6 m²/L/second coat | 30 min | 4 h | On normal plastered surface by brushing |
| Acrylic plastic emulsion (water based) | 9 to 11 m²/L/first coat 5.5 to 6 m²/L/second coat | 30 min | 4 h | On normal plastered surface by brushing |
| Emulsion 100% Acrylic All Colours | 8 to 10 m²/L/coat | 20–30 min | 4–6 h | On normal plastered surface by brushing |
| **Metal paints** | | | | |
| Steel primer (zinc chromate/red oxide) | 14 to 18 m²/L/coat | 6–8 h | 8–10 h | Depending on surface and surface preparation |
| Enamel paint (Solvent based)- Glossy finish | 18 to 22 m²/L/first coat 12 to 14 m²/L/second coat | 3 h | 6–8 h | On smooth primed mild steel surface by brushing |
| Synthetic Enamel paint (Solvent based) | 18 to 22 m²/L/first coat 12 to 14 m²/L/second coat | 3 h | 8 h | On smooth primed mild steel surface by brushing |
| Chlorinated Rubber Paint | 8–10 m²/L/coat | 15 min | 24 h | On smooth primed mild steel surface by brushing |
| Aluminium paint | 20 to 22 m²/L/coat | 30 min | 8 h | Spray, brush/roller in small areas |

(*Contd*)

**Table 18.5** (*Contd*)

| Type of paint | Covering capacity | Drying time | Recoating time for final finish (at 25°C and 50% RH) | Remark |
|---|---|---|---|---|
| **Preservatives and varnishes, wood paints** | | | | |
| Wood preservative | 15 to 16 m²/L/coat | 6–12 h | 24 h | On smooth sanded wood surface by brushing |
| Wood sealer | 9 to 10 m²/L/coat | 2 h | 24 h | On smooth sanded wood surface by brushing |
| Wood finish (Touch wood)-Solvent based | 9 to 11 m²/L/coat | 1 h & tack free time-4 h | 24 h | On smooth sanded wood surface by brushing |
| Wood finish (melamine)-Solvent based | 9 to 10 m²/L/coat | 20 min & tack free time-2 h | 8–10-brushing 6-spraying | On smooth sanded wood surface by spraying |
| Varnish/ French polish | 12 to 14 m²/L/first coat 8 to 10 m²/L/second coat | 10–15 h | 24 h or more | On smooth sanded wood surface by brushing |

## 18.17 Defects in Painting

Painting defects may arise due to a variety of reasons but some of the main reasons are the incorrect choice of paint for a particular surface, improper surface preparation before painting, application of paint on damp surface, poor workmanship, and ingress of moisture to painted surface. Even if the paint is properly selected, surface well-prepared, and the paint applied properly, it may gradually deteriorate and eventually fail due to weathering. When failure is identified after the painting has cured, the paint in the affected areas has to be removed to the substrate and the painting has to be redone after preparing the substrate. Some common types of paint defects, the reasons for such defects, and possible remedies are provided in Table 18.6. More information on paint defects and their remedy may be found in CAMTECH Guide (2007) and Wicks, Jr., et al. (2007). It is to be noted that when the defect is restricted to the paint surface only (involving colour, gloss, or texture), it can be corrected by applying another coat of paint.

**Table 18.6** Common defects in painting, possible causes, and remedies

| Defects | Causes | Remedies |
|---|---|---|
| Blistering | Trapped oil or gas forming bubbles resulting in localized loss of adhesion between one or more layers of coating (likely in enamels), or between primer and parent surface. Owing to (a) heat, (b) insufficient drying time between coats, and (c) porous paint that allows moisture to pass through. | Allow surface to be fully dry before painting, and avoid painting under direct sunlight. Scrap off blisters, sand the surface smooth, prime bare places, and then repaint it. |
| Blooming or blushing | Dull patches on finished paint surfaces due to: (a) applying paint under high humidity or in a poorly ventilated room and (b) use of improper thinner. | Clean the affected area and repaint avoiding bloom activating conditions (moist or fume-laden atmosphere). |
| Crawling | Failure of a new coat of paint to form a continuous film over the preceding coat. | Do not apply latex paint over high-gloss enamel and do not apply paint on concrete or masonry treated with a silicone water repellent. |

(Contd)

**Table 18.6** (*Contd*)

| Defects | Causes | Remedies |
|---------|--------|----------|
| Chalking (formation of a white, chalky powder on the surface of paint ) | All paints will show some chalking when subjected to outdoor exposure. Prevalent with flat white paints or very light-coloured paints that contain high levels of titanium dioxide and extenders or due to insufficient oil in primer. The 'chalk' is the powder (other than dirt) that comes off on finger when rubbed, and is the pigment used. This may be due to (a) applying paint over surfaces with dust or rust,(b) not sanding previous coat of gloss paint, and (c) weathering of paint and slow degrading of the binder by sunshine and moisture<br>***Prevention:*** Proper surface preparation | Soft paints (those based on linseed oil), enamel paints or lower gloss acrylics are likely to be chalky. Avoided by (a) using good quality of paint and mixing it well before use, (b) storing paint in dust free atmosphere, and (c) removing the chalk and repainting the surface. |
| Checking and Cracking | These are breaks in a coating after paint becomes hard and brittle. Checking is a mild form of cracking. Hair cracks appearing in smaller area it is called *crazing*. If appearing in a larger area called *alligatoring* or *crocodiling*. *Cracking* is wider and extends through to the substrate. Checking occurs when (a) paint film lacks tensile strength, (b) painting done during very cold weather, and (c) insufficient drying of undercoat. Cracking is due to (a) cracks in the plaster or masonry (b) painting on glossy surface, (b) applying top coat before previous coat dries, (c) defective surface (e.g., improperly seasoned wood), and (d) excessive use of putty/ thick paint coat. | Cracking normally found on wood, than on other substrates, due to its grain. Avoided by (a) regulating the use of putty and paint, (b) complete removal of the defective coats, and repainting after preparing and priming of exposed strata, (c) when small cracks do not enlarge with time, scrubbing top coat with emery and applying a fresh coat of paint, and (d) taking proper precautions during painting when there are temperature variations or rainfall. |
| Fading | Due to (a) improper paint formulation, (b) pigment not properly mixed, (c) use of incompatible pigment, and (d) paint not mixed well before application | Avoided by (a) using good quality of paint, and mixing it well before use, (b) good surface preparation, and (c) good workmanship. |
| Flaking | The moisture penetration through cracks on the coatings resulting in the loss of bond between surface and paint film. This is due to (a) poor surface preparation, (b) residue left after cleaning with solvent, (c) paint with poor flexibility and adhesion, and (d) paint film is not having elasticity. | Avoided by (a) use of plastic emulsion paints or paint with good elastic/flexible properly, (b) ensuring proper surface preparation and solvent cleaning, and (c) removing all dirt and dust on the surface prior to painting and rubbing the surface with emery paper before applying fresh coat. |
| Peeling | Inadequate bonding of the topcoat with the undercoat or the underlying surface; due mainly to inadequate surface preparation. May also be caused by (a) moist, oily, waxy, greasy, or glossy surface, or trapped gases between the painted surface and the paint film, which expand under the influence of heat, (b) poor adhesive quality of paint, and (d) paint applied on wet surfaces. | Emulsion paints provide a porous coating and allow the moisture to pass through. Hence, (a) use of approved quality of paint, (b) prepare the surface well before painting-all glossy surfaces must be sanded, (c) mix paint well before application, and (d) ensure moisture free surface before painting. |

(*Contd*)

**Table 18.6** (*Contd*)

| Defects | Causes | Remedies |
|---|---|---|
| Sagging or rundown | Due to (a) excessive thick coat, (b) spraying too close or at a high pressure, (c) poor workmanship, (d) poor surface preparation, and (e) less viscosity of the paint or use of wrong thinner.<br>*Prevention:* Apply finish-coat using recommended thinner. | Care must be taken while brushing (a) squeeze out paint carried away with brush in the container, (b) use correct viscosity paint, (c) maintain a distance of about 300 mm between spray gun and the surface, (d) if possible, avoid keeping the painted objects in vertical position during drying, (e) spread paint finally with vertical strokes, and (f) sand surface to smooth the sagged areas and repaint. |
| Wrinkling | Due to (a) poor workmanship, (b) Poor adhesive property of paint, (c) applying paint on a greasy surface, (d) applying very thick coating, (e) presence of heavy pigments, and (f) exposure of the wet film to rapid drying conditions | Care must be exercised while brushing: (a) ensure proper surface preparation, (b) use paint with good adhesive and with recommended viscosity, (c) use curing agents for drying, and (d) in spray painting keep the gun in constant motion. |
| Yellowing | Paint film subjected to too much sunlight.<br>*Prevention:* (a) Use anti-yellowing water based enamel, and (b) improve ventilation for inside job. | Surface should be cleaned and repainted. |

If shrinkage is expected on surfaces, especially on wood and concrete, painting should be delayed. Special paints should be used in conditions where thermal movements may occur. When painting in warm weather, a latex paint may dry very quickly. Owing to this film formation or curing may be incomplete, affecting long-term durability. When painting exteriors, avoiding the following conditions, occurring simultaneously, will result in better painted surfaces: (a) painting under very hot conditions, above 35°C, (b) painting in direct sunlight, especially when the paint has dark colour, (c) the surface being painted is hot, (d) very porous surface, and (e) breezy or windy conditions. To minimize the effects of summer heat, painting may be done early in the morning, when it is cooler. Exterior painting in the late afternoon should be avoided as the surfaces might have heated up and may be too hot to apply paint. It is always better to keep the brushes/rollers loaded with paint, work in smaller areas, and apply paint at a consistent, steady pace to get a good, uniform finish.

## 18.18 Harmful Effects of Paints

Lead, formerly a common additive in paint, has been gradually phased out. Even though Australia, the European Union, Canada, the USA, the UK, and other countries have banned lead paint as early as 1977, lead paint is not prohibited in India. A 2015 study found that over 31% of household paints in India had lead concentration of more than 10,000 parts per million (ppm), which far exceeds the BIS standard of 90 ppm for lead in paint. Titanium white (based on the pigment titanium dioxide) and zinc whites have replaced lead in paints. Cases of lead poisoning still occur in children who have been in direct contact with old, flaking paintwork, or with some toys. But the problem does not stop with lead. Epoxy paints and varnishes contain synthetic resins and phenols, which may be carcinogenic and have been banned in many countries (Pearson, 1998).

Labels on solvent-based paints and varnishes usually contain a warning that these products should be applied with windows and doors kept wide open – a clear indication of the toxicity of the vapours released into the atmosphere (Pearson, 1998). Most paints, varnish, and thinner solvents contain VOCs such as xylene, epoxy, toluene, benzene, hexane, MEK, as well as toxic heavy metals. These off-gas during application and while drying, and some even after they become dry

(Pearson, 1998). Urethane varnish and latex paints may also contain additives, such as insecticides and fungicides, that off-gas indefinitely. All isothiazolinone derivatives used in paints as preservatives may lead to allergic contact dermatitis and hence their use is now restricted to a maximum of 15 ppm (Lundov et al., 2014). In addition to the aforementioned, synthetic paints are serious fire hazards; once alight, they emit toxic gases and dense smoke that may prove fatal (Pearson, 1998). When working, during stripping or clean-up operations with solvent-based paints, varnishes, or stains, it is desirable to wear air masks. When sanding the painted surfaces, proper masks should be worn to prevent harmful dust particles entering lungs. It is better to avoid sleeping in the freshly painted residences for at least 2 days.

## SUMMARY

- Paints and varnishes are applied over surfaces to give decorative finish and to provide protection to the base material, from weathering actions. Paint is essentially a mixture of (a) a binder, (b) pigments, and (c) solvents. Various additives and extenders are also added by chemical engineers to enhance certain properties.
- Paints are classified based on the solvent, function, pigment used, sheen of paint, and also the appearance of the painted surface.
- Paints are mainly differentiated as water-based (environment friendly, easily applied, and cleaned) and oil-based. They are usually applied by brushes and rollers and sometimes by spraying.
- Good quality of paints is judged by their ease of application, quick drying, high opacity, good flow out, quality of protective film, durability, water resistance, flexibility, and ease of clean-up.
- The PVC controls factors such as gloss, washability, adhesion, hiding power, permeability, and durability.
- The three basic types of paint primers are: oil-based, latex based, and pigmented shellac primers. Primers ensure that the surface to be painted has an ideal and uniform texture for the paint to adhere effectively.
- In cement paints, such as *Snowcem* and *Supercem*, white cement is used as the vehicle or binder. Cement paints require wetting of the surfaces before painting and also curing of the painted surface.
- Distempers are the most economical water-based paints and their quality is better than white or colour wash.
- Plastic emulsion paints consists of pigments and polymeric particles dispersed in water, as an emulsion. They are easy to apply, quick drying, and do not have objectionable odour.
- Latex paint is a water-based paint that has synthetic polymers, such as acrylic, PVA, or styrene acrylic, as binders. As latex paint with 100% acrylic resins costs more, paint manufacturers substitute acrylic with vinyl resins.
- Solvent/oil-based paints such as lustre paints, enamel paints, and oil paints, take longer time to dry and produce strong odours. Most exterior paints are acrylic paints, which can also be made antifungal, anti-algae, corrosion resistant, etc.
- Varnish is similar to paint but provides a harder and transparent film. It has oil, resins, and solvents, but in most of the cases will not contain any pigment.
- Before painting/varnishing, the surfaces should be cleaned and properly prepared.
- Paints and varnishes may contain harmful substances such as lead and VOCs, which may off-gas during their life time. Hence, they require proper precautions while handling.

## EXERCISES

### Multiple-choice Questions

1. The function of base pigment in paint is
   - (a) to provide a protective film on surface
   - (b) to reduce imperfections on the surface
   - (c) to increase the strength of the surface
   - (d) to bring down the overall cost
2. Which of the following is not a vehicle in paints?
   - (a) Alkyd polymer
   - (b) Tung oil
   - (c) Turpentine
   - (d) Soybean oil
3. Which of the following is not used as a thinner?
   - (a) Turpentine
   - (b) Naptha
   - (c) Xylene
   - (d) Linseed oil
4. Which of the following is not a drier in paints?
   - (a) Lead oxide
   - (b) Lead acetate
   - (c) Red lead
   - (d) Silica

5. Match List-I with List-II.

| List-I | List-II |
|--------|---------|
| (a) Pigment | (i) Talc |
| (b) Drier | (ii) Titanium oxide |
| (c) Thinner | (iii) cobalt octoate |
| (d) Extender | (iv) Turpentine |

6. Which of the following is not correct with respect to water-based paints?
   (a) They do not dry fast
   (b) They are environment friendly
   (c) They can withstand movements
   (d) They can be cleaned using soap and water

7. Which of the following is not correct with regard to oil-based paints?
   (a) They are cheaper than latex paint
   (b) They take longer time to dry
   (c) Require chemical solvent for clean-up
   (d) They release VOCs

8. Paints can be applied using
   (a) brush
   (b) roller
   (c) airless spraying
   (d) all of these

9. Match the following types of sheen with the usually adopted location.

| List-I | List-II |
|--------|---------|
| (a) Gloss | (i) Ceiling |
| (b) Semi-gloss | (ii) bed rooms |
| (c) Eggshell | (iii) Kitchen |
| (d) Flat | (iv) Cabinets |

10. Match paint with its PVC.

| (a) Gloss | (i) 10 |
|-----------|--------|
| (b) Semi-gloss | (ii) 55–80 |
| (c) Varnish | (iii) 30–45 |
| (d) Flat | (iv) 25–35 |

11. Which of the following increases by adding colouring pigments in paints?
    (a) Washability
    (b) Adhesion
    (c) Durability
    (d) Pigment volume concentration

12. In the following, which is *not* correct with regard to cement paints?
    (a) They should not be applied over whitewash, colour wash, or dry distemper.
    (b) They should not be applied on damp walls.
    (c) They require only less skill for application.
    (d) They require water curing.

13. Distemper is used to coat
    (a) external concrete surfaces
    (b) interior surfaces not exposed to weather
    (c) wood work
    (d) compound walls

14. Acrylic resins are used for interior and exterior applications, because
    (a) they offer better washability
    (b) they have better adhesion
    (c) they have resistance to alkali cleaners
    (d) all of these
    (e) none of these

15. Putty for oil-based paints consists of
    (a) white lead and turpentine
    (b) whiting powder and raw linseed oil
    (c) red lead and linseed oil
    (d) zinc oxide and boiled linseed oil

16. Approximate coverage of two coats of whitewash is
    (a) 2.5–3.0 m$^2$/kg
    (b) 4.0–4.5 m$^2$/kg
    (c) 6.0–7.0 m$^2$/kg
    (d) 7.0–8.0 m$^2$/kg

17. Oil-varnish is a homogeneous solution of resin in
    (a) kerosene oil
    (b) alcohol
    (c) naptha
    (d) linseed oil

18. Which of following is an example of spirit varnish?
    (a) French polish
    (b) Asphalt varnish
    (c) Oil varnish
    (d) Spar varnish

19. In synthetic varnishes, thinner used is
    (a) oil
    (b) spirit
    (c) water
    (d) petroleum-based

20. Coverage of one coat of *Snowcem* is
    (a) 1 to 2 m$^2$/kg
    (b) 2 to 3 m$^2$/kg
    (c) 2 to 4 m$^2$/kg
    (d) 3 to 5 m$^2$/kg

21. Coverage of one coat of water-based enamel semigloss paint is
    (a) 8 to 10 m$^2$/kg
    (b) 10 to 13 m$^2$/kg
    (c) 12 to 15 m$^2$/kg
    (d) 15 to 16 m$^2$/kg

22. Which of the following is not a defect in paints?
    (a) Blistering
    (b) Crawling
    (c) Flaking
    (d) Efflorescence

23. Blistering in paints is
    (a) mild cracking of paint film
    (b) swelling of paint film
    (c) detachment of paint film
    (d) delamination of paint film in layers

## Review Questions

1. Explain the difference between decorative and industrial paint industry.
2. (a) What are the various components of paints? (b) State the functions of each of them.
3. (a) What are the functions of additives and extenders? (b) List any two additives that may act as anti-corrosive pigments. (c) Give two examples of extender pigments. (d) What is the necessity to add driers to paints?
4. How are paints packaged?

5. How are paints classified?
6. Describe briefly the two general types of paints. Why are latex paints often used in residences?
7. What are the disadvantages of alkyd or oil-based paints? In what circumstances are they considered the best choice?
8. How are paints applied?
9. Compare the performance of good quality latex paint and oil-based paint.
10. What are the considerations while choosing paint for a particular project?
11. What are the sheens recognized by IS 1303? Explain each of them briefly.
12. Should cost alone be considered while selecting paint?
13. What is pigment volume concentration and critical pigment volume concentration?
14. (a) What is a primer? (b) What are its functions? (c) What are the three basic types of primers?
15. (a) What is cement paint? (b) What are its ingredients? (c) How is it manufactured?
16. Describe the procedure of applying cement paint. What precautions are necessary while applying cement paints?
17. What are the advantages and disadvantages of cement paints?
18. What are distempers? Describe the application procedure for soft distemper.
19. What is oil-bound distemper? What are its advantages?
20. What are the four classes of plastic emulsion paint? Describe their common characteristics. Why are they popular?
21. What are latex emulsion paints? Why are acrylic resins considered better for interior and exterior applications?
22. Differentiate vinyl emulsion paint from acrylic emulsion. What are the three main types of vinyl emulsions?
23. What are enamel paints? How are they applied?
24. What is the difference between ordinary enamel and synthetic enamel paint?
25. What are exterior paints? Explain their use in buildings.
26. What is the coverage of one coat of whitewash as compared to colour wash?
27. Describe the procedure to be followed while doing the following activities:
    (a) Painting of new wood work     (b) Painting of a new iron work
    (c) Painting of an old iron work     (d) Painting of a plastered surface
    (e) Painting old plastered surfaces
28. What are the objects of varnishing a surface? Where will you prefer varnish over paint?
29. (a) What are the ingredients of varnish? (b) Classify different types of varnish and briefly describe them.
30. Write short notes on (a) acrylic varnish, (b) alkyd varnish, (c) lacquer, (d) shellac, and (c) spar varnish.
31. What are stains? How are they differentiated from varnish? What are the different types of stains?
32. How is varnish applied on wooden surfaces?
33. Give short notes on (a) aluminium paints, (b) anti-corrosive paints, (c) cellulose paints, (d) epoxy paints, (e) intumescent coatings, (f) rubber-based paints, and (g) powder coatings.
34. What is bitumen paint? How does it differ from normal paint? Where is it used?
35. What is meant by the covering capacity of paints?
36. Discuss the reasons for the causes of defects in painting work. Describe briefly the different types of defects in paints, their causes, and remedial measures.
37. What safety precautions should be exercised while painting?

## ANSWERS

**Multiple-choice Questions**

| | | | | |
|---|---|---|---|---|
| **1.** (a) | **2.** (c) | **3.** (d) | **4.** (d) | **5.** (a)-(ii), (b)-(iii), (c)-(iv), (d)-(i) |
| **6.** (a) | **7.** (a) | **8.** (d) | **9.** (a)-(iv), (b)-(iii), (c)-(ii), (d)-(i) | |
| **10.** (a)-(iv), (b)-(iii), (c)-(i), (d)-(ii) | **11.** (d) | | **12.** (b) | **13.** (b)  **14.** (d) |
| **15.** (b) | **16.** (b) | **17.** (d) | **18.** (a) | **19.** (d)  **20.** (c) |
| **21.** (c) | **22.** (d) | **23.** (b) | | |

# CHAPTER 19
# ASPHALT, BITUMEN, AND TAR

## 19.1 Introduction

Asphalt, bitumen, and tar are essentially hydrocarbons, and collectively referred to as *bituminous materials*. It has been discovered that bituminous materials have been used in Iran and Egypt, during 7000 to 6000 B.C., as mortar in the construction of palaces, temples, terraces, for roadway coating, waterproofing containers, and in wooden boats. The first British patent for the use of asphalt/bitumen was in 1834. Asphalt and bitumen are the products derived from petroleum whereas tar is obtained from destructive distillation (heating with a minimal exposure to air) of organic substances such as coal or wood. The terms asphalt and bitumen are often used interchangeably and hence may indicate the same product; however, in the USA, asphalt may contain bitumen as binder and some aggregates. Bitumen has replaced tar in road construction, because of its easy availability in different forms. Tar is used as preservative for wood and as an ingredient in waterproof paints. Different grades are specified for bitumen, and the current IS 73:2013 code specification is based on viscosity grading. Historically, asphalt concrete has been used for a variety of purposes, including bulletproofing British warships in the early 1940s. Bituminous materials are now used extensively for roadway construction, primarily because of their excellent binding characteristics, waterproofing properties, and relatively low cost. In addition, they are used in roofing, insulation, varnishes, and bases for coal tar paints. IS 73 also specifies seven different tests for determining the quality of bitumen (see Section 25.23 of Chapter 25, for the details of some of these tests).

## 19.2 Asphalt

The terms 'asphalt' and 'bitumen' are often used interchangeably to mean both natural and manufactured forms of the substance. Though the term bitumen is used universally, in the USA the term asphalt is used to represent a mixture of bitumen and aggregates. Thus, *asphalt* may be defined as a mixture of bitumen, which is a sticky and viscous solid or semi-solid cementitious material, obtained mostly by refining petroleum. It has a black or brownish black colour and remains solid in normal temperatures and gradually liquefies when heated between 50 and 100°C. The following are the different types of asphalt.

## 19.2.1 Types of Asphalt

Asphalt is classified into two categories: (a) natural asphalt and (b) petroleum or refined asphalt.

**Natural asphalt**   This kind of asphalt is found to occur in nature. Depending on the source of occurrence, it may be divided into two classes: (a) lake asphalt and (b) rock asphalt. *Lake asphalt* is found to occur naturally in lakes such as Pitch Lake in Trinidad, the La Brea Tar Pit in Los Angeles, and the Lake Bermudez in Venezuela, at depths ranging from 3 to 60 m. It may contain about 40 to 70% pure asphalt. *Rock asphalt* is obtained from rocks, such as sandstone or limestone, impregnated naturally with asphalt and found in oil sands of Alberta, Canada, and tar sands of Utah, USA (The Athabasca oil sands are the largest bitumen deposits in Canada.). The chemical composition and consistency of these deposits vary significantly. It may contain about 4–20% of pure bitumen by volume.

**Petroleum or residual asphalt**   Known as artificial asphalt, it is obtained by the fractional distillation of crude petroleum oil. When crude oil is separated in distillation towers at a refinery, the heaviest hydrocarbons with the highest boiling points settle at the bottom and contain asphalt.

## 19.2.2 Uses of Asphalt

Up to 80–85% of asphalt/bitumen is used in road construction, where it is used as the binder and mixed with aggregate particles to create asphalt concrete. Another 10% is used as bituminous waterproofing products such as roofing felts and waterproof papers, as damp-proof course for preventing dampness in walls, for lining walls of tanks and swimming pools, as pipeline coatings, water repellent layer over flat roofs and basements, and in electrical insulation products. The remaining is used in paints, sealants, and joint fillers.

## 19.2.3 Forms of Asphalt-Paving Materials

Some basic terminology applicable to asphalt-paving materials (note that many are nearly synonymous, depending on the specifying agency and may change over time) (AI, 2007), are as follows:

**Asphalt cement**   It is the asphalt derived from crude oil that is used directly in paving. In this form, it is heated to make it flowing and then reverted to a semisolid state as it cools. It is used as the binder in hot-mix asphalt, warm-mix asphalt, open-graded asphalt, stone-mastic asphalt (SMA), and as a tack coat. It is also used to produce asphalt emulsion, polymer-modified asphalt, rubberized asphalt, and asphalt cutback.

**Asphalt concrete**   Commonly called asphalt, tarmac, or black top, *asphalt concrete* is a composite material used to surface roads, parking lots, airports, as well as the core of embankment dams. It is a mixture of aggregate materials bound together with asphalt, laid in layers, and compacted. In 1901, Edgar Purnell Hooley patented a material called 'tarmacadam', which became the forerunner to modern asphalt concrete. The shortened name of Hooley's material, 'tarmac', is often used to refer to asphalt concrete even though modern asphalt concrete does not contain tar.

**Asphalt cutback**   It is made when asphalt cement is dissolved in a petroleum-based solvent. Solvents include gasoline or naphtha (rapid-curing cutback), kerosene (medium-curing cutback), or low-volatility oils (slow-curing cutback). In the dissolved state, this asphalt is less viscous and the mix is easy to work and compact. After the mix is applied as a coating, the lighter solvent evaporates. Owing to concerns of pollution from the emission of volatile organic compounds, cut-back asphalt has been largely replaced by asphalt emulsion. Asphalt cutback can be mixed with aggregate at an asphalt mixing plant to create cold-mix asphalt.

**Asphalt emulsion**   It is made by shearing asphalt cement into microscopic droplets (0.5 to 10 microns), which are mixed with water (typically, in ratio of asphalt: water between 40:60 and 60:40) and an emulsifying agent such as soap (very small percentages) that keeps the drops in suspension in the water. The asphalt reverts to the semisolid state when the emulsifying agent is neutralized or 'breaks', allowing the particles to join together, which is followed by the evaporation of water. Asphalt emulsions, polymer-modified asphalt, and rubberized asphalt emulsions are used extensively for the surface treatment of pavements. Emulsions are also used as foundation coatings, sound absorbers, roof coatings, insulation coatings, masonry coatings, in black curing compound (for use on fresh concrete that will be subsequently coated with bituminous mastic compounds), as well as to seal patching work on asphalt. Asphalt emulsion can be mixed with aggregate at an asphalt mixing plant to create cold-mix asphalt. As asphalt emulsion is water based, fire hazard for the applicator is eliminated. In general, asphalt emulsions will dry faster than the solvent-applied coating of comparable thickness.

**Tack coat**    It is an asphalt cement, asphalt emulsion, or asphalt cutback sprayed onto a paved surface to assist in bonding asphalt concrete layers together during construction.

**Mastic asphalt/bitumen**    It is a dense mixture consisting of coarse aggregate, and/or sand, and/or limestone fine aggregate, and/or filler and bitumen, which may contain additives (for example, polymers, waxes). It is produced by heating asphalt in a mixer for several hours before adding the aggregate (see IS 1195:2002 and IS 1196:1978, for the specifications). The aggregates are added after the bitumen has become a viscous liquid, and the mixture is heated for another 6 to 8 h. It is then transported to the job site for use. The mixture is usually designed to be of low-void content. It requires no compaction at site. Mastic asphalt is used in Europe since 1890. It is used in road constructions [with a thickness of 25 mm (normal traffic) to 40 mm (heavy traffic)]. Mastic asphalt is placed manually or mechanically, with specially designed spreading machines, at the desired thickness. On the other hand, *asphalt/bitumen mastic* (AM) is a term used in Europe to describe a mix of sand (that is, without aggregates > 2 mm), and/or limestone fine aggregate, and/or filler and bitumen that is used specially for water/damp proofing in a variety of applications (see also IS 3037:1986, IS 4365:1967, IS 5871:1987, IS 7198:1974).

**Stone mastic asphalt**    Known as *stone-matrix asphalt* as well, it was developed in Germany in the 1960s. SMA has a high coarse aggregate content that interlocks to form a stone skeleton that resists permanent deformation. The bitumen content in mastic mixtures is typically double that found in concrete asphalt, which helps to bind the particles more closely together to keep water out. Typical SMA composition consists of 70–80% coarse aggregate, 8–12% filler, 6–7% binder, and 0.3% fibre. It provides a deformation-resistant, durable surfacing material, suitable for heavily trafficked roads.

**Crumb rubber modifier**    It is created by grinding recycled tire rubber after stripping out steel reinforcement in the tires. The CRM can be mixed with asphalt cement, natural rubber, and other ingredients to produce rubberized asphalt (As per the American Society for Testing and Materials—ASTM—rubberized asphalt should have minimum 15% recycled rubber by mass).

**Modified asphalt**    Different additives may be added to modify asphalt depending on mix specifications. Some of the most common additives are ethyl vinyl acetate (EVA), a polymer which makes the mix more workable, especially in colder conditions, Sasobit, a wax which allows the asphalt to be mixed at lower temperatures, and styrene-butadiene-styrene (SBS), a thermoplastic rubber which offers tensile strength and strain recovery.

## 19.2.4 Mixture Formulations

Asphalt may be mixed with aggregates in one of the following ways:

**Hot-mix asphalt concrete**    It is produced by heating the asphalt binder to about 135 to 163°C to decrease its viscosity, and drying the aggregate, prior to mixing, to remove moisture from it. Mixing is generally performed with the aggregate at about 150°C for virgin asphalt and 166°C for polymer-modified asphalt, and the asphalt is applied to the roadway at about 110°C. Paving and compaction is performed, when the asphalt is sufficiently hot. In many countries, paving is restricted to summer months because in winter the compacted base will cool the asphalt in such a way that it cannot be packed to the required density. The HMAC is commonly used on high traffic pavements such as those on major highways and airfields. It is also used as an environmental liner for landfills and reservoirs.

**Warm mix asphalt concrete** The development of *WMA concrete* was initiated in Europe in the late 1990s primarily to reduce greenhouse gas emissions under the Kyoto Protocol. It represents a broad range of technologies used with asphalt concrete that allow the mixture to stay workable and compactable at lower temperatures (104 to 135°C) than typical HMA (135 to 163°C). Technologies such as Aspha-min, WAM Foam, and Sasobit were developed in Europe in the 1990s. Now, there are about 20 technologies to produce WMA, including several technologies developed in the USA such as Evotherm, Rediset WMX, REVIX, Thiopave, Aquablack, AquaFoam, Ultrafoam GX, LEA (Low Emission Asphalt), and Double Barrel Green.

WMA technologies can be classified broadly as follows:

1. Those that use water which causes the hot asphalt binder to foam (e.g., zeolite, WAM Foam, Low Energy Asphalt, and Double Barrel Green)
2. Those that use some type of organic additive or wax (e.g., Sasobit)
3. Those that use chemical additives or surfactants (e.g., Evotherm, Rediset WMX, and REVIX)

WMA has a wide range of production temperatures ranging from slightly over 100°C to about 20 to 30°C below typical HMA temperatures.

Water expands in volume by a factor of 1673, when turns into steam at atmospheric pressure; this fact is utilized in the technologies that suggest to add small amounts of water to hot asphalt binder (Kandhal, 2010). The use of these technologies increases the volume of asphalt binder many folds, which help to coat the aggregate easily and also lowers the apparent viscosity of the mix. Water can be introduced into the asphalt mix by using foaming nozzles, hydrophilic material such as zeolite or damp aggregates. However, asphalt binder temperature is typically kept the same as that used for hot-mix asphalt. The technologies where organic additives or waxes are used, lower the viscosity of asphalt binder above their melting point. Their melting point should be more than the in-service pavement temperature in hot summer, in order to avoid problems such as permanent deformation or rutting. The technologies where chemical additive and/ or surfactants are used, a variety of different mechanisms occur to coat the aggregate and allow compaction at lower temperatures (Kandhal, 2010). Details of these technologies may be found in Kandhal (2010).

WMA can be used to reduce the mixing temperature and facilitates paving in cooler weather, and also allows longer transportation distances (Kandhal, 2011). Utilization of WMA technology can reduce compaction temperatures by approximately 14 to 25°C (FHWA, 2016). The amount of reduction depends on the WMA technology used and the characteristics of the mix, plant, climate, lift thicknesses, and hauling distance. Because less energy is needed to heat the asphalt mix, less fuel is needed to produce WMA. Fuel consumption during WMA manufacturing is typically reduced by 20%, thus resulting in releasing less carbon dioxide, aerosols, and vapours. Because WMA makes compaction easier, cost savings are achieved by reducing time and labour spent to compact the mix. The working conditions are also improved. In addition, the lower laying-temperature allows the finished surface to be used rapidly, which is important for construction sites with critical time schedules. WMA is also versatile and has been used successfully in a range of pavement thicknesses. WMA has been used in all types of asphalt concrete: dense-graded, stone matrix, porous, and mastic asphalt. It was found to be durable enough to withstand high traffic demands. Figure 19.1(a) shows a roadway using WMA.

(a)                                                    (b)

**Fig. 19.1** Asphalt pavements (a) Road made with warm mix asphalt
(b) Sealing cracks with flexible rubberized asphalt

**Cold mix asphalt concrete**   It is used as a storable patching material. Most often, it uses cutback asphalt and/or asphalt emulsion mixed with aggregate and/or recycled asphalt pavement (RAP). In its emulsified state, this asphalt is less viscous and the mixture is easy to work and compact at temperatures of 15.5°C.

## Case Study   Porous asphalt

Developed in the mid-1970s, porous asphalt is an environmentally friendly tool for storm-water management. The use of porous asphalt conserves water, reduces runoff, and promotes infiltration resulting in cleansing of storm-water, replenishing of aquifers, and protecting streams. A typical porous pavement has a distinctive 'open-graded' surface layer over an underlying 'open-graded' stone base (see Fig. 19.2). When it rains, the water drains through the 65–100 mm thick porous asphalt into the 200–300 mm thick stone base, then, slowly, infiltrates into the soil. Many contaminants are removed as the storm-water passes through the porous asphalt, stone recharge base, and soils through filtration and microbial action. Compared to impervious asphalt, porous asphalt drains better, and has better traction and visibility in wet conditions, and produces less glare and vehicular noise. It has be noted that porous asphalt pavements should never be seal-coated or crack sealed. More details about porous asphalt pavements may be obtained from FHWA-HIF-15-009 and www.porouspavement.org.

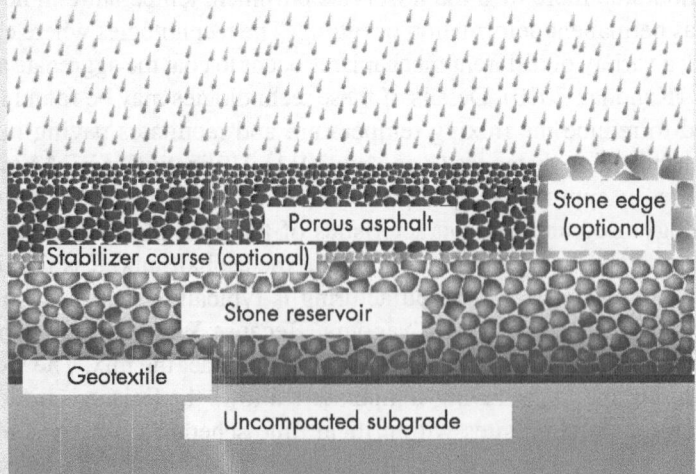

**Fig. 19.2** Cross-section of typical porous asphalt pavement
(*Source*: FHWA-HIF-15-009)

## 19.3 Bitumen

It is manufactured from crude oil. In petroleum refineries, components such as LPG, naphtha, kerosene, and diesel are separated through the process of fractional distillation. The heaviest material obtained from this process is bitumen, which is further treated and blended to make different grades of paving grade bitumen. Of the multitude of crude oils commercially available, only a limited number are considered suitable for producing bitumen/asphalt binder of the required quality in commercial quantities (FHWA, 2016). Bitumen is also found as a natural deposit or as a component of naturally occurring asphalt. Bitumen is defined as per IS 334:2002 as a black or dark brown non-crystalline solid or viscous material, derived from petroleum either by natural or refinery processes and substantially soluble in carbon disulphide and trichloroethylene. It is highly waterproof, resilient, and ductile, has adhesive properties, is chemically inert and resistant to atmospheric exposure and the effects of dilute acids and alkalis. Because of these properties, it is used in a variety of applications from road making to joint sealing compounds.

For engineering and construction purposes, the three physical properties of bitumen, namely, consistency, purity, and safety are important. Consistency is the term used to measure its degree of stiffness, i.e., its ability to flow. Bitumen is basically a thermoplastic material; it becomes a liquid when heated and solidifies when cooled. Hence, its stiffness or its ability to flow is temperature sensitive. Consistency of bitumen can be judged by some empirical tests measuring properties such as penetration, softening point, and ductility and also by testing the fundamental property of bitumen such as viscosity (*Viscosity is defined as inverse of fluidity. Viscosity thus defines the fluid property of bituminous material*). As per recent specifications, bitumen is graded as per its consistency (viscosity/stiffness).

Pure bitumen is completely soluble in solvents such as carbon disulphide and carbon tetrachloride. Hence, any impurity in bitumen in the form of inert minerals or carbon could be quantitatively analysed by dissolving the samples of bitumen in any of the aforementioned solvents. Bitumen, when heated, gives out volatile gases at certain temperatures depending upon their grade. These volatile gases catch fire causing a flash. The definition of flash and fire points as per Indian standards are as follows: the *flash point* of a material is the lowest temperature at which the vapour of a substance momentarily takes fire in the form of a flash under specified condition of test. The *fire point* is the lowest temperature at which the material gets ignited and burns under specified conditions of test.

### 19.3.1 Types of Bitumen

The various types of bitumen are available, based on the type of application, and are briefly discussed as follows (see also IS 334:2002):

**Straight-run bitumen**    It is the bitumen obtained as the end product or residue from the refining of crude petroleum under direct distillation.

**Steam refined bitumen**    It is the residue from the distillation of crude petroleum processed further with the injection of steam to a specified viscosity or penetration.

**Hot-bitumen**    The viscosity of the straight-run bitumen has to be reduced to make it workable in any application. In hot-bitumen, the viscosity is reduced by heating it to 135 to 163°C.

**Industrial bitumen**    Industrial grade bitumen is also known as *blown bitumen* (R-grade). This is obtained by blowing air under pressure into hot liquid bitumen at high temperatures (more than 180°C), which drives out volatile components and results in the modification of physical and rheological properties (this process is sometimes incorrectly referred to as oxidation). The effectiveness of air-blowing process may be enhanced by the use of catalysts, which further enhance the properties of bitumen. Blown bitumen has rubbery consistency and has higher softening point and very low penetration number.

Six grades of industrial bitumen are specified in IS 702:1988, as 85/40, 85/25, 90/15, 115/15, 135/10, and 155/6. The two numbers given in the grade designation denote the softening point and penetration point. For example, grade 85/40 means that this grade of bitumen has a softening point of 85°C and a penetration of 40 (measured in one tenths of 1 mm). Industrial grade bitumen is used for the manufacture and fixing of roofing and damp-proofing felts, plastic bitumen for leak stops, waterproof packing paper, pipeline coating, joint filler boards and joint sealing compounds, bituminous filling compounds for cable boxes, and for sealing batteries. It is also used as an insulation material for buildings, refrigeration, and cold storage equipment, in acid/alkali resistant flooring, and in the preparation of bitumen mastic.

**Cutback bitumen**  It is the bitumen dissolved in volatile solvent, usually a petroleum distillate, such as petrol, kerosene, or fuel oil to reduce the viscosity. The solvent evaporates after the application of material and the bitumen will bind the aggregate. Cutback bitumen is used for cold weather bituminous road construction and maintenance. As per IS 217:1988, four grades of rapid-curing type, which are blended with a naphtha type distillate (RC 70, RC 250, RC 800, RC 3000), five grades of medium-curing type, which are blended with a kerosene-type distillate (MC 30, MC 70, MC 250, MC 800, MC 3000), and four grades of slow-curing type, which are blended with high boiling oils or containing a higher viscous oil than in medium or rapid curing cutback (SC 70, SC 250, SC 800, SC 3000) are specified. In all these grades, the numbers after RC/MC/SC indicate kinematic viscosity at 60°C. Rapid-curing type should be used with aggregates containing practically no fine aggregate passing through 2.36-mm sieve. Medium-curing type should be used with aggregates containing less than 20% of fine aggregate passing through 2.36-mm sieve. Slow-curing type should be used with aggregates containing more than 20% of fine aggregate passing through 2.36-mm sieve. Other requirements of these grades and types are provided in IS 217:1988. RC grade is recommended for surface dressing and patchwork, MC grade for premix with less quantity of fine aggregates, and SC grade is used for premix with appreciable quantity of fine aggregates.

**Bitumen emulsion**  It is a liquid product in which a substantial amount of bitumen is dispersed in the form of finely divided droplets, in an aqueous medium containing an emulsifier and a stabilizer. In India, cationic type emulsions, in which the cation of the emulsifier is at the interface with the bitumen particles which are positively charged and the aqueous phase is acidic, are normally used (IS 8887:2004). Whereas, *anionic emulsion* is one in which the anion of the emulsifier is at the interface with the bitumen particles which are negatively charged and the aqueous phase is alkaline (IS 3117:2004). The bitumen content in the emulsion is around 60% and the remaining is water. When the emulsion is applied on the road, it breaks down resulting in the release of water and the mix starts to set. The time of setting depends upon the grade of bitumen. As per IS 8887:2004, three types of bituminous emulsions are available, which are Rapid setting (grades RS-1 and RS-2), Medium setting (grade MS), and Slow setting (grades SS-1 and SS-2). The physical and chemical requirements of the five grades of emulsions should satisfy the requirements of IS 8887. *Rapid setting emulsions* are used for surface treatment, penetration macadam, and seal coating. *Medium setting emulsions* are used for plant or road mixes with fine aggregates between 5 and 20% retained on 2.36-mm sieve, and used in open graded premix work, patch repair work, and bituminous macadam. *Slow setting emulsions* are used for plant or road mixes with graded fine aggregates greater than 20%, passing through 2.36-mm sieve and a portion passing through 75-μm sieve, and used in slurry seals, tack coats, and soil stabilization. Guidance for the selection of emulsion type is given in Table 19.1. Usually, the type of aggregate and climate dictates the type of bitumen emulsion to be used. The various bitumen emulsions are ideal binders for hill road construction, where the heating of bitumen or aggregates is difficult. Slow setting emulsions are preferred in rainy season. The international status for emulsion usage is about 10% of total bitumen usage. Out of the present emulsion consumption, nearly 60–65% of emulsion is used in tack coat applications. These emulsions are always stored in airtight drums.

**Table 19.1** Guidance for selection of emulsion type

| Application | Rapid setting | Medium setting | Slow setting |
|---|---|---|---|
| Penetration Macadam (MS) | – | √ | – |
| Prime Coat (SS-1) | – | √ | √ |
| Tack Coat(RS-1) | √ | – | – |
| Slurry Seal (SS-1 and SS-2) | – | √ | √ |
| Fog Seal (SS-1) | – | – | √ |
| Soil Stabilization (SS-1 and SS-2) | – | √ | √ |
| Crack Filler (SS-2) | – | – | √ |
| Cold Mix (MS and SS-1) | – | √ | √ |

It has to be noted that bitumen is always applied hot, cutbacks are applied either hot or cold, and emulsions are always applied cold.

**Macadam bitumen**   It is a road surface made of compacted layers of open graded mixture of high quality even-sized aggregate bound together with the designed proportion of bitumen hot-mixed and hot-laid and rolled. It is named after Scottish engineer John Loudon McAdam (1756–1836), who invented this type of construction.

**Modified bitumen**   The properties of straight run bitumen may be improved (typically, with enhanced high temperature performance characteristics) by the addition certain polymers, such as SBS, EVA, or polyethylene (PE), or by a blend of additives called *bitumen modifiers*. When polymers are added, it is called *polymer-modified bitumen* (PMB). *Crumb rubber modified bitumen* (CRMB) is produced by grinding bitumen with recycled tire rubber, after stripping out steel reinforcement. CRM can be mixed with asphalt cement, natural rubber, and other ingredients to produce rubberized bitumen/asphalt (ASTM specifies that *rubberized asphalt* should have a minimum 15% of recycled rubber by mass). Rubberized asphalt is used in different types of asphalt-aggregate mixtures for structural and surface layers, and for chip seals. CRM is also used with polymers to produce blended rubberized asphalt, but with no required minimum CRM content and more finely ground particles (FHWA, 2016). Use of rubberized asphalt as a pavement material was pioneered by the city of Phoenix, Arizona, in the 1960s due to its high durability. Greater interest was then evinced in rubberized asphalt because of its ability to reduce road noise. Approximately 1500 tires are used for every lane-mile of rubberized paving.

Specifications for modified bitumen are provided in IRC SP 53:2010 and IS 15462:2004. Modified bitumen is transported and applied hot. IS 15462:2004 classifies modified bitumen into the following four types: (a) PMB(P)-Plastomeric thermoplastic based, (b) PMB(E)-Elastomeric thermoplastic based, (c) NRMB- Natural rubber and SBR latex based, and (d) CRMB-Crumb rubber/treated crumb rubber based. These are further divided into three grades based on the penetration value/softening point value. It must be noted that the performance of PMB and CRMB is dependent on strict control on temperature during construction. The advantages of using modified bitumen are as follows (IRC SP 53:2010):

1. Lower susceptibility to daily and seasonal temperature variations
2. Higher resistance to deformation at high pavement temperature

3. Better age resistance properties
4. Higher fatigue life for mixes
5. Better adhesion between aggregates and binder
6. Delay of cracking and reflective cracking
7. Overall improved performance

### 19.3.2 Specifications of Paving Bitumen

Bituminous mixes prepared with binders having high wax content have tendencies to become brittle in cold weather and bleed in hot weather. Hence, in the second revision of IS 73:1992, paving bitumen for use in roadways or runways is classified as follows:

1. Type 1 Paving bitumen from non-waxy crude
2. Type 2 Paving bitumen from waxy crude

As per this specification, there were two types of bitumen: Type 1 (assigned with the letter 'S' with six different grades) and Type 2 (assigned with the letter 'A' with four different grades), respectively. These grades were designated as per their empirical penetration limits. Thus, as per IS 73:1992, grade A35 means paving bitumen having high wax content and a penetration value in the range of 30 to 40. Similarly, S35 means paving bitumen having low wax content and a penetration value in the range of 30 to 40. Thirteen performance tests have also been prescribed.

However, from the third revision of IS 73:2006, the grading of bitumen was changed from penetration grade to *viscosity grade*, at 60 and 135°C, with four grades designated with the letters 'VG', as shown in Table 19.2 (the number after VG is the average value of absolute viscosity at 60°C divided by 100). This specification of bitumen is based on viscosity at 60°C, instead of the specification based on penetration at 25°C, accounts for variability in performance at high temperatures. This specification with another requirement of minimum viscosity at 135°C, minimized the mix problems in the field. In addition, the number of total tests were also reduced to seven (see Table 19.2), thus reducing the cost of testing paving bitumen. Furthermore, empirical tests/parameters such as, penetration ratio, paraffin wax content, and Fraass breaking point were also eliminated, without compromising the quality of bitumen.

In the fourth revision of IS 73 released in 2013, viscosity measurement at 60°C is given more importance. For each grade of bitumen, a range of viscosity values at 60°C and the minimum value of penetration at the annual average pavement temperature of 25°C are specified (see Table 19.2). Such a revision rationalized the binder selection process and now the choice of the grade depends upon the design maximum air temperature of the location where the binder is used (see Table 19.3). Ductility test is no longer mandatory. The paving bitumen binder should be homogenous and should not foam when heated to 175°C. The higher the grade, the stiffer will be the bitumen. It has to be noted that in the viscosity grading, viscosity tests are specified at 60 and 135°C, which represent the temperature of road surface during summer, and mixing temperature, respectively.

**Table 19.2** Requirements for paving bitumen as per IS 73:2013

| S. no. | Characteristics | Requirements for grades | | | | To be tested as per |
|--------|-----------------|-------|-------|-------|-------|----------------------|
| | | **VG 10** | **VG 20** | **VG 30** | **VG 40** | |
| 1 | Absolute viscosity at 60°C, Poises | 800–1200 | 1600–2400 | 2400–3600 | 3200–4800 | IS 1206 (Part 2) |
| 2 | Kinematic viscosity at 135°C, cSt, Minimum | 250 | 300 | 350 | 400 | IS 1206 (Part 3) |

*(Contd)*

**Table 19.2** *(Contd)*

| S. no. | Characteristics | Requirements for grades | | | | To be tested as per |
|--------|-----------------|------|------|------|------|------------------|
| | | **VG 10** | **VG 20** | **VG 30** | **VG 40** | |
| 3 | Penetration at 25°C, 100 g, 5 Sec., 0.1 mm, minimum | 80 | 60 | 45 | 35 | IS 1203 |
| 4 | Softening point (Ring & Ball test)°C, minimum | 40 | 45 | 47 | 50 | IS 1205 |
| 5 | Flash point (Cleveland open cup test), °C, Minimum | 220 | 220 | 220 | 220 | IS 1448 (Part 69) |
| 6 | Solubility in trichloroethylene, percent, Min | 99.0 | 99.0 | 99.0 | 99.0 | IS 1216 |
| 7 | Tests on residue from rolling thin film oven test: | | | | | |
| | (a) Viscosity ratio at 60°C, Maximum | 4.0 | 4.0 | 4.0 | 4.0 | IS 1206 (Part 2) |
| | (b) Ductility at 25°C, cm, Minimum | 75 | 50 | 40 | 25 | IS 1208 |

**Note:** 1 cSt = $1 \times 10^{(-6)}$ m²/s

**Table 19.3** Suitability of different grades for the maximum air temperature, as per IS 73:2013

| Grade | Suitable for 7-day average maximum air temperature °C |
|-------|------------------------------------------------------|
| VG 10 | <30 |
| VG 20 | 30–38 |
| VG 30 | 38–45 |
| VG 40 | >45 |

*Note*: This is the 7-day average maximum air temperature for a period not less than 5 years from the start of the design period.

Research by Shell International Petroleum Company has resulted in multi-grade bitumen, which is purely crude-based and without any additive/chemical (see IS 15808:2008, for details). This multi-grade bitumen has been found to perform better over a wider temperature range compared to the conventional bitumen, as specified in IS 73:2013.

## 19.3.3 Tar Paper and Bitumen Felts

**Tar paper**   It is a heavy-duty paper (Virgin Kraft paper), impregnated with tar or glass fibre mat with tar (since they are not made of paper, they are known as *roofing felt*) producing a waterproof material useful for roof construction, and produced as per ASTM D227 or ASTM D4990. Tar paper is different from bitumen felt, which is essentially impregnated with bitumen instead of tar, but these two products serve the same purpose and used in the same way. Tar paper has been in use for centuries. Originally, felt was made from recycled rags but today felts are made of recycled paper products (typically cardboard) and sawdust. Nowadays, tar paper is far less common than bitumen felt and is used for waterproofing roofs to prevent ingress of moisture. Since tar paper is not wind or sun resistant, it is placed between sheathings (the supporting wood, plywood, slats) and the outer roof covering (shingle, gravel, etc.). It is sold in rolls of various widths, lengths, and thicknesses – 0.91-m wide rolls, 15 or 30-m long, and 7 and 14 kg weights are common in the US – often marked with chalk lines at certain intervals to help laying it on roofs with proper overlap (more overlap for flatter roofs). It can be installed in several ways, such as staples or roofing nails, but it is also sometimes applied in several layers with hot asphalt, cold asphalt (adhesive), or non-asphaltic adhesives.

**Bitumen felts** Felt is a general term used to describe roll-roofing materials, consisting of mat of organic or inorganic fibres soaked with bitumen to create a waterproof product (see also IS 1346:1991). Fiberglass-based felts are long-lasting and not prone to tearing; they also hold up well against extreme weather (IS 7193:2013). Recently, polyester fibres have been developed as the base fleece material for roofing felt. Polyester is also highly resistant to tearing and able to withstand extreme weather.

As per IS 1322:1993, the following three types of bitumen felts are specified:

1. Type 1- Saturated felt for underlay with fibre base
2. Type 2- Self-finished felt with fibre base (for waterproofing)
3. Type 3 - Grade 1- Self-finished felt with hessian base (for waterproofing) and Grade 2- Self-finished felt with hessian base (for damp proofing)

The base fabric for fibre-base felts should consist of a suitable blend of vegetable and/or animal fibres. The weight of ash on incineration of the fabric shall not exceed 10% of its original weight. The fabric for the hessian base felts should conform to Type H hessian conforming to IS 1218(Part 2):1971. In the manufacture of self-finished felts, the saturated felt is treated by a uniform bituminous coating. The resultant coated felts are given a superficial application of fine mineral powder, such as mica, talc, or slate, with the power material passing through 600-micron IS Sieve. The bitumen in all types of felts should conform to IS 73:2013, with a penetration of not less than 80 at 25°C. The weight of the bitumen should be not less than 110% of the weight of the untreated felt for fibre-base felts and not less than 60% of the weight of the untreated fabric for hessian base felts subject to the minimum weight given in IS 1322:1993. The bitumen felts are supplied in width of 900 or 1000 mm and in lengths of 10 or 20 m. Bitumen felts Types 1 and 2 should be subjected to tests such as breaking strength test, and pliability test and Type 3 felt to additional tests such as storage sticking test, pressure head test, heat resistance test, and water absorption test as per IS 13826 (Parts 1 to 7):1993. For other specifications of bitumen felts, reference should be made to IS 1322:1933. Damp-proofing treatment using bitumen felt is covered in IS 1609:1991.

Owing to the advancements in technology, we now have thermoplastic polyolefin (TPO) membrane as the underlayment, which is used in most of the modern roofs. It is used like roofing felt, but is lighter, more resistance to punctures, and even stronger than polyester roofing felt. It is important to note that heavier paper is not always better. With a proper roofing job, the felt will be completely protected by the shingles.

## 19.3.4 Commercial Roofing Systems

Bitumen is still the best-known waterproofing material. However, it has poor aging characteristics, becomes brittle in cold weather, is fluid in hot weather, and has little resistance to fatigue. Elastomeric compounds such as APP (Atactic-Polypropylene) and SBS are added with bitumen to substantially improve these properties. Commercial roof systems and materials generally are divided into generic classifications: low slope and steep slope. Low-slope roofing includes water impermeable, or weatherproof, types of roof membranes installed on slopes less than or equal to 3:12 (14°). Steep-slope roofing includes water shedding types of roof coverings installed on slopes exceeding 3:12 (14°).

Most low-slope roof membranes have the following three principal components:

*Weatherproofing layer or layers* The weatherproofing component is the most important element because it keeps water from entering a roof assembly.

*Reinforcement* It adds strength, puncture resistance, and dimensional stability to a membrane.

*Surfacing* It is the component that protects the weatherproofing and reinforcement from sunlight and weather. Some surfacing provide other benefits such as increased fire resistance, improved traffic and hail resistance, and increased solar reflectivity.

In some roof membranes, a component may perform more than one function.

There are the following four basic types of low-slope roof membranes or systems:

1. Built-up roof (BUR) membranes
2. Modified bitumen system
3. Single-ply membranes
4. Spray polyurethane foam-based (SPF) roof systems

These low-slope membrane systems are discussed briefly here. Note that the aforementioned third and fourth systems are not asphalt based. However, they are briefly discussed here for completeness.

## Built-up Roof System

These are also known as tar and gravel-roof systems. These are reliable and have been used in the USA and other countries for more than 100 years. A finished product of this membrane typically consists of alternating layers of bitumen (applied hot by mopping) and reinforcing fabrics (usually, three to five layers are used). These reinforcing fabrics are also referred to as *roofing felts* or *ply sheets*. Recommended application temperature and maximum heating temperature for mopping grade roofing bitumen ranges from 166 to 229°C and from 246 to 274°C, depending on the type of bitumen used. The lower surfacing sheet is a heavy sheet of felt, generally used as first ply to protect the roof elements. The upper surface layer is the protecting surface with a gravel or slag embedded in a heavy top coating of hot bitumen, to provide UV protection; it can also be a granule-surfaced cap sheet, or having weather-resistant coating. The bitumen used is often asphalt, coal tar, or a cold-applied adhesive. Roofing felts are reinforced with either inorganic mats or organic mats, such as rag fibre (hessian) or cellulose fibres (wood or jute). It has to be noted that organic felt papers have become obsolete. Inorganic base products are polyester, fiberglass, and historically, asbestos mat (which is banned in several countries). Fibreglass bitumen felt is discussed in IS 7193:2013 (see also IS 9918:1981). Polyester mat is weaker and less chemically stable than fiberglass but used because it is cheap. Inorganic felts are lighter, more tear-resistant, more fire-resistant, and do not absorb water. Another type of felt paper is perforated for use in built-up roofing and is not for use as a water-resistant underlay. Felts are produced in a standard width of about one meter and length of 10 to 45 m. Built-up roofing systems exhibit exceptional resistance to the conduction of heat between the exterior and interior of a building, resulting in noticeable reductions in heating and cooling costs. They can be coated with UV reflective material (typically, with an aluminium coating) to have a highly reflective *cool roof*. BUR systems have now been replaced by modified bitumen and single ply membrane systems.

## Modified Bitumen System

These are bitumen roofs similar to built-up roofing, but the bitumen is chemically modified to give it a rubber-like or plastic consistency and durability (IS 15462: 2004). The most common chemical agents used to modify the bitumen are atactic polypropylene (APP) and SBS. Use of APP typically results in plastic-like quality to bitumen and SBS in a rubber-like quality. Addition of 30% of APP to bitumen can make it stretch up to 50% of its original length and that with 10–15% of SBS, it can stretch up to six times its original length (see also Table 19.4, which compares the properties of APP and SBS modified bitumen). In addition, bitumen with SBS will return to its original size when allowed to relax. APP-modified bitumen systems provide enhanced tensile strength and resistance to foot traffic. SBS membranes have improved elongation and provide increased resistance to brittleness at cold temperatures and fatigue resistance. APP membranes are provided with a polyester mat (weighing 0.16 kg/m²), whereas SBS membranes with polyester or fiberglass mats (weighing 0.09 kg/m²). Both APP and SBS membranes are available in an assortment of colours. MB roof materials are available in rolls and have a wide variety of surfacing options (such as granular, aluminium, copper, aggregate, etc.). Specifications for SBS membrane with polyester reinforcement and APP membranes with glass or polyester reinforcement are provided in 16525:2017, IS 16526:2017, IS 16532:2017, and IS 16570:2017.

There are several ways of connecting these membranes. APP membranes are applied using a propane-fired torch or specially designed hot-air welders. The heat is applied only as needed to soften the bitumen in order to bond it with the substrate. Since, it may involve fire hazard, it is banned in several places. SBS membranes are applied with mopped hot asphalt or by using gas flame torch. The hot-mopped application is similar to that used in conventional built-up roofs. In a heat application process, the seams are heated to melt the asphalt together and create a seal. Recently, cold-applied adhesives and self-adhesive membranes have been introduced. An MB roof system can be a one-ply, two-ply, or three-ply system; in 2–3 ply applications, the thickness ranges from 220 to 300 mils. These roofing systems contain materials that resist expansion and contraction, and reflect much of the UV rays. As these roofing systems are attached effectively to the top of a building, the need for providing gravel is eliminated. In addition, it is easy to locate any breaks or leaks and can be patched relatively easily.

**Table 19.4** A comparison of physical properties of APP and SBS systems

| Property | APP | SBS |
|---|---|---|
| Ultimate elongation, % | 300 | 2200 |
| Tensile strength at break, MPa | 0.09 | 0.78 |
| Softening point, °C | 149 | 116 |
| Cold flexibility, °C | –5.0 | –35 |
| Fatigue failure, 1000 cycles | 700 | 10,000 |
| Permanent set, % | >300 | <10 |
| Application | T | H, T, C |
| H-Hot bitumen, T-Torch application, C-Cold application | | |

## Single-ply Roof Membranes

These membranes are factory-made and classified into two groups: thermoplastic materials (which include poly vinyl chloride [PVC] and thermoplastic polyolefin [TPO] materials) and thermoset (synthetic rubber) materials (which include Ethylene Propylene Diene Monomer [EPDM], CSPE, CR, and ECR). There are five subcategories of thermoplastics, but the most common are PVC and TPO materials. Both are highly reflective roof membranes that provide excellent weathering and are highly resistant to UV light, tears, punctures, and most chemicals. Thermoplastic membranes include a reinforcement layer that provides more strength and stability. Three application methods (mechanically attached, fully adhered, and point affixed) are available for thermoplastic systems. Single ply membranes are applied in one layer of 40–80 mils thickness.

The EPDM is a single-ply rubber roofing membrane that is resistant to a high degree of ozone, UV, weathering, and abrasion damage. Because of its tolerance to a wide array of weather conditions including wind and hail, the EPDM is considered the most sustainable and environmental friendly material. EDPM sheets are bonded together at the seams to form one continuous membrane. The finished roof's thickness is usually between 30 and 60 mils.

### Spray Polyurethane Foam Roofing

This material is foam-based and features a two-part mixture, which is applied as the system's base layer. Thereafter, a surfacing layer is applied on top, which can be a membrane but often a roof coating is used. The base layer of spray polyurethane foam insulation is applied at different levels of thickness for greater insulation value, or R-value. The surfacing layer serves as the protection against damaging agents such as harmful ultraviolet rays or powerful storms.

### Asphalt Shingles

Asphalt shingles, as shown in Fig. 19.3, are used in roof coverings installed on slopes exceeding 3:12 (14°). Two types of base materials are used to make asphalt shingles: organic and fiberglass. Organic shingles are made with an organic material such as paper, cellulose, or wood fibre saturated with asphalt to make it waterproof. A top coating of adhesive asphalt is then applied over it and ceramic granules are embedded. Organic shingles contain about 40% more asphalt than similar fiberglass shingles.

**Fig. 19.3** Asphalt Shingle roofing (a) Typical shingle (b) Roof with asphalt shingles (*Photo*: Dr N. Subramanian)

Some organic shingles, produced before the early 1980s, contained asbestos.

Fiberglass shingles have a base layer of glass fibre-reinforcing mat coated with asphalt containing mineral fillers, thus making the shingle waterproof. Fiberglass shingles have better fire rating, and are more flexible and stronger than organic shingles. They are also cheaper and easier to construct. Both types of shingles are made in a similar manner with the top surface covered with ceramic granules and the back side treated with sand, talc, or mica to prevent the shingles from sticking to each other before use. The top surface protects the shingles from the deteriorating effect of UV rays of the Sun, and could be provided with different colours. Some shingles have copper or other materials added to the surface to help prevent algae growth. Self-sealing strips prevent the shingles from being blown off by high winds.

Some manufacturers use a fabric backing known as a 'scrim' on the back side of shingles to make them more impact resistant. The shapes and textures of asphalt shingles include: three tab, jet, signature cut, and Art-Loc. Special locking asphalt shingles are designed to lock together and called tie lock or T lock. Architectural (laminated) shingles are multi-layer, laminated shingles, which give more varied, contoured visual effect to a roof surface and add more resistance for water. Solar reflecting shingles help reduce air conditioning costs in hot climates by providing a better reflective surface. The asphalt shingles are affixed to the roof substrate by means of mechanical fasteners or cold-applied adhesives; hot mopping asphalt is not used. They are typically installed over an underlayment membrane of roofing felt or fibre-reinforced roofing felt.

## 19.3.5 Health, Safety, and Environmental Aspects of Bitumen

Bitumen presents a low order of potential hazard provided adequate safety precautions are taken while handling bitumen. In case of accidental contact with hot bitumen, the affected part should be immediately plunged in water. Ice pack can also be given. However, no attempt should be made to remove firmly adhered bitumen from the skin. It can be allowed to fall off gradually or can be removed by medicinal paraffin. In all cases, the affected person should be taken to qualified doctor immediately. During the mixing of bitumen with aggregate, fumes are emitted. The compositions of these vapours vary depending on the temperature, manufacturing process, presence of additives and modifiers, and work practices. These fumes contain particulate matter, hydrocarbons, and compounds of sulphur, oxygen, and nitrogen. When exposed to bitumen fumes and vapours, workers have reported symptoms of irritation of the eyes, nose, and throat. Bitumen also contains low levels of polycyclic aromatic hydrocarbons, especially benzo(a)pyrene, which are carcinogenic, and may lead to lung cancer, if there is prolonged exposure.

### 19.3.6 Other Bituminous Products

Bitumen is also used in paints, sealants, joint fillers and primers. In addition, bitumen is used as an adhesive in electrical laminates and as a base for synthetic turf.

**Water-based bitumen paints**  It is an emulsion used as a damp-proof membrane and protective coating to prevent the corrosion of metals. Unlike regular paint, bitumen paint is extremely durable, waterproof, and can be used both above and below ground. It also dries to the touch in just four hours. Water-based bitumen paint is UV-resistant, and maintains its black colour for years. It can also be used to paint metal pipes, storage tanks, and masonry walls. Coverage on non-porous surfaces is about 10 $m^2$/L, and on porous surfaces 8 $m^2$/L.

**Bitumen primer**  These are quick drying solvent based low viscosity bitumen primers (IS 3384:1986). They are used as a primer and sealer coat on masonry and concrete substrates to improve the adhesion of bitumen-based waterproofing membranes and coatings. They can also be used as a curing compound on freshly cast concrete.

**Bitumen-based sealants**  Water is the most destructive element to asphalt pavements. Water entering roadway through cracks accelerates the deterioration of the roadway. Sealing pavement cracks to prevent water from entering the base and sub-base will extend the pavement life from 3 to 5 years. Flexible rubberized asphalt is found to bond with the crack walls and move with the pavement thus preventing water intrusion [see Fig. 19.1(b)]. Both crack preparation and sealant application are important for the success of the repair. Cracks must be prepared by crews using compressed air (0.69 MPa minimum) and a simple blowpipe to receive hot pour sealants (at 190°C), which are effectively applied through a delivery hose and wand. Some sealants are available as heavy-duty aluminium foils coated with thick layer of bitumen. They are used as a self-adhesive material for permanent seal under all weather conditions.

**Bitumen joint fillers**  They are composed of a blend of asphalts and mineral fillers formed under heat and pressure between two asphalt-saturated liners. They are waterproof, permanent, flexible, and self-sealing. They are ideally suited for expansion joints in driveways, streets, and floor slabs (see also IS 1838(Part 1):1983 and IS 1834:1984). Owing to their unique self-sealing characteristics, no subsequent joint sealing is required.

## 19.4 Tar/Pitch

*Tar* is a dark brown or black viscous liquid of hydrocarbons and free carbon, obtained from a wide variety of organic materials through destructive distillation. It is chemically distinct from bitumen. Tar can be produced from coal, wood, or peat. Its main use was in preserving wooden sailing vessels against rot. The largest user in the past was the Royal Navy. Demand for tar declined with the advent of iron and steel ships. Naturally occurring 'tar pits' (such as the La Brea Tar Pits in Los Angeles) actually contain asphalt, not tar. The terms *tar* and *pitch* are often used interchangeably. However, pitch is considered more solid, whereas tar is more liquid. There are four different types of tar, which are as follows:

**Coal tar**   It is a tar produced by the destructive distillation of coal. Coal tar pitch serves as a valuable ingredient in the production of a number of waterproofing, protective, and binding compounds [see IS 13758 (Parts 1 to 72): 1993, IS 14695:1999, IS 14948:2001, IS 15337: 2003]. It is also used for waterproofing concrete structures, as a saturant for roofing felts, for damp-proof courses, flooring mastics, and as a base for coal tar paints (IS 212:1983, IS 216:2006, and IS 9912:2008). Tar made from coal or petroleum is considered toxic and carcinogenic (cancer producing) because of its high benzene content, although coal tar in low concentrations is used in medicine. Coal and petroleum tar have a pungent odour.

**Wood tar**   The dry distilling of pine wood causes tar and pitch to drip away from the wood and leave behind charcoal. Earlier, it was used as a water repellent coating on wooden boats and roofs. The by-products of wood tar are turpentine and charcoal. It is also used as an additive in flavouring candy and other foods and has a pleasant odour.

**Road tar**   It is obtained by blending pitch, anthracene oil, and creosote oil (IS 218:1983) in such a manner that it conforms to specifications suitable for road use (IS 215:1995).

**Mineral tar**   Mineral tar is obtained by the distillation of bituminous shales. It has less volatile matter than the wood tar.

Tar was attractive as a paving material because it is not soluble in petroleum-derived fuels or lubricants and thus will not degrade in parking areas, where it may be exposed to fuel or lubricant leaks or spills. As a result, it is still sometimes used as a surface sealant for asphalt parking lots and driveways, even though there are human health concerns associated with its use.

Mixing tar with linseed oil varnish produces tar paint. Tar paint has a translucent brownish hue and can be used to saturate and tone wood and protect it from weathering. Tar paint can also be toned with various pigments, producing translucent colours and used to preserve wood texture. Wet tar should not be touched with bare skin, because when it dries, it leaves a stain. However, the stain can be removed by applying paint thinner.

## 19.5 General Properties of Bituminous Products

All bituminous materials are best-known waterproofing materials. In general, the hot applied material and those constituted with solvent (not emulsions) are more likely to withstand pressure. Emulsions, however, are satisfactory when the surfaces are vertical or nearly vertical.

**Effect of heat**   Though bituminous materials are solid in normal temperatures, they will revert to plastic state upon heating. When used on roofs, they are liable to soften due to solar heating, especially if insulation is provided between the surface of the structure and the covering. Similar softening will occur, near hot air exhausts, near radiators, or other heat sources. In addition, on roofs, blisters may form due to solar heating, when felts are not attached uniformly to the substructure and the air entrapped below the felt expands upon heating. This defect may be overcome by laying the felts on mopped hot bitumen. This effect can be further reduced by providing a light-coloured reflective coating on the top surface.

**Resistance to fire**   Mastic asphalt and pitch mastic are not readily combustible and may reduce fire hazard.

**Durability**   Bituminous materials are durable – some asphalt mastic roofs are more than 100 years old. However, their durability may be affected, if they are directly exposed to sunlight and also to acids and fats, or subjected to mechanical damage. Usually, bitumen felts are covered by granular surface or gravel to minimize exposure to direct sunlight. Periodic maintenance may help in identifying the effects due to natural weathering. Damage due to acids or vibration may be minimized by choosing the correct type and grade of the material.

**Appearance** The appearance of floors covered with bituminous material is reasonably good, though they are black in colour. However, they need to be well maintained. Bituminous paints are also available in comparatively lighter colours.

**Effect of physical loading** When used under load, bituminous materials tend to flow. This effect may be seen as in squeezing out of damp proof course and indentation in floors by pointed objects such as chair and table legs. Hence, it is important to choose the correct grade of material.

**Effect of acids/fats and oils** Bituminous materials used in acid-resisting construction are usually efficient. However, the nature of the acid and working temperatures are to be considered, while selecting the bitumen (see IS 9510:1980). Products based on tar are superior in resisting the effects of fats and oils compared to those based on bitumen.

**Effect of paints** Bituminous materials should not be painted with any paint having a strong binder, since it will result in crazing of the surface. They should not be painted with oil or other paints which are light in colour, since bitumen will always bleed though. Light-coloured bituminous paints, however, are satisfactory.

A comparison between asphalt, bitumen, and tar is provided in Table 19.5.

**Table 19.5** Comparison between asphalt, bitumen, and tar

| Property | Asphalt | Bitumen | Tar |
|---|---|---|---|
| Colour | Blackish brown | Dark black | Deep black |
| Manufacture | Fractional distillation of petroleum | Fractional distillation of petroleum | Destructive distillation of organic material |
| Carbon content | Low | Moderate | High |
| State | Solid or semi-solid | Usually solid at normal temperature | Viscous liquid |
| Heating effect | Burns with smoke and becomes plastic at 250°C | Becomes liquid on heating | Becomes less viscous on heating |
| Specific gravity | 0.92–1.02 | 1.0–1.07 | 1.08–1.24 |
| Setting time | Less | Less | More |
| Adhesive property | Less | Good | Very good |
| Acid and water resistance | Good | Good | Less |
| Solubility | Soluble in $CS_2$ | Soluble in $CS_2$ | Insoluble in $CS_2$ |
| Use | Road works, damp-proofing, and roofing felt | Roofing felt and damp-proofing | Preservative for wood, waterproofing paint |

## 19.6 Sustainability of Bituminous Materials

Most of the bitumen comes from petroleum residues from the distillation of crude oils, which are being depleted at a faster rate and may last another 50–60 years only. Hence, it is important to look at the sustainability of bituminous products. Asphalt concrete is 100% recyclable and is the most widely reused construction material in the world. Less than 1% of asphalt concrete is found to have disposed of in landfills. Lot of changes have occurred in the use of bitumen-based materials in the recent years. Reclaimed asphalt pavement and recycled asphalt shingles are being used increasingly to preserve virgin binder and aggregates. In addition, more polymerization and rubber are used to create binders that are not only suitable to modern pavements, but also preserve natural resources. In addition, these specialized mixtures have better structural support, enhance safety, and also reduce traffic noise. Multiple approaches are used to increase sustainability and extend life of pavements, which include the following:

1. Reducing virgin binder and virgin aggregate content in hot-mix and warm-mix asphalt mixtures,
2. Reducing the energy and $CO_2$ emissions during the production of asphalt mixtures,
3. Using alternate binders,
4. Extending the life of asphalt pavements,
5. Reducing the impacts of materials transportation,
6. Extending the life of seal coats, and
7. Increasing surface reflectivity (where required) (FHWA, 2015).

More information on these strategies may be found in FHWA (2015) and FHWA (2016). The effects of bitumen and bitumen emissions on the health of humans are summarized in IARC Monograph 103(2013).

## SUMMARY

- Asphalt, bitumen, and tar provide economic solutions to road construction and waterproofing of buildings.
- Bitumen (called in the USA as asphalt) is derived from the fractional distillation of petroleum, whereas tar is obtained from the destructive distillation of organic materials.
- Asphalt may be mixed with aggregates as a hot mix, warm mix, or cold mix. Warm mix results in many economic advantages and is energy efficient.
- Bitumen, which is the binder in asphalt, is available in different forms such as straight-run bitumen, steam-refined bitumen, hot bitumen, industrial bitumen, cutback bitumen (with solvents such as petrol, kerosene, or fuel oil), bitumen emulsion (dispersed in water), macadam bitumen, and modified bitumen (with plastics).
- Current Indian standards specify bitumen based on viscosity, though some still use the penetration grade used in the earlier versions of IS 73.
- Tar paper (not used extensively now) and bitumen felts are used as roof waterproofing materials. Three different types of felts are available based on the supporting mat.
- Several commercial roofing systems such as built-up roof (BUR) membranes, modified bitumen system, single-ply membranes, and spray polyurethane foam-based (SPF) roof systems are available for low-slope roofs. Asphalt shingles are used on roof slopes exceeding 14°.
- When hot bitumen is used, workers should be protected against the health hazards.
- Bitumen is also used in paints, sealants, joint fillers, and primers.
- To test the quality of bitumen, various tests are specified such as viscosity test, penetration test, softening point test, ductility test, etc., and are explained in Section 25.23 of Chapter 25.
- Tar is available as coal tar, wood tar, road tar, and mineral tar. Tar is not soluble in petroleum and has better fire resistance. However, its use is restricted due to its cancer producing effects.
- Recycled and reclaimed bitumen, the use of alternate binders, emulsions and plastics, and the technologies like warm-mix asphalt, etc., are some of the strategies adopted to increase the sustainability of bitumen.

## EXERCISES

### Multiple-choice Questions

1. Hot-mix asphalt is heated to a temperature of about
   (a) 135 to 163°C
   (b) 104 to 135°C
   (c) <100°C
   (d) >200°C

2. Warm-mix asphalt is heated to a temperature of about
   (a) 135 to 163°C
   (b) 104 to 135°C
   (c) <100°C
   (d) >200°C

3. Petroleum bitumen is obtained from
   (a) destructive distillation
   (b) extraction
   (c) atmospheric-vacuum distillation
   (d) fractional distillation

4. Industrial bitumen, as per IS 702, is specified based on
   (a) softening temperature and kinematic viscosity
   (b) softening temperature and penetration
   (c) kinematic viscosity and penetration
   (d) kinematic viscosity

5. Cutback bitumen, as per IS 217, is specified based on
   (a) softening temperature and penetration
   (b) softening temperature and kinematic viscosity
   (c) kinematic viscosity and penetration
   (d) kinematic viscosity

6. Industrial bitumen is one in which
   (a) bitumen is dissolved in a solvent such as petrol, kerosene, or fuel oil
   (b) bitumen is dispersed as droplets in water
   (c) air is blown under pressure into liquid bitumen at high temperatures
   (d) bitumen is added with polymers

7. Bitumen emulsion is one in which
   (a) bitumen is dissolved in a solvent such as petrol, kerosene, or fuel oil
   (b) bitumen is dispersed as droplets in water
   (c) air is blown under pressure into liquid bitumen at high temperatures
   (d) bitumen is added with polymers

8. Cutback bitumen is one in which
   (a) bitumen is dissolved in a solvent such as petrol, kerosene, or fuel oil
   (b) bitumen is dispersed as droplets in water
   (c) air is blown under pressure into liquid bitumen at high temperatures
   (d) bitumen is added with polymers

9. Modified bitumen is one in which
   (a) bitumen is dissolved in a solvent such as petrol, kerosene, or fuel oil

   (b) bitumen is dispersed as droplets in water
   (c) air is blown under pressure into liquid bitumen at high temperatures
   (d) bitumen is added with polymers

10. Paving bitumen as per IS 73:2013 is specified based on
    (a) softening temperature and penetration
    (b) softening temperature and absolute viscosity at $60°C$
    (c) absolute viscosity at $60°C$ and penetration
    (d) absolute viscosity at $60°C$

11. The minimum ductility value, specified by IS 73:2013, for VG 40 grade bitumen is
    (a) 50 cm          (c) 75 cm
    (b) 25 cm          (d) 100 cm

12. The absolute viscosity at $60°C$ measured in Poises, specified by IS 73:2013, for VG 20 grade bitumen should not be more than
    (a) 1200           (c) 3600
    (b) 2400           (d) 4800

13. As per IS 73:2013, VG 30 grade bitumen is suitable for use in 7-day average maximum air temperature of
    (a) $<30°C$        (c) $38–45°C$
    (b) $30–38°C$      (d) $>45°C$

14. The bitumen in all types of felts should have a penetration of not less than
    (a) 40 mm at $25°C$    (c) 70 mm at $25°C$
    (b) 60 mm at $25°C$    (d) 80 mm at $25°C$

15. Bitumen is made sustainable by
    (a) reclaiming asphalt pavements
    (b) using warm-mix asphalt with reduced binder content
    (c) extending the life of asphalt pavements
    (d) all of these

## Review Questions

1. Differentiate between asphalt, bitumen, and tar.
2. What is asphalt? What are the different classifications of asphalt?
3. List and describe four different forms of asphalt-paving materials?
4. What are the different uses of asphalt?
5. What is the difference between hot-mix and warm-mix asphalt concrete?
6. What is bitumen and how is it obtained?
7. What are the various types of bitumen and what are their uses?
8. Explain the following briefly: (a) straight-run bitumen, (b) blown bitumen, (c) cutback bitumen, and (d) bitumen emulsion.
9. What are the four types of cutback bitumen as per IS 217? Describe them briefly.
10. What are the three different types of bituminous emulsions as per IS 8887? Describe them briefly.
11. What is modified bitumen? Describe its applications. What are the four types of modified bitumen as per IS 15462? What are the advantages of using modified bitumen?
12. How were grades of bitumen designated in the second, 1992 version of IS 73? How are the grades designated as per the current fourth revision of IS 73?
13. Write short notes on (a) tar paper and (b) bitumen felts.
14. What are the three types of bitumen felts as per IS 1322?

15. What are the four basic types of low-slope roof systems? Explain them briefly.
16. Describe briefly the built-up-roof (BUR) system. How is it installed?
17. Differentiate between APP-modified and SBS-modified bitumen systems.
18. What are the two types of base materials used in asphalt shingles?
19. What are the health and safety hazards of bitumen?
20. List the four other bituminous products and explain them briefly.
21. What is tar and how does it chemically differ from bitumen?
22. List three different types of tar and explain them briefly.
23. Discuss the general properties of bituminous materials.
24. How are asphalt/bitumen made sustainable?

# ANSWERS

## Multiple-choice Questions

| | | | | | |
|---|---|---|---|---|---|
| **1.** (a) | **2.** (b) | **3.** (d) | **4.** (b) | **5.** (d) | **6.** (c) |
| **7.** (b) | **8.** (a) | **9.** (d) | **10.** (c) | **11.** (b) | **12.** (b) |
| **13.** (c) | **14.** (d) | **15.** (d) | | | |

# CHAPTER 20
# THERMAL INSULATING MATERIALS

## 20.1 Introduction

India has extraordinary climatic regions that range from tropical in south to alpine climate in the Himalayas. The ambient temperature in summer varies from 37.5 to 45°C and in some places, it even touches 50°C. Temperature during winter varies from 5 to 17.5°C and in some places even sub-zero temperatures, up to –7.5°C, are reached. India also has high humidity that varies from 44 to 81%. Providing thermal comfort without excess air conditioning costs is one of the primary requirements of buildings, and an important architectural requirement of all buildings. Thermal comfort is determined by the room's temperature, humidity, and air speed. It has been found from experience that temperature ranging from 20 to 25°C, together with relative humidity of 30 to 65% provides thermal comfort for most of the people. It has to be noted that the thermal comfort temperature varies from person to person and also depends on factors such as activity level, clothing, and humidity. The ASHRAE guidelines recommend 20 to 23.33°C in the winter and 22.22 to 26.67°C in the summer [Thermal comfort calculations according to ANSI/ASHRAE Standard 55 can be freely performed with the CBE Thermal Comfort Tool using the *psychrometric chart* (http://comfort.cbe.berkeley.edu/)]. Understanding the concepts of heat transfer and thermal properties of building materials and assemblies is important for assessing energy use, thermal comfort, thermal movements, durability, and moisture problems.

Heat energy always flows from a warmer area to a cooler area; thus, heat flows out of the building during winter and into the building during summer. Heat flow can be a transient or a steady process. When it is transient, temperature and/or heat flow vary with time. Steady-state heat flow occurs when the temperature and heat flow reach a stable equilibrium condition that does not vary with time. In reality, heat flow occurs in three dimensions but, for practical considerations, it is simplified and assumed to be one-dimensional. Thermal performance is usually concerned with reducing the heat transmission (outwards or inwards) through enclosures comprising walls, roofs, or floors. Thus, reduction of heat flow in buildings requires insulation layers with few thermal bridges, an effective air barrier system, good control of solar radiation, and management of interior heat generation. It has to be noted that insulation can be in the form of *thermal insulation* to reduce the rate of heat transfer, *acoustic insulation* to reduce the intensity of sound, *vibration insulation* to reduce the impact of vibration, or *fire insulation* to reduce the effect of fire. In 2017, about 40% (or about 40 EJ) of total US energy consumption was consumed by the residential and commercial sectors (www.eia.gov). The largest energy use in any building is heating and cooling (about 40%). Insulation not only provides thermal comfort to the occupants, but also reduces the energy required to heat/cool the building.

### 20.1.1 General Principles of Thermal Insulation

A thermal insulator is a poor conductor of heat and should have low thermal conductivity. Thermal resistance of an insulating material depends on the type of material and the thickness. Provision of an air gap is very important and useful insulating agent. Orientation of the building has important role in thermal resistance. The building should be designed in such a way that the entry of solar energy should be minimum during summer and maximum during winter.

### 20.1.2 Choice of Insulating Materials

The choice of thermal insulating materials depends on several factors such as thermal resistance, thermal conductivity, thermal transmittance, stability, density of the material, long life, fire resistance, lack of odour, low chemical activity, ability to cope with moisture, and availability at reasonable cost (SP 7:2016). Cost of energy consumed for heating and cooling, reasonable fire resistance, good moisture permeability, not liable to undergo deformation, and insect resistance are some of the factors to be considered before making the choice of materials. The insulation materials used for various purposes are cellulose, fiberglass, rock wool, polystyrene, urethane foam, and vermiculite. Other insulation materials are rock wool, mineral wool, gypsum board, foam glass, cork sheet, and foam plastic.

## 20.2 Importance of Thermal Insulation in Buildings

The primary requirement of thermal insulation of the building envelope is to provide thermal comfort to its occupants and reduce energy requirements. This results in healthy living environments and provides better productivity at workplaces. It may also reduce unwanted heat loss or heat gain through a building envelope. As a result, the energy demand for cooling and heating of buildings can be reduced, ultimately resulting in the reduction of greenhouse gas emissions. Passive solar design strategy of thermal insulation should take into account all possible measures to reduce energy consumption. These may include location, orientation and vegetation around the building, prevailing weather conditions, thermal mass, wind direction, height of ceiling, ventilation provided by the chimney effect, cross-ventilation, and glazed area of doors and windows (Kibert, 2005). For example, it is possible to have a plan shape of the building and orient it in such a way that it will collect as much sun as possible in winter, and yet avoid collecting much sun in summer [see also IS 7662(Part 1):1974]. As shown in Fig. 20.1, if the building in the northern hemisphere is elongated on the east-west axis, with a relatively longer southern wall, it will have a larger collecting surface to the sun's radiation in winter. During summer, the sun will be much higher in the sky and hence the radiation on the longer southern wall will be at an oblique angle; hence, the collected radiation will be less (see also IS 11907:1986). In temperate or cold climates, the side of the building facing the sun should have more windows. More details of such planning may be found in Roaf et al. (2007).

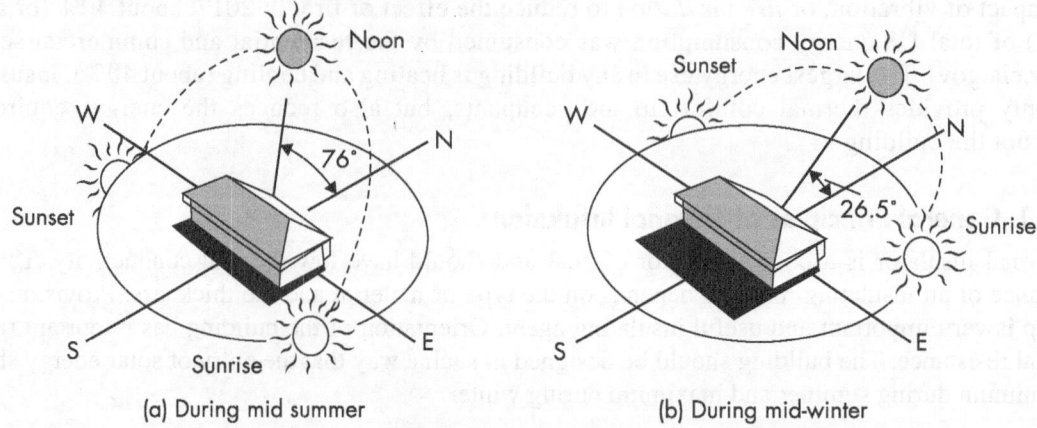

(a) During mid summer
(b) During mid-winter

**Fig. 20.1** Solar influence on plan shape and orientation

Thermal insulation ensures that the temperature of the internal surfaces of exterior building walls does not fall below a critical level so that damage from condensation and the formation of mould and associated health risks can be avoided. Vapour barriers are usually installed to prevent moisture penetration into insulation.

Thermal insulation guarantees sufficiently high surface temperature on the inside of external building walls during the winter, thus allowing the same level of comfort with lower room air temperatures resulting in less energy consumption. Efficient operation of certain sensitive equipment may require effective thermal insulation. Usual building material provides some resistance to heat flow and thermal insulation, if provided, will enhance this resistance.

Using thermal insulation in buildings may help to reduce the reliance on mechanical/electrical systems and at the same time conserve energy and associated natural resources. As the energy cost is an operating cost, greater energy savings can be achieved in the long run with minimum capital investments. Use of thermal insulation also reduces disturbing noise from the outside and enhances the acoustical comfort of buildings.

High temperature changes may result in undesirable thermal movements, which could damage the structure and contents of any building. Reducing the temperature fluctuations with insulations will help to preserve the integrity of building structures and contents and even increase the lifespan of a building. Suitable insulation material can help retard the heat and prevent the spreading of flames into the building.

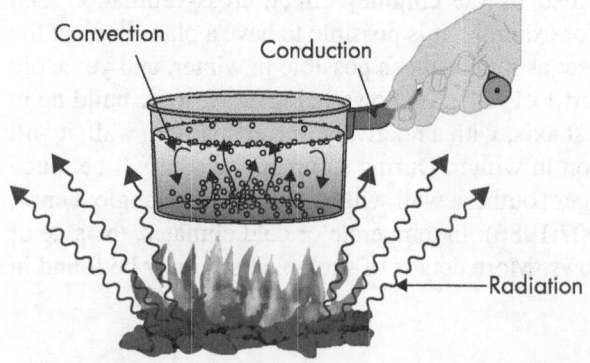

**Fig. 20.2** Conduction, convection, and radiation

## 20.3 Heat Transfer Fundamentals

Whenever a temperature difference exists in a medium, heat transfer must occur. Heat transfer is transit of thermal energy due to a spatial temperature difference. Heat transfer occurs by three primary mechanisms, acting alone or in some combination: conduction, convection, and radiation (see Fig. 20.2). Changes in moisture conditions may also be considered important, as these changes in moisture absorb and release heat energy, i.e., latent heat.

## 20.3.1 Conduction

It is the direct flow of heat through a material resulting from physical contact, as shown in Fig. 20.2. The transfer of heat by conduction is caused by molecular motion, in which molecules in a material transfer their energy to adjoining molecules, and increase their temperature. It is the most important heat transfer mode for solids, and occurs sometimes in liquids, and occasionally in gases. In conduction, transfer of heat energy occurs from a high-temperature object to a lower-temperature object. Thus, in conduction, the energy moves and the matter do not move. The following factors influence conduction: temperature gradient, cross-section of the material, length of the travel path, and physical material properties. When thermal equilibrium between the two objects is reached, the thermal transfer stops. This mode of heat transfer is important in buildings where there is a temperature difference between the inside and outside of a building, e.g., in a heated building during winter. Through conduction the internal heating or cooling will be lost to the outside, resulting in high operating costs, high carbon emissions, and occupant discomfort. Conduction in different materials occurs in different rates. The coefficient of thermal conductivity of a material will indicate its conductivity. For example, metals such as steel and aluminium are good conductors of heat. Thus, a steel framing or aluminium cladding in an exterior wall will perform poorly from the standpoint of energy efficiency. Common types of insulation, such as fiberglass, cellulose, and urethane foam are poor conductors of heat. Conduction heat transfer problems relevant to buildings include (Kusuda, 1977): (a) exterior wall conduction: transient heat transfer due to climatic effects, such as temperature fluctuation, solar radiation, wind and precipitation; thermal storage, and cold-bridge effects, (b) interior mass conduction: heat gain/loss due to cooling and heating loads, and (c) ground heat loss from slab-on-grade floor and basement walls.

## 20.3.2 Convection

In *convection*, unlike in conduction, the matter moves. Thus, as shown in Fig. 20.2, when a fluid or gas, such as air or water, is heated, it expands, becomes less dense, and rises to higher level and carries the thermal energy along with it. This effect causes displacement of the fluid/gas. As the heated liquid/air rises, it pushes denser, colder liquid/air down. A series of such events causes convection currents to form. This is natural convection as opposed to forced convection made by using mechanical fans. The operation of space heater is a classic example of convection. When hot water is circulated in a space heater, it heats the air surrounding it near the floor. Thus, the heated air expands and rises to the top of the room. This will force the cooler air to be pushed down, which in turn will be heated, thus creating convection current. It is to be noted that the movement of wind powered by the heat of the sun is an example of convection found in nature. Convection is the primary way of heat transfer through gases and liquids. Convective heat transfer problems relevant to buildings include (Kusuda, 1977): (a) heat transfer at the exterior surface considering both wind and surface roughness characteristics, (b) convection in and through the cavity walls, (c) convection between the window glass panes, (d) inter- and intra-space air motion due to temperature and pressure gradient, (e) convection heat transfer due to air leakage through exterior walls, and (f) convection heat transfer within the porous insulating structure.

## 20.3.3 Radiation

*Thermal radiation* is the heat energy emitted as electromagnetic waves (see Fig. 20.2). These waves carry the heat energy away from the emitting object. The radiation can occur through vacuum or any other medium of solid or fluid. All objects which are hotter than 0 K emit thermal radiation, with the warmer objects radiating more energy than cooler objects. They also absorb thermal radiation emitted by their surroundings. Radiated energy travels in straight lines through space or vacuum. Only when the radiation from a hot object strikes the surface of another cooler object, it is converted into heat energy. The sun is an example of an object that radiates and transfers heat across the solar system. Solar energy is composed of ultraviolet (UV) rays, visible light, and infrared energy, each reaching the earth in different percentages: 5% of solar energy is in the UV spectrum, including the type of rays responsible for sunburn; 43% of solar energy is visible light, in colours ranging from violet to red; and the remaining 52% of solar energy is infrared, felt as heat (USEPA, 2008). Energy in all of these wavelengths contributes to the *urban heat island effect*. Everyday examples of radiation include heat from a light bulb, fire, or electric heater. Radiation is of practical importance for the heat transfer between solids, within highly porous solids, and occasionally between high-temperature gases.

Radiation heat transfer is very important in building application in the following areas (Kusuda, 1977):

1. Short-wavelength radiation: (a) solar heat absorption on opaque exterior surfaces, (b) solar heat transmission through transparent surfaces, (c) solar heat absorption and reflection by interior building surfaces, (d) absorption and reflection of solar heat by window glasses
2. Long-wavelength radiation: (a) heat emission by the exterior surfaces to the sky, (b) heat exchange among interior surfaces, (c) heat exchange between interior surfaces and occupants, (d) heat exchange between the lighting fixtures and interior surfaces

## 20.4 Thermal Properties of Insulating Materials

A thermal insulator is a poor conductor of heat and has low thermal conductivity. The upper and lower temperature limit within which the material retains all its useful properties is a very important consideration in choosing insulation materials. Guidance for heat insulation may be found in IS 3792:1978; it also provides definition of terms, symbols, and units of quantities used in heat insulation. Some of these terms are briefly explained in this section.

**Specific heat**   It is the amount of heat energy per unit mass required to raise the temperature by 1°C. The specific heat of materials is constant at a particular temperature. The specific heat of water is 4,176 J/kg/°C, which is higher than any other common substance. By comparison, it only takes 385 J of heat to raise 1 kg of copper by 1°C.

**Thermal capacity**   It is the amount of heat required to raise the temperature of a mass of material by 1°C. The thermal capacity per unit area of surface of roof, wall, or floor is the sum of the products of the mass per unit area of each individual material in the component multiplied by its individual specific heat. The unit of thermal capacity is J/kgK (Joule per kilogram-Kelvin).

**Thermal conductivity ($k$ or $\lambda$)**   It is defined as the rate at which heat is transferred by conduction through unit cross-section area of a material, when a temperature gradient exits perpendicular to the area. Thermal conductivity is the characteristic property of a material and its value depends on a number of factors such as density, porosity, moisture content, fibre diameter, pore size, type of gas in the material, mean temperature, and outside temperature range (SP 7:2016). It is expressed in the units of W/(mK) (Watt per meter-degree Kelvin). Its value ranges from 0.03 W/mK for insulators to 400 W/mK for metals. Heat transfer occurs at a higher rate across materials of high thermal conductivity than those of low thermal conductivity. The reciprocal of thermal conductivity is called thermal resistivity. Metals with high thermal conductivity, e.g., copper, aluminium, and silver exhibit high electrical conductivity, and are used in electrical applications. On the other hand, materials with low thermal conductivity, such as polystyrene, are used as thermal insulators in building construction or in furnaces. Thermal conductivity of insulation materials could be determined by using *heat flow meter* or *water calorimeter* (also see IS 3346:1980, IS 9489:1980, and IS 9490:1980). Thermal conductivity of common materials is given in Table 20.1 (see also Appendix C of IS 3792).

**Table 20.1** Thermal conductivity, $k$, of few building materials W/(mK) at room temperature

| Material/Substance | Thermal conductivity, $k$ | Material/Substance | Thermal conductivity, $k$ |
|---|---|---|---|
| Air | 0.024 | Cork slab | 0.044 (density 192 kg/m³) |
| Aluminium | 238 | Fiberglass, loose-fill | 0.043 |
| Argon (gas) | 0.016 | Soda-lime-glass | 0.96–1.7 |
| Asbestos-cement sheets | 0.166 | Iron | 80 |
| Asphalt | 0.52–0.75 | Plywood | 0.13 |
| Brick, dense | 1.4 | Polyethylene | 0.33 |
| Brick, insulating | 0.15 | Expanded polystyrene | 0.035–0.038 |
| Brickwork | 0.6–1.0 | Rock wool (unbounded) | 0.047 |
| Cast iron | 58 | Steel, carbon 1% | 47 |
| Cement mortar | 0.72 | Stainless steel | 16 |
| Concrete, lightweight | 0.19 (density 840 kg/m³) | Timber, oak | 0.17 |
| Concrete | 1.63 (density 2300 kg/m³) | Wood wool slab | 0.10 |
| Foam concrete | 0.084 (density 400 kg/m³) | Zinc | 116 |
| Copper | 401 | Urethane foam | 0.021 |

**Thermal conductance (C)**    It is a measure of the thermal transmission per unit area through the total thickness of the material. Thus, thermal conductance is the quantity of heat that passes in unit time through a plate of particular area and thickness when its opposite faces differ in temperature by one Kelvin. The value of the thermal conductance can be calculated by dividing the thermal conductivity with the thickness of the material, and is expressed in W/(m²K) (Watt per square meter per degree Kelvin). Thus, $C = \dfrac{k}{L} = \dfrac{1}{R}$.

**Thermal emissivity (E)**    It is the rate at which radiant energy is given off by a material per unit area per unit time (often in W/m²). Whereas, *emittance* is the ratio of the emissivity of a material to the emissivity of a perfectly-emitting and absorbing (theoretical) black body. Thus, emittance is always less than 1.0. During hot summer days, the lower the roof emissivity, the higher will be the surface temperature of the roof and hence there will be increased heat conduction into the building. In air-conditioned buildings, this will lead to a higher cooling energy use. On the other hand, during the winter when heating is required, due to the lower emissivity, there will be lower heat loss from the roof of the building. This would lead to a lower heating energy. The *urban heat island effect is* created by heat absorbent roofs and black asphalt pavements, which result in temperatures in cities to be 3 to 5.6°C warmer than nearby rural areas (Kibert, 2007). Emissivity of some materials is given in Table 20.2.

**Table 20.2** Emittance of some materials

| Material | Emissivity | Material | Emissivity |
|---|---|---|---|
| Aluminium foil | 0.03 | Copper, polished | 0.04 |
| Aluminium, anodized | 0.77 | Glass, uncoated | 0.95 |
| Asphalt | 0.88 | Polished marble | 0.89–0.92 |
| Brick | 0.90 | White paint | 0.90 |
| Concrete | 0.85 | Polished silver | 0.02 |

**Thermal reflectivity** Thermal or solar reflectance, or *albedo*, is the percentage of solar energy reflected by the surface of roofing material. Traditional roofing materials have low solar reflectance. However, cool roof materials will have a high solar reflectance of more than 65% and hence transfer only 35% or less of the energy into the building. These materials reflect radiation across the entire solar spectrum, especially in the visible and infrared (heat) wavelengths (USEPA, 2008). Thus, cool reflective roof surfaces help to keep a building cool during the summer, decreasing the cooling load, and saving energy. The greatest benefit of reflectivity is achieved in warm climates, such as those in Chennai, where cooling is required throughout the year. Reflective roofs are not recommended for cold climates where there is no need to cool the buildings. Reflectivity of some common materials is given in Table 20.3. Reflectivity of the roofing surfaces may be increased by using some special paints (www.roofcoatings.org). A database of cool materials is available at https://heatisland.lbl.gov/resources/cool-roofing-materials-database.

**Table 20.3** Reflectivity of some common materials

| Material | Reflectivity (%) | Material | Reflectivity (%) |
|---|---|---|---|
| Fresh snow | 80–90 | Bare soil | 17–20 |
| New concrete with white Portland cement | 70–80 | Vegetation | 15–20 |
| White acrylic paint | 80 | Wood shingle | 17 |
| Aluminium | 65–75 | Water/ocean | 5–6 |
| Plaster | 40–45 | Worn asphalt | 12 |
| Concrete (traditional) | 40–55 | Fresh asphalt | 4 |

**Solar reflectance index** The coolness of a roof can be judged by comparing its surface temperature on a sunny day to that of a reference black roof or to that of a reference white roof. Such a judgment is provided by the *solar reflectance index* (SRI), which assigns a coolness of 0 to the reference black roof (solar reflectance $R = 0.05$, thermal emittance $E = 0.90$) and a coolness of 100 to the reference white roof ($R = 0.80$, $E = 0.90$) (https://heatisland.lbl.gov/). Most roofing materials have an SRI (coolness rating) between 0 and 100, though values can be below 0 (hotter than reference black) or above 100 (cooler than reference white). The higher the SRI, the cooler will be the surface.

**Thermal transmittance or U-value** It is defined as thermal transmission through unit area of the building component divided by the difference in temperature across the thickness of the component, in steady-state conditions, and expressed in the units of W/m²K. Losses due to thermal radiation, thermal convection, and thermal conduction are considered in the U-value. The U-value is a measure of the heat transfer through a building envelope due to a temperature difference between the indoors and outdoors. It will indicate the heat loss in a building component such as a wall, floor, or roof. Higher U-values show poor thermal performance of the building envelope and lower U-values usually indicate high levels of insulation. The U-value of a material is usually used to select an insulating material.

Thermal transmittance differs from thermal conductance in that in the former case temperatures in the surrounding air of the building component on the two sides are measured, whereas in the latter case the temperatures on the two surfaces of the building component are measured. Thermal conductance is a characteristic of the building component whereas thermal transmittance depends on the conductance and surface coefficients of the building component under the prevailing conditions.

The U-value of roof or wall is calculated as the reciprocal of all the resistances of the component materials found in the building element (see Fig. 20.3).

$$U = \frac{1}{R_{si} + R_{so} + R_A + R_1 + R_2 + \ldots}$$

(20.1)

where $R_{si}$ is the thermal resistance of internal surface, $R_{so}$ is the thermal resistance of outside surface, $R_A$ is the thermal resistance of unvented thermal cavities, and $R_1$, $R_2$, etc., are the thermal resistance of insulating boards/materials. Common values for the thermal resistance of internal and external surfaces are 0.12 m²K/W and 0.06 m²K/W, respectively. It has to be noted that the relationship between U-value and R-value is not always exactly the inverse and therefore R-value cannot be precisely extrapolated for a material of different thickness. However, assuming an inverse relationship may be adequate in most of the cases.

$$U = \frac{1}{R_{si} + R_{so} + R_A + R_1 + R_2 + \ldots}$$

$R_{si}$ is the thermal resistance of internal surface

$R_{so}$ is the thermal resistance of outside surface

$R_A$ is the thermal resistance of unvented air cavities

$R_1$, $R_2$ are the thermal resistance of insulating boards/materials

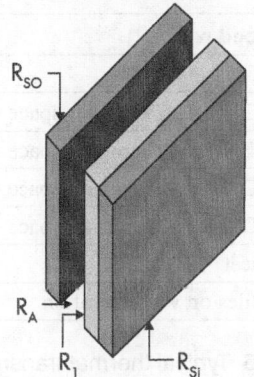

**Fig. 20.3** Calculation of U-values

U-values are determined from the default tables in Appendix D, Table 11.4.3 of the energy conservation building code of India (ECBC:2007) or determined from data or procedures contained in the ASHRAE Fundamentals (2005). The U-values of some typical flat roof construction as per IS 3792:1978 and SP 7:2016 are shown in Table 20.4 and sloped roof construction in Table 20.5. Typical thermal transmittance values of some common building elements are provided in Table 20.6 (also see Table 15.5 of Chapter 15).

**Table 20.4** Thermal performance of flat roof and wall constructions (Extracted from IS 3792)

| Type of flat roof | U-value [W/(m²K)] |
| --- | --- |
| 100-mm RCC | 3.59 |
| 100-mm RCC + 50-mm foam concrete + waterproofing | 1.08 |
| 50-mm RCC + 25-mm expanded polystyrene | 1.08 |
| 50-mm RCC + 25-mm expanded polystyrene + waterproofing | 0.62 |
| 100-mm RCC + 25-mm extruded polystyrene- 36 kg/m³ | 0.749 |
| 100-mm RCC + 75-mm extruded polystyrene- 36 kg/m³ | 0.312 |
| 100-mm RCC + 25-mm expanded polystyrene- 24 kg/m³ | 0.931 |
| 100-mm RCC + 75-mm expanded polystyrene- 24 kg/m³ | 0.409 |
| 100-mm RCC + 25-mm phenolic foam- 32 kg/m³ | 0.725 |
| 100-mm RCC + 75-mm phenolic foam- 32 kg/m³ | 0.301 |
| 100-mm RCC + 25-mm polyurethane spray- 42 + 2 kg/m³ | 0.664 |
| 100-mm RCC + 75-mm polyurethane spray- 42 + 2 kg/m³ | 0.259 |

*(Contd)*

**Table 20.4** (*Contd*)

| Type of flat roof | U-value [W/(m²K)] |
|---|---|
| 100-mm RCC + 25-mm polyisocyanurate spray- 42 + 2 kg/m³ | 0.664 |
| Uninsulated cavity wall | 1.60 |
| 225-mm solid brick wall | 2.23 |
| 225-mm solid brick wall + 50-mm of PUR insulation | 0.40 |

**Table 20.5** Thermal performance of sloped roof constructions (Extracted from IS 3792)

| Type of sloped roof | U-value [W/(m²K)] |
|---|---|
| 6.25-mm AC sheet | 5.47 |
| 6.25-mm AC sheet +25-mm air space + insulating board | 2.44 |
| 6.25-mm AC sheet +25-mm air space + 50-mm fibreglass + 6.25-mm hard board | 1.40 |
| 6.25-mm AC sheet +25-mm air space + 25-mm sandwich of fibreboard/ expanded polystyrene | 1.22 |
| 6.25-mm AC sheet +25-mm air space + 50-mm sandwich of fibreboard/ expanded polystyrene | 0.65 |
| 3-mm GI sheet | 6.16 |
| Mangalore tiles on wooden rafters | 4.07 |

**Table 20.6** Typical thermal transmittance values of some common building elements (*Source*: Wikipedia)

| Type of sloped roof | U-value [W/(m²K)] |
|---|---|
| Single glazing | 5.7 |
| Single-glazed windows, allowing for frames | 4.5 |
| Doubled-glazed windows, allowing for frames | 3.3 |
| Double-glazed windows with advanced coatings | 2.2 |
| Double-glazed windows with advanced coatings and frames | 1.2 |
| Triple-glazed windows, allowing for frames | 1.8 |
| Triple-glazed windows with advanced coatings and frames | 0.8 |
| Well-insulated roofs | 0.15 |
| Poorly-insulated roofs | 1.0 |
| Well-insulated walls | 0.25 |
| Poorly-insulated walls | 1.5 |
| Well-insulated floors | 0.2 |
| Poorly-insulated floors | 1.0 |

*Total isolation value* or *K-value* of a building is obtained by multiplying the form factor of the building (equals the total inner surface of the outer walls divided by the total volume of the building), with the average U-value of the outward walls of the building. Thus, the K-value is expressed as [W/(Km³)].

**Thermal resistance (R)** It is usually considered as the reciprocal of thermal transmittance (U-value). For a material, the thermal resistance is given by thickness (*d*) divided by the thermal conductivity and expressed in the unit of m²K/W (square meter Kelvin per Watt). Thus, $R = d/k$. The R-value of an element, also called as thermal insulation, is the direct measure of its resistance to transferring energy or heat. Higher R-value indicates higher insulating properties (see Table 20.7). The greater the material thickness, greater will be the thermal resistance. R-value is the resistance of each layer to the flow of heat. A 130-mm thick low density or a 100-mm thick medium density glass wool batt can achieve an R of 2.5. For materials composed of different layers, the individual thermal resistances can simply be added to get the total thermal resistance of the material.

**Table 20.7** R-values of insulation materials (m²K/W)

| Category | Insulation type | Approximate R/25 mm |
|---|---|---|
| Loose-fill insulation | Vermiculite (grey or brown metallic granules)* | 0.41–0.48 |
| | Cellulose blown | 0.53–0.63 |
| | Loose fiberglass or rock wool | 0.30 |
| Fiberglass/Mineral wool batts/boards or blown | Low-density pink, yellow, or green batts installed perfectly without gaps | 0.39–0.46 |
| | High density stiffer batts installed without gaps | 0.53–0.77 |
| Rigid foam | Expanded polystyrene-EPS (white 'beadboard') | 0.67–0.77 |
| | Extruded polystyrene-XPS (blue, pink, or green board) | 0.88 |
| | Polyisocyanurate board (stiff yellow or tan foam with foil facing) | 1.14–1.40 |
| Sprayed foam | Urea-formaldehyde (Sprayed)* | 0.53–0.70 |
| | Open-cell polyurethane foam (yellow, green, or white, soft foam) | 0.53–0.60 |
| | Closed-cell polyurethane foam (usually yellow or green, tough foam) | 1.05 |

R-values may vary depending on type, density, and the quality of installation. *Not used now

Unfortunately, heat moves in and out of a building in several different ways and R-values take into account only the conduction and do not include either convection or radiation. Hence, it may be better to use the U-value, which takes into account all the different mechanisms of heat loss.

## 20.5 Criteria for Selection of Insulation Materials

The factors that are considered while selecting a thermal insulating material in any application include the following: energy required for production, cost, low thermal conductivity, thermal transmittance, stability, density, durability, fire resistance and flammability, lack of odour, low chemical activity, moisture resistance, good thermal resistance, compressive strength, heat storage capacity, acoustic properties, and ease of application.

Air infiltration should be considered while insulting a building. If a well-insulated building has poorly fitted windows and doors, air leaks through them and also through electrical openings and wall penetrations will reduce the efficiency of the whole system. Thus, proper construction techniques are necessary to take full advantage of the building insulation.

The thermal resistance of insulation materials is the most important property that is of interest when considering thermal performance and energy conservation issues. It has to be noted that brick, stone, concrete, and slate are poor insulators, although they may have large thermal mass. Thus, when they are placed inside an insulated area, will act as stabilizers of inside temperature. However, when they are used as exterior finishes, they will not aid the insulation of the building.

**Thermal performance**   The lower the density of a material, the lower will be its conductivity (K) and hence it will insulate better. Air has the lowest density of all the materials and hence the best insulator is one in which the air can be made immobile, in order to prevent heat transfer by convection (e.g., walls with unvented air cavities). Mineral wool, cellulose fibre, foamed plastic, and foamed concrete work on the principle of trapped air spaces. The higher the number of air cells and the smaller their size, the lower will be the thermal conductivity of such insulating material. These cells should not be interlinked, as this will allow the convection of heat. As indicated earlier, higher R-values indicate higher insulating properties. The materials should have thermal bridging capabilities (see Section 20.5.1).

**Moisture absorption**   Absorbed moisture will reduce the R-value of insulation, as it will fill the air spaces of insulation and conduct heat far better than air. It has been found that only 5% of humidity within the insulation will reduce its efficiency by 50%. In addition, water and ice can damage the insulation, and wet insulation may corrode metals or feed water to insects and microorganisms which will rot organic building materials. Materials with low R-value combined with water absorption will create great moisture problems. Vapour diffusion can penetrate through pervious materials like drywall and masonry. Hence, fibrous insulation products are always protected against exterior humidity by using a breather membrane and against interior humidity by using a vapour barrier (see Section 20.8). It is also better to select an insulation material, which will not absorb moisture.

**Structural strength**   Although the higher void content of a material may increase the insulating property, it will reduce the structural strength. Strength is important only when the insulation supports loads. Materials having low compressive strength could be used as fillers.

**Chemical reaction and resistance**   When used in areas where flammable chemicals are present, potential fire hazards of the material should be considered. Corrosion resistance must also be considered, when used with steel structures. When the atmosphere is salt or chemical laden or when leaks in pipes are anticipated, chemical resistance should assume importance.

**Fire resistance**   Combustibility, flame spread, and smoke development rating of the insulating material are important in many situations. Usually, porous insulating materials have good fire resistance and will also resist the spread of flames. In potential fire hazard zones, the toxicity of the fumes must also be considered.

**Shrinkage, compaction, and settlement**   Some insulating materials may suffer dimensional instability while in use. This can be anticipated and overcome by careful design and installation procedures.

**Economy and ease of installation**   The cost of insulation is based on its durability, effectiveness, and ease of installation. Durability of insulation is governed by its resistance to compression, moisture, and degradation.

**Weather resistance**   When used in outdoor applications, stability against environmental influence and resistance to UV light should be considered. In order to obtain an optimum solution and cater to varying climate conditions, a combination of materials may be required. Some products are available as a combination of different types of insulations.

## 20.5.1 Thermal Bridging

It is the occurrence of rapid heat transfer through more conductive (or poorly insulating) building materials like wood, steel, and aluminium, when combined with insulating materials. For example, thermal bridging may occur in wall studs. Suppose that the wall is insulated with 150-mm thick fibre-glass batts (with R-value of 0.42/25 mm) and has wooden studs of size 50 × 100 mm at 400 mm centres (see Fig. 20.4). The heat flows through the wood studs at a rate that is three times faster than the heat flow through the surrounding insulation. Hence, the effective R-value of the assembly will be lower than the fibre-glass batts' R-value of 2.5. Several strategies could be adopted to reduce thermal bridging. For example, the walls can be made of SIP panels (see Section 17.9.7 of Chapter 17) or advanced framing techniques with reduced the number of wall studs could be used. Another strategy involves minimizing the thickness of framing materials and applying strips of insulation over the wood studs to provide a *thermal break*. In airtight and insulated homes, thermal bridges can account for heat loss of up to 30%. Insulating steel framing in cold climates is a challenging

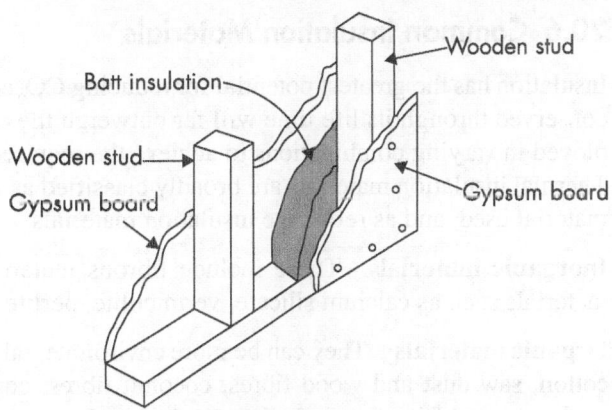

**Fig. 20.4** Insulation batts installed between wooden studs

job. Without insulated sheathing, the steel studs can cause condensation in cavities resulting in wetting of the surrounding materials and lowering whole wall R-value.

### 20.5.2 Convection and Air Leakage

Convective loops within insulated building cavities may increase heat transmission. Wall and ceiling cavities can pump heat out of a building even if there is no direct air leakage from indoors to outdoors. Air convection can also occur around the insulation's edge gaps-between batts and framing members. Edge gaps of about 4% of the insulated surface area may result in 30% loss in effective R-value of ceiling insulation. In addition, if there is leakage of air from inside or outside of a building into an insulated cavity, there can be a reduction of 15 to 50% in the effectiveness of insulation (Krigger & Dorsi, 2004). It has to be noted that air can flow even through fibrous insulating material such as loosely installed fiberglass. In cold climates, the installed density of the insulation is also an important factor.

Roof products are applied to either low-slope or steep-slope roofs, such as roof coatings and single-ply membranes, must meet the requirements given in Table 20.8

**Table 20.8** Efficiency recommendation for cool roofing products (www.energystar.gov and ECBC:2007)

| Roof slope | Recommended solar reflectance | | Best available solar reflectance | | As per ECBC:2007 |
|---|---|---|---|---|---|
| | Initial | 3 years after installation | Initial | 3 years after installation | For slopes < 20 degrees: initial solar reflectance ≥ 0.70 and initial emittance ≥ 0.75 |
| Low slope (9.46° or less) | ≥0.65 | ≥0.5 | 0.87 | 0.85 | |
| Steep slope (greater than 9.46°) | ≥0.25 | ≥0.15 | 0.77 | 0.60 | |

Maximum assembly U-factor requirement (W/m²K) for some building elements in hot and dry climate zone as per ECBC:2017 are as follows:

1. Roof assembly: 0.33
2. Opaque external walls: 0.40
3. Sky lights: 4.25 max
4. Vertical fenestration: 3.00

For other types of buildings in different climatic zones, reference should be made to ECBC:2017.

## 20.6 Common Insulation Materials

Insulation has the greatest potential for reducing $CO_2$ emissions during the life time of buildings. Energy conserved through its life time will far outweigh the energy used in its manufacture. They may be employed in varying combinations to achieve the required thermal comfort with low energy consumption. Thermal insulation materials are broadly classified as inorganic, organic materials, according to the raw material used, and as reflective insulation materials.

**Inorganic materials**   These include fibrous materials such as glass, rock, and slag wool, cellular materials such as calcium silicate, vermiculite, perlite and pumice, gypsum boards, and aerogels.

**Organic materials**   They can be more environmental friendly natural materials such as cellulose fibres, cotton, saw dust and wood fibres, coconut fibres, cork boards, sheep's wool and hemp fibres or less environmental friendly synthetic materials such as polyurethane rigid foam and spray-in place foams.

**Reflective insulation materials**   These are highly reflective metallic or metalized reflective membranes (e.g., aluminium foil), which retard the heat transfer up to 97%. They are used mainly in warm climates and are nontoxic, thin, and act as vapour barrier also. They usually require some backing material.

   Insulation materials may also be classified as bulky fibrous materials (fiberglass, rock and slag wool, cellulose, and natural fibres), cellular materials (glass or foamed plastic such as polystyrene [closed-cell] and polyisocyanurate), granular material (calcium silicate, expanded vermiculate, perlite, expanded polystyrene, and cellulose), rigid foam boards, and foils. They can also be classified as follows:

1. Batt blanket and matt insulation (fiberglass, mineral/rock and slag wool, polyester fibre batts)
2. Loose fill insulation (exfoliated vermiculate, mineral wool, polystyrene or cork granules)
3. Blown insulation (glass, mineral or cellulose fibres)
4. Rigid board insulation [vermiculite, gypsum, calcium silicate, expanded polystyrene (EPS), extruded polystyrene (XPS), polyurethane and polyisocyanurate, phenolic foam, expanded silica, or perlite]
5. Natural fibre insulation (wood fibre and cellulose, wool, hemp, cotton and flax)
6. Spray foam insulations (polyurethane and isocyanate foams, cementitious foam)

### 20.6.1 Inorganic Materials

**Mineral wool**   (IS 3144:1992, IS 8183:1993) It may represent glass wool, rock wool, or slag wool. Mineral wool is the commonly used insulation made from melting glass, rock, and slag and drawing the molten material into fibres. The fibres can be used as loose-fill insulation in places like attics. The fibres could be bonded together using organic resins into products like batts and blankets (which are flexible), and also as rigid boards. The majority of mineral wool in the USA is slag wool, produced using the slag from steel mills. Mineral wool products are light and have very good thermal insulation properties. Mineral wood products are not combustible, and absorb very little water. They may be used in conjunction with other fire resistant insulation, to cover large areas (see also Tables 20.7 and 20.9). Specification for sprayed mineral wool thermal insulation is provided in IS 9742:1993.

**Glass wool (IS 3690:1974)**   The glass fibres are made from 70% recycled glass, silica sand, limestone, and boron. The manufacture of fiberglass is explained in Section 15.9 of Chapter 15. The fibres could be bonded using phenol formaldehyde or acrylic resin and made as batts and blankets, which have facing of kraft or foil-kraft paper. The density of the material can be varied through pressure and binder content. The fibreglass itself is noncombustible, but the binder and facings are combustible and even flammable. Batts are commonly installed in cavities of buildings. Their thermal performance is dependent on proper installation. They are available in widths of 400 or 600 mm to fit between the framing members (see Figs 20.4 and 20.5). Blankets are used as insulation in metal buildings, ducts, and tanks, as well as sound insulation. Blankets

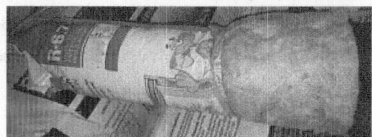

**Fig. 20.5** Unfaced fiberglass insulation (R-value shown is for inches)

are available in 1 to 2 m rolls of various lengths. The main advantages of fibreglass are low cost and easy availability. The R-value/25 mm ranges from 0.42 $m^2K/W$ for a density of 10 kg/$m^3$ to 0.81 $m^2K/W$ for a density of 50 kg/$m^3$.

**Rock and slag wool (IS 3677:1985)** Rock wool is made from natural basalt rock at a temperature of about 1600°C. It is a lightweight and fire-resistant product, with high embodied energy due to the high temperature involved in its production. Rock wool however may contain 10% recycled content from building waste or 15% slag from the steel industry. They are available as loose filling material, as shown in Fig. 20.6, used in bulk

**Fig. 20.6** Unfaced rock wool insulation (R-value shown is for inches)

insulation (done by hand packing), or as stitched mats. The mats may be faced with a confining media such as metallic cage (wire netting, expanded metal, and metallic lath), hessian cloth, scrim cloth, kraft paper, or glass tissue (IS 3677). The mats are available in 25, 40, 50, 60, 75, 90, and 100-mm thicknesses and usually in widths of 900 ± 50 mm. It is denser than glass wool and hence its R-value is also higher than glass wool. Density varies from 100 to 200 kg/$m^3$. Thermal conductivity (k value) at 50°C ranges from 0.50 W/(mK) for an apparent density of 100 kg/$m^3$ to 0.42 W/(mK) for an apparent density of 100 kg/$m^3$ (IS 3677). It has short fibres, very high mechanical resistance, maximum working temperature of 700°C, high melting temperature of about 1000°C and is noncombustible. It can also absorb sound energy.

**Vermiculite** It is a naturally occurring flaky mineral, resembling mica in appearance, and usually light-brown, grey, or gold in colour. When vermiculite is heated to 800–1100°C, it expands like popcorn to a porous mass of density 70 to 120 kg/$m^3$. Expanded vermiculite is lightweight, fire-resistant, odourless, nontoxic, and has an R-value of about R-0.35 per 25 mm (Popular brand name in USA Zonolite). It can be used as loose insulation or as aggregate in lightweight concrete (vermiculate: cement in the ratio of 6:1). Prefabricated slabs of thickness 15 to 100 mm can be made and used in high temperature applica-

**Fig. 20.7** Spray protection of steel members for fire-resistance (*Photo:* NS)

tions. When used in wall cavity, it has to be compressed properly, to counter its settlement. It can absorb lot of moisture. Vermiculate boards can be fixed around structural steel members for fire protection (see Chapter 16 of Subramanian [2008]). Sprays containing vermiculate or mineral fibre in cement or gypsum binder can also be applied to the required thickness on structural steel members, as shown in Fig. 20.7 (Subramanian, 2008). Although not all vermiculite may contain asbestos, some products that were made till 1990s were contaminated with asbestos (especially those mined near Libby, Montana, USA). Vermiculite can be bonded with bitumen and used as a composite thermal insulation/waterproofing material.

**Perlite and pumice products**   These are natural glasses of volcanic origin. When perlite is pulverized and heated in kilns to 800–1150°C, it expands 10–30 times of its original volume. Its density may range between 80 and 240 kg/m³ (Celik, et al., 2016). It has properties such as low thermal conductivity, sound insulation, high porosity, lightweight, non-flammability, and nontoxicity. It can be used either as loose fill insulation or as aggregate in mortar and lightweight concrete blocks. As perlite can absorb moisture, it may be necessary to add bitumen or silicon. Perlite with pumice yields lightweight insulation with good thermal resistance.

**Calcium silicate**   **(IS 8154–1993, IS 9428–1993)** It is used to insulate high-temperature pipes and equipment and for fire-endurance applications. It is manufactured and sold in three different forms: pre-formed block, preformed pipe, and board. It has high compressive strength, corrosion-inhibiting properties, and high-temperature structural integrity. Its density ranges from 220 to 280 kg/m³ (IS 9428:1993). Thermal conductivity ranges between 0.080 W/(mK) at 200°C and 0.148 W/(mK) at 500°C. The moisture content of the material should not exceed 7.5%. Specification for preformed calcium silicate for temperatures up to 650°C is provided in IS 8154:1993 and up to 950°C in IS 9428–1993. For standard sizes of preformed products refer IS 9428:1993. Up to 1972, calcium silicate insulation was reinforced with asbestos fibres. Now, glass fibre, plant fibres, cotton linters, or rayon are used (see also Section 3.10.2 of Chapter 3). Specification for spray applied hydrated calcium silicate thermal insulation is provided in IS 11128:1994.

**Gypsum boards**   It has a relatively low R-value (R-0.08 for 12.5-mm thickness) and hence could not be used as a standalone thermal insulation material. It has to be combined with other insulation materials to provide an acceptable overall R-value. Brick, stone, and concrete are examples of other basic materials that also require supplemental insulation. Gypsum board is usually used to cover foam plastic materials as codes usually prohibit the installation of exposed foam plastic in occupied areas of most buildings.

**Aerogels**   It was invented by chemist Samuel Stephens Kistler in 1931. This is a synthetic porous ultra-light material derived from a gel, in which the liquid component is replaced with a gas. It is a solid with extremely low density (5 to 200 kg/m³), low thermal conductivity (0.03 W/(mK) in atmospheric pressure), and high R-value (R-1.76 per 25 mm) compared to the commonly used plastic foams such as XPS (R-0.88) and EPS (R-0.70). However, it is very expensive. Its melting point is 1200°C. Micro-porous silica, micro-porous glass, and zeolites are some examples of aerogels – silica aerogel is the most common type. Aerogel is fire resistant, lightweight, nontoxic, and water repellent. People handling aerogel for extended periods should wear gloves to protect their skin. Transparent silica aerogels, when used as thermal insulation material for windows, can significantly limit thermal losses (www.aerogel.org). Some types of aerogel provide 39 times more insulation than fiberglass.

## 20.6.2 Organic Materials

Organic materials include different types of plastics, cellulose, and wood products, sheep's wool, and hemp boards (plastic insulation materials are discussed in Section 20.7). Wood chips, sawdust, hemlock fibre, etc., are used rarely, as they absorb water, reducing their effectiveness as a thermal insulator. In addition, wood is susceptible to mould, mildew, and rot attack, in the presence of moisture. Masonry coatings containing acrylic adhesives are also available, and are applied in multiple coatings.

They adhere to the polystyrene insulation; after it is attached to exterior walls of a building (see www. eima.com/eifs).

**Cellulose fibres**   These are made from different types of cellulosic materials such as waste/recycled paper, card boards, old newspapers, cotton, straw, saw dust, and hemp. They have low embodied energy and are inexpensive. They have to be treated with chemicals like borax, or boric acid to have fire resistance. Cellulose insulation, like rock wool, is denser and more resistant to airflow than fibreglass. It is used as loose fill insulation of roofs, floors, and walls. They can be reused and recycled but should not be incinerated (see also Table 20.9). Cellulose is a respiratory irritant due to the inclusion of fire-retardant chemicals. In addition, it produces considerable dust during installation and hence during its installation it is wise to protect the skin and eyes and wear respirator mask.

**Cotton**   It is as environmentally friendly product made from both natural fibres and industrial waste of cotton textiles with some polyester fibre added to it. It is nontoxic, requires less energy for manufacture, and also is a good sound insulator. Cotton insulation is available in batts and costs about 15 to 20% more than fiberglass batt insulation. It is available in 87.5 and 137.5 mm thick unfaced batts. Rated insulating values are R-2.28 for the thinner batts and either R-3.34 or R-3.70 for the thicker batt (depending on density).

**Saw dust and wood fibres**   *Wood fibre* insulation was introduced during 1920s using timber industry wastes. These wastes are ground, pressed under pressure to the desired thickness, dried and cut to the desired sizes. In addition to wood fibres, sugar cane fibres are also used. They can be used to insulate walls and ceilings. They are renewable, chemically inert, and nontoxic, but are not fire resistant. Pulverized clay could be added to improve the fire resistance. Products ranging in density between 50 and 180 kg/m³ are available, with 'k value' in the range of 0.038–0.043 W/mK. Specification for wood wool building slabs is provided in IS 3308:1981.

**Coconut fibres**   These are obtained from the outer layers of coconut and are hard and elastic. These fibres are usually sandwiched between two Teflon (polytetrafluoroethylene: PTFE) sheets to prevent the produced boards from sticking onto the hot plates during hot-pressing process. When hot pressed in a slab-like thermal insulation batt, the density of these boards is in the range of 250–350 kg/m³ (Panyakaew & Fotios, 2011). The thermal conductivity values are in the range of 0.046–0.068 W/mK. More details of these boards may be found in Panyakaew and Fotios (2011).

**Cork-board**   *Cork* is obtained from cork-oak tree. The bark of the oak tree is stripped, after preliminary drying and is used to produce a high-void elastic material. The material is moulded in the form of boards. Typical thermal conductivity values for cork are between 0.035 and 0.043 W/mK. Cork is naturally fire retardant and can be used as acoustic and thermal insulation in walls, floors, and ceilings. Because cork is a cellular material, its water vapour resistance is higher than that of other renewable materials. Cork board may be faced with aluminium sheet on both sides for improving its thermal properties.

**Sheep's wool**   Wool insulation is made from *sheep wool fibres* that are bonded together using adhesives to form insulating batts and rolls. It is made with recyclable and renewable source, nontoxic, and biodegradable. Sheep wool insulation requires less energy to produce than glass fibre insulation. As wool contains moisture, it is fire resistant, and extinguishes itself when the source of flame is removed. It is safe and easy to handle and does not require protective clothing or special breathing apparatus during installation. It is ideally suited to timber frame structures as it has a natural synergy with wood. It has a disadvantage of absorbing volatile organic compounds (VOCs).

**Hemp fibres** They are produced from hemp plant, which grows up to a height of nearly 4 m within a period of 100–120 days. The product is composed of 85% hemp fibre and the rest polyester binding, and 3–5% soda added for fire proofing. Hemp is a natural material and has high thermal resistance. In addition, it has the ability to absorb and release moisture without affecting thermal performance. Hemp fibre products are available as semi-rigid panels looking like mineral wool. It is designed for use in wood-framed construction and available in sheets 380 and 580 mm wide by 1200-mm long and in thicknesses of 87.5 mm for 50 × 100 mm frame construction (R-2.30), 137.5 mm for 50 × 150 framing (R-3.5), and 200 mm. The density is 35 kg/m³. The insulation is not available in rolled form, like fiberglass.

## 20.7 Plastic Insulation Products

Plastic insulation products can be (a) rigid foam board or (b) spray-in place foams. Rigid foam boards are of three types: (a) expanded polystyrene, (b) extruded polystyrene, and (c) isocyanurate boards. Similarly, there are three types of sprayed-in place foams: (a) urea-formaldehyde, (b) open-cell spray polyurethane foam, and (c) closed-cell spray polyurethane foam. Foam insulation reduces thermal bridging through wooden or steel studs, because it serves as thermal break (Krigger & Dorsi, 2004). The rigidity of rigid foam boards may provide structural strength when combined with wood sheathing. Plastic foam insulation may require metal or masonry covering, because it may deteriorate due to UV solar radiation.

### 20.7.1 Rigid Foam Boards

The three different types of rigid foam boards are briefly described here.

**Expanded polystyrene (IS 4671:1984, IS 14164:2008)** Made from polystyrene, a colourless, transparent thermoplastic, expanded polystyrene (EPS) is composed of small plastic beads that are fused together. The most versatile of the three rigid insulation options (see Table 20.7). It can be used to insulate roof, wall, floor, and also used in structural geofoam applications. EPS foam is often used in insulating concrete forms (ICFs), and structural insulated panels (SIPs) (see Section 17.9.7, for more details). EPS foam is the cheapest, does not retain water, and hence is approved for ground contact and below-grade applications. EPS is available faced or unfaced. Faced products are considered vapour retardant and their R-value does not degrade over time. Specifications for expanded polystyrene in the form of finished boards and blocks, for use in building applications in the temperature range from −150 to 80°C is provided in IS 4671:1984. The density of the material, calculated at nominal thickness, excluding facing, is in the range 15 to 35 kg/m³. Its thermal conductivity ranges from 0.031 to 0.037 W/(mK) depending on the density of the material. The size of finished boards is 1.0 × 0.5 m and they are available in thicknesses of 15, 20, 25, 40, 50, 60, 75, and 100 mm. For other requirements of EPS, such as compressive strength, moisture absorption, water-vapour permeability, etc., and for the tests to determine these properties, reference should be made to IS 4671.

**Fig. 20.8** Extruded polystyrene board

**Extruded polystyrene** Known as XPS, it is made by melting polystyrene and then pressing it to form sheets (see Fig. 20.8). It consists of closed cells and offers improved surface roughness, higher stiffness, and reduced thermal conductivity. XPS is commonly used as foam board insulation. Over time, the R-value of XPS may reduce slightly as some of the intercellular gas is replaced by air. The R-value of polystyrene foam board depends on its density. Recognized by its blue, green, or pink colour, XPS is used in walls or below-grade applications and is recyclable. XPS may be available as unfaced or with a number of different plastic facings. Thicker and faced XPS is stronger and can have a lower permeability rating, but is not a vapour barrier – it has to be noted that EPS absorbs more moisture than XPS products (www. xpsa.com).

**Preformed rigid polyurethane (PUR) and polyisocyanurate (PIR) boards (IS 12436:1988)**  Polyisocyanurate insulation is available as a liquid, sprayed foam, and rigid foam board. Laminated insulation panels, with a variety of facings, are also available. Foamed-in-place application of PIR insulation is cheaper than PIR foam boards, and may perform better. In the first 2 years of installation, the R-value of PIR insulation may reduce slightly as some of the intercellular gas is replaced by air. PIR foam boards with foil/plastic facings may stabilize the R-value. Properly installed reflective foil, and facing an open air space, can also act as a radiant barrier. Wall panels made PIR foam are typically 89-mm thick. Ceiling panels are up to 190-mm thick. These panels, although more expensive, are more fire and water vapour-diffusion resistant than EPS. They also insulate 30 to 40% better. Specification for preformed rigid polyurethane and polyisocyanurate foams, used in the temperature range of −180 to +140°C, are provided in IS 12436:1988. They may have density in the range of 32 to 80 kg/m$^3$ and thermal conductivity in the range of 0.017 to 0.025 W/(mK) at 50°C. They should be protected from fires. The application of polyurethane insulation by *in-situ* pouring is explained in IS 13205:1991. Specification for rigid phenolic foams is provided in IS 13204:1991.

## 20.7.2 Sprayed In-place Plastic Foams

Sprayed polyurethane foam (SPF) is an alternative to traditional insulation materials such as fiberglass and sprayed in place using a spray-gun. A two-component mixture of polyurethane and isocyanate is pumped together with a blowing agent at the tip of a gun, and is then sprayed as expanding foam. A blowing agent agitates the plastics-causing bubbles to form, and the material hardens within about 1 min. This expanding foam is sprayed onto roof tiles, concrete slabs, into wall cavities, or through holes drilled in a cavity of a finished wall. It has to be noted that cementitious foam is applied in a similar manner but does not expand. Buildings treated with spray foam insulation insulate about 50% better than traditional insulation products. The advantages are as follows:

1. The foam blocks airflow by expanding and seals all leaks, gaps, and penetrations.
2. It serves as vapour barrier, with a better permeability rating than plastic sheeting, thus reducing the build-up of moisture, which can cause mould growth.
3. It provides acoustical insulation.

The disadvantages are as follows:

1. They are expensive than traditional insulation.
2. Most foams, with the exception of cementitious foams, emit toxic fumes when they burn.
3. Although chlorofluorocarbon (CFC) is no longer used as blowing agent, many use hydro-chlorofluorocarbon (HCFC) or hydro-fluorocarbon (HFC) gas as blowing agents. Both are potential greenhouse gases and HCFCs also have ozone depletion potential.
4. Most foams require protection such as drywall on the interior of a house.
5. Most foams require protection from sunlight and solvents.
6. During their application, protective mask or goggles should be worn, else will lead to temporary impairment of vision.

The three different types of sprayed in-place plastic foams are briefly described here.

**Urea-formaldehyde foam** It is a spray foam used prior to 1980s. It is banned now due to the potentially toxic emission of formaldehyde gas. As the gas emissions fade with time, any such foam found now may not be hazardous. However, if the foam is disturbed, it may become powdery dust and hence should be handled with caution.

**Closed-cell polyurethane** It is more common than open-cell polyurethane. They are usually dense and strong and once placed look like solids foam. Thus, in addition to insulating the buildings, they also strengthen the walls on which it is applied. Owing to this, they provide better insulation and are also more resistant to water degradation or penetration (see Table 20.7, for their R-values). The advantages of closed-cell foam over open-cell foam are as follows:

1. As it is non-porous and has better moisture impermeability, it acts as a better vapour barrier.
2. It is a superior insulator – only about half the thickness of open-cell foam is required to attain the required R-rating.
3. It structurally reinforces the walls.
4. As it rarely requires any trimming, it results in little or no waste. However, it is expensive than open-cell foam.

**Open-cell polyurethane** It is not as common as closed-cell foam, but is cheaper. However, its installation is similar to that of closed-cell foam, but the effects are different. When sprayed, open cell foam expands and hence insulates hard to reach places, nooks, and corners. Hence, wall and ceiling cavities, which are difficult to seal with closed-cell foams, could be insulated. The expansion feature will hermetically seal the entire area and provide proper insulation. Open-cell foam is ideally suited for walls which have 87.5-mm cavities, and in the attic space. However, the R-value of open-cell polyurethane spray foam is lower (see Table 20.7). Hence, heat-resistance level will be lower, causing insulation failures during extreme temperatures. Even though they provide adequate insulation during normal weather, they do not have adequate water resistance, and their lower strength may provide less support to walls. Open cell foams will allow timber to breathe, more permeable to vapour and air, and effective as a sound barrier.

Properties of some thermal insulation materials are compared in Table 20.9.

**Table 20.9** Comparison of properties of some thermal insulation materials

| Type of insulation | Thermal conductivity W/(mK) | Embodied energy GJ/t | Moisture resistance | Chemical properties | Toxicity | CFC emissions /recyclability |
|---|---|---|---|---|---|---|
| Cellular glass | 0.042 | 27 | 0.2% by volume; no capillary action | Totally inorganic, impervious to common acids except hydrofluoric acid; may release hydrogen sulphide and CO in fire | None | Does not use CFCs or HCFCs; reclaimable on demolition |
| Cellulose | 0.033 | 0.63 to 1.25 | Absorbs much more water than fiberglass, mineral wool, or polystyrene | Treated with inorganic salts for fire protection | None, fully biodegradable | Does not use CFCs, HCFCs, or VOCs; 100% recyclable. |

*(Contd)*

**Table 20.9** (*Contd*)

| Type of insulation | Thermal conductivity W/(mK) | Embodied energy GJ/t | Moisture resistance | Chemical properties | Toxicity | CFC emissions /recyclability |
|---|---|---|---|---|---|---|
| Cork | 0.037 | Can be harvested every 9–12 years during the tree's 160–200 year life time | Water repellent with zero capillarity; relatively high rate of vapour transmission | Unaffected by water, alkalis, and organic solvents | None | None; can be recycled |
| Rock mineral wool | 0.034 to 0.036 | 25 | Does not absorb moisture and no capillary action | May need isolating board under asphalt. | None | Does not use CFCs or HCFCs; can be recycled |
| Expanded Polystyrene | 0.033–0.038 | 120 | Not a vapour barrier. Low water vapour transmission, no capillary action and high resistance to moisture absorption | Resistant to diluted acids and alkalis; not resistant to organic solvents; can chemically interact with membranes such as PVC | Thermal decomposition products are not toxic than that of wood | Does not use CFCs, HCFCs, or $CO_2$; easily melted and reformed |

## 20.8 Vapour Barriers

Vapour barriers resist water-vapour diffusion though the surface of the building envelope. The use of vapour barrier in an insulated building is necessary when the winter temperature is likely to fall below 4°C. When there is low temperature outside and warm inside temperature, there is a possibility of reaching dew point in some places inside the structure, which will result in water collecting and freezing. This moisture will eventually cause the insulation to lose its effectiveness and may cause problems to surface finish and structural member decay. The vapour barrier in such situations should be placed on the heated side of the structure, as normal moisture flow will be from this side towards the colder side. It is very dangerous to have internal and external surfaces of a cavity wall both completely watertight, because if the water does get into the wall, it cannot get out again, causing mould growth. This problem can be seen in the majority of American timber stud homes (Roaf et al., 2007).

When a vapour barrier is applied over insulation, it should cover all exposed edges of the insulation and bonded to the surface of element, on which the insulation is applied. In warm climates, with air conditioning and heating, moisture comes from outside to inside during summer and from inside to outside during winter. It may be best not to install vapour barriers in such cases (Krigger & Dorsi, 2004).

The following materials may be used as vapour seals (IS 14164:2008):

*Foils* These include (a) aluminium foil with a minimum thickness of 0.05 mm, and (b) aluminium foil laminated to kraft paper of minimum 60 g/m².

***Bituminous and resinous mastics*** Two coats of bitumen (conforming to fully blown type of IS 702) or resinous mastics having a water vapour permeability of not more than 2800 ng/(m²·s·Pa)

*Plastic sheets* These include polyester, polyethylene, polyisobutylene, and PVC coated fabrics. These sheets should require additional protection.

Any material that is relatively impermeable to water vapour such as oil-based primers and vapour barrier paints.

## 20.9 Low-E™ Reflective Insulation

**Fig. 20.9** Reflective insulation

Low-E™ Reflective Insulation is a new pin perforated product with a pure aluminium facing that is heat bonded to one or both sides of closed-cell polyethylene foam (see Fig. 20.9). As this system effectively seals cracks/gaps in the external sheathing, it acts as a superior air infiltration barrier. In addition, it blocks up to 97% of radiant heat thus providing better insulation to the exterior of the building, and better year round thermal efficiency and comfort levels. Furthermore, Low-E™ reflective insulation is totally nontoxic, and its performance is not affected by initial exposure to the atmospheric conditions.

## 20.10 Other Products

Other products that are used to provide thermal insulation include the following:

1. Autoclaved aerated (cellular) concrete blocks (see Section 9.6.1 of Chapter 9)
2. Preformed foam cellular concrete blocks (see Section 9.6.2 of Chapter 9 and IS 6598–1972)
3. Hollow concrete blocks (see Section 3.10.4 of Chapter 3)
4. Cavity wall construction
5. Sprayed asbestos fibres (which are banned in several countries due to the risk of lung cancer)
6. Thermal insulating cement (It can be of four types: hard-setting compositions, self-setting cements, gypsum plaster compositions, and fire-proof finishing cements – more details of these cements may be found in IS 7509:1993, IS 7510:1974, IS 9350:1980, and IS 9743:1990)
7. Straw bales

Cavity wall is constructed using two layers of masonry (using ordinary or fly ash concrete blocks) separated by a hollow space that provides thermal as well as sound insulation. This hollow space could be filled with insulation materials to increase thermal comfort. Although *rat-trap bond* construction technique could provide thermal comfort as well as saving about 25% of bricks used, it will occupy more floor area.

## 20.11 Fenestration

It refers to the openings in the building envelope, including the installation of windows, doors, and skylights. Windows are often the major source of solar heat gain in hot climates or summer and heat loss in cold climate or winter. Windows, doors, skylights can gain and lose heat through: (a) direct conduction through the glass or glazing, frame, and/or door, (b) the solar radiation of heat into a house and out of a house from objects, such as people, furniture, and interior walls, and (c) air leakage through and around them.

Window glass transmits from 20 to 84% of sun's heat depending on the type of glass used. The glasses that are used include clear glass, low e-glass, reflective glass, or heat-absorbing glass. The surface area of the windows and their R-values determine the level of comfort and extent of energy problems associated with windows. Windows should be selected based on three factors: U-value, solar heat gain coefficient (SHGC), and visible transmittance (VT).

**U-value** It indicates the rate of heat flow due to conduction, convection, and radiation through a window as a result of a temperature difference between the inside and outside. It is identical to the U-value discussed earlier. Lower U-values are more energy conserving than higher U-values (also see Section 15.8.6 of Chapter 15).

**Solar heat gain coefficient** SHGC is the ratio of solar heat passing through the window to solar heat striking the window. As the SHGC increases, solar energy transmitted through the window increases. The climate, orientation, and external shading of the building will determine the optimal SHGC for a particular window. The SHGC is a ratio between 0 and 1. Windows with lower SHGC values are desirable where high air-conditioning loads are expected while windows with higher SHGC values may be better where passive solar heating is needed (www.energy.gov).

**Visible transmittance** VT indicates how much visible light is transmitted through the window glass.

Table 20.10 provides energy characteristics of different window glass options in metal frames (Carmody & Haglund, 2012). As seen in this table, the thermal performance of windows can be improved by (a) multiple layers of glass or plastic film, (b) high-performance low-e or solar control coatings, (c) low-conductance gas fills, (d) warm edge spacers, and (e) high-performance frames. For superior energy performance, windows with a lower U-factor (≤0.40), lower SHGC (≤0.4), where heavy air conditioning is required, and higher VT (>0.70) to maximize daylight and view should be used. Windows with *Energy Star* certification provide higher energy performance.

**Table 20.10** Comparison of properties of different window glasses in metal frame (extracted from Carmody & Haglund, 2012)

| Type of Glazing | U-Value W/m²K | R-Value m²K/W | SHGC | VT | |
|---|---|---|---|---|---|
| Single glazing (standard clear) | 7.32 | 0.137 | 0.73 | 0.69 | New frame materials and designs |
| Double glazing (standard clear) | 4.71 | 0.212 | 0.65 | 0.63 | Low-emittance and/or solar control coating |
| Double glazing tinted | 4.71 | 0.212 | 0.54 | 0.47 | Low-conductance gas fill |
| Double Low-e, high SHGC, with argon gas fill | 3.69 | 0.271 | 0.58 | 0.61 | Warm edge spacer between glazings |
| Double Low-e, medium SHGC, with argon gas fill | 3.63 | 0.275 | 0.38 | 0.56 | |
| Double Low-e, low SHGC, with argon gas fill | 3.58 | 0.279 | 0.26 | 0.49 | Improved weatherstripping |
| Triple Low-e, low SHGC, with argon, improved | 1.07 | 0.935 | 0.18 | 0.37 | |

# 20.12 Insulation and Ventilation

Insulation and ventilation are inseparable and considerably influence the indoor air quality. If the building is sealed perfectly with good insulation but is not allowed adequate ventilation, the building will become stale and moisture-laden, leading to odours and mould and mildew growth. Many new, well-insulated buildings incorporate an air exchanger such as *heat recovery ventilator*, which ensures that the indoor air is exchanged regularly with outdoor air. It also allows part of the heat from the outgoing indoor air is transferred to the incoming outdoor air in order to boost energy efficiency (see https://zehnderamerica.com/heat-recovery-ventilator/ for more details).

## SUMMARY

- Insulation not only provides thermal comfort to the occupants, but also reduces the energy required to heat/cool buildings.
- Insulation materials are used to manage the three basic forms of heat transfer, i.e., radiation, conduction, and convection.
- The thermal properties of insulating materials include: specific heat, thermal capacity, thermal conductivity, thermal conductance, thermal emissivity, thermal reflectivity, solar reflectance index, thermal transmittance or U-value, and related thermal resistance or R-value.
- While selecting a thermal insulating material, the following factors should be considered: energy required for production, cost, low thermal conductivity, good thermal resistance, thermal transmittance, stability, density, durability, fire resistance and flammability, lack of odour, low chemical activity, moisture resistance, compressive strength, heat storage capacity, acoustic properties, and ease of application.
- Inorganic thermal insulation materials include fibrous materials such as glass, rock, and slag wool, cellular materials such as calcium silicate, vermiculite, perlite and pumice, gypsum boards, and aerogels.
- Organic thermal insulation materials consists of natural materials (such as cellulose fibres, cotton, saw dust and wood fibres, coconut fibres, cork boards, sheep's wool, and hemp fibres) or synthetic materials (such as rigid foam board and spray in-place foams).
- Rigid foam boards are of three types: (a) axpanded polystyrene, (b) extruded polystyrene, and (c) isocyanurate boards.
- The three types of sprayed-in place foams: (a) urea-formaldehyde, (b) open-cell spray polyurethane foam, and (c) closed-cell spray polyurethane foam. Vapour barriers such as foils, bituminous and resinous mastics, and plastic sheets resist water-vapour diffusion through the surface of the building envelope.
- Low-E™ reflective insulation can effectively seal cracks/gaps in the external sheathing and block up to 97% of radiant heat.
- The thermal performance of windows can be improved by (a) multiple layers of glass or plastic film, (b) high-performance low-e or solar control coatings, (c) low-conductance gas fills, (d) warm edge spacers, and (e) high-performance frames. Insulation and ventilation should be considered together to provide better indoor air quality.
- Several tests to determine properties such as compressive strength at 10% deformation, water absorption, water-vapour permeability, thermal stability, flammability, cross-breaking strength, etc., are prescribed in IS codes such as IS 4671, IS 11129, IS 11239 (Parts 1 to 13), and IS 13286.

## EXERCISES

### Multiple-choice Questions

1. Thermal comfort for most is achieved when
   - (a) temperature is between 20 and 25°C
   - (b) RH is between 30 and 65%
   - (c) temperature is between 15 and 20°C and RH is between 30 and 40%
   - (d) temperature is between 20 and 25°C and RH is between 30 and 65%

2. Heat energy transferred by displacement and circulation is called
   - (a) conduction
   - (b) convection
   - (c) radiation
   - (d) mechanical transfer

3. Unit for specific heat is
   - (a) J/Kg /°C
   - (b) Kg /°C/W
   - (c) J/K /m
   - (d) K W/m

4. Unit for thermal conductivity is
   - (a) mK/W
   - (b) W/m²K
   - (c) W/mK
   - (d) J/kgK

5. Least thermal conductivity material is
   - (a) wood
   - (b) air
   - (c) glass
   - (d) concrete

6. U-value of 100 mm RCC
   - (a) 0.749
   - (b) 1.08
   - (c) 2.23
   - (d) 3.59

7. Which one of the following is an inorganic thermal insulation material?
   - (a) Plastic fibre
   - (b) Vermiculite
   - (c) Cellulose
   - (d) Hump fibre

8. A good thermal insulation material should not possess
   - (a) high density
   - (b) high temperature range
   - (c) high volume
   - (d) high moisture content
9. Identify the non-vapour barrier.
   - (a) Aluminium foil
   - (b) Mineral wool
   - (c) Plastic sheets
   - (d) Bituminous mastics
10. Which is not considered while selecting a window for thermal insulation?
    - (a) U-value
    - (b) Solar heat gain coefficient
    - (c) Thermal capacity
    - (d) Visible transmittance

## Review Questions

1. Explain how location, orientation, and vegetation around the building affect the thermal performance.
2. What are the three basic mechanisms of heat energy transfer?
3. List any five basic thermal properties that are considered while selecting thermal insulation materials.
4. Define thermal conductivity (K), thermal transmittance (U), and thermal resistance (R).
5. What are the basic factors that are considered while selecting thermal insulating materials?
6. What is meant by thermal bridging? What strategies could be adopted to reduce thermal bridging?
7. How are insulation materials classified?
8. What are reflective insulation materials? How do they function? Where will you recommend them?
9. Write short notes on the following insulating materials: (a) glass and rock/slag wool, (b) vermiculite, (c) calcium silicate, and (d) aerogels.
10. Describe some of these organic insulation materials: (a) cellulose fibres, (b) cork-board, (c) cotton, and (d) sheep's wool.
11. Explain any three naturally available organic insulation materials used in construction.
12. What are the three different types of rigid foam boards? Explain each of them.
13. What are the three different types of sprayed-in place foams? Explain each of them.
14. Why vapour barriers are considered important for insulation? What are three normally used vapour barriers?
15. Will rat-trap bond of brickwork provide insulation? What is its advantage and drawback?
16. How windows, doors, and skylights affect the thermal insulation of buildings? What are the three factors while considering window glass selection? Explain them briefly.
17. How does ventilation affect insulation?
18. List the various eco-friendly insulating materials used in construction. List three insulation materials which are injurious to health.

## ANSWERS

### Multiple-choice Questions

| | | | | | |
|---|---|---|---|---|---|
| **1.** (d) | **2.** (b) | **3.** (a) | **4.** (c) | **5.** (b) | **6.** (d) |
| **7.** (b) | **8.** (d) | **9.** (b) | **10.** (c) | | |

# CHAPTER 21
# SOUND INSULATING MATERIALS

## 21.1 Introduction

Rapid urbanization in India has led to various public health challenges, including environmental pollution. Noise pollution is one of the major environmental concerns in India and sadly many are not even aware of the hazards it can cause. According to the World Health Organization (WHO) guidelines, for a sound sleep, the noise in bed rooms should be less than 30 dB. Similarly, to have better concentration in any classroom, the noise level should not exceed 35 dB (WHO, 2011). Noise levels at airports and railway stations could reach levels ranging from 70 to 100 dB. According to the US Bureau of Transportation Statistics, noise pollution from aviation and interstate highways in the USA are at levels below 50 dB. Whereas, the situation in India is alarming; for example, the recorded noise level in Mumbai during Ganeshutsav was 123.2 dB in 2013 (reduced to 116.4 dB in 2016), with an average of 71.25 dB (www.hindustantimes.com).

Constant exposure to noise results in uncomfortable living conditions as well as serious illnesses such as poor concentration, insomnia, hearing loss, increased stress, behavioural and mental problems, heart ailments, and hypertension (Jamir et al., 2014). Realizing this, the central government came up with Noise Pollution (Regulation and Control) Rules 2000 (see Table 21.1).

**Table 21.1** Central board for pollution control norms for noise (*Source*: http://envfor.nic.in)

| Area | Day, dB (A) | Night, dB (A) |
| --- | --- | --- |
| Residential | 55 | 45 |
| Commercial | 65 | 55 |
| Industrial | 75 | 70 |
| Silence zone | 50 | 40 |
| Day time: 6 am to 9 pm and Night time: 9 pm to 6 am | | |

The aforementioned underlines the importance of keeping the noise levels within acceptable limits inside the buildings, by adopting sound isolation measures and using sound insulation materials in buildings. As windows are responsible for the noise problem in 90% of the cases, proper attention should be paid while selecting the glazed surfaces of buildings, such that quite working and living conditions are achieved.

### 21.1.1 Brief History

The Greeks knew how sound propagates and used this knowledge while building the semicircular outdoor amphitheatre at Epidaurus, in the 5th century B.C., where one can hear a whisper from as far as 60 m away. The Romans could design interiors with ideal acoustics, as proved in the Pantheon in Rome. The Middle Age builders of Europe also built cathedrals for maximum acoustical effect. Sir Christopher Wren and other 18th century architects designed concert halls with optimized listening experience at any seat. However, only Sir Isaac Newton (1643–1727) scientifically demonstrated that sound waves travel through any medium—solid, liquid, or gaseous—and that the speed with which they propagate depends upon the elasticity and density of the medium. In 1866, the German scientist, Charles Kundt, using a primitive device, could calculate with reasonable accuracy the speed of sound in air to be 343 m/s (Using the present-day precision electronic instruments, the speed of sound at 20°C air and at normal atmospheric pressure is calculated as 344.4 m/s). Now, noise can be measured using sound level meter, noise dosimeter, or impulse-sound level meter.

## 21.2 Fundamentals of Acoustics

To understand the basics of noise control measures to be undertaken in buildings, it is essential to understand the fundamental properties of sound such as: (a) frequency (pitch), (b) wavelength, and (c) amplitude (loudness). Once these fundamental properties of sound waves are understood, effective noise control measures could be implemented.

### 21.2.1 Sound and Noise

The science of sound is known as *acoustics* (IS 9736:1981). *Sound* is a form of energy which can be transmitted over a distance from its source through a medium, such as air or solid element of construction such as wall or floor, as waves. For example, when a door is slammed, the door vibrates, sending sound waves through the air, which produces irregular and harsh sound. The loud wave at the start become soft (small) wave as it travels a distance as shown in Fig. 21.1(a). Similarly, when a guitar string is struck, it sends a sound wave as shown in Fig. 21.1(b), resulting in a continuous, regular series of repeated cycles, which we hear as a musical tone. This waveform also makes the same transition from loud to soft waves as in the case of sound produced by door slamming.

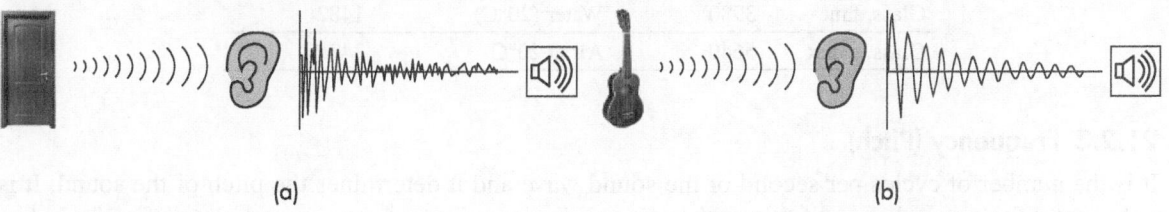

(a)                                                      (b)

**Fig. 21.1** Sound waves (a) Sound due to slamming of door (b) Sound due to playing of guitar

Thus, when a musical instrument, as shown in Fig. 21.2, is played, it generates pressure waves in a medium such as air. These air pressure changes travel as waves through the air and the sound produced are heard by humans using ears. As shown in Fig. 21.2, a sound wave consists of alternating compressions and rarefactions, or regions of high and low pressure, moving at a certain speed.

Noise is loud sound and unwanted sound that is disturbing, irritating, annoying, and even painful. As already mentioned, noise may be hazardous to health and interfere with verbal communications. Noise is composed of many frequencies and hence it may affect one person more than another. The basic

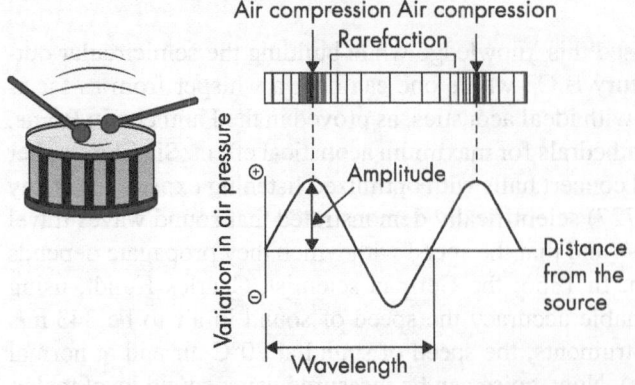

**Fig. 21.2** Graphic representation of a sound wave

problems associated with the measurement of noise and evaluation of its effects on humans are outlined in IS 9876:1981 (see also IS 9989:1981).

### 21.2.2 Speed of Sound

It depends on the type of medium and the temperature of the medium.

Speed of sound in gases, $c = \sqrt{\gamma R T}$ m/s (21.1)

where $\gamma$ is the ratio of specific heats (1.4 for air), $R$ is the gas constant (286.9 J/kg·K for air), and $T$ is the absolute temperature (273.15 +°C).

Thus, for air at 20°C, $c = \sqrt{1.4 \times 286.9 \times (273.15 + 20)} = 343.1$ m/s

Speed of sound in solids, $c = \sqrt{E/\rho}$ m/s (21.2)

where, $E$ is the modulus of elasticity and $\rho$ is the density of the material.

Table 21.2 shows the speed of sound in various materials. It has to be noted that the denser the medium, the faster sound travels in that medium.

**Table 21.2** Speed of sound in various materials

| Material | Speed, m/s | Material | Speed, m/s |
| --- | --- | --- | --- |
| Aluminium | 6420 | Granite | 5950 |
| Brick | 3650 | Gypsum board | 6800 |
| Copper | 3560 | Iron | 5130 |
| Concrete | 3100 | Rubber, vulcanized | 54 |
| Cork | 500 | Steel, mild | 5050 |
| Glass, crown | 5100 | Wood, oak | 3850 |
| Glass, flint | 3980 | Water (20°C) | 1482 |
| Glass, pyrex | 5640 | Air at 20°C | 343.1 |

### 21.2.3 Frequency (Pitch)

It is the number of cycles per second of the sound wave and it determines the pitch of the sound. It is only useful for musical sounds, where there is regular waveform. A pure sound wave of a single frequency takes the shape of a sine wave, as shown in Fig. 21.3. The time taken by a wave to complete one oscillation is called the *time period*. Thus, frequency [denoted by $f$ and having a unit of hertz (Hz)] is

Frequency, $f = c/\lambda$ (21.3)

Time period, $T = \dfrac{1}{f} = \dfrac{\lambda}{c}$ (21.4)

where $c$ is the wave speed in m/s, and $\lambda$ is the wave length in m.

A frequency of 1 Hz means one wave cycle per second. A frequency of 10 Hz means 10 wave cycles per second, where the cycles are much shorter and closer together [Fig. 21.3(b) and (d)].

**Fig. 21.3** Frequency of sound (a) Low frequency (b) High frequency
(c) Frequency of 1 Hz (d) Frequency of 10 Hz

Most of the sounds we normally hear consist of a combination of many different frequencies, depending on the nature of the sound generator. For example, the sound of a bell in a temple has a mix of predominantly high frequencies and the sound of a bass drum or large truck has a mix of predominantly lower frequencies. A healthy young person can hear frequencies in the range of 20 to 20,000 Hz, which deceases at middle age to about 70 to 14,000 Hz. However, the human ear is less sensitive to lower frequencies of sounds below 100 Hz. The sound at a frequency near about 4000 Hz is considered by us a loud. Frequencies which fall below 20 Hz are known as *infrasound* and frequencies above 20,000 Hz are called *ultrasound*.

For purposes of noise control, acousticians divide the audible sound spectrum into '*octaves*', and express them in *octave bands*, which are their centre frequencies, as shown in Table 21.3. It has to be noted that each centre frequency shown in this table is twice that of the one before it.

**Table 21.3** Octave band and band limits

| Octave band centre frequencies, Hz | Band limits |
|---|---|
| 32 | 22–45 |
| 64 | 45–89 |
| 125 | 89–178 |
| 250 | 178–355 |
| 500 | 354–709 |
| 1000 | 707–1414 |
| 2000 | 1411–2822 |
| 4000 | 2815–5630 |
| 8000 | 5617–11234 |

## 21.2.4 Wavelength

The wavelength of a sound wave is the length in space of one complete cycle of a sound wave as shown in Fig. 21.2. Numerically, it will be equal to the speed of sound divided by the frequency of the sound wave. For example, the wavelength of a 100 Hz sound in air at room temperature is 344.4 (m/sec)/ 100 Hz = 3.44 m.

## 21.2.5 Amplitude (Loudness)

Amplitude is the maximum displacement of a sound wave, on either side of its mean position and shows how loud the sound is (see Fig. 21.4). It has to be noted that the sounds having the same wavelength (equal frequency)

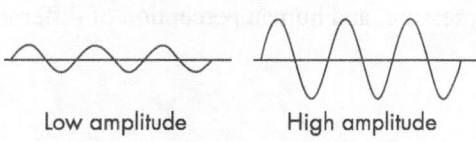

Low amplitude          High amplitude

**Fig. 21.4** Low and high amplitude

may have differing loudness. Loudness of a sound is measured by the logarithm of the intensity. Sound intensity level is defined by

$$\beta = 10 \log_{10} \left( \frac{I}{I_0} \right) dB \qquad (21.5)$$

where $I$ is the intensity of sound and $I_0$ is the reference intensity, which is the threshold of hearing intensity of $10^{-12}$ W/m². The intensity of sound at a location varies inversely to the square of the distance of source.

A decibel (dB) is one tenth of a bel (B) (Bel was originally called transmission unit or TU but was renamed in 1924 in the honour of Alexander Graham Bell).

**Example 21.1** The sound level at 25 m from a loudspeaker is 68 dB. What is the rate at which sound energy is being produced by the loudspeaker, assuming it to be an isotropic source?

**Solution:** Given: $\beta = 10 \log_{10} \left( \frac{I}{I_0} \right) = 68 \, dB$

Thus, $\log_{10} \left( \frac{I}{I_0} \right) = 6.8$, $\left( \frac{I}{I_0} \right) = 10^{6.8}$

Hence, $I = I_0 \times 10^{6.8} = 10^{-12} \times 10^{6.8} = 6.31 \times 10^{-6}$ W/m²

*Note*: The intensity of an isotropic source is defined by: $I = \frac{P}{4\pi r^2}$ or $P = I \left( 4\pi r^2 \right)$ $\qquad (21.6)$

Thus, $P = 6.31 \times 10^{-6} \times 4\pi \times 25^2 = 0.0496$ Watt

## 21.2.6 Sound Pressure Level

Sound intensity is difficult to measure. Sound pressure detected by human ear is usually the easiest to measure in acoustics and measured in the units of Pa (N/m²). For this reason, sound level is usually defined in terms of *Sound Pressure Level* (SPL). The SPL is actually a ratio of the absolute sound pressure and a reference level (usually, the threshold of hearing or the lowest intensity of sound that can be heard by most people). The SPL is measured in decibels (dB), because of the broad range of intensities we can hear. Note that the SPL can be measured with a sound pressure level meter. The commonly used reference sound pressure (threshold of human hearing) in air is 20 μPa (beyond 200 Pa, we experience pain). Most sound level measurements are made relative to this reference, meaning 1 Pa will equal an the SPL of 94 dB. Thus, the SPL can be expressed using a logarithmic scale, with units of dB, and ranges from 0 to 150 dB (Everest and Pohlmann, 2014).

$$SPL = 10 \log_{10} \left( \frac{p^2}{p_{ref}^2} \right) = 20 \log_{10} \left( \frac{p}{20 \mu \, Pa} \right) dB \qquad (21.7)$$

where the SPL is sound pressure level (dB), $p$ is the acoustic pressure (μPa) , and $p_{ref}$ is the sound pressure at the threshold of hearing (0.00002 Pa). Table 21.4 shows the approximate intensity, sound pressure, and human perception of different sources of noise.

**Table 21.4** Sound levels and loudness sensation

| Noise source | Approximate intensity (dB) | Approximate sound pressure (µPa) | Human perception of sound |
|---|---|---|---|
| Hearing threshold | 0 | 20 | – |
| Rustle of leaves, ticking clock, whisper | 10–20 | 60–200 | Whisper |
| Quiet home or private office, empty auditorium, library | 20–40 | 200–2000 | Faint |
| Noisy home, average radio, conversation of two people standing 1 m apart | 40–60 | 2000–20,000 | Moderately loud |
| Highway traffic noise at 15 m from the highway, cocktail party, printing press, loud TV/radio, noisy restaurant | 70–80 | 80,000–200,000 | Loud |
| Loud street noise, noisy factory, police whistle, power lawn mover | 80–100 | 2,00,000 to 2,000,000 | Very loud |
| Ambulance siren, artillery, auto horn, thunder, elevated train, discotheque, hammer drill, impact wrench, fire crackers | 100–120 | 2,000,000–20,000,000 | Deafening |
| Jet aircraft at 30 m, bass drum at 1 m, or auto horn at 1 m | 130–140 | >100,000,000 | Pain threshold |
| Rocket launch | 180 | | Dangerous |

A reduction of sound from 65 to 55 dB reduces the loudness of the sound by one half, while a reduction of sound from 65 to 45 dB results in a loudness reduction of one quarter. The formula for the sum of the sound pressure levels of $n$ random radiating sources is

$$SPL = \sum SPL_i = 10 log_{10}\left(\frac{p_1^2 + p_2^2 + \ldots + p_n^2}{p_{ref}^2}\right) = 10 log_{10}\left[\left(\frac{p_1^2}{p_{ref}^2}\right) + \left(\frac{p_2^2}{p_{ref}^2}\right) + \ldots + \left(\frac{p_n^2}{p_{ref}^2}\right)\right]$$

Realizing that $\left(\dfrac{p_1^2}{p_{ref}^2}\right) = 10^{\frac{SPL_i}{10}}$, $i = 1,2,\ldots,n$, we get

$$\sum SPL_i = 10 log_{10}\left(10^{\frac{SPL_1}{10}} + 10^{\frac{SPL_2}{10}} + \ldots 10^{\frac{SPL_3}{10}}\right) \tag{21.8}$$

**Example 21.2**   What is the resulting sound level when there are two concurrent sources of sound, each measuring 70 dB(A)?

**Solution:** When there are two equal concurrent sources of sound, the resulting sound level will not be 140 dB.

Combined dB $= 10 log_{10}\left(10^{\frac{70}{10}} + 10^{\frac{70}{10}}\right) = 10 log_{10}\left(20 \times 10^6\right) = 10 \times 7.301 = 73.01 = $ dB(A)

*Note:* The (A) after dB refers to a weighting scale, built into the sound measuring instruments, that approximates the manner in which we hear higher frequencies better than lower frequencies.

## 21.2.7 Airborne and Structure-borne Sound

The origin of noise may be outdoor or indoor. Outdoor noises are due to traffic (roadways and honking of car horns, railways, and airplanes), blaring loudspeakers used during festivals and by politicians, construction activities, fireworks, children playing, hawkers, services deliveries, building operations such as lifts, and various types of machinery in the neighbourhood of buildings (e.g., grass-mowing equipment). Whereas, indoor noise include all noise emitted from domestic appliances such as television, radio, musical instruments, refrigerator, juicer-mixer-grinder, vacuum cleaners, and washing machine, operation of the cistern and water-closets, cooling and ventilation machinery, conversation/shouting of the occupants, footsteps, banging of doors, movement of furniture, and children dancing on the upper floor. These may emanate from an adjacent room or an adjacent building. Noise conditions vary from time to time; a noise which may be tolerable and not objectionable during the day time may become annoying during the night time, when one is sleeping or taking rest. Quiet conditions are essential, especially in bed rooms and study rooms. As mentioned earlier, sound waves can travel through any media – air, water, wood, masonry, or metal. However, sound cannot travel through empty or vacuum space. Depending on how sound is transferred, it is classified as airborne or structure-borne (see Fig. 21.5).

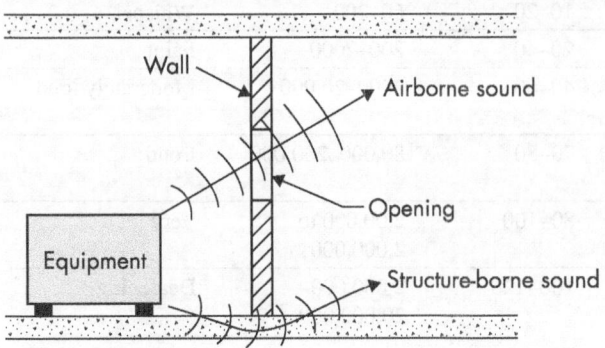

**Fig. 21.5** Airborne and structure-borne sound

**Airborne sound**   It is the sound that radiates from a source directly into the room through the air. It may travel into rooms through doors, windows, ventilators, ventilating ducts, and other openings, or even through holes and cracks (see Fig. 21.5). The sound of traffic from the road, the sound of music, voices from the next room, and the noise from a low-flying aircraft – all travel as airborne sound. Air-tightness is essential and important to insulate against airborne noises. Reduction of airborne noise requires the use of rigid and massive walls, or acoustically designed dry walls, without any openings. All doors and windows should be properly gasketed where a high degree of sound insulation is desired. Airborne sound can make a wall to vibrate as a diaphragm; due to this, the wall can transmit the sound through the interconnected solid structure (IS 1950:1962). Ventilating ducts or air transfer openings should be designed to minimize transmission of noise.

**Structure-borne noise**   This sound is usually caused by impact, travels through solid materials in direct mechanical contact with the sound source, and can be transmitted as airborne noise (see Fig. 21.5). Examples are heavy footsteps, falling of objects, and dragging of furniture on the floor above, knock at the door, and vibration due to equipment. A motor bolted to a floor or a slammed door can cause the structure to vibrate significantly, which travel great distances through solid structure with little loss (As per Table 21.2, sound travels in steel about 15 times faster than air). Air-conditioner noises can be transmitted to a room by the air in the ducts, by the metal of the ducts, or by both. Water pipes and plumbing fixtures have excellent sound-carrying capacities (Everest and Pohlmann, 2014). All structure-borne sound is heard by us when they are airborne. Otherwise, they are felt as vibrations in a material. Impact sound travels more readily in denser materials (IS 1950:1962).

Structure-borne noise cannot be attenuated by constructing walls or barriers; it requires isolating the vibration – by using rubber/neoprene pads or springs below the sound producing machine, or by providing gaps in the walls of the structure such that the vibration cannot pass through it. During noise control, both airborne and structure-borne sound must be considered.

### 21.2.8 Transmission Loss

When airborne noise is transmitted through a structure, there will be a reduction in sound intensity. This reduction in sound is termed as *transmission loss* (TL) and is measured in decibels. The efficiency of any sound insulation material is expressed in terms of this TL of airborne sound. Thus, if 80 and 45 dB are the sound levels measured on either side of a wall, the TL or 'sound insulation of that wall' is 80 – 45 = 35 dB. Sound insulation offered by a wall, floor, ceiling, or insulating material depends on the materials used, their thickness, porosity, and density. It has to be noted this value also varies with the frequency; it will be high for high frequencies and low for the medium and low frequencies. Hence, the performance of a sound insulator cannot be exactly determined based on the TL at one frequency. Hence, in practice, the average of such values in the range of 250 to 2,000 Hz is considered.

## 21.3 Maximum Acceptable Noise Levels

The maximum acceptable noise level inside buildings from the point of view of comfort, economy, practical considerations, and as per Indian conditions is prescribed in IS 1950:1962 and is reproduced in Table 21.5. The desirable sound insulation required to be provided for the various types of buildings is provided in IS 1950:1962.

**Table 21.5** Maximum acceptable noise levels

| Type of building | Accepted noise level (dB) | |
|---|---|---|
| | As per NBC, Part 8, Section 4, 2016 | As per ASHRAE Handbook |
| Large offices, banks, and stores | 45–50 | 25–45 |
| Residences, hotel rooms | 35–40 | 25–35 |
| Class rooms | 40–45 | 25–40 |
| Hospitals, small offices, and libraries | 35–40 | 25–40 |

## 21.4 Principles of Soundproofing

While considering home acoustics, it should be realized that minimizing sound and sound proofing are not the same. In general usage, soundproofing a room refers to is sound-attenuation, which is easier than to make it totally soundproof. Soundproofing is a complicated process, and is rarely done in an ordinary house building.

The following six principles of soundproofing will be useful while solving any soundproofing problem: (a) mass, (b) absorption, (c) conduction, (d) resonance, (e) mechanical decoupling or isolation, and (f) damping.

**Mass**　Sound is transferred through vibration. Hence, a heavier partition is more difficult to vibrate than a lightweight partition. Thus, low-frequency isolation can be obtained by adding some mass. However, for middle- or high-frequency isolation, very large addition of mass is required. For example, doubling the weight of a concrete wall (without an air cavity), will increase the TL only by about 5 dB (see also Table VII of IS 1950:1962). Similarly, in single wood stud walls, doubling the number of drywall layers yields 4–5 dB of improvement. However, it will be more effective when the second layer of drywall is coupled with a heavier soundproofing material, such as mass-loaded vinyl.

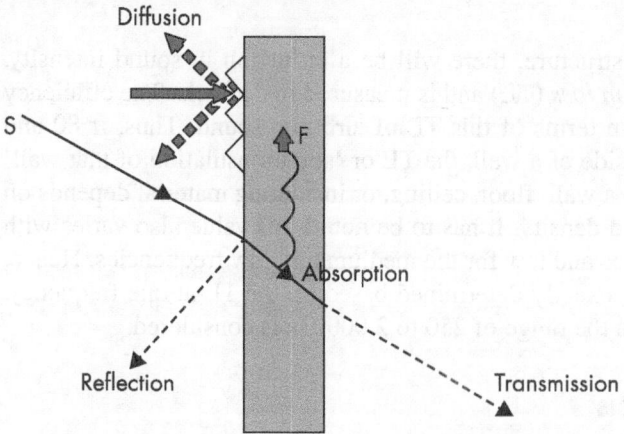

**Fig. 21.6** Absorption, reflection, and transmission of sound

**Absorption** When sound wave hits a wall, part of the sound energy will be absorbed within the material, some of it will be reflected, and some is transmitted through it, as shown in Fig. 21.6. The direction of travel of refracted component will be downwards because the wall material is denser than air. A fraction of the sound energy is lost as heat (as shown by F in Fig. 21.6) due to the friction offered by the material to the vibration of air particles.

The proportion of sound that is reflected, absorbed, or transmitted depends on the type of construction hit by the sound wave, and the frequency of the sound. All materials absorb some acoustical energy. Denser materials, such as gypsum board, wood, concrete, brick, and tile, are fairly reflective, do not absorb much sound, and are considered *sound barrier materials*. On the other hand, fibrous or soft materials, such as carpet, foam padding, fibreglass insulation, are *sound absorbers* and are poor barriers. In general, low-frequency sounds are very difficult to absorb because of their long wavelength. Absorption will help increase isolation in the mid-to-high frequency ranges, but will not help to isolate low frequencies and footfall impact noise. The value of absorption is increased significantly when the assembly is decoupled.

**Noise reduction coefficient** The sound-absorbing quality of a material is described by an *absorption coefficient*, $\alpha$, which is defined as the ratio of the total energy incident on a surface minus the energy reflected from the surface, to the energy incident upon the surface. The absorption coefficient also called the *noise reduction coefficient* (NRC) has a value between zero and one. An NRC of 0 indicates perfect reflection and NRC of 1 indicates perfect absorption. This single number NRC rating of a material is the average, rounded to the nearest multiple of 0.05, of the absorption coefficients of a material determined at the octave band centre frequencies of 250 Hz, 500 Hz, 1 kHz, and 2 kHz (see Table 21.3). A material is considered a sound absorber if it has an NRC value greater than 0.35. A carpet on rubber underlay may have a NRC of about 0.4, whereas a glass window will have an NRC of about 0.05 only. ASTM C423–09 has replaced NRC by *sound absorption average* (SAA). The SAA is a single-number rating of sound absorption properties of a material similar to NRC, except that it is taken as an average of 12 one-third octave bands from 200 to 2500 Hz, and rounded to the nearest 0.01. [see also the Indian standard method of rating of sound insulation in IS 11050 (Parts 1–3):1984].

**Conduction** While sound can travel through air pockets, such as ductwork, stud and ceiling and joist cavities, vibrations from sound can also be conducted along wood framing, plumbing pipes, sheet metal, concrete, and glass. This *flanking* transmission is structure-borne sound transmission that bypasses the separating wall and travels through other building elements, such as the floor, ceiling, or walls, abutting the separating wall or ceiling (see Fig. 21.7). Flanking transmission can also occur through joints or penetrations in the assembly. Sound transmission through conduction may be broken by either decoupling or damping the materials. Both these approaches can be accomplished by a number available of products.

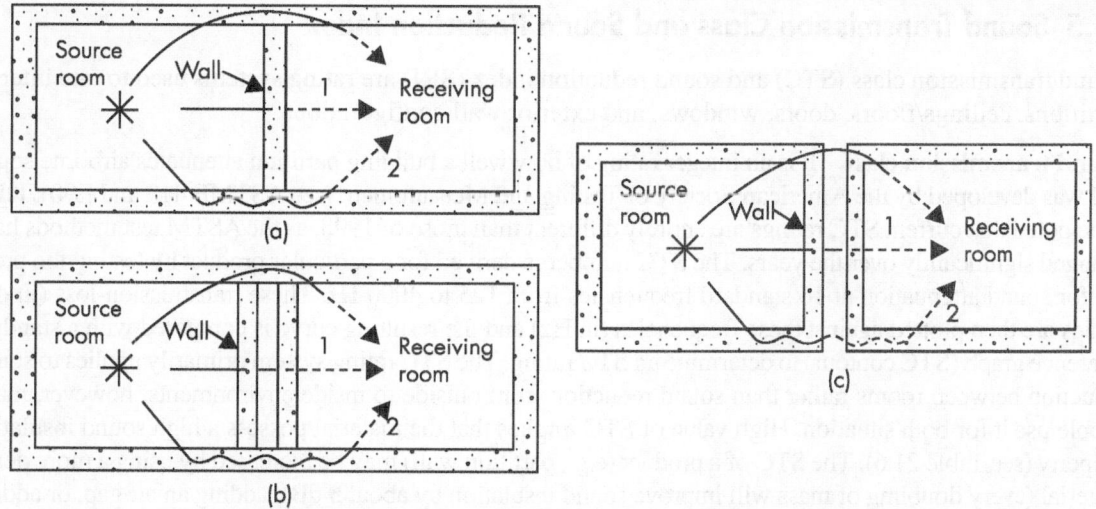

**Fig. 21.7** Schematic illustration of flanking paths of sound (adapted from UFC 3–450–01, 2003) (a) Single wall (b) Double wall (c) Isolated structure

**Resonance**    It occurs when natural frequency and forcing frequency coincides causing maximum amplitude. Each building material will have a specific resonance frequency. This frequency will vary depending on the type of product and the mass of that product. At resonance frequencies, even a massive decoupled wall with insulation will vibrate. Resonance increases the ease with which sound is transmitted. Resonance is avoided by damping or lowering the resonance frequency of an assembly. It is better to avoid creating more than one airspace in any assembly and insulate that one airspace with the proper amount of insulation. Other suggestion to avoid resonance is: if one side of the assembly has 12-mm drywall, provide the other side with 16-mm drywall. If multiple layers of drywall or wood are added to one side of the assembly then use a different thickness for each layer.

**Mechanical decoupling**    Decoupling is not effective at all frequencies. In fact, decoupling a wall results in reduced low frequency performance around its resonance point, and is only effective at frequencies much higher than resonance. Decoupling a wall means mechanically separating the two sides of a wall to make it harder for sound to pass through the wall. Decoupling may be done by providing air cavity in the wall or providing insulation material in the air cavity. Other means of providing decoupling include adding 'whisper clips' to a surface (which mechanically separate two sides of a wall), double studding the walls, or even staggering the locations of the studs.

**Damping**    Sound-dampening materials can be used to dissipate sound and reduce the amplitude and energy of the noise. They come in the form of mats, rolls, and viscoelastic compounds (such as Green glue®) and can be used to soundproof floors, walls, and ceilings. The concept of damping is to convert sound vibrations to non-perceptible amounts of heat. When glass is used for a wall or a window, damping can be increased by using laminated glass, in which a plastic sheet is sandwiched between two layers of glass (see also Section 15.8.1). Damping is most effective against low-frequency noises such as the noise from construction machinery or booming bass beats of music.

## 21.5 Sound Transmission Class and Sound Reduction Index

Sound transmission class (STC) and sound reduction index (SRI) are rating systems used to rate interior partitions, ceilings/floors, doors, windows, and exterior wall configurations.

**Sound transmission class**   It is an integer rating of how well a building partition attenuates airborne sound and was developed by the American Society of Testing and Measurement (see ASTME 413 and E90). It has to be noted that current STC ratings are entirely different than those of 1998, as the ASTM test methods have changed significantly over the years. The STC number is derived for a particular product by testing the product for sound attenuation at 16 standard frequencies from 125 to 4000 Hz. These transmission-loss (in dB) values are then plotted against these frequencies (in Hz) and the resulting curve is compared with a standard reference graph (STC contour) to determine an STC rating. The STC rating system primarily applies to sound reduction between rooms rather than sound reduction from outside to inside environments; however, many people use it for both situation. High value of STC implies that the material possess a high sound insulating property (see Table 21.6). The STC of a product (e.g., partition wall) may be increased by using a more dense material (every doubling of mass will improve sound insulation by about 5 dB), adding an air gap, or adding sound-absorbing material. More information on STC may be found in Everest and Pohlmann (2014).

**Table 21.6** Sound transmission class (STC) and effectiveness of noise control

| STC rating | Speech audibility | Effectiveness/Rating |
|---|---|---|
| 25 | Normal speech easy to understood through the wall | Poor |
| 30 | Loud speech can be easily understood, normal speech heard but not understood | Poor |
| 35 | Loud speech audible but not intelligible | Average |
| 40 | Loud speech just audible (onset of privacy) | Fair |
| 45 | Loud speech faintly heard but not understood | Good |
| 50 | Loud speech usually not heard | Very good |
| 60 | Most sounds inaudible (superior sound proofing) | Excellent |

**Note:** Assumes background noise of 30 dB on the listening side

Thus, NRC is a measurement of how well a product makes the room quieter, whereas STC measures how well a product keeps sound from escaping the room.

**Sound reduction index**   In countries outside the USA, the SRI is used and is expressed in decibels as per ISO 16283 (Parts 1–3). Even though the basic method, the actual measurements, and the mathematical calculations of both STC and SRI are similar, they differ in the detail and numerical results. In general, the apparent SRI is given as (ISO 16283–1:2014)

$$SRI = D + 10 log_{10}\left(\frac{S}{A}\right) dB \tag{21.9}$$

where $D$ is the difference in the energy (average sound pressure levels between the source and receiving room), $S$ is the area of the common partition, in $m^2$, and $A$ is the equivalent absorption area of the receiving room, in $m^2$. Like STC, SRI values are highly dependent on sound frequencies.

## 21.6 Sound-absorbing Materials

Most of the thermal insulation materials described in Chapter 20 could be used for sound insulation. The desirable characteristics of an sound insulating material are acoustically efficient, environment friendly, resistance to deterioration, incombustibility, resistance to moisture, lightweight, easy to maintain, and aesthetically pleasing appearance. To be an effective sound absorber, a material must have interconnecting cells. For example, fibreglass insulation, due to its interconnecting air cells, is a very good sound absorber. Other effective sound absorbers, called *resonators*, will have small perforations or slots, through which sound can enter but cannot escape out easily. Wood slat panels and slotted concrete masonry units are examples of such resonators. The *Helmholtz Resonator* is another type of sound absorber, which is a chamber with a small orifice, and is similar to a bottle; the sound entering the chamber is refracted within it and cannot escape from it (see Everest and Pohlmann [2014], for more discussions on Helmholtz resonators). Most resonators are effective only in a very narrow frequency range (Certain-Teed Report, 2018). Membranes or diaphragms stretched tightly over rigid perforated materials are also effective sound absorbers. Measurements of sound absorption and sound insulation are described in IS 8225:1987, IS 10420:1982, and IS 9901:1981(Parts 1 to 9).

A wide range of sound-absorbing materials exist; their effectiveness dependents upon the thickness, density, composition, surface finish, and method of mounting. These materials absorb sound waves, soften echo and reverberate, thus reducing the sound in a room. They may be soft, semi-hard, and hard. Soft materials have sufficient porosity and have good sound-absorbing capacity and the examples include glass fibres, glass or mineral wool, and fabrics. Semi-hard materials are stiff; examples are all boards made of fibrous materials and polyurethane or melamine boards. Hard materials are very stiff and made porous during manufacture; they may be used as protective surfaces. Acoustic tiles, masonry, stone, and concrete are examples of hard materials – mass and rigidity of these materials make them highly noise resistant. Neither the non-porous rigid materials nor the porous materials alone provide the desired insulation in a reasonable thickness. For example, about 100-mm thick brick wall (1950 kg/m$^3$) or about 850-mm thick rock wool (80 kg/m$^3$ density) would be required to secure an insulation of about 60 dB (IS 1950:1962). A double wall construction with an intervening airspace of 100 mm is however more effective than a single partition of the same weight.

Wood is a light material and hence its sound insulation performance is not particularly good. A sufficient level of sound insulation can be achieved in wooden buildings by using multi-layered plywood and providing air gap or fibrous insulation material between the layers.

**Porous materials**   Porous absorbing materials can be classified as cellular, fibrous, or granular. Materials made from open-celled polyurethane and foams are the examples of cellular materials. Fibrous materials are those made from natural or synthetic fibres. Granular absorbing materials include some kinds of asphalt, porous concrete, granular clays, sands, gravel, and soils (Arenas and Crocker, 2010).

The fibres used can be natural or synthetic. The natural fibres include vegetable fibres such as cotton, kenaf, hemp, flax, and wood, animal fibres (wool, fur felt), or mineral fibres (e.g., asbestos—it is however carcinogenic and is banned in several countries). Synthetic fibres can be cellulose (e.g., bamboo fibre), mineral (fibreglass, mineral wool, glass wool, graphite, and ceramic), or polymer (polyester, polypropylene, or Kevlar) (Arenas and Crocker, 2010).

As the production of synthetic fibres involves high-temperature and extrusion of petrochemical-based materials, their carbon footprint is significant. They also pose health issues while handling. On the other hand, natural fibres are biodegradable and do not require any precautions while handling them. Fibrous materials are used in the form of boards, foams, fabrics, carpets, and cushions. Usually, they are employed as rolls or in slabs with different thermal, acoustical, and mechanical properties. If fibres are too closely packed, there will be little energy lost as heat. In addition, if they are packed densely, sound cannot penetrate them, and the air motion cannot generate enough friction to be effective. The effectiveness of these fibrous materials depends on their thickness, the airspace, and density (Everest and Pohlmann, 2014). Carpet and padding on the floor are highly effective in reducing the transmission of impact noise to the room below. Acoustic panels are constructed with mineral wool and a solid wood frame. They are finished with jute fabric cover to provide good appearance. Owens Corning 703 fibreglass boards are commonly used for making acoustic panels.

Porous absorbing materials can also be solid and sound waves enter through their cavities, channels, or interstices. Thus, coarse concrete blocks can absorb sound, as it will have tiny surface pores, dissipating sound energy. However, painting these blocks will fill the surface pores and reduce their sound penetration and absorption. However, if spray painted with a non-bridging paint, the absorption may be reduced modestly (Everest and Pohlmann, 2014).

**Porous liners with perforated panel facings**   When mechanical protection is needed for a porous liner, it may be covered using perforated wood, plastic, or metal panels. The performance of these panels can be improved by mounting them with airspace (equal to the thickness of the panel) between the material and the hard backing wall. The perforations in metal perforated panels are usually holes or slots, and porous material is usually included in the airspace to introduce damping into the system. The porous liner has to be protected from contamination by moisture, dust, oil, or chemicals. In practice, such a protection is provided by wrapping them using a thin, limp, impervious blanket made of polyurethane, polyester, aluminium, or PVC (Bies and Hansen, 2009). If such a porous liner wrapped in an impervious membrane is used, the facing and liner should be separated by 12-mm square wire- mesh spacer.

**Foam material**   Polyurethane and melamine foams are the most used cellular porous sound-absorbing materials. *Auralex* is a well-known acoustic foam brand that has dozens of foam shapes and sizes and *Studiofoam* is the most popular product (www.auralex.com). It has a distinctive wedge or pyramid shape that is highly effective at absorbing sound. These foams are attached to walls as panels where noise reflection is a problem, or hung from ceilings as baffles to cut down echo and reverberation. *Pro Studio* foam gives superior performance compared to the cheap "egg crate" variety foam.

Other types of foams have been developed for use where heat or corrosion resistance is required. Although aluminium foams have been used traditionally, other metals, such as nickel, steel, titanium, and copper are being used now. Such metal foams exhibit high stiffness, low weight, fire resistance, and low moisture absorption. These metal foams are usually expensive but can be recycled. *Open-celled metal foams* utilize polyurethane foams as the base in the matrix, and have porosity of 75–95%. These foams with cell sizes of 0.2–0.5 mm are used in high-temperature applications. *Closed-celled metal foams* have a large number of small closed pores of size 1–8 mm; they are used as impact absorbers and as cores in sandwich panels. These closed-cell metal foams are made by injecting a gas (air, nitrogen, argon, carbon dioxide, or carbon monoxide) or by adding a foaming agent (titanium hydride or calcium carbonate) into the molten metal (Arenas and Crocker, 2010). As closed-celled metal foams do not absorb sound well, they are made absorptive by either rolling or drilling holes. When rolled, the faces of some cells break and form tiny sharp-edged cracks which can absorb sound by viscous flow across the cracks, thermo-elastic damping, and the cells acting as Helmholtz resonators. Better sound absorption can be obtained by placing the metal foam at a suitable distance from a rigid wall.

Ceramic foams, normally used in high-temperature applications, are made in a similar way to polyurethane foams with open cell walls forming reticulated channels through the foams. They are usually made of silicon based elements (Arenas and Crocker, 2010).

**Aerogels**   Silicon aerogels, though very expensive, are emerging as sound-absorbing materials. They have unique properties such as high specific surface area (500–1200 $m^2$/g), high porosity (80–99.8%), low density (~5 to 200 kg/$m^3$), low thermal conductivity [0.005–0.1 W/(mK)], large sound absorption and low sound velocity (damping > 50 dB, $v_1 \approx 100$ m/s), and low index of refraction (~1.05). They are highly lightweight and translucent and have average pore size of 30 to 40 nm and melting point of 1200°C. Composite aerogels with varying concentrations of silica and poly-dimethylsiloxane, with pore sizes between 5 and 20 nm, have been found to exhibit better sound-absorbing properties than those of commercial fibreglass. Cost-effective silica-cellulose aerogels have also been developed from recycled cellulose fibres and methoxytrimethylsilane. The sound absorption coefficient of 100-mm thick silica-cellulose aerogels is about 0.39–0.50, which is better than those of cellulose aerogels (0.30–0.40) and commercial polystyrene foams [Feng, et al. (2016) and Maleki et al. (2014)]. The production of metal foams, ceramic foams, and aerogels, however, can contribute to greenhouse gas emissions.

Values of sound reduction for typical types of (a) continuous construction, (b) semi-discontinuous constructions, and (c) discontinuous construction are given Appendix A of IS 1950:1962. Some sound reduction values of continuous construction are provided in Table 21.7.

**Table 21.7** Materials and sound reduction (Based on IS 1950:1962)

| Material | Surface mass, kg/m² | Average sound reduction, dB |
|---|---|---|
| 12.5-mm fibreboard | 3.75 | 20 |
| 12.5-mm plasterboard | 11.8 | 31 |
| 18-mm wood particle board | 8 | 27 |
| 16 to 19-mm plywood | 3.0–4.5 | 24 |
| 20-mm plasterboard, plastered 16 mm each side | 40 | 35 |
| 75-mm concrete block, plastered | 160 | 40 |
| 100-mm brickwork or concrete, plastered | 192 | 45 |
| 200-mm brickwork, plastered | 384 | 50 |
| 400-mm brickwork, plastered | 768 | 55 |
| 3-mm sheet glass | 7 | 25 |
| 6.5-mm plate glass | 15 | 30 |

**Mass loaded vinyl**    Called as 'Limp Mass Barrier', it is a heavy, vinyl sheeting material impregnated with metal particles to increase its mass. It is used for adding mass and to soundproof walls, floors, and ceilings.

**Acoustic spray**    It is made primarily from recycled cellulose with water-based polymer binders. It is applied directly to most ceiling substrates. As it has an open-cell structure, it creates a porous sound-absorbing layer. It is sprayed typically to a thickness of 15–35 mm (NRC of 0.7 to 0.95) depending on the absorption requirement.

**Acoustical plaster**    It is a plaster-containing primarily recycled natural cellulose fibre with water-based polymer binders. It is easy to apply even on curved surfaces and is fireproof. It is applied on acoustic board to a thicknesses of 2 to 5 mm. Acoustic spray and acoustic plaster can be applied directly on either concrete, plasterboard, or metal deck ceilings. They can be used to get lightly textured to smooth finish. Acoustical plaster is completely fireproof, and LEED rated. Most acoustical plasters have a NRC of 0.50–1.00.

**Floor underlayment**    Soundproofing a hardwood or tiled floor requires the decoupling of the flooring surface and the subfloor to reduce the noise transmission. Cork, felt, and polymers are commonly used as underlayment materials.

*Noise control curtains*    These are industrial soundproof curtain products specially designed to block sound by incorporating a very heavy layer inside. These curtains are made of quilted fibreglass or rock wool layers, sandwiched over mass loaded vinyl.

**Green glue compound and sealant**  Green glue noise proofing compound is a unique damping product that is ideal for new construction and renovation projects. When used between two rigid layers of drywall of minimum 12-mm thickness, the compound works to isolate the noise travelling from one room to the other. It is applied using a standard quart-sized calking gun. This product is ideal for dissipating low-frequency noise. The compound reaches its peak performance 30 days after application.

## 21.7 Techniques used in Noise Reduction/Mitigation

Basically, wherever air can enter a room, sound also can enter. Planning against noise should be an integral part of town and country planning proposals, ranging from detailed zoning, building layouts, and roads within built-up areas (see also IS 4954:1968). In addition, housing colonies should be adequately set back from busy airports, state and national highways, factories, and railway lines. Reduction of noise in a house requires considerations during the planning stage itself. The best method to reduce the sound in a room is to properly seal it of any opening or cracks. Overall interior noise levels can be reduced by the extensive use of thick, heavy carpeting, drapes, wall hangings, and acoustical ceiling tiles. These materials absorb sound. They cannot prevent noise from coming through the walls, but they can reduce overall sound levels by reducing sound reverberations. Sound control could be achieved by using any of the sound-absorbing materials (see Section 21.6) and by special construction techniques. In general, sound transmission can be controlled by mass absorption, separation, resiliency, and dampening. Owing to the complex nature of noise and noise control engineering, it will often be necessary to engage an acoustic consultant, whenever soundproofing is required in recording studios, theatres, auditorium, etc. Some guidance is also provided in IS 2526:1963 and IS 3483:1965.

### 21.7.1 Planning of Buildings

During the site planning stage itself, appropriate measures should be taken to reduce/mitigate noise levels. Building layouts should be done in such a way that noisy areas are located in one part of a building and quiet areas in a different part of the building. Some of the other measures include locating houses in appropriate building zone and also in relation to the noise source(s), positioning and orientation of buildings on site (buildings can be arranged so that they form a natural acoustic barrier against noise sources), leaving enough space on the road side, reduction of noise at source (e.g., quieter road surfacing), shielding of noises using dense foliage, earth berms, barriers or screens, proper room arrangement (e.g., corridors, kitchens, bathrooms, stairs/lift rooms, and balconies/verandas may be located on the noisy side and bedrooms and living space on the quieter side), choosing better room sizes [Ancient Greeks used the famous golden ratio of 1 (height) × 1.6 (width) × 2.56 (length) for better acoustics in rooms. Modern sound engineers use a ratio of 0.62 × 1 × 1.62], use of solid walls on noisy side, keeping windows and doors away from the noisy side, and increasing the sound insulation of the building envelope (e.g., constructing the façade facing a noise source with suitable built-in acoustic mitigation measures). More details of these aspects may be found in Part 8, Section 4 of the National Building Code of India.

### 21.7.2 Sound Insulation of Walls

Normally, walls are considered to be sound barriers, but absorbent materials could be used on the walls to reduce noise levels. Sound insulation of walls could be made by using the following techniques:

*Mass absorption*  As discussed in Section 21.4, mass absorption is based on the principle that heavier a wall more difficult it is to vibrate and transmit sound; however, the cost of construction may limit the feasibility of large increases in wall mass [Fig. 21.8(a)].

*Separation*   Separation by means of airspace makes a more effective sound insulator than a single wall of equal weight, leading to cost savings [Fig. 21.8(b)]; when two separate partitions are involved, it will be better to mount them on independent supports to avoid passing on the vibrations. Airspace of 75 mm provides significant noise reduction, but increasing the spacing to 150 mm can reduce noise levels by an additional 5 dB (www.fhwa.dot.gov). However, extremely wide airspaces are difficult to design and will consume more space. As per Doelle (1972), in single stud wooden wall, increasing the stud spacing from 400 to 600 mm will give 2–5 dB increase in STC [Fig. 21.8(c)]. Sound transmission can be reduced by staggering the studs as shown in Fig. 21.8(d).

*Resiliency*   Nails severely reduce the ability of walls to reduce noise. Resilient and flexible materials such as glass fibre board, resilient clips/ channels, and semi-resilient attachments could be used to in-crease STC rating from 2–5 dB [Fig. 21.8(e)] (Doelle, 1972). A *resilient channel* or isolation rail is a thin metal channel attached to the framework of the wall through sound deadening pads, which isolates the drywall from the framing and weakens the sound waves (www.soundproofing.org). Further, these materials are relatively inexpensive and simple to insert. Resilient construction using suspended ceiling tiles, mineral wool, and carpeted floors result in loss of sound as they convert some sound into heat.

*Damping*   As discussed earlier, the use of dissimilar materials (such as gypsum wall board and fibre-glass) and/or two different thicknesses will improve damping [Fig. 21.8(f)].

*Acoustical blankets*   These are also known as isolation blankets. They can increase sound attenuation when placed in the airspace between two leaves of walls [Fig. 21.8(g)]. They are generally made of sound-absorbing materials such as mineral or rock wool, fibreglass, or wood fibres. The use of these blankets can attenuate noise as much as 10 dB.

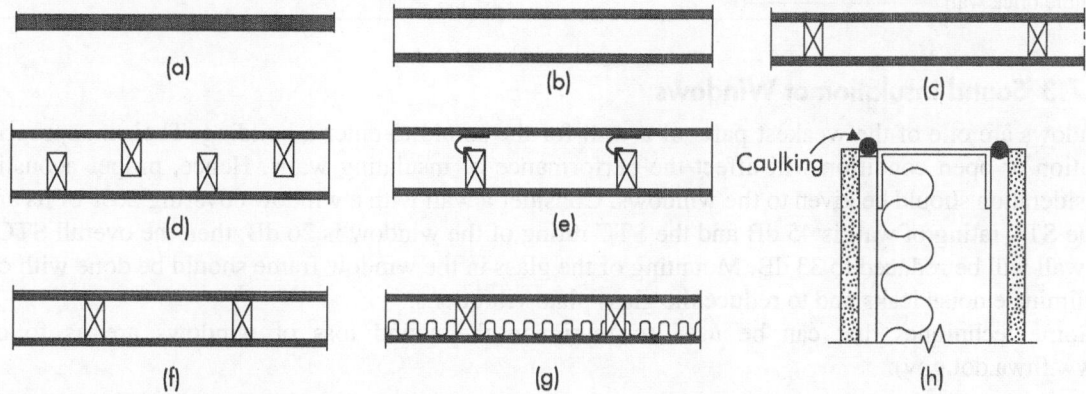

**Fig. 21.8** Improving acoustical properties of walls (www.fhwa.dot.gov) (a) Increased mass (b) Use of air-space (c) Wide spacing between studs (d) Staggered studs (e) Use of resilient attachment (f) Dissimilar panels (g) Sound absorbing blanket in air-space (h) Well sealed cracks and edges

*Proper sealing*   As sound can sneak into cracks and holes, the walls must be well sealed at the perime-ter [Fig. 21.8(h)]. A 25-mm square hole or a 400-mm long and 1.6-mm wide crack will reduce a 50 STC wall to 40 STC (www.fhwa.dot.gov).

*Noise-reducing drywall*   Recently, a single-piece of noise reducing drywall has been introduced (e.g., SilentFX® QuickCut drywall). Generally, the noise-reducing drywall consists of two dense gypsum cores separated by a layer of viscoelastic polymer.

Table 21.8 shows wall types and their sound insulation values. The relative approximate cost of these walls (as per rates in the USA) is also provided for comparison of cost effectiveness.

**Table 21.8** Walls and their sound insulation values (adapted from www.fhwa.dot.gov)

| Type of wall | STC value | Approximate relative cost (%) |
|---|---|---|
| Common stud wall | 35 | 100 |
| Staggered stud wall | 39 | 129 |
| 100-mm brick wall | 40 | 230 |
| Staggered stud wall with absorbent blanket | 43 | 144 |
| 230-mm brick wall | 52 | 290 |
| 175-mm concrete wall | 52 | 226 |
| Double brick wall | 53 | 322 |

## 21.7.3 Sound Insulation of Windows

Windows are one of the weakest parts of a wall for the sound to enter a building. Their inappropriate location or open condition will affect the performance of insulating walls. Hence, proper acoustical consideration should be given to the windows. Consider a wall with a window covering 20% of its area. If the STC rating of wall is 45 dB and the STC rating of the window is 26 dB, then the overall STC of the wall will be reduced to 33 dB. Mounting of the glass in the window frame should be done with care to eliminate noise leaks and to reduce the glass plate vibrations.

Some techniques that can be used to minimize the sound loss of windows are as follows (www.fhwa.dot.gov).

*Close windows*   Closing windows and sealing them permanently is the best solution. However, openable acoustical windows are now available and are reasonably effective in reducing sound. Such permanent sealing of windows require air-conditioning systems, which also may provide some masking of noise (see 'masking').

*Reduce window size*   Although reducing the window size may reduce the cost and provide some relief from sound, it may have only marginal effect on sound reduction (A reduction in window area by a half may reduce sound levels by 3 dB.). Moreover, several building codes stipulate minimum window to wall size ratio.

*Increase glass thickness*   If sealing is not sufficient to reduce the noise, thicker glass can be used. In addition, the glass can be laminated with a transparent plastic, which is both noise and shatter resistant and mounted on resilient material such as rubber, cork, or felt. A 12-mm glass may have a STC rating of 35 dB as compared to 25 dB rating for 4.7-mm glass. However, increasing glass thickness may be practical only to a certain extent.

*Install double-glazed windows*   A double-glazed window having two 4.7 mm-thick panes separated by airspace will have an STC up to 49 and can cost less than a window with thicker single glass pane (Everest and Pohlmann, 2014). The acoustic performance of these windows may be improved further by: (a) increasing the width of airspace up to 100 mm, (b) increasing glass thickness, (c) using better sealing, (d) using differing pane thicknesses (preferably 30%), (e) using slightly non-parallel panes (used in control rooms of television studios), or (f) filling the space between panes with nontoxic, odour-less, and colourless gas such as argon or krypton; the cheaper argon is used to fill 12-mm gaps, whereas the better insulating krypton is used to fill thinner 6 to 9-mm gaps.

*Masking*   It is a technique in which the noise is drowned out with a background noise. Thus, unwanted sound is masked with white noise, which is a combination of all of the different frequencies of sound (*white noise* gets its name from 'white light'). White noise masking machines are popular in commercial systems (produced by air conditioning and heating systems, soft music, or free smartphone apps) and among new parents.

Sound reduction values obtainable with the various types of windows are given in Appendix A of IS1950:1962.

### 21.7.4 Sound Insulation of Doors

Acoustically, doors are very difficult to handle than windows. The common hollow core panel door has a STC rating of less than 20 dB. If the door occupies 20% area of the wall, the STC of wall will be reduced from 48 to 24 dB. Replacing a hollow core door by a solid core door will be a relatively inexpensive solution. A solid core door with vinyl seal around the edges on a carpeted floor will reduce the same 48 STC wall to only 33 dB (www.fhwa.dot.gov). The sound insulation can be further increased by installing drop-bar threshold closers or gasket-stops at the bottom edge of the door. If the gap at the bottom of door is over 6 mm, a door sweep seal (a metal strip with a rubber flap) could be screwed to the door (www.soundproofing.org). The doors should not be located directly facing the noise; in such cases, some shielding should be provided for the doors.

### 21.7.5 Sound Insulation of Floors and Ceilings

Both airborne and structure-borne sounds may be transmitted through floors and ceilings. Acoustical treatment of ceilings is usually not necessary unless the noise problem is extremely severe. Where the ceilings and floors are RCC slabs, they have sufficient weight and rigidity to provide adequate insulation for airborne sounds, but offer poor insulation for structure-borne or impact sounds. Acoustical baffles with absorptive materials or perforated tiles are used where sound is a major concern (see Fig. 21.9). They are designed to be unobtrusive and visually nondescript so that the aesthetics of a room is not altered drastically.

**Fig. 21.9** Use of acoustic baffles
(*Courtesy*: Asona India Pvt Ltd)

Insulation against impact sounds may be made in the following three ways:

**By using a resilient material**   Resilient materials such as linoleum (green product), vinyl (synthetic product based on PVC resin), cork, asphalt mastic, and carpet are highly effective in reducing the transmission of impact noise to the room below. Carpets (and also vinyl or linoleum) serve the dual purposes of floor covering and noise reduction. Noise reduction is achieved in two ways: carpets absorb the incident sound energy; and sliding and shuffling movements on carpets produce less noise than on bare floors (www.carpet-rug.org). A 150-mm thick concrete floor with 12-mm thick carpet will provide sufficient airborne and impact sound insulation.

**By using floating floors**   A 50-mm thick floating RC floor, constructed over the existing concrete floor, and separated by resilient materials (such as glass-wool), will isolate impacts and vibrations to be transmitted to the room below (see Fig. 21.10). It will also improve the insulation of airborne sounds. The waterproof paper and lapping of both the quilt and the paper should be properly done to separate the two concrete slabs. A floating floor gives the greatest amount of sound and vibration insulation; however, it is extremely expensive.

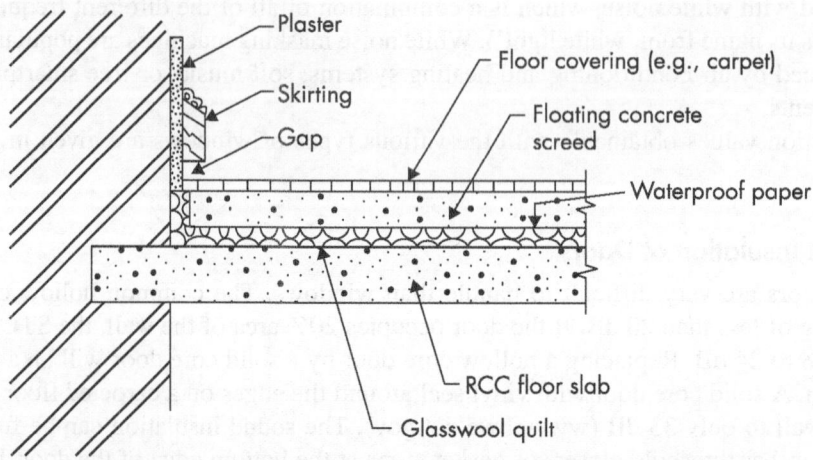

**Fig. 21.10** Concrete floating floor construction (Based on IS 1950)

In the case of wooden floors, isolation can be achieved using mineral or glass-wool quilt. Resilient mountings provide even more satisfactory results. Further improvements can be obtained by providing a 'pugging' or a 'deadening' material (sand with a load of at least 100 kg/m²) in the airspace between the wooden joists, as shown in Fig. 21.11 (see also IS 1950, for more details). Wooden floating floors are not nailed into the subfloor like typical flooring. Instead, they are installed using special adhesive. Without the use of nails in a floor joist, floating floors can eliminate creaks and will help prevent sound from travelling between different levels of the house (www.certainteed.com). Squeaking floors can be fixed with joist tape. It will be better to use joist tape before installing the walls and flooring, or else part of the wall/floor has to be removed to use the tape.

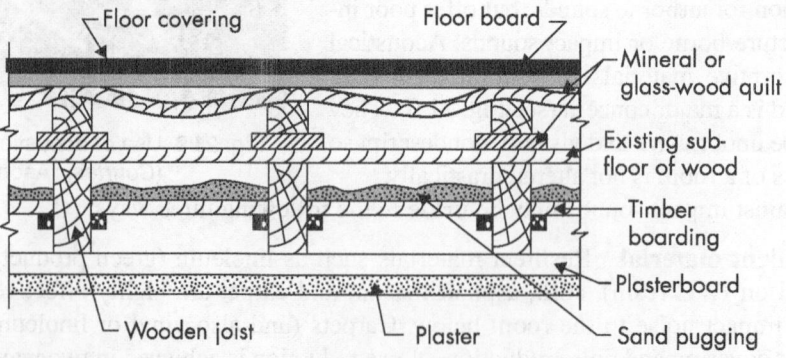

**Fig. 21.11** Floating construction in timber floors (based on IS 1950)

**By using a suspended ceiling** Suspended ceiling with an airspace will improve the insulation of both airborne and structure-borne sounds by attenuating and isolating them from the room below (Fig. 21.12). Suspended ceilings are most effective as noise reducers but are the most expensive.

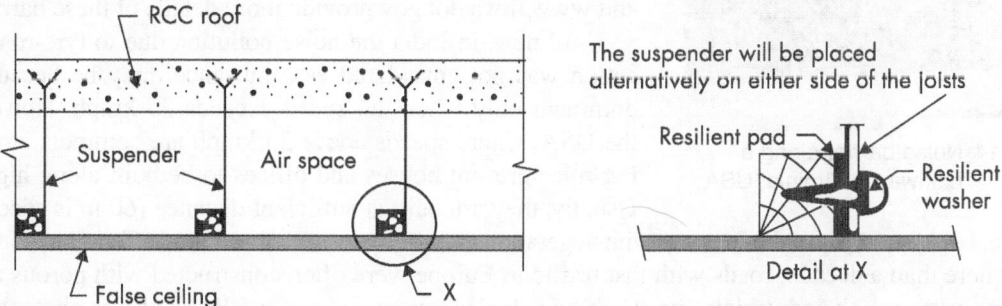

**Fig. 21.12** Floor with suspended ceiling (Based on IS 1950)

Foam or neoprene padding can be put under bass speakers or vibrating items to isolate them from the floor, reducing the transmission of sound and vibration through the building structure. Sound insulation values of typical timber and other floors are given in Appendix A of IS 1950:1962.

## 21.8 Barriers and Noise Reduction at Source

*Barriers* can be placed in between noise sensitive areas and the source of noise to reduce the sound. Different types of barriers are possible, such as (a) walls and fences made of different materials including precast concrete, wood, metal, and plastic (concrete walls are usually built with attractive surface designs), (b) sloping mounds of earth, called berms, (c) dense planting of trees and shrubs, and (d) a combination of these techniques. The available space, cost, safety, aesthetics, and the desired level of sound reduction should be considered while selecting the barrier. Noise barriers reduce the sound by either by absorbing, transmitting, or reflecting back the sound. When the barrier is tall enough to break the line-of-sight, it can provide noise reduction up to 5 dB; every additional height of 1 m can reduce sound by about 1.5 dB, as shown in Fig. 21.13. Noise barriers have been built in the USA since the mid-20th century and Fig. 21.14 shows such a wall constructed along a highway. To be most effective, a barrier must be long and continuous to prevent noise from passing around the ends. It must also be solid, with few openings. It must be designed to withstand wind pressure.

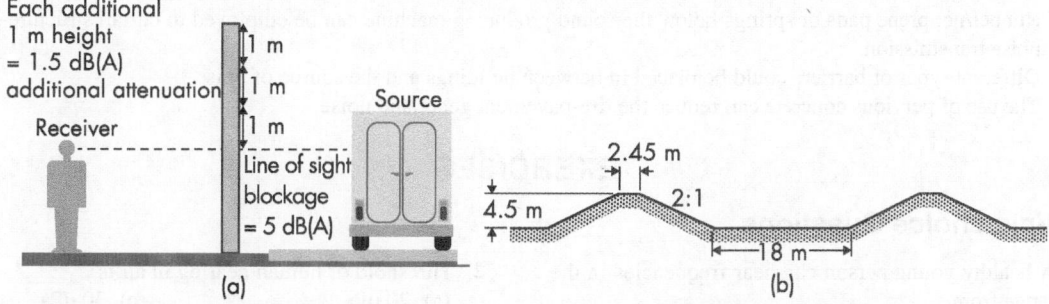

**Fig. 21.13** Typical sound barriers (a) Effectiveness of wall as barrier (b) Cross-section of berm

Earth berms have a natural appearance and reduce noise by about 3 dB more than vertical walls of the same height. However, earth berms require considerable land and cost to construct, especially if they are very tall. Walls require less space, but are limited in height to 8 m for structural and aesthetic reasons. In general,

**Fig. 21.14** Noise barrier along a highway in Virginia, USA

plantings by themselves do not provide much sound attenuation. It is better to use plantings in conjunction with other types of barriers and for aesthetic enhancement. Kotzen and English (2009) and www.fhwa.dot.gov provide more details of these barriers.

Until now, in India the noise pollution due to tyre-road interaction was not considered as a nuisance, may be because it is dominant only when the speed exceeds 35 kmph. However, in the USA, where speeds above 35 kmph are common, strict zoning rules prevent houses and offices to be built along highways. Usually, they are built at sufficient distance (60 m is specified in Virginia, USA) away from the freeway with vegetation in buffer zones, so that sound effects are controlled.

For more than a decade, roads with fast traffic in Europe were often constructed with porous asphalt, to reduce water splash and vehicle spray behind vehicles during heavy rainfall and also to abate the noise pollution (Bendtsen, et al., 2008). The field-measured tire-pavement noise showed that porous concrete (also called *pervious concrete*) reduced the tire-pavement generated noise by 4 to 8 dB (Tian et al., 2014). Generally, the thicker the porous surface, the lower will be the peak frequency (Arenas and Crocker, 2010). However, porous concrete and porous asphalt have to be maintained properly; otherwise, there may be problems such as voids silting up in due course and slipperiness in winter (Subramanian, 2016).

## SUMMARY

- Controlling noise pollution is important for maintaining good health.
- There are two types of noises, viz., airborne and structure-borne.
- The six principles of soundproofing in practice are: (a) mass, (b) absorption, (c) conduction, (d) resonance, (e) mechanical decoupling, and (f) damping.
- To reduce the intensity of airborne noise, sound absorbent materials may be used. An absorbent material may be applied to walls, floors, ceilings or used as furnishings to reduce the sound level by absorption. The selected materials should also have adequate fire resistance.
- A wide range of sound-absorbing materials exist; their effectiveness dependents upon the thickness, density, composition, surface finish, and method of mounting. These materials absorb sound waves, softening echo and reverberation, and reduce the sound in a room.
- To reduce the transmission of airborne noise, sound insulating materials may be used, which reduce the noise by virtue of their mass and physical properties. Heavy panels or double panel construction with thin sheet materials and an intervening air cavity filled with loose mineral/fibreglass wool, are effective.
- Efforts to mitigate noise levels should start at the planning stage itself. Sound insulation can be done to walls, windows, doors, floors, and ceilings.
- Rubber/neoprene pads or springs below the sound producing machine can be employed to curtail structure-borne noise transmission.
- Different types of barriers could be placed in between buildings and the source of noise.
- The use of pervious concrete can reduce the tire-pavement generated noise.

## EXERCISES

### Multiple-choice Questions

1. A healthy young person can hear frequencies in the range from
   (a) 20 to 20,000 Hz
   (b) 70 to 14,000 Hz
   (c) 100 to 4000 Hz
   (d) 2000 to 10,000 Hz

2. Threshold of human hearing in air is
   (a) 20 μPa          (c) 30 μPa
   (b) 25 μPa          (d) 40 μPa

3. The maximum acceptable noise level in residences as per IS 1950:1962 is
   (a) 20–25 dB        (c) 35–40 dB
   (b) 25–35 dB        (d) 40–45 dB

4. NRC rating is the average of the absorption coefficients of a material determined at the octave band
   centre frequencies
   (a) 200 to 2500 Hz      (c) 250 to 2000 Hz
   (b) 125 to 4000 Hz      (d) 500 to 4000 Hz

5. The STC number is derived by testing the product for
   sound attenuation at standard frequencies of
   (a) 200 to 2500 Hz      (c) 250 to 2000 Hz
   (b) 125 to 4000 Hz      (d) 500 to 4000 Hz

6. At what STC rating, loud speech is not heard usually?
   (a) 35      (c) 50
   (b) 40      (d) 60

7. Open-celled metal foams have a porosity of
   (a) 60–70%      (c) 80–85%
   (b) 75–80%      (d) 75–95%

8. Though ancient Greeks used the famous golden ratio of 1 (height):1.6 (width):2.56 (length) for better
   acoustics, now the following ratio is used
   (a) 1:1.5:2.5
   (b) 0.62:1:2.5
   (c) 0.5:1.2:2.0
   (d) 0.62:1:1.62

9. The common hollow core panel door has a STC
   rating of about
   (a) 20 dB      (c) 30 dB
   (b) 25 dB      (d) 35 dB

10. When the barrier is tall enough to break the
    line-of-sight, it can provide noise reduction of
    (a) 1.5 dB      (c) 3.25 dB
    (b) 2.5 dB      (d) 5.0 dB

## Review Questions

1. What are the two major types of noises? What are the causes of these types of noises and their consequences?
2. Define the terms sound, noise, frequency of sound, wavelength, amplitude, and sound pressure.
3. Differentiate between airborne and structure-borne sound.
4. How structure-borne noise can be minimized in a building?
5. Write short notes on: transmission loss, noise reduction coefficient, sound transmission class (STC), and sound reduction index (SRI).
6. What are the six basic principles of soundproofing? Explain each one of them.
7. How resonances can be avoided in buildings?
8. Describe briefly how mechanical de-coupling can be done in a wall.
9. What are the characteristics of an efficient sound-insulating material?
10. How is the sound-absorbing materials generally classified? Name some of these materials.
11. What are the different porous absorbing materials that could be used?
12. Briefly explain about the foam materials used for sound-absorbing?
13. Can aerogel be used for soundproofing? What are its advantages and drawbacks?
14. How noise levels could be mitigated during the planning stage?
15. Explain the techniques used in the sound insulation of walls?
16. How can we sound insulate windows?
17. Is it possible to sound insulate doors? How can it be done?
18. How are floors and ceilings sound insulated?
19. What are the different types of sound barriers that can be used? Explain them.
20. How are sounds created by tires on pavements reduced?

## ANSWERS

### Multiple-choice Questions

| | | | | | |
|---|---|---|---|---|---|
| **1.** (a) | **2.** (a) | **3.** (c) | **4.** (c) | **5.** (b) | **6.** (c) |
| **7.** (d) | **8.** (d) | **9.** (a) | **10.** (d) | | |

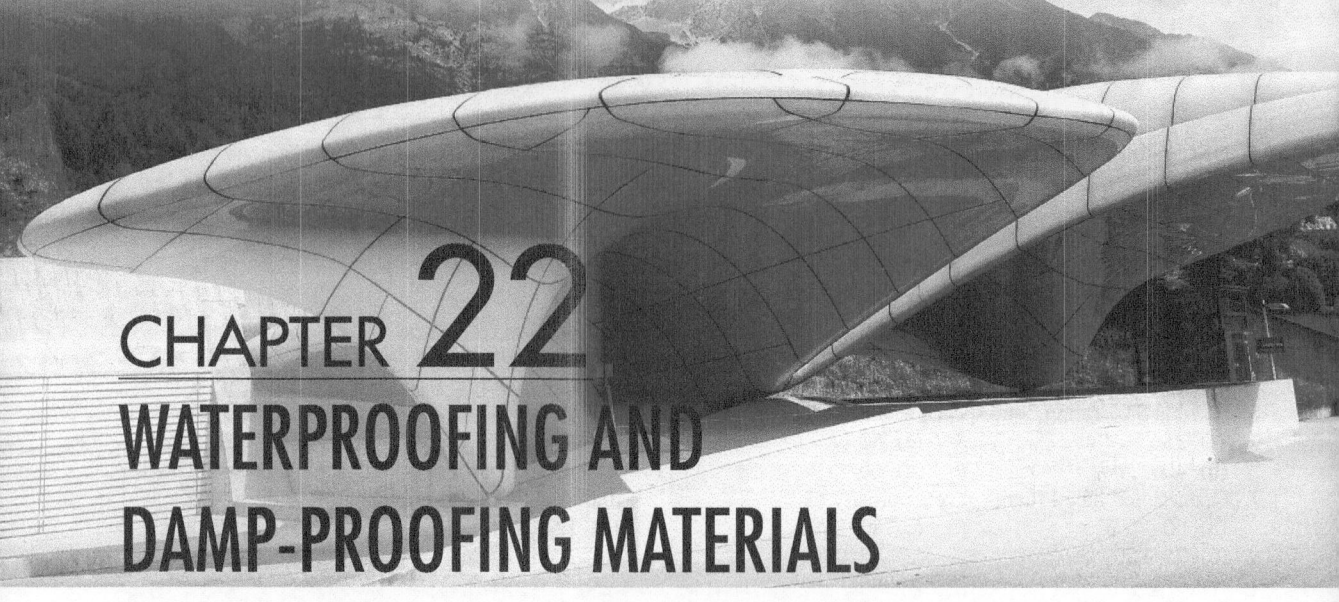

# CHAPTER 22
# WATERPROOFING AND DAMP-PROOFING MATERIALS

## 22.1 Introduction

Water is detrimental to practically all construction materials. Since most structures are subject to moisture conditions in varying degrees, it is necessary to protect these materials, against moisture/water. Waterproofing is the treatment given to any surface or structure to prevent the passage of water under hydrostatic pressure. On the other hand, damp-proofing is the treatment given to any surface or structure to stop the rise of water by capillary action. A waterproofing treatment may be as simple as a single brush or spray application to a surface, or as complex as a multi-coat, trowel-and-spray system involving cement, bitumen, or even synthetic materials. Bitumen and tar-based waterproofing is the oldest type, which is still in use, but has been improved over the years for better performance. Products involving silicon, urethane, and other synthetic elastomer can be incorporated in one-coat, continuous film with very good expansion and durability. The same materials can be used for waterproofing and damp-proofing. In this chapter, only the materials used in waterproofing are discussed. The methods used in construction while using them may be found in books on building construction. Waterproofing of major works like tunnels is also beyond the scope of this book.

## 22.2 Causes and Effects of Dampness in Building

*Dampness* is the presence of unwanted moisture on the surfaces of a building, either as a result of seepage from outside or due to condensation from within the building. The various causes of dampness/leakage in a building are as follows:

**Moisture rising up the walls from the ground** All the buildings are founded on soil, and their foundations are embedded into them. Owing to moisture present in the soils, it may rise up into the walls/floor through capillary action.

**Rainwater entering from top floors** Efficient roof drainage is an important prerequisite for the economic maintenance/waterproofing of a building. For efficient drainage, drainage slope should not be flatter than 1 in 80 and should preferably be 1 in 40. In addition, the size of rainwater pipes should be fixed as per Table 3 of IS 2527:1984, depending upon the rainfall intensities, ranging from 3.3 to 85.4 m². Installation of rainwater gutters and pipes requires a careful attention to their capacity, alignment, water tightness, and firmness of support. There should not be any undulations in the roof surface, which will result in the accumulation of rainwater, leading to leakage. If wall tops are not properly sealed, water will enter through these walls and travel down.

**Rain beating entering through external walls**  If the external faces of walls are not properly plastered, water may enter during the pounding of heavy rain on these walls, causing dampness on the interior.

**Condensation**  Owing to condensation of atmospheric moisture, water may be deposited on walls, floors, and ceilings, causing dampness. Condensation is caused when moisture-laden air from bathrooms, kitchens, laundry, and even breathing of persons, comes into contact with cold surfaces. Condensation can be controlled by proper ventilation, proper insulation of the building envelope, increasing the temperature, or lowering the humidity of the air circulated in the building.

**Inadequate waterproofing**  This may be due to the presence of air bubbles between roofing material and the waterproofing layer, inadequate overlaps of waterproofing layers, and improper details at parapet wall and slab junctions.

**Imperfect joint detailing and treatments**  Unless expansion/contraction and construction joints are sealed with flexible sealant, waterproofed, and maintained properly, they lead to ingress of water.

**Other problems**  Water may also seep from roofs due to shrinkage of concrete resulting in micro cracks, imperfect joint detailing and treatments, poor quality of concrete or poor workmanship, and plumbing leaks.

Continued dampness causes secondary damage to a building. Plaster and paint deteriorate and wallpaper loosens. Stains from salts (called *efflorescence*) and moulds may show up as patches and mar surfaces. Rusting of steel and iron fasteners and even embedded reinforcement in concrete may occur. In extreme cases, mortar or plaster may fall down from the affected areas. Timber in contact with damp areas will be deteriorated and result in warping, buckling, and dry rotting. It may also cause poor indoor air quality and respiratory illness to the occupants. It may provide favourable conditions for dust mite populations to grow, which can affect asthma sufferers. The unwanted moisture facilitates the growth of various fungi (mould) and bacteria that can cause respiratory problems and/or allergic reactions and may eventually lead to sick building syndrome. It can also result in odours in poorly ventilated spaces because of fungal growth. Dampness along with warmth and darkness breeds germs, which may cause dangerous diseases. Dampness also promotes and accelerates the growth of termites.

A number of waterproofing systems and materials are available in the market. Hence, selecting the appropriate product for one's needs is often challenging, especially when information regarding the environment in which they are applicable and their durability are not readily available. In addition, some products work well initially but prone to deterioration after a period of exposure to aggressive environments.

## 22.2.1 Requirements of Waterproofing Systems

Ideally, a waterproofing or damp-proofing material should be watertight, durable, strong (capable of resisting super-imposed loads), flexible, tear-resistant, and elastic so that it can stretch to cover cracks and also move with the building. When it is exposed to the sun, it should be resistant to ultraviolet (UV) radiation, microbial attacks, and fungus growth. In addition, the other desirable characteristics are bondability, breathability, providing insulation, and cost effectiveness. However, it is almost impossible to have all the aforementioned properties in one material and hence it may be necessary to use judicious combination of materials appropriate to the requirements (Surlaker, 2005).

## 22.3 Waterproofing Systems and their Applications

The classification of different types of systems and materials available for waterproofing of buildings is shown in Fig. 22.1. The waterproofing systems that are commonly used can be grouped as: (a) flexible membranes that are available as sheets or can be applied as a liquid, (b) coatings that are polymer based (without and with cementitious component), (c) modified cement mortars (made watertight by the addition of chemical and mineral admixtures), and (d) polymer modified cementitious mortars. It may be of interest to note that many of these materials are similar to those used in repair. These systems are discussed briefly.

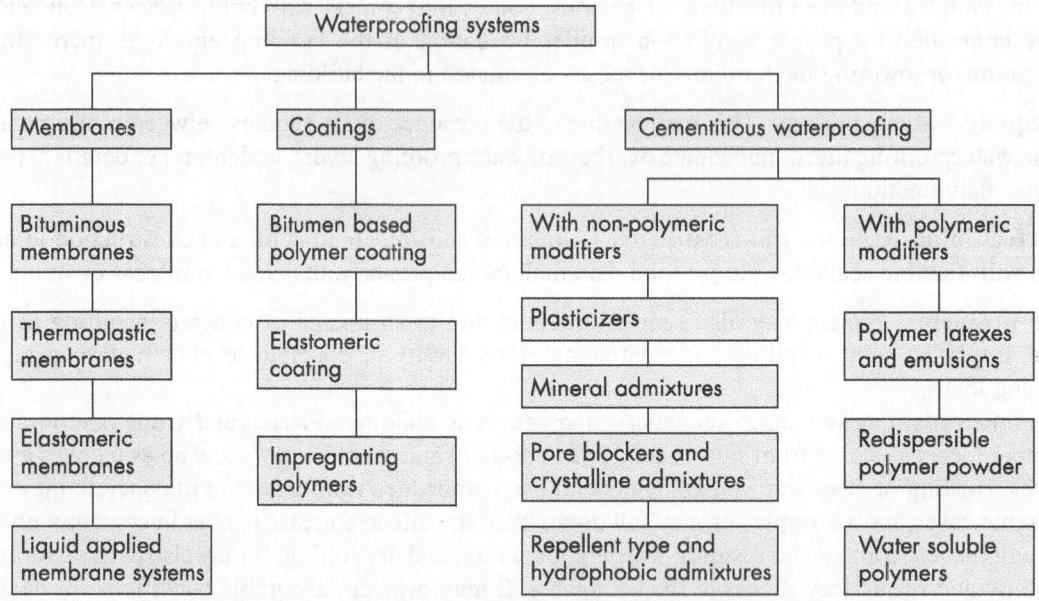

**Fig. 22.1** Different types of waterproofing systems and materials
(*Source*: Adapted from Nair and Gettu, 2016)

### 22.3.1 Bitumen Primer (IS 3384:1986)

Bitumen primer is commonly used for priming concrete and masonry surfaces prior to the application of the first mopping coat of melted bitumen in laying built-up roofing or membrane waterproofing, to promote the bonding of bitumen with the concrete roof or masonry surface (see also Section 19.3.6 of Chapter 19).

## 22.4 Sheet Membranes

A waterproofing membrane is a thin layer of watertight material that is laid over a surface. Since this creates a continuous layer, water will not pass through it. For example, when a waterproofing membrane is laid in between the RCC slab and weathering tiles, water will not be able to pass through the RCC slab. The membrane should be flexible enough to take any shape as it is laid over. These membranes are composed of thin layers of waterproofing material with thickness of 2 to 4 mm. Usually, they are placed on the positive side of the basement (same side of hydro-static pressure) to prevent ingress of water from the roof/soil or the sides (Nair and Gettu, 2016).

There are essentially two types of membranes, sheet-based membranes and liquid applied membranes. Sheet-based membranes can be bituminous, thermoplastic, or elastomeric. Many of these formulations, both bitumen and synthetic, are similar if not identical to those used for roofing (see Section 19.3.4 of Chapter 19). It has to be noted that bituminous membranes are not suitable when expose to sunlight. They become brittle and fragile when exposed longer to sunlight. Plastic polymer products have good waterproofing properties and are lightweight, elastic, and non-porous. However, they are not sustainable. The evaluation of these membranes are done as per ASTM D5, ASTM D41, ASTM D146, ASTM D882, ASTM D412, ASTM D471, and ASTM G155 (see also Nair and Gettu, 2016).

### 22.4.1 Bituminous Materials and Membranes

The most common exterior membrane is made of petroleum-based bitumen (also called as asphalt) and is extremely resistant to moisture penetration. It is very sticky and holds well to vertical concrete surfaces. A blend of vegetable and/or animal fibre or hessian-based tar felt was made in India earlier, by impregnating hessian mat with bitumen to form impermeable sheets.

**Bitumen felts**   *Bitumen felts* are classified as Types 1, 2 (fibre base), or 3 (hessian base) depending on the type of base used in their manufacture and their intended use (IS 1322:1993, see also Section 19.3.3 of Chapter 19). The Type 1 felt is manufactured by saturating the dried base fabric (fibre or hessian) by immersing it in molten bitumen and the surplus saturant is removed. The self-finished felts are obtained by treating the Type 1 felt by passing it through bituminous coatant and applying it uniformly and then giving a superficial application of fine mineral powder. The mineral granules, added on top of the felt, will protect the base from ultraviolet degradation, and also decrease its vulnerability to fire. In addition, a thin, transparent film may be added to the base of the felt during the manufacture on all 'torch-on' products. This film prevents the felt from sticking to itself when rolled up for packaging. The glass fibre-based bitumen felt, available in two grades, consists of a continuous mat of resin-bonded glass fibres treated with bitumen (IS 7193:1994).

As per Table 1 of IS 1322, the minimum weight of bitumen felt range from 7.6 to 37.1 kg/10 m². The bituminous 'saturant' conforming to IS 73:2013 should be used and the penetration should be greater than 80 at 25°C. As per IS 1322:1993, the weight of bitumen should be greater than 110% of the minimum weight of fibre base and greater than 60% of hessian base, respectively. The bituminous 'coatant' conforming to IS 702: 1988 should be used for Types 2 and 3 felts, and the softening point should be greater than 105°C when tested by ring and ball method and the penetration should be greater than 8 at 25°C (see also Sections 19.3.2 and 19.3.3 of Chapter 19).

The felts may be subjected to the following tests: (a) breaking strength test as per IS 13826 (Part 1), (b) pliability test as per IS 13826 (Part 2), (c) storage sticking test as per IS 13826 (Part 3), (d) pressure head test as per IS 13826 (Part 4), (e) heat resistant test as per IS 13826 (Part 5), and (f) water absorption test as per IS 13826 (Part 6) [see also Table 2 of IS 1322].

**Fig. 22.2** Fixing of bitumen felts using blow-torches (*Courtesy*: Findlay & Evans Waterproofing, Australia, https://www.waterproofingfew.com.au)

The sheet-based membranes normally arrive at the site in the form of rolls (in widths of 900 or 1000 mm and in lengths of 10 or 20 m). These are then unfurled and laid on a firm surface. This type of membrane is stuck to the substrate with a hot bitumen-based adhesive using 'blow-torches' (see Fig. 22.2). Since these membranes are made in the factory-excepting the joints, they are consistent in quality.

The bonding material between the felt and the roof surface and between the successive felts should be industrial blown type bitumen of Grade 85/25 or 90/15 conforming to IS 702:1988 (see also

Section 19.3.1 of Chapter 19). For topdressing, bitumen used should be industrial blown type with allowable penetration not more than 40 when tested in accordance with IS 1203:1978. For vertical surfaces up to 1 m height, blown type bitumen of grade 85/25 or 90/15 and above 1-m height 115/15 grade are recommended, as per IS 1346:1991.

On flat or sloping concrete and masonry roofs, five (normal treatment) to nine courses (extra-heavy treatment) are to be applied after preparatory work as per IS 3067:1988. For example, the normal treatment for moderate conditions consists of the following five courses (IS 1346:1991):

1. Primer conforming to IS 3384:1986 at a minimum rate of 0.27 L/m².
2. Hot applied bitumen at a minimum rate of 1.2 kg/m².
3. Hessian-base self-finished felt, Type 3- Grade 1 or glass fibre base Type 2-Grade-1.
4. Hot applied bitumen at a minimum rate of 1.2 kg/m².
5. Pea-sized gravel devoid of fine sand at the rate of 0.006 m³/m².

When the waterproofing treatment is to be isolated from the roof structure, an additional layer of bitumen saturated felt is used between the roof surface and the first layer of self-finished bitumen felt. In such cases, it is called a floating treatment (IS 1346).

Joints between adjacent membranes are also made with the same hot adhesive. The sheets are overlapped by about 100 mm to form a waterproof joint. All overlaps are firmly bonded with hot bitumen. The second layer of felt is laid with their joints staggered with the first layer. With this type of membrane, joints between sheets are critical, and must be done perfectly to avoid leakage. For the overlaps of the adjacent layers of felt to offer minimum obstruction to the flow-off of water, the felt is laid with their lengths at right angles to the direction of the run-off gradient. Usually, they are laid commencing at the lowest level and working up to the crest. It is essential to apply a cement mortar fillet of 1:4, along the junction of parapet wall and floor.

On the flashings and at the drain mouths, the gravel layer can be omitted and instead two coats of bituminous paint at the minimum rate of 0.1 L/m² per coat or a single coat of bituminous emulsion at the rate of 0.5 L/m² are applied. When the roof surface is subjected to foot traffic, or to provide fire protection to the roof surface, an additional top layer of cement concrete flooring tiles (conforming to IS 1237:2012) or a burnt clay flat terracing tiles (conforming to IS 2690:1992) is provided. Expansion joint may be covered with zinc or lead sheet or two layers of bitumen felt, with a top dressing of gravel. For heat-reflecting surface or aesthetic reasons, bitumen-based aluminium paints or coloured bituminous emulsions may be used. For the treatment of sloping timber roof, reference should be made to IS 1346:1991.

**Bitumen mastic (IS 5871:1987)**   It is manufactured as follows: Limestone powder passing 75-micron IS Sieve and with greater than 80% calcium carbonate content by weight (filler) and fine aggregate are mixed together and heated to a temperature of 190 to 205°C. The required quantity of bitumen is separately heated to 170 to 180°C and added to the aggregate. These are again mixed and cooked in a mechanically agitated mastic cooker for about 30 min until the materials are thoroughly mixed. Then, the mastic is cast into blocks weighing about 25 kg. It is used in waterproofing of roofs (IS 3037:1986, IS 4365:1967) and damp-proofing of foundations (IS 5871:1987 and IS 7198:1974).

**Polymer-modified bitumen**   This is often polymer- and rubber-modified bitumen and manufactured by mixing thermoplastics, ordinary, or chemically treated crump rubber powder and elastomers, in fully automated mobile plants with high shear mixing facility. The different modifiers which are used to modify bitumen, and the details of tests conducted for separation, penetration, softening point, and elastic recovery may be found in IS 15462:2004. Depending on the materials used they are classified as follows (see also Section 19.3.1 of Chapter 19):

1. Type A PMB(P) Plastomeric thermoplastics-based.
2. Type B PMB(E) Elastomeric thermoplastics-based.
3. Type C NRMB Natural rubber and styrene butadiene rubber (SBR) latex-based.
4. Type D CRMB Crumb rubber/treated crumb rubber-based (for more details, refer to IS 15462:2004)

Two types of modifiers are commonly used: (a) atactic polypropylene (APP) and (b) styrene-butadiene-styrene (SBS). APP polymer-modified bitumen membrane systems are ideal for waterproofing under rafts/basements along with PVC and water-stops. This system, when employed for waterproofing roofs, is found to be better than SBS-modified bituminous membranes (Nair and Gettu, 2016). In these systems, dimensionally stable non-woven polyester/fibre glass mats are coated with polymerized bitumen to obtain high tensile strength, tear, and puncture resistance. More details about these systems along with their application in concrete bridge decks may be found in NCHRP Synthesis 425 (2012).

**Bituminous compound**   It is suitable for cold application and has several uses. As per IS 1580:1991, it is available in two grades: Grade 1- semi-stiff, smooth, and homogenous paste suitable for application by spreading with hand, trowel, or spatula; and Grade 2-homogenous paste with light consistency suitable for application by putty knife. Rubberized bitumen is also used as sheet membrane. It is also available in liquid/semi-solid state and can be applied hot or cold. Adjustments to viscosity allow it to be applied with a trowel, roller, or sprayer. More discussions on bitumen roofing systems may be found in Section 19.3.4 of Chapter 19.

**Problems with bituminous products**   The objections/problems of using bituminous roofing products are discussed here. Bitumen is mostly produced from crude oil and hence is not a sustainable building product. The products which require 'blow-torches' to melt and struck them to the concrete surface may affect the concrete surface. Pipes, wiring, timber, and other items may be damaged due to flame/heat applied while fixing the waterproofing membrane to the surface, and may become a fire hazard. In addition, the bitumen fumes generated at work sites during hot application contain known carcinogens, which can also cause skin and respiratory problems. Moreover, in most of the places in India, where the surface temperature exceeds 25°C, bitumen felts crack due to the effect of solar radiation, heat and atmospheric oxidation, and volatility of solvents. Thus, they become brittle and allow moisture to enter. Entrapping of air during lying of tar felt, and the difference in the values of thermal expansion of concrete and bitumen felt may result in blister formation. Moisture and rainwater may penetrate through such blisters and cause cracks and subsequent leakage. In addition, exposure to extreme heat and UV radiation can drastically decrease the lifespan of bitumen felts.

**Blisters in bituminous membranes**   Blisters are the most common problem with built-up roofs (BUR). They are pockets of entrapped air and/or moisture in roof membranes. They can occur between the cap and the base sheets and/or between the base sheet and the substrate (see Fig. 22.3). Blisters are not only unsightly, but also shorten the service life of a roof, as they increase the vulnerability of membrane to physical and chemical degradation. Their sloping sides can also change the direction of water flow on a roof and cause ponding. Blisters occur more frequently in modified

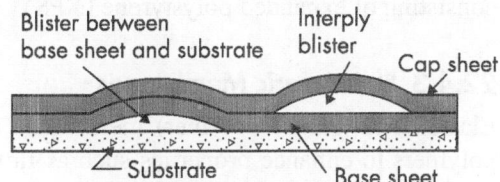

**Fig. 22.3** Blisters in modified bituminous membranes

bituminous roofs. Field research indicates that blistering is more common with SBS than APP-modified membranes (Liu et al., 2000). Good interply bonding is important to resist blister growth.

Roof blisters can be prevented by (a) using dry materials in dry settings, (b) ensuring that the materials have firm contact between them for the adhesion to work properly. Other recommendations to reduce blistering in SBS-modified bituminous roofs are given by Liu et al. (2000). The blisters can be repaired by cutting away the membrane that has lost adhesion, replacing any wet insulation under them with similar material, and then replacing the membrane with overlapping edges by at least 150 mm.

### 22.4.2 Thermoplastic Membranes

The most common thermoplastic roof membranes are polyvinyl chloride (PVC), thermoplastic olefin (TPO), discussed in ASTM D4434 and ASTM D687, respectively (see also Section 19.3.4 of Chapter 19). *PVC membrane* systems are well known and may be non-reinforced or reinforced with polyester or glass-fibre mats. PVC sheets contain additives such as plasticizers and stabilizers, to impart flexibility and achieve other desired physical properties. They may be attached to the concrete surface using adhesives, mortar, tape, straps, etc. Some PVC membrane systems are backed up with geo-textile layer for additional support. Seams are sealed by manual/automatic hot air or chemical welding, forming a monolithic, continuous sheet membrane. They are tested for leaks on the joints since they are not usually bonded to the concrete surface (Nair and Gettu, 2016). As no adhesive or primer is required, they can be used even against hydrostatic heads. PVC membranes are produced in numerous colours, though grey and white are the most common. The advantages of this membrane system include good low-temperature flexibility, high-temperature tolerance, high puncture and impact resistance, and fire resistance (https://usa.sarnafil.sika.com). Numerous tunnels in Switzerland are lined with PVC, polyethylene (PE), or other polymeric sheets.

*TPO sheets* have a blend of polypropylene and ethylene propylene polymers and are usually reinforced with polyester (see also Section 19.3.4 of Chapter 19). This single-ply membrane is composed of three layers: TPO polymer base, polyester-reinforced fabric centre (scrim), and thermoplastic polyolefin compounded top ply. Common fillers used in its manufacturing include: fibreglass, carbon fibre, and metal oxy sulphate. They are available in white, light grey, and black reflective colours. TPO roof is as UV-resistant and as heat-resistant as ethylene-propylene-diene monomer (EPDM), and as heat-weldable as PVC. TPO offers energy efficiency similar to PVC, but at a lower cost. TPO sheets can resist mould growth, dirt accumulation, tears, impact, and punctures. Reinforced TPO membranes can accommodate thermal expansion and contractions of buildings more effectively than other single ply roofing products. However, they may have accelerated weathering when subjected to high thermal or solar loading (www.tporoofing.org).

*High-density polyethylene (HDPE) liners*, with rubber bitumen adhesive, also serve as an impervious barrier in water-retaining structures. To make them more reliable, dual HDPE/bentonite membranes are produced now, which are heat-weldable. By welding the seams, the HDPE liner acts as a continuous waterproofing barrier. These membranes, when used in roofing applications, contain a backing layer consisting of expanded polystyrene (XPS).

### 22.4.3 Elastomeric Membranes

Elastomeric (elastic polymer) membranes are manufactured by adding elastomers to thermoplastic polymers to enhance properties such as flexibility, elasticity, and tensile strength. Examples are SBS copolymer, EPDM, and PE. Commercially available *SBS* self-adhesive membranes are provided with adhesive on one side, so that they may be bonded easily to the concrete surface. SBS is a rubber-type modifier that gives bitumen the ability to stretch and resist damage, and improves its cold-temperature flexibility. The advantages of these systems are that they are highly flexible, do not require any heating for the installation, and are more durable than bituminous membranes (have a life of about 12–15 years). However, pin holes and blisters may affect their effectiveness.

*EPDM* is an extremely durable synthetic rubber roofing membrane, and covered in ASTM D4637 (see also Section 19.3.4 of Chapter 19). Its two primary ingredients are ethylene and propylene, and are derived from oil and natural gas. EPDM is available in both black and white colours. They are sold in a variety of widths, ranging from 2.3 to 15 m, and in two thicknesses, 45 and 60 mils (1.15 and 1.50 mm). White colour is achieved by adding titanium dioxide that reflects UV rays and reduces air-conditioning costs. EPDM can be installed fully adhered, mechanically attached, or ballasted, with the seams of the system sealed with liquid adhesives or specially formulated tape (www.epdmroofs.org). The membrane can be fixed directly to wood and structural concrete decks, or mechanically anchored to steel decks.

*Polyethylene film* (IS 2508:2016) can be used to waterproof concrete slabs with bitumen and/or bituminous compositions. The various steps involved as per IS 7290:1979 are as follows (see Fig. 22.4):

1. Primer is first applied to properly impregnate the prepared surface at 0.3 to 0.5 kg/m$^2$.
2. Over this, hot bitumen (conforming to IS 73:2013) is applied at a rate of 0.70 kg/m$^2$.
3. The polyethylene film with cold cutback adhesive is then struck in overlaps.
4. The 100-g brown Kraft paper is then laminated *in situ* over the film with semi-hot layer of straight-run bitumen.
5. Straight-run grade semi-hot bitumen, dusted with fine sand, is applied over it at the rate of 0.7 kg/m$^2$.
6. Finally, the finishing layer of tiles or stones may be laid.

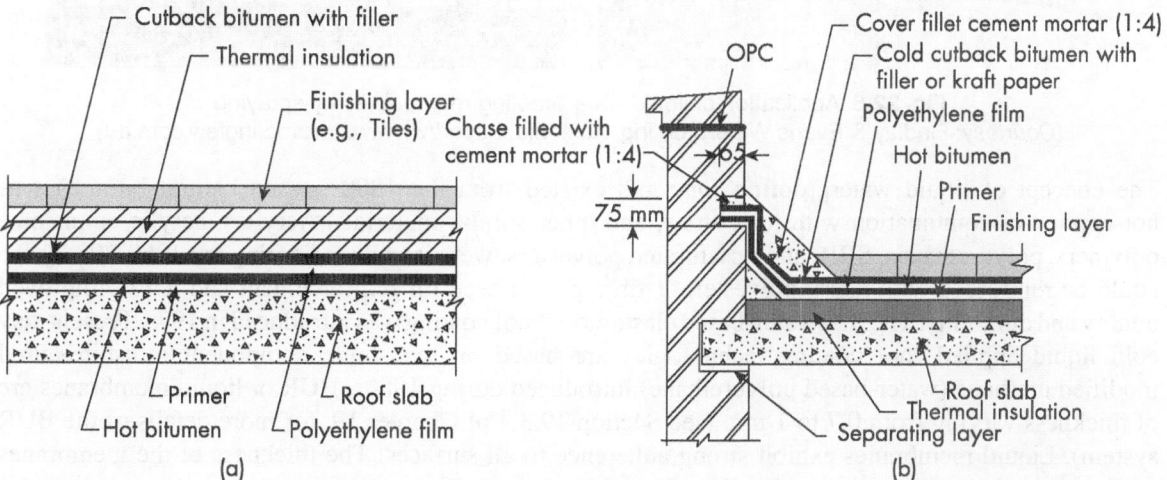

**Fig. 22.4** Use of polyethylene film in waterproofing as per IS 7290:1979 (a) Treatment laid below thermal insulation (b) Waterproofing treatment when roof abuts parapet wall

It has to be noted that several sheet membranes such as those having urethanes, vinyl, and epoxy-based materials, are high in volatile organic chemicals.

**Roof gardens**　Where roof gardens are desired, the waterproofing treatment should be made as per IS 1609:1991. As far as possible, plants should be planted in containers to avoid root penetration into the membrane and roof under it.

## 22.4.4 Liquid Applied Membranes

Liquid applied waterproofing membranes in liquid form are either applied manually by brush and roller or sprayed onto the surface (see Fig. 22.5). The applied liquid cures in the air to form a seamless, joint-free membrane. Since the application procedure is very quick, large areas could be waterproofed in a single day without the danger of cold joints. However, when a very large area is encountered, a new membrane can be overlapped over the old one, as it will stick to itself readily. Liquid-applied membranes are seamless, semi-flexible, easy to apply, detail, maintain, and repair, and hence are considered better than sheet membranes. However, careful supervision and control is needed while applying, to ensure proper curing, consistent thickness, and uniform application. If the applied membrane is too thin, it can tear or break. In addition, the adhesion of the membrane to concrete must be good. When concrete screed is provided over the membrane, the membrane has to be made rough by throwing a thin layer of sand over the wet membrane (before it is cured).

**Fig. 22.5** Application of liquid waterproofing membranes by spraying
(*Courtesy*: Findlay & Evans Waterproofing, Australia, https://www.waterproofingfew.com.au)

The concept of liquid waterproofing for roofs existed from the 1800s, when natural bitumen was hot applied in combination with jute, straw, and other similar materials. Acrylics, acrylic emulsions, polymers, polyurethanes, SBR, and unsaturated polyesters, were introduced during the 1960-70s. These could be reinforced with glass, polyester, or fibre glass fleece for reinforcement and led to improved quality and durability. The first water-based elastomeric roof coating was introduced in 1975. Present-day cold liquid applied roof coating technologies are based on the single component moisture-cured modified urethane (water-based polyurethane) introduced during 1980s. BUR or liquid membranes are of thickness varying from 0.7 to 1 mm (see Section 19.3.4 of Chapter 19, for more details of the BUR system). Liquid membranes exhibit strong adherence to all surfaces. The thickness of the membranes can be controlled by the spray application. Sprayed waterproofing is a rapid process and can cover more than three times the area than conventional brush/roller type application.

*Acrylic polymer-based liquid* applied membranes are mostly single component, flexible, and ready to use systems. The coating does not require any primer and is odourless, dries tan and can be painted with any colour. Their use has the following advantages: water vapour breathability, resistance to corrosive environment, UV resistance, high solar reflection, and low absorption. The solar reflectance property of the material results in 'cool roofs' and hence reduces both the heat-island effect and the energy consumption for cooling of buildings. They also reduce thermal stresses on the roof and on the material itself [for example, Sika® SolaRoof® MTC 15 has an initial solar reflectance of 0.85 (after 3-years-0.71), thermal emittance of 0.9 (after 3-years-0.88) and solar reflectance index (SRI) of 107 (after 3-years-87)]. They are in conformity with LEED 2009 (SSc 7.2)/Version 4 (SSc 5). These membranes are suitable for all surfaces – concrete, brickwork, metal, and timber – and can be applied for above and below ground waterproofing. Acrylic and epoxy-based polymer coating systems penetrate surface pores and can bridge only hairline cracks. Methods of test of acrylic-based polymer waterproofing materials are covered in IS 13435 (Parts 1 to 4).

One-component *polyurethane-based liquid* applied membranes and roof coatings were invented in 1960s. Owing to their high mechanical properties and flexibility, especially at lower temperatures, and their capability to cure under a wide range of conditions they are used in climatically challenging environments. They also have excellent adhesion, high elastic and crack-bridging ability. They are UV stable and resistant to yellowing with high SRI. Moisture-triggered chemistry polyurethane systems use atmospheric moisture to trigger the curing process. This means the waterproofing membranes are capable of curing in a wide range of conditions including extreme temperature ranges and humidity variations. Unlike traditional polyurethane systems, they do not release $CO_2$, which often causes gassing, and application is not delayed by adverse weather conditions (example: Sikalastic®-612). The i-Cure hardener-based liquid applied membranes have reduced solvent content and reduced odour emission during and after curing and hence are ideal for hospitals, schools, food, and pharmaceutical industries (example: Sikalastic®-631 BC).

Two-component hot spray applied roofing membranes are based on a *hybrid of polyurethane/polyurea*. In these systems, the high mechanical strength and abrasion resistance of polyurea is complimented by the flexibility and low modulus of elasticity of polyurethane. These are solvent free with solid content of 100%, resulting in efficient consumption rates without any material loss due to the evaporation of water or solvents. This technology is found to be very efficient and economic, especially for large surface areas. Owing to the fast curing of this membrane, pedestrian traffic could be allowed and an UV protective top coat be applied after a few minutes (example: Sikalastic®-851 R).

As liquid-applied membranes are not very elastic, they do not bridge over cracks and gaps well. Hence, it will be a good practice to fix a lax bitumen membrane over expansion and movement joints, as shown in Fig. 22.6.

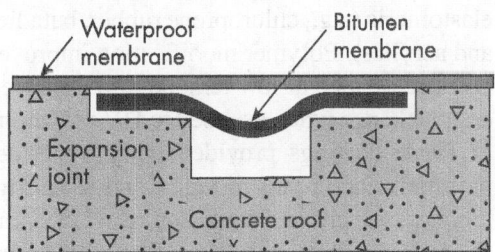

**Fig. 22.6** Detail at expansion and movement joints of concrete roof

## 22.5 Waterproofing Coatings

Coatings were employed earlier for aesthetic reasons but now are used to protect reinforced concrete against penetration of water, $CO_2$, and other aggressive environments and to protect corrosion of reinforcements. They are applied after cleaning and applying a primer on the concrete surface, using a brush or roller. Coatings have advantages such as good adhesion to substrate, application even on the negative side, impermeability to chemicals and moisture, permeability to vapour, some crack bridging ability (up to 0.5-mm width), resistance to microbial growth, and low dirt pick-up. Several coating systems based on acrylic resin, acrylic rubber, bitumen, epoxy, polyester, silicone, urethane, and polymer-modified cement, and mortar are available (Nair and Gettu, 2016). The coatings may be classified as high-viscosity surface coatings (such as acrylic, epoxy, and urethane) that do not penetrate capillary pores and low-viscosity monomers (such as methyl methacrylate and styrene) that penetrate capillary pores by means of special drying and vacuum process (Bassi and Roy, 2002).

It is difficult to control the thickness of the coating during the application. Hence, two coats are applied perpendicular to each other to ensure an even thickness over the applied area. On the other hand, if the coating is too thick, it may break and cracks may appear. In addition, the coatings cannot take abrasion and can be damaged when exposed to direct sunlight. Hence, a protection screed should be applied as soon as the curing process is completed.

### 22.5.1 Bitumen-based Waterproof Coating

Standard *bitumen emulsions* are normally considered to be of the oil-in-water (O/W) type and contain 40 to 75% bitumen, 0.1 to 2.5% emulsifier, 25 to 60% water, and some minor components. The bitumen droplets range from 0.1 to 20 micron in diameter (TRC E-C102, 2006). Bitumen emulsion is obtained by mixing hot bitumen with water-containing emulsifying agents and agitating it sufficiently so that the bitumen is broken into droplets. Bitumen emulsions are classified according to the sign of the charge on the droplets and according to their reactivity (rapid/medium/slow setting). *Cationic emulsions* have droplets which carry a positive charge. *Anionic emulsions* have negatively charged droplets (see also ASTM D977 and D2397). Emulsified bitumen has considerably lower viscosity than asphalt itself, allowing it to be used at lower temperatures and thus have reduced emissions and energy consumption. However, they cannot meet the increasing requirements of flexibility, strength, crack resistance, and high temperature performance. Hence, bitumen-based waterproof coatings are not extensively used now.

### 22.5.2 Bitumen-based Polymer Coating

These are a mixture of bitumen, mineral fillers, and polymers. Polymers commonly used are either elastomeric (e.g., chloroprene rubber, butadiene rubber, SBS, or natural rubber), or plastomeric (e.g., APP and acrylics). Polymer modification improves the properties such as mechanical strength, flexibility, and crack resistance at low temperatures, and resistance to flow at high temperatures. Its sealing properties do not change when exposed to UV radiation and can be used in the temperature range of –25 to +100°C.

These coatings provide highly impermeable barrier against ingress of water and hence used in pavements and roads, bridges, and repairing of old bituminous roofs and tunnels. Since polymers also increase solvent and corrosion resistance, the system can be applied on damp surfaces by brush, roller, or spray and forms a uniform film after drying. It has a wide range of applications, including application in underwater conditions. On aging, bitumen-based polymer coating remains flexible, nontoxic and eco-friendly (Nair and Gettu, 2016).

### 22.5.3 Elastomeric Coating

These are special coatings and their elastomeric membrane forming feature distinguishes them from ordinary paints. Elastomeric coatings are higher volume solids (45–60%) than conventional paints, and, when applied on surfaces, result in a dry film thickness of 10–20 mils (0.25 to 0.5 mm) per coat (as against conventional paints with a dry film thickness of 2–3 mils only). They have excellent waterproofing properties and can tolerate some substrate movement. Their stretchiness (150–400% greater elongation without breaking throughout their service life) allows them to fill or bridge even moving hairline cracks. To bridge 1-mm crack-width, a minimum film thickness of about 0.1 mm and, for a 2-mm crack, a film thickness of 0.25 mm may be required (Varghese, 2005). It is also possible to introduce very fine fibreglass mat as reinforcement along with these coatings to enhance the crack-bridging capacity. Elastomeric coatings can be used on walls as well as roofs and on surfaces such as stucco, concrete, and masonry. They are UV and heat resistant and help improve the energy efficiency of building considerably. However, elastomeric coatings achieve very little penetration of the substrate. It may be essential to repair any bigger-than-hairline cracks and fill any voids prior to the coating application. A primary coating may also be necessary.

Elastomeric coatings may contain acrylic resin solutions, water-repellent silicon resins, acrylics, and polyurethanes. Acrylic ester waterproof coating is obtained with acrylate or methacrylate, as main monomers, and copolymerized with other monomer containing vinyl groups, and adding appropriate fillers and additives. Polyurethane waterproof coating is a kind of polymer waterproof material with polyurethane resin as the main film-forming material.

Exterior coatings, based on styrene acrylic copolymer, after application cure to form durable, protective, and waterproof barriers. Acrylic polymer-modified elastomeric coating has excellent flexibility, breathing properties, water resistance, crack-bridging properties, weathering durability, and resistance to growth of microbes. *Acrylic coatings* are more popular and available in India (e.g., Sika® CemCrete and Brushbond of Fosroc).

Water-based *silicone coating* systems can be applied with a roller, brush, or spray. For better performance, a minimum of two coats are necessary. They have the advantages of stability under weathering conditions, resistance to sunlight, ozone, rain, snow, and temperature extremes, protection against cracking, chalking, peeling, and blistering. These systems are breathable, resist fungus, mould, and mildew, and bridge hairline cracks. Their stretchiness (600% elongation) can withstand normal, seasonal expansion and contraction (Nair and Gettu, 2016). Pre-treatment of the substrate is necessary to prevent leaching of chemicals. Alkali (sodium, potassium, or lithium) silicates are promoted as water repellents, but their efficiency has not been conclusively proven yet (ACI 515.2 R-13).

*Polyurethane coating*, also known as urethane coatings, can be easily applied by a brush, spray, or roller. After curing, it forms a fairly thick waterproof film due to the chemical reaction between its components. Rough or porous surfaces may require two coats. Polyurethane coatings can be one- or two-part systems. They are formed through the reaction of a polyol with an isocyanate (ACI 515.2 R-13). When cured with the ambient moisture in the air, they should be applied on dry surfaces and protected from moisture to prevent blistering during the curing period. An epoxy coating may be applied prior to applying polyurethane coating to close the pores in the surface. *Aliphatic urethanes* have very good abrasion resistance, colour, and gloss stability, and resistance to UV light but are more expensive than *aromatic-based polyurethanes* (ACI 515.2 R-13). With little solvent in the coating, the volume of the film shrinks slightly. This type of coating is preferred due to its high strength, good bonding, tear, puncture, and wear resistance, extensibility and tensile strength, weather and chemical resistance, and high adaptability to environmental temperature differences and base-layer deformation. It is a synthetic high molecular waterproof coating with excellent performances. Some of the disadvantages are its certain toxicity, minimal fire resistance, and high cost. Recently, *waterborne polyurethane coating* has been developed. It is nontoxic and nonflammable, and has advantages of no pollution, easy storage, and convenient to use.

## 22.5.4 Impregnating Polymer Coatings

Precast concrete products can be impregnated with low viscosity monomer-activator systems (such as MMA or Styrene with AIBN) and then the monomer be polymerized by using high temperature or microwaves. Owing to this, a protective coating is formed on the surface of concrete, which improves its mechanical properties and also protects it from freeze-thaw cycles and from severe environments. Silicon resin-based water repellent systems can also be used to impregnate concrete, though they have long reaction period to form a silicone-resin network. They have low viscosity and hence penetrate into the concrete and improve the UV and weathering resistance of concrete. But they have no effect on the carbonation of concrete (Nair and Gettu, 2016).

## 22.5.5 Epoxy Coating

Epoxy resin was first patented in Germany by Paul Schlack (Bayer) in 1934 and by Pierre Castan of Switzerland in 1943 (licensed to Ciba Geigyand from 2003 to Huntsman Corporation of the USA). It is obtained as a result of thoroughly mixing thermosetting epoxide polymer with a hardening agent and curing the mixture in air, involving the cross-linking of polymer molecules (see also Section 17.9 and Table 17.2 of Chapter 17). Epoxies are known for their excellent durability, adhesion, low porosity, chemical and heat resistance, good-to-excellent mechanical properties, and very good electrical insulating properties. Epoxy materials are used for the structural repair of concrete as well as for waterproofing terraces and sunken toilet slabs (http://csscorp.net).

### 22.5.6 Testing to Evaluate Waterproof Coatings

The rapid chloride permeability test is widely used indirect method for determining the permeability of concrete (ASTM C1202). Resistance to chloride ion penetration can also be determined by ponding chloride solution on a concrete surface and determining the chloride content of the concrete after 90 days, at particular depths (AASHTO T259). Other tests are stipulated in ASTM D412, ASTM C 836, ASTM E96, ASTM D2243, and E 1062 (see also Nair and Gettu, 2016).

## 22.6 Cementitious Waterproofing

In India, *cementitious waterproofing* systems are preferred over other waterproofing systems, as they are compatible with concrete. Concrete having appropriate cement content (as per Table 5 of IS 456) and w/cm ratio less than 0.45, when properly mixed, placed, and cured, usually results in a very good impermeable concrete. In order to improve workability with the aforementioned w/cm ratio, and to finally get a concrete with better impermeability, plasticizers or superplasticizers can be added. It has to be remembered that proper consolidation and curing is also important to get crack free concrete (see also Chapter 8).

The modifiers used with cement-based systems can be classified as non-polymeric and polymeric materials (see Fig. 22.1). Non-polymeric integral waterproofing materials are in the form of dry powders or liquids (see also IS 2645:2003). When added to concrete or mortar, at the time of mixing (at the rate of 1 kg per bag of cement), they function as plasticizers, mineral admixtures, pore blockers, repellents, or crystalline admixtures. Their main requirement is to reduce the required quantity of water and at the same time to increase workability. The water reduction reduces the permeability of concrete. As seen in Section 8.2.1 of Chapter 8, plasticizers, such as modified lignosulphonates, are used to reduce the w/c ratio and at the same time increase the fluidity of concrete, thus making the effective placing and compaction of concrete possible. As discussed in Section 9.3 of Chapter 9, when mineral admixtures, such as fly ash, silica fume, blast furnace slag, or reactive rice-husk ash, are used in concrete, in conjunction with low w/cm ratio and appropriate superplasticizer, the resulting concrete will have less permeability and higher durability, in addition to increased strength. These supplementary cementing materials reduce permeability through hydration and pozzolanic-reaction process. Repellent-type admixtures react with calcium hydroxide present in the hydrated cement paste, forming water-repellent salts that prevent the ingress of water (Nair and Gettu, 2016). The correct dosage rate and directions for use of each of these admixtures can be obtained from the manufacturers.

### 22.6.1 Integral Waterproofing

A number of proprietary integral waterproofing compounds for mortar and concrete exist in the market [e.g., Accoproof (ACC India), Conplast WL (Fosroc), Dichtament DM (MC-Bauchemie), Pidiproof LW (Pidilite), and Pudlo CWP (David Ball Group)]. They may be added to cement to make the mortar or concrete waterproof (the quantity to be added should be according to the manufacturers' recommendations, and generally will not exceed 3% by weight of cement or about 200 mL per 50 kg bag of cement). The permeability of the standard cylindrical specimen prepared with the recommended proportions of waterproofing compound should be less than half of similar specimen prepared without the addition of the compound, as per the permeability test described in IS 2645:2003. This code also prescribes other requirements of integral waterproofing compounds.

## 22.6.2 Crystalline Admixtures

In 1943, a Danish chemist, Lauritz Jensen, patented a revolutionary crystalline-active concrete waterproofing method. In 1946, he founded his first company in Denmark and called it VANDEX ('water out'). Owing to its many advantages and distinct benefits, the VANDEX method soon succeeded in the Scandinavian market and subsequently used throughout Europe. Such *pore blockers*, including those that are called as *crystalline admixtures*, reduce the permeability and sorptivity of concrete considerably. The hydrophilic nature of these materials causes them to react with calcium hydroxide and water and increase the density of calcium silicate hydrate and/or generate pore-blocking deposits that resist water penetration (ACI 212.3R-16). These crystalline deposits develop throughout the depth of the concrete and become a permanent part of the concrete mass. Such admixtures, designed for use in high-cement content and low-w/cm ratio concretes, contain a *hydrophobic pore-blocking ingredient* (*HPI*). The HPI has to be added to the concrete at the rate of 30 L/m$^3$ of concrete (this is 5 to 12 times the normal dosage of ordinary water repellents). HPI has two principal components – reactive aliphatic fatty acids and aqueous emulsion of polymeric and aromatic globules. Each has distinctive actions (Aldred, 1988), which are as follows:

1. During mixing, the fatty acids react with the initial hydration products and create a water and moisture repelling lining on the capillaries and concrete surfaces.
2. When hydrostatic pressure is applied, the globules are pushed in front of the water until they jam together plugging the capillary, preventing any further penetration.

In the absence of moisture, these chemicals remain dormant in the pores of concrete. But when they are subjected to water pressure, they come into action. It is interesting to note that the effectiveness of crystalline admixtures is yet to be established scientifically.

## 22.6.3 Repellent Type Admixtures

Repellent type or *hydrophobic admixtures* are materials including soaps and vegetable oils (tallows, soya-based), long-chain fatty acid derivatives (such as stearates, oleates, and caprylates), petroleum (mineral oil, paraffin waxes, and bitumen emulsions), and fine particle fillers (e.g., silicates, bentonite, talc). These materials provide a water-repellent layer along pores in the concrete, but the pores remain physically open (ACI 212.3R-16). They react with the calcium hydroxide in the hydrated cement paste to form water-repellent salts in the pores of concrete, thus preventing water from penetrating into the surface of concrete. As per ACI 212.3R-16, these materials can be used only where no or small amounts of hydrostatic pressure is applied.

Other permeability-reducing admixtures may contain finely divided solids such as inert and chemically active fillers (talc, bentonite, silicious powders, clay, hydrocarbon resins, and coal tar pitches) and other chemically active fillers (lime, silicates, and colloidal silica). These fine solids make the concrete denser, thus physically restricting the water to pass through the pores (ACI 212.3R-16).

## 22.6.4 Polymeric Modifiers

*Polymer-based waterproofing systems* are effective because they bond well to the base concrete, have low permeability, better workability and strength, and have low-shrinkage. They are compatible with concrete when the mortar containing these materials is applied over the concrete base (see also Section 9.8 of Chapter 9). Polymer-modified cement mortars may be cost effective than resin and silicone-based systems (Ohama, 1998). When polymer is incorporated in the cement mortar or concrete, both cement hydration and polymer film formation (i.e., coalescence of polymer particles and the polymerization of resins) occur. Owing to this, the aggregates are bound by such polymer-modified cement paste and the properties are enhanced. Polymer content (by weight) should be at least 10% for continuous film formation. The co-matrix phase formation is explained in Ohama (1998). Wet curing period of 3–7 days is required for cement hydration and film formation.

*Polymer latexes* and emulsions are polymer particles dispersed in an aqueous medium and are prepared by emulsion polymerization. The latexes used in cement mortars include elastomeric latexes (e.g., SBR and natural rubber), thermoplastic latexes [e.g., polyacrylic ester (PAE), polyethylene-vinylacetate (EVA), polystyrene-acrylic ester (SAE)], thermosetting latexes (e.g., epoxy resin), and bituminous latexes. In India, SBR and SAE are more widely used due to their lower cost. The use of latexes increases the workability and decreases the w/cm ratio for the same workability. Higher percentage of SBR, of about 15% by weight of cement, reduces the porosity and shrinkage and increases the tensile and flexural strengths (Nair and Gettu, 2016). The addition of SEA increases the silicate polymerization, thus improving the stability of the C-S-H gel. Acrylic polymers fill and seal the pores in the co-matrix phase resulting in reduced water absorption and permeation and reduced transmission of gases such as $CO_2$ and water vapour. Polyvinyl acetate and EVA have also been used, both as latex and redispersible powders to improve the fracture toughness, impermeability, and bond strength of cement mortars. However, the performance of EVA-modified concrete immersed in water has been found to deteriorate due to swelling. In addition, their use may result in leaching leading to hairline cracks affecting the permeability. This may the reason why EVA and PVA are not very commonly used (Nair and Gettu, 2016).

### 22.6.5 Redispersible Polymer Powders

These are polymer latexes in the form of powders with particle sizes of 1 to 10 μm. They should be dry blended with cement before mixing with water. Polymers that are used include EVA, SAE, PAE, PVA, and vinyl acetate and versatate copolymer (VA/VeoVa) (Wang and Wang, 2011). Although they may be used in applications similar to those of using latex-modified cement mortars, they are more popular as cementitious tile adhesives.

### 22.6.6 Water-soluble Polymers

Commonly used water-soluble polymers are biodegradable cellulose derivatives such as polysuccinimide, polyvinyl alcohol (PVAl), and polyacrylamide (PAN) (Jalalvandi and Shavandi, 2018, www.gantrade. com). They are added to the cement mortar or concrete either in the powdered form or as aqueous solutions during mixing. The problem of leaching of polyvinyl alcohol can be decreased by incorporating borax in the cement mortar. PAN-modified cement concrete results in improved flexural strength, bonding strength, dynamic impact resistance, and fatigue life of concrete, though the compressive strength may be slightly reduced (Nair and Gettu, 2016).

### 22.7 Bentonite Membrane Systems

Bentonite clay active membranes are sheets of sodium bentonite clay sandwiched between two layers of geo-composite liners. The system consists of a polyethylene honeycomb liner where the sodium bentonite powder is held and a rain-protection fleece, which are mechanically stitched together to form a strong composite membrane (www.bpa-waterproofing.com and www.mineralstech.com). The bentonite used may be of two forms: (a) dry bentonite, which will subsequently hydrate and swell, thereby bonding together and forming an impermeable barrier, and (b) pre-hydrated bentonite, which has been activated in the factory to provide an immediate impermeable barrier. Both systems can swell to many times their original thickness, potentially sealing any surface cracks.

## 22.8 Brickbat Coba with Lime

It is an old system, but still in use in certain parts of India. It involves laying mortar with broken brick as aggregates and brick ground with lime or cement as binding matrix. In this process, lime mortar is prepared separately by mixing one part of slaked lime and two parts of *surkhi* by volume. Then, the lime concrete is made by mixing one part of lime mortar with 2.5 parts of brick bats (prior to using, these brick bats should be soaked in water for a minimum of 6 h) and is laid over RC roof for a thickness of about 75 mm. A solution is made by soaking 600 g of Myrobalan (ink nut tree – '*Kadukkai*' in Tamil) and 200 g of *Gur* (jaggery) in 40 L of water (for 10 $m^2$ of work) and brewed for 12 to 24 h (see IS 3036:1992, for more details). This liquid is poured onto the surface of lime concrete and manually beaten with wooden mallets for 3 days to make it firm and weatherproof. The lime concrete after compaction has to be cured for a minimum of 10 days. After it is set, one to two layers of burnt clay flat terracing tiles (conforming to IS 2690) should be laid over it over a thin layer of lime mortar. The tiles should be joined with non-shrinking impervious mortar by adding suitable integral waterproofing admixtures (IS 3036). Instead of flat roofing tiles China mosaic chips in polymer bedding may be laid. All along the junction of the roof surface with the masonry of parapet wall, a strip of lime concrete fillet should be laid and finished smooth. With skilled labour, this system has proved in the past to give better waterproofing. The system may be made cost effective by using waste brick bats generated during construction or demolition.

This system may be used in combination with other waterproofing systems to be more effective. The thick mass of brick bats also provides thermal insulation to the roof of the building. The main disadvantage of this system is that it increases the thickness and dead load, as compared to other water-proofing systems. In addition, this system is laborious (skilled and experienced workers are a must) and time consuming. In addition, brickbat coba system is rigid and cannot accommodate any movements due to thermal stresses and hence may lead to the development of cracks, which are difficult to repair. The bricks used in the system are porous and when water enters, these bricks readily absorb and may hold large amounts of water creating further leakages.

## 22.9 Damp-proofing of Walls and Basements

There is a distinct difference between damp-proofing and waterproofing. Damp-proofing is done to prevent the capillary rise of water from the soil with no hydrostatic pressure, whereas waterproofing is a treatment that resists the passage of water under pressure. Though less frequently encountered in modern buildings, damp-proof course (DPC) is still an acceptable form of treatment in many situations. The DPC may be horizontal or vertical. A DPC layer is usually laid below all masonry walls, regardless of whether the wall is a load bearing wall or a partition wall. The damp-proofing work should be taken up only when the subsoil water level is at its lowest, i.e., during dry season.

Materials widely used for damp-proofing include the following:

1. Flexible materials such as butyl rubber, hot bitumen, plastic sheets, bituminous felts (IS 1609:1991), and sheets of copper or lead.
2. (b)Semi-rigid materials such as bitumen mastic (conforming to IS5871:1987).
3. Rigid materials such as impervious bricks, stones, slates, cement mortar, or cement concrete painted with bitumen.
4. Mortar with waterproofing compounds.

Damp-proofing admixtures include certain soaps, stearates, and petroleum products. They do not reduce the permeability of concretes that have low cement contents, high w/c ratios, or reduced fines in the aggregate.

Damp-proofing is accomplished several ways, which are as follows:

1. Using bitumen felt in such a way that it provides an effective barrier to prevent capillary rise of water in the wall as shown in Fig. 22.7(a). Two or more layers of bitumen felt, (conforming to 1322:1982 and IS 7193:2013) are used with overlapping joints and are bonded together with blown bitumen conforming to IS 702:1988, of grade of 85/25 or 90/15. These coatings are normally black in appearance. If a bituminous primer is recommended, about 0.2 to 0.4 L/m² of primer is first brushed over the surface and allowed to dry. The method of application of felt is as follows (IS 1609:1991): on floors, hot blown bitumen is first applied at the rate of 1.5 kg/m², followed by hessian-based self-finished felt (Type 3-Grade 2) or glass fibre-based felt (Type 2-Grade 2), and then hot blown bitumen is again applied at the rate of 1.5 kg/m². For walls, one or two layers of hessian-based self-finished felt (Type 3-Grade 2) or glass fibre-based felt (Type 2-Grade 2) is applied. Each layer of the felt may provide an expected life of up to 10 years.

2. Applying horizontal as well as vertical DPC outside the surface of RCC foundation/basement wall as shown in Fig. 22.7(b). At least, three layers on bitumen felts/mastic are used with thickness of 300 mm (see also IS 5871:1987). Half-brick thick outer protecting wall is provided to protect the DPC. Joints in successive coats of bitumen mastic are staggered at least 150 mm for horizontal and 75 mm for vertical work (IS 7198:1974). It has to be noted that such a damp-proofing may fail after a few years due to cracks forming in the DPC. Hence, it will be better to waterproof the walls and base slab from the inside face – this could be done even in old buildings.

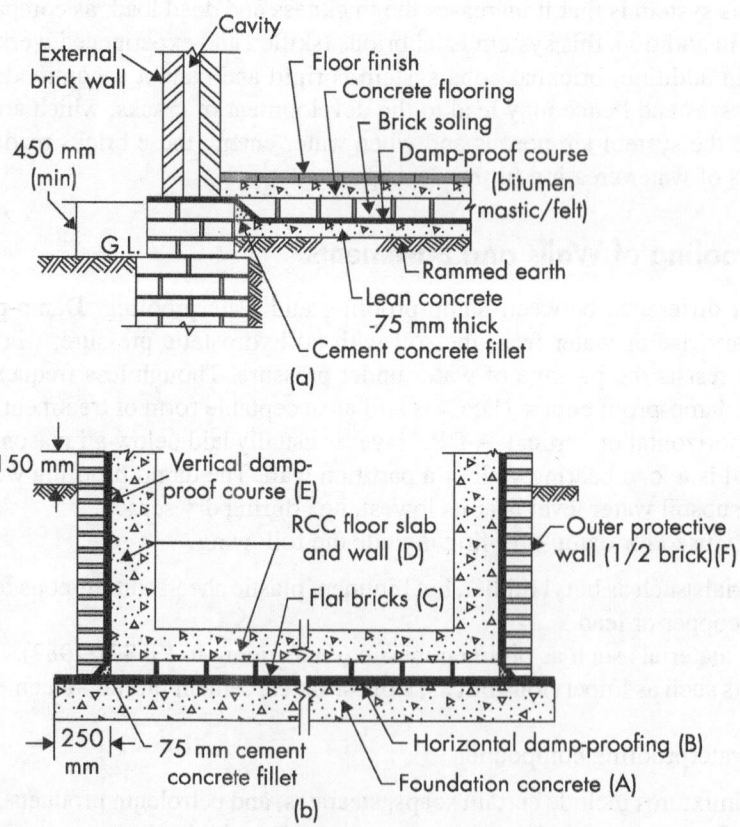

**Fig. 22.7** Damp-proofing of walls and basement in damp soil (a) Damp-proofing for flooring with high ground water level (b) External asphalt tanking for basement

3. Providing a damp-proof membrane (DPM), such as polyethylene sheeting, under a concrete slab to prevent capillary action occurring through the concrete. A DPM may be also be used to protect DPC.
4. Integral damp-proofing of concrete involving adding materials to the concrete mix itself to make the concrete impermeable as discussed in Section 22.6.
5. Surface coating with thin waterproof materials, which can resist non-pressurized moisture or applying shotcrete, which can resist water under pressure.
6. Cavity wall construction such as rat-trap wall, where the interior walls are separated from the exterior walls by a cavity.
7. Pressure grouting cracks and joints in masonry materials.

For a dry basement, both damp-proofing and waterproofing are necessary, because foundations settle over time, which could lead to minor cracks in the coating.

When basement rests on soils which are not properly drained, hydrostatic pressure will be exerted on the floors and walls, resulting in water oozing out of them. In such situations, it is necessary to provide a trench all around up to the foundation level and fill it with 0.6-m deep pervious material such as gravel, as shown in Fig. 22.8. The water may be collected using perforated drain tile pipes placed at 50 mm above foundation level. Drain pipes embedded in gravel bed may also be provided in the foundation concrete, at suitable intervals as shown in Fig. 22.8. All these drain pipes should be provided with suitable longitudinal slope such that the collected water is drained in a catch drain and pumped out. Various fibreglass or plastic drainage mats could be used instead of gravel filled in trenches. These mats are about 12.5-mm thick and are very porous. When they are placed next to a foundation wall, any water gets nearer to them is quickly drained. Guidelines for the protection of below-ground structures against water from the ground are provided in IS 16471:2017.

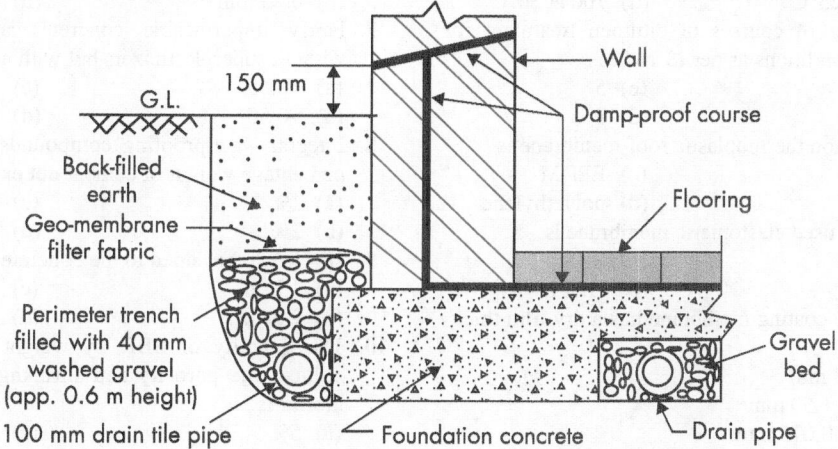

**Fig. 22.8** Damp-proofing treatment for basement on undrained soils

It is also better to keep water away from the foundation by extending downspouts away from the building, and by keeping a down slope of the ground, preventing rainwater or snow melt drain down into the soil near the foundation. For waterproofing of underground water reservoirs and swimming pools reference should be made to IS 6494:1988.

## SUMMARY

- Waterproofing is the treatment given to any surface or structure to prevent the passage of water under hydrostatic pressure. On the other hand, damp-proofing is the treatment given to any surface or structure to stop the rise of water from the foundation soil by capillary action.
- There are several causes of dampness in buildings. Continued dampness may cause rusting of steel, rotting of timber, and may even cause odours and illness.
- A waterproofing material should be watertight, durable, strong, flexible, tear-resistant, be able to crack-bridge, resistant to UV radiation, microbial attacks, and fungus growth.
- The waterproofing systems may be categorized as membranes, coatings, and cementitious waterproofing (with non-polymeric or polymeric modifiers).
- Membranes may consist of bituminous, thermoplastic, elastomeric, or liquid applied membranes.
- Coatings consist of bitumen-based polymer, elastomeric, and impregnating polymers.
- Non-polymeric cementitious modifiers include plasticizers and mineral admixtures, pore blockers and crystalline admixtures, repellent type and hydrophobic admixtures. Polymeric cementitious modifiers are found as polymer latexes and emulsions, redispersible polymer powders, and water-soluble polymers.
- Several companies such as BASF, FOSROC, Mc-Bauchemie, Pidilite, Roffe Construction Chemicals, and Sika sell their proprietary products, which have to be selected for any particular use, after carefully studying their specifications.

## EXERCISES

### Multiple-choice Questions

1. The bituminous saturant used in felts should have a penetration of not be less than
   (a) 80 at 25°C
   (b) 100 at 25°C
   (c) 80 at 30°C
   (d) 100 at 30°C

2. The number of courses of bitumen treatment for moderate conditions as per IS 1346 is
   (a) 3
   (b) 4
   (c) 5
   (d) 9

3. Most common thermoplastic roof membrane is
   (a) PVC
   (b) HDPE
   (c) EPDM
   (d) polyethylene

4. Commonly used elastomeric membrane is
   (a) PVC
   (b) TPO
   (c) SBS
   (d) HDPE

5. Elastomeric coating membrane has a dry film thickness of
   (a) 2.5 to 5 mm
   (b) 1.15 to 1.50 mm
   (c) 0.05 to 0.075 mm
   (d) none of these

6. To bridge 1-mm crack-width, the required minimum film thickness is
   (a) 0.1 mm
   (b) 0.25 mm
   (c) 1 mm
   (d) none of these

7. Fairly impermeable concrete can be produced without superplasticizers but with a w/c ratio of
   (a) <0.45
   (b) >0.50
   (c) <0.30
   (d) none of these

8. Integral waterproofing compounds are added as a percentage weight of cement not exceeding
   (a) 1%
   (b) 2%
   (c) 3%
   (d) 5%

9. HPI has to be added to the concrete at the rate of
   (a) 1 L/m$^3$
   (b) 2 L/m$^3$
   (c) 10 L/m$^3$
   (d) 30 L/m$^3$

10. The quantity of SBR by weight of cement that reduces the porosity and shrinkage of the cement mortar is
    (a) 5%
    (b) 10%
    (c) 15%
    (d) 25%

### Review Questions

1. What are the reasons and effects of dampness in buildings? What are the possible remedies?
2. What are the requirements of a good waterproofing system?
3. Draw a sketch showing the different types of systems used for waterproofing.
4. What are sheet membranes? How are they applied on roofs?
5. How are bituminous materials and membranes used to waterproof buildings? What are their main problems?
6. How is bitumen mastic manufactured?
7. What are the four types of polymer-modified bitumen membranes as per IS 15462:2004?

8. What are roof blisters? Why are they considered undesirable? How can they be prevented and repaired?
9. What are the objections/problems of using bituminous roofing products?
10. What are the two most common thermoplastic roof membranes? Briefly discuss about them.
11. Write short notes on SBS and EPDM elastomeric membranes. What properties are enhanced when these membranes are used?
12. What are the various steps involved in laying polyethylene film waterproofing as per IS 7290:1979?
13. What are the advantages and precautions to be adopted when using liquid applied waterproofing membranes for waterproofing?
14. State the advantages of using acrylic polymer-based liquid applied membranes?
15. What materials are used as waterproof coatings? What are their advantages? What are the two classifications of these coatings?
16. Write short notes on (a) bitumen-based waterproof coating, (b) bitumen-based polymer coating, (c) elastomeric coating, and (d) impregnating polymer coatings.
17. What are the two types of cement-based waterproofing systems?
18. What is meant by integral waterproofing?
19. Explain the action of crystalline admixtures, containing HPI, in reducing the permeability of concrete.
20. Write short notes on (a) repellent type or hydrophobic admixtures, (b) polymer-based waterproofing systems, (c) water-soluble polymers, and (d) bentonite clay active membranes,
21. Explain the system of waterproofing using brickbat coba with lime. Why is it not used now?
22. What is the difference between waterproofing and damp-proofing? What are the materials used for damp-proofing?

## ANSWERS

**Multiple-choice Questions**

| | | | | | |
|---|---|---|---|---|---|
| **1.** (a) | **2.** (c) | **3.** (a) | **4.** (c) | **5.** (b) | **6.** (a) |
| **7.** (a) | **8.** (c) | **9.** (d) | **10.** (c) | | |

# CHAPTER 23
# MISCELLANEOUS MATERIALS

## 23.1 Introduction

In addition to the various materials discussed in the previous chapters, we also require some materials like adhesives (which bonds two different materials), caulks, and sealants (to bridge the gaps between materials and prevent ingress of water). Bamboo, which is a renewable plant material, is gaining importance as a building material. Composite material is one in which two different materials are advantageously combined to supplement the weakness of the other (e.g., reinforced concrete). Polymer composites or *fibre composite materials* are multiphase materials produced by combining polymer resins such as polyester, vinylester, and epoxy, with fillers and reinforcing fibres to produce a bulk material with properties better than those of the individual base materials. Asbestos is a naturally occurring mineral and an effective insulator. It has been mixed with cement, plastic, and other materials to produce different products – but it is banned in many countries as it causes diseases such as lung cancer and mesothelioma. Recent research on nanotechnology has resulted in nanomaterials, in which a single unit is sized between 1 and 100 nm. These nanomaterials include nano-silica, titanium dioxide and carbon nanotubes, and are used to improve the properties of several materials such as concrete, steel, glass, paints, and coatings. Smart materials are those which are engineered to respond to cracks, excessive stresses, or environmental effects and include shape-memory alloys. 3D printing involves special materials that are used by special printers to create computer controlled optimized objects. All these materials are considered briefly in this chapter.

## 23.2 Adhesives

An *adhesive* (also called as *glue*) is used when it is required to bond two or more parts together so that they act as a single unit. Adhesives are one of the most important raw materials used in the plywood, woodworking, and joinery industry. Several types of adhesives are available in the market for different types of applications. Some adhesives require sophisticated application techniques whereas others require simple application by a brush, spray, or trowel.

Modern adhesives offer many advantages over traditional mechanical fastening methods. These advantages include the ability to bind different materials together, speedy assembly, distribution of stress across the entire bonded joint, cost effectiveness, aesthetic appearance, and increased design flexibility. However, they may have disadvantages such as decreased stability at high temperatures, and difficulty of bonding objects with small bonding surface area. Epoxies, acrylics, and urethanes all have high strength but differ in cure times, solvent resistance, and service temperature (www.adhesives.com).

They may be classified in a variety of ways depending on the materials used (e.g., organic, synthetic, and rubber based), their form (e.g., paste, liquid, film, pellets, and tape), their type (e.g., hot melt, reactive hot melt, thermosetting, pressure sensitive, or contact), or their load-carrying capability (e.g., structural, semi-structural, or non-structural).

Based on the material used, they are classified into the following two types of adhesives:

*Natural adhesives*　These are made from natural resources such as plants or animals, starch, casein glues, or natural rubber. Until Second World War, all the glues were of natural origin.

*Synthetic adhesives*　Synthetic adhesives are based on thermoplastics, thermosetting plastics, and chemically modified natural resins. The properties of synthetic resins can vary widely depending upon their basic raw materials, proportions, and conditions of manufacture.

Thermosetting adhesives (which will not soften with heat) are mixtures of urea, melamine, phenol, and resorcinol with formaldehyde. (Resorcinol is a phenol formaldehyde resin, black in colour, which is marginally less toxic, performs better in heat but more expensive than urea-formaldehyde [UF] glue.) Polyvinyl acetate emulsions (white glue) and hot-melt glues are thermoplastic adhesives and can be softened by heating (see IS 848:2006, IS 851:1978, IS 1508:1972, and IS 4835:1979). It has to be noted that adhesives with formaldehyde (which is human carcinogen and not renewable) emit toxic vapours during application and drying. In many countries, there is an exposure limit to formaldehyde ranging from 0.3 to 2.0 ppm. Products with formaldehyde may be substituted by isocyanate-based products, though their toxicity is also high. One method of lowering formaldehyde emission levels is to chemically modify UF resins with polyamines. Formaldehyde-free melamine glyoxylic acid/glyoxal resins are under research (e.g., Acrodur® by BASF).

Synthetic adhesives could be 1-component or 2-component adhesives as follows:

*1-component adhesives*　These are available in a single container, such as moisture-curing polyurethane adhesive, cyanoacrylates, silicones, moisture curing adhesives, and modified silanes.

*2-component adhesives*　These are available in two different containers; it is necessary to mix the two in the correct proportion to begin the process of curing, e.g., 2-component epoxy, polyurethane, and acrylic adhesives.

Based on the type of curing that occurs in the adhesive, they are designated as physical curing and chemical curing adhesives.

*Physical curing adhesives*　They can be contact adhesives and pressure sensitive adhesives (PSA) adhesives. *Contact adhesives* are applied to both the surfaces to be joined and allowed to dry (sometimes even up to 24 h) before the two surfaces are pressed together. It is not necessary to apply pressure for a long time and hence clamps are not necessary. PSAs form a bond only when the parts being bonded are given a light pressure.

*Chemical curing adhesives*　These include polyurethane, epoxy, and acrylic adhesives.

More information on adhesives may be found in Pizzi and Mittal (2017). Let us now briefly discusses about the uses of adhesive in construction.

## 23.2.1 Adhesives used in Timber Constructions

Adhesives based on phenolic and aminoplastic synthetic resins are used in woodwork (IS 848:2006 and IS 851:1978). UF resins are the most widely used thermosetting resin for wood. In the manufacture of particleboard, fibreboard, and oriented strandboard (OSB), pieces of wood are bonded using UF, melamine-formaldehyde, melamine-urea-formaldehyde, or phenol-formaldehyde (PF) resins (UF resins comprise about 80% of the amino resins produced worldwide). Owing to its lack of resistance to moisture, UF resins are usually used for the manufacture of products intended for interior use only. Melamine resins are used primarily to improve the moisture resistance of urea-resin adhesives. Phenol-formaldehyde resins are widely used to produce softwood plywood for severe service conditions. It has to be noted that UF resin has the highest emission rate of VOCs, and PF resin the lowest (Goyer et al., 2006).

Depending upon their degree of resistance, synthetic resin adhesives used in plywood are of the following three types (IS 848:2006):

*Boiling water proof*   They are also called BWP and are highly resistant to weather, cold, boiling water, steam, and dry heat.

*Boiling water resistant*   They are also called BWR and are inferior to BWP but can withstand cold water indefinitely but boiling water for a limited period.

*Moisture resistant*   They are inferior to BWR; can withstand cold water for a long period and hot water for a limited time but fail in boiling water.

Polyvinyl acetate adhesives (e.g., *Fevicol*) are widely used in the furniture industry, due to their competitive cost. They do not melt, but soften over a temperature range, and are unaffected by sunlight, ultraviolet light, and air.

## 23.2.2 Adhesives for Fixing Ceramic Tiles and Mosaics

The suitability of the tile adhesive depends on two factors: (a) the porosity/density of the tiles to be fixed and (b) the nature, surface, and location where the tiles are fixed. According to IS 15477:2004, Type 1 (polymer-modified adhesive) adhesive can be used for most ceramic (i.e., non-vitrified) tiles (with porosity greater than 3%) and the majority of porous stones and less demanding backgrounds. Type 2 adhesive (highly polymer-modified adhesives) has to be used for less porous/denser tiles (porosity less than or equal to 3%) in demanding backgrounds. In practice, vast majority of tiles and situations can be used with Type 1 adhesive but where either fully vitrified tiles/large dense tiles and/or demanding backgrounds occur, Type 2 adhesive has to be used.

## 23.2.3 Adhesives for Joining Concrete

Liquid epoxy adhesives are used to bond other materials with existing concrete surfaces. Surface preparation is absolutely essential to ensure proper bonding (ASTM D5258). This involves intensive cleaning, usually by pressure washing, or vacuum shot blasting followed by the removal of dust. For very good bonding, acid etching of the concrete surface with 15% by weight hydrochloric acid solution, followed by thorough flushing with a high pressure stream of water is preferable.

Usually, a two-part epoxy resin concrete adhesive is used, which is mixed together at site thoroughly and applied uniformly (15 mils minimum thickness) over the existing concrete surface. Lightweight concrete or other porous substrates may require a second coat. Ambient curing epoxies have short working times and cure quickly in hot weather and cure slowly in cold weather. Epoxy adhesives should not be applied in the rain or in the presence of standing water. When used to bond two layers of concrete together, it also acts as moisture vapour barrier between the two layers. Epoxies are often expensive and overexposure to them can trigger allergies.

## 23.3 Caulks and Sealants

These have been in use for many hundreds of years. The Tower of Babel was reportedly built with mortar and tar or pitch as a sealant. Naturally occurring bitumen and asphalt materials have been widely used as sealants for many centuries. Glazing putty was first used in the 17th century for sealing window glass into the panes. Development of polymer-based sealants started in the 1930s, in the form of acrylic and butyl rubber sealants. Polysulphides and polyurethane were introduced in the 1950s and 1960s, silicones in the 1970s and silyl-terminated polyether (MS Polymer) in the 1990s, respectively.

Caulks have lower quality and limited service life (3–5 years) than sealants. Caulks are of three types: oil-and alkyd, butyl rubber, and acrylic latex based. Acrylic latex based caulks are better than other two types, having greater life and better performance, but may cost twice as much as oil-based caulks.

Joint sealants are used in construction to prevent water, moisture, and other substances from passing through material surfaces, joints, or openings. They can also prevent the passage of air, sound, dust, insects, etc., as well as acting as vapour retarder. While some sealants have adhesive qualities, they differ from conventional adhesives as they have lower strength and higher elongation. Joint sealants are used in doors, window perimeters, expansion joints, glazing, roofing terminations, skylights, and also in airport pavement runways and aprons, and bridge and highway joints.

The main job of caulks and sealants is to bridge the gaps between materials used in buildings, especially the exterior surfaces, thus preventing water or moisture entering through these gaps. In order to perform satisfactorily, these 'water stoppers' must be able to bridge openings of various shapes, adhere to diverse materials, remain flexible enough to allow the expansion and contraction of adjoining components, and yet harden sufficiently to resist degradation by sun, water, and other chemicals. As it is difficult to fulfil all these requirements in a single product, several types of caulks and sealants have been developed to meet these differing demands.

Caulks and sealants resemble paints and other coatings, since they are composed of pigments and binders. The pigments, however, are mainly fillers, or extender pigments, which have higher percentage than in paints. This, in addition to the low concentration of solvents, prevents caulks and sealants from sagging and running in wider openings. In general, caulks can effectively seal openings in the range of 1.5 to 4 mm and sealants in the range of 3 to 20 mm. They have a wide range of properties such as strength, flexibility, hardness (ASTM C661), adhesion (ASTM C794), movement capability (ASTM C719), tack free time (ASTM C 679), appearance, and permanence (see also ASTM C920 and ISO 11600:2002). Some sealants may have high viscosity, so that they do not flow from where they are applied, and some have low viscosity, allowing them to penetrate into a substrate. Acrylic sealants can cure in the absence of air, whereas other sealants require air to cure. Sealants are generally more elastic than caulks, thus allowing a greater movement of adjoining materials without breaking.

Sealants are basically of two types: elastomeric, solvent release, and polymerizing or chemical cure. They are used when wider openings are to be sealed, more stretching is required, where more severe conditions of exposure are expected, and a longer life is desired than caulks. Sealants are naturally costlier than caulks. Some of the most common types of sealants include: acrylic sealants, butyl rubber, Neoprene® sealants, polysulphide sealants [IS 12118 (Parts 1 & 2):1987 and IS 11433 (Parts 1 & 2): 1985], polyurethane sealants, silicone sealants, silyl-terminated polyether (STPE or MS polymer), and silyl-modified polyurethanes (SPUR). A comparison of the characteristics of these sealants is provided in Table 23.1 (see also Panek and Cook, 1992).

**Table 23.1** Characteristics of sealants

| Type | Setting by | Life expectancy (years) | Elongation | Advantages | Disadvantages |
|---|---|---|---|---|---|
| Acrylic | Solvent release | 20 | 250% at 7°C | Excellent adhesion to most surfaces, wide choice of colours, self-resealing | Some odour during cure, slow cure (14 days), heat required for proper application |

*(Contd)*

**Table 23.1** (*Contd*)

| Type | Setting by | Life expectancy (years) | Elongation | Advantages | Disadvantages |
|------|------------|-------------------------|------------|------------|---------------|
| Butyl Rubber | Solvent release | 10–20 | 25–50% | Low cost, easy to apply and cleanup. Bond excellent to most substrates. May be used in curtain walls where adhesion to rubber compounds is required. | Some formulations may shrink (up to 45%), relatively low elongation, poor weathering and recovery after stretch. |
| Neoprene® | Solvent release | 30 | Up to 500% | Fast set (4 h), weather resistant, good adhesion to bitumen and metals | Slow cure (3 to 7 days), up to 40% shrinkage, available in black only. |
| Polysulphides | Chemical | 20 | 150–300% | Poor recovery after elongation, excellent weather resistance, good in submerged applications | Surfaces must be very clean, require priming. |
| Polyurethanes | Chemical | 20 | Up to 500% | Good abrasion resistance, high strength, excellent weather resistance, good UV resistance, used with several substrates | Primer normally required, relatively slow cure (5–10 days) |
| Silicone | Chemical | 5–20 | Over 500% | Viscosity not affected by temperature, sets rapidly (1 hour), good weather, UV, and heat resistance, good adhesion | May require primed surfaces, strong odour |
| Silyl-Terminated Polyether (STPE or MS Polymer) and silyl-modified polyurethanes (SPUR) | Chemical | 15–25 | 150–350% | Environment friendly(solvent and isocyanate-free), good weather and stain resistance, high strength, elastic recovery > 70%, applied on variety of substrates, used without primers, superior adhesion, quick curing | Deterioration of bond strength when over exposed to sunlight or UV, used in temperature < 85°C |

Problems in sealants may be due to (a) poor surface preparation, (b) choosing a wrong sealant for an application, (c) improper priming, (d) improper installation, and (e) contamination.

## 23.4 Bamboo

The use of bamboo dates back to over 3000 years by the Chinese. Bamboo, belonging to the grass family *Poaceae*, is among the fastest growing plants on earth, and can grow about 0.9 m per day (most commonly at 30–100 mm per day) under optimum climate and growing conditions, and can grow over 30-m tall (common height is 4.5–12 m), and can be as large as 88 to 120 mm in diameter. About 1575 species of bamboo in 111 different genera are found to exist. About 100 species are used commercially, of which 20 are identified as priority species (Bhowmick, 2017). Most bamboo species are native to warm and moist tropical and warm temperate climates, and found in countries like China, India, Japan, Korea, Australia, Central America, Mexico, and South-eastern United States. India is the second largest producer of bamboo in the world (with 30% of world's bamboo resources). Bamboo has long been used as scaffolding in India and China (China has now banned its use as scaffolding for buildings over six stories). Large consumption of traditional slow growing hardwoods such as mahogany, cedar, and rosewood, which can take 50 years or more to reach maturity, has resulted in some of these species disappearing. Hence, there is a renewed interest in using bamboo as building material.

## Do You Know?

Bamboo has extraordinary resilience – it was the only plant to survive the radiation of the atomic bombings in Hiroshima, Japan, in 1945, and grows in the Himalayas withstanding temperatures well below –20°C. Soft 'bamboo' towels, T-shirts, socks, and other textiles are produced through a chemical process similar to those used to make rayon. (The U.S. Federal Trade Commission requires most bamboo-based textiles to be labelled as rayon.)

Bamboo, being globally available and rapidly renewable, is a good example of a sustainable material. Bamboo is a crucial element in the balance of oxygen and carbon dioxide in the atmosphere. Bamboo releases 35% more oxygen than an equivalent tree and takes four times as much carbon as some trees. Bamboo can be cultivated with little to no fertilizer, pesticides, heavy harvesting machinery or irrigation, and bamboo root systems may even protect steep banks from erosion. However, it has to be noted that the glues that bind the strips into planks may contain formaldehyde, posing health risks to workers/consumers. Dyes used in them may contain heavy metals. Structural and engineered bamboo products are comparatively low-energy-intensive materials with structural properties sufficient for the demands of modern construction (Gatóo et al., 2014). Construction professionals around the world are not yet adequately familiar with the structural design of bamboo, although a few codes and standards already exist (IS 10145:1982, IS 13958:1994, IS 14588:1999, IS 15476:2004, IS 15972:2012, and IS 16073:2013). Structural design using bamboo is covered in IS 15912:2018. Tests for bamboo are stipulated in IS 6874:2008 and IS 8242:1976.

**Fig. 23.1** (a) Bamboo (b) Culms of different bamboo species (c) Prototype of ZERI bamboo pavilion used in the EXPO 2000 in Hannover, Colombia (by Sebastian Kaminski of Arup) (*Source*: Bhowmick, 2017)

The raw material is a giant grass consisting of a hollow culm having longitudinal fibres aligned within a lignin matrix, divided by nodes (solid diaphragms) along the culm length, as shown in Fig. 23.1. Bamboo can be cut and laminated into sheets and planks. The bamboo culm is split, planed flat, processed (boiling and drying the strips in kilns), laminated by gluing, and pressed to form the board product. (Shreds of bamboo can also be soaked in adhesive and pressed into blocks or panels.) Some of these products retain their original blond colour, some are steamed to a darker shade, and some are dyed. Preservatives are also added for preventing fungus attack and durability (IS 1902:2006 and IS 9096:2006). First used in China and Japan, laminated bamboo flooring was introduced in the West during the mid-1990s. Flooring, cabinetry, furniture, and even decorations, made of bamboo laminates are currently popular. The annual value of the bamboo industry has grown 500-fold since 1981, to $32 billion. The hardness of bamboo flooring depends on the maturity of the plant when it was harvested (ideal age is 5 to 6 years) and the variation of the manufacturing process. Some floors are softer than a typical hardwood floor, but others are much harder (almost three times harder than oak).

Bamboo is considered anisotropic as its mechanical properties vary in the longitudinal, radial and transverse directions. However, engineered bamboo composites have standard shapes and relatively low variable material properties. Bamboo scrimber and laminated bamboo are two examples of engineered bamboo (Sharma et al., 2015a & 2015b). Bamboo scrimber, also called strand woven or parallel strand bamboo, is made by crushing fibre bundles, saturating them in resin and then compressing the mass into a dense block. This process is efficient because it utilizes almost 80% of the raw materials, and the product could be used in deck flooring (Sharma et al., 2015a). The longitudinal direction of the bamboo fibres is maintained in this process and the resin matrix connects the fibre bundles. However, laminated bamboo plywood is made from flat rectangular bamboo strips that are cut from the bamboo stem. These strips or 'slats' are then glued together either horizontally (plain pressed) or vertically (side pressed, with a thickness of 20 mm). In plain pressed bamboo plywood, the bamboo strips are pressed together facing up. In this 1-ply plywood, the pattern and characteristic of the bamboo nodes are clearly visible, and have a thickness of 5 mm. In 3-ply and 5-ply plywood panels, different layers are glued together (usually crosswise). In order to make thicker panels or beams of side pressed bamboo plywood, several layers are glued together. Because laminated bamboo panels consist of individual strips, they show less swelling and shrinkage than solid wood panels (see www.bambooimport.com, for the way in which they are made). The final products may use only 30% of raw material input due to large losses of material during the production, when they are planned (Sharma et al., 2015a).

Experimentally measured properties of structural bamboo and comparable natural bamboo and timber products are shown in Table 23.2.

**Table 23.2** Material properties for structural bamboo and comparable natural bamboo and timber product (Adapted from Sharma et al., 2015a)

| Material | Density | Compression | | Tension | | Shear | Flexural | |
|---|---|---|---|---|---|---|---|---|
| | $\rho$, kg/m³ | $f_c^{\parallel}$, MPa | $f_c^{\perp}$, MPa | $f_t^{\parallel}$, MPa | $f_t^{\perp}$, MPa | $\tau^{\parallel}$, MPa | $f_b$, MPa | $E_b$, GPa |
| Raw Bamboo | 666 | 53 | – | 153 | – | 16 | 135 | 9 |
| Laminated bamboo | 686 | 77 | 22 | 90 | 2 | 16 | 77–83 | 11–13 |
| Bamboo Scrimber | 1163 | 86 | 37 | 120 | 3 | 15 | 119 | 13 |
| Douglas-fir | 520 | 57 | – | 49 | – | 11 | 68 | 13 |
| Sitka spruce | 383 | 36 | – | 59 | – | 9 | 67 | 8 |

Bamboo used in construction should be treated (with a mixture of borax and boric acid) to resist insects and rot. Alternatively, cut bamboo may be boiled to remove the starches that attract insects. Bamboo has been used as reinforcement for concrete but its effectiveness is not yet confirmed positively. Although bamboo has the necessary strength, when untreated it will swell by absorbing water from the concrete, causing it to crack. Several procedures must be followed to overcome this shortcoming (Sharma et al., 2015b). Bamboo behaves in a similar way to timber in fire. It chars at a slow and predictable rate and is also a poor conductor of heat. The charring rates are similar to those for timber (e.g., 0.6 mm/min).

Bamboo buildings have historically performed well in earthquakes, mainly because of their lightweight (high strength-to-weight ratio), and secondly because of their ability to absorb energy at connections, especially when nails are used for connection (Bhowmick, 2017). Bamboo was used for the structural members of the India pavilion at Expo 2010 in Shanghai. The pavilion is the world's largest bamboo dome; about 34 m in diameter, with bamboo beams/members overlaid with a ferro-concrete slab, waterproofing, copper plate, solar PV panels, a small windmill, and live plants (Soni, 2010).

## 23.5 Composite Materials

These are distinguished from alloys, which consist of two or more components, and are formed through processes such as casting (e.g., alloys of steel). Wood is a natural composite and consists of long cellulose fibres (a polymer) held together by a much weaker lignin. An early example of composite is bricks made of mixing mud and straw. In general, composites are classified according to their matrix materials. A well-known example of a composite material is concrete, which is a structural composite obtained by combining cement, sand, gravel, and optionally other ingredients such as mineral and chemical admixtures. Reinforced concrete is a composite of concrete and rebars; fibre-reinforced concrete is a composite of concrete and different kinds of fibres (see Chapters 8 and 9, for more details of the various types of concrete). Although cement-matrix composites are the most widely used structural materials, there are other classes of composites such as polymer-matrix, metal-matrix, carbon-matrix, and ceramic-matrix composites (Chung, 2010 & 2017). Polymer-matrix and cement-matrix composites are the most common, due to the low cost of fabrication. They basically consist of a matrix and reinforcements. Sandwich structures are also considered as composites (see Section 17.10.7 of Chapter 17, for more details of structural insulated panels).

The first *polymer-matrix composite material* was *fibreglass*. It is still used today for making boat hulls, sports equipment, building panels, and many car bodies. In fibreglass, the matrix is plastic and the reinforcement is glass that is made into fine threads and woven like a cloth. Some advanced composites are now made using carbon fibres instead of glass. These materials are lighter and stronger than fibreglass but are more expensive to produce. These lightweight composites are cheaper to make and easier to transport and install. Carbon nanotubes have also been used successfully to make new composites (see Section 23.7). These are even lighter and stronger than composites made with ordinary carbon fibres, but are extremely expensive. However, they provide opportunities to build lightweight, and fuel efficient cars and airplanes.

*Fibre-reinforced plastics* (FRP) are composites comprising high-strength continuous fibres embedded in a polymer matrix. The fibres provide the main reinforcing elements whereas the polymer matrix (epoxy resins) acts as a binder, protecting the fibres and transfers loads to and between the fibres (see also Section 17.12 of Chapter 17). The reinforcement provides strength, stiffness, rigidity, and other mechanical properties to the composite. The reinforcement may be in the form of steel, glass, carbon, aramid, or aluminium fibres. As a general rule, the stiffness and strength of a composite will increase in proportion to the amount of fibres present [up to 60–65% of fibre volume fraction, the tensile strength will increase]. The fibres in the matrix can be oriented in four different ways: (a) unidirectional, (b) bi-directional, (c) random or discontinuous, and (d) woven (see Fig. 23.2). Carbon fibre based composite has superior mechanical properties and higher tensile strength, stiffness, and durability compared to glass fibre based composite. The inherent corrosion resistance of FRP composites not only increases their durability, but also reduces the need for their maintenance. Such FRP composites have been used in bridge decks, cable stayed bridge, electrical transmission towers, wrapping freeway support columns, piers, seawalls, dock fenders, and GRP pipes (IS 12709:1994 and IS 13916:1994).

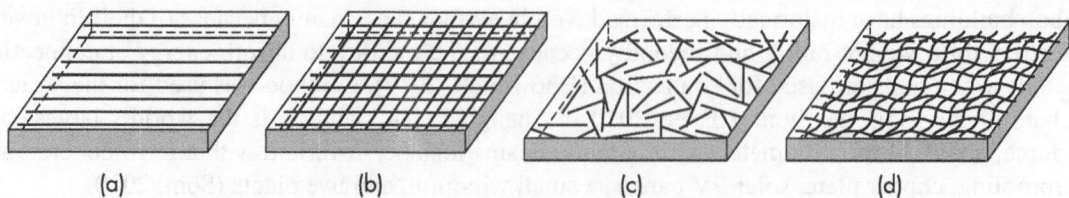

**Fig. 23.2** Various types of fibre-reinforced composite laminates (a) Unidirectional (b) Bi-directional (c) Discontinuous (d) Woven

Carbon fibre polymer-matrix composites have the following attractive properties (Chung, 2010):

1. Low density (lower than aluminium)
2. High strength (as strong as high-strength steels)
3. High stiffness (stiffer than titanium, yet much lower in density)
4. Good fatigue resistance
5. Good creep resistance
6. Low friction coefficient and good wear resistance
7. Toughness and damage tolerance (as enabled by using appropriate fibre orientations)
8. Chemical resistance (chemical resistance controlled by the polymer)
9. Corrosion resistance
10. Dimensional stability (can be designed for zero coefficient of thermal expansion)
11. Vibration damping ability
12. Low electrical resistivity, and
13. High electromagnetic interference shielding effectiveness

*Carbon-matrix composites (CMC)* are comprising carbon or graphite fibres embedded in a carbon or graphite matrix. The carbon fibre reinforcement makes these composites stronger, tougher, and more resistant to thermal shock than conventional graphite. The advantages of carbon-carbon composites include lightweight, low density, high strength, high stiffness, high thermal conductivity, low coefficient of thermal expansion, high fracture toughness, and good fatigue and creep resistance (Chung, 2017). The combination of low expansion and high thermal conductivity makes them resistant to thermal shock. The uniqueness of these composites is that their strength increases with increasing temperature up to about 2000°C. However, carbon-matrix composites suffer from their tendency to be oxidized, thereby becoming vapour. Another drawback is their high manufacturing costs. About 63% by volume of these composites produced is used in aircraft braking systems (first fitted to Concorde in 1973). CMCs have been used in the nose cone and leading edges of the US space shuttle. They are also used as wall tiles of fusion reactors (Taylor et al., 2017).

*Metal-matrix composites* (MMCs) are made by dispersing reinforcing material into a metal matrix. The surface of the reinforcement may be coated to prevent any chemical reaction with the matrix. The matrix usually consist of a lighter metal such as aluminium (most widely used), magnesium, copper, nickel, tin alloys, silver-copper, and lead alloys, and titanium. In high-temperature applications, cobalt and cobalt-nickel alloys are used. The reinforcement can be either continuous, or discontinuous (Chung, 2010). There are two types of discontinuous reinforcement for MMCs: particulate and whiskers. The common types of particulate are alumina, boron carbide, silicon carbide, titanium carbide, and tungsten carbide. The common type of whisker is silicon carbide, but whiskers of alumina and silicon nitride have also been used. Whiskers generally cost more than particulate. Continuous reinforcement of carbon or ceramic fibres is also used. In order for the diffusion bonding of the fibre/matrix interface to occur, MMCs are usually fabricated at elevated temperatures. When they are cooled down to ambient temperature, residual stresses are generated in the composite. The reinforcement of MMCs serves to reduce the coefficient of thermal expansion and increase the strength and modulus. MMCs can operate in wider range of temperatures, do not absorb moisture, have better electrical and thermal conductivity, greater wear resistance, are resistant to radiation damage, and do not release any harmful gases. However, MMCs tend to be more expensive, have lower ductility and toughness, and are difficult to fabricate. Applications are found mostly in aircraft components, space shuttle, military hardware, cutting tools, diesel engine pistons (Toyota), and high-end sports equipment. Other applications include high-temperature applications (nickel), bicycle frames (carbide-aluminium), automobile cylinder liners (aluminium), electronics (copper-silver), electronic heat sinks and substrates (aluminium and copper), soldering and bearings (tin alloys), and brazing (silver-copper) (Chung, 2010).

*Ceramic-matrix composites* consist of ceramic fibres embedded in a ceramic matrix. The fibres used may include carbon, special silicon carbide, alumina, and mullite (a rare silicate mineral). These reinforcements enhance the fracture toughness, thermal shock resistance, resistance to crack propagation, and avoid abrupt brittle failure. They are used in high-temperature applications (e.g., aerospace and engine components), slide bearings of pumps, and disk brakes of cars and airplanes. They can also be used in applications, which employ conventional ceramics or in which metal components have limited life due to corrosion or high temperatures.

### 23.5.1 Concrete Canvas

Concrete Canvas® (CC) [also called Concrete Cloth® or Geosynthetic Cementitious Composite Mat (GCCM)] is a cement impregnated flexible fabric, developed by the British company Concrete Canvas. CC consists of a 3D fibre matrix, made up of a specially formulated dry concrete mix sandwiched between waterproof PVC backing layer at the bottom and a top layer of hydrophilic fibres (polyethylene and polypropylene yarns), which aids hydration by drawing water into the mixture (see Fig. 23.3). CC can be hydrated either by spraying or fully immersing it in water. It hardens when hydrated (to 80% strength within 24 h) to form a thin, durable, water, and fire proof concrete layer. CC is available in three types of thickness-

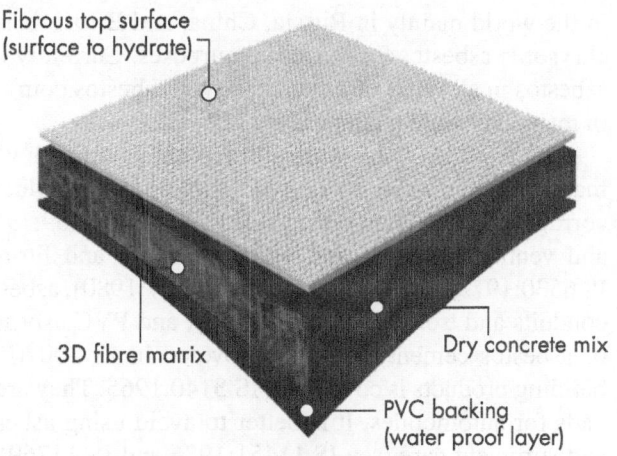

Fibrous top surface (surface to hydrate)

3D fibre matrix

Dry concrete mix

PVC backing (water proof layer)

**Fig. 23.3** Concrete Canvas®
(*Source*: www.ccportal.co.uk)

es: CC5 (5 mm), CC8 (8 mm), and CC13 (13 mm). CC is supplied to the site in the form of rolls of width 1 to 1.1 m, which can be spread on the required place and watered. Thus, it does not require any plant or mixing equipment and can be rapidly laid by three persons at the rate of 200 m²/h, which may be 10 times faster than conventional concrete solutions. It can be easily nailed, stapled through, or coated with an adhesive for easy attachment to other surfaces. Once set (initial set 120 min and final set 240 min), the fibres reinforce the concrete, preventing crack propagation and providing a safe plastic failure mode. CC has a 10-day compressive failure stress of 40 MPa (tested as per ASTM C 473–07). CC may be twice as abrasion resistant as standard OPC concrete, has excellent chemical resistance, good weathering performance, and will not degrade in UV rays.

CC is often used as an alternative to shotcrete due to its speed and ease of installation, material and costs savings, and long service life of over 10 years. CC is used in a variety of civil infrastructure applications, such as ditch lining, slope protection, and capping secondary containment bunds. Compared to traditional concrete solutions, CC is faster, easier, and more cost effective to install and has the additional benefit of reducing the environmental impact of concreting works by up to 95%. More information of CC may be found in Srinivas and Ravinder (2012) and in www.ccportal.co.uk.

## 23.6 Asbestos Cement Products

*Asbestos* is a naturally occurring silicate fibrous mineral. Asbestos fibres are soft and flexible. They have excellent tensile strength, poor heat conduction, and resistance to chemical attack. Asbestos has been used as an effective insulator for several years. Six types of asbestos have been recognized, which are classified into the following two groups (IS 11707:1986).

1. Serpentine asbestos: chrysotile (white)
2. Amphibole asbestos: crocidolite (blue), amosite (brown), anthophyllite, tremolite, and actinolite

More than 55 countries around the world, including Japan, Australia, and the European Union, have banned the use of asbestos as they are feared to cause diseases such as lung cancer and mesothelioma (Asbestos-related illnesses often take 20–50 years to develop). However, 2 MT of asbestos is produced in the world mainly in Russia, China, and Brazil. It is not banned in India as the Indian industry uses chrysotile asbestos for industrial purposes. Currently in the USA, it is legal to include more than 1% of asbestos in all types of products (www.asbestos.com). Plastic materials are replacing asbestos products in many developing countries.

About 15% of asbestos fibres can be mixed with cement to produce very strong composite material called *asbestos cement*. It is used to produce several products such as asbestos cement flat, corrugated, and semi-corrugated sheets (IS 459:1992, IS 2096:1992, and IS 13008:1990), waste and ventilating pipes, and rainwater pipes and fittings [IS 1592:2003, IS 1626(Parts 1 to 3):1994, IS 6530:1972, IS 6908:1991, and IS 9627:1980], asbestos cement building boards (IS 2098:1997), cable conduits and troughs (IS 8870:1978), and PVC asbestos floor tiles (IS 3461:1980). Methods of laying of asbestos cement sheets are covered in IS 3007(Parts 1 & 2):1999 and painting asbestos cement building products is covered in IS 3140:1965. They are also used in clutches, brake linings, gaskets, and pads for automobiles. It is better to avoid using asbestos products or use them with abundant caution and sufficient care (see IS 11451:1986 and IS 11769(Parts 1–3):1987 for the guidelines). In any case, it is advisable to apply a coat of white paint to the underside of asbestos sheets, when they are used in roofing sheets. Several test methods are also prescribed in Indian codes (IS 3632:1969, IS 4844:1968, IS 5328:1969, IS 5748:1969, and IS 5913:2003).

*Asbestos cement products* are popular due to the following advantages offered by them:

1. They are very good thermal insulators and may increase the energy efficiency of the building.
2. They are highly resistant to fire and do not burn easily (melting point 1200 to 1500°C).
3. They are inexpensive and cost effective (AC pipes are very much cheaper than PVC pipes).
4. They are highly durable and weather proof.
5. They are resistant to damage from termites and rust free.

6. They have high tensile strength along the fibre.
7. They are very easy to clean and maintain but difficult to repair.

## 23.7 Nanomaterials

*Nanotechnology* is the technology of using very small particles of material, of size in the order of 10–100 nanometre (1 nm = $10^{-9}$m) (approximately, 100,000 times smaller than the diameter of a human hair), or by manipulating them to create large-scale materials. Materials engineered to such a small scale are often referred to as *engineered nanomaterials* or *nanomaterials* (ENMs or NMs). Nanomaterials offer unique and much improved performance properties for adhesives, concrete, coatings, flooring, glass, and other construction products. Already, numerous products such as sunscreens, cosmetics, sporting goods, stain-resistant clothing, tires, and electronics are manufactured using ENMs. They are also used in medical diagnosis, imaging, drug delivery, and in environmental remediation (www.niehs.nih.gov). Nanomaterials which could be used in construction include: nano-silca, titanium dioxide, carbon nanotubes, and various nano-coatings. A comparison of nanofillers with supplementary cementitious materials is shown in Fig. 23.4.

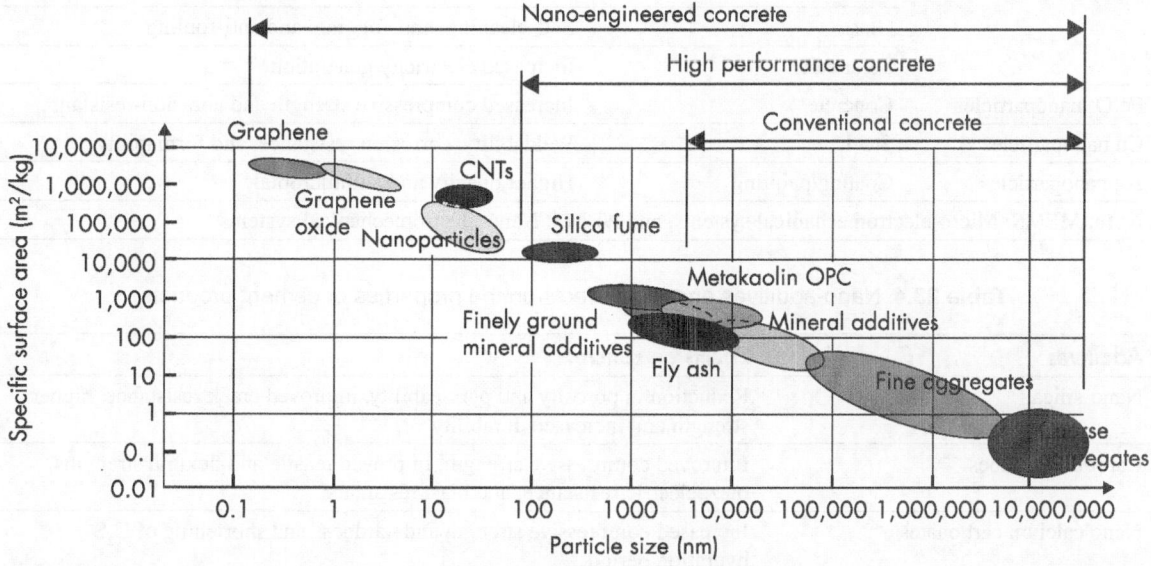

**Fig. 23.4** Comparison of nanofillers with supplementary cementitious materials and aggregates in concrete (Adapted from Sobolev and Gutiérrez, 2005)

Nano-silica particles can be obtained by the sol-gel process from the hydrolysis of tetraethoxysilane in ethanol using ammonia as base catalyst (Chatterjee, 2017). Like micro-silica, (silica fume) nano-silica is also primarily composed of $SiO_2$, but with finer, 10 to 150 nm amorphous particles available as dry powder or in a stabilized suspension. Resent research has shown that nano-silica also affects concrete durability on a chemical and physical level, albeit slightly differently from micro-silica due to its much finer size. Its improved particle size distribution and packing in the concrete results in better strength, durability, and impermeability (Ghafoori, et al., 2016). Chatterjee (2017) and Lee et al. (2010) provide a review of the beneficial aspects of nanotechnology in cement and concrete industry (see Tables 23.3 and 23.4).

**Table 23.3** Examples of ENMs used in construction

| MNMs | Construction materials | Expected benefits |
|---|---|---|
| Carbon nanotubes | Concrete | Mechanical strength, increased durability, and crack prevention |
| | Ceramic matrix composites | Enhanced crack and shatter resistance, improvement of strength, thermal, and optoelectric properties |
| | NEMS/MEMS | Real-time structural health monitoring |
| | Solar cells | High carrier mobility and high stability |
| $SiO_2$ nanoparticles | Concrete | Higher densification of the matrix leading to higher compressive strength and increased durability |
| | Ceramics | Improvement in thermo-mechanical, electro-magnetic, and optical properties |
| | Glass | Fire-resistant and anti-reflection |
| $TiO_2$ nanoparticles | Cement | Accelerated C-S-H gel formation and upgradation of durability and strength |
| | Glass | Self-cleaning, anti-fogging, and anti-fouling |
| | Solar cells | Increased electricity generation |
| $Fe_2O_3$ nanoparticles | Concrete | Increased compressive strength and abrasion-resistant |
| Cu nanoparticles | Steel | Weldability, corrosion resistance, and formability |
| Ag nanoparticles | Coating/painting | Higher opacity and antimicrobial |

**Note:** MEMS: Micro-electromechanical systems, and NEMS: Nano-electromechanical systems

**Table 23.4** Nano-additives and their effects on the properties of cement products

| Additives | Effects on concrete |
|---|---|
| Nano-silica | Reduction in porosity and permeability, improved crack resistance, higher strength and increased durability |
| Carbon nanotubes | Improved compressive strength, Improved tensile and flexural strengths, piezoelectric resistance, and blast resistance |
| Nano calcium carbonates | Increased compressive strength and hardness and shortening of $C_3S$ hydration period |
| Nano alumina | Improved elastic modulus, compressive strength, enhanced heat resistance, and thermal shock resistance |
| Nano cellulose | Increased bond strength, modulus of elasticity, and reduced moisture absorption |
| Nano zinc oxide | Increased compressive strength and aging resistance |

Although engineered nanomaterials provide greater benefits, their potential effects on human health and the environment is not fully studied (possible exposure scenarios are shown in Fig. 23.5). Nano-sized particles can enter the human body through inhalation, ingestion, or through the skin. Fibrous nanomaterials made of carbon have been found to induce inflammation in the lungs, similar to asbestos (www.niehs.nih.gov).

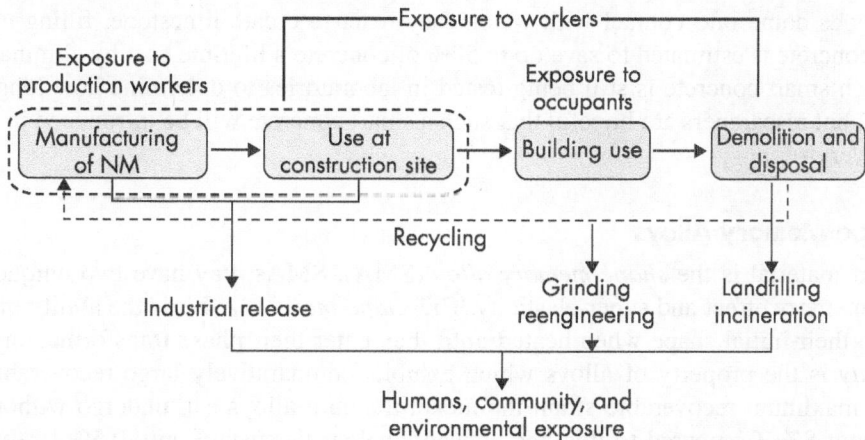

**Fig. 23.5** Possible exposure scenarios during the lifecycle of MNMs used in construction

## 23.7.1 Graphene

Carbon is the second most abundant mass within the human body and the fourth most abundant element in the universe (by mass), after hydrogen, helium, and oxygen. This makes graphene, which is made of carbon atoms, the ecologically friendly, sustainable solution for an almost limitless number of applications. Prof. Andre Geim and Prof. Konstantin Novoselov of the University of Manchester, England, developed a technique to produce graphene and published their findings in 2004 (Novoselov, et al. 2004) and subsequently received the Nobel Prize in physics in 2010. In simple terms, graphene is a single, tightly packed layer of carbon atoms that are bonded together in a repeating lattice pattern of hexagonal honeycombs. It is the thinnest compound and the lightest material (with 1 $m^2$ at around 0.77 mg), but the strongest (about 200 times stronger than steel and with a tensile strength of 1,034,213 MPa), and at the same time is flexible (bendable and stretchable). It can stretch 20–25% of its original length without breaking. It is the best conductor of heat (with thermal conductivity of 2000–4000 W/mK at room temperature) and also the best conductor of electricity. In addition, graphene absorbs 2.3% of the visible light that hits it and hence one can see through it without any glare (www.graphenea.com).

Some of the biggest emerging applications are: solar cells (very much thinner and lighter than those based on silicon), transistors, transparent screens, and bendable electronics. Graphene could also have applications for camera sensors, DNA sequencing, gas sensing, material strengthening, and water desalination. Another use for graphene is in paint. Graphene is highly inert and hence can act as a corrosion barrier between oxygen and water diffusion. Hence, future vehicles can be made corrosion resistant, as graphene can be made to grow on any metal surface (given the right conditions). Owing to its strength, graphene is seen as a potential replacement for Kevlar in protective clothing and could be used in vehicle manufacture. Carbon fibres can be made with higher strength and stiffness but with less weight. *Carbon nanotubes* are made of rolled graphene. These are used in bikes, tennis rackets, and even living tissue engineering. If the 20th century was the age of plastics, the 21st century may be called as the age of graphene.

## 23.8 Smart Materials

Smart materials are those which are engineered to respond to cracks, excessive stresses, or environmental effects such as temperature, pressure, and the presence of oxygen. A new type of smart concrete, called *self-healing concrete* contains dormant bacteria spores and calcium lactate in self-contained pods.

When these pods come into contact with water, they react to create limestone, filling up the cracks. Self-healing concrete is estimated to save up to 50% of concrete's lifetime cost by eliminating the need for repair. Such smart concrete is still being tested in laboratories to determine how long the bacteria sustains itself, but researchers are hopeful that such a smart concrete will be introduced in the construction industry very soon.

### 23.8.1 Shape-Memory Alloys

Another smart material is the *shape-memory alloy* (SMA). SMAs may have two unique characteristics: a shape-memory effect and super-elasticity. The *shape-memory effect* is the ability of the alloys to revert back to their initial shape when heated until they enter their phase transformation temperature. *Super-elasticity* is the property of alloys which exhibits comparatively large recoverable strain. For example, the maximum recoverable strain the nickel-titanium alloys can undergo without permanent damage is about 8%. Compared to this, the conventional steels sustain only 0.5% strain (Chang and Araki, 2016).

SMAs have been developed since the early 1960s. The three main types of SMA are copper-zinc-aluminium, copper-aluminium-nickel, and nickel-titanium (NiTi) alloys. Although iron-based and copper-based SMAs, such as Fe-Mn-Si, Cu-Zn-Al, and Cu-Al-Ni, are also commercially available and cheaper than NiTi, SMAs using NiTi are preferable due to their stability, practicability, and superior thermo-mechanic performance. NiTi alloys were first developed in 1961–1963 by William J. Buehler, a researcher at the US Naval Ordnance Laboratory, in White Oak, Maryland. It was commercialized under the trade name Nitinol (an acronym for Nickel Titanium Naval Ordnance Laboratories). Its remarkable properties were discovered accidentally. When a sample, that was bent out of shape many times, was presented at a meeting, one of the associate directors, Dr David S. Muzzey, just applied heat from his pipe lighter to the sample. To everyone's amazement, the sample stretched back to its original shape.

In civil structures, SMAs can perform as passive, semi-active, or active components to reduce damage caused by earthquakes. The passive structural control may utilize SMA's damping property to reduce the response and consequent plastic deformation subjected to severe loadings. SMAs can be used effectively via two mechanisms: ground isolation system and energy dissipation system (Song et al., 2006). Active tuning of structural natural frequency using martensite SMA wires for vibration suppression is an example of semi-active control of civil structures. Structural self-rehabilitation using stranded martensite SMA wires for post-tensioning is an example of active structural control. By monitoring the electric resistance change of the shape memory alloy wires, the strain distribution inside the concrete can be obtained. When cracks appear due to explosions or earthquakes, the SMA wire strands can be heated electrically, thus contracting them to reduce the cracks. This self-rehabilitation can handle macro-sized cracks. The concrete structure is considered intelligent since it has the ability to sense and to self-rehabilitate (Song et al., 2006). Companies like Shape Change Technologies LLC (www.shapechange.com) produce products of SMAs. SMAs have been successfully used for medical, robotic, aerospace, building, and automobile applications. More details about the use of SMAs in construction may be found in Chang and Araki (2016).

### 23.8.2 3D Printing of Materials

*3D printing*, also called *additive manufacturing*, is a process in which material is joined or solidified under computer control using a special printer to create 3D objects (Bos, et al., 2016). 3D-printers build 3D objects using a computer-aided design (CAD) model or *AMF file*, usually by successively adding material layer by layer. The technology used by most 3D printers is *fused deposition modelling*, a special application of plastic extrusion, developed in 1988 by S. Scott Crump and commercialized by his company Stratasys, which marketed its first FDM machine in 1992.

A large number of additive processes are available. The main differences between processes are in the way the layers are deposited to create parts and in the materials that are used. ISO/ASTM52900–15 defines seven categories of *Additive Manufacturing* (AM) processes. AM systems are employed in industries including aerospace, architecture, automotive, defence, and medical replacements. For example, General Electric uses the high-end model to build parts for turbines. In these systems, inside the printer, the material used is melted and distributed in layers until the desired thickness, texture, and patterns are made.

The materials used for 3D printing include the following:

1. Sintered powders such as polyamide (Nylon) and alumide (a mix of polyamide and grey aluminium)
2. Metals, such as stainless steel, bronze, nickel, gold, aluminium, and titanium
3. Carbon fibre and other composites
4. Carbon nanotubes and graphene embedded in plastics
5. Nitinol (alloy of nickel and titanium)
6. Water-absorbing plastic
7. Conductive carbomorph (carbon black plus plastic)
8. Paper
9. Cement mortar and concrete
10. Stem cells

The three main 3D printing processes used in building systems are contour crafting, concrete printing, and D-shape. Comparison and details of these processes are given by Perkins and Skitmore (2015). In building applications, typically a mortar with high cement content and maximum particle size in the order of 2 to 3 mm is used (larger aggregates have also been used). The shape of the extrusion varies and is either circular, ovular, or rectangular, and linear rates of extrusion are in the range of 50 mm/s to 500 mm/s (Buswell et al., 2018). In 2014, the Chinese company WinSun printed 10 houses in 24 h. In 2015, it built a six-storey apartment building at Suzhou Industrial Park, Jiangsu Province, China. The printer used by the company is 6-m high, 10-m wide, and 40-m long, which can be used to create significantly large objects, using a combination of recycled concrete, fibreglass, sand, and a hardening agent. The Shanghai-based company claims that they can save 60% of the materials typically needed to construct a home of the same size, can build it 70% faster, and with 80% less labour. Such a process can minimize the environmental footprint of buildings, as well as speed up the construction process, and greatly reduce the chance of on-site injury.

The first 3D concrete printed (3DCP-fused deposition modelling printed) bicycle prestressed concrete bridge was designed, printed, tested (both in the laboratory and in situ), constructed, and opened in October 2017, in the village Gemert in the Netherlands. It also includes the innovative use of cable reinforcement in the print filament (Salet et al., 2018).

The toxicity from the emissions of 3D printing depends on the material used and also on the size, chemical properties, and quantity of emitted particles. Excessive exposure to VOCs can lead to irritation of the eyes, nose, and throat, headache, loss of coordination, and nausea and sometimes may lead to asthma (NIOSH Report, 2016). Wojtyła et al. (2017) investigated commercially available thermoplastic filaments [such as acrylonitrile-butadiene-styrene (ABS), polylactic acid (PLA), polyethylene terephthalate (PET), and nylon] and found that ABS is significantly more toxic than PLA. Based on animal studies, it was found that carbon nanotubes and carbon nanofibres, sometimes used in fused filament printing, can cause pulmonary effects including inflammation, granulomas, and pulmonary fibrosis.

# SUMMARY

- Adhesive or glues are used when they are required to bond two or more parts together so that they act as a single unit. They are used in the plywood, and woodworking and joinery industry. They can be natural or synthetic.
- Caulks and sealants are used to seal openings in joinery. Caulks are of three types: oil-and alkyd, butyl rubber, and acrylic latex based. Joint sealants are used in construction to prevent water, moisture, and other substances from passing through material surfaces, joints, or openings. Sealants are basically of two types: elastomeric and polymerizing.
- Bamboo, being globally available and rapidly renewable, is a good example of a sustainable material. Bamboo used in construction should be treated (with a mixture of borax and boric acid) to resist insects and rot.
- Composite materials are those in which two different materials are advantageously combined to supplement the weakness of the other. Composites include cement-matrix composites, polymer-matrix, metal-matrix, carbon-matrix, and ceramic-matrix composites.
- In fibreglass, the matrix is plastic and the reinforcement is glass that is made into fine threads and woven like a cloth.
- FRP are composites comprising high-strength continuous fibres embedded in a polymer matrix. Carbon fibre based composite has superior mechanical properties and higher tensile strength, stiffness, and durability compared to glass fibre based composite.
- Polymer composites are produced by combining polymer resins such as polyester, vinylester and epoxy, and reinforcing fibres. CMCs comprise carbon or graphite fibres embedded in a carbon or graphite matrix. MMCs are made by dispersing reinforcing material into a metal matrix. Ceramic-matrix composites consist of ceramic fibres embedded in a ceramic matrix.
- Asbestos cement products are produced by mixing cement with asbestos fibres – they are banned in several countries as inhaling asbestos fibres may cause lung cancer and mesothelioma.
- Nanomaterials, which have sizes in the range of 1 to 100 nm, include nano-silica, titanium dioxide, and carbon nanotubes. These are used to improve the properties of several materials such as concrete, steel, glass, paints, and coatings.
- Graphene, which is made of carbon atoms, is an ecologically friendly, sustainable solution for an almost limitless number of applications.
- Smart materials are engineered to respond to cracks, excessive stresses, or environmental effects and include shape-memory alloys.
- 3D printing involves using special printers and materials to create computer controlled objects of optimized shape.

# EXERCISES

## Multiple-choice Questions

1. Caulks have a service life of
   (a) 1–2 years    (c) 5–8 years
   (b) 3–5 years    (d) 8–10 years
2. Sealants can be used to close openings in the range
   (a) 1.5 to 4 mm    (c) 3 to 10 mm
   (b) 3 to 4 mm    (d) 3 to 20 mm
3. Polyurethanes sealants can elongate up to
   (a) 100%    (c) 150–300%
   (b) 250%    (d) 500%
4. Common growth of bamboo per day is
   (a) 5 to 15 mm    (c) 300 to 900 mm
   (b) 30 to 100 mm    (d) None of these

5. The asbestos fibre content in asbestos cement is about
   (a) 10%    (c) 20%
   (b) 15%    (d) 25%
6. The strength of graphene is greater than that of steel by
   (a) 10 times
   (b) 25 times
   (c) 100 times
   (d) 200 times
7. The maximum recoverable strain the NiTi alloy can undergo without damage is
   (a) 0.5%    (c) 8%
   (b) 2%    (d) 10%

## Review Questions

1. Distinguish between adhesives and sealants.
2. Where adhesives are used in building construction? What are the advantages of using adhesives over traditional fastenings? What are the two basic types of adhesives?
3. Write short notes on adhesives used in (a) timber constructions, (b) fixing ceramic tiles, and (c) joining concrete.
4. What are caulks? How do they differ from sealants? What are the two basic types of sealants?

5. State the advantages of using bamboo as a building material. How is bamboo processed? What are the current uses of bamboo?
6. Write short notes on (a) polymer-matrix, (b) metal-matrix, (c) carbon-matrix, and (d) ceramic-matrix composites. State at least five advantageous properties of carbon fibre polymer-matrix composites.
7. What is asbestos cement? What products are made in India using asbestos cement? Why is it banned in several countries?
8. What are nanomaterials? Give four examples of nanomaterials and state the advantages of using them over traditional materials.
9. Write short notes on (a) graphene, (b) shape-memory alloy, (c) smart materials, and (d) 3D printing materials.

## ANSWERS

**Multiple-choice Questions**

**1.** (b)   **2.** (d)   **3.** (d)   **4.** (b)   **5.** (b)   **6.** (d)   **7.** (c)

# CHAPTER 24

# DEFORMATION AND FRACTURE OF MATERIALS

## 24.1 Introduction

In order to make sure that a structure and its components are safe, durable, and have assured intended performance, it is necessary that they do not deform excessively, and do not crack during the intended life, which may lead to complete fracture or failure. Mechanical behaviour of materials is concerned with the study of deformation and fracture of materials. One aspect of this study is the physical testing of specimens, which is done by applying different forces and deformations. This chapter provides a brief information on the types of material failures including in tension, bending, torsion, creep, and fatigue. An introduction to fracture mechanics and fracture toughness testing of different materials is also provided. The various tests conducted to evaluate the properties of materials discussed in the previous chapters are discussed in Chapter 25.

## 24.2 Types of Material Failures

*Deformation failure* may be defined as the change in the physical dimensions or shape of an element that makes it dysfunctional. If such deformations result in cracking and breakage of the element into two or more pieces, it is called *fracture*. If there is loss of material such as steel due to chemical action, it is called *corrosion*. *Wear* is the surface damage and progressive loss of material of solid surfaces in contact and subjected to relative motion. All wear processes involve one or a combination of wear mechanisms including abrasion, adhesion, fatigue, and oxidation or other chemical actions. If wear is caused by fluids, it is called as *erosion*. Although corrosion and wear are also of great importance, only deformation and fracture are considered in this chapter. Material failure may be classified as either due to deformation or fracture as shown in Fig. 24.1.

### 24.2.1 Elastic and Plastic Deformation

Deformation that is seen as soon as the load is applied can be classified as either elastic deformation or plastic deformation, as shown in Fig. 24.2. *Elastic deformation* is recovered immediately upon unloading. When there is only elastic deformation, stress and strain are usually proportional. For axial loading, the constant of proportionality is the modulus of elasticity, *E*, as defined in Eq. (24.1). The graph between engineering stress and strain, as shown in Fig. 24.2(b) and (c), is called the *stress-strain curve*. Thus, we have the following:

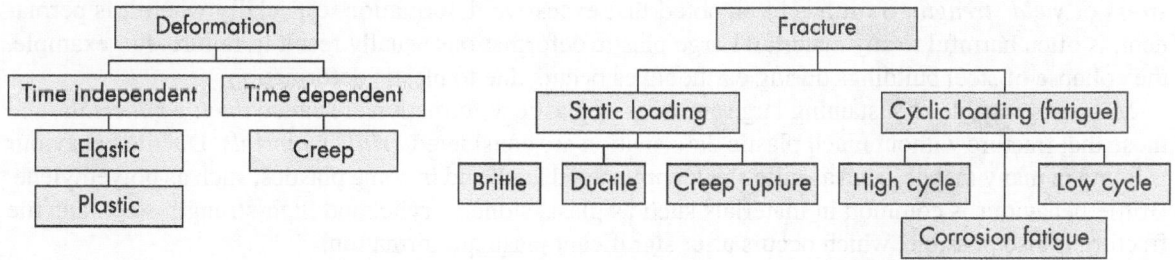

**Fig. 24.1** Basic types of deformation and fracture

$$\text{Engineering stress, or nominal stress} = \frac{\text{Load}}{\text{Original area}} \text{ (MPa)} \tag{24.1}$$

$$\text{Engineering strain, or nominal strain} = \frac{\text{Change in length}}{\text{Original length}} = \frac{\Delta L}{L_o} \tag{24.2}$$

where, $L_o$ is the initial gauge length and $\Delta L$ is the change in gauge length.

$$\text{Young's modulus or modulus of elasticity, } E = \frac{\text{Stress}}{\text{Strain}} \text{ (MPa)} \tag{24.3}$$

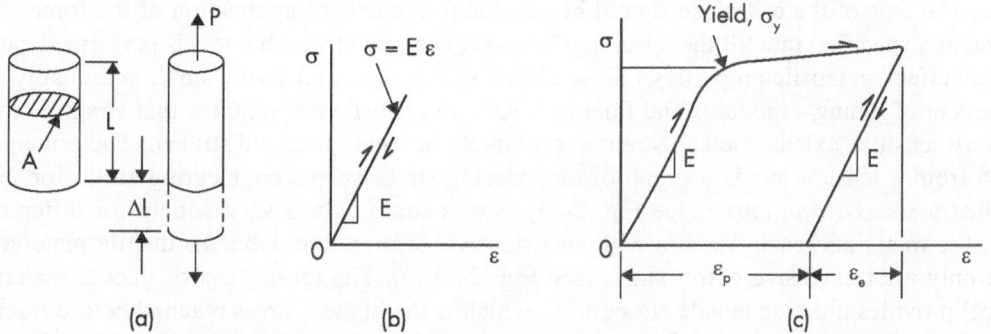

**Fig. 24.2** Tensile loading resulting in elastic and elasto-plastic deformation (a) Member subjected to axial tension (b) Elastic deformation only (c) Both elastic and plastic deformation

Brittle materials absorb relatively little energy prior to fracture, even those with high strength. Breaking is often accompanied by a snapping sound. The point of stress up to which the material behaves elastic is called the *elastic limit*. Elastomers, rubber, shape memory metals such as nitinol exhibit large elastic deformation ranges. However, elasticity is nonlinear in these materials. The behaviour of concrete in tension will be similar to that shown in Fig. 24.2(b). Most metals and ceramics show linear elasticity and a smaller elastic range.

Unlike elastic deformation, *plastic deformation* is not recovered upon unloading and hence is permanent. Elastic-plastic deformation of some materials is shown in Fig. 24.2(c). It can be easily seen from this figure that when the material is loaded, the stress-strain relation remains linear up to the elastic limit. When the load is increased further, plastic deformation begins, and even a small increase in load causes a relatively large additional deformation [Note that the term 'plastic deformation' does not mean that the deformed material is a plastic (a polymeric material)]. The load at which the plastic deformation starts is called the *yield load* and the process of is called *yielding*. The value of stress is called the *yield stress* or *yield strength*, $\sigma_y$. It has to be noted that excessive deformation, especially when it is permanent, is often harmful to any material. Large plastic deformations usually result in failure. For example, the collapse of steel buildings during earthquakes occurs due to plastic deformation.

Materials capable of sustaining large amounts of plastic deformation are called *ductile materials*, and those that fracture without much plastic deformation are considered *brittle materials*. Ductile behaviour is found in many metals, such as mild steel, copper, and lead, and in some plastics, such as polyethylene. Brittle behaviour is common in materials such as glass, stone, acrylic, and high-strength steel. Ductile fracture is also possible, which occurs after significant plastic deformation.

*Buckling* is deformation due to compressive stress that causes large changes in the alignment of columns or plates, leading to collapse. Either elastic or plastic deformation, or a combination of both, can dominate the behaviour. More details of buckling behaviour may be found in Subramanian (2008).

### 24.2.2 Tension Tests

Tension tests, as shown in Fig. 24.3, are often used to assess the strength and ductility of materials. Such a test is done by slowly stretching a bar or 'coupon' of the material in tension until it breaks (fractures). It has enlarged ends or shoulders for gripping in testing machines. Specifications like Indian Standards stipulate the dimensions and shape of the test specimens (see, for example, IS 1608:2005). The cross-section of the test pieces may be circular, square, rectangular, annular, or, in special cases, of some other shape. The gripped ends of the specimen may be of any shape to suit the grips of the testing machine. The axis of the test piece should be parallel to the axis of application of the force. The load is applied at a specified rate till the specimen breaks. (The rate at which a test is performed can have a significant effect on tensile properties.) Since elastic strains are usually very small, reasonably accurate measurement of Young's modulus and Poisson's ratio in a tensile test requires that strain be measured with a very sensitive extensometer. (Strain gages should be used for lateral strains.) The principal result obtained from a tension test is a graph of engineering stress versus engineering strain for the entire test, called a *stress-strain curve* (see Fig. 24.3). Stress-strain curves vary widely for different materials. Brittle materials reach fracture without extensive deformation, whereas ductile materials reach fracture only after extensive deformation [see Fig. 24.3(c)]. The tensile test of ductile materials like mild-steel provides ultimate tensile strength, $f_u$, which is the highest stress reached before fracture, the yield strength, $f_y$, and the strain at fracture, $\varepsilon_u$ [see Fig. 24.3(c)]. Materials having high values of both $f_u$ and $\varepsilon_u$ are said to be tough, and tough materials are generally desirable in building construction.

*Ultimate tensile strength* or stress is called the *engineering ultimate tensile stress* and is different from the true or absolute ultimate stress (see also Section 25.20 of Chapter 25). For ductile failure, the area of cross-section of the specimen reduces at the plastic stage forming a neck [see Fig. 24.3(a)], and the true stress is obtained by considering the actual area of cross-section at that instant instead of the original area of cross-section. Thus, it is calculated by the following formula:

$$\text{True stress = Ultimate tensile load/Actual area of cross-section} \tag{24.4}$$

This stress will be larger than the engineering stress, as seen in Fig. 24.3(c). The percentage elongation is a measure of ductility of the material and is calculated as follows:

$$\text{Percentage elongation at fracture} = \frac{\left(\text{Final length} - \text{Initial length}\right)}{\text{Initial length}} \times 100 \tag{24.5}$$

For ductile steel, the percentage elongation may be 20–24% or more.

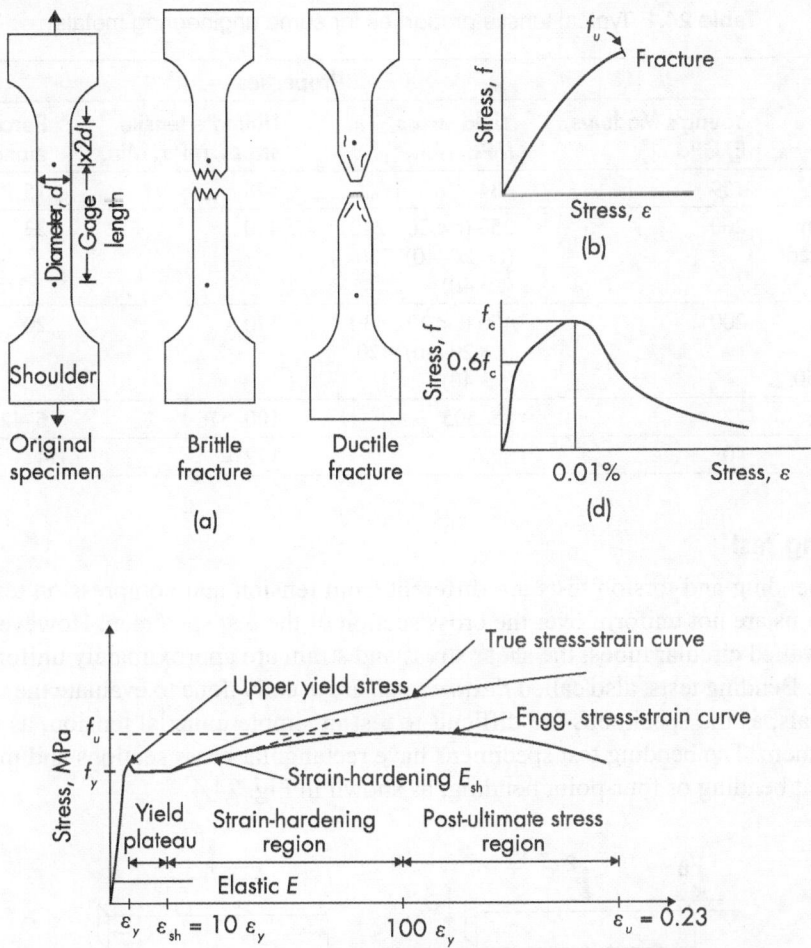

**Fig. 24.3** Tension test showing brittle and ductile behaviour (a) Tensile testing of specimen (b) Stress-strain curve of brittle material (c) Stress-strain curve of mild-steel (ductile material) (d) Stress-strain curve of quasi-brittle material (e.g., concrete)

**Strain hardening**   Under tensile stress, plastic deformation is characterized by *strain hardening region*, necking region, and finally, *fractur*e [see Fig. 24.3(a) and (c)]. Necking begins after the ultimate strength is reached. During necking, the material can no longer withstand the maximum stress and hence the strain in the specimen increases rapidly. Plastic deformation ends with the fracture of the material. The rise in the stress-strain curve after yielding is termed as *strain hardening*, since the resistance increases with increasing strain. The degree of strain hardening is usually indicated by the ratio of the ultimate tensile strength to the yield strength. Hence, we define the strain hardening ratio = $f_u/f_y$. Values of this ratio above about 1.4 are considered relatively high for metals, and those below 1.2 relatively low (Dowling, 2012). Typical tensile properties of some engineering metals are given in Table 24.1.

**Table 24.1** Typical tensile properties for some engineering metals

| Material | Properties | | | |
|---|---|---|---|---|
| | Young's Modulus, E, GPa | Yield stress MPa, Min. | Ultimate tensile stress, MPa, Min. | Percentage elongation, Min. |
| Ductile cast iron | 159 | 334 | 448 | 15 |
| Hot rolled medium steel (IS 2062) grade E 250 | 200 | 250 (t < 20), 240 (t = 20–40), 230 (t > 40) | 410 | 23 |
| Hot rolled High tensile Steel (IS 2062) Grade E 450 | 200 | 450 (t < 20), 430 (t = 20–40), 420 (t > 40) | 570 | 20 |
| Aluminium alloys | 72 | 85–505 | 100–570 | 6–42 |
| Magnesium | 40 | 89.6 | 172.4 | 4 |

## 24.2.3 Bending Test

The results of bending and torsion tests are different from tension and compression tests, because the stresses and strains are not uniform over the cross section of the test specimen. However, in the torsion testing of thin-walled circular tubes, the shear stress and strain are approximately uniform if the wall is sufficiently thin. Bending tests, also called *flexure tests*, are usually done to evaluate the tensile strengths of brittle materials, as the specimens are difficult to test in simple uniaxial tension, as they may crack while gripping them. The bending test specimens have rectangular cross sections and may be loaded in either three-point bending or four-point bending, as shown in Fig. 24.4.

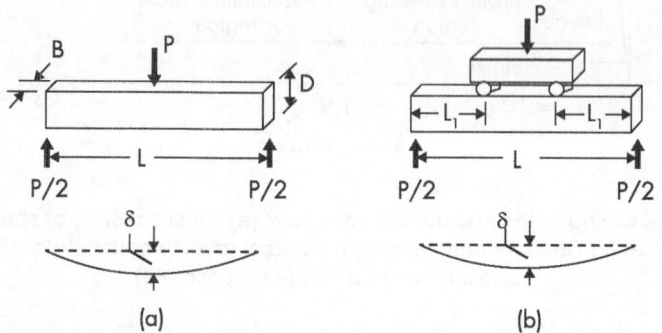

**Fig 24.4** Bending tests (a) Three-point bending (b) Four-point bending

Assuming the linear behaviour, the stress at fracture may be estimated as follows:

$$f = \frac{M}{I}\left(\frac{D}{2}\right) \tag{24.6}$$

where, $M$ is the bending moment. For a rectangular cross section with depth $D$ and breadth $B$, the moment of inertia $I$ about the neutral axis is $BD^3/12$. Considering the case of three-point bending of length $L$ due to the central load of $P$ [see Fig. 24.4(a)], the maximum bending moment at mid-span is $M = PL/4$. Thus, using Eq. (24.6), we get the following:

$$f_{fb} = \frac{3}{2}\frac{P_f L}{BD^2} \tag{24.7}$$

where $P_f$ is the force causing fracture in the bending test and $f_{fb}$ is the calculated fracture stress. Similar expressions can be derived for the four-point bending shown in Fig. 24.4(b). Usually, brittle materials will be stronger in compression than in tension and hence the maximum tension stress causes failure in bending.

## 24.2.4 Torsion Test

It is relatively easy to conduct test on round bars loaded in simple torsion and unlike tension tests, necking of specimen does not occur. Torsion test is not as common as tension test, although the test specimen may be similar in both tests. The state of stress and strain in a torsion test on a round bar corresponds to pure shear. In the torsion test, a circular metallic rod is twisted to a specified degree, until the material fails in torsion (see Fig. 24.5). Thus, a torque 'T' is applied to one end of the rod while the other end is held fixed in a stationary grip. Torsion machines use an electrical motor and gear-drive to apply the torque. Troptometers or commercially available digital inclinometers can be used to measure the angle of twist. Torsion tests are used for the direct evaluation of the shear modulus, as well as strength and ductility in shear.

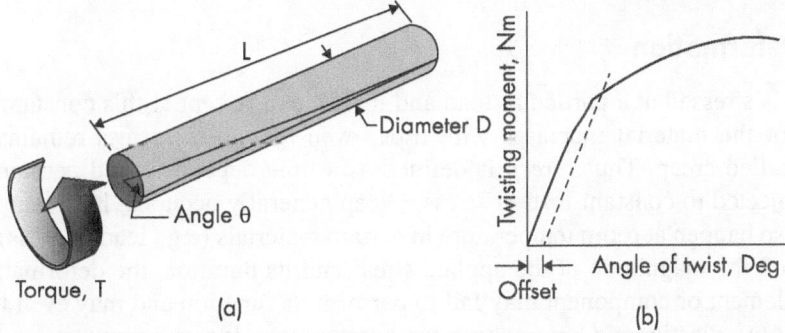

**Fig. 24.5** Torsion test (a) Round bar in torsion (b) Torque vs angle of twist plot

Within elastic behaviour, the shear stress $\tau$ is related to the torque $T$ by,

$$T = \frac{\tau J}{r} \qquad (24.8)$$

where $r$ is the radius of the rod and $J$ is the polar moment of inertia $= \dfrac{\pi D^4}{32}$

Thus, the shear stress at fracture $\tau_f$ is calculated as follows:

$$\tau_f = \frac{16 T_f}{\pi D^3} \qquad (24.9)$$

where, $T_f$ is the torque at fracture, $D$ is the diameter of the rod.

Similarly, for a hollow section, with inner radius $r_1$ and outer radius $r_2$, the shear stress at fracture $\tau_f$ can be derived as follows:

$$\tau_f = \frac{2 T_f r_2}{\pi \left( r_2^4 - r_1^4 \right)} \qquad (24.10)$$

The angle of twist measured during the test is usually plotted against the applied torque (see Fig. 24.5), from which the shear modulus could be evaluated in the elastic region, as follows:

$$G = \frac{L}{J} \left( \frac{T}{\theta} \right) \qquad (24.11)$$

When subjected to torsion, a ductile specimen breaks along a plane of maximum shear, i.e., a plane perpendicular to the shaft axis. However, brittle materials fracture on planes of maximum tension stress, i.e., at 45° to the axis of the specimen. As the fracture wraps around the circumference, maintaining 45° to the specimen axis and causes a helical fracture pattern. For more information on torsion, refer to Subramanian (2008).

### 24.2.5 Quasi-brittle Fracture

Cement-based materials are traditionally regarded as being brittle, but in reality exhibit a far more sophisticated response. They have moderate strain hardening prior to the attainment of their ultimate tensile capacity, and are characterized by an increase in deformation with decreasing tension carrying capacity past the ultimate strength [see Fig. 24.3(d)]. Such a response is called *tension softening*. These materials that exhibit moderate strain hardening prior to the attainment of ultimate tensile strength and tension softening thereafter are called *quasi-brittle*. Quasi-brittle failures are typical of reinforced concrete structures. They are also typical of certain kinds of rock, ice, modern tough ceramics, and various composites. For more details on fracture of quasi-brittle materials refer to Bažant and Planas (1997). *Brittle fracture* of steel has already been covered in Section 13.9.4 of Chapter 13.

## 24.3 Creep Deformation

When a material is stressed at a particular load and if this load is kept at this constant value for a long time, the strain of the material increases with time, even though the stress remains the same. This phenomenon is called creep. Thus, *creep* is defined as a time-dependent and permanent deformation of a material, subjected to constant load or stress. Creep generally occurs at high temperature (thermal creep), but can also happen at room temperature in certain materials (e.g., lead or glass), at much slower rate. Depending on the magnitude of the applied stress and its duration, the deformation may become so large that an element or component may fail to perform its function and may even lead to failure, as shown in Fig. 24.6(a). Plastics and low-melting-temperature metals may creep at room temperature, and virtually any material will creep upon reaching its melting temperature. Creep is thus often an important consideration in engineering design, especially in applications involving high temperature, and also in prestressed concrete structures. Buckling may also occur in a time-dependent manner due to creep deformation. Concrete creeps significantly at all stresses and for a long time. Creep deflection can easily be observed in long concrete cantilever beams, if they are not designed for creep.

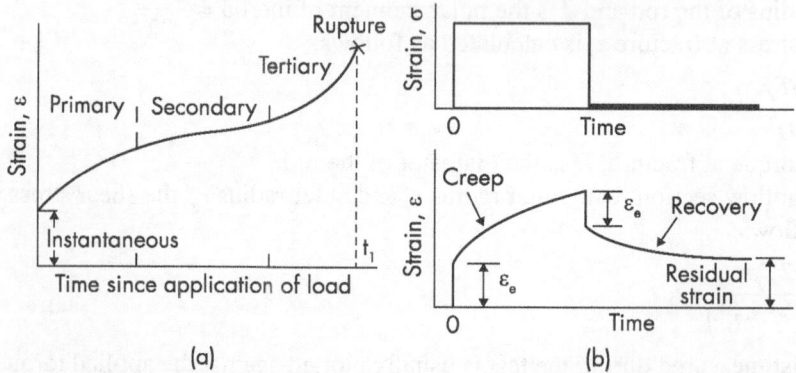

**Fig. 24.6** Creep deformations (a) Schematic creep curve (b) Creep recovery

The rate of deformation is called the *creep rate*. The slope of the line in the creep strain vs. time curve is shown in Fig. 24.6(a). There is an initial elastic deformation $\varepsilon_e$, and following this, the strain slowly increases as long as the stress is maintained. The three stages of creep deformation are as follows:

1. *Primary creep* starts at a rapid rate and slows with time.
2. *Secondary creep* has a relatively uniform rate.
3. *Tertiary creep* has an accelerated creep rate and terminates when the material breaks or ruptures. It is associated with both necking and formation of grain boundary voids in the material.

The creep mechanism is often different for metals, plastics, rubber, and concrete. The critical temperature for creep in metals is at 40% of the melting temperature $(t_m)$, i.e., if $t > 0.40 \, t_m$, creep is likely to occur (Bažant and Zhou, 2002).

If the applied load is released before the *creep rupture* occurs, there is an immediate elastic recovery equal to the elastic deformation, followed by a period of slow recovery [see Fig. 24.6(b)]. The material in most cases does not recover to the original shape and a permanent deformation remains. The magnitude of the permanent deformation depends on the length of time, amount of stress applied, and temperature.

## Case Study  Collapse of World Trade Center towers at New York, USA

The 1 & 2 World Trade Center towers with heights of 417 m and 415.1 m respectively, were the tallest buildings in the world, when they were completed in 1972–73. The architect Minoru Yamasaki and the structural engineering firm Worthington, Skilling, Helle & Jackson used the frame tube structural system, originally developed by Fazlur Rahman Khan, for these towers. On 11 September 2001, terrorists hijacked aircrafts and crashed them into these towers. Due to the impact of the planes and the ensuing fire, the two towers collapsed resulting in the death of 2,996 people. The collapse is attributed to the creep buckling of the perimeter and core steel columns.

### 24.3.1 Creep Testing

Conceptually, conducting a creep test is rather simple – a constant force has to be applied on the test specimen and its dimensional change has to be measured over time with exposure to a relatively high temperature. Although conceptually quite simple, creep tests in practice are more complicated. Temperature control is critical and resolution and stability of the extensometer is an important factor. The usual creep test consists of applying constant stress to a tensile specimen, often by using suspended weights as shown in Fig. 24.7. A thermostatically controlled furnace is also used, whose temperature is controlled by a thermocouple attached to the gauge length of the specimen (see Fig. 24.7). The extension of the specimen is measured by very sensitive extensometers, since the actual amount of deformation before failure may be only 2 or 3%.

The creep test has the objective of precisely measuring the rate at which secondary or steady state creep occurs. Increasing the stress or temperature has the effect of increasing the slope of the line, i.e., the amount of deformation in a given time increment. The results are presented as the amount of strain (deformation), generally expressed as a percentage,

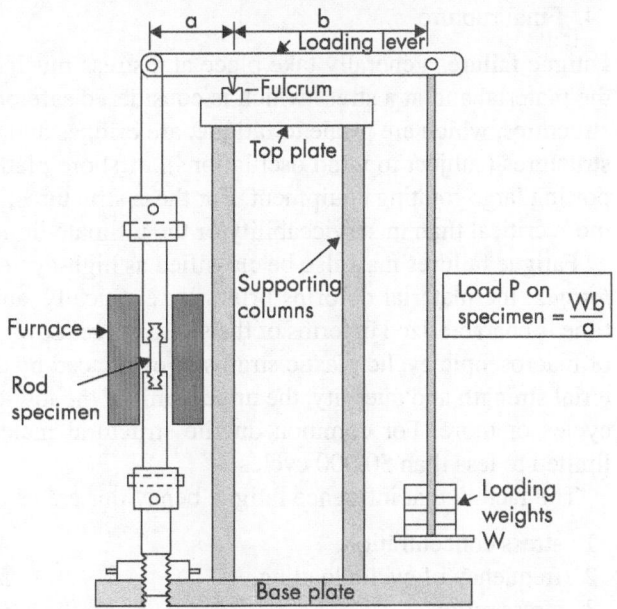

**Fig. 24.7** Schematic of a creep testing machine

produced by applying a specified load for a specified time and temperature, e.g., 1% strain in 90,000 h at 30 MPa and 450°C. This enables the designer to calculate how the component will change in shape during service and hence can specify its design creep life. This is of particular importance where dimensional control is crucial, as in the case of gas turbines (www.twi-global.com).

Two additional tests that use the same equipment and test specimen are the *creep rupture test* and the *stress rupture test*. Both of these tests are continued until the specimen fails. In the creep rupture test, the amount of creep that has occurred at the point of failure is recorded. The test results are expressed as percentage strain, time and temperature, e.g., rupture occurring at 2% strain at 400°C in 75,000 h. The stress rupture test gives the time of rupture at a given stress and temperature, e.g., 45 MPa causing failure at 450°C in 80,000 h. This data, if properly interpreted, are useful to specify the design life of components since they give a measure of the load carrying capacity of a material as a function of time (www.twi-global.com). More details of these tests may be found in BS 3500, BS EN 10291, and ASTM E139.

Creep can be prevented in metals by the following three ways:

1. Using metals with higher melting point
2. Using materials with greater grain size
3. Using alloying (Super-alloys based on copper, nickel, and steel can be engineered to be highly resistant to creep).

## 24.4 Fatigue under Cyclic Loading

A common cause of fracture is *fatigue*, which is failure due to repeated loading. The fatigue failure is progressive in nature and occurs in the following four stages:

1. Crack initiation at points of stress concentration
2. Crack growth
3. Crack propagation
4. Final rupture

Fatigue failures generally take place at a stress much lower than the ultimate strength (yield stress) of the material and at a stress which is considered safe on the basis of static failure analysis. Examples of structures, which are prone to fatigue, are bridges and gantry girders, cranes, slender towers like (open) structures (subject to wind oscillations), offshore platforms (subject to wave load), and structures supporting large rotating equipment. For these structures, verifications in the limit state of fatigue are often more critical than in serviceability or the ultimate limit state (Subramanian, 2008).

Fatigue failures may also be classified as high-cycle and low-cycle fatigue failures. Under *high-cycle fatigue*, the material deforms primarily elastically, and the number of cycles for failure or the failure time is characterized in terms of the stress range. *Low-cycle fatigue* can be characterized by the presence of macroscopic cyclic plastic strains as evidenced by a stress-strain hysteresis loop. Depending on material strength and ductility, the upper limit of the low-cycle fatigue regime may be from 100 to 100,000 cycles or more. For common ductile structural materials, the low-cycle fatigue regime is generally limited to less than 50,000 cycles.

The factors that influence fatigue behaviour are as follows:

1. stress concentration,
2. frequency of cyclic loading,
3. stress ratio,
4. connection details,
5. residual stress in the steel,
6. plate thickness,

7. material strength,
8. imperfections,
9. post weld treatment,

10. service temperature of the steel, and
11. environment (important in the case of corrosion fatigue).

More details of these factors and the use of S-N methods for estimating the fatigue life of structures may be found in Subramanian (2008).

## 24.5 Background to Fracture Mechanics

The field of study of the propagation of cracks in solid materials is termed as *fracture mechanics*. In this study, analytical solid mechanics methods are used to calculate the force required for developing cracks and experimental work is done to study the extent of resistance offered by the material to fracture.

Under static loading, rapid fracture can occur but will not vary with time. If such a fracture occurs with minimum plastic deformation, it is called a *brittle fracture*. Such a fracture occurs in glass and other similar materials. Brittle fracture is more likely to occur under impact loading, where the loading is applied rapidly. Brittle fracture can occur even in materials that deform plastically by large amounts (e.g., ductile steels or aluminium alloys), when cracks or sharp flaws are present. (A crack is typically a 'sharp' void in a material, which acts like a *stress concentrator* or amplifier.) The three factors that have a profound influence on the nature of fracture are: (a) temperature, (b) strain rate, and (c) the state of stress. At low temperature, failure can occur in steel by brittle fracture suddenly and without warning. Thus, a right combination of low temperature, an abrupt change in section size (notch effect) or a flaw/crack, and also the presence of tensile stress can initiate a brittle fracture (see Section 13.9.4 of Chapter 13, for more information on brittle fracture). The material property called the *fracture toughness* indicates the ability of any material to resist brittle fracture.

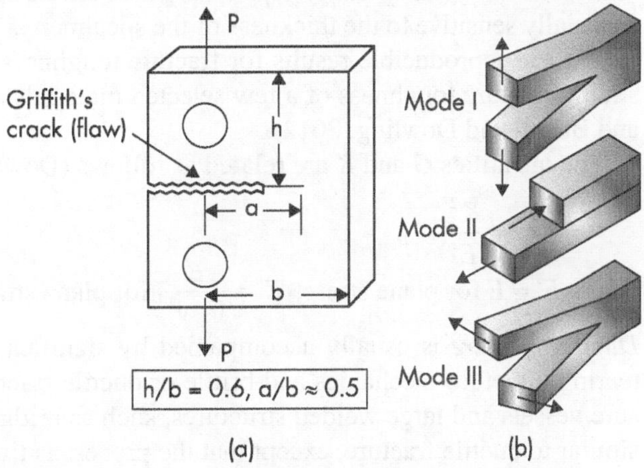

British aeronautical engineer A. A. Griffith was the first to develop a theory of fracture mechanics, using elastic strain energy concepts, which was applicable to elastic materials (Griffith, 1921). Griffith suggested that the low fracture strength in glass is due to the presence of microscopic flaws and, to verify this hypothesis, he introduced an artificial flaw in his experimental glass specimens [as shown in Fig. 24.8(a)]. He found that product of the square root of the flaw length ($a$) and the stress at fracture ($f_f$) was nearly constant. (The importance of Griffith's work in fracture was largely unrecognized until the 1950s.)

**Fig. 24.8** Fracture-toughness test and fracture modes (a) Fracture-toughness test (b) The three fracture modes

$$f_f = \sqrt{\frac{2E\gamma_s}{\pi a}} \text{ and } K_c = f_f \sqrt{\pi a} \ (a \ll b) \tag{24.12}$$

where $f_f$ is the fracture stress, $E$ is the Young's modulus, $a$ is the crack length in m, $\gamma_s$ is the surface energy density of the material (J/m²). The Griffith equation is strongly dependent on the crack size $a$, and is applicable only to ideally brittle materials like glass. In addition, the direct application of Griffith's criterion is seldom practical because of the difficulties of determining $\gamma_s$.

While the energy-balance approach provides great insight into the fracture process, an alternative method that examines the stress state near the tip of a sharp crack directly is found to be more useful in engineering practice. Three types of cracks, termed modes I, II, and III as illustrated in Fig. 24.8(b) are usually defined. Mode I is the normal-opening mode which is of importance, while modes II and III are shear sliding and tearing modes, respectively, which are not usually considered.

When the material exhibits more ductility, such as steel, consideration of the surface energy alone fails to provide an accurate model for fracture. For such materials, where there are plastic deformations, Irwin of the US Naval Research Laboratory (Irwin, 1957) and Orowan (Orowan, 1955) independently proposed the following equation:

$$f_f = \sqrt{\frac{EG_c}{\pi a}} \text{ with } G_c = 2\left(\gamma_s + \gamma_p\right) \tag{24.13}$$

With fracture toughness $K_I = Yf_f\sqrt{\pi a}$ (24.14)

where, where $\gamma_s$ is the surface energy, $\gamma_p$ is the plastic dissipation per unit area of crack growth ($\gamma_s$ will be in the range of 1 to 2 J/m$^2$ whereas $\gamma_p$ will be in the range of $10^2$ to $10^3$ J/m$^2$), and the quantity $Y$ is a function of $a/b$ [see Fig. 24.8(a)]. Y = 1 for centre double-ended cracked plate ($a/b \leq 0.4$), $Y = 1.12$ for a single-edge-cracked plate ($a/b \leq 0.13$), and $Y = 2/\pi$ for an embedded penny shaped circular crack of radius $a$ in an infinite plate. More accurate expressions for $Y$ are available in Dowling (2012). Materials with high strength generally have low fracture toughness, and vice versa. $K_I$ may be considered a material property (like yield stress) and can be determined for different materials using standard testing methods. However, in reality, $K_I$ also depends on the specimen geometry and loading conditions, and is especially sensitive to the thickness of the specimen. $K_I$ is also a microstructure sensitive property. Consistent and reproducible results for fracture toughness can only be obtained under conditions of plane strain. Fracture toughness of a few selected materials is provided in Table 24.2 (also see Subramaniam and Balani and Dowling, 2012).

The quantities $G$ and $K$ are related as follows (Dowling, 2012):

$$G = \frac{K^2}{E'} \tag{24.15}$$

where $E' = E$ for plane stress, $E' = \dfrac{E}{1-v^2}$ for plane strain, and $v$ = Poisson's ratio

*Ductile fracture* is usually accompanied by significant plastic deformation and may include gradual tearing. Fracture mechanics and brittle or ductile fracture have to be considered while designing pressure vessels and large welded structures, such as bridges and ships. Creep rupture (see Section 24.3) is similar to ductile fracture, except that the process is time dependent.

Sometimes, fracture occurs as a result of a combination of stress and chemical effects. This kind of fracture is usually observed in structures of chemical industry and is called *stress-corrosion cracking*. In addition, some low-strength steels may crack in the presence of chemicals such as sodium hydroxide. In prestressed concrete, atomic hydrogen may be liberated as a result of the action of acids, which can penetrate the steel surface, making it brittle and fracture prone when subjected to tensile loads. Even a small amount of hydrogen can cause considerable damage to the tensile strength of high-tensile steel wires of prestressed concrete. Use of high-alumina cement and blast furnace slag cement, which are rich in sulphides can also cause *hydrogen embrittlement*. Use of dissimilar metals such as aluminium and zinc for sheaths to house high-tensile steel wires also results in hydrogen embrittlement. Even minute traces of sulphur, when in contact with high-tensile steel wires in the presence of moisture, can reduce their strength considerably. To prevent hydrogen embrittlement, steel should be properly protected from the action of acids. Protective coverings such as bituminous crepe-paper should be used during transport to reduce the chances of contamination. The wires should always be stored in dry conditions.

**Table 24.2** Fracture toughness of some selected materials

| Material | Toughness, $K_{lc}$ (MPa $\sqrt{m}$) | Material | Toughness, $K_{lc}$ (MPa $\sqrt{m}$) | Material | Toughness, $K_{lc}$ (MPa $\sqrt{m}$) |
|---|---|---|---|---|---|
| Cast iron | 33 | Low-carbon steel | 77 | Stainless steel | 220 |
| Aluminium-titanium alloy Al 2024-T351 | 34 | ABS polymer | 3 | PVC | 2.4 |
| Soda-lime glass | 0.76 | Concrete | 1.19 | | |

**Example 24.1** Estimate the fracture strength of a brittle material (meaning that we can ignore plastic yield) with properties, $E = 100$ GPa, $\gamma_s = 1$ J/m², and crack length = 2.5 μm.

**Solution**: The stress at fracture as Griffith's energy balance equation is

$$f_f = \sqrt{\frac{2E\gamma_s}{\pi a}} = \sqrt{\frac{2 \times 100 \times 10^9 \times 1}{\pi \times 2.5 \times 10^{-6}}} = 159.58 \ MPa$$

## 24.5.1 Fracture Toughness Testing – Different Materials

There are several types of tests used to determine fracture toughness of materials, which generally utilize a notched specimen with various configurations. The linear-elastic plane strain fracture toughness test (or $K_{lc}$ value), described in Section 24.3, is used to determine the fracture toughness of metallic materials, across a range of temperatures to determine design life and crack growth, or remaining life. Similar tests are also done for other materials such as polymers and ceramics, which are described in ASTM D5045 and ASTM C1421, respectively. In these materials, the $K_Q$ value (similar to $K_{lc}$) decreases with increasing specimen thickness. More details of these tests may be found from Dowling (2012).

Another widely used test is the three point beam bending test, in which a thin crack is pre-set into the test specimen before applying the load (see Fig. 24.9). Growth of the crack is detected by plotting the force versus displacement (P-v) behaviour. A deviation from the linearity of the P-v plot identifies the point $P_Q$ corresponding to the early stage of cracking. The value of $K$ is calculated for this point.

Crack tip opening displacement, or *CTOD test* is used to measure the physical opening of the fatigue crack tip at the point of failure. It is typically used to measure the toughness of lower strength materials ($f < 1400$ MPa), which exhibit a small amount of plastic deformation before failure and tear before failure. All CTOD tests are carried out in the temperature range of −129 to 200°C.

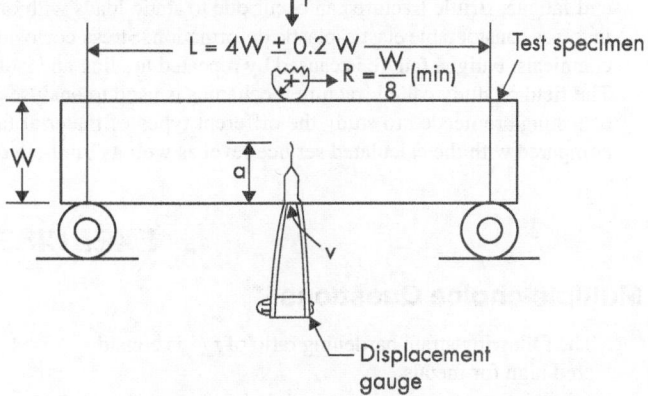

**Fig. 24.9** Fracture toughness test on a bend specimen

Unlike $K_{Ic}$, a material's $J_{IC}$ (J-integral) value can be used to determine toughness using the energy required to fracture the sample. It is also used for lower strength materials, exhibiting small amount of plastic deformation before failure. A related value, $JR$, uses one or more samples to determine the initial point of tearing. More details of these testing can be found in Zhu and Joyce (2012).

A widely utilized standardized test method is the Charpy impact test, in which a standard rectangular simply supported beam having a V-notch or a U-notch at mid-length is subjected to an impact load by a swinging pendulum, and the energy absorbed by the specimen is measured (see Section 25.20.2 of Chapter 25, for more details of this test).

### 24.5.2 Structural Integrity Assessment Procedure and Fracture Mechanics

*Structural integrity* is a relatively new scientific and engineering discipline, which includes stress analysis and diagnostics of the behaviour and weakening, and lifetime assessment and revitalization of structures (Sedmak and Radaković, 2004). Besides common situations in which structural integrity is accessed when non-destructive testing detects a defect, stress analysis of crack-free structures is done by using the finite element method [software such as SINTAP (Structural Integrity Assessment Procedure)/FITNET (European Fitness-for-Service Network) are also available]. The purpose of structural integrity assessment is to determine the significance, in terms of fracture and plastic collapse, of flaws present in structures and components. Owing to such an analysis, an accurate and detailed distribution of displacements, strains, and stresses is obtained, making it possible to identify 'weak' locations in a structure, even before a crack appears. This approach is very important for structures exposed to operating conditions, which will initiate crack, such as fatigue, creep, and corrosion. It is important to note that this approach is not intended to supersede existing methods, but to serve in conjunction with them throughout the lifetime of a structure, to guarantee safety.

## SUMMARY

- The study of the deformation and fracture of materials is termed as mechanical behaviour of materials.
- Materials are often tested to evaluate their strength and resistance to failure to determine properties such as yield strength or fracture toughness.
- The failure of materials may be due to elastic, plastic, and creep deformation. Elastic deformation is permitted as it will recover immediately upon unloading, whereas plastic deformation is a permanent, and hence is not allowed. In certain materials, creep deformation occurs, which accumulates over time; this may lead to creep rupture, which is usually a ductile fracture.
- Other material failures may be due to cracking, which can lead to brittle or ductile fracture, stress-corrosion cracking, and fatigue. Brittle fracture can occur due to static loads with only small elastic deformation, whereas ductile fracture involves considerable elastic-plastic deformation. Stress-corrosion cracking is caused by constant exposure to moisture/chemicals. Fatigue failure is caused by repeated loading and is due to the gradual development and growth of cracks.
- The field of study called fracture mechanics is used to analyse cracks in engineering components. Different methods of testing are needed to study the different types of material failures. In design, the strength evaluated by testing is compared with the calculated service level as well as limit-state stresses of components to assure safety and stability.

## EXERCISES

### Multiple-choice Questions

1. The following strain hardening ratio of $f_u/f_y$ is considered high for metals
   - (a) 1.1
   - (b) 1.2
   - (c) 1.3
   - (d) 1.4

2. Troptometers are used to measure
   - (a) deformation
   - (b) stress
   - (c) angle of twist
   - (d) strain

3. The percentage of melting temperature that is considered critical temperature for creep in metals is
   (a) 30%    (c) 50%
   (b) 40%    (d) 60%

4. Creep can be prevented in metals by
   (a) using metals with higher melting point
   (b) using materials with greater grain size
   (c) using alloying
   (d) none of these
   (e) all of these

5. The factors that have influence on the nature of fracture are
   (a) temperature
   (b) strain rate
   (c) the state of stress
   (d) none of these
   (e) all of these

6. The fracture toughness of low-carbon steel is approximately
   (a) 33 MPa√m    (c) 77 MPa√m
   (b) 50 MPa√m    (d) 210 MPa√m

## Review Questions

1. What are the basic types of deformation and fracture?
2. Sketch the elastic and elastic-plastic stress-strain diagram for axial loading and explain the behaviour.
3. Explain how tension test is conducted on a steel specimen.
4. What is meant by strain hardening and necking?
5. Why are bending tests conducted? Explain the three-point bending test and how the fracture stress is estimated.
6. Explain how torsion test is conducted on a steel specimen.
7. What is meant by quasi-brittle fracture?
8. Define creep and discuss its importance in materials engineering. Explain the classical creep curve. Detail experimental ways to determine creep. How can creep be prevented in metals?
9. Define fatigue. What are the four stages of fatigue failure? List at least five factors that may influence fatigue behaviour.
10. Distinguish between brittle fracture and ductile fracture. What are the three factors that influence the nature of fracture?
11. How the two fracture mechanics theories proposed Griffith and Irwin differ?
12. What is hydrogen embrittlement and how is it caused?
13. Describe the three point beam bending test used to determine fracture toughness.
14. Explain how structural integrity assessment procedures could be developed based on fracture mechanics.

## ANSWERS

### Multiple-choice Questions

**1.** (d)    **2.** (c)    **3.** (b)    **4.** (e)    **5.** (e)    **6.** (c)

# CHAPTER 25

# TESTING AND EVALUATION OF BUILDING MATERIALS

## Part I

## 25.1 Introduction

Before using the various materials discussed in this book, they have to be tested in order to evaluate whether they have the same properties, as assumed in the design. Samples of these materials, called *specimens*, are often subjected to a variety of mechanical tests to measure their tensile or compressive strength, toughness, hardness, and other properties of interest. Some of the common forms of test specimen and the form of loading are shown in Fig. 25.1. In the tension test, the sample shown in Fig. 25.1(a) is tested by applying a tensile force. Concrete cylinders or cubes are usually loaded in compression up to failure [Fig. 25.1(b)]. Bending tests as shown in Fig. 25.1(c) on prisms are employed in concrete to find the modulus of rupture. Cylindrical rods or tubes are used in torsion tests [Fig. 25.1(d)]. Hardness of the material is found by measuring the penetration of a hard ball on the surface of the specimen [see Fig. 25.1(e)]. Usual specimens have smooth surfaces. Notched or pre-cracked specimens, as shown in Fig. 25.1(f-g), are used to find the fatigue strength.

Tension, compression, flexural, shear, and torsion tests are conducted using *universal testing machines* (UTM), which became available from 1900s. Two types of machines are generally used: (a) mechanical system, in which two large screws are used to apply the load, and (b) hydraulic system, which uses the pressure of oil pumped into a hydraulic piston (Dowling, 2012). A typical UTM is shown in Fig. 25.2. The earlier UTM machines that were analog with a chart recorder have been replaced by modern UTMs with digital controls and PC-driven software. The specimen is placed in the machine between the grips and modern machines can automatically record the change in gauge length during the test. The applied load is measured by the *load-cell* in the testing machine. The machine produces the load/deformation curve with force on the Y-axis and deformation on the X-axis. More information on UTMs may be found in the product manuals of these machines and also from Dowling (2012).

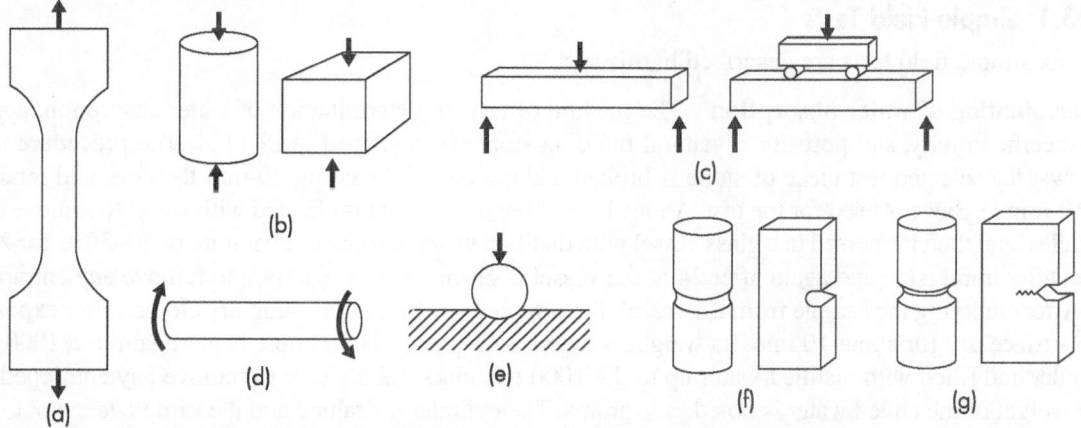

**Fig. 25.1** Various types of test specimen and their loading (a) Tension (b) Compression (c) Center-point and third-point loading (d) Torsion (e) Indentation hardness (f) Notched specimen (g) Pre-cracked specimen

In this chapter, the various tests that are conducted to determine the mechanical properties of various materials are briefly discussed.

## 25.2 Sampling of Specimen

Sampling is of fundamental importance for estimating the quality of a lot or ascertaining its conformity to the requirements of any specification. The economy, reliability, and practicability of the sampling procedures are indispensable in most of the applications. However, the reliability of the conclusions drawn on the basis of the sample depends on its representativeness and the method of its selection. Basic procedures for the selection of a random sample under diverse situations are

**Fig. 25.2** Hydraulic UTM (*Courtesy*: Prof. G. Mohan Ganesh, VIT, Vellore)

provided in IS 4905:2014. These procedures ensure a truly random and representative sample leading to sound and satisfactory estimation of lot quality. In addition, several standards also stipulate the number of samples to be tested for a particular material, in order to get reliable results.

## 25.3 Tests Conducted on Stones

The tests that are carried out on stones to determine their strength and other properties, durability, and quality are (a) absorption test, (b) Smith's test, (c) toughness test, (d) hardness test, (e) acid test, (f) crystallization test, (g) attrition test, (h) crushing test, (i) freezing and thawing test, (j) impact test, and (k) microscopic test. Out of these tests, the first six tests are simple tests and can be carried out at the building site. The next six tests are carried out in the laboratory. These tests are briefly described as follows:

## 25.3.1 Simple Field Tests

The six simple field tests are described in this section.

**Determination of water absorption** The method of test for determination of water absorption, apparent specific gravity, and porosity of natural building stones is described in IS 1124. The procedure is as follows: the selected test piece of stone is broken and the material passing 20-mm IS Sieve and retained on 10-mm IS Sieve is used for the test. About 1 kg of the test sample is cleaned with water to remove dust particles, and then immersed in a glass vessel with distilled water at room temperature of 20–30°C for 24 h. Soon after immersion and again after 24 h, the vessel is given a gentle agitation to remove any entrapped air. After emptying the sample from the vessel, the sample is surface dried using dry cloths. After exposing it to surface dry for about 10 min, its weight is noted as B grams. The sample is put again in a 1000 mL cylinder and filled with distilled water up to the 1000 mL mark, taking care to remove any entrapped air. The weight of the added water is noted as C grams. The cylinder is drained and the sample taken out. The test sample is again dried in an oven at 100–110°C for about 24 h, cooled in a desiccator to bring it to room temperature, and its weight is again measured as A grams. The room temperature is also recorded.

$$\text{Apparent specific gravity, at room temperature} = \frac{A}{1000 - C} \tag{25.1}$$

$$\text{Water absorption} = \frac{B - A}{A} \times 100 \tag{25.2}$$

$$\text{Apparent porosity} = \frac{B - A}{1000 - C} \times 100 \tag{25.3}$$

The true porosity may be calculated from the following formula:

$$\text{True porosity} = \frac{\text{True specific gravity} - \text{Apparent specific gravity}}{\text{True specific gravity}} \tag{25.4}$$

$$\text{Saturation coefficient} = \text{Water absorption/Total porosity} \tag{25.5}$$

where $A$ is the weight of oven-dry test piece (g), $B$ is the weight of saturated surface-dry (SSD) test piece (g), and $C$ is the quantity of water added in 1000 mL jar containing the test piece (g).

**Smith's test** This test is performed to determine the presence of soluble matter in a sample of stone. A sample of stone is placed in distilled water in a glass vessel. After about an hour, the vessel is vigorously stirred or shaken. If the water turns muddy, the stone contains earthy matter. If water remains clear, stone will be durable and free from any soluble matter.

**Toughness test** This test is performed by breaking the stone with a hammer. Toughness is indicated by its resistance to hammering.

**Hardness test** As discussed in Section 2.2 of Chapter 2, this is a simple test, in which the stone is scratched with a penknife, and the hardness determined based on Table 2.2 of Chapter 2.

**Acid test** In this test, a sample of stone weighing about 50 to 100 g is taken. It is placed in a solution of hydrochloric acid having strength of 1% for 7 days. The solution is agitated at intervals. A good building stone should maintain its sharp edges and keep its surface free from powdering at the end of this period. If the edges are broken and the surface is powdered, it indicates the presence of calcium carbonate and such a stone will have poor weathering quality. This test is usually carried out on sandstones.

**Crystallization test (IS 1126:2013)** Three test pieces of 50 mm diameter and 50 mm height are dried for 24 h and are weighed ($W_1$). The specimens are placed in 14% sodium sulphate solution (density 1.055 kg/m$^3$)

for 16 to 18 h at room temperature (20 to 30°C). The specimens are then taken out of the solution and kept in air for half an hour, oven dried at a temperature of $105 \pm 5°C$ for 4 h, and then cooled at room temperature. This process is repeated for 30 cycles. The specimens are weighed ($W_2$) after every five cycles and the difference in weight is found. The change in weight indicates the degree of decay of stone. There should not be any visible defect in the stones and the loss in weight should be minimal.

## 25.3.2 Laboratory Tests

The six laboratory tests conducted on stones are briefly described here.

**Attrition test [IS 2386 (Part 4):1963]**  This test is carried out in Deval's testing machine to find the abrasion resistance of aggregates. A sample of stones is broken into pieces of about 60-mm size. Such samples, weighing 5 kg, are put in both the cylinders of Deval's attrition test machine, which has a diameter and length of 200 and 340 mm, respectively. Their axes make an angle of 30° with the horizontal. The cylinders are closed and rotated about the horizontal axis for 5 h at the rate of 33 rpm. After this period, the contents are taken out from the cylinders and they are passed through a sieve of 1.5-mm mesh. The material, which is retained on the sieve, is weighed. Percentage wear is calculated as follows:

$$\text{Percentage wear} = \frac{\text{Loss in weight}}{\text{Initial weight}} \times 100 \tag{25.6}$$

**Crushing test**  This test consists of finding the compressive strength of a stone cube of size 40 mm in a compression testing machine. At least three specimens are tested. The specimens are immersed in water for 72 h in room temperature. The surface dry specimens are placed in the compression testing machine, and the load is applied at the rate of 14 $N/mm^2$ per minute. Crushing strength of the stone per unit area is the maximum load at which the sample crushes or fails divided by the area of the bearing face of the specimen.

**Freezing and thawing test**  This test is conducted in regions where the temperature can go beyond the freezing point. The freezing and thawing resistance is estimated by keeping the stone specimen in water for 24 h and then freezing it to −12°C for 24 h. It is then thawed. This cycle is repeated at least seven times, after which, the specimens are carefully inspected for any damage.

**Hardness test**  In the laboratory, the Dorry's testing machine is used. For this test, a 25-mm diameter cylindrical stone specimen of 25-mm height is prepared and weighed. The specimen is then placed in Dorry's testing machine and subjected to a pressure of 25 $g/cm^2$. The annular steel disc of the machine is then rotated at a speed of 28 rpm. During the rotation of the disc, standard coarse sand (used as an abrasive) is sprinkled on the top of disc. After 1000 revolutions, the specimen is taken out and weighed. The coefficient of hardness is calculated as follows:

$$\text{Coefficient of hardness} = 20 - (\text{Loss of weight in g/3}) \tag{25.7}$$

**Impact test**  For determining the toughness of stone, it is subjected to impact test in a Page Impact Test machine as follows: A cylinder of diameter 25 mm and height 25 mm is taken out from the sample of stones. It is then placed on the cast iron anvil of machine. A steel hammer of weight 2 kg is allowed to fall axially in a vertical direction over the specimen. With the height of first blow as 10 mm, the height of each subsequent blow is increased by 10 mm, and the blow at which specimen breaks is noted. If it is $n^{th}$ blow, '$n$' represents the toughness index of stone.

**Microscopic test**  In this test, thin sections of the stone are taken and placed under microscope to study its grain size, mineral constituents, and also the presence of any harmful substances.

## 25.4 Tests for Burnt Clay Bricks

The properties of bricks that may affect their performance during their service life are (a) compressive strength, (b) water absorption, (c) efflorescence, (d) dimensional tolerance, (e) hardness, (f) soundness, and (g) structure. These properties are depended upon the chemical composition of the clay, the quality of manufacturing process, and the degree of firing of bricks. To perform various tests to determine these properties, a sample of bricks should be chosen as per IS 5454:1978 (20 bricks per 50,000 bricks or more for class 10 and above). Bricks should be tested for the following physical requirements to ascertain their suitability for construction work.

### 25.4.1 Compressive Strength

The compressive strength of any individual brick, when tested as per IS 3495 (Part 1), should not be less than the minimum average compressive strength as specified in Table 3.3 of Chapter 3 for the corresponding class of brick, by more than 15%. The method of testing of solid and perforated bricks for compressive strength as per IS 3495 (Part 1), is briefly explained here.

**Solid bricks**   Five brick specimens are immersed in water at $27 \pm 2°C$ for 24 h. They are removed from the water and the frogs are filled and flushed with the face of the brick with 1:3 cement-sand mortar ratio. The samples are then cured for four days (1 day under damp jute bags and 3 days in clean water). Then, they are surface dried and placed in a compression testing machine with the flat faces horizontal and the mortar filled face upwards. The load is applied at the rate of 14 N/mm$^2$ per minute until the brick specimen fails. The maximum load at which the specimen fails is divided by the average net area of the two faces of the brick to get the compressive strength. The average of five values is considered.

**Perforated bricks**   Five randomly taken specimens are immersed in water at room temperature for 24 h. Then, they are surface dried and kept in a compression testing machine with perforated faces horizontal. The load is applied at the rate of 14 N/mm$^2$ per minute until the brick specimen fails. The maximum load in which the specimen fails is divided by the average net area of the two faces of the brick to get the compressive strength. The average of five values is considered.

### 25.4.2 Water Absorption

As mentioned in Table 3.3 of Chapter 3, bricks should not absorb water more than 20% by weight up to class 12.5 and 15% by weight for higher classes, when soaked in cold water for a period of 24 h. There are two types of water absorption tests specified by IS 3495 (Part 2) and described as follows:

**Twenty-four hour immersion cold water test**   Five brick specimens are taken at random. They are dried in an oven at $110 \pm 5°C$, for about 48 h. They are allowed to cool for 4–6 h without a fan and another 2–3 h with a fan and then weighed ($W_1$). They are then immersed in cold water for 24 h, surface dried, and weighed again ($W_2$). The difference in weight indicates the amount of water absorbed by the brick and from which the percentage of water absorption is determined.

$$\text{Percentage absorption} = \frac{W_2 - W_1}{W_1} \times 100 \tag{25.8}$$

The average of five tests is considered.

**Five hour boiling water test**   Five brick specimens are taken at random. They are dried in an oven at $110 \pm 5°C$, for about 48 h. They are allowed to cool in room temperature and weighed. The dried bricks are then immersed in water. The water is boiled for about 6 h and then allowed to cool down to room temperature. The weight of the brick is again taken. The percentage of water absorption is found out by using these weights. The average of five tests is considered.

### 25.4.3 Potential to Efflorescence

Efflorescence is a crystalline white deposit of water-soluble salts that can form on the surface of brickwork. The salts usually come from ground water or out of the mortar, but may also come from within the bricks themselves. Salts leading to efflorescence are mostly sulphates of calcium, magnesium, aluminium, sodium, and potassium, with carbonates occurring to a lesser extent. Dampness is favourable to its formation. The principal objection is an unsightly appearance, though it typically is not harmful to brick. According to IS 1077:1992, the rating for efflorescence should not be more than ''moderate' up to Class 12.5 and 'slight' for higher classes, when tested as per IS 3495 (Part 3), as described here.

This test should be conducted in a well-ventilated room at 18–30°C. Average value of five random samples should be taken. The brick is kept vertically in a vessel of size 300 × 200 mm, which is filled with distilled water for a depth of at least 25 mm. The water is allowed to be absorbed by the brick and evaporated through it. When the brick looks dry, a similar quantity of water is poured into the dish and allowed to evaporate again. The brick is examined after the second evaporation and reported as follows:

1. Nil: when there is no perceptible deposit of salts on the brick
2. Slight: when the brick is not covered by salt for not more than 10% of the area
3. Moderate: when there is heavy deposit of salt covering up to 50% of the area of brick but there is no powdering or flaking of the surface
4. Heavy: when there is heavy deposit of salt covering more than 50% of the area of brick and there is powdering or flaking of the surface
5. Serious: when there is heavy deposit of salt accompanied by powdering and/or flaking of the surface and this deposition increases with repeated wetting

More information on the causes and the prevention of efflorescence may be found in BIA TN23a (2006).

### 25.4.4 Warpage

For certain classes of bricks, such as the acid-resistant brick, the maximum permissible warpage should be less than 2.5 mm at any point. The warpage of the brick is measured with the help of a flat steel surface and measuring ruler graduated in 0.5-mm divisions or wedge of steel 60 × 15 × 15 mm, as shown in Fig. 25.3. To measure warpage, 10 random bricks from a lot are chosen.

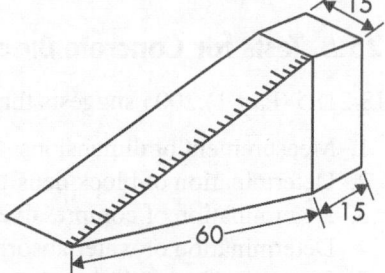

**Fig. 25.3** Measuring metallic wedge

### 25.4.5 Dimensional Tolerance

As per IS 1077:1992, 20 whole bricks are selected at random to check measurement of length, width, and height. They are arranged on a level surface successively as indicated in Fig. 25.4(a–c) in contact with each other and in a straight line. The overall length of the assembled bricks in one lot of 20 or two lots of 10 is measured. Variations in dimensions are allowed only within a limit of ±3% for Class I bricks and ±8% for other classes.

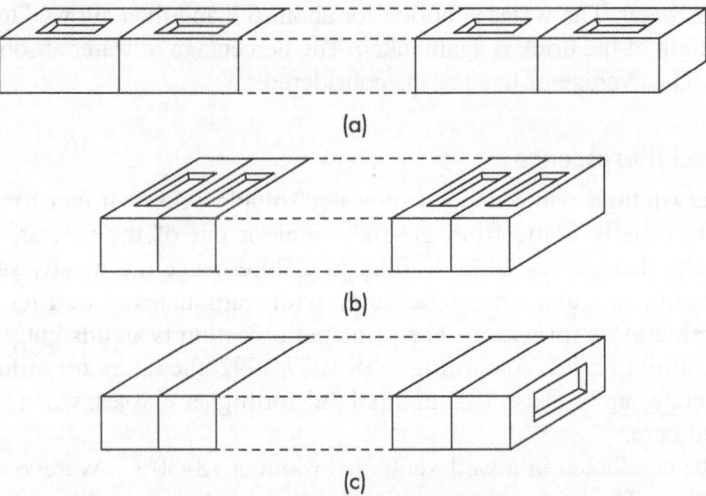

**Fig. 25.4** Dimensional tolerance of bricks (20 numbers used) (a) Measurement of length
(b) Measurement of width (c) Measurement of height

### 25.4.6 Hardness, Soundness, and Structure

To find the hardness, a scratch is made on the brick surface with finger nails. If no impression is left on the surface, the brick is assumed to be sufficiently hard.

To determine soundness, two random bricks are taken, one in each hand, and struck lightly with each other. They should not break and a clear ringing sound should be heard. If they break or a dull sound is heard, the bricks are considered unsound. Alternatively, a random brick is dropped from a height of 1 m to fall flat on a hard ground. If the brick does not break, it is considered sound.

For examining the internal structure, a random brick is broken and the structure examined. The broken surface should show homogeneous and compact texture and should not contain defects like holes. Otherwise, the brick is not considered good.

## 25.5 Tests for Concrete Blocks

IS 2185 (Part 1):2005 suggests that the following tests should be conducted on concrete blocks:

1. Measurement of dimensions
2. Determination of block density
3. Determination of compressive strength
4. Determination of water absorption
5. Determination of drying shrinkage
6. Determination of moisture movement

These tests are similar to those conducted on bricks. More details of these tests may be found in IS 2185.

## 25.6 Tests for Building Limes

Lime is a reactive material and constantly undergoes chemical changes on exposure to the atmosphere. Even during manufacture, there are chances of variability in the quality. It is therefore necessary to check its quality at various stages such as after burning, on slaking, during storage, and before actual use.

Several simple field tests have been provided in IS 1624:1986, which can give quick and fairly reliable results about the quality of lime. The laboratory tests on lime are specified in IS 6932 (Parts 1–11):1973.

## 25.6.1 Field Tests

The field tests as per IS 1624 are as follows:

**Visual examination** A sample of lime is examined for its colour and for the state of aggregation, namely, lumpy, powdery, soft, hard, etc.: (a) white colour indicates fat or pure lime (Classes C & D limes), and (b) lumps of lime indicate quicklime or unburnt limestone.

**Hydrochloric acid test** The procedure for doing the test is as follows:

1. Place about 10 cc powder lime in a test tube.
2. Add about 100 mL of hydrochloric acid (50% strength) to the test tube.
3. Stir the contents with a glass rod, until there is no effervescence.
4. Leave the sample standing for about 24 h.
5. The bubbling reaction will indicate the presence of lime and the liberation of carbon dioxide.
6. The volume of insoluble residue (IR) at the bottom of the cylinder compared with the original volume of lime will indicate the proportion of inert material (adulteration) in the lime.
7. A good thick gel formation which does not flow when the tube is inverted indicates Class A lime, while a gel which flows indicates Class B lime. No gel formation indicates Classes C, D, or F lime.

**Ball test**

1. Make a ball of 50-mm diameter (about the size of an egg), with enough water to give a stiff paste, and store it undisturbed for 6 h.
2. Then, place it in a basin of water.
3. If it expands and disintegrates within a few minutes, it indicates Class C or D lime. If there is little expansion with numerous cracking, it can be classified as Class B or E. No adverse effects indicate Class A lime.

**Impurity test**

1. Mix a known quantity of lime with water in a beaker and stir the contents well and allow it to settle for 2 h.
2. Pass the milk of lime with addition of water, if necessary, through 850 μm IS sieve.
3. Transfer the residue to a metal tray and allow it to settle and decant off the water from the tray.
4. Dry the residue for 8 h in hot sun and weigh.
5. If the residue is less than 10%, the lime Classes B and F. If it is less than 5%, it is considered as Class C and D.

**Plasticity test**

1. Mix the lime with water to a thick paste and leave it overnight.
2. Spread it on a blotting paper with a knife to check its plasticity. If it spreads nicely, it shows that the lime has good plasticity.

## Workability test

1. Make a 1:3 lime sand mortar with enough water.
2. Throw a trowel of mortar onto a brick wall, just as a mason does while plastering.
3. If the mortar sticks well on the wall, it is considered to be having good workability.

IS 1624:1986 describes another test also. In this test, a frustum of a cone (of diameter 100 mm at bottom, 70 mm at top and of height 50 mm) is placed on a square metal plate of size 120 mm. The mortar is filled in the frustum of a cone, by tapping it with a wooden rod of 16-mm diameter. The mortar at the top is levelled with a mason's trowel. The cone along with the plate is raised to a height of 300 mm. The metal plate at the bottom of cone is slid horizontally and the mortar is allowed to fall freely onto another graduated square plate of size 300 mm, placed below horizontally on the ground. The spread of the mortar on the plate is taken at two or three perpendicular locations and the average value is noted. A spread of about 150 to 160 mm indicates a good workable mortar. IS 6932 (Part 8) also describes a workability test conducted on a standard flow table using the truncated conical mould.

### 25.6.2 Laboratory Tests

Some of the laboratory tests on lime are specified in IS 6932 (Parts 1–11):1973 are briefly described as follows:

**Loss on ignition test as per IS 6932 (Part 1)**

1. Place 1 g of the prepared sample in a weighed platinum crucible and cover with a lid.
2. Ignite it at a temperature not less than 1000°C to constant mass.
3. Weigh the residue. The difference between the original and the final mass of the sample represents the loss on ignition. This is expressed as a percentage of the mass of the sample taken.

$$\text{Loss on ignition (LOI), percent by mass} = 100\frac{M_o - M_i}{M_o} \tag{25.9}$$

where $M_o$ = mass of the original sample, and $M_i$ = mass of the sample after ignition.

**Fineness test as per IS 6932 (Part 4)**
The procedure is as follows:

1. Arrange the IS sieves one above the other with the coarser sieves at the top and the finer sieves at the bottom.
2. Place 100 g of hydrated lime on the top sieve and wash it through the sieves with a moderate jet of water.
3. Do the sieving by holding the sieves in both hands with a gentle wrist motion.
4. Dry the residue on each sieve at $100 \pm 10°C$ to constant mass and weigh.
5. Express the result as a percentage of the original mass of hydrated lime.

**Soundness test as per IS 6932 (Part 9)**   This test is conducted using Le-Chatelier apparatus; conforming to IS 5514:1996 (The apparatus is shown in Fig. 25.6). The procedure is as follows:

1. Dry mix a suitable amount of cement, hydrated lime, and standard sand at a ratio of 1:3:12.
2. Gauge the above mix, add with 12% mass of water (calculated based on the dry mixture weight) and mix thoroughly.
3. Take three Le-Chatelier moulds, greased internally, and place them on a small non-porous plate and fill them with the mortar.
4. Cover the moulds with non-porous plates and place small weights over them. Leave all the three moulds undisturbed for 1 h.
5. Measure the distance between the indicator pointers. Keep the apparatus in damp air for 48 h.
6. Subject the moulds to steam at atmospheric pressure for 3 h, without immersing them in water.

7. Now cool the samples to room temperature and measure the distance between the pointers is again.
8. The difference in the two measurements should not be more than 10 mm after deducting of 1 mm from the measured expansion, to allow for the expansion of added cement. This figure is recorded as the net expansion due to lime.

**Compressive strength test (IS 6932-Part 7)**   The test procedure is as follows:

1. Prepare 12 cubes with side 50 mm from standard lime-sand mortar mix of 1:3.
2. Keep them undisturbed for 72 h in an atmosphere of at least 90% relative humidity and at temperature $27 \pm 2°C$.
3. Take the specimens out of the mould and keep them in the air for 4 days.
4. Cure six cubes in water (kept at $27 \pm 2°C$) for 7 days, take them out of the water, wipe the surface dry, and test them in compression testing machine, to give the strength at 14 days. Place the cubes in such a way that the faces that are loaded are those that were in contact with the sides of the mould, when moulded. Conduct the test by increasing the load steadily and uniformly at the rate of 150 N/min.
5. Cure the remaining six cubes in water for 21 days and test them similarly to give the strength at 28 days.
6. The load divided by the area of the side of the cube gives the compressive strength of the mortar. The average of the six specimens gives the compressive strength of the mortar.

**Popping and pitting test (IS 6932-Part 10)**   The test procedure is as follows:

1. Take 70 g of hydrated lime and mix it with 70 mL of clean water at a temperature of $27 \pm 3°C$ and keep it soaked for 2 h.
2. Mix the lime putty so obtained thoroughly and 'knock' it up with a trowel, if necessary, with a small amount of clean water, in order to obtain a plastic mass.
3. Spread it over a non-porous surface, add 10 g of Plaster of Paris spread evenly over it and mix the whole mass rapidly and thoroughly for 2 min.
4. Gauge the material and press small quantities of it into a greased ring mould 100 mm in diameter and 5-mm deep, using a broad palette knife, avoiding air bubbles.
5. Smooth off the top surface, in level with the top edge of the ring, with not more than 12 strokes of knife. This entire process should be completed within 5 min from the time of adding Plaster of Paris.
6. Form four such pats and leave them to set for 30 min.
7. Transfer them to a drying oven maintained at a temperature between 35–45°C and keep them for 4 to 16 h (4 h may be sufficient for most limes). Reject any test pat containing cracks.
8. Now, place the test pats horizontally in a steam boiler and subject them to the action of saturated steam for 3 h.
9. Examine the pats for disintegration, popping, or pitting. If any of these occur the lime is considered to be unsound.

## 25.7 Sampling and Testing of Cement

The usual tests carried out for cement are for chemical and physical requirements. They are given in relevant Indian Standards. IS 4031 (Parts 1–11) and IS 4032. The chemical tests on cement are concerned with permissible limits for IR, LOI, and other compounds and impurities such as magnesium oxide and sulphate. The physical tests are to determine fineness, soundness, setting time, and compressive strength. Most of these tests are conducted at a laboratory (Neville, 2012).

## 25.7.1 Methods of Sampling

It has been found that it is not practical to make large quantities of cement without any variation in quality. Hence, it is important for the user to ascertain the quality of purchased cement, especially in important projects. The methods for sampling and the criteria for conformity of hydraulic cements from bags, bulk storage (silos), ship's hold, wagons, and conveyors are given in IS 3535:1986. Broad outlines with regard to the controls to be exercised during the manufacturing process have also been indicated in this code.

After the cement is delivered at the site, samples for laboratory tests should be taken within 1 week, and sent to laboratories for physical and chemical testing within 1 week of sampling. As the selected samples represent the quality of a particular lot, they should be selected carefully. It may be possible to select a bag of cement randomly and send for testing. If there is a delay in the testing of the samples, they should be packed and stored in airtight containers. Only 10 kg of cement is required for one set of laboratory tests. Routine testing of cement is mandatory in large construction projects, to determine the strength as well as the fineness of cement. As fineness testing is absolutely necessary when the cement is stored for more than 3 months.

## 25.7.2 Test for Fineness

The particle size distribution is critical for controlling the rate at which cement sets and gains strength. There must be a certain amount of small particles to ensure that the cement sets in a reasonable amount of time; but, if there are too many small particles, the cement will set too quickly, leaving no time for mixing and placing (A particle that has a diameter of 1 µm will react completely within the 1st day, whereas a particle with a diameter of 10 µm will react completely in about 1 month. Particles larger than about 50 µm will probably never become fully reacted, even if there is a sufficient source of water). Fortunately, the grinding process has a natural tendency to produce a wide range of particle sizes and hence will not pose any problem. A universally recognized standard method for characterizing the complete particle size distribution of cement particles does not currently exist. A better parameter for describing the fineness of the cement (at least in terms of knowing how reactive it will be at early times) is the specific surface area, because most of the surface area comes from the smallest particles. The most common method is the Blaine air permeability test (covered by ASTM C 204–16 and IS 4031-Part 2). The Blaine and laser diffraction test tests assume that the cement particles are spheres, which is obviously not true and thus these methods only estimate the surface area of the cement particles per unit mass (m²/kg). IS 4031-Part 1 suggests actual dry sieving to find the fineness. Other methods of determining fineness of cement are Wagner Turbidimeter test (ASTM C115–10), Laser Diffraction Spectrometry test (most commonly used by the cement industry to quantify the particle size distribution (PSD) of a powder and is less time consuming than the Blaine test and can be automated for efficient measurement), Nitrogen Brunauer–Emmett–Teller method (It is the most fundamental measurement of specific surface area because it makes no assumption about the shape of the particles), and X-ray computed tomography. Most cement standards have a minimum limit on fineness [ranges from 225 m²/kg (OPC) to 400 m²/kg (supersulphated cement)].

**Sieving method**    In the *sieving method*, as per IS 4031-Part 1: 1996, the fineness of cement is measured by sieving it on a standard sieve.

In this procedure, the proportion of cement whose grain size is larger than 90-µm IS sieve is determined. In addition to the sieve, we may require a balance capable of weighing 10 g to the nearest 10 mg and a nylon brush with 25 to 40 mm bristle, for cleaning the sieve.

The procedure to determine fineness of cement by sieving is as follows:

1. About 10 g of cement should be weighed to the nearest 10 mg and placed on the sieve.
2. The sieve is agitated in swirling, planetary, and linear movements for about 15 min, until no more fine material passes through it.

3. The residue is weighed and expressed as a percentage of the quantity first placed on the sieve ($R_1$) to the nearest 0.1%.
4. The base of the sieve is gently brushed off to get rid of all the fine material.
5. The aforementioned procedure is repeated by using a fresh 10 g sample to obtain $R_2$. Now, the value of $R$ is calculated as the mean of $R_1$ and $R_2$. If these values differ by more than 1% absolute, a third sieving is done and the mean of the three values is taken as the value of $R$, i.e., the residue on the IS 90 μm sieve for the tested cement. According to IS 4031, this value should not be greater than 10%.

It has to be noted that the sieve test is rarely used.

**Blaine air-permeability test**    In the Blaine's air-permeability test, as per IS 4031 (Part 2): 1999, the following equipment/materials are required:

1. Variable flow type air permeability apparatus (Blaine Type) and the accessories conforming to IS 5516:1996 and shown in Fig. 25.5
2. A balance (range 3 g and accuracy 1 mg)
3. A timer (range 30 min and accuracy 0.2 s)
4. Manometer liquid such as dibutyl phthalate or light mineral oil
5. Pyknometer for determining the density of cement
6. Mercury of reagent grade
7. Light oil, to prevent formation of mercury amalgam on the inner surface of the cell
8. Circular discs of filter paper of medium porosity (mean pore diameter 7 μm)
9. Light grease, for ensuring an airtight joint between cell and manometer, and in the stopcock

The test should be carried out at a temperature of 27 ± 2°C and a relative humidity not exceeding 65%.

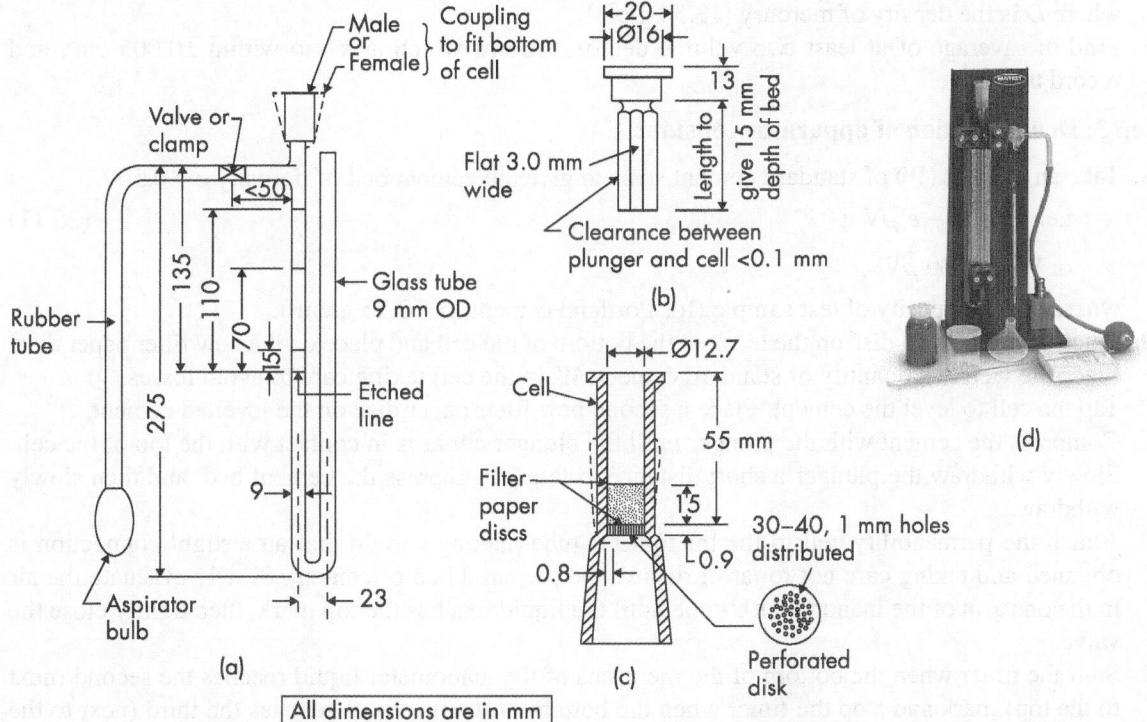

**Fig. 25.5** Variable flow type air permeability apparatus (Blaine Type) as per IS 5516:1996
(a) Manometer (b) Plunger (c) Cell (d) Equipment (*Photo Courtesy*: Matest S.p.A. Unipersonale, Italy)

The procedure for finding the fineness of cement as per this method consists of the following steps:

### Step 1: Determine the density of cement
The density of cement may be determined as per the procedure described in Section 25.7.9.

### Step 2: Determination of the bed volume
The volume of the compacted cement bed is established for the given cell-plunger clearance, as per the following procedure:

1. Apply a very thin film of light mineral oil to the cell interior. Place the perforated disc on the ledge in the cell. Place two new filter paper discs on the perforated disc.
2. Fill the cell with mercury, removing air bubbles with a clear dry rod. Level the mercury with the top of the cell by lightly pressing a small glass plate against the mercury surface until the glass is flush to the surface of the mercury and rim of the cell, being sure that no bubble or void exists between the mercury surface and the glass plate.
3. Remove the mercury from the cell and weigh to the nearest 0.01 g ($M_1$) and record the temperature.
4. Remove the top filter paper from the cell. Using a trial quantity of 2.80 g of cement, compress the cement with one filter disk above and one below the sample.
5. Refill the remaining space in the cell above the filter paper with mercury. Level the mercury to the top of the cell with a glass plate and remove mercury from the cell and weigh it to get $M_2$.
6. Calculate the bulk volume occupied by the cement bed to the nearest 0.005 cm³ from the following equation:

$$V = \frac{(M_1 - M_2)}{D} \ \text{(cm}^3\text{)} \tag{25.10}$$

   where $D$ is the density of mercury (13.54 g/cm³).
7. Find the average of at least two volume determinations, which agree to within ±0.005 cm³, and record this value.

### Step 3: Determination of apparatus constant

8. Take an amount ($W$) of standard cement so as to give the cement bed of porosity $e = 0.500$.

   i.e., $W = (1 - e)\rho V$ g $\tag{25.11}$

   or W = 0.500 $\rho$V

   where $\rho$ is the density of test sample (for Portland cement, $\rho = 3.15$ g/cm³).
9. Place the perforated disc on the ledge at the bottom of the cell and place on it a new filter paper disc. Place the weighed quantity of standard cement, $W$, in the cell taking care to avoid losses.
10. Tap the cell to level the cement. Place a second new filter paper disc on the levelled cement.
11. Compress the cement with the plunger until the plunger collar is in contact with the top of the cell. Slowly withdraw the plunger a short distance, rotate 90°, repress the cement bed, and then slowly withdraw.
12. Attach the permeability cell to the manometer tube making certain that an airtight connection is obtained and taking care not to jar or disturb the prepared bed of cement. Slowly evacuate the air in the one arm of the manometer U-tube until the liquid reaches the top mark, then tightly close the valve.
13. Start the timer when the bottom of the meniscus of the manometer liquid reaches the second (next to the top) mark and stop the timer when the bottom of the meniscus reaches the third (next to the bottom) mark. Record the time $t$ and temperature of test.

14. Repeat the whole procedure on two further samples of the same reference cement. Calculate the average time of the three determinations. Then, calculate the apparatus constant using the following formula.

$$K = 1.414 S_0 \rho_0 \frac{\sqrt{0.1 \eta_0}}{\sqrt{t_0}} \qquad (25.12)$$

where $K$ is apparatus constant, $S_0$ is specific surface of reference cement, $\rho_0$ is density of reference cement, $t_0$ is mean of three measured times, and $\eta_0$ is air viscosity at the mean of the three temperatures [ranges from 0.00001788 (at 16°C) to 0.00001876 (at 34°C)].

**Step 4: Determination of fineness**

15. Repeat the steps (9 to 15) as done in determination of apparatus constant, but this time use the cement whose fineness is to be calculated.
16. Calculate fineness of cement using following formula, with the specified porosity of $e = 0.500$ and temperature at $27 \pm 2°C$.

$$S = \frac{521.08 K \sqrt{t}}{\rho} \text{ cm}^2/\text{g} \qquad (25.13)$$

where $S$ is specific surface area, $K$ is apparatus constant, $\rho$ is density of cement, and $t$ is time.

The various available methods for the determination of the particle size distribution of cement are listed and compared by Ferraris et al. (2002). Lea and Nurse Air permeability apparatus, as described in IS 5536:1969, can also be used.

## 25.7.3 Tests for Soundness

As discussed in Section 5.6 of Chapter 5, volume expansion and unsoundness in cement mortar or concrete are caused by the presence of excess lime (CaO), MgO, or $CaSO_4$. Soundness of cement may be determined by Le-Chatelier's method and Autoclave tests as per IS 4031 (Part 3):1988.

**Le Chatelier's method** By this method, we can only find out the existence of unburnt lime (CaO). The apparatus required are:

1. The Le Chatelier Apparatus (conforming to IS 5514:1996) and shown in Fig. 25.6,
2. A balance, whose permissible variation at a load of 1000 g should be +1.0 g, and
3. A water bath.

The Le Chatelier's apparatus (IS 5514:1996) consists of a small split cylinder of spring brass 0.5-mm thickness forming a mould 30-mm internal diameter and 30-mm high. Two parallel indicating arms (made of 2-mm diameter brass wire) with pointed ends are fixed on either side of the split, as shown in Fig. 25.6(a), such that the distance of these ends from the centre of the cylinder is 165 mm. A pair of plane glass base and cover plates, of size 35 × 35 mm each, are provided for the mould. The cover plate should weigh at least 75 g; an additional small mass may be placed on the cover plate to satisfy this requirement. Two loops of suitable material and strength are soldered to the upper half of the mould on each side of the central split, as shown in Fig. 25.6(b), to facilitate de-moulding of the hardened paste specimen after test.

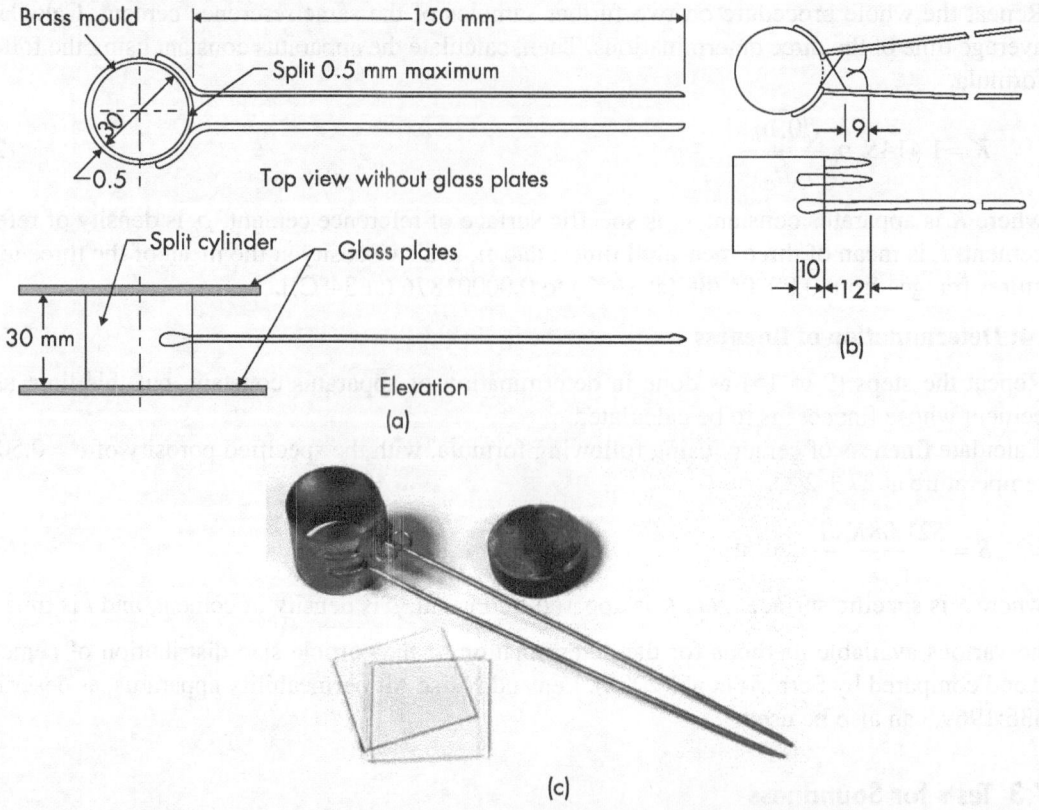

**Fig. 25.6** Le Chatelier's apparatus (a) Le-chatelier apparatus (b) Arrangement of loops for demoulding (c) The apparaus, glass plates, and lead weight (*Photo Courtesy*: ELE International, UK)

The procedure to determine soundness of cement as per Le Chatelier's method is as follows:

1. Make cement paste by mixing it with 0.78 times water required for the standard consistency.
2. Place the lightly oiled mould on a lightly oiled glass sheet and filled it with the cement paste.
3. Cover the mould with another piece of lightly oiled glass sheet, put a small weight on this covering glass sheet, and immediately submerge the whole assembly in the water bath at a temperature of $27 \pm 2°C$. Keep it this arrangement for 24 h.
4. Remove it from the water bath and measure the separating distance between the indicator points to the nearest 0.5 mm ($d_1$).
5. Submerge the mould again in the water bath and bring the temperature of water to boiling point in 25–30 min. Keep it boiling for 3 h.
6. After 3 h, allow the mould to cool down to room temperature and take it out from the water bath.
7. Finally, measure the distance separating the two indicator points to the nearest 0.5 mm ($d_2$).
8. ($d_2 - d_1$) represents the expansion of cement.

If the test results does not satisfy the requirement of 10-mm expansion for most of the cements and 5 mm for super sulphated and high-alumina cement, a further test should be made from another portion of the same sample in manner described earlier, but after aeration (done by spreading out in a layer of 75-mm thickness at $27 \pm 2°C$ and at a relative humidity of 50 to 80%, for a period of 7 days). The expansion of this aerated cement test should not be more than 5 mm.

**The autoclave test** The autoclave test is another test used for the soundness of cement (IS 4031 (Part 3):1988. It is sensitive to both lime and magnesia. All the cements containing more than 3% magnesia content should be tested for soundness using this test with unaerated cement. It has to be noted that no satisfactory test is yet available for the deduction of unsoundness due to excess of calcium sulphate; however, it can be easily determined by chemical analysis. The autoclave test consists of heating bars made of cement paste with water of normal consistency and measuring its expansion. Higher pressure and temperature are used in this test to accelerate the reactions.

The apparatus required for the autoclave test are as follows:

1. An Autoclave- The autoclave shall consist of a high pressure steam boiler (capable of maintaining a pressure of 2.1 ± 0.1 MPa with a temperature of 215.7 ± 1.7°C), equipped with suitable safety device.
2. A balance, whose permissible variation at a load of 1000 g should be +1.0 g.
3. Moulds- These should be of 25 × 25 mm size and 282 mm internal length and other accessories conforming to IS: 10086:1982.
4. Graduated glass cylinders of 150 mL capacity.
5. Length comparator- These are used to measure the changes in length of test specimen.

The procedure to determine soundness of cement as per autoclave method is as follows:

1. Thinly cover the mould with mineral oil. Then, attach the stainless steel reference inserts with knurl heads so as to get an effective gauge length of 250 mm.
2. Take 500 g of cement and mix with sufficient water to give a paste of standard consistency.
3. After mixing, immediately fill the mould in one or two layers, each layer compacted by pressing the paste into corners by thumb. Smoothen the top layer by trowel. (During the operations of mixing and moulding, the hand shall be protected by rubber gloves.)
4. After preparing the mould, store it in a moist room for a period of 24 h.
5. After 24 ± ½ h after moulding, remove the specimen from the moist atmosphere, measure its length ($L_1$) and place it in the autoclave at room temperature in a rack so that the four sides of each specimen are exposed to saturated steam vapour during the entire period of test.
6. To permit air to escape from the autoclave during the early portion of the heating period, leave the vent valve open until steam begins to escape.
7. Then, close the vent valve and raise the temp of the autoclave at such a rate that the gauge pressure of the steam is brought to 2.1 MPa in 60 to 75 min from the time heat was turned on. This pressure has to be maintained for 3 h.
8. After 3 h switch off the autoclave, and let it be cooled at the rate to bring the pressure down to less than 0.1 MPa in 1 h. The slowly bring it to atmospheric pressure by opening the vent valve.
9. Now, remove the specimen from the autoclave and place it in water maintained at a temperature of 90°C. Then, cool the water to 27 ± 2°C in 15 min. Dry the surface of the specimen and measure its length ($L_2$).
10. The soundness of cement is equal to $L_1 - L_2$ mm.

If the test results does not satisfy the requirement of 0.8% expansion for most of the cements except super sulphated and high-alumina cement, a further test should be made from another portion of the same sample in manner described earlier, but after aeration (done by spreading out in a layer of 75-mm thickness at 27 ± 2°C and at a relative humidity of 50 to 80%, for a period of 7 days). The expansion of this aerated cement test should not be more than 0.6%.

### 25.7.4 Test for Consistency

The basic aim is to find out the water content required to prepare a plastic cement paste of standard consistency, as per IS 4031 (Part 4). This code defines it as that consistency at which the Vicat plunger penetrates to a point 5–7 mm from the bottom of the Vicat mould. It is important to note that the water requirement for various tests of cement depends on the standard consistency of cement, which itself depends upon the compound composition and fineness of the cement.

The following apparatus are required for the test:

1. Vicat apparatus, as per IS 5513:1996 and shown in Fig. 25.7
2. A balance, whose permissible variation at a load of 1000 g should be ±1.0 g
3. Gauging trowel conforming to IS 10086:1982

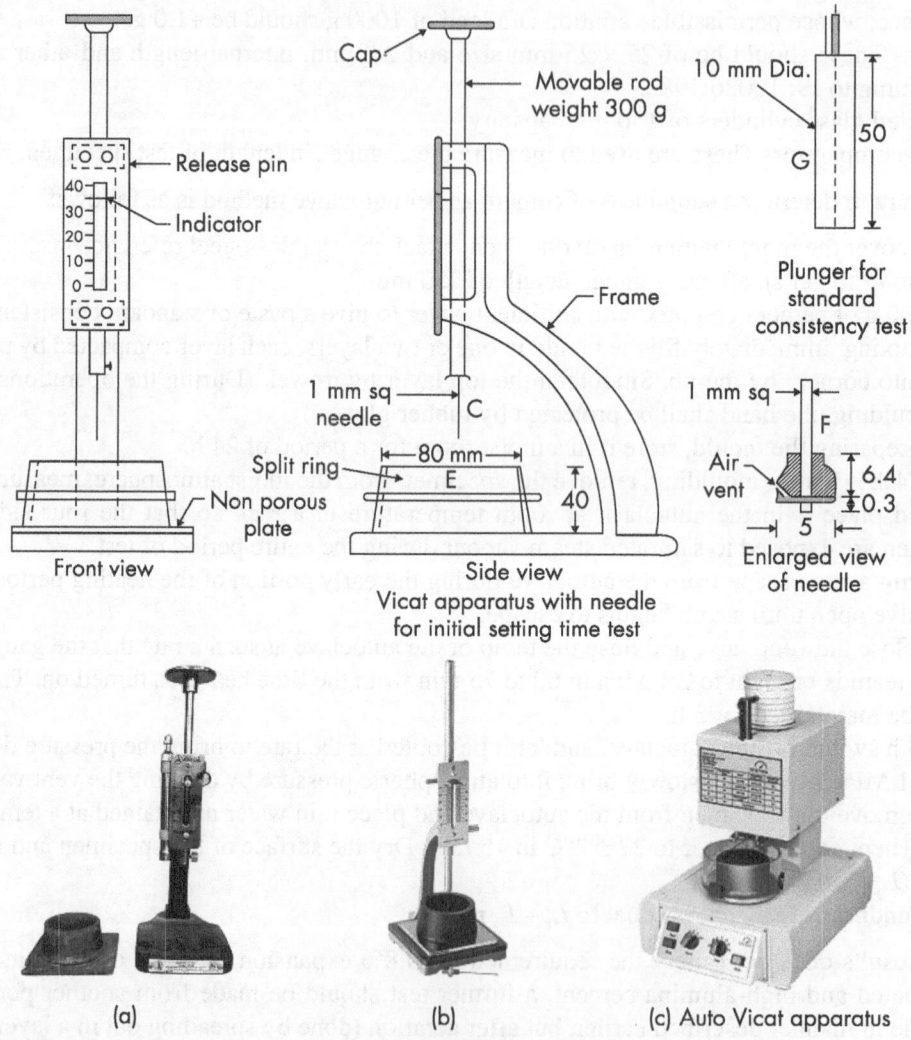

**Fig. 25.7** Vicat apparatus as per IS 5513:1996 (*Photos Courtesy*: (a) Prof. G. Mohan Ganesh, VIT, Vellore, (b) and (c) Mr. Robin Bailey, Humboldt Mfg Co.)

The procedure to determine the consistency of cement is as follows:

1. Weigh about 400 g of cement and mix it with about 20% by weight of cement of potable distilled water. The mixing time should be monitored carefully so that it is not less than 3 min and that gauging is counted from the time of adding water to the dry cement.
2. Fill the paste in the Vicat mould, with the mould resting on non-porous plate. Trim off the excess paste and level the paste mould with the trowel. Slightly shake the mould to expel the air bubbles.
3. Immediately after filling the mould, place the mould, together with the non-porous resting plate, under the rod bearing the plunger.
4. Lower the plunger (G) gently to touch the surface of the test block, and quickly release, allowing it to sink into the paste.
5. Note the reading on the gauge.
6. Repeat the procedure, taking fresh samples of cement paste with different quantities of water, until it penetrates 5–7 mm from the bottom.

Express the amount of water as a percentage by mass of the dry cement to the first place of decimal.

## 25.7.5 Test for Initial and Final Setting using Vicat Apparatus

The initial and final setting time of cement is measured, as per IS 4031 (Part 5): 1988. The following apparatus are required:

1. Vicat apparatus with different penetrating attachments as per IS 5513:1996 and shown in Fig. 25.7
2. A balance
3. Gauging trowel

The procedure to determine initial and final setting time of cement is as follows:

1. Prepare a neat cement paste by mixing the cement with 0.85 times the water required to give a paste of standard consistency. Use potable or distilled water while preparing the paste.
2. Start a stop-watch, the moment water is added to the cement.
3. Fill the Vicat mould completely with a cement paste, with the mould resting on a nonporous plate. Smooth off the surface of the paste making it level with the top of the mould. The cement block thus prepared in the mould is the test block.

**Determination of initial setting time**  This can be done as follows:

1. Place the test block, resting on the non-porous plate, under the rod bearing the needle.
2. Lower the needle (C) gently until it comes in contact with the surface of the test block and quickly release, allowing it to penetrate into the test block. In the beginning, the needle will completely pierce the test block.
3. Repeat this procedure until the needle fails to pierce the block beyond 5.0 ± 0.5 mm measured from the bottom of the mould.
4. The period elapsing between the time when water was added to the cement and the time at which the needle fails to pierce the test block, as in step 3, is considered as the initial setting time.

The initial setting time of OPC is 30 min. Note that setting time decreases with rise in temperature; the setting time of cement can be increased by adding some admixtures.

**Determination of final setting time**  This can be done as follows:

1. Replace the needle of the Vicat apparatus by the needle with an annular attachment F.
2. Lower the needle gently until it comes in contact with the surface of the test block.

3. The cement should be considered as finally set when, upon applying the needle gently to the surface of the test block, the needle makes an impression thereon, while the annular attachment fails to do so.
4. The period elapsing between this time and the time water was added to the cement is considered as the final setting time.

The final setting time of OPC is 600 min. See Table 5.4 of Chapter 5, for the initial and final setting times of the various kinds of cements.

## 25.7.6 Test for Compressive Strength

The compressive strength of cement is the most important of all the properties. The compressive strength is found using a cement-sand mortar (ratio of cement to sand is 1:3) cube of size 70.6 mm, as per IS 4031-Part 6:1988. The apparatus required are as follows:

1. Vibration machine-It should be conforming to IS 10080:1982
2. A gauging trowel
3. A balance
4. Graduated glass cylinders of 150 to 200 mL capacity
5. Cube moulds of 70.6-mm size conforming to IS 10086:1982 and a poking rod. The temperature of moulding room, dry materials, and water should be maintained at $27 \pm 2°C$. The relative humidity of the laboratory shall be $65 \pm 5\%$. The moist closet or moist room should be maintained at $27 \pm 2°C$ and at a relative humidity of not less than 90%.

The procedure to determine the compressive strength of cement is as follows:

1. Take 200 g of cement and 600 g of standard sand and mix them thoroughly in dry state.
2. Add water [$(P/4 + 3)\%$ of combined mass of cement and sand (where $P$ is the percentage of water required for preparing a paste of standard consistency, as per IS 4031-Part 4)] to the dry mix of cement and sand and mix thoroughly for a minimum of 3 min and maximum of 4 min to obtain a mix of uniform colour. If even after 4 min uniform colour of the mix is not obtained, reject the mix and mix fresh quantities of cement, sand, and water to obtain a mix of uniform colour.
3. Assemble the moulds and cover the joints with thin film of petroleum jelly in order to ensure that no water escapes during vibration. Treat the interior faces of the mould with a thin coating of mould oil.
4. Place the mould on the vibrating machine and hold it in position by clamps provided on the machine for the purpose.
5. Fill the mould with mortar in the first layer using a suitable hopper attached to the top of the mould (to facilitate filling). The hopper should be removed only after the completion of vibration. The mortar should be prodded with the poking rod 20 times in about 8 s to ensure the elimination of entrained air and honeycombing. Place the remaining quantity of mortar in the hopper of the cube mould and prod again as mentioned for the first layer.
6. Compact the mortar by vibration. The period of vibration should be 2 min at a specified speed of $12,000 \pm 400$ vibrations per minute to achieve full compaction.
7. Remove the mould, together with the base plate, from the vibrating machine and finish the top surface using a trowel.
8. Cure the specimen by keeping it in a moist place, with temperature of $27 \pm 2°C$ and relative humidity of 90%, for 24 h.
9. At the end of 24 h, remove the cube from the mould and immediately submerge it in fresh clean water kept at a temperature of $27 \pm 2°C$. The cube should be taken out of the water only at the time of testing.

10. Prepare at least six cubes in the aforementioned manner.
11. Place the test cubes, on their sides, without any packing between the cube and the plattens of the compression testing machine.
12. Apply the load steadily and uniformly, starting from zero at a rate of 35 N/mm$^2$/min.
13. The measured compressive strength of the cube is calculated by dividing the maximum load by the cross-sectional area of cube and expressed to the nearest 0.5 MPa.

The minimum 3$^{rd}$, 7$^{th}$, and 28$^{th}$ day compressive strength of cubes for different cements as specified in IS codes is given in Table 5.4 of Chapter 5. The compressive strength of masonry cement should be determined as per IS 4031 (Part 7): 1988.

### 25.7.7 Test for Tensile Strength

The briquette test was once used to find the tensile strength of cement, but now used only for rapid hardening cement, but is used for determining the tensile strength of mortars. This test is explained in Section 25.11.4.

### 25.7.8 Test for Heat of Hydration

Depending on the relative quantities of the clinker compounds, heat is evolved during hydration of cement. The evolution of heat causes an increase in the temperature of concrete, and is the greatest for mass concreting. In mass concrete, the cooling can take place only from the surfaces exposed to the atmosphere; hence, the temperature will be higher in the interior (can rise up to 55°C) than the outer surfaces. Owing to this differential temperature, shrinkage cracks may also occur. (As a rule of thumb, the maximum temperature differential between the interior and exterior of concrete should not exceed 20°C to avoid crack development.) Moreover, there will be a rapid increase in the interior strength. In such cases, we may have to use low-heat cements or adopt methods of cooling concrete. The heat of hydration may be determined in accordance with IS 4031 (Part 9):1988 using calorimeter.

The apparatus required are as follows:

1. A calorimeter, conforming to IS 11262:1985 and shown in Fig. 25.8
2. Mortar and pestle (approximately, 200 mm in diameter) for grinding partially hydrated samples
3. Glass/plastic vials having the dimension approximately 80 × 20 mm with tight fitting stoppers or caps
4. Stopwatch or timer accurate to 0.5 s or less for time interval up to 60 s and to 1% or less for time intervals of 60 to 300 s
5. IS Sieve - 150 and 850 μm conforming to IS 460 (Part 1):1985
6. Muffle furnace capable of maintaining a temperature of 900 to 950°C
7. Analytical balance-It should be capable of reproducing results within 0.0002 g with an accuracy of ±0.0002 g
8. Weighing bottles
9. A camel hair brush

Nitric acid of 2.00 ± 0.05 N strength, hydrofluoric acid around 40% (w/w), zinc oxide, paraffin wax, and distilled water are also required for this test.

The procedure for determining the heat of hydration as per IS 4031-Part 9:1988 is as follows: hand mix 60 g of cement using 24 mL of distilled water for 4 min at temperature between 15 and 25°C. Fill three specimen glass vials with this mixture, cork, and then seal them with wax. Store these vials with the mixture in a vertical position at 27 ± 2°C.

**Determination of heat of solution** This is done as follows:

For determining the heat of solution of unhydrated cement, weigh a sample of about 3 g. At the same time, weigh about 7 g of cement for finding the loss on ignition, both the weights should be correct to the nearest 0.001 g. The heat of solution is (kJ/kg) is calculated as follows:

Heat of solution (kJ/kg) of unhydrated cement

$$= \frac{\text{Heat capacity} \times \text{Corrected temperature rise}}{\text{Weight of sample corrected for ignition loss}} - 0.8(\varphi_o - \varphi) \tag{25.14}$$

where 0.8 is the specific heat of unhydrated cement.

1. For determining heat of solution of the hydrated cement, break open one of the glass vials and the remove the adherent wax and glass from the cement.
2. Grind the cement (rapidly to avoid carbonation) to pass an 850-μm IS sieve. From this, weigh out 4.2 and 7 g of cement samples for heat of solution and loss on ignition, respectively.
3. To obtain the ignition loss, place the sample in a cool furnace and raise the temperature of the furnace to 900°C over a period of 1 h.
4. Keep the sample at 900 ± 50°C for 3–4 h and then cool it in a desiccator containing anhydrous calcium chloride.
5. Weigh after half an hour, to the nearest milligram. The difference in the two weights gives the loss on ignition.

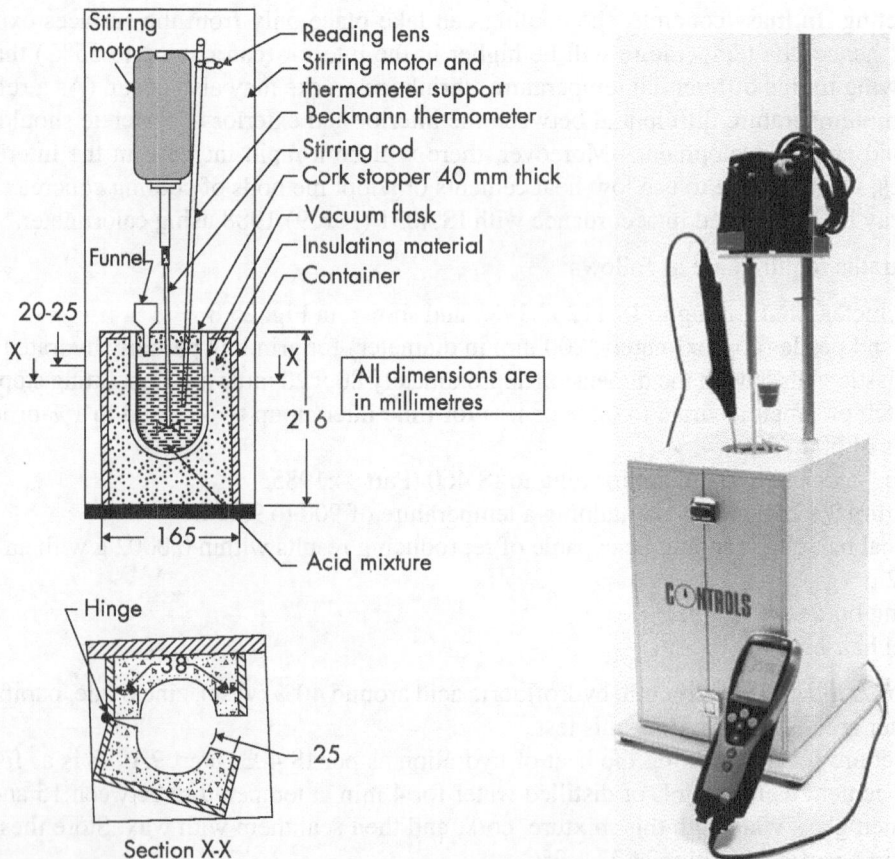

**Fig. 25.8** Calorimeter as per IS 11262:1985 (*Source*: http://www.controls-group.com/eng/cement-testing-equipment/calorimeters.php)

**Determination of heat capacity**     This can be done as follows:

1. To determine the heat capacity, ignite sufficient quantity of zinc oxide for 1 h at 900–950°C.
2. Cool in a desiccator containing anhydrous calcium chloride and grind it to pass a 150-μm IS sieve. For each determination, about 7 g of this ignited oxide is heated again to 900 to 950°C for 5 min and then cooled for about 2½ h (not more than 5 h) in the desiccator containing anhydrous calcium chloride before weighing accurately.
3. Now, assemble the calorimeter and run the stirrer for at least 5 min to allow the temperature to become uniform.
4. Take temperature reading correct to 0.001°C every minute, for 5 min, to determine the initial heating or cooling correction.
5. Then, introduce the zinc oxide using the funnel steadily for a period of 1–2 min. Now, clean the funnel with camel-hair brush.
6. Take temperature readings at 1 min intervals until the solution is complete, as indicated by a steady rate of heating or cooling of the calorimeter. The solution period should not exceed 20 min. Continue the readings for next 5 min to determine the final heating or cooling correction.
7. Plot the initial and final heating or cooling rates against the corresponding calorimeter temperature, namely the Beckmann readings at the beginning of the solution period and at the end, respectively.
8. Join the two points by a straight line, as shown in Fig. 25.9, which shows a typical measurement. From this graph, read off the corrections, for each temperature reading during the solution period. These corrections should be summed and the total added or subtracted as appropriate to the observed temperature-rise.

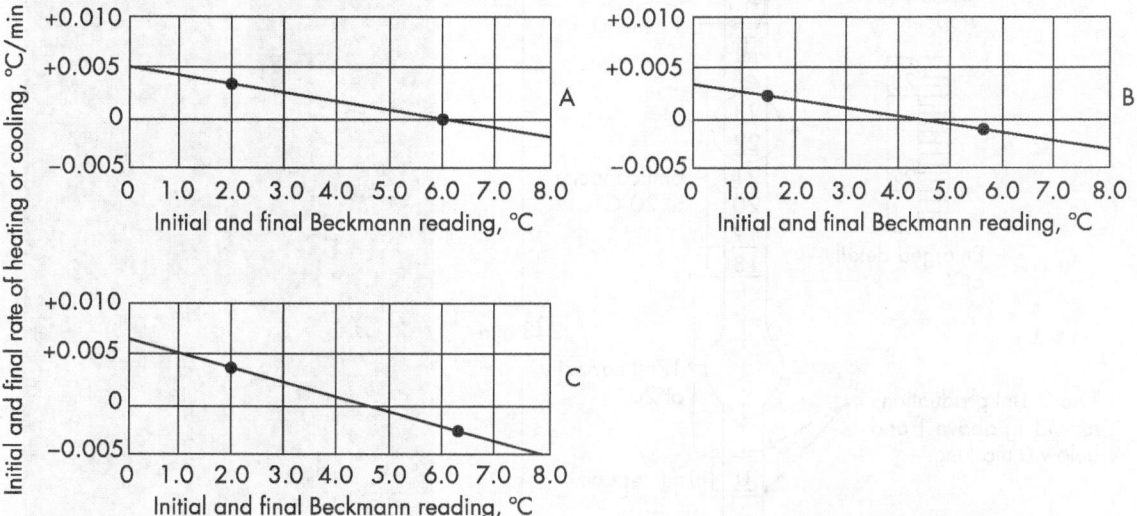

**Fig. 25.9** Typical heating or cooling correction as per IS 4031 (Part 9):1988

Heat capacity is calculated from the following expression:

$$\text{Heat capacity (J/°C)} = \frac{\text{Weight of ZnO}(g)}{\text{Corrected temperature rise}}\left[1072 + 0.4(30 - \varphi) + 0.5(\varphi_o - \varphi)\right] \quad (25.15)$$

where 1072 is the heat of solution of zinc oxide at 30°C, 0.4 is the negative temperature coefficient of the heat of solution, $\varphi$ is the final temperature of the calorimeter and contents in °C, 0.5 is the specific heat of zinc oxide, and $\varphi_o$ is the room temperature in °C.

Equation 25.15 can be simplified to:

$$\text{Heat capacity (J/°C)} = \frac{\text{Weight of ZnO(g)}}{\text{Corrected temperature rise}}(1084 + 0.9\varphi + 0.5\varphi_o) \qquad (25.16)$$

The heat of solution of hydrated cement (kJ/g, ignited weight)

$$= \frac{\text{Heat capacity} \times \text{Corrected temperature rise}}{\text{Weight of sample corrected to ignition loss}} - 1.7(\varphi_o - \varphi) \qquad (25.17)$$

where 1.7 is the specific heat of hydrated cement. The mean of three determinations on separate vials should be taken. More details about this test may be found in IS 4031 (Part 9).

### 25.7.9 Specific Gravity Test

Specific gravity is defined as the ratio between the weight of a given volume of cement and weight of an equal volume of water. The equipment required for this test are:

1. Le-Chatelier flask as shown in Fig. 25.10
2. A balance
3. A constant temperature water bath

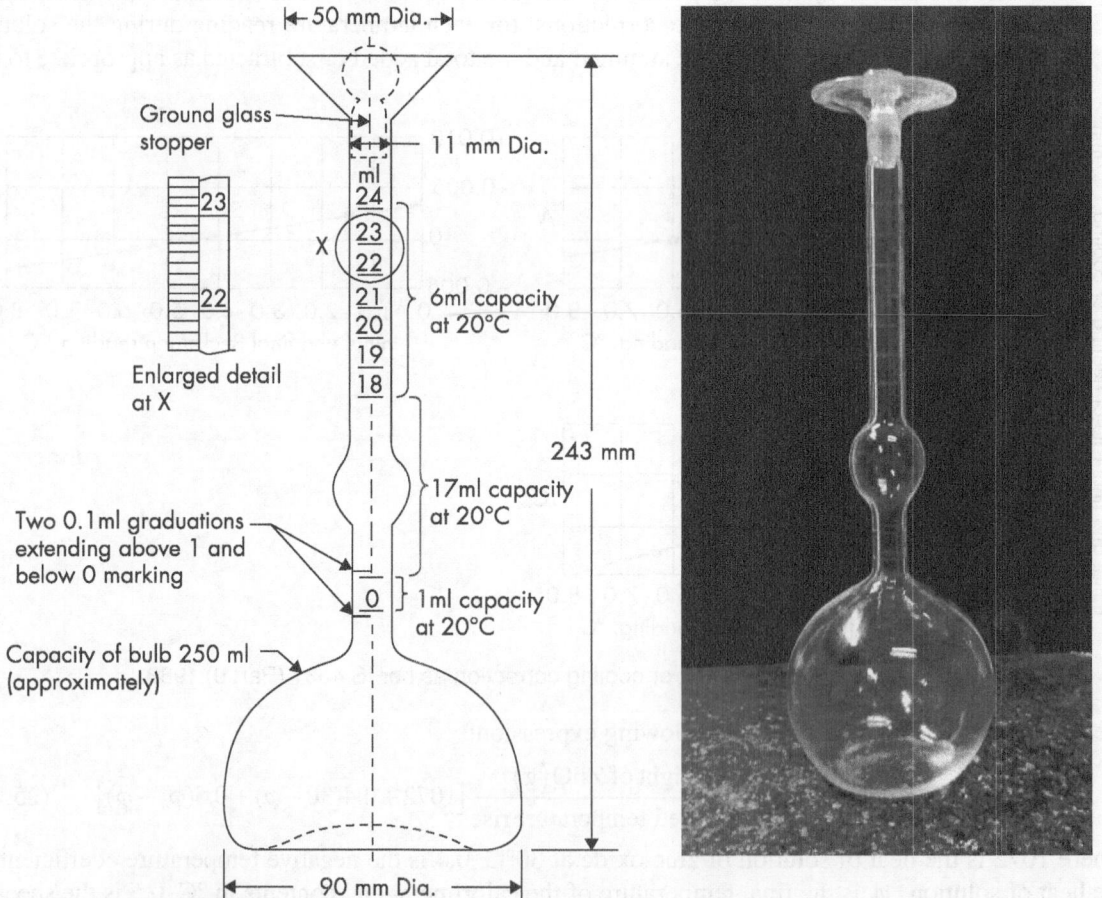

**Fig. 25.10** Standard Le-Chatelier flask as per IS 4031 (Part 11) 1988 (*Photo courtesy*: Er Dar Adil of IITD)

The procedure to find the specific gravity of cement, as per IS 4031 (Part 11):1988, consists of the following steps:

1. Dry the Le-Chatelier flask and fill with either kerosene oil or Naphtha (both should be free of water and having a specific gravity not less than 0.7313) to a point on the stem between 0 and 1-mL mark.
2. Dry the inside of the flask above the level of the liquid, if necessary.
3. Immerse the flask in a constant temperature water bath maintained at room temperature for sufficient time.
4. Record the level of the kerosene oil in the flask as initial reading.
5. Introduce about 64 g of cement into the flask, in small amounts and at the same temperature of the liquid so that the level of kerosene rises to about say 22 mL mark. Splashing should be avoided and cement should not be allowed to adhere to the sides of the flask above the liquid.
6. Insert the glass stopper into the flask and roll it gently in an inclined position to free the cement from air until no further air bubble rises to the surface of the liquid.
7. Keep the flask again in constant temperature water bath and note down the new liquid level as final reading.
8. The difference between the first and final readings represents the volume of liquid displaced by the mass of cement used in test.
9. Calculate the density as follows to the second place of decimal.

$$\text{Density} = \frac{\text{Mass of cement, g}}{\text{Displaced volume, cm}^3} \tag{25.18}$$

10. Two tests should be carried out and the average should be considered. If the difference between the two values differs by more than 0.03, the test should be repeated.

**Web-based software for virtual testing of cement**   The web-based computer software called Virtual Cement and Concrete Testing Laboratory has been developed by scientists at NIST, USA, which can be used to explore the properties of cement paste and concrete materials. This software may be found at http://www.nist.gov/el/building_materials/evcctl.cfm.

## 25.7.10 Test Certificate from Cement Manufacturer

Every cement company is continuously testing the cements they are manufacturing. They keep a record of both physical and chemical properties of the cement, based on a batch number, which indicates the date, month, and year. Based on this, they can issue a test certificate. The purchaser can demand to get a copy of this test certificate.

## 25.8 Field Tests on Cement

Some tests can be performed at the site to ascertain the quality of cement. Although these tests may not be very accurate, they may provide some idea about the quality of cement. A few of these tests are as follows (http://civiltoday.com):

**Date of manufacturing**   The date of manufacturing of cement may give some idea about its quality, as the strength of cement reduces with age.

**Cement colour**   The colour of cement may provide some idea about the addition of cementitious materials. OPC will have uniform-grey colour with a light greenish shade. Of course, the addition of cementitious materials may change the colour. The higher tetracalcium alumino-ferrite content will make cement darker in colour and lighter colour of the pozzolana will make it lighter in shade. However, note that we cannot gauge the quality based on its colour.

**Presence of hard lumps**   In cement, this will indicate the absorption of moisture from the atmosphere. Such a cement should be discarded.

**Temperature inside cement bag**   When one inserts a hand in a bag of cement and feels warmth instead of coolness, the possible hydration of cement should be suspected.

**Smoothness test**   When cement is rubbed between fingers and does not give smooth feeling, it may indicate adulteration with sand or other materials.

**Water sinking test**   Small quantity of cement, when thrown inside a mug of water, should float for a while before sinking. Otherwise, the quality should be suspected.

**Stickiness of cement paste**   A paste made with cement and water should feel sticky. If it gives an earthy smell, adulteration with clay or silt should be suspected.

**Glass plate test**   Cement is capable of setting even under water. Hence, when a cake with sharp edges is made with a thick paste of 100 g of cement, put on a glass plate, and immersed in water, its shape should not change while settling. Such a cake made with good quality cement will set under water and attain strength.

## 25.9 Chemical Tests on Cement

Chemical tests are done to determine lime saturation factor, alumina iron ratio, LOI, IR, and the amount of sulphuric anhydride ($SO_3$), magnesia (MgO), alkalies, and chlorides in cement. The procedure for the determination of these ingredients is provided by IS 4032:1985. Only a few important tests are described here.

For conducting the chemical tests, prepare a sample as follows:

1. Keep about 10-g sample in a large dish and spread it evenly into a thin layer.
2. Put it on oven maintained at 105 + 5°C for an hour to remove any superficially absorbed moisture.
3. Keep it in a desiccator (a glass container holding a drying agent for removing moisture from specimens and protecting them from water vapour in the air) and cool it to room temperature for chemical analysis.

### 25.9.1 Loss on Ignition

Loss on ignition (LOI) represents the percentage weight loss suffered by a sample of cement after heating to 1000°C. The higher the LOI, the less strength the cement will develop. The following steps are required to obtain the % loss on ignition as per IS 4032:1985:

1. Heat about 1.0 g of the oven-dried sample, $m_1$, for 15 min in a weighed and covered platinum crucible of 20 to 25 mL capacity by placing it in a muffle furnace at a temperature of 950 + 25°C.
2. Cool it and weigh. Check the loss in weight by a second heating for 5 min and reweigh ($m_2$).

$$\text{The\% loss on ignition} = \frac{(m_1 - m_2)}{m_1} \times 100 \tag{25.19}$$

Round off the calculated value and report up to 1 decimal place. The percentage loss on ignition should not exceed 4%. A high loss on ignition can indicate prehydration and carbonation, which may be caused by improper and prolonged storage or adulteration during transport or transfer.

## 25.9.2 Silica

Silica ($SiO_2$) imparts strength to the cement due to the formation of dicalcium and tricalcium silicates. If silica is present in excess quantity, the strength of cement increases while it's setting time is prolonged. The following steps are required to obtain the% silica content in the cement as per IS 4032:1985:

1. Take 0.5–1.0 g (denote this weight as $W$) of oven-dried cement sample in an evaporating dish and moisten it with 10 mL of distilled water at room temperature to prevent lumping.
2. To this add 5 to 10 mL of hydrochloric acid, and digest with the aid of gentle heat and agitation until the sample is completely dissolved. You may also apply light pressure with the flattened end of a glass rod to aid the dissolution.
3. Evaporate the solution to dryness on a steam-bath.
4. Without heating the residue any further, treat it with 5 to 10 mL of hydrochloric acid and then with an equal amount of water.
5. Now, cover the dish and digest for 10 min using a hot-plate. Dilute the solution with an equal volume of hot water, and immediately filter through an ashless filter paper.
6. Wash the separated silica ($SiO_2$) thoroughly with hot water and preserve the residue.
7. Evaporate the filtrate again to dryness, baking the residue in an oven for 1 h at 105 to 110°C.
8. Now, add 10 to 15 mL of hydrochloric acid (1:1) and heat the solution using a hot-plate.
9. Dilute the solution with an equal volume of hot water and wash the small amount of silica it contains on another filter paper. The filtrate and washings may be preserved for the determination of combined alumina and ferric oxide.
10. Transfer the papers containing the residues to a weighed platinum crucible. Dry and ignite the papers, first at a low heat until the carbon of the filter paper is completely consumed without inflaming, and finally at 1100 to 1200°C until the weight remains constant (say $W_1$).
11. Add to the ignited residue thus obtained (which will contain small amounts of impurities), a few drops of distilled water and two drops of sulphuric acid and evaporate cautiously to dryness.
12. Finally, heat the small residue at 1050 to 1100°C for a minute or two; cool and find the weight (say $W_2$). The difference between this weight and the weight of ignited sample represents the amounts of silica:

$$\text{Silica (\%)} = 100 \, (W_1 - W_2)/W \tag{25.20}$$

## 25.9.3 Sulphuric Anhydride

Sulphuric anhydride ($SO_3$) is an indirect measure of the amount of gypsum or calcium sulphate ($CaSO_4$) in the cement. The following steps are required to obtain the % sulphuric anhydride ($SO_3$) content in the cement as per IS 4032:1985.

1. Add 25 mL of cold water to 1 g of the cement sample, and while the mixture is stirred vigorously add 5 mL of hydrochloric acid. If necessary, heat the solution and crush the material with flattened end of a glass rod until it is evident that the decomposition of the cement is complete.
2. Dilute the solution to 50 mL and digest for 15 min at a temperature just below boiling.
3. Filter and wash the residue thoroughly with hot water. Set aside the filter paper with the residue.
4. Dilute the filtrate to 250 mL and heat to boiling. Add slowly drop-by-drop, 10 mL of hot barium chloride (100 g/L) solution and continue the boiling until the precipitate is well formed.

5. Digest the solution on a steam-bath for 4 h or preferably overnight.
6. Filter the precipitate through a filter paper and wash the precipitate thoroughly.
7. Place the filter paper and the contents in a weighed platinum or porcelain crucible and slowly incinerate the paper without inflaming. Then, ignite at 800 to 900°C, cool in a desiccator, and weigh the barium sulphate obtained.
8. Calculate the percentage of sulphuric anhydride ($SO_3$) content of the sample as follows:

$$SO_3 (\%) = W \times 34.3 \qquad\qquad (25.21)$$

where $W$ is the weight of residue ($BaSO_4$) in g and 34.3 is the molecular ratio of $SO_3$ to $BaSO_4$ (0.343), multiplied by 100.

## 25.9.4 Insoluble Residue

IR is a non-cementing material which is present in Portland cement. This residue material affects the properties of cement, especially its compressive strength. The following steps are required to obtain the quantity of IR in cement as per IS 4032:1985:

1. Digest the filter paper containing the residue of Sulphuric Anhydride test ($W_4$) in 30 mL of hot water and 30 mL of 2 N sodium carbonate solution maintaining constant volume, the solution being held for 10 min at a temperature just short of boiling.
2. Filter and wash with dilute hydrochloric acid (1:99) and finally with hot water until the residue is free from chlorides.
3. Ignite the residue in a crucible at 900 to 1000°C, cool in a desiccator, and find out the weight. The IRs should not exceed 1.5%.

IS 4032 also provides procedures to find the quantity of other chemicals such as CaO, MgO, $Fe_2O_3$, and $Al_2O_3$ in ordinary and low-heat Portland cement, rapid hardening Portland cement, Portland slag cement, and Portland pozzolana cement.

## 25.9.5 X-Ray Fluorescence Spectrometer Test

In this method, the sample (converted into a suitable tablet form by using either a pressed pellet or fused bead technique) is irradiated by X-ray beam from an X-ray source. These X-rays are absorbed by the elements present in the sample which in turn emit X-rays called secondary or fluorescent X-rays. These X-rays are characteristic of the elements present in the sample in terms of their wavelength (or energy) by the way of their origin, i.e., transitions among various energy states. Their intensities are directly proportional to the concentration of emitting element in the sample. The fluorescent X-rays emitted by the elements are analysed by using a set of collimators, dispersing crystals, detectors, and intensity measuring system. A calibration is carried out using a set of suitable reference standards with varying ranges of oxide concentration. Concentration of the elements is determined from the calibration curves. The advantage of this technique is its rapidity of analysis and its suitability as 'on-line' as well as 'off-line' system. Availability of quick data is extremely useful for correcting, proportioning, and controlling the raw mix to ultimately achieve the desired quality of clinker and cement. More details about this method may be found in IS 12803:1989.

## 25.9.6 Atomic Absorption Spectrophotometer Method

The conventional methods of chemical analysis which are generally practised, such as gravimetric and volumetric methods, though accurate and precise, are time consuming and hence may delay the necessary corrective actions. In addition to the conventional methods given in IS 4032:1985, the technique of atomic absorption spectrophotometer analysis may be used for routine quality control purposes. The advantages of atomic absorption technique over the conventional analytical methods are its rapidity, relative freedom from interferences (which affords a high degree of selectivity), and a high degree of sensitivity for over 60 elements. Application of such rapid analytical methods for the analysis of major and minor constituents of cement for the routine control purposes will be immensely beneficial. This method may be suitably used for the analysis of clinker as well as raw materials and raw mix used in cement manufacture.

In this method, the sample is fused with lithium metaborate in a crucible. The fused bead is dissolved in dilute nitric acid and the solution is diluted to that concentration, which meets optimum requirements for atomic absorption analysis. The standard solutions are prepared from standard reference materials and, after necessary dilutions to the desired analytical range, their absorbance is measured and is plotted against respective concentration to get a calibration curve. The concentration of element in the unknown sample is determined from the absorbance by means of calibration curve. IS 12813:1989 may be consulted for more details about atomic absorption spectrophotometer analysis.

The personnel doing this test should adhere to the manufacturer's recommended practice for igniting and extinguishing the burner on the atomic absorption spectrophotometer to avoid an explosion, which could cause physical injury.

### 25.9.7 Calorimetric Analysis

As mentioned already, the general practice of following the conventional gravimetric and volumetric methods described in IS 4032:1985, for the chemical analysis of hydraulic cement, is tedious and time consuming. The use of instruments based upon electromagnetic radiation has brought speed and accuracy to analytical methods. The use of this method will result in a rapid analysis of major and minor constituents of cement for quality control purposes by the cement industry. The colorimetric methods of analysis of hydraulic cement has been described in IS 12423:1988 as an as alternate rapid method of analysis; however, this code suggests that in case of any dispute, gravimetric methods described in IS 4032:1985 should be taken as the reference method. IS 12423:1988 should be consulted for the procedure of conducting calorimetric analysis of major and minor constituents of the different varieties of hydraulic cement.

## 25.10 Tests for Aggregates

As explained in Chapter 6, the size, shape, grading of aggregate, and their surface moisture affect directly the workability and strength of concrete whereas soundness, alkali-aggregate reaction, and the presence of deleterious substances adversely affect the soundness and durability of concrete. Hence, the following tests are conducted to ensure satisfactory performance of aggregates.

### 25.10.1 Flakiness and Elongation Index

The flakiness index of aggregates is the percentage by weight of particles whose least dimension (thickness) is less than three-fifths (0.6 times) of their mean dimension. The elongation index of an aggregate is the percentage by weight of particles whose greatest dimension (length) is greater than nine-fifths (1.8 times) their mean dimension. Both the tests are not applicable for sizes smaller than 6.3 mm.

The apparatus to measure the flakiness or elongation index of coarse aggregate is a standard length gauge with a serious of holes of various sizes as shown in Fig. 25.11. In order to conduct the test, select randomly a minimum of 200 pieces of any fraction to be tested.

The test procedure, as per IS 2386 (Part 1):1963, is as follows:

1. Take some quantity of aggregate, sufficient to provide a minimum number of 200 pieces of any fraction to be tested.
2. Sieve the sample through sieves of various sizes, as described in Section 6.8 of Chapter 6.

3. Try to pass the particles of each size in the direction of its length (maximum dimension) between the corresponding bars of the thickness gauge.
4. Weigh the total amount of aggregates passing through gauge holes to an accuracy of 0.1% of the weight of the sample.
5. The elongation index is the total weight of the material retained on the various slots of the thickness gauge, expressed as a percentage of the total weight of the sample gauged.
6. The flakiness index is the total weight of the material passing through the various slots of the thickness gauge, expressed as a percentage of the total weight of the sample gauged.

As per British codes, the flakiness index should be less than 50 for natural aggregates and 40 for crushed stone aggregates.

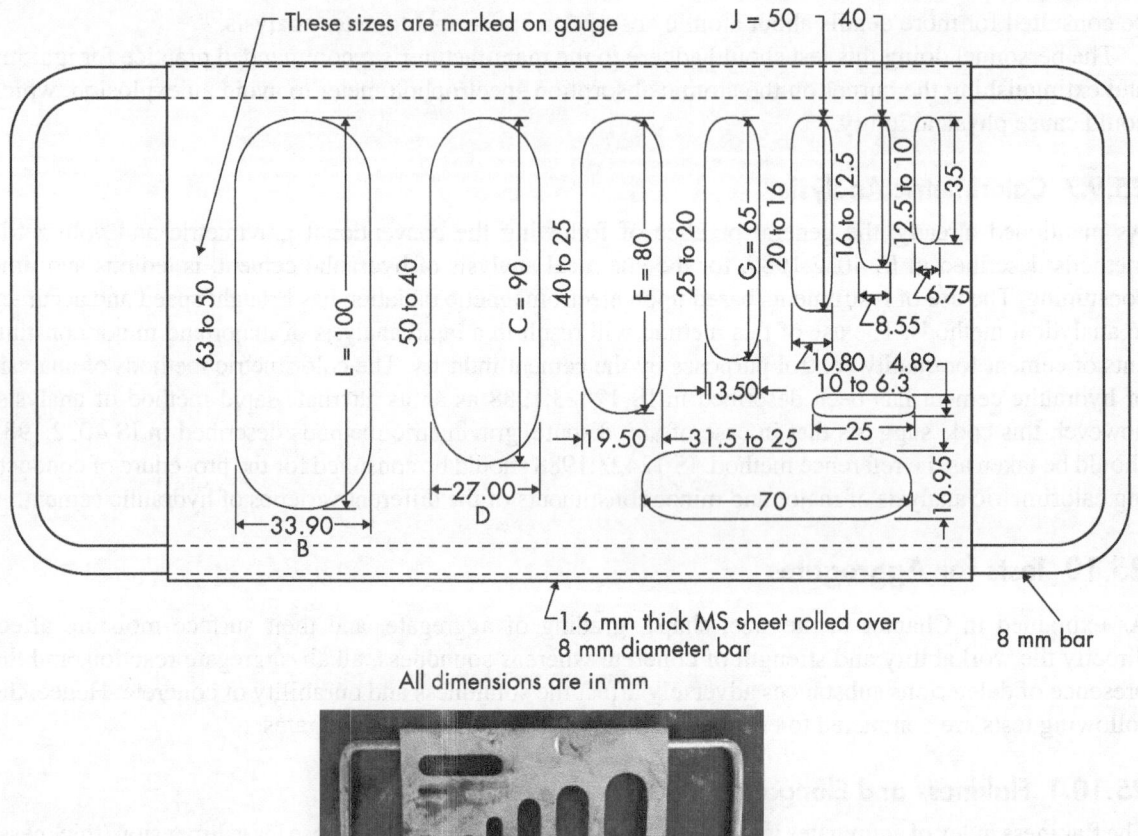

**Fig. 25.11** Thickness gauge for measuring the flakiness of coarse aggregate as per IS 2386 (Part 1):1963
(*Photo Courtesy*: Er Dar Adil of IITM)

## 25.10.2 Limits of Deleterious Materials

IS 2386 (Part 2):1963 describes test methods for the determination of clay lumps, determination of clay, fine silt and fine dust, determination of lightweight pieces (coal and lignite), determination of soft particles, and the estimation of organic impurities. Only the test for the determination of clay, fine silt and fine dust, and the test for the estimation of organic impurities are described in this section. For other tests, the reader is advised to refer to IS 2386 (Part 2):1963.

**Determination of clay, fine silt, and fine dust**   Sedimentation method is used to determine clay, fine silt, and dust, including particles up to 20 μm. The test procedure as per IS 2386 (Part 2):1963 consists of the following:

1. Weigh approximately 300 g of the air-dried sample, passing the 4.75-mm IS sieve, and place it in the screw-topped glass jar, together with 300 mL of diluted sodium oxalate solution (as a dispensing agent for silt and clay), made with 0.8 g of sodium oxalate per litre.
2. Rotate this glass jar about its long axis (with this axis horizontal) at a speed of 80 ± 20 rpm for 15 min. The clay and sand will go into suspension.
3. Pour the suspension into 1000 mL measuring cylinder. Wash the residue by gentle swirling and decantation of successive 150 mL portions of sodium oxalate solution, and pour the washings also into the cylinder until the volume is made up to 1000 mL. Go to step 5 for sand.
4. For coarse aggregate, place the sample in a suitable container, covered with a measured volume of sodium oxalate solution (0.8 g/L). Agitate vigorously and transfer the liquid suspension to the 1000 mL measuring cylinder. Repeat this process until all clayey material is transferred to the cylinder. Make up the volume up to 1000 mL with sodium oxalate solution.
5. Now use a sedimentation pipette of the Andreason type as shown in Fig. 25.12. This consists mainly of a pipette fitted at the top with a two-way tap and held rigidly in a clamp, which can be raised or lowered as required, and which is fitted with a scale from which the changes in height of the pipette can be read. First, determine the volume of the pipette A, including the connecting bore of the tap B, by filling it with distilled water and then by draining the water by reversing the tap, into a bottle, and weighing it.
6. Shake the cylinder containing the suspension again thoroughly and place the tube and contents under the pipette.
7. Gently lower the pipette until its tip touches the surface of the liquid. Now, lower it further into the liquid by 100 mm.
8. Three minutes after placing the tube in position, fill the pipette with the suspension by applying gentle suction at C.
9. Now remove the pipette from the measuring cylinder and run its contents into a weighed container, by washing any adherent solids into the container by distilled water.
10. Dry the contents of the container at 100 to 110°C for some time, cool it, and weigh it.
11. Determine the percentage of fine clay, silt, and dust, by the following formula:

$$\text{\% of clay and fine silt or fine dust} = \left(\frac{100}{W_1}\right)\left(\frac{1000W_2}{V} - 0.8\right) \tag{25.22}$$

where $W_1$ is the weight of original total sample (g), $W_2$ is the weight of dried sample in the pipette (g), $V$ is the volume of the sample taken in the pipette (mL), and $0.8$ is the weight of sodium oxalate (g) in 1 L of the diluted solution.

**Test for organic impurities**   This is an important test for dirty sands. The test procedure is as follows:

1. Fill a 350 mL graduated glass bottle with 75 mL of 3% solution of sodium hydroxide.
2. Add sand to this solution until 125 mL mark is reached. The volume is made up to 200 mL by adding more solution of sodium hydroxide.
3. Close the bottle with a screwed lid and shake it vigorously and then make the solution to stand for 24 h.
4. Now the colour of the solution will indicate whether or not there is a dangerous amount of organic matter. A colourless or straw colour will indicate some organic matter, which is not objectionable.

5. On the other hand, a dark colour indicates an objectionable quantity of organic matter. In such a case, the sand should be washed before use, and a retest has to be done to check whether the washing is alright.

6. The colour of the liquid can be compared to the colour of a solution made as follows. Add 2.5 mL of 2% solution of tannic acid in 10% alcohol to 97.5 mL of a 3% sodium hydroxide solution, thus making it 100 mL. Place it in a 350 mL bottle and shake it vigorously and allow it to stand for 24 h. The colour of this solution is compared with the liquid above the sand.

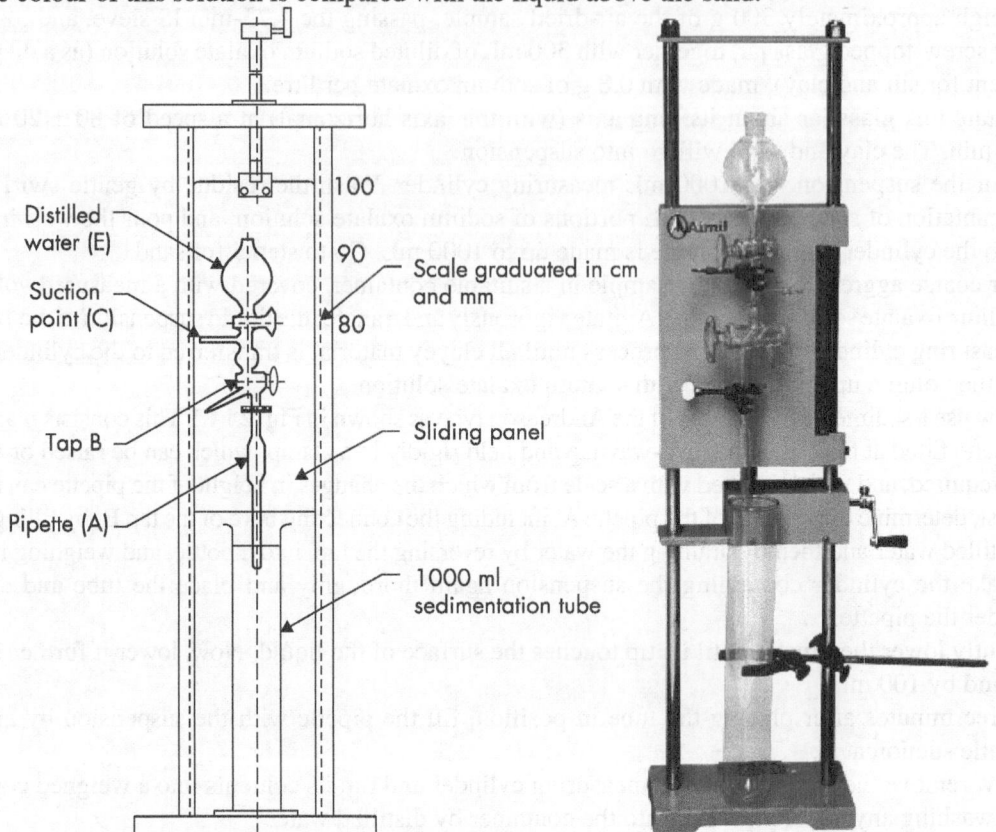

Distilled water (E)

Suction point (C)

Tap B

Pipette (A)

100

90

80

Scale graduated in cm and mm

Sliding panel

1000 ml sedimentation tube

**Fig. 25.12** Sedimentation pipette of the Andreason type as per IS 2386 (Part 2):1963
(*Photo Courtesy*: Aimil Ltd, New Delhi)

## 25.10.3 Specific Gravity, Density, Voids, Absorption, and Bulking

Test methods for determining specific gravity of coarse and fine aggregates are described in IS 2386 (Part 3):1963. The relative density of an aggregate may be determined on an oven-dry basis or a SSD basis. Both the oven-dry and SSD relative densities may be used in concrete mixture proportioning calculations.

The test as per IS 2386 (Part 3):1963 consists of immersing a basket containing about 2000 g of sample of stone aggregate in distilled water at temperature between 22 and 32°C and removing the entrapped air. The sample is kept immersed in water for 24 ± ½ h. The basket containing the sample is weighed in water (immersed weight A). The aggregates are then surface dried and weighed in air (Weight B). Now the aggregates are oven dried at a temperature of 100 to 110°C and maintained at this temperature for 24 ± ½ h, and then cooled, and weighed (Weight C) in air. The specific gravity and water absorption are calculated as follows:

$$\text{Specific gravity} = \frac{C}{B - A} \tag{25.23}$$

$$\text{Apparent specific gravity} = \frac{C}{C - A} \tag{25.24}$$

$$\text{Water absorption (percent of dry weight)} = \left( \frac{B - C}{C} \right) \times 100 \tag{25.25}$$

where $A$ is the weight of the saturated surface dry aggregate in water, $B$ is the weight of the saturated surface dry aggregate in air, and $C$ is the weight of oven-dried aggregate in air. Most natural aggregates have specific gravity ranging from 2.4 to 2.9 with corresponding mass densities of 2400 and 2900 kg/m³ (Kosmatka et al., 2011).

The easy test to determine the moisture content of aggregates is to weigh the aggregates first and then heat them in an open pan in the field. After the water is dried, the weight of the sample may be found. The difference between the two weights gives the water content.

The bulk density may be determined as described in IS 2386 (Part 3):1963 by placing three layers of oven-dry aggregate in a container of known volume, rodding each layer 25 times with a tamping rod, and determining the weight of aggregates after deducting the weight of container.

$$\text{Bulk density}, \gamma = \frac{\text{Weight}}{\text{Volume}} \tag{25.26}$$

A simple method of determining percentage bulking of sands at sites has been explained in Section 6.12.3 of Chapter 6. For more details about these tests, IS 2386 (Part 3) should be referred to.

### 25.10.4 Mechanical Properties: Aggregate Crushing Value, Impact Value, and Abrasion Value

Several tests for evaluating the mechanical properties of aggregates are described in IS 2386 (Part 4). The aggregate crushing value test is generally used for all aggregates and the 10% fines value test is performed on soft aggregates. Aggregate impact test and aggregate abrasion value tests are required for special applications involving impact and abrasive actions.

**Aggregate crushing value test**   The aggregate should be tested in a surface-dry condition. If dried by heating, the period of drying shall not exceed 4 h, the temperature should be 100 to 110°C and the aggregate should be cooled to room temperature before testing. The test procedure as per IS 2386 (Part 4):1963 is described as follows:

1. Take about 6.5 kg of aggregate passing 12.5-mm IS Sieve and retained on a 10-mm IS Sieve. The quantity of aggregate should be such that the depth of material in the cylinder, after tamping as described below, should be 100 mm. Compact it in the standard cylinder used for this test in three layers; each layer should be compacted with the rounded end of a tamping rod with 25 strokes. Level off the top layer, using the tamping rod as a straight-edge.
2. Weigh the sample and record the weight (Weight A). The same weight of sample should be taken for the repeat test also.
3. Place the cylinder in position on the base-plate in the compression testing machine between the platens of the testing machine. Level the surface of the aggregates and place the no-friction plunger so that it rests horizontally on the surface of the aggregates, taking care to ensure that the plunger does not jam in the cylinder (see Fig. 25.13).
4. Apply the load at a uniform a rate so that the load of 40 tonnes is reached in 10 min.

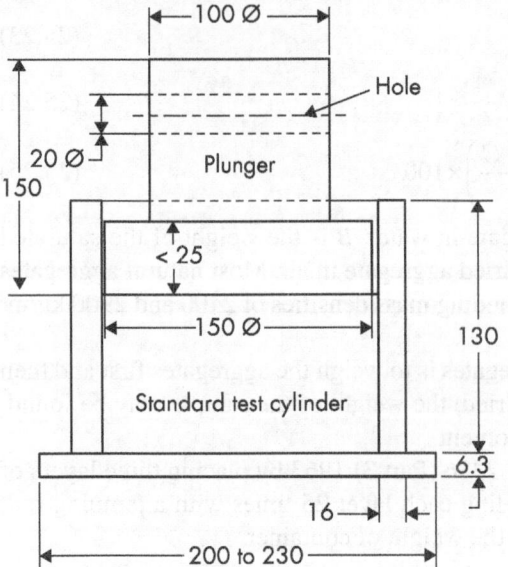

**Fig. 25.13** Apparatus for aggregate crushing value and ten percent fines value test

5. Release the load and remove the whole of the material from the cylinder. Sieve it on a 2.36-mm IS Sieve for the standard test, or the appropriate sieve as given in IS 2386 (Part 4):1963. Weigh the fraction passing the sieve (Weight B).

6. The ratio of the weight of fines formed to the total sample weight in each test is expressed as a percentage, as follows:

$$\text{Aggregate crushing value} = \frac{B}{A} \times 100 \qquad (25.27)$$

where $B$ is the weight of fraction passing the appropriate sieve, and $A$ is the weight of surface-dry sample. It has to be noted that two tests should be made and the mean value should be taken. Typical values of ACV for river gravel, crushed granite, and slag are 38, 25, and 22.5, respectively.

**Ten percent fines value test** The apparatus that is used for the ten percent fines value (TFV) test is the same as that used for the ACV test. Sample preparation for this test is also similar to the aggregate crushing value test, i.e., perform steps 1 to 3 of aggregate crushing value test first. Then, apply the load at a uniform a rate so that the plunger penetration is as follows in 10 min:

1. 15.0 mm for rounded or partially rounded aggregate (for example, uncrushed gravels),
2. 20.0 mm for normal crushed aggregate, and
3. 24.0 mm for honeycombed aggregate (for example, expanded shale and slags).

After reaching the necessary penetration, release the load and sieve the material through 2.36-mm IS Sieve. The percentage of the fines passing through the aforementioned sieve is expressed as the percentage of the weight of the test sample. Repeat the test with another sample. This TFV should be in the range of 7.5 to 12.5% (i.e., about 10%) but if it does not, do another test at a load calculated as follows:

$$\text{Load for 10 percent fines} = \left(\frac{14}{y+4}\right)x \qquad (25.28)$$

where $x$ is the load in tones for causing 7.5 to 12.5% fines and $y$ is the mean of the percentage of fines from two tests at $x$ tonnes load. This load should be reported to the nearest 0.5 tonne. The recommended values are:

1. For normal concrete: >5 tonnes
2. For concrete in wearing surfaces such as pavements: >10 tonnes
3. For granolithic concrete in factories where high wear resistance is required: >15 tonnes

Typical values of TFV for river gravel, crushed granite, and slag are 10, 15, and 18 tonnes, respectively. The British Transport Research Laboratory in its Overseas Road Note 19 (2002) proposed a correlation between ACV and TFV in the following form:

   $\text{ACV} = 38 - 0.08 \times 10\%\text{TFV}$

This relationship is said to be valid in the strengths of 14 to 30 ACV and 100 to 300 kN TFV. A TFV value of 160 kN equates approximately to an ACV of 25 using this relationship.

**Aggregate impact value test**   This test is required for aggregates which will be used in concrete that may be subjected to impact as in runways in airports. The procedure for conducting this test as per IS 2386 (Part 4):1963 is as follows:

1. Take a sample of aggregates passing through the 12.5-mm IS Sieve; however, it is retained on a 10-mm IS Sieve. These aggregate should be dried in an oven for a period of 4 h at a temperature of 100 to 110°C and cooled.
2. Compact it in the standard cylinder, 102-mm internal diameter and 50-mm deep, in three layers; each layer should be compacted with the rounded end of a tamping rod with 25 strokes. Level off the top of cylinder, using the tamping rod as a straight-edge.
3. Determine the net weight of aggregate in the cylinder to the nearest gram (Weight A); the same weight of aggregate should be used for the duplicate test.
4. Firmly fix the cylinder in position on the base of an aggregate impact test machine (see IS 2386 (Part 4), for details) and compact the sample placed in it by a single tamping of 25 strokes of the tamping rod.
5. Raise the hammer, weighing 14 kg, until its lower face is 380 mm above the upper surface of the aggregate in the cylinder, and allow it to fall freely on to the aggregate, 15 times, at an interval of not less than 1 s.
6. Remove the crushed aggregate from the cylinder and sieve the whole of it using a 2.36-mm IS sieve until no further significant amount passes in one minute.
7. Weigh the fraction passing the sieve to an accuracy of 0.1 g (Weight B). Weigh also the fraction retained on the sieve (Weight C). If the total weight (B + C) is less than the initial weight (A) by more than 1 g, discard the result and do a fresh test. Do two such tests.
8. Record the ratio of the weight of fines formed to the total weight of sample in each test as a percentage, to the first decimal place by:

$$\text{Aggregate impact value} = \left(\frac{B}{A}\right) \times 100 \tag{25.29}$$

where $B$ is the weight of fraction passing 2.36-mm sieve and $A$ is the weight of oven-dried sample. As per Overseas Road Note 19, London (2002), the aggregate Impact Value should be less than 25.

**Aggregate abrasion value test**   The aggregates used for roads and pavements should be hard enough to resist abrasion. The abrasion value of coarse aggregate may be determined by either Deval Machine or by Los Angeles machine. Only the test using Los Angeles machine is described here. For the details of the method using Deval Machine, refer to Section 25.3.2.

The Los Angeles abrasion testing machine consists of a hollow steel cylinder, closed at both ends, and has an inside diameter of 700 mm and an inside length of 500 mm (see Fig. 25.14). The cylinder is mounted in such a manner so that it can be rotated about its axis in a horizontal position. An opening is provided in the cylinder for inserting the test sample. The opening can be closed dust-tight with a removable cover, which is bolted in place. A removable steel shelf, projecting radially 88 mm into the cylinder and extending its full length, is mounted along one element of the interior surface of the cylinder. The thickness of the shelf is chosen to make it firm and rigid. The position of the shelf is such that the distance from the shelf to the opening, measured along the circumference of the cylinder in the direction of rotation, is not less than 1250 mm. The test procedure as per IS 2386 (Part 4) is as follows:

1. Depending upon the maximum size of aggregates, take about 5000 to 10,000 g of test sample consisting of clean aggregates of specified grading (A-G). Dry the sample in an oven at 105–110°C for a period of 4 h and cool it.
2. Place the test sample and the abrasive charge in the rotating drum of the Los Angeles testing machine. The charge consists of 6 to 12 numbers cast iron spheres or steel spheres (depending on the grading) approximately 48 mm in diameter and each weighing about 417 g.

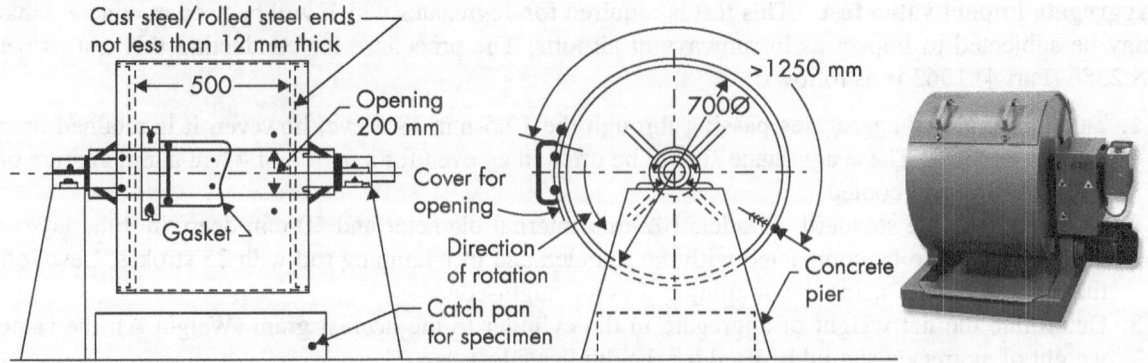

**Fig. 25.14** The Los Angeles abrasion testing machine, as per IS 2386 (Part 4)
(*Photo Courtesy*: M/s Controls USA, Inc)

3. The shelf in the drum lifts and drops the aggregates and the steel balls, while the machine is rotated at a speed of 20 to 33 revolutions/min. For grading A, B, C, and D, the machine should be rotated for 500 revolutions and for grading E, F, and G, it should be rotated for 1000 revolutions. The machine should be so driven and so counter-balanced as to maintain a substantially uniform peripheral speed.
4. At the end of the test, discharge the material from the machine and sieve it through l.70-mm IS Sieve.
5. Wash the material coarser than the 1.70-mm IS Sieve, dry it in an oven at 105 to 110°C, and find the weight.
6. Calculate the percentage of wear from the original weight and the final weight of the test sample, expressed as a percentage of the original weight. This percentage wear is called the Los Angeles abrasion value.

As per Overseas Road Note 19, London (2002), the Los Angeles abrasion value should be less than 30 for wearing course and less than 35 for binder course of roads.

### 25.10.5 Soundness of Aggregate

The soundness of aggregates can be determined as per IS 2386 (Part 5) as follows:

**Table 25.1** Size of sieve hole and yield

| Size (mm), square hole sieve | Yield (g) |
| --- | --- |
| 10 to 4.75 | 300 |
| 20 to 10 (Consisting of) | 1000 |
| 12.5 to 10 (33%) | |
| 20–12.5 (67%) | |
| 40 to 20 (Consisting of) | 1500 |
| 25–20 (33%) | |
| 40–25 (67%) | |
| 63 to 40 (Consisting of) | 3000 |
| 50–40 (50%) | |
| 63–50 (50%) | |
| 80 or more | 3000 |

1. Wash the sample of coarse aggregates and dry at 105 to 110°C and separate to different size ranges as given in step 2.
2. Pass a sample of fine aggregate through different IS sieves such that it yields not less than the amounts in different sizes as shown in Table 25.1, after removing sizes finer than 4.75 mm.
3. In case of fine aggregate, these are thoroughly washed and dried at 105 to 110°C and separated into different sizes by the set of sieves ranging from 10 mm to 600 µm, to yield at least 100 g in each.
4. Immerse the aggregate sample for 16 to 18 h in the prepared solution of either sodium sulphate or magnesium sulphate solution such that the sample is covered by the solution to a depth of at least 150 mm. Maintain the sample at 27 ± 1°C, by covering the container to reduce evaporation.

5. Remove the aggregate sample from the solution and drain it for $15 \pm 5$ min. Then, dry in an oven at 105–110°C to constant weight for 4 h. Allow it to cool to room temperature. The process of immersion and drying is repeated for as many numbers of cycles as agreed to between the purchaser and vendor.

6. Now subject the sample for quantitative and qualitative examinations.

*Quantitative examination* After the completion of final cycle, wash the cooled sample with barium chloride to free it from the sodium sulphate or magnesium sulphate. Dry each sample to constant weight at 105–110°C and weigh. Fine aggregate are sieved over the same sieve on which it was retained before the test, and coarse aggregate over the sieves shown in Table 25.2.

**Table 25.2** Size of aggregates and sieve size used

| Size of aggregate (mm) | Sieve size used to determine loss (mm) |
|---|---|
| 10 to 4.75 | 4 |
| 20 to 10 | 8 |
| 40 to 20 | 16 |
| 63 to 40 | 31.5 |

*Qualitative examination* Examine each fraction of samples coarser than 20 mm qualitatively for the evidence of disintegration of grains, splitting, crumbling, cracking, or flaking.

## 25.10.6 Mortar-making Properties of Fine Aggregates

IS 2386 (Part 6) covers the test procedure for measuring the mortar-making properties of fine aggregate for concrete by means of flow test and compression test on specimens made from a mortar of a plastic consistency and gauged to a definite water-cement ratio.

The flow table procedure as per IS 2386-Part 5 is as follows:

1. Place cement and water in quantities that will give a water-cement ratio of 0.6 by weight in a vessel and permit the cement to absorb water for 1 min. Mix the materials into a smooth paste with a spoon.

2. Add a known weight of the sample of sand (in SSD condition) and mix it thoroughly, until the mortar appears to be of desired consistency (flow $100 \pm 5$). Continue the mixing for 30 s.

3. Carefully wipe the flow table top clean and dry, and place the flow mould at the centre (flow table, as specified in IS 1727:1967, is specially designed equipment that can be made to repeatedly raise and drop and shown later in the chapter in Fig. 25.22).

4. Place a layer of about 25-mm thickness mortar in the mould and tamp it 20 times with the tamping rod. Fill the mould again with mortar and tamp as specified for the first layer. Cut off the mortar to a plane surface, flush with the top of the mould, using the trowel.

5. Wipe the table top clean and dry. Lift the mould away from the mortar 1 min after completing the mixing operation. Immediately, drop the flow table through a height of 12.5 mm, 10 times in 6 s.

6. The flow is the resulting increase in average diameter of the mortar mass, measured on at least four diameters at approximately equal angles, expressed as a percentage of the original diameter. For most mortars, the required flow is $100 \pm 5\%$.

7. The flow test is repeated, using a fresh batch of mortar each time, until the desired flow is achieved. The quantity of water needed to achieve flow is recorded, and this mortar is then tested for compressive strength.

**Test for compressive strength** Once the proper flow is achieved, place the mortar and compact it in 50-mm cube moulds, in two layers, each layer rodded with 25 strokes of the tamping rod. Finish the surface of each cube flat using a trowel, and place the moulds into a moist curing cabinet. (The temperature of the mixing water, moist closet, and storage tank should be maintained at $27 \pm 2$°C.) After 24 h of curing, strip the moulds from the cube specimens. Determine the compressive strength using a compression testing machine at specified curing intervals, usually 1 or 3 days, 7 days, 28 days, and 56 days (in case fly ash is added to cement). Calculate the average crushing strength of three specimen cubes. Visit www. caer.uky.edu/kyasheducation/testing-mortar.shtml, for more details on tests on mortars.

## 25.10.7 Alkali Aggregate Reactivity

**Table 25.3** Grading requirements as per IS 2386 (Part 7)

| Sieve size | | Percentage by weight |
|---|---|---|
| Passing | Retained on | |
| 4.75 mm | 2.36 mm | 10 |
| 2.36 mm | 1.18 mm | 25 |
| 1.18 mm | 600 μm | 25 |
| 600 μm | 300 μm | 25 |
| 300 μm | 150 μm | 15 |

IS 2386 (Part 7) describes two tests for determining the potential reactivity of aggregates: mortar bar method [also called as the *concrete prism test* (CPT) in ASTM C1293] and the chemical method [also called as *accelerated mortar bar test* (AMBT) in ASTM C1260 and ASTM C1567]. In the former method, aggregates with a specified gradation (see Table 25.3) are used in making 25 × 25 × 250 mm mortar bars (at least four numbers should be made). The mortar is made using one part of cement to 2.25 parts of graded aggregates by weight. The amount of mixing water, measured in millilitres should be such as to produce a flow of 105 to 120%. After 24 ± 2 h, remove the test specimen from the mould and measure the length at a temperature of 27 ± 2°C. Then, place the specimen on end in a metal or plastic container maintained at 38 ± 2°C. Keep the specimens in the container over, but not in contact with, water. When length measurements are to be made at subsequent periods, remove the container holding the specimens from the 38 ± 2°C storage and place in a room at a temperature of 27 ± 2°C for at least 16 h prior to measuring the specimens. Measure the length of the specimens periodically. Readings at ages of 1, 2, 3, 6, 9, and 12 months and, if necessary, at least every 6 months thereafter, are suggested. Report the average of the expansions of the four specimens of a given cement-aggregate combination to the nearest 0.01% as the expansion for the combination for a given period. As per IS 383:2016, the permissible limits for mortar bar expansion at 38°C are 0.05% at 90 days and 0.10% at 180 days, respectively.

In the *chemical method* (also called as *accelerated mortar bar test*), aggregates with a specified gradation are used in making 25 × 25 × 250 mm mortar bars that are cured for 48 h, and then stored in a 1 N sodium hydroxide (NaOH) solution at 80°C (1 N means 1 gram-equivalent weight per litre of solution). This test is found to be especially suitable for slowly reactive aggregate. As per IS 383:2016, mortar bars that expand in this harsh environment by less than 0.10%, at 16 days after casting, are indicative of innocuous behaviour in most cases; expansions of more than 0.20% at 16 days after casting are indicative of potentially deleterious expansion; and expansions between 0.10 and 0.20% at 16 days after casting include both aggregate that are known to be innocuous and deleterious in field performance. For these aggregates, additional petrographic analyses should be done to confirm that the expansion is due to alkali aggregate reactivity (ASR). However, in the absence of such information, the aggregate should conservatively be considered reactive and measures to mitigate ASR reaction should be undertaken (IS 549.2–2011). The chemical method has a significant practical advantage in its speed, as the mortar bar method, which correlates better with field observations, takes 1 year to perform when evaluating aggregate reactivity, or 2 years when evaluating the efficacy of supplementary cementitious materials (SCMs) to mitigate deleterious expansion. More details about these tests and the effects of some SCMs (fly ash, natural pozzolan, and natural zeolite powder) on ASR expansions can be found in Andiç-Çakir et al. (2009), ACI 221.1R-98, and IS 549.2–2011.

IS 383:2016 also cautions that a few locations in India have dolomitic and limestone aggregates. In such cases, it suggests concrete prism test over mortar bar test. The test should cover measurement of length change of concrete prisms, the susceptibility of cement-aggregate combinations to expansive alkali-carbonate reaction involving hydroxide ions associated with alkalis (sodium and potassium) on certain calcitic dolomites and dolomitic limestone.

### 25.10.8 Petrographic Examination

Potentially reactive components of an aggregate can be identified and quantified through petrographic examination when performed by an experienced petrographer. The petrographic examination is generally done according to procedures outlined in IS 2386 (Part 8) or ASTM C295 M-03. The Indian code specifies two methods of petrographic examination of aggregates for concrete. Method I may be adopted for routine purposes and Method II for detailed investigations. These methods require hand lens, stereoscopic microscope, petrographic microscope, and auxiliary equipment necessary for adequate petrographic examination and identification of rocks and minerals.

A qualified petrographer, who is familiar with the problems of concrete, should examine each of the size fractions, considered in the sample, as per IS 2386 (Part 8), individually, to establish whether the particles are coated with mineral substances (such as opal, calcium carbonate, or gypsum), silt, or clay. If organic coatings are present, they should be identified and evaluated by petrographic or chemical techniques. The coated aggregates should be washed to remove the coatings. The coarse aggregate fractions should be examined particle-by-particle, by a hand lens or microscope. After the completion of such an analysis, the petrographer should prepare a report summarizing the observations and conclusions regarding the suitability of the aggregate under the anticipated environmental conditions, and indicating any restrictions of its use, or the need for any special tests to elucidate the significance of some properties.

### 25.11 Tests for Mortars

The quality of mortars is checked by performing the following tests: compressive strength, consistency, water retention, tensile strength, adhesion, and brickwork in compression.

### 25.11.1 Compression Strength

The compressive strength of mortar should be determined as per Appendix A of IS 2250:1981. Accordingly, the dry ingredients are placed in the mixing bowl and mixed for 30 s (An electrically driven mechanical mixer as per IS 2259:1981 is to be used.). In the next 30 s, while mixing, water is poured at a uniform rate into the bowl and the mixing is continued for 60 s. Trial mortars are to be made with specified proportion of dry ingredients and adding different percentages of water until the specified flow of 110 to 115 mm is obtained [The flow should be measured as per IS 5512:1983 (see also Section 25.12.4)]. The bowl is covered with a damp cloth and the mortar is allowed to stand for a period of 10 min, and then remixed for 60 s. When using lime putty, similar procedure is used, except that the sand and lime putty are premixed by hand/mixer until the lime is uniformly distributed.

Specimens of 50-mm mortar cubes are made using metal moulds. After spraying the interior faces of the moulds with mineral oil or light grease, the moulds are filled with the mortar in two layers. Each layer should be compacted properly, using the tamping rod. The specimens are then stored by covering them by damp material, at a temperature of $27 \pm 2°C$, for 1 to 3 days. Though lime mortar cubes can be cured at the same temperature in air, cement mortar cubes are cured in clean water tanks. The cubes are tested on the 28th day. The bearing surfaces of the testing machine and that of the cube are wiped clean and tested such that the load is applied to opposite sides of the cube as cast. The load is applied at a uniform rate of 2 to 6 N/mm² per minute, until failure occurs. The compressive strength (N/mm²) is calculated as the failure load divided by the area of cross section of specimen; the average of three test results is taken as the compressive strength of the mortar.

## 25.11.2 Consistency

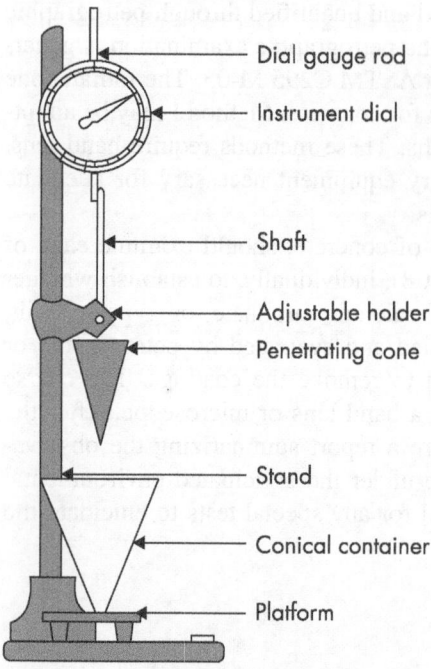

Dial gauge rod

Instrument dial

Shaft

Adjustable holder

Penetrating cone

Stand

Conical container

Platform

**Fig. 25.15** Standard cone apparatus for consistency test

The working consistency of the mortar is usually judged by the mason during application of the mortar. The quantity of water needed for maintaining consistency will also depend upon the masonry to which the mortar is used. Thinner joints will require greater fluidity but bed joints subject to heavy pressure require stiffer mortar. In addition, the mortar should be able to hold the water against suction by the masonry unit, particularly in the case of burnt clay and concrete blocks.

The consistency of mortars is measured as per Appendix B of IS 2250:1981. Accordingly, a conical container (150 mm in height, and diameter at base of 75 mm) is filled with mortar mix, to a level that is 10-mm below its rim. Mortar mix should be placed in the mould in one continuous operation and compacted by tamping rod. The mould filled with mortar mix is bumped five or six times over a flow table. The container is then placed over the base below the penetration cone of the apparatus, as shown in Fig. 25.15. The apex of the penetrating cone is brought first in contact with the surface of the mortar and the cone is clamped in position. The instrument dial is now in contact with the cone. The cone is then released and allowed to sink into the mortar mix. After the cone has stopped penetrating into the mortar, the dial is once more set to record the position of the cone and the difference between the two noted dial readings gives the depth of penetration of the cone into the mortar. The average of two determinations is considered as the consistency of the mortar.

In the field, the aforementioned procedure may be further simplified. The shaft of the cone is held by hand in a perpendicular position so as to be in contact with the surface of the mortar. It is then gently released to sink into the mortar taking care that the shaft remains vertical during penetration. The depth of penetration is computed from the measurement of the wetted depth along the surface of the cone.

The following values of depth of penetration are recommended by IS 2250:1981.

1. For laying walls with solid bricks: 90 to 130 mm
2. For laying perforated bricks: 70 to 80 mm
3. For filling cavities: 130 to 150 mm

## 25.11.3 Water Retention Test

Measurement of water retentivity may be necessary only in the case where mortar is to be used with masonry unit, which has got high suction characteristics. Water retentivity can be determined as per Appendix C of IS 2250:1981. The apparatus used for this test is shown in Fig. 25.16.

The complete unit consists of an aspirator pump (not shown in Fig. 25.16), vacuum regulator and vacuum gauge system, three-way stopcock, flask, rubber gasket, brass funnel, perforated brass dish, filter paper, and hardwood stand (see photo). Older models had a mercury manometer and a relief column in the place of vacuum regulator and vacuum gauge system. The vacuum regulator and vacuum gauge system are used to regulate the supplied vacuum pressure and are connected to a three-way stopcock to a funnel upon which rests a perforated dish. The perforated brass dish is used, since it will not be affected by masonry mortars. The metal dish has a diameter of about 154–156 mm, thickness of 1.7–1.9 mm, and has perforations as shown in Fig. 25.16(b). The bore of the stopcock as well as the inside diameter of connecting glass tubing is 4-mm diameter. A synthetic rubber gasket is permanently sealed to the top of the funnel. During the testing, this gasket is lightly coated with petrolatum/grease to ensure a seal between the funnel and the brass dish. While applying the grease, it should be ensured that it is not clogging the holes in the perforated brass dish. Hardened filter paper of grade Whatman No. 50 or equivalent should be used. The filter paper should lie flat and completely cover the bottom of the brass dish.

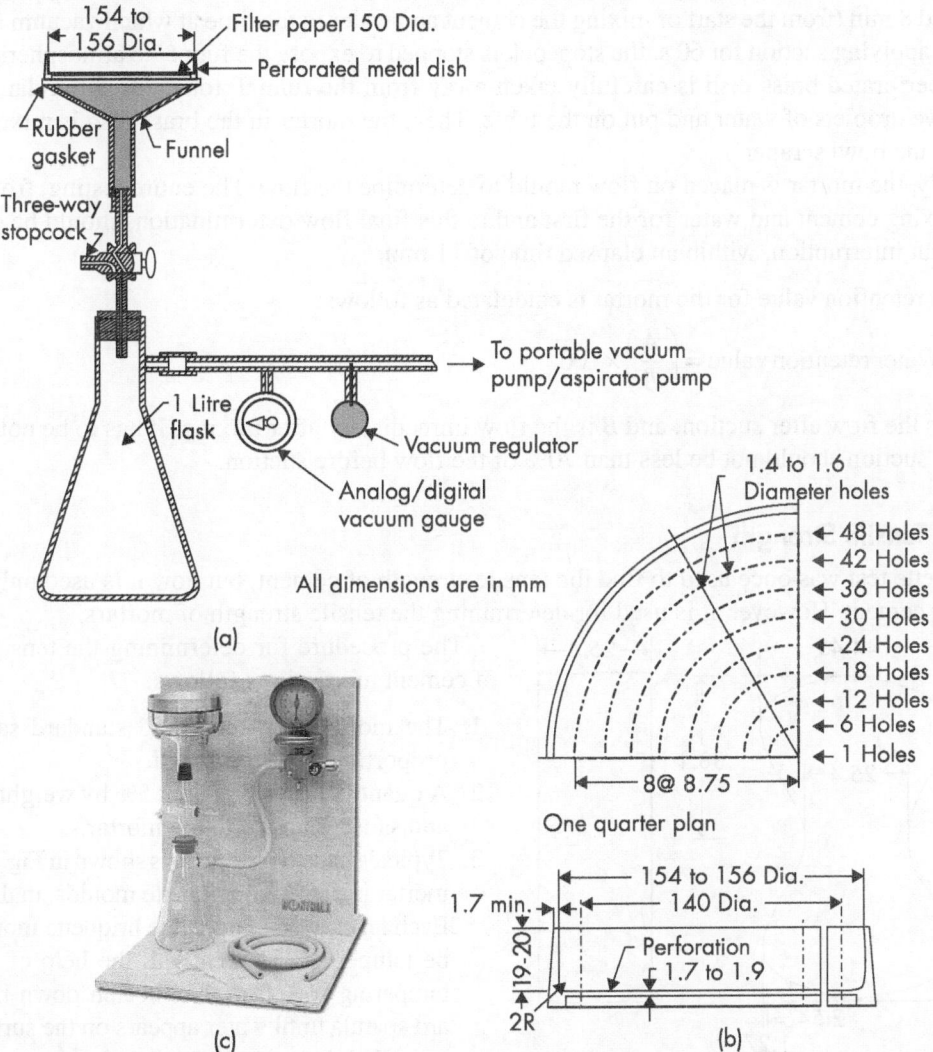

**Fig. 25.16** Apparatus assembly for water retention test (a) Test apparatus (b) Details of perforated metal dish (c) Photo of apparatus (*Courtesy*: Controls USA, Inc.)

The following procedure is adopted while conducting the test:

1. The vacuum regulator is adjusted such that it maintains a vacuum of 50 mm as measured on the vacuum gauge.
2. The perforated brass dish is then placed on top of the greased gasket of the funnel.
3. A wet filter paper is now placed at the bottom of the brass dish. The stopcock is turned to apply vacuum to the funnel: the apparatus has to be checked for any leaks to ensure that the required suction is applied. The stopcock is turned to shut off the vacuum.
4. The mortar is mixed to a consistency such that it will give a flow of 110 to 115 mm [The flow is determined as per IS 5512:1969 (see also Section 25.12.4).].
5. The perforated brass dish is filled with the mortar to a level slightly above the rim. The mortar is then tamped 15 times, at uniform spacing, with the tamper, and cut-off to a plane surface flush with the rim of the dish.
6. Now the stopcock is turned to apply vacuum to the funnel. Note that the time elapsed should not exceed 8 min (from the start of mixing the cement and water to the time at which vacuum is applied).
7. After applying suction for 60 s, the stopcock is stopped to expose the funnel to atmospheric pressure.
8. The perforated brass dish is carefully taken away from the funnel, touched with a damp cloth to remove droplets of water and put on the table. Then, the mortar in the brass dish is mixed for 15 s, using the bowl scraper.
9. Finally, the mortar is placed on flow mould to determine the flow. The entire testing, from the start of mixing cement and water for the first and to this final flow determination, should be carried out without interruption, within an elapsed time of 11 min.

The water retention value for the mortar is calculated as follows:

$$\text{Water retention value} = \left(\frac{A}{B}\right) \times 100 \qquad (25.30)$$

where $A$ is the flow after suction, and $B$ is the flow immediately after mixing. It has to be noted that the flow after suction should not be less than 70% of the flow before suction.

### 25.11.4 Tensile Strength

The briquette test was once used to find the tensile strength of cement, but now it is used only for rapid hardening cement. However, it is used for determining the tensile strength of mortars.

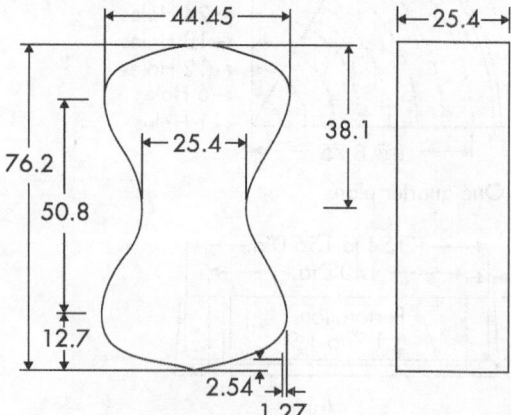

**Fig. 25.17** Standard cement-mortar briquette

The procedure for determining the tensile strength of cement mortar is as follows:

1. The mortar of cement and standard sand, in the proportion 1:3 is prepared.
2. A quantity of water, about 5% by weight of cement and sand, is added to the mortar.
3. Typical shape of briquette is shown in Fig. 25.17. The mortar is placed in briquette moulds, in three layers. Each layer, when laid in the briquette mould, should be tampered 25 times with the help of a standard tampering rod. It is then beaten down by a standard spatula until water appears on the surface. Same procedure is repeated for the other face of briquette. Twelve such standard briquettes are prepared.

4. The briquettes are then kept in a damp cabin for 24 h, for the initial and final setting to take place. The briquettes are then carefully removed from the moulds and submerged in clean water for curing.
5. The briquettes are tested in tensile testing machine at the end of 3 days and 7 days. Six briquettes are tested in each test and the average is found out. During the test, the load is applied uniformly at the rate of 3.50 N/mm$^2$.
6. It may be noted that cross-sectional area of briquette at its least section is 645 mm$^2$. Hence, the ultimate tensile stress of cement paste is obtained as follows:

   Ultimate tensile stress = Failure load/645

7. The tensile stress at the end of 3 days should not be less than 2 N/mm$^2$ and that at the end of 7 days should not be less than 2.50 N/mm$^2$

### 25.11.5 Test for Adhesion

Bond strength of masonry may be defined as 'adhesion perpendicular to the load between the masonry mortar and the masonry unit'. Bond strength is required to withstand tensile forces due to wind, structural and other applied forces, movement of the masonry units, and temperature changes. In general, the higher the cement content, the greater will be the bond strength. Excessively high air contents reduce bond at the brick interface.

**Couplet tests**   This test consists of two brick units connected by only one mortar joint [Fig. 25.18(a)]. The test specimen is tested in universal testing machines after curing it for 28 days. It is the simplest specimen and at the same time, the most economic, taking into account the number of bricks required. Several variations of this method are discussed in Almeida et al. (2002). A similar test, called the *crossed brick couplet test,* is included in the American Standard C952-02, which is performed by applying compressive loads in the upright bars as shown in Fig 25.18(b). The bond strength of the couplet is calculated based on the failure load and bond area between brick and mortar (Konthesingha et al., 2007).

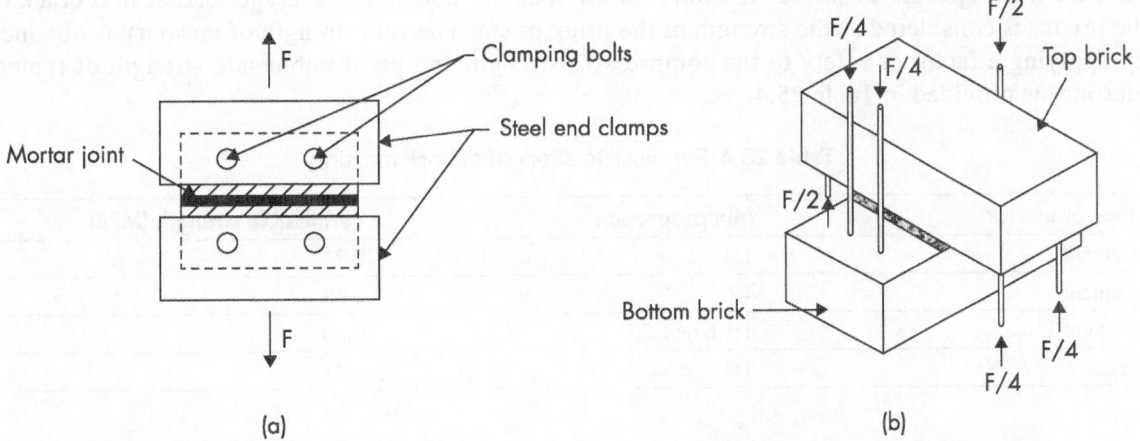

**Fig. 25.18** Couplet tests for adhesiveness of mortar (a) Couplet test using special clamps
(b) Crossed brick couple test

More information on various tests conducted on mortars may be found in Cemex Guide (2010), BS EN 1015 (Parts 1–21), IS 3085:1965, and BS 4551 (Parts 1 and 2).

### 25.11.6 Test for Brickwork in Compression

If a brick wall made from bricks and mortar is tested under compression, the wall will not behave like a brick subjected to compression. It is because the bricks bedded on the mortar behave like a beam on elastic foundation. Thus, the failure of the wall will be due to the failure of bricks in tension, which act like small beams resting on the mortar bed. Thus, the wall cracks at a stress level equal to the tensile strength of bricks, which is approximately one-tenth of the compression strength of bricks. This clearly shows that the required strength of mortar is only to bond the bricks together. In general, it should not be greater than the strength of bricks used in the construction. In fact, even if the strength of mortar is greater than the strength of bricks, the compressive strength of brickwork will not be more than one-tenth of the strength of bricks. Mortars stronger than masonry units tend to cause concentrated and wider shrinkage and settlement cracks, whereas weaker mortars will result in distributed joint cracks. According to European standard EC:8, mortars used in the construction of plain and confined masonry, situated in earthquake zones, should have a minimum compressive strength of 5 MPa.

In order to find the compressive strength of masonry, prisms of brickwork (minimum three) should be made. The thickness of the prisms should be the same as the thickness of the masonry part of the wall in the structure. The length of the prism should be equal to or greater than the thickness of the prism. The height of the prism should be at least twice the thickness but not more than 5 (If the *h/t* ratio of the prisms tested is less than 5 in case of brickwork and more than 2 in case of blockwork, compressive strength values indicated by the tests should be corrected by multiplying with the factor as per Table 12 of IS 1905:1987), contain at least two mortar joints, and be a minimum of 400 mm. The mortar joints should be 10 mm in thickness, and the head joints should be fully filled with mortar. Tests carried out by Fishburn (1961), Francis et al. (1970), and Thaickavil and Thomas (2018) showed that the prism compressive strength is a function of average joint thickness and reduces as the joint thickness increases. After curing the specimens for 28 days, the prisms should be tested in compression testing machine at the rate of 350 to 700 kN/m. The load at which the first crack appears as well as the final failure load are noted. The average load at first crack of the prisms is considered as the strength of the brick prism. The safe strength of masonry is obtained by applying a factor of safety to the compressive strength and permissible safe strength of typical masonry is provided in Table 25.4.

**Table 25.4** Permissible strength of brick masonry

| Type of mortar | Mix proportion | Permissible strength (MPa) |
| --- | --- | --- |
| Cement | 1:3 | 0.75 |
| Cement | 1:6 | 0.45 |
| Cement-Lime | 1:1:6 or 1:2:9 | 0.50 |
| Lime | 1:3 | 0.45 |

# Part II

## 25.12 TESTS ON FRESH CONCRETE

The workability of concrete should be checked frequently by one of the standard tests (slump, compacting factor, Vee-Bee consistency, or flow table) as described in IS 1199 (Parts 1–7):2018

### 25.12.1 Slump Test

Although it does not measure all factors contributing to workability, slump test is the most commonly used method to measure the consistency of the concrete, because of its simplicity. This test is carried out using an open ended cone, called the *Abrams cone or slump mould*. The following procedure is used to determine the slump of fresh concrete:

1. The internal surface of the Abrams cone (Fig. 25.19) is thoroughly cleaned and freed from superfluous moisture and any set concrete before commencing the test.
2. This cone is placed on a smooth, horizontal, rigid, and non-absorbent surface.
3. Then, it is filled with fresh concrete in four layers, each approximately one fourth of the height of the mould.
4. Each layer of concrete is tamped 25 times with the rounded end of the 0.6-m long tamping rod of 16-mm diameter. The strokes are distributed uniformly over the cross section and for the second and subsequent layers should penetrate into the underlying layer. The bottom layer is tamped throughout its depth.
5. At the end of the fourth stage, the concrete is struck off level with a trowel at the top of the mould, so that the mould is exactly filled.
6. Now, the mould is carefully lifted vertically upwards, without disturbing the concrete in the cone, thereby allowing the concrete to subside.
7. The *slump* is measured immediately by determining the difference between the height of the mould and that of the highest point of the subsided concrete (see Fig. 25.19). The slump measured is recorded in terms of millimetres of subsidence of the specimen.

Figure 25.19 shows the slump testing mould, measurement of slump, and types of slumps. If a shear slump (indicates concrete is non-cohesive) or collapse slump (indicates a high workability mix) is achieved, a fresh sample should be taken and the test repeated. A slump of about 50–100 mm is used for normal reinforced concrete (see Table 8.5 of Chapter 8). Too high or too low a slump gives immediate warning and enables the mix operator to remedy the situation. It has to be noted that the slump test is unreliable for lean mixes.

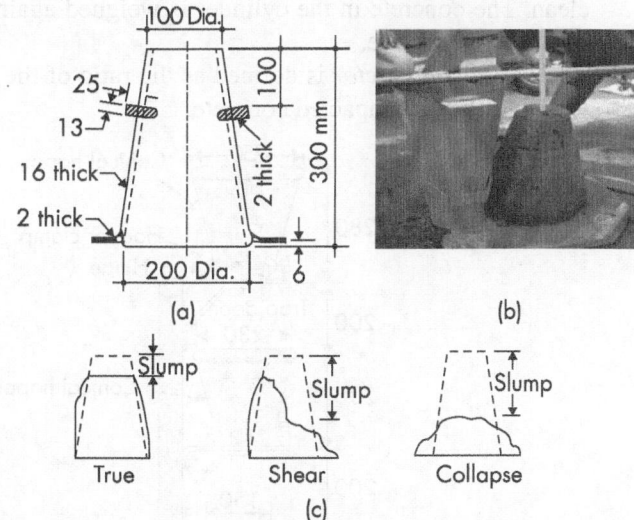

**Fig. 25.19** Slump testing (Subramanian, 2013) (a) Typical mould for slump test (b) Measuring slump (c) Types of slump

## 25.12.2 Compacting Factor Test

This test is more precise and sensitive than the slump test. It is particularly useful for concrete mixes of very low workability and is normally used when concrete is to be compacted by vibration; such dry concretes are ill-suited to slump test. This test is designed primarily for use in the laboratory, but can also be used in the field. When concrete contains aggregates greater than 38-mm size, the concrete should be wet sieved through 1.5 in screen to exclude aggregate particles bigger than 38 mm.

The apparatus used for compaction factor test is shown in Fig. 25.20. The following procedure is used to determine workability of fresh concrete using this apparatus (IS 5515:1983):

1. The sample of concrete to be tested is placed gently in the upper hopper using a hand scoop, and filled up to the brim. The top hopper has a top internal diameter of 250 mm, bottom internal diameter of 125 mm, and an internal height of 280 mm.
2. The trap-door is opened to allow the concrete to fall into the lower hopper, which has a top internal diameter of 230 mm, bottom internal diameter of 125 mm, and an internal height of 230 mm. Certain types of mixes may have the tendency to stick in one or both of the hoppers. If this happens, the concrete may be pushed gently using a rod. During this process, the bottom cylinder having an internal diameter of 150 mm and internal height of 300 mm should be covered.
3. Immediately after the concrete has come to rest, the cylinder is uncovered, and the trap-door of the lower hopper is opened, and the concrete is allowed to fall into the cylinder.
4. The excess of concrete remaining above the top level of the cylinder is then cut off using trowels.
5. The weight of the concrete in the cylinder is then determined to the nearest 10 g. This is known as weight or partially compacted concrete.
6. The cylinder is then refilled with concrete from the same sample in layers of approximately 50 mm, and preferably vibrated to obtain full compaction. The top surface of the fully compacted concrete is carefully struck off level with the top of the cylinder. The outside of the cylinder is then wiped clean. The concrete in the cylinder is weighed again. This weight is known as the weight of fully compacted concrete.
7. The *compacting factor* is defined as the ratio of the weight of partially compacted concrete to the weight of fully compacted concrete.

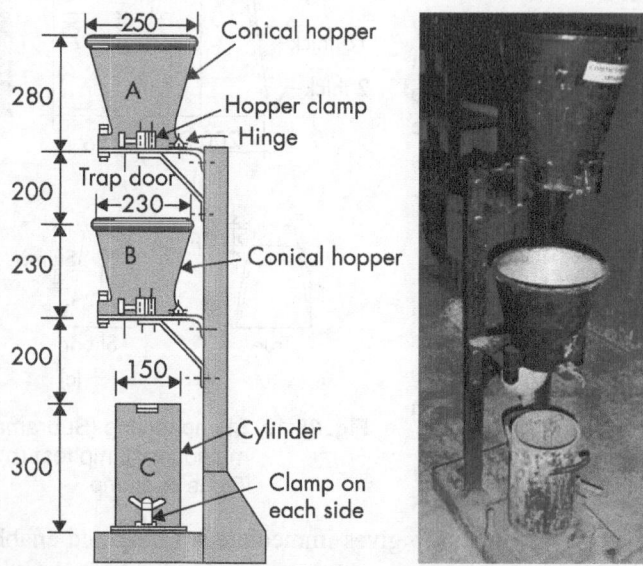

**Fig. 25.20** Compacting factor apparatus (*Photo Courtesy*: Prof. Manu Santhanam, IITM)

### 25.12.3 Vee-Bee Consistometer Test

The consistency of concrete may be determined using a *Vee-Bee Consistometer* (see Fig. 25.21), which determines the time required for transforming, by vibration, a concrete specimen in the shape of a conical frustum into a cylinder. Vee-Bee test is used for concrete with low workability. The Vee-Bee consistometer consists of a 380-mm long and 260-mm wide vibrator table (G) that is supported on rubber shock absorbers at a height of 300 mm above floor level (IS 10510:1983). The table is mounted on a base (K), which rests on three rubber feet and is equipped with an electrically operated vibrometer mounted under it, operating on either 65 or 220 volts three phase, 50 cycles alternating current. A sheet metal slump cone (B) open at both ends is placed in the metal container (A) and the metal container is fixed on to the vibrator table by means of two wing-nuts (H). The sheet metal slump cone is 300-mm high with a bottom and top diameter of 200 mm and 100 mm, respectively. A swivel arm holder (M) is fixed to the base and another telescopic swivel arm (N) is fixed into it by means of a set screw (F). A funnel (D) and a guide-sleeve (E) are connected to the telescopic swivel arm (N). The swivel arm can be readily detached from the vibrator table. The graduated rod (J) is fixed on to the swivel arm (N) and at the end of the graduated arm a glass disc (C) is screwed. The division of the scale on the rod records the slump of the concrete cone in centimetres and the volume of concrete after vibration of the cone in the metal container. The standard iron rod is 20 mm in diameter and 500 mm in length.

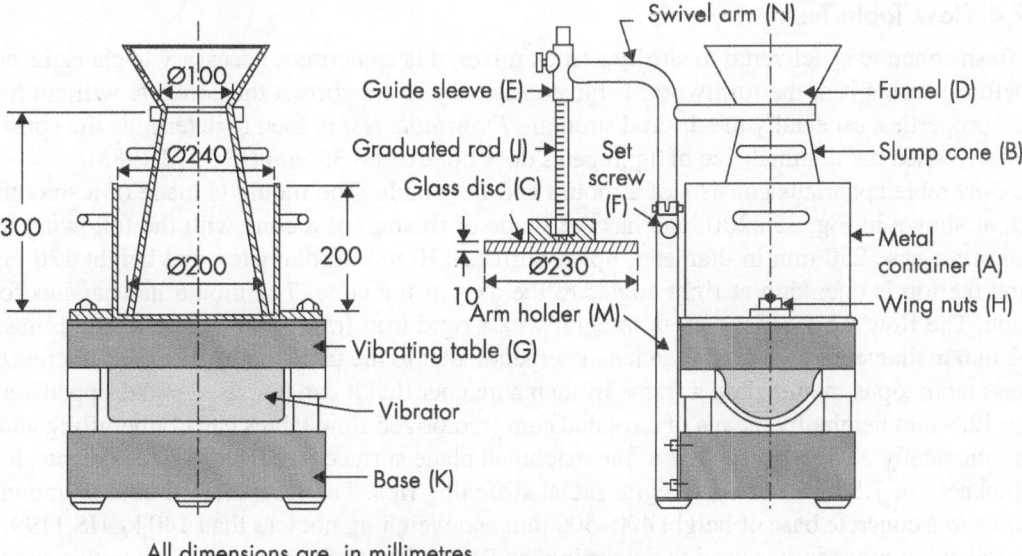

All dimensions are in millimetres

**Fig. 25.21** Vee-Bee consistometer as per IS 10510:1983 (*Photo Courtesy*: Prof. Manu Santhanam, IITM)

The following procedure is used to determine workability of fresh concrete using this apparatus:

1. A conventional slump test is performed, placing the slump cone inside the cylindrical metal container of the consistometer.
2. The glass disc (C) attached to the swivel arm is turned and placed just on the top of the slump cone in the metal container and before the cone is lifted up, the position of the concrete cone is noted by adjusting the glass disc attached to the swivel arm.
3. The cone is lifted up and the slump is noted on the graduated rod by lowering the glass disc on top of the concrete cone.
4. The electrical vibrator is switched on and the concrete is allowed to spread out in the metal container. At the same time, a stopwatch is started simultaneously.
5. The vibration is continued until the conical shape of the concrete disappears and the concrete assumes a cylindrical shape.
6. When the concrete fully assumes cylindrical shape, the stopwatch is switched off and the time taken is noted; the time is recorded in seconds. The consistency of the concrete is expressed in VB degree, which is equal to the recorded time in seconds. The relation between workability, slump, compaction factor, and VB degree is given in Table 8.3 of Chapter 8.

### 25.12.4 Flow Table Test

When fresh concrete is delivered to site by a truck mixer, it is sometimes necessary to check its consistency before pouring it in the formwork. If the consistency is not correct the concrete will not have the required properties, especially the desired strength. *Flow table test* is used to determine the consistency of concrete, when the nominal size of aggregates does not exceed 38 mm (IS 5512:1983).

The flow table apparatus consists of a mould and flow table. The mould is made of a smooth brass casting, as shown in Fig. 25.22(a), and has the shape of frustum of a cone with the following internal dimensions: a base 250 mm in diameter, upper surface 170 mm in diameter, and height 120 mm; the base and the top is open and at right angles to the axis of the cone. The mould has handles for easy operation. The flow table consist of an integrally cast rigid iron frame and a circular rigid brass table top 762 mm in diameter, with a shaft attached perpendicular to the table top by means of a screw thread. The brass table top is mounted on a frame in such a manner that it can be raised and dropped vertically through 12.5-mm height, by means of a rotated cam (motorized flow tables can do the lifting and dropping automatically). The table top has a fine machined plane surface free from surface defects; it has an edge thickness of 7.5 mm and six integral radial stiffening ribs. The flow table should be mounted on and bolted to a concrete base of height 400–500 mm and weighing not less than 140 kg (IS 1199:2018).

The following procedure is used to determine the flowability of fresh concrete using this apparatus:

1. Before commencing the test, the table top and inside of the mould is wetted and cleaned of all gritty material and the excess water is to be removed with a rubber squeezer.
2. The mould is held firmly at the centre of the table and filled with concrete in two layers, each approximately one-half the volume of the mould. Each layer is rodded with 25 strokes with a 16-mm diameter tamping rod (having a length of 60 mm and rounded at the lower tamping end). The strokes are distributed in a uniform manner over the cross section of the mould and should penetrate into the underlying layer. The bottom layer should be rodded throughout its depth.
3. After the top layer has been rodded, the surface of the concrete is to be struck off with a trowel so that the mould is exactly filled. The overflowed excess concrete on the table should be removed and the entire table outside the mould should be cleaned again.
4. The mould is then removed from the concrete by a steady upward pull.

5. The table is then raised and dropped by 12.5 mm, 15 times in about 15 s.
6. The diameter of the spread concrete is measured at six symmetrically distributed locations and the average of these measurements, to the nearest 5 mm, is taken as the spread.

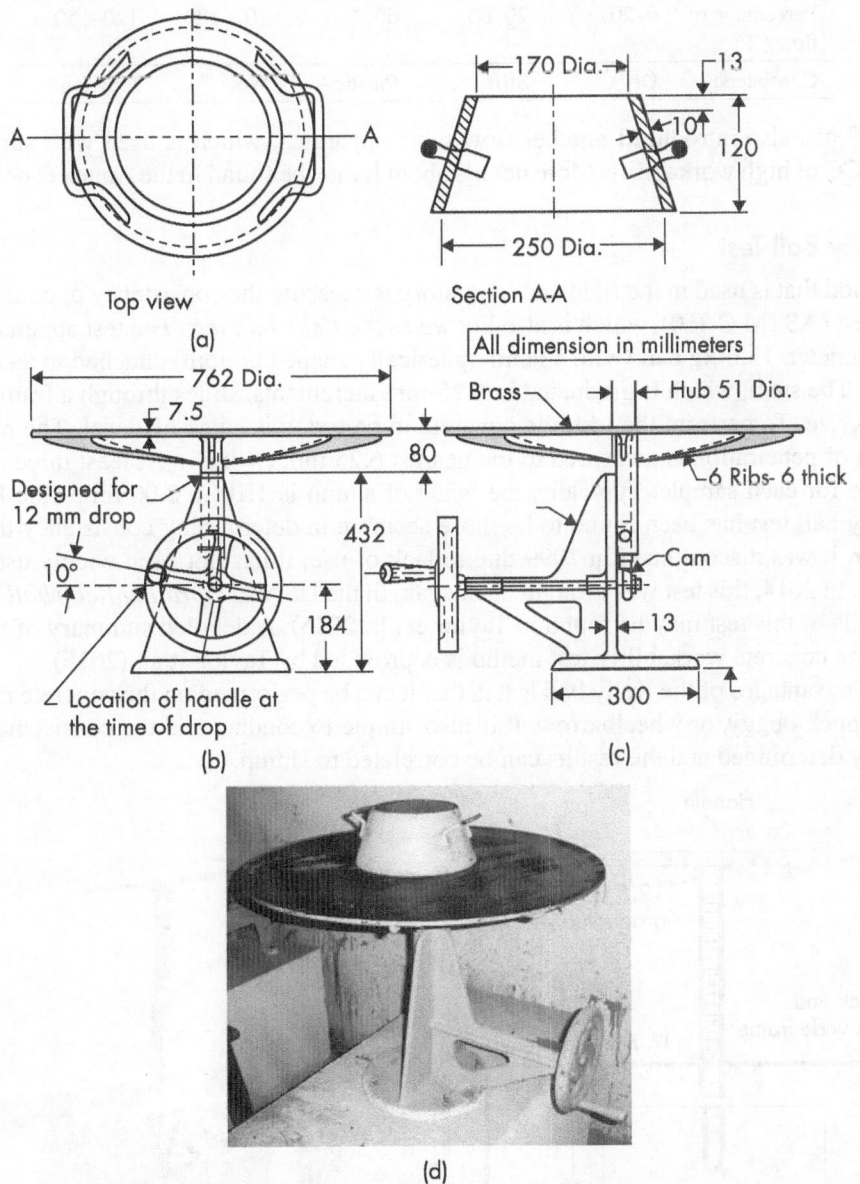

Top view

(a)

170 Dia.

13

10

120

250 Dia.

Section A-A

All dimension in millimeters

762 Dia.

7.5

Designed for 12 mm drop

10°

Location of handle at the time of drop

80

432

184

(b)

Brass

Hub 51 Dia.

6 Ribs- 6 thick

Cam

13

300

(c)

(d)

**Fig. 25.22** Flow table apparatus as per IS 1199:2018 (*Photo Courtesy*: Prof. Manu Santhanam, IITM)
(a) Mould for flow-table test (b) Front view of flow-table (c) Side view of flow-table (d) Photo

The flow of the concrete is now calculated as the percentage increase in diameter of the spread concrete over the base diameter of 250 mm. It is calculated from the following formula:

$$\text{Flow} \, (\%) = \frac{\text{Spread diameter in mm} - 250}{250} \times 100 \qquad (25.31)$$

Table 25.5 shows the relation between the percentage of flow and consistency of concrete.

**Table 25.5** Relation between percentage flow and consistency

| Percentage of flow (%) | 0–20 | 20–60 | 60–100 | 100–120 | 120–150 |
|---|---|---|---|---|---|
| Consistency | Dry | Stiff | Plastic | Wet | Flowing |

Note that BIS has also introduced another flow table apparatus, which is used with super-plasticized concrete or SCC of high workability. More details about it may be found in the Annex C of IS 9103:1999.

## 25.12.5 Kelly Ball Test

Another method that is used in the field and laboratory to measure the consistency of concrete is the ball penetration test (ASTM C 360), which is also known as the *Kelly ball test*. The test apparatus consists of a 150-mm diameter, 13.6-kg ball (with a hemi-spherically shaped bottom) attached to a stem, as shown in Fig. 25.23. The stem, which is graduated in 6.25-mm increments, slides through a frame that rests on the fresh concrete. To perform the test, the concrete to be tested is stuck off level. The ball is released and the depth of penetration is measured to the nearest 6.25 mm (¼ inch). At least three measurements must be made for each sample. Typically, the value of slump is 1.10 to 2.00 times the Kelly ball test reading. Kelly ball test has been found to be more accurate in determining consistency than the slump test. However, it was discontinued in 1999 due to lack of use; it has not been widely used outside the United States. In 2014, this test was brought back again in the USA as *Vibrating Kelly ball* test or VKelly Test; the details of this test may be found in Taylor et al. (2015). A detailed summary of the features of several existing concrete workability test methods is provided by Taylor et al. (2015).

One of the advantages of the Kelly ball test is that it can be performed on the concrete in any container such as hopper, buggy, or wheelbarrow. It is also simple to conduct and the consistency of concrete can be rapidly determined and the results can be correlated to slump.

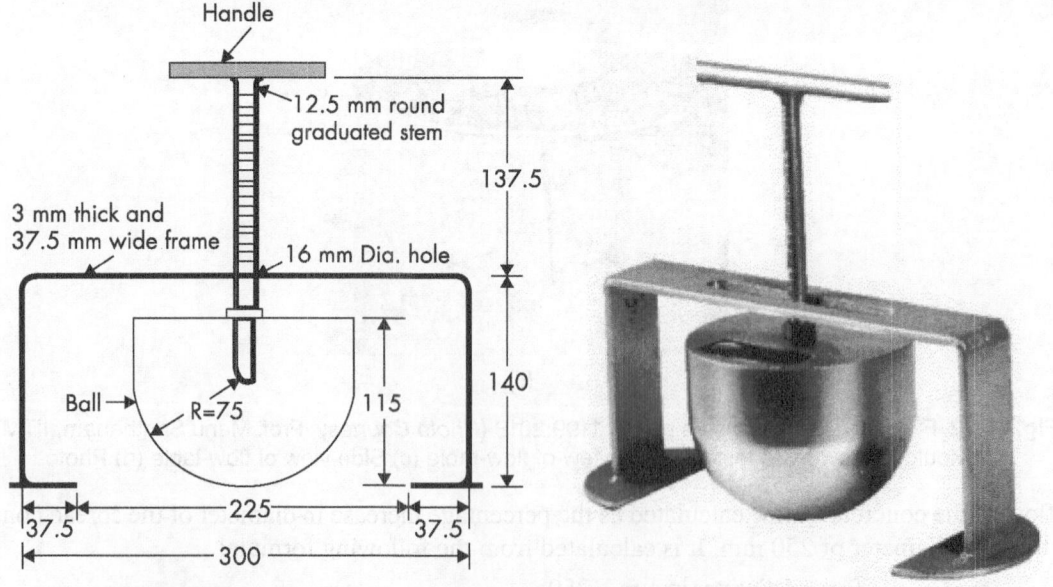

**Fig. 25.23** Kelly ball test apparatus (*Image Courtesy*: Humboldt Mfg Co.)

### 25.12.6 Other Tests on Freshly Mixed Concrete

IS 1199:2018 also describes the procedures for the determination of the proportions of the constituents, determination of the weight per cubic meter, determination of the air content of freshly mixed concrete by the pressure method (also see IS 9799:1981).

## 25.13 Tests on Hardened Concrete

Tests may be conducted on hardened concrete for determining the compressive strength, flexural strength, and tensile strength. These tests are described in this section (IS 516 (Part 1–4):2018).

### 25.13.1 Compression Tests (on Cylinder/Cube or Core Samples)

In India, the UK, and several European countries, the characteristic compressive strength of concrete (denoted by $f_{ck}$) is determined by testing to failure 28-day old concrete cube specimens of size $150 \times 150 \times 150$ mm, as per IS 516:2018). When the largest nominal size of aggregate does not exceed 20 mm, 100-mm cubes may also be used. However, in the USA, Canada, Australia, and New Zealand, the compressive strength of concrete (denoted by $f_c'$) is determined by testing to failure 28-day old concrete cylinder specimens of size 150-mm diameter and 300-mm long. Recently, 70-mm cube or 75-mm cylinder high strength concrete (HSC)/ultra-high strength concrete (UHSC) specimen is being recommended for situations in which machine capacity may be exceeded (Graybeal and Davis, 2008). Compression tests can also be carried out on cylindrical cores taken from actual concrete structures.

The concrete is poured in the mould in layers of 50 mm and compacted properly either by hand or vibrator, so that there are no voids. The top surface of these specimens are made even and smooth, by putting cement paste and spreading smoothly on whole area of specimen. The test specimens are then stored in moist air of at least 90% relative humidity and at a temperature of $27 \pm 2°C$ for 24 h. After this period, the specimens are marked and removed from the moulds and kept submerged in clear fresh water, maintained at a temperature of $27 \pm 2°C$ until they are tested (the water should be renewed every seven days). The making and curing of test specimen at site is similar (see also Clause 3.0 of IS 516).

These specimens are tested by compression testing machine after 7 or 28 days curing. Specimens stored in water shall be tested immediately on removal from the water, and while they are still in the wet condition. Load should be applied gradually at the rate of 14 N/mm$^2$ per minute until the specimen fails. Load at the failure divided by the area of specimen gives the compressive strength of concrete. Minimum three specimens, preferably from different batches, should be tested at each selected age. If strength of any specimen varies by more than ±15% of average strength, results of such specimen should be rejected (see Clause 15.4 of IS 456). The average of *three* specimens gives the compressive strength of concrete. Sampling and acceptance criteria for concrete strength, as per IS 456 are provided in Section 25.16 (In the USA, the evaluation of concrete strength tests is done as per ACI 214R-11). Figure 25.24 shows the cube testing and various failure modes of concrete cubes. The ideal failure mode, with almost vertical cracks [see Fig. 25.24(b)] is rarely achieved due to the rough contact surface between the concrete cube and the plate of testing machine. When the stress level reaches about 75–90% of the maximum, internal cracks are initiated in the mortar throughout the concrete mass, parallel to the direction of the applied load. The concrete tends to expand laterally due to Poisson's effect, and the cube finally fails leaving two truncated pyramids one over the other [see Fig. 25.24(b)]. Sometimes, the failure may be explosive, especially in cubes of HSC; to avoid injuries, proper precautions should be taken to contain the debris using high resistance and transparent polycarbonate or steel mesh shields around the testing machine.

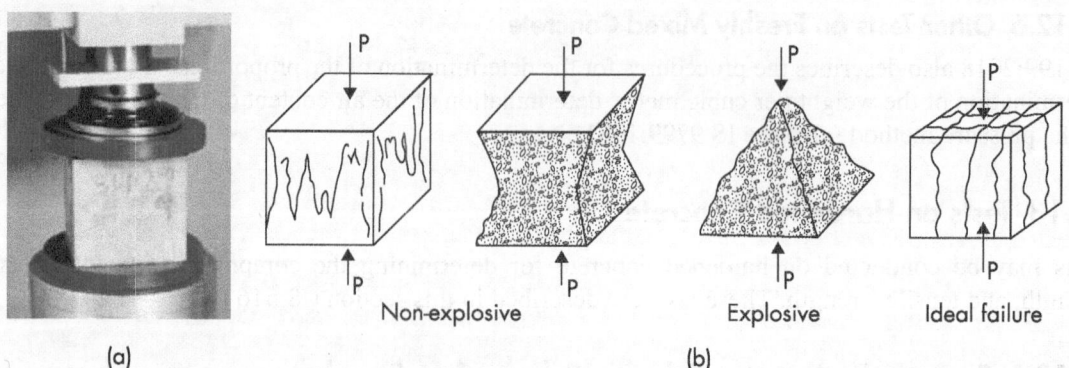

**Fig. 25.24** Cube testing and failure of concrete cubes (Subramanian, 2013) (a) Cubes in testing machine (b) Failure of concrete cubes

It has to be noted that when cubes are tested as per the standard compressive test, a lateral frictional restraint exists between the loading platen and the end of the specimen, and this restraint increases proportionally as the axially applied monotonic compressive load goes up. This frictional restraint will act as a lateral confinement at the specimen end, thus unavoidably enhancing the 'compressive strength' of the target specimen. Owing to the existence of such an end friction, the typical compressive failure of the standard specimen exhibits an obvious cone failure characteristic, and only a small central portion of the cylinder is in true uniaxial compression, the remainder being in a state of triaxial stress [see Fig. 25.24(b)]. Consequently, the standard specimen test tends to overestimate the compressive strength of the concrete cube/cylinder. To obtain true uniaxial properties of the concrete, end friction-reducing measures such as using brush platen, or introducing sandwich of thin teflon sheet, bearing grease, and aluminium foil between the loading platen and the specimen must be undertaken.

## 25.13.2 Flexure Test (Equivalent Cube Method)

Tensile strength is one of the basic and important properties of concrete because concrete structures are prone to tensile cracking due to the applied loads or other environmental effects. Tensile strength is required for the design of concrete structural elements subject to transverse shear, torsion, shrinkage, and temperature effects. Its value is also used in the design of prestressed concrete structures, liquid retaining structures, roadways, and runway slabs. It has to be realized that tensile strength of concrete is very low compared to its compressive strength. Hence, it is difficult to conduct uniaxial tension test on concrete specimens. Owing to this, indirect test methods such as (a) split cylinder test or (b) flexural test are used to determine the tensile strength of concrete. The split cylinder test is discussed in Section 25.13.3.

As previously mentioned, the flexural test is performed to determine the tensile strength of concrete. The test has to be performed as per IS 516:2018 and IS 9399:1979. Any reliable compression machine can be used. The bed of the testing machine is provided with two steel rollers, 38 mm in diameter, on which the specimen will be supported. These rollers are mounted in such a way that the distance from centre to centre is 600 mm for 150-mm specimens or 400 mm for 100-mm specimens. The load is applied through two similar rollers mounted at the third points of the supporting span, i.e., spaced at 200 or 133 mm centre-to-centre. The load should be divided equally between the two loading rollers, and all rollers should be mounted in such a manner that the load is applied axially and without subjecting the specimen to any torsional stresses or restraints. One such arrangement is shown in Fig. 25.25.

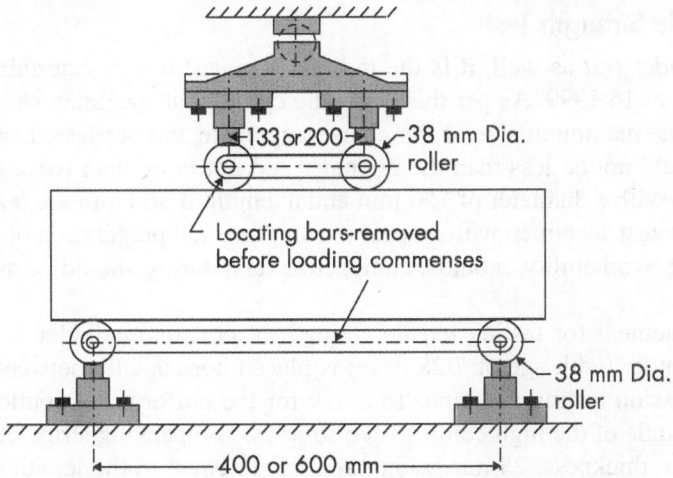

**Fig. 25.25** Test setup for flexural test

Normal standard size of concrete beam specimen is $150 \times 150 \times 750$ mm. The span is usually fixed as four times its depth. Before testing, test specimens are to be stored in water at a temperature of 24–30°C for 48 h, and tested immediately on removal from the water whilst they are still in a wet condition, but wiped to remove surface moisture. The procedure of testing is as follows:

1. Place the specimen in the testing machine as shown in Fig. 25.25, i.e., the specimen should be placed in such a manner that the load is applied to the uppermost surface as cast in the mould, along two lines spaced 133 to 200 mm. The axis of the specimen should be carefully aligned with the axis of the loading device. No packing should be used between the bearing surfaces of the specimen and the rollers.

2. Apply the load without shock and increasing continuously at a rate that increases the extreme fibre stress approximately at 0.7 N/mm²/min, i.e., at a rate of loading of 4 kN/min for the 150-mm specimens and at a rate of 1.8 kN/min for the 100-mm specimens. The load has to be increased until the specimen fails, and peak load at failure is noted.

3. Calculate the flexural strength of the specimen, expressed as the modulus of rupture $f_b$, as follows:
   (a) When '$a$' is greater than 200 mm for 150-mm specimen, or greater than 133 mm for a 100-mm specimen,

$$f_b = \frac{Pl}{bd^2} \qquad (25.32a)$$

where $P$ is the maximum load at failure, $l$ is the span length, $b$ and $d$ are the width and depth of the beam respectively, and '$a$' equals the distance between the line of fracture and the nearer support, measured on the centre line of the tensile side of the specimen, in mm.

(b) When '$a$' is less than 200 mm but greater than 170 mm for 150-mm specimen, or less than 133 mm but greater than 110 mm for a 100-mm specimen,

$$f_b = \frac{3Pa}{bd^2} \qquad (25.32b)$$

If '$a$' is less than 170 mm for a 150-mm specimen, or less than 110 mm for a 100-mm specimen, discard the results of the test.

### 25.13.3 Split Tensile Strength Test

Named the *split cylinder test* as well, it is the indirect standard test to determine the tensile strength of concrete, as per IS 5816:1999. As per this code, the cylindrical specimen should have diameter not less than four times the maximum size of the coarse aggregate and not less than 150 mm. The length of the specimens should not be less than the diameter and not more than twice the diameter. Usually, cylindrical specimens with a diameter of 150 mm and a length of 300 mm are tested. The procedure of making and curing the test specimen with respect to sampling and preparation of materials, proportioning, weighing, mixing, workability, moulds, compacting, and curing should be as per the requirements of IS 516:2018.

The general arrangement for testing tensile strength of concrete cylinder is shown in Fig. 25.26. The test concrete cylinder (with age of 7/28 days) is placed horizontally between the loading surfaces of a standard compression testing machine. To allow for the uniform distribution of the applied load and reduce the magnitude of the high compressive stresses, two packing strips of tempered hardboard/steel of nominal 4-mm thickness, 25-mm width, and length equal to the length of cylinder are placed between the specimen and loading platens of the testing machine. The load is applied without shock and increased continuously at a uniform rate [1.2 to 2.4 N/mm²/min] until the specimen splits, and no greater load is sustained. Concrete cylinders split into two halves along the vertical plane due to indirect tensile stress generated by Poisson's effect. The maximum load applied to the specimen is recorded. At least three specimens shall be tested for each age of tests.

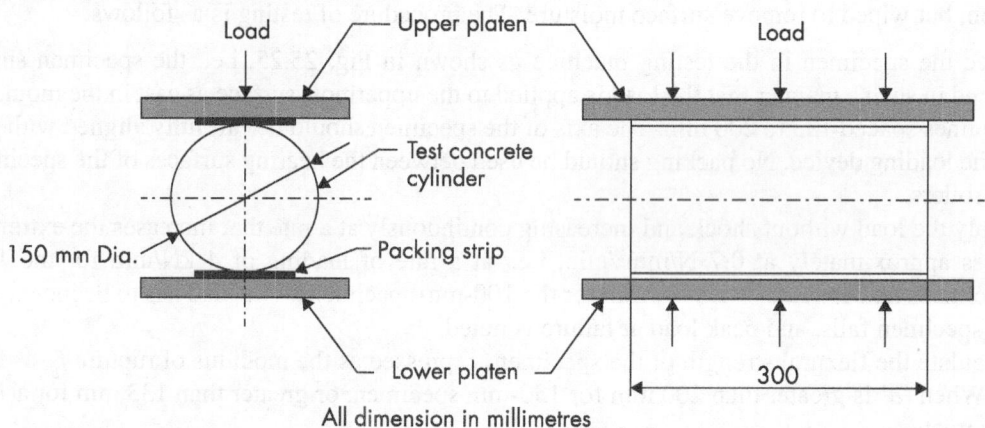

Load — Upper platen — Load

Test concrete cylinder

150 mm Dia. — Packing strip

Lower platen — 300

All dimension in millimetres

**Fig. 25.26** Split tensile strength test

The measured splitting tensile stress $f_{ct}$ of the specimen is calculated using the following formula:

$$f_{ct} = \frac{2P}{\pi l d} \tag{25.33}$$

where $P$ is the maximum load in Newton applied to the specimen, $l$ is the length of the cylinder, and $d$ is the diameter of the cylinder. It has to be noted that the tensile strength found out from this test varies from 1/8th to 1/12th of the cube compressive strength.

### 25.13.4 Testing for Elastic Modulus

The test specimens shall consist of concrete cylinders 150 mm in diameter and 300-mm long (IS 516:2018). The test specimens should be prepared as described for the cube test (see Section 25.13.1) and should be stored in water at a temperature of 24 to 30°C for at least 48 h before testing. At least three specimens should be tested. They should be tested when they reach the age of 28 days. A 2000 kN compression testing machine is required. Two extensometers are required each having a gauge length of not less than 102 mm and not more than half the length of the specimen. They should be capable of measuring strains to an accuracy of $2 \times 10^{-6}$. The procedure for this test is as follows:

**Fig. 25.27** Compressometer (*Photo Courtesy*: Humboldt Mfg. Co)

1. Test three cylindrical specimens for compressive strength as described in Section 25.13.1 and record the average compressive strength $f_{cu}$ to the nearest 0.5 MPa.
2. Use a compressometer to determine the deformation and strains on the concrete cylinder specimen. It consists of a frame with a bottom ring and a top ring with hardened and tapered tightening screws to firmly clamp the compressometer to the cylinder specimen (see Fig. 25.27). Two spacers hold the two frames in position. A dial gauge with a range of 12 mm and readable to 0.002 mm is mounted on the upper ring and the tip of the dial gauge should rest on the anvil.
3. Apply proper capping to the top of cylinders.
4. Place the concrete cylinder upright. Unscrew the contact screws of the compressometer until the points are flush with the inside surface of the rings.
5. Place the compressometer over the wet concrete specimen locating the specimen at the centre of the ring so that the tightening screws of the bottom and top frame are at equal distance from the two ends.
6. Tighten the screws so that the compressometer is held on the specimen.
7. Remove the spacers by unscrewing the spacer screws. Follow any other instruction supplied by the manufacturer of the compressometer.
8. Place the specimen with attached compressometer on the lower platen of the compression testing machine and centre it.
9. Apply load continuously without stock at a rate of 14 N/mm²/min until an average stress of $(C + 0.5)$ N/mm² is reached where $C$ is one third of the average compressive strength of cubes calculated to the nearest 0.5 N/mm².
10. Maintain the load at this stress for at least 1 min and reduce it gradually to an average stress of 0.15 N/mm² and note the readings of the compressometer.
11. Apply the load again a second time at the same rate until an average stress of $(C + 0.15)$ N/mm² is reached. Note the readings of the compressometer while the load is maintained at this level. The load shall again be reduced gradually and readings again taken at 0.15 N/mm².
12. Apply the load for the third time and note the readings of the compressometer at 10 approximately equal increments of stress up to an average stress of $(C + 0.15)$ N/mm².
13. Calculate the stress and strains for each cylinder as follows:

$$\text{Stress}, f = P/A \tag{25.34a}$$

where $P$ is the applied load and $A$ is the area of cross section of the cylindrical specimen.

$$\text{Stain}, \varepsilon = d/L_o \tag{25.34b}$$

where $d$ is the longitudinal deformation of specimen obtained from the dial gauge readings and $L_o$ is the gauge length, the distance between rings, and typically equal to 200 mm for 150-mm diameter specimens.

14. After loading to 35% of ultimate compressive cube strength and recording the load versus displacement data, unload the specimen. Remove the compressometer.
15. Perform an unconfined compression test in accordance with the codal provisions.
16. Plot the stress-strain curve (stress on the ordinate and strain on the abscissa).
17. Calculate the *secant elastic modulus, E,* at 35% of cube strength as (see also Section 8.8.6 of Chapter 8).

$$E = \frac{f_2 - f_1}{\varepsilon_2 - 0.00005} \tag{25.35}$$

where $f_2$ is the stress corresponding to 35% of ultimate load, $f_1$ is the stress corresponding to a strain of 0.00005, and $\varepsilon_2$ is the strain at stress $f_2$.

## 25.14 Non-destructive Testing

*Non-destructive tests* (NDT) are used to find the strength of existing concrete elements. A number of non-destructive testing methods are available, which can be broadly classified as those which measure the overall quality of concrete (e.g., dynamic or vibration methods such as resonance frequency and ultrasonic pulse velocity tests); and those which involve the measurement of parameters such as surface hardness, rebound, penetration, and pullout strength, which are considered to be indirectly related to the compressive strength of concrete. In addition to these methods we can have radiographic, radiometric, nuclear, magnetic, and electrical methods. The measurements of NDTs are influenced by materials, mix and environmental factors. In addition, the data on the materials and mix proportions used in the construction are usually not available. Hence, it is necessary that the interpretation of the results is evaluated by experts.

**Advantages of non-destructive methods**   These have the following advantages over the destructive methods of testing:

1. NDTs can be conducted directly on concrete structures or elements, thus eliminating representative samples. On the other hand, in destructive testing, a large number of specimens are required and tested for destruction at various ages. It is because the change in strength of concrete is usually studied over a long period of time (normally, for 28 to 56 days); sometimes, it may be necessary to study the effect of curing or deterioration due to certain causes, which will extend over several months. Moreover, as the quality of all the specimens may not be the same, the results obtained may not be reliable.
2. Using NDTs, the variation in the strength of concrete at any age can be studied easily; most of the NDT equipment are portable and hence it is easy to carry out the tests.
3. The structure or elements are not loaded to destruction in NDT. The quality of concrete is judged by measuring some of its physical properties, which are related to its quality.
4. There is no wastage of material as in destructive methods of testing.

Some of these methods are as follows:

*Schmidt/Rebound hammer test [ASTM C 805M-2013, IS 13311 (Part 2):1992]*   It is used to evaluate the surface hardness of concrete.

*Ultrasonic pulse velocity test [ASTM C 597-2009, IS 13311 (Part 1):1992]*   It is mainly used to measure the sound velocity of the concrete and hence the compressive strength of the concrete.

*Penetration resistance test (ASTM C 803M-2010)*   A specially designed gun is used to drive a hardened steel probe into the concrete and the amount of penetration of the probe is used to indicate the strength of concrete (commercial system is known as the Windsor Probe). Essentially, it measures the surface hardness and hence the strength of the surface and near surface layers of the concrete.

*Pullout test (ASTM C 900-2015)*   It measures the maximum force required to pull an embedded metal insert with an enlarged head from a concrete specimen or structure.

*Maturity method (ASTM C 1074-2011)*   It is a technique for determining strength gain of concrete based on the measured temperature history during curing.

*The break-off test (ASTM C 1150-1996)*   It measures the force required to break off a cylindrical core from a larger concrete mass. The measured force and a pre-established strength relationship are used to estimate the in-place compressive strength.

*Half-cell electrical potential method (ASTM C 876-2009)*   It is used to detect the corrosion potential of reinforcing bars in concrete.

*Permeability test (IS 3085:1965)*   In this test, the flow of water through the concrete is measured to calculate the permeability of a concrete specimen.

*Covermeter test*   It is used to measure the distance of steel reinforcing bars beneath the surface of the concrete and also to measure the diameter of the reinforcing bars.

*Carbonation depth measurement test*   It is used to determine whether moisture has reached the depth of the reinforcing bars and hence the possibility of corrosion.

*Radiographic test*   It is used to detect voids in the concrete and the position of pre-stressing ducts.

*Tomographic modelling*   It uses the data from ultrasonic transmission tests in two or more directions to detect voids in concrete.

*Impact echo testing*   It is used to detect voids, delamination, and other anomalies in concrete.

*Ground penetrating radar or impulse radar testing*   It is used to detect the position of reinforcing bars or stressing ducts.

*Infrared thermography*   It is used to detect voids, delamination, and other anomalies in concrete and also to detect water entry points in buildings.

IS 516:2018 gives the procedure for determining the elastic modulus of concrete by an electro-dynamic method, i.e., by measuring the natural frequency of the fundamental mode of longitudinal vibration of concrete prisms. As this is a NDT, the same specimens may subsequently be used for determining the flexural strength, as discussed in Section 25.13.2.

Only a few of the aforementioned methods, which are more often used, are described in Sections 25.14.1–25.14.5. The details of other tests may be found in ACI 228.1R-03 and Malhotra and Carino (2003).

## 25.14.1 Rebound Hammer Test

Developed in 1948 by Swiss engineer Ernst Schmidt, the device measures the hardness of concrete surfaces using the rebound principle. The device is often referred to as a *Swiss Hammer*. It works on the principle that the rebound of an elastic mass depends on the hardness of the surface against which the mass impinges. There is little apparent theoretical correlation between the compressive strength of concrete and the rebound number of the hammer. However, within limits, empirical correlations have been established between strength properties and the rebound number. It weighs about 1.8 kg and is suitable for both laboratory and field work.

The Schmidt hammer, consists of a spring-loaded hammer mass that slides on plunger within a tubular housing, as shown in Fig 25.28 [IS 13311 (Part 2):1992]. During the test, hold the plunger perpendicular to the concrete surface to be tested and slowly push the hammer towards the surface. As the hammer is pushed towards the concrete surface, the main spring connecting the hammer mass to the plunger is stretched, as shown in Fig. 25.29(b). When the hammer is pushed to the limit, the latch is automatically released. Now the energy stored in the spring propels the hammer mass towards the concrete member, as shown in Fig. 25.29(c). The mass impacts the shoulder area of the plunger rod and rebounds as shown in Fig. 25.29(d). During the rebound, the slide indicator travels with the hammer mass and records the rebound distance, which is called as the *rebound number*. A button on the side of the hammer is pushed to lock the plunger in the retraced position, and the rebound number is read from the scale. The rebound distance is indicated by a pointer on a scale graduated from 0 to 100; the rebound reading is termed as 'R-value'. Take 10 readings from each test area. No two impact tests should be closer than 25 mm. Examine the impression made by the hammer on the surface of concrete after the impact; if the impact crushes or breaks through a near-surface air void, discard the reading and take another reading. If a reading differs by more than seven units from the average, that reading should be discarded and a new average should be computed based on the remaining readings. If more than two readings differ from the average by seven units, the entire set of readings is discarded.

The compressive strength of concrete is determined by referring to the conversion chart, provided by the manufacturer. Prior to testing, the Schmidt hammer should be calibrated using a calibration test anvil supplied by the manufacturer for that purpose. Twelve readings should be taken, and after dropping the highest and lowest, the average of the 10 remaining reading should be considered. When conducting the test the hammer should be held at right angles to the surface which in turn should be flat and smooth. The rebound reading will be affected by the orientation of the hammer, when used in a vertical position. (For example, when used on the underside of a slab, gravity will increase the rebound distance of the mass.) Because the rebound number test probes only the near-surface layer of concrete, the rebound number may not actually represent the interior concrete. The results are affected by factors such as smoothness of surface, size and shape of specimen, moisture condition of the concrete, type of cement, and coarse aggregate and extent of carbonation of the surface of concrete. Thus, though the rebound hammer test is easy to perform, there are many factors (other than concrete strength) that may influence the test results (ACI 228.1R-03).

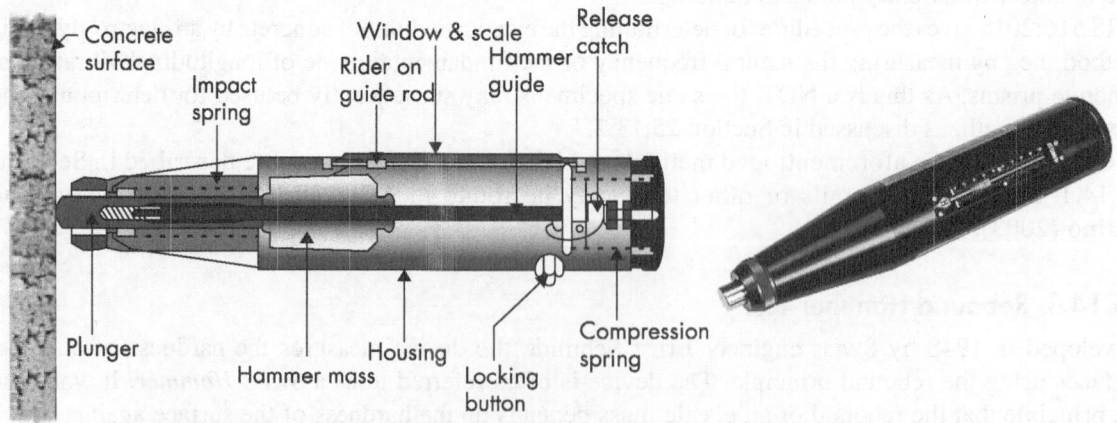

**Fig. 25.28** Rebound hammer (*Source*: www.spectro.in/Rebound-Hammer-Test.html and *Photo Courtesy*: Humboldt Mfg Co.)

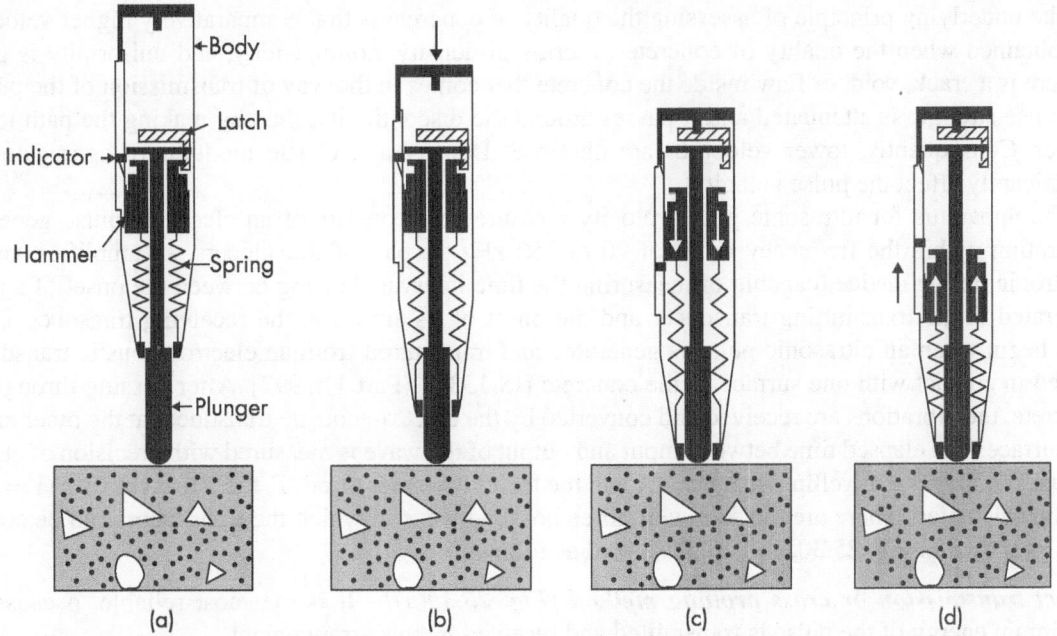

**Fig. 25.29** Schematic to illustrate operation of the rebound hammer (reprinted with permission from ACI 228.1 R-95) (a) Instrument ready for test (b) Body pushed towards test object (c) Hammer is released (d) Hammer rebounds

The Schmidt hammer provides an inexpensive, simple, and quick method of obtaining an indication of concrete strength, but an accuracy of about ±20% is possible only for specimens cast, cured, and tested under conditions in which calibration curves have been established.

### 25.14.2 Ultrasonic Pulse Velocity Test

The *ultrasonic pulse velocity method* involves measuring the travel time over a known path distance of a pulse of ultrasonic waves. This test method is applicable to assess the uniformity and relative quality of concrete, indicate the presence of voids and cracks, and evaluate the effectiveness of crack repairs. The pulse velocity method is a truly non-destructive method and the same sample can be tested again and again, which is very useful for testing concrete undergoing internal structure changes over a long period of time (such as deterioration due to aggressive chemical environment, cracking, and changes due to freezing and thawing) or in cases where the early age strength has to be determined. Using the pulse velocity test, it is also possible to obtain the dynamic elastic modulus, Poisson's ratio, thickness of concrete slabs, and estimate the strength of concrete test specimens as well as in-place concrete (Malhotra and Carino, 2003).

When the pulse is induced into the concrete from a transducer, a complex system of stress waves is developed, which includes longitudinal or compression waves, transverse or shear waves, and Rayleigh or surface waves. These stress waves travelling through an elastic medium are similar to the sound waves travelling through air. The receiving transducer is used to detect the onset of the longitudinal waves, which are the fastest (compression waves are followed by shear and surface waves). The velocity of the pulses is independent of the geometry of the material through which they pass and depends only on its elastic properties.

The underlying principle of assessing the quality of concrete is that comparatively higher velocities are obtained when the quality of concrete in terms of density, homogeneity, and uniformity is good. If there is a crack, void, or flaw inside the concrete that comes in the way of transmission of the pulses, the pulse strength is attenuated and it passes around the discontinuity, thereby making the path length longer. Consequently, lower velocities are obtained. Density and elastic modulus of aggregate also significantly affect the pulse velocity.

The apparatus for ultrasonic pulse velocity measurement consists of an electrical pulse generator (operating within the frequency range of 20 to 150 kHz), a pair of transducers, an amplifier, and an electronic timing device (capable of measuring the time interval elapsing between the onset of a pulse generated at the transmitting transducer and the onset of its arrival at the receiving transducer). The tests begin when an ultrasonic pulse is generated and transmitted from an electro-acoustic transducer, placed in contact with one surface of the concrete [IS 13311 (Part 1):1992]. After passing through the concrete, the vibrations are received and converted by the electro-acoustic transducer at the other end of the surface. The elapsed time between input and output of the wave is measured with precision of at least 0.1 μs. With known travelling distance, $D$, and the travel time measured, $T$, the pulse velocity ($V = D/T$) can be calculated. There are the following three possible ways in which the transducers can be configured as shown in Fig. 25.30, and are discussed as follows:

***Direct transmission or cross probing method [Fig. 25.30(a)]*** It is the most reliable, because the maximum energy of the pulse is transmitted and received by this arrangement.

***Semi-direct transmission method [Fig. 25.30(b)]*** This method can be used in situations where the two opposite faces of the structural member may not be accessible for measurements. However, care has to be taken to keep the transducers not too far apart so that the transmitted pulse is not attenuated and hence not received properly. This method has been used in situations of heavy reinforcing steel.

***Indirect transmission or surface probing method [Fig. 25.30(c)]*** This method can also be used when the two opposite faces of the structural member are not accessible – in this method, the receiving transducer is also placed on the same face of the concrete member. But it is least accurate because the signal produced at the receiving transducer has amplitude of only 2 to 3% of that produced by cross probing. This method is also prone to errors and may require special procedures for determining pulse velocity (Malhotra and Carino, 2003).

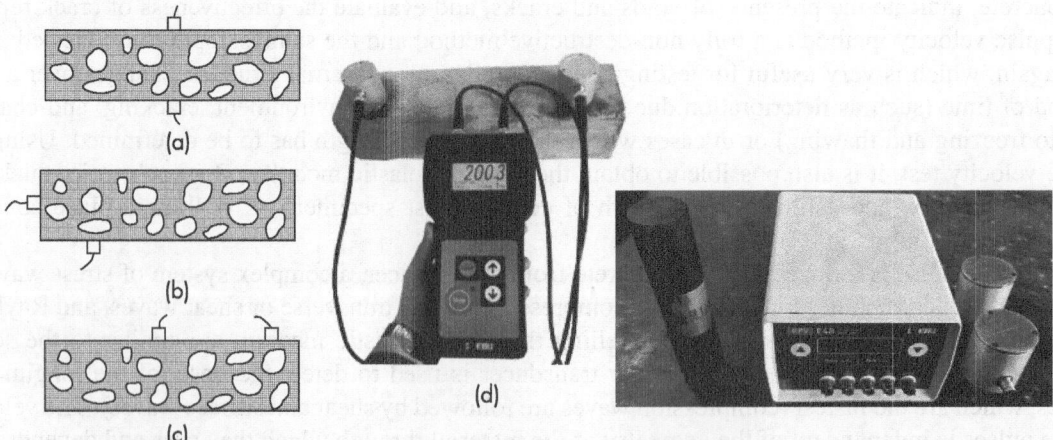

**Fig. 25.30** Methods of pulse-velocity measurement (*Source*: http://www.controls-group.com and *Courtesy*: Er Dar Adil) (a) Direct method (b) Semi-direct method (c) Indirect method (d) Testing on specimen

For the ultrasonic pulses to be generated and then detected properly at the other end, the transducers must be in full contact with the surface of concrete; otherwise, air pocket between the transducer and the test specimen will introduce an error in the indicated transit time. Hence, it is necessary to have adequate acoustical coupling between the concrete and the face of each transducer. Many *couplants* are available, such as grease, petroleum jelly, liquid soap, and kaolin glycerol paste, of which petroleum jelly has proven to be more effective. It has to be noted that the couplant should be spread as thin as possible, in order that it does not affect the reading. Repeated readings at a particular location should be taken until a minimum value of transit time is obtained. If the concrete surface is very rough, where the transducers are located, the surface should be ground smooth or the surface should be made smooth by the application of a thin layer of Plaster of Paris or a suitable quick-setting mortar or paste.

The factors contributing to the variability of ultrasonic pulse velocity methods as applied to concrete are aggregate properties, cement type, water-cement ratio, admixtures, degree of compaction, curing, age of concrete, moist condition (the pulse velocity of saturated concrete may be up to 2% higher than that of similar dry concrete), and stress level in concrete. Additionally, embedded reinforcement in the pulse path may have a significant effect on the measurements of pulse velocity (the pulse velocity in steel is 1.2 to 1.9 times the velocity in plain concrete). IS 13311 (Part 1):1992 recommends a minimum path length of 150 mm for the direct transmission method involving one unmoulded surface and a minimum of 400 mm for the surface probing method along an unmoulded surface. High frequency transducers are preferable for short path lengths and low frequency transducers for long path lengths. (Transducers with a frequency of 50 to 60 kHz may be used for most of the applications.) Since size of aggregates influences the pulse velocity measurement, IS 13311 (Part 1):1992 recommends a minimum path length of 100 mm for concrete with a nominal maximum size of aggregate up to 20 mm and a path length of 150 mm for concrete with a nominal maximum size of aggregate of 20–40 mm. By taking these factors into account, ultrasonic pulse velocity method may prove to be a simple and inexpensive method for the determination of the properties of concrete.

Table 25.6 may be used to have a correlation between the pulse velocity and concrete quality. It may be useful to access the level of workmanship employed and indicate the presence or absence of internal flaws, cracks, and segregation of concrete.

**Table 25.6** Pulse velocity and concrete quality

| Pulse velocity, km/s | Quality of concrete |
| --- | --- |
| Greater than 4.5 | Very good to excellent |
| 3.5–4.5 | Good to very good |
| 3.0 to 3.5 | Medium and Satisfactory |
| Less than 3.0 | Poor and doubtful |

## 25.14.3 Impact-echo Test

It is a recent development of ultrasonic methods. The test involves the measuring of concrete thickness and integrity using one surface. (Since the early 1970s, this method has been successfully used for the evaluation of concrete piles.) The test is also used to determine the location of cracking, voids and delamination. It is based on monitoring the surface P-waves or echoes of concrete resulting from a short-duration mechanical impact. Specifically, the test measures the amplitude of reflected shock waves to detect flaws in concrete. The success of this method depends on using the correct impact. The method was invented by Nicholas J. Carino and Mary J. Sansalone at the U.S. National Bureau of Standards (now called National Institute of Standards and Technology-NIST) in the 1980s and later developed by Mary J. Sansalone at the Cornell University, Ithaca, New York, from 1987–1997 (Carino, 2001).

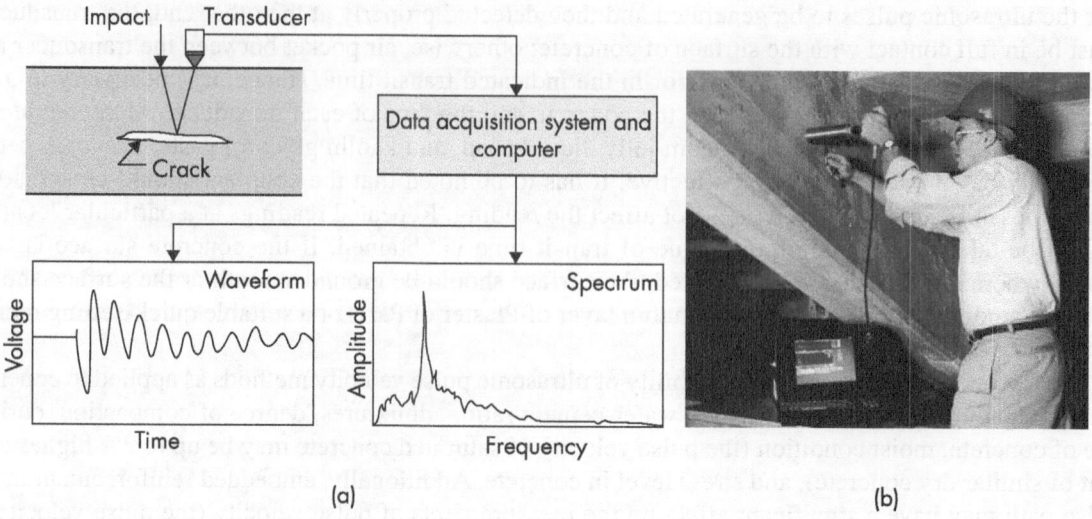

**Fig. 25.31** The impact-echo test system (*Source*: Carino, 2001) (a) Simplified diagram of the impact-echo method (b) Example of PC-based impact-echo test equipment

The impact-echo system consists of three components as shown in Fig. 25.31. The setup has an impact source (hammers are used in piles, spring-loaded spherically tipped impactors, which produce shorter duration impacts of 20 to 60 µs, are used in slabs and walls having thickness from 0.15 to 1 m), a receiving transducer, and a waveform analyser or digital processing oscilloscope (which captures the output of the transducer, stores the digitized waveform, and performs the signal analysis). The fundamental equation of impact-echo is as follows:

$$d = C/(2f) \tag{25.36}$$

where $d$ is the depth from which the stress waves are reflected (the depth of a flaw or the thickness of a solid structure), $C$ is the wave speed, and $f$ is the dominant frequency of the signal. The distance of the reflecting interface can be determined by knowing the speed of the stress wave. The standard guideline on the application and interpretation of the impact-echo method is ASTM C 1383–15. A simplified diagram of the impact-echo test is shown in Fig. 25.31(a) and a system manufactured by Impact-Echo Instruments, LLC, is shown in Fig. 25.31(b).

The factors that affect the detection of a flaw within concrete are: the type of the flaw and its orientation, the depth of the flaw, and the contact time of the impact. The impact-echo method is a reliable method for locating a variety of defects in concrete structures and the equipment is lightweight, portable, easy to operate, and requires access to only one side of the structure. As with most methods for flaw detection in concrete, experience is required to interpret impact-echo test results.

## 25.14.4 Maturity Method

It is a NDT technique for determining the strength gain of concrete at early ages, generally less than 14 days. The method accounts for the combined effects of time and temperature on concrete strength development. Since the degree of cement hydration depends on both time and temperature, the strength of concrete may be evaluated from the concept of *maturity*, which is expressed as a function of the curing time and internal concrete temperature. It is assumed that batches of the same concrete mixtures of same maturity will attain the same strength regardless of the time temperature combinations leading to that maturity. The maturity method has various applications in concrete construction such as formwork removal and post-tensioning.

In the maturity method, the thermal history of the concrete and a maturity function are used to calculate a *maturity index* that quantifies the combined effects of time and temperature. The maturity function is a mathematical expression that converts the temperature history of the concrete to a maturity index. Several such functions have been proposed and reviewed in Malhotra and Carino (2003). Two expressions have found widespread usage. In the first one developed by Saul in 1951, the maturity is computed based on temperature history using the following equation:

$$M(t) = \sum_0^t (T_a - T_0)\Delta t \tag{25.37}$$

where $M(t)$ is the maturity at age $t$ (°C·h), $\Delta t$ is a time interval (hours), $T_a$ is the average concrete temperature during time interval $\Delta t$ (°C), and $T_0$ is the datum temperature (°C). Equation (25.37) is known as *Nurse-Saul function*. Traditionally, a value of −10°C (14°F) has been used as the datum temperature when using the Nurse-Saul function.

In the second expression, the equivalent age maturity function assumes that the rate of strength gain varies exponentially with time. This function includes an activation energy term, similar to equations that describe chemical reactions. This exponential function is used to compute an equivalent age of the concrete at some specified temperature as follows:

$$M = \sum_0^t e^{-\left(\frac{E}{R}\right)\left(\frac{1}{273+T} - \frac{1}{273+T_r}\right)} \Delta t \tag{25.38}$$

where $M$ is the Arrhenius maturity index at a reference temperature $T_r$ (hours), $E$ is the activation energy in J/mol (For $T \geq 20$°C $\rightarrow E = 33500$ J/mol, and for $T < 20$°C $\rightarrow E = 33500 + 1470(20\text{-}T)$ J/mol, $R$ is the universal gas constant which is equal to 8.3144 J/(mol· K), $T$ is the average temperature of concrete during time interval $\Delta t$, (°C), $T_r$ is the reference temperature, (°C), and $\Delta t$ is the time interval (hours). Equation (25.38) is known as *Arrhenius equation* (or *equivalent age maturity function*) and was developed in Europe by Freiesleben Hansen and Pederson (1977). Research has shown that the equivalent age maturity function (Eq. [25.38]) accounts for temperature more accurately over a wider temperature range than the temperature-time factor. Tikalsky et al. (2001) summarizes the scientific basis for both these functions (Eqs. [25.37] and [25.38]).

The correlation curve equation can be based on any function that accurately describes the data. Two of the more common relationships are the logarithmic and hyperbolic functions. The form of the logarithmic function is as follows (Plowman, 1956):

$$S = A + B \cdot log(M) \tag{25.39}$$

where $S$ is the estimated strength of concrete at a given maturity (a variable), $A$ and $B$ are the regression constants, and $M$ is the maturity index (a variable).

The form of the hyperbolic function is as follows (Carino, 1981):

$$S = S_u \frac{K(M - M_o)}{1 + K(M - M_o)} \tag{25.40}$$

where $S$ is the estimated strength of concrete at a given maturity (a variable), $S_u$ is the regression constant analogous to the ultimate strength that the concrete will attain, $M$ is the maturity index (a variable), $M_o$ is the regression constant analogous to the maturity when strength gain begins, and $K$ is the regression constant analogous to a rate constant.

The standard guideline on the testing and interpretation of the maturity method is given in ASTM C 1074–17. To estimate the concrete strength at any construction project, the following general steps are followed:

1. Develop the strength-maturity relationship of the concrete mixture using laboratory tests. [At least 15 cylindrical/cube concrete specimens should be used. The mix proportions and constituents of the concrete should be similar to those of the concrete whose strength will be estimated in actual structure. Embed temperature sensors at the centres of at least two specimens. Connect sensors to maturity instruments. Moist cure the specimens in a water bath. Perform compression tests at ages of 1, 3, 7, 14, and 28 days. Test two cylinders at each age and compute the average strength. At each test age, record the average maturity index for the instrumented specimens. On graph paper, plot the average compressive strength as a function of the average value of the maturity index. Draw the best-fit curve through the data (see Fig. 25.32). The resulting curve is the strength-maturity relationship to be used for estimating the strength of the concrete.]

2. Prior to concrete placement in the actual structure, install thermocouples inside the concrete at locations deemed critical due to the strength requirements or the rate of hydration.

3. Record temperature data as soon as possible after concrete placement.

4. Use the temperature data to calculate the maturity index during aging. A *maturity meter* (e.g., Humboldt 4101 4-channel maturity meter) can be used to calculate the maturity and then output the values.

5. Using this maturity index, estimate the in-place concrete strength from the pre-established maturity-strength curve from laboratory tests (or the correlation equation) as illustrated in Fig. 25.32. Alternatively, use the correlation equation to determine the maturity of the concrete at a desired strength level and monitor the maturity until that level is achieved.

It has to be noted that the traditional approach of using field-cured cylinders/cubes does not replicate the same temperature profile of the in-place concrete and hence does not estimate its in-place strength as accurately as the maturity method.

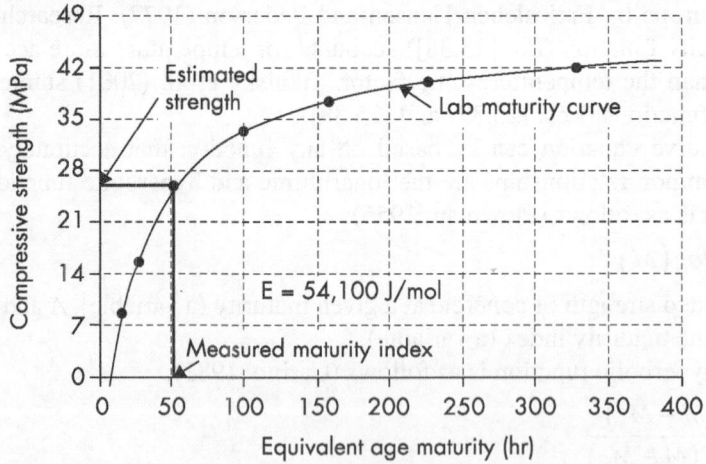

**Fig. 25.32** Laboratory-developed compressive strength versus maturity curve (Source: www.fhwa.dot.gov)

The factors that lead to variability in testing are aggregate properties, cement properties, water-cement ratio, and curing temperature. Before attempting to estimate in-situ strength of concrete, laboratory testing on concrete samples of similar characteristics must be performed in order to develop the correct maturity function in order to minimize the effect of the these factors. Temperature probe locations must be carefully selected to measure a representative temperature of the entire concrete section. More information about the maturity method and its use in high-strength concretes can be found in Malhotra and Carino (2003).

**Example 25.1**  In a concrete wall construction, the contractor can remove the forms once the compressive strength of 27.5 MPa is reached. Estimate the approximate length of time that the contractor has to wait before removing the forms. The strength-maturity relationship for the concrete mix used in the project is as given in Fig. 25.32. Assume the cement hydration will cease at −10°C and the temperature of the concrete at the time of removing forms is 20°C.

## Solution:
From Fig. 25.32 for strength of 27.5 MPa, the maturity index is 50 h.

$$M = \sum_{0}^{t} e^{-\left(\frac{E}{R}\right)\left(\frac{1}{273+T} - \frac{1}{273+T_r}\right)} \Delta t$$

$$50 = e^{-\frac{33500}{8.3144}\left(\frac{1}{273+20} - \frac{1}{273+(-10)}\right)} \Delta t$$

$$50 - e^{1.5686} \Delta t$$

or, $\Delta t = 10.42$ h

The contractor has to wait at least 10 h and 25 min after the concrete is placed before removing the forms.

## 25.14.5 Permeability Test

In this test, a concrete specimen of known dimensions is kept in a specially designed cell and is subjected to a known hydrostatic pressure from one side; the quantity of water percolating through it is measured during a given interval of time and the coefficient of permeability is computed based on the measured quantity of water (IS 3085:1965).

The apparatus for the test consists of a permeability cell, water reservoir, and pressure lines (consisting of heavy-duty rubber hose or suitable metal tubing with watertight joints). In addition, it requires a supply of compressed air at 0.5 to 1.5 N/mm² to the permeability cell (with suitable regulating valves and pressure gauges), and supply of de-aired water (water may be easily de-aired for this purpose by boiling and cooling). Figure 25.33 shows the details of permeability cell and schematic test setup.

The permeability cell consists of a metal cylinder with a ledge at the bottom for retaining the specimen, a flange at the top, a removable cover plate, and a sheet metal funnel, which can be securely bolted to the cell. The cell and the cover plate should be fabricated using any corrosion-resistant metal such as aluminium, and the thickness of the cell should be designed to safely withstand the maximum test pressure. A rubber or neoprene O-ring gasket, seated in matching grooves, should be used between the cell and the cover plate to obtain a watertight joint. The dimensions A, B, and C of the cell as shown in Fig. 25.33(a) depend on the diameter of the specimen (for example, for 150-mm diameter of specimen, the dimensions of A, B, and C are 170, 120, and 160 mm, respectively). A metal pipe (50 to 100 mm in diameter and about 500 mm long) may be used as the water reservoir and should be fitted with a graduated side arm glass-gauge, and the necessary fittings and valves for administrating water and compressed air and for draining, bleeding and connection to the permeability cell, as shown in Fig. 25.33(b).

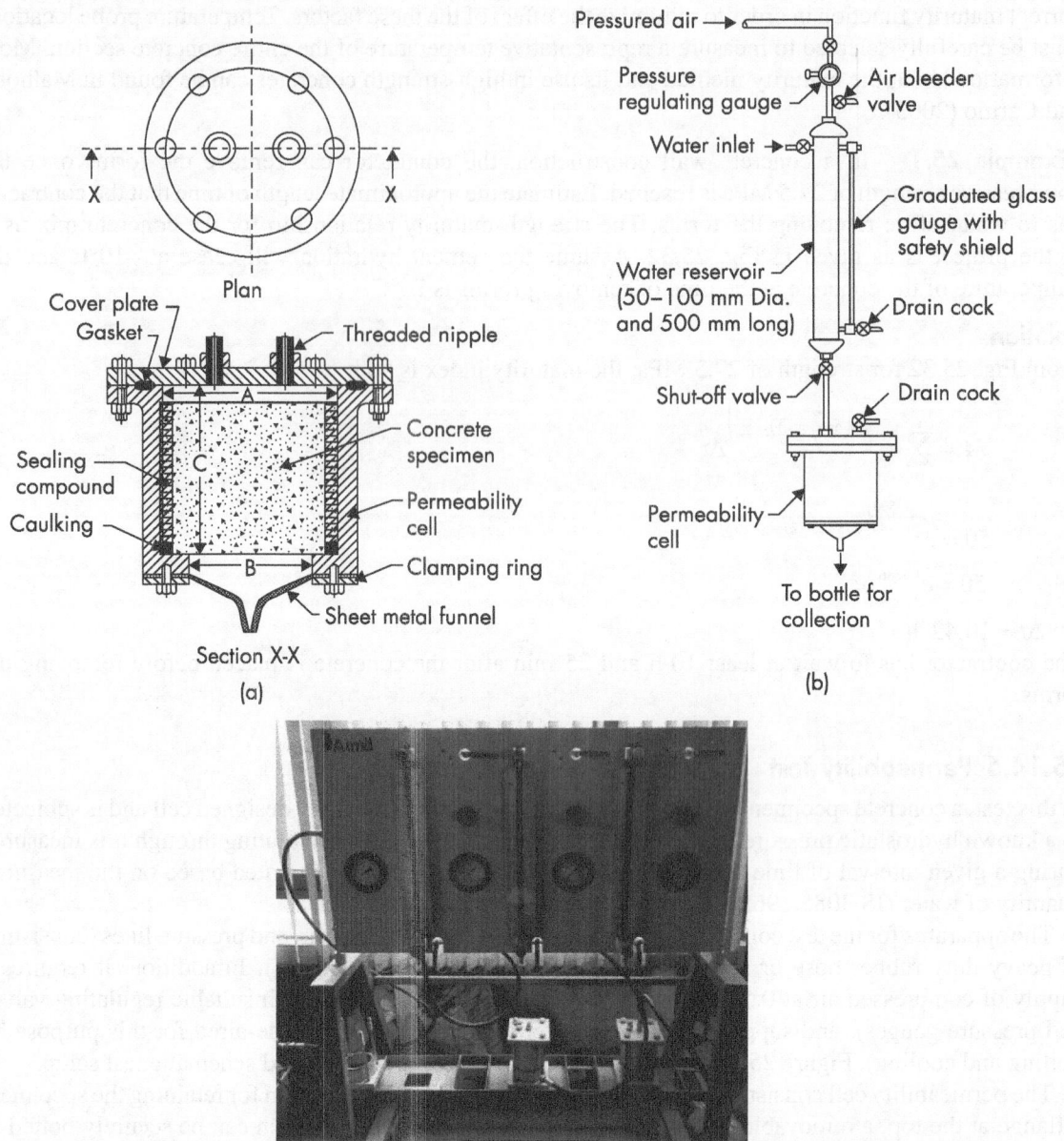

**Fig. 25.33** Permeability test as per IS 3085:1965 (*Photo Courtesy*: Er Dar Adil of IITD)
(a) Details of permeability cell (b) Schematic set-up of permeability test

The normal size of specimen is 150-mm diameter (and height). When specimens are made with aggregates not exceeding 20 mm, the diameter (and the height) of the specimen may be reduced to 100 mm. The specimen should be cast and cured as per the normal norms of casting and curing of standard cylinders.

The test procedure is as follows:

1. Calibrate the reservoir under the operating pressures of 0.5 to 1.5 N/mm².
2. Clean the specimen thoroughly and the sand blast the end faces. Also, surface-dry the specimen and measure its dimensions.

3. Place it at the centre of the cell, with the lower end resting on the ledge.

4. Caulk the annular space between the specimen and the cell tightly to a depth of about 10 mm using cotton or hemp cord soaked in a suitable molten sealing compound. Fill the rest of the space carefully with the molten sealing compound, up to the top level of the specimen (A mixture of bees-wax and rosin, applied smoking hot, forms an effective seal). It is important to check the seal for water tightness.

5. After sealing the specimen properly, fix the funnel in position and connect the cell assembly to the water reservoir, as shown in Fig. 25.32(b).

6. Now check whether the air bleeder valve, the valve between the reservoir and the cell, and the drain-cock are opened, and then allow the de-aired water to enter into the reservoir.

7. When water passes freely through the drain-cock, close it, and fill the water reservoir fully. Then, close the reservoir water inlet and air bleeder valves.

8. When the system is completely filled with water, apply the desired test pressure to the water reservoir and record the initial reading of the gauge. At the same time, weigh and place a clean collection bottle under the funnel to collect the water percolating through the specimen.

9. Record the quantity of percolated water and the gauge-glass readings at periodic intervals.

10. In the beginning, the rate of water intake will be larger than the rate of outflow. As the steady state of flow is approached, the two rates tend to become equal and the outflow reaches a maximum and stabilizes. (The steady state of flow is defined as the stage at which the outflow and inflow of water become equal for the first time.) With further passage of time, both the inflow and outflow generally register a gradual drop. Continue the permeability test for about 100 h after the steady state of flow has been reached. Consider the outflow as the average of all the outflows measured during this period of 100 h.

It has to be noted that the test has to be carried out at a temperature of $27 \pm 2°C$. For each 5°C increase/decrease of temperature, apply a correction of 10% increase/decrease in the coefficient of permeability.

The coefficient of permeability is calculated using the following formula:

$$K = \frac{Q}{AT\left(\dfrac{H}{L}\right)} \tag{25.41}$$

where $K$ is the coefficient of permeability in cm/s, $Q$ is the quantity of water in millilitres percolating over the entire period of test after the steady state has been reached, $A$ is the area of the specimen face in cm$^2$, $T$ is the time in seconds over which $Q$ is measured, and $H/L$ is the ratio of the pressure head to thickness of specimen, both expressed in the same units.

ASTM C1585–13 and ASTM C642–13 are the most used standards for determining water absorption of concrete.

## 25.15 Partial Destructive Tests

A few tests can be classified as partial destructive tests, as they are not fully non-destructive and damage the concrete specimen to some extent. Two such tests are discussed in this section.

### 25.15.1 Penetration Resistance Method (Windsor Probe)

In this method, the depth of penetration of a hardened steel probe is measured after it is driven into the hardened concrete using a power-actuated driving device. The probe penetration technique involves the use of a specially designed gun to drive a hardened steel probe (of diameter 6.5 mm and length 80 mm)

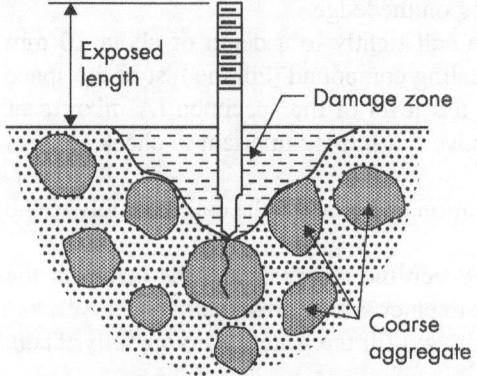

**Fig. 25.34** Approximate shape of failure zone in concrete during probe penetration test

into the concrete. (The commercial test system is known as the *Windsor Probe*.) In these tests, the amount of penetration of the probe is assumed to indicate the strength of concrete. This method is similar to the rebound hammer test, except that the probe impacts the concrete with much higher energy than the plunger of the rebound hammer.

The test is based on the initial kinetic energy of the probe and energy absorption by the concrete. The probe penetrates into the concrete until its initial kinetic energy is absorbed. The penetration of the Windsor probe creates dynamic stresses that lead to the crushing and fracturing of the near-surface concrete (Fig. 25.34). A cone-shaped zone develops upon penetration, which encompasses fracturing and is resisted by the compression of the adjacent concrete. Although calibration charts are provided by the manufacturer, the instrument should be calibrated for the type of concrete and the type and size of aggregate used. An essential requirement of this test is that the probes should have a consistent value of initial kinetic energy as per ASTM C 803M-2018.

**Limitations and advantages** The hardness of the coarse aggregate can have a significant effect on the penetration depth and reinforcing steel must be avoided, especially when there are deep penetrations (ACI 228.1R-03). Though penetration-resistance tests are easier to perform than the other alternatives, they are generally not as accurate. Owing to the insignificant effect of the penetration resistance methods on the structural integrity of the probed sample, the tests are considered to be non-destructive despite the disturbance of the concrete during penetration (structural members can be tested in-situ with only minor patching of holes on exposed faces).

### 25.15.2 The Pullout Test

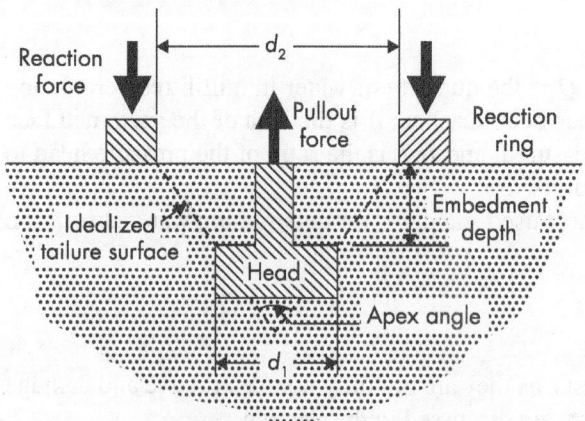

**Fig. 25.35** Schematic diagram of typical pullout test

It measures the maximum force required to pull an embedded metal insert with an enlarged head from a concrete specimen or structure. In such tests conducted per ASTM C 900–2015, a metal insert with an enlarged head (Fig. 25.35) is embedded about 25 to 30 mm into the fresh concrete or post-installed in hardened concrete. The most commonly used pullout test method is the *LOK test* developed in 1962 by Kierkegaard-Hansen (Kierkegaard-Hansen, 1975). The pullout force on the insert is applied using a device (e.g., tension jack) that bears against the concrete surface through a reaction ring that is concentric with the insert. As the insert is pulled out, a cone-shaped fragment of the concrete is also extracted. The large diameter of the conic fragment $d_2$ is determined by the inner diameter of the reaction ring, and the small diameter $d_1$ is determined by the insert-head diameter. Because the test is conducted before the formwork is removed, access to the concrete and insert for the testing device must be provided in the formwork.

The average value for the coefficient of variation for the pullout test has been found to be around 8%. The factors that affect the results of pullout teat are the maximum aggregate size, cement mortar percentage, the type of insert, ring dimensions, and depth of embedment. Hence, it is important to use the same type of testing apparatus and embedment during correlation and testing in the actual structure. More details about the pullout test may be found in ACI 228.1R-03.

## 25.16 Sampling and Acceptance Criteria of Concrete

Quality assurance of concrete has to be ensured by proper design of the concrete mix, use of proper materials (good quality aggregates, cement, fly ash and other cementitious materials, water, and admixtures), proper workmanship in the execution (mixing, transporting, placing, vibrating, and curing) at site by the contractor, and proper care during the use of the structure including timely maintenance and repair by the owner. Hence, all the parties involved in the making and using of a concrete structure, i.e., the designer, the concrete supplier, the contractor, and also the owner should take equal responsibility for providing good quality assurance (see also Clause 10 of IS 456, for more information). To assure good quality, Clause 15 of IS 456 suggests that samples from fresh concrete should be taken as per IS 1199, and cubes made, cured, and tested at 28 days as per IS 516. This clause stresses that the 28 day compressive strength as specified in Table 25.7 alone should be the criterion for acceptance or rejection of concrete (for the criteria of flexural strength, consult Clause 16.2 of IS 456). It has to be noted that the *characteristic strength* of any material is defined as the value of strength below which not more than 5% of test results are expected to fall.

**Table 25.7** Criterion for acceptance of characteristic compressive strength

| Specified grade | Mean of the group of 4 non-overlapping consecutive test results, N/mm² | Individual test results, N/mm² |
|---|---|---|
| Any grade | $\geq f_{ck} + 0.825 \times s$ or $f_{ck} + 3$ N/mm², whichever is greater, where $s$ is the established value of standard deviation | $\geq f_{ck} - 3$ |

**Note**: In the absence of established value of standard deviation, a value of 3.5 N/mm² may be assumed for M15 grade, 4 N/mm² for M20 and M25 grades, 5 N/mm² for M30 to M60 grades, and 6 N/mm² for M65 to M80 grades (see also Tables 1 and 2 of IS 10262:2019).

The target mean strength of the concrete mix should be equal to characteristic compressive strength plus 1.65 times the standard deviation (IS 10262). The minimum size of the sample needs to be about 50 for normal distribution, but Clause 9.2.4.1 of IS 456 accepts a size of 30 samples. When insufficient test results (less than 30 samples) are available, an assumed value of standard deviation, $s$, has to be used; the code suggests some values (see Table 25.7). The standard deviation of the results from the mean value is regarded as an index of the scatter and hence may reveal the site control. A small value of standard deviation will result in a curve with dominant peak, while a larger value of standard deviation will result in a flatter curve as shown in Fig. 25.36. Three test specimens form a sample. The minimum frequency of sampling depends on the quantity of concrete, as shown in Table 25.8. Note that IS 456 specifies both compressive and flexural strength tests (Clauses 16.1 and 16.2); flexural strength test is considered only in some special situations, where the tensile strength of concrete plays an important role.

**Table 25.8** Sampling frequency

| Quantity of concrete involved, m³ | Required number of samples |
|---|---|
| 1–5 | 1 |
| 6–15 | 2 |
| 16–30 | 3 |
| 31–50 | 4 |
| >51 | 4 + 1 additional for each additional 50 m³ |

**Note:** A minimum of one sample in each shift is required.

In most constructions, it may be difficult to obtain 30 samples and hence four consecutive non-overlapping samples (it means that the test results of one group should not be taken in another group also) are considered to be practicable size to test the acceptability of concrete. As this size is smaller than the minimum population size, the following expression, which satisfies 5% acceptability criterion, may be used (SP 24: 1983).

$$f_{ml} = f_{ck} + 1.65\ s\ (1 - 1/\sqrt{n}) \tag{25.42a}$$

where $n$ is the size of the sample and $f_{ml}$ is the mean value of smaller size sample. The above equation for four non-overlapping samples reduces to the following:

$$f_{ml} = f_{ck} + 0.825\ s \tag{25.42b}$$

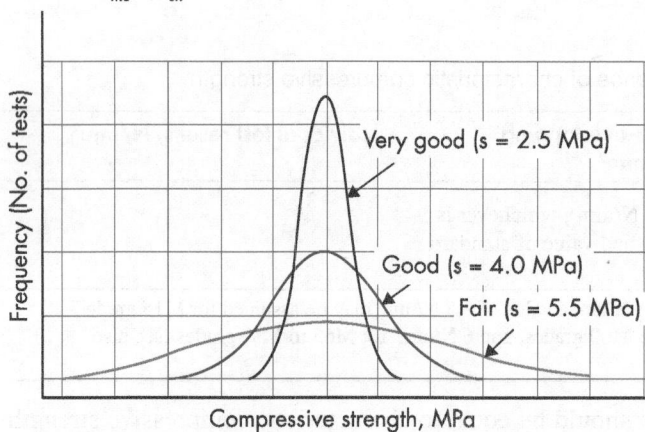

**Fig. 25.36** Typical normal frequency curves for different levels of control

The *acceptability criteria*, as given in Table 25.7, are supposed to be valid for concretes up to grade M60. It means that even if the concrete is of M60 grade, a mean value of the order of 63 MPa of four consecutive samples is acceptable! Hence, this criterion is liberal for HSC. However, the criterion in the last column of Table 25.7 (i.e., strength $\geq f_{ck} - 3$) is more stringent for HSC, as the margin for individual sample strength is small.

As per ACI 318:18, the concrete may be accepted if the strength is larger than (1) $0.8f_{ck} + 1.34\ s$ and (2) $0.8f_{ck} + 2.33\ s - 3.45$ for concrete of grade less than M44 (with cylinder strength converted to cube strength). If concrete strength is greater than M44, the strength should be larger of (1) $0.8f_{ck} + 1.34\ s$ and (2) $0.72f_{ck} + 2.33\ s$, where $s$ is the standard deviation (note that the ACI values have been converted from cylinder to cube strength to compare with IS code values). As per Clause 5.6.3.3 of ACI, the concrete strength is considered to be satisfactory as long as averages of any three consecutive strength tests remain above the specified strength and no individual strength test falls below the specified strength by more than 3.5 MPa if $f_{ck}$ is 44 MPa or less, or falls below $f_{ck}$ by more than 10% if $f_{ck}$ is over 44 MPa.

It has to be noted that the above strengths denote 28th day strength. Concrete made with Portland cement attain about 85 to 90% of strength on the 28th day; however in Portland-Pozzolana Concrete (PPC) using pozzolonic materials like fly ash, such a percentage of strength will be attained only after 56th day.

**Example 25.2**   In a concrete work, concrete of grade M 25 is to be used. The standard deviation has been established as 4.0 N/mm². In the course of testing concrete cubes, the following results were obtained (average strength of three specimens tested at 28 days, in each case expressed in N/mm²): 29.8, 32.0, 33.6, 28.6, 23.0, 27.7, and 22 N/mm². Determine whether the concrete is acceptable.

**Solution**:

1. The first four tests are straightaway accepted, the sample strength being greater than the characteristic strength of 25 MPa in each case.
2. The fifth test result of 23.0 N/mm² is less than the characteristic strength of 25 MPa. But greater than $f_{ck} - 3$ N/mm², or 22 N/mm². Average strength of the samples $= (29.8 + 32.0 + 33.6 + 28.6 + 23.0)/5 = 29.4$, which is greater than $f_{ck} + 1.65$ s $(1 - 1/\sqrt{n}) = 25 + 1.65 \times 4$ $(1 - 1/\sqrt{5}) = 25 + 0.912 \times 4 = 28.65$ N/mm² and greater than $f_{ck} + 3 = 28$ N/mm². Hence, acceptable.
3. The sixth result is also acceptable being greater than the characteristic strength of 25 MPa.
4. The seventh one is equal to 22 N/mm². The average strength of all seven samples is: $(29.8 + 32.0 + 33.6 + 28.6 + 23.0 + 27.7 + 22)/7 = 28.1$, which is less than $f_{ck} + 1.65$ s $(1 - 1/\sqrt{n}) = 25 + 1.65 \times 4$ $(1 - 1/\sqrt{7}) = 25 + 1.026 \times 4.1 = 29.1$ N/mm² but greater than $f_{ck} + 3 = 28$ N/mm². The seventh sample thus does not comply with all the requirements given in Table 11 of IS 456. However, the acceptance will depend upon the discretion of the site engineer. In this case, as only one specimen fails to meet single criteria, the concrete may be accepted.

## 25.17 Tests on Self-Compacting Concrete

Several new tests have been evolved for testing the suitability of SCC (see Fig. 25.37 and Table 25.9). They essentially involve testing the following: (a) flowability (slump flow test), (b) filling ability (slump flow test, V-funnel, and Orimet), note that in the slump flow test the average spread of flattened concrete is measured horizontally, unlike the conventional slump test, where vertical slump is measured, (c) passing ability (L-box, J-ring which is a simpler substitute for U-box), (d) robustness, and (e) segregation resistance or stability (simple column box test, Sieve stability test). The details of these test methods may be found in Okamura and Ouchi (2003) and Hwang et al. (2006).

**Table 25.9** Acceptability of SCC based on test results

| S. No. | Test | Unit | Acceptable range of values |
|---|---|---|---|
| 1 | (a) Slump flow | mm | 650–800 |
| | (b) $T_{50}$ Slump flow time | sec.* | 2–5 |
| 2 | (a) V-funnel-$T_0$ | sec. | 8–12+ |
| | (b) V-funnel-$T_5$ | Increase over 2(a) in sec. | 0–3 |
| 3 | L- box ($h_2/h_1$) | Ratio | 0.8–1.0 |
| 4 | U-box ($h_1$-$h_2$) | mm | < 30 |
| 5 | J-ring | mm | 0–10 |
| 6 | Orimet | sec. | 0–5 |
| 7 | Fill-box | % | 90–100 |

**Notes:**
*It is the needed time for the concrete to spread by 500 mm (mean diameter).
+In the V-funnel test result, if the time is less than 8 s, w/cm ratio has to be increased, and if it is more than 12 s, w/cm has to be decreased.

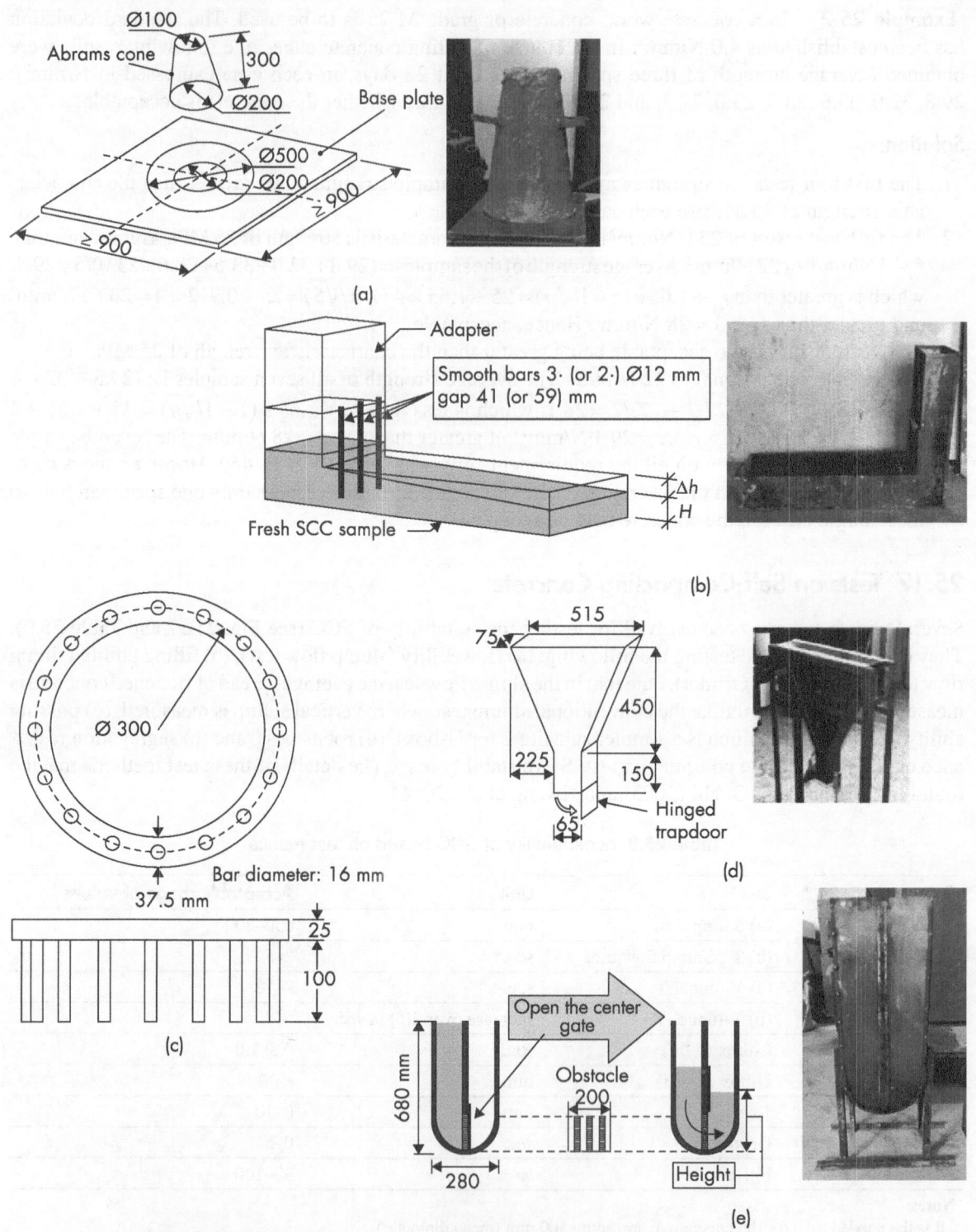

**Fig. 25.37** Tests on self-consolidating concrete [*Source*: Okamra and Ouchi (2003); Reprinted with permission from JCI; *Photos Courtesy*: Er Dar Adil of IITD] (a) Slump flow test (b) L-box (c) J-ring (d) V-funnel (e) U-flow test

SCC has been used in a number of bridges and precast projects in Japan, Europe, and the USA (Okamura and Ouchi, 2003). Recently, SCC has been used in a flyover construction in Mumbai, India (ICJ, Aug. 2009). The various developments in SCC undertaken in India may be found in ICJ, 2004 & 2009. An amendment (No. 3, Aug. 2007) in the form of Annex J was added to IS 456, which prescribes the following for SCC:

1. Minimum slump flow: 600 mm
2. Amount of fines (<0.125 mm) in the range of 400 to 600 kg/m³, which may be achieved by having sand content more than 38% and using mineral admixture to the order of 25 to 50% by mass of cementitious materials.
3. Use of high range water reducing admixtures and viscosity modifying agents.

## 25.18 Testing of Gypsum Products

Methods of test for gypsum plaster and concrete products are stipulated in IS 2542 (Part 1/Sec. 1–12): 1978, and IS 2542 (Part 2/Sec. 1–8):1981.

## 25.19 Testing of Timber and Timber Products

Many tests are specified in Indian codes to determine the following properties of timber: (a) moisture content, (b) specific gravity, (c) volumetric shrinkage, (d) radial and tangential shrinkage, and fibre saturation point, (e) static and impact bending strength, (f) compressive strength parallel and perpendicular to grain, (g) hardness, (h) shear strength parallel to grain, (i) tensile strength parallel and perpendicular to grain, (j) torsional strength, and (k) brittleness by Izod or Charpy impact. These tests are described in IS 1708 (Parts 1–18). Determination of moisture content test, which is generally required by government departments, is explained in Section 25.19.1. Sampling and several methods of tests for plywood are specified in IS 7638:1999 and IS 1734 (Parts 1–20):1983, respectively (also see IS 12077:1987). Methods of test for wood particle boards and boards from other lignocellulosic materials are stipulated in IS 2380 (Parts 1–21):1977. Tests for timber in structural sizes are described in IS 2408:1963, wood poles in IS 1900:1974, and for bamboo in IS 6874:2008.

### 25.19.1 Determination of Moisture Content

The moisture content of timber may be determined using the methods described in IS 1708 (Part 1):1986 or IS 11215:1991. In the *oven-dry method*, as per IS 11215:1991, test specimens having a length of 15 to 20 mm in the direction of the grain, free from all defects, are cut from each sample (At least three samples from lots containing up to 200 pieces). If weighing is not done immediately, the test specimens are cut from a point at least 450 mm from one end of the sample or from its centre, and weighed accurately. Then, they are oven dried for about 12–18 h at $103 \pm 2°C$, and weighed again. The moisture content is determined as follows:

$$MC = \frac{W_1 - W_0}{W_0} \times 100 \qquad (25.43)$$

where $MC$ is the moisture content in percent, $W_1$ is the initial weight, and $W_0$ is the oven-dried weight.

$MC$ can also be measured using electrical moisture meters, which are of two general types: *electrical resistance meters* (conductance-type) and *capacitance meters* (dielectric-type). As per IS 11215, the resistance meters are preferable as their readings are affected to a lesser extent by the natural variation of density within the particular species being tested. Conductance-type meters with pin-type electrodes

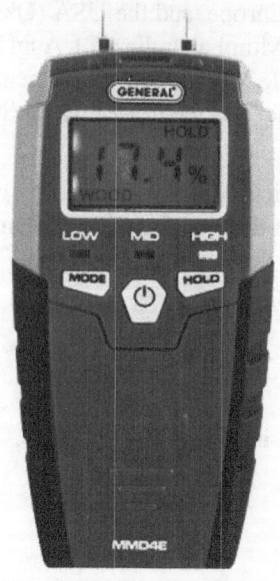

**Fig. 25.38** Conductance-type moisture meter

are widely used because of their simplicity (see Fig. 25.38). The nail-like pins are driven into the timber to measure the moisture content. The electrodes should be oriented in such a way that the current flows parallel to the grain. Electrode pins of varying lengths are generally provided for use with timber of different range of thicknesses. Dielectric-type meters use surface electrodes that do not puncture the wood surface, and can measure the moisture content of relatively dry wood. The electrical resistance meters are capable of reading moisture content to an accuracy of ±1.5 to 3%, for moisture contents in the range 8 to 24% (which covers the usual range of seasoned timber). The meter readings are taken at three sections along the length in the middle width of each face of every sample, and the average is taken. For timber over 50 mm in thickness, readings are taken on all the faces of each sample. The temperature of the wood (not air temperature) may significantly influence the readings of electrical resistance meters (James, 1998).

The oven-drying method has been the most universally accepted method for determining moisture content, but it is slow and necessitates cutting the wood. In addition, certain wood species like deodar, containing volatile oils, may give values slightly greater than true moisture content by the oven-drying method. The electrical method is rapid, does not require cutting the wood, and can be used on timber in place in a structure. However, considerable care must be taken to use and interpret the results correctly (Simpson, 2010). For species like deodar, containing volatile oils, the moisture content has to be determined using the distillation method described in IS 11215:1991.

## 25.20 Tests on Metals/Steel

Several tests are conducted on metals/steel to determine their properties. Only tension test, the Charpy V-notch (CVN) test, and the tests to determine the hardness are explained here.

### 25.20.1 Tension Test

Tension test is an important test for metals and used to evaluate the ultimate tensile strength (UTS), yield strength, stiffness (elastic modulus, $E$), Poisson's ratio, ductility, and toughness. Ultimate strength is the minimum guaranteed UTS at which the metal would fail. It is obtained from tensile test on a standard specimen, generally called *coupon*. A typical specimen as per IS 1608:2005 is shown in Fig. 25.39. In this test, the *gauge length* '$L_g$' and the initial cross-sectional area $A_o$ are important parameters. The dimensions of the specimens are established to ensure that failure occurs within the designated gauge length. The test coupons are actually cut out from a specified portion of the member for which the tensile strength is required. The initial gauge length is taken as $5.65\sqrt{A_o}$ in the case of specimen with rectangular cross section. For circular specimen, ASTM specifies $L_g/d = 4$ but other standards specify it as $L_g/d = 5$.

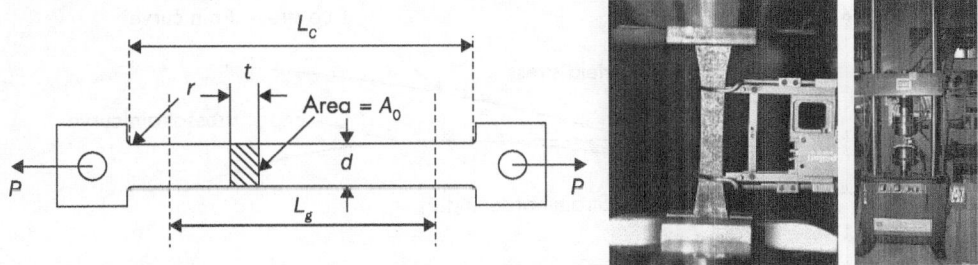

**Fig. 25.39** Tension test (*Photos courtesy:* Er Dar Adil of IITD) (a) Standard tensile test specimen as per IS 1608 (b) Close-up of the tension steel coupon (c) Steel coupon test

The coupon is fixed in a tensile testing machine, with specified distances between the grips and tested under uniaxial tension. The loads are applied, at specific rate of loading as per IS 1608, through the threaded ends. Typical stress-strain curve of two grades of steel specimen subjected to gradually increasing tensile load are shown in Fig. 25.40(a) and enlargement of initial part of stress-strain curve of mild-steel specimen is shown in Fig. 25.40(b).

The UTS is the highest stress at which a tensile specimen fails by fracture and is given by the following formula:

$$\text{UTS} = \text{Ultimate tensile load/Original area of cross section} \qquad (25.44)$$

UTS or stress is called the *engineering ultimate tensile stress* and is different from the *true or absolute ultimate stress*. As a specimen is subjected to tension, the area of cross section of the specimen decreases due to the Poisson's effect, and the true stress is obtained by considering the actual area of cross section at that instant instead of the original area of cross section. Thus, it is calculated by the following formula:

$$\text{True stress} = \text{Ultimate tensile load/Actual area of cross section} \qquad (25.45)$$

In most of the engineering structures, the stress in the members at service condition will not cause any appreciable change in the cross section. Moreover, it is extremely difficult to determine the area of cross section at various stages of loading. Hence, in all the design calculations, only the original area of cross section is used.

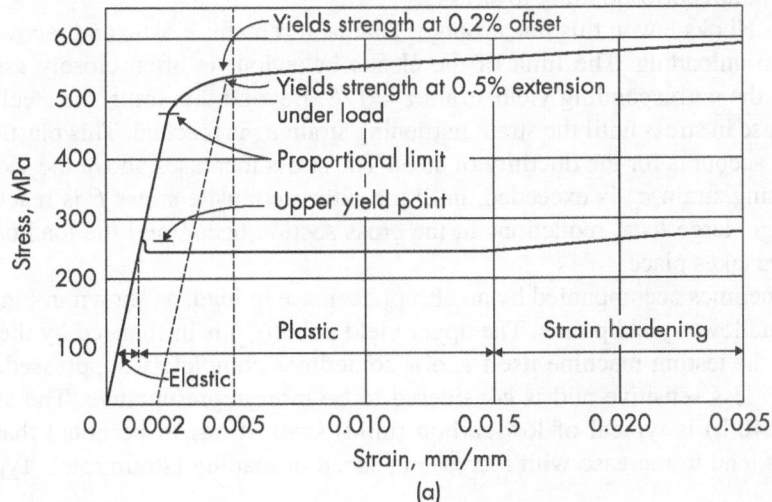

(a)

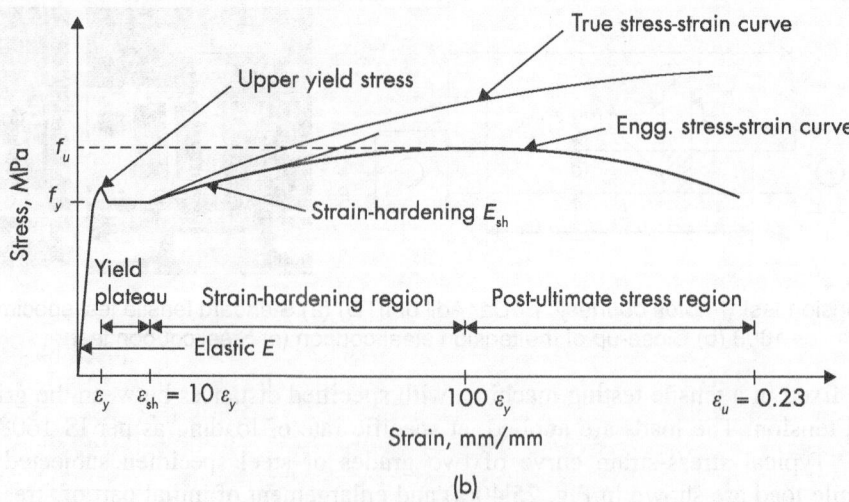

**Fig. 25.40** Typical stress-strain curves (a) Stress-strain curves of ordinary and high strength steel (b) Stress-strain curve of a mild steel specimen

As seen in Fig. 25.40(a), after reaching the ultimate tensile stress, a localized reduction in area, called *necking*, begins, and elongation continues with diminishing load until the specimen breaks. After failure, the fractural surface of the two pieces is found to form a cup and cone arrangement. This cup and cone fracture is considered as an indication of *ductile fracture*.

As shown in Fig. 25.40(a), initially the steel has a linear stress-strain curve whose slope is the Young's elastic modulus, $E$. Thus, the modulus of elasticity is defined as follows:

$$\text{Elastic modulus} = \text{Stress within proportional limit/Strain} \tag{25.46}$$

It can be written as follows:

$$E = f/\varepsilon \tag{25.47}$$

where $E$ is the elastic modulus also called Young's modulus, $f$ is the uniaxial stress below the proportional limit, and $\varepsilon$ is the strain corresponding to stress $f$.

The steel obeys Hooks law in this linear range. That is, it remains elastic and recovers to the original shape perfectly on unloading. The limit of the elastic behaviour is often closely associated with the *yield stress* $f_y$ and the corresponding yield strain $\varepsilon_y = f_y/E$. Beyond this limit, the steel flows plastically without any increase in stress until the strain hardening strain $\varepsilon_{sh}$ is reached. This plastic range is usually considerable, and accounts for the ductility of steel. The stress increases above the yield stress $f_y$, when the tensile hardening strain $\varepsilon_{sh}$ is exceeded, until the ultimate tensile stress $f_u$ is reached. As indicated earlier, at this stage, large local reductions in the cross section occur, and the load capacity decreases until tensile failure takes place.

Yielding is sometimes accompanied by an abrupt decrease in load, as shown in Fig. 25.40(a), which results in upper and lower yield points. The upper yield point ($f_{yu}$) is influenced by the shape of the test specimen and by the testing machine itself and is sometimes completely suppressed. The lower yield point ($f_{yl}$) is much less sensitive and is considered to be more representative. The stress-strain curve shown in Fig. 25.40(b) is typical of low-carbon (mild) steel. It has to be noted that both upper and lower yield points tend to increase with increase in speed of loading (strain rate). Typical value of the ratio $f_{yu}/f_{yl}$ for normal structural steel is about 1.05 to 1.10. The term yield stress is commonly used to mean either yield point or yield strength when it is not necessary to make the distinction. Steel in compression has the same elastic modulus as in tension. The lower yield stress is also the same and there is about the same length of level yielding (contraction).

**Poisson's ratio**   It, denoted by $\nu$, can also be determined from tension test by measuring transverse strains during elastic behaviour. A strain gauge may be used for this purpose. Then, the Poisson's ratio can be determined by using the following formula:

$$\nu = -\frac{\text{Transverse strain}}{\text{Longitudinal strain}} = -\frac{\varepsilon_y}{\varepsilon_x}$$  (25.48)

Since $\varepsilon_y$ is of opposite sign of $\varepsilon_x$, positive $\nu$ is obtained. Poisson's ratio of steel is often around 0.3.

**Ductility**   Measurement of ductility is obtained from the tension test specimen by determining the percentage elongation (comparing final and original lengths over a specified gauge distance).

%  Elongation = [(Elongated length between gauge point – Gauge length)
/Gauge length] × 100  (25.49)

The measured elongation is influenced by the gauge length, strain rate of test, and failure within or outside the gauge length. Minimum elongation of Grade Fe 415D steel specimen on a gauge length of $5.65\sqrt{(A_o)}$ is 20% (see also Table 13.9 of Chapter 13).

**Toughness**   In structural steel design, toughness is a measure of the ability of steel to resist fracture under impact loading, i.e., the capacity to absorb large amounts of energy. The area under the stress-strain curve is a measure of toughness (see Fig 13.11 of Chapter 13).

## 25.20.2 The Charpy Impact test

It is also known as *the CVN -test*. It is a standardized high strain-rate test that determines the amount of energy absorbed by a material during fracture (IS 1757). In this test, a standard rectangular simply supported beam having a V-notch at mid-length is subjected to an impact load by a swinging pendulum, and the energy absorbed by the specimen is measured (see Fig. 25.41). The larger the impact value, the tougher the material. The temperature of the specimen is varied and the energy absorbed by each specimen is recorded, and energy – temperature curve is plotted as shown in Fig. 25.42. From this curve, a *transition temperature* corresponding to some level of energy absorption (usually, 20 or 27 J) is selected. The transition temperature is the temperature below which fractures are mostly brittle and above which fractures are mostly ductile. A value of 20 J is gen-

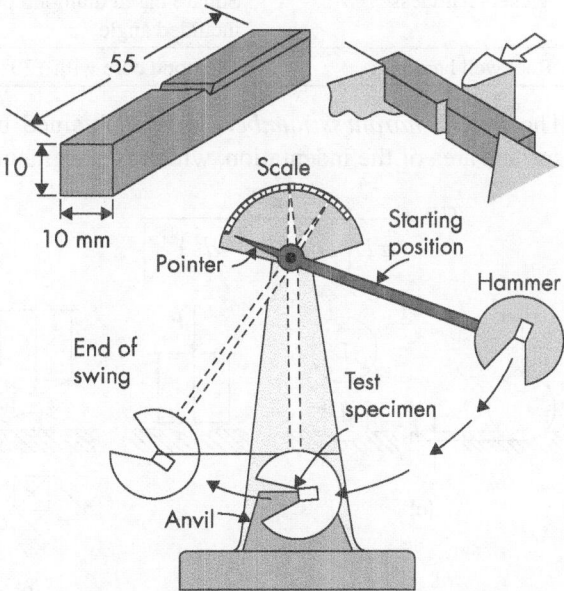

**Fig. 25.41** Charpy impact test as per IS 1757

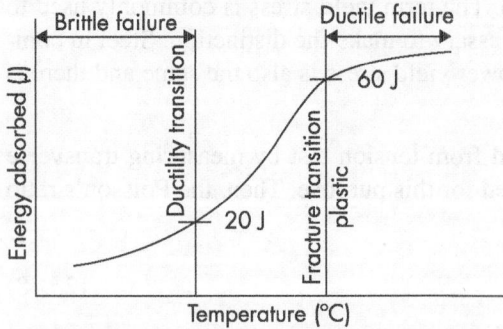

**Fig. 25.42** Energy-temperature curve of steel

erally used to define the lower transition temperature and 60 J has sometimes been specified for the higher one (Gaylord, Jr., et al., 1992). If the temperature is more than the ductility transition temperature, there will be appreciable plastic flow at the root of a notch before cracking begins.

Though the CVN test has been a common means of determining notch toughness, other tests such as drop-weight, nil-ductility test (NDT), dynamic tear test, etc., have also been developed. Details of these tests may be found in Barsom and Rolfe (1987).

### 25.20.3 Tests to Determine Hardness

Several methods are available to determine the hardness of steel and other metals. In all these methods, an 'indentor' is forced on the surface of the specimen. On removal, the size of indentation is measured using a microscope. Based on the size of the indentation, the hardness of the specimen is determined. The various methods used to determine hardness are listed in Table 25.10 (also see Fig. 25.43). Rockwell hardness testing is not normally used for structural steel (Peterson, 1990).

**Table 25.10** Various testing methods used to determine hardness of metals

| Testing methods | Indentor | Typical values for steel |
|---|---|---|
| Brinell hardness | Hard steel ball | 150–190 |
| Vickers hardness | Square based diamond pyramids of 135° included angle | 157–190 |
| Rockwell hardness | Diamond core with 120° included angle | 80–105 |

The *Brinell hardness number*, HB, is obtained by dividing the applied force $P$ (kg), by the curved surface area of the indentation, which is a segment of a sphere, as follows:

$$HB = \frac{2P}{\pi D \left[ D - \left( D^2 - d^2 \right)^{0.5} \right]} \qquad (25.50)$$

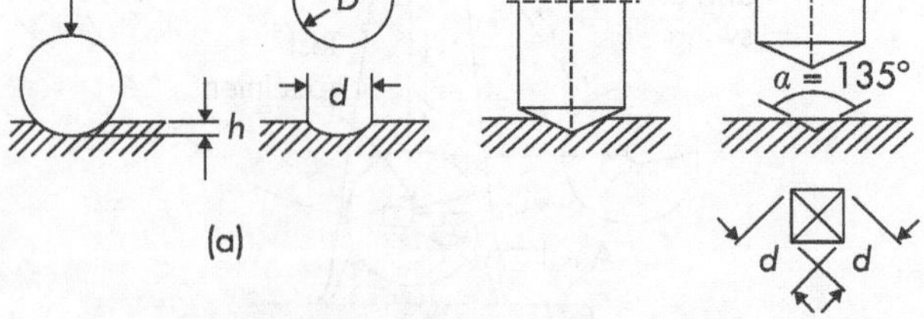

where $D$ is the diameter of the ball and $d$ is the diameter of the indentation [see Fig. 25.43(a)], both measured in mm.

The *Vickers harness number*, HV, is obtained by dividing the applied force $P$ (kg) by the surface area of the pyramidal depression (mm²) [see Fig. 25.43(b)] as follows:

$$HV = \frac{2P}{d^2} \sin\left( \frac{\alpha}{2} \right) \qquad (25.51)$$

**Fig. 25.43** Illustration of methods to determine hardness

In Rockwell test, a diamond point (a cone with an included angle of 120° and a slightly rounded end) or a steel ball (ranging in diameter from 1.6 to 12.7 mm) is employed as the indenter. Rockwell test differs from other hardness tests in that the depth of the indentation is measured, rather than the size. A small initial force of about 10 kg is first applied and the penetration ($h_1$) is measured. Then, the major load (60 to 150 kg) is applied and the additional penetration due to the major load ($h_2$) is measured. The Rockwell hardness number is calculated as follows:

$$HRX = M - \frac{\Delta h}{0.002} \tag{25.52}$$

where $\Delta h = h_2 - h_1$ in mm and $M$ is the upper limit of the scale. For regular Rockwell hardness, $M = 100$ for all scales using the diamond point indenter (A, C, D scales) and $M = 130$ for all scales using ball indenter (B, E, M, R scales). In *HRX*, the *X* denotes the scale and has to be specified, such as 60 HRC (i.e., 60 points in C scale). Usually, instead of using Eq. (25.52), the hardness is read directly from the dial of the machine.

It is possible to approximately estimate the UTS, $f_u$, for low- and medium-strength carbon alloy steels from the Brinell hardness, by using the following relation:

$$f_u = 0.35(HB) MPa \tag{25.53}$$

ASTM E140: 02 provides conversion tables showing the relationship among Brinell hardness, Vickers hardness, and Rockwell hardness for different metals.

## 25.21 Testing for Quality of Glass

Specifications for flat transparent glass used for glazing are provided in IS 2835:1987. The following are to be tested:

*Quality*   Glass should be flat, transparent, and clear. It should be free from cracks, scratches, bubbles, etc., that impair the visibility (see also Table 2 of IS 2835, for the distribution of allowable defects).

*Tolerance of size*   The following tolerances are allowed in the length and width of the sheet glass: ±1.5 mm for glass of thickness, 1 to 3 mm, ±2 mm for thickness of 3.5 to 6.3 mm, ±3 mm for thickness of 8 to 12 mm, ±4 mm for thickness of 15 to 19 mm, and ±5 mm for thickness of 25 to 32 mm.

*Waviness*   The sheet glass should not show any distortion of light when tested according to the standard test specified in Appendix A of IS 2835 for waviness.

## 25.22 Testing of Clay Tiles

Tests for water absorption, permeability, flexural strength, breaking strength, and impact resistance are specified in IS 2960, IS 654, IS 1464, and IS 1478 for clay tiles.

### 25.22.1 Water Absorption Test

Six tiles are selected randomly and dried in an oven at a temperature of 105 to 110°C until they attain constant weight and then cooled and weighed. These tiles are then immersed completely in clean water at 24 to 30°C for 24 h, surface water wiped off, and weighed again. The percentage of water absorption is calculated as follows:

$$\text{Percentage absorption} = \left( \frac{B - A}{A} \right) \times 100 \tag{25.54}$$

where $A$ is the weight of dry specimen and $B$ is the weight of the specimen after 24 h immersion in cold water.

## 25.22.2 Permeability Test

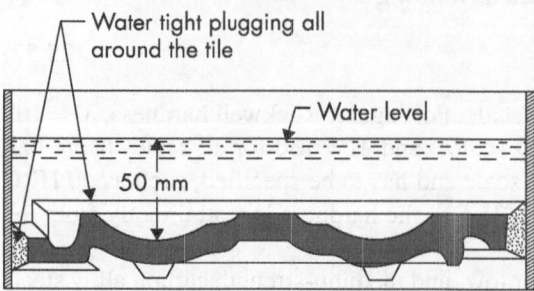

**Fig. 25.44** Arrangement for permeability test

The test is conducted in a rectangular trough (see Fig. 25.44), which is open at bottom, the dimensions at bottom being equal to the size of the Mangalore Pattern tile. The test is conducted at temperature of $27 \pm 2°C$ and relative humidity of $65 \pm 5\%$. The tile specimen is fitted at the bottom of the trough and the space between the tile and the sides of the trough is plugged watertight with a suitable material like wax or bitumen. Water is poured into the mould so that it stands over the lowest tile surface to a height of 50 mm, and allowed to stand for a period of 6 h. The bottom of the tile is carefully examined to see whether the water has seeped through the tile.

## 25.22.3 Test for Flexural strength

As per IS 2690 and IS 654, six tiles are chosen randomly and immersed in water at a temperature of $27 \pm 2°C$ for 24 h. Each of these wet but surface wiped tiles is supported on parallel steel bearers (A and B), with a span equal to three-fourths the dimensions of the tile, in the compression testing machine as shown in Fig. 25.45(a). Rectangular tiles shall be supported along longer face. The load is applied perpendicular to the span, through a third steel bearer placed midway between the supports, at a uniform rate of 450 to 550 N/min. The average breaking load of the six tiles should not be less than that specified in the code. The flexural strength is calculated as follows:

$$\text{Flexural strength} = \frac{15WS}{bt^2} \text{ N/mm}^2 \tag{25.55}$$

where $W$ is the breaking load in N, $S$ is the span (3/4$^{th}$ tile length) in mm, $b$ is the width of tile in mm, and $t$ is the thickness of tile in mm.

## 25.22.4 Test for Breaking Strength

Six ridge tiles, selected at random for every 5000 tiles, are tested as per IS 1464:1992. They are first soaked in water for 24 h, and then wiped dry. The two longitudinal edges of the ridge tile are kept in the normal position, over two strips of 25-mm thick rubber sheets placed on the compression testing machine. Then, the load is applied on the ridge of the tile, at a uniform rate of 2.7 kN/min, through a rubber-lined wooden block of size $75 \times 100 \times 300$ mm as shown in Fig. 25.45(b). The breaking strength is obtained by dividing the breaking load by the length of the tile and the average of the values, in N/mm is considered.

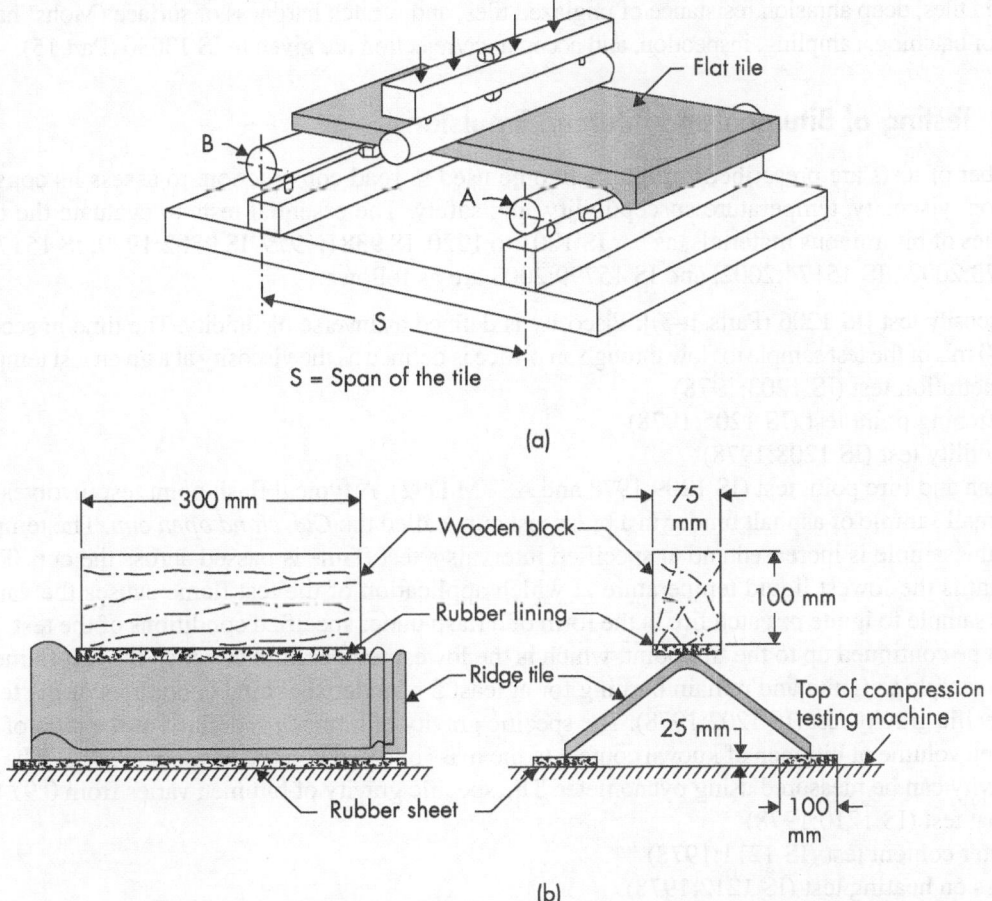

**Fig 25.45** Testing arrangement for flexural strength (a) Plain or flat tile (b) Ridge tile

## 25.22.5 Test for Impact Resistance

The apparatus for the impact test as per IS 1478 consists of an upright stand fixed to a heavy base, with a clamp holding between its jaws a steel ball of 35-mm diameter and a mass of 170 g. Three tiles selected at random are dried in an oven at temperature of 100 to 110°C to a constant weight and cooled to room temperature. Then, the tile is placed horizontally, with its face upwards over a 25-mm thick rubber sheet, which in turn be placed over a rigid horizontal surface. The tile is positioned in such a way that when the ball is released from the clamp it falls vertically at its centre. Starting from a height of 75 mm, the steel ball is released in different heights increasing in steps of 75 mm, until the tile fractures, and the average load that causes the fracture is noted.

## 25.22.6 Tests for Ceramic Tiles

Several tests are specified for the determination of properties of ceramic tiles in IS 13630 (Parts 1–14), which include: dimensions and surface quality, water absorption and bulk density, moisture expansion (using boiling water), linear thermal expansion, resistance to thermal shock, modulus of rupture and breaking strength, frost and impact resistance, chemical resistance of glazed/unglazed tiles, crazing and surface abrasion resistance of glazed tiles, deep abrasion resistance of unglazed tiles, and scratch hardness of surface (Mohs' hardness). Rules for batching, sampling, inspection, and acceptance/rejection are given in IS 13630 (Part 15).

## 25.23 Testing of Bitumen and Bitumen Emulsions

A number of tests are prescribed for bitumen to be used in road construction, to assess its consistency, gradation, viscosity, temperature susceptibility, and safety. The essential tests to evaluate the different properties of bituminous materials, as per IS 1201 to 1220, IS 9381:1979, IS 9382:1979, IS 15172:2002, IS 15173:2002 , IS 15174:2002, and IS 15799:2008 are as follows:

1. Viscosity test [IS 1206 (Parts 1–3)]. Viscosity is defined as inverse of fluidity. The time in seconds for a 50 mL of the test sample to flow through an orifice is defined as the viscosity at a given test temperature.
2. Penetration test (IS 1203:1978)
3. Softening point test (IS 1205:1978)
4. Ductility test (IS 1208:1978)
5. Flash and Fire point test (IS 1209:1978 and ASTM D92). A typical flash point test involves heating a small sample of asphalt binder in a brass test cup, called the *Cleveland open cup*. The temperature of the sample is increased and at specified intervals a test flame is passed across the cup. The flash point is the lowest liquid temperature at which application of the test flame causes the vapours of the sample to ignite *or* catch fire in the form of a flash under specified conditions of the test. The test can be continued up to the fire point, which is the lowest temperature at which the test flame causes the sample to ignite and remain burning for at least 5 s, under specified conditions of the test.
6. Specific gravity test (IS 1202:1978). The specific gravity of bitumen is defined as the ratio of mass of given volume of bitumen of known content to the mass of equal volume of water at 27°C. The specific gravity can be measured using pycnometer. The specific gravity of bitumen varies from 0.97 to 1.02.
7. Float test (IS 1210:1978)
8. Water content test (IS 1211:1978)
9. Loss on heating test (IS 1212:1978)
10. Durability test (IS 15799:2008)

Only the first four tests are briefly described here. For other tests, the corresponding IS code has to be consulted.

### 25.23.1 Viscosity Test

The viscosity test at 60°C is conducted by using the absolute viscosity test by vacuum capillary viscometer [IS 1206 (Part 2) and ASTM D2171], which measures the viscosity in units of poise. The viscosity test at 135°C is conducted by using the kinematic viscosity test [IS 1206 (Part 3) ASTM D2170], which measures the kinematic viscosity in units of Stokes or centi-Stokes. To convert from kinematic viscosity (in units of Stokes) to absolute viscosity (in units of poises), one simply multiplies the number of Stokes by the density in units of g/cm³.

To find the absolute viscosity, a bitumen sample is heated to a temperature not more than 90°C, i.e., above its approximate softening point, until it becomes sufficiently fluid. About 20 mL of the solution is stirred in a container maintained at a temperature of 135 ± 5.5°C to drive away entrapped air. Then, the hot bitumen is poured in the Cannon–Manning vacuum viscometer through the larger diameter filling tube (see Fig. 25.46). The viscometer is then placed in an oven or bath maintained at 135 ± 5.5°C for a period of 10 ± 2 min to allow larger air bubbles to escape. The test bath is maintained at a temperature of 60 ± 0.1°C, and viscometer is placed vertically in the test bath. A vacuum of 30 ± 0.05 cm of mercury is established in the vacuum system and the viscometer is connected to it with the valve closed. After the viscometer has been in the bath for 30 ± 5 min, the valve is opened and the bitumen is allowed to flow inside the viscometer. The time required for the leading edge of the meniscus to pass between successive pairs of timing marks is measured. The first flow time, which exceeds 60 s between a pair of timing marks, is reported.

The absolute viscosity is calculated as follows:

Viscosity in poises = $K\,t$

where K is the calibration factor in poise per second supplied with the viscometer tube for the pair of timing marks where the flow time exceeded 60 s, and $t$ is the flow time in seconds.

For finding the kinematic viscosity at 135°C, the same oil bath as used for absolute viscosity can be used, but the viscometer tube will be different. At this temperature, bitumen flows readily so no vacuum needs to be applied.

### 25.23.2 Penetration Test

This test measures the hardness of bitumen, as per IS 1203:1978, by measuring the depth in tenths of a millimetre to which a standard loaded needle will penetrate vertically in 5 s. The penetrometer consists of a needle assembly, as shown in Fig. 25.47, with a total weight of 100 g and a device for releasing and locking in any position. The material is softened to a pouring consistency between 60 (tars and pitches) and 90°C (bitumen), stirred thoroughly, and poured into a container with a depth at least 10–15 mm in excess of the expected penetration. The container is cooled in atmosphere at a temperature of 15 to 30°C, for 1.5 to 2 h. Then, it is placed in temperature controlled water bath in the penetrometer, at a temperature of 25°C, for a period of 1.5 to 2 h, and then the test is conducted at a specified temperature of 25°C. It may be noted that the pouring temperature, size of the needle, weight placed on the needle and the test temperature will affect the penetration value. In hot climates, a lower penetration grade is preferable.

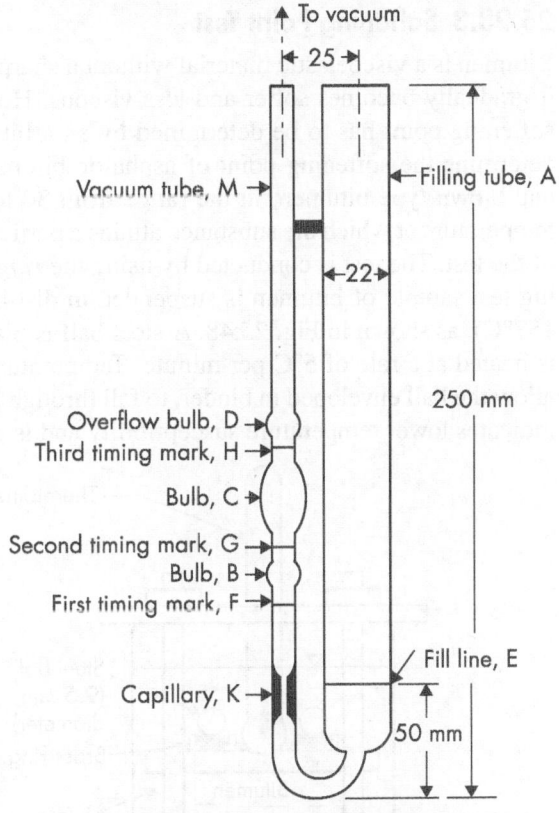

All dimensions are in millimetres.

**Fig. 25.46** Cannon–Manning vacuum viscometer based on IS 1206 (Part 2)

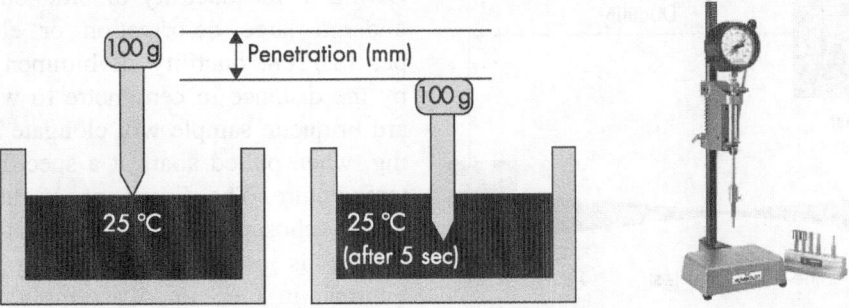

**Fig 25.47** Schematic of test setup for the measurement of penetration
(*Photo Courtesy*: Humboldt Mfg Co.)

### 25.23.3 Softening Point Test

Bitumen is a viscoelastic material without a sharply defined melting point. As the temperature is raised, it gradually becomes softer and less viscous. Hence, for obtaining consistent value of the results, the softening point has to be determined by an arbitrary and closely defined method. This test is done to determine the softening point of asphaltic bitumen and fluxed native asphalt, road tar, coal tar pitch, and blown-type bitumen, in the range from 30 to 157°C, as per IS 1205:1978. Softening point is the temperature at which the substance attains a particular degree of softening under the specified conditions of the test. The test is conducted by using the *ring-and-ball apparatus*. In this test, a brass ring containing test sample of bitumen is suspended in distilled water (30 to 80°C) or USP glycerin (above 80 to 157°C), as shown in Fig. 25.48. A steel ball is placed upon the bitumen sample and the liquid medium is heated at a rate of 5°C per minute. Temperature is noted when the test sample is sufficiently soft to allow the ball enveloped in binder, to fall through a height of 25 mm. Generally, a higher softening point indicates lower temperature susceptibility and is preferred in hot climates.

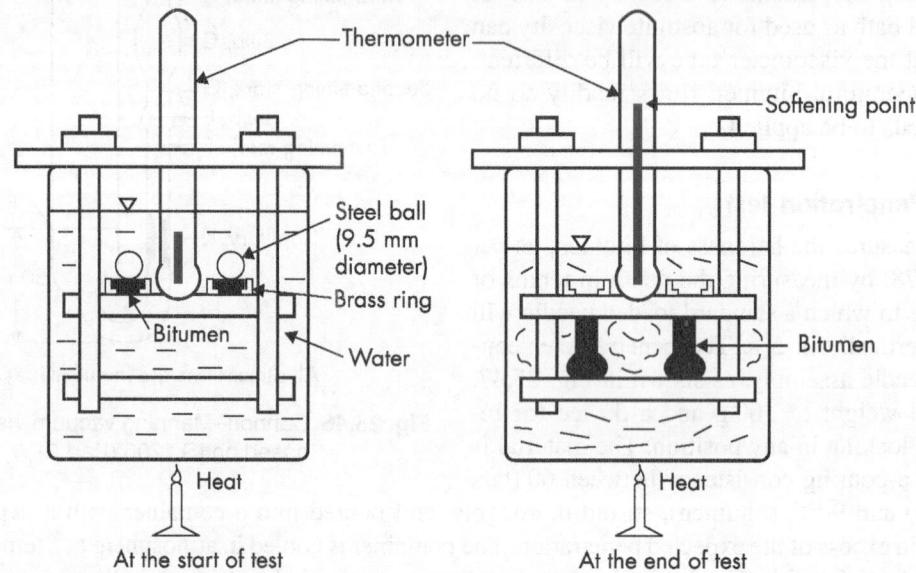

**Fig. 25.48** Ring-and-ball test for determining softening point temperature of bitumen

### 25.23.4 Ductility Test

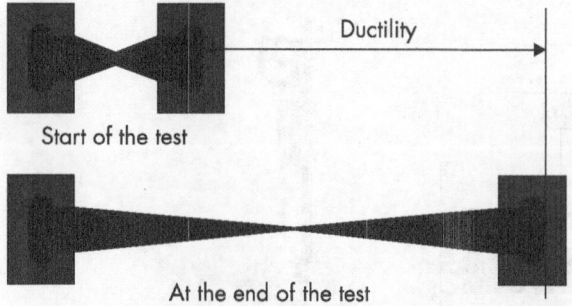

Ductility

Start of the test

At the end of the test

**Fig. 25.49** Ductility test

Owing to the ductility of bitumen it is able to undergo large deformation or elongation. As per IS 1208, ductility of bitumen is measured by the distance in centimetre to which a standard briquette sample will elongate before breaking, when pulled apart at a specified speed and temperature. The dimension of the briquette is usually chosen as 1 cm². This bitumen briquette sample is prepared by heating and pouring bitumen in the mould assembly, as shown in

Fig. 25.49, which is placed on a plate. This sample, along with the mould, is cooled in air and then in a water bath at a temperature of 27°C. The excess bitumen protruding the mould is cut and the surface is levelled using a hot knife. This mould containing the sample is kept in a water bath for about 90 min. The sample is then hooked on to tensile testing machine, after removing the mould on the sides, and elongated as shown in Fig. 25.49, until the specimen breaks. The distance up to which it elongates, before breaking, is recorded as the ductility value in centimetre. It has to be noted that factors such as pouring temperature, testing temperature, and rate of pulling, may affect the ductility of bitumen. Refer to Table 19.2 of Chapter 19 for the minimum ductility values specified by the BIS, for the various grades of bitumen.

More details on other tests performed on asphalt and bitumen may be found in Khanna et al. (2013).

## 25.24 Testing of Polymers

The physical and mechanical testing of polymers is an important part of the product development and production process. Physical testing is done to determine density, hardness, and scratch resistance. The mechanical properties of polymeric material that are of interest are the strength (tensile, flexural, shear, and compressive strength), elasticity, viscoelasticity, and anisotropy. Typically, polymeric materials are characterized as elastomers, plastics, or rigid polymers depending their mechanical properties (see Fig. 25.50). Polymers exhibit a wide variation of behaviour in stress-strain tests, ranging from hard and brittle to ductile, including yield and cold drawing. The utility of stress-strain tests for design with polymeric materials can be greatly enhanced if tests are carried out over a wide range of temperatures and strain rates (see Fig. 25.51: in this figure, $T_g$ is glass transition temperature and $T_m$ is the melting temperature). Anisotropy plays a large role in the mechanical properties of crystalline or semi-crystalline polymers. The degree of crystallinity also affects the physical properties of a polymer. Differential scanning calorimetry can be used as a rapid method for determining polymer crystallinity based on the heat required to melt the polymer. Tensile testing is used to measure tensile strength, yield strength, and Young's modulus. The actual deformation of the polymer chains during tensile testing is very complex, but some simple models are often used to understand the deformation mechanisms during tensile elongation. Prior to the application of the tensile load, the crystalline lamellae are closely connected by the amorphous regions between the tightly packed polymer chains. When the specimens of amorphous and semi-crystalline polymers begin to deform, the stress causes the polymer chains to elongate and uncoil. If the stress is applied in the direction of chain alignment, the polymer will exhibit a higher yield stress and strength. The amount of crosslinking (crosslink density) and the molecular weight between the crosslinks govern the tensile stress-strain curves. However, if the stress is applied normal to the direction of chain alignment, the yield stress will decrease.

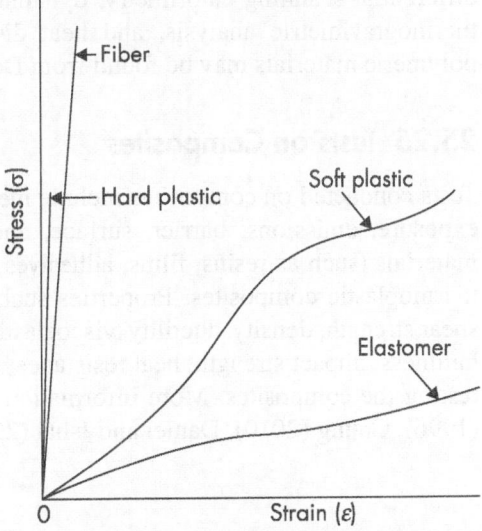

**Fig. 25.50** Stress-strain curve of polymers

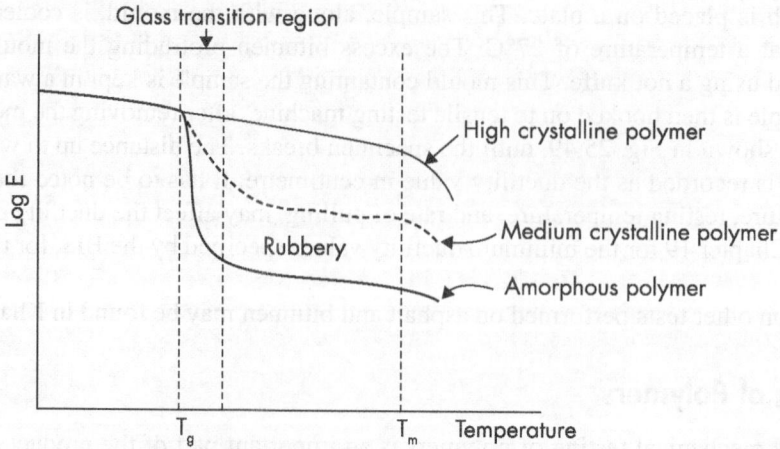

**Fig. 25.51** Effect of temperature on polymers

There are many properties of polymeric materials that influence their mechanical properties. As the degree of polymerization goes up, so does the polymer's strength, as longer chains have higher interactions and chain entanglements. *Crazes* are small cracks that form in a polymer matrix, which can increase the strength and decrease the brittleness of a polymer by allowing the small cracks to absorb higher stress and strain without leading to fracture. These crazes can be seen with transmission electron microscopy and scanning electron microscopy. Crosslinking, typically seen in thermoset polymers, can also increase the modulus, yield stress, and yield strength of a polymer (Dowling, 2012).

Dynamic mechanical analysis is the most common technique used to characterize viscoelastic behaviour, common in many polymeric systems. Other tests include ultra-violet/visible, spectrophotometry, differential scanning calorimetry, dynamic mechanical thermal analysis, thermomechanical analysis, thermogravimetric analysis, and heat distortion temperature. More information on the testing of polymeric materials may be found from Dowling (2012).

## 25.25 Tests on Composites

Tests conducted on composites include mechanical, physical, electrical, optical, thermal, flammability, exposure, emissions, barrier, surface, and chemical, which identify the characteristics of the raw materials (such as resins, films, adhesives, and fillers) or laminates such as thermoset composites and thermoplastic composites. Properties such as elastic modulus, yield strength, UTS, compressive and shear strength, density, ductility, viscoelasticity, fatigue resistance, creep, abrasion resistance, toughness, hardness, impact strength, heat resistance, resistance to cracking, and flame resistance are found out by testing the composites. More information on testing of composite materials may be found in Adams (1996), Chung (2010), Daniel and Ishai (2006), and MIL-HDBK-1701F (2002).

# SUMMARY

## PART I

- Before using any material, it is important to test whether it has the assumed or required properties.
- Samples of these materials, called specimens, are often subjected to a variety of mechanical tests to measure their tensile or compressive strength, toughness, hardness, and other properties. The number of samples to be tested is usually specified in codes based on probability and statistical theories.
- Natural stones are subjected to the following tests: (a) absorption test, (b) Smith's test, (c) toughness test, (d) hardness test, (e) acid test, (f) crystallization test, (g) attrition test, (h) crushing test, (i) freezing and thawing test, (j) impact test, and (k) microscopic test.
- Burnt clay bricks are tested for (a) compressive strength, (b) water absorption, (c) efflorescence, (d) dimensional tolerance, (e) hardness, (f) soundness, and (g) structure.
- Concrete blocks are tested for (a) measurement of dimensions, (b) block density, (c) compressive strength, (d) water absorption, (e) drying shrinkage, and (f) moisture movement.
- The laboratory tests on lime include (a) the LOI, (b) fineness, (c) soundness, (d) compressive strength, and (e) popping and pitting test. In addition, several field tests could be conducted on lime as per IS 1624.
- The chemical tests on cement are concerned with permissible limits for the IR, the LOI and other compounds, and impurities such as magnesium oxide and sulphate, and tests such as X-ray fluorescence spectrometer test, atomic absorption spectrophotometer method, and calorimetric analysis.
- The physical tests on cement are to determine (a) fineness (Blaine air-permeability test), (b) soundness (Le Chatelier's method and autoclave test), (c) consistency (test with Vicat apparatus), (d) setting time (with vicat apparatus), (e) compressive strength (on cement cubes of 70.6-mm size), (f) heat of hydration (using calorimeter), and (g) specific gravity (using Le-Chatelier flask).
- Tests conducted on aggregates include the determination of (a) flakiness and elongation index, (b) limits of deleterious materials (using sedimentation pipette of the Andreason type), (c) specific gravity and density, (d) voids, absorption and bulking, and (d) mechanical properties such as aggregate crushing value, impact value, and abrasion value (using the Los Angeles abrasion testing machine). Tests are also conducted on aggregate to find its soundness, mortar making properties, and alkali aggregate reactivity, and for petrographic examination.
- Mortars are subjected to tests such as: compressive strength, consistency, water retention, tensile strength, adhesion, and brickwork in compression.

## PART II

- Tests on concrete are done when it is green for workability-slump and compacting factors test, and for consistency (Vee-Bee consistency or flow table, and Kelly ball test) and when it is hardened to determine its properties. The tests on hardened concrete include compression tests on cylinder/cube or core samples, tensile strength (through flexure or split cylinder test), and the elastic modulus (compression tests on cylinders using compressometer).
- Several NDTs can also be conducted on concrete and include (a) Schmidt/rebound hammer test, (b) ultrasonic pulse velocity test, (c) impact-echo test, (d) maturity method, and (e) permeability test. There are also a few partial destructive tests such as Windsor probe test (penetration resistance method) and the pullout test.
- IS 456 suggests an equation and some values of standard deviation to determine the target mean strength of concrete, based on compressive tests conducted on limited number of samples.
- Tests conducted on self-compacting concrete (SCC) include (a) flowability (slump flow test), (b) filling ability (slump flow test, V-funnel, and Orimet), (c) passing ability (L-box, J-ring, and U-box), (d) robustness, and (e) segregation resistance or stability (simple column box test and Sieve stability test).
- IS 1708 (Parts 1–18) specify the following tests to determine the properties of timber: (a) moisture content, (b) specific gravity, (c) volumetric shrinkage, (d) radial and tangential shrinkage, and fibre saturation point, (e) static and impact bending strength, (f) compressive strength parallel and perpendicular to grain, (g) hardness, (h) shear strength parallel to grain, (i) tensile strength parallel and perpendicular to grain, (j) torsional strength, and (k) brittleness by Izod or Charpy impact.
- Tests conducted on steel include tension test, Charpy V-notch test, hardness test (Brinell/Vickers/Rockwell), fracture toughness test, creep test, and fatigue tests.

- Tests on clay tiles include (a) water absorption, (b) permeability, (c) flexural strength, (d) breaking strength, and (e) impact resistance.
- Various tests on bitumen and bitumen emulsions are conducted and include tests for (a) viscosity test, (b) penetration, (c) softening point, and (d) ductility.
- Several tests are also prescribed for polymers and composites. The Indian standard codes listed in the references of this chapter in the ORC should be consulted before attempting to do the prescribed tests for each material.

## EXERCISES

## Multiple-choice Questions

### PART I

1. Which of the following test is used to determine the rate of wear of stones?
   - (a) Crushing test
   - (b) Abrasion test
   - (c) Attrition test
   - (d) Impact test

2. Mohs' scale is used on stone to determine the following:
   - (a) Toughness
   - (b) Hardness
   - (c) Flakiness index
   - (d) Durability

3. Which of the following tests on stones are correctly matched?
   - (a) Absorption test
   - (i) Resistance to impact
   - (b) Attrition test
   - (ii) Resistance to abrasion
   - (c) Crushing test
   - (iii) Resistance to freezing and thawing
   - (d) Dorry's testing machine
   - (iv) Resistance to load

4. During hydrochloric acid test on limes, a gel is formed. If the gel is flowing freely, it indicates
   - (a) Class A lime
   - (b) Class C lime
   - (c) Class B lime
   - (d) Class F lime

5. During the ball test of lime, if it expands and disintegrates within a few minutes, it indicates
   - (a) Class A lime
   - (b) Class B lime
   - (c) Class C lime
   - (d) Class F lime

6. During the impurity test of lime, if the residue is less than 5%, it indicates
   - (a) Class A lime
   - (b) Class B lime
   - (c) Class C lime
   - (d) Class F lime

7. During the workability test, the following spread indicates a good workable mortar
   - (a) 100–110 mm
   - (b) 120–130 mm
   - (c) 130–150 mm
   - (d) 150–160 mm

8. Match the following tests and apparatus correctly.

   | (Cement test) | (Apparatus) |
   |---|---|
   | (a) Fineness | (i) Vicat |
   | (b) Consistency | (ii) Le-Chatelier |
   | (c) Soundness | (iii) Calorimeter |
   | (d) Heat of hydration | (iv) Blaine Type |

### PART II

9. The 28-day target mean compressive strength of concrete as per IS 10262 is:
   - (a) $f_{ck}' = f_{ck} + 1.65 \times s$
   - (b) $f_{ck}' = f_{ck} + 1.6 \times s$
   - (c) $f_{ck}' = f_{ck} + 1.5 \times s$
   - (d) None of these

10. IS 456 assumes the following standard deviation for M20 concrete
    - (a) 3.0 N/mm$^2$
    - (b) 3.5 N/mm$^2$
    - (c) 4.0 N/mm$^2$
    - (d) 4.0 N/mm$^2$

11. The criteria for acceptance in IS 456 for compressive strength of concrete is (with s for the standard deviation)

   - (a) $\geq f_{ck} + 1.34 \times s$ or $f_{ck} + 7$ N/mm$^2$, whichever is greater for mean, and $f_{ck} - 3$ N/mm$^2$ for individual test
   - (b) $\geq f_{ck} + 0.825 \times s$ or $f_{ck} + 5$ N/mm$^2$, whichever is greater for mean, and $f_{ck} - 4$ N/mm$^2$ for individual test
   - (c) $\geq f_{ck} + 0.825 \times s$ or $f_{ck} + 3$ N/mm$^2$, whichever is greater for mean, and $f_{ck} - 3$ N/mm$^2$ for individual test
   - (d) None of these

12. If the quantity of concrete involved is 20 m$^3$, the required number of samples to be tested for quality control is
    - (a) 2
    - (b) 3
    - (c) 4
    - (d) 5

13. Minimum elongation of Grade FE 415D steel specimen on the gauge length is
    (a) 10%          (c) 20%
    (b) 15%          (d) 12.5%

14. The energy absorption at lower and higher transition temperatures specified in Charpy V-notch test are
    (a) 20 and 60 J      (c) 20 and 50 J
    (b) 15 and 40 J      (d) 25 and 60 J

15. In the test to determine the flexural strength of tiles, the load is applied in compression testing machine at an uniform rate of
    (a) 250 to 300 N/min
    (b) 350 to 450 N/min
    (c) 450 to 550 N/min
    (d) 500 to 550 N/min

16. To test the breaking strength of ridge tiles, for every 5000 tiles a sample of
    (a) 4 tiles are tested
    (b) 6 tiles are tested
    (c) 8 tiles are tested
    (d) 10 tiles are tested

17. To test the impact resistance of tiles, a steel ball of the following size and mass is dropped on them
    (a) 25-mm diameter and 200-g mass
    (b) 35-mm diameter and 170-g mass
    (c) 40-mm diameter and 200-g mass
    (d) 35-mm diameter and 150-g mass

18. Ring-and-ball apparatus is used for finding the following property of bitumen
    (a) penetration        (c) softening point
    (b) viscosity          (d) ductility

## Review Questions

### PART I

1. What are the simple field tests that are carried out to determine the suitability of stones and determine their quality?
2. Briefly explain the laboratory tests that are carried out to determine the quality of stones.
3. Briefly describe the following tests of stones:
    (a) Attrition test          (b) Crushing test          (c) Page impact test
    (d) Hardness test using Dorry's testing machine, Also, state the significance of each of them.
4. List the tests that are conducted on bricks to determine their properties.
5. Explain a test to determine the water absorption of a brick.
6. What is efflorescence and how to determine whether a brick will not have it?
7. Name any three field tests for lime.
8. Name any three laboratory tests for lime.
9. Describe briefly the following tests of lime:
    (a) Impurity test          (b) Workability test          (c) Fineness test          (d) Soundness test
    (e) Popping and pitting test  (f) Compressive strength test
10. Write short notes on the following tests on cement:
    (a) Fineness test          (b) Soundness test          (c) Consistency test
    (d) Initial and final setting test                      (e) Compressive strength test
11. What is the effect of fine grinding of cement? Describe the method of determining fineness by air-permeability method of testing.
12. Describe the test for determining the heat of hydration of cements.
13. How is the specific gravity of cement determined?
14. How are the following determined using chemical tests of cement?
    (a) Loss on ignition       (b) Silica content       (c) Sulphuric anhydride   (d) Insoluble residue
15. Briefly describe the following tests:
    (a) Specific gravity       (b) Aggregate crushing value   (c) Aggregate impact value
    (d) Aggregate abrasion value   (e) Ten percent fines value   (f) Petrographic examination
16. Briefly describe the tests to determine the following:
    (a) Deleterious materials   (b) Organic impurities
    (c) Soundness               (d) Fineness modulus
17. Describe the two tests conducted to determine alkali aggregate reactivity.
18. List any four tests carried out to check the quality of mortars.
19. Describe the procedure for determining the compressive strength of mortar.
20. How is the consistency of mortar determined as per IS 2250:1981?
21. How is the water retention of mortar determined as per IS 2250:1981?

22. How is the tensile strength of mortar determined?
23. How is the adhesion of mortar determined by couplet tests?
24. How does brickwork fail under compression? How is the compressive strength of masonry determined?

# PART II

25. Name any three tests conducted on fresh concrete.
26. What is workability of concrete? What is the most commonly used test to measure workability? Describe slump test.
27. Describe the procedure used to determine workability of fresh concrete using compacting factor test.
28. Describe the procedure used to determine consistency of fresh concrete using Vee-Bee consistometer test.
29. Describe the procedure used to determine the flowability of fresh concrete using flow table apparatus.
30. How is Kelly ball test used to measure the consistency of concrete?
31. Name any three tests conducted on hardened concrete.
32. How is compressive strength of concrete determined?
33. Describe flexure test of hardened concrete. What is measured using this test?
34. Explain the split tensile strength test conducted on concrete cylinders.
35. Describe the test conducted to determine elastic modulus of concrete.
36. Name any five non-destructive tests performed on concrete.
37. State the advantages of non-destructive methods.
38. What is Schmidt hammer? How is it used to determine the compressive strength of concrete?
39. Explain the ultrasonic pulse velocity method of testing of concrete.
40. How is the impact-echo test used to determine the integrity of concrete?
41. What is the principle behind the maturity method? Write down the two expressions that are commonly used in the maturity method. How can the concrete strength be determined using the maturity method?
42. Describe the permeability test.
43. What are the two partial destructive tests conducted on concrete?
44. Explain the penetration resistance method.
45. Describe the pullout test.
46. How is quality assurance of concrete ensured?
47. What is meant by the characteristic strength of concrete?
48. State the two criteria specified in IS 456 for accepting the compression strength of test cubes?
49. What are the various tests performed to test the suitability of timber?
50. Describe the methods to determine moisture content in timber. Are the results by electrical moisture meters reliable?
51. Describe the specimen and test to predict the tensile and yield strength of steel.
52. Sketch the typical stress-strain curve of steel, indicating the three important regions.
53. What is Charpy V-notch test? Explain the procedure of this test.
54. Describe the tests used to determine hardness of metals.
55. How are clay tiles tested for water absorption, permeability, flexural strength, breaking strength, and impact resistance?
56. Describe the tests on bitumen to determine the following:
    (a) Viscosity
    (b) Penetration
    (c) Softening point
    (d) Ductility

# ANSWERS

**Multiple-choice Questions**

|  |  |  |  |  |  |  |
|---|---|---|---|---|---|---|
| **1.** (c) | **2.** (b) | **3.** (b)-(ii) | **4.** (b) | **5.** (c) | **6.** (c) | **7.** (d) |
| **8.** (a)-(iv), (b)-(i), (c)-(ii), (d)-(iii) | **9.** (a) | **10.** (c) | **11.** (c) | **12.** (b) | **13.** (c) |
| **14.** (a) | **15.** (c) | **16.** (b) | **17.** (b) | **18.** (c) |  |  |

# BIBLIOGRAPHY

ACI 211.1-91 (Reapproved 2009), *Standard Practice for Selecting Proportions for Normal, Heavyweight and Mass Concrete*, American Concrete Institute, Farmington Hills, MI, 38 pp.

ACI 211.4R-08, *Guide for Selecting Proportions for High-Strength Concrete with Portland Cement and Fly Ash*, American Concrete Institute, Farmington Hills, Michigan, 1993, 13 pp.

ACI 213R-14, *Guide for Structural Lightweight-Aggregate Concrete*, American Concrete Institute, Farmington Hills, Michigan, 2003, 38 pp.

ACI 301M:10, *Specifications for Structural Concrete*, American Concrete Institute, Farmington Hills, Michigan, 2010, 75 pp.

ACI 530-13/530.1–13 *Building Code Requirements and Specification for Masonry Structures and Commentaries*, American Concrete Institute, Farmington Hills, MI, 2013, 380 pp.

ACI E1-16, *Aggregates for Concrete*, Developed by ACI Committee E-701, American Concrete Institute, Farmington Hills, MI, 30 pp.

ACI SP-219-04, *Recycling Concrete and other Materials for Sustainable Development*, 2004, American Concrete Institute, Farmington Hills, MI, 202 pp.

Adams, D.F., *Test Methods for Composite Materials*, CRC Press, Boca Raton, FL., 1996.

AI-2007. MS-4: *The Asphalt Handbook*. Seventh Edition, Asphalt Institute, Lexington, KY, USA, 2007, 832 pp.

Archer, K. and Lebow, S.T., *Wood Preservation, in Primary Wood Processing, Principals and Practice*, Walker, J.C.F. Ed., Second edition, Springer, The Netherlands, 2006, 596 pp.

ASTM C 33 (2003), *Standard Specification for Concrete Aggregates*, American Society for Testing and Materials, Philadelphia, PA.

Aïtcin, P.-C., *High Performance Concrete*, CRC Press, Boca Raton, FL., 1998, 624 pp.

Bapat, J.D., 2012, *Mineral Admixtures in Cement and Concrete*, CRC Press, Boca Raton, FL., 310 pp.

Barsom, J.M., and S.T. Rolfe, *Fracture and Fatigue Control in Structures, Applications of Fracture Mechanics*, 3rd Edition, Butterworth-Heinemann, Woburn, MA, 1999, 548 pp.

Bassi, R., and S.K. Roy (Ed.), *Handbook of Coatings for Concrete*, Whittles Publishers, UK, 2002, 256 pp.

Bažant, Z.P. and J. Planas, *Fracture and Size Effect in Concrete and Other Quasi-brittle Materials*, CRC Press, Boca Raton, FL., 1997, 640 pp.

Bentur, A., and Mindess, S., *Fibre Reinforced Cementitious Composites*, 2nd Edition, Taylor and Francis, Oxon, U.K., 2007, 624 pp.

Bies, D.A. and Hansen, C.H. (2009), *Engineering Noise Control: Theory and Practice*, 4th Edition, CRC Press, London.

Boynton, R.S., *Chemistry and Technology of Lime and Limestone*, 2nd Edition, John Wiley & Sons, Inc., New York, 1980, 592 pp.

Bungey, J.H., Millard, S.G., and Grantham, M.G., *Testing of Concrete in Structures*, 4th Edition, Taylor & Francis, London & New York, 2006, 340 pp.

Calkins, M., *Materials for Sustainable Sites: A Complete Guide to the Evaluation, Selection, and Use of Sustainable Construction Materials*, John Wiley & Sons, Hoboken, NJ, 2009, 457 pp.

Cary, H.B., *Modern Welding Technology*, 5th Edition, Prentice Hall, Upper Saddle River, New Jersey, 2002, 801 pp.

CC & AA 2006b, Compaction of Concrete, *Data Sheet of Cement Concrete & Aggregates*, Cement Concrete & Aggregates Australia, Sydney, June, 7 pp. (www.ccaa.com.au)

CC & AA Report, *Use of Recycled Aggregates in Construction*, Cement & Concrete Aggregates Australia, May 2008, 25 pp.

Cemex Guide, *Educational Guide to Introduction to Mortar*, Cemex Corporation, May 2008, 11 pp. [www.cemex.co.uk]

Chanda, M., *Plastics Technology Handbook*, 5th Edition, CRC Press, Boca Raton, FL., 2017.

Chandra, S., and L. Berntsson, *Lightweight Aggregate Concrete- Science, Technology, and Applications*, William Andrew Publishing, Norwich, New York, 2002, 430 pp.

Christine, B., *Masonry Design and Detailing: For Architects and Contractors*, 5th Edition. Mc-Graw Hill, New York, 2004.

Chung, D.L.L., *Carbon Composites-Composites with Carbon Fibers, Nanofibers, and Nanotubes*, 2nd Edition, Butterworth-Heinemann (Elsevier), 2017, 706 pp.

Chung, D.L.L., *Composite Materials-Science and Applications*, 2nd Edition, Springer-Verlag London Limited, 2010, 349 pp.

Daniel, I.M., and Ishai, O., *Engineering Mechanics of Composite Materials*, Oxford University Press, New York, 2nd Edition, 2006.

DK Pocket Genius- *Rocks and Minerals*, DK Publishing, New York, NY, 2012, 156 pp.

Doelle, L.T. *Environmental Acoustics*, McGraw-Hill Book Company, New York, 1972, pp. 246 pp.

Domone, P.L.J. and Jefferis, S.A.(Editors), *Structural Grouts*, Blackie Academic & Professional (An imprint of Chapman & Hall), London,1994, 236 pp.

Dowling, N.E., *Mechanical Behaviour of Materials-Engineering Methods for Deformation, Fracture, and Fatigue*, 4th Edition, Pearson Education Limited, Essex, England, 2012, 954 pp.

Drysdale R.G., A.A. Hamid, and L.R. Baker, *Masonry Structures: Behavior and Design*, 2nd Edition, Pearson, 1993, 800 pp.

ECBC:2017, *Energy Conservation Building Code of India*, Ministry of Power, Govt. of India, New Delhi, India, 2007, 75 pp. (also *Energy Conservation Building Code 2017*, Bureau of Energy Efficiency, New Delhi, India, 2017, 200 pp.).

Euro Inox, European Stainless Steel Development Group, *Design Manual for Structural Stainless Steel*, Nickel Development Institute, Toronto, Canada, 1994.

Everest, A.F. and Pohlmann, K.C., *Master Handbook of Acoustics,* 6th Edition, The McGraw-Hill Companies, Inc., New York, 2014, 640 pp.

FHWA-2015, *Towards Sustainable Pavement Systems: A Reference Document,* FHWA-HIF-15-002. Federal Highway Administration, Washington, D.C., USA, Jan. 2015, 458 pp.

Freed, E.C., *Green Building & Remodeling for Dummies*, Wiley Publishing, Inc., Hoboken, NJ, 2008, 361 pp.

Gambhir, M.L., 2013, *Concrete Technology—Theory and Practice*, 5th Edition, McGraw Hill Education (India) Pvt. Ltd, New Delhi, 788 pp.

Ganapathy, C., *Modern Construction Materials*, Eswar Press, Chennai, 2015, 409 pp.

GangaRao, H.V.S., N. Taly, and P.N. Vijay, *Reinforced Concrete Design with FRP Composites*, CRC Press, Boca Raton, FL, 2007, 382 pp.

Gupta, T.N., ed., *Building Materials in India: 50 years – A Commemorative Volume*, Building Materials and Technology Promotion Council, New Delhi, 1998, 560 pp.

Harley, B., *Insulate and Weatherize-for Energy Efficiency at Home*, The Taunton Press, Newtown, CT, 2012, 233 pp.

Holmes, S. and M. Wingate, *Building with Lime: A Practical Introduction,* Revised Edition, Practical Action Publishing Ltd., Rugby, UK, 2002, 310 pp.

Horvath, J.S., *Geofoam Geosynthetic: A Monograph, Horvath Engineering*, Scarsdale, NY, 1995, 217 pp. (also see http://www.michiganfoam.com/docs/cgt-2001-1.pdf).

Hurd, M. K., 2005, *Formwork for Concrete*, SP-4, 7th Edition, American Concrete Institute, Farmington Hills, Michigan, 500 pp.

*ICI Handbook on Recycling, Use and Management of C&D Wastes*, Indian Concrete Institute, Chennai, 2016.

IS 10262: 2019, *Concrete Mix Proportioning-Guidelines*, Second Revision, Bureau of Indian Standards, New Delhi.

Jackson, R., T. Luthi, and I. Boyle, 'Mass Timber: Knowing Your Options', *STRUCTURE Magazine*, ASCE, Jan. 2017, pp. 22–25.

Jenkins, A., *300 Tips for Painting & Decorating-Tips, Techniques & Trade Secrets*, Firefly Books, Ontario, Canada, 2014, 176 pp.

Kibert, C.J., *Sustainable Construction; Green Building Design and Delivery*, 2nd Edition, John Wiley & Sons, Inc, Hoboken, NJ, 2007, 432 pp.

Kissell, J.R. and R.L. Ferry, *Aluminum Structures-A Guide to Their Specifications and Design*, Second Edition, John Wiley & Sons, New York, 2002, 532 pp.

Kosmatka, S.H., B. Kerkhoff, and W.C. Panarese, 2016, *Design and Control of Concrete Mixtures*, 16th Edition, Portland Cement Association, Skokie, 616 pp.

Kubal, M.T., *Construction Waterproofing Handbook*, 2nd Edition, McGraw-Hill Education, New Delhi, 2008, 576 pp.

Kustin, K., J. Costa-Pessoa, and D.C. Crans (editors), *Vanadium: The Versatile Metal*, Oxford University Press, 2007, 492 pp.

Lambourne, R. and T.A. Strivens (ed.), *Paint and Surface Coatings-Theory and Practice*, 2nd Edition, Woodhead Publishing Ltd, Cambridge, England, 1999, 784 pp.

Lawson, B., *Embodied Energy of Building materials, Environment Design Guide*, Royal Australian Institute of Architects, Melbourne, 2006. www.environmentdesignguide.com.au

Lea, F. M., 1971, *The Chemistry of Cement and Concrete*, 3rd Edition, Chemical Publishing Co., Inc., New York, 740 pp.

Llewellyn, D.T. and Hudd, R.C., *Steels: Metallurgy and Application*, Butterworths–Heineman, 3rd Edition, 1998, 400 pp.

MacDermott, C.P. and Shenoy, A.V., *Selecting Thermoplastics for Engineering Applications*, 2nd Edition, CRC Press, Boca Raton, FL, 1997, 328 pp.

Malhotra, V.M. and N.J. Carino, 2003, *Handbook on Nondestructive Testing of Concrete*, 2nd Edition, CRC Press, Boca Raton, FL, 384 pp.

Mamlouk, M. S. and Zaniewski, J,P., *Materials for Civil and Construction Engineers*, 4th Edition, Pearson Prentice Hall, Upper Saddle River, New Jersey, 2016, 672 pp.

Marotta, T., J.C. Coffey, C. LaFleur-Brown, and C. LaPlante, *Basic Construction Materials*, 8th Edition, Pearson Prentice Hall, Upper Saddle River, New Jersey, 2010, 336 pp.

*Masonry Designer's Guide*, 7th Edition, Masonry Society, 2013, 479 pp.[ www.masonrysociety.org]

Mazzolani, F.M., *Aluminium Alloy Structures*, 2nd Edition, E & FN SPON, An imprint of Chapman & Hall, London, 1994.

Mehta, P.K. and P.J.M. Monteiro, 2006, *Concrete -Microstructure, Properties and Materials*, 3rd Edition, McGraw-Hill, New York, 659 pp. (Also contains a CD-ROM with numerous PowerPoint slides, videos, and additional material)

Mindess, S., J.F. Young, and D. Darwin, 2003, *Concrete*, 2nd Edition, Prentice Hall & Pearson Education, Harlow, England, 644 pp.

Minke, G., *Building with Earth-Design and Technology of a Sustainable Architecture*, Birkhäuser – Publishers for Architecture, Basel, Switzerland, 2006, 199 pp.

Mommertz, E., *Acoustics and Sound Insulation: Principles, Planning, Examples,* Birkhäuser, 2009

Morely, M., *Building with Structural Insulated Panels: Strength & Energy Efficiency through Structural Panel Construction*, The Taunton Press, New Town, CT, 2000, 186 pp.

Neville A.M., *Properties of Concrete*, 5th Edition, Pearson Education Limited, Essex, England, 2011.

Neville, A.M., and J.J. Brooks, 2010, *Concrete Technology*, 2nd Edition, Prentice Hall & Pearson Education, Harlow, England, 442 pp.

Ohama, Y., *Handbook of Polymer-Modified Concrete and Mortars- Properties and Process Technology*, Elsevier, 1995, 246 pp.

Ohji, T., and M. Singh (Editors), *Engineered Ceramics: Current Status and Future Products,* John Wiley & Sons, Inc., Hoboken, NJ, 2016, 536 pp.

Pearson, D., *The New Natural House Book-Creating a Healthy, Harmonious, and Ecologically Sound Home*, Fireside-Simon & Schuster, New York, 1998, 304 pp.

Peterson, Ch., *Stahlbau-Grundlagen der Berechnung und baulichen Ausbuildung von Stahlbauten*, Second Edition, Vieweg Verlag, Braunschweig, 1990, 1413 pp.

Pfaender, H.G. (Ed.), *Schott Guide to Glass*, 2nd Edition, Springer, Netherlands, 1996, 224 pp.

Pizzi, A. and K.L. Mittal, *Handbook of Adhesive Technology*, 3rd Edition, CRC Press, Boca Raton, FL., 2017, 658 pp.

Popovics, S., 1988, *Strength and Related Properties of Concrete*, John Wiley, New York, 552 pp.

Porter, F.C., *Corrosion Resistance of Zinc and Zinc Alloys*, CRC Press, Boca Raton, FL, 1994, 536 pp.

Powel, F. J. and S. L. Matthews (ed.), *Thermal Insulation: Materials and Systems,* ASTM, Philadelphia, PA, 1987, 735 pp.

Ramachandran, V.S., (Ed.) 1995, *Concrete Admixtures Handbook—Properties, Science and Technology*, 2nd Edition, Noyes Publications, Park Ridge, NJ, Indian edition by Standard Publishers, New Delhi, India, 2002, 1183 pp.

Richerson, D.W. and W.E. Lee, *Modern Ceramic Engineering: Properties, Processing, and Use in Design,* 4th Edition, CRC Press, Boca Raton, FL., 2018, 840 pp.

Rindel, J.H., *Sound Insulation in Buildings,* CRC Press, Boca Raton, FL, 2018, 450 pp.

Rixom, R. and N.P. Mailvaganam, 1999, *Chemical Admixtures for Concrete*, 3rd Edition, CRC Press, Boca Raton, FL, 456 pp.

Roaf, S., M. Fuentes, and S. Thomas, *Ecohouse: A Design Guide*, Third Edition, Architectural Press & Elsevier, Amsterdam, 2007, 479 pp.

Ross, R.J. (ed.), *Wood Handbook-Wood as an Engineering Material*, Centennial Edition, General Technical Report FPL-GTR-190, Forest Products Laboratory, US Department of Agriculture, Forest Service, Madison, Wisconsin, April 2010, 509 pp.

Santhakumar, A.R., 2018, *Concrete Technology*, 2nd Edition, Oxford University Press, New Delhi, 712 pp.

Schmidt, M., E. Fehling, C. Glotzbach, S. Fröhlich, and S. Piotrowski. (Eds.), *Proceedings of the 3rd International Symposium on UHPC and Nanotechnology for High Performance Construction Materials*, University of Kassel, Germany, March 7-9, 2012, Kassel University Press, 1058 pp. http://www.uni-kassel.de/upress/online/frei/978-3-86219-264-9.volltext.frei.pdf

Shetty, M.S., *Concrete Technology—Theory and Practice*, 7th Revised Edition, S. Chand & Company Ltd., New Delhi, 2013, 656 pp.

Siddique, R., *Waste Materials and By-Products in Concrete*, *Springer Verlag*, Berlin, 2008, 414 pp.

Siegesmund, S. and R. Snethlage (Eds.), *Stone in Architecture, Properties and Durability*, Springer Verlag, Germany, 2014, 550 pp.

SP 21:2005 *Summaries of Indian Standards for Building Materials*, Bureau of Indian Standards, New Delhi.

SP 33:1986 *Handbook on Timber Engineering,* Bureau of Indian Standards, New Delhi, 1986.

SP 41(S&T): 1987 *Handbook on Functional Requirements of Buildings (Other Than Industrial Buildings)*

SP 7-2016, *National Building Code of India 2016* (NBC 2016), 2 volumes, Bureau of Indian Standards, New Delhi (note Part 11- *Approach To Sustainability*).

SP 20(S&T):1991 *Handbook on Masonry Design and Construction*, First Revision, Bureau of Indian Standards, New Delhi, 1991.

Subramanian, N., *Principles of Space Structures*, Second Edition, Wheeler Publishing, Allahabad, 1999, 666 pp.

Subramanian, N., 'Sustainability-Challenges and Solutions', *The Indian Concrete Journal*, Vol. 81, No.12, Dec 2007, pp. 39–50.

Subramanian, N., *Space Structures: Principles and Practice*, 2 Volumes, Multi-Science Publishing Co., Essex, U.K., 2007, 820 pp.

Subramanian, N., 'Pervious Concrete—A Green Material that Helps Reduce Water Run-off and Pollution', *The Indian Concrete Journal*, Vol. 82, No. 12, Dec. 2008, pp. 16–34.

Subramanian, N., *Design of Steel Structures*, 1st Edition, Oxford University Press, New Delhi, 2008, 1240 pp.

Subramanian, N., The Principles of Sustainable Building Design, Chapter in the book *Green Building with Concrete: Sustainable Design and Construction,* 2nd Edition, Sabnis, G.M., Ed., CRC Press, Boca Raton, FL., 2016, 441 pp.

Subramanian, N., *Design of Reinforced Concrete Structures*, Oxford University Press, New Delhi, 2013, 857 pp.

Subramanian, N. and M. Mota, *Sustainability of Steel Reinforcement, in Green Buildings with Concrete,* 2nd Edition, (Sabnis, G.M., ed.), CRC Press, Boca Raton, USA, 2016, pp. 373–394.

Taylor, H. F. W., *Cement Chemistry*, Thomas Telford Publishing, London, 1997, 477 pp.

Trechsel, H.R. and Bomberg, M.T. (Ed.), *Moisture Control in Buildings: The Key Factor in Mold Prevention,* 2nd Edition, ASTM International, West Conshohocken, PA, 2009, 620 pp.

Turner, W.C., and J.F. Malloy, *Thermal Insulation Handbook,* 2nd Edition, R.E. Krieger Publishing Co, Malabar, FL., 1981, 627 pp.

Varghese, P.C., *Building Materials*, Prentice-Hall of India Pvt. Ltd., New Delhi, 2006, 267 pp.

Wicks, Jr., Z.W., F.N. Jones, S.P. Pappas, and D.A. Wicks, *Organic Coatings- Science and Technology*, 3rd Edition, Wiley Interscience, John Wiley & Sons Inc., Hoboken, New Jersey, 2007, 722 pp.

# INDEX

# ABOUT THE AUTHOR

Dr N. Subramanian is an award winning author, consulting engineer, researcher, and mentor now living in Maryland, USA, and former Chief Executive of *Computer Design Consultants* (CDC), Chennai. He had a brilliant academic career and earned his PhD from the *Indian Institute of Technology, Madras*. He also worked as a Post Doctoral Fellow for 2 years at IITM and then worked in the then West Germany as the prestigious *Alexander von Humboldt Fellow* with Prof. J. Lindner and Prof. Ch. Peterson. After his return from West Germany, he worked for a brief period in Anna University and then started his own company, CDC, during 1982.

He has more than 45 years of professional experience which include research, teaching, and consultancy. He has served as a consultant to several leading organizations in India and designed several noteworthy structures involving RCC, structural steel, and cold-formed steel in India and abroad. More than 600 microwave/transmission towers, ranging in height from 20 m to 120 m have been designed by him. One Defense project earned the certificate of merit from the Consultancy Development Centre, DSIR, Govt of India in 1992.

Dr N. Subramanian also developed several special-purpose software packages for the analysis and design of different types of structures. Being a member/fellow of several professional bodies, he took very active part in these organizations, and is the past All India Vice president of the Association of Consulting Engineers (India) and also the Indian Concrete Institute. He organized several conferences/seminars for ACCE(I) and ICI.

Dr Subramanian has authored more than 25 wide selling books and more than 260 technical papers, published in national and international journals/conferences. He is the author of *Design of Steel Structures – Limit States Method*, 2nd Edition (OUP), *Steel Structures: Design and Practice* (OUP), *Design of Reinforced Concrete Structures* (OUP), 8 books in the series *Access to Computer Education* (OUP), and *Space Structures: Principles and Practice* (Multi-Science, UK). Three of these books earned the coveted *ACCE-Nagadi Award* and are prescribed as text books in several universities.

Dr Subramanian is in the editorial/review boards of prestigious International journals such as International Journal of Space Structures, Structural Journal of the American Concrete Institute, Journal of Indian Concrete Institute and the Indian Concrete Journal. He is also honored with several awards, which include the *Tamilnadu Scientist Award, Lifetime Achievement Award* given jointly by the Indian Concrete Institute and L & T, Life member of the American Society of Civil Engineers, and Distinguished Alumnus Award of College of Engineering, Guindy (Anna University).

He is active as mentor to young engineers through several online forums such as the *Structural Engineering Forum of India* (has more than 23,400 members), Quora, and LinkedIn. He also served as the President of the Rotary Club of Madras Metro.

# RELATED TITLES

**Concrete Technology (2e):**
*A.R. Santhakumar*
(9780199458523)

This second edition of *Concrete Technology* provides a comprehensive coverage of the theoretical and practical aspects of the subject and includes the latest developments in the field of concrete construction. It incorporates the latest Indian Standard specifications and codes regulating concrete construction.

**New to this Edition:**

- A new chapter on rheological models of concrete focuses on the flow characteristics of fresh concrete
- Exclusive chapters on properties of fresh and hardened concrete
- A new chapter provides improved treatment of the different stages of manufacture of concrete

**Design of Steel Structures – Limit States Method:** *N. Subramanian*
(9780199460915)

*Design of Steel Structures: Limit States Method* is a comprehensive textbook designed to cater to the undergraduate students of civil and structural engineering. It will also prove useful for postgraduate students and serve as an invaluable reference for practising engineers unfamiliar with the limit state design of steel structures.

**Key Features:**

- Based on the limit state method (IS 800:2007) that reflects the latest developments and is state of the art
- Case studies on structural failures, applications, new developments in structural design, and more new examples
- More than 200 examples, 950 figures, 200 tables, and 1000 references to facilitate a wholesome understanding of concepts
- Flowcharts for design of tension members, compression members, beams, plate girders, and gantry girders for systematic and methodical explanation of the design process

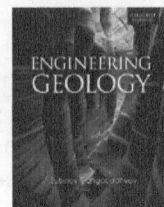

**Engineering Geology:** *Subinoy Gangopadhyay*
(9780198086352)

Engineering Geology is designed as a textbook for the undergraduate and postgraduate students of engineering geology, applied geology, mining, and civil engineering. It will also serve as a reference text for civil engineers and professional geologists.

**Key Features:**

- Provides examples of various geological sites in India and closely follows the work done by the Geological Survey of India
- Explains concepts with numerous well-illustrated figures and on-the-spot pictures of geological project sites from across the country
- Includes a chapter that gives guidelines for writing geological reports which will be immensely valuable for practising civil engineers and geologists

**Strength of Materials (3e):**
*R. Subramanian*
(9780199464739)

This new revised edition of Strength of Materials is a rich and comprehensive textbook for all undergraduate and diploma courses in civil as well as mechanical engineering disciplines. An invaluable resource for professionals and practitioners alike, the book offers crisp and precise explanation of concepts underlying the analysis and design of structures.

**New to this Edition:**

- Two new chapters—Chapter 13: Bending of Curved Bars and Beams, Chapter 14: Stresses due to Rotation
- Problems and multiple choice questions from previous UPSC examinations solved as ready reference for UPSC aspirants
- New application-based solved examples with different load conditions and improved illustrations for better readability

## Other Related Titles

| | | | |
|---|---|---|---|
| 9780195694833 | Subir K. Sarkar & Subhajit Saraswati | : | *Construction Technology* |
| 9780198083535 | Satish Chandra & M.M. Agarwal | : | *Railway Engineering* |
| 9780198083528 | S.K. Duggal | : | *Earthquake-resistant Design of Structures* |
| 9780198069188 | T.S. Thandavamoorthy | : | *Structural Analysis* |
| 9780199456772 | Mohd. Kaleem Khan | : | *Fluid Mechanics and Machinery* |
| 9780195690385 | Rajesh Srivastava | : | *Flow Through Open Channels* |
| 9780198086949 | N. Subramanian | : | *Design of Reinforced Concrete Structures* |